ENVIRONMENTAL
SCIENCE

TENTH EDITION

ENVIRONMENTAL SCIENCE

DANIEL D. CHIRAS
Visiting Professor, Colorado College
Director, The Evergreen Institute
Center for Energy and Green
Building

JONES & BARTLETT
LEARNING

World Headquarters
Jones & Bartlett Learning
5 Wall Street
Burlington, MA 01803
978-443-5000
info@jblearning.com
www.jblearning.com

Jones & Bartlett Learning books and products are available through most bookstores and online booksellers. To contact Jones & Bartlett Learning directly, call 800-832-0034, fax 978-443-8000, or visit our website, www.jblearning.com.

Substantial discounts on bulk quantities of Jones & Bartlett Learning publications are available to corporations, professional associations, and other qualified organizations. For details and specific discount information, contact the special sales department at Jones & Bartlett Learning via the above contact information or send an email to specialsales@jblearning.com.

05781-2

Production Credits

Chief Executive Officer: Ty Field
President: James Homer
Chief Product Officer: Eduardo Moura
Executive Publisher: William Brottmiller
Publisher: Cathy L. Esperti
Senior Acquisitions Editor: Matthew Kane
Editorial Assistant: Audrey Schwinn
Production Manager: Louis C. Bruno, Jr.
Production Assistant: Talia Adry
Marketing Manager: Lindsay White

Art Development Editor: Joanna Lundeen
Art Development Assistant: Shannon Sheehan
Manufacturing and Inventory Control Supervisor: Amy Bacus
Composition: Cenveo Publisher Services
Cover Design: Kristin E. Parker
Manager of Photo Research, Rights & Permissions: Lauren Miller
Cover Image: © Antonio Jorge Nunes/Shutterstock, Inc.
Printing and Binding: Manufactured in the United States by RR Donnelley

Library of Congress Cataloging-in-Publication Data
Chiras, Daniel D.
  Environmental science / Daniel D. Chiras. — Tenth edition.
    pages cm.
  ISBN 978-1-284-05705-8 (alk. paper)
  1. Environmental sciences. I. Title.GE140.C48 2015
  GE140.C48 20152014013408
  363.7—dc23   2014013408

6048

Printed in the United States of America
18 17 16 15   10 9 8 7 6 5 4 3 2

TO MY FATHER WHO STARTED ME ON THE CONSERVATION PATH MANY YEARS AGO.

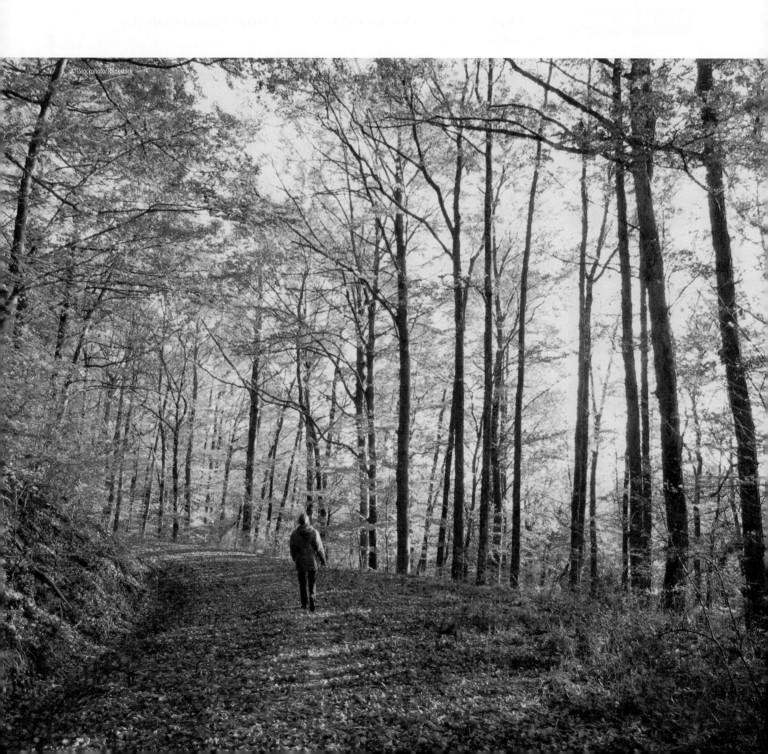

BRIEF CONTENTS

TABLE OF CONTENTS

© Antonio Jorge Nunes/Shutterstock, Inc.

PREFACE

Environmental science courses vary widely. Some professors prefer to discuss basic science before they delve into environmental issues such as air pollution, water pollution, global climate change, species, and extinction. Other professors focus closely on the main issues. In between these two approaches are a wide range of options.

This book is meant to serve the diverse approaches of professors. This book presents the basic science needed to understand environmental issues and offers an in-depth look at all of the environmental problems facing the world. This book discusses solutions to environmental issues, including the more traditional ones and a new generation of policies and actions that promise to achieve lasting results. They are sustainable solutions that involve all stakeholders: businesses, governments, and individuals.

During the nearly four decades I have spent working on this book, it has evolved dramatically in response to profound changes in my thinking. Perhaps the most important change has been the shift toward a more comprehensive and more systematic way of viewing environmental problems. One of the useful discoveries I have made over many years is that most environmental problems spring from a common set of root causes. Unlike other environmental science textbooks, this one discusses the common root causes and ways to address them.

Over time, I have also concluded that we've been addressing environmental problems superficially. I've realized that most efforts are stopgap measures—solutions that treat the symptoms while neglecting the underlying root causes. To create lasting solutions we must address their root causes. One root cause is that the systems we depend on—for food, energy, materials, water, waste treatment, and other services—are unsustainable. To solve environmental problems and create an enduring human presence (a sustainable future), we must rethink and restructure basic human systems, such as transportation, energy, waste management, water supply, industry, and housing. This book shows the reader how this can be done—and how such changes can have profound positive effects on the environment.

Another important realization I have made is that to be sustainable, solutions to environmental problems must make sense from social and economic perspectives, not just an environmental one. One-sided solutions don't work. This shift to solutions that make sense from all three perspectives will take enormous creative energy on our part in the years to come.

This edition reflects a pressing need for sustainable solutions—measures that confront the root causes of the environmental crisis and create a society that meets its needs without bankrupting the Earth. This book includes discussions meant to help students understand why human systems need to be revamped and how we go about this challenging task, creating a clean, healthy environment.

In this new edition, I have added new material, updated statistics, and polished the writing. As in the first nine editions, I worked diligently to ensure that this text remains user-friendly. My goal continues to be to present the most important facts and concepts in a clear and exciting fashion. As always, I have strived to present both sides of the issues in the text and in Point/Counterpoint essays that deal with controversial environmental issues. Even though this text presents a strong case for sustainability and systems reform, it is left to you, the reader, to make up your own mind as to the need and desirability of such an approach. My efforts to make this book as unbiased as possible support my objective of letting students make up their own minds about our predicament and ways to extricate ourselves from it. Critical thinking skills presented in the book also help students learn to analyze issues and make up their own minds.

Themes

All textbooks have a central theme or, in some cases, a set of themes that shape the presentation. This book is molded by six central ideas: focus on key principles, sustainability, addressing root causes, systems reform, critical thinking, and action.

FOCUS ON KEY PRINCIPLES Environmental science is a vast field, requiring many years to master. Most students, however, can become proficient in the subject much more quickly when they are taught key concepts. Vital principles are highlighted in special Key Concept boxes at the end of each section. They encapsulate the material covered in each section and provide an excellent tool for studying.

SUSTAINABILITY The main theme of this book is that the long-term well-being of this planet and its inhabitants is in jeopardy and that to create an enduring human presence we must make a massive course change. We must transition to a sustainable society. A sustainable society seeks balance between human and ecological needs. Its economic systems serve people and the planet. Creating a sustainable society may be our only realistic hope for surviving on a finite planet, but it will not evolve without foresight, planning, and action.

ADDRESSING ROOT CAUSES AND SYSTEMS REFORM

Creating a sustainable future requires serious efforts to understand and confront the root causes of the environmental crisis. This book outlines those causes and shows ways to address them. Part of this struggle will involve efforts to revamp human systems—make sustainable systems to provide energy, transportation, and waste management, for example.

CRITICAL THINKING This text also stresses critical thinking skills—learning to think critically about issues, a task that is essential to create a sustainable society.

ACTION Finally, this text emphasizes an often overlooked point, that building a sustainable future requires actions by all of us. Air pollution is not just caused by inadequate laws or corporate wrongdoings, it is a result of our own often wasteful lifestyles. Because we are all part of the problem, we must all be part of the solution. Individual action is as essential as responsible corporate and government policies and practices. This book offers even more advice than previous editions on ways we can all go green.

Expanded Basic Science

In recent editions, I have expanded the coverage of basic science, especially information to help students understand global climate and energy. You'll find detailed discussion, for instance, of weather, ocean currents, aquatic ecology, the laws of thermodynamics, and other important facts that help you better understand the world we live in.

Global Focus and Expanded Coverage of International Issues and Solutions

Environmental problems affect us locally, in our own homes and towns and cities. They affect larger regions, too, often covering many states or crossing international borders. And, of course, they often span the globe. As in previous editions, I present a wealth of information on local, regional, and global problems and solutions. Special efforts were made to include examples from Canada as well as countries in Europe, Africa, and Asia.

Organization

This book is divided into six parts.

PART I introduces students to four of the principal themes of this book: sustainability, addressing root causes, restructuring human systems, and critical thinking. It lays an important foundation for the study of environmental science.

PART II introduces the student to basic principles of ecology necessary for understanding environmental issues. These chapters elaborate on six operating principles of sustainability—ideas that will help us revamp modern society one system at a time.

PART III begins the discussion of environmental issues, dealing with one of the most pressing of all; the population crisis. This part examines the impact of rapid population growth—one of the root causes of modern environmental problems—and explores culturally acceptable ways of slowing it down.

PART IV deals with a variety of resource issues, such as wildlife extinction and energy, and outlines strategies for solving them sustainably. In this material, I attempt to show how we can overhaul some of the vital human systems such as energy and waste management.

PART V discusses pollution and legal, technical, and personal solutions for it, including both traditional and sustainable strategies.

PART VI the capstone of the book, attempts to place the population, resource, and pollution crises in a social context. It reexamines ethics and explores economics and government in more detail.

New to This Edition

The *Tenth Edition* has been thoroughly updated with new discoveries, new concepts, and the most recent statistics on resources, population, and pollution. I have added information on changes in environmental laws and regulations. I have also included important new information on peak oil and natural gas, renewable energy options, global climate change, environmentally friendly transportation, environmentally friendly building, ecosystem protection, tropical rainforests, species extinction, and more.

The Student Experience

Over the years, this book has undergone some dramatic changes to help make the study of environmental science more interesting, meaningful, and memorable. Below is a list of features students should find helpful:

Study Skills

Immediately following the preface is a brief section on study skills. The study skills section includes numerous simple but effective tips that help students improve their memory, note-taking skills, reading abilities, test-taking abilities, and much more. Study skills can help all students, even A students, become more efficient learners. Skills learned here will carry over to virtually every other course students take and will be helpful throughout life.

Learning Objectives

Every chapter opens with a comprehensive list of Learning Objectives. Students should review this list before diving into the chapter to help guide their focus. As they progress through the chapter, they should periodically flip back to the Learning Objectives to ensure they are fully grasping that chapter's key environmental science concepts. This practice will encourage students to think critically about environmental issues and solutions.

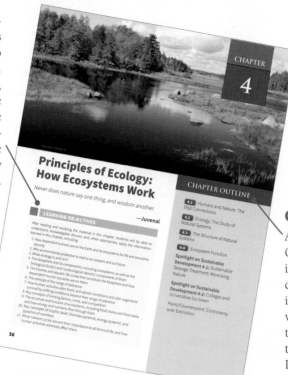

Chapter Outlines

Along with the Learning Objectives, a Chapter Outline is included at the start of every chapter. This list succinctly identifies the sections students will encounter as they progress through a chapter, including the Spotlight on Sustainable Development boxes and the Point/Counterpoint topic. It provides students with a "big picture" understanding of the overarching environmental principles for each chapter.

Critical Thinking

As I mentioned earlier, critical thinking is one of the central themes of this book. It enables students to analyze complex issues and make rational decisions.

A number of important critical thinking rules are discussed in Chapter 1. Each chapter also begins with a brief critical thinking exercise, which asks students to critically analyze an issue, a research finding, or an assertion.

Students are also asked to exercise critical thinking skills after Point/Counterpoints, and many of the discussion questions at the end of each chapter call on students to put their knowledge and their critical thinking skills to use.

Key Concepts

Key concepts are highlighted in boxes at the end of each section. These brief encapsulations are designed to help students understand the concepts that form the foundation of environmental science. They serve as a great tool for reviewing chapters as well as studying for tests. Read them before you read each chapter, after completing a chapter, and while you're studying for tests. They're the glue that holds this book together.

Green Living Tips

In this edition, I've added even more green living tips, labeled *Go Green*. These simple, inexpensive, and practical tips will help you become a part of the solution and help all of us build a sustainable future.

Foundation Tools

This book offers several Foundation Tools, conceptual models that help students understand how the world works and how solutions can be applied. These models are designed to encourage systems thinking and help students organize facts into a solid conceptual framework. Below is a brief description of each model.

POPULATION, RESOURCES, AND POLLUTION This model shows how populations of organisms, like ourselves, affect their environment and how their actions, in turn, affect the populations themselves.

IMPACT ANALYSIS This model shows the impacts that humans have on various components of the environment. It helps to clarify a confusing topic.

RISK ANALYSIS This model presents an overview of a process called *risk assessment*. This process helps us assess risks in a meaningful way.

ROOT CAUSES This model shows the many factors that contribute to the crisis of unsustainability and helps students identify key leverage points for change.

HUMAN SYSTEMS This model shows the complex systems that support our lives and how they are functionally related to one another.

SUSTAINABLE PRINCIPLE APPLICATION This model demonstrates how the ecological principles of sustainability can be applied to complex human systems.

Examining Both Sides of the Issue

In keeping with my long-standing belief that students must examine all sides of an issue, I have continued to present information from opposing viewpoints. You will find this approach exercised in the text itself, where I examine contradictory positions, and in the Point/ Counterpoints.

POINT/COUNTERPOINTS AND VIEWPOINTS

Complex environmental issues often result in hotly contested debates. These editorials can stimulate individual thinking as well as classroom discussion on complex problems. They are also a perfect avenue for developing one's critical thinking skills.

SPOTLIGHT on Sustainable Development

1-1 The Netherlands' Green Plan Revolutionizes the Way Industries Function

The Netherlands is a tiny country, about one-fourth the size of the state of Illinois (**FIGURE 1**). It is, however, a giant among nations. Why this glowing praise for such a small country?

In 1989, the country adopted a National Environmental Policy Plan (NEPP) commonly called its Green Plan, after widespread public alarm over many environmental trends. An even tougher plan came into existence several years later. Known as the NEPP 2, it gave the Dutch government more power to write and enforce environmental laws and to earmark substantial sums of money to research and redesign industrial processes. Since then, it has been twice upgraded.

The most exciting aspect of the Green Plan is that it allows for a new approach—a set of voluntary agreements with industry and other sectors of society responsible for producing the bulk of the nation's pollution. Why was all of this necessary?

Although the Netherlands has some of the most stringent environmental laws in the world, they weren't working well. Because of this, the government took a bold step: It admitted that there are limitations to regulating complex and interdependent environmental problems on an issue-by-issue basis. In place of this unworkable system of rules and regulations, the government decided to meet with the key players (industries and citizen groups, for example) to reach agreement on establishing bold new targets and timetables for drastically reducing pollution. The government let the companies select ways to reduce pollution; all that it asked of industry was a commitment to meet ambitious government targets and timetables. Industry groups happily signed the agreements, recognizing

FIGURE 1 The Netherlands is leading the world in sustainable development.
© fdeestock.

Spotlights on Sustainable Development

To give students further insight into sustainable solutions, I've included numerous case studies throughout the text called Spotlights on Sustainable Development. They show what cities, towns, businesses, and even professional sports teams are doing to create a sustainable future. They provide guidance and hope.

FIGURE 4-15 A simplified view of the nitrogen cycle. Nitrogen in the atmosphere is converted into ammonium ions by bacteria in the soil. Ammonia is converted to nitrates and taken up by plants.

CRITICAL THINKING

Exercise Analysis

It is difficult for me to know the list of misconceptions you came up with, but here are a few that might have arisen along the way. The exact meaning of the term ecology may have been one. Many people use it very loosely. You may not have understood the term niche, either. Like many people, you may have thought of it as a physical place an organism occupied, its habitat.

I suspect that your sense of the word environment has shifted. Now you can see that it includes biotic and abiotic components. The environment of an organism is quite complex.

I suspect that the discussion of the range of tolerance may have helped you to understand how our activities and natural events alter an organism's chances of survival.

What about energy? Has your understanding of energy changed? Do you see food chains and food webs now in a different light? Rather than being a simple matter of one organism feeding on another, food chains and food webs are elaborate pathways for the flow of nutrients and energy.

You may have gained new perspective on nutrient cycles. You may already have known that they existed, but did you know that humans are altering them in major ways?

Detailed Illustrations

The text features numerous detailed illustrations to demonstrate clearly important environmental processes and cycles, such as the carbon cycle and the nitrogen cycle, pictured here. The illustrations serve as powerful study tools for students in preparation for lectures and exams.

Critical Thinking Analysis

Each chapter begins with a brief critical thinking exercise, which asks students to critically analyze an issue, a research finding, or an assertion. The author responds to the exercise at the end of each chapter and walks the reader through the steps to thinking critically about the environmental issue under discussion, with reference back to the chapter material.

Critical Thinking and Concept Review Questions

Every chapter concludes with Critical Thinking and Concept Review Questions. These questions challenge students to apply the information from the chapter, ensuring they thoroughly grasp the important environmental science concepts. Many of the questions encourage students to think critically about a specific environmental issue and push them to think deeply about both sides of the matter.

Key Terms

A list of Key Terms is included at the end of every chapter. Furthermore, the key terms in the chapter appear in bold to draw the reader's attention. Students should refer to the Key Terms as part of their study to assess their understanding of chapter material.

Additional Online Study Tools

Additional resources such as practice activities, links to relevant websites, flashcards, and an interactive glossary are available online to help students study.

Focus on Sustainability and Systems Reform

While the main focus of the book is environment, not development, I attempt to show how development must be revamped to become sustainable. I describe basic human systems and why they are unsustainable. Moreover, I outline a variety of ways that we can make them environmentally, socially, and economically sustainable.

Teaching Tools

The **IMAGE BANK** provides the illustrations, photographs, and tables (to which Jones & Bartlett Learning holds the copyright or has permission to reprint digitally). These images are not for sale or distribution but may be used to enhance your existing slides, tests, and quizzes, or other classroom material.

The **LECTURE OUTLINES** in PowerPoint format provide lecture notes and images for each chapter of *Environmental Science, Tenth Edition.* Instructors with Microsoft PowerPoint software can customize the outlines, art, and order of presentation.

The **INSTRUCTOR'S MANUAL** is provided as a text file. The Instructor's Manual has been updated and expanded for this edition. It contains chapter outlines, objectives, and key terms as well as suggestions for presenting each chapter. We have also included sample syllabi to demonstrate a few of the many approaches that can be taken to an environmental science course.

A **TESTBANK** is available with hundreds of questions in a variety of formats.

WEB LINKS are available to direct instructors and students to outside sources where they can learn more about topics discussed in the book and environmental issues specific to their own areas.

The Advanced Industrial Age

❖ In recent times, industrialization has grown rapidly.

❖ Resource demand and environmental destruction have reached unsustainable levels.

Figure 7.5: Industrialization on the farm. Modern agriculture depends heavily on machinery, energy, and additional resources.
© Jack Dagley/ShutterStock, Inc.

Copyright © 2015 by Jones & Bartlett Learning, LLC an Ascend Learning Company
www.jblearning.com

Where Do Acids Come From?

❖ Acid precursors come from natural and anthropogenic sources, the latter being the most important.

Figure 20.9A: Sources of sulfur dioxide. Volcanic eruptions, such as the Mount St. Helens blast shown here, is one natural source of sulfur dioxide.
© Weyerhaeuser. Used with Permission.

Copyright © 2015 by Jones & Bartlett Learning, LLC an Ascend Learning Company
www.jblearning.com

Acknowledgments

Although I have spent thousands of hours researching and writing it, this book is really the product of thousands of researchers and scholars in anthropology, biology, chemistry, demography, natural resources, political science, economics, ecology, and dozens of other disciplines. Their findings and their thoughts form the foundation on which this book rests. To them, a world of thanks!

A genuine thanks also to the staff at Jones & Bartlett. A very special thanks to my editors, Erin O'Connor and Matt Kane, who offered moral support and encouragement throughout the project. Thanks also to Lou Bruno, Audrey Schwinn, Kristin Parker, and Lauren Miller for their diligence and hard work on the manuscript, design, and photo research. Their creativity, bright ideas, and hard work have been vital to the success of this book. As always, it is has been a pleasure to work with the Jones & Bartlett Learning staff in this complex and sometimes tedious production. Also a special thanks to my research assistant, Linda Stuart, who helped update the book. Her persistence and attention to detail were much appreciated. Updating statistics can be tiring, frustrating, and at times overwhelming, and I'm thankful for her kind assistance. Also, a word of thanks to the many people in government agencies and nonprofit groups who shared articles, reports, and data with us. Although they are too numerous to mention here, their part in this project is much appreciated. The author and publisher would also like to thank Stacy Zell of Carroll Community College for her assistance with the preparation of the supplemental material for instructors.

Finally, many manuscript reviewers and survey respondents provided helpful and constructive criticism on all editions of this book. I am very thankful for their helpful comments.

David Armstrong, University of Colorado, Boulder

Robert Auckerman, Colorado State University

Terry Audesirk, University of Colorado, Denver

Nancy Bain, Ohio University

Michael Bass, Mary Washington College

Richard Beckwitt, Framingham State College

Mark C. Belk, Brigham Young University

Charles Blalack, Kilgore College

Bayard H. Brattstrom, California State University, Fullerton

Lester Brown, Worldwatch Institute

Anya Butt, Central College

Bonnie Callan, Des Moines Area Community College

Ann S. Causey, Auburn University

Jay Clymer, Marywood University

Sheree E. Cohn, Saint Cloud State University

Donald Collins, Montana State University

Peter Colverson, Mohawk Valley Community College

Anne Cummings, Pikes Peak Community College

Craig B. Davis, Ohio State University

Lola Deets, Penn State Erie, The Behrend College

JodyLee Estrada Duek, Pima Community College

Sally DeGroot, St. Petersburg Junior College

Jessica Dutton, Adelphi University

Joseph Farynairz, Mattatuck Community College

Eric J. Fitch, Marietta College

David Fluharty, University of Washington

Gary J. Galbreath, Northwestern University

Ted Georgian, St. Bonaventure University

Lisa Akers Greenwood, Rochester Institute of Technology

Sigurdur Greipsson, Troy State University

James H. Grosklags, Northern Illinois University

Richard Haas, California State University, Fresno

Zac Hansom, San Diego State University

William S. Hardenbergh, Southern Illinois University

John P. Harley, Eastern Kentucky University

John N. Hoefer, University of Wisconsin, La Crosse

James Hornig, Dartmouth College

William Hoyt, Saint Joseph's College of Maine

Gary James, Orange Coast College

Pat Hilliard Johnson, Palm Beach Community College

John A. Jones, Miami Dade Community College

Alan R. P. Journet, Southeast Missouri State University

Beverly W. Juett, Midway College

Thomas L. Keefe, Eastern Kentucky University

Suzanne Kelly, Scottsdale Community College

Thomas G. Kinsey, State University College at Buffalo

Ronald Kolbash, Ohio University

Mary Hurn Korte, Concordia University of Wisconsin

Kip Kruse, Eastern Illinois University

Steve LaDochy, California State University, Los Angeles

David Lovejoy, Westfield State College

Timothy F. Lyons, Ball State University

Ted L. Maguder, St. Petersburg Junior College, Clearwater

Ravindra Malik, Albany State University

Heidi Marcum, Baylor University

Richard Meyer, University of Arkansas

Myles Miller, Philadelphia University

Sheila Miracle, Southeast Kentucky Community and Technical College

Glenn P. Moffat, Foothill College

Charles Mohler, Cornell University

Joseph Moore, California State University, Northridge

Bryan C. Myres, Cypress College

Paul E. Nowack, University of Michigan

Natalie Osterhoudt, Broward Community College

Nancy Ostiguy, California State University

John H. Peck, St. Cloud State University

Charles R. Peebles, Michigan State University

Michael Picker, Western Director for the National Toxics Campaign

David Pimental, Cornell University

Michael J. Plewa, University of Illinois, Urbana-Champaign

Michael Priano, Westchester Community College

Joseph Priest, Miami University

Martha W. Rees, Baylor University

Robert J. Robel, Kansas State University

Jack Schelein, York College, City University of New York

Michael P. Shields, Southern Illinois University

Cynthia Simon, University of New England

Doris Shoemaker, Dalton College

Rocky Smith, Colorado Environmental Coalition

Steven Soldheim, University of Wisconsin, Madison

Stephen Sousa, Andover High School

Laura Tamber, Nassau Community College

Roger E. Thibault, Bowling Green State University

J. Rafael Toledo, University of North Texas

Leland Van Fossen, DeAnza College

Bruce Webb, Animal Protection Institute

Ross W. Westover, Canada College

Jeffrey R. White, Indiana University

Ray E. Williams, Rio Hondo College

Larry Wilson, Miami Dade Community College

Stephen W. Wilson, Central Missouri State University

Susan Wilson, Miami Dade Community College

Timothy Winter, Louisiana State University, Shreveport

Robert Wiseman, Lakewood Community College

Richard J. Wright, Valencia Community College

Paul A. Yambert, Southern Illinois University

Astatkie Zikarge, Texas Southern University

Daniel D. Chiras
The Evergreen Institute
Gerald, Missouri

ABOUT THE AUTHOR

Courtesy of Dan Chiras

Dan Chiras received a Ph.D. in reproductive physiology from the University of Kansas Medical School. In 1976, he accepted a teaching and research position at the University of Colorado, Denver, where he taught a variety of courses, including general biology, cell biology, histology, reproductive biology, and endocrinology. Over the years, Dr. Chiras pursued his interest in environmental issues and developed a number of courses on the environment, including a graduate course on global environmental issues and the environmental and health effects of pollution. He has also taught undergraduate science modules on air pollution, nuclear power, noise pollution, impacts of coal development, and strategies for sustainability.

While at the university, Dr. Chiras worked on several EPA projects, including a study of the health impacts of chlorinated organics from wastewater treatment and the impacts of coal mining in the West. He also prepared an assessment of the impacts of shale oil development in Colorado for the U.S. Department of Energy.

In 1981, Dr. Chiras resigned his full-time position at the university to pursue a writing career. Since that time, he has published nearly 350 articles in journals, magazines, encyclopedias, and other publications, including *Environment, American Biology Teacher, The Amicus Journal, Solar Today, Natural Home, Mother Earth News, Sustainable Futures, Colorado Outdoors, Home Power,* and *Environmental Carcino-genesis and Ecotoxicology Reviews.*

Dr. Chiras wrote the environment section for World Book Encyclopedia's Science Year from 1993 to 2010. He wrote an extensive article titled "The Population Explosion" in Science Year 1998. Dr. Chiras has written articles on ecology, the environment, environmental issues, air pollution, and population articles for *Encyclopedia Americana.*

Dr. Chiras has published 30 books on ecological design, environmental science, sustainable systems design, green building, and residential renewable energy. Included in this list are a college-level textbook titled *Natural Resource Conservation* (with John P. Reganold) and a high-school textbook titled *Environmental Science: Framework for Decision Making.* Dr. Chi-ras has also published two college-level biology textbooks: *Human Biology* and *Biology: The Web of Life.*

Dr. Chiras has written several books for a more general audience, *Beyond the Fray: Reshaping America's Environmental Response,* which offers advice on building a truly effective response to the environmental crisis, and *Lessons from Nature: Learning to Live Sustainably on the Earth,* which outlines ways to build a sustainable future. He has also assembled an anthology titled *Voices for the Earth: Vital Ideas from Amer-ica's Best Environmental Books.* This book includes 14 essays summarizing the key ideas of books nominated by a national committee as America's best environmental books.

Dr. Chiras's more recent books include *Power from the Wind, Power from the Sun, Wind Energy Basics, Solar Electricity Basics, Green Home Improvement, Green Transportation Basics, Solar Home Heating Basics, The Natural House: A Complete Guide to Healthy, Energy-Efficient, Environmental Homes; The Solar House; The New Ecological Home; Superbia!; 31 Ways to Create Sustainable Neighborhoods; The Homeowner's Guide to Renewable Energy; The Natural Plasters Book;* and *EcoKids: Raising Children Who Care for the Earth.*

Dr. Chiras is a visiting professor in the environmental science program at Colorado College, where he has taught courses on renewable energy, green building, and climate change. Dr. Chiras has also served as a visiting professor at the University of Washington in Seattle, where he taught the introductory environmental science course.

Besides teaching and writing, Dr. Chiras has played an active role in the environmental movement. He has served on the Board of Directors of the Colorado Environmental Coalition for five years and was president of this coalition of 40 environmental groups for two years. In 1988, he co-founded Friends of Curbside Recycling, which was instrumental in convincing the city of Denver to begin a curbside recycling program. In 1989, he cofounded Speakers for a Sustainable Future, which offered slide programs on recycling, water conservation, and sustainability. In 1993, Dr. Chiras cofounded another nonprofit organization, the Sustainable Futures Society. He served on the board of directors of the Advanced Technology Environmental Education Library, which produced a website that serves as a repository for environmental information.

Dr. Chiras has spoken to a wide range of audiences, including the National Association of County Agricultural Agents; the American Society of Interior Designers; the American Home Economics Association; Architects, Designers, Planners for Social Responsibility; the Colorado Renewable Energy Society; the American Solar Energy Society; and many others.

In 2009, Dr. Chiras formed The Evergreen Institute: Center for Renewable Energy and Green Building in east-central Missouri (www.evergreeninstitute.org), where he teaches classes on solar electricity, wind energy, passive solar heating and cooling, green building, natural building, and more. He is currently turning the 60-acre farm and educational center into a net-zero energy facility by implementing efficiency measures and installing solar electric, solar hot water, passive solar, and wind energy systems that generate all the energy the facility consumes.

In addition to his active scientific pursuits, Dr. Chiras is an organic gardener and bicyclist. He raises homing pigeons and all natural grass-fed cattle, Belted Galloways.

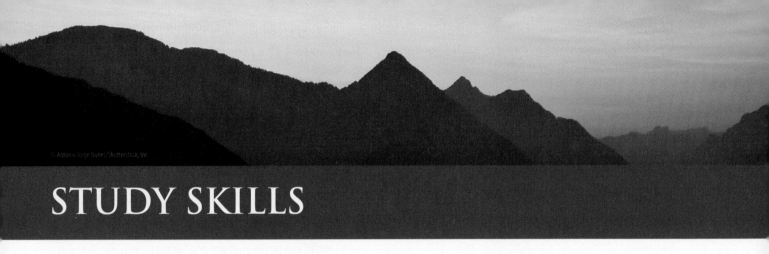

© Antonio Jorge Nunes/Shutterstock, Inc.

STUDY SKILLS

College can be an extremely demanding time. For many students, term papers, tests, reading assignments, and classes require a new and higher level of commitment to their education. At times, the workload can become overwhelming.

Fortunately, there are many ways to lighten the load and make time spent in college more profitable. This section offers some helpful tips on ways to enhance your study skills. It teachers you how to improve your memory, how to become a better note taker, and how to get the most out of what you read. It also helps you prepare better for tests and become a better test taker.

Mastering these study skills will require some work, mostly to break old, inefficient habits. In the long run, though, the additional time you spend now learning to become a better learner will pay huge dividends. Over the long haul, improved study skills will save you lots of time and help you improve your knowledge of facts and concepts. That will, no doubt, lead to better grades and very likely a more fruitful life.

General Study Skills

- Study in a quiet, well-lighted space. Avoid noisy, distracting environments.
- Turn off televisions, radios, and cell phones.
- Work at a desk or table. Don't lie on a couch or bed.
- Establish a specific time each day to study, and stick to your schedule.
- Study when you are most alert. Many students find that they retain more if they study in the evening a few hours before bedtime.
- Take frequent breaks—one every hour or so. Exercise or move around during your study breaks to help you stay alert.
- Reward yourself after a study session with a mental pat on the back or a snack.
- Study each subject every day to avoid cramming for tests. Some courses may require more hours than others, so adjust your schedule accordingly.
- Look up new terms or words whose meaning is unclear to you in the glossaries in your textbooks or in a dictionary.

Improving Your Memory

You can improve your memory by following the PMC method. The PMC method involves three simple learning steps: (1) paying attention, (2) making information memorable, (3) correlating new information with facts you already know.

STEP 1. Paying attention means taking an active role in your education—taking your mind out of neutral. Eliminate distractions when you study. Review what you already know, and formulate questions about what you are going to learn before a lecture or before you read a chapter in the text. Reviewing and questioning help prime the mind.

STEP 2. Making information memorable means finding ways to help you retain information in your memory. Repetition, mnemonics, and rhymes are three helpful tools.

- Repetition can help you remember things. The more you hear or read something, the more likely you are to remember it, especially if you're paying attention. Jot down important ideas and facts while you read or study to help involve all of the senses.
- Mnemonics are useful learning tools to help remember lists of things. I use the mnemonic CAR-RRP to remember the biological principles of sustainability: conservation, adaptability, recycling, renewable resources, restoration, and population control.
- Rhymes and sayings can also be helpful when trying to remember lists of facts.
- If you're having trouble remembering key terms, look up their roots in the dictionary. This often helps you remember their meaning.
- You can also draw pictures and diagrams of processes to help remember them.

STEP 3. Correlating new information with the facts and concepts you already know helps tie facts together, making sense out of the bits and pieces you are learning.

- Instead of filling your mind with disjointed facts and figures, try to see how they relate with what you already know. When studying new concepts, spend some time trying information together to get a view of the big picture.

- After studying your notes or reading your textbook, go back and review the main points. Ask yourself how this new information affects your view of life or critical issues and how you may be able to use it.

Becoming a Better Note Taker

- Spend 5 to 10 minutes before each lecture reviewing the material you learned in the previous lecture. This is extremely important!
- Know the topic of each lecture before you enter the class and spend a few minutes reflecting on facts you already know about the subject about to be discussed.
- If possible, read the text before each lecture. If not, at least look over the main headings in the chapter, read the topic sentence of each paragraph, and study the figures. If your chapter has a summary, read it too.
- Develop a shorthand system of your own to facilitate note taking. Symbols such as = (equals), > (greater than), < (less than), w/ (with), and w/o (without) can save lots of time so you don't miss the main points or key facts.
- Develop special abbreviations to cut down on writing time. E might stand for energy, AP might be used for air pollution, and AR could be used to signify acid rain.
- Omit vowels, abbreviate words, and use special symbols to decrease writing time (for example: omt vwls & abbrvte wrds to ↓ wrtng tme). This will take some practice.
- Don't take down every word your professor says, but be sure your notes contain the main points, supporting information, and important terms.
- Watch for signals from your professor indicating important material that might show up on the next test (for example, "This is an extremely important point…").
- If possible, sit near the front of the class to avoid distractions.
- Review your notes soon after the lecture is over while they're still fresh in your mind. Be sure to leave room in your notes written during class so you can add material you missed. If you have time, copy your notes after each lecture.
- Compare your notes with those of your classmates to be sure you understood everything and did not miss any important information.
- Attend all lectures.
- Use a tape recorder if you have trouble catching important points.
- If your professor talks too quickly, politely ask him or her to slow down.
- If you are unclear about a point, ask during class. Chances are other students are confused as well. If you are too shy, go up after the lecture and ask, or visit your professor during his or her office hours.

How to Get the Most Out of What You Read

- Before you read a chapter or other assigned readings, preview the material by reading the main headings or chapter outline to see how the material is organized.
- Pause over each heading and ask a question about it.
- Next, read the first sentence of each paragraph. When you have finished, turn back to the beginning of the chapter and read it thoroughly.
- Take notes in the margin or on a separate sheet of paper.
- Underline or highlight key points.
- Don't skip terms that are confusing to you. Look them up in the glossary or in a dictionary. Make sure you understand each term before you move on.
- Use the study aids in your textbook, including summaries and end-of-chapter questions. Don't just look over the questions and say, "Yeah, I know that." Write out the answer to each question as if you were turning it in for a grade, and save your answers for later study. Look up answers to questions that confuse you. This book has questions that test your understanding of facts and concepts. Critical thinking questions are also included to sharpen your skills.

Preparing for Tests

- Don't fall behind on your reading assignments.
- Review lecture notes as often as possible.
- If you have the time, you may want to outline your notes and assigned readings.
- Space your study to avoid cramming. One week before your exam, go over all of your notes. Study for two nights, then take a day off that subject. Study again for a couple of days. Take another day off from that subject. Then make one final push before the exam, being sure to study not only the facts and concepts but also how the facts are related. Unlike cramming, which puts a lot of information into your brain for a short time, spacing will help you retain information for the test and for the rest of your life.
- Be certain you can define all terms and give examples of how they are used.
- You may find it useful to write flash cards to review terms and concepts.
- After you have studied your notes and learned the material, look at the big picture—the importance of the knowledge and how the various parts fit together.
- You may want to form a study group to discuss what you are learning and to test one another.
- Attend review sessions offered by your instructor or by your teaching assistant, and study before the session and go to the session with questions.
- See your professor or class teaching assistant with questions as they arise.

- Take advantage of free or low-cost tutoring offered by your school, or, if necessary, hire a private tutor to help you through difficult material. Get help quickly, though. Don't wait until you are hopelessly lost. Remember that learning is a two-way street. A tutor won't help unless you are putting in the time.
- If you are stuck on a concept, it may be that you have missed an important point in earlier material. Look back over your notes or ask your tutor or professor what facts might be missing and causing you to be confused.
- Write and take your own tests. Include all types of questions.
- Study tests from previous years, if they are available legally.
- Determine how much of a test will come from lecture notes and how much will come from the textbook.
- Purchase a study guide, if one is available, and use it to review material and test our knowledge.
- Check out your instructor's website or the online resources that accompany this text. Offered are valuable study aids, such as review questions and practice quizzes.

Taking Tests

- Eat well and get plenty of exercise and sleep before tests.
- Remain calm during the test by deep breathing.
- Arrive at the exam on time or early.
- If you have questions about the wording of a question, ask your professor.
- Skip questions you can't answer right away, and come back to them at the end of the session if you have time.
- Read each question carefully and be sure you understand its full meaning before answering it.
- For essay questions and definitions, organize your thoughts first on the back of the test before you start writing. A sample outline may help you get your thoughts straight and allow you to write a coherent answer.

Now take a few moments to go back over the list. Check off those things you already do. Then, mark the new ideas you want to incorporate into your study habits. Make a separate list, if necessary, and post it by your desk or on the wall and keep track of your progress.

PART 1

INTRODUCTION TO ENVIRONMENTAL PROTECTION

SUSTAINABILITY, SCIENCE, AND SYSTEMS

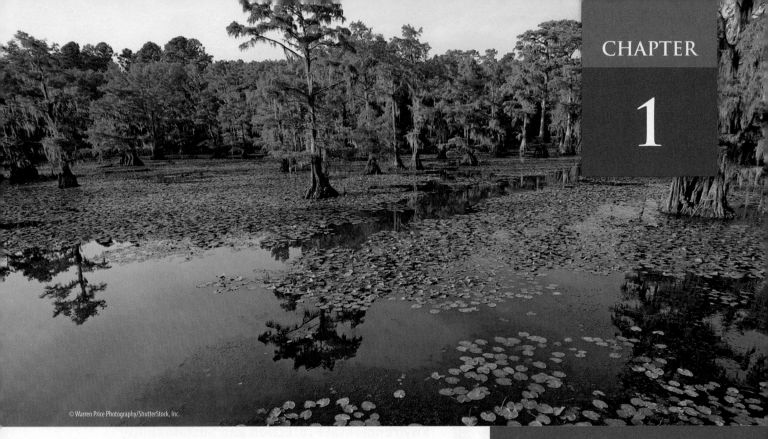
© Warren Price Photography/ShutterStock, Inc.

Environmental Science, Sustainability, and Critical Thinking

A problem well stated is a problem half solved.

—Charles Kettering

LEARNING OBJECTIVES

After reading and studying the material in this chapter, students will be able to understand, knowledgably discuss, and, when appropriate, apply the information learned in this chapter, including:

1. The encouraging signs like green plans and EPA Green Lights Program that human society is moving toward a more sustainable future
2. The negative trends that need to be corrected for human society to build a sustainable future
3. The terms *environmental science* and *sustainable development* and why environmental protection requires sustainable development
4. Scientific method and its components including experimentation and hypothesis formation and how we engage in this process almost every day of our lives
5. Hypotheses, theories, and paradigms
6. How science informs values
7. Rules of critical thinking

CRITICAL THINKING

Exercise

The term *critical thinking* has become the latest buzzword in colleges and universities. This term means different things to different people. Some college textbooks, for example, contain critical thinking questions designed to encourage students to think a little deeper—that is, to assimilate facts and concepts. Other books ask critical thinking questions that call on students to use their ability to reason. This calls on students to weigh evidence.

What do you think the term *critical thinking* might mean? Hint: You might begin by looking up the word *critical* in the dictionary. Ask friends and teachers. After you have completed your inquiry, write a definition, and then check the definition given in this chapter. What rule of critical thinking would you propose based on this exercise?

The author's analysis of this exercise is located at the end of the chapter.

I f you are like most people, you've probably heard a lot of conflicting information about the environment and environmental issues. Teachers, friends, news reporters, writers, and scientists, for example, may have warned you of serious problems. Others have offered an opposite view, saying things are not nearly as bad as some people would have you believe or that some of the problems don't exist. What is the truth?

In this book, you will find ample scientific evidence to suggest that the problems are real. In fact, a survey of U.S. environmental trends that I conducted covering 30 years showed that the vast majority (nearly 70%) of environmental indicators are pointing in the wrong direction. They indicate movement away from a healthy environment. Although there are some very encouraging signs of improvement—about 20% of the trends showed improving conditions—there is still cause for serious concern and action.

As you read this book, I invite you to explore the issues with an open mind, looking carefully at all aspects of each one. We'll examine the scientific evidence behind the issues and hear what the experts from all sides of the issues have to say. We'll explore traditional solutions and a new brand of responses that could help us find lasting solutions to pressing problems—solutions that are good for people, the

environment, *and* the economy. These are part of a new strategy known as *sustainability*. Let's begin with some of the encouraging trends.

 KEY CONCEPTS

Although there are some signs of improvement, the vast majority of the environmental trends show signs of movement away from sustainability.

1.1 Encouraging Signs/ Continuing Challenges

All across the world, change is underway. It's not ordinary change, either. It is profound change in the ways societies conduct their everyday affairs—ways that are leading to a new wave of environmental protection with lasting impacts. In the Netherlands (Holland), for example, industries banded together to produce a nationwide Green Plan. Drafted in conjunction with several key government agencies, the Green Plan calls on industries to reduce toxic emissions drastically and pollutants voluntarily (see **Spotlight on Sustainable Development 1-1**). This bold new plan will, if its goals are reached, practically eliminate hazardous wastes and a host of additional pollutants. To date, over 70% of the plan's goals have been met. The plan does not require regulations and fines, which governments typically use to get companies to improve their environmental performance. Rather, companies are left largely to their own devices to find economical ways to meet national goals. The government is there to lend a hand, providing financial resources and technical assistance. If companies do not meet goals, fines may be imposed. Because of Holland's success, many other nations, including Bhutan, New Zealand, Nepal, Mexico, and Singapore, have launched similar programs.

In the United States, impressive change is also in the works. For example, a number of programs have been established by the **U.S. Environmental Protection Agency (EPA)**, a federal agency charged with writing many environmental regulations and enforcing environmental laws passed by the Congress. One program, EPA's Green Lights program, was designed to reduce electrical consumption by businesses and government agencies. So far, thousands of the nation's largest corporations, including Boeing, and many governmental agencies have signed up to participate in this program (**FIGURE 1-1**).

Participating companies work with the EPA to cut electrical demand through lighting efficiency measures. Companies can cut their lighting electrical demand by 50% or more. Their economic return on investment is typically 30% or higher, making this an excellent investment. All efforts are voluntary. The EPA provides technical assistance.

Many businesses have also voluntarily taken on energy efficiency and renewable energy projects to lower

FIGURE 1-1 EPA green lights program.
Boeing Corporation, based in Seattle, Washington, joined the EPA Green Lights program (now part of the EPA's Energy Star Program) and saves over $12 million per year by reducing electrical energy consumption. Every eight years, this would buy one Boeing 757 jet.

© Doug Scott/age footstock.

© hugolacasse/ShutterStock, Inc.

1-1 The Netherlands' Green Plan Revolutionizes the Way Industries Function

The Netherlands is a tiny country, about one-fourth the size of the state of Illinois (**FIGURE 1**). It is, however, a giant among nations. Why this glowing praise for such a small country?

In 1989, the country adopted a National Environmental Policy Plan (NEPP) commonly called its Green Plan, after widespread public alarm over many environmental trends. An even tougher plan came into existence several years later. Known as the NEPP 2, it gave the Dutch government more power to write and enforce environmental laws and to earmark substantial sums of money to research and redesign industrial processes. Since then, it has been twice upgraded.

The most exciting aspect of the Green Plan is that it allows for a new approach—a set of voluntary agreements with industry and other sectors of society responsible for producing the bulk of the nation's pollution. Why was all of this necessary?

Although the Netherlands has some of the most stringent environmental laws in the world, they weren't working well. Because of this, the government took a bold step: It admitted that there are limitations to regulating complex and interdependent environmental problems on an issue-by-issue basis. In place of this unworkable system of rules and regulations, the government decided to meet with the key players (industries and citizen groups, for example) to reach agreement on establishing bold new targets and timetables for drastically reducing pollution. The government let the companies select ways to reduce pollution; all that it asked of industry was a commitment to meet ambitious government targets and timetables. Industry groups happily signed the agreements, recognizing

FIGURE 1 The Netherlands is leading the world in sustainable development.
© AbleStock.

that the responsibility of meeting the targets was theirs and that failure to meet the goals would result in stiff penalties. The Dutch, therefore, combined the carrot (incentive) and stick (penalty) approaches in hopes of restructuring industry for sustainability.

This approach was widely hailed by many major Dutch industrialists. They liked the agreement because it gave them free rein to find solutions that made the most sense to their businesses. The net effect is a kind of customized implementation that avoids the one-size-fits-all solutions typically employed in industrial nations' environmental control policies. Such approaches have met with considerable resistance in the United States because they often imposed prohibitive and sometimes unnecessary costs on businesses.

The voluntary approach was also popular because it helped companies become more efficient and more profitable. In addition, the Netherlands has become an exporter of technologies created by the country's industries to meet the goals of the nation's ambitious Green Plan. As other nations see the need for this shift, the Dutch pollution prevention and sustainable technology industry could profit enormously.

The Netherlands' Green Plan relies on four basic principles. First is life cycle management, which makes companies responsible for products throughout their life cycle—from production to consumption to disposal. Under this principle, producers are responsible for disposing of products after their useful life span has expired. This has spawned interest in manufacturing products that are easily recycled.

Another essential component of the strategy is energy efficiency. By using energy more efficiently, companies can save huge sums of money—and gain a competitive edge in the market by producing goods and services more cheaply than international competitors.

Sustainable technologies are also an important component of this plan. The government has launched a major program to develop sustainable technologies and make existing technologies more readily available to people and industries.

Finally, success also hinges on improving public awareness. The government launched a media blitz that echoes the important theme: "A better environment begins with you."

Today, over 250,000 Dutch businesses participate in the Green Plan, focusing on ways to conserve energy, produce more environmentally friendly products, and reduce pollution emissions. Seventy percent of the original goals have already been achieved according to the Resource Renewal Institute, which tracks green plans. (You can learn more about the Netherland's Green Plan on their website, www.rri.org/green-plans-netherlands.php.)

The successes of this ambitious program have been impressive: ozone-depleting substances have been phased out. Industry has reduced its waste disposal by 60%. Recycling has increased to 80%, and sulfur dioxide emissions have been reduced by 70%. In the 1980s, the Rhine river was so polluted the few remaining fish were too toxic to eat. Today, children fish and swim in the river thanks to the Netherlands Green Plan and international agreements that reduce pollutants released into the river. The country has been able to achieve economic growth while reducing its impact on the environment (except for CO_2 emissions).

Not all environmental problems facing this small country were the result of industry. Realizing this, the Dutch government has revised its plan to address the citizens and enlist their aid in reducing consumption. They're even working on land use and transportation issues with a greater emphasis on consumer responsibility. In addition, after 2010 sustainability has become a major criterion for all government purchases, which stimulates businesses to produce even more green products. All in all, this tiny, progressive country stands as a model for the rest of the world, proving that economic prosperity and environmental protection are compatible goals.

their carbon footprint, that is, to reduce the amount of greenhouse gas (carbon dioxide) they emit. Johnson and Johnson, for example, has invested $100 million in energy efficiency and renewable energy projects at their many facilities—and are realizing a substantial return on their investment of around 15%. Wal-Mart has pledged to makes its new stores even more energy efficient than their already energy-efficient stores. They've pledged to cut waste and make their transportation fleet substantially more efficient within 10 years, saving millions of dollars each year while helping reduce greenhouse gas emissions. Google and Microsoft have both recently installed huge solar electric arrays at their facilities on the West Coast. Even the buildings that replace the Twin Towers of the World Trade Center in Manhattan were built green—using environmentally friendly materials. These energy-efficient buildings and others around them will be powered entirely by renewable energy.

What motivates companies to use energy more efficiently? Many adopted these programs to become better environmental neighbors, but they also did it to save money, sometimes millions of dollars a year on energy.

What has brought about all of the changes discussed previously? Quite simply, these and a great many other changes underway have been brought about by a growing recognition that the current paths many nations are following are not sustainable. That is to say, many individuals have realized we can't go on as we have been without serious social, economic, and environmental repercussions. In my analysis of environmental trends spanning a period of 30 years, mentioned earlier, I found that nearly 70% of U.S. indicators, including energy consumption, population growth, waste production, and wildlife habitat, showed decline. About 20% of the trends showed some improvement. The rest of the conditions were neither improving nor deteriorating (**FIGURE 1-2**).

Worldwide, similar trends have caused concern among many political leaders, activists, citizens, scientists,

and business leaders. The world population continues to increase. Forests are being cut down. Species are disappearing, and global pollution is increasing. Although global statistics can become overwhelming, consider what will happen in the next 24 hours. During this period, nearly 230,000 people will be added to the world population. In the next 24 hours, a huge amount of tropical rain forest will be leveled to make room for farms, roads, dams, mines, and towns. The daily loss comes to more than 800 square kilometers (312 square miles)—that's equivalent to a swath 150 miles long and 2 miles wide. Unfortunately, very few of these trees will ever be replanted. In this same period, 330 km² square kilometers (127 square miles) of land will turn to desert, in large part because of poor farming practices, overgrazing, and a warming trend that may be caused by pollutants—largely from industrial societies—that trap heat in the Earth's atmosphere. The list goes on.

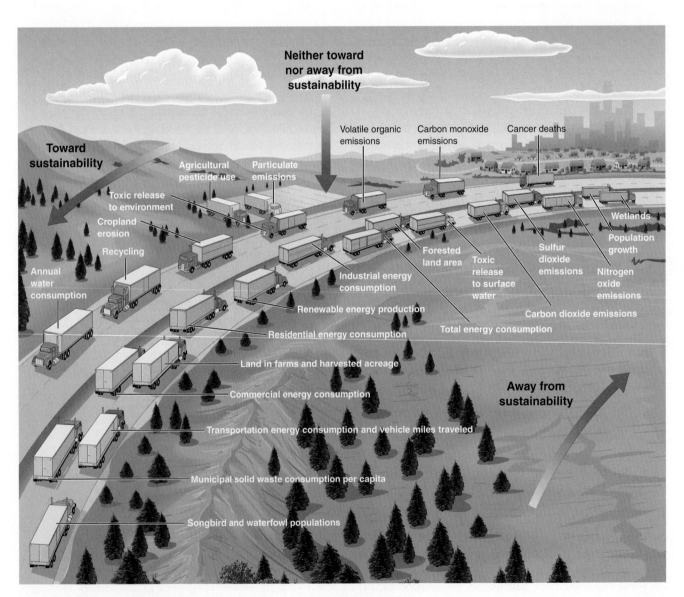

FIGURE 1-2 Environmental trends. Summary of 25 environmental trends studied by the author.

These and a host of other environmental trends presented throughout this book suggest to many observers that the current course of virtually all nations of the world is unsustainable. This is not to say that we are doomed, however. Stating we're on an unsustainable course is simply another way of asserting that things can't go on as they have been. To avoid problems—potentially serious problems—and to create a better future, experts contend that we will need to change our ways. Thankfully, many of the world's people are realizing this. You should not give up hope. There are many very favorable trends underway and new and promising developments. New technologies are also helping solve problems, as are individual actions (**FIGURE 1-3**). Another encouraging sign is that more and more people have come to realize that there are a number of fundamental driving forces—or root causes—that have created the present conditions. Two of the most important are continuing growth in both economic output and population.

Go Green

Use compact fluorescent light bulbs in your dorm room or apartment to save energy and reduce pollution. These bulbs use 75% less energy than a standard light bulb and last a lot longer.

FIGURE 1-3 Solar electricity. Photovoltaic (PV) modules produce electricity from sunlight. On these homes in Brazil, PVs provide electricity for two indoor compact fluorescent lights. Photo by Roger Taylor/NREL.

Encouragingly, many nations have responded swiftly to the population challenge. They have developed educational programs and other voluntary measures that are helping to slow the growth of their populations.

Many countries have begun to take steps to create economies that are better for the environment. Important agreements have been reached to curb global pollution,

SPOTLIGHT on Sustainable Development

1-2 The Philadelphia Eagles Go Green!

"The Philadelphia Eagles may not have a winning record on the field," writes NBC's Laura Marquez, "but they are setting the record in the NFL as the greenest team in football." How can a football team become a leader in green?

Like individuals, businesses, and governments, the Philadelphia Eagles football team has adopted a number of policies and practices to help ease their impact on the environment. For example, they now print their programs and tickets on recycled paper. They also require the use of plastic made from corn, a renewable resource, for the beer cups sold at home games. Plates used in the executive suites are made from plant products as well.

Even the lights in their stadium are green—powered by renewable energy. Overall, one-third of the energy the team consumes derives from renewable resources, mostly wind power. The team even subsidizes renewable energy purchases by employees.

The team minimizes the use of potentially toxic chemicals and fertilizes their playing field with organic fertilizer. They also recycle much of the mountain of waste left behind after each home game.

The team is also helping to educate its fans on the importance of things they can do to reduce their impact on the environment. They do this by sharing facts about their individual contributions to the problems and the solutions.

Although the Eagles are leading the charge, they're not alone. The St. Louis Rams are also taking steps to become a greener team. Their efforts have spurred interest in greening high school and college teams, not just in football, but in other sports as well. You can learn more about the Eagle's and Rams' efforts to go green by logging on to their websites.

protect forests, and protect wild species and in recent years, some nations have taken steps to combat global climate change. Although there's a lot more to do, the world seems to be moving in the right direction.

KEY CONCEPTS

There are many positive changes in perception and policy that have led to real progress in solving local, regional, and global environmental problems. However, the problems still outweigh the solutions.

1.2 What Is Environmental Science?

This book focuses on the environmental problems that we face—issues in three principal areas: pollution, natural resources, and population. However, it also focuses on solutions, that is, what individuals, businesses, and governments can do about our many environmental problems. The topics covered in this book fall within the domain of a relatively new science, known as *environmental science*. What is environmental science?

Environmental science is, in broadest terms, a branch of science that seeks to understand the many ways that we affect our environment and the many ways that we can address these issues. Environmental science draws on the research and expertise of specialists from numerous traditional sciences, including biology, ecology, chemistry, geology, engineering, and physics. The study of environmental issues and solutions also draws heavily on the humanities: economics, political science, anthropology, history, law, sociology, and even psychology. Thus, environmental science is a multidisciplinary field of endeavor.

To many observers, myself included, a full understanding of issues and solutions requires this broad, multidisciplinary approach. We won't solve the complex environmental issues that we face today through science and technology alone. We must understand and address the human element.

This book takes a multidisciplinary approach. It discusses the science behind the issues, technologies that will help us address problems, but also draws on a wide range of other disciplines to help you develop a deeper understanding of issues and solutions.

Environmental science, like engineering, is an applied science. Environmental scientists seek knowledge to expand our body of knowledge, but also to enable us to develop solutions to our problems. Those solutions may involve new technologies, but they may also involve new ways of thinking, new behaviors, and new laws and regulations. Today, numerous environmental scientists realize that many solutions proposed since the 1960s were stop-gap measures—that is, they solved the problems but only temporarily. This realization has led many to rethink previous ideas and propose more lasting, preventive measures—solutions that are good for people, the economy, and the environment. We call these sustainable solutions. In this book, I present traditional, stop-gap–type solutions and the newer set of sustainable solutions. As you shall see, these solutions are part of a new movement called **sustainable development**. You'll see the connection in the next chapter.

KEY CONCEPTS

Environmental science is a multidisciplinary endeavor that seeks to understand and solve a wide range of environmental problems. In recent years, environmental scientists have begun to propose more sustainable solutions that address the root causes of environmental problem.

1.3 Science and Scientific Method: Keys to Understanding Environmental Issues and Creating Lasting Solutions

The term *science* comes from the Latin word *scientia*, which means "to know or to discern." Technically, the term if refers to a body of knowledge derived from observation, measurement, and experimentation—and to the process used to gain such knowledge.

Throughout this book you will learn both fundamental principles of science and many of the scientific concepts and facts behind environmental issues. Some of the most important scientific principles come from the science of **ecology**, the study of ecosystems. An **ecosystem** is a system composed of living things (animals, plants, microorganisms) and the interrelated physical and chemical environment. To understand science, let's begin by seeing how scientific information is acquired.

KEY CONCEPTS

Science is a body of knowledge and a method of acquiring knowledge about the world around us.

Scientific Method: The Basis of Good Science

Scientific exploration generally occurs in an orderly fashion. As a rule, scientific discovery begins with observations and measurements (**FIGURE 1-4**). From them, scientists often formulate tentative explanations called **hypotheses**. Consider an example: Based on observations of wild rattlesnakes, scientists hypothesize that rattlesnakes shake their rattles to warn large animals to stay away. It is not an aggressive act, scientists have observed, but a

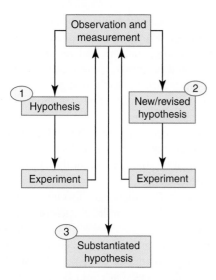

FIGURE 1-4 **The scientific method.** This drawing shows a simplified view of the scientific method, beginning with observation and measurement. In some cases, the original hypothesis may have to be revised many times.

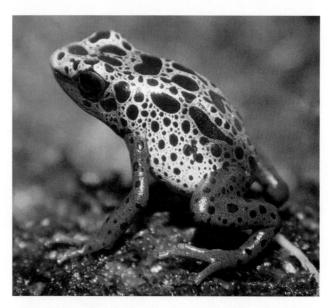

FIGURE 1-5 **Predator beware.** Many brightly colored reptiles and amphibians are poisonous, like this poison dart frog from the tropical rain forest. Predators learn to avoid them.

© John Arnold/ShutterStock, Inc.

protective measure that reduces the chances of a snake's being stepped on and killed. Hypotheses (generalizations) such as this one that are based on observation and measurement are derived by inductive reasoning. **Inductive reasoning** occurs any time a person uses facts to arrive at generalizations, that is, conclusions or hypotheses. This is very common in scientific study. Charles Darwin based his theory of natural selection on observations. He was engaged in inductive reasoning.

Inductive reasoning is the opposite of **deductive reasoning**, in which one arrives at a specific conclusion drawn from general principles. Deductive reasoning is common in mathematics, philosophy, politics, and ethics. General principles are used to examine and evaluate decisions. Consider an example: Scientists have found that brightly colored frogs of the tropics are generally poisonous, a fact established by scientific research (based on inductive reasoning). Brightly colored snakes are also often poisonous. Studies suggest that this is a kind of warning coloration. Wary predators know to avoid such colorful meals. If you were to encounter a brightly colored snake or frog, you might assume that it is poisonous (**FIGURE 1-5**). You would have used deductive reasoning to do this. You would have used a general principle to derive a conclusion about a specific organism.

KEY CONCEPTS

Scientific knowledge is often gained through a deliberate process of observation and measurement, which leads scientists to formulate hypotheses that can then be tested through experiments. This is a form of inductive reasoning. Deductive reasoning is the process of formulating a conclusion based on general principles.

Experimentation: Testing Hypotheses

After a hypothesis is made, the scientist must determine if it is valid. This is usually achieved by experimentation. Properly performed, experiments provide information that either supports or refutes hypotheses under question. As shown in Figure 1-4, if a hypothesis is refuted, a modified or new one is generally substituted. The new hypothesis must then be tested.

Scientific method, then, consists of observations and measurements that result in hypotheses. These are tested through experimentation and refined as needed. Surprisingly, most people use scientific method nearly every day of their lives. Suppose, for example, that every time you went to a friend's house you became ill. To determine what made you sick, you'd probably start by making a mental list of all of the activities you engaged in at your friend's house during your visits. After making the list and weeding out activities that seem irrelevant, you might conclude that your problems began when you petted your friend's cat. That's your hypothesis.

To test this hypothesis, you could perform an experiment. For instance, you might ask your friend to keep the cat outdoors when you were there. If you come home feeling sick again, you might alter your hypothesis. Maybe it's the new carpet in her apartment. Of course, scientific study is much more involved than this example, but the process is much the same.

Many scientific experiments are carried out on animals, plants, and other living organisms. Studies such as these are often designed to test the effects of a chemical substance—for example, the effect of a pollutant on an animal's lungs. In experiments of this nature, it is necessary to establish two distinct groups of organisms, an experimental group and a control group. The **experimental group** is the one that you treat or "experiment on." In a study of the effects of air pollution on laboratory mice, for example, the experimental group would be exposed to certain amounts of air pollution. The **control group** is treated identically, except that it is not exposed to the pollution. By setting up experiments in this manner, scientists can test the effect of one and only one variable, which is known as the **independent variable**. Any observed differences in the two groups should result from the treatment.

In experiments outside the laboratory, it is not always easy to isolate one variable. Scientists studying areas in which acidic air pollutants are being deposited in rain, for example, have found many biological effects, such as damage to crops and forests. However, other pollutants such as ozone may be responsible for some of the effects. Sorting out the specific pollutants involved can be quite complicated.

KEY CONCEPTS

Experiments enable scientists to test hypotheses and gain new knowledge, but experiments must be carried out very carefully. Good experiments often require control and experimental groups.

Studies on Human Health

Studying the effects of pollution on humans is much more difficult than analyzing the effects of chemicals on laboratory animals described previously. Why? First, ethics generally prohibit experimentation on humans such as those performed on laboratory rats or mice. Second, unlike laboratory mice, which are genetically identical, humans represent a genetically diverse group of organisms. Setting up control and experimental groups containing large numbers of genetically similar humans who have been exposed to the same conditions throughout life is impossible. Consequently, scientists often rely on **epidemiological** (EP-eh-DEEM-ee-ah-LO-jeh-cul) studies to study the effects of chemical pollutants on people.

Literally translated, **epidemiology** (EP-eh-DEEM-ee-OL-oh-gee) is the study of epidemics caused by infectious disease organisms. Today, however, epidemiology is a branch of science that also studies the health effects of environmental pollutants, food additives, radiation, and a variety of other agents. Epidemiological studies generally rely on statistical analysis of data on populations that have been exposed to various potentially harmful substances. Consider an example: Suppose a scientist wanted to know the cause of a kind of cancer in humans. If she were an epidemiologist, she would generally begin by searching through death certificates from hospitals to identify patients who had succumbed to the cancer in question. She would then interview their next of kin to learn about the deceased's occupation, personal habits, lifestyle, and possible exposures. She would then put all of the information into a computer and sort through it to see if

she could find any commonalities. For example, a large percentage of the people might have been employed in a chemical factory. The job is not finished yet, though. Before any conclusions could be drawn, the researcher would assemble a control group as identical to the first group as possible. The control group would consist of people of the same age and sex, except that these people would have died from causes other than the cancer under study. By analyzing their life histories and comparing the results with those from the cancer group, she might be able to pinpoint the cause of the cancer in the experimental group.

Epidemiological studies may also involve comparisons of living subjects. For example, a scientist might compare the health records of workers exposed to a certain pollutant in a factory to workers at the same company who are not exposed to the pollutant (for example, office workers). This may turn up differences that could be attributed to workplace exposure.

Science requires a lot of experimentation to validate hypotheses. No one study can be taken as the final word. However, by repeating experiments and achieving the same results each time, scientists can gain confidence in their hypotheses. (Such repetition may seem like overkill to some people, but it is essential to the scientific process.)

KEY CONCEPTS

Studies of the effects of chemicals on humans are often performed by comparing carefully selected groups of individuals exposed at work or at home to individuals of the same sex and age who were not exposed. These are known as epidemiological studies.

Scientific Theories and Paradigms

Scientific knowledge grows as facts accumulate from observation and experimentation. As researchers gain an understanding of how things work, they often formulate **theories**, explanations that account for many different facts, observations, and hypotheses. Theories are scientific principles, broad generalizations regarding natural, physical, and chemical phenomena. Unlike hypotheses, theories generally cannot be tested by single experiments because they encompass many different pieces of information. Atomic theory, for instance, explains the structure of the atom and fits numerous observations.

Science is not static, however. It constantly grows and evolves. Thus, scientists sometimes find that new interpretations emerge, replacing entrenched ideas. This sometimes forces scientists to alter or abandon well-established theories. Consequently, scientists must be open-minded about theories; they must be willing to replace their most cherished theories with new ones, as more information accumulates.

Perhaps the best known example of a changing theory is the **Copernican revolution** (coe-PURR-nick-in) in astronomy. The Greek astronomer Ptolemy (TOLL-eh-me) hypothesized in A.D. 140 that the Earth was at the center of the solar system. This is called the *geocentric view*. The moon, the stars, and the sun, he argued, all revolve around the Earth. In 1580, however, Nicolaus Copernicus (coe-PURR-nick-us) showed that the many observations of astronomers of the time were better explained by assuming that the sun was the center of the solar system. Copernicus was not the first to suggest this *heliocentric* view. Early Greek astronomers had proposed the idea, but it gained little attention until his time.

Over time, scientists have created cognitive frameworks within which theories, laws, and generalizations are formulated. These frameworks that underlie all scientific disciplines are called *paradigms*, a term popularized by the philosopher and science historian Thomas Kuhn. A **paradigm** is a basic model of reality in science. It is how scientists think about their discipline, for example, evolution is a paradigm of the biological sciences. All biologists interpret findings with evolution as a key organizing principle. It is a way of thinking about biological processes.

Paradigms govern the way scientists think, form theories, and interpret the results of experiments. They also govern the way nonscientists think. Once a paradigm is accepted, it is rarely questioned. New observations are generally interpreted according to the paradigm; those that are inconsistent with it are often ignored or disputed. In some cases, though, observations that fail to fit the paradigm may amass to a point at which they can no longer be ignored, causing scientists to rethink their most cherished beliefs and, sometimes, discard them. This is called a **paradigm shift**.

The term *paradigm* is commonly used today in a more general way, referring simply to the way we see the world. Many observers, for instance, speak about the **dominant social paradigm** of modern society when they refer to the set of beliefs that govern our lives. Our attitudes and behaviors often flow from our beliefs.

Paradigms also shape our language, thought, and perceptions. Slogans and common sayings reflect our paradigms and are repeated and reinforced day after day. Perhaps the most obvious expression of the dominant social paradigm is the growth-is-essential philosophy. You hear it in the business world every day when businesspeople proclaim, "A company must grow or die." Politicians echo a similar sentiment. Systems analyst Donella Meadows wrote in her book *The Global Citizen*, "Your paradigm is so intrinsic to your mental process that you are hardly aware of its existence, until you try to communicate with someone with a different paradigm."

Perhaps the most difficult problem is that people often get attached to their paradigms (to their ways of thinking) and fiercely resist any effort to change them. Because of the importance of our paradigms, success in building a sustainable future may require concerted efforts to overcome this resistance.

Although this description of science and scientific method shows that they require analytical skills and reasoning, they are also creative and fascinating endeavors. Scientific research may sound dull and boring to many people, but the truth is that, in many cases, it is extremely exciting. Far from being the left-brained, analytical types depicted in many movies and books, scientists are often highly creative and—believe it or not—intuitive people (**FIGURE 1-6**). They see hidden connections in the chaos of experimental data and make leaps of logic that, upon testing, often turn out to be true.

KEY CONCEPTS

As scientific knowledge accumulates, scientists are able to formulate theories that explain their observations. Overarching theories or paradigms emerge and profoundly shape scientific thought and our perceptions of reality.

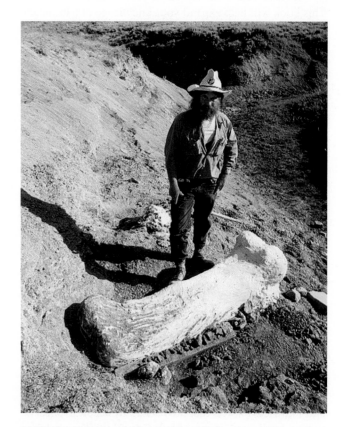

FIGURE 1-6 Science is fun for the mind! Robert Bakker, one of America's most creative scientists, is a man of great energy, enthusiasm, and creativity.
© Francois Gohier/Ardea London.

Science and Values

Science is all about facts and figures that explain our world. Scientists endeavor to remain objective, letting the data determine their hypotheses and theories. Ultimately, scientists hope to find objective truths—assertions about the world around us that can be backed up by data they've acquired through careful experimentation and observation. Because of this, most scientists argue that science is amoral—outside the sphere in which moral judgments are made; that is, it is not about right and wrong.

Scientists bring human qualities to their work, including their own values and preconceived notions that can taint their observations. Thus, scientists are not always as objective as one might hope. Today, scientists can be found to argue both sides of controversial issues. They may look at the same data and come away with markedly different conclusions because their biases often get in the way of objective analysis. Numerous Point/Counterpoints in this book illustrate this situation.

Just as human values affect scientific interpretation of data, science can also affect public values. A good understanding of ecological science, for instance, can help reshape human values concerning nature and our part in the natural world. In fact, your study of environmental science will very likely alter your value system.

Societal values can be based on scientific knowledge. This book will help expand your knowledge of natural and human systems, helping you see things more clearly. This may help you to rethink your values.

One value-shaping role of science rests in its ability to help us see hidden connections—for example, the role of forests in maintaining atmospheric carbon dioxide levels and global climate. This book shows the many ways that natural systems support human life. Knowing how important the planet is to our health, our well-being, even our economic prosperity cannot help but change our attitudes and, possibly, our actions.

KEY CONCEPTS

Scientists seek to be objective in their work so that science is fact based and value free. However, biases based on values can influence scientific analysis at times. Scientific knowledge can also affect human values.

 1.4 ## Critical Thinking Skills

One of the skills a good scientist (and many a great thinker) has is called *critical thinking*. **Critical thinking** is the capacity to distinguish between beliefs (what we think is true) and knowledge (facts that are backed by accurate observation and valid experimentation). Critical thinking helps us separate judgment from facts. It is the most ordered kind of thinking of which people are capable. Critical thinking involves subjecting facts and conclusions to careful analysis, looking for weaknesses in logic and other errors of reasoning. Critical thinking skills are essential in analyzing a wide range of environmental problems, issues, and information.

So how does one think critically? Clearly, there is no single formula. However, most critical thinkers would agree that several steps contribute to this process. These steps or principles of critical thinking, summarized in **TABLE 1-1**, will help you analyze arguments, research findings, issues, and statements you read.

KEY CONCEPTS

Critical thinking is an acquired skill that helps us analyze issues and distinguish between beliefs and accurate facts.

Gather All Information

The first requirement of critical thinking is to gather all of the information you can before you make a decision. In order to understand most issues, it is often necessary to

TABLE 1-1

Summary of Critical Thinking Rules

Gather all information.
- Dig deeper.
- Learn all you can before you decide.
- Don't mistake ignorance for perspective.

Understand all terms.
- Define all terms you use.
- Be sure you understand terms and concepts others use.

Question how information/facts were derived.
- Were they derived from scientific studies?
- Were the studies well conceived and carried out?
- Were there an adequate number of subjects?
- Was there a control group and an experimental group?
- Has the study been repeated successfully?
- Is the information anecdotal?

Question the source of the information.
- Is the source invested in the outcome of the issue?
- Is the source biased?
- Do underlying assumptions affect the viewpoint of the source?

Question the conclusions.
- Do the facts support the conclusion?
- Correlation does not necessarily mean causation.

Expect and tolerate uncertainty.
- Hard-and-fast answers aren't always possible.
- Learn to be comfortable with not knowing.

Examine the big picture.
- Study the whole system.
- Look for hidden causes and effects.
- Avoid simplistic thinking.
- Avoid dualistic thinking.

acquire additional information. When you do, you often find a slightly different picture of reality. This is especially true in environmental debates, in which advocates tend to simplify issues or to present information that supports their case while ignoring contradictory findings. Although you'll rarely have all the information you need to make a decision, you can always get more. So get in the habit of learning all you can before you make a decision. Be sure not to mistake ignorance for perspective.

When gathering facts, seek out statistics. Watch out for anecdotes—or stories—that may sound convincing but don't accurately represent the truth. Humans prefer stories to statistics and often reach erroneous conclusions as a result.

 KEY CONCEPTS

Critical thinking requires one to know as much information about an issue as possible before rendering an opinion or making a decision.

Understand All Terms

The second rule of critical thinking is that when analyzing any issue, solving any problem, or judging the accuracy of someone's statements, you must understand all terms. In some instances, you will find that the people presenting a case have inaccurate or incomplete understanding. A good example is the term *sustainability*, which is currently being used by people who have a poor understanding of it. They even use the word to construct oxymorons, such as *sustainable economic growth*. In a finite system, as you will soon see, growth cannot be sustained.

 KEY CONCEPTS

To think critically about an issue, one must understand the terms and concepts related to it.

Question the Methods

The third principle of critical thinking is to question the methods by which information has been acquired. Were the facts derived from experimentation? Was the experiment performed correctly? Did the experimenters use an adequate number of subjects? Did the experiment have a control group? Were the control and experimental groups treated identically except for the experimental variable?

 KEY CONCEPTS

Critical thinking requires that we know how information has been acquired and that we question the methods by which it was derived.

Question the Source

A fourth requirement of critical thinking is to question the source of "facts." Ask yourself. Who is the source? When business owners say that pollution from their factories isn't causing any harm, beware. Environmentalists are not immune to bias and are also fair game for your critical thinking skills.

Be on the lookout for bias and underlying assumptions, even in this text. Some of the most well-entrenched assumptions and myths are listed in **TABLE 1-2**. Study them and be aware of them. Also, look at the Individual Actions Count! table. It will help you examine your own values and assumptions for bias.

 KEY CONCEPTS

Critical thinking requires one to question the source of information, which often requires us to search for biases and assumptions that influence individuals.

Question the Conclusions

The fifth requirement of critical thinking is to question the conclusions derived from facts. Do the facts support the

TABLE 1-2
Sixteen Myths and Assumptions of Modern Society

1. People do not shape their future; it happens to them.
2. Individual actions don't count.
3. People care only about themselves and money; they can't be counted on to take action on the part of a good cause unless they'll gain.
4. Conservation is sacrifice.
5. For every problem there is only one solution; find it, correct the problem, and all will be well.
6. For every cause, there is one effect.
7. Technology can solve all problems.
8. Environmental protection is bad for the economy.
9. People are apart from nature.
10. The key to success is through the control of nature.
11. The natural world is here to serve our needs.
12. All growth is unqualifiably good.
13. Favorable economics justifies all actions; if it's economical it's all right.
14. The systems in place today were always here and will always remain.
15. Results can be measured by the amount of money spent on a problem.
16. Slowing the rate of environmental destruction and pollution solves the problems.

Adapted from Donella Meadows (1991). *The Global Citizen*. Washington, DC: Island Press; and Daniel D. Chiras (1990). *Beyond the Fray: Reshaping America's Environmental Response*. Boulder, CO: Johnson Books.

conclusions? Are other interpretations possible? One of the earliest studies on lung cancer, for instance, showed a correlation between lung cancer and sugar consumption. The researchers who performed the study found that people with the highest rate of lung cancer ate the most sugar. However, upon reexamination it turned out that cigarette smokers consumed more sugar than nonsmokers. The real cause of lung cancer was cigarette smoking.

This example illustrates a key scientific principle worth remembering: Correlation doesn't necessarily mean causation. A correlation is an apparent connection between two variables.

 KEY CONCEPTS

Critical thinking requires us to question the conclusions drawn from facts to see if other interpretations might be possible.

Tolerate Uncertainty

The sixth rule of critical thinking is to tolerate uncertainty. Although this may seem contradictory at first, it really isn't. As noted earlier, science is a dynamic process. Scientific knowledge is constantly evolving. What we know, however, is dwarfed by our ignorance. Although knowledge of poisons and hazardous materials is immense, in actuality very little is known about the toxic effects of chemicals on humans. How do they interact? Are impacts affected by diet?

Some current debates on environmental issues involve a fair amount of uncertainty. Hard-and-fast answers just aren't available yet. Sometimes we have to make decisions with limited data.

 KEY CONCEPTS

Our knowledge of the world around us is evolving, and, thus, it is necessary to accept uncertainty as an inevitable fact of life and make decisions with the best information possible.

Examine the Big Picture

The seventh rule of critical thinking is always to examine the big picture. Don't get trapped in simplistic thinking. Dualistic thinking is one of the most common forms of simplistic thinking. Dualistic thinking is reasoning that is black-and-white, right-or-wrong oriented. Dualistic thinking sees two options for every issue and nothing in between.

This book presents several conceptual models that will help you understand the big picture. These learning tools organize a great deal of information into a simple format.

To understand the big picture, we must often look at whole systems, both human and natural. For example, it is important to understand how value systems influence the systems (infrastructure) that serve our needs.

For example, agricultural systems in many countries are a reflection of the view that the key to success is through domination and control of nature. We level grasslands, plant acre after acre of the same crop, and then control the insects that arise to take advantage of this bonanza with an arsenal of potentially harmful pesticides (**FIGURE 1-7**).

One of the goals of this book is to help you become a better systems thinker. This is one of the most valuable skills architects of a sustainable future must possess.

 KEY CONCEPTS

To become a critical thinker it is necessary to examine the big picture—relationships and entire systems.

To be ignorant of one's ignorance is the malady of the ignorant.
—**Bronson Alcott**

FIGURE 1-7 The control of nature. A belief that humans are superior to other living things and that domination and control of natural systems are the key to success profoundly influences how we design human systems such as agriculture. Sustainability calls for systems to be ecologically intelligent.

Photo by Jeff Vanuga/NRCS USDA.

CRITICAL THINKING

Exercise Analysis

Critical thinking requires students to think carefully and objectively and to assimilate facts and concepts. At the heart of critical thinking, however, is something much more important. Your dictionary probably offered several definitions of the word critical. For example, it may have defined it as "finding fault," which is clearly not appropriate in this instance. The dictionary may have also defined critical as "being characterized by careful analysis."

Careful analysis requires that we examine issues very thoughtfully, that we study the facts that back up our opinions. It also means that we look for loopholes in people's logic. Critical thinking helps us separate beliefs that may have little factual basis from facts.

One of the rules of critical thinking illustrated by this exercise is that to be a critical thinker you must understand terms you encounter. Be sure a speaker explains the terms he or she is using. Don't be timid about asking for an explanation. Also, be sure you understand new words you're reading. Don't launch into an argument until you really understand what the other person is saying. One thing you might try when you're talking with someone is a process called active listening—repeating what you think someone is trying to say. This technique is very simple and is one way you can be sure you truly understand another person's point before you propose your own ideas.

CRITICAL THINKING AND CONCEPT REVIEW

1. Companies throughout the world are recognizing that good environmental practices are good for business. Make a list of reasons why this may be true.

2. In what ways can science help us solve problems sustainably? What are the limitations of science in problem solving?

3. Describe the scientific method. What are the steps, and why is each one necessary?

4. Describe the components of a good scientific experiment.

5. A graduate student interested in studying the effects of air pollutants on lichens (a fungus that grows on rocks) designed an experiment as follows: He proposed to bring rocks with lichens into the laboratory, where he would expose them to urban air pollution. (His university was located downtown in a major metropolitan area.) He proposed circulating outside air into the laboratory and measuring the effects. What is wrong with this experimental design? How would you design a better experiment?

6. What is the difference between a theory and a hypothesis? How is the term *theory* used incorrectly?

7. Numerous studies on the natural toxin known as aflatoxin, which is found in peanuts, show that it is mutagenic and carcinogenic; that is, it causes mutations in the genetic material and cancer in some organisms. Drawing on these studies, scientists have labeled aflatoxin as a potentially dangerous substance. What kind of reasoning is this, inductive or deductive?

8. "From the facts of the case," said Sherlock Holmes, famous investigator of fictional repute, "I can deduce that the butler did it." Is Holmes using the word *deduce* correctly in this context?

9. How can science affect our values? Give some examples from your life.

10. Make a list of the critical thinking skills outlined in this chapter. Describe and give an example of each one. Over the next week, make a list of all the times you use these rules.

11. Study the list of myths and assumptions in Table 1-2. Which ones do you agree with? Which ones do you disagree with?

KEY TERMS

control group
Copernican revolution
critical thinking
deductive reasoning
dominant social paradigm
ecology
ecosystem

epidemiological
epidemiology
experimental group
hypotheses
independent variable
inductive reasoning
paradigm

paradigm shift
sustainable development
theories
U.S. Environmental Protection
  Agency (EPA)

© Matej Hudovernik/Shutterstock, Inc.

Environmental Protection and Sustainability

There are no passengers on spaceship earth. We are all crew.

—Marshall McLuhan

 LEARNING OBJECTIVES

After reading and studying the material in this chapter, students will be able to understand, knowledgably discuss, and, when appropriate, apply the information learned in this chapter, including:

1. The history of environmental protection in the United States and the important changes that have occurred over time to create more comprehensive and lasting solutions
2. Ways we can meet our needs while protecting, even enhancing, the environment
3. Sustainable development and ways we can meet social, economic, and environmental needs simultaneously
4. The social, economic, and environmental principles of sustainable development
5. The important differences between growth and development
6. The human-made and natural systems that human society depends on and why the human-made systems are unsustainable
7. How the principles of ecological sustainability can be applied to human systems to promote lasting environmental solutions

17

CRITICAL THINKING

Exercise

In our society, people like to name and categorize things around them. In other words, we like to label everything. Of course, labeling has many benefits. It makes our lives easier to organize and talk about. However, labeling has its downfalls, too. One of those is seen when we categorize (label) people. Someone might be labeled an "environmentalist," "radical environmentalist," or "tree hugger." To be sure, environmentalists also label others. A corporate executive might be labeled an "opponent" or an "exploiter," with the clear implication that he greedily seeks to take resources for his own economic advantage.

What are the effects of labels? Does categorization (labeling) hinder debate? How can it be avoided?

Human civilization has experienced two major revolutions—changes in the ways people lived on the Earth. The first was the **Agricultural Revolution**, a long, progressive shift from hunting and gathering to subsistence-level farming then to a system of highly mechanized farming. The Agricultural Revolution started around 10,000 B.C. The second major revolution was the **Industrial Revolution**, the shift from manufacturing goods by hand to machine manufacturing powered by coal—and later by other fossil fuels such as oil and natural gas. This period started in the late 1700s in England and in the 1800s in the United States.

The successes of the Agricultural and Industrial Revolutions are many: economic prosperity, an impressive list of goods and services, and a better way of life, to name a few. The gains made during these periods have not been without serious environmental impacts, several of which threaten our long-term future. Today, people are recognizing that changes are needed to reverse these trends. We may, in fact, be entering a new cultural revolution, a **Sustainable Revolution**: a long period in which human systems that evolved during the previous 12,000 years are redesigned to create an enduring human presence—a sustainable future. This chapter describes the history of environmental protection and why environmental protection is essential to create a sustainable future. It also outlines the rationale for systems reform, gives an overview of the key systems that support

our lives, and ends with a discussion of ways to revamp systems for sustainability. As you will see, this chapter outlines the need for a sustainable approach to environmental protection—solutions that could result in fundamental changes in the ways we meet our needs for resources, deal with waste, and run our economies.

2.1 Environmental Protection and Sustainable Development

In this book, we'll examine dozens of important environmental issues and look at many ways we can protect the environment. We'll study numerous laws and pollution control devices, for example, and will look at innovative ways to protect forests and farmlands. We'll examine ways to help protect species from becoming extinct and ways to lower the impact of our fossil fuel-based economy. But unlike most other textbooks, this one takes a broader look. Rather than view environmental protection as an isolated set of actions, I view it as a part of every decision we make about our society and how it develops. In short, we will look at how we can meet human needs while protecting the environment. We call this process **sustainable development.** Clearly, environmental protection via sustainable measures is crucial to sustainable development.

KEY CONCEPTS

Increasingly, environmental protection is being incorporated more broadly into all human actions and into the process of development. Meeting our needs while protecting the environment is called *sustainable development*.

To understand the need for this broader approach and the importance of learning about sustainable development in a course on the environment, let's take a look at the evolution of environmental protection efforts in the United States and other developed nations.

The Evolution of Environmental Protection

FIGURE 2-1 shows the stages through which environmental protection has progressed in the United States and other industrial nations. As illustrated, environmental protection in the United States began in the early 1950s in a piecemeal fashion, as cities and towns began to address local environmental problems. In the early stages, most of the problems revolved around lakes and rivers that were polluted by human sewage or industrial waste. The response to these problems was largely technological. Most of the solutions involved the installation of **pollution control devices**, equipment that removed pollutants from the waste stream. However, local officials soon discovered that pollution control devices

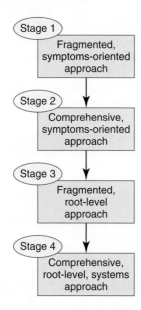

FIGURE 2-1 The evolution of environmental protection.
Environmental protection has evolved considerably over the past four decades. Now efforts are under way to incorporate it into sustainable development—that is, make it part of the design of new technologies, new homes, new vehicles, and so on. This effort could dramatically reshape basic human systems.

were inadequate. Why? Even though a factory or factories within its jurisdiction were required to clean up, pollution from sewage treatment plants or factories upstream rendered their efforts meaningless. It soon became clear that neighboring "upstream" cities and towns had to take action as well. However, they weren't always willing to enact the necessary regulations. Consequently, environmental protection efforts soon graduated to the state level. States began to enact laws requiring statewide compliance.

The first stage of environmental protection in the United States involved cities, towns, and later state governments. Important as these efforts were, they focused on a limited number of problems such as water and air pollution. Efforts were uncoordinated. Each municipality and state had its own regulations and ways of solving problems. The effort was also *symptoms oriented*; that is, solutions were primarily designed to treat symptoms of the problems (not underlying root causes). Pollution control devices, commonly called **end-of-pipe controls**, were placed on factories and power plants to remove air pollutants from smokestacks. But they, in turn, produced vast amounts of waste that had to be disposed of.

In time, the states found that their efforts were not enough. They failed to halt pollution entering from neighboring states either in the air or in rivers that crossed state boundaries. The federal government was then drawn into the fray, and Congress enacted numerous laws starting in the 1960s to ensure uniform environmental protection in all states. The federal government created programs to monitor pollution emissions and levels of pollutants in the air and water and to encourage compliance among businesses (including penalties). In the 1960s and 1970s, a fairly

comprehensive nationwide program of environmental protection emerged. Although it was not complete, it covered many problems.

In recent years, environmental protection has escalated to an international level, for the same reason that it moved from local to state to national governments: Pollution does not honor national boundaries. International protection efforts became essential as people began to realize that what happens to the environment in one part of the world may have serious impacts elsewhere. In the 1980s and 1990s, numerous multilateral and international agreements were enacted to address global problems.

The involvement of the federal government, followed by international accords involving many nations, constitutes the second stage of environmental protection. It was during this phase that the United States established uniform national standards and federal inducements to encourage states to undertake environmental protection. Over time, additional environmental issues were addressed.

Stage two represents a major improvement in environmental protection. It was much more comprehensive. More issues were addressed, and more players entered the scene. Although these changes represent a major step forward in the evolution of environmental protection, the main focus remained on treating pollution after the fact, with end-of-pipe controls. In other words, despite the fact that the scope of environmental protection had broadened considerably, the second stage remained largely a symptoms-level approach.

KEY CONCEPTS

Environmental protection has evolved from piecemeal local efforts to a much more comprehensive global strategy involving high levels of cooperation among states and nations covering a wide assortment of environmental problems. These efforts, however, primarily address symptoms through end-of-pipe controls, not the root causes of environmental problems.

The Next Generation of Environmental Protection Efforts

The second phase of environmental protection resulted in many gains. In recent years, however, people have come to realize that many environmental protection measures, although effective in the short term, are inadequate over the long haul. End-of-pipe controls have slowed the rate of environmental destruction, but they have not stopped it. In fact, after years of improvement many problems have begun to worsen. That's because reductions in pollution from automobiles, power plants, and factories are being offset by increases in population size and economic activity. They naturally result in more automobiles, more power plants, and more factories, and that, in turn, is resulting in an increase in pollution and environmental damage.

Recognizing the shortcomings of the environmental response, many nations have entered into the third stage of environmental protection. They are developing solutions that address what many environmental scientists believe are the root causes of the environmental crisis: population growth, inefficiency in resource use, heavy dependence on fossil fuels, and our "throw-away" mentality. In the United States, the Environmental Protection Agency launched bold experiments in environmental protection. The EPA's 33/50 program, for instance, enlisted major corporations in hazardous waste reduction. Participants slashed their output of hazardous wastes by 50% or more through fairly simple and cost-effective measures. The EPA's Green Lights program promotes corporate energy efficiency in lighting, and its Energy Star program promotes energy efficiency in more than 50 products, including computers, printers, monitors, and televisions. Similar programs have been instituted to promote more energy-efficient housing and appliances (**FIGURE 2-2**).

Other nations have launched similar programs, some much bolder than those under way in the United States. Clearly, then, a handful of nations are on the threshold

Go Green

When shopping for electronic devices like stereos, select an Energy Star–qualified model. They're the most energy efficient on the market.

of the third stage of environmental protection shown in Figure 2-1. By addressing the root causes, these and numerous other measures help prevent problems and create lasting solutions. Nevertheless, root-level efforts are in their infancy. We have a long way to go to be meaningfully engaged in the third stage of environmental protection.

KEY CONCEPTS

Although environmental policy and environmental protection efforts have evolved dramatically in the past six decades, most solutions dealt with symptoms. Efforts are now under way to address the root causes of the problems, which promise to create lasting solutions.

Creating Sustainable Solutions

Important as it is, entry into the third state of environmental protection probably won't ensure an enduring human future. To do so, we must move into a fourth stage (Figure 2-1). The fourth stage would be comprehensive in nature, involving actions at all levels: local, state, national, and international. Moreover, it would entail activities by all players—not just government, but businesses and citizens as well. Although it would address root causes, as in stage three, its emphasis would be on restructuring human systems—among them energy, industry, agriculture, transportation, housing, and waste management. These efforts could eliminate environmental problems that threaten our long-term social, economic, and environmental health. Why is this necessary?

The environmental problems facing the world are largely the result of population pressure. In this chapter, you will see that environmental problems are also caused by "flaws" in the systems that were designed to meet basic human needs. Systems of agriculture, industry, housing, transportation, energy production, waste management, and water supply were designed when the human population was small, resource supplies were vast, and our understanding of environmental issues was still growing. But things have changed. The human population has exceeded 7 billion, and the systems that have served us well are spawning problems of global impact. (You can watch at grow by logging onto www.census.gov/popclock.) In this book, I discuss ways that we can revamp these systems to reduce or eliminate environmental problems by making human systems sustainable.

Although ample evidence suggests that we are not on a sustainable course and that we are undermining our future, we are not doomed. We have the brainpower and know-how to invent machines and systems that are sustainable— that produce what we need without destroying the planet. The challenge will be immense, however. The transition to a sustainable society will require fundamental redesign and fundamental changes in the way we think, deep changes that may take a century or more to put into effect. We're just on the threshold of that change now.

The fourth stage involves integrating environmental protection into every decision we make. Environmental

FIGURE 2-2 EPA's Energy Star program. The U.S. EPA enrolled many manufacturers of computers, monitors, printers, televisions, and appliances in a voluntary program to design and manufacture energy-efficient products. The most efficient products come with an Energy Star label (lower left on control panel and inset).

Courtesy of Electrolux. Used with permission. Inset © Jones & Bartlett Learning.

considerations become a key part of the development path we choose. End-of-pipe controls are still important; they're a kind of first aid. Ultimately, however, pollution control devices must be replaced by strategies that meet our needs yet produce little or no pollution at all. Gas-powered automobiles, for example, may be replaced by mass transit and electric cars powered by electricity generated from wind and solar energy. This book discusses all solutions—traditional and new—showing the role each has to play in building a sustainable future.

Despite substantial progress in sustainable development, in recent years the United States has begun to back pedal on important environmental issues, refusing to support international efforts to curb global climate change, loosening air pollution standards, and weakening a host of other hard-won environmental protections. Hard economic times often spur efforts to undermine environmental protections. Doing so, however, will only make conditions worse and increase costs to society created by unresolved environmental issues.

KEY CONCEPTS

All environmental problems result from the fact that human systems such as energy production and agriculture are unsustainable. Making them sustainable will greatly reduce or eliminate costly environmental problems.

2.2 Meeting Human Needs While Protecting the Environment

In this section, we take a closer look at sustainable development and the principles that underlie this promising idea. Our focus will be on ways that environmental protection can be integrated into human activity or development.

What Is Sustainable Development?

Sustainable development is a term rich with meaning. Although dozens of definitions have been proposed over the past few years, one of the most widely used was published in *Our Common Future*, a book written by the World Commission on Environment and Development (WCED). The WCED was established by the United Nations in the 1980s to examine the relationship between the environment and development and to suggest ways to make the two compatible. The WCED defined *sustainable development* as "development that meets the needs of the present without compromising the ability of future generations to meet their needs." Because the environment is the source of all our resources and our economic well-being, meeting our needs while ensuring future generations the ability to meet their needs means we must protect the environment.

To understand sustainable development, we begin by dissecting the term into its parts, beginning with the word *sustainable*. According to *Webster's New World Dictionary*, sustainable means "able to be sustained." To sustain means "to keep in existence, to maintain, and endure." According to Webster's, *development* is "a step or a stage in advancement or improvement." In short, *sustainable development is a way of improving or advancing our culture in a way that can be maintained over the long haul.*

Then Vice President Al Gore noted at a national conference in 1993 that the term *sustainable development* implies that "there's something called *unsustainable development.*" He and many others would agree that our current development path fits that bill. You will see evidence of this throughout this book.

Sustainable development, although something of a newcomer on the political scene, is not a new idea by any stretch of the imagination. In fact, more than 100 years ago, U.S. President Theodore Roosevelt alluded to the concept in his annual message to Congress, a speech that has since become known as the State of the Union address. Roosevelt noted, "To waste our natural resources, to skin and exhaust the land instead of using it (so) as to increase its usefulness, will [undermine the prospects] of our children." He went on to talk about "the very prosperity which we ought by right to hand down to [future generations] amplified and developed." In other words, we ought to meet our needs while improving environmental conditions for future generations. Roosevelt was not the first to hold such a view, either. Native American cultures and indigenous peoples the world over have espoused a similar view and lived accordingly for thousands of years before Roosevelt's time.

Sustainable development, then, refers to improvements or advancements in human well-being that are enduring. Advocates of sustainable development concern themselves with strategies designed to meet all human needs—not just the need for a clean, healthy environment, but also needs for respectable work, good pay, recreation, peace, freedom from harm, and a host of other factors. Individuals clearly differ in the goals they set for human development, but several goals seem to be almost universal. These include a long and healthy life, good education, and a decent standard of living. The list also includes political freedom and a guarantee of human rights. Freedom from violence and meeting survival needs for food, shelter, water, and clothing are also basic goals of human development. The challenge today is to meet these and other needs while protecting—even enhancing—the quality of the environment. A good example would be agricultural methods that provide healthy food while enhancing the fertility of the soil.

Sustainable development strategies seek ways to forge lasting relationships between humans and the environment, ones that protect and restore ecosystems rather than destroy and deplete them. Environmental protection is, therefore, vital to the economic success of any state, local, and national development plans. Sustainable development also seeks ways to create enduring relationships among people (for example, cooperative rather than adversarial) and a new economic system that is kind to the Earth and to all individuals, all genders, and all races. These measures are as important as sound environmental policies.

FIGURE 2-3 is a concept map that encapsulates much of the thinking on the social, economic, and environmental conditions essential to creating a humane, sustainable society. Take a few moments to study it. We won't deal with all the social aspects of sustainable development in this book, but concentrate primarily on economics and the environment. It is important to remember, though, that creating a sustainable future requires much more than efforts to protect the environment.

KEY CONCEPTS

Sustainable development is a means of meeting present needs in ways that do not impair future generations—and other species—from meeting their needs. Because the environment is essential to satisfying the needs of present and future generations, environmental protection and improvement are keys to its success.

Satisfying the Triple Bottom Line

For many years, social, economic, and environmental policies have been devised in isolation. Economic decisions were made with little regard for their environmental impact. Environmental policies were made with little regard for economics. Sustainable development requires an integration of social, economic, and environmental goals. As I like to tell my students, all policy is environmental; that is, to be successful, all social and economic policy must be crafted in a way that ensures environmental protection. By the same token, environmental policy must make sense from social and economic perspectives. **FIGURE 2-4** illustrates this important concept. The three circles on the left side of the figure show three realms—social, economic, and environmental. In traditional government policymaking, solutions in each realm have long been devised without regard to the others. In contrast, sustainable development calls for solutions that make sense from all three perspectives simultaneously, as shown on the right side of Figure 2-4. The area of intersection in

SUSTAINABLE DEVELOPMENT: CREATING A NEW RELATIONSHIP

Sustainable development calls for policies and actions that foster enduring relationships among people and between people and the planet

that result in the creation of

Sustainable communities, states, and nations

which require

Social, economic, and environmental conditions conducive to harmony and long-term survival

including

Social Conditions
- Adequate food, shelter, and other basic needs, creating a high quality of life
- Freedom from oppression
- Freedom from physical harm
- Democratically based decision making
- Participation
- Cooperation
- Intergenerational and intragenerational equity (environmental justice)

Environmental Conditions
- Waste output that does not exceed assimilative capacity of environment
- Use of renewable resources that does not exceed Earth's capacity to regenerate or renew them
- Use of nonrenewable resources that does not exceed the rate of their replacement by renewable substitutes
- Clean air and water
- Biodiversity

Economic Conditions
- Broad prosperity
- Greater local/regional self-reliance
- Ecologically sound economics
- Well-paying jobs for all who want them
- Healthy, stable economies

FIGURE 2-3 Conditions created by sustainable development. Sustainable development is a process that seeks to promote conditions essential for an enduring human presence. This will require changes in environmental as well as social and economic aspects of our lives. Some people think that profound changes will have to occur in all three realms to create a sustainable society.

| Traditional decision making | Decision making in a sustainable society |

FIGURE 2-4 Sustainable solutions. For many years, social, economic, and environmental issues were addressed independently. Solutions made in one sector often had adverse impacts in the others. Sustainable development calls for solutions that make sense from all three perspectives simultaneously.

this drawing represents a new way of thinking and making policy. However, sustainable solutions are not compromises that only partially satisfy human needs for a safe, healthy environment and social and economic conditions conducive to a long, healthy life. Rather, sustainable solutions strike out in new directions to achieve the social, economic, and environmental conditions vital to a sustainable future. By satisfying the triple bottom line (that is, economic, environmental, and social goals), sustainable development could conceivably reshape human civilization. Numerous examples of sustainable solutions will be presented in this book.

Sustainable development is having a profound effect on people throughout the world. It is a common theme of conferences and workshops. It has also been adopted by numerous political entities, agencies, and organizations as a central focus. Included in this rather impressive list are the United Nations General Assembly, the U.S. Agency for International Development, the United Nations Development Program, and the World Bank. Environmentalists—and a growing number of business leaders and economists—have embraced the notion and have begun to explore and implement policies and actions needed to create a much more enduring human way of life. How do we create a sustainable society?

> **KEY CONCEPTS**
>
> Sustainable development requires strategies that satisfy social, economic, and environmental goals simultaneously.

Principles of Sustainable Development

Sustainable development is based on 10 key principles, which derive from three areas: ecology, social thought and ethics, and economics. In this section, we examine each one very briefly and reflect on ways they will help ensure a clean, healthy environment, a sustainable society, and ultimately a better future.

Ecological Principles of Sustainable Development

Four ecological principles are central to the concept of sustainable development: dependence, biophysical limits, living within the carrying capacity of the environment, and interdependence.

Principle 1: Dependence Many people think of the environment as something only birdwatchers and hikers care about, something that is of little worldly interest to the average person. Nothing could be further from the truth. Humans depend on the Earth and its ecosystems for a wide array of goods and services. These goods and services are vital to our personal well-being and also contribute significantly to the global economy. In the words of the economist Robert Goodland, the environment is the source of all material inputs, feeding the economic system, and it is the sink for all its wastes. In fact, so important is the environment that many economists refer to the Earth's natural systems (ecosystems) as the **biological infrastructure**—or **infra-infrastructure**—of modern society because the Earth and the natural systems it contains make possible all of the **physical infrastructure** of our society—the buildings, dams, ports, highways, and so on (**FIGURE 2-5**). The Earth is the source of all wealth, too. It provides us with these goods and services free of charge. However, even though these services are free, they are infinitely costly to replace.

If one recognizes human dependence on the Earth and on ecosystems, it becomes evident that environmental protection and environmentally sustainable development are forms of self-protection—an insurance policy for long-term human prosperity. In short, planet care is the ultimate form of self-care.

> **KEY CONCEPTS**
>
> Humans depend on the environment for countless goods and services that are essential to our lives and our economy.

Principle 2: Biophysical Limits The Earth and its ecosystems provide numerous vital resources. Some resources such as soil and forests are **renewable**—capable of being regenerated. Others such as iron ore are **nonrenewable**—not able to be regenerated. Both types of resources have limits. For example, there are limits to the amount of oil, a nonrenewable resource, that we can extract, and there are limits to the number of fish, a renewable resource, that can be caught without depleting their population.

Go Green

Although consuming cold-water fish like salmon is good for your health, choose wisely. Many fish populations are being overfished. Obtain a list of fish to avoid and fish to eat from the Marine Conservation Society at www.fishonline.org. It will help you identify which fish are from well-managed sources and/or caught using methods that minimize damage to marine wildlife and habitats.

BIOLOGICAL INFRASTRUCTURE
(Infra-infrastructure)

HUMAN INFRASTRUCTURE

Wood and paper

Iron, aluminum, and copper

Water

Food

Influx of materials into the city

FIGURE 2-5 Infrastructure and infra-infrastructure. The Earth and the environment provide fuel for the economy and provide the materials needed to create our infrastructure—our highways, buildings, water systems, and so on. The Earth and the environment are the infra-infrastructure of our society. Damaging and depleting the environment detract from our possibilities in the long run. The main goal of sustainable development is to create a society and an infrastructure, a set of systems, that protects and even enhances the infrastructure.

A growing body of evidence shows that humankind has transcended certain biological and physical limits and stands perilously close to others. An example of our transgression of limits: Overfishing in the Atlantic Ocean has already eliminated more than 25 commercially important fisheries (**FIGURE 2-6**).

As evidence that we stand on the brink of transcending other limits: According to a 2012 report from the United Nation's Food and Agriculture Organization, 57% of the world's marine fish stocks are fully exploited—that is, they are at or very close to their maximum sustainable production limit—that is up from 52% in their 2005 report. Twenty-nine percent are overexploited—up from 16% in the 2005 report. One percent is recovering from depletion.

In addition to these examples of biological limits, a growing body of scientific evidence indicates that chemical limits are being exceeded by pollution generated by human society. In both the industrial and nonindustrial nations of the world, many polluted rivers no longer support fish or support a fraction of their original populations. The buildup of carbon dioxide (a pollutant produced by the combustion

of oil, coal, and natural gas) in the atmosphere and the changes it is causing to the climate of planet Earth are other examples. The partial destruction of the ozone layer by chlorofluorocarbons is another example of limits that modern society has transcended.

KEY CONCEPTS

The renewable and nonrenewable resources that support our lives are limited, as is the Earth's ability to assimilate and detoxify wastes produced by human society. Many signs that we have exceeded limits of the planet's resources and pollution assimilation capacity are evident.

Principle 3: Living Within Limits Although there are limits to resources and the waste assimilation capacity of the Earth, we can live and prosper. To do so, we must address rapid population growth and redesign systems so that they operate within limits. Living within limits requires us to learn to live within the carrying capacity of the environment.

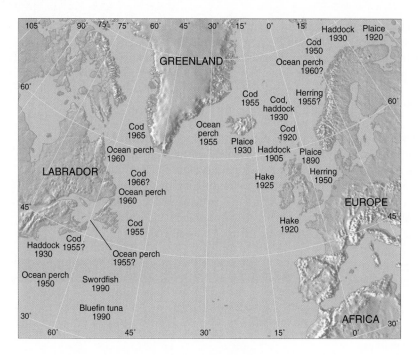

FIGURE 2-6 **Ocean in peril.** This map of North Atlantic fisheries shows the dates when major fisheries were depleted and thus no longer commercially fishable.

Carrying capacity is defined as the number of organisms an environment can support indefinitely and is determined by two elements: resource supplies and the capacity of the environment to absorb, assimilate, and/or detoxify wastes. Proponents of sustainable development call for policies and practices that honor the very real limits in the Earth's carrying capacity—limits to the amount of arable land, fish in the sea, and oil in the Earth's crust, for example. You'll learn more about this daunting task in upcoming chapters.

 KEY CONCEPTS

Living sustainably means finding ways of prospering within limits of resources and the waste assimilation capacity of the Earth's ecosystems.

Principle 4: Interdependence The fourth ecological principle is interdependence. *Humans and the Earth exist in a precarious interdependence.* Clearly, just as we humans are dependent on the environment, the fate of the global environment is dependent on us. Our influence has grown so great—and continues to grow with incredible speed—that we humans have become custodians of an entire planet. What befalls the Earth befalls humanity.

 KEY CONCEPTS

We depend mightily on the Earth and its ecosystems, but the Earth depends on our actions.

Social/Ethical Principles of Sustainable Development

Living sustainably on the Earth is a biological challenge faced by all species. As noted in the previous section, living sustainably in the environment requires species to thrive within the limits of the resource supplies and waste assimilation capacity of the Earth—the source and sink functions. Although the challenge of living sustainably on the planet is basically the same for all species, for humans, the task is complicated by a variety of factors, among them ethics, economics, and politics. This section describes three key social or ethical principles: intergenerational equity, intragenerational equity, and ecological justice. Keep them in mind as you consider various solutions to environmental problems.

Principle 5: Intergenerational Equity One of the most important principles of sustainable development is **intergenerational equity**—fairness to future generations. The doctrine of intergenerational equity calls on us to live in a way that does not rob future generations opportunities to prosper. By managing resources well, for example, we ensure that there are ample supplies for future generations. Adhering to a sense of fairness—practicing intergenerational equity—could profoundly influence the technologies we develop and the ways human systems are designed and operated. As a simple example: Honoring intergenerational equity might dictate houses powered by sunlight and wind and made from many recycled materials because it helps to ensure future generations the resources they will need to build homes.

 KEY CONCEPTS

Intergenerational equity calls on us to conduct all human activities in ways that honor the needs of future generations, ensuring them the resources and conditions they need to life well.

Principle 6: Intragenerational Equity In the international arena, debates over global environmental protection have drawn attention to another important ethical principle, the notion of **intragenerational equity**. The doctrine of intragenerational equity calls on us to act in ways that satisfy our needs while safeguarding the welfare of all who are alive today. What we do is tempered by a concern for potential impacts on other people—our neighbors, our fellow citizens, and global citizens as well.

Intragenerational equity is based on a realization that all members of the world community share the same air and water—and that abuse of these resources is often felt far from the site of impact. Soil erosion on a Kansas farm, for instance, may lead to siltation in a downstream reservoir in Arkansas and flooding along the banks of the Mississippi River. Impacts of human actions can even extend globally. Ozone depletion caused by the release of CFCs (chlorofluorocarbons) from old refrigerators in Canada, for instance, may lead to skin cancer in Andean peasants who have never even seen a refrigerator.

The lines of impact are reciprocal. That is to say, the developed nations can be profoundly distressed by resource abuse in the developing nations. Deforestation of tropical rain forests in Africa to accommodate population growth, for example, could have a profound impact on the climate of European nations by altering the movement of moist air masses that affect precipitation patterns hundreds of miles northward.

KEY CONCEPTS

Intragenerational equity calls on us to act in ways that honor the rights and needs of all people alive today.

Principle 7: Ecological Justice A small but growing number of proponents of sustainable development support the idea of **ecological justice**, the notion that the Earth is the rightful property of all species—not just humans. According to this view, humans have an obligation not just to other people, but to all living things, present and future.

KEY CONCEPTS

The notion of ecological justice says that all species have a right to a clean environment and adequate resources—and current generations have a responsibility to ensure their rights.

Political Principles of Sustainable Development

Numerous principles of a political nature underpin the current thinking on sustainable development; three of the most important are the need for widespread participation, the need for global cooperation, and the need to address the root causes of problems.

Principle 8: Participation A sustainable future will require much more than new laws and regulations. It will also require the efforts of businesses and broad citizen support and action. How can citizens were you help? Citizens are an integral part of the economic system as consumers of goods and services. By becoming conscientious consumers—buying green (environmentally responsible) products, but also buying less—individuals can help stimulate the transition to sustainability. Citizens can take hundreds of additional actions to promote a sustainable future. Driving fuel-efficient vehicles, carpooling, riding buses, bicycling, or walking—all make significant contributions. Recycling is very important, too. In sum, the transition to a sustainable society requires the participation of all of us. As the opening quote of this chapter reminds us, "There are no passengers on spaceship earth. We are all crew."

KEY CONCEPTS

Building a sustainable society will require participation by governments, businesses, and individuals like you.

Principle 9: Cooperation Ensuring an enduring human presence will require a level of cooperation unwitnessed in human history. Regional agreements among nations, such as the multilateral treaties designed to clean up the Baltic Sea, are examples of the kind of cooperation needed to create a sustainable global human presence. International accords signed in the past decade, such as the global warming agreement and the ozone treaty, are additional examples of cooperation. Improvements in these and other treaties, as well as new agreements, are needed for further progress. Cooperation is also needed among others, including nonprofit organizations, governments, citizens, businesses, and even opposing political parties.

KEY CONCEPTS

Environmental protection and sustainable development will require cooperation of all participants.

Principle 10: Addressing the Root Causes One emerging realization in business and government is that many environmental solutions enacted over the past six decades have addressed the symptoms of environmental problems while overlooking fundamental root causes. Treating symptoms may help in the short term, but it won't solve the problems. A patient who is under enormous stress might be relieved by medicine, but if the patient doesn't learn to reduce stress, he or she may never get better. Treating symptoms is a trait or tendency found in virtually all areas of human endeavor from medicine to crime fighting to educational reform. In the arena of environmental protection, the examples of our short-sighted approach are many: smokestack scrubbers, catalytic converters on automobiles, hazardous waste incinerators, sewage treatment plants, and so on. Over the years, as noted earlier in this chapter, it has become clear to many observers that these and other measures are inadequate for several reasons. Most importantly, they address problems such as pollution and hazardous waste too late—after they are produced, at the end of the pipe. A pollutant attacked at the point of origin can be eliminated, but after it is produced, it is too late: It becomes a social, economic, and environmental liability. Unfortunately, most nations have based their environmental protection strategy on policies that attack pollutants well after they've been generated.

This conventional approach to environmental protection—treating the symptoms—has several additional shortcomings. One is that it results in short-term gains that may be offset by other forces—among them continuing population growth, increasing economic activity, and increasing per capita consumption. This is especially evident in the United States and other Western nations. Take for example, the catalytic converter, a device used on automobiles to reduce the emissions of certain pollutants (notably, hydrocarbons and carbon monoxide) spewing from tail pipes. Catalytic converters work well and are responsible for a dramatic decline in the pollution our cars produce. They're even responsible for a general cleansing of the air in many urban centers. However, since their introduction, continuing expansion of the U.S. population, dramatic increases in the number of vehicle miles traveled per year especially in expanding urban areas, and the popularity of large, energy-inefficient vehicles (notably SUVs) are starting to overwhelm the gains that resulted from the use of catalytic converters (**FIGURE 2-7**).

Another problem is that the symptoms approach has resulted in a shifting of pollutants from one medium to another. **Smokestack scrubbers**, a type of pollution control device that captures sulfur oxide gases from coal-fired power plants, convert air pollutants into a toxic sludge. This waste is typically dumped into landfills, where toxic substances contained in it may leak into groundwater.

Other solutions have been much too narrow—that is, they address only a small portion of the problem. For example, although catalytic converters reduce hydrocarbon and carbon monoxide emissions, they do nothing to reduce emissions of carbon dioxide—a pollutant that scientific evidence shows is causing the Earth's atmosphere to heat up.

Fortunately, a growing number of policymakers are beginning to realize that lasting solutions require policies and practices that confront the root causes of the problems we face. Even prominent business leaders recognize the importance of root-level solutions such as energy efficiency as a way to comply with or even exceed environmental performance standards, and many recognize that such an approach is not detrimental to business but can be highly profitable. For a quick review of the principles discussed in this section, see **TABLE 2-1**.

KEY CONCEPTS

To create a sustainable society, we must focus on strategies that address the root causes of environmental problems.

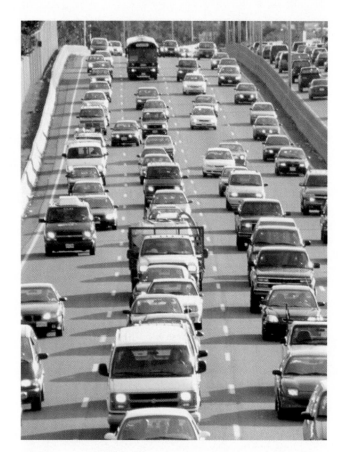

FIGURE 2-7 Rush hour? Although substantial gains have been made in reducing pollution from individual automobiles, the growth in the automobile fleet and the dramatic annual increase in the number of miles we travel threaten to overwhelm progress resulting from increased vehicle efficiency and pollution control devices.
© AbleStock.

Growth and Development: Understanding the Differences

One of the key issues that must be addressed to protect the environment in the coming years is growth—economic growth and population growth. Can we reach a sustainable state if population and the economy continue to grow? There are two schools of thought with widely differing views on the issue of economic growth: progrowth and antigrowth.

We will address economic growth here. The antigrowth school questions policies that promote continued, unlimited economic growth. Their "growth principle" might read something like this: Continued economic growth within a finite system is unsustainable. To create a sustainable human presence, economies must eventually reach a stable state. Growth—expansion of the economy—must come to an end. Rather than continued growth, we should focus on ways to improve the quality of our lives to reduce continued resource extraction and pollution.

TABLE 2-1

Ecological Principles

Dependence—Humans are dependent on a clean, healthy environment for many goods and services vital to our personal and economic well-being.

Biophysical Limits—Very real biophysical limits exist in the planet's ability to supply resources and absorb wastes from human civilization.

Living within the Carrying Capacity of the Environment—Living sustainably on the Earth will require steps to live within the Earth's biophysical limits.

Interdependence—The fate of the environment is in our hands. What we do or don't do will profoundly affect the natural environment in ways that could have serious consequences for virtually all living things, including ourselves.

Social/Ethical Principles

Intergenerational Equity—Present generations have an obligation to meet their needs in ways that do not foreclose on future generations.

Intragenerational Equity—Present generations also have an obligation to act in ways that do not prevent or impair others who are alive today from meeting their needs.

Ecological Justice—Human actions should not endanger other species, which also have an inherent right to the resources they need to survive.

Political Principles

Participation—Building a sustainable future will require unprecedented participation from all sectors of society.

Cooperation—Creating a sustainable society will require cooperation among many different participants.

Addressing the Root Causes—Successful solutions will require efforts that address the root causes of the problems.

Making buildings throughout the world more energy efficient is one example. It helps reduce our demand for energy and all the impacts that come with it, while saving money and making us more comfortable. They refer to this as *development*.

In contrast, the proponents of growth believe that continued economic growth is absolutely essential to building a sustainable future. Continued economic growth will, they say, promote an improvement in the standard of living needed to eliminate poverty and slow the growth of the human population, especially in the developing nations. How else can we improve the economic well-being of the world's people if we cannot continue to grow economically?

Before we explore these diametrically opposed views, a few definitions are in order. First, **economic growth** is defined as an expansion of human economic activity. One barometer of economic growth is the consumption of resources such as energy. Growth strategies typically result in an expansion of infrastructure—additional buildings, power plants, roads, water projects, and so on. This expansion, in turn, requires more mineral extraction and energy consumption. It results in the additional harvest of natural resources (forests, for example) and additional pollution, all of which can contribute to economic growth and environmental degradation.

Advocates of continued economic growth view it as an essential means of raising the standard of living of the world's vast and growing population of the poor. They see continued economic growth as a desirable path for the wealthy nations of the world, as well as for the less fortunate developing nations. The rationale is that economic growth in countries such as the United States will benefit all the world's people. Economic growth in the developing nations would benefit many people within those nations.

Development, as used in this book, implies a qualitative improvement. It is measured by a betterment in both the quality of our lives and the condition of the environment. Development involves measures that ensure adequate housing and work space. It is aimed at providing more efficient transportation, sufficient energy, and education. It often includes steps to improve cultural opportunities, health care, and freedom from physical danger. Sustainable development seeks to achieve these goals while safeguarding the environment. Development is consequently not a simple numerical definition, but a complex quality-of-life issue.

Antigrowth or, more appropriately, development advocates recognize the importance of reducing poverty but argue that continued economic growth in nations such as the United States will cause further unsustainable deterioration of the environment and won't help the people of the developing nations very much, if at all. Economic expansion in the developing nations is essential, as long as the benefits are truly distributed among the population.

The belief in the benefits of economic growth enjoys such widespread acceptance that it is difficult for many to consider alternative economic strategies. This book presents some of these alternatives to stimulate your thinking. Nevertheless, the controversy over growth is bound to endure for decades. In the meantime, a compromise position seems to be emerging: Most people agree that economic growth, which is inevitable, should maximize the quality of people's lives but minimize—or, better yet, avoid—undesirable environmental and social impacts. This strategy calls on us to develop and deploy the least environmentally disruptive technologies and systems designs possible. You will have to develop your own opinions about growth. It is not an issue that is easy to resolve. I encourage you to gather facts and give this thoughtful consideration as you read the rest of this book and ponder environmental problems and solutions.

KEY CONCEPTS

Growth and development are fundamentally different goals. Growth results in an increase in material production and consumption that may be unsustainable in the long run; development is a strategy that calls for improvements in culture that do not necessarily require further increases in resource consumption, pollution, and environmental destruction.

2.3 Human Settlements: Networks of (Unsustainable) Systems

Your life and mine are supported by a handful of systems that operate behind the scenes to provide us with a mind-boggling range of goods and services. A few moments of reflection on these systems may prove helpful in understanding just how dependent we are on them. Let's begin with your breakfast. If you ate at home, you probably cooked your breakfast on a stove or in a microwave oven supplied with electricity from a power plant. The power plant, in turn, is probably fueled by coal from a coal mine. The transmission lines that supply your energy, the power plants, and the mines are all part of an extensive **energy supply system**. Where did the food come from? Most likely from a grocery store. Before that, though, the food was processed and packed in a food processing plant, possibly from corn or wheat grown in an Iowa or Kansas farm field. All of these components—the farms, food processing facilities, and grocery stores—are part of an elaborate **food production system**.

Food production, however, is just one of many interdependent systems that provide for us. As illustrated in **FIGURE 2-8**, we are served by an extensive waste management system, consisting of landfills, recycling facilities, incinerators, compost facilities, and trucks that move waste from its source to its destinations. All countries have extensive systems for providing water and housing. Of course, there are also **manufacturing systems** for producing goods and a **transportation system** for conveying people and goods.

Gluing all of this together are an **economic system** and a **system of government**. Increasingly, an informational system has emerged to convey information on the way the other systems are working. An **educational system** provides the necessary training (at least, we hope) for people to function in this complex web of systems.

Supporting these human systems is the Earth and environment, the **ecological system**. The Earth and environment are really the infra-infrastructure of society, for they provide the goods and services that make all of human society possible. The ecological system also serves as a sink for our wastes.

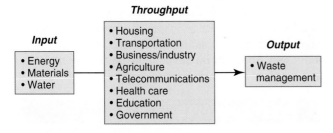

FIGURE 2-8 Human systems. Our lives depend on a variety of systems, such as waste management, transportation, energy, and agriculture. Some systems provide inputs needed to make others run. All systems produce waste, handled by our waste management system.

Human settlements are made up of many interdependent systems. They usually function so well that we don't have to think about them. In fact, most of us take the systems for granted. This is to be expected, for they have been designed to make our lives easier so that we can go about whatever we do unburdened by such concerns as growing our own food or repairing the roads so that we can get back and forth to work.

The invisibility of the systems is a benefit to most of us, but it also has a downside. Because the systems are invisible, it is difficult for many people to think in terms of systems. It is also difficult for some people to understand the central assertion of this book—that the fundamental problem facing the world's people is that the systems that support our lives are unsustainable. You may be having some trouble with this idea. After all, every time you flick on a light switch, there's electricity. Every time you go to the grocery store, there's an abundance of food on the shelves. Every time you turn on a faucet, there's clean, safe water to drink or wash with. How can someone assert that these systems are unsustainable?

KEY CONCEPTS

Human settlements are made up of many interdependent systems, such as transportation, energy, and waste management. Growing evidence shows that these systems as designed and operated are not sustainable.

2.4 Why Are Human Systems Unsustainable?

Just because a system seems to be reliably supplying goods and services does not mean it is sustainable. On the surface, many systems seem to be performing well because they provide us with what we need and at a relatively decent price. It is what's taking place behind the scenes—what's happening to the environment, for instance—that makes them unsustainable. Consider the U.S. transportation system.

People and goods are transported via an extensive network of flight paths, highways, railways, and navigable waterways covering many millions of miles. They are transported in all sorts of vehicles: jets, planes, cars, buses, trucks, motorcycles, bicycles, trains, and boats. Together, the components of the transportation system represent an investment of trillions of dollars. A brand new Boeing 757 costs more than $100 million. The transportation industry employs millions of people.

The transportation system has made our lives infinitely easier. Those who can afford it can jet around the globe in the time it used to take a person on horseback to ride a few hundred miles. Unfortunately, this system destroys open space and wildlife habitats. It produces millions of tons of harmful pollutants that are imperiling human health and the health of the entire planet. Each year, for example, the combustion of gasoline and other liquid fuels in the global transportation system produces billions of tons of the gaseous pollutant known as carbon dioxide (CO_2). Much of this enormous quantity of carbon dioxide cannot be absorbed by the oceans and plant life. Because carbon dioxide captures heat like a pane of glass in a greenhouse, its accumulation in the atmosphere is causing the Earth's atmosphere to heat up. This heating could accelerate in years to come, as many scientists predict, causing a serious threat to all life on the planet. This warming of the Earth's atmosphere, referred to as **global warming**, could have devastating effects on human populations.

Supplying energy to the transportation system has also resulted in significant pollution of the world's oceans with oil. Oil drilling in pristine areas is having many adverse effects on the local environment and local species. Supplies are also limited. In the long run, the Earth's oil reserves are bound to run out, again making this resource and the system that depends on oil unsustainable. Many examples of the unsustainable impacts of our systems are provided in this book.

A study of human systems suggests that they are unsustainable for three basic reasons: (1) They produce levels of pollution that exceed the local, regional, and even global capacity to absorb and render them harmless; (2) they deplete finite nonrenewable resources; and (3) they use renewable resources such as forests faster than they can naturally regenerate. In short, human systems exceed the carrying capacity of the planet to supply resources and deal with our wastes. Ultimately, they erode the planet's ability to support life. Signs of these three problems can be seen in cities and towns and nations across the planet.

(a)

(b)

KEY CONCEPTS

Although human systems may provide us with a steady stream of goods and services, they are systematically reducing the carrying capacity of the planet. Human systems (1) produce pollution in excess of the planet's ability to absorb and detoxify wastes, (2) deplete nonrenewable resources, and (3) use renewable resources faster than they can be regenerated.

The Challenges Ahead

This book's main theme is that current systems are unsustainable and that to live sustainably on the planet we must revamp the systems that support our lives. The enormous challenge of making human systems sustainable can be distilled into two basic challenges. The first part of the challenge is to modify or retrofit what already exists—for example, making existing homes and buildings more efficient. The second challenge occurs when building anew. As opportunities present themselves to design new components of the systems we've discussed, we must create new, sustainable components that incorporate the best materials, technologies, and designs to minimize resource demand and impact. New houses provide an opportunity to create structures with a fraction of the impact on the environment of existing homes. They could, for example, be made from rapidly renewable and abundant materials such as straw bales (**FIGURE 2-9A**) or recycled waste materials such as used automobile tires (Figure 2-9B). Some builders are creating energy-efficient homes that generate as

FIGURE 2-9 Alternative housing. (a) Straw bale house and **(b)** packed tire home made from discarded automobile tires and dirt from the site, described in Spotlight on Sustainable Development 2-1.

Courtesy of Dan Chiras.

much energy as they consume. The author's new home on his farm and educational center is a good example (**FIGURE 2-10**). Instead of building new highways or expanding existing highways, we can lay down rail lines on highway medians to move much larger numbers of people to and from their work with minimum environmental impact. Although this requires the use of resources and results in the production of pollution, it is far more desirable than the alternative. The goal here is to achieve ways to change or grow that produce sustainable impacts and ensure a strong, vibrant economy and a better future. See **Spotlight on Sustainable Development 2-1** for an example of sustainable housing.

KEY CONCEPTS

To create a sustainable future, we must revamp the systems that support our lives. The challenge is twofold: to retrofit existing systems to make them as sustainable as possible and to build new systems using the principles of sustainable design.

FIGURE 2-10 Energy-efficient building. The author's home creates as much energy as it consumes via solar and wind.

Courtesy of Dan Chiras.

2.5 Applying the Principles of Sustainable Development

The principles of sustainable development outlined in this chapter and summarized in Table 2-1 provide a philosophical foundation for sustainable development, a strategy vital to solve the many environmental problems facing the world today. Think of them as directive or guiding principles. Although they provide some general direction, they do not suggest specific actions needed to create a sustainable future. It is one thing to say that we should recognize that the world has limits or promote intergenerational equity, for instance, but it is quite another to figure out how we actualize these goals. The question that arises then is, how do we put ideals into effect? In other words, how do we *operationalize* the guiding principles?

Operating Principles

This section offers some suggestions for living well while protecting, even enhancing, the environment—that is, making sustainable development happen. Students and instructors are encouraged to add their own ideas to this list of seven operating principles. To be sustainable:

1. We must **stabilize our population**. Our population cannot continue to grow indefinitely.
2. We must **better manage how we grow**. In other words, we cannot spread willy-nilly across the landscape, usurping prime agricultural land needed to feed people and provide other valuable services such as recreation.
3. We must **use resources much more efficiently**. All resources, from fossil fuels to building materials to drinking water, must be used more efficiently. Resource efficiency provides a great economic benefit to businesses and clearly helps societies ensure adequate supplies for future generations.
4. We must turn to **clean, renewable energy supplies**, such as solar and wind energy. They provide the fuel we need at a fraction of the environmental impact.

5. We must **recycle** everything we can. Waste is intolerable in a world of limited resources, and recycling can make good economic sense. It also helps to ensure a supply of resources for future generations. Not only must we recycle, we must **manufacture a large portion of our goods with recycled materials**.
6. We must **restore natural systems** that have been damaged in years past. Moreover, we cannot continue to exploit the land, deplete natural systems, and then move on, as we have through much of history. To support people and the other living things that share this planet with us, we must replant forests, rebuild wetlands, and restore farmlands, grasslands, and pastures.
7. We must **manage resources sustainably**. That is to say, we must do a much better job of managing forests, grasslands, and other natural systems. Short-term profit at the expense of long-term productivity only leads us further from the path of sustainability.

KEY CONCEPTS

Creating a society that is sustainable requires a number of steps: (1) population stabilization, (2) growth management, (3) efficiency, (4) renewable energy use, (5) recycling, (6) restoration, and (7) sustainable resource management.

Applying the Operating Principles to Human Systems

FIGURE 2-11 shows a simple yet effective foundational tool I developed to help students apply the operating principles of sustainable development to the human systems to create a clean, healthy environment and to help humankind prosper in the long term. This simple tool can help citizens, business owners, and government officials concentrate their efforts and optimize limited budgets.

Figure 2-11 lists the systems (for example, energy) in the first column. The top row lists the operating principles of sustainability (such as conservation). Each box in this table can be filled with ideas to put that principle into action within a given system. Dozens of ideas could be included. The ideas could include policies and practices that operationalize the principles of sustainability outlined earlier.

The initial work with the table should be considered brainstorming and thus should not be subject to critical thinking, criticism, or prioritization. After a full set of ideas has been generated, the next step is to think critically about them. Discard those ideas that don't make sense and choose the highest priority items—those that are most effective and affordable. Remember that they should make sense from social, economic, and environmental perspectives—a fundamental prerequisite of sustainable development.

© hugolacasse/ShutterStock, Inc.

2-1 Treading Lightly on the Earth: Creating Sustainable Homes for the 21st Century and Beyond

The home featured in **FIGURE 1** is my home. It uses only a fraction of the natural resources of a typical home, yet it offers the amenities most of us have grown accustomed to. It's part of a new wave of *sustainable housing*. Why?

In designing and building the house, I tried to incorporate four of the operational principles of sustainable development discussed in Section 2.5: conservation (efficiency), recycling, renewable resource use, and restoration. As you read through this description, try to pinpoint the specific principle featured in each description.

First of all, this home generated all of the electricity needed by me, my two sons, and an in-home office—via a small wind generator and solar modules, mounted on the roof. This system and superefficient appliances and lighting made it unnecessary for me to hook up to the electrical grid, saving a thousand dollars or so in hookup costs and a thousand dollars or more a year in electrical bills.

The home is also an extremely efficient user of electricity. Every appliance and electronic device in the house operates on a fraction of the electricity required in most conventional appliances and consumer electronics. The lighting system, for instance, requires only about a quarter of the electrical energy of a typical house. The television uses half of the energy of most TVs. The refrigerator uses about one-tenth of the energy of a typical fridge at the time I purchased it.

This house is dependent on the sun for wintertime heating. South-facing windows permit ample sunlight to enter the house in the fall, winter, and early spring (**FIGURE 2**). The tile floors and thick interior mass walls of the house store the solar heat and radiate it back evenly, keeping the house at a fairly constant temperature during the winter. A superinsulated ceiling also helps hold the heat in during cold winter nights. Most of the north side of the house is sheltered by an earth berm, which also helps maintain a constant temperature. The roof over the bedrooms and my office is completely underground. It's planted in native vegetation and is called a living roof. Native vegetation on the roof helps it blend in with the surrounding meadow and helped us return more of the site to nature.

During the summer, the house is naturally cooled by the Earth. Earth sheltering helps keep the house naturally cool, like most basements. But it isn't a dark, dank basement. The house is very well lighted by south-facing windows (Figure 2).

Unlike most homes in the region, this one has no well. Rainwater and snowmelt are captured by the roof, stored in a 5,000-gallon underground tank, and pumped into the house as needed. Drinking water is purified by a filtration system. Gray water—water from the washing machine—is isolated from black water (water from the

FIGURE 1 The author's sustainable home.

Courtesy of Dan Chiras.

FIGURE 2 Solar heating. Sunlight streams through the windows, warming the house even on the coldest winter days.

Courtesy of Dan Chiras.

toilets) and used to supply an extensive set of planters in the house and an outside garden.

Further adding to its sustainability, the home is built largely of recycled products, including discarded automobile tires. Yes, 800 discarded tires. As shown in **FIGURE 3**, the back walls and many of the interior walls of this house are made of automobile tires, packed with dirt and piled on top of one another like bricks. After the tire walls were erected, we covered them with stucco and earthen plasters, creating attractive thermal mass walls that absorb heat from sunlight and radiate it back into the rooms at night. By using tires, I greatly reduced the amount of timber needed to build this house. In fact, by our estimates we used about 50 to 60% less lumber than a typical house of this size.

The house is insulated with 100% recycled cellulose insulation made from old newspaper. The roof is made of steel that can be recycled when its useful life span is over—approximately 50 years. The floor tile is made from recycled feldspar mine waste, and the carpeting is made from recycled plastic pop bottles (PET plastic). The carpet pad is made from cloth scraps from a clothing factory. The cabinets and vanities are made from

old barn wood from a 150-year-old barn that was slated for demolition (**FIGURE 4**). The wood was refinished and stained. The paint on the walls is 50% reclaimed paint—paint that was discarded by homeowners during the county's monthly hazardous-waste roundup. When repainting, we used nontoxic paints.

So, you ask, how much did it cost? Surprisingly, this house cost just a little less than most other homes being built in the area—about $95 per square foot. New homes were going for $100 to $110 per square foot in this area at the time this house was built. Interestingly, virtually all of the recycled building materials were competitively priced. Tires came to us free of charge. The slightly lower price resulted from the fact that I acted as the general contractor and did a lot of the work myself.

Living sustainably also requires personal actions that all of us can engage in. My boys and I, for example, recycled virtually all of our household waste. We also composted all organic material in an outdoor compost facility. Because of these actions, we produced only about one garbage bag of trash every month. We used cloth bags when we went shopping and purchased recycled products (for example, toilet paper). We were careful in how we used water and energy. The superefficient showerhead I installed use one gallon per minute, compared to a normal shower head which might use 5 to 10 gallons per minute. The faucets in the bathrooms have been retrofitted with 0.7-gallon-per-minute aerators.

For the 14 years I lived in the house, I worked at home, so I didn't have to commute. When I had to travel to meetings or other events and to access

(a)

(b)

FIGURE 3 Tire walls. Interior walls are made of tires packed with dirt. Stucco covers the tires, creating an aesthetic look. (a) Close-up and (b) a view of the entire home.
Photos a and b courtesy of Dan Chiras.

FIGURE 4 Barn wood cabinets. These cabinets are made of wood salvaged from a 150-year-old barn. The wood is remilled, stained, and treated with a water-based (nontoxic) finish.
Courtesy of Dan Chiras.

skiing or hiking, I drove an energy-efficient Toyota hybrid Prius Gen II, which gets around 50 miles per gallon year round. I offset carbon dioxide emissions from our wood stove, the car, travel, and other activities by purchasing renewable energy credits. They help fund wind farms owned by companies owned by Native Americans. We ate low on the food chain. We're not vegetarians but had a lot of meatless meals.

I lived in the house for nearly 15 years and found that it worked very well. We installed insulated window shades. We ran out of water periodically, so we installed a more efficient washing machine, which helped us cut down on demand. We never ran out of water again. Most people aren't able to build a sustainable home from scratch. Many things can be done to create a more sustainable living space, though.

Next, you can devise an action plan. The plan may include a timetable, a list of key players (for example, business leaders and activists), and resources. It might include education programs through schools, businesses, and government. It might also contain some indicators to plot progress in achieving goals. The action plan might list anticipated obstacles and address them.

As an example, consider the box in Figure 2-11 that relates the conservation principle to housing/buildings. Remember, however, that sustainability is a twofold challenge: One goal is to revamp existing buildings; the second is to make sure that all new buildings use resources efficiently.

The necessary policies and actions that a town might include to make this happen could involve changes in local building codes to improve insulation in new homes and remodeling projects, energy-efficient lighting designs, and educational programs for builders.

The systems approach outlined here is a way to streamline sustainable development process and prevent environmental problems. It gives people a framework to organize and address complicated issues in a systematic fashion. It can be used by cities, towns, state governments, and even national governments to meet needs while protecting the environment and saving money. One important

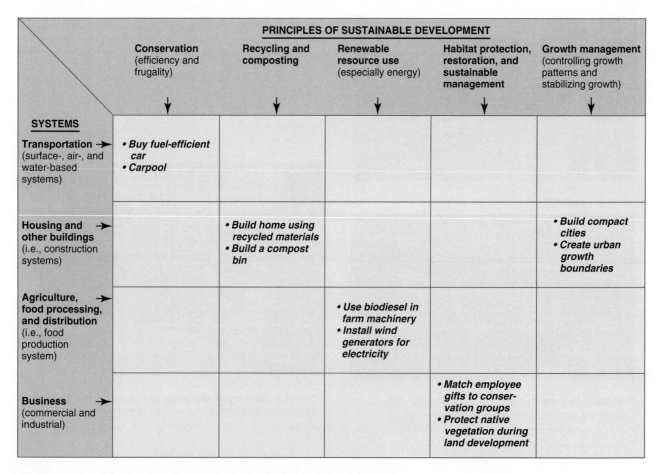

FIGURE 2-11 Matrix showing the systems and principles of sustainability.

piece that's missing from the discussion is how to address connections among systems, which are often interdependent. Energy, for instance, is a component of all other systems. It is as vital to the transportation sector as it is to industry, housing, and food production. Thus, decisions in one system often affect others, so certain systems might be best addressed in conjunction with others. For example, creating a sustainable transportation system cannot be done without discussing housing (at least the placement of subdivisions) and business development (notably the location of new businesses). To streamline the process, people involved in drafting sustainable development plans are encouraged to group interrelated systems and tackle them together.

You may have noted that this approach omits some familiar environmental issues—for example, loss of open space and air quality. The reason for this is quite simple, although sometimes difficult to grasp: Air pollution and loss of open space are *symptoms* of unsustainable systems (**FIGURE 2-12**). Air pollution is best addressed by creating sustainable systems of transportation, housing, and industry—for the same reason that symptoms of a disease are best addressed by finding out what is fundamentally wrong and treating the person for that. As an example, severe neckaches and eyestrain in a computer operator are best addressed by changing the position he or she sits in while at the computer, not by giving the person a larger dose of pain reliever.

Air pollution might be sustainably addressed in a number of ways, for example, by encouraging alternative forms of transportation such as buses and carpools, creating energy-efficient houses and factories, using renewable energy resources, and emphasizing more compact forms of development so that people have shorter commutes.

Sustainability won't occur until cities and towns, states, nations, and businesses get serious about restructuring human systems. It won't occur until we recognize that the problems many communities face are symptoms of deeper problems—unsustainable systems that erode the life-support systems of the planet. But keep reading this book and decide for yourself.

Sustainable development will require root-level solutions. This concept can be applied to all problems—social, economic, and environmental. Unfortunately, most political leaders and citizens don't give much thought to the systems that undergird their communities and the ways these systems could be restructured to create a lasting future. Also, many people may find the assertion that human systems are unsustainable difficult to grasp, much less believe.

To help overcome these barriers, we need leaders who can convince others that a sustainable approach is required and that it is not just a matter of doing more of the same. Leaders need not be government officials—citizens, nonprofit organizations, and progressive-minded business owners all can play a role.

To cherish what remains of the Earth and foster its renewal is our only legitimate hope of survival.

—Wendell Berry

FIGURE 2-12 It's the systems! Air pollution, waste, and species extinction are symptoms of overpopulation and unsustainable human systems. Redesigning these systems could help reduce many of the symptoms of the environmental crisis and put us back on a sustainable course.

CRITICAL THINKING

Exercise Analysis

Labels are valuable to human beings. They help us communicate about the world around us. They also help us create an identity. ("I'm a scientific writer, a father, and a musician.") Labels can be a form of thought stopper, though. A thought-stopping label hinders expression and thought in one of several ways. Labels can put people into categories that may or may not be accurate. To be introduced as an "avid environmentalist" at a dinner held by a nonprofit environmental organization, for instance, might have a very different effect than, say, being introduced similarly at a Chamber of Commerce luncheon. A label can be a barrier to communication. You might have a perfectly reasonable viewpoint, but many people in the Chamber audience might not listen.

Labels also simplify things and may mischaracterize the person who is being labeled. To say that a person is an "environmentalist" is sometimes taken to mean that he or she cares only about the environment. To say that someone is a "corporate type" may wrongly imply that the person cares only about business. Few of us have such a narrow realm of interest.

The lesson in all this is to be careful about the labels you use and the way you interpret the labels others use. Assume that people are more than their labels may imply. Labels may create inaccurate pictures. It is better to ask someone what he or she believes.

CRITICAL THINKING AND CONCEPT REVIEW

1. Define the term *sustainable development*. Why is it vital to environmental protection? Make a list of practices that you think would contribute positively to sustainable development efforts.

2. Describe the evolution of environmental protection in the United States. What forces have caused environmental protection efforts to change over the past 60 years? What changes are needed to make environmental protection more effective? Why is sustainable development essential to environmental protection?

3. List the 10 principles of sustainable development outlined in this chapter and briefly describe each one. Do you agree with all of these principles? Why or why not?

4. What is meant by an end-of-pipe control? Give reasons why you think such an approach may have been favored in the past. Why have such controls proved to be inadequate?

5. Summarize the views of the progrowth and antigrowth advocates. Why is growth so important to the issue of environmental protection?

6. Do you believe that economies and populations can continue to grow indefinitely? Why or why not?

7. Make a list of five of the major activities you were involved in today—for example, riding the bus to school, eating meals, and so on. Then make a list of the major systems that made these activities possible—for example, transportation system, food production system, and communication system. Describe how human systems are interdependent.

8. This chapter argues that human systems are unsustainable. Why? Cite evidence. Can you think of other reasons?

9. What is meant by the statement that the invisibility of systems makes it difficult for people to understand that they are unsustainable? How does this invisibility contribute to our natural inclination to focus on symptoms of problems rather than on the systems themselves? Why is this a problem?

10. Restructuring human systems (physical infrastructure and social and economic systems) will require two broad types of actions outlined in this chapter. What are they? Give some examples of each.

11. This chapter presents a technique to help people apply the principles of sustainable development to systems reform. Draw a diagram of the matrix and explain how it is used. What are the advantages of this approach? What are the disadvantages?

12. Why is it necessary to understand and address the root causes of unsustainability to build a sustainable society?

13. List the principles of sustainable development and describe each one. Then give an example of a policy or practice that would operationalize each principle in a system of your choice—for example, housing or waste management.

14. Using the matrix presented in this chapter (Figure 2-11), take a system such as energy production and apply the various principles to it. Come up with as many ideas as you can. Pick out your best ideas. What criteria helped you determine which ideas were better than others?

KEY TERMS

Agricultural Revolution
biological infrastructure
biophysical limits
carrying capacity
dependence
development
ecological justice
ecological system
economic growth
economic system
educational system

end-of-pipe controls
energy supply system
food production system
generational equity
global warming
Industrial Revolution
infra-infrastructure
interdependence
intergenerational equity
intragenerational equity
living within limits

manufacturing systems
nonrenewable
physical infrastructure
pollution control devices
renewable
smokestack scrubbers
sustainable development
Sustainable Revolution
system of government
transportation system

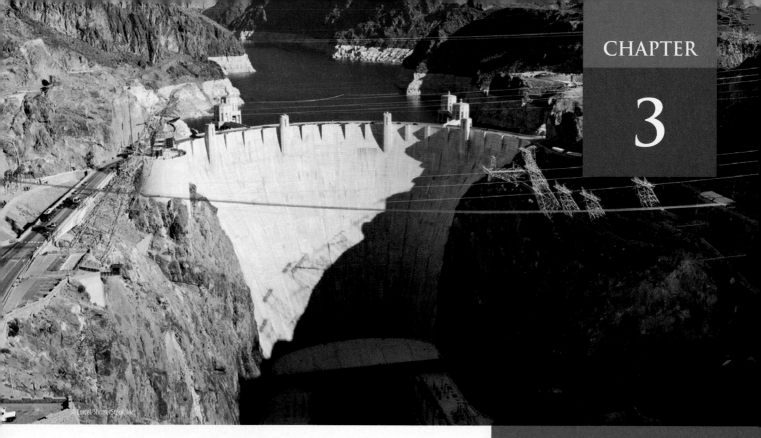

© Corcel/ShutterStock, Inc.

Understanding the Root Causes of the Environmental Crisis

There are a thousand hacking at the branches of evil to one who is striking at the root.

—Henry David Thoreau

LEARNING OBJECTIVES

After reading and studying the material in this chapter, students will be able to understand, knowledgably discuss, and, when appropriate, apply the information learned in this chapter, including:

1. The current thinking on the many root causes of the environmental crisis, or crisis of unsustainability
2. Ways that we can address the root causes of the environmental crisis and the benefits they will have on the environment and human future
3. Our current unsustainable system of ethics (or values)
4. Key principles of a sustainable ethical system that could help us address root causes of the environmental crisis and create lasting solutions that allow human society to protect, even enhance, the environment

Exercise

One of the most important skills of critical thinkers is to be able to uncover hidden bias or preconceived notions that we and others have. Below is a list of statements that you may agree or disagree with. They are designed to help you uncover some of your biases and beliefs to help you become a better critical thinker. Read each statement and decide whether you agree with it.

Statements to Ponder

- Technology is the answer to environmental problems. Whatever is wrong in the environmental arena can be fixed by the application of proper technologies.
- Humans are superior to other living things.
- Human needs should generally come first.
- All people care about is money. If something makes economic sense, people will do it.
- Continued economic growth is a desirable goal.

After reading and thinking about each statement, take a few moments to determine where your views come from. Finally, take some time to consider whether your view or value is well founded.

I t seems that just about everyone has an opinion on the causes of environmental issues. To some people, the current environmental dilemma is largely the result of overpopulation—too many people. To others, it is the fault of those corporations whose concern for the environment ranks too far below their interest in making money. Others may attribute it to inept or corrupt government officials. Many think that the environmental problems of the world result primarily from overconsumption and waste. Modern, environmentally destructive technology is seen as the culprit by some. The list goes on. I'm sure you have your own thoughts on the subject.

If the truth be known, the current environmental crisis or, more appropriately, the **crisis of unsustainability** is the result of many factors. Population is a big player. So is consumption. Corporate policies play a part, too. There are many other factors as well. This chapter describes the major reasons for environmental unsustainability and presents a diagram of the complex web of cause and effect to help you understand how we arrived at our current situation.

The chapter's exploration of the roots of the environmental crisis is intended to help you not only comprehend things better but also understand key leverage points where change can be effected. The main theme of this chapter is that if we know the root causes and address them, we have a far better chance of creating lasting solutions and a better future.

3.1 Roots of the Environmental Crisis

Books and articles on the root causes of the environmental crisis reveal many interesting ideas. This chapter combines the insights of historians, social scientists, economists, psychologists, anthropologists, theologians, and others into a coherent picture. It also discusses my own biological/evolutionary hypothesis. We study the ideas of these thinkers one at a time, and then combine them to show a map of cause and effect. Remember as you read this material that books could be written on each subject. I present the main ideas *and opposing ideas as well*, to give you as balanced a view as possible. As you read the various hypotheses, imagine them as pieces of a larger puzzle, all leading to the problem we call the crisis of unsustainability.

KEY CONCEPTS

Many different hypotheses have been presented to explain the root causes of the environmental crisis. When combined, many of these hypotheses do a good job of explaining environmental troubles.

Religious Roots

An early, influential hypothesis on the roots of the environmental crisis was presented by UCLA historian Lynn White in the 1960s. He argued that the roots of our environmental crisis are "largely religious." In his classic paper entitled "The Historical Roots of Our Ecological Crisis," White argued that the emergence of science and technology four generations ago in western Europe and North America spawned an era of massive environmental manipulation that was paralleled by enormous environmental problems. However, White argued that Western-style science and technology "got their start, acquired their character, and achieved world dominance in the Middle Ages." The most important influence on science and technology at this time was Christianity's view of human dominance over nature.

White noted the familiar passage in Genesis 1:28 that directs humankind to "fill the earth and subdue it, and have dominion over the fish of the sea and over the birds of the air and over every living thing." In Christian teaching, White contended, no item in physical creation had any purpose except to serve man's purposes.

White also noted that during the Middle Ages Christianity promoted a sense of dualism, a view of humanity as apart from nature. In a paganistic society, humans viewed themselves as a part of the whole.

White contended that the disastrous environmental problems facing the world could not be solved by simply applying more Western-style science and technology to our problems, which, in his view, are rooted in Christian domination and duality. Professor White wrote that "we are not, in our hearts, part of the natural process. We are superior to nature, contemptuous of it, willing to use it for our slightest whim."

Widely circulated in academic circles in the 1960s and 1970s, White's ideas were embraced by many scholars, lay people, and even some theologians. Not all theologians and scholars, such as René Dubos and Gagriel Fackre, however, agreed with his hypothesis. They cited biblical verses that promote stewardship and protection. Another critic of White's premise was Lewis Moncrief, a professor at Michigan State University. Moncrief finds no fault in White's contention that human ecology is deeply conditioned by religious belief, "but to argue that it is the primary conditioner of human behavior toward the environment is much more than the data that [White] cites to support this proposition will bear." Moncrief points out that White himself notes in his classic paper that humans have drastically altered the environment since antiquity and cites several examples of pre-Christian environmental destruction that undermine his argument.

René Dubos pointed out that in both ancient Eastern and Western civilizations, environmental destruction was, and still is, rampant (**FIGURE 3-1**). The deforestation, soil erosion, and overgrazing prevalent in some cultures suggest that Christian beliefs have had no monopoly on ecological damage. In other words, exploitation of the natural world may be a universal human trait. If environmental destruction has occurred since antiquity, White's thesis must be questioned. White answers his critics by pointing out that although the Bible proclaimed it humans' duty to protect the Earth, little heed was paid to such instructions.

What can we make of the debate? Obviously, opinions vary. While some are skeptical of White's ideas there are those who contend that religion may have played a role—and may still be playing a role—in shaping our thoughts in ways that are detrimental to the environment and our long-term future. A Gallup poll found that one-third of Americans believe the Bible is literally true. Why is this significant? Because roughly 50 million of these Americans are expecting the Apocalypse soon. There are those who believe efforts to address serious long-term environmental issues is pointless. Some oppose actions on the grounds that environmental destruction will help hasten the Apocalypse.

(a)

(b)

FIGURE 3-1 Destruction in the East. Environmental damage is not a Western phenomenon. All cultures have had a profound effect on the environment, even those whose philosophy embraces the protection of nature as in the case of the Three Rivers Gorge Dam in China. **(a)** The river before dam construction began. **(b)** The dam under construction.

Photo a © Geoffrey Morgan/Alamy Images; photo b © China Photos/Reuters/Landov.

Some scholars think that early Christian teaching shaped many people's attitudes toward nature, which in turn fostered the creation of exploitive systems of science and technology that are largely responsible for the destruction of the environment. Others argue that humans have a long history of environmental destruction dating back long before the advent of Christianity.

Some Cultural Roots: Democracy, Industrialization, and Frontierism

Lewis Moncrief (mentioned previously) attributes the environmental crisis to two dominant forces: the spread of democracy and industrialism. Democracy began to take hold after the French Revolution. Prior to that time, land and resources in the Western world were owned primarily by royalty or the Roman Catholic Church. They bestowed control of the land resources on the nobility and others who pledged their allegiance. Much of the land was protected as royal property. However, the French Revolution marked the unraveling of these feudal systems of land tenure. As a result, land became vested in the hands of more and more individuals.

At about the same time, but over a more extended period, another kind of revolution was taking place that would dramatically change the world. Known as the **Industrial Revolution,** this shifted labor from manual to energy-intensive machine production. It resulted from the application of science and technology to the service of humankind. With this change, the productive capacity of each worker greatly increased. It became possible to mass-produce goods (**FIGURE 3-2**).

Moncrief contends that the spread of democracy and industrialization resulted in a more equitable distribution of wealth among the human population and, more important, rising affluence and consumption. Pollution and environmental decay followed (**FIGURE 3-3**).

Moncrief asserts that the colonists of North America were quick to adopt the "technical and social innovations" taking place in Europe. In America, national policy was designed to transmit both ownership of the land and many of the nation's natural resources from the federal government to private individuals. Huge parcels of land were given to settlers. In the 1800s, mining laws were designed to transfer land to those willing to make the effort to exploit it. (Incidentally, these mining laws are still in effect today!) In short, much of the land and the natural resources of the nation came to be controlled by citizens and businesses. Decisions that affect the environment were made by millions of private owners. Industrialization and the spread of wealth in the United States further added to the destruction of the environment, as it did in Europe.

Moncrief also touches on an important dimension of the environmental crisis, frontierism. **Frontierism** is a way of life that is based on an assumption of plenty. To most frontierspeople—and particularly to early farmers—many of the natural resources we now fight to protect were originally perceived as obstacles rather than as assets. Forests needed to be cleared to permit farming. Marshes needed to be drained.

Part of the reason for this obstacle mentality was a false perception of the inexhaustibility of the vast resources of the new nation, a frontier ethic. "After all," observes Moncrief, "if a section of timber was put to the torch to clear it for farming, it made little difference because there was still plenty to be had." Frontier thinking continues today as a major factor in the ever-worsening environmental crisis (**FIGURE 3-4**).

In the frontier society, the relationship of humans to nature was indifferent at best, and often antagonistic. Antagonism and indifference prevailed in large part because of a lack of knowledge on the part of settlers about the long-term implications of the type of changes being

FIGURE 3-2 Industrial production. Industrialization made it cheaper to produce goods and made them more accessible to a large number of people, thus increasing our demand for resources and our impact on the environment.

© RedTC/ShutterStock, Inc.

FIGURE 3-3 Industrial pollution from factories. Scenes like this are common in many less developed nations and were very common in the developed nations until the 1960s and 1970s.

© Eric Gevaert/ShutterStock, Inc.

FIGURE 3-4 Frontierism continues. Tropical rain forests are cut down in record number each year to supply wood and make way for farms. The belief that there are plenty more makes this resource highly vulnerable.

© Photographer's Choice/age footstock.

made to the environment. Lacking an understanding of ecology and the consequences of abusing the life-support systems of the planet, frontier societies plundered one ecosystem after another. Ignorance, another dynamic in the complex web of causation, is alive and well today. Although our understanding of ecology and the impacts of humans on the environment has reached an all-time high, many people remain ignorant of the impacts, and this ignorance drives many environmentally unsound practices—building energy-inefficient homes, for instance. Ignorance, therefore, occupies a prominent place in the network of causation.

KEY CONCEPTS

Some scholars believe that the spread of democracy, which put land ownership and wealth in the hands of many, and the Industrial Revolution, which made mass production of goods possible and spread wealth throughout society, are at the root of the environmental crisis. Frontierism, a belief in the inexhaustibility of resources, may have also been a root cause of the environmental crisis.

Biological and Evolutionary Roots

Philosopher Bertrand Russell once wrote that "every living thing is a sort of imperialist, seeking to transform as much as possible of the environment into itself and its seed." Although the vast majority of the world's species do not intentionally seek to convert the resources of the environment into their kind, populations do naturally expand in proportion to available resources—water, soil, sunlight, and so on. Organisms are biological imperialists. In this chapter, I use the term **biological imperialism** to describe an expansionist tendency.

Countless examples of biological imperialism exist in the natural world. This characteristic of living things is

amply illustrated by the spread of a vine known as *kudzu* (pronounced CUD-zoo), which grows rampant in the southeastern United States, smothering farm fields, forests, and abandoned buildings (**FIGURE 3-5A**).

The water hyacinth (HIGH-ah-sinth) is another biological imperialist extraordinaire (**FIGURE 3-5B**). Introduced to the United States in 1884 from South America as an aquatic ornamental by a private collector, the water hyacinth escaped from the pond where it grew. After entering the warm, nutrient-rich waterways of Florida, it spread throughout the extensive network of canals and rivers that lace the state. Aided by a remarkable reproductive capacity—10 plants can multiply to 600,000 in 8 months—the water hyacinth today chokes many of Florida's waterways. The plant has crowded out native species and made navigation virtually impossible in some water bodies. From Florida, the plant has spread throughout much of the southern United States and today covers nearly 2 million acres of rivers and lakes along the southern tier of states from Florida to California.

Biological imperialism refers to a fundamental tendency for populations of organisms to proliferate in response to available resources. Bacteria in a petri dish reproduce

(a)

(b)

FIGURE 3-5 Biological imperialism. Two supercolonizers, **(a)** the kudzu and **(b)** the water hyacinth.

Photo a © Danny E. Hooks/ShutterStock, Inc.; photo b © sittitap/iStock/Thinkstock.

and colonize. Organisms don't choose to be imperialists. It is simply a biological fact of life: When conditions are favorable, organisms prosper and reproduce successfully—and populations expand. Biological imperialism is aided by the availability of nutrients, sunlight, moisture, and a host of other factors.

In nature, checks and balances tend to prevent many populations from reaching their full potential. However, some species fare better than others, especially when placed in foreign environments where the natural checks and balances that evolved over millions of years are not present. *High reproductive rates and a host of other special characteristics make some organisms superb imperialists.* The success of any one species depends on the structural, functional, and behavioral characteristics (called *adaptations*) it arrives with or acquires through evolution.

Humans are superior colonizers. They have special features such as brainpower and technology to remake the planet—to create conditions and expand food supplies in ways that enable them to prosper and expand.

My hypothesis is that the major environmental transgressions of the past and the present owe their origin in large part to a sort of human biological imperialism. From its meager origins in Africa, humankind has expanded widely. Today, human civilization in one form or another inhabits virtually every biome on the Earth and makes use of many aquatic life zones. Hardly a square foot of the planet has been spared the human imprint.

Early environmental transgressions in North America—for example, the extinction of many species of mammals—may have occurred as the first settlers spread across the continent about 10,000 years ago (**FIGURE 3-6**). According to some scientists, the extinction of these species can be attributed in large part to our successful planetary colonization (human biological imperialism) aided by the use of very primitive technology, such as fire and early weapons. Human populations expanded as a result, migrating throughout the world, often wreaking havoc along the way.

As technology became more advanced, the pace and severity of environmental destruction quickened. The past 60 years have seen a further acceleration of the Earth's destruction as technology and per capita consumption, particularly in the Western world, have increased dramatically.

Over the years, many technologies have helped us to overcome natural forces that might otherwise have controlled human population expansion. *The bottom line is that technologies have permitted us to expand almost uncontrollably.* Agricultural technologies have dramatically increased food production. Medical technology has greatly reduced outbreaks of infectious diseases, such as the plague, that once took a massive toll on human populations. Together, these and other advances led to an explosion in the human population. As our numbers increase, so does our impact.

Technological liberation from biological limits has unleashed our imperialistic tendencies, creating a global environmental crisis. *In short, technological liberation has translated a natural biological urge into a global environmental disaster.*

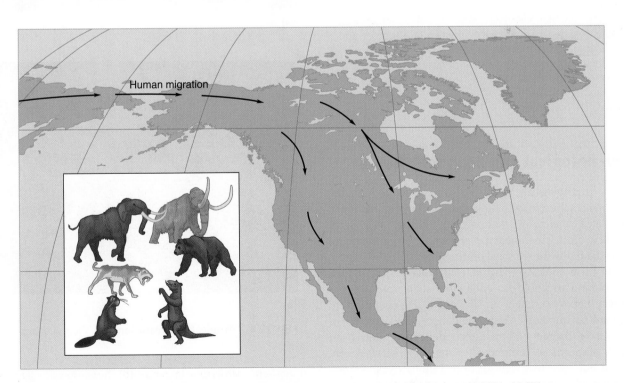

FIGURE 3-6 Early inhabitants of North America, who entered by a land connection with Siberia about 10,000 to 12,000 years ago, may have wiped out many species of megafauna like the saber-tooth tiger, giant sloth, mastodon, and wooly mammoth shown here.

Technological development was made possible by two evolutionary developments unique to humans: the evolution of the brain and the evolution of manual dexterity. From an environmental perspective, one of the most important developments in human nervous system evolution was the emergence of the frontal lobe. The frontal lobe, which is part of the forebrain, houses several functions, among them (1) ideation (the ability to create ideas) and (2) the ability to plan and think about the future. Both functions are vital to technological development. Of equal importance to the development of technology were refinements of the brain that gave humans the ability to communicate and build tools and weapons.

The ability to create ideas, plan for the future, invent tools, communicate ideas, and manipulate objects were all essential to the development of technology. These uniquely human abilities today give us a power to do many great things—to till the Earth to grow crops, for instance.

Appreciated by many for expanding our capacity to improve on nature, these abilities are also responsible, in large part, for much destruction. In some cases, the destruction wrought by technology is intentional, as manifest in tropical deforestation. In others, it is unintentional, as manifest in ozone depletion and acid deposition. From an environmental standpoint, the distinction is irrelevant. Either way, this damage reduces the Earth's carrying capacity for humans and the millions of species that share this planet with us.

Psychological and Economic Roots

Many psychological factors also contribute to the environmental crisis by aiding and abetting our biological imperialist tendencies. Although psychology may seem far afield from environmental science, it is important to understand if we are to solve problems.

In their book, *New World, New Mind*, ecologist Paul Ehrlich and psychologist Robert Ornstein note that the human nervous system is "designed" by evolution to respond to immediate physical danger. The survival of early cave dwellers, they assert, depended largely on quick reactions to imminent threats. This survival tool served our ancestors well for over 3 million years—99% of our evolutionary history. Ehrlich and Ornstein argue, though, that the human nervous system today is rather

ill equipped to respond to trends that threaten our long-term future.

The human brain may be something of a double-edged sword. The frontal lobe, discussed earlier, helps us plan for the future, and engineer and design technologies. However, it may not equip us with a very adequate response to dangers unless they are immediate and life-threatening. Although some environmental problems create an imminent threat to us, most pose long-term, less obvious threats to the health and welfare of humans— among them global warming, acid deposition, and tropical deforestation (**FIGURE 3-7**). To build a sustainable future, we must learn to respond to long-term dangers.

Denial is another psychological factor that contributes to our inability to respond to the environmental crisis. Many environmental controversies have been marked by long-standing denial that has hindered progress toward lasting solutions.

(a)

(b)

FIGURE 3-7 Tropical deforestation. (a) The loss of tropical rain forests creates local environmental problems of enormous magnitude, but for most of the world's people the impacts are long term, such as climate change. **(b)** The rain forests influence climate hundreds of miles away.

Photo a © iStockphoto/Thinkstock; photo b © Photodisc.

Denial is often prevalent among those who have a vested interest in maintaining the status quo, as well as those of limited perspective and knowledge. Such individuals often find it difficult to believe that human activities could endanger the biosphere and humankind. Individuals who have difficulty accepting the reality of a problem cannot be counted on to take responsibility or action. In many cases, they will actively obstruct efforts to solve crucial problems.

Denial can be confused with healthy skepticism. Scientists are often naturally skeptical about ideas. They require adequate data to support new views.

Another powerful influence in the environmental crisis is the seemingly near obsession with economic growth that is prevalent in many of the more developed countries. Most Western industrial nations, which produce the bulk of the world's goods and services, are predicated on ever-increasing material production and consumption. With the growth-is-good (indeed essential) logic so deeply imbedded in many societies, modern Western culture has evolved what many think is a lopsided view of human progress based on producing and consuming more and more, regardless of the cost to the planet or to future generations.

Continued human expansion and economic growth are impossible in a finite system. In his book *The Poverty of Affluence*, Paul Wachtel observes that growth is "an omnipresent symbol of the good." With this mindset, it is "very difficult for us to accept any idea of a limit to growth as implying anything other than stagnation. Our emphasis on growth leads us to equate contentment with complacency." Why is this penchant for growth so prevalent?

Numerous authors have studied the psychological roots of the economic growth philosophy and overconsumption. Andrew Bard Schmookler, a leading thinker in this area, contends that the modern culture of mass consumption is developed around a core of unfulfilled longing. People buy to make themselves feel better—to satisfy internal longings. Modern advertisers have capitalized on our sense of inadequacy. According to Schmookler, advertisers promise that the goods we purchase will make us happy or successful.

In addition, the many ads that bombard people in consumer societies suggest that success is measured by what and how much we own. Most of us, however, can never fully match the beautiful or handsome models portrayed in ads. There's always some area that we can improve, but our efforts to placate an unsatisfiable longing for adequacy are like trying to fill a sieve with sand. From a marketer's standpoint, the effect is remarkable: We never stop consuming in our vain attempts to fill the emptiness. The economy expands as a result. Unfortunately, the more we acquire to offset the sense of personal inadequacy, the more environmental damage we cause.

Psychologists tell us that the roots of the sense of inadequacy can be traced to other factors. The lack of self-esteem that advertisers feed off in our culture results, in large part, from inadequate family systems and deficiencies in child rearing. We may feel inadequate because of the messages we received as children. Inadequate child rearing results in part from the lack of parental education on the subject.

Advertisers also play off a powerful human desire to conform. At the root of conformity is the obvious and pervasive need to belong. In order to belong, a person must have the right car, the right clothes, the right cologne or perfume. In other words, to belong, one must consume. When styles change and one's sense of conformity and belonging are threatened, people buy more. In such a culture, it is better to be in debt to the hilt than to be with*out*.

Underlying the need for conformity and belonging may be the erosion of community. *Homo sapiens* is by nature highly social. Our early evolutionary success can be traced in part to social groups that permitted cooperative hunting and food-gathering ventures and joint efforts in warding off predators. As our communities dissolve into frenetic megalopolises, belonging at a deep and genuine level is replaced by superficial membership in society facilitated by conformity with transient fashions orchestrated by advertising firms.

Consumption also stems from the joy of novelty and the usefulness and convenience of many new products. In addition, some consumption results from the relative costs of mass production versus labor for repair. How many of us have had to throw away a $250 stereo component because it would cost $200 to replace a faulty part?

Go Green

Rather than giving gifts to friends and loved ones, consider giving a little of yourself—for example, cooking a dinner for friends or loved ones, or taking them on a hike in a nearby park, or even volunteering to help them clean their apartment or home.

The growth-oriented economic system is also shaped by an intense competitive drive in humans. In modern society, competitiveness is culturally propagated through advertisements, education, and the public utterances of our political and business leaders. The telecommunications industry, for instance, sells its wares on the value of the competitive edge gained by such innovations as overnight delivery, faxes, e-mail, cell phones, and other forms of "instantaneous" communication—features that give one company a leg up on the competition. Improvements in the American educational system have been deemed important primarily for the competitive edge they could give the U.S. economy in the global marketplace.

Acquisitiveness and greed also drive economic systems that exploit people and the planet with detrimental consequences. Garrett Hardin discussed the consequences of greed in a classic paper entitled "The Tragedy of the Commons." In England, Hardin noted that cattle growers grazed their livestock freely on the commons, land open to all. The commons fell into ruin as the users became caught in a blind cycle of greed. To increase their personal wealth, individuals expanded their herds. Each additional cow meant more income at only a small cost. Hardin notes that the cattle growers were rewarded for doing wrong. Although they may have realized that increasing their herd might lead to overgrazing and deterioration of the pasture, they also knew that the negative effects of overgrazing would be shared by all members of the community. Each herdsman had more to gain than to lose by expanding his herd. This shortsighted thinking (another root cause) resulted in a spiraling decay of the commons. Hardin summarized the situation as follows: "Each man is locked into a system that compels him to increase his herd without limit in a world that is limited." As each pursues what is best for himself, the whole system is pushed toward ruin.

Tragedies of the commons date back to the days of ancient Greece, when Aristotle recognized that property shared freely by many people often received the least care. Early civilizations toppled forests and carelessly overgrazed their cattle on rangelands, just as the tribespeople of the Sahel (the region bordering the southern Sahara) do today. History, however, shows that early civilizations paid dearly for their disregard. The skeletons of buildings from ancient cities still stand in deserts that were the once-rich forests and grasslands of the Fertile Crescent (**FIGURE 3-8**). Much of Iran and Iraq, now barren desert, once supported cattle, farms, and forests. Greece and Rome fell, historians believe, partly because of the misuse of their lands. The tragedy of the commons is the cumulative result of the self-serving tendencies of humans (greed), as well as of ignorance, competitiveness, and lack of regulation.

Hardin's analysis of the tragedy of the commons, although important, gives the impression that common resources are alone in being mismanaged. In many places private lands are no better cared for because of short-term profit or poor management (in the case of former communist nations).

All of the psychological factors discussed so far influence the way we conduct business, government, and our personal lives throughout the world. They influence the laws that govern economic activity. They affect the way we have settled the land and made our place on the Earth. Beliefs and values are, in essence, the philosophical underpinnings of all human systems—from law to the economy to government. These psychological factors have either contributed to our biological imperialist tendencies or failed to create mechanisms to temper them.

KEY CONCEPTS

Human attitudes and beliefs are also responsible for many unsustainable practices. Denial, apathy, inability to respond to subtle threats, greed, acquisitiveness, and other factors influence our economic systems, laws, and way of life in profound ways. In short, they worsen our biological imperialist tendencies.

Putting It All Together

It should be clear from the forgoing discussion that many factors contribute to global environmental problems. People are complex, as are the systems that support our lives. So are the issues. An analysis of thinking on root causes shows that we can't point to a single root cause, or even a couple of root causes. There are many. To understand how the factors fit together, let's build a model, one piece at a time, drawing on the information you've just read.

FIGURE 3-9 shows the environmental crisis and its link to human systems designed to meet our many needs. Ultimately, it is these systems that are responsible for the problems. As you will learn in other chapters, these systems have many characteristics that

Socioeconomic system characterized by overconsumption, inefficiency, linearity, fossil fuel dependence, and failure to manage and restore ecosystems properly

↓

Environmental crisis

FIGURE 3-9 It's the systems. Unsustainable systems are largely to blame for environmental ills. But the "shape" of these systems is influenced by many other factors.

FIGURE 3-8 Remnant of past action. This abandoned city in the desert was once a thriving region, but environmental abuses caught up with residents and destroyed their homeland.

© P. Narayan/age footstock.

contribute to environmental problems. They are dependent on fossil fuels, for example, and many are highly inefficient.

Human systems that enhance our survival unleash population growth and bring us power unimagined even 100 years ago. Put another way, they unleash our biological imperialism. **FIGURE 3-10** draws the connections. Democratization and the spread of wealth through industrialization, shown in Figure 3-10, are also responsible for the evolution of the modern systems.

FIGURE 3-11 shows how our inability to respond to long-term trends, based on economic self-interest, disbelief, denial, and ignorance, influence the socioeconomic system. **FIGURE 3-12** illustrates how the obsession with economic growth contributes to the socioeconomic system that is responsible for the crisis. Many factors contribute to this obsession, as described earlier. Among them are our sense of inadequacy, competitiveness, acquisitiveness, frontier ethics, and laws and regulations. These factors, in turn, are influenced by a host of other factors.

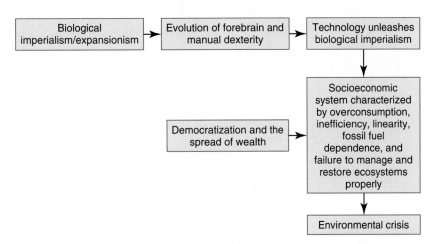

FIGURE 3-10 Biological imperialism, technology, democratization, and capitalism. Technology unleashes our biological imperialism and partly is responsible for overpopulation and the vast, unsustainable systems that support our lives. Democracy, industrialism, and the spread of wealth, while generally viewed as positive advances, may contribute to excessive resource demand and pollution production of human society as explained in the text.

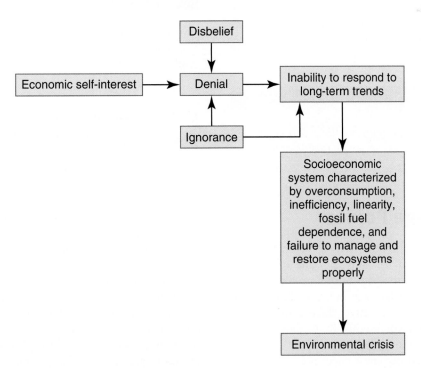

FIGURE 3-11 Psychological factors. Humans seem unable to respond to long-term trends or warnings of problems. Many factors contribute to this inability.

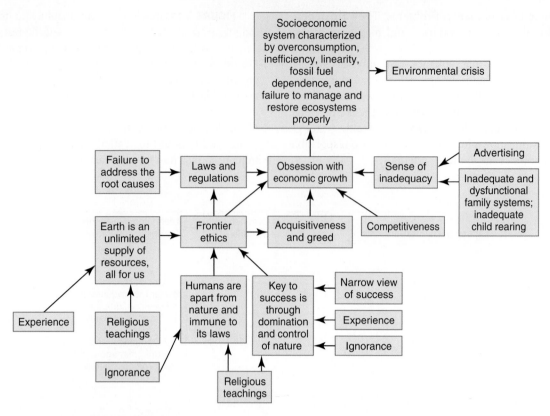

FIGURE 3-12 More psychology and ethics. This section of the map showing the causes of the environmental crisis attempts to map out roots of our obsession with economic growth. As you can see, many psychological and ethical factors come into play. Religious teachings may also influence our thinking.

If all of the components are placed together, the model shown in **FIGURE 3-13** emerges. It helps illustrate how complex the problems really are. But if you understand each one, then it is not that difficult to see how they all fit together.

3.2 Leverage Points

Charles Kettering once wrote that a "problem well understood is a problem half solved." One of the key benefits of analyzing the root causes of a problem is that it helps us recognize key leverage points—places where actions would prove most effective. Take a moment to study Figure 3-13; while you do this, think about areas in which action would be most effective.

One area that would become a major point of focus in our efforts to transition to a more sustainable human way of life is technology.

Making Human Systems and Technologies Sustainable

If the environmental crisis could be distilled to a single sentence, it would be this: *The world's environmental woes result from the fact that human technologies and human systems are fundamentally unsustainable.* Put another way, the environmental problems we face result from the fact that the systems that supply our many needs were designed in ways that do not respect the limits of the Earth's carrying capacity. Human systems may have been sufficient in a world of 2 billion people, but today, with the world's population well over the 7 billion mark and growing by approximately about 85 million per year, evidence strongly suggests that the ways we produce goods and services, house our people, handle our waste, and so on are fundamentally unsustainable. Just as hazardous waste and pollution are a sign of failed technology, the world's many environmental problems are symptoms of

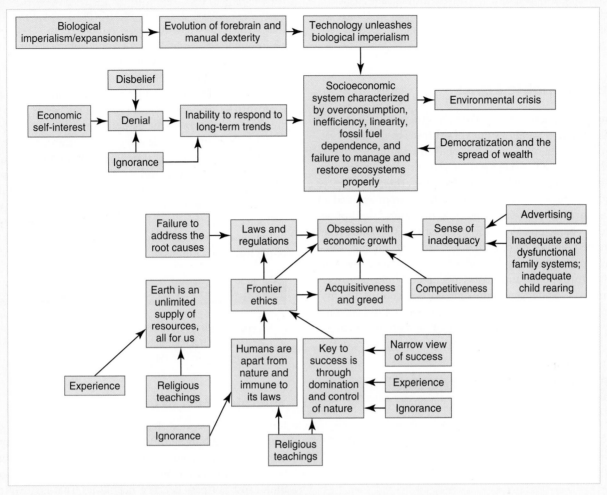

FIGURE 3-13 Many factors contribute to the environmental crisis. They interact in many ways to produce a socioeconomic system that threatens its own existence.

systems that are inadequately designed for a world of limits. In the next section, I present an alternative worldview, a sustainable ethic to ponder as you study this subject. This ethic may be one of the most fundamental leverage points for change.

 KEY CONCEPTS

Unsustainable human systems are a result of many factors. Building a sustainable society will require a restructuring of all human systems. Changes in technology will be essential to this transition.

A New Worldview: Changing Our Perceptions, Values, and Beliefs

The material in the previous sections indicates the importance of our perceptions, values, and beliefs. Science can shape our perceptions—how we see the world and the nature of the problems facing us—and our understanding of many issues. This book could change your values and beliefs. Creating the broad cultural support for sustainability, however, will very likely require much more education. One recommendation for changing these important factors is to institute environmental courses at all levels of the educational system—from kindergarten through college—throughout the world. These could help play a role in educating all people. They could help develop a more accurate picture of the state of affairs—how serious the environmental crisis is—and could help people understand the many options we have to forge sustainable lifestyles and a sustainable economy. Further ideas on changing our values are outlined throughout this text. Many people believe that fundamental changes in our ethics are required.

Unsustainable Ethics

Many people agree that building a sustainable future will require a new system of values or **ethics.** The system of values of a person, religion, group, or even a nation determines how it acts. The prevalent system of Earth values in Western culture, as well as others, is the **frontier ethic.** It embraces a rather narrow view of humans in the environment and an even narrower view of the purpose of nature. The frontier ethic is characterized by three tenets: (1) The Earth is an unlimited supply of resources for exclusive human use—there's always more, and it's all for us; (2) humans are apart from nature and immune to its laws; and (3) human prosperity and well-being result from our efforts to control the environment. In other words, nature is something to overcome or conquer. We get ahead by reshaping natural systems to our liking. Let's consider each of these tenets in more detail.

The First Tenet of Unsustainable Ethics

The view of the Earth as an unlimited supply of resources for exclusive human use no doubt evolved in prehistoric time when human numbers were small and the Earth's resources did indeed appear inexhaustible. The massive increase in economic activity and the upsurge in population growth in the last 200 years, however, have brought us face to face with the limits of the planet. Since 1920, dozens of ocean fisheries have been depleted. Dozens more are now threatened. In the past 100 years, half of the world's rain forests have been destroyed, along with countless species of plants and animals. Within most of our lifetimes, tropical rain forests could be wiped out. Oil and natural gas supplies are also likely to run out.

Despite growing evidence that our frontiers have vanished, many people still persist in viewing the Earth as an unlimited supply of resources.

The Second Tenet of Unsustainable Ethics

Humankind positioned itself outside the realm of nature for thousands of years. Today, many people continue to view humans as separate from nature, and they persist in thinking that we can do what we please without harm. The French philosopher and writer Albert Camus summed our philosophy of separation best when he wrote, "Man is the only creature that refuses to be what he is." In reality, says ecologist Raymond Dasmann, "A human apart from environment is an abstraction—in reality no such thing exists." Our lives are intricately linked to the natural world. How?

A huge portion of the oxygen we breathe each year is replenished by plants and algae. All of the food we eat comes from plants, soil, water, and air. Even plastic, which comes from oil, comes from organisms (phytoplankton) that lived on the Earth several hundred million years ago. They died and sank to the the ocean floor where they were covered by sediment and turned into oil. To think of ourselves outside of the realm of nature is not only foolish, it is suicidal.

The Third Tenet of Unsustainable Ethics

Industrial nations view nature as a force that must be conquered and subjugated. Matthew Arnold, a 19th-century British poet, summed up the modern view best when he wrote: "Nature and man can never be fast friends. Fool, if thou canst not pass her, rest her slave!" That is, if we can't subdue nature, we will become a slave to natural forces.

Believing in the need to reign supreme, humankind has conquered rivers and subdued the wilderness. In the past four decades, it has become clear to many that our acts of conquest have not come without a cost. In fact, many of the environmental problems we face today stem from our efforts to control and dominate nature. Such impacts are often called *ecological backlashes.*

A good example of ecological backlash can be seen on today's chemical-intensive farms. To increase crop production, many farmers use an assortment of pesticides, chemicals used to kill organisms that farmers view as pests. However, research has shown that pesticide use may be a Pyrrhic victory—one that comes at a high cost. Why? Since World War II, chemical pesticide application has increased 10-fold, while pest damage has doubled. Today, 584 species of insects are resistant to at least one chemical pesticide, and 20 insect pests are resistant to every known pesticide, according to the Arthropod Pesticide Resistance database maintained by Michigan State University.

Even bedbugs, which bite us when we sleep, creating painful sores, are showing signs of pesticide resistance; according to professor Susan Jones at Ohio State University in Columbus, Ohio.

KEY CONCEPTS

The third tenet of sustainable ethics is that success comes through domination and control of nature.

Some Impacts of Frontier Thinking

Frontier ethics profoundly affects how people act and has led to numerous environmental problems. It also affects our attitudes about the seriousness of environmental problems. Projections of declining fossil fuel supplies, for instance, generally spark little interest among people convinced that supplies will never run out—or that if they do, scientists will find substitutes.

The frontier ethic also influences how people solve environmental problems. The solution to impending oil shortages, for example, is to increase our search for new supplies, even if it takes us into pristine wildernesses like Alaska's Arctic National Wildlife Refuge. Our faith in unlimited resources prevents many among us from finding ways to use resources more efficiently. Why be efficient if resources are unlimited?

The frontier mentality also permeates our own lives, influencing personal goals and expectations. Most of us tend to make our buying decisions on the basis of what we can afford. Very rarely do we ask whether a decision to purchase a product helps or hinders the environment. Rarely do we think about the product's effects on the long-term habitability of the planet. If the world is an unlimited supply of resources, why should we?

Over the years, the unquestioning acceptance of frontier ethics has spawned a way of life that, while successful in many ways, threatens our own future. Truly, frontier ethics has been a fundamental driving force in the creation of human systems that are out of sync with the natural systems that support them.

Societies that are out of sync with natural systems are unsustainable. They grow crops at the expense of their soils. They may also destroy wetlands, which are a vital habitat for many species of fish—which in turn are an important source of human food. Many other examples of this phenomenon are given in this book.

In contrast, sustainable societies satisfy their needs for food, shelter, and other resources without destroying the natural systems upon which they depend. For example, they grow crops without damaging their soils. They harvest trees while protecting forests and without clogging nearby streams and lakes with sediment eroded from the land. Such communities exist in harmony with nature (see **Spotlight on Sustainable Development 3-1**).

A sustainable society does not merely prevent destruction; it also seeks ways to enhance natural systems.

Sustainable farming, for example, increases the organic content of the soil, which makes land more productive. Ultimately, creating a sustainable society means fitting human societies within the economy of nature. It is about creating a better future, a stronger nation, and broad prosperity.

KEY CONCEPTS

Frontier thinking influences how we design and operate all human systems, from the economy to government to education to waste management, to agriculture.

Toward a Sustainable Ethic

Creating a sustainable society that lives within the Earth's means may require the world's people to adopt a new, sustainable ethic. **Sustainable ethics** contrast sharply with the frontier ethics described earlier. They assert that (1) the Earth has a limited supply of resources, and they're not all for us; (2) humans are a part of nature, subject to its laws; and (3) success stems from efforts to cooperate with the forces of nature. I like to add a fourth element of sustainable ethics to remind us of the importance of the natural world. It is that our future depends on creating and maintaining a healthy, well-functioning global ecosystem. An **ecosystem** consists of a community of organisms and all of the interactions between them and their physical environment.

As you can see, the principles of sustainable ethics are the opposite of the principles of frontier ethics. Just as frontier ethics leads us to exploitive behavior, sustainable ethics could lead to a less exploitive human presence, one that could endure for thousands of years and greatly improve our chances of long-term success. Sustainability is not about limiting our future but ensuring it—creating a better world.

KEY CONCEPTS

Creating a sustainable society that protects the environment will require a new value system—one that respects limits, sees humans as a part of the natural world, and recognizes the need to cooperate with nature, not dominate it.

Planet care is the ultimate form of self-care.

—Dan Chiras

SPOTLIGHT on Sustainable Development

3-1 Ashland, Oregon Shows That Water Conservation Works

Ashland is a small city in Oregon. Like a growing number of cities and towns, Ashland is showing the value of conservation as a means of saving money and providing for the future needs of people, with little or no environmental impact. Ashland's story is an inspiring example of critical thinking—more specifically, of questioning conventional ideas and conventional means of supplying resources.

This story begins in 1989, when city officials started discussions about a potentially severe problem: future water shortages that would be caused when the city lost badly needed water rights in just 8 years. This would reduce its supply of drinking water considerably. Accordingly, the officials contacted a consulting firm that, after reviewing the problem, proposed the construction of an $11 million dam on nearby Ashland Creek (for which the city held water rights). According to the consultants, the dam and reservoir would supply 2 million liters (500,000 gallons) of water a day for the city and prevent potential water shortages.

Having adopted progressive energy conservation standards 8 years earlier, the city officials thought that conservation might be a better way of making up for the potential loss. City officials also realized that their town of around 20,000 would be financially hard-pressed to pay for an $11 million dam. Consequently, they contacted another consulting firm and asked them to look into the feasibility of making up for the lost water by using conservation—a program to install water-efficient showerheads and faucets and low-flush toilets. After some study, the company found that at one-twelfth the cost of the dam—or $825,000—the city could phase in a water conservation program that could save the required 2 million liters (500,000 gallons) of water a day relatively painlessly. The net effect: They wouldn't need to dam the creek. They'd also save more than $10 million! Their solution made sense from economic, social, and environmental perspectives—and is a tribute to the value of critical thinking: questioning traditional ways.

For years, water shortages have been addressed by building new dams. But supplies of water and many other resources can be stretched considerably by using what we already have more efficiently. Such efforts often cost a fraction of traditional approaches and are an important part of the strategy for sustainable development.

CRITICAL THINKING

Exercise Analysis

This exercise is very personal, and I won't tell you what you should or should not think. That's up to you, and I guarantee it will probably change as you learn more about the environment. I can assure you that the more you think about what you believe, the better off you'll be. The more open you are to new ideas and alternative viewpoints, the more apt you are to make good judgments. Be patient, though. It may take a long time to arrive at the truth. (This is another rule of critical thinking!)

I will, however, spend a few moments talking about where your ideas and values come from. Most of the ideas expressed in this exercise are expressions of values. These come from many sources. Our parents, for example, often help forge our values. Sometimes they teach us through example—positive and negative. Sometimes they lecture us in an attempt to instill values. We also acquire values from teachers, ministers, friends, relatives, government leaders, speakers, books, documentaries, movies, and advertisements. Almost everything we read or see or hear has some value statement behind it. Values often shape our ideas and actions.

Science can also shape values. You will be learning a great deal about the science of the environment and sustainability in this book, facts that may change both your thinking and your answers to the questions posed during this exercise.

1. Everyone seems to have a personal opinion on the root causes of the environmental crisis—or even if there is a crisis in the first place. Why do you think this is so?

2. Describe the views of Professor Lynn White on the religious roots of the environmental crisis. Do Christian teachings about dominance and about humans being apart from nature (duality) influence the way you think? How might science and technology be influenced by these Christian teachings?

3. Describe some of the criticisms of Lynn White's hypothesis. Do you see any flaws in his arguments? Does this mean that his hypothesis is invalid or may be just part of the picture?

4. Describe the way in which some scholars think that the spread of democracy and industrialization, especially mass production, may have contributed to the environmental crisis.

5. I use the term *biological imperialism* in this chapter to describe a natural biological tendency of species. What is this tendency? How has this tendency in humans contributed to the environmental crisis?

What factors make humans superior biological imperialists? How do values and beliefs affect our actions? *(Hint: You can think of both positive and negative influences on human beliefs.)*

6. If technology is so central to the environmental crisis, what can be done to make technology work for us, instead of undermining our future?

7. Make a list of the psychological factors discussed in this chapter and explain each one. Then list ways these psychological factors affect our laws, government, businesses, and personal lives.

8. Make a list of the systems that underlie your community and describe why you think they might be unsustainable in the long run. To help you think about this, you might consider their impacts on the local, regional, and global economy, society, and environment.

9. In your own words, describe the frontier ethic and the problems it can create.

10. Describe the sustainable ethic and the change it could have on the way we live our lives.

KEY TERMS

biological imperialism
crisis of unsustainability
ecosystem

ethics
frontier ethic
frontierism

Industrial Revolution
sustainable ethics

PART 2

NATURAL SYSTEMS/ HUMAN SYSTEMS

SEARCHING FOR A SUSTAINABLE RELATIONSHIP

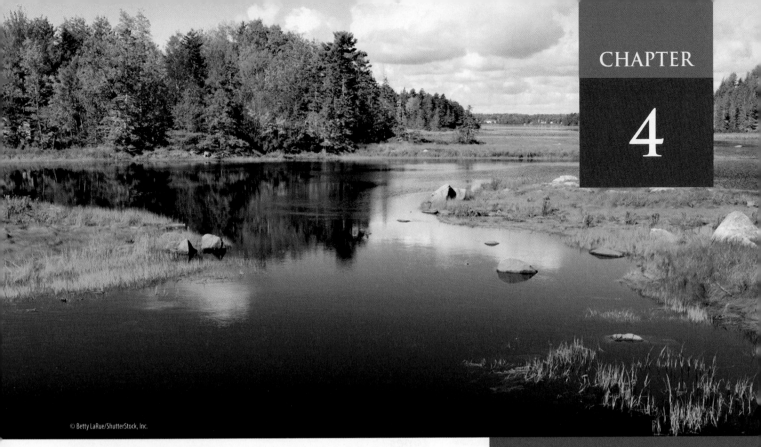

© Betty LaRue/ShutterStock, Inc.

Principles of Ecology: How Ecosystems Work

Never does nature say one thing, and wisdom another.

—Juvenal

 LEARNING OBJECTIVES

After reading and studying the material in this chapter, students will be able to understand, knowledgably discuss, and, when appropriate, apply the information learned in this chapter, including:

1. How dependent humans are on the Earth and its ecosystems for life and economic security
2. Why environmental protection is vital to our present and our future
3. What ecology is and isn't
4. The biosphere and its components, including ecosystems, as well as the biological (biotic) and nonbiological (abiotic) components of them
5. The biomes and aquatic life zones that constitute the biosphere and how dependent human societies are on them
6. The concept of the range of tolerance
7. How human activities alter biotic and abiotic conditions and alter organisms' survival by shifting conditions beyond their range of tolerance
8. Key concepts of limiting factors, niche, and competition
9. The structure and function of ecosystems, including food chains and food webs and how energy and nutrients flow through them
10. Key concepts of trophic level, biomass pyramid, energy pyramid, and pyramid of numbers
11. What nutrient cycles are and their importance to all forms of life, and how human activities adversely affect them

CRITICAL THINKING

The information gained from various fields of science such as ecology is often loosely translated in the public arena. Terms are sometimes misinterpreted. Facts are taken out of context. New findings are given more credence than they deserve, and old, disproved ideas remain in the popular thinking for a long time. As you read this chapter, make a list of terms, ideas, concepts, and facts you encounter that contradict what you thought was true.

Most of us live our lives seemingly apart from nature. We make our homes in cities and towns, surround ourselves with concrete and steel, and drown out the songs of birds with noise. The closest many of us get to nature is a romp with the family dog on the grass in the backyard. A lucky few come in much closer contact with the great outdoors through hiking, camping, canoeing, and kayaking. For many of these people, though, nature is still viewed as something apart from humans—a thing to protect to preserve a few pristine places for people to enjoy.

4.1 Humans and Nature: The Vital Connections

Hard as it may be for many people to accept, human beings are part of the fabric of life. We are a part of nature. We are dependent on the Earth and natural systems in thousands of ways. Consider our dependence first by taking a look around the room in which you are sitting. Everything in that room comes from the Earth or a natural system. The clothes you wear, your morning tea or coffee, and even the cornflakes you ate for breakfast are products of the Earth—the soil, water, air, and plants.

Like all other species, humans depend on the soil, air, water, sun, and a host of living organisms to survive. Each year, in fact, human beings (and other animals) consume enormous quantities of oxygen, which is used in the cells of our bodies to break down food molecules to generate energy. Oxygen is produced by plants and algae. Without these organisms, humans and other animals could not survive. Trees, grasses, and other plants also provide a host of additional free services. For example, plants protect the watersheds near our homes, preventing flooding and erosion. Swamps purify the water in streams and lakes—water many of us drink. Birds help to control insect populations.

Clearly, nature serves us well. Although many of us have isolated ourselves from nature, we still depend on nature in many ways. We have not emancipated ourselves from it at all. We also influence natural cycles, and therefore, not only do we depend on nature, the fate of natural systems depends on us.

 KEY CONCEPTS

Humans are a part of nature, dependent on natural systems for a variety of economically important resources and ecological services essential to our survival and long-term prosperity.

4.2 Ecology: The Study of Natural Systems

This chapter explores *ecology*, the study of living organisms and the web of relationships that binds all of us together in nature. The late professor Garrett Hardin, a world-renowned ecologist, wrote that ecology takes as its domain the entire living world. Ecologists study how organisms interact with one another and how they interact with the *abiotic*, or nonliving, components of the environment (sunlight, for example).

Throughout this chapter, we explore our connections to the living world. We examine the ways in which human systems depend on natural systems and the ways in which humans affect them. One of the goals of this chapter is to help you understand how nature works and how we can work better with nature to create a sustainable future. You will find that a great many of the lessons learned from the study of ecology can be applied to human society. Before we begin our journey, however, let me say a few words about the term *ecology*.

Ecology is probably one of the most misused words in the English language. Banners proclaim, "Save Our Ecology." Speakers argue that "our ecology is in danger," and others talk about the "ecological movement." These common uses of the word *ecology* are incorrect. Why?

Ecology is a branch of science. It describes and quantifies the web of interactions in the environment. But ecology is not synonymous with the word *environment*. Thus, we can save our ecology department and ecology textbooks, but we cannot save our ecology. Our ecology is not in danger, but our environment is. You cannot join the ecology movement, but you would be a welcome addition to the environmental movement.

Go Green

Check out current efforts to green your campus and become an active participant! If your college or university isn't actively pursuing ways to go green, consider organizing an effort with key student organizations, faculty members, and administrators. Chances are there are plenty of environmentally active faculty and staff that would love to help out.

Ecology is a field of science that seeks to describe relationships between organisms and their chemical and physical environment.

4.3 The Structure of Natural Systems

To understand ecology, you must study the structure of natural systems. We begin with the biosphere.

The Biosphere

The science of ecology focuses much of its attention on biological systems, examining their components and interactions. The largest biological system is the **biosphere** (BI-oh-sfere), the skin of life on planet Earth. As shown in **FIGURE 4-1**, the biosphere forms at the intersection of air, water, and land. In fact, all living organisms consist of components derived from these three realms. The carbon atoms in body proteins, for example, come from carbon dioxide in the atmosphere. Carbon dioxide is captured by plants and made into food molecules by a process known as **photosynthesis.** Animals eat plants; the food molecules then become the building blocks of proteins and other important molecules in animals. The minerals in the bones of animals come from the soil, again through plants. Water comes directly from streams and lakes and plant matter.

The biosphere extends from the bottom of the ocean, approximately 11,000 meters (36,000 feet) below the surface, to the tops of the highest mountains, about 9,000 meters (30,000 feet) above sea level. Although that may seem like a long way, it's really not. In fact, if the Earth were the size of an apple, the biosphere would be about the thickness of its skin. Although life exists throughout the biosphere, it is rare at the extremes, where conditions for survival are less than optimum. Most living things are concentrated in a narrow band extending from less than 200 meters (600 feet) below the surface of the ocean to about 6,000 meters (20,000 feet) above sea level.

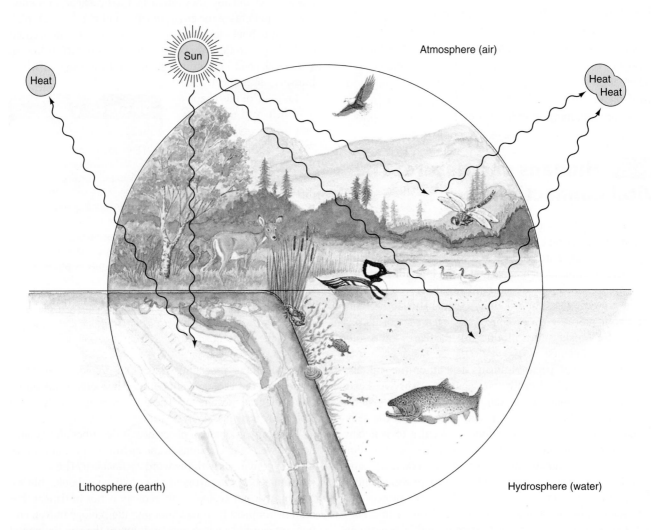

FIGURE 4-1 The biosphere. The biosphere exists at the intersection of land, air, and water. Organisms derive essential minerals and other substances from these three spheres.

The biosphere is a **closed system,** much like a sealed terrarium. By definition, a closed system receives no materials from the outside. The only outside contribution is sunlight, which is vital to the health and well-being of virtually all life.

Sunlight powers almost all life on the planet. Even the energy released by the combustion of coal, oil, and natural gas, which we use to power our homes and factories, comes from sunlight that fell on the Earth several hundred million years ago.

Because the Earth is a closed system, all materials necessary for life must be recycled. The carbon dioxide you exhale, for instance, may be used by a rice plant during photosynthesis next month in Indonesia. Those carbon dioxide molecules will be incorporated into carbohydrate produced by the plant and stored in the seed. Consumed by an Indonesian boy, the carbohydrate will be broken back down during cellular energy production. The carbon dioxide molecules are released into the atmosphere and made available to other plants. Without this and dozens of other recycling processes, all life on the planet would grind to a

The biosphere is an enormous biological system, spanning the entire planet. The materials within this closed system are recycled over and over in order for life to be sustained. The only outside contribution to the biosphere is sunlight, which provides energy for virtually all living things.

halt. Protecting the environment, then, helps to protect global recycling systems on which we all depend.

Biomes and Aquatic Life Zones

The biosphere consists of terrestrial and aquatic systems. Viewed from outer space, the Earth—the terrestrial portion of the biosphere—resembles a giant jigsaw puzzle, consisting of large landmasses among vast expanses of ocean. The landmasses, or continents, can be divided into fairly large regions called *biomes* (**FIGURE 4-2**). A **biome** is a terrestrial

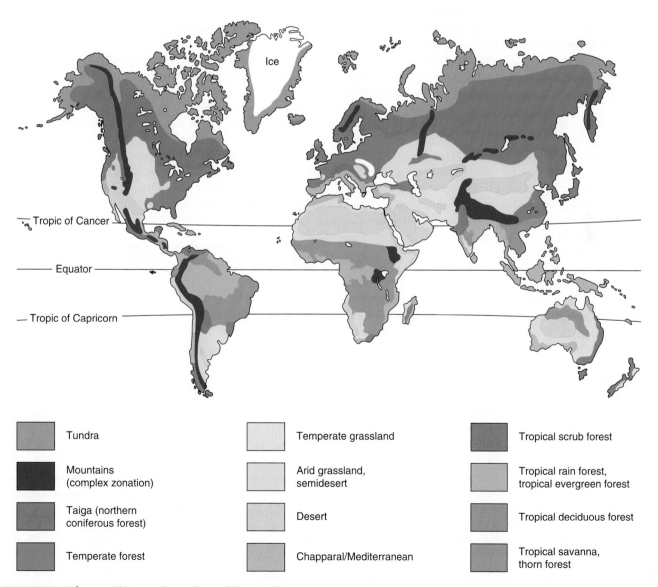

FIGURE 4-2 Biomes. This map shows the world's major biomes.

portion of the biosphere characterized by a distinct climate and a particular assemblage of plants and animals adapted to it.[1] This section provides an overview of biomes.

In a biome, abiotic conditions—such as soil type, temperature, and rainfall—determine the plant communities that can survive. These, in turn, determine which animals can subsist. As illustrated in Figure 4-2, the North American continent contains seven major biomes, five of which are discussed here. Starting in the north is the **tundra** (TON-dra), a region of long, cold winters and rather short growing seasons (**FIGURE 4-3A**). The rolling terrain of the tundra supports grasses, mosses, lichens, wolves, musk oxen, and other animals adapted to the bitter winter cold. Trees cannot grow on the tundra because of the short growing season and because the subsoil (called *permafrost*) remains frozen year round, preventing the deep root growth necessary for trees.

Immediately south of the tundra lies the **taiga** (TIE-ga), also known as the **northern coniferous** or **boreal forest.** The taiga's milder climate and longer growing season result in a greater diversity and abundance of plant and animal life than exists on the tundra. Evergreen trees, bears, wolverines, and moose are characteristic species (**FIGURE 4-3B**).

East of the Mississippi River lies the **temperate deciduous forest biome,** characterized by an even warmer climate and more abundant rainfall (**FIGURE 4-3C**). Broad-leaved trees make their home in this biome. Opossums, black bears, squirrels, and foxes are characteristic animal species.

West of the Mississippi lies the **grassland biome** (**FIGURE 4-3D**). Inadequate rainfall and periodic drought prevent trees from growing on the grasslands, except near rivers, streams, and human habitation. Over the years, deep-rooted grasses have evolved on the plains. These grasses can withstand fire, drought, and grazing even in arid grasslands. Coyotes, hawks, and voles are characteristic animal species.

In the Southwest, where even less rain falls, is the **desert biome** (**FIGURE 4-3E**). In contrast to what many people think, the desert often contains a rich assortment of plants and animals uniquely adapted to aridity and heat. Cacti, mesquite trees, rattlesnakes, and a variety of lizards all make their home in this seemingly inhospitable environment. The scorching hot deserts of Saudi Arabia receive less moisture than the desert around Tucson, Arizona, and therefore contain far fewer plants and animals.

The oceans can also be divided into distinct zones, known as **aquatic life zones**. The aquatic equivalent of biomes, each of these regions has a distinct environment and characteristic plant and animal life adapted to conditions of the zone. Both freshwater and saltwater (marine) systems exist. Freshwater systems include lakes, rivers, ponds, and marshes. The four major marine aquatic life zones are coral reefs, estuaries (the mouths of rivers, where fresh and salt water mix), the deep ocean, and the continental shelf.

Humans inhabit all biomes on Earth but are concentrated in those in which conditions are mildest and most conducive to growing food. Within many biomes is a wide

assortment of natural resources such as minerals and fuel. Although they are vital to our economies and well-being, these resources are but a fraction of the benefit we gain from biomes. Plants and animals, for example, provide food and great enjoyment to many. Microorganisms in the soils of biomes detoxify wastes and recycle nutrients, keeping the soil rich and productive. The soil itself serves as a growing medium for all plants—crops as well as forests. Besides providing timber, trees provide oxygen, remove air pollutants, and protect the soil from erosion, thus reducing sediment pollution in surface waters. Vegetation also reduces flooding.

KEY CONCEPTS

The biosphere consists of distinct regions called *biomes* and aquatic life zones, each with its own chemical and physical conditions and unique assemblage of organisms. Humans inhabit all biomes, but are most prevalent in those with the mildest climates.

What Is an Ecosystem?

The biosphere is a chemical, physical, and biological system that encompasses the entire surface of the planet. Therefore, the biosphere is often referred to as a global **ecological system,** or **ecosystem**. If the biosphere is a global ecosystem, biomes are regional ecosystems. For the sake of convenience, ecologists often limit their studies to smaller portions of a biome—individual forests, ponds, or meadows. All of these ecosystems, no matter how big or small, consist of two components: the living or **biotic** components and the nonliving or **abiotic** components. Numerous interactions exist among these various components.

KEY CONCEPTS

Ecosystems are biological systems consisting of organisms and their environment.

Abiotic Components of Ecosystems and the Range of Tolerance The abiotic components of an ecosystem are the physical and chemical factors necessary for life—sunlight, precipitation, temperature, and nutrients. In most ecosystems, the abiotic conditions vary during the day and often shift from one season to the next. To live in most ecosystems, then, organisms must be able to survive a range of conditions. The range of conditions to which an organism is adapted is called its **range of tolerance**. As **FIGURE 4-4** shows, organisms do best in the **optimum range**. Outside of that are the **zones of physiological stress,** where survival and reproduction are possible but not optimal. Outside of these zones are the **zones of intolerance,** where life for that organism is implausible. As a rule, organisms that have a wide range of tolerance are more widely distributed than organisms with a narrow range of tolerance.

---

[1] Although each biome has its specific climate, there can be considerable climatic variation within a given biome.

FIGURE 4-3 North America's five major biomes. (a) Tundra, **(b)** taiga, **(c)** temperate deciduous forest, **(d)** grassland, and **(e)** desert.

Photo a courtesy of U.S. Fish & Wildlife Services; photos b and c © Photos.com; photo d courtesy of Tim McCabe/NRCS USDA; photo e © Anthony Ricci/ShutterStock, Inc.

Many examples of the range of tolerance can be given. Think of your own tolerance for temperature. Most people are comfortable around 23°C (70°F). When the temperature becomes much warmer, you enter the zone of physiological stress. If it increases even more, death occurs if you can't find a way to cool yourself down. This is the zone of intolerance. The same occurs at the lower end of the temperature scale.

It is important to note that the tolerance of any given species for an environmental factor may vary with the age of the organism. Newly hatched salmon, for instance, may

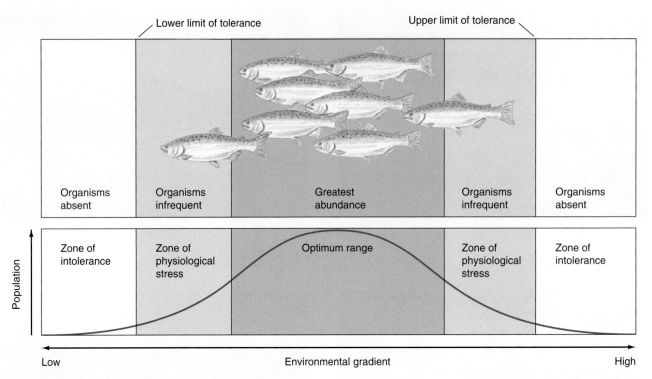

FIGURE 4-4 **Range of tolerance.** Organisms can live within a range of conditions, called the range of tolerance, but they thrive in the optimum range.

be much more vulnerable to pollutants in the water than adult salmon are. The tolerance of individuals in any population also varies. Thus, some humans are more tolerant than others to heat or cold.

Humans often alter the abiotic components of their environment. Dams, for example, create lakes in streambeds, altering water temperature and water flow, with devastating effects on native fish populations. Dams also seriously alter downstream conditions. In the Colorado River, for instance, native razorback suckers and other species that once thrived in the warm waters of the river are now **endangered species** (in danger of extinction) because of the large dams that release extremely cold water from the bottom of the reservoir (**FIGURE 4-5A**). This water is too cold for the sucker and other native fish species (**FIGURE 4-5B**). (The Point/Counterpoint in this chapter gives two opposing views on human-caused extinction.)

Natural events can also alter the biotic and abiotic conditions of the environment. Floods, tropical storms, and volcanic eruptions all change conditions for varying periods, with sometimes dramatic effects on native species.

 KEY CONCEPTS

Organisms thrive within a range of abiotic conditions; altering those conditions can have severe consequences and can cause extinction.

Limiting Factors

Although species are sensitive to all of the abiotic factors in their environment, the one factor that is in short supply, called a **limiting factor,** tends to regulate population size. In freshwater lakes and rivers, dissolved phosphate is a limiting factor. Phosphate is needed by plants and algae for growth, but phosphate concentrations are naturally low. As a result, plant and algal growth is held in check. When phosphate is added to a body of water—say, by detergents released in the effluent of a sewage treatment plant— plants and algae proliferate. Algae often form dense surface mats, blocking sunlight (**FIGURE 4-6**). Plants that are rooted on the bottom of the water body may perish for lack of sunlight. Because these plants produce oxygen, their demise often causes oxygen levels in the deeper waters to decline, killing fish and other aquatic organisms.

On land, precipitation tends to be the limiting factor. At any given temperature, the more moisture that falls, the richer the plant and animal life.

 KEY CONCEPTS

Organisms require many different abiotic factors to survive, but one factor—*the limiting factor*—tends to be critical to survival and growth of a population. Increasing the concentrations of a limiting factor can result in dramatic increase in population growth with many adverse effects on organisms.

Biotic Components

Ecosystems consist of numerous organisms, including bacteria, plants, fungi, and animals. These are the biotic components of the ecosystem. Organisms of the same species within a biome or aquatic life zone usually occupy a specific region. This group of organisms is called a **population.**

(a)

(b)

FIGURE 4-5 Upsetting the ecological balance. (a) Dams like the mighty Glen Canyon Dam in Arizona not only inundate upstream areas, they often change the water temperature in the downstream section. **(b)** The cool water flowing out of the bottom of the reservoir created by Glen Canyon Dam has endangered the razorback sucker and several other native fish species.

Photo a © gary 718/ShutterStock, Inc.; photo b © Ken Lucas/Visuals Unlimited.

FIGURE 4-6 ALGAL BLOOM. This pond is choked with algae due to the abundance of plant nutrients from human sources.

© Josh Meyer/ShutterStock, Inc.

In any given ecosystem, populations of different organisms exist together in an interdependent **biological community**.

All organisms in a community, including humans, are part of the web of life. Organisms interact in many ways. Some organisms prey on others. Others are preyed upon. Still others compete with fellow members of the community of life.

KEY CONCEPTS

Organisms are the biotic components of ecosystems; they form an interdependent community of life.

Niche and Competition If asked to give a brief description of yourself, you would probably begin by describing the place where you live. You might discuss the college or university you attend, your job, the friends you have, and other important relationships that describe your place in human society. Now that you're discovering your part in nature, you may even describe your place in the community of life.

A biologist would do much the same when describing an organism. He or she would start with a description of the place where an organism lives—that is, its **habitat.** Next, he or she would describe how the organism fits into the ecosystem, its **ecological niche,** or simply **niche.** An organism's niche consists of its relationships with its environment—both the abiotic and biotic components. This niche includes what an organism eats, what eats it, its range of tolerance for various environmental factors, and other important elements. The niche of an organism is its functional role compared with the habitat, which is its "address."

Organisms in a community occupy the same habitat, but most of them have quite different niches, a phenomenon that minimizes competition. The fact that organisms occupy separate niches provides for a wider use of an ecosystem's resources, especially food.

Niches do overlap somewhat. For example, two species may feed on some of the same foods. Coyotes and foxes, for example, both feed on rabbits and mice. Coyotes also feed on larger prey, however—even deer. Foxes tend to feed on smaller prey, including reptiles and amphibians.

The more two species' niches overlap, the more they compete. When niches overlap considerably, competition becomes intense, and one species usually suffers. If two species occupy identical niches, competition will eliminate one of them. As a result, two species cannot occupy the same niche for long. This rule is called the **competitive exclusion principle.**

The concept of the niche is very important to the sustainable management of natural resources. For example, successful control of an insect pest on crops is best achieved through an understanding of the species' niche. An analysis of an insect's niche might show that certain birds or insects feed on the pest. By encouraging these beneficial species—say, by providing trees for the birds—farmers can reduce pest populations naturally. They could also save enormous sums of money on chemical pesticides and help protect the environment.

Humans: Competitors Extraordinaire Humans, like other organisms, compete with one another for a variety of resources. People also compete with the many other species that share this planet with us. For example, we compete with sea otters and seals over salmon. When we graze a cow on a pasture, our livestock compete with woodchucks and rabbits for food.

The competition between people and other species is fairly lopsided. Humans possess a marked advantage over most other species. This edge on the competition stems primarily from our technological prowess—high-powered rifles, bulldozers, and chain saws being three examples. Fishing nets and sonar give commercial fishers a marked advantage over their two major competitors, seals and sea otters.

As the human population grows, as our demand for food and resources climbs, and as our technological prowess increases, our competitive advantage will only increase, but the effect may not be advantageous to human civilization in the long run. Already, commercial overfishing has depleted dozens of the world's fisheries. Many other fisheries are now in danger. Overfishing has had a ripple effect on other species, reducing populations of seals and other fish-eating animals.

On every continent and in every nation, humans are outcompeting other species. According to various estimates, 40 to 100 species become extinct every day, largely because of tropical deforestation. Unless we do something, hundreds of thousands of species will become extinct in the next decade. Many scientists warn that none of us will be immune to the impacts of such widespread biological impoverishment. Cutting down the rain forests, for example, could alter global climate. Why? Rain forests absorb massive amounts of carbon dioxide and thus help control atmospheric levels of this important greenhouse gas. As forests are destroyed, carbon dioxide levels increase, and the Earth's atmosphere will very likely become warmer. We could feel the effect in higher utility bills to cool our homes because of noticeably hotter summers. If predictions come true, we could face much higher food costs as hotter summers reduce crop production. If conditions became bad enough, massive food shortages could occur. Although all this human intrusion is rather depressing, there is hope. We can find ways of coexisting with nature. Understanding how natural systems operate and patterning human systems after sustainable natural systems could help us redesign our infrastructure. See **Spotlight on Sustainable Development 4-1** for an example.

KEY CONCEPTS

The place an organism or population of organisms lives is known as its habitat. Its ecological niche is all its relationships with the biotic and abiotic components of its environment. Competition occurs between species occupying the same habitat if their niches overlap. Understanding an organism's niche is vital to sustainable management of wildlife and other natural resources.

KEY CONCEPTS

Humans are a major competitive force in nature. Our advanced technologies and massive population size, however, enable us to outcompete many species causing their demise and ecological backlashes that affect humans as well.

Evolution is the process of change in gene pools through time. When one gene pool becomes reproductively independent, a new species has formed. Such speciation generates species; extinction takes them away. Simply put, extinction is a failure to adapt to change, the termination of a gene pool, the end of an evolutionary line.

Extinction is a natural process. Most species that ever lived are now extinct. The 3 to 30 million species on Earth today are no more than 1 to 10% of the species that have evolved since life began about 3.5 billion years ago. So why are thoughtful people concerned about endangered species? History makes it clear that, given enough time, each species will become extinct.

The concern is that today the natural process of extinction proceeds at an unnatural rate. Let us estimate by how much human activity has accelerated rates of extinction. The average life span of species is 1 to 10 million years. Assume (to be conservative) that the average longevity of species of higher vertebrates is 1 million years. In round numbers, there are 13,000 species of birds and mammals. So, on average, one species ought to go extinct each century. However, between 1,600 and 2,000, at least 36 species of mammals and 94 species of birds, became extinct—32 species per year, or 32 times the natural rate.

What does it mean to increase a rate by 32 times? Exceed the 55 mph speed limit by 32-fold, and you are moving 1,760 miles per hour, over twice the speed of sound. The difference between natural rates of extinction and present, human-influenced rates is analogous to the difference between a casual drive and Mach 2! Is that a problem? You decide: Concern is a moral construct, not a scientific one.

Several human activities have contributed—mostly inadvertently— to accelerating rates of extinction. The dodo and the passenger pigeon were extinguished by overhunting. Wolves and grizzly bears were exterminated over much of their ranges as threats to livestock. The black-footed ferret was driven to the verge of extinction because prairie dogs, its staple food, were poisoned as an agricultural pest.

Habitat change is the most important cause of endangerment and extinction. Clearing forests for agriculture has decimated the lemurs of Madagascar. Pesticides led to the decline of the peregrine falcon. Introducing exotic species (like goats on the Galapagos and mongooses in Hawaii) displaces native animals and plants. Developing the Amazon Basin is a habitat alteration, and a cause of extinction, on an unprecedented scale.

Many urge saving species for their aesthetic value. Whooping cranes are beautiful, and part of their beauty

Humans Are Accelerating Extinction

David M. Armstrong

Dr. David M. Armstrong is a mammologist who taught environmental biology at the University of Colorado at Boulder and has written several books on mammals and ecology of the Rocky Mountain region.

is that they are products of a marvelous evolutionary process. Most concern about accelerated extinction, however, stresses economic value. A tiny fraction of seed plants are used commercially. An obscure plant like jojoba eventually could be a source of oil more reliable than the Middle East. Wild grasses have yielded genes that have improved disease resistance in wheat. Some animals like musk ox, kudu, and whales could contribute protein to our diet. Species may have medical value; penicillin, after all, was once an obscure mold on citrus fruit. Some sensitive species are monitors of environmental quality, and the presence of healthy populations of many species may promote greater stability or resilience of ecosystems. Naturalist Aldo Leopold noted that humans have a way of "tinkering" with the ecosphere. But he suggested that the first rule of tinkering ought to be that one never throws away any of the parts.

"Extinction is forever" and impoverishes both ecosystems and the potential richness of human life. Borrowing again from Aldo Leopold, I believe our concern about unnaturally rapid extinction is part of a "right relationship" between people and the landscapes that nurture and inspire them.

Biologist Sir Julian Huxley noted that "we humans find ourselves, for better or worse, business agents for the cosmic process of evolution." We hold power over the future of the biosphere, the power to destroy or preserve. German philosopher Georg Hegel noted that freedom implies responsibility. I agree. The power to destroy species implies a responsibility to preserve them. The question of human-accelerated extinction boils down to a simple ethical question, "Does posterity matter?" Some of us have ethics that are human centered. We ask simply, "Do my children deserve a life as rich, with as much opportunity, as mine?" If they do, then we have the responsibility to choose restraint. (Perhaps my life has been rich enough without the dodo, but I am reluctant to make that judgment for future generations.)

Extinction Is the Course of Nature

Norman D. Levine

Norman D. Levine was a professor emeritus at the College of Veterinary Medicine and Agricultural Experiment Station, University of Illinois at Urbana. His research interests included parasitology, protozoology, and human ecology. This article is provided by the American Institute of Biological Sciences (© 1989).

Courtesy of University of Illinois, Urbana-Champaign

Evolution is the formation of new species from preexisting ones by a process of adaptation to the environment. Evolution began long ago and is still going on. During evolution those species better adapted to the environment replaced the less well-adapted species. It is this process, repeated year after year for millennia, that has produced the present mixture of wild species. Perhaps 95% of the species that once existed no longer exist.

Human activities have eliminated many wild species. The dodo is gone, and so is the passenger pigeon. The whooping crane, the California condor, and many other species are on the way out. The bison is still with us because it is protected, and small herds are raised in semicaptivity. The Pacific salmon remains because we provide fish ladders around our dams so it can reach its breeding places. The mountain goat survives because it lives in inaccessible places. But some thousands of other animal species, to say nothing of plants, are extinct, or soon will be.

Some nature lovers weep at this passing and collect money to save species. They make lists of animals and plants that are in danger of extinction and sponsor legislation to save them.

I don't. What the species preservers are trying to do is stop the clock. It cannot and should not be done. Extinction is an inevitable fact of evolution, and it is needed for progress. New species continually arise, and they are better adapted to their environment than those that have died out.

Extinction comes from failure to adapt to a changing environment. The passenger pigeon did not disappear because of hunting alone, but because its food trees were destroyed by land clearing and farming. The prairie chicken cannot find enough of the proper food and nesting places in the cultivated fields that once were prairies.

And you cannot necessarily introduce a new species, even by breeding it in tremendous numbers and putting it out into the wild. Thousands of pheasants were bred and set out year after year in southern Illinois, but in the spring of each year there were none left. Another bird, the capercaillie, is a fine, large game bird in Scandinavia, but every attempt to introduce it into the United States has failed. An introduced species cannot survive unless it is preadapted to its new environment.

A few introduced species are preadapted, and some make spectacular gains. The United States has received the English sparrow, the starling, and the house mouse from Europe, and also the gypsy moth, the European corn borer, the Mediterranean fruit fly, and the Japanese beetle. The United States gave Europe the gray squirrel and the muskrat, among others. The rabbit took over in Australia, at least for a time.

The rabbit and the squirrel were successful on new continents because their requirements are not as narrow as those of other species that failed. Today, adjustment to human-made environments may be just as difficult as adjustment to new continents. The rabbit and the squirrel have succeeded in adjusting to the backyard habitat, but most wild animals have disappeared.

Human-made environments are artificial. People replace mixed grasses, shrubs, and trees with rows of clean-cultivated corn, soybeans, wheat, oats, or alfalfa. Variety has turned into uniform monotony, and the number of species of small vertebrates and invertebrates that can find the proper food to survive has become markedly reduced. But some species have multiplied in these environments and have assumed economic importance; the European corn borer in this country is an example.

Would it improve Earth if even half of the species that have died out were to return? A few starving, shipwrecked sailors might be better off if the dodo were to return, but I would not be. The smallpox virus has been eliminated, except for a few strains in medical laboratories. Should it be brought back? Should we bring panthers back into the eastern states? Think of all the horses that the automobile and tractors have replaced, and of all the streets and roads that have been paved and the wild animals and plants killed as a consequence. Before people arrived in America about 10,000 years ago, the animal–plant situation was quite different. What should we do?

Should we all commit suicide? Evolution exists, and it goes on continually. People are here because of it, but people may be replaced someday. It is neither possible nor desirable to stop it, and that is what we are trying to do when we try to preserve species on their way out. It can be done, I think, but should we do it to them all? Or to just a few, as we are doing now?

Critical Thinking Questions

1. Summarize and critically analyze the key points of both authors. Do you see any flaws in their reasoning?
2. Which viewpoint do you adhere to? Why?

4-1 Sustainable Sewage Treatment: Mimicking Nature

Since the advent of cities and towns, dealing with human waste has proved to be a huge challenge. Even in the rich, industrialized countries of the world, modern waste treatment practices leave something to be desired. The liquid waste is chemically treated and dumped into streams and lakes in most cases. The solid material or sludge that remains after treatment is trucked off to landfills.

Traditional waste treatment methods not only pollute our environment, they also waste valuable nutrients, which come from the foods we eat and ultimately the soil on which crops are grown. In nature, plant and animal waste is returned to the soil, where it nourishes new plant life.

Over the years, scientists and others have sought more environmentally compatible—and sustainable— ways of dealing with human waste. Biologist John Todd has led this effort through the invention of waste disposal systems that mimic nature's ways. Todd designed and built his first solarpowered sewage treatment plant at Sugarbush ski resort in Vermont. In this system, known as an Eco Machine, raw sewage enters a solar-heated greenhouse and then flows into cylinders, where naturally occurring bacteria convert the ammonia in the sewage into nitrate, a plant nutrient. The effluent then enters special channels where algae consume the nitrates. The algae are part of an artificial ecosystem containing freshwater shrimp (that feed on algae) and fish (that feed on the shrimp). Snails in the system consume the sludge (organic wastes that fall to the bottom) and also serve as food for fish.

At the far end of the greenhouse is a small marsh containing organisms that remove additional impurities before the water is released into a nearby stream (**FIGURE 1**). In this artificial marsh are many plants that absorb toxic substances.

In Providence, Rhode Island, Todd installed a much larger system that handles up to 16,000 gallons of raw sewage per day flowing through a greenhouse containing 1,200 aquariums, which house a variety of organisms to purify the water. This system, which costs one-third as much as an equivalent sewage treatment plant, has negligible environmental impact. To date, Todd's company, which is now operated by his son, has installed dozens of systems in 11 countries. To learn

FIGURE 1 Proponents of this new technology point out that malfunctions at conventional plants can be serious and sometimes require the evacuation of nearby residents to avoid poisonous chlorine gas emitted from them. A malfunction in a solar-aquatic plant may kill off some organisms, but it poses no threat to people.
Courtesy of John Todd, University of Vermont.

more you can log on to Todd's website, www.toddecological.com.

Another pioneer in ecologically sound waste disposal is environmental engineer Bill Wolverton. In the mid-1970s, while working at NASA, Wolverton began experiments to purify wastewater from NASA facilities using lagoons filled with prolific, nutrient-hungry water hyacinths, which have become a "pest" in lakes, streams, and rivers in the southern United States. One system he designed for a 4,000-person NASA facility in Mississippi has saved the agency millions of dollars in sewage fees. Since he began, Wolverton has designed more than a hundred systems in the southern United States. The effectiveness of these systems is illustrated in Figure 2. This graph shows permitted discharges of three pollutants from a conventional sewage treatment plant (STP) and levels achieved at a Walnut Cove wastewater treatment facility, designed by Wolverton (**FIGURE 2**).

Researchers are also experimenting with smaller constructed wetlands for biological treatment that we can use in our own backyards. Many are designed so waste water percolates through a bed of gravel under a layer of soil and plants, and thus, there is no way sewage could reach the surface.

Another pioneer in biological waste treatment is Tom Watson, who lives in New Mexico. Watson specializes in household waste treatment centers using inexpensive plastic infiltrators in a pumice bed (**FIGURE 3**). Plants growing in the overlying soil send their roots into pumice, where they feed on water and nutrients from household waste, including toilet water. Bacteria in the pumice decompose organic matter, releasing nutrients for plants.

Biological treatment systems have pros and cons. Those being built by John Todd can operate in virtually any climate. These systems can also be scaled up to any size simply by adding more greenhouses. Watson's systems and constructed wetlands can also function well in a variety of climates, but because they're not protected from the weather, they may have problems in extremely cold climates.

Biological treatment systems, both large and small, are well suited to developing countries because they offer a low-cost option for treating waste. In fact, some developing nations are currently considering installing biological systems to upgrade existing facilities. They're finding that the necessary upgrades can be made at moderate cost. Moreover, the facilities do not require as much management or use as much energy as conventional systems. They even produce cleaner treated water than their high-tech counterparts—and at a lower cost.

One concern is that in large biological treatment facilities plants may absorb heavy metals such as mercury and lead, which would need to be disposed of in a way that was safe. Todd's colleague, Alan Liss, notes, "It's better to have a small amount of highly toxic plants than a huge amount of moderately toxic sludge." Eliminating such wastes from the waste stream through preventive measures would easily solve this problem.

Larger biological systems such as those being built by John Todd and Bill Wolverton require a level of biological knowledge that few sewage treatment engineers have. They represent a radically different approach for society, even though natural systems have been purifying water for millennia. This appearance of being different makes it difficult to convince city officials to give them a try when they know that existing technology will do the job. Some facilities produce enormous amounts of plant waste, sometimes containing hazardous substances including heavy metals that must be safely disposed of.

Despite their drawbacks, biological systems that mimic nature can be adapted for most or all situations, and their use will undoubtedly rise in the years to come as cities and towns look for more sustainable ways of treating waste.

FIGURE 2 Graph of pollutants released from a conventional water treatment facility (STP) and Walnut Cove wastewater treatment. [BODS is biological oxygen demand, a measure of organic waste. TSS is suspended solids, and NH₃-N is nitrogen in the form of ammonia.]

Data compiled from Walnut Cove wastewater treatment-monitoring data.

FIGURE 3 **The Watson Wick filter.** In this simple, effective set up wastewater enters the pumice bed, where bacteria go to work on waste. Plant roots absorb nutrients and water. Water coming out of the system is relatively free of nutrients. Tests so far have shown them to be more effective than conventional sewage treatment methods.

4.4 Ecosystem Function

Life on land and in the Earth's waters is possible principally because of the existence of the **producers,** organisms such as the algae and plants. These organisms absorb sunlight and use its energy to synthesize organic foodstuffs from atmospheric carbon dioxide and water via photosynthesis. These organic molecules are used by the producers themselves, but they also provide nourishment for all the other organisms. Producers, therefore, form the foundation of the living world.

Another large group of organisms is the **consumers**. Ecologists place consumers into four general categories, depending on the type of food they eat. Some, such as deer, elk, and cattle, feed directly on plants and are called **herbivores** (ERB-ah-voors). Others, such as wolves, feed on herbivores and other animals and are known as **carnivores** (CAR-neh-voors). Humans and a great many other animal species subsist on a mixed diet of plants and animals and are known as **omnivores** (OM-neh-voors). Another group feeds on animal waste or the remains of plants and animals and are called **detritivores** (dee-TREH-teh-voors) or **decomposers**. This group includes many bacteria, fungi, and insects.

KEY CONCEPTS

Photosynthetic organisms such as plants and algae produce food within ecosystems and are known as producers. Their well-being is essential to the survival and well-being of all other organisms including herbivores, carnivores, and detritivores.

Food Chains and Food Webs

Biological communities consist of numerous food chains. A **food chain** is a series of organisms, each one feeding on the organism preceding it (**FIGURE 4-7**). All organisms in the community are members of one or more food chains.

FIGURE 4-7 Simplified grazer food chains. These drawings show terrestrial (land-based) and aquatic grazer food chains.

Biologists recognize two general types of food chains: grazer and decomposer. **Grazer food chains** begin with plants and algae. These organisms are consumed by herbivores, or grazers. Herbivores, in turn, may be eaten by carnivores.

Decomposer food chains begin with dead material—either animal wastes (feces) or the remains of plants and animals. These are consumed by decomposers including insects, worms, and a host of microorganisms such as bacteria. These organisms are responsible for the decomposition of the waste and the return of its nutrients to the environment for reuse (recycling).

In ecosystems, decomposer and grazer food chains are tightly linked (**FIGURE 4-8**). Thus, waste from the grazer food chain enters the decomposer food chain. Nutrients liberated by the decomposer food chain enter the soil and water and are reincorporated into plants at the base of the grazer food chain.

Food chains exist only on the pages of textbooks; in a community of living organisms, food chains are part of a much more complex network of feeding interactions, **food webs** (**FIGURE 4-9**). Food webs present a complete picture of the feeding relationships in any given ecosystem.

As with so many other topics in ecology, an understanding of food chains and food webs is essential to living sustainably on the planet. For instance, efforts to protect the ozone layer, which shields the Earth from harmful ultraviolet radiation, are important to protect people from cancer, but they are also important because they protect phytoplankton in the world's oceans. Phytoplankton, small microscopic photosynthesizers, form the base of aquatic food chains. Ultraviolet radiation can kill phytoplankton and thus cause a collapse of aquatic food chains.

KEY CONCEPTS

Food chains are feeding relationships that are part of much larger food webs in ecosystems. Two types of food chains exist: grazer and decomposer.

The Flow of Energy and Nutrients Through Food Webs

Energy and nutrients both flow through food webs, but in very different ways. Let us begin with energy. As you just learned, solar energy is first captured by plants and then used to produce organic food molecules. Energy from the sun is then stored in these molecules. In the food chain, organic molecules pass from plants to animals. In both plants and animals, the molecules are broken down. This process releases stored solar energy, which is used to power numerous cellular activities.

During cellular energy release, a good portion of the energy stored in organic food molecules is lost as heat. Heat escaping from plants and animals is radiated into the atmosphere and then into outer space. It cannot be recaptured and reused by plants or animals. Because all solar energy is eventually converted to heat, energy is said to flow unidirectionally through food chains and food webs. Put another way, energy cannot be recycled.

In contrast, nutrients flow cyclically; that is, they are recycled. Nutrients in the soil, air, and water are first incorporated into plants and algae and are then passed from plants to animals in various food chains. Nutrients in the food chain eventually reenter the environment through waste or the decomposition of dead organisms.

Every time you exhale, for example, you release carbon dioxide, a waste product of cellular energy production. Carbon dioxide reenters the atmosphere for reuse. Thus, through the act of breathing you play an important role in the global recycling system that makes life possible. Other wastes must be decomposed before releasing their nutrients. The feces of a rhinoceros, for example, are broken down by bacteria, which liberate carbon dioxide, nitrogen, and minerals.

Nutrients also reenter the environment through the decomposition of dead organisms. When a plant or animal dies, bacteria and fungi devour its organic remains. This process is called **decomposition**.

Although microorganisms utilize nutrients released during this process, some nutrients escape and enter the soil and water for reuse. (Of course, when a bacterium dies, it also breaks down, releasing nutrients into its environment.)

One way or another, nutrients eventually make their way back to the environment for reuse. New generations of organisms, therefore, rely on the recycling of nutrients in the biosphere. Every atom in your body has been recycled since the beginning of life on Earth. Perhaps some of those atoms were in the very first cells.

KEY CONCEPTS

Food chains are biological avenues for the flow of energy and the cycling of nutrients in the environment. Energy flows in one direction through food chains, but nutrients are recycled.

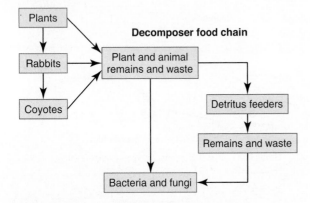

FIGURE 4-8 A grazer food chain and a decomposer food chain, showing the connection between the two.

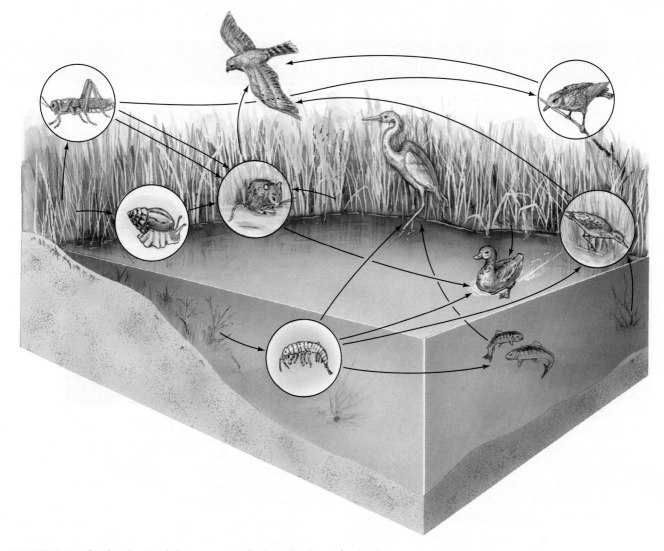

FIGURE 4-9 A food web. Food chains are actually threads in larger food webs.

Trophic Levels

Ecologists classify the organisms in a food chain according to their position, or **trophic level** (literally, "feeding" level). The producers are the base of the grazer food chain and belong to the first trophic level. The grazers are part of the second trophic level. Carnivores that feed on grazers are in the third trophic level, and so on.

Most terrestrial food chains are limited to three or four trophic levels. Longer terrestrial food chains are rare because food chains generally do not have a large enough producer base to support many levels of consumers. Why not?

Plants absorb only a small portion of the sunlight that strikes the Earth (only 1 to 2%), which they use to produce organic matter, or biomass. Technically, **biomass** is the dry weight of living material in an ecosystem. The biomass at the first trophic level is the raw material for the second trophic level. The biomass at the second trophic level is the raw material for the third trophic level, and so on.

As shown at the top of **FIGURE 4-10**, not all of the biomass produced by plants is converted to grazer biomass. At least

three reasons account for the incomplete transfer of biomass from one trophic level to the next. First, some of the plant material, such as the roots, is not eaten. Second, not all of the material that the grazers eat is digested. Third, some of the digested material is broken down to produce energy and heat and therefore cannot be used to build biomass in the grazers (Figure 4-10). As a rule, only 5 to 20% of the biomass at any trophic level is passed to the next level. (The amount varies depending on the organisms involved in the food chain.)

When plotted on graph paper, the biomass at the various trophic levels forms a pyramid, the **biomass pyramid** (**FIGURE 4-11**). Because biomass contains energy (stored in the bonds between atoms), the biomass pyramid can be converted into a graph of the chemical energy in the various trophic levels. This graph is called an **energy pyramid**.

In most food chains, the number of organisms also decreases with each trophic level, forming a **pyramid of numbers**.

Knowledge of ecological pyramids helps us understand why people in many developing countries subsist on a diet of grains (corn, rice, or wheat) rather than meat. As shown in **FIGURE 4-12**, in the grain → human food chain on the

© hugolacasse/ShutterStock, Inc.

4-2 Colleges and Universities Go Green

All across North America, colleges and universities are taking steps to green their campuses. Campus green is not a new color. Many colleges and universities have been recycling waste and taking other measures to reduce their impact on the environment for many years.

Today's green movement is much deeper and greener, however, and is often designed to help create a sustainable future. Over 700 colleges and universities in the United States, for instance, have joined the Campus Climate Challenge, aimed at reducing their contribution to global warming. They're buying renewable energy and implementing energy-efficiency measures that dramatically lower their carbon emissions. In fact, colleges and universities are the third largest purchasers of renewable energy in the United States.

Many colleges and universities are also building new classrooms and other facilities to much higher, more energy-efficient standards using green building materials—often thanks to student insistence (**FIGURE 1**). The University of Vermont and the University of Denver, for instance, built green buildings to house their law schools. The Colorado College, where I teach, built new green science and art buildings. Dozens of others universities and colleges have followed suit.

The challenge of greening college campus can be enormous. In addition to dorms and classroom buildings, institutions of higher learning also maintain and operate office buildings, laundry service, food service, vehicle repair and maintenance, healthcare facilities, bookstores, restaurants, and even their own power plants. They're like a small city! Leadership varies too. At most schools, students have taken the lead. They can often create change that would be nearly impossible for faculty and administration. At other schools, the administration and faculty have led the charge. Some colleges and universities have even appointed sustainability specialists who coordinate all sustainability activities on campus. Sustainability specialists are helpful to ensure continuity because the most avid supporters of such movements, the students, are on campus for a rather short period—either two or four years—and the task of greening a campus can take many years to accomplish.

Some governing bodies that are in charge of entire university systems, like the University of

FIGURE 1 The Tutt Science building at Colorado College, where the author is a visiting professor, is LEED-certified with many green features, including energy-efficient design, energyefficient lighting, water-efficient bathrooms, and a host of environmentally friendly building materials.

Courtesy of the Tutt Science Center at Colorado College.

California's Regents, have also taken a leadership role in campus greening. The UC Regents, for instance, established a Green Building Policy and Clean Energy Standard for all of the schools in the university system.

Financing green initiatives can be a challenge. With rising tuition costs, the administrations of many schools are reluctant to raise fees. Students, however, have actively pursued this avenue, pushing for small increases in fees to fund renewable energy and energy efficiency. Such campaigns are almost always overwhelmingly supported by students.

Students and faculty interested in learning more can find direction through the National Wildlife Federation's Campus Ecology Program. Another excellent resource is the Campus Sustainability Assessment Framework, which lists 170 social, environmental, political, and economic indicators to assess the sustainability of a college or university campus.

Although greening a campus can be a challenge, it does help colleges walk their talk. And it provides good hands-on experience for students who will someday work for corporations that are looking for ways to green their operations, too.

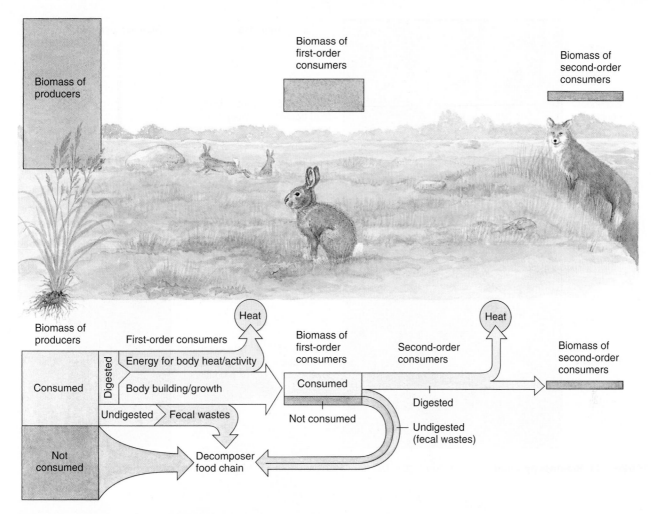

FIGURE 4-10 Flow of energy and biomass through a food chain. Not all biomass from one trophic level ends up in the next level, for reasons shown in the diagram and discussed in the text. Note that all energy is eventually lost as heat.

right, 20,000 kilocalories of grain can feed 10 people for a day. (A kilocalorie is the same unit as a calorie.) If that amount of grain is fed to a steer and the beef is then fed to humans, however, only one person can subsist on the original 20,000 kilocalories. Why? In the grain → steer → human food chain, the 20,000 kilocalories fed to the cow that day produce only 2,000 kilocalories of food, barely enough to feed one person for a day (assuming a 10% transfer of biomass). Although people don't eat meat-only diets, this simplified example does illustrate a key point: the shorter the food chain, the more food is available to top-level consumers. (This is not a criticism of the cattle industry. Cattle often feed on grasses that grow on land too poor to support crops, although they are often fattened on grain for a year or so before being slaughtered.)

This simple rule has profound implications for the human race. The human population increases by about 83 million people a year. Feeding these people poses an enormous challenge. How can new residents be fed most efficiently?

The most efficient food source will be crops such as corn, rice, and wheat that are fed directly to people. It is far less efficient to feed corn and other grains to cattle and other livestock that are slaughtered for human consumption. Vegetarianism, say proponents, is not only good for your health but good for the environment because it requires less grain production than a nonvegetarian diet. It should be pointed out, however, that vegetarianism is not environmentally benign. Plowing former grasslands or forests to grow food for people has a profound impact on the environment. Fertilizer and pesticide use can have an enormous negative effect as well. Many wild animals are killed to grow grains.

Eating lower on the food chain is not always possible or advisable. In the case of grain-fed cattle, it may be both; but range-fed beef or elk are quite different. They feed on plants that are, of course, not edible for humans, and they live on land that would be poor farmland. Consuming meat from these sources may be a better option from an environmental perspective. Gain-fed beef is also much healthier than corn-fed beef.

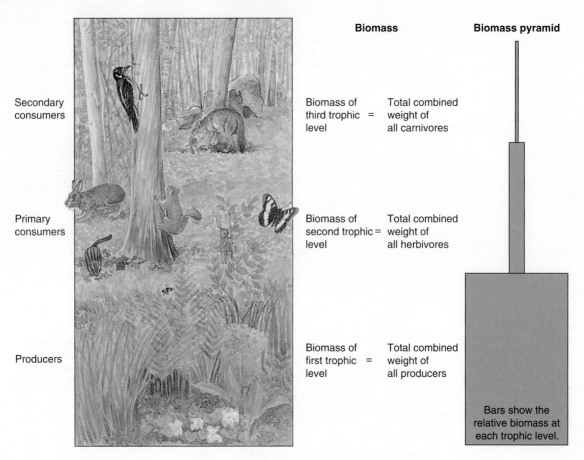

Biomass

Biomass of
third trophic =
level

Total combined
weight of
all carnivores

Biomass of
second trophic =
level

Total combined
weight of
all herbivores

Biomass of
first trophic =
level

Total combined
weight of
all producers

Biomass pyramid

Bars show the
relative biomass at
each trophic level.

Secondary
consumers

Primary
consumers

Producers

FIGURE 4-11 Biomass pyramid. In most food chains, biomass decreases from one trophic level to the next higher one.

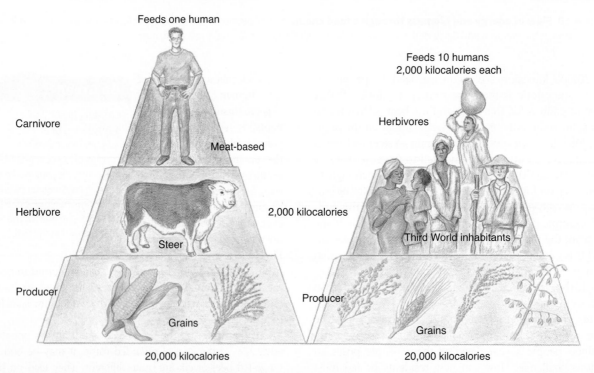

Feeds one human

Carnivore

Meat-based

Herbivore

2,000 kilocalories

Steer

Producer

Grains

20,000 kilocalories

Feeds 10 humans
2,000 kilocalories each

Herbivores

Third World inhabitants

Producer

Grains

20,000 kilocalories

FIGURE 4-12 Energy pyramids in two food chains. (a) The typical meat-based diet. The 20,000 kilocalories of corn fed to cows produces only 2,000 kilocalories of meat. An adult needs only about 2,000 calories per day. **(b)** In a shorter food chain, 20,000 kilocalories can feed 10 people directly. This is the reason many people in developing nations subsist primarily on a vegetarian diet. Although few, if any, people eat a meat-only diet, this example does illustrate an important point: more food is available to those societies that eat lower on the food chain.

The position of an organism in a food chain is called its *trophic level*. Producers are on the first trophic level. Herbivores are on the second level. Carnivores are on the third level. The length of a food chain is limited by the loss of energy from one trophic level to another. The largest number of organisms is generally supported by the base of the food chain, the producers.

Nutrient Cycles

In ecosystems, nutrients flow from the environment through food webs and then are released back into the environment for reuse.

The term **nutrients** is used here to refer to all ions (charged atoms) and molecules used by living organisms.

This circular flow constitutes a **nutrient cycle,** also known as a **biogeochemical cycle**.

Nutrient cycles consist of two parts: the environmental and the organismic (**FIGURE 4-13**). In the **environmental phase**, a nutrient exists in the air, water, or soil, or sometimes in two or more of them simultaneously. In the **organismic phase**, nutrients are found in plants, animals, and microorganisms.

Dozens of global nutrient cycles operate continuously to ensure the availability of chemicals vital to all living things, present and future. Unfortunately, however, a great many human activities disrupt nutrient cycles. Disruptions can profoundly influence the survival of species, including our own.

This section examines two of the most important nutrient cycles, the carbon and nitrogen cycles, and the ways they are being altered.

Nutrients are recycled in global nutrient cycles. In these cycles, nutrients alternate between organisms and the environment. Humans can disrupt nutrient cycles in many ways, with profound impacts on ecosystems and our own future.

The Carbon Cycle

The carbon cycle is shown in **FIGURE 4-14** in a slightly simplified form. To understand how it operates, we begin with carbon dioxide. In the environmental phase of the cycle, carbon dioxide resides in two reservoirs, or sinks: the atmosphere and the surface water (oceans, lakes, and rivers). As illustrated, atmospheric carbon dioxide is absorbed by plants and other photosynthetic organisms in terrestrial ecosystems, thus entering the organismic phase of the cycle.

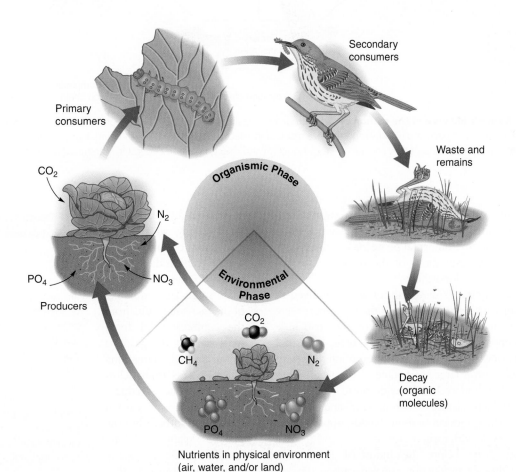

FIGURE 4-13 Nutrient cycle. Nutrients exist in organisms and their abiotic environment; they cycle back and forth between these two components of the ecosystem. CO_2 = carbon dioxide, N_2 = nitrogen, PO_4 = phosphate, NO_3 = nitrate, and CH_4 = methane.

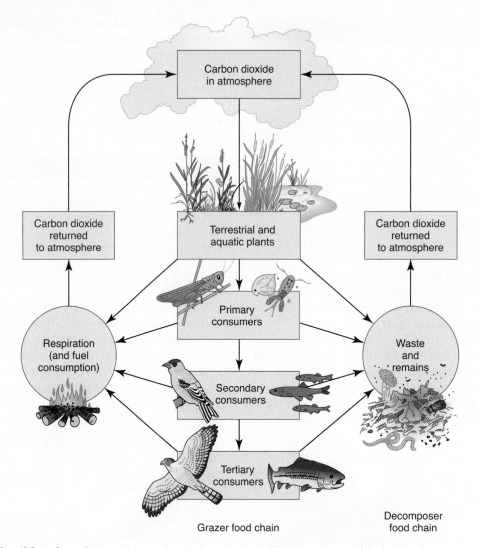

FIGURE 4-14 A simplified view of the carbon cycle. Carbon dioxide in the atmosphere is absorbed by plants and passed through the food chain. It is released back into the environment as a result of the decomposition of the waste and dead remains of plants, animals, and other organisms. It is also released by cellular energy production and the combustion of organic materials such as coal, oil, gasoline, and wood.

These organisms convert carbon dioxide into organic food materials, which are passed along the food chain. Carbon dioxide reenters the environmental phase via cellular energy production (cellular respiration) of the organisms in the grazer and decomposer food chains.

For tens of thousands of years, our ancestors lived in relative harmony with nature. Because their numbers were small and their technology fairly primitive, they had little impact on the environment. With the advent of the Industrial Revolution, however, human beings began to interfere with natural processes on a large scale. One of the victims has been the global carbon cycle. The widespread combustion of fossil fuels (which releases carbon dioxide) and rampant deforestation (which reduces carbon dioxide uptake) have overloaded the cycle with carbon dioxide.

For many years before the Industrial Revolution, global carbon dioxide production equaled carbon dioxide absorption by plants and algae. Today, well over 8 billion tons of carbon in the form of carbon dioxide are added to the atmosphere each year. Three-quarters of the increase results from the combustion of fossil fuels such as the gasoline in our cars; the remaining quarter stems from deforestation, which reduces the amount of carbon dioxide absorbed by the planet's plants. Making matters worse, many forests are burned after

GO GREEN

In the summer, turn the thermostat setting up a little at night and when you are away from your home or apartment, for instance, from 78 to 82°F. In the winter, turn the heat down a little when you are sleeping or away, for instance, from 68 to 60–62°F. Doing so can make a substantial dent in your heating and cooling costs and reduce carbon dioxide emissions, helping to combat global warming.

cutting, further adding to the carbon dioxide levels in the atmosphere.

In the past 130 years, global atmospheric carbon dioxide levels have increased by around 35%. In the atmosphere, carbon dioxide traps heat escaping from Earth and reradiates it to the Earth's surface. As carbon dioxide levels increase, global temperatures rise. Such a rise could shift rainfall patterns, destroy agricultural production in many regions, and wipe out thousands of species. The increase in global temperature is already causing glaciers and the polar ice caps to melt, raising the sea level and flooding many low-lying coastal regions. Fortunately, there are many cost-effective strategies to reduce our dependence on fossil fuel. The most notable are energy efficiency and the use of renewable fuels—solar energy and wind, for example. Transitioning to these clean, reliable, and cost-effective fuels could help us ensure a healthy economy, continued prosperity, and a better future.

KEY CONCEPTS

The carbon cycle is vital to the survival of the Earth's many species. It is the basis of food and energy production in the living world. It is also vital to maintaining global temperature. The carbon cycle is currently being flooded with excess carbon dioxide as a result of the combustion of fossil fuels and deforestation, which could have devastating effects on climate and ecosystems.

The Nitrogen Cycle

Nitrogen is an element that is essential to many important biological molecules, including amino acids, DNA, and RNA. The Earth's atmosphere contains enormous amounts of it, but atmospheric nitrogen is in the form of nitrogen gas (N_2), which is unusable to all but a few organisms. As a result, atmospheric nitrogen must first be converted to a usable form, either nitrate or ammonia.

The conversion of nitrogen to ammonia is known as **nitrogen fixation**. It occurs in terrestrial and aquatic environments. As **FIGURE 4-15** shows, the roots of leguminous plants (peas, beans, clover, alfalfa, vetch, and others) contain small swellings called *root nodules*. Inside the nodules are bacteria that convert atmospheric nitrogen to ammonia (ammonium ions). Ammonia is also produced by bacteria called *cyanobacteria* that live in the soil. Once ammonia is produced, other soil bacteria convert it to nitrite and then to nitrate. Nitrates are incorporated by plants and used to make

amino acids and nucleic acids. All consumers ultimately receive the nitrogen they require from plants.

Nitrate in soil also comes indirectly from the decay of animal waste and the remains of plants and animals. As shown on the right side of Figure 4-15, decomposition returns ammonia to the soil for reuse. Ammonia is converted to nitrite, then to nitrate and reused. Some nitrate, however, may be converted to nitrite and then to nitrous oxide (N_2O) by denitrifying bacteria, as illustrated in Figure 4-15. Nitrous oxide is converted to nitrogen and released into the atmosphere.

Humans alter the nitrogen cycle in at least four ways: (1) by applying excess nitrogen-containing fertilizer on farmland, much of which ends up in waterways; (2) by disposing of nitrogen-rich municipal sewage in waterways; (3) by raising cattle in feedlots and pastures adjacent to waterways; and (4) by burning fossil fuels, which release a class of chemicals known as nitrogen oxides into the atmosphere. The first three activities increase the concentration of nitrogen in the soil or water, upsetting the ecological balance. Nitrogen oxides released into the atmosphere by power plants, automobiles, and other sources are converted to nitric acid, which falls with rain or snow. Besides changing the pH (acidity) of soil and aquatic ecosystems, nitric acid also adds nitrogen to surface waters and may be responsible for 25% of the nitrogen pollution in some coastal waters in the United States.

Nitrogen, like phosphorus, is a plant nutrient. It stimulates the growth of aquatic plants and causes rivers and lakes to become congested with dense mats of vegetation, making them unnavigable. Sunlight penetration to deeper levels is also impaired by the growth of plants, causing oxygen levels in deeper waters to decline. In the autumn, when aquatic plants die and decay, oxygen levels may fall further, killing aquatic life.

Nothing can survive on the planet unless it is a cooperative part of a larger global life.

—**Barry Commoner**

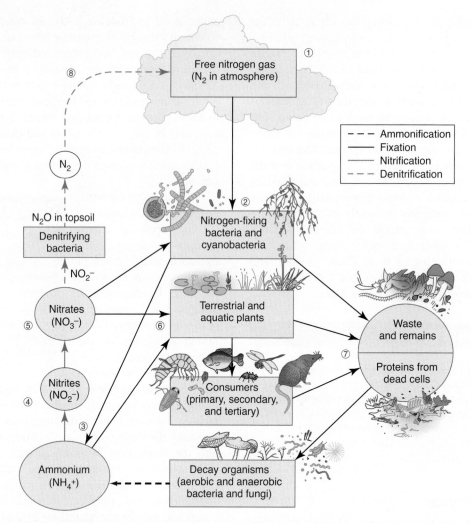

FIGURE 4-15 A simplified view of the nitrogen cycle. Nitrogen in the atmosphere is converted into ammonium ions by bacteria in the soil. Ammonia is converted to nitrates and taken up by plants.

CRITICAL THINKING

Exercise Analysis

It is difficult for me to know the list of misconceptions you came up with, but here are a few that might have arisen along the way. The exact meaning of the term *ecology* may have been one. Many people use it very loosely. You may not have understood the term *niche*, either. Like many people, you may have thought of it as a physical place an organism occupied, its *habitat*.

I suspect that your sense of the word *environment* has shifted. Now you can see that it includes biotic and abiotic components. The environment of an organism is quite complex.

I suspect that the discussion of the range of tolerance may have helped you to understand how our activities and natural events alter an organism's chances of survival.

What about energy? Has your understanding of energy changed? Do you see food chains and food webs now in a different light? Rather than being a simple matter of one organism feeding on another, food chains and food webs are elaborate pathways for the flow of nutrients and energy.

You may have gained new perspective on nutrient cycles. You may already have known that they existed, but did you know that humans are altering them in major ways?

CRITICAL THINKING AND CONCEPT REVIEW

1. "Humans are a part of nature." Do you agree or disagree with this statement? Support your answer.
2. Define the term *ecology* and give examples of its proper and improper use.
3. The Earth is a closed system. What does this mean, and what are the implications of this fact?
4. Define the following terms: *biosphere, biome, aquatic life zone*, and *ecosystem*.
5. Define the term *range of tolerance*. Using your knowledge of ecology and nutrient cycles, give some examples of ways in which humans alter the abiotic and biotic conditions of certain organisms. Describe the potential consequences of such actions.
6. Describe ways in which humans alter conditions within their own range of tolerance.
7. What is a limiting factor? Give some examples.
8. Define the following terms: *habitat, niche, producer, consumer, trophic level, food chain*, and *food web*.
9. A hunting advocate in your state is proposing the introduction of a foreign species, one very similar to deer, that he encountered in Russia on a hunting expedition. He thinks the introduced species will provide additional hunting opportunities and additional tax revenue for the state, which will be good for the economy. The governor is in favor of the proposal. Write a letter to the governor explaining what needs to be known about this species before it should be considered for introduction.
10. Explain why the biomass at one trophic level is less than the biomass at the next lower trophic level.
11. Outline the flow of carbon dioxide through the carbon cycle, and describe ways in which humans adversely influence the carbon cycle. How can you reduce negative impacts on the carbon cycle?
12. Using what you have learned about ecology, describe why it is important to protect natural ecosystems.
13. With the knowledge you have gained, explain why it is beneficial to set aside habitat to protect an endangered species.
14. Looking back over the principles you have learned in this chapter, write a set of guidelines for human society that would help us live sustainably on the Earth.
15. Reread the Point/Counterpoint in this chapter. Which view do you agree with? Why? Is your support of one view based on values or science, or both?

KEY TERMS

abiotic
aquatic life zones
biogeochemical cycle
biological community
biomass
biomass pyramid
biome
biosphere
biotic
carnivores
closed system
competitive exclusion principle
consumers
decomposer food chains
decomposers
decomposition
desert biome
detritivores

ecological niche
ecological system
ecology
ecosystem
endangered species
energy pyramid
environmental phase
food chain
food webs
grassland biome
grazer food chains
habitat
herbivores
limiting factor
niche
nitrogen fixation
northern boreal forest
northern coniferous forest

nutrient cycle
nutrients
omnivores
optimum range
organismic phase
photosynthesis
population
producers
pyramid of numbers
range of tolerance
taiga
temperate deciduous forest biome
trophic level
tundra
zones of intolerance
zones of physiological stress

© Kletr/ShutterStock, Inc.

Principles of Ecology: Biomes and Aquatic Life Zones

Let us permit nature to have her way: she understands her business better than we do.

—Montaigne

LEARNING OBJECTIVES

After reading and studying the material in this chapter, students will be able to understand, knowledgably discuss, and, when appropriate, apply the information learned in this chapter, including:

1. The difference between weather and climate
2. The major factors that determine the weather and climate on planet Earth
3. The climatic and biological characteristics of biomes and aquatic life zones
4. The valuable resources we acquire from the Earth's biomes and aquatic life zones
5. Ways humans are affecting the Earth's biomes and aquatic life zones
6. The importance of protecting the Earth's biomes and aquatic life zones

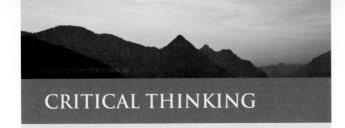

CRITICAL THINKING

Exercise

An article in *Science News* reports on studies by a scientist who examined sonar measurements of ice thickness in the Arctic Sea near the north pole. He studied assessments made by two British submarines that took similar routes across the Arctic in 1976 and 1987. During the first trip, the average ice thickness was 5.3 meters (17.5 feet), and on the second trip, it was 4.5 meters (14.85 feet).

To some scientists, these data suggest that the Arctic ice is melting, a sign that global warming may be occurring. What questions would you ask to determine the validity of this conclusion? Write them down. What additional information might be helpful in analyzing the possibility that the polar ice caps are shrinking?

The Earth formed approximately 4.6 billion years ago. At first, the planet was a barren mass of rock and ice. Over billions of years, however, life evolved. Today, the Earth's surface is carpeted with a rich and diverse array of life forms existing in a complex web of life. As anyone who has traveled across any major continent knows, however, the biosphere varies from one region to the next. That is to say, the biosphere is divided into distinct regions, which ecologists call **biomes**. Biomes differ from one another in climate—that is, in rainfall, sunlight, temperature, and other abiotic factors. These differences, in turn, lead to marked variations in the types and abundance of species.

The Earth's surface is also covered by massive bodies of water. These, too, have distinct regions, known as **aquatic life zones**. The life forms that exist in these zones are determined by different abiotic factors.

This chapter describes the major biomes and aquatic life zones and outlines some threats to them. It also offers suggestions for protecting the world's rich biological legacy and points out how important protection is for the survival of people and the many other species that share this planet with us. Before we discuss these topics, however, let's review some facts about weather and climate, which determine the biotic community of a region.

5.1 Weather and Climate: An Introduction

Most of us have heard the terms *weather* and *climate* many times in our lives, but have you ever stopped to consider what they mean and what factors determine what the weather and climate are like?

Weather refers to the daily conditions in our surroundings, including temperature and rainfall. Weather changes constantly from week to week, day to day, and even hour to hour. **Climate** is the average weather over a long period—approximately 30 years. As one wind energy expert told me, climate is what you expect, weather is what you get.

The climate of a region determines what plants can live in an area. Dry climates such as those found in the desert are home to cacti and other plants that are adapted to survive with little rainfall. Wetter climates such as those of the tropical rain forest are home to a wide variety of plants adapted to abundant rain. Plant life in each biome and aquatic life zone forms the base of grazer food chains and is the source of nutrients and energy for all other species.

KEY CONCEPTS

Weather refers to daily conditions such as rainfall and temperature. Climate is the average weather over a long period. Climate determines the plant and animal life of a region.

Major Factors that Determine Weather and Climate

Several factors determine a region's weather patterns and, ultimately, its climate. First is the amount of light and heat that strike different parts of the Earth. These come from the sun. As shown in **FIGURE 5-1**, the Earth can be divided into three climatic zones: tropics, the temperate zones, and the polar regions. The **tropics** lie on either side of the equator between 30° north and 30° south latitudes. They receive the most light and heat energy from the sun and are the warmest. The **temperate zones** lie between 30° and 60°, both north and south latitudes. They receive less heat and sunlight and are therefore cooler. The **poles** receive the least amount of solar energy and are the coolest.

The unequal heating of the Earth—and the three major climatic zones that occur as a result—are the product of the Earth's tilt (Figure 5-1). Tropical regions receive the most direct sunlight and are, therefore, the hottest. Temperate and polar regions receive sunlight that filters through more atmosphere. This reduces solar energy including heat. Besides influencing climate, the unequal heating of the Earth's surface creates air and water currents that profoundly influence the climate of the Earth. Let's begin with the wind currents.

When air is heated, it expands and becomes less dense. As shown in **FIGURE 5-2a**, hot air in the tropics rises and

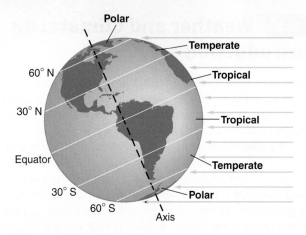

FIGURE 5-1 Climate zones. The Earth's surface can be roughly divided into the three major climate zones shown here.

moves toward the poles, transporting heat and moisture from the equator to the poles. Cold air from the poles tends to move southward in the northern hemisphere toward the equator. These phenomena create a general air circulation pattern between the equator and the poles that tends to distribute excess heat from the equator throughout the globe.

KEY CONCEPTS

The Earth is unequally heated, which creates three major climatic zones: tropical, temperate, and polar. Air tends to flow from the equator to the poles.

The Coriolis Effect and Topography

Air circulation is actually a bit more complicated than the previous discussion might lead you to believe. In the northern hemisphere, for example, warm air from the equator rises and moves northward. As it moves northward, much of its moisture is lost as rainfall over the tropics. As the air

moves northward it also cools, and some of it sinks back to the Earth's surface (**FIGURE 5-2b**). This air flows back toward the equator, creating winds that blow toward the equator called the **trade winds** (**FIGURE 5-2c**).

The rest of the equatorial air continues northward over the temperate zone. This air is relatively dry, having lost much of its moisture in the tropics. As Figure 5-2b shows, the air circulating toward the north pole splits into higher and lower level winds—both flowing generally northward. However, as the lower-level air travels northward it picks up moisture from the temperate zone, which it deposits as rainfall and snow. When the equatorial air reaches the poles, it is cold and very dry. It then begins its way back to the equator (Figure 5-2b).

This general circulation distributes heat and moisture throughout the planet. It makes the poles warmer than they might otherwise be and makes the tropics a bit cooler. The distribution of heat and moisture caused by air flow also helps determine weather. However, this distribution is complicated a bit by the fact that Earth spins on its axis. This, in turn, shifts the wind direction and helps to create distinct climatic zones within the major regions.

Figure 5-2c shows three distinct wind patterns, one in each zone of the northern hemisphere. The trade winds blow toward the equator, but they don't blow directly south. In the northern hemisphere, they blow from northeast to southwest. In the temperate zone, the low-level winds generally blow from south to north (as shown in Figure 5-2b). However, because they are deflected a bit, they actually blow from the southwest to northeast. They are often refered to as **westerlies** or **south westerlies**. In the polar regions, winds are deflected so they blow from the northeast to the southwest. They are called the **polar easterlies**.

Scientists call the deflection of wind currents by the spin of the Earth the **Coriolis** (CORE-ee-ole-iss) **effect.** To understand this phenomenon, suppose that a punter kicks a football 50 yards south; the ball is in the air for 5 seconds, and it flies in a perfectly straight course. If this were so, you would

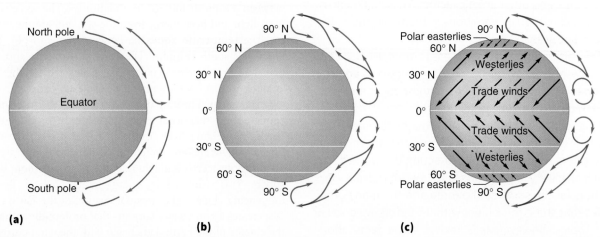

(a)   **(b)**   **(c)**

FIGURE 5-2 Global air circulation. (a) Warm tropical air rises and flows toward the poles. Cold polar air tends to flow toward the equator. **(b)** Air circulation actually is more complicated than shown in part (a). Warm tropical air loses some of its heat and sinks, creating a circulation pattern that brings air back toward the equator. This creates the trade winds. **(c)** Wind patterns caused by the spin of the earth.

expect it to land in a perfectly straight line from the kicker. It does not. In actuality, it lands a little less than half an inch off target, just west of the straight line. The reason for this is the Earth's spin. At the equator the Earth is spinning 1,000 miles per hour! While the ball is in the air in a straight path, the Earth is spinning underneath it. The path of the ball thus appears to be deflected. Air flowing along the Earth's surface is also deflected because of the Earth's spin. Thus, the trade winds, the westerlies, and the easterlies are produced by the Earth's rotation. Wind currents, instead of flowing straight north or south (as they would if the Earth did not spin), actually flow across continents. Why is this worth noting?

As noted previously, wind currents help to determine weather patterns. In the United States, moist air tends to flow southwesterly from the Pacific Ocean across the continent toward the northeast. Topography then comes into play. Mountain ranges along the West Coast and in the interior, for example, rob some of the moisture carried in this air, creating deserts or semiarid lands on their downwind side. This occurs because mountains thrust warm, moist air upward. As it rises it cools, and the moisture condenses and falls as precipitation. When the air comes down the other side of the mountain range, it is drier. Little precipitation falls, and the area is either desert or semidesert (**FIGURE 5-3**). This phenomenon is called the *rain shadow effect*. As the air flows across the Great Plains, it picks up and deposits moisture downwind. Rainfall increases as one travels eastward. This increase in precipitation, in turn, explains why the prairie due east of the Rocky Mountains starts off as short grass, which requires little rain, but gradually becomes tallgrass prairie and then deciduous forest from eastern Kansas eastward.

KEY CONCEPTS

Weather within the major climatic zones is determined in large part by wind flow patterns, which are profoundly influenced by the spin of the Earth. Weather is also affected by topography, especially mountain ranges.

Ocean Currents

Climate is also influenced by ocean currents. At the equator, water warmed by the sun rises and tends to drift toward both poles, creating huge currents. The Gulf Stream, shown in **FIGURE 5-4**, is a good example. The warm water of the Gulf Stream flows northward from equatorial regions, affecting the climate of landmasses that it passes near. The Gulf Stream, for instance, warms England and the rest of Europe and produces much milder winters than would be expected otherwise. Northern Japan and Alaska are also much warmer than one would predict based on their location because of a warm-water current traveling up from the equator (North Equatorial current). Should these currents cease, they could bring about massive climate changes (cooling) in these areas. Atmospheric scientists believe that global warming could drastically alter ocean currents, bringing about massive cooling in areas now warmed by ocean currents that originate near the equator.

As the warm-water currents move northward, however, they begin to cool. As the waters cool, they sink and then flow back toward the equator, in much the same way that cold Arctic air tends to flow toward the equator. Cold water returning to the equator creates broad, deep currents in the opposite direction of the warm-water surface currents. As shown in Figure 5-4, the cold water Humboldt current in the South Pacific Ocean flows up from the south pole, bringing cold, deep waters toward the equator. This water eventually emerges at the surface in regions called **upwellings.** Because it is rich in nutrients, it supports a diverse aquatic

Go Green

To help combat climate change, plant trees. You can join groups that are planting trees in national forests or state-owned lands or even cities and towns. A few shade trees planted around your home or your parents' home will reduce cooling costs and help combat global warming.

FIGURE 5-3 Rain shadow effect. This shows the effect of mountain ranges on rainfall.

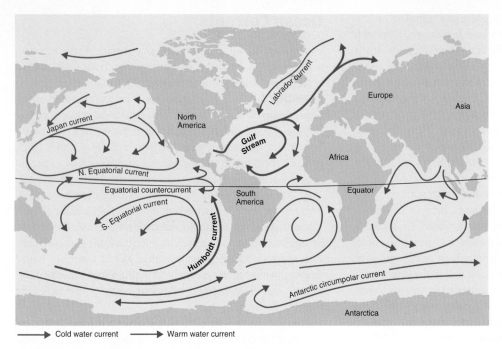

FIGURE 5-4 Major ocean currents.

→ Cold water current → Warm water current

community of great ecological and economic value. Every 10 years or so, the Humboldt current turns warm, a phenomenon called **El Niño.** This creates heavy rain and rough seas, which can have a devastating effect on coastal communities and often affects weather worldwide. (Today, El Niños are occurring more frequently, quite possibly as a result of global climate change.)

Weather is much more complicated than discussed here, but this introduction gives you an overview that may help you understand why particular biomes are located where they are. It will also help you to understand some major global environmental issues, such as global climate change.

5.2 The Biomes

We begin our journey through the Earth's diverse biomes (**FIGURE 5-5**) in the north and work our way south. On this journey, you will learn about the climate of several key biomes and the unique biological characteristics (or adaptations) that permit plants and animals to thrive within them. As you will see, each biome is characterized by a dominant form of vegetation. However, within any given biome, regional differences in climate (and other factors such as soil) can alter the composition and abundance of species.

The study of biomes underscores an important lesson, Life can adapt to a wide range of conditions. In the frozen tundra, for example, some organisms survive in air temperatures as low as –70°C (–90°F). In contrast, some desert species can tolerate 50°C (120°F).

The Tundra

Stretching across the northernmost portions of North America, Europe, and Asia is the Arctic **tundra**. The Arctic tundra is the northern limit to plant growth. Covering about 10% of the landmass of the Earth, the tundra lies between a region of perpetual ice and snow to the north and a band of coniferous forests to the south. The tundra is a treeless tract characterized by grasses, shrubs, and mat-like vegetation (mosses and lichens) adapted to the harsh climate.

The Arctic tundra receives very little precipitation (less than 25 centimeters, or 10 inches, a year), and most precipitation occurs in the summer. During the long, cold winters, the mean average temperature remains well below zero for months on end. Because Arctic summers are short and winters are so cold, the deeper layers of soil remain frozen throughout the year and are called **permafrost.**

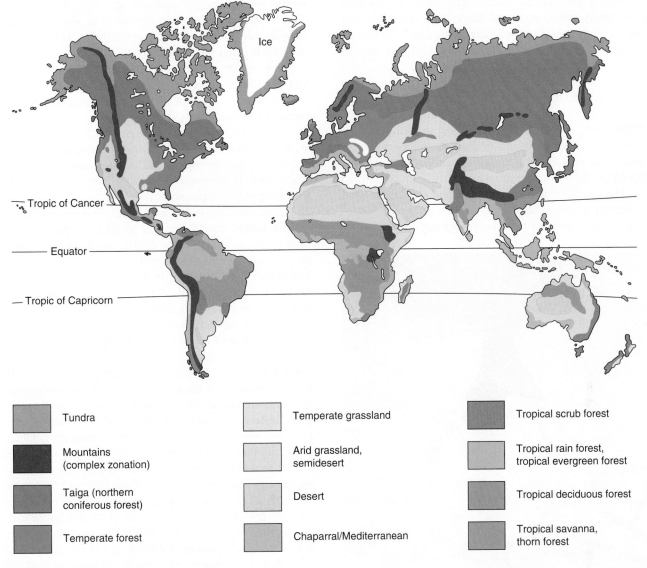

FIGURE 5-5 **The Earth's biomes.**

Legend:
- Tundra
- Mountains (complex zonation)
- Taiga (northern coniferous forest)
- Temperate forest
- Temperate grassland
- Arid grassland, semidesert
- Desert
- Chaparral/Mediterranean
- Tropical scrub forest
- Tropical rain forest, tropical evergreen forest
- Tropical deciduous forest
- Tropical savanna, thorn forest

The Tundra

The permafrost and the harsh winters of the Arctic tundra prevent deep-rooted plants such as trees from growing there. In the brief Arctic summer, however, the long days and relatively warm temperatures permit the superficial layers of soil to melt. Because evaporation is low and because water from melted snow and ice cannot percolate downward into the soil in the summer (due to the permafrost), the tundra becomes dotted with shallow ponds, lakes, and bogs (**FIGURE 5-6**).

FIGURE 5-6 **The tundra.** In the summer, water collects on the surface in the tundra because evaporation is low and water percolation into the ground is prevented by the permafrost.
Courtesy of U.S. Fish & Wildlife Services.

During the summer months, the tundra comes alive with insects (mosquitoes and black flies) and birds that migrate north to nest on the rolling plains. The birds feed on the swarms of insects, raise their young, and then migrate south with their offspring, often in great flocks.

Despite the harsh conditions, a variety of animals live year round on the tundra. Ptarmigan, musk oxen, and arctic hares, for example, are adapted to the extreme cold (**FIGURE 5-7**). Animals survive by living in burrows or by having thick layers of insulation or a large body size (which retains heat well). Some animals, like caribou, are migratory, moving southward when winter comes.

The tundra is a fragile environment, easily damaged by human actions. The short growing season provides little time for vegetation to recover from damage caused by mining, oil and gas development, and other human activities. Tire tracks that destroy vegetation on the tundra take many decades to heal. Damage caused by spills of oil or hazardous waste may take much longer.

Nowhere is the threat to the tundra greater than in northern Alaska. Alaska's northern coastline is approximately 1,760 kilometers (1,100 miles) long, and all of it except a 184-kilometer (115-mile) stretch, which is part of the Arctic National Wildlife Refuge (ANWR), one of the last great wilderness areas on the planet, is open to oil and gas exploration and development. Efforts are under way to open that section to development as well, despite almost overwhelming public opposition.

Many conservationists and citizens fear that the region will be turned into a network of roads, airports, power plants, oil platforms, waste ponds, buildings, and gravel pits—to supply oil to make fuel for cars and other motorized vehicles (**FIGURE 5-8**). Proponents of oil development in the ANWR believe that wildlife and oil development can peacefully coexist despite estimates by biologists suggesting that full-scale oil development will result in a 20 to 40% decline in the large caribou herds that spend their summers in the region. Biologists also predict that full-scale oil development will wipe out half of the existing musk ox population. Populations of grizzly bears, polar bears, wolverines, and other animals are likely to decline, too. Oil development in the region would also result in air pollution, water pollution, and noise. Spills of oil and toxic wastes are common in nearby Prudhoe Bay Oil fields. According to the Alaska Department of Environmental Conservation, 400 spills were reported, on average, every year from 1996 to 2002 (2002 being the date of the most recent data). Since 1973 there have been more than 17,000 spills of oil and other toxic chemicals. Proponents are currently planning on developing only a small portion of ANWR, but critics are worried that after development commences, proponents will call for further development that could be devastating to this pristine area.

(a)

(b)

FIGURE 5-7 Tundra species. (a) The ptarmigan and **(b)** musk ox are both well adapted for life in the cold Arctic tundra and live there year round.

FIGURE 5-8 Arctic oil. Oil development on Alaska's North Slope can cause significant damage to the delicate tundra that, because of the short growing season, may never be repaired by natural processes.

The Taiga

Just south of the tundra is a wide band of cone-bearing, or coniferous, trees (pine, fir, and spruce). It extends across Canada, part of Europe, and Asia and constitutes the **taiga** (TIE-ga), or **northern coniferous forest biome** (Figure 5-5).

The taiga, which a student of mine once defined as a "carnivorous forest" on his test, has a longer growing season than its northern neighbor, the tundra. It also receives far more precipitation.

In the summer, the subsoil of the taiga thaws, permitting deep-rooted plants such as trees to live. But the summer growing season is still short in comparison with that in the southern biomes (grassland and temperate deciduous forest). The rather cold, snowy winters and limited growing season of the taiga have resulted in numerous adaptations. Few organisms illustrate these adaptations as well as **conifers,** trees that remain green throughout the year. Conifers have narrow, pointed leaves called *needles*. Needles contain photosynthetic cells and are retained throughout the year, allowing conifers to continue to photosynthesize (albeit slowly) throughout the winter. The presence of needles throughout the year also permits the plant to make the most of the growing season. Unlike the deciduous hardwoods of the south, which lose their leaves in the winter and must develop new ones each growing season, conifers can take immediate advantage of the sunlight, moisture, and warmer weather of spring.

Retaining needles may help conifers survive in the taiga's short growing season, but it also creates a problem: The needles tend to capture snow. Instead of breaking under the weight of snow, though, the branches of conifers are adapted to bend and shed snow. Another adaptation of great importance to conifers is the waxy coating found on their needles. This layer greatly reduces evaporation. Because water transport up the tree from the ground is restricted during the winter, the waxy coating prevents the needles from drying out and helps the tree survive.

As anyone who has flown east to west over Canada will attest, the dense coniferous forests of the taiga extend for miles on end. These forests are interspersed with meadows,

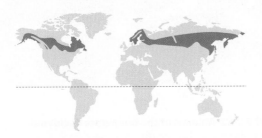

Northern Coniferous Forest (Taiga)

lakes, ponds, and rivers. Together, they support a variety of species such as bears, moose, and deer. Many smaller mammals also make their home there, including foxes, wolverines, and snowshoe hares. Because of the relatively cold winter and short growing season, however, species diversity in the taiga is fairly low compared with that in biomes to the south.

Along the west coast of North America, the forests of the taiga are bathed in moisture from the Pacific Ocean. Because the ocean moderates temperature, this region enjoys a long growing season. Consequently, forests in this area contain some of the world's largest trees. Heavy pressure from timber companies and consumers, however, has destroyed many of these magnificent trees and the species that live in these forests (**FIGURE 5-9**).

Efforts are under way to reduce the overcutting, but in many areas of Canada and the United States, powerful logging companies are resisting. The U.S. Congress, however, reduced cutting in the Tongass National Forest of Alaska and set aside a large tract of the forest for permanent protection.

The Temperate Deciduous Forest Biome

The temperate deciduous forest biome is located in the eastern United States, Europe, and northeast China. In the United States, this biome is home to about half of the human population. Characterized by abundant precipitation and a long growing season (5 to 6 months), this biome supports a

FIGURE 5-9 Clear-cutting. This forest in the taiga has been clear-cut. It once contained trees 500 to 700 years old. Old-growth forests such as these contain many animal species that cannot survive in younger forests.
© Photodisc/Getty Images

wide variety of plants and animals.

The dominant plants of the temperate deciduous forest biome are **deciduous trees**—the broad-leaved trees that shed their leaves each year in the fall. Maple, oak, black cherry, and beech trees are examples (**FIGURE 5-10**). The loss of leaves in the fall is believed to be an adaptation that greatly reduces evaporation at a time when the supply of liquid water is limited.

In the spring, new leaves develop from buds. In the relatively brief period after the ground has thawed and before the leaves have fully developed, numerous species of wildflowers sprout on the sunny forest floor. The shade created by

Go Green

During your summer vacation, work with your parents to create some wildlife habitat on their property—that is, a flower garden or small pond to promote birds and butterflies and other small animals. For advice, contact the National Wildlife Federation's Certified Wildlife Habitat program at www.nwf.org/backyard/.

deciduous trees, however, greatly limits plant growth on the forest floor throughout most of the rest of the growing season.

The temperate deciduous forest biome is characterized by a deep, rich soil. The richness of the soil results from mineral and organic nutrients. The nutrients are drawn up through the roots and incorporated in the leaves. During the fall, the leaves detach and fall to the ground. There, they decay and release their nutrients into the soil.

The fertile soil and abundant plant life of the forest support a rich and varied population of insects, microorganisms, birds, reptiles, amphibians, and mammals (**FIGURE 5-11**). Common mammals include the raccoon, white-tailed deer, red fox, and black bear.

Ernest Hemingway once wrote that "a continent ages quickly once we come." In fact, few biomes have been so heavily altered by human activities as the temperate deciduous forest. Large-scale destruction of the forests began when colonists first settled the continent. Early settlers cleared the forests to grow crops. They also cut trees to make room for

Temperate Deciduous Forest

FIGURE 5-10 Temperate deciduous forest. Broad-leaved trees in this biome lose their leaves each fall. Years of leaf litter have produced rich soils that humans have long used for farming.
© Snehit/ShutterStock, Inc.

(a)

(b)

FIGURE 5-11 Animals of the temperate deciduous forest. **(a)** Black bear and **(b)** white-tailed deer.
Photos a and b © Photodisc.

homes, towns, orchards, and roads. In the United States, only about 10% of the land east of the Mississippi is still forested, and only about 0.1% of the original forests remains. Much of the existing forest consists of young trees that have regrown after successive harvests.

Unfortunately, many early settlers failed to practice soil conservation on their farms. Heavy rains often washed the soil away. Moreover, farmers usually did little to replenish soil nutrients. Many farmers depleted the land, destroyed the soil that had taken centuries to form, and then moved westward.

The Grassland Biome

Grasslands exist in temperate and tropical regions in areas that receive intermediate levels of precipitation—that is, less precipitation than forested regions but more than deserts (**TABLE 5-1**). Grasslands are found in North America, South America, Africa, Europe, Asia, and Australia.

In North America, grasslands form a continuous, broad zone extending from the Gulf of Mexico northward through Canada to the taiga (Figure 5-5). A small area of grassland is found in the Great Basin, between the Cascade Mountains on the west and the Rockies on the east.

Grasslands

All grasslands bear a remarkable similarity (**FIGURE 5-12**). Most are on flat or slightly rolling terrain. Carpeted in thick grasses, the soils are probably the richest in the world as a result of thousands of years of plant growth and decay.

In the United States, the grasslands begin just east of the Rocky Mountains. The grasslands of eastern Colorado, however, are not as rich as those further east because of lower rainfall, as explained earlier. The grasslands of the eastern plains of Colorado, Wyoming, and Montana are known as the **shortgrass prairie.** Further east, as rain and snowfall increase, the shortgrass prairie gives way to the **tallgrass prairie** of the Dakotas, Nebraska, and eastern Kansas. Today, most of the tallgrass prairie has been destroyed by farming so that only small patches remain.

Grasslands are usually devoid of trees because they lack sufficient annual precipitation and experience periodic drought. Thus, the only trees found in the Great Plains of the United States are those that have been planted around homes and farms or those that live along streams and rivers.

Grasses, however, require less water than trees, and many species of grass send their roots deep into the Earth, drawing on reliable moisture supplies far below the surface. Grasses are also well adapted to periodic fires ignited by lightning. Fires burn the plant above the ground but do not harm the roots, from which new life can spring.

Because the soils of the grassland biome are so rich, they have been heavily exploited by people, primarily for

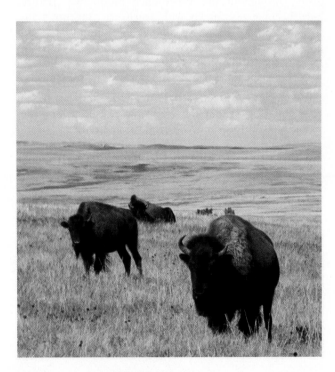

FIGURE 5-12 Grassland biome. This biome does not receive enough moisture to support trees. The only trees found in the grasslands, therefore, were planted by humans or live along streams.

© MediaImages/Photodisc/age footstock.

TABLE 5-1

Precipitation in Major Biomes

Biome	Annual Precipitation Centimeters (Inches)
Tundra	Under 25 (10)
Taiga	38 to 100 (15 to 40)
Temperate deciduous	75 to 150 (30 to 60)
Grassland	25 to 75 (10 to 30)
Desert	Under 25 (10)
Tropical rain forest	150 to 400 (60 to 160)

agriculture, a practice that profoundly alters the plant and animal life. The grassland biome of North America, for example, once supported an enormous population of bison—perhaps as many as 60 million animals. Roaming the prairies, bison herds were so large that it would take men traveling on horseback 3 to 4 days to pass through a single herd. These animals were killed for food and sport and squeezed out of their habitat by the spread of farms.

The grassland biome of North America also supported grizzly bears and elk. As humans settled the grasslands and began to farm and raise cattle, grizzlies were exterminated. Elk herds fled to the protection of the Rocky Mountains, where they remain today. Many species—including deer, pronghorn antelope, badgers, and coyotes—still live on the grasslands of North America, sharing habitat with cattle, horses, sheep, and people.

In the grasslands, as in the temperate deciduous biome, farming has probably had the most significant impact of all human actions. Today, many of the world's tallgrass prairies have been plowed under to plant corn, wheat, soybeans, and other crops. The shortgrass prairie, being drier and less productive, has fared much better. Except where irrigation water has allowed farmers to cultivate the land, much of the shortgrass prairie still exists. This land, however, is often overgrazed by cattle. Continual overgrazing eliminates many hardy grasses, and in the semiarid grasslands of North America (in Colorado, for example), it creates dry soil conditions suitable for the growth of weedy species such as sagebrush. Because cattle will not eat it, sagebrush thrives in overgrazed fields, eventually taking over. Throughout the West, once productive pastures have become overgrown with sagebrush (**FIGURE 5-13**). As you travel in eastern Wyoming and Colorado, look for fields of sagebrush; chances are that they are the result of overgrazing. With a little care, such an ecological tragedy could have been avoided.

Poor agricultural practices on farms and ranches in grasslands throughout the world have resulted in widespread soil erosion and desertification in the grassland biome. The most dramatic evidence of abuse came in the 1930s in the western and midwestern United States. An extended drought, combined with fencerow-to-fencerow planting, spawned one of the most significant environmental disasters of human history: the **dust bowl.** Millions of tons of topsoil were lost from U.S. farms in huge dust storms.

The dust bowl stimulated a rash of conservation efforts to help protect farmland. Perhaps the most significant step was the planting of long rows of trees alongside fields to reduce wind erosion (**FIGURE 5-14**). Today, farmers sometimes leave wheat stubble and residues from crops on the ground over the winter to protect soils. Unfortunately, many of the improvements in soil management made in the immediate post–dust bowl era are being lost. Many farmers are once again planting from fencerow to fencerow in an effort to increase their output. Not surprisingly, devastating dust storms are becoming more and more common in some states, such as Texas and California.

The Desert Biome

Deserts exist throughout the world. Some cover vast regions. The Sahara, for example, stretches across northern Africa and is about the size of the United States. In North America, deserts exist primarily on the downwind side of mountain ranges.

Deserts, although dry, are not devoid of precipitation (Table 5-1). Rain often comes in violent downpours, however, causing flash flooding and severe erosion that have

FIGURE 5-13 Sagebrush. Overgrazed western rangeland is often overrun with this hardy species, the sagebrush, which is edible only to a limited number of species such as pronghorn antelopes.
© Photodisc.

FIGURE 5-14 Shelterbelts. Rows of trees planted along the margins of farm fields help reduce wind erosion and drying.
Courtesy of Erwin Cole/NRCS USDA.

Desert

sculpted the magnificent canyons of the desert Southwest. Many species of desert wildflowers are well adapted to the infrequent but intense spring rains. These species have an accelerated life cycle, in which they grow and produce flowers within a few days of a thunderstorm, turning the desert into a colorful garden almost overnight (**FIGURE 5-15**).

Plants that live in the desert must also be adapted to tolerate a wide range of temperatures. In the desert, temperatures may reach 50°C (120°F) during the day and then drop to near freezing at night. The temperature in the desert fluctuates so much in large part because of the absence of moisture in the atmosphere, which reduces heat retention by the nighttime sky. Thus, even though the sun warms up the desert floor during the day, the heat escapes quickly at night.

Plants in the desert are adapted to low soil moisture. Many desert plants, such as cacti, have shallow root systems that extend laterally from the plant. Running just below the surface, these roots absorb rain and melted snow and then transport the water to the main body of the plant, where it is stored. Other desert plants, such as the mesquite tree, have deep taproots that extend downward, sometimes over 30 to 60 meters (100 to 200 feet), to moist soil. Taproots anchor the plant and tap into a relatively reliable source of water.

Water absorbed by the roots of many desert plants is often stored in succulent, water-retaining tissues, giving the plant an ample supply on which to draw during the rainless months. In most desert plants, water supplies are protected by the presence of thick outer layers and waxy coats that reduce moisture loss.

Desert plants are often widely spaced on the desert floor, which reduces competition for water and ensures an adequate supply. How do plants space themselves? Some plants release growth-inhibiting chemicals into the soil that deter competitors from taking root in a region around them.

The thorns of cacti are yet another adaptation that reduces water loss. Thorns protect cacti from being eaten, but they also give some degree of protection from the sun by providing shade and reflecting some sunlight from the plant (**FIGURE 5-16a**). White, fluffy hairs found on some species of cacti also reflect sunlight and provide shade (**FIGURE 5-16b**).

Many reptiles and other animals also make the desert biome their home. Like the plants, the animals are well adapted to desert conditions. The thick scales of snakes and lizards, for example, minimize water loss, permitting these creatures to thrive in the dry, hot conditions (**FIGURE 5-17a**). Lizards and other species, such as mice, also survive by avoiding daytime heat. They rest in caves or burrows and venture forth only at night. The ringtail, for example, sleeps all day and comes out at night to find its food (**FIGURE 5-17b**).

(a)

(b)

FIGURE 5-16 The cactus. (a) Spines protect cacti from hungry animals and also provide shade. **(b)** Some cacti have white hairs that protect them from the sun.

Photo a © Photos.com; photo b © Kim Pin Tan/ShutterStock, Inc.

FIGURE 5-15 Desert garden. Sudden spring rains cause desert flowers to bloom.

© SuperStock/age footstock.

(a)

(b)

FIGURE 5-17 Animals of the desert. (a) Gila monster and **(b)** ringtail. To survive in deserts, both animals are active primarily at night.

Photo a © Photos.com; photo b © Arco Images GmbH/Alamy Images.

Because water is a rare commodity in the desert, many species acquire the moisture they need from cellular energy production. Water released during this process is called *metabolic water*. In some species, metabolic water provides nearly all of the water needed to survive. The kangaroo rat receives *all* of the moisture it needs from this process and from the plants and insects it eats. Although cellular energy production does not produce large amounts of water, it can be enough if species possess additional adaptations that help them conserve body water. The kangaroo rat, for example, excretes a highly concentrated urine, a physiologic adaptation to the hot, dry conditions of the desert. Snakes and lizards also excrete a highly concentrated urine that reduces water loss.

Deserts can turn cold. Winter in the desert often brings freezing temperatures and snow for several days and sometimes weeks on end. Thus desert plants must also be able to tolerate cold.

Large cities have sprung up in many of the world's deserts. To supply inhabitants, food is trucked in from farms hundreds of miles away. Water is often pumped from deep aquifers or is transported from distant sources via extensive pipelines. Phoenix, Arizona, for example, receives much of its water from the Colorado River, several hundred miles to the northwest, through a gigantic (and costly) canal and pipeline.

The continuing expansion of cities in the deserts of the world and the growing water demand have created serious problems, however. Southeast of Phoenix, for instance, over 300 square kilometers (120 square miles) of land has subsided (sunk) more than 2 meters (6 feet) because of intensive groundwater withdrawal. Cracks in the Earth's surface have developed in subsided areas. Some of the largest cracks are 3 meters (10 feet) wide and 3 meters deep and run for 300 meters (990 feet), three times the length of a football field. Subsidence and cracks in the Earth's surface can break pipelines and can destroy homes and highways. Depletion of groundwater can also eliminate the supply of water for native plant species.

Each year, millions of acres of new desert form on semiarid grasslands. Research suggests that deserts are expanding principally because of human actions. Livestock overgrazing destroys grasses in semiarid regions. The loss of vegetation may reduce rainfall, creating desertlike conditions. Climate change, both natural and human induced, may also be contributing to the spread of deserts. Stopping desertification will require dramatic improvements in land management, especially grazing, and reductions in greenhouse gases.

KEY CONCEPTS

The desert biome is characterized by dry, hot conditions, but often abounds with plants and animals adapted to the heat and lack of moisture. Unfortunately, deserts of the world are expanding as a result of human activities such as overgrazing livestock and the production of greenhouse gases.

The Tropical Rain Forest Biome

Heading south to the equator, we find one of the most endangered biomes on Earth, the tropical rain forest (**FIGURE 5-18**). Tropical rain forests exist near the equator in South and Central America, Africa, and Asia (Figure 5-5). By far the most complex and diverse of all the Earth's biomes, tropical rain forests support a wealth of plants, animals, and micro-organisms. A small tropical island, for example, may have as many butterfly species (500 to 600) as the entire United States. Two hundred fifty different tree species exist in a single hectare (2.5 acres) of rain forest; in the temperate deciduous forest biome of the United States, a similar tract may have 20 to 30 species.

(a)

Tropical rain forest

(b)

FIGURE 15-18 Tropical rain forest. This endangered biome covers vast areas and is rich in species.

Photos a and b © Photodisc.

The tropics contain a variety of different ecosystems, of which the rain forest is the largest and best known. With 200 to 400 centimeters (60 to 160 inches) of rain falling per year, trees of the tropical rain forest grow to heights of 60 meters (200 feet) or more. The tops of the tallest trees in a tropical rain forest form a dense canopy that blocks much of the incoming sunlight. Shorter trees form a lower canopy that intercepts most of the remaining light. As a result, only about 1% of the incoming sunlight reaches the ground. Because so little light strikes the forest floor, ground-level vegetation is sparse. Only those species adapted to low light (for example, African violets, philodendrons, and ferns) can survive in such conditions. The lack of ground-level vegetation relegates most life to the treetops. Dwelling in this zone are monkeys, birds, and countless species of insects.

The tropical rain forest is a paradox to the untrained observer. Although it is the richest and most diverse biome on Earth, most of its soils are thin and nutrient poor. If a section of forest growing on such poor soil is cleared to make room for crops or cattle ranches, it will generally produce for only 4 or 5 years before its nutrients are depleted. Large-scale farming and ranching on many rain forest soils are generally doomed to fail.

An understanding of basic ecology explains the paradox of the tropical rain forest, in which the world's richest biome grows on some of the world's poorest soils. To begin, we ask: Why is the soil so poor? In tropical rain forests, dead trees, leaves, and other forms of biomass (dead animals) that fall to the ground are consumed by countless insects. The material that the insects do not eat is rapidly decomposed by bacteria, fungi, and other organisms. Nutrients released into the soil by bacterial decay are quickly absorbed by the roots of trees located just beneath the ground's surface. Thus, virtually all of the nutrients are locked up in plants and animals. In essence, then, tropical soils are so poor because life is so rich.

The lush vegetation of the tropics protects the soil from erosion. Thus, clearing the land for agriculture can be a chancy proposition. Heavy rainfall in the wet season erodes soils no longer protected by trees and other vegetation, rendering the new farmland useless and choking nearby streams and rivers with sediment. The loss of the thin topsoil also makes recovery from damage more difficult and protracted. Some forestry scientists believe that recovery from logging could take large forest clearings 500 years or more.

Many rain forest soils are also useless for farming and ranching because they contain large amounts of iron. These soils are called **lateritic** (LA-ter-it-ick) **soils** (*later* is Latin for "brick"). When cleared and exposed to sunlight, lateritic soils bake as hard as bricks. The impenetrable crust that develops 1 to 2 years after clearing makes the soils virtually impossible to cultivate.

Tropical forests are the lungs of the planet. They "breathe in" vast amounts of carbon dioxide and release oxygen vital to animals. By absorbing carbon dioxide, tropical rain forests help reduce global warming.

Once covering a region approximately the size of the United States, only about half of the rain forest remains. Despite worldwide concern for the fate of tropical rain forests, timber cutting continues. According to a 2010 report by the United Nations Food and Agricultural Organization (FAO), from 2000 to 2010, an average of 13 million hectares or 32 million acres of forest, mostly tropical rainforest, were destroyed each year. Thanks to aggressive replanting efforts, the FAO notes that the annual average net loss each year comes to about 5.2 million hectares or nearly 13 million acres, an area about the size of Costa Rica or half the size of Tennessee. As rainforests vanish, tens of thousands of species, including many potentially beneficial plant species, will vanish. Widespread destruction of tropical forests could upset global rainfall patterns and could affect global climate by reducing the amount of carbon dioxide removed from the atmosphere each year.

Altitudinal Biomes

Temperature and precipitation (rain and snow) profoundly influence life in biomes, affecting species diversity and the abundance of life. The relationship between climate and life is dramatically illustrated in mountainous terrain. If you walked from the top of a mountain to its base, you would very likely progress through several distinct life zones called **altitudinal biomes** (**FIGURE 5-19**).

Altitudinal biomes mirror the latitudinal biomes described in this chapter and are a reflection of differences in temperature and precipitation. Starting at the top of the mountain, for example, one encounters a cold, treeless region similar to the Arctic tundra. Known as the **alpine tundra,** this area is characterized by a short growing season and long, cold winters (**FIGURE 5-20**). But unlike the Arctic tundra, the alpine tundra receives lots of moisture. Most of the precipitation comes in the form of snow, and most blows away, ending up at the next lower level.

Climbing downward, one encounters a band of conifers akin to the taiga. In the Rocky Mountains, this zone borders on grassland. On the west slope of the Cascade Mountains, however, the taigalike region borders deciduous forests.

KEY CONCEPTS

The tropical rain forest is the richest and most diverse biome on Earth because of its abundant rainfall and warm climate. Paradoxically, tropical rain forests have the poorest soils in the world. About half of the world's rain forest has been destroyed; huge tracts could be eliminated in the near future, with devastating effects on climate and plants and wildlife, if current trends continue.

KEY CONCEPTS

Climate varies with altitude, as does the distribution and abundance of life.

FIGURE 5-19 Altitudinal biomes.

FIGURE 5-20 Alpine tundra. This biome is, in many ways, like the Arctic tundra, except that it receives far more moisture. Because the growing season is short, plants remain small.
© Harald Høiland Tjøstheim/ShutterStock, Inc.

5.3 Aquatic Life Zones

On land, precipitation and temperature are the chief determinants of the distribution and abundance of life. In aquatic life zones, however, water is abundant and temperature relatively constant. The abundance and diversity of life forms are determined principally by energy and nutrients.

Aquatic life zones may be either freshwater or saltwater. In both freshwater and saltwater aquatic life zones, many food chains begin with a group of producer organisms, the **phytoplankton** (FIE-toe-plank-ton). Phytoplankton are microscopic, free-floating, photosynthetic organisms, mostly algae and diatoms (**FIGURE 5-21**). These organisms capture solar energy, using it to produce carbohydrates from carbon dioxide dissolved in the water.

Phytoplankton are consumed by microscopic **zooplankton** (ZOE-oh-plank-ton), single-celled organisms called protozoans, and multicellular crustaceans —organisms with a hard outer shell like shrimp and lobsters (**FIGURE 5-22**). Zooplankton form the second trophic level of many aquatic food chains. Zooplankton, in turn, are consumed by small fish, which are a food source for larger fish and other organisms.

KEY CONCEPTS

Aquatic systems are divided into distinct regions, known as *aquatic life zones*, which may be freshwater or saltwater. The abundance of life is determined by energy and nutrient levels. Phytoplankton form the base of aquatic food chains.

Freshwater Lakes

Freshwater aquatic life zones include lakes, ponds, rivers, and streams. We'll begin by looking at ponds and lakes.

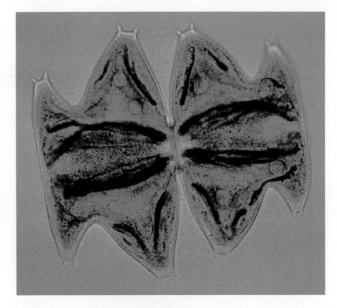

FIGURE 5-21 Phytoplankton. Unicellular algae like these are called phytoplankton and form the base of many aquatic food chains.
© T.E. Adams/Visuals Unlimited.

FIGURE 5-22 Zooplankton. These organisms feed on algae and are consumed by small fish.
© John D. Cunningham/Visuals Unlimited.

Ponds are relatively small, shallow bodies of water. Because they are shallow, sunlight often penetrates to the bottom, providing plenty of energy for all aquatic life. Lakes are deeper than ponds and generally contain four distinct zones. The first is the **littoral** (LIT-tore-el) **zone**, from the Latin word *litoralis*, meaning "seashore" or "coast" (**FIGURE 5-23**). The littoral zone consists of the shallow waters at the margin of a lake, where rooted vegetation often grows. It contains abundant phytoplankton and rooted vegetation. Some rooted vegetation never breaks the surface; it is submerged throughout its life. Others—for example, cattails—extend above the water's surface. Still other plants—such as water lilies—have leaves and flowers that float on the water's surface.

As Figure 5-23 shows, the **limnetic** (limb-NET-tick) **zone** is a region commonly called "open water." Extending

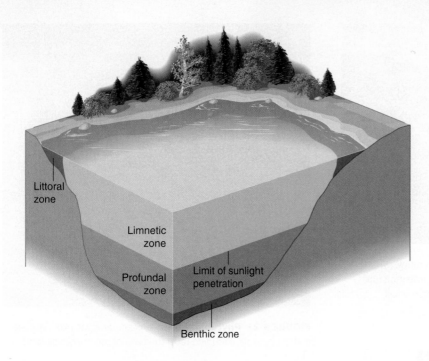

FIGURE 5-23 **Zones of a lake.**

downward to the point at which light no longer penetrates, the limnetic zone is the main photosynthetic body of a lake. In many lakes, the limnetic zone supports abundant phytoplankton—so many, in fact, that phytoplankton biomass may exceed the biomass of rooted vegetation along the shoreline. The limnetic zone is therefore something of a biological factory, producing the food that supports most other aquatic life forms. It is also a major source of oxygen, required by zooplankton, many bacteria, and animals such as fishes.

The **profundal zone** lies beneath the limnetic zone in deeper lakes and extends to the bottom (Figure 5-23). Because no sunlight penetrates the profundal zone, conditions are not favorable for plant and algal growth. Without sunlight and plants, oxygen levels remain fairly low. Fishes can survive in the region, but they rely on food produced in the limnetic and littoral zones.

The bottom of a lake is the **benthic zone** (Figure 5-23). It is home to organisms that tolerate cool temperatures and low oxygen levels—such as snails, clams, crayfish, various aquatic worms, and insect larvae (including those of the mayfly, dragonfly, and damselfly). The larvae emerge during the spring and become free-flying insects. Mayflies, for example, generally breed within a day of emergence and then die. Many of them are eaten by fish and are therefore an important source of food in freshwater aquatic ecosystems.

KEY CONCEPTS

Lakes are divided into four regions—the littoral zone, the limnetic zone, the profundal zone, and the benthic zone—with very different conditions and, consequently, very different life forms.

Lake Turnover

During the summer in temperate climates (such as North America), the water of most lakes forms three distinct layers

characterized by different temperatures (**FIGURE 5-24**). The warm surface water of a lake is called the **epilimnion** (ep-eh-LIM-knee-on) (**FIGURE 5-24a**). Just beneath the epilimnion is a region of abrupt temperature change, the **thermocline** (thur-moe-CLINE). Swimmers can experience the thermocline by diving toward the bottom of a lake. Below the thermocline is a region of fairly uniform temperature, the **hypolimnion** (high-poe-LIM-knee-on). In the summer, the hypolimnion contains dense, cold water.

As Figure 5-24a shows, the three summertime layers also differ in oxygen levels. The oxygen concentration is highest in the epilimnion, where light penetration and photosynthesis are greatest, and falls rather rapidly from the surface to the bottom as photosynthesis declines.

The thermal layering of lakes is not static. During the fall, for example, the surface waters gradually cool, and the water temperature of a lake eventually becomes fairly uniform from top to bottom (**FIGURE 5-24b**). In other words, thermal stratification disappears. In the fall, winds often churn the water, causing a thorough mixing of surface and deep water. As a result, oxygen levels also become fairly uniform from top to bottom. This mixing of surface and bottom waters is known as the **fall overturn.**

In late fall and early winter, the surface waters of a lake cool even more. Ice may form on the lake, often covering the entire surface. Because ice is less dense than liquid water, it floats. At this time, water oxygen levels are fairly uniform from top to bottom, although levels are lowest in the deepest waters and at the bottom (**FIGURE 5-24C**).

During the winter, many species such as turtles burrow in the mud, where they hibernate. Fishes, however, remain active and continue to feed throughout the winter, living on oxygen produced during the summer months. Fortunately, cold water reduces their metabolism and thus their demand for oxygen. Algae, which are present in reduced number, replenish the oxygen so long as sunlight can penetrate the ice.

In the spring the ice melts, and the water temperature becomes uniform once again. Winds agitate the water, causing a mixing known as the **spring overturn** (**FIGURE 5-24d**). As the days get longer and warmer, however, the surface waters begin to warm, and the lake again becomes thermally stratified.

Rivers and Streams

Rivers and streams are complex ecosystems. As with lakes, no two streams are alike. In many areas, streams begin in mountains or hilly terrain, collecting water that falls to the

KEY CONCEPTS

During the summer in temperate climates, lakes consist of three zones: the upper layer, the epilimnion; the deepest layer, the thermocline, and the lowest layer, the hypolimnion. They differ with respect to temperature, oxygen content, and sunlight penetration. Stratification disappears in the fall and returns again as summer approaches. Photosynthesis of algae and plants is a major source of oxygen.

Earth as rain or snow. The region drained by a stream is called a **watershed.** Small streams join to form rivers, and rivers flow downhill to the sea.

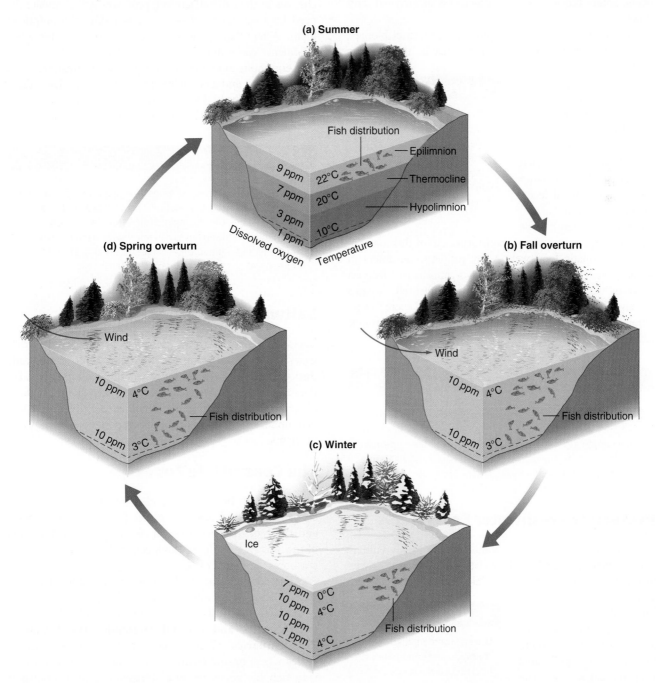

FIGURE 5-24 Thermal stratification and turnover in lakes. Lakes in the temperate zone are thermally stratified, but waters turn over in the spring and fall, resulting in a mixing of waters and a distribution of oxygen to deeper levels.

Streams and rivers are generally well oxygenated because they have a relatively large surface area (relative to their water volume) to absorb oxygen from the air. Current also facilitates oxygenation. Current velocity is determined by the gradient—that is, the steepness of the terrain. Fast-moving currents of mountain streams produce waves and rapids as the water collides with rocks or drops over ledges. This agitation greatly increases oxygenation. For these and other reasons, photosynthesis is a less important source of oxygen in rivers and streams than it is in ponds and lakes. Certain fish are adapted to different currents. Trout, for example, generally inhabit cold, oxygen-rich mountain streams where the current is quite rapid. Black bass live in rivers, where the waters are warmer, less oxygenated, and slower.

Unlike many lakes and ponds and terrestrial biomes, streams are rather open ecosystems; that is, they receive a great many nutrients from bordering ecosystems. In some streams, much of the biologically available energy actually comes from nearby terrestrial vegetation, such as leaves that fall into the stream. Animal feces, insects, stems, nuts, and other biomass may also be washed into streams during rainstorms. All of this material feeds the aquatic food web. Many primary consumers in streams, therefore, are detritivores—organisms that feed on waste or remains of plants and animals (detritus). Streams do have their own producers, mostly algae and rooted vegetation, but in some cases, these organisms play only a minor role in providing food.

Go Green

Join groups that are cleaning up rivers and streams in your area. Participate in their programs. If there aren't any programs, you might want to start your own.

Protecting Freshwater Ecosystems

Lakes are repositories for pollutants and are highly vulnerable to them, especially if their waters are replaced slowly. Streams generally fare better than lakes because of their flow, which tends to whisk pollutants away. Despite this natural purging, streams and rivers are also quite vulnerable to pollutants from human sources. The Rhine River of Europe, for example, once supported

Go Green

Use environmentally safe shampoo and conditioner to help reduce water pollution. You can purchase them at natural grocers.

150 species of fishes. Today, only about 15 species remain, in large part because of pollutants from the hundreds of factories, towns, farms, and sewage treatment plants dotting its banks and in the watershed in the seven countries through which it flows.

Protecting rivers and streams requires measures to control pollution through a river's watershed, that is, to reduce the release of pollutants into water bodies. One way they are achieving this goal is by waste minimization. **Waste minimization,** or **waste reduction,** involves measures that greatly reduce or eliminate waste. Waste minimization is a preventive measure with great benefit to people and the planet. Many companies are finding that simple, often inexpensive changes in their manufacturing processes will eliminate the need for toxic chemicals—or greatly reduce the output of them. Where there is no waste, there is no need for waste treatment. Waste minimization, therefore, can save companies enormous amounts of money in waste treatment and disposal costs, and it eliminates future liability for harm done by pollution. **Spotlight on Sustainable Development: 5–1** shows ways Christmas tree farmers can reduce pollutants.

Saltwater Life Zones

The oceans cover over 70% of the Earth's surface. Like freshwater systems, the ocean can be divided into ecologically distinct life zones. The sections that follow discuss several of the most important ones. In the ocean, as in all other bodies of water, the distribution and abundance of life are dependent on many factors, but the most important are energy and nutrients.

The Coastal Life Zones

The ocean can be crudely divided into two groups of life zones, those lying near the coasts (the coastal life zones) and those of the deeper ocean. We begin our exploration of the seas by examining three coastal life zones: estuaries, seashores, and coral reefs.

The coastlines are highly productive waters characterized by abundant sunlight and a rich supply of nutrients, both of which contribute to an abundance of life forms.

Estuaries and Coastal Wetlands: The Estuarine Zone Estuaries (ES-stu-air-ees) are the mouths of rivers, places where freshwater mixes with saltwater. Estuaries are very rich life zones because streams and rivers transport many nutrients washed from the land and incoming tides carry nutrients into them from the ocean. These nutrients support abundant plant and algal growth and sizable populations of fish and molluscs (clams, oysters, and scallops).

Most estuaries are located near **coastal wetlands**—salt marshes, mangrove swamps, and mud flats (**FIGURE 5-25**). Together, estuaries and coastal wetlands form the **estuarine zone.** For most people, coastal wetlands are seen as muddy, smelly places of little value. However, studies show that approximately two-thirds of the world's commercially valuable fishes and molluscs (shellfish) depend on the estuarine zones at some point in their life cycle. For example, many marine fishes spawn in the estuarine zone. Their eggs are laid on the bottom and are extremely susceptible to the

SPOTLIGHT on Sustainable Development

5–1 Dreaming of a Green Christmas

Each year, Americans buy an estimated 24.5 million Christmas trees, which are grown on tree farms throughout the country. They are worth slightly more than $1 billion, according to the University of Illinois Extension service. Trees are grown in fields cleared from native forest (**FIGURE 1**). Although the industry helps to boost many local economies, tree farms displace native wildlife. Because of the heavy use of chemical pesticides, they can also threaten waterways and species, including humans, living nearby.

In 1992, residents of Boone, North Carolina, in the Appalachian Mountains, found that the heavy use of chemicals to control both insects and weeds on local Christmas tree farms was polluting local water supplies and might be linked to an elevated rate of childhood leukemia in the region. Although health officials could not say with certainty that there was a correlation between pesticide use and leukemia, hundreds of farmers in Watauga County decided to take action—drastically slashing the amount of pesticide they used. In some instances, farmers were able to eliminate certain pesticides.

Weeds and pests still persist, but they're being controlled by natural biological control methods. The results have been encouraging. Groundwater supplies are improving. Wildlife are rebounding in fields and nearby woodlands and streams.

This effort to help improve both human and environmental health was successful partly because the county agricultural extension office developed a series of workshops for local farmers who wanted to learn alternative ways of controlling pests. Some solutions have been quite simple. In years past, farmers sprayed fields with diazinon or chlordane to kill grubs that dine on the roots of fir trees. Wooly aphids, another pest, were controlled with yet another potent pesticide.

To control weeds, farmers sprayed fields in the late winter with herbicides. This wipes out weeds before they sprout. They then mowed the grass around the trees. It turns out, however, that grubs prefer short grass for laying eggs. Today, many farmers are forgoing the early herbicide treatment and letting the grass grow longer. Grubs have been effectively controlled. Farmers save on mowing costs and herbicide use.

FIGURE 1 Tree farm in North Carolina.
© Jason Tench/ShutterStock, Inc.

FIGURE 5-25 Estuarine zone. Aerial view of an estuary and mangrove swamp (coastal wetland), collectively forming an estuarine zone.
Courtesy of NOAA.

deposition of sediment. The larvae that hatch from the eggs feed off phytoplankton and zooplankton that flourish in the nutrient-rich waters of the estuary. The estuarine zone also provides food, shelter, and breeding grounds for millions of waterfowl and fur-bearing animals such as muskrats.

Unfortunately, the estuarine zone is one of the most altered ecosystems on Earth. Dams built on rivers—to control flooding, supply drinking or irrigation water, or provide recreational activities—reduce the flow of freshwater into estuaries. Reservoirs behind dams also capture sediment that would otherwise travel to the estuary. Sediment contains many nutrients vital to estuarine phytoplankton. Dams, therefore, literally starve the organisms at the base of the estuarine food chain.

The impact of dams is aptly illustrated by the Aswan High Dam, built in the early 1960s on the Nile River in Egypt to provide irrigation water for farms. Although the dam provides many benefits, it has nearly halted the flow of nutrients into the river's estuary. This in turn caused a collapse in the sardine fishery in the Mediterranean Sea, as phytoplankton perished. The commercial sardine catch in the Mediterranean dropped from 18,000 tons per year before the dam was built to about 500 tons per year today. Some improvement was recorded in the early 1990s, but in appears that the sardine catch has pretty much plateaued.

Coastal wetlands have been drained and filled with dirt to build homes, highways, recreational facilities, and factories. These activities destroy habitat for fish and other species. In the United States, over 40% of the coastal wetlands have been destroyed. In California, an estimated 90% have been destroyed.

KEY CONCEPTS

Estuaries are nutrient-rich zones at the mouths of rivers, often associated with coastal wetlands; together they form the estuarine zone. The estuarine zone is highly productive and of great value to humans and other species. Human activities severely threaten this important biological asset.

The Shoreline

The Earth's coastlines are generally rocky or sandy regions that support a variety of specially adapted organisms. Abundant sunlight and nutrients account for much of the biological diversity in this zone. Rocky coastlines are home to seaweed (algae), sea urchins, barnacles, and sea stars (**FIGURE 5-26**). Many of these organisms anchor themselves to the rocks and are thus able to withstand turbulence generated by waves. Organisms that do not attach themselves, such as sea urchins, live in rocky crevices protected from water currents.

The coastline is also home to a variety of shorebirds that dine on insects, crustaceans, and other organisms. In addition, the coastlines serve as nesting sites for turtles and are home to breeding colonies of seals and walruses. Popular recreation and building sites, many of the world's coastlines have been severely altered. Turtle nesting sites have been destroyed in many places. Coastlines are often badly contaminated by oil from nearby spills. Protecting the coastlines from offshore oil spills and human encroachment is extremely important.

KEY CONCEPTS

The shorelines of the world are rocky or sandy regions that are home to a surprising number of organisms adapted to the tides and the turbulence created by wave action.

Coral Reefs

Coral reefs are found in relatively warm and shallow waters in the tropics or nearby regions (subtropics). A **coral reef** is created by corals, marine organisms that live in compact colonies. Stony corals, one type of coral, secretes calcium carbonate forming a hard skeleton that creates reefs. Most corals contain algae that produces, energy and nutrients via photosynthesis. These corals form coral reefs in shallow

FIGURE 5-26 Tidal pool inhabitants. Numerous species live along the rocky shorelines of the world's oceans.
Courtesy of NOAA.

water. (**FIGURE 5-27**). The stiff calcium carbonate outer skeletons persist long after the death of these, creating a base on which other corals grow. Coral reefs take thousands of years to form.

Coral reefs are the aquatic equivalent of the tropical rain forest. They are home to a dazzling variety of organisms, many of which are colorful beyond imagination (**FIGURE 5-28**).

Like estuaries and coastal wetlands, coral reefs are also highly vulnerable and experiencing considerable damage. Ships running aground on the coral reefs of Florida, for example, are damaging these fragile structures. Each year, thousands of divers visit these coral reefs. Careless divers break delicate coral with their swim fins. Over the years, many of the most popular reefs have been severely damaged.

Sediment is a particularly troublesome pollutant in coral reefs. It clouds seawater, which reduces photosynthesis in a group of microscopic organisms (called *dinoflagellates*) that live in a mutualistic relationship with many species of corals. Decreasing photosynthesis decreases the amount of food available to coral, and reefs grow more slowly. Particularly heavy sedimentation can bury a reef, choking the life out of it. Several large reefs near Honolulu, in fact, have been destroyed by sediment washed from construction sites, highways, and farms. Housing development on the Florida Keys is also increasing sedimentation, threatening nearby coral reefs. Many of the world's coral reefs are dead or dying as a result of warming ocean waters associated with global climate change brought on by human activities.

KEY CONCEPTS

Coral reefs are found in warm, shallow waters of the world's oceans. Rich in life, they are considered the aquatic equivalent of the tropical rain forests. Coral reefs are being damaged by ships, divers, sediment pollution, and rising water temperatures brought on by global warming.

FIGURE 5-27 Coral reef. The coral reef is bathed in sunlight and found in warm water. These features and ample nutrients make it the richest of all marine aquatic zones.

Courtesy of Mr. Mohammed Al Momany, Aqaba, Jordan/NOAA.

(a)

(b)

FIGURE 5-28 Coral reef inhabitants. (a) Coral reefs house some of the most spectacular fish known to science. **(b)** Brightly colored anemones filter food from the water.

Photo a © krishnacreationz/ShutterStock, Inc.; photo b © Photos.com.

The Marine Ecosystem

FIGURE 5-29 shows a cross-section of the ocean floor. As illustrated, the ocean floor slopes downward away from landmasses and then drops more steeply. The gradually sloping region is called the **continental shelf**. The more steeply falling region is the **continental slope**, and the bottom of the deep ocean is called the **abyssal plain**.

As Figure 5-29 shows, the ocean is divided into four *ecologically distinct* life zones: the neritic, euphotic, bathyal, and abyssal. The **neritic** (neh-RIT-ick) **zone** is equivalent to the littoral zone of lakes and lies above the continental shelf, which varies in width from 10 to 200 miles away from land. The neritic zone contains relatively shallow water and consequently receives abundant sunshine. Its waters are relatively warm and well oxygenated.

Nutrients in the neritic zone come from streams and rivers that flow into the ocean. They are also supplied by upwellings, the transport of nutrient-rich water from the floor of the ocean up the continental slope. These nutrients

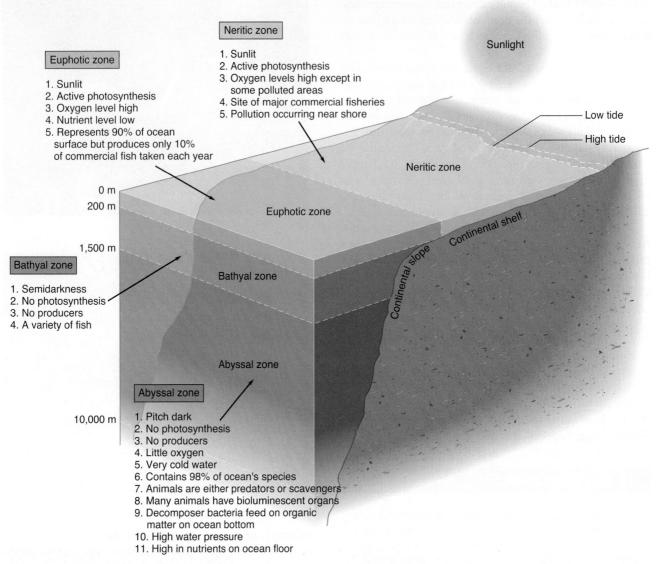

Euphotic zone

1. Sunlit
2. Active photosynthesis
3. Oxygen level high
4. Nutrient level low
5. Represents 90% of ocean surface but produces only 10% of commercial fish taken each year

Neritic zone

1. Sunlit
2. Active photosynthesis
3. Oxygen levels high except in some polluted areas
4. Site of major commercial fisheries
5. Pollution occurring near shore

Sunlight

Low tide

High tide

Neritic zone

0 m
200 m

Euphotic zone

1,500 m

Bathyal zone

1. Semidarkness
2. No photosynthesis
3. No producers
4. A variety of fish

Bathyal zone

Continental slope
Continental shelf

Abyssal zone

Abyssal zone

10,000 m

1. Pitch dark
2. No photosynthesis
3. No producers
4. Little oxygen
5. Very cold water
6. Contains 98% of ocean's species
7. Animals are either predators or scavengers
8. Many animals have bioluminescent organs
9. Decomposer bacteria feed on organic matter on ocean bottom
10. High water pressure
11. High in nutrients on ocean floor

FIGURE 5-29 Ocean zones.

help support abundant phytoplankton, zooplankton, and fishes.

In the neritic zone, sunlight normally penetrates to the ocean bottom. Ample sunlight supports large populations of algae and rooted plants, which, in turn, support a great many other species. In fact, most commercial fishing operations in the ocean concentrate their operations in the neritic zone.

The **euphotic** or **photic** (you-FO-tic or FO-tic) **zone** is the oceanic equivalent of the limnetic zone of lakes. It is the open-water region that extends to the lower limits of sunlight penetration, or about 200 meters (650 feet) below the ocean's surface. Abundant sunlight supports numerous species of phytoplankton. Phytoplankton support a variety of zooplankton, mostly minute crustaceans.

Phytoplankton also produce plenty of oxygen. Given the high levels of oxygen and ample supply of sunlight, one might think that the euphotic zone would be extremely productive. Unfortunately, the waters are not very rich in nutrients, and the zone is therefore not very productive. Thus,

even though the euphotic zone covers 90% of the ocean's surface, it produces only about 10% of the commercial fish taken each year.

Beneath the euphotic zone is a region of semidarkness, the **bathyal** (BATH-ee-el) **zone**. The bathyal zone is too dark to support photosynthesis. It is therefore characterized by a lack of photosynthetic organisms and low oxygen levels. Despite these factors, the bathyal zone is home to a variety of fish and other organisms, such as shrimp and squid, which feed on organisms "raining down" from above.

Beneath the bathyal zone is the abyssal zone. The **abyssal** (ah-BISS-el) **zone** is a region of complete darkness. It contains no photosynthetic organisms and is characterized by low oxygen levels. Most animals that live there are either predators or scavengers. To live in this zone, an animal must be adapted to extremely cold water, high water pressure, low oxygen, and complete darkness.

Sediment in the abyssal zone is often rich in nutrients. The deep-ocean floor is populated by a variety of bizarre-looking creatures that make use of these nutrients. Some species of fishes have evolved luminescent organs that shine in the dark, presumably helping them attract food and mates.

KEY CONCEPTS

The marine ecosystem consists of four ecologically distinct life zones, similar to those found in lakes.

Nature's laws affirm instead of prohibit. If you violate her laws you are your own prosecuting attorney, judge, jury and hangman.

—**Luther Burbank**

CRITICAL THINKING

Exercise Analysis

Scientists have asked many questions about these studies. First, could the differences in ice thickness have been the result of normal variation? (Possibly.) Were the measurements of the two submarines taken during the same time of year? (No.) Did the submarines follow the same path? (Actually, they followed slightly different routes.) Could winds have altered ice buildup between the two periods? (Possibly. Winds do alter ice buildup.) Have other studies shown thinning? (Yes. A study of ice measurements by two submarines on a virtually identical route across the Canada Basin during 1958 and 1970 showed that the mean ice cover was thinner in 1970 by 0.69 meter, about 2 feet.)

What other information/data might be useful in analyzing the hypothesis that the world's polar ice caps are shrinking? One bit of information that would be very helpful would be the extent of the ice sheets—that is, the square kilometers of ice sheet, plotted over time. Recent studies of this phenomenon show a marked decrease in the extent of the Arctic ice sheet, valuable habitat of the polar bear.

CRITICAL THINKING AND CONCEPT REVIEW

1. Why are environmental conditions profoundly different in tropical, temperate, and polar regions.
2. How are global winds generated? What factors influence the direction in which they flow?
3. In what way do global air circulation patterns affect land surface temperatures throughout the world?
4. How does global water circulation in the oceans affect climate?
5. What is a biome? Why is one biome so different from another? Give some examples.
6. Using what you have learned, critically analyze the following comment. "Environmentalists speak about ecosystems as being fragile; yet look at the tundra. The conditions there are as harsh as any on Earth."
7. Design an experiment to determine how long it would take for the Arctic tundra to recover from human

impact caused by truck traffic on the unprotected vegetation.
8. Animals are generally well adapted to the biome in which they live. Give some examples of plant and animal adaptations that enable them to live and reproduce in their biomes.
9. Explain why there are regional differences in vegetation and animal life within a biome.
10. Which biome do you live in? Describe it. How has it been altered?
11. List the pros and cons of oil development in the ANWR. Which side do you take on the issue? Why?
12. Critically analyze the following comment. "The United States has created a rich agricultural industry by plowing grasslands and clearing and plowing temperate forests. There is no reason to believe that countries in the tropics can't follow the same

13. Why do many deserts form downwind from mountain ranges?

14. Explain why the desert biome is spreading.

15. What are the primary determinants of species composition and abundance in terrestrial and aquatic habitats?

16. Describe each of the four zones of a lake (littoral, limnetic, profundal, and benthic) in terms of light availability, nutrient levels, oxygen concentration, and life forms.

17. Describe thermal stratification in deep lakes. Why does it occur during the summer but disappear throughout most of the rest of the year?

18. Why is the current so important in determining the abundance and type of life in a river?

19. Explain how streams and rivers recover from pollution and why natural recovery methods are so often overwhelmed.

20. What is an upwelling? How is it similar to an estuary? How is it different?

21. Describe ways in which human activities affect coral reefs and the estuarine zone.

KEY TERMS

abyssal plain
abyssal zone
alpine tundra
altitudinal biomes
aquatic life zones
bathyal zone
benthic zone
biome
climate
coastal wetlands
conifers
continental shelf
continental slope
coral reef
Coriolis effect
deciduous trees
dust bowl
El Niño

epilimnion
estuarine zone
estuaries
euphotic (or photic) zone
fall overturn
grasslands
hypolimnion
lateritic soil
limnetic zone
littoral zone
neritic zone
northern coniferous forest biome
permafrost
phytoplankton
polar easterlies
poles
profundal zone
shortgrass prairie

south westerlies
spring overturn
taiga
tallgrass prairie
temperate zones
thermocline
trade winds
tropics
tundra
upwellings
waste minimization
waste reduction
watershed
weather
westerlies
zooplankton

Principles of Ecology: Self-Sustaining Mechanisms in Ecosystems

And this our life, exempt from public haunt, finds tongues in trees, books in running brooks, sermons in stones, and good in everything.

—**Shakespeare**

CHAPTER OUTLINE

LEARNING OBJECTIVES

After reading and studying the material in this chapter, students will be able to understand, knowledgably discuss, and, when appropriate, apply the information learned in this chapter, including:

1. How ecosystems function and mechanisms that allow them to sustain themselves over time
2. The concept of homeostasis and why it is vital to all organisms and ecosystems
3. Factors that help create ecosystem balance
4. Why ecosystems are resilient and how resilience allows ecosystems to resist some of the changes brought about by human activities
5. How ecosystems regulate population and why population control is so vital to ecosystem sustainability
6. The process of natural succession and the difference between primary and secondary succession
7. The theory of evolution, the end products of evolution, and how evolution occurs
8. How humans alter the environment by influencing abiotic and biotic factors

CRITICAL THINKING

Exercise

You are a policy analyst for your state's department of natural resources and are asked to comment on a proposal to import a species of deer from Siberia into one of the state's wildlands. This area is devoid of large grazers such as deer. The lands are well vegetated and appear to be prime deer habitat. Proponents of this proposal think that it would enhance hunting in the state and generate substantial economic benefits to local communities as well as the state. Evaluate the proposal. What information would you need? What problems would you anticipate? After you have completed the exercise, list the critical thinking rules you used.

FIGURE 6-1 Beautiful but obnoxious The water hyacinth was intentionally introduced into southern Florida but was accidentally released into its waterways. Since its release, the plant has proliferated wildly and today chokes canals and streams throughout the state and much of the southern United States.

Courtesy of Nancy Mayberry/U.S. Army Corps of Engineers.

I n 1884, the water hyacinth was introduced into Florida from South America as an ornamental plant. Housed in a private pond, this lovely flowering plant, which grows on the water's surface, was accidentally released into a nearby river. From there, it spread throughout the state's extensive network of canals and rivers. Aided by a remarkable ability to reproduce—10 plants can multiply to 600,000 in 8 months—the hyacinth now chokes waterways throughout much of the state. This hearty colonizer crowds out native plants and grows so thick that it has made navigation impossible in some rivers (**FIGURE 6-1**).

From Florida, the water hyacinth continued to spread, and it now inhabits the rivers and lakes of much of the southern United States. At one time, the plant occupied nearly 800,000 hectares (2 million acres) of rivers and lakes from Florida to California. Thanks to extensive control efforts, the water hyacinth's range has been seriously reduced. In Louisiana, for example, the water hyacinth's distribution has been slashed from 400,000 (988,000 acres) to 80,000 hectares (197,000 acres). In Florida, another state where the water hyacinth has proliferated, its range has been dramatically reduced. The state spends about $5.8 million a year to control the invasive species.

The story of the water hyacinth illustrates a common phenomenon: the disruption of a natural ecosystem by a seemingly harmless human action—the importation of an alien species. It also serves as a warning to humans that seemingly harmless actions can have severe ecological repercussions. The economic cost of such acts can also be

staggering. See the Point/Counterpoint in this chapter for a debate on related issues.

This chapter continues the discussion of ecosystems by exploring mechanisms in natural systems that have permitted them to persist for millions of years. An understanding of these mechanisms can help us manage natural systems—and ourselves—better. This material also permits us to explore the principles of sustainability: conservation, recycling, restoration, population control, and adaptability.

6.1 Homeostasis: Maintaining the Balance

Life is a balancing act. Take the human body as an example. Inside each of us are hundreds of mechanisms that evolved to help maintain fairly constant internal conditions—that is, to keep our bodies in balance. These mechanisms help us maintain body temperature, blood

One of the key realizations of ecology is that the Earth is a kind of living system governed by many self-regulating (homeostatic) mechanisms. The Earth is in a state of equilibrium. If pushed too far in any one direction, the self-regulating mechanisms can become overloaded and break down, resulting in radical changes.

In its scientific aspects, ecology seems to offer an extraordinary broadening of our understanding of life on the planet. Yet with its transfer into the public domain, it has become little more than a sophisticated way of saying, "We don't want any more progress." Somehow this exciting discipline has been translated into a very conservative social doctrine. People have often waved the flag of "ecology" as a new way of saying that nature must be preserved and human activity minimized. Ecology is sometimes viewed as "subversive" to technological progress. It supposedly tells us that our ignorance of natural systems is too great for us to proceed any further with human enterprise. Just as nationalistic conservatives always try to throw a veil of reverence around such concepts as "patriotism" and "national tradition," so environmentalists try to maintain the same indefinable quality around ecosystems.

The lesson environmentalists drive home is that because we do not understand ecosystems in their entirety, and never will, we dare not touch them. Our knowledge is too limited, and nothing should be done until we understand more fully the implications of our actions.

To say that ecology is the science that does not yet grasp the complex interrelationships of organisms is like trying to define medicine as the science that does not yet know how to cure cancer. Environmentalists emphasize the negative parts of the discipline because it fits their concept that we have already had enough technological progress.

The lessons of ecology tell us many things. They tell us that organisms cannot go on reproducing uncontrollably. But they also tell us that many organisms have developed behavioral systems that keep their populations from exploding. The laws of ecology tell us that we cannot throw things away into the environment without having them come back to haunt us. But they also tell us that nature evolved intricate ways of recycling wastes long before human beings appeared, and ecosystems are not as fragile as they seem.

In fact, the whole notion of "fragile ecosystem" is somewhat contradictory. If these systems are so fragile, how could they have survived this long? If ecology teaches us anything, it enhances our appreciation of

Environmentalists Are Subversive to Progress

William Tucker
The author, a critic of the environmental movement, has written numerous articles on environmental issues. His book *Progress and Privilege* is a thought-provoking and controversial discussion of environmentalism.

how resilient nature is and how tenaciously creatures cling to life in the most severe circumstances. This, of course, should not serve as an invitation for us to see how efficiently we can wipe them out. But it does suggest that the rumors of our powers for destruction may be exaggerated.

The environmentalist interpretation of ecology has been that ecosystems have somehow perfectly evolved and that human intervention always leads to degradation. It should be clear that even if a particular ecosystem did represent biological perfection, that is not reason in and of itself to preserve it at the expense of human utility. Our ethical position cannot be one of completely detached aesthetic appreciation. We must first be human beings in making our ethical judgments. We cannot be completely on the side of nature.

We are not a group of imbeciles aimlessly poking into the backs of watches or tossing rocks into the gears of Creation. There is purpose to what we do, and it is essentially the same as nature's. We are trying to rearrange the elements of nature for our own survival, comfort, and welfare. We can certainly act stupidly, but we can also act out of wisdom. It is foolish to argue that everything is already perfect and must be left alone. To portray humans as meddling outsiders in an already perfected world is nonsense. In going to this extreme to reaffirm nature, we only deny that we are a part of it.

Environmental writers suggest that we practice an "ecological ethic," extending our moral concerns to other animals, plants, ecosystems, and the entire biosphere. I would accept this proposal, with one important qualification: that is, that our ethical concerns still retain a hierarchy of interest. We should extend our moral concerns to plants, trees, and animals, but not at the expense of human beings. Our first obligation is to humanity. We should avoid actions that are destructive to the biosphere, but we must recognize that at some point our interests are going to impinge upon other living things.

(continued)

Environmentalists Support Sustainable Progress

Daniel D. Chiras

Contrary to what some might have you think, the environmental movement is not composed of a homogeneous group of people who uniformly object to all human progress. The environmental movement attracts a following linked by a common interest in protecting the environment with different political leanings.

My experience within the movement, however, shows that environmental protection means many things to many people. People sympathetic with the plight of the natural world range from those who call for "complete preservation," which tolerates little human intervention, to the "preserve-it-in-parks" folks, who seek to protect small pieces of the environment for future generations to enjoy, while developing most of the rest for human use.

William Tucker's essay in this Point/Counterpoint casts environmentalists as narrow-minded obstructionists who condemn all human progress and technological development. No doubt, many businesspeople share this view.

Progress is not one thing to all people. In fact, much of the environmental debate hinges, quite precariously, on the difference in how people view progress.

To most environmentalists, progress has a meaning much broader than economic growth, faster jets, and new gadgets to make life a little easier. Progress means prospering within the limits of nature, living on Earth while protecting the air, water, and soil on which our lives depend. Progress means finding a way of living sustainably on the planet.

Most environmentalists embrace progress, but no human endeavor can be counted a success if it destroys the Earth's life-support systems.

Tucker's assertion that environmentalists don't want any more progress misses the point. Environmentalists want lots of progress. We want progress in recycling and tapping into renewable energy in a big way. We want progress in using energy and other resources more efficiently. We want progress in restoring ecosystems that we depend on and in controlling human numbers.

As for Tucker's assertion that environmentalists think that nature must be preserved and human activity minimized, my experience in environmental protection shows that most environmentalists are working to keep society from destroying the Earth's life-support systems. The knowledge and skills necessary to fit into the web of life and succeed are readily available. Restraint and proper design are necessary, but stopping all human activity is not.

Ecology has taught environmentalists to look to the future with their eyes wide open, rather than narrowly focused on material welfare, economic wealth, and convenience. In attempting to rearrange the elements of nature to satisfy our needs, we must tread carefully, finding ways to meet our needs without impairing the ability of future generations—and other species—to meet theirs.

As for the claim that environmentalists see society as a stupid meddler whose every activity leads to degradation, let me say this. History teaches us that human activities have had profound, even devastating, effects on the natural world and human society. The dust bowl days, the great deserts spreading through Africa, and destruction of the once rich Fertile Crescent are blatant reminders of the severity of our intrusions. More recent studies show that our impact continues on a grand scale, as witnessed by the threats of deforestation, species extinction, global warming, and stratospheric ozone depletion.

Human intervention doesn't always lead to degradation, but it is also not a force to be trivialized. Ecosystems are not as fragile as some would have you believe, but they can be ruined on a massive scale. Witness the widespread loss of wetlands and rain forests throughout the world. Nature may be resilient, but against widespread human intervention, it is often powerless.

The crux of the disagreement between environmentalists and "developmentalists" lies in each group's view of progress. In my view, many environmentalists have a more sustainable view of progress. Environmentalism is not subversive. It advocates an alternative form of progress that seeks to foster a sustainable relationship with the planet. Our interests are likely to impinge on the interests of other species, but our challenge is to find ways to minimize these conflicts. By adopting the principles of sustainability outlined in this book, we can create more efficient and effective government, a better future, broader prosperity, and a stronger country. Through mutual responsibility, we can live well while protecting, even enhancing the quality of the air, water, and soil. Let's stop fighting over the environment and find a new way of life and worldview that serve humans and the environment simultaneously.

Critical Thinking Questions

1. Summarize the main points in each essay. Were arguments adequately supported by facts?
2. Use your critical thinking skills to analyze and comment on these views.
3. Whose views most closely represent yours? Why?

sugar levels, hormone levels, water concentration, and many other things.

The body's state of internal constancy is called **homeostasis** (HOE-me-oh-STAY-siss). The word *homeostasis* comes from two Greek words, *homeo*, which means the "same," and *stasis*, meaning "standing." Literally translated, *homeostasis* means "staying the same." Homeostasis, however, is not to be misconstrued as unwavering constancy. Internal conditions do not stay the same from one minute to the next. Scientists refer to homeostasis as a state of **dynamic equilibrium**. This means that conditions are dynamic—ever changing—but they stay more or less the same over long periods. Consider an example.

Blood sugar (glucose) levels oscillate a bit around a set point of approximately 100 milligrams per deciliter. Blood sugar levels increase after a meal as sugars enter the bloodstream, but within an hour or so, they return to the set point. As we increase activity, which burns sugar, levels drop. Then internal storage depots release sugar so that levels return to stay fairly constant. Blood sugar is said to be in a *dynamic equilibrium*. In sum, then, homeostasis isn't about maintaining absolutely constant conditions. It is about maintaining conditions within a range that is required for survival.

KEY CONCEPTS

Homeostasis is a state of relative constancy vital to the survival of all organisms.

Homeostasis in Natural Systems

Ecosystems also possess a number of homeostatic mechanisms that help to maintain relatively constant conditions. Predation is one such mechanism. Predators are animals that consume other organisms, their prey. Wolves, for example, are predators of deer. They help to hold their prey populations in check. Diseases and extreme weather also help to control deer populations. All three factors help to ensure homeostasis in ecosystems. Predators and diseases are beneficial to prey populations, for they help to maintain their number within the carrying capacity of the environment.

Ecosystem homeostasis is readily observable to those who study tropical rain forests and other ecosystems, such as coral reefs, where conditions remain more or less constant year round. It is less obvious in temperate climates such as the deciduous forest biome, where seasonal climate changes result in dramatic shifts in conditions. It is important to note that even though climatic conditions shift, ecosystems in the temperate climate remain fairly constant from one year to the next. Thus, if you studied a forest ecosystem near your home each spring for an extended period, you would find that (1) the same species were present each year and (2) the population size of

each species was approximately the same from one year to the next.

This is not to say that conditions are static (unchanging). In fact, things do change. Just like the human body, ecosystems exist in a state of dynamic equilibrium. Over the course of a year, some organisms die and are replaced by newborns. Some organisms may migrate out; others move in. Population sizes may increase a bit if the weather is favorable, or they may decline if the weather turns bad for a long time. Despite these changes, the system remains more or less the same over the long haul.

KEY CONCEPTS

Like organisms, ecosystems possess many mechanisms that help to keep them in a state of relative constancy.

Factors That Contribute to Ecosystem Homeostasis

Numerous factors affect the organisms in any ecosystem. These factors fall into two groups: those that tend to increase population size, which I call **growth factors**, and those that tend to decrease it, **reduction factors** (**FIGURE 6-2**). Factors that increase or decrease population size of any given organism can be biological or biotic (as in the case of predators) or nonbiological or abiotic (as in the case of rainfall or temperature). At any given moment, the population size of a given organism is determined by the interplay of numerous biotic and abiotic growth and reduction factors. Because ecosystems contain many species, the balance within an ecosystem can be crudely related to the sum of the individual population balances.

To understand this concept, let's examine an elk population in the foothills of the Rocky Mountains. Numerous biotic factors contribute to the growth of elk populations—for example, an abundance of food. Favorable abiotic conditions also stimulate increases in the elk population size. Ample sunshine, mild temperatures, and an abundance of spring rain, for example, promote the growth of the elk's primary food source: the rich grasses that grow in the meadows dotting the foothills. This, in turn, enhances survival and reproduction of elk.

In nature, growth is generally opposed by a number of abiotic and biotic factors. Ecologists describe these opposing factors (reduction factors) collectively as **environmental resistance**. In the elk populations of Colorado, predators, disease, parasites, and competition tend to reduce the elk population size, as do unfavorable weather and lack of food and water. In general, when environmental resistance is low, organisms such as the elk proliferate.

In most relatively undisturbed ecosystems, the interplay of growth and reduction factors creates a state of homeostasis. Consider another example.

REDUCTION FACTORS	ECOSYSTEM BALANCE	GROWTH FACTORS

REDUCTION FACTORS

Biotic
Predators
Disease
Parasites
Competitors
Lack of food
Lack of suitable habitat

Abiotic
Unfavorable weather
Lack of water
Alterations in chemical environment

ECOSYSTEM BALANCE

GROWTH FACTORS

Biotic
High reproductive rate
Ability to adapt to environmental change
Ability to migrate to new habitats
Ability to compete
Ability to hide
Ability to defend
Ability to find food
Adequate food supply

Abiotic
Favorable light
Favorable temperature
Favorable chemical environment

FIGURE 6-2 Ecosystem balance. Ecosystem homeostasis is affected by both forces that tend to increase population size and forces that tend to decrease it. Growth and reduction factors consist of both abiotic and biotic components.

In the grasslands of Kansas lives a mouse-like rodent known as the prairie vole (**FIGURE 6-3**). Its population size depends on many factors, among them light, rainfall, available food supply, predation, temperature, and disease. As an undergraduate student, I conducted laboratory experiments on prairie voles, and found that the optimum conditions for reproduction are low temperature (slightly above freezing), long days (14 hours of light a day), and plenty of water and food. Under these conditions a female gives birth to a litter of seven pups every 3 weeks for several years. For the captive-raised vole, motherhood is no picnic!

Fortunately, in the wild, optimum conditions rarely occur simultaneously. In the summer, for instance, when food, water, and day length are optimal, the temperature is much too warm for optimum reproduction. In the winter, when the temperature is optimal for breeding, the days are short and the food supply is low. Reproduction is less than optimum, and the population is held in check.

FIGURE 6-3 The prairie vole. The prairie vole is a small, mouse-like rodent that lives on the grasslands of much of the midwestern United States. Like all other species, its population is regulated by a myriad of biotic and abiotic factors.

Courtesy of Ryan L. Rehmeier, Department of Biology and Environmental Science, Simpson College.

KEY CONCEPTS

Numerous biotic and abiotic factors influence the growth of populations. Some stimulate growth; others deter growth. Ecosystem homeostasis is the result of the interaction of these factors.

The Resilience of Ecosystems

In ecosystems, normal fluctuations in environmental conditions such as rainfall and temperature influence all of the species, from the tiniest bacterium to the largest carnivore. Thus, abiotic changes result in biotic shifts. Fortunately, natural systems are resilient. Consider the prairie vole again.

Among prairie voles, an unusually cool summer results in an increase in reproduction, resulting in an increase in the population size. As the population of prairie voles increases, though, populations of hawks and coyotes, which prey on voles, expand. If conditions favorable to vole reproduction are sustained, a new balance might be achieved—with substantially larger populations of voles *and* their predators.

If, as is often the case, normal weather returns, reproduction among the voles would decrease, causing the population to decline. The hawk and coyote populations, which have increased in previous years, would continue to feed on voles, resulting in further declines. But as the vole population declined, the predators themselves would decline, bringing the system back into equilibrium.

The ability of an ecosystem to recover from temporary changes in conditions is referred to as **resilience**. It is essential to the maintenance of life on Earth.

Resisting Changes from Human Activities

Human impact on the environment may vary from relatively minor to severe. As a rule, minor damage can be repaired by homeostatic mechanisms. Sewage accidentally dumped into a stream, for example, adds organic chemicals to the water (**FIGURE 6-4**). These substances are food for naturally occurring bacteria. In a healthy stream, the population of these bacteria is normally small. If organic food supplies increase, however, the bacterial population expands. Because the bacteria consume oxygen as they devour organic waste, the level of dissolved oxygen in the stream plummets.

Go Green

Never dump hazardous wastes like paints or solvents down a sink or sewer or discard them with your trash. Collect hazardous wastes in their original containers and either donate them to community reuse programs or dispose of them through local household hazardous materials programs.

This, in turn, may kill aquatic organisms and fish (or causes them to migrate to better waters).

The impact of a spill can be substantial because of the chain reaction of effects. However, if the spill is a one-time event, the system will cleanse itself. The flow of water in the stream will purge it of pollutants. Bacteria will destroy the organic matter. These two factors will cause the organic nutrient levels to decline. As food supplies decline, the bacteria will die off. As the bacteria decline, the levels of dissolved oxygen will begin to increase (because of oxygen uptake at the surface). Populations of fish and other aquatic organisms will recover.

Problems arise when human activities push ecosystems too far. Continued pollution of the stream, for example, may strain the limits of resilience. In some cases, damage may be irreversible. Deforestation in the tropics, for instance, can also result in long-term damage. Although the system may recover if left alone for long periods, the recovery takes hundreds of years.

KEY CONCEPTS

The ability of ecosystems to recover from small changes minimizes and sometimes negates the impacts of human actions. In many instances, though, human actions can overwhelm the recuperative capacity of natural systems.

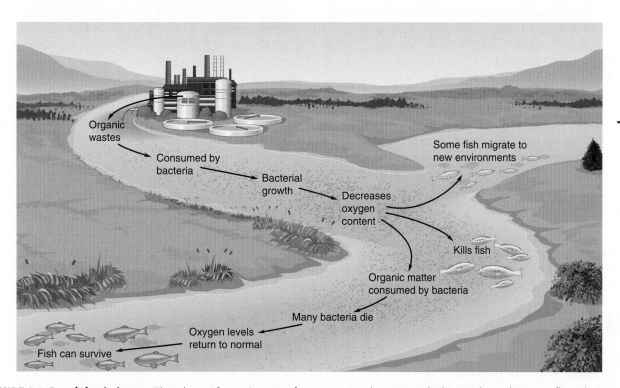

FIGURE 6-4 Regaining balance. The release of organic wastes from a sewage plant causes the bacterial populations to flourish. As they increase, oxygen levels decline, causing troubles for fish and other aquatic organisms. Oxygen levels and fish populations return to normal in the stream, thanks both to naturally occurring bacteria that decompose the organic matter in the sewage and to the water flow. Continued dumping or dumping by a downstream sewage treatment plant, however, could seriously hinder the stream's recovery.

Population Control and Sustainability

Population control is essential to the sustainability of ecosystems. It is not surprising then to find numerous population control mechanisms at work in ecosystems. These mechanisms, which are part of environmental resistance, help to maintain populations within an ecosystem's carrying capacity.

In most cases, population control mechanisms are extrinsic—that is, they are factors that influence an organism from the outside. Predation, disease, and adverse weather, for example, are extrinsic factors that help to regulate population size. In many cases, these factors are a consequence of changes in populations themselves. Population growth, for instance, causes an increase in population density, which may result in food shortages or more rapid spread of disease.

Intrinsic factors are rare. They arise from behavioral mechanisms *within* a species. Consider an example. Wolves live in packs that occupy a specific habitat—their territory. For reasons not well understood, only the dominant male and female of the pack breed and produce young. This intrinsic mechanism helps to control the size of the wolf pack within its habitat.

Humans are one of nature's many species and are also subject to natural extrinsic controls. Epidemics (outbreaks) of disease, for instance, can reduce populations. Adverse weather can take its toll as well, a phenomenon witnessed in the sweltering summer of 2003 when 40,000 people died in Europe as a result of intense heat. Humans also employ intrinsic methods to control population such as abstinence and contraceptives, which we use to reduce our numbers voluntarily.

In nature, organisms are held within the carrying capacity of their environment. Even though humans are subject to the laws of nature, especially population control, we can break nature's laws. We have, for instance, found ways to combat, even eliminate, disease and expand food production. But our ability to expand the Earth's carrying capacity in some areas often has serious consequences—for example, species extinction, pollution, and climatic disturbance. Most experts agree that to create a sustainable human society, we humans must learn to live within the Earth's carrying capacity. We must respect limits and live within them.

First and foremost on the list of things to do is reduce population growth. Robert Gilman, a proponent of sustainability, argues that the challenge is to "set our own limits or have limits disastrously imposed on us."

KEY CONCEPTS

Balance in natural systems is achieved in large part by intrinsic and extrinsic mechanisms that help to maintain populations within the carrying capacity of the environment. To create a sustainable society, many experts believe that humans must also control population size.

Species Diversity and Stability

Ecosystem stability may also be enhanced by species diversity. Roughly speaking, **species diversity** is a measure of the number of species living in a community. The more species, the more diversity. As a general rule, ecologists have found that the higher the species diversity, the greater an ecosystem's stability. This conclusion is based on observations that extremely complex ecosystems, such as rain forests, remain unchanged almost indefinitely if undisturbed. A stable system has fairly constant numbers of organisms and species. In stable systems, small changes such as the harvest of fruits and nuts by native peoples or the killing of an animal for food have little impact on the forest.

In contrast, ecosystems with fewer species such as the Arctic tundra tend to be less stable. They experience sudden, drastic shifts in population size. Other greatly simplified ecosystems such as fields of wheat and corn are also extremely vulnerable to change, and they deteriorate rapidly if abiotic or biotic factors shift very much.

FIGURE 6-5 illustrates the reason some ecologists think there is a connection between stability and diversity. This diagram shows two food webs, one in a simple (less diverse) ecosystem and the other in a more complex or diverse ecosystem. Because the number of species in a complex ecosystem is greater, the food webs are more complex. In a complex ecosystem, the elimination of one species would have little effect on the rest of the food web. In contrast, because the number of species in the food web of a simple ecosystem is small, the elimination of one species would very likely have major repercussions.

This principle actually provides an opportunity to exercise our critical thinking skills. Some ecologists point out that the stability of the tropical rain forest is not the result of species diversity but of constant climate. On the relatively simple and unstable tundra, the climate shifts dramatically from season to season. These shifts may be responsible for the tundra's relative instability.

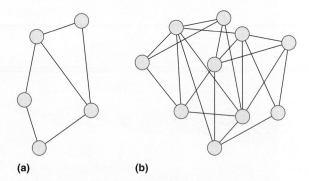

(a)   (b)

FIGURE 6-5 Food Webs. (a) Food web in a simple ecosystem. Circles represent organisms. Note the dearth of links in the simplified web. **(b)** food web in a complex ecosystem. Many ecologists contend that complex ecosystems are more stable because the increased number of links reduces the importance of any one species.

FIGURE 6-6 supports this interpretation. It illustrates that species diversity among mammals varies with latitude in North and Central America. In the frozen northern regions of Canada and Alaska, species diversity is low. Heading south, diversity increases until one reaches the tropics of Central America, where the highest diversity is found. The relationship between species diversity and latitude is also found in plants and virtually all other kingdoms (major groups of organisms including plants, animals, and fungi). Latitude, therefore, is an important factor affecting species diversity. The connection between latitude and species diversity is climate. Quite clearly, the milder the climate, the more species that live there. The more stable the climate, the more stable the ecosystem. Digging a little deeper helps us get a more accurate picture.

Truthfully, no one knows for sure whether species diversity fosters stability. However, we do know that simplifying ecosystems by reducing species diversity makes systems less stable and more vulnerable to outside influences. This is discussed later in the chapter.

6.2 Natural Succession: Establishing Life on Lifeless Ground

As noted earlier in the chapter, small shifts in the biotic and abiotic factors in an ecosystem, whether caused by natural forces or humans, are very common and readily correctable. Drastic shifts also occur. Although such changes can cause serious ecological disruption, they may be repairable. The process of repair is called *secondary succession*; it is one type of natural succession. The other type is called *primary succession*.

Natural succession is a process in which a biotic community forms on a lifeless or relatively lifeless piece of ground. (It also occurs in aquatic systems.) To understand it, let's begin with a look at primary succession.

KEY CONCEPTS

Diverse ecosystems tend to be more stable than less-diverse ecosystems. Species diversity is related to climate, however. The warmer the climate, the more diverse and stable the ecosystem is. Diversity and stability may both be related to stability of the climate. Reductions in species diversity can destabilize ecosystems.

KEY CONCEPTS

Ecosystems can form on barren or relatively lifeless ground by a process called *natural succession*.

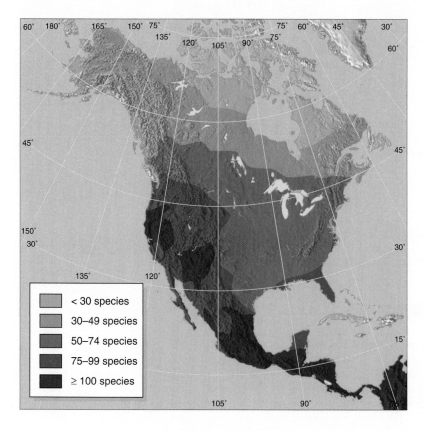

	< 30 species
	30–49 species
	50–74 species
	75–99 species
	≥ 100 species

FIGURE 6-6 Diversity vs. latitude. The number of mammals (shown here) and most other species varies considerably with latitude. The highest diversity is found in the tropics; the lowest is found in the Arctic tundra. (After Simpson, 1964)

Primary Succession: Starting from Scratch

Primary succession occurs when ecosystems form on previously barren terrain, for example, after the retreat of glaciers. As a glacier begins to melt, it exposes a barren, lifeless terrain composed largely of rocks. Over time, biological communities form on this barren surface, transforming the landscape into a verdant garden. The process that leads to the formation of a mature ecosystem on lifeless land is called **primary succession.** Consider an example.

In North America when the great glaciers began to retreat 15,000 years ago, large areas of barren land and rock were exposed. The exposed rock first became populated with lichens. Lichens are primitive organisms, actually a type of fungi, that house photosynthetic bacteria in their bodies, which permit them to survive, grow, and reproduce on barren rock (**FIGURE 6-7A**). Clinging to a rock's surface, lichens and their algae life partners live off moisture from the rain and carbon dioxide from the air, which they use to produce organic nutrients via photosynthesis. Lichens also secreted (release) carbonic acid, a weak acid that dissolves rock.

(a)

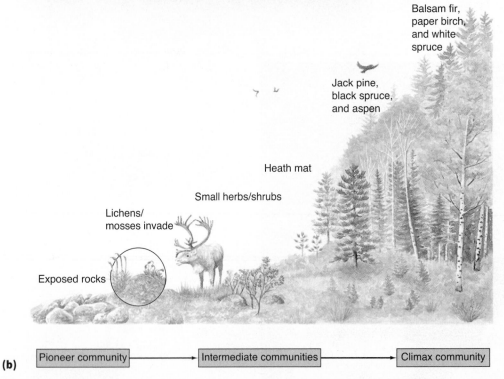

(b)

Balsam fir, paper birch, and white spruce

Jack pine, black spruce, and aspen

Heath mat

Small herbs/shrubs

Lichens/ mosses invade

Exposed rocks

| Pioneer community | → | Intermediate communities | → | Climax community |

FIGURE 6-7 Primary succession. (a) Lichens are fungi that grow on rocks and slowly erode their surface. In this photo taken in Missouri, the dark blotches on the rocks are lichens. The green plants are mosses that have established a foothold on the rock. **(b)** A representation of primary succession on Isle Royale, located in Lake Superior. Rock exposed by the retreat of glaciers is colonized by lichens and then mosses, followed by other plant communities and leading to a climax community. One biotic community replaces another until a mature community is formed. During succession, the plants of each community alter their habitat so drastically that conditions become more suitable for other species.

Photo courtesy of Dan Chiras.

Carbonic acid liberates nutrients from the rock, which lichens use. It also forms small mineral particles that will, over time, become part of the soil. Tiny insects join the lichens. Together, the insects and lichens formed a **pioneer community,** the first community to become established on the barren land (**FIGURE 6-7B**). Lichens and insects of the pioneer community thrived for a long time. Over time, they captured windblown dirt particles. Dead lichens crumbled. Insects died, too. Together, the particles of dirt and the decaying bodies of lichens began to form soil. In time, enough soil developed so that mosses took root. Eventually, the mosses overran the lichens' habitat, displacing them. Together, the mosses, new insects, and bacteria combined to form a new **intermediate community**.

These communities persisted for a long period, building soil very slowly. Eventually, however, the mosses and their cohorts were replaced by herbs and small shrubs, and yet another intermediate community was formed. As the soil depth increased, larger shrubs were able to take hold. Eventually, a forest began to grow. The forest is believed to be the final stage in this process and is therefore referred to as a *climax community*. **Climax communities** are relatively stable ecosystems believed to be the end point of natural succession.

Primary succession also occurs on newly formed volcanic islands. The Hawaiian Islands, for example, arose from lava flows from deep within the Earth. After the lava cooled, rock formed. This rocky landscape was colonized by lichens, and the process of succession began. As soil began to form, plants took root. The first plants to colonize the islands most likely came from seeds deposited in the feces of seabirds that happened upon the islands. Other plants are thought to have come from neighboring continents and islands. How did they get there? Biologists hypothesize that uprooted vegetation may have drifted to the newly formed islands over many thousands of years. Some of them took root there, turning the once-barren mass of rock into a rich tropical garden.

KEY CONCEPTS

Primary succession is a process in which mature ecosystems form on barren ground where none previously existed. Invaders create a community of organisms that changes slowly over time, creating more and more diversity.

Secondary Succession: Natural Ecosystem Restoration

Natural succession also occurs in the wake of ecological disturbance, ranging from mild to severe. This process is called **secondary succession.** Secondary succession is the sequential development of biotic communities after the complete or partial destruction of an existing community. In other words, secondary succession is nature's way of restoring damaged ecosystems.

Biological communities may be destroyed by natural events—for example, volcanic eruptions, floods, droughts, and fires. The eruption of Mount St. Helens in 1980 and the devastating fires in Yellowstone in 1988 are examples (**FIGURE 6-8**). Biological communities are also commonly destroyed by human activities such as agriculture, intentional flooding, fire, or mining.

Secondary succession generally takes place more rapidly than primary succession. Why? The development of soil that takes place in primary succession is unnecessary because the soil was not destroyed. Thus, an important component of life on Earth is present. The soil is generally able to accept seed—or already contains enough seeds—so that the process of succession can occur much more rapidly.

Abandoned farm fields provide an excellent opportunity to observe secondary succession. As illustrated in **FIGURE 6-9** (top left), abandoned farmland is first invaded by hardy species such as crabgrass or broom sedge, depending on the area. These plants are well adapted to survive in bare, sunbaked soil. In the eastern United States, crabgrass, insects, and mice invade abandoned fields, forming the pioneer community. Crabgrass is soon joined by tall grasses and other plants as shown in Figure 6-9 (left center). The newcomers, however, eventually produce so much shade that they eliminate the sun-loving crabgrass. Tall grasses and other plants dominate the ecosystem for a few years and become home to mice, woodchucks, rabbits, insects, and birds.

In time, pine seeds settle in the area, and seedlings begin to spring up in the open field (bottom left). Like crabgrass, the pine trees flourish in the sunny fields. Unlike crabgrass, they have the ability to rise above their competitors, the grasses. Over the next three decades, pines begin to shade out the grasses and other plants (top right). As a forest forms, animals such as woodchucks that feed on grasses move on to more hospitable environments. Squirrels and chipmunks, which prefer a wooded habitat, colonize the new ecosystem.

Shade from the pines, however, gradually creates an inhospitable environment for pine seedlings and a favorable environment for shade-tolerant hardwood trees such as maple and oak (middle right). As a result, hardwoods begin to take root. Over time, they begin to tower over the pines, and their shade gradually kills many of their predecessors. The climax stage consists primarily of hardwoods (bottom right).

KEY CONCEPTS

Secondary succession is a repair process that takes place after an ecosystem is destroyed by natural or human causes. It occurs more rapidly than primary succession because there's generally no need to form soil.

(a)

(b)

(d)

(c)

FIGURE 6-8 Mount St. Helens. (a) This volcano in southwest Washington was surrounded by a scenic northern coniferous forest prior to its eruption in 1980. **(b)** The aftermath of the eruption. One hundred and fifty thousand acres of forest were flattened in just a few minutes. The ridge in the lower left was once densely covered; the side facing the mountain was blasted away down to bedrock. **(c)** More than 9 years later, vegetation was just starting to return in areas less than 6 miles from the mountain. **(d)** The Mount St. Helens area today (photo taken in 2013).

Photos a, b, and c copyright © Weyerhaeuser. Used with permission. Photo d © jtstewartphoto/iStock/Thinkstock.

Changes During Succession: An Overview

Succession is a biological race to make optimal use of available resources—especially sunlight, soil nutrients, and water. As noted previously, the success of pioneer communities and many intermediate communities is transient. In both primary and secondary succession, organisms in the pioneer communities create conditions conducive to the establishment of other species. The species of the intermediate community, in turn, create conditions conducive to others. This occurs until a climax community forms.

During succession, animal populations shift with the changing plant communities. The early stages of succession are characterized by low species diversity (**TABLE 6-1**).

FIGURE 6-9 Secondary succession. Here, abandoned eastern U.S. farmland is gradually replaced by crabgrass, which in turn gives way to other herbaceous plants. Pine trees move in, and over time a mature hardwood ecosystem is formed. Note that many nonplant species (not shown here) also appear in these communities. In the early stages, for example, insects, mice, woodchucks, and seed-eating birds would be found. In the later stages, as trees come to dominate, squirrels and chipmunks (which prefer a wooded habitat) invade the new ecosystem.

TABLE 6-1

Characteristics of Mature and Immature Ecosystems

Characteristic	Immature Ecosystem	Mature Ecosystem
Food chains	Linear, predominantly grazer	Weblike, predominantly detritus
Net productivity	High	Low
Species diversity	Low	High
Niche specialization	Broad	Narrow
Nutrient cycles	Open	Closed
Nutrient conservation	Poor	Good
Stability	Low	Higher

Data from E. Odum (1969). "The Strategy of Ecosystem Development," *Science* 164: 262–270. Copyright 1969 by the American Association for the Advancement of Science.

Because there are fewer species early on, food webs tend to be simple, and populations tend to be volatile. In contrast, mature biotic communities have a high species diversity and relatively stable populations.

As ecosystems become increasingly complex, the food chains become woven into more complex food webs. In immature and intermediate stages, grazer food chains account for the bulk of the biomass flow. In climax ecosystems, however, most of the biomass flows through the detritus food chain. In fact, in a mature forest less than 10% of the net primary productivity is consumed by grazers.

Interestingly, in mature ecosystems nutrients are recycled more efficiently than in immature or intermediate ecosystems, both of which tend to lose a considerable amount of their nutrients because of erosion and other factors. This fact is extremely important to those who manage forests. Leaving a mature stand of trees on a hillside on the banks of a stream, for example, will minimize soil erosion and could protect the stream from filling with sediment.

Environmental resistance also changes over time during natural succession in each community as shown in **FIGURE 6-10**. In pioneer communities, for example, there is little environmental resistance to early colonizers such as crabgrass. Over time, environmental resistance increases. This increase is caused in large part by increasing competition from other species that become established in the evolving landscape. In an abandoned farm field, for example, crabgrass and the grasses that invade later compete for sunlight. The grasses, which can grow taller, eventually win the struggle, shading out the crabgrass.

Climax communities that emerge from secondary succession contain organisms that exist in a complex web of life controlled by a variety of growth and reduction factors. These communities tend to stay in relative balance over long periods if undisturbed. These relatively stable assemblages of organisms are sustained in large part because of their efficient use of resources, recycling of nutrients, and dependence on a reliable, renewable resource base. They also sustain themselves because they are capable of restoring damage and maintaining population balance through a variety of population control mechanisms. These principles can be used as guidelines for managing natural resources and restructuring our own society for sustainability.

Those of you who will go on to study ecology will learn that the view of secondary succession just presented is the classical interpretation. New research, however, has thrown into question a few of the key tenets of this theory. Research, for example, shows that succession varies considerably in different regions and with different types of vegetation. In the Rocky Mountains in New Mexico, Colorado, Wyoming, and Montana, for example, avalanches sometimes destroy stands of conifers. The opening in the forest created by an avalanche first fills with grasses, wildflowers, and aspen trees. Over the years, dense stands of aspen grow. Next, shade-tolerant fir and spruce grow up in the forests, eventually reestablishing the climax community. All of this complies with the theory. However, in this example, species diversity in the intermediate community—the aspen forest—actually exceeds that of the mature system. Aspen forests support grasses, herbs, wildflowers, and a great many birds and mammals, such as deer and elk. As the climax community becomes established, species diversity falls because the dense forests inhibit growth of grasses and other sun-loving plants that provide food for many animals. The needle-covered floors of a fir and spruce forest support relatively few species.

New research also shows that the sequential replacement of one community by another is not always as clear-cut as was once thought. In the succession of an eastern farm field to forest, for example, pine trees are often present early in the vegetational development. Because they grow slowly and are often browsed (grazed), however, their presence is not noticeable until later stages.

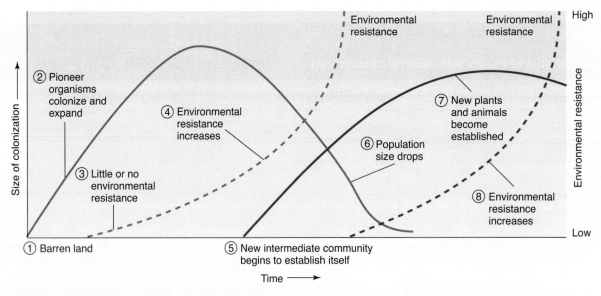

FIGURE 6-10 Changing conditions during succession. A graphic representation of the population growth and decline of a pioneer community (solid black line) and the rise of an intermediate community (red line) during succession. Environmental resistance (dashed black line) is created as the pioneer community becomes established. This, in turn, makes conditions conducive to growth of the intermediate community. Follow numbers to understand the progression of events. Environmental resistance will increase as the intermediate community becomes established, too (dashed red line).

KEY CONCEPTS

During succession, biotic and abiotic conditions change. Pioneer and intermediate communities alter conditions so much that they promote the growth of new communities that eventually replace them. During succession, two of the most notable changes are an increase in species diversity and ecosystem stability.

6.3 Evolution: Responding to Change

Most people think of evolution as a process in which life formed and new life-forms emerged, which is, of course, an accurate view. They overlook an important point, though: Evolution is a life-sustaining process. Put another way, life on Earth has sustained itself, in part, because populations of plants, animals, and other organisms have changed or evolved as conditions changed, often dramatically. Evolutionary change provided organisms with new means of adapting to their environment—new ways of eking out an existence or adjusting to changes in biotic and abiotic conditions.

Evolution has two widely recognized outcomes. In some cases, it results in modifications in existing species that make them better suited to their environment—that is, better able to survive and reproduce and thus sustain their kind. Changes in the coloration of a species of butterfly, for example, might make it better able to avoid being eaten by predatory birds. Evolution also results in the formation of new species. Birds, for example, evolved from reptiles (**FIGURE 6-11**). Fish gave rise to the first amphibians.

Much of what scientists know about evolution comes from fossils—preserved bones, imprints of organisms captured in rock, or footprints of early animals. By examining fossils embedded in rock strata whose age can be determined through various means, scientists have been able to construct a biological history of the Earth (Figure 6-11).

KEY CONCEPTS

The ability of species to evolve in response to changes in the biotic and abiotic conditions has contributed to the sustainability of life on Earth.

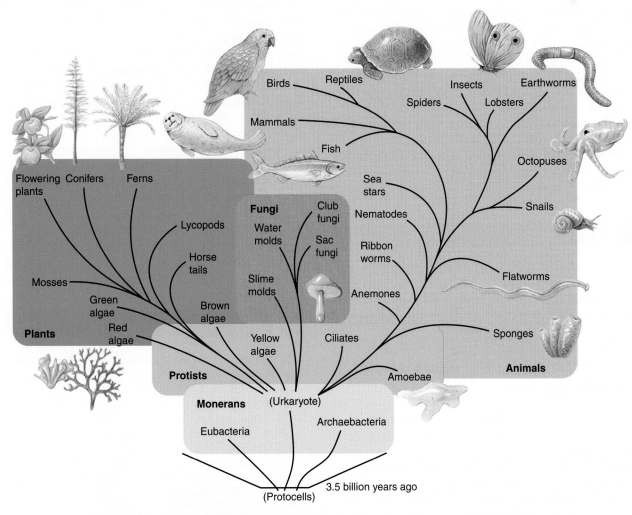

FIGURE 6-11 The evolution of life on earth. This drawing shows the general evolutionary progression of life on Earth.

Evolution by Natural Selection: An Overview

The theory of evolution is often attributed to Charles Darwin, a 19th-century British naturalist. Interestingly, Darwin did not originate the idea; it dates back to ancient Greece. In fact, the theory was already widely discussed in Darwin's time. What Darwin and another scientist Alfred Russell Wallace did (working independently) was to propose a mechanism by which evolutionary changes occurred—the **theory of evolution** by natural selection. This theory states that species evolve as a result of natural selection.

Natural selection is a kind of weeding out that takes place in populations of organisms. Within any given population, some members are better adapted to their environment than others; that is, they possess structural, functional, and behavioral characteristics that give them an advantage over other members of the population. These characteristics are called **adaptations**. Why are some members of a population better adapted than others?

The reason for the structural, functional, and behavioral differences is genetic. In most instances, adaptations increase an individual's chances of surviving and reproducing. For instance, predators that are faster than others in their population are more likely to acquire the food they need to survive. They are likely to be stronger and more successful in reproduction. Others that are less fortunate in their genetic endowment produce fewer offspring. Consequently, the predators that outhunt their cohorts will leave more surviving offspring in the population. Future generations will have a higher percentage of individuals possessing the favorable adaptation.

Like many other great ideas, natural selection took many years to be understood and appreciated. Not until the 1940s, nearly a hundred years after they developed the concept, did Darwin and Wallace's ideas on natural selection become widely accepted.

KEY CONCEPTS

Natural selection is the driving force behind evolution. It consists of natural forces that select for those members of a population that are superior in one or more features—advantages that increase their chances of surviving and reproducing.

Genetic Variation: The Raw Material of Evolution

As the previous section showed, evolution occurs as a result of the presence of genetically determined traits (adaptations) that give organisms an advantage over others and natural selection. Adaptations result from **genetic variation,** naturally occurring differences in the genetic composition of organisms in a population that may give some members of a population an advantage over others. Genetic variation comes from several different sources, three of which are described here.

In sexually reproducing species, genetic variation results from mutations in the genetic material, or DNA, of the germ cells (ova in females and sperm in males). DNA is a storehouse of information that controls the structure and function of the cells of the body. DNA consists of many segments called **genes,** each of which plays a specific role in regulating cell structure and function. Mutations in the genes may be caused by ultraviolet radiation, chemicals in the environment, and cosmic rays from the sun. Some may occur randomly. Some germ cell mutations are neutral—that is, they have no effect on an organism's offspring. Others are harmful. If not repaired, they can be passed on to an organism's offspring and may result in birth defects, cancer, or, in some instances, death. Mutations that result in beneficial adaptations in the offspring increase their survivability—for example, by improving their ability to escape predators, tolerate cold temperatures, or find food. Increased survivability, in turn, increases their chances of breeding and passing on this favorable trait.

Genetic variation may also arise during cell divisions that take place in the production of germ cells, or **gametes.** During this process, genetic material may be transferred from one chromosome (long strand of DNA with associated protein) to another. This process, called **crossing over,** results in new and sometimes favorable genetic combinations.

Another source of new genetic combinations is sexual reproduction. **Sexual reproduction** occurs when offspring are produced by a union of sperm from males and ova from females. Each offspring contains half of each parent's genetic information. Offspring, therefore, represent a new genetic combination that may provide benefits.

Genetic variation within a population results in structural, functional, and behavioral variation among the individuals of that population. As noted earlier, this results in some members being better adapted to environmental conditions than others. For these reasons, biologists refer to variation as the raw material of evolution. Organisms with favorable adaptations are said to possess a **selective advantage**. They are more likely to survive and reproduce.

KEY CONCEPTS

Genetic differences in organisms of a population, called *genetic variation,* are the raw material of evolution. *Genetic variation* results in naturally occurring differences in the genetic composition of organisms in a population that may give some members of a population an advantage over others. Genetic variation comes from mutations, sexual reproduction, and a process known as crossing over, which occurs during the formation of sperm and ova. Genetic variation produces variation in populations in structure, function, and behavior.

Natural Selection: Nature's Editor

Sociologist Andrew Schmookler once wrote that evolution employs no author but only an extremely patient editor. By that he meant that although evolution may appear to move in certain directions, it is not consciously directed. Variation occurs naturally. Beneficial genetically based adaptations tend to persist because they are selected for by the abiotic and biotic conditions of the environment. In other words, the environment preserves (*selects*) those organisms in a population with traits that confer some advantage over the rest. Ultimately, natural selection results in organisms better adapted to their environment.

Natural selection is really about the survival and reproduction of the fittest. **Fitness** is commonly thought of as a measure of strength. To a biologist, though, fitness is actually a measure of reproductive success—and thus the genetic influence an individual has on future generations. By definition, the fittest individuals leave the largest number of descendants in subsequent generations. Because they make up a higher proportion of future generations, their influence on the genetic makeup of those generations is greater than that of less fit individuals.

It is important to remember that fitness is not necessarily dependent on superior strength. Greater fitness may be conferred by species' ability to hide or digest food better than others. One important measure of fitness is the efficiency with which organisms use resources. Adaptations that make an organism's offspring more efficient in their use of water in a desert environment, for instance, confer a tremendous advantage on its offspring. Numerous adaptations also contribute to energy conservation in many species, especially those that live in harsh environments such as the Arctic tundra. In a world of growing human population and dwindling resources, human survival will also very likely require the evolution of similar efficiency measures. But evolution in human society will be a conscious or deliberate process, a kind of cultural evolution, whereas evolution in the nonhuman world is not a thought-driven or directed process.

KEY CONCEPTS

Natural selection weeds out the less fit organisms of a population, leaving behind the fittest. The fittest individuals are those best adapted to environmental conditions and most likely to reproduce, thus contributing more offspring to future generations.

Speciation: How New Species Form

The foregoing discussion shows why the evolutionary process helps to change—and ultimately sustain—species in an ever-changing environment. This section discusses another important process, **speciation:** how new species arise.

Speciation occurs most commonly when members of a species are separated from one another. In scientific language, this is referred to as **geographic isolation.** Geographic isolation may occur when a mountain range or a river forms in an organism's habitat. If isolated by impenetrable physical barriers and exposed to different environmental conditions, two new populations may evolve. If the populations are separated long enough, their members may lose their ability to interbreed, a phenomenon called **reproductive isolation.**

When geographic isolation leads to reproductive isolation, new species have formed. Scientists call the formation of new species in different regions **allopatric speciation** (*allopatric* is derived from the Greek for "other" and "fatherland"). New species may also form without geographical isolation, though. This process is common in plants and is termed **sympatric speciation** ("same fatherland"), a subject beyond the scope of this book.

KEY CONCEPTS

Evolution may result in the formation of new species. When members of a population are separated by a physical barrier, they may evolve along different lines, forming separate species.

Coevolution and Ecosystem Balance

The discussion of natural selection to this point has focused primarily on abiotic factors as the chief agents of evolutionary change. Numerous studies, however, show that organisms themselves can also be important agents of natural selection. Predators, for example, can put pressure on their prey, which results in changes in the prey. Consider a hypothetical example.

Barn owls hunt for mice at night. If a population of barn owls evolves better mechanisms to detect prey—for example, improved night vision—it would put additional pressure on the mice. This change would first be felt by the slower mice, who would fall victim to the owls. The fastest mice would survive, and they and their offspring would form a larger and larger portion of the mouse population. The second result of the owl's improved performance would therefore be a new, genetically superior population of mice that was less apt to be snatched by the owls with improved night vision. Faced with a faster population of mice, the owls might evolve even greater hunting skills. So not only can one organism affect the evolutionary course of another, but the relationship can be reciprocal.

Biologists refer to this process as **coevolution**. During coevolution, each species serves as an agent of natural selection that affects the evolution of the other. Coevolution can be likened to an arms race between predator and prey because improvements in a predator's ability are matched by an improvement in its prey—much as countries try to outmatch each other during arms races.

Coevolution may also occur between plants and the animals that feed on them. This process prevents one species from destroying another. It is, therefore, an important element of sustainability in natural systems. Without it, species might destroy their own food supply.

Understanding evolution is important to understanding environmental issues, especially species extinction. Reductions in populations threaten the survivability of a species in part because they can reduce genetic variation. They may, for example, eliminate organisms that could survive other changes wrought by human society.

Perhaps the most important lesson in the study of evolution is that to sustain itself, a species must change in response to changing realities. Organisms generally have no choice; either they evolve, or they die out. For humans, all of the signs suggest that change is needed. But our change must be cultural in nature and purposefully directed. This book is largely about the kinds of cultural changes we need to make to create an enduring human presence on planet Earth.

Courtesy of NSSDC/NASA

foundation tools

6.4 Human Impact on Ecosystems

Environmental science can be overwhelming at times. There are many problems and lots of solutions from which to choose. To help you understand environmental issues, I've created a model, shown in **FIGURE 6-12** that illustrates how numerous factors affect the environment. As shown, humans alter the environment via two basic routes: changes in abiotic and biotic factors of the natural world.

FOUNDATION TOOL

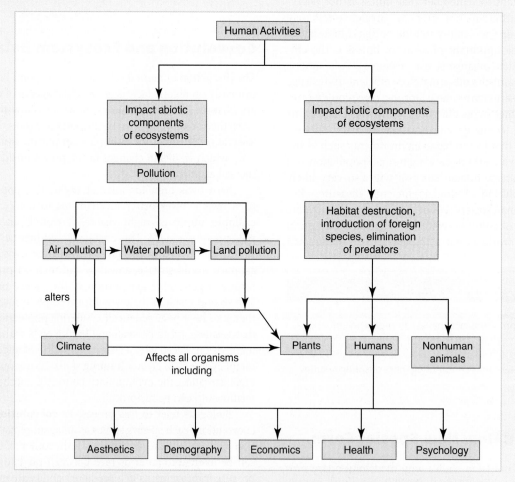

FIGURE 6-12 Impact Analysis model. The concept model shows the range of impacts caused by human activities. Note that human activities impact two broad areas, abiotic and biotic conditions. Also note that alterations to the abiotic conditions (air, water, land) can also impact biotic conditions.

Altering Abiotic Factors

Let's begin with the abiotic factors. Be sure to refer to Figure 6-12. As illustrated, some human activities produce pollutants that contaminate the air, water, and land, changing the chemical composition of the environment. Some air pollutants alter the climate. Changes to the chemical environment and the climate have profound impacts on other species as indicated by the lines connecting air pollution, water pollution, land pollution, and climate to plants, humans, and nonhuman animals. These may stress organisms, making survival problematic. They may also impair reproduction, which lessens a population's chances of survival, or they may kill organisms outright. **Spotlight on Sustainable Development 6-1** shows some positive influences humans can have on natural systems.

Some chemical pollutants may be identical to naturally occurring substances found in the environment. If their concentrations exceed an organism's range of tolerance it may suffer. Nitrates and phosphates released from sewage treatment plants are good examples. Other chemical pollutants from human sources are entirely foreign to natural systems. These are synthetics such as pesticides. Because they are not naturally occurring, organisms generally have no natural defenses against them. That is, they have no way of detoxifying them. These substances can cause tumors, reduce reproduction, or make other changes. One substance found in detergents and pesticides, called nonylphenols, when released into the waterways in the Pacific Northwest may be partly responsible for the decline in salmon populations. This substance, research suggests, alters the hormonal system of young salmon, making it impossible to adapt to salt water as they migrate from rivers to the ocean.

Pollution from human activities can also affect the physical conditions of ecosystems. Chlorofluorocarbons, for instance, deplete the ozone layer and increase the amount of ultraviolet radiation striking the Earth. Power plants release hot water that changes a stream's temperature, with adverse effects on aquatic species. Carbon dioxide may alter global climate, resulting in an increase in the Earth's temperature with potentially serious effects on human health.

Go Green

To help reduce pollution, including carbon dioxide emissions, which are responsible for global climate change, turn off the lights and your computer and other electronic devices when you leave your dorm room or apartment.

Altering Biotic Factors

As illustrated in Figure 6-12, humans also alter biotic components of the natural world. As noted in this figure, we destroy habitat or severely alter it when we build homes and highways. We also introduce foreign species such as the water hyacinth to areas where there are no natural controls to hold them in check. These new species may outcompete native species, eliminating them entirely. In some instances, we eliminate predators such as mountain lions and wolves, which may cause prey populations to skyrocket, or we introduce predators for which prey species have no natural defense. We also accidentally introduce disease organisms. These changes affect plants and animals and humans as well, as illustrated.

A few examples demonstrate the severity of these seemingly innocuous alterations that have environmental and economic impacts. First, consider the Africanized honeybee. Africanized honeybees, commonly known as *killer bees*, were introduced to South America in 1956 by a well-meaning geneticist, Warwick Kerr. He imported the bees to Brazil to develop a successful stock of honey producers to replace the docile European honeybees, which had fared poorly in the tropical climate. Kerr hoped that interbreeding the two might yield a more successful tropical strain.

Knowing their aggressiveness, Kerr isolated the bees in screened-in hives. However, in 1957 a visitor unwittingly lifted the screen, allowing 26 queens and their entourages to escape. Trouble soon began. The killer bee population quickly spread, moving a remarkable 350 to 500 kilometers (200 to 300 miles) a year. As they spread, they began interbreeding with local honeybees, eventually destroying the honey industry in Brazil and other countries. The bees also viciously attacked people, horses, and livestock that crossed their paths, stinging them severely and sometimes killing their victims.

The venom of killer bees is not any more toxic than that of native honey bees. It is just that the bees attack in great numbers and pursue enemies for greater distances than native honey bees. In addition, after a hive has been disturbed, the bees remain agitated for 24 hours, attacking people and animals within a quarter of a mile from their hive.

Killer bees arrived in Texas in 1990 and have spread into New Mexico, Arizona, Nevada, and California. In 2009, they arrived in Utah. They have also moved eastward into Oklahoma, Lousiana, southern Arkansas, Georgia, and Florida (**FIGURE 6-13**). According to one source, 27 people were killed by the bees in the United States between 1990 and 2009. Although there is no accurate accounting of

6-1 Restoration Ecology: Sustaining the Biosphere

One of the most often overlooked solutions to living sustainably is restoring the damage humans have created over the past several hundred years. There's much talk about solar energy, energy efficiency, ways of controlling growth, and preventing pollution, but very little attention has been given to the need to restore the lands and waters that have been so badly damaged over the past 200 to 300 years.

Restoring natural systems, farmland, waterways, and lakes is important, not only to protect wild species but also to ensure an adequate food supply for the world's people. Such efforts are good for the economy and help create a better future. These systems also provide many ecological benefits, such as oxygen production and flood control with many economic and noneconomic benefits.

Fortunately, the scientific community has been at work on this subject for a long time. In 1985, scientist Ed Garbisch began the long and costly process of restoring an ancient freshwater wetland in New Jersey. In a $4 million project, Garbisch and his workers seeded marsh grass, cut channels to restore water flow, and built knolls for nesting sites for waterfowl. Despite initial skepticism, Garbisch and his coworkers restored the swamp to its previous condition.

Garbisch and others like him have given rise to a new field of ecology, called **restoration ecology**, the scientific study of how ecosystem recovery occurs and how it can be facilitated. Restoration ecology is one branch of science known as **conservation biology**, which seeks to understand natural ecosystems and ways to protect and restore them.

In many respects, restoration ecology is akin to rehabilitative medicine. Some of the earliest restoration projects were designed to reclaim land that had been surface mined for coal and other minerals. Today, however, ecologists and others are working to restore marshes, tropical rain forests, streams, and prairies throughout the world to help create a sustainable future (**FIGURE 1**).

Ecologist John Berger has spearheaded the movement in the United States to restore battered ecosystems. Berger founded a nonprofit group called Restoring the Earth, dedicated to repairing damaged ecosystems.

One of the most ambitious projects currently under way is that of Professor Daniel Janzen, a University of Pennsylvania biologist. With others, Janzen hopes to restore an entire dry tropical forest in northwest Costa Rica that covers 120,000 hectares (296,400 acres). Janzen's project will take 200 years or so to complete, one-fifth of the time it would take for nature to reestablish a forest that has been cut and burned over the years.

Ecological restoration offers economic, environmental, and aesthetic benefits; for example, protecting an eroding shoreline with concrete or rocks can cost five times more than planting a 7-meter (21-foot) wide strip of salt marsh to protect the shoreline. This can result in a huge savings to owners of shoreline property. Plants can also turn a barren beachfront into a dense, lush marsh that attracts a variety of colorful birds.

Creating a marsh costs $5,000 to $50,000 per hectare ($2,000 to $20,000 per acre). Although a marsh's value cannot easily be calculated, biologists in Louisiana estimate that a hectare of salt marsh is worth about $210,000 solely for its ability to reduce water pollutants such as sediment. Moreover, marshes are one of the world's most productive ecosystems. Today, approximately one half of the U.S. salt marshes have been lost. When a salt marsh disappears, shorelines erode rapidly,

FIGURE 1 Restored wetland in the southeastern United States.

Courtesy of Gary Kramer/ NRCS USDA

fish populations collapse, birds vanish, wildlife retreats, and some of nature's remarkable plant communities are destroyed.

Some environmentalists worry that restoration legitimizes further environmental destruction. The Sierra Club, for instance, points out that developers who want to build on existing wetlands frequently offer to replace them with wetlands "created" elsewhere.

A closer look at the issue—so essential to critical thinking—suggests several reasons that support the conservation group's position. First, natural wetlands are complex ecosystems. A flooded field planted with some swamp vegetation is a far cry from the natural system it replaced. Although it may come to resemble a true swamp in a decade, it will very likely have much lower species diversity. Second, rebuilding wetlands or other ecosystems is a complex and costly task that requires expert attention and follow up. All in all, many think that it is best to protect what we have and not rely on strategies designed to compensate for losses by creating new wetlands.

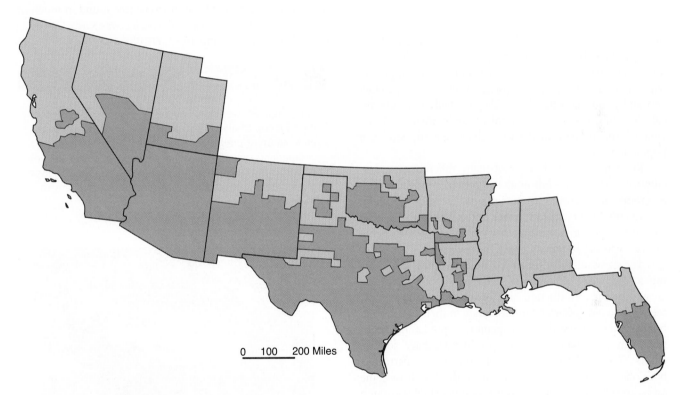

FIGURE 6-13 The range of Africanized honeybees in the United States. The areas indicated in gold show the range of the bee as of March 2011.

Reproduced from Spread of Africanized honey bees by year, by county/ARS/USDA.

deaths from killer bees, since 2011 the death toll is estimated to be about 40 per year. Scientists fear that if the bee spreads into northern climates and breeds with its tamer cousin, thousands of hives would have to be destroyed. This would cripple the $140-million-a-year honey industry. It will also be devastating to farmers, for America's honeybees pollinate 130 crops worth an estimated $20 billion a year.

Many biologists hope that the northward spread of the killer bee would be halted by the colder climate. A study published in 1988, however, showed that killer bees could survive at 0°C (32°F) for 6 months. Some scientists now fear that the bees could migrate as far north as Canada, causing widespread damage to the honey industry and to bee-pollinated crops.

Another example is the zebra mussel, accidentally introduced into North America by tankers arriving from Europe (**FIGURE 6-14**). Already widespread in Russia, the zebra mussel has invaded many streams from New York to Minnesota, where it sets up home in water pipes coming from power

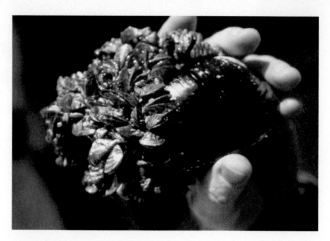

FIGURE 6-14 **Zebra mussel.** The zebra mussel is an alien species that has wreaked havoc in many lakes in the northeastern United States.

Courtesy of U.S. Fish & Wildlife Services.

plants and factories. There the mussel proliferates and can greatly reduce the flow of water. To keep their facilities running, plant managers must treat their pipes more frequently with chemicals that kill the mussels, adding to their product cost and polluting waterways with potentially toxic chemicals. The mussels also feed on microscopic phytoplankton, which form the base of the food chain that supports fish populations in lakes.

Russian scientists have tried many different tactics to eliminate the mussel, but none has proved very successful. A group of scientists from Virginia recently proposed introducing blue crabs (which feed on zebra mussels) from Chesapeake Bay, in hopes that they would help control zebra mussels without becoming a pest themselves.

The sagas of the Africanized honeybee and the zebra mussel are two of a number of severe biological disruptions created by the introduction of a foreign species into a new region. Foreign species often proliferate in their new homes because there are few if any competitors or predators to control them. Fortunately, not all introductions have adverse effects. The ring-necked pheasant and chukar partridge, both aliens in the United States, have done well in some areas. In other cases, alien species have perished without a trace. Hardy species such as the killer bee, however, are the ones that demand our attention and remind us to be vigilant.

Another example of human alteration of biotic components of the environment is the elimination of predators. Predators have never fared well in human societies. Early hunters and gatherers killed them for food and because they viewed them as competition for their prey. Modern societies have carried on the tradition, killing bears, eagles, hawks, wolves, coyotes, and mountain lions with a vengeance, often with serious ecological consequences.

Problems also arise when predators are introduced into a new habitat. The mosquito fish, a native of the southeastern United States, for example, has been introduced into many subtropical regions throughout the world because it eats the larvae of mosquitoes and thus helps to control malaria, a mosquito-borne disease. Unfortunately, the mosquito fish

also feeds heavily on zooplankton. When zooplankton populations decline, algae proliferates and forms thick mats that reduce light penetration and plant growth in waterways.

The recent outbreaks of the deadly ebola virus in Africa, Europe, and the United States illustrate the dangers of spreading disease organisms, or **pathogens.** Pathogens are harmful microorganisms, mostly viruses and bacteria, that are a natural part of ecosystems. As a rule, they are held in check by natural forces. When introduced into new environments where there are no natural controls, though, they can spread quickly, killing people and other species.

One of the classic tragedies of pathogen introduction began in the late 1800s, when a fungus that infects Chinese chestnuts was introduced accidentally into the United States. The fungus arrived in a shipment of Chinese chestnut trees purchased by the New York Zoological Park. Chinese chestnuts are immune to the fungus, but the American chestnuts are not. Once a valuable commercial tree found in much of the eastern United States, this species was virtually eliminated between 1910 and 1940 (**FIGURE 6-15**).

KEY CONCEPTS

Many human activities such as habitat destruction, introduction of alien species, and the introduction or the elimination of predators have a direct effect on the biotic components of ecosystems. Introduction of foreign species is particularly troublesome because these species may proliferate wildly, causing major economic and environmental damage.

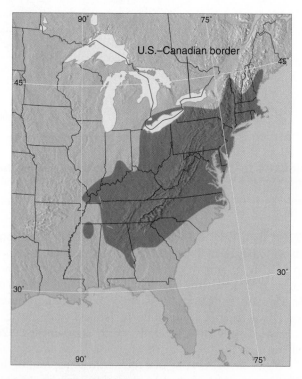

FIGURE 6-15 **Decline of the American chestnut.** Former range of the American chestnut, a species nearly wiped out by the accidental introduction of a harmful fungus.

Simplifying Ecosystems

When humans change the biotic and abiotic conditions of an ecosystem, species diversity is often a victim. Ecologists refer to the loss of species diversity as **ecosystem simplification**. One example of this phenomenon is the conversion of forests and grasslands to farmland. Grassland contains many species of plants and animals. When plowed under and planted in one crop, called a **monoculture,** the field becomes highly simplified. It contains few species—the crop and a few varieties of weed. Although farm fields are vital to food production they are highly vulnerable to insects, diseases, drought, and wind.

The reasons for this susceptibility are many. Consider the vulnerability of such monocultures to insects and plant pathogens. Monocultures generally stretch for miles, providing a virtually unlimited food source for insects and plant pathogens, especially viruses and fungi. As crops grow, food supplies for the pests increase dramatically, favoring rapid growth and spread of their populations. Viruses, fungi, and insects can become major pests. A more diverse farm field, one containing a number of different crops, reduces the food supply for insects and tends to hold populations in check.

Monocultures also tend to be genetically homogeneous. They contain one crop species with the same genetic makeup. If it is not resistant to pests or disease, the crop is highly vulnerable. Given an unlimited food supply with little resistance, insects and diseases can spread rapidly.

Protecting monocultures from harm leads to many environmentally harmful practices—for example, the application of chemical pesticides to control insects, a practice that may have severe environmental impacts.

Why Study Impacts?

Figure 6-12, described in the preceding paragraphs, helps us understand and perhaps even predict the impacts of human activities on entire systems. It also alerts us to the potential dangers of different options. In our frontier society, impact analysis is generally used to identify existing or potential impacts so that we can find ways to **mitigate** (offset) damage. For example, hydropower projects flood recreational streams. The loss of recreation opportunities such as kayaking, rafting, and fishing resulting from the construction of a dam might be compensated for by opening a previously inaccessible stream to recreational users.

In the United States, all major construction projects on federal land (for example, U.S. Forest Service land) or supported by federal dollars require analyses of potential impacts that are then reported in an **Environmental Impact Statement (EIS).** The EIS discusses the environmental consequences of proposed actions, as well as the potential social, cultural, and archeological impacts. Many states require similar analyses for projects funded by state money or projects on state-owned land.

As it stands, the EIS requires a review of other options. In most cases the options are ignored, and efforts are concentrated on ways to mitigate damage created by the project. After completion, the EIS is subject to public review. The federal agencies responsible for the project then either approve or deny it. Permits that are granted for the work often include provisions to mitigate damage.

Although this system may sound good, it is not a sustainable approach because, as mentioned, most EISs focus on ways to mitigate, not prevent, the impact of projects. To promote sustainable development, some critics argue that the EIS should require proponents to analyze several options and select the option that provides the services at the very least economic *and* environmental cost. Instead of a new dam, for instance, a massive water-conservation program might be the recommended choice. Water conservation costs less and can yield more water than a new dam. The environmental impact of water conservation would be minimal compared to the huge impact of building a massive dam to flood a stream or river. This alternative system is called **least impact analysis** or **sustainability analysis**. It is preventive in nature and may help us shift to much more sustainable development alternatives that are good for people, the economy, *and* the environment.

For the first time in the whole history of evolution, responsibility for the continued unfolding of evolution has been placed upon the evolutionary material itself. . . . Whether we like it or not, we are now the custodians of the evolutionary process on earth. Within our own hands—or rather, within our own minds—lies the evolutionary future of the planet.

—**Peter Russell**

CRITICAL THINKING

Exercise Analysis

The first step in evaluating this proposal would require you to gather additional information. You might find it helpful to examine potential ecosystem effects of introducing a foreign species. Has this species been introduced into similar habitats before? What were the effects? Is the deer adapted to the region into which it will be introduced? Is its native habitat similar to its new habitat? Would it have enough food?

Second, assuming conditions are appropriate, you may want to estimate how the transplanted deer population would grow. Would numbers increase so much that the deer would destroy its new home? Could hunters keep the population under control to prevent it from destroying the range? Would predators need to be introduced to control numbers? What would their effect be?

Third, you would have to study native grazers to determine if the introduction of deer would affect them. In addition, it would be wise to determine if the introduced deer carries diseases that could affect native or domestic animals.

Fourth, you may want to determine if the deer would remain in the region or if they would migrate into other areas, affecting other large grazers such as white-tailed deer and elk. Can you think of other concerns?

As this example shows, there are many factors to take into consideration besides the possible increase in income from hunting licenses and other incidental expenses. Critical thinking requires us to gather as much information as possible and to take a careful look at all the possibilities—that is, to examine the big picture. In Montana and Wyoming, for example, red deer from central Europe were introduced by game ranchers. Some have escaped and formed wild populations that are now interbreeding with native elk populations. Although the idea of introducing the deer seemed good at the outset, many proponents are having second thoughts.

CRITICAL THINKING AND CONCEPT REVIEW

1. Define the term *ecosystem homeostasis*.
2. If you were to examine a mature, undisturbed ecosystem over the course of 30 years at the same time each year, would you expect the number of species and the population size of each species to be the same from year to year? Why or why not?
3. What is environmental resistance? What role does it play in maintaining ecosystem balance?
4. Give evidence that species diversity affects ecosystem stability. Is there any evidence to contradict this idea? What is it?
5. What is a mature ecosystem? What are its major features? How does it differ from an immature ecosystem?
6. Do you think that human communities resemble mature or immature ecosystems? If immature, are there any dangers one should be aware of?
7. Describe temporary imbalances caused in ecosystems you are familiar with. Indicate how the ecosystems respond to them.
8. The process of secondary succession is nature's way of restoring itself. In some cases, restoration may be protracted or impossible. Why?
9. What is the difference between primary and secondary succession? Why does secondary succession generally occur more rapidly than primary succession?
10. Explain why organisms in the pioneer and intermediate communities are replaced by others.
11. Describe how introducing alien species into an ecosystem can affect ecosystem stability. Give some examples.
12. Give some examples of ways humans tamper with abiotic components of ecosystems.

13. Discuss why simplified ecosystems (monocultures) are highly susceptible to pests.
14. Define the following terms: *evolution, natural selection, genetic variation.*
15. Describe how geographical isolation results in speciation.
16. In what ways has the study of ecology broadened your view of life? Has it made you reconsider any of your views?
17. Make a list of the major principles you have learned in this chapter. Then note how each one could be useful in managing natural resources. Describe what applications they might have for designing sustainable human systems.

KEY TERMS

adaptations
allopatric speciation
climax community
coevolution
conservation biology
crossing over
dynamic equilibrium
ecosystem simplification
Environmental Impact Statement (EIS)
environmental resistance
fitness
gametes

genes
genetic variation
geographic isolation
growth factors
homeostasis
intermediate community
least impact analysis
mitigate
monoculture
natural selection
natural succession
pathogens
pioneer community

primary succession
reduction factors
reproductive isolation
resilience
restoration ecology
secondary succession
selective advantage
sexual reproduction
speciation
species diversity
sustainability analysis
sympatric speciation
theory of evolution

Human Ecology: Our Changing Relationship with the Environment

For 200 years we've been conquering nature. Now, we're beating it to death.

—Tim McMillan

LEARNING OBJECTIVES

After reading and studying the material in this chapter, students will be able to understand, knowledgably discuss, and, when appropriate, apply the information learned in this chapter, including:

1. The biological evolution of humankind and the changes that have contributed to the environmental crisis
2. The stages of human cultural evolution and how the human-environment relationship has transformed over time
3. How changes in human culture have contributed to an environmentally destructive and unsustainable way of life
4. The importance of creating a cultural shift toward a sustainable society and a sustainable relationship with the natural environment

CRITICAL THINKING

Exercise

In a speech before the United Nations in the 1960s, Adlai Stevenson called our home "Spaceship Earth," likening the planet to a spacecraft with everything required to survive on board. Using your critical thinking skills, analyze the appropriateness of this analogy. You may want to begin by making a list of all the features of a spaceship, say the space shuttle, that you know about. In particular, you will want to know how life is sustained on such a craft. You may find that you will have to do some research in encyclopedias, in books on spacecraft, or on the web. After you have gathered this information, compare the spaceship to the planet Earth. What similarities and what differences do you see? Is the analogy a good one?

In 1948, the noted British astronomer Sir Frederick Hoyle predicted that "once a photograph of the Earth, taken from the outside, is available ... a new idea as powerful as any in history will let loose." It was not too many years later that the first photograph of Earth from outer space came to us, and Hoyle's prediction bore fruit. Standing in sharp contrast to the darkness of space, the brilliant sphere with its gossamer veil of clouds was breathtaking and yet disturbing—for it showed our home, which we had always seen as inexhaustible, as a tiny body alone in a vast universe. Sparkling in the sun's rays, the Earth seemed exquisite, fragile, and highly vulnerable.

We know now the Earth is indeed vulnerable. Its climate is being severely altered. Its life-support systems—which provide food, clean water, and oxygen—are being destroyed at a record pace, dooming tens of thousands of species to extinction and threatening our own well-being, too. How did we end up in such a precarious situation? To answer this question we journey back in time to trace human cultural evolution. This brief survey of several million years of our life on Earth provides important insights into the modern environmental crisis. It will conclude with two conceptual models designed to improve your understanding of human impacts and your systems thinking abilities, skills vital to building a sustainable future. *Systems thinking* refers to one's capacity to grasp interactions and the complexity of systems—for example, human systems. It helps us see how systems. It helps us see how systems interact and how our systems affect natural systems. As you read this material, take time to think about the changes in human systems—how they have moved away from sustainable designs—and how they can be revamped to create a sustainable future.

7.1 Human Biological Evolution: Some Highlights

Will Cuppy once quipped that "all modern men are descended from a wormlike creature, but it shows more in some people." As best we can tell from the fossil record, humans did not evolve from worms. They evolved from a group of insect-eating mammals that resembled the modern-day tree shrew of Southeast Asia (**FIGURE 7-1**). During evolution, they gave rise to yet another group of animals, the prosimians or premonkeys. The prosimians were the very first primates. They gave rise to monkeys, apes, and our earliest ancestors.

Unlike the dinosaurs, our earliest ancestors did not live in habitats that were conducive to fossil preservation. Whereas the mud in swamps or along rivers preserved the bones of many a dinosaur, the forests and grasslands in which primates lived were not environments where bones of a dead animal would be covered with sediment and preserved. In addition, most of them were probably eaten by predators or scavengers. Their bones were scattered widely and eaten. Because of this, the fossil record of our early ancestors is less than complete, so the origin of modern humans is not entirely clear. Even so, new discoveries in the past 20 to 30 years have helped researchers chart the course of early human evolution.

For years, archaeological evidence suggested that the first hominids, the first humanlike primates, belonged to the genus ***Australopithecus*** (awe-STRAY-loh-PITH-eh-cuss; meaning "southern apeman"). They lived about 3 million years ago. More recent work, however, has uncovered a much earlier ancestor, known as *Sahelanthropus tchandensis*. This apelike creature lived 6 to 7 million years ago in the Sahel region (southern range of the Sahara) in northern

FIGURE 7-1 Primates very likely evolved from insect-eating mammals that resembled this creature, a tree shrew native to Southeast Asia.
© kajornyot/Shutterstock, Inc.

Chad (in Africa), just after humans split from the great apes. It may represent the earliest human ancestor. They gave rise to other species that eventually formed modern humans. Homo sapiens, the species to which we belong, emerged about 400,000 years ago.

The evolutionary history of humans is marked by a number of significant developments vital to our understanding of the modern human predicament. One very important development was the evolution of **bipedal** (two-legged) locomotion, which requires less energy than quadrupedal (four-legged) locomotion and may have given our early ancestors a competitive advantage over other animals. Many scientists believe that because the front appendages were no longer needed for walking, the hands improved over time through natural selection. This improvement in manual dexterity clearly made it easier for our predecessors to make and use tools and weapons and to manipulate objects.

Bipedalism and changes in the physical structure of the hand were not solely responsible for improvements in manual dexterity. During the long evolutionary history of humans, brain size also increased. This in turn improved hand–eye coordination and enhanced our ability to make tools and shape our environment. These developments obviously set the stage for human technological development, which has contributed in many ways to human progress. The development of technology has been a boon to humans that has helped to overcome environmental resistance. Irrigation practiced by early residents of the Tigris–Euphrates valleys (the Fertile Crescent), for example, reduced the threat of food shortages. Improvements in sanitation and medical science have controlled many infectious diseases, such as the plague, that once killed millions of people. Expanded food production and other benefits of technology fueled a massive increase in the human population in the past 200 years.

Humans also developed the ability to create and communicate ideas. This attribute—when combined with the ability to design technologies, plan for the future, and manipulate the environment—gave humans a power unequaled in the biological world, a power that has grown ever stronger in recent decades. Seen by many as a means of expanding the Earth's carrying capacity and enhancing our survival and well-being, these evolutionary developments have also contributed to many of the current environmental problems facing humankind. Our goal today is to learn to redirect our thinking to create a way of life and new technologies that promote a sustainable way of life.

KEY CONCEPTS

Humans evolved from primates and, over time, have evolved remarkable abilities that give us a distinct advantage over many other species. These advantages have permitted us to colonize much of the world and reshape the environment, sometimes to our own detriment.

7.2 Human Cultural Evolution: Our Changing Relationship with Nature

Humans have evolved biologically over a period of 3.5 million years. During this period, social systems also changed. The earliest humanlike creatures were hunters and gatherers. Their societies were largely replaced by agricultural societies, which have in many places been supplanted by industrial societies (**TABLE 7-1**). All three exist today, but hunting and gathering and agricultural societies are on the decline.

As the following material will show, each distinct social structure used various forms of technology to manipulate the environment, but the impacts of each stage in cultural development vary considerably. The most primitive forms of technology, and the least damaging, existed in hunting and gathering societies; the most advanced and Earth-threatening are found in industrial societies. The following sections highlight important changes in the human–environment relationship over our long cultural evolution—especially changes in how we view and subsequently treat the natural world.

KEY CONCEPTS

Humans have evolved culturally and technologically through three distinct phases: hunting and gathering, agricultural, and industrial. During this time, changes in technology altered our impacts on the environment moving us away from a sustainable relationship with nature.

Hunting and Gathering Societies

Hunting and gathering societies were the dominant form of social organization throughout most of our evolutionary history. In fact, for 99% of the time we have spent on Earth, humans have gathered fruits, seeds, and berries and have hunted animals. Growing evidence suggests that hunters and gatherers probably gained a substantial amount of their meat by scavenging—picking the bones of animals that had been killed by others or that had died from natural causes. Only in the last few fleeting moments of geological time have we turned to agriculture, then to industry.

Studies of present-day hunters and gatherers suggest that members of these societies had a profound knowledge of the environment. They knew where to find water, edible plants, and animals; how to predict the weather; and what plants had medicinal properties. Studies of present-day hunter-gatherers also suggest that, contrary to popular conception, in many regions of the world, especially the warmer climates, our early ancestors did not live with the constant threat of starvation, nor did they spend the greater part of their lives in search of food. Studies also suggest that they were healthy, well nourished, and suffered from few diseases.

Many hunter–gatherer scavengers were nomads who foraged for plants and captured a variety of animals by using primitive traps and weapons. Hunter-gatherers fashioned

TABLE 7-1

Classification of Human Social Systems

Social System	Features
Hunting and gathering	1. The people were nomadic or semipermanent. 2. They benefited from their intelligence and ability to manipulate tools and weapons. 3. They were knowledgeable about the environment and skilled at finding food and water. 4. On the whole, they were generally exploitive of their resources. 5. The environmental impact was generally small because of low population density and lack of advanced technology. 6. They lived healthy lives, were well fed, and experienced low disease rates. 7. Their widespread use of fire may have caused significant environmental damage in some locations.
Agriculture	1. Farmers were generally either subsistence level or urban based. 2. They benefited from new technologies to enhance crops and resource acquisition needed for their survival. 3. They were knowledgeable about domestic crops and animals. 4. They were highly exploitive of their resources. 5. The impact of subsistence-level farming was significant, but because population size was small, damage was minimized. The impact of urban-based agriculture was much larger because of new technologies, trade in food products, increasing population, and lack of good land-management practices. 6. Disease was more common among city dwellers because of increased population density. 7. Poor agriculture, overgrazing, and excessive timber cutting caused widespread environmental damage.
Industry	1. Industry includes early and advanced forms. 2. It relies on new technologies, energy, energy-intensive forms of transportation, tremendous input of materials, reduced number of workers, and, recently, biotechnology. 3. Mass production and modern technology are transferred to the farm. 4. Industry is highly exploitive, more so than earlier societies, and devoted to maximum material output and consumption. 5. Impact is enormous and includes pollution, species extinction, waste production, and dehumanization. 6. Humans become subject to infectious disease and new industrial-age diseases including ulcers, heart disease, and mental illness. 7. Widespread environmental damage results from industry, agriculture, and population growth.

tools from sticks, stones, and bones. Because their technology was primitive, they did not enjoy a great advantage over other species, and their populations never grew very large.

Judging from existing hunting and gathering societies, anthropologists believe that hunter-gatherers had a deep reverence for the environment and the plants and animals on which they depended. These people understood that they were a part of the Earth and dependent on it.

Some hunter-gatherers developed semipermanent lifestyles, setting up homes near hunting or fishing grounds that could provide a year-round supply of food. These groups were more likely to cause noticeable damage. New research on hunting and gathering societies also suggests that many groups also grew their own food and raised animals to feed themselves. Food may have been traded with other groups, a step that marked the advent of the first system of commerce. Cave dwellers in Europe 28,000 to 10,000 years ago, for example, probably participated in extensive networks to trade food and other commodities.

On balance, the hunter-gatherers had little impact on the environment, and their way of life was generally sustainable, as witnessed by their long history. Their existence was sustainable because their numbers were held in check by natural forces and their demands for resources were small. Because they generally lived nomadic lifestyles, the damage they created was easily repaired. Moreover, they lived off the land's renewable resource base and recycled their waste, so they fit well within the economy of nature. Additionally, they seem to have exhibited a reverence for the Earth. In short, they had the lifestyle and the ethics necessary for sustainability.

It is important to point out that although hunter-gatherers were relatively benign, they did cause considerable damage in some areas. In North America, hunter-gatherers may have been responsible for the extinction of many species of large animals after the last ice age. Species lost during this period include the cave bear, giant sloth, mammoth, giant bison, mastodon, saber-toothed tiger, and giant beaver (**FIGURE 7-2**). These animals may have been killed directly, driven out of their preferred habitats, or wiped out as their prey were destroyed. In the Great Plains region of North America, some Indian tribes ignited grass fires to drive buffalo over cliffs for slaughter, killing many more than were needed. Such a wasteful action does not fit the image of a wise steward of the land. Compared with today's societies, though, hunter-gatherers were relatively benign.

FIGURE 7-2 Extinct mammals of North America. These species became extinct during the Pleistocene, possibly as a result of overhunting by humans.

Throughout most of human evolutionary history, our ancestors were hunter-gatherers who lived in relative harmony with the Earth. Primitive technology, small population size, a nomadic lifestyle, reliance on renewable resources, and recycling all waste were the features of these societies that held human impact to sustainable levels.

Agricultural Societies

Anthropologists believe that agricultural societies emerged between 10,000 and 6,000 B.C., although evidence shows that some hunter-gatherers were tending to crops long before this time. The roots of agriculture are generally traced to the rain forests of Southeast Asia (**FIGURE 7-3**, green circle), where early agriculturalists practiced slash-and-burn or *swidden* agriculture. In the **slash-and-burn agriculture** of this region, farmers cleared small sections of jungle to plant crops. Because rain forest soils are generally nutrient poor, crops failed after several years, and plots were abandoned. Farmers moved on to clear neighboring forests. In the abandoned fields, native species invaded and returned the land to its original state. (For a discussion of this practice in modern times, see **Spotlight on Sustainable Development 7-1**.) The early agricultural societies of Southeast Asia also domesticated many animals, such as pigs and fowl. These became vital food sources, greatly supplementing crops.

Seed crops originated in a wide region extending from China to eastern Africa (Figure 7-3). The first farmers cleared rich woodlands, but agricultural efforts were fairly limited until the development of the plow. Coming into use in the Middle East around 3,000 B.C., the first plows were nothing more than tree limbs with branches that cut through the topsoil. In subsequent years, more elaborate plows were developed. Pulled by oxen, these plows enabled farmers to cultivate grassland soils whose thick sod was difficult to work with previous implements. This dramatically increased crop production. As in Southeast Asia, a variety of domesticated animals such as goats, sheep, and cows supplemented the human diet.

FIGURE 7-3 Roots of agriculture. This map shows the areas where tropical planting and seed crop agriculture originated. The Mayans of Mexico were probably the first people in the Western hemisphere to grow seed crops.

The plow gave agricultural societies the means to increase dramatically the productivity of the land. As a result, farmers achieved a greater degree of control over their destiny and could expand beyond the limits previously set by the natural food supply.

Before the development of the plow, most farmers fed themselves and their families. Farming was a subsistence activity. As food became mass-produced, two major changes occurred. First, the human population began to swell. Since more food was available, survival became less problematic. Second, because fewer people were needed to work to provide food, many migrated from farms to villages and cities, where they took up crafts and small-scale manufacturing. As cities and towns formed, they became centers of trade, commerce, government, and religion.

The growth of population, cities and towns, and small-scale industry placed greater demands on the environment for resources—wood, metals, and stone. Heightened exploitation accompanied by poor land management resulted in widespread destruction of the natural environment. Many fertile areas were destroyed by overgrazing, excessive timber cutting, and poor agricultural practices.

The shift to mass-produced food had a potentially more harmful effect: It severed, in large part, the link to nature. Unlike hunter-gatherers and early subsistence farmers—who depended on a wide variety of plants

Go Green

Watch how much you eat. Overconsumption is not only unhealthy, it results in an increase in agricultural production with many impacts on the environment.

SPOTLIGHT on Sustainable Development

7-1 Agroforestry: Tapping Ancient Wisdom

The destruction of tropical forests to make way for highways, towns, and farms is an issue of worldwide concern. Not only does it wipe out thousands of wild species, it also destroys vital economic assets, increases soil erosion, and contributes to global warming.

Population growth is a key factor in the loss of rain forest, as is economic development. About 25% of the rain forest cut down each year can be attributed to the search for new agricultural land to accommodate population growth. Migration to deforested regions is often viewed as a means of relieving overcrowding in urban settings and landlessness in agricultural regions.

Government policies in the industrial and nonindustrial nations of the world also contribute significantly to the rapid deforestation. In the developing countries, for example, governments seeking to raise foreign exchange earnings or to finance economic development programs often turn to their forests to grow cash crops such as tea and coffee that they can export to the industrialized nations.

As a result, rain forests are often used for brief periods, exhausted, and then abandoned. The inhabitants move on, starting a whole new cycle of exploitation. To curtail deforestation, many experts believe that we need to find ways to use tropical rain forests sustainably—so that they can continue to be productive for many years. This could stop the cycle of exploitation and abandonment. One candidate is **agroforestry**, a sustainable management system that combines agriculture with tree crops or other forest plants (**FIGURE 1**). These activities occur on the same parcel of land, either simultaneously

or sequentially. Agroforestry maintains or improves the environmental quality of the land, provides food and income, and reduces the need to exploit new land. It can be practiced on forestland that is already degraded and cleared. It, therefore, helps restore soils. It is good for society, the environment, and the economy.

Traditional agroforestry systems have evolved over many years. In fact, forests have been prudently managed by indigenous peoples for many generations without any apparent loss of species diversity or deterioration of soil quality.

FIGURE 1 Agroforestry. By planting trees and crops on the same plot, farmers can create a sustainable system of food and fiber production. It provides for human needs and eliminates further cutting of tropical rain forests.
© Wendy Stone/Corbis.

One form of agroforestry is swidden agriculture, briefly described in this chapter. In swidden agriculture, people clear small plots of rain forest in which they plant food and tree crops. When the soil loses its fertility, the plots are abandoned and allowed to regrow. This restores the soil in time, making it useful once again.

The Bora, an Amazonian tribe native to northeastern Peru, practice swidden agriculture. Some scientists believe that large areas of the Amazon forest may have been managed in this manner by considerably larger populations in the past.

The Kenyah people, who live in a remote section of Indonesia, have practiced swidden agriculture for at least 200 years. Over time, they have converted much of the virgin forest into a mosaic of secondary forest of different ages, but they have retained uncut reserves in which they harvest products that are rare or nonexistent in secondary forest.

Many of the agroforestry systems studied to date have been practiced by relatively remote tribes, suggesting that these systems are suitable only for subsistence production. Recent studies, however, prove that they can also be important cash producers near urban centers. In Tanshiqacu, Peru, for example, a nontribal group of 2,000 people of mixed Amazonian and European ancestry produce food, fiber, handicraft materials, and charcoal from the forests. Their products appear in local markets as well as in markets in the Peruvian capital, Lima. Most households maintain several agroforestry fields and sell a variety of products, a tactic that ensures a year-round income.

In Java, farmers clear forests and plant teak trees, whose wood is exported to many developed nations. Among the teak trees, they plant rice and corn. This system of agroforestry has been modified to include horticultural species, animal food, and fuelwood crops that provide additional income.

In the Philippines, the rural poor venture into steep, mountainous terrain to clear forests to make room for crops. To prevent widespread destruction of these regions, governmental agencies constructed numerous demonstration plots during the 1980s to introduce agroforestry practices to the rural people. Because deforestation of steep hillsides leads to soil erosion, the first step in these projects was to stabilize the soil by laying bamboo poles along the contour of the land. The bamboo barriers capture soil, preventing it from washing down the slope. Fast-growing trees were then planted along the contours. The trees provide wood for fuel and leaves for animal feed and soil enrichment. The seeds of these trees are used to produce traditional medicines. Fruit trees such as mango and food crops such as pineapple and cassava were planted in the fields. Today, the agroforestry plots provide a year-round food and fuel supply for the people. By stabilizing the soil, they reduce the need to cut more forest.

As these examples illustrate, agroforestry provides many benefits. It helps protect virgin forest and provides a sustainable supply of food, fuel, and fiber. It also reduces soil erosion and land destruction and often improved soil fertility, restaring the land. Agroforestry benefits the rural poor, providing (in addition to food) income and employment opportunities. Agroforestry is also being practiced in developed nations such as the United States. Agroforestry is not a panacea; it's just one of many strategies for living sustainably on the planet.

and animals for food and who had an impressive working knowledge of the environment—the new generation of farmers relied on a relatively small number of plants and animals. For many people, especially city dwellers, a profound knowledge of the environment was no longer necessary. The emergence of technological agriculture also ushered in dramatic changes in the way people treated the environment. Most important, domination of nature replaced the harmonious ways of earlier peoples.

The environmental impact of agricultural societies and the large urban centers they supported were enormous. Archeological and historical records show that overgrazing, widespread destruction of forests, and poor farming practices changed many once-productive regions into barren landscapes. Crop failures and the invasion of displaced people caused the demise of numerous ancient civilizations. These occurrences were especially evident in the Middle East, North Africa, and the Mediterranean from 5,000 B.C. to A.D. 200. The Babylonian Empire, for example, once occupied most of what is now Iran and Iraq. At the outset, this land was covered with productive forests and grasslands. Huge herds of cattle, goats, and sheep overgrazed the grasslands and eventually destroyed the natural vegetation. Forests were cut to provide timber and create pasture. The loss of grassland and forests decreased rainfall and eventually parched the land. Sediment washed from the barren soils, robbing them of nutrients and filling irrigation canals. These changes and a succession of invading armies eventually destroyed this once great empire.

In summary, agricultural societies were relatively benign at first, but as technologies improved, agriculture became a major source of environmental damage. The success of agriculture resulted in an increase in population size and ushered in the age of commerce. As commerce began to flourish, material demands increased, and waste from shops and small factories began to accumulate outside cities. Local resources began to decline. Finally, as commerce and technologically based agriculture took hold, humans began to see the natural world in a much different way—more as a source of wealth and a force to dominate and control. Agriculturalists lost the physical and spiritual connection with nature that characterized their hunting and gathering predecessors. The human population began to increase, and the systems that supported humans—agriculture and the economy—became less sustainable.

KEY CONCEPTS

Agriculture started as a subsistence activity, but the plow and other forms of technology gave our ancestors the ability to produce excess food. This, in turn, displaced farm workers, who took up crafts and trades in cities and towns, and it fostered a dramatic increase in human population. The growth of human population and the emergence of cities and towns as centers of commerce had a significant negative impact on the environment.

The Industrial Society

The **industrial society** resulted from a major shift in manufacturing, from small-scale production by hand to large-scale production by machine. Historians refer to this epic change as the **Industrial Revolution** (**FIGURE 7-4**). Industrial societies emerged in England in the 1700s and in the United States in the 1800s.

The Industrial Revolution was made possible by new technologies *and* by an abundant supply of fuel, notably coal. Industrialization led to the mass production of products.

This meant that more goods were available to more people. As the demand for goods rose and industries grew, the influx of fuel, food, minerals, and timber into the city rose sharply.

Manufacturing technologies that made the Industrial Revolution possible were the outcome of many scientific and engineering advances. Although the technologies improved the economic output of cities and towns, they often made work meaningless and boring. They also produced large quantities of smoke, ash, and other toxic wastes that were released into the environment. The shift to machine production, therefore, changed the workplace, the city, and the surrounding countryside that supplied the resources.

Mechanization also swept through farms during the Industrial Revolution (**FIGURE 7-5**). Technological advances such as Jethro Wood's cast-iron plow with interchangeable parts and Cyrus McCormick's reaper helped produce a rapid increase in agricultural production. One of the most significant advances was the invention of the internal combustion engine, which made horse-drawn implements obsolete. The motor-powered tractor could plow as much land in a week as one of our forebears could work in a lifetime using hand tools. Another significant advance was the development of fertilizers, which allowed an increase in agricultural productivity (output per hectare). Plant breeding produced higher yield crops that also contributed to a rise in productivity. Because of more efficient farming methods, fewer farm workers were needed. Unemployed workers migrated to cities, swelling their populations.

New medicines and better control of infectious disease through improved sanitation also occurred during the Industrial Revolution. These important developments enhanced human survival. Because the death rate declined and people began to live longer, the human population began another upward turn.

(a)

(b)

FIGURE 7-4 Changing ways. (a) Carpets being made by hand in Nepal. **(b)** A factory can turn out thousands of carpets per year using the latest technology. The Industrial Revolution made goods more readily available. Although they improved people's lives, they also increased the demand on resources and resulted in widespread pollution and habitat destruction.

Photo a © Khoo Si Lin/ShutterStock, Inc.; photo b © AbleStock.

FIGURE 7-5 Industrialization on the farm. Modern agriculture depends heavily on machinery, oil, fertilizers, pesticides, and other resources. Large fields are worked to achieve maximum output.

 © Jack Dagley/ShutterStock, Inc.

Like the Agricultural Revolution, the shift to an industrial society further changed our relationship with nature. Fewer and fewer people were in contact with nature as urban populations swelled. In addition, control over nature became increasingly important. Industrial people came to view themselves as more and more apart from nature. The prevailing attitude is best summarized by the 17th-century English philosopher John Locke, who wrote, "The negation of nature is the way toward happiness." Locke argued that people must become "emancipated from the bonds of nature." These notions have been passed from generation to generation throughout the industrial age. Locke also preached unlimited economic growth and expansion, with the belief that individual wealth was socially important for a harmonious society, an idea that persists today.

It must be pointed out that the Industrial Revolution provided many benefits. On the whole, people ate better, lived better, and began to live longer. Many new products were available, and there was more leisure time, but there was a price to be paid in environmental deterioration for the unsustainable way of life that emerged. Populations increased, and systems for providing human needs became less and less sustainable.

KEY CONCEPTS

The Industrial Revolution resulted in many changes that enhanced peoples lives but also exacted a huge toll. An increase in population size, resource demand, pollution, and environmental destruction dramatically altered the human–environment interaction. Attitudes toward nature shifted even farther away from stewardship.

The Advanced Industrial Age

Industrialization, well established by World War II, expanded rapidly after the end of the war, creating an **Advanced Industrial Society.** Several features distinguish this new phase. First and foremost is a marked rise in production and consumption, which have resulted in huge increases in energy and resource demand by virtually all sectors of our society. Consider energy. In hunting and gathering societies, individuals require only about 2,000 to 5,000 kilocalories of energy per day. A kilocalorie (KILL-oh-CAL-er-ee) is a measure of energy. One kilocalorie is equal to 1,000 calories, which is the amount of energy needed to raise one gram of water one degree Celsius. Kilocalorie and calorie, a measure used by dieters, are equivalent. The 2,000 to 5,000 kilocalories used by hunter-gatherers provided metabolic energy and additional heat to cook food. Early agricultural societies, however, required twice that amount per person to grow crops—about 10,000 kilocalories. Advanced agricultural societies required twice as much again, or about 20,000 kilocalories per person per day. Increased energy use resulted from improvements in agriculture and the growth of crafts, commerce, and cities. In industrial societies per capita energy use climbed again, increasing to 60,000 kilocalories per day per person. Modern industrial society, however—with its mechanized farming, energy-intensive industries, and affluent lifestyles—has doubled that amount, raising energy consumption to about 120,000 kilocalories per person per day. Some nations such as the United States and Canada consume, on average, even more: 250,000 kilocalories per person per day.

The problem with this is that fossil fuel combustion is the source of most air pollution. The production of fossil fuels can also result in significant amounts of water pollution, as witnessed by the 1989 Exxon Valdez oil spill off the coast of Alaska. Another problem with the energy we use is that it produces carbon dioxide. Carbon dioxide gas released in large quantities now threatens to change the global climate in ways that could disrupt agriculture and cause severe economic problem.

Go Green

If you own a car, keep it well tuned and keep the tires properly inflated. Drive the speed limit and avoid rapid acceleration. All these efforts will improve mileage and reduce pollution, especially carbon dioxide, a major greenhouse gas.

A second characteristic of the advanced industrial society is a shift toward synthetics, such as plastics, and non-renewable resources, such as oil and metal. Synthetics create problems in nature because bacteria in soil and water, which decompose naturally occurring substances, are frequently unable to break them down. Consequently, synthetics may persist in the environment for decades. The persistent insecticide DDT, for example, can accumulate in the fatty tissues of birds and disrupt reproduction. Plastic pollution in water has proven to be a problem for some aquatic organisms.

Yet another feature of the advanced industrialized society is disposability. In the name of convenience, more and more products are designed to be discarded after use. Disposable pens, diapers, contacts, razors, and cameras are just a few of the products we use and then throw away. Throwaways increase consumption, energy demand, and pollution.

The advanced industrial society seems to be caught up in an ever-escalating production–consumption cycle that is, say critics, highly unsustainable. Domination of nature continues as the central theme of modern industrial societies, and economic growth retains its commanding allure—despite evidence that both threaten the long-term future of our planet and our long-term well-being.

In this intense period of economic and population growth, the unsustainability of virtually all human systems—from agriculture to energy production to water supply to waste management—is becoming clear. The challenge today is to restructure those systems to ensure a better future. Environmental protection efforts must be integrated with development efforts to create lasting change and an enduring human presence.

7.3 The Population, Resources, and Pollution Model

We have seen how humans have an impact on the environment. As our culture has shifted away from hunting and gathering to agriculture and then to industrial societies, the impact on these conditions has grown to a level that many scientists believe is unsustainable.

This section provides an ecological model of human activity that shows how impacts arise. It is called the **Population, Resources, and Pollution (PRP) model** (**FIGURE 7-6**). The PRP model illustrates several important relationships between humans and their environment and offers a big-picture view of the human–environment interaction. Not only is this vital to becoming a critical thinker, but also, it helps promote systems thinking, which is vital to sustainable development.

The PRP model (Figure 7-6) shows that humans, like all other organisms, acquire resources from the environment. The acquisition of resources—for example, through coal mining—alters both the abiotic and biotic conditions of ecosystems. Surface coal mines, for example, disrupt wildlife habitat. They can also lead to soil erosion that pollutes nearby streams—hence, the arrow between resource acquisition and pollution in the model.

The resources we extract from the Earth are put to use. After being mined, for example, coal is burned in power plants. Minerals extracted from the Earth are fashioned into finished products such as automobiles. Such activities can produce substantial amounts of pollution, indicated in the model by an arrow connecting resource use to pollution.

As we ponder society's contribution to building a sustainable society, it is important to remember our personal role, too. Although we rarely think of it, everything we purchase comes from a mine, forest, field, stream, or the ocean. The acquisition of the raw materials needed to make the products we purchase creates some physical damage. The conversion of those raw materials into finished goods results in pollution of air, water, and soil.

Some pollutants are **biodegradable**—that is, broken down by bacteria and other organisms in air, water, and soil. Others are **nonbiodegradable**—not broken down by organisms. Both forms can significantly alter abiotic and biotic conditions of ecosystems.

Scientists also classify pollutants by the medium they contaminate—for example, air, water, and land. Thus we have air pollution and water pollution. In recent years, scientists have found that pollution readily crosses the boundaries between these media, a phenomenon called **cross-media contamination.** Air pollution, for example, washes from the sky and is deposited in lakes and forests. Some water pollutants evaporate from lakes and streams and enter the atmosphere.

The negative impacts of development on the human population are indicated on the PRP model by minus signs on the arrows that lead from pollution and resource use to population. This response is an example of a **negative feedback loop,** in which one factor leads to a decrease in a second factor. Negative feedback loops are the chief means of controlling biological systems and regulating homeostasis. Many human systems are controlled by negative feedback. A familiar example is the furnace in your home, which is regulated by a thermostat. When the temperature drops below the setting on the thermostat, the furnace is turned on. Heat from the furnace then warms the room air until the air temperature reaches the desired level. When it does, the furnace shuts off.

The negative feedback loops shown on the PRP model are of great concern to scientists. They could in fact limit future growth catastrophically. Continued soil erosion from the world's farmland resulting from poor agricultural practices, for example, could cause a marked decline in food production that could have a devastating effect on the global human population.

As the previous discussion of cultural evolution pointed out, resource acquisition and use (especially that permitted by technology) also have a positive effect. Put another way, they enhance survival and promote human population growth. The efficient harvest of food and fiber, for example, has made it possible for large numbers of people to inhabit the Earth. This relationship is indicated in the model by the plus signs on the arrows leading from resource use to population. It is an example of a positive feedback loop, in which one factor leads to the growth of a second factor, which in turn stimulates the first one in a repetitive cycle.

Positive feedback mechanisms are rare in nature; in the human–environment system, they may cause serious problems, for they can create devastating cycles of depletion and environmental destruction. For example, although

FIGURE 7-6 Population, resources, and pollution model. This model applies to all organisms. It outlines how organisms interact with their environment. The + signs by the arrows indicate a positive feedback loop, in which one activity enhances another. The – signs indicate a negative feedback loop, in which one activity adversely affects another.

fossil fuel energy has increased our capacity to produce food, it is creating a dangerous rise in global temperature and a massive shift in rainfall patterns that might devastate crop production on much of the existing farmland.

The PRP model represents a fundamental ecological relationship true to all living organisms. It's as relevant to humans as it is to black bears. What is more, it provides some insight into the human–environment interaction. It allows us to predict the impacts of human actions. If you change one variable in the model, the rest change. If you add more people, for example, resource use is bound to rise. Acquire more resources, and pollution is likely to climb.

The model also provides insight into sustainable solutions. To solve rising pollution levels, resource demand might be cut. To control resource depletion, population growth control measures might be desirable.

KEY CONCEPTS

Human populations acquire and use resources from the environment to produce goods and services that provide us with many benefits. These activities, however, degrade the environment by altering abiotic and biotic conditions.

7.4 The Sustainable Society: The Next Step

For most of our evolutionary past, humankind has lived in a sustainable relationship with nature. In the past few centuries, though, human civilization has undergone many changes that have shifted it off its previous sustainable course. These changes in both attitude and way of life arose during the immensely successful Agricultural and Industrial Revolutions.

One thing that this book makes clear is that major changes are needed to create a sustainable future. New systems of transportation, housing, waste management, industrial production, and the like are needed to protect Earth's systems, which provide us with the goods and services on which we depend.

Fortunately, many of the necessary changes are already under way. Many examples are given in this book. The Spotlights on Sustainable Development present uplifting examples (see **Spotlight on Sustainable Development 7-2**). Although change is in its infancy, the signs are encouraging, leading some observers to believe that human society may indeed be in the very early stages of a **Sustainable Revolution,** the next step in human cultural evolution. These changes designed to restore and protect natural systems and to undo the mistakes of the past may take decades, perhaps centuries to complete. Change is occurring all around you, though.

Local governments are drafting sustainable development programs. Businesses are finding ways to eliminate hazardous waste and slash energy consumption. Individuals are taking steps, as well, to reduce consumption, recycle waste, and reuse products. Books could be written on the changes under way. Efforts to build a sustainable society could have many benefits, among them stronger, more secure nations; reduced political tension; stronger economies; a cleaner environment; and healthier people. In short, sustainability could create a better future for all.

KEY CONCEPTS

Many steps are under way to create a more enduring way of life. These changes may be the early signs of a new cultural shift—a Sustainable Revolution—designed to restructure human systems to honor the limits of the natural systems on which all living things depend.

> The danger is not what nature will do with man, but what man will do with nature.
>
> **—Evan Esar**

© hugolacasse/ShutterStock, Inc.

SPOTLIGHT on Sustainable Development

7-2 Monfort Boys' School: Mining Human Wastes to Feed People

Fish are a popular source of protein in many countries. Today, in fact, the world's fishing fleets catch over 100 million metric tons of fish from the world's oceans each year. This is made possible because of advanced technologies such as sonar, which help fishing fleets find and capture fish.

Unfortunately, commercial overfishing has depleted several dozen of the world's richest fisheries—just in the Atlantic Ocean alone—areas of the ocean once served up huge fish catches. Many other fisheries are now in danger. In fact, of the world's 15 major oceanic fisheries, 11 are in decline. So heavily harvested are many of these fisheries that if we stopped fishing immediately, it would take them from 5 to 20 years to recover.

Overfishing has had a ripple effect on other species, reducing populations of seals and other fish-eating animals. Thus, not only are we threatening our own food supply, we're eliminating the supply of the many species that share this planet with us. Unfortunately, the story does not end here. Each year, millions of tons of fish netted by commercial fishing operations are thrown overboard because they are the wrong species or sex or are too small. International fishing regulations require this to protect various species and populations.

We can also turn to fish farming—growing fish and shellfish—to increase food production. Fish and shellfish can be raised in artificial environments such as ponds or in natural bodies of water (for example, in bays, where they are contained by nets) then harvested for food when they've reached market size. Worldwide, fish farms produce about 22 million metric tons of food a year.

Important as it is, fish farming is not without its problems. Fish need to be fed, for example, and fish food needs to be grown. That takes resources and a considerable amount of energy. Moreover, fish in ponds produce lots of waste that can pollute the waters into which their ponds are emptied, waters that we may eventually drink or swim in or waters upon which species depend.

There is environmentally sustainable way to raise fish for human consumption, however. Consider the work of a group called ZERI, an acronym that stands for Zero Emissions Research and Initiatives, founded by Gunter Pauli, a European businessman with a firm commitment to the environment. Pauli's organization is pioneering an ecosystems approach to raising fish—and many other foods as well—for human consumption in ways that do not harm the environment or, preferably, enhance it. His group helps people in less developing nations use waste from one activity, for example, brewing beer, to produce fish, fruit, vegetables, and animal food.

At the Monfort Boys' School, a Catholic technical school in the island nation of Fiji, located in the Pacific north of New Zealand, officials use waste grain from a local commercial brewery to feed an entire human-made ecosystem, reaping a huge profit in the process. The waste grain begins its travel through this human-made food chain as a substrate for a commercial mushroom venture. It could be fed directly to livestock such as pigs, but the waste grain is not well digested as is. Enzymes released by the mushrooms, however, render the waste more digestible. Thus, after the school has harvested the mushrooms, the biologically processed waste grain is mixed with pig food and fed to the hogs. Waste grain is also mixed with duck and chicken feed, cutting down the need to purchase expensive feeds for them as well.

Hogs are occasionally slaughtered, and the meat can be sold or eaten by the school's 100 students. However, that's not the end of the story. In fact, it is only the beginning. Waste from the pigpens is washed out twice daily and drains into a huge settling tank. Here, the waste is broken down by anaerobic bacteria, microorganisms that break down organic material in the absence of oxygen. One byproduct of this process is methane, a combustible gas that can be burned to produce light and to cook food. The farm produces enough methane for a family of six.

The sludge, a liquid waste is drawn off by bucket, is used to fertilize crops—bananas and various vegetables—that are then sold for food or consumed by the students. Liquid waste from the settling tank is then fed into a series of settling ponds where naturally occurring algae consume much of the nitrogen and phosphorus in the waste. The algae-rich water is then delivered to a fish pond containing six different species of fish. The fish consume the algae, living entirely off this free food source. At the end of each year, the fish are harvested and sold for food. Interestingly, ducks and chickens, which are fed waste grain, are raised in coops suspended over the fish ponds. Their waste freely drops into the fish ponds, adding additional nutrients to the waters. In this ecological production system, all waste is food for something else.

At one time, waste from the hog farm polluted nearby mangrove swamps, killing aquatic life. Today, when the water leaves this "facility," it is so clean that the nearby mangrove swamps are showing promising signs of recovery. Species of fish and crabs wiped out by pollution are now returning.

Besides producing food from waste, this farm also generates a fair amount of income. This system, in fact, can produce an annual income of as much as $60,000 for a family of six...get this...working a half a day a week each! Compare that to the typical income of a worker in the less developed world—averaging about $2,000 a year. Needless to say, this system is being seriously considered by a great many other people in other countries.

What can you and your family do to help save endangered fish populations?

One thing that will help immediately is to buy fish whose populations are not in danger. See the Monterey Aquarium's website (at www.mbayaq.org) for a list of fish (Seafood Watch) that you can eat without guilt.

Another important step you can take is to cut down on shrimp consumption or avoid it altogether. Shrimp harvesting is one of the most environmentally damaging activities in which humans engage. Trawlers drag huge nets along the bottom of the ocean, scooping up shrimp and a host of other sea creatures. As noted earlier, for every pound of shrimp that is harvested, there are 5 pounds of other sea life that are swept up. Most are dead before they are thrown back overboard. Shrimp nets also scrape up the bottom of the ocean, removing plants and leaving the ocean floor in a state of ecological disarray.

Another thing you can do is to purchase fish raised on the farm, fish farm, that is.

Yet another option is for us to eat lower on the food chain. Most Americans eat way more meat protein than they need. By reducing fish and other meats from the diet, or eliminating it, we can reduce our demand on fish farms and natural fisheries. A good vegetarian diet can provide all of the protein one needs.

Adapted with permission from Dan Chiras, *EcoKids: Raising Children Who Care for the Earth*. Gabriola Island, B.C.: New Society Publishers, 2004.

CRITICAL THINKING

If you are like me, you probably had to do a little research to answer this question well. Gathering more information is one of the most important rules of critical thinking. Without adequate information, it is difficult to answer a question accurately or to form reliable opinions. Unfortunately, as you may have discovered, all too many people volunteer their opinions or make crucial decisions without sufficient information. Remember the critical thinking rule: Don't mistake ignorance for perspective.

Your research into the Spaceship Earth analogy probably revealed many similarities and differences between a spaceship and the Earth. Like the Earth, the spaceship is a fairly self-contained unit. It floats around in outer space, and everything the occupants need is aboard. I found, though, that the similarities pretty much ended there.

I knew that a spaceship carries its food from another source—the Earth. Food is not generated on board; it is contained in packages. The Earth, however, produces its own food, from plants that take sunlight, carbon dioxide, and water to make food molecules (via photosynthesis).

Another difference is that a spaceship contains a special filter that captures carbon dioxide from the respiration of the astronauts. This carbon dioxide comes from the breakdown of food molecules in the cells of their bodies. On Earth, inhabitants release carbon dioxide, but it is absorbed by plants, which use it to produce food. The food molecules are consumed by animals, and the carbon dioxide is released once again. So on Earth, carbon dioxide is recycled. In a spaceship, it is not.

Oxygen is consumed by animal life both on spaceships and on Earth. The oxygen on the spaceship, however, is released from huge storage tanks filled before the spaceship blasts off. On Earth, oxygen is supplied by plants, which generate oxygen during photosynthesis. Thus, oxygen is recycled on Earth but not on the shuttle.

Water is consumed by animal life, both on Earth and in the spaceship. It is also released as we exhale. On a spaceship, water is collected and discarded. Drinking water is supplied in foods and drinks. On Earth, water is recycled. It enters the atmosphere and rains down upon the land, where it is absorbed by plants.

Finally, on the spaceship, waste matter (feces and urine) are collected. Urine is jettisoned into space, but solid matter is stored in huge tanks that are emptied when the astronauts return to Earth. On Earth, human waste (in modern technological societies) is typically collected and treated. Some of the material is returned to the land—that is, recycled by applying it to pasture and soils.

Thus, it turns out that the Earth as a spaceship has some validity, but it is really a rather poor analogy. It makes for a great image but falls short of scientific accuracy.

CRITICAL THINKING AND CONCEPT REVIEW

1. What developments occurring during human biological evolution have contributed to the modern environmental crisis? In your opinion, do these developments mean that we are doomed to continue living unsustainably? How can they be used to our advantage?

2. Describe each of the major steps in human cultural evolution. How are they similar? How are they different? How did the human–environment interaction shift over time? How did human systems shift over time away from sustainability?

3. In your view, what general changes are needed in modern society to restructure systems that evolved during the Agricultural and Industrial Revolutions to create a sustainable future?

4. Describe the Population, Resources, and Pollution model. Illustrate through example some of the positive and negative feedback loops.

5. How have the elements of the PRP model shifted as human culture evolved?

6. Has the PRP model presented in this chapter clarified your thinking on environmental problems and solutions? How? Can you see how using it would help expedite efforts to build a sustainable society?

7. One reviewer commented that the material in this chapter gives the impression that any way of life "above" that of the hunter-gatherer was unsustainable. Do you agree with this assessment? Can human society be made sustainable? What will it take to create a sustainable way of life?

8. Before one can devise strategies to create a sustainable society, he or she must ask, "What are we trying to sustain?" How would you answer this question, given the material you have encountered so far in this book and in your course?

9. Using your critical thinking skills, debate the statement, "Efforts under way to create a sustainable society are bound to negate the progress of the past 200 years and should be stopped."

10. Draw a diagram of all the components of the Monfort Boys' School's production system, showing how the components are related to one another. What features of their system make it a model of sustainability?

11. Draw a diagram of the college's or university's (where you attend) food system. How could it be modified to make it more sustainable?

KEY TERMS

Advanced Industrial Society
agroforestry
Australopithecus
biodegradable
bipedal

cross-media contamination
Industrial Revolution
industrial society
negative feedback loop
nonbiodegradable

Population, Resources, and Pollution (PRP) model
slash-and-burn agriculture
Sustainable Revolution

3

THE POPULATION CHALLENGE
SOLVING THE WORLD'S MOST PRESSING PROBLEM SUSTAINABLY

© Antonio Jorge Nunes/Shutterstock, Inc.

© Sam Chadwick/Shutterstock, Inc.

Population: Measuring Growth and Its Impact

One generation passeth away, and another generation cometh, but the earth abideth forever.

—Ecclesiastes 1:4

LEARNING OBJECTIVES

After reading and studying the material in this chapter, students will be able to understand, knowledgably discuss, and, when appropriate, apply the information learned in this chapter, including:

1. The history of human population growth
2. Factors that have contributed to the rapid growth of the human population in the past few decades
3. The concept of carrying capacity
4. Why it is difficult to estimate the Earth's carrying capacity for human beings
5. Factors that suggest that the human population has grown beyond the Earth's carrying capacity
6. Key concepts and terms related to population growth, including crude birth rate, crude death rate, doubling time, total fertility rate, and replacement-level fertility
7. How migration affects local and regional population growth
8. Predictive tools that help determine whether populations will grow—notably the population histogram
9. Why we should be concerned by population growth in less developed and more developed countries
10. U.S. and global projections of population growth

CRITICAL THINKING

A U.S. senator speaking on the population issue noted that 90% of the current growth in the world population is taking place in less developed countries such as India and China. In order to protect the environment, he said, population growth in these countries needs to be stabilized—and soon. When asked about population growth in the United States and other more developed nations, he replied, "The U.S. population is growing at a slow rate, only about 1% per year. The developed nations, on the whole, are growing at an even slower pace, about 0.2% per year—much, much slower than the growth rate in the developing nations. Therefore, population growth is not a problem in the United States or other developed countries."

Do you think his emphasis on stabilizing population growth in the less developed countries and ignoring growth in the United States is justified? Why or why not?

After you have completed the exercise, answer these questions: What data did you have to call on to make your argument? What critical thinking rules did you use to judge this assertion?

In the 60 seconds it took you to find this book on your bookshelf and turn to this page, the human population of planet Earth increased by approximately 155 people. That's the equivalent of about 2.6 new world residents every second! At this remarkable rate of growth, nearly 1.6 million people join the human population every week and nearly 83 million people join every year. In just under 4 years, the human population increases by an amount equivalent to the entire U.S. population.

In 2013, world population reached an all-time high of nearly 7.2 billion. By the year 2025 it could reach 8.1 billion, according to a report by the United Nations. By 2050, world population could reach 9.6 billion, according to the same report. Ninety percent of this growth takes place in the poorer nations, where over 80% of the world's people currently live. To see how fast the world population is increasing you can log onto the online population clock sponsored by the U.S. Census Bureau.

8.1 The Growing Human Population

Throughout most of our history the human population has been small (**FIGURE 8-1**). Two thousand years ago, in fact, there were about 200 to 300 million people in the world. By 1850, though, world population had climbed to 1 billion (**TABLE 8-1**). But then things began to change, and rapidly. In the next 80 years the population doubled. Forty-five years later it had doubled again, reaching 4 billion. What caused the rapid upsurge in population growth?

KEY CONCEPTS

The human population has grown slowly for most of human history; the rapid increase in population size is a relatively recent phenomenon.

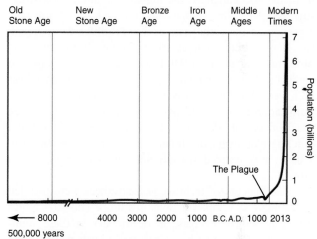

FIGURE 8-1 World population growth. This graph dramatically illustrates the growth curve of the world population. Human numbers were held in check by famine, disease, war, and primitive technological development for most of human history. Advances in technology and health care have caused a massive increase in human numbers in the last 200 years.

TABLE 8-1

Global Population Growth

Population Size	Year	Time Required to Double
1 billion	1850	All of human history
2 billion	1930	80 years
4 billion	1975	45 years
8.2 billion (projected)	2025	50 years

Why Has the Human Population Grown So Large?

Hunting and gathering and early agricultural practices could not support large populations. Disease, famine, and war also caused numbers to remain low. In the course of human history, however, several important changes occurred that altered our prospects. Among them were the development of technology (tools and weapons), the Agricultural Revolution, the Industrial Revolution, and the opening of the New World for settlement. Each advance helped to unleash human population from the constraints of environmental resistance. Improvements in agriculture, such as the plow and irrigation during the Agricultural Revolution, for example, increased food supplies, which spurred human population growth. The most dramatic influence of technology on food supplies, however, occurred during the Industrial Revolution as new forms of machinery were introduced. Many new technologies improved food production, boosted food supplies, and permitted the population to increase. Modern medicine also enhanced the human condition. New drugs such as penicillin dramatically lowered the death rate, especially in the early years of life (**FIGURE 8-2**). Although we later learned that it caused serious environmental problems, the pesticide DDT was used in the tropics to combat malaria-carrying mosquitoes, saving millions of lives each year. Finally, improvements in sanitation and water purification also boosted population growth by decreasing the incidence of infectious diseases. As a result of these and other developments, death rates in many countries dropped. This, in turn, resulted in a substantial increase in population growth.

The benefits of technological advances were first felt in the more developed nations in Europe and later North America. But the decrease in death rates in these countries was not matched by a decrease in birth rates. It took over 150 years for most countries to bring the two in line. The result was a rapid increase in population in the developed countries during this period.

The benefits of the Agricultural and Industrial Revolutions came later to the less developed countries (LDCs) of Asia, Latin America, and Africa—in fact, not until after World War II. Prior to this time, high growth rates and high death rates were commonplace in these nations. The introduction of modern medicine, advanced agricultural techniques, and sanitation, however, caused a sudden decrease in death rates in LDCs after World War II. Because birth rates remained high, growth took off and the populations of the LDCs began to expand.

KEY CONCEPTS

The human population has skyrocketed in the last 200 years primarily because of a worldwide lowering of death rates without a corresponding decrease in birth rates. Death rates plummeted primarily as a result of increases in food supply and better medicine and sanitation.

Expanding the Earth's Carrying Capacity: An Ecological Perspective

Each of the advances cited previously here—technology, agriculture, medicine, and others—has helped alter the delicate balance between population growth and population reduction factors. The net effect has been an increase in those factors that stimulate growth and a decrease in those that hold growth in check. This, in turn, has caused a dramatic increase in the Earth's carrying capacity for human beings.

The **carrying capacity** of the Earth is the number of organisms (in this case, humans) the Earth can support over the long haul. The carrying capacity of the Earth is determined by two major factors: resource availability, including food supplies, and the environment's capacity to safely absorb and detoxify wastes.

All ecosystems have a specific carrying capacity for each population. The carrying capacity is not static, however. In other words, it fluctuates from year to year, depending on climate and other factors. As a rule, most species have little effect on the carrying capacity of their surroundings. Humans, however, have a remarkable ability to expand the Earth's human carrying capacity. Although such changes enhance the survival and reproduction of our kind, the gains we make often occur at the expense of other organisms. The more fish we harvest from the sea through advanced fishing techniques, the fewer there are for sea mammals.

Expansion of the carrying capacity affects humans as well. Importing water via pipelines to desert communities, for example, increases the regional carrying capacity. However, dramatic increases in the human population may result in sizable increases in pollution. At a certain point, the amount of waste produced cannot be assimilated by natural mechanisms. People begin to suffer.

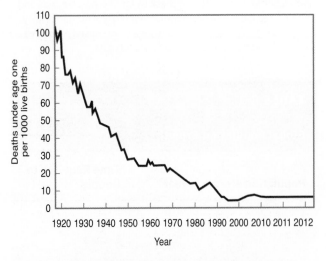

FIGURE 8-2 Infant mortality in the United States.
The infant mortality rate declined since 1915. This decline has contributed to a marked increase in average life expectancy.

Technological advances lower environmental resistance and promote population growth, increasing the carrying capacity for humans. This, in turn, generally decreases the prospects of other species and may cause adverse effects in human populations as well.

Determining the Earth's carrying capacity for humans is a task fraught with difficulty. Some people think that we have not reached the carrying capacity. Others believe that the human population already exceeds the Earth's long-term carrying capacity and that further growth will only further undermine the ecological health of the planet and our own prosperity and well-being.

What Is the Earth's Carrying Capacity for Humans?

Much of the debate over the environment often revolves around arguments concerning the Earth's carrying capacity for human beings. Current estimates range from 500,000 to 1,000 billion people—140 times the present population. The reason for this vast discrepancy lies in the assumptions one makes about the standard of living and the level of technological development. If everyone lived frugally, recycled resources, drove solar-powered cars, lived in small solar homes, used energy efficiently, and so on, the Earth could support a fairly large population. But in the absence of environmental lifestyles and substantial changes in industrial production, the Earth's carrying capacity would be much smaller.

Where are we today? Many scientists believe that the 2013 human population of nearly 7.2 billion already exceeds the carrying capacity. Polluted air, depleted ocean fisheries, species extinction, denuded landscapes, deforestation, widespread starvation, and poverty are signs that humanity has pushed beyond the carrying capacity of the Earth. While recognizing these problems, critics say that new laws and improvements in technology could reduce these problems and perhaps even allow the Earth to support many more people. By using energy more efficiently or converting to solar and wind energy, for example, they say that many more people could live on the Earth. Is this realistic?

As you study the material presented in this book and in your course, you will see that the problems faced by the world's people are massive. Many will be difficult to solve. Many are growing worse. Adding more people could worsen the problems. A larger human population, even one that uses resources more efficiently, will require massive amounts of food, wood, water, fuel, and building materials. According to one scientific estimate, the human population currently consumes about 40% of the Earth's primary terrestrial productivity—that is, 40% of the plant matter produced on land. A doubling of human population *without any increase in the standard of living* would virtually turn the entire planet into a farm to support humans, an outcome that would cause widespread extinction of plants and animals and further loss of the ecological services essential to human survival. A doubling of human population would also generate huge amounts of pollution and waste. (For a debate on the pros and cons of further population growth, see the Point/Counterpoint in this chapter.)

Too Many People, Reproducing Too Quickly?

The impacts of the human population can be felt everywhere—in both cities and rural lands. In fact, every environmental problem discussed in this book is rooted in population, although not everyone agrees with this assertion. See what you think after you study the chapter and look at both sides of the issue.

Scientists who study populations use the term **overpopulation** to refer to a condition in which a population of organisms—any organism—exceeds the carrying capacity of its environment. In short, the population exceeds the capacity of the environment to supply resources to all and to assimilate waste. Based on this definition, they say that humans suffer from overpopulation. Is this true? Let's take a look at the issue.

Overpopulation occurs when any organism exceeds the carrying capacity of the environment. Overpopulation is at the root of all environmental problems.

Overpopulation: Urban and Rural Despair In 1819, the British poet Shelley wrote, "Hell is a city much like London." If he were alive today, the poet might have put it differently: "Hell is a city much like Cairo, Calcutta, Shanghai, Bangkok, London, Los Angeles, Mexico City, and dozens of others." In both the rich and poor nations of the world, many people live in deplorable conditions without adequate water and sanitation. Social problems run rampant. Crowding in urban centers has been implicated in a variety of social, mental, and physical diseases. Many social psychologists assert that divorce, mental illness, drug and alcohol abuse, and social unrest result, in part, from stress caused by crowding. Prenatal death and rising crime rates may also be attributed partly to overcrowding. They label this malaise the **inner city syndrome.** Research on animals supports the contention that crowding is unhealthy.

In addition to the social problems, urban centers frequently exceed the capacity of the environment to assimilate wastes. Cities throughout the world are centers of intense pollution from automobiles, factories, power plants, and sewage treatment plants. Even pollution from homes contributes to urban air pollution. Pollution affects people's health and has significant effects on crops and the natural environment. In China, for example, studies suggest that air pollution from urban and rural sources reduces crop production by 5 to 30%, depending on the level of pollution. By cleaning the air, the nation might eliminate its need for imported grain.

Large urban populations also place considerable demands on the outlying countryside for resources such as fuel, water, and food. Also, as cities spread, they expand onto farmland and open space, literally devouring the resources they need to survive and prosper.

A surplus of people strains many rural areas, too. Rural Bangladesh, Kenya, Ethiopia, Mexico, and other less developed nations, for example, reel under the burden of massive populations. Overpopulation in these and other countries results in unsanitary living conditions, water shortages, food shortages, and disease. In Africa, trees around rural villages are cut down for firewood, creating a barren landscape that extends many miles from the villages. Grasses are overgrazed by livestock, and once-fertile land is turned into desert (**FIGURE 8-3**).

The problems caused by overpopulation are especially evident in Bangladesh, which lies just east of India along the Bay of Bengal. No larger than Wisconsin, Bangladesh houses 25 times as many people, about 157 million in 2013, according to the Population Reference Bureau, an international nonprofit organization that tracks populations of all countries. Seventy-five percent of the population lives outside of cities.

Rural and poor, Bangladesh is the most densely populated nation in the world. Over the years, cattle have severely overgrazed the land. Desperate peasants have stripped away the vegetation in search of food, fuel, and shelter. When the rains come, floods follow because denuded hillsides cannot hold the rain back. Soil from these badly abused lands washes into streams. Sediment-choked streams flow to the sea and deposit their silt in river deltas. In search of farmland and a place to live, people flock to the deltas by the tens of thousands. As almost any ecologist will tell you, the deltas belong to rivers and the sea, not to people. When the heavy rains and hurricanes come, as they do most years, high waters flood the deltas, driving people away and destroying their farms and homes. Thousands of lives are lost each year from storms. Catastrophic cyclones (hurricanes) in 1970, 1985, 1988, 1991, and 2007 destroyed the homes of millions of people and killed hundreds of thousands of people. In 1991, a powerful storm killed an estimated 250,000 people.

On the surface, the disasters in Bangladesh look like severe natural disasters. On closer examination, however, it is clear that much of the blame can be pinned on an underlying human problem—overpopulation, which forces people onto marginal lands unsuitable for human habitation. Changing climate, which results in more powerful storms, is also a likely contributor.

Rural overpopulation feeds the urban crisis. Throughout the world, dismayed peasants and their families who are unable to survive on their farms migrate in large numbers to the cities in search of jobs and security. In Mexico City, thousands of rural farmers and their families arrive each year. Unfortunately, what awaits them and their counterparts in other countries is intense crowding, skyrocketing unemployment, poverty, crime, inadequate food, and pollution.

KEY CONCEPTS

The enormous size of the human population causes massive environmental problems in urban and rural areas. These include shortages of resources, environmental pollution, and a host of social problems. Crowding may also result in massive loss of life, for example, from severe weather, such as hurricanes.

FIGURE 8-3 Rural despair. Overpopulation, desertification, and drought have diminished the resources available to these people in northern Africa. Many rural poor living at or beyond the carrying capacity of their environment face similar problems.

© Patryk Kosmider/ShutterStock, Inc.

To get at the heart of the question "Is more better?" study the daily flow of water over Niagara Falls. You will find that twice as much water flows over the falls during each daylight hour as during each nighttime hour. There in a nutshell you have the population problem.

Puzzled? You should be. The connection between Niagara Falls and population is not obvious. Before we can understand it we need to review a little biology.

For every nonhuman species there is an upper limit to the size of a population. Near the maximum, individuals are not so well off as they are at lower densities. Starvation appears. Crowded animals often fight among themselves and kill their offspring. Wildlife managers and advocates agree that the maximum is not the optimum.

What about humans? Will we be happiest if our population is the absolute maximum the Earth can support? Few people say so explicitly, but some argue that "more is better!"

Admittedly, we need quite a few people to maintain our complex civilization. A population the size of Monaco's, with about 30,000 people, could never have enough workers for an automobile assembly line. But Sweden, with nearly 9 million people, turns out two excellent automobiles. Nine million is a long way from 6.2 billion, the population of the world today.

Some say, "More people—more geniuses." But is the number of practicing geniuses directly proportional to population size? England today is 12 times as populous as it was in Shakespeare's day, but does it now boast 12 Shakespeares? For that matter does it have even one?

Consider Athens in classical times. A city of only 40,000 free inhabitants produced what many regard as the most brilliant roster of intellectuals ever: Solon, Socrates, Plato, Aristotle, Euripides, Sophocles—the list goes on and on. What city of 40,000 in our time produces even a tenth as much brilliance?

Of course, the free populace of Athens was served by 10 times as many slaves and other nonfree classes. This left the 40,000 free people to apply themselves to intellectual and artistic matters. Modern peoples are given creative freedom by labor-saving machines, certainly a more desirable form of slavery. But where are our geniuses?

Business economists are keenly aware of "economies of scale," which reduce costs per unit as the number of units manufactured goes up. Communication and transportation, however, suffer from diseconomies of scale. The larger the city, the higher the monthly phone bill. Crimes per capita increase with city size. So do the costs of crime control. All these suggest that more may not be better.

Is More Always Better?

Garrett Hardin
The author was a renowned environmentalist, writer, and lecturer. He taught at several universities and served as chairman of the board and CEO of the Environmental Fund. He wrote numerous books and articles on environmental ethics and is best known for his article "The Tragedy of the Commons."

Courtesy of Garrett Hardin.

Democracy requires effective communication between citizens and legislators. In 1790, each U.S. senator represented 120,000 people; in 2004, the figure was 4 million. At which time was representation closer to the ideal of democracy? To communicate with his or her constituents each senator now has an average of 60 paid assistants. President Franklin D. Roosevelt had a staff of 37 in 1933, when the population was 125 million. Ronald Reagan had a staff of 1,700 in 1981, when the population was 230 million. We have to ask whether democracy can survive unrestrained population growth.

Let's look at another aspect of the more-is-better argument. An animal population is limited by the resources available to it. With humans, a complication arises. Though the quantities of minerals on Earth are fixed, improved technology periodically increases the quantity of resources available to us. In the beginning copper ores we mined contained 20% copper; now we are using ores with less than 1%. Available copper has increased but not the total amount of copper on Earth.

Let's return to Niagara Falls. Less water flows over the falls at night because more water is diverted to generate electricity when people aren't looking at the falls. It would be possible to use all the water to generate electricity, but then there would be no falls for us to look at. As a compromise, the volume of water "wasted" falling over the falls is reduced only at night. Therefore, the turbines and generators are not fully used 24 hours of the day, which means that local electricity costs are just a bit higher.

If the population continues to grow, the day may come when electricity is so scarce and expensive that the public will demand that Niagara Falls be shut down so that all the water can be used to generate electricity. Similar dangers face every aesthetic resource. Wild rivers can be dammed to produce more electricity, and estuaries can be filled in to make more building sites for homes and factories. The maximum is never the optimum. With human populations, quantity (of people) and quality (of life) are trade-offs. Which should we choose—the maximum or the optimum?

The Case for More People

Julian L. Simon

The author was professor of business administration at the University of Mayland at College Park. He wrote several important books on population and economics, including *The Ultimate Resource* and *Population Matters.*

Courtesy of Julian L. Simon.

Classical economic theory holds that population growth reduces the standard of living: the more people, the lower the per capita income, all else being equal. However, many statistical studies conclude that population growth does not have a negative effect on economic growth. The most plausible explanation is the positive effect additional people have on productivity by creating and applying new knowledge.

Because technological improvements come from people, it seems reasonable to assume that the amount of improvement depends in large measure on the number of people available. Data for developed countries show clearly that the larger the population, the greater the number of scientists and the larger the amount of scientific knowledge produced.

There is other evidence of the relationship between population increase and long-term economic growth: An industry, or the economy of an entire country, can grow because population is growing, because per capita income is growing, or both. Some industries in some countries grow faster than the same industries in other countries or than other industries in the same country. Comparisons show that in the faster growing industries the rate of increase of technological practice is higher. This suggests that faster population growth, which causes faster growing industries, leads to faster growth of productivity.

The phenomenon economists call "economy of scale"— greater efficiency of larger-scale production where the market is larger—is inextricably intertwined with the creation of knowledge and technological change, along with the ability to use larger and more efficient machinery and greater division of labor. A large population implies a bigger market. A bigger market is likely to bring bigger manufacturing plants, which may be more efficient than smaller ones and may produce less expensive goods.

A bigger population also makes profitable many major social investments that would not otherwise be profitable—railroads, irrigation systems, and ports. For instance, if an Australian farmer were to clear a piece of land for farming a distance away from neighboring farms, he might have no way to ship his produce to market. He might also have trouble finding workers and supplies. When more farms are established nearby, however, roads will be built that link him with markets in which to buy and sell.

We often hear that if additional people have a positive effect on per capita income and output, it is offset by negative impacts such as pollution, resource shortages, and other problems. These trends are myths. The only meaningful measure of scarcity is the economic cost of goods. In almost every case the cost of natural resources has declined throughout human history relative to our income.

Conventional wisdom has it that resources are finite. But there is no support for this view. There is little doubt in my mind that we will continue to find new ore deposits, invent better production methods, and discover new substitutes, bounded only by our imagination and the exercise of educated skills. The only constraint on our capacity to enjoy unlimited raw materials at acceptable prices is knowledge. People generate that knowledge. The more people there are, the better off the world will be.

Critical Thinking Questions

1. Outline the main points and supporting evidence given by both authors. Which views do you agree with? Why? Are your reasons based on feelings or facts?
2. Is population growth an issue of concern? To answer this question, make two lists. On one, list all of the things that you would expect to be better if the population doubled; on the other, list all of the things that would be worse.
3. On which list would you put the availability of wilderness? What about the noise level? Amount of democracy? Amount of pollution? Availability of parking spaces? Personal freedom?
4. When you are through, compare your list with those of your friends. What value judgments account for the differences? Can these differences be reconciled? How?

Reproducing Too Quickly

Hunger, starvation, disease, poverty, illiteracy, pollution, unemployment, and barren landscapes are, to many observers, signs that the human population already exceeds the Earth's carrying capacity. Although efforts are being mounted to solve these problems, continued growth makes it difficult to keep even with them, let alone improve conditions. Nowhere is this problem more evident than in Africa.

The population of Africa is currently growing at a rate of 2.6% per year—or doubling every 27 years. In order to keep up with this growth rate, the amount of food must be doubled every 27 years—so must the number of doctors, nurses, and teachers, as well as water supplies and transportation facilities. To improve the already deplorable conditions, the growth in facilities and services would have to increase faster than the population.

Rapid population growth makes social, economic, and environmental problems more difficult to solve and increases the impact we have on the planet. Rapid population growth, for example, causes hunger, starvation, poverty, and illiteracy to spread. It causes local, regional, and global pollution and species extinction to increase and it causes further shortages in resources.

Although global population growth has slowed and certain countries have shown dramatic declines in growth—some are even shrinking—growth continues at a rapid pace. As noted in the introduction, the human population is expanding by about 82 million people per year. While the decline in growth may be comforting to some critics, it is still a potentially dangerous force.

Population growth is occurring at different rates in different parts of the world. The world's fastest-growing areas are, in decreasing order, Africa, Latin America, and Asia. These regions account for nearly 90% of the annual growth in the human population (**FIGURE 8-4**). This leads many observers to focus their attention on measures to control growth in these regions. However, growth in the more developed countries, while less robust, has major impacts on the condition of the planet.

Robert McNamara, president of the World Bank for 14 years, maintains that "short of nuclear war itself, population growth is the gravest issue the world faces. . . . If we do not act, the problem will be solved by famine, riot, insurrection, and war." Solving the population, pollution, and resource "trilemma" say many experts will require a substantial and immediate humane decrease in the population growth. To see what one country is doing, see **Spotlight on Sustainable Development 8-1** in this chapter.

KEY CONCEPTS

All of the social, economic, and environmental problems of cities and rural areas are aggravated by rapid population growth. The large size of many populations makes it difficult for governments to meet current demands and continued rapid growth makes it nearly impossible to improve conditions and create a sustainable human presence.

8.2 Understanding Populations and Population Growth

Newspapers and magazines sometimes present a confusing array of statistics on population, making it difficult for the public to understand this issue. With a little effort you can learn enough demography, or population science, to weed out the fallacies from the facts of the population debate.

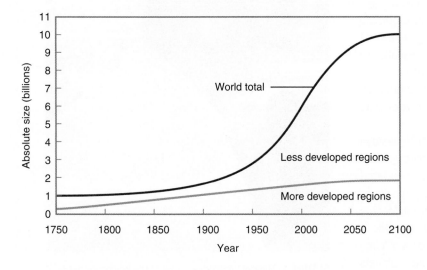

FIGURE 8-4 Global population growth. Note that the vast majority of growth is occurring in the less developed nations, a trend which began after World War II in large part because of the introduction of modern medicines and other measures that lowered death rates. Birth rates continued at high levels, resulting in massive growth. Measures are now being taken to reduce birth rates and slow the rate of growth.

8-1 Thailand's Family Planning Success Story

Paul and Anne Ehrlich noted in their book *The Population Explosion,* "The population/resource/environment predicament was created by human actions, and it can be solved by human actions. All that is required is the political and social will." Thailand is a stellar example of what can happen when people decide to take action.

In 1971, Thailand adopted a national population policy. By 1986, the country's population growth rate plummeted from 3.2 to 1.6%. During that period, the use of contraceptives by married couples rose from 15 to 70%. Thailand's success has continued for another decade. In 2010, the population growth rate had dropped to 0.6%. In 2013, it was 0.4%.

Thailand's success stems from many factors. The first is religious in nature. Ninety-five percent of the Thai people are Buddhists, and Buddhist scripture warns that "many children make you poor." In addition, Buddhism embraces family planning. In an effort to support population control, Buddhist monks distribute contraceptives with cards that remind the Thai people that "many births cause suffering."

Another factor is cultural openness. The Thai people are open to new ideas, and relationships between men and women are egalitarian. Consequently, women have equal say in decisions about family matters.

A third reason for the success in family planning is political. The government of Thailand encourages nationwide family planning and offers considerable financial support. Over the years, Thailand's government has made a wide range of contraceptives available to the public and has worked cooperatively with an influential nonprofit agency, the Population and Community Development Association of Thailand or PDA, to elevate family planning to a national goal.

PDA was founded in 1974 by Mechai Viravaidya (**FIGURE 1**). Mechai's creative and high-profile ways of promoting family planning in Thailand are considered by some as the single most important factor in the country's remarkable success in population control. Mechai promotes condom use by handing them out in public at any opportunity he can find. He even sponsors a "cops and rubbers" program, during which police officers hand out condoms on New Year's Eve. Today, condoms are affectionately known in Thailand as "mechais."

Because of Mechai's efforts, condoms, birth control pills, IUDs, and spermicidal foams are all available in bus terminals. In addition, his organization has opened vasectomy clinics across the country and celebrates the King's birthday each year by offering free vasectomies. Sterilization is the most widely used form of contraception in Thailand.

Mechai has also launched an economic development program. As a former government economist, he realized that income-generating alternatives could help change his people's attitude toward large families. With German financing, the PDA introduced a revolving loan scheme to support efforts to develop clean drinking water supplies. However, loans were only available to those individuals who participated in family planning programs. Money provided by the fund helped Thai people install thousands of toilets and containers to catch rainwater for drinking. The PDA also initiated a loan program to support farmers who are participating in family planning programs. Loans were offered at rates far better than those traditionally available to farmers.

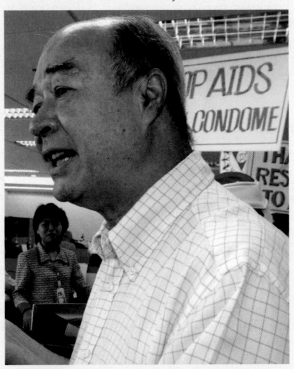

FIGURE 1 Mechai Viravaidya is one of the key proponents of family planning in Thailand. His dedication has helped make Thailand a family planning success story.
© Barbara Walton/EPA/Landov.

The government of Thailand also sponsors a loan program to individuals in various communities. Loan funds, in fact, are apportioned on the basis of a village's use of contraceptives. The total amount of the loan fund increases as the level of contraceptive use goes up.

Today, widespread commitment of the Thai people to family planning is believed by many to be associated with both the PDA's and the government's economic development programs. The PDA's programs have been so successful in Thailand that the nation now offers a 3-week training course to representatives from other less developed countries. Two of the many participants are Bangladesh, where family planning is considered inappropriate, and the Philippines, which is predominantly Catholic. The training course is designed to allow participants to develop ways to adapt the PDA's strategies for use in their countries.

Thailand's success in family planning has been paralleled by economic success. Economic progress has allowed the government to expand its family planning services and medical facilities throughout the country. Economic improvements, in turn, have meant that families live better and have a wider variety of choices.

Efforts to put the brakes on population growth have been remarkable, proving that when people recognize the problems of overpopulation and decide to take action, they can make a big difference.

Measuring Population Growth— Growth Rates and Death Rates

One of the most important population measurements is the growth rate. What does it mean when demographers report that a population is growing at a rate of 3% per year? Is this something to be concerned about? To answer this question, let's first see how a population's growth rate is determined.

The growth rate of the world population, also known as the **natural increase,** is calculated by subtracting the number of deaths in a population from the number of births in any given year, as shown below:

Growth Rate (%) = Crude Birth Rate - Crude Death Rate × 100

Crude birth rate is the number of births per 1,000 people in a population. Crude death rate is the number of deaths per 1,000 people. Worldwide, the crude birth rate is currently 20/1,000, and the crude death rate is 8/1,000. The difference between the birth and death rates is the population growth rate, in this case 12/1,000. This means that every year 12 people join the world population for every 1,000 people in the population.

To convert this to a percentage, simply multiply by 100, as shown below:

Growth Rate = 12/1,000 × 100 = 1.2%

The relationship between birth rates and death rates determines whether the world's population is growing, shrinking, or staying the same. Some important growth rates are listed in **TABLE 8-2**. This table also shows the doubling times, which are discussed shortly.

Growth rates are influenced by a variety of factors affecting birth and death rates. The birth rate in a given population, for instance, depends on (1) the age at which men and women get married, (2) their educational level, (3) whether a woman works after marriage, (4) whether a couple uses reliable contraceptives, (5) the number of children a woman and her husband want, (6) cultural values, and (7) religious beliefs.

These factors can also be viewed as leverage points for controlling human numbers. For example, efforts to reduce population growth often rely on improving educational and employment opportunities for men and women, as well as making contraceptives available.

Death rates are also influenced by a variety of factors and are equally important in determining population growth. As you learned earlier, food supplies, proper sanitation, water purification, and modern medicine have a profound effect on death rates.

TABLE 8-2

Growth Rate and Doubling Time (2013)

Region	Growth Rate(%)	Doubling Time (years)[1]
World	1.2	58
More Developed Countries	0.1	700
Less Developed Countries	1.4	50
Less Developed Countries (excluding China)	1.7	41
Africa	2.6	27
Asia	1.1	64
North America	0.6	117
Latin America/Caribbean	1.3	54
Europe	0.0	—
Oceania[2]	1.1	64

[1]Discrepancies in doubling time are the result of rounding off growth rates.
[2]Australian, New Zealand, Fiji, Guam, etc.
Data from *2013 World Population Data Sheet*, Washington, DC: Population Reference Bureau, 2013.

Doubling Time

Another important measurement of population dynamics is the **doubling time,** the time it takes for a population to double. The following formula is used to determine doubling time:

Doubling Time = 70/Growth Rate (%)

In this equation, 70 is a demographic constant. Using this equation, you can quickly convert growth rates into doubling times. A 2% growth rate would yield a doubling time of 35 years.

Growth Rates and Doubling Times The world is crudely divided into three categories: (1) the more developed countries, (2) the less developed (or developing) countries, and (3) the least developed countries (**TABLE 8-3**). The **more developed countries,** such as the United States, Canada, Australia, and Japan are industrial nations with a strong economic base. Many have strong agricultural economies as well, which are highly mechanized. The population growth rates of the more developed countries have decreased dramatically over the years so that most developed nations are growing at fairly slow rates—on the average around 0.1% per year in 2013. The doubling time of the more developed

TABLE 8-3

Comparison of More Developed and Less Developed Countries (2013)

Feature	More Developed	Less Developed	Least Developed
Standard of living	High	Low	Extremely low
Per capita food intake	High (3,100–3,500 cal/day)	Low (1,500–2,700 cal/day)	Low (2,000 cal/day)
Crude birth rate	Low (11/1,000 population)	High (22/1,000 population)	Very high (34/1,000) population
Crude death rate	Low (10/1,000 population)	Low (7/1,000 population)	Low (10/1,000 population)
Growth rate	Low (less than 0.1%)	High (1.4%)	High (2.5%)
Doubling time	High (809 years)	Low (40 years)	Very low (28 years)
Infant mortality	Low (5/1,000 births)	High (44/1,000 births)	Extremely high (66/1,000 births)
Total fertility rate	At or below replacement level (1.6)	High (2.6)	Extremely high (4.4)
Life expectancy at birth	High (78 years)	Lower (69 years)	Lowest (61 years)
Urban population	High (76%)	Low (47%)	Lowest (28%)
Wealth (per capita GNP) (2012 U.S. dollars)	High ($35,800)	Low ($6,600)	Extremely low (1,430)
Industrialization	High	Low	Lowest
Percent of married women (15–49) using contraception	High (71%)	Moderate (60%)	Low (33%)

Selected data from *World Population Data Sheet.* Washington, DC: Population Reference Bureau, 2013.

nations is around 700 years. Averages can be deceiving, however. Some more developed countries such as Canada, the United States, and Australia have relatively fast growth rates of 0.4, 0.5, and 1.1%, respectively per year, respectively. Note: These growth rate figures do not take into account legal and illegal immigration. Others, such as Denmark and Belgium, have extremely slow growth rates, 0.1 and 0.2%, respectively, per year. Some, such as Austria, have stopped growing altogether. A number of others, such as Greece, Italy, Germany, and Russia are shrinking, or, in the language of demographers, are experiencing "negative growth."

In contrast, the **less developed nations** are growing much more rapidly—at an average of 1.4% per year in 2013 (doubling time = 50 years). Many Asian and South American countries fit into this category.

The third category, a new one called the **least developed countries**, are the poorest nations with growth rates averaging 2.5% per year. They also experience high death rates, especially among infants. Nigeria, for instance, is a country of more than 176 million people. Its annual growth rate is 2.8%, and its infant mortality rate is 97/1,000. The U.S. infant mortality rate 6/1,000 as a reference.

KEY CONCEPTS

Growth rates in the more developed countries are relatively low. In the less developed nations, growth rates are generally much higher, and are higher still in the least developed countries.

The Total Fertility Rate and Replacement-Level Fertility

The growth rate and doubling time of a population are vital to understanding its future. Another important measure of human population, which helps us understand trends in growth, is the total fertility rate. The **total fertility rate** (TFR) is the number of children women in a population are expected to have in their lifetimes based on current trends in childbearing. In the United States, for example, the total fertility rate was 1.9 in 2013; this means that, on average, each woman in the reproductive age group of 15 to 45 is expected to have, on average, 1.9 children over her lifetime. Canada's TFR is 1.6. In India, a much poorer nation, women have a TFR of 2.4.

Women in many countries have achieved **replacement-level fertility,** a total fertility rate at which couples produce exactly the number of children needed to replace themselves. In the United States and other more developed countries, replacement-level fertility is a TFR of 2.1 children. That means that 10 women must have 21 children to replace themselves and their partners. The reason for the additional child is that in these countries 1 of every 21 children dies before reaching reproductive age. Replacement-level fertility is higher in less developed nations (around 2.3) because death rates for children are higher.

Today, 104 more developed and less developed nations (out of 200) have reached or fallen below replacement-level fertility. Most of these nations are in Europe. However, even though a nation's population has reached replacement-level fertility, that does not mean that it has stopped growing. If a country has a large number of women in their reproductive years, their population can continue to grow. It is only when the death rate equals the birth rate and when the net migration is zero—that is, when the number of people entering the country equals the number leaving it—that population growth stops. Demographers refer to this state as **zero population growth,** or ZPG.

In the United States, the TFR fell below replacement level in 1972, thanks in large part to modern contraception, rising affluence, and widespread public awareness of the population dilemma (**FIGURE 8-5**). Even though the TFR has remained at or below replacement-level fertility, demographers project that the U.S. population will continue to grow well into the future. By 2050, the U.S. population is expected to reach 400 million, up from 316 million in 2013, according to the Population Reference Bureau.

The U.S. population continues to grow for two reasons: First, each year numerous people move to the United States from other countries. Some of these people come legally— about 1.03 million in 2013; others arrive illegally (**FIGURE 8-6**). In 2011 and 2012, the number of illegal immigrants in the United States totaled just under 12 million; net migration from Mexico stopped and maybe even reversed. As a result legal and illegal immigration currently add about 1.0 million people each year to the U.S. population and is responsible for slightly more than half of the annual growth.

The rest of the growth rate comes from within. To understand why, we begin with Figure 8-5, which shows

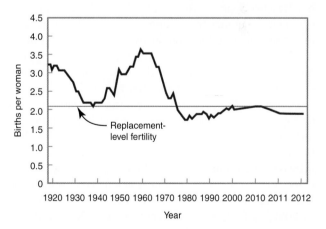

FIGURE 8-5 Total fertility rate in the United States. Total fertility rate (TFR) is the number of children women are expected to have, based on the current age-specific fertility rates (fertility rates in each age group). The TFR in the United States has fluctuated widely with changing economic conditions. In the decade of the Great Depression, women had a low TFR. In the several decades following World War II, the TFR shot up; in 1972 it fell below the replacement level. It has remained at or slightly below 2.1 since then, and in 2013 it was 1.9.

Selected data from *World Population Data Sheet.* Washington, DC: Population Reference Bureau, 2013.

FIGURE 8-6 **The flow of legal immigrants (or legal permanent residents) to the United States from 1900 to 2006.**

Adapted from Office of Immigration Statistics. *Annual Flow Report, March 2007*. Washington, DC: U.S. Department of Homeland Security, 2007.

the total fertility rate for the United States since 1920. After World War II, the total fertility rate climbed dramatically and remained high for several decades. This increase resulted in a **postwar baby boom,** a rapid increase in childbirth that in turn resulted in a long-term increase in the U.S. population. The children of the baby-boom era form something of a tidal wave in the American population. Today, many of these individuals are having their own children. Even though they are having smaller families than their parents—on average, about two children each—the number of women giving birth is still on the rise. In other words, more women are having children. This delayed effect is called the **lag effect.**

Because of this lag effect and because of immigration (discussed shortly), populations can grow long after they have reached replacement-level fertility. Stabilization is possible *only* if a number of conditions are met: (1) the number of women entering the reproductive age group levels off or declines, (2) the total fertility rate remains below replacement level, and (3) immigration is held in check. Small changes in any of these factors can have drastic effects on a population. As testimony to the effect of relatively small changes, in the late 1980s, the U.S. Census Bureau projected that the U.S. population would stabilize at about 300 million by 2050. As a result of changes in immigration policy that increased the number of legal immigrants and because of higher than expected fertility rates among U.S. women, the U.S. population is now expected to reach nearly 400 million by 2050, up from 317 million at the end of 2013, with stabilization nowhere in sight.

Migration

The previous sections have dealt primarily with global population growth, which is determined by birth rates and death rates. To calculate the growth of specific regions—cities, towns, states, and countries—we must take into account the number of people moving out of and into the region to set up residence—that is, **migration.**

Migration has two forms—immigration and emigration. The term **immigration** refers to the movement of people into a country; **emigration** refers to movement out. **Net migration** is the difference between the two. Population growth in a country will stabilize if the growth rate and net migration are zero.

KEY CONCEPTS

The growth of a town, city, state, or region is determined by two factors: growth rate (natural increase) and migration—the movement of people into and out of the population.

The Pros and Cons of Immigration Immigration is one of the hottest topics today in the United States, in part because legal and illegal immigration into the United States accounts for over half of the annual growth. Proponents of growth controls argue that efforts to stabilize the U.S. population are doomed unless the country reduces all forms of immigration.

Public opinion polls in the United States in the late 1980s showed that a majority favors reducing immigration of all kinds. Accordingly, Congress passed a law in 1986 levying penalties on those who knowingly employ illegal immigrants and requiring job seekers to provide proof of citizenship or legal immigration status. Four years later, though, a new Congress switched its position and passed the 1990 Immigration Act, which increased the number of legal immigrants by 35%—increasing from 530,000 in 1991 to nearly 700,000 a year from 1992 to 1994. In 2012, the latest data for which statistics were available, about 1,030,000 legal immigrants entered the United

Go Green

Talk to friends and family members to help them understand the seriousness of exponential growth in the human population and resource consumption.

States. Most of them were from Asia (primarily China) and Mexico (Figure 8-6).

Opponents of liberal immigration policies argue that such policies have many negative effects. They say that it increases the tax burden on already struggling cities such as New York, Miami, and Los Angeles, where the bulk of the immigrants end up. Social programs in these and other cities are designed to support low-income families, and immigrants often require assistance. Opponents argue that immigration also worsens crowding and internal strife in racially torn neighborhoods. It increases resource demand and puts additional stress on school districts that are already coping with severe budget problems. It also increases competition for jobs.

Proponents of more liberal immigration policies argue that immigration offers many benefits to a country. It increases cultural diversity and provides workers for low-income jobs, thus keeping the cost of running businesses down, which means cheaper products for consumers. It also helps immigrants escape political persecution, squalid living conditions, and poverty and enables families to improve their economic standing and educational opportunities.

Internal migration, the movement of people within a country from one region to another, is also of concern to scientists and environmentalists. Such movements may have enormous impacts on local economies, cultures, and environments. In the United States, for example, in the 1970s through the 1990s, Americans migrated en masse from the northeastern, north central, and midwestern regions to the South (mostly Florida) and West (Arizona, Wyoming, Utah, Alaska, Idaho, Colorado, New Mexico, Texas, and California). This movement, referred to by many as the "sunning of America," was caused by (1) the expansion of industries in the West and Southwest; (2) a decline in industries of the northeastern and north central states, especially auto and steel manufacturing; (3) a desire for a warmer climate; (4) a desire for a lower cost of living; (5) a preference for abundant recreation; (6) an aspiration for a less hectic, less crowded lifestyle; and (7) the growth of retirement communities.

The massive migration to the South and West brought many important economic benefits to these regions. Builders, restaurant owners, bankers, and other members of the service sector of the economy all benefited economically from the growth. Ironically, the influx of people into the Sunbelt destroyed many of the values the migrants sought. With growth rates in the range of 3 to 5% per year, many western cities and states were swamped by new people. Air pollution in Sunbelt cities has often worsened with increasing population; the clean air that many came for has disappeared. Traffic has become congested. The rapid rise in demand for housing sent the cost of homes skyward. The West now has the most expensive new homes in the nation. Spreading cities engulfed smaller outlying communities and changed the tempo of life. The exodus of retirement-age individuals to warmer climates also put strains on existing healthcare facilities.

Internal migrants also affected the places they came from. Since many of the outmigrants were young, educated, and skilled workers, they created a significant drain on human resources in the areas they left.

KEY CONCEPTS

Internal migration, the movement of people from one region of a country to another, affects regional population growth and has positive and negative social, economic, and environmental impacts.

Population Histograms

Demographers use growth rates, fertility rates, doubling times, and net migration to explain the dynamics of the human population—where it is and where it is going. However, few techniques shed as much light on a subject as a nicely drawn graph.

One particularly helpful graphic tool is the population **histogram,** a bar graph that offers a profile of a population (**FIGURE 8-7**). The population histogram displays the age and sex composition of a population as shown in Figure 8-7. The area of each horizontal bar on the histogram represents the size of a certain age and gender group.

As illustrated in Figure 8-7, three general profiles exist: expansive, constrictive, and stationary. Kenya and Mexico are two countries that are expanding. Their population histograms resemble the top graph. Both countries have a large number of young people. If they produce more offspring than their parents did, the population will continue to expand at the base. If, on the other hand, family size decreases, the base of the histogram will begin to constrict. This is what is happening to the Austrian population; its population histogram is constrictive like the middle graph. Sweden presents an entirely different picture. For many years Swedish couples have been having the same number of children as their parents had; as a result, Sweden's population is nearly stationary, and their population histogram is similar to the lowest graph.

Population histograms can be used to predict population size and age structure—information that's useful for planning schools, hospitals, retirement homes, and so on. For example, government officials looking at an expansive population such as that in Kenya would be wise to consider special efforts to expand food production and educational facilities to accommodate new children. Officials in Austria, with its constrictive population profile, would be well advised to plan for an increasing number of retirees.

A population histogram of the world is expansive. Surprisingly, nearly 26% of the world's people today are under the age of 15 years. If we're going to create a sustainable future, many experts agree that something must be done to curb childbearing in this group.

(a)

(b)

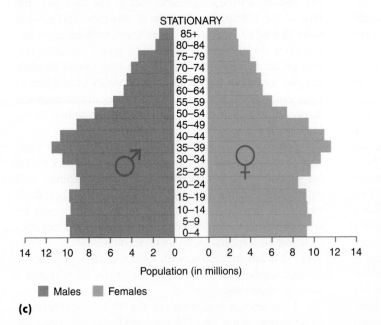

■ Males ■ Females

(c)

FIGURE 8-7 Three population profiles. (a) Expansive populations have a large percentage of young people and are expanding at the base. **(b)** Constrictive populations have a tapering base, resulting from lower total fertility. **(c)** Stationary or nearly stationary populations show no expansion or constriction. This results from couples having only the number of children that will replace them. The world population histogram is expansive.

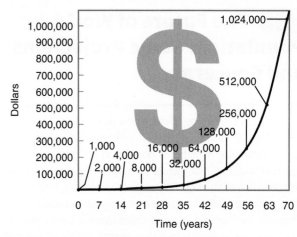

FIGURE 8-8 Exponential growth in savings. This graph plots the growth of a savings account with an initial deposit of $1,000 and an interest rate of 10% per year. Notice how slowly the bank account grows until it "rounds the bend" of the exponential growth curve.

Exponential Growth

So far you've learned that many populations are large and growing and that the size and growth rate have many implications to people, the economy, and the environment. You've learned how demographers measure growth and predict the future. To comprehend how serious population growth is, you must understand a phenomenon called *exponential growth*. **Exponential growth** occurs when something increases by a fixed percentage every year. For example, a savings account with 5% compounded interest grows exponentially. The remarkable characteristic of all exponential growth is that early growth in absolute numbers is quite slow. Once growth "rounds the bend," however, the item being measured—whether money in a bank account or number of people in a population—begins to increase more and more rapidly.

To appreciate the true nature of exponential growth, consider an example. Suppose that your parents had opened a $1,000 savings account in your name on the day you were born. Also suppose that this account earned 10% interest and that all of the earned interest was applied to the balance, where it also earned interest. At this rate of growth, your account will double every 7 years. Consequently, when you were 7 years old, your account would have been worth $2,000 (**FIGURE 8-8**). By the age of 14 years, it would have increased to $4,000. By the age of 42 years, you would have $64,000. If you left the money in a little longer, you'd find it growing faster and faster. At the age of 49 years the account would be worth $128,000; at the age of 56 years you'd have a quarter of a million dollars. In 7 more years, you'd have half a million dollars. But if you waited until you were 70 years old, your account would be worth over $1 million.

Looking back over your account records, you'd find that during the first 49 years it grew from $1,000 to $128,000, but that during the last 21 years it increased by nearly $900,000! The rate of growth was constant over the entire period, and the account doubled every 7 years; yet it was not until the balance in your account "rounded the bend" of the growth curve that the growth seemed to take off.

How did this rapid upturn occur, even though the rate of growth was the same? The reason for this is simple: Once the base amount becomes extremely large, even small percentage increases result in substantial absolute increases. For instance, the 1.2% increase in a world of 6.7 billion people results in a net increase of about nearly 80 million people a year! At what appears to be a seemingly slow rate of growth (1.2%), world population could double in 58 years.

The problem isn't just that the population is increasing exponentially. It's that population, resource use, pollution, and environmental damage are all increasing at exponential rates. If you accept the argument that the human population has exceeded the Earth's carrying capacity, further rapid growth only pushes us past the point of sustainability. Systems analysts refer to this phenomenon as an **overshoot.** As you will see later, an overshoot could result in a collapse of the system. Although population growth is slowing, many scientists expect social, economic, and environmental conditions to deteriorate as the population continues to expand.

8.3 The Future of World Population: Some Projections and Concerns

Perhaps the questions most often asked of demographers are "What is the future of the world population?" and "How big will it get?" As the previous discussion has shown, many factors determine whether a population will grow or shrink and how quickly. The factors that determine future population growth—such as birth rates, fertility, and death rates—can change dramatically. Several less developed countries such as Thailand, for example, have mounted massive family planning programs that caused population growth to fall dramatically. Consequently, demographers can only guess about the future of the human population. Population projections, they like to remind us, are mathematical calculations based on assumptions about current levels and future trends. Although demographers may not know for sure how large the population will be in the year 2050, they do know that it will very likely be a lot larger than it is today. **FIGURE 8-9** presents several possible scenarios for the world and U.S. populations. Let's start with global population. As illustrated in Figure 8-9a, estimates for global population size vary from a decline to 6 billion by 2100 to a high of about 16 billion. The middle estimate places world population at 10 billion.

The U.S. population is currently 317 million, but could increase substantially by 2050. How much it will grow largely depends on legal and illegal immigration (Figure 8-9b). According to the U.S. Census Bureau, the U.S. population could grow to 458 million with high levels of immigration (nearly 2.5 times what they are today). If migration remains the same, the U.S. population, they project, would reach 399 million by 2050.

Projected increases in the U.S. population have an impact on wildlife, open space, air quality, waste production, traffic, energy demand, and water pollution.

(a)

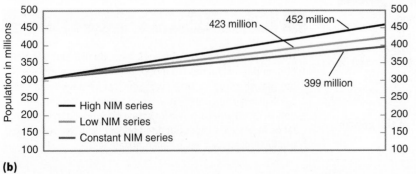

(b)

FIGURE 8-9 Projected population growth.
(a) Future world population, based on a variety of assumptions. **(b)** The future U.S. population depends on future total fertility rate and annual net migration.

Part a data from United Nations, 2010; part b data from U.S. Census Bureau.

We will also need to find new ways of meeting human needs—ways that are far less damaging to the environment. The mandate for redesigning basic human systems, seems all the more pressing given this possibility. Changes in systems alone, though, may not be enough to build a sustainable society. Most experts agree that population growth must be slowed, if not stopped. Some believe that the human population must actually decrease over the long run. Without a halt in growth, gains made in efficiency and other areas such as recycling and restoration will very likely be offset by rising population size.

KEY CONCEPTS

Predicting the future size of the United States' and the world's population is fraught with difficulty. It is likely, though, that the population of the United States and the world will increase dramatically before it stabilizes. This increase could bring about massive changes in the environment.

Population Growth in the Less Developed World: Why Should We Worry?

Many people in the more developed nations consider population growth in the less developed nations an issue of little importance to them. What difference does it make to the average North American how fast the less developed countries are growing?

A survey of the impacts of growth in the less developed countries clearly shows that residents of the industrial North do indeed have something to worry about. For example, rapid population growth in Latin America has resulted in a sharp increase in the size of its workforce, but many workers cannot find jobs at home and thus enter the United States illegally in search of work. Although this cultural infusion is seen as a positive impact by some, these individuals compete with U.S. workers for jobs. New residents also strain U.S. schools and hospitals, which are not well equipped to accommodate Spanish-speaking immigrants.

Illegal immigrants also strain the natural environment. An estimated 11.6 million illegal workers and their families were employed in 2012 in the United States, each needing water, food, shelter, and other goods and services. This places additional burdens on already stressed land, air, and water.

Growing population abroad increases the production of food for export. Although that may be good for farmers, it also places additional stresses on agricultural soils. It also means additional pesticide use, energy consumption, and demand for irrigation water—all of which impact our environment.

Expanding population abroad also results in the destruction of forests, fields, and wetlands, as well as the many species that live in them. The loss of species has numerous economic implications. Many new drugs, for instance, could come from plants in the rain forests that are now being cut down to make way for more people.

Population growth in the less developed nations also threatens global climate. The loss of rain forests, to accommodate growing populations in tropical nations, for example, is partly responsible for rising carbon dioxide levels that contribute to global warming. Deforestation of rain forests in central Africa may have a profound effect on rainfall patterns in Europe. And as these countries strive for economic prosperity, they will produce more and more pollution that contributes to global problems. China, home to nearly 1.4 billion people, is the second leading producer of carbon dioxide, a major contributor to global climate change. Air pollution from China also drifts across the Pacific Ocean affecting air quality in the United States.

Rapid population growth in the less developed nations contributes to the vicious cycle of poverty gripping many nations, making it difficult for them to pay back loans to more developed countries. Today, billions of dollars in international debt remain unpaid, in part because countries cannot cope with their population growth. When countries fail to pay back loans to U.S. banks, consumers suffer. Banking services invariably become more costly. Savings and loan failures in the United States in the early 1990s resulted in part from loans made to less developed nations that were unable to make payments.

Why worry? In our highly interconnected world, population growth in the less developed nations creates problems everywhere. There are few boundaries that this problem does not cross.

KEY CONCEPTS

Population growth in the less developed nations of the world has many social, economic, and environmental impacts—serious impacts that affect all the world's people.

A World of Possibilities

The human population will likely grow larger in the near term. Over the long haul, most experts believe that growth cannot continue. Population size will rise to a certain level and then stabilize or, very likely, fall. The curves in **FIGURE 8-10** show some of the potential patterns.

No doubt all three patterns will be seen in individual countries.

In Figure 8-10a, the population transitions smoothly and rather gradually to a stable size. Such curves are said to be sigmoidal or S-shaped. This is the most optimistic of all possibilities. World population may follow a domed pattern as shown in Figure 8-10b, climbing for a while and then falling to a more sustainable level. A third and less desirable pattern, known as an *irruptive pattern,* is shown in Figure 8-10c. In this scenario, population grows until it goes so far beyond the carrying capacity that it crashes. It could either fall precipitously or fall to a lower level and then rise and fall in a dangerous cycle.

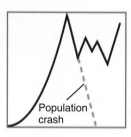

(a) S-shaped or sigmoidal **(b) Domed**   **(c) Irruptive**

FIGURE 8-10 Patterns of population growth. (a) Smooth transition to a stable population size. **(b)** Gradual drop-off caused by population exceeding the carrying capacity. **(c)** Population crash caused by irreparable change to the ecosystem.

No one knows what will happen to world population. The only opinion on which the experts approach agreement is that world population cannot grow indefinitely. Some countries will very likely make a smooth transition to a stable state; others may experience periodic crashes caused by epidemics in crowded urban populations. Still others may fall to low population levels because of starvation, disease, and widespread environment damage.

H. G. Wells once wrote that "human history becomes more and more a race between education and catastrophe." The current population/resource/pollution bind clearly illustrates this fact. What we decide today will have farreaching effects that determine the kind of lives our children and theirs will have. What we do and don't do will literally shape the future of the world and determine the future of many species that live on this planet with us.

KEY CONCEPTS

Most experts agree that the human population cannot grow indefinitely. Some countries will very likely make a smooth transition to a stable population size; others may experience periodic crashes that will eliminate large numbers of people. Still others may overshoot the carrying capacity and destroy their ability to support people so drastically that their populations may fall to much lower levels.

> You cannot escape the responsibility of tomorrow by evading it today.
>
> **—Abraham Lincoln**

CRITICAL THINKING

Exercise Analysis

Stabilizing population growth in the less developed nations is extremely important. Local conditions in many countries are deplorable and environmental damage is significant. However, critical thinking skills you've learned suggest the importance of examining the big picture—that is, looking at issues in their entirety. This is one of the facts you will find: Although population growth in less developed nations accounts for 90% of the annual increase, an increase of about 76 million people, residents of the industrial nations actually have a far greater impact on the environment. The United States, for instance, houses about 5% of the world's population but consumes 25% of the world's energy. All told, the more developed nations, which represent 30% of the world's people, consume 75% of the world's resources! Our numbers are few, but our impact is enormous.

In general, people living in the industrial nations of the world consume 20 to 40 times more resources per capita and produce 20 to 40 times as much pollution per capita as their counterparts in the less developed nations. Consequently, the 8.4 million new residents of the industrial world added to the world population have the environmental impact of 170 to 340 million people in the less developed nations. Because of this, many experts argue that measures to slow growth in the industrial nations are just as important, if not more important, than efforts to slow growth in the less developed nations. Thoughtful analysts realize the importance of growth stabilization everywhere—in the rich as well as the poor nations of the world.

One of the critical thinking rules you've learned is to question conclusions. In this case, and in others, it is necessary to scrutinize the conclusions of the speaker, which sound valid at first, but fall apart when analyzed more carefully. They may not be based on adequate information.

CRITICAL THINKING AND CONCEPT REVIEW

1. What is the population of the world? What is the population of the United States?
2. Using your critical thinking skills, analyze the following statement: "Population growth is not the cause of environmental problems; technology is."
3. Analyze the following statement using your critical thinking skills and knowledge you have gained in your readings and in lecture presentations: "Population growth in the United States is of little concern to the future of the planet. It is population growth in the developing world that is out of control and needs to be addressed."
4. You are asked to give a lecture on population growth in the United States and why it should be of concern. Write an outline with sufficient detail to make your case. Be sure to include projections of future growth and the impact they could have on wildlife, open space, pollution, and resource demand.
5. How many years did it take the world to reach a population of 1 billion people? How quickly did we reach the second, third, and fourth billion?
6. What factors kept world population in check for so many years? Discuss the advances that have unleashed population growth in the last 200 years.
7. Some individuals think that technology, especially energy-efficient machines and appliances, will be sufficient to solve many of the growing environmental problems. Substantial reductions in energy consumption, for instance, could dramatically cut air pollution. Analyze this point of view. Hint: You may want to do some research on the gains that can be made through energy conservation and compare them to rising demand.
8. Define the term *exponential growth rate*.
9. Using the knowledge you have gained in this course, discuss this statement: "The world population is growing at a relatively slow rate—only about 1.2% per year. Growth at this rate is of little concern. If I had a bank account growing at this rate, I'd be frightened at how slowly it was growing."
10. What is a population histogram? Describe the three general profiles. Why are histograms useful?
11. How is the world population growth rate calculated?
12. How is doubling time calculated? What is the difference in doubling times of populations growing at 0.4% per year and 0.2% per year?
13. Define *replacement-level fertility* and *zero population growth*.
14. Discuss the pros and cons of a lenient policy toward legal and illegal immigration in the United States. Do you favor strong immigration quotas?

KEY TERMS

carrying capacity
doubling time
emigration
exponential growth
histogram
immigration
inner city syndrome

internal migration
lag effect
less developed nations
migration
more developed countries
natural increase
net migration

overpopulation
overshoot
postwar baby boom
replacement-level fertility
total fertility rate
zero population growth

© TonyV3112/Shutterstock, Inc.

Stabilizing the Human Population: Strategies for Sustainability

To rebuild our civilization we must first rebuild ourselves according to the pattern laid down by life.

—Alex Carrel

LEARNING OBJECTIVES

After reading and studying the material in this chapter, students will be able to understand, knowledgably discuss, and, when appropriate, apply the information learned in this chapter, including:

1. Why many observers believe it is vital to creating a sustainable society and a sustainable relationship to the environment to stop the growth of human population and humanely reduce population size
2. How economic development can lead to population stabilization
3. How family planning can assist in reducing population growth
4. How small-scale, sustainable economic development may be the most sustainable form of economic development in less developed nations
5. Why per capita consumption and resource use per capita affect the sustainability populations in more developed countries
6. Ways more developed nations can help create sustainable populations in less developed and more developed nations
7. Barriers to controlling population growth, especially in less developed nations
8. How many countries have begun to slow human population growth

Exercise

Many advocates of sustainable development put their faith in technological improvements: new energy-efficient homes, new high-mileage vehicles, pollution-free factories, water-efficient fixtures, energy-efficient appliances, and others. They contend that such innovations can allow both our economy and our population to grow and still permit us to reach a sustainable state. They often use the term *sustainable growth*, especially in an economic arena, to convey their belief in this idea. Their argument boils down to this: By conserving and managing resources properly, we can provide access to energy and food indefinitely. In short, all consumption patterns can be made sustainable simply through conservation.

Using your critical thinking skills and the knowledge you have gained so far, analyze this viewpoint. Is it valid? If not, what principles of sustainability does it violate? Provide some hard evidence to back up your arguments.

Hint: To explore this issue, you will very likely want to gather some additional information on the subject. Two sources that may be of assistance are "Using Less and Still Running Out" by Mark W. Nowak, published in the *National Association of Environmental Professionals News* (November/December 1995) and "Can Technology Spare the Earth?" by Jesse Ausubel, published in *American Scientist* (March/April 1996).

Throughout the world, many countries are taking steps to slow their population growth. For example, in Kenya, which has one of the fastest growing populations in the world, large families have been the norm for many years. Because economic conditions are worsening, however, many parents are worried about supporting large families and are taking action. Between 1984 and 2013, the average number of children a Kenyan woman bore dropped from 7.7 to 4.5, according to the Population Reference Bureau. Forty-six percent of all married women from age 15 to 49 are using some form of contraception.

Even more encouraging news comes from the northern African nation of Morocco. In 1980, women in this country had an average of seven children. In 1995, that number dropped to four. By 2012, it had dropped to 2.3 where it remains today. In Morocco the use of **contraceptive measures,** devices, or techniques that reduce the chance of fertilization, has increased to 63% of all married couples. This progress is even more encouraging when one considers Morocco's social system, in which women are relegated to an inferior role and gain status primarily by bearing children.

9.1 Achieving a Sustainable Human Population: The Challenges

Achieving a sustainable human population involves two basic challenges. The most immediate is finding acceptable means of slowing the growth of the human population, in order to create a stable population size. Many proponents of sustainability, however, argue that slowing growth and reaching a stable number may be insufficient in the long run. These individuals maintain that the human population already exceeds the Earth's carrying capacity. Further additions would result in even more severe environmental deterioration, especially if people sought higher standards of living. The second, long-term challenge, then, may be to find ways for the human population to decrease in size naturally.

Is the human population already unsustainable? So far, you have learned many facts that demonstrate that humans are living beyond the carrying capacity of the Earth. In texts on natural resources and pollution you may encounter numerous additonal facts that support this assertion.

Stopping population growth, reducing population size, and pursuing sustainable economic plans, all discussed in this chapter, can help break the vicious cycle of poverty and environmental destruction. However, the road ahead will be long and difficult. Even if we could miraculously reach replacement-level fertility today, the world population would continue to swell, adding 2 to 4 billion people before it stopped growing, in large part because of the lag effect. Human conditions would very likely deteriorate, as would the condition of the environment.

KEY CONCEPTS

Most people consider the main challenge of achieving a sustainable human population to be finding acceptable means of reducing the population growth rate, to allow the world population to stabilize. Many others, however, argue that in order to live sustainably on the Earth, we must eventually reduce human numbers through humane, socially acceptable means, notably family planning.

Stabilizing the Human Population: Some Strategies

Over the last few years, experts on population have come to realize that slowing growth and achieving a stable population will require more than education on birth control and easy access to contraceptives and family planning. Most experts agree that we must apply a number of additional remedies; and the closer they are to the roots of the problem, such as poverty, lack of education, the inequality of women, and poor health care, the better.

KEY CONCEPTS

Stabilizing the human population will require a number of measures besides access to contraception and family planning that attack the root causes of rampant population growth.

Economic Development and the Demographic Transition

For many years, economic growth was viewed as one of the most powerful forces for slowing or even stopping population growth. How does economic development slow growth? Economic growth creates jobs and increases personal wealth.

People can then afford decent housing, food, and education. Poverty and disease are often reduced or nearly eliminated. As incomes rise, fertility rates decline. This scenario has been repeated dozens of times over the past couple hundred years in what are now the more developed countries (MDCs) of the world. In all of these countries, population growth was brought under control as economic conditions improved. This phenomenon is called the **demographic transition.**

The demographic transition takes place in four stages (**FIGURE 9-1**). The present-day industrial nations have progressed through all four. In Stage 1, birth rates and death rates are high. In this phase, the population is stable because high birth rates and death rates cancel each other out. In Stage 2, improvements in health care and sanitation (brought about by technological development, including advances in medicine and improving economic conditions) cause death rates to begin to fall. However, birth rates in Stage 2 tend to remain high. The discrepancy between the birth rates and death rates results in a period of rapid population growth. In Stage 3, as the country continues to develop economically, birth rates begin to decrease, and population growth slows. Finally, over time, birth rates and death rates come into balance. Population growth is stopped in Stage 4.

The decrease in birth rates in Stage 3 can be attributed to several factors. Perhaps the most important is the shift in people's attitudes toward children. Preindustrial farmers view children as an asset because they help on the farm and often support their parents in old age. With industrialization

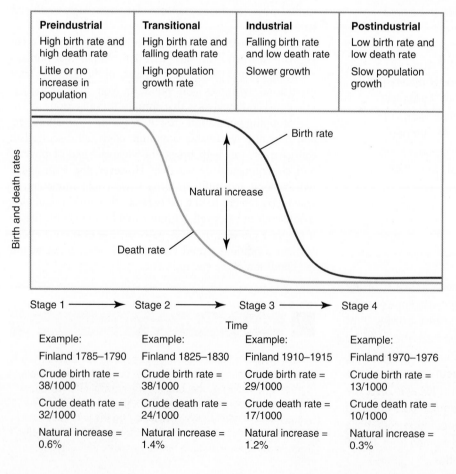

Preindustrial	Transitional	Industrial	Postindustrial
High birth rate and high death rate	High birth rate and falling death rate	Falling birth rate and low death rate	Low birth rate and low death rate
Little or no increase in population	High population growth rate	Slower growth	Slow population growth

Stage 1 ⟶ Stage 2 ⟶ Stage 3 ⟶ Stage 4

Time

Example:	Example:	Example:	Example:
Finland 1785–1790	Finland 1825–1830	Finland 1910–1915	Finland 1970–1976
Crude birth rate = 38/1000	Crude birth rate = 38/1000	Crude birth rate = 29/1000	Crude birth rate = 13/1000
Crude death rate = 32/1000	Crude death rate = 24/1000	Crude death rate = 17/1000	Crude death rate = 10/1000
Natural increase = 0.6%	Natural increase = 1.4%	Natural increase = 1.2%	Natural increase = 0.3%

FIGURE 9-1 Demographic transition. This graph of birth rates and death rates in Finland illustrates the demographic transition. As you can see, the transition results in a shift from high birth and death rates to low birth and death rates, a change that is brought about by industrialization and an overall increase in wealth of people in a given population. Industrial nations have all experienced the demographic transition.

and the inevitable migration of families to the city, however, children become an economic liability.

Because of competition for living space, each child means that more money has to be devoted to necessities. If the children do not work, they create an additional financial drain on the family. As a result, smaller families generally prevail.

If economic development has brought about a demographic transition in so many countries, helping them to stabilize or reach a state of extremely slow growth, it should also work for less developed nations in Asia, Africa, and Latin America. Right?

Unfortunately, the evidence suggests that it will not. At least four reasons for the failure of economic development to effect a timely demographic transition become apparent.

First, the economic resources of many of the less developed countries (LDCs) are too limited to build the type of industry needed for a demographic transition. In some instances, resources were heavily exploited during the era of colonialism, which contributed to the wealth and demographic transition of the more developed nations.

Second, the demographic transition in the more developed nations of the world did not take place overnight. For example, it took Finland over 200 years to approach a balance between birth rate and death rate (Figure 9-1). The less developed countries, with rapid doubling times, do not have the luxury of this kind of time.

Third, population growth in many LDCs outstrips economic growth. Recent studies show that a 1% growth in the labor force requires a 3% economic growth rate. Think of the economic growth needed to sustain populations growing at 2.5 to 3% per year. For many LDCs it has been hard enough to keep up with population growth; getting ahead has proven to be impossible.

The fourth reason is that the fossil fuel energy sources that were essential to the demographic transition in the more developed countries are diminishing and becoming ever more costly. Without the rich mines of England or the great oil deposits of Arabia and North America, the less developed nations will probably never experience similar economic growth.

Many of the world's less developed nations are in Stage 2 or early Stage 3 of the demographic transition, with high birth rates and low death rates. Because a demographic transition brought on by industrialization is unlikely, the transition to stable populations must come about in other ways, most likely through family planning and small-scale sustainable economic development.

KEY CONCEPTS

Economic development can be a powerful force for reducing population growth. Although economic development caused a shift in population growth in the more developed nations, this change took many decades and substantial resources, which the less developed nations of the world do not have. Because of this, the transion to a stable population must come about through family planning and sustainable economic development.

Family Planning and Population Stabilization

To help usher a transition to Stage 4, most population experts believe that less developed nations must institute aggressive family planning programs. **Family planning** allows couples to determine the number and spacing of offspring. For these countries, family planning programs can accelerate the decline in the birth rate that occurs slowly through economic growth and the demographic transition. Some researchers believe that family planning programs are responsible for a 40 to 50% decline in fertility rates in the less developed nations since the 1960s.

Family planning programs offer various means of birth control. **Birth control** includes any device, method, or chemical designed to reduce births in a population. Generally, birth control measures fall into three broad categories: (1) **abstinence,** refraining from intercourse; (2) **contraceptives,** chemicals, devices, or methods that prevent sperm and egg from uniting; and (3) the most controversial, **induced abortion,** the intentional interruption of pregnancy through surgical means or drug treatments.

Family planning programs vary considerably from one nation to another. Most of them offer education on birth control along with clinics that dispense contraceptives or perform abortions (**FIGURE 9-2**). Family planning programs may be either privately run or state sponsored. Planned Parenthood in the United States, for example, is a private, nonprofit organization with clinics in many cities. It offers low-cost medical care, contraceptives, and abortions.

State-sponsored family planning programs lie on a continuum, from the purely voluntary to the compulsory. **Voluntary programs** as a rule make birth control information and methods available to the public at low cost. There is no promotion on the part of the government. People choose the type of birth control and family size they want.

Family planning programs promoted by governments are called **extended voluntary programs.** In these programs, governmental agencies may hand out information on birth control and sterilization or sponsor posters, newspaper ads, television and radio announcements, and billboards (**FIGURE 9-3**). In Egypt, for instance, a song promoting birth control was played on a government-sponsored commercial. The song was so popular that it became a national hit.

In extended voluntary programs, payments or other incentives are often offered by governments to couples practicing birth control or undergoing sterilization. Transistor radios, for example, were once handed out in

Go Green

If you are contemplating marriage and children, consider delaying motherhood or fatherhood for a few years and consider having only one or two children to help reduce population growth. Be sure to raise your children to take care of the environment.

FIGURE 9-3 Family planning poster in China. This poster, extolling the virtues of the one-child family, is a familiar sight in China, a nation of nearly by 1.4 billion people. China's government has a strong interest in reducing the nation's birth rate and eventually reducing the overall population size in an effort to temper resource demand and reduce environmental destruction. Special healthcare workers travel to villages offering advice on birth control. Similar posters are found in India, Thailand, and other Asian countries.

© Adrian Bradshaw/EPA/Landov.

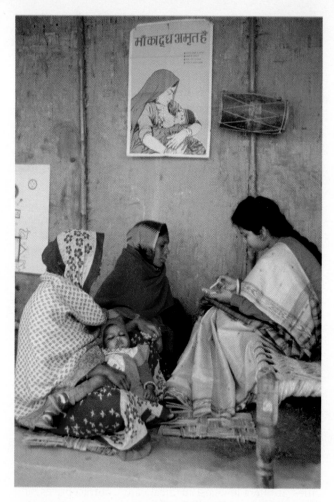

FIGURE 9-2 Family planning in India. A doctor lectures to a group of Indian women on various methods of birth control at a family planning clinic in New Delhi. These and other clinics throughout the world are vital to efforts to educate women on family planning.

© Tim Graham/Alamy Images.

rural Egypt to couples who adopted some form of birth control. Special loan programs were used in Thailand. Some government programs offer reminders. In Jakarta, Indonesia, for example, music extolling the virtues of small families plays at intersections. In Bali, church bells ring at 5 P.M., reminding women to take their birth control pills.

Governmental programs often attempt to change people's thinking about family size. Posters in Vietnam, for instance, extol the virtues of a one-child family. Murals and posters in India feature a smiling family of four with the slogan "Two or three children is enough." These efforts strike at one of the roots of the population problem: people's mindsets. Clearly, as Varinda Vittachi, a writer from Sri Lanka, notes, "The world's population problem will be solved in the mind and not in the uterus."

Informational campaigns may also be designed to change stereotypical sex roles. They may try to persuade men that masculinity and self-importance are not related to the number of children their partners have, or they may try to convince men and women alike that childbearing is not the only way for a woman to be a valuable community member.

Forced family planning programs involve governmental limitations on family size and punishment for those who exceed quotas. They are extremely rare. They may impose sterilization after a family reaches the allotted size, limit food rations for "excess" children, or tax couples who exceed the allowed number of children. China, which started a forced family planning program in the 1980s, enacted a one-child policy designed to reduce its population from over 1 billion at the time to 700 or 800 million because of concerns over water supplies, food, and energy. This effort obviously failed because China's population is now around 1.4 billion. Female workers met regularly in small groups at work to discuss birth control. Government workers reportedly urged men and women with one child to undergo sterilization. According to unsubstantiated reports, pregnant women who already had one child were forced to have abortions. The Chinese government also supplemented the monthly income of couples who pledged to have only one child. If a couple that pledged to have one child had a second one, however, the government required repayment of the monthly bonus and cut the monthly salary by 15%. It even imposed hefty fines.

In recent years, China's one-child family planning policy has been relaxed. Concerned about potential problems that could be caused by China's aging population, the coastal city of Shanghai enacted a two-child family policy. This measure was taken to encourage families to have more children to ensure an adequate workforce as current workers retired. For a discussion of another aspect of population growth control, see **Spotlight on Sustainable Development 9-1**.

Small-Scale, Sustainable Economic Development, Jobs for Women, Better Health Care, and Improvements in the Status of Women

Although less developed nations may lack time, funds, and resources needed to promote economic development to reduce fertility, many now understand that small-scale, community-level economic development can raise individual wealth. Increasing economic well-being often reduces population growth and can protect the environment if carried out in a sustainable manner.

Recognizing that family planning works only if people want fewer children, the **United Nations Fund for Population Activities (UNFPA),** a leader in promoting family planning throughout the world, invests money in economic programs, especially those that provide employment for women. In Egypt, for instance, the UNFPA has made substantial investments in clothing factories that employ Egyptian women. The logic behind these and other activities is that working women often delay marriage and childbearing—and thus have fewer children.

In 1994, the United Nations sponsored the International Conference on Population and Development (ICPD), which explored many ways to reduce population growth, including steps to improve the status of women and health care. Although controversial, many efforts are underway to raise the social status of women, give them more say in family decisions, and decouple self-worth and child production. Improvements in health care for women are also being

© hugolacasse/ShutterStock, Inc.

SPOTLIGHT on Sustainable Development

9-1 Controlling Urban Sprawl: The Other Kind of Population Control

Controlling population means more than slowing the rate of growth. It also entails efforts to halt the spread of human settlements onto farmland, forest, fields, and wildlife habitat—valuable resources vital to create a sustainable future. This goal is as vital in the more developed nations of the world as it is in the less developed nations.

A world leader in growth management is the state of Oregon. In the early 1970s, Oregon's legislature passed a law that required all cities and towns to establish urban growth boundaries designed to protect open space, forests, wildlife habitat, and farmland (**FIGURE 1**). Within the growth zone, city officials were permitted to plan for an annual population growth of 2% over a period of 20 years.

Urban boundaries have had many positive effects in Oregon. They have helped reduce the leapfrogging of subdivisions, a leading cause of the loss of open space and agricultural land. Furthermore, the concentrating effect of urban growth boundaries has helped to reduce taxes by reducing the cost of providing basic services such as roads, sewer lines, and schools to new housing developments. Urban sprawl, on the other hand, makes services much more costly to provide.

By concentrating growth in a more confined region, urban growth boundaries can enhance the success of mass transit such as light rail systems. In Portland, the number of passengers who rode the light rail system in its first year was twice what planners had projected. In 1995, the light rail system logged about 24,000 trips per day. Because of the popularity, light rail has been greatly expanded. By 2012, it averaged 130,000 trips per day.

Concentrating growth has been good for business and the environment. Downtown Portland, for example, has seen its share of the regional retail market increase from 7% to nearly 30%, and has added 30,000 jobs without any increase in the number of cars. The number of days that air quality doesn't meet health standards has dropped from about 100 per year to zero.

Since the late 1970s, Portland's urban growth boundry has been moved numerous times, although most adjustments have been small—about 20 acres or less. Much larger additions have been made to provide housing. Even so, the urban boundary has been an effective means of containing growth and could be implemented in many other cities.

Growth management has been successfully implemented in several other states, including Maine, Florida, Vermont, New Jersey, Rhode Island, Washington, Hawaii, and Georgia. Worldwide, many cities and nations have launched programs to control urban growth, among them Belgium, France, the Netherlands, and Germany. Several Canadian cities, including Toronto and Vancouver, have taken the lead in providing compact urban centers with outlying, high-density suburban areas. Because of efforts to concentrate housing and business, public transportation became a feasible way of linking the areas, decreasing automobile usage, energy consumption, and pollution. Canadians walk, cycle, or use local public transportation within a

FIGURE 1 View of Portland, Oregon's suburb, showing the developed and protected lands side by side.
© Fary Braasch/Corbis.

given area and rely on rapid light rail or express buses to reach other areas. One of the world's best success stories is that of Curitiba, Brazil.

widely promoted. Part of the reason behind this strategy is that improvements in health care may help lower infant death rates and thus will reduce the desire to have more children.

KEY CONCEPTS

Small-scale sustainable economic development, jobs for women, efforts to promote equality, and improvements in health care for women are all vital components of a global strategy to reduce fertility and population growth.

Sustainable Populations in the More Developed Countries

Many people think that curbing population growth applies only to the less developed countries, where 90% of the present growth occurs. Observers point out, however, that the high per capita consumption of the MDCs puts enormous strains on the Earth's environment. The environmental impact of a nation is often approximated by this equation:

$$\text{Enivronmental Imapct} = \text{Population Size} \times \text{per Capital Consumption} \times \text{Pollution and Resource Use per Unit of Consumption}$$

Population size is a major determinant of impact. If lifestyles are similar in two populations, 1 with 10 million and 1 with 1 million people, the larger population will have 10 times the environmental impact of the smaller one. The larger population uses more resources and creates more pollution.

Per capita consumption is also a major determinant in environmental impact. In the MDCs, especially the United States and Canada, the per capita consumption (the amount of resources each of us uses) far exceeds consumption in the less developed world. In fact, a single American or Canadian uses 20 to 40 times more resources than a citizen of the less developed world and, thus, has 20 to 40 times the environmental impact.

To understand the implications of this statistic, let's look at it in another way. Each year, approximately 82 million people join the world population, 76 million in the LDCs and 6 million in the MDCs. The paltry 6 million new residents of the more developed nations, however, will consume as many resources and produce as much waste as 120 to 240 million residents of the less developed nations. Widespread pollution, species extinction, and global resource depletion are three results of the more developed nations' higher standards of living. Because of this, population stabilization is as important (perhaps even more important) in the MDCs as it is in the less developed nations. Population stabilization strikes at the root of environmental destruction.

The third element of the equation, pollution and resource use per unit of consumption, also affects the impact of a given population. The amount of energy required to meet human needs varies. Providing electrical energy from wind turbines and solar electric systems uses far fewer resources and produces much less pollution than burning coal. To understand this component of the equation, imagine two populations, each with 1 million people. If population A relies on strategies outlined in this book such as recycling, renewable energy use, mass transit, energy efficiency, and so on, its use of resources and production of waste could be 50 to 90% lower than population B, which is on the traditional, unsustainable path.

The impact of a population depends on many factors, most importantly, the size of the population, per capita consumption (how much citizens consume on average), and the resources used and pollution produced to meet needs. Most of the attention on curbing population growth is focused on the less developed nations—the largest sector of the global population—however, population controls in the more developed nations are even more important because of their high level of per capita consumption.

Sharing Knowledge: Another Role to Play By using fewer resources (being frugal), using resources more efficiently, developing technologies and methods that reduce or eliminate pollution, and reducing population growth, the more developed countries can also make enormous progress toward a sustainable future.

Frugality and efficiency are often viewed as ways of freeing up resources for the less fortunate people of the world. The logic behind this belief is that if the wealthy nations use less, there would be more for people of the LDCs. This, in turn, would greatly reduce current disparities and growing tensions between the haves and have-nots. The late Garrett Hardin, author of numerous books and noted environmental thinker, however, contended that global sharing of resources is not the answer to global sustainability. First, the residents of the less developed nations generally lack the financial resources to purchase surpluses. Hardin argued that the rate of growth and needs of the residents of the less developed nations far exceed the capacity of the more developed nations to help—food donations could not feed the world's hungry. Even a 10% reduction in the demand for food in the United States, if it could be effected, would not come close to meeting the needs of the world's hungry and poor.

According to Hardin, the MDCs have an important role to play in bringing about population stabilization. They can assist the LDCs by sharing their knowledge of birth control, sustainable agriculture, health care, and sustainable development. Financial assistance to help achieve sustainable development might also go a long way. TABLE 9-1 lists additional measures that more developed nations can take to help reduce population growth.

The industrial nations can do many things to help build a sustainable future—for example, reducing their own growth, use of resources, and pollution. They can also assist less developed nations through financial aid, especially for family planning and sustainable development, and by sharing information and technical expertise, on family planning, sustainable agriculture, and sustainable economic growth.

TABLE 9-1

Population Control Strategies for More Developed Countries

Strategy	Rationale	Benefits
Stabilize population growth by restricting immigration and by spending more money and time on sex education and population awareness in public schools.	High use of resources taxes the environment. Immigrants create a serious strain on the economy and create social tension in conditions of high unemployment. Education helps citizens realize the importance of population control.	Limiting resource use leaves more for future generations and less developed countries.
Provide financial assistance to less developed countries for agriculture and appropriate industry. Aid should come from government and private sources.	Economic growth in less developed countries will raise the standard of living and aid in population control.	The rich–poor gap would narrow. A decrease in sociopolitical tension and resource shortages would result.
Provide assistance to population control programs.	Better funded population programs can afford the increased technical assistance and community outreach programs necessary to provide information to the public.	This could result in faster decrease in population growth.
Make trade with less developed countries equitable and freer.	Freer trade will increase per capita income and raise standards of living with little effect on home economy.	A higher standard of living and increased job opportunities could result.
Concentrate research on social, cultural, and psychological aspects of reproduction.	Techniques available today are effective and reliable. What is needed is more motivation for population control, especially among poor countries.	Money will be better spent; research of this nature may help facilitate family planning in less developed countries.

9-2 Cleaning Products Go Green

In January 2008, the Clorox Company announced the introduction of GREEN WORKS® all-purpose cleaners, a line of natural cleaning products that are as effective as conventional cleaners but made from plant-based ingredients (**FIGURE 1**).

According to the company, GREEN WORKS® all-purpose cleaners are at least 99% natural and made from chemicals derived from coconuts and lemon oil. The cleaning agents are biodegradable, nonallergenic, and were not tested on animals. The GREEN WORKS® all-purpose cleaners line includes natural cleaning products for use in the home, kitchen, and bathroom: an all-purpose cleaner, a glass and surface cleaner, a toilet bowl cleaner, and a bathroom cleaner. The company also packages their cleaning agents in recyclable (number 1) plastic.

In lab and blind consumer in-home testing, GREEN WORKS® all-purpose cleaners performed as well as, or better than leading conventional cleaners, according to the company. (I tried a couple before the product was released and was extremely impressed!)

GREEN WORKS® all-purpose cleaners are the first line of natural cleaners developed by a major consumer products company and are affordable alternatives to natural cleaners. The company has even teamed up with the Sierra Club. "One of the Sierra Club's primary goals is to foster vibrant, healthy communities with clean water and air that are free from pollution," said Sierra Club Executive Director Carl Pope. "Products like GREEN WORKS® all-purpose cleaners help to achieve this goal in the home. We're looking forward to working with Clorox and the GREEN WORKS® all-purpose cleaners team to promote a line of natural cleaning products for consumers who are moving toward a greener lifestyle."

GREEN WORKS® all-purpose cleaners are available at grocery stores, drug stores, and mass retail outlets. To learn more, log on to http://greenworkscleaners.com.

Clorox is not the only company to go green. In February, 2008, SC Johnson, a leader in corporate green, began to print a Greenlist™ logo on its eco-friendly products, starting with an old standby, Windex®, a window cleaner. Their Greenlist logo indicates that it is a cleaning agents made with lower concentrations of volatile organic chemicals (VOCs) that can be hazardous to human health and contribute to photochemical smog.

The company has also reformulated Pledge®, a furniture polish, to increase its biodegradability while increasing cleaning power, and Fantastik Orange Action®, a cleaner with no VOCs.

In 2006, SC Johnson received the Presidential Green Chemistry Challenge Award and the Ron Brown Award for Corporate Leadership for its work on the Greenlist process. SC Johnson has also moved toward greener energy to reduce its emissions.

More and more North Americans are looking for ways to live greener lifestyles, and there are hundreds of things they can do to realize their goals. Green cleaning products like those from Clorox and SC Johnson are two simple, inexpensive steps we can take to create a healthier, more sustainable world.

FIGURE 1 You can help create a sustainable future by using environmental- and people-friendly cleaning products like these.

GREEN WORKS® is a registered trademark of The Clorox Company. Used with permission. © 2014 The Clorox Company. Reprinted with permission.

Creating Sustainable Populations in the Less Developed Countries

Most of the LDCs recognize the need for population control and have implemented policies to reduce population growth. Support for these programs comes from a variety of sources. One of the major players in the international effort to promote family planning in the less developed nations is the International Planned Parenthood Federation (IPPF), a nonprofit organization established in the 1950s. The IPPF disseminates information on family planning and provides various kinds of assistance to countries.

Another major source of financial support for family planning is the UNFPA. The UNFPA spent about $181 million in 2012 in less developed nations for reproductive health, including family planning programs.

Foreign assistance programs sponsored by more developed countries also play a key role in meeting population goals. Germany, Japan, the United States, and other countries all donate huge sums of money for various programs, including sustainable development and family planning. The United States donates money through the **U.S. Agency for International Development (USAID).** USAID currently spends about $660 million on family and reproductive health programs, which includes family planning.

Even international lending agencies help out. For example, since the 1960s, the World Bank, which is largely financed by the United States, has offered financial aid to LDCs with population control policies; it has recently stepped up its support.

Although many hundreds of millions of dollars are provided for family planning, it is far from sufficient. In addition, spending on family planning has actually fallen in the past decade or so because of inflation. Today, many countries with active family planning programs provide most of their own financing. Few of these governments, though, spend more than 1% of their national budget on family planning services—far below what is needed, according to some proponents.

Despite the obstacles facing the LDCs, family planning makes good sense from environmental and economic standpoints. Depending on the country and the program, each birth averted by family planning yields a savings of between $15 and $200 per year in social services. Estimates in the United States suggest that every dollar invested in family planning saves the LDCs at least $3 in health and welfare costs in the first year. **TABLE 9-2** lists population control strategies for less developed countries.

9.3 Overcoming Barriers

Achieving a sustainable human population will not be easy. Many barriers lie in the way. This section describes three major obstacles: (1) psychological and cultural, (2) educational, and (3) religious barriers.

> **KEY CONCEPTS**
>
> Three primary barriers lie in the way of achieving a sustainable human population: psychological and cultural, educational, and religious.

Psychological/Cultural Barriers

In the less developed countries, children are generally valued by their parents, especially in rural areas, in part because they assist with chores and provide a degree of social security for parents in their old age. Because mortality rates are high in most LDCs, having many children also increases the likelihood that some will survive, but with decreases in death rates as a result of improvements in health care such as vaccination, nutrition, and sanitation, the reason for childbearing is becoming less significant.

Traditional views of family size often change slowly, even in light of changing conditions. In India, for example, larger families remained the norm even in economically developed regions for many years. Although India's government officially promoted families of two or three children, many parents still saw the ideal family as two sons and a daughter. The problem is that in trying to reach this goal, couples had, on average, 3.4 children. "One son is no sons," one Indian argues. "To be sterilized is to tempt fate," another asserts In recent years, however, total fertility dropped to 2.4, which is replacement-level fertility. If the fertility rate remains at this level, India's massive population of nearly 1.3 billion could some day stabilize.

Having children is a psychologically enriching activity for many people the world over. Sociologists report that in many LDCs men and women have long been admired for the number of children they have. As a result, social acceptance and other psychological factors result in the birth of many children that may never have adequate food, clothing, shelter, and education.

Children are psychologically enriching in the MDCs too. However, many citizens of MDCs, even those who

> **KEY CONCEPTS**
>
> Numerous private and governmental organizations such as the International Planned Parenthood Federation and the United Nations Fund for Population Activities spend millions of dollars a year to support programs to slow population growth in LDCs through family planning and sustainable economic development.

TABLE 9-2

Population Control Strategies for Less Developed Countries

Strategy	Rationale	Benefits
Develop effective national plan to ensure better dissemination of information and availability of contraception and other methods of population control. Do not rely on one type of control.	Each country better understands its people and thus can design better programs to spread population control information and devices.	More effective dissemination of information and, probably, a higher rate of success.
Finance education in rural regions, emphasizing population control and benefits of reduced population growth.	Education can help make population control a reality.	Slower population growth, more effective use of contraceptives, and more incentive.
Seek to change cultural taboos against birth control and cultural incentives for large families.	Changes in culture and psychology may be needed to make population control programs effective.	Such changes will help programs succeed.
Develop appropriate industry and agriculture, especially in rural areas to reduce or eliminate the movement of people from the country to the city.	Appropriate agriculture and industry will create jobs and better economic conditions for families. A higher standard of living could translate into better health care and greater survival of young, thus destroying the need for large families.	This will result in a higher standard of living, better health care, and impetus for control of family size.
Seek programs of development that attain a maximum spread of wealth among the people.	Development must not just help a select few because benefits may not trickle down to the needy.	Plans of this nature yield good distribution of income and help the needy rather than a select few.
Integrate population policy with economic, resource, food, and land-use policy to achieve a stable state.	Finite resources require wise allocation and use; success in the long run depends on attempts to achieve a sustainable future.	Longevity and permanence are attainable if policies are integrated and take into account the requirements of a sustainable society.
Seek funding from the United Nations and more developed countries.	More developed countries have a stake in stabilizing world population growth.	More developed countries could provide significant financial support.

love children and put stock in family life, realize that children are an economic drain. In 2013, the U.S. Department of Agriculture estimated that it will cost a family in the United States earning $60,640 to $105,000 per year about $241,000 to raise that child until he or she is 18 years old. Families making $105,000 a year or more will spend nearly $500,000 to raise a child to age 18.

KEY CONCEPTS

In the less developed countries of the world, children are often seen as an asset to their parents, and childbearing enhances a woman's social status. Having many children seems desirable because mortality tends to be high, too. These psychological factors contribute to larger family size and rapid growth. In more developed nations, children are valued but simultaneously viewed as an economic drain, and a woman's status is not so heavily dependent on childbearing.

Educational Barriers

Lack of education is also a barrier to reducing population growth. As shown in **FIGURE 9-4**, the more education a woman receives, the fewer children she has. Education influences fertility in all nations, rich and poor, in one of several ways. First of all, because education takes considerable time and energy, women often wait till they are finished to get married. The longer they postpone marriage, the fewer children they will have. Second, educated men and women often pursue careers. This also causes couples to postpone marriage and childbearing. In most women, the childbearing years occur between ages 15 and 44; thus, a woman who graduates from college at age 21, marries, but delays having children until she is 30, has decreased her childbearing years by half. She will have half the number of years to produce offspring. Men and women who lack educational opportunities or who choose not to go to school generally marry younger and pursue careers that do not interfere with childbearing. Thus, the period of childbearing is much longer than for couples that pursue higher levels of

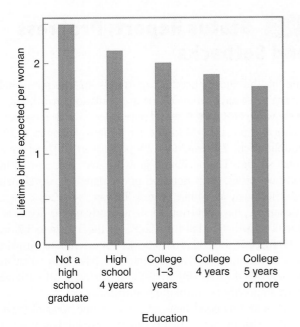

FIGURE 9-4 Total fertility rates of U.S. women, by level of education. In the United States and abroad, education opens up a world of employment opportunities for men and women and lowers fertility rates. As such, education is vital to all efforts to reduce fertility rates and is needed to build a sustainable world community.

education. Third, educational programs can also help raise the literacy rate, making it easier for people to understand the benefits of family planning as well as the instructions that accompany contraceptives. Fourth, education improves economic opportunities and, as noted previously in this chapter, shifts attitudes toward larger families. Because of these and other factors, improvements in education worldwide may be an essential component of a successful population stabilization program.

KEY CONCEPTS

As a general rule, the higher the level of education in a population, the lower its fertility rate. Education and careers that follow decrease the number of childbearing years and open up many options besides childbearing.

Religious Barriers

Religion may also be a powerful force in reproduction. Buddhism promotes family planning in Thailand. The Catholic religion, on the other hand, forbids all "unnatural" methods of birth control, such as the pill, the condom, the diaphragm, and abortion. Theoretically, the Catholic Church guides the sexual practices of approximately 1.2 billion people. Recent surveys show, however, that the use of contraceptives by Catholic women, especially in Western nations, is nearly as high as that among non-Catholics. In Latin America, many priests speak out against the Vatican's official policy.

Birth control is a generally undiscussed subject among other religions. However, in many of the Eastern and Middle Eastern religions, which compete with one another for followers, birth control is not advocated. The total fertility rate in Iraq was 4.3 in 2013.

KEY CONCEPTS

Religions influence population both positively and negatively. Some religions openly denounce efforts to control population growth; others openly support smaller families. Some have no official view.

If family planning programs are going to work, they must address these three obstacles. Consider some examples of ways to surmount the psychological and cultural obstacles. In many Middle Eastern countries where family planning is culturally unacceptable, birth control can be promoted as a means of spacing births to improve maternal and child health. This new direction appeals to more people and broadens the base of support for family planning programs. Thus, a couple who may not relate to lofty goals of slowing the growth rate of their country can understand the benefits family planning brings to them in improved economic conditions and maternal and infant health.

Go Green

Help support nonprofit organizations such as the International Planned Parenthood Federation that are working to promote family planning and improve maternal health care in developing countries.

Attitudes toward family size and childbearing can be changed, particularly from within the culture by women themselves. Especially important are ways to improve the status of women and to give women a say in the economic decisions of a village, described earlier.

KEY CONCEPTS

To reduce population growth, we must overcome psychological, educational, and religious barriers.

9.4 Ethics and Population Stabilization

To many people, reproduction is a basic human right. To deny the right to reproduce is to deny one of the most fundamental and important of all human activities. To other people, population control measures violate deep religious beliefs or cultural norms. For various ethnic groups, family planning may even have overtones of genocide. To help you sort out the debate and make up your own mind about the ethics of population stabilization, this section addresses an important ethical question: Is reproduction a personal right? Arguments from both sides of the issue are given to help you formulate your own view. We begin with advocates of family planning.

Some proponents argue that the right to reproduce at will should be curtailed when the rights of the individual interfere with the welfare of society—that is, the **collective rights.** More people means more suffering for everyone and more environmental destruction that has many adverse effects on people and the species that share this planet with us.

Others who argue in favor of population control focus on the welfare of the unborn. Ecologist Paul Ehrlich, for example, notes that we "must take the side of the hungry billions of living human beings today and tomorrow, not the side of potential human beings. . . . If those potential human beings are born, they will at best lead miserable lives and die young." He argued that we cannot let humanity be destroyed by a doctrine of individual freedoms conceived in isolation from the biological facts of life. Along these lines, the late Garrett Hardin argued that the integrity of the biosphere and the Earth's carrying capacity should be the guiding principle in the debate. We must recognize that the condition of the planet determines the well-being of all living things, including humans. Hardin suggested that we have a moral obligation to future generations to protect the biosphere, the life support system of the planet. Human population control is viewed as a way of honoring this obligation.

Proponents of this viewpoint do not ask whether it is ethical to control population growth, but rather whether it is ethical to let the human population continue to grow unabated. If we permit population to grow, they argue, it will very likely cause a massive deterioration of the environment, which adversely affects present and future generations. It will also cause the extinction of many species.

Opponents say that individual rights are paramount. To limit family size takes away personal freedom. No one has a right to tinker with basic freedoms. Moreover, to counter arguments about suffering and deterioration of conditions, they say that we need to find ways to accommodate new people—improve economies, health care, food supplies, and other areas to make conditions better. We can, they say, have more people and not ruin the planet. We need to grow smarter.

Although debate over population stabilization continues today, most countries favor the sustainable viewpoint—that population stabilization is necessary and that the collective rights of present and future generations take precedence over individual rights. While few countries put strict limitations on family size, most encourage smaller families, leaving the decision to the parents. Acting out of this belief, many nations have mounted programs to promote family planning and make contraceptives available to those who want them—even those who cannot afford them. Many of the countries that promote family planning are striving to slow the rate of growth, reach a stable state, or, in some instances, decrease their population size.

KEY CONCEPTS

Many people think that the right to have children is a fundamental personal freedom. Others believe that the collective rights of present and future generations to a clean, healthy environment and adequate food, shelter, and clothing supersede individual rights.

9.5 Status Report: Progress and Setbacks

One of the most encouraging trends in environmental science these days is the decline in population growth rates. As shown in **FIGURE 9-5**, world population growth rates have declined impressively since the early 1960s, falling from 2% (doubling time 35 years) to 1.2% per year in 2007, where it remains today. The decline in growth rates has occurred in many countries, both rich and poor. Many less developed countries such as China, Korea, Taiwan, and Costa Rica, for example, have contributed significantly to this decrease. China's growth rate fell from 2.5% in the 1960s to 1% in 1999. By 2007, it had fallen to 0.5%, its current rate. Growth rates in the more developed countries have also fallen in the past two decades (Figure 9-5). Today, must MDCs have nearly stationary or shrinking populations.

Despite this good news, there are some areas of grave concern. One of them is that population growth has resulted in a massive buildup of people in urban areas, who become susceptible to the spread of infectious disease. This is especially troublesome in less developed nations. Poor air quality in LDCs may lower immune system function and unsanitary conditions increase the risk of exposure to new and deadly disease-causing organisms. Increased mobility, largely via jet airliners, may also serve as a means of spreading disease internationally.

Another discouraging trend is the increase in death rates. Researchers estimate that approximately one-third of the decline in the annual global population growth rate is the result of rising death rates, largely from AIDS, acquired immune deficiency syndrome. AIDS is a disease caused by the sexual transmission of a virus (HIV) that attacks the immune system. Although the HIV infection rate among adults age 15 to 49 is 0.8% in the more developed nations, the incidence in populations in sub-Saharan Africa is 4.9%. In many countries rates are in the 15% to 24% range, according to data from the Population Reference Bureau. Further declines in the growth rate are projected by the United Nations because of AIDS, but also because of declining groundwater and food supplies, which are setting the stage for a potentially massive increase in death rates.

Another discouraging bit of news is that today, nearly one quarter of the world's population is under the age of 15 and is soon to enter its reproductive years. Worldwide, contraceptive use is substantial, but demand still greatly exceeds availability (**FIGURE 9-6**). In addition, population growth in Africa, Latin America, and Asia is still fairly rapid. Even though population growth has declined, a 1.2% growth rate still results in the addition of about 83 million people every year. Most of the new growth in the coming decades will occur in the less developed nations, which are ill equipped to feed and house the new residents. Widespread hunger, poverty, and environmental decay are likely to occur.

Human civilization need not end in an explosion of people, but many experts agree that we can improve only

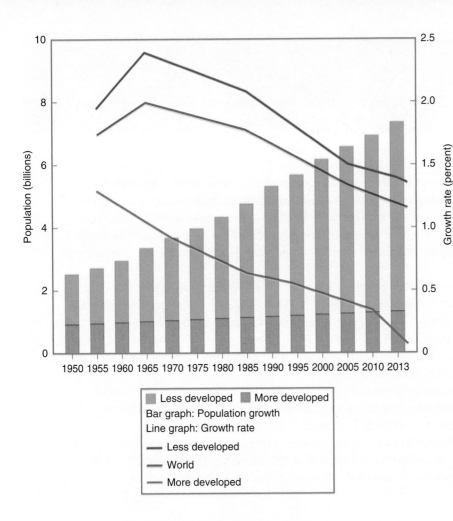

Less developed ■ More developed
Bar graph: Population growth
Line graph: Growth rate
—— Less developed
—— World
—— More developed

FIGURE 9-5 Encouraging trends. The growth rate of the human population has been declining steadily. Although it has declined, the population continues to increase.

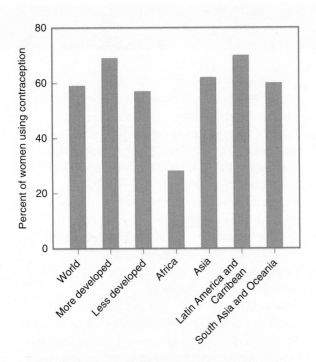

FIGURE 9-6 Contraceptive use. This graph shows the percentage of women using contraception in the world, in developed versus developing countries, and on various continents.

if we hold our numbers within the carrying capacity of the Earth. This means stopping the growth of the human population and very likely achieving a smaller population size. To do so will require a multitude of approaches from family planning to sustainable economic development to improvements in education, the status of women, and health care. By attacking the problem at all fronts, we have a better chance of achieving success and creating a sustainable future.

I cannot believe that the principal objective of humanity is to establish experimentally how many human beings the planet can just barely sustain. But I can imagine a remarkable world in which a limited population can live in abundance, free to explore the full extent of man's imagination and spirit.

—Philip Handler

CRITICAL THINKING

Exercise Analysis

Population growth is one of the driving forces of environmental deterioration. In other words, it is a root cause. Even with new ways of doing things, more people generally means more pollution, more resource use, and more environmental disturbance. No matter how carefully we conduct our affairs or how efficiently we use resources, human society will always have an impact on the environment. The more of us there are, the greater the impact. However, the impact depends on per capita consumption, resource demand per unit of production, and pollution output per unit of production.

Here's where the issue gets confusing to many people. Being more efficient is absolutely essential to creating a sustainable society. So are efforts to reduce our output of pollution. Technological innovation can have a major beneficial impact on our future. These savings, however, can be easily offset by population growth. Energy efficiency measures can substantially cut energy demand, but those gains can be offset by growth. Recall from the last chapter that the U.S. population may increase by nearly 50% by the year 2050. So, if we cut energy demand and the output of pollution by 50%, population growth would result in no net gain.

This kind of reasoning helps us understand the need for sustainable technologies coupled with population stabilization—and perhaps, as some advocates suggest, efforts to eventually reduce the size of the human population by means that are acceptable and, of course, humane.

Is the notion that all consumption patterns can be made sustainable through conservation and other means valid? Probably not. What principles of sustainability does it ignore? It ignores the fact that the Earth is a finite source of resources. Limits are very real and must be reckoned with, if we are to create a sustainable existence.

CRITICAL THINKING AND CONCEPT REVIEW

1. Critically evaluate this statement: "The world cannot support the people it currently has at a decent standard of living. Thus, efforts should focus on helping the less developed nations industrialize. Population growth will decline as a result, so family planning programs are not necessary."

2. Define *family planning*. Make a list of the three major types of family planning programs. Give examples of each. What are the weaknesses and strengths of each?

3. The United Nations appoints you as head of family planning programs. Your first assignment is to devise a family planning program for a less developed country with rapid population growth, high illiteracy, widespread poverty, and a predominantly rural population. Outline your program in detail, justifying each major feature. What problems might you expect to encounter?

4. Describe ways in which more developed countries might aid less developed countries in stabilizing population growth.

5. Discuss attitudes about the value of children in less developed countries. How do these views differ from those in more developed countries?

6. How are children viewed among your friends, relatives, and parents? Are large family sizes viewed favorably by them? Are small families viewed favorably? Why or why not?

7. If you are considering having children, what factors influence the desirable size of your future family?

8. Discuss reasons why the total fertility rate tends to be lower among more educated women.

9. Discuss general ways to ensure a high rate of success in family planning programs.

10. Do we have the right to have as many children as we want? Should that right be curtailed? Explain your answer.

11. Make a list of the pros and cons of population growth. In other words, what are the benefits, and what are the adverse impacts? You may want to look at U.S. population growth and global population growth separately. You may also want to go online to find what other people think.

KEY TERMS

abstinence
birth control
collective rights
contraceptive measures
contraceptives
demographic transition

extended voluntary programs
family planning
forced family planning programs
induced abortion
United Nations Fund for Population
  Activities (UNFPA)

U.S. Agency for International
  Development (USAID)
voluntary programs

Creating a Sustainable System of Agriculture to Feed the World's People

Civilization itself rests upon the soil.

—Thomas Jefferson

LEARNING OBJECTIVES

After reading and studying the material in this chapter, students will be able to understand, knowledgably discuss, and, when appropriate, apply the information learned in this chapter, including:

1. The dimensions of human hunger and malnutrition
2. How hunger and malnutrition affect people, the environment, and the economic prospects of nations
3. The importance and formation of top soil
4. Factors that contribute to the unsustainability of modern agriculture and reduce the prospects of human population in the long term, including soil erosion, desertification, farmland conversion, declines in irrigated farmland, waterlogging, salinization, and declining genetic diversity among livestock and crops
5. How government affects the nature of modern farming for better and worse
6. The many ways to build a sustainable system of agriculture to reduce and eliminate problems listed in number 4
7. Ways we can enhance soil quantity and quality of the world's cropland, including minimum tillage, contour farming, strip cropping, crop rotation, and organic fertilizers
8. Limits to expanding the amount of land in agricultural production
9. Ways to expand food supplies through new strains, protecting genetic relatives of modern crop species, genetic engineering, alternative crops and livestock, and eating lower on the food chain

CRITICAL THINKING

Exercise

Some agricultural interest groups object to the phrase *sustainable agriculture*. They argue that it implies that current agricultural practices are not sustainable. In support of their case, they point out that agricultural productivity (the output of food per hectare or acre) has been rising steadily in many countries—such as Canada, Australia, and the United States—thanks to widespread use of insecticides, herbicides, fertilizer, irrigation, and other modern practices. They also point out that farmers in many modern agricultural countries are feeding more and more people every year. Is there anything wrong with this line of reasoning? What critical thinking rules are essential to analyze this issue?

Modern farmers typically plant huge expanses of corn, wheat, and soybeans to produce food in quantities sufficient to meet human needs. Unfortunately, large fields containing a single crop tend to be highly vulnerable to insects, plant diseases, and hail. In response to this problem, some progressive Nebraskan farmers have broken with the traditions of modern agriculture. Instead of planting their fields in one crop, they plant two crops (such as corn and soybeans) on the same land but in alternating strips. This simple technique called *intercropping* increases the productivity of soybeans and corn dramatically. Why? In one study, researchers found that in this system the corn protects the soybeans from the drying effects of the wind. This increases soybean output by 11%. They also found that the stands of corn are less dense than in a field planted from fencerow to fencerow. This results in better sunlight penetration, which increases corn production by a remarkable 150%. Intercropping not only increases production, it reduces the need for chemical pesticides, for reasons explained later. Reductions in pesticide use save farmers considerable amounts of money, reduce pesticide exposure to farmers and their families, and reduce environmental problems.

This is one example of many efforts aimed at building a more sustainable system of agriculture. **Sustainable agriculture** is a system that produces high-quality foods while maintaining or improving the soil and protecting the environment—the air, water, soil, and wealth of wild species. It is a system that can endure, providing benefits for centuries. This chapter tackles the subject of food and agriculture, beginning with a look at hunger and malnutrition.

10.1 Hunger, Malnutrition, Food Supplies, and the Environment

Hunger and starvation are two consequences of overpopulation. According to estimates from the United Nations (UN) Food and Agriculture Organization, about 12% of the world's population — around 840 million people—in less developed countries (LDCs) are chronically undernourished—that is, they fail to get enough protein, calories, or both. Hunger is greatest in sub-Saharan Africa and southern Asia, notably India and Bangladesh. In sub-Saharan Africa, for instance, about 25% of the population is chronically undernourished, according to the UN Task Force on Hunger. Many of them are children.

Scientists recognize two types of undernutrition. The first, **kwashiorkor** (KWASH-ee-OAR-core), results from a lack of protein. The second, **marasmus** (meh-RAZ-mess), results from an insufficient intake of protein and calories (food that provides energy). In reality, kwashiorkor and marasmus are two extremes of protein–calorie deficiency and most individuals who are malnourished exhibit symptoms of both. People may also suffer from specific deficiencies such as a lack of vitamin A, which may cause serious damage to the eyes, leading to blindness.

FIGURE 10-1 shows children suffering from kwashiorkor, extreme protein deficiency. As illustrated, the legs and arms of victims of this disease are thin because the lack of

FIGURE 10-1 Kwashiorkor. This protein deficiency leads to swelling of the abdomen. The loss of muscle protein results in thin arms and legs. Children are physically and mentally stunted, apathetic, and anemic.

Courtesy of Dr. Lyle Conrad/CDC.

protein in the diet results in a decrease in muscle mass. Notice also that their abdomens are swollen with fluids. Victims are weak and passive. Kwashiorkor is most common in children 1 to 3 years of age and generally begins after they are weaned (thus losing the protein-rich milk of their mothers) and have been switched to a low-protein, starchy adult diet.

Victims of marasmus are thin and wasted (**FIGURE 10-2**). Their ribs stick out through wrinkled skin. They often suck on their hands and clothes to appease a gnawing hunger. Unlike victims of kwashiorkor, however, children suffering from marasmus are alert and active. Marasmus often occurs in infants separated from their mothers during breastfeeding as a result of maternal death, a failure of milk production (lactation), or the use of milk substitutes (baby formulas) often provided for free by hospitals. For example, the distribution of baby formula by food manufacturers to LDCs to introduce mothers to this product can have disastrous results in areas where safe drinking water is unavailable or when poverty-stricken mothers dilute the formula to make it go farther, thus, reducing their infants' intake of protein and calories. A mother's milk production will cease after a baby begin feeding from a bottle, so women are forced to continue using baby formula.

Public outcry led to measures to curb such activities. In 1981, the World Health Organization convened over 200 members that adopted an international code for marketing

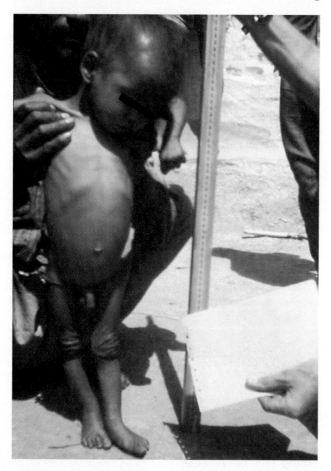

FIGURE 10-2 Marasmus. Victims of marasmus (protein and calorie deficiency) are thin but alert and active. Survivors of malnutrition, however, may be left with stunted bodies and minds.
Courtesy of Dr. Edward Brink/CDC.

of breast milk substitute that included strict restrictions on advertising in LDCs and a prohibition of giving free samples. To date, however, only 65 of 199 nations (32%) have adopted most or all of the recommendations. In the remaining countries, the recommendations are ignored by governments and formula manufacturers.

Sadly, for every clinically diagnosed case of marasmus and kwashiorkor in LDCs, there are hundreds of children with mild to moderate forms of undernutrition, a condition much more difficult to detect. Undernourishment is not restricted to the LDCs of the world, however. In the United States, for example, an estimated 10 to 15% of the population is undernourished. Hunger is most prevalent in Mississippi, Arkansas, Alabama, New Mexico, and the District of Columbia.

Although many people suffer from undernourishment—insufficient protein and caloric intake—experts note that 1 of every 3 people in the LDCs (nearly 2 billion people) receive inadequate vitamins and minerals. That is, although they may consume a sufficient number of calories and enough protein, their diets may lack essential vitamins. Vitamin A deficiency, for instance, is quite prevalent in LDCs where it causes blindness.

Undernutrition causes considerable human suffering. Imagine the pain of going to bed hungry night after night. Malnutrition can lead to death. For example, mild cases make people more susceptible to **infectious diseases,** ailments caused by bacteria and viruses that can be spread from one person to another. A person weakened by a lack of food is more likely to die from an ordinarily nonfatal disease than a person who is well nourished. In fact, undernutrition so weakens the immune system that normally nonthreatening diseases such as measles and diarrhea can be fatal among children. Making matters worse, crowding in urban centers facilitates the spread of disease. According to the United Nations, nearly 8 million people die prematurely each year from undernutrition, malnutrition, and nonfatal diseases worsened by poor nutrition. Many of the victims—1.5 million—are children. That's over 20,000 per day!

KEY CONCEPTS

A large segment of the world's people (most of whom live in Asia, Africa, and Latin America) either do not get enough to eat or fail to get all of the nutrients and vitamins they need—or both. Nutritional deficiencies make people more susceptible to infectious disease and, if they are severe enough, can cause death. Nutritional deficiency is even a problem among the poor in more developed countries such as the United States.

Hunger, Poverty, and Environmental Decay

Many students ask why they should study malnutrition in a course on environmental issues. One reason is that food problems are a direct result of population growth, climate change, and loss of farmland caused by growing human population—all environmental issues. But there's another reason. Nutritional deficiencies early in life often lead to mental and physical retardation. The more severe the deficiency, the more severe the impairment. Mental retardation occurs because 80% of the brain's growth occurs before the age of two. Malnourished children who survive to adulthood remain mentally impaired. Typically plagued by malnutrition their whole lives, they are often prone to infectious diseases and provide little hope for improving their own or their nation's prospects. Hunger and malnutrition, therefore, may contribute to growing poverty, rising population, and worsening environmental conditions.

 KEY CONCEPTS

Hunger and malnutrition cause mental and physical retardation that may contribute to widespread poverty and population growth, which contribute to further environmental destruction.

Declining Food Supplies

Malnutrition abounds. But what are the prospects for the future? Will we be able to feed the world's people and accommodate the growing population, which expands by nearly 230,000 people per day?

For many years, hope has been pinned on our ability to grow more food. The use of modern agricultural methods and increases in farmland caused grain production per capita to rise approximately 30% from 1950 to 1970. This resulted in a substantial improvement in the diet of many of the world's people. From 1971 to 1984, however, world grain production barely kept pace with population growth. Between 1984 and 2012, grain production per capita fell 7%, according to the Worldwatch Institute.

The decline in food production per capita over the past decade results from numerous factors, among them the warming global climate, population growth, soil erosion, and soil deterioration (from causes described in the next section). Many experts think that food production per capita will continue to decline throughout the next decade as a result of these problems. If global warming, population growth, soil erosion, and other problems worsen, then hunger, poverty, and environmental destruction will become even more widespread.

Because of declining per capita food production, many countries have lost the ability to feed their people and have become dependent on imports from the more developed countries such as Canada, Australia, and the United States.

Some experts believe that the safety net provided by major food-producing nations is in danger. One of the most serious threats is global warming. Droughts, record-breaking heat, and the spread of deserts—all caused by global warming—could dramatically reduce global food production, making it difficult to meet current demand and future demand caused by growing populations. Many experts believe that unless decisive steps are taken—and soon—millions of people in the less developed nations could perish from hunger and diseases.

 KEY CONCEPTS

Grain production per capita has been on the decline for two decades, a trend that bodes poorly for those suffering from hunger and malnutrition—as well as for those trying to provide food for the ever-growing human population.

The Challenge Facing World Agriculture: Feeding People/ Protecting the Planet

Three interrelated challenges face the world today: First, we must find ways to feed the malnourished people alive today (an immediate challenge). Second, we must find ways to meet future needs for food (a long-term challenge). Third, we have to find ways to produce food to meet present and future needs while protecting and improving the soil and water upon which agriculture depends (both an immediate and a long-term challenge). In other words, we have to ensure that current agricultural practices are sustainable.

 KEY CONCEPTS

The challenge today is to find sustainable ways to feed current and future world residents.

10.2 Understanding Soils

Malnutrition is only one of a handful of serious food and agricultural problems facing the world today. The next section outlines many problems, such as soil erosion, that are responsible for the deteriorating condition of the world's cropland and the decline in food production. Before we examine this set of challenges and discuss ways to build a sustainable agriculture, however, a brief study of soils is in order. This information will provide you with some of the scientific knowledge you need to assess various solutions and their sustainability.

What Is Soil?

High-quality soils promote plant growth, both in natural ecosystems and on rangeland and farmland. Because of this, soils are vital to our long-term health and our economic well-being.

Soil is a complex mixture of inorganic and organic materials with variable amounts of air and water. The inorganic material includes clay, silt, sand, gravel, and rocks. The organic component consists of living and nonliving plant and animal materials. Living organisms include insects, earthworms, and microorganisms. The nonliving matter includes plant and animal waste and residues (remains of dead bodies) in various stages of decomposition.

Soils are described according to six general features: texture, structure, acidity, gas content, water content, and biotic composition. These components and characteristics combine to form many different soil types throughout the world. A detailed discussion of each soil type is well beyond the scope of this book. As you shall soon see, some soils are better suited for agriculture than others.

KEY CONCEPTS

Soils consist of four components: inorganic materials, organic matter, air, and water.

How Is Soil Formed?

Soil formation is a complex and slow process, even under the best of conditions. It results from an interaction between the **parent material,** the underlying substrate from which soil is formed, and the organisms. The time it takes soil to develop depends partly on the type of parent material. To form 2.5 centimeters (1 inch) of topsoil from hard rock may take 200 to 1,200 years, depending on the climate. Softer parent materials such as shale, volcanic ash, sandstone, sand dunes, and gravel beds are converted to soil at a faster rate—in 20 years or so under very favorable conditions. Because of the time it takes to replenish soil and because soils are so important to society, they should be protected and carefully managed.

As discussed later in this chapter, the cornerstone of sustainable agriculture is prevention—measures that preserve and protect topsoil so that it can remain productive indefinitely. The food we grow on soils is much like interest from a bank account. We can draw on it forever, as long as we do not deplete the soil itself. Soil is, therefore, one form of **natural capital** on which society depends.

Four factors are responsible for the type of soil that forms, the thickness of the soil layer, and the rate of soil formation. They are climate, parent material, biological organisms, and topography.

The **climate** is the average weather conditions, notably temperature and precipitation. The parent material, the underlying rock or sand or gravel from which the mineral matter is derived, is acted on by climate and biological organisms and converted into soil. For example, heating and cooling (both elements of climate) cause barren rock to split and fragment, especially in the desert biome, where daily tempera-tures vary widely. Water entering cracks in rocks expands when it freezes, causing the rock to fragment further. The roots of trees and large plants reach into small cracks and fracture the rock. Over time, rock fragments produced by these processes are slowly pulverized by streams or landslides, by hooves of animals, or by wind and rain.

Soil formation is facilitated by numerous organisms. Lichens erode the rock surface by secreting carbonic acid. Lichens also capture dust, seeds, excrement, and dead plant matter, which help form soil. The roots of plants also help to build soil by serving as nutrient pumps, drawing up inorganic nutrients from deeper soil layers. These chemicals are first used to make leaves and branches, which can fall and decay; thus, they become part of the uppermost layer of soil, the **topsoil.**

Grazing animals drop excrement on the ground, adding to the soil's organic matter. The white rhinoceros, for example, produces about 27 metric tons (24 tons) of manure each year, which is deposited in its habitat. A variety of insects and other creatures, such as earthworms, also participate in soil formation.

Topography, the final soil-forming force, is the shape or contour of the land surface. It determines how water moves and how quickly soil erodes. Steeply sloping land surfaces, for instance, are more susceptible to soil erosion. As a result, thin soils tend to form on them. Relatively flat terrain, on the other hand, suffers little from erosion. Soils tend to be thicker in such regions. Valley floors benefit from their flatness and their proximity to steeper terrain whose soil washes away and is deposited on the floor. Many a valley, even in the arid West, has been converted to valuable cropland because of the rich soil that has built up in them over many years. Today, however, valley floors are also desirable land for development, and farmers and ranchers are squeezed out as the population expands and new subdivisions pop up.

KEY CONCEPTS

Soil formation is a complex process involving an interaction among climate; the parent material, which contributes the mineral components of soil; biological organisms; and topography. Because soil is so valuable and because it takes so long to form, we should take care to protect and manage soils carefully.

The Soil Profile

Most of us have seen a road cut or excavation for a building and noticed that soil is composed of layers. The layers, called *horizons,* differ in color and composition. Not all horizons are present in all soils; in some, the layering may be missing altogether.

Soil scientists recognize five major horizons (**FIGURE 10-3**). The uppermost region of the soil is the **O horizon,** or **litter layer.** This relatively thin layer of organic waste from animals and detritus is the zone of decomposition and is characterized by a dark, rich color. Plowing mixes it in with the next layer.

The **A horizon,** or topsoil, is the next layer. It varies in thickness from 2.5 centimeters (1 inch) in some regions to 60 centimeters (2 feet) in the rich farmland of Iowa. Topsoil is generally rich in inorganic and organic materials and is important because it supports crops. It is darker and looser than the deeper layers. The organic matter of topsoil, called **humus,** acts like a sponge, holding moisture. Grasslands have the deepest A horizons. The thinnest A horizons are found in deserts, coniferous forests, tundra, and tropical rain forests. As you shall soon see, most of the land that is viewed as potentially arable lies in the tropics, where the A horizon is practically nonexistent.

O horizon (litter)

A horizon (litter and topsoil)

B horizon (subsoil)

C horizon (transition zone)

D horizon (parent material)

Organic matter

Dark, rich in humus

Light colored

Varied

Rock or gravel

FIGURE 10-3 Soil profile. Typical soils consist of five distinct layers. The topsoil is the most important to agriculture because it contains organic matter and nutrients essential for plant growth.

The **B horizon,** or **subsoil,** is also known as the *zone of accumulation* because it collects minerals and nutrients leached from above. This layer is lightly colored and much denser than the topsoil because it lacks organic matter. The next layer, the **C horizon,** is a transition zone between the parent material below and the soil layers above. The **D horizon** is the parent material from which soils are derived.

Soil scientists have identified 11 major soil types or orders. These are defined primarily by their *diagnostic horizons*—that is, as having a specific soil profile. Soil profiles tell soil scientists whether land is best for agriculture, wildlife habitat, forestry, pasture, rangeland, or recreation. They also tell us how suitable soil might be for various other uses such as home building, landfills, and highway construction.

The **soil profile** is determined by the climate (especially rainfall and temperature), parent material, biology, and topography—that is, the same soil-building forces discussed in the previous section. Soil profiles are histories of the interaction among these factors.

KEY CONCEPTS

Soils are typically arranged in layers. For agriculture, the most important are the upper two layers: the O horizon, which accumulates organic waste from plants and animals, and the A horizon, the topsoil.

10.3 Barriers to a Sustainable Agricultural System

With this introduction to soils, we now turn our attention to problems facing world agriculture. We examine the challenges that lie ahead as we attempt to forge a sustainable system of agriculture, beginning with unsustainable rates of soil erosion.

As already mentioned, as the world population continues to expand, per capita food supplies—the amount of food available per person—are on the decline. This problem is especially acute in the less developed nations. In the more developed nations, in contrast, food surpluses are common and obesity is on the rise.

Soil Erosion

Thomas Jefferson wrote that "civilization itself rests upon the soil." The first towns, early empires, and powerful nations can all trace their origins to the deliberate use of the soil for agriculture. Agricultural expert R. Neil Sampson wrote that in most places on Earth, "We stand only 6 inches from desolation, for that is the thickness of the topsoil layer upon which the entire life of the planet depends."

In many places, soil is being washed or blown away—that is, eroded by wind and water. **Erosion** occurs when rock and soil particles are detached by wind or water, transported away, and deposited in another location, often in lakes and streams. **Soil erosion,** the loss of soil from land, is one of the most critical problems facing agriculture today. It is a problem in the MDCs as well as the LDCs.

Soil erosion is classified as either natural or accelerated. As the name implies, **natural erosion** occurs in areas in the absence of human intervention. It generally occurs at such a slow rate that new soil is generated fast enough to replace what is lost. In other words, natural erosion generally occurs at a sustainable rate. In contrast, **accelerated erosion** largely results from human activities such as overgrazing, and it occurs at a rate that outstrips the formation of new soil. Accelerated erosion is dangerous not just because it removes productive topsoil, but because it decreases soil fertility that may cause declines in productivity. Studies on corn and wheat indicate that each inch of topsoil lost to erosion results in a 6% decline in productivity. Severe soil erosion may also result in the formation of deep gullies that make farmland unworkable (**FIGURE 10-4**).

Soil erosion is such a pressing problem because soil washes away quickly but forms very slowly. Soil erosion also has a number of serious environmental impacts. For instance, pesticides attach to soil particles. Transported to nearby waterways, pesticide-laden particles may be ingested by fish and other aquatic organisms. They may

Go Green

If you or your parents own land that suffers from erosion, take steps to revegetate eroded areas or hire a professional to do the work.

FIGURE 10-4 Soil erosion on farmland. All soil erosion above replacement level bodes poorly for farmers and the world's people. This field is suffering from extreme erosion, which not only robs the land of topsoil but also greatly reduces its productive capacity.

Courtesy of Lynn Betts/NRCS USDA.

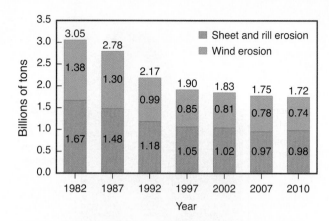

FIGURE 10-5 Soil erosion on the decline. Over the past four decades soil erosion from wind and water has declined dramatically. Although it still exceeds replacement level in many areas, progress has been substantial.

Data from USDA, Natural Resources Conservation Service, *2010 Natural Resources Inventory.*

then be passed to birds and human consumers in the food chain. Another problem is that sediment deposited in waterways fills streams and rivers and reduces their capacity to hold water. This, in turn, increases flooding. Furthermore, sediment destroys breeding grounds of fish and other wildlife and increases the need for dredging harbors and rivers. The World Resources Institute estimates that the off-site damage from soil erosion in the United States is over $10 billion a year.

Since agriculture began in the United States, one-third of the nation's topsoil has been lost to erosion, according to the Soil Conservation Service. Unfortunately, soil erosion continues today. According to U.S. Department of Agriculture estimates, about 1.5 billion metric tons (1.7 billion tons) of topsoil were lost annually from U.S. farmland in 2010 (**FIGURE 10-5**). Thanks to efforts of the U.S. government and farmers, soil erosion has declined dramatically since 1982. Between 1982 and 2007, U.S. cropland erosion fell by nearly 43%. Although soil erosion is down from earlier years (Figure 10-5), the average rate of erosion on U.S. farmland is approximately seven times greater than soil formation, a situation that is clearly unsustainable. Should erosion continue, the U.S. agricultural system could experience substantial declines in productivity. **FIGURE 10-6** shows where wind and water erosion are concentrated.

Unfortunately, little information is available on soil erosion rates throughout the world. Scientists currently estimate that approximately one-third of the world's cropland topsoil is being eroded faster than it is regenerated. Soil erosion is especially rapid in many of the less developed nations. In China, for example, the Yellow River annually transports 1.6 billion tons of soil from badly eroded farmland to the sea. The Ganges in India carries twice that amount. The Worldwatch Institute (a nonprofit organization based in Washington, D.C.) estimates that 22.5 billion metric tons

(25 billion tons) of topsoil is eroded from the world's croplands each year! At this rate, the world loses about 7% of its cropland topsoil every 10 years! To put this into perspective: If erosion continues at this rate, 225 billion metric tons will be lost in the next decade; this is equivalent to more than half of the topsoil on U.S. farms.

Soil erosion above the natural rate of soil formation is unsustainable. Year after year, it depletes a valuable resource needed to feed the world's people. It also will make it more difficult to feed the growing population. Without efforts to halt soil erosion, malnutrition and starvation could increase in the coming century.

KEY CONCEPTS

Soil is vital to the success of a nation, indeed the world, but agricultural soils are being lost at record rates in many countries—a trend that is clearly unsustainable.

Desertification: Turning Cropland into Desert

Each year millions of acres of cropland, pasture, and rangeland are becoming too arid to be farmed. This phenomenon, called **desertification,** occurs most often in highly vulnerable semiarid lands—that is, lands that are already fairly dry. In such areas, even small changes in rainfall or in agricultural practices can have a profound effect on the ability of land to support livestock and crops.

Numerous factors can be blamed for this problem. One of the leading causes is drought: long, dry periods. Drought may result from both natural climatic changes and changes brought about by humans. Global warming, overgrazing, and deforestation are three human causes. Overcropping semiarid lands—that is, planting them too frequently—also contributes to desertification.

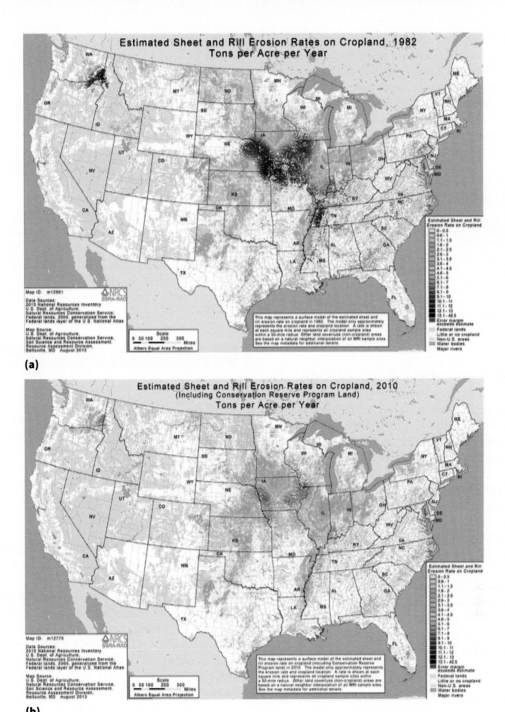

FIGURE 10-6 Soil erosion in the U.S. (a) Soil erosion in 1982. Notice where the most significant erosion is occurring. **(b)** This map of erosion in 2010 shows the decline in soil erosion in the U.S. thanks to improvements in land management.

Parts a and b are from USDA, Natural Resources Conservation Service, *2010 Natural Resources Inventory*.

How do global warming, overgrazing, and deforestation cause drought? Global warming is caused by certain pollutants, one of the most significant being carbon dioxide. These gases trap heat in the atmosphere. Rising temperature increases evaporation rates and thus tends to dry the soil in various regions. Global warming also appears to be changing climate, especially rainfall patterns. Some areas, scientists predict, will become hotter and drier—that is, experience less precipitation. Many atmospheric scientists believe that

if global warming is not stopped, changes in rainfall and average daily temperature could be severe enough to render entire agricultural regions such as the midwestern United States unsuitable for farming.

Removing vegetation from the land also tends to alter the local climate. Deforestation, for example, may decrease rainfall downwind from a site. Why? A well-vegetated surface acts like a sponge, absorbing moisture that supports plants and replenishes groundwater. Some water evaporates, only to fall on downwind sites, creating a cycle of precipitation and evaporation that continues on down the line. In denuded areas, however, water tends to run off the surface of the land. The drier the landscape, the less water there is to evaporate. This, in turn, tends to decrease rainfall downwind.

The loss of farmland to desertification is a serious problem, especially when combined with other factors, including soil erosion and climate change.

KEY CONCEPTS

Throughout the world, cropland, rangeland, and pastures especially in semiarid regions are becoming too dry to use because of climate change (natural and human-induced) and poor land management practices such as overgrazing and deforestation. This phenomenon is called *desertification*. Desertification destroys millions of hectares of farmland and rangeland each year, further decreasing our ability to produce food.

How Serious Is the Problem? Desertification is a problem rooted in overpopulation, global climate change, and unsustainable land-use patterns. It afflicts numerous countries and regions, including the United States, Africa, Australia, Brazil, Iran, Afghanistan, China, and India. According to the United Nations Environment Programme (UNEP), desertification is occurring on 73% of all the world's rangeland. Desertification is also occurring on a large percentage of the world's cropland. Desertification is worst in Africa, Asia, and Latin America. According to the United Nations, 12 million hectares (nearly 30 million acres) are lost to desertification worldwide every year—that's 57 acres per minute!

Desertification is not just an environmental problem, it causes extreme hardship currently. An estimates 1.5 billion people are directly affected by it, according to the United Nations. Many of them may soon need to leave their homes because of the loss of cropland and rangeland they depend on.

Desertification is not new to humankind. In the ancient Middle East, for instance, the destruction of forests, overgrazing, and poor agricultural practices caused a reduction in the water-absorbing capacity of the land and reduced the amount of rainfall. Coupled with a long-term regional warming trend, the decline in rainfall turned once-productive pastureland and farmland in much of the Fertile Crescent (where agriculture had its roots) into desert.

A more recent example occurred in the United States in the infamous dust bowl era of the 1930s. This disaster resulted from prolonged drought combined with fencepost-to-fencepost cultivation of fields in part to supply Europe with food in the early years of World War II. During the prolonged drought, crops withered and died. Field after field turned into an arid tract of dry dirt. Winds swept the parched topsoil into huge dust storms and carried the topsoil away. Only through extensive conservation measures in the postwar years were farmers able to slowly rebuild

their soils. Today, however, some of these gains have been lost as farmers attempt to raise food production to increase their earnings. Small dust bowls are occurring in southern California, Colorado, and Texas.

Desertification is especially severe in parts of Africa, especially the sub-Saharan region known as the Sahel. Beginning in 1968, a long-term drought in the Sahel (coupled with overpopulation, overgrazing, and poor land management) began the rapid southward expansion of the desert in Ethiopia, Mauritania, Mali, Niger, Chad, and Sudan. The Sahara is also spreading northward.

Desertification and erosion are taking a huge toll on world food production. In Africa, a continent straining under the pressures of 1.1 billion people in 2013, one of every three people do not have enough food to eat. In Chad, Mozambique, Somalia, and Uganda, 4 of every 10 people are malnourished. Food supplies are declining in Latin America as well.

KEY CONCEPTS

Desertification and soil erosion are destroying agricultural land worldwide, contributing to present-day food shortages and reducing our ability to meet future demands caused by the continuously expanding human population.

Farmland Conversion

Besides soil erosion and desertification, valuable farmland is being lost by the conversion to other nonagricultural uses, a phenomenon called **farmland conversion.** Expanding cities, new highways, shopping malls, and other nonfarm uses consume millions of acres of farmland each year in the United States, Canada, and abroad (**FIGURE 10-7**). Farmland lost to development is often been replaced by lower quality farmland. According to the Natural Resources Conservation Service, between 1982 and 2007, the nation's cropland acreage declined from 170 million hectares (420 million acres) to 144.5 million hectares (357 million acres), a loss of slightly more than 1 million hectares (over 2.6 million acres) a year. About 46% of the losses were forest; most of the remainder were cropland and pasture. If this continues, one-third of the United States' rural farmland will be gone within the next 100 years. Although the number of acres under cultivation increased slightly (2 million acres or 0.8 million hectares) from 2007 to 2010, it remains to be seen if this trend will continue.

Farmland conversion is a worldwide phenomenon in both MDCs and LDCs. Little is known about the rate of agricultural land conversion in the LDCs of the world, but it is believed to be substantial. The loss of productive farmland is

Go Green

When you graduate, consider buying an existing home within city limits rather than a new home in the suburbs, most likely built over former farmland or pasture.

FIGURE 10-7 Farmland conversion. Urban sprawl, as shown here in Des Moines, Iowa, swallows up farmland at an alarming rate throughout the world. Although home construction and sales contributes to the economy, once houses and other structures are built, the land is lost from agricultural production, a trend with serious consequences.

Courtesy of Lynn Betts/NRCS USDA.

clearly an unsustainable trend that is made all the more troubling by the continual expansion of the world population and losses from soil erosion and desertification.

 KEY CONCEPTS

Each year, millions of hectares of productive farmland are lost to human development in the United States and abroad—roads, airports, shopping centers, sub-divisions, and so on—a phenomenon called *farmland conversion*.

Declines in Irrigated Cropland per Capita

Water is as essential to agriculture as soil and sunlight. Plants need water to grow. Water plays an important role in photosynthesis—the sunlight-driven conversion of carbon dioxide into organic food molecules.

Most of the world's cropland is nourished by rainfall. However, a growing percentage of the world's cropland is irrigated—supplied with water from streams and lakes (surface waters) or from wells that draw water from the ground (groundwater). In the United States, 17% of all cropland is now irrigated. This land produces approximately one-third of the nation's food. Globally, about 20% of the world's cropland is irrigated, and that farmland produces about 40% of the world's food.

According to the Worldwatch Institute, further increases in irrigation are likely to be modest. Groundwater depletion and rising demand for water in growing urban centers are the main reasons for such a prediction. Groundwater is already being pumped faster than it can be replaced in several major agricultural areas, including regions of China, India, the Middle East, Northern Africa, and the United States. Water shortages are also evident in many other areas.

KEY CONCEPTS

Irrigated cropland supplies enormous amounts of food to the world's people, but the amount of irrigated cropland per capita is on the decline—a trend that bodes poorly for world food production. Measures that increase the efficiency of water use will help ensure an adequate supply of irrigation water.

Although such trends bode poorly for the long-term prospects of world agriculture, there is hope. Numerous sustainable solutions exist, the most important being water conservation (a topic discussed in the next section).

Waterlogging and Salinization

Although irrigation greatly increases food production, in many semiarid regions, it has created some serious problems that could also significantly reduce food production in the coming decades, further adding to declines caused by soil erosion, desertification, and farmland conversion. Two such problems are waterlogging and salinization.

Waterlogging occurs when too much water is applied to farmland. Irrigating poorly drained fields, for example, often raises the **water table,** the upper level of the groundwater (**FIGURE 10-8**). If the water table rises too near the surface, it can fill the air spaces in the soil and suffocate the roots of plants. It also makes soil difficult to cultivate. Worldwide, about one-tenth of the irrigated cropland (an area slightly smaller than Idaho) suffers from waterlogging. Productivity on this land has fallen approximately 20%.

Salinization occurs when irrigation water that has accumulated in the soil evaporates, leaving behind salts and

FIGURE 10-8 **Salinization and waterlogging.** Salts and other minerals accumulate in the upper layers of poorly drained soil (salinization) when irrigation waters raise the water table and water begins to evaporate through the surface. The rising water table also saturates the soil and kills plant roots (waterlogging). Lower arrows indicate the movement of water from groundwater into the topsoil. Upper arrows indicate evaporation.

minerals once dissolved in it (Figure 10-8). If not flushed from the soil, enormous quantities of salts and minerals can accumulate in the soil, greatly reducing crop production and making some soils impenetrable.

Worldwide, about one-tenth of the irrigated farmland is seriously affected by salinization. Another 33% is affected to some degree. In the United States, the productivity of cropland suffering from salinization is believed to be reduced up to 25%. Moreover, when saline buildup reaches a critical level, soil becomes unproductive and must be abandoned. One expert estimates that 1.5 million hectares (2.5 to 3.75 million acres) of land are abandoned worldwide each year because of salinization and waterlogging.

 KEY CONCEPTS

Irrigation can cause waterlogging, the buildup of excess water in the soil, which suffocates plants. It may also cause salinization, the deposition of salts that are toxic to most plants. Waterlogging and salinization affect many millions of hectares of land worldwide.

Declining Genetic Diversity in Crops and Livestock

Before the advent of modern agriculture, grains and vegetables existed in thousands of varieties. Now, only a few of these varieties are commonly used. In Sri Lanka, for example, farmers once planted 2,000 varieties of rice; today, however, only 5 varieties are in use. In India, 30,000 strains of rice were once grown; today, 10 varieties are responsible for about 75% of the nation's rice production.

In most cases, new varieties are chosen because they are more suitable for machine harvesting and because they respond favorably to fertilizer and irrigation water. They also produce higher yields.

A similar trend is occurring on ranches throughout the world as ranchers adopt livestock breeds developed for maximum yield. The problem, as you shall soon see, is that huge expanses planted in one species are extremely vulnerable to pests, disease, and adverse weather. To combat disease and pests, farmers often turn to potentially harmful chemical pesticides.

 KEY CONCEPTS

The number of species of cultivated plants and domestic animals has declined dramatically. Reducing diversity results in huge monocultures of genetically similar plants, which make crops more susceptible to disease, adverse weather, insects, and other pests and more dependent on chemical pesticides.

The Green Revolution The trend toward reduced genetic diversity began with development of high-yield varieties in the 1960s as part of a worldwide agricultural movement called the **Green Revolution.** Research began in 1944, when the Rockefeller Foundation and the Mexican government established a plant-breeding station in northwestern Mexico. The program was headed by Norman Borlaug, a University of Minnesota plant geneticist who developed a high-yield wheat plant for which he was later awarded a Nobel prize. Before the program began, Mexico imported half of the wheat it consumed each year. By 1956, it had become self-sufficient in wheat production, and by 1964 it was exporting half a million tons per year (**FIGURE 10-9**).

The success in Mexico led to the establishment of a second plant-breeding center in the Philippines. High-yielding rice strains were developed at this center and introduced into India in the mid-1960s. Again, the results were spectacular. India more than doubled its rice production in less than a decade.

Important as it was, the Green Revolution contributed greatly to the decrease in species diversity in cultivated crops. One of the most important concerns was that the new crops were not as resistant to diseases and insects. Local varieties of plants are acclimated to their environment. New varieties, on the other hand, often have little such resistance.

Moreover, as diversity dwindles, huge monocultures become more and more common. Expansive fields of one genetic strain facilitate the spread of disease and insects. The potato famine in Ireland in the 1840s is one example of the effects of reducing crop diversity. At that time, only a few varieties of potatoes were planted in Ireland. When a fungus (*Phytophthora infestans*) began to spread among the plants, there was little to stop it. Within a few years, 2 million Irish perished from hunger and disease, and another 2 million left the country.

FIGURE 10-9 The Green Revolution continues. Genetic research and conventional cross-breeding have allowed scientists to transform wheat to high-yielding lines shown here. Land and water scarcities, rising temperatures, and emerging crop diseases challenge breeders to continue improving the crop, and wheat landraces and wild grasses offer valuable sources of new traits for that purpose.

Courtesy of International Center for Maize and Wheat Improvement (CIMMYT).

The decline in genetic diversity resulting from the Green Revolution and other developments adds to the unsustainability of modern agriculture. But it also presents a focal point for change, a way we can reverse the trend and ensure a more environmentally sustainable system of food production.

 KEY CONCEPTS

The Green Revolution was a worldwide effort to improve the productivity of important food crops: wheat and rice. It succeeded in its primary objectives but created a substantial decline in genetic diversity, which makes world food crops more vulnerable to insects, plant pathogens, and other factors.

Habitat Destruction: Contributing to the Loss of Genetic Diversity The loss of genetic diversity among crop species is paralleled by an equally troublesome extinction of wild species throughout the world, especially in the tropics, as a result of habitat destruction. The loss of species

in the tropics could adversely affect modern agriculture. Why? Many modern crops came from tropical regions. Many of their relatives remain there today, growing as they have for centuries. These wild relatives (as well as other species) serve as a source of new genetic information for domestic crops to combat drought, disease, and insects. Keeping modern crops vital and successful, therefore, depends on keeping their relatives alive and well.

 KEY CONCEPTS

The loss of wild plant species that gave rise to modern crop species throughout the world, especially in the tropics, is eroding our capacity to improve crops and make them more resistant to pests, disease, and drought.

Politics, Agriculture, and Sustainability

Governments also add to agriculture's growing list of problems, sometimes fostering unsustainable practices. Consider some examples. **Subsidies,** payments made to farmers from their governments, can contribute to an unsustainable system of farming. In the United States, for example, the federal government subsidizes farmers through a variety of programs. In one program, farmers are given a guaranteed price for certain crops such as corn, cotton, wheat, rice, and soybeans. If the market price of their crop falls below the federal price guarantee, the government makes up the difference. However, this program has many unforeseen consequences. First, it encourages farmers to plant crops that are "insured" by the federal government through price support. As a result, farmers tend to plant one or a few crops. It also encourages farmers to plant all of the land they can, even marginal land that may be easily eroded by wind and water. This practice encourages huge monocultures that are susceptible to insects and other pests. To combat them, farmers rely on an arsenal of toxic insecticides and other chemicals. Many of these chemicals end up in groundwater and in lakes and rivers, where they poison many species.

Government lending policies can also encourage unsustainable practices. In Mexico, for instance, most credit for irrigation systems and roads is given to farmers who produce cash crops such as tomatoes and cattle, both for export to the United States. Cash crop farms and cattle pastures usurp farmland once used to produce crops for domestic consumption.

Governments may also dictate unsustainable agricultural policy. In Ethiopia, farmers have traditionally left semi-arid lands fallow for 7-year periods so that nutrients from the highly weathered, poor soil could be replenished by natural processes. This practice, however, has been condemned by the Ethiopian government, which is interested in increasing farm production. If land is not cultivated within 3 years, it is confiscated. Unfortunately, bypassing the fallow period results in rapid soil deterioration.

Numerous other examples could be cited here. The important point in all of this is that to create a sustainable system of agriculture, laws and policies must be systematically reexamined and carefully revised with sustainability in mind.

KEY CONCEPTS

The problems facing world agriculture are not all technical. Some result from inadequate or self-defeating laws and regulation. Lawmakers throughout the world have unwittingly facilitated the creation of an unsustainable system of agriculture.

10.4 Solutions: Building a Sustainable Agricultural System

Providing food for the growing human population will require a variety of policies and actions—both private and governmental. As is the case with other environmental issues, one of the most important solutions to famine is family planning to slow the growth and perhaps reduce the size of the human population. Efforts must also be made to increase food supplies, and such measures must be sustainable. That is, they must protect and improve the soil, water, and other resources upon which farming is dependent. Most analysts recommend a multifaceted approach that includes efforts to (1) protect existing soil and water resources; (2) increase the amount of land in production; (3) sustainably raise output per hectare of farmland—that is, increase productivity; (4) develop alternative foods; (5) eat lower on the food chain; (6) reduce food losses to pests; (7) increase the agricultural self-sufficiency of less developed nations; (8) enact legislation and policies that ensure a better distribution of food and more sustainable agriculture; and (9) end wars. This section describes each part of this multifaceted strategy.

KEY CONCEPTS

A sustainable system of agriculture consists of practices that produce high-quality food in ways that protect and enhance the long-term health and productivity of soils. Creating such a system will require a multifaceted approach, including measures to slow and perhaps stop the growth of the human population and, ultimately, reduce population size.

Protecting Existing Soil and Water Resources

The old adage that "an ounce of prevention is worth a pound of cure" applies to many aspects of our lives. It also applies to the task of building a sustainable system of agriculture. In fact, few measures are as important to creating a sustainable agricultural system as prevention—notably, preventing soil erosion, desertification, salinization, waterlogging, and farmland conversion.

KEY CONCEPTS

Prevention is vital to creating a sustainable system of agriculture. Protecting soil and water resources from deterioration is the first line of defense in meeting present and future needs for food.

Soil Conservation: Reducing Soil Erosion Protecting soil from erosion by water and wind is one of the most important steps we can take to ensure adequate food supplies both now and in the future—and protect the environment, too. Fortunately, soil erosion can be minimized and even halted by a variety of simple, often cost-effective techniques. This section describes six strategies: minimum tillage, contour farming, strip cropping, terracing, gully reclamation, and shelterbelts.

KEY CONCEPTS

One of the highest priorities in making the transition to a sustainable system of agriculture is putting an end to excessive soil erosion. Fortunately, there are many simple yet effective measures that can help farmers achieve this goal.

Minimum Tillage Farmers typically plow their fields before planting new crops turning crop residue in the soil. They then break up the clumps of soil with a device called a *disc,* making the soil suitable for sowing seeds. In areas where soil is too wet in the spring, farmers plow and disc in the fall, leaving their land barren and subject to wind erosion during the winter.

With special implements, however, farmers can forgo these costly, time-consuming, and energy-intensive steps and plant right over the previous year's crop residue (**FIGURE 10-10**). This technique is one form of **minimum tillage,** or **conservation tillage,** a strategy that reduces the physical disruption of the soil. According to the U.S. Department of Agriculture in 2008, minimum tillage was practiced on about 46 million hectares (113.8 million acres), 41.5% of all U.S. farmland (**FIGURE 10-11**).[1] Unfortunately, this practice is not widely used in other countries.

Because fields are protected much of the year by crops or crop residues, soil erosion can be decreased substantially—in some cases by as much as 90%. Minimum tillage also reduces energy consumption by as much as 80%, saving farmers money, and conserves soil moisture by reducing evaporation from plowed feeds. Crop residues can increase habitat for insects that prey on pests, reducing the need for pesticides and subsequent contamination of the environment. Curtailing the use of heavy machinery on farm fields also has the benefit of reducing soil compaction, which makes soils harder, increases surface runoff, and impairs

---

[1] The Natural Resource Conservation service is no longer collecting data on acres in conservation tillage.

FIGURE 10-10 Minimum tillage planter. This device is designed to dig furrows in the presence of crop residue, avoiding plowing and discing. Seeds are sown at the same time. Leaving the previous crop residue on the land over the fallow period greatly reduces soil erosion.

Courtesy of Ken Hammond/USDA.

root growth. Thus, by cutting back on machinery use, farmers may find that crops actually grow more rapidly and produce more food.

Despite its benefits, minimum tillage has several drawbacks. For example, **herbicides,** chemicals that kill weeds, are often used in place of mechanical cultivation to control weeds. In addition, crop residues may harbor insects that damage crops. Minimum tillage also requires new and costly farm equipment. Already strapped for cash, many farmers can't afford the new equipment. In Canada, this problem has been partially solved by the provincial government of Manitoba, which purchased or leased conservation seeding equipment, then made it available to farmers for a nominal fee. As a result, hundreds of thousands of acres have been brought into minimum tillage.

 KEY CONCEPTS

Reducing the amount of tillage protects the soil from the erosive forces of wind and rain and from compaction. This technique, although effective in reducing energy demand, erosion and labor, often requires additional chemical herbicides to control weeds.

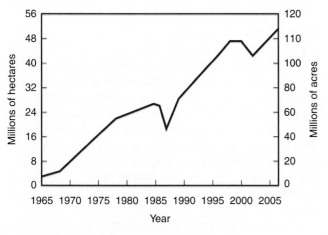

FIGURE 10-11 Growth in minimum tillage in the United States.

Data from USDA Natural Resource Conservation Service. Data no longer being collected.

Contour Farming   On hilly terrain, crops can be planted along lines that follow the contour of the land, a technique called **contour farming** (**FIGURE 10-12**). Planting crops across the direction of water flow on hilly terrain reduces the rate at which water flows across the land, resulting in a 50 to 80% reduction in soil erosion and a marked increase in water retention. This technique, which is widely practiced on farms in hilly terrain, also reduces demand for irrigation water and sediment pollution in streams, rivers, and lakes.

KEY CONCEPTS

Planting crops perpendicular to the slope—that is, along the land contour lines— dramatically reduces soil erosion and increases water retention.

Strip Cropping   As **FIGURE 10-13** shows, **strip cropping** is a measure in which farmers alternate strips of two or more crops on hilly terrain. Strip cropping reduces wind

FIGURE 10-12 Contour farming. This land is farmed along the contour lines to reduce soil erosion and surface runoff, thus saving soil and moisture.

© Clint Farlinger/Alamy.

Chapter 10: Creating a Sustainable System of Agriculture to Feed the World's People   **197**

FIGURE 10-13 Strip cropping. Alternating strips of alfalfa with corn on the contour protects this crop field in northeast Iowa from soil erosion. It can be practiced on hilly terrain to reduce erosion and increase production.

Courtesy of Tim McCabe/NRCS USDA.

and water erosion and increases productivity. For example, farmers may alternate row crops such as corn with cover crops such as alfalfa on hilly terrain. Water flows more easily through row crops and begins to gain momentum, but when it reaches the cover crop, its flow is nearly stopped. Strip cropping can be combined with contour farming to further reduce erosion. As noted in the introduction of this chapter, it can also be used on flat terrain to increase production.

 KEY CONCEPTS

Crops can be planted in alternating strips, a practice called *strip cropping*. When combined with contour farming, this technique greatly reduces soil erosion.

Terracing For thousands of years many peoples have grown crops in mountainous regions using **terraces,** small earthen embankments placed across the slope to check water flow and minimize erosion. Terraces have also been used in the United States for many years on land with less pronounced slope (**FIGURE 10-14A** and **B**).

A University of Nebraska study showed that terracing and contour farming each reduce erosion by 50%. Together, they can reduce soil loss by 75%. Minimum tillage is even more effective. By itself it can reduce erosion rates by 90%. When it is combined with terracing and contour farming, soil erosion rates can be reduced by 98%. Unfortunately, terraces are expensive to construct and may interfere with the operation of large farm equipment. Cost-sharing programs with the U.S. Department of Agriculture can help make them more affordable.

 KEY CONCEPTS

Terraces, small earthen embankments that run across the slope of the land, greatly reduce soil erosion. Contour farming, minimum tillage, and terracing combined then can nearly stop all erosion.

Gully Reclamation Gullies are a danger sign indicating rapid soil erosion. In fact, some gullies can work their way up hills at a rate of 4.5 meters (15 feet) a year. Gullies also make it difficult to farm land by impairing the use of machinery.

To prevent gullies from forming, farmers must dramatically reduce water flow over their land. Contour farming, strip crops, and terraces all help. In existing gullies, small earthen dams can be built across them to reduce water flow, retain moisture for plant growth, and capture sediment, which will eventually support vegetation as erosion is reduced. Seeding also helps to ensure revegetation. Too often, land with severe gullies is abandoned or haphazardly reclaimed, only to suffer worse erosion.

 KEY CONCEPTS

Gullies form quickly on poorly managed hilly terrain and grow rapidly. Preventing them is absolutely essential to protect farmland. If they form, small earthen or rock dams can be built across gullies to stop erosion. They can also be replanted with fast-growing species to prevent further expansion.

Shelterbelts Wind erosion is a major problem, accounting for considerable amounts of soil erosion in many parts of the world. Wind erosion can be controlled by minimum tillage. Shelterbelts, rows of trees planted along the perimeter of crops, are another effective measure aimed at reducing wind erosion. In 1935, the U.S. government mounted a campaign to prevent the recurrence of the disastrous dust bowl days. This program involved planting long rows of trees as windbreaks, or **shelterbelts,** on farms in the Great Plains to slow the winds that carry soil away (**FIGURE 10-15**). Thousands of kilometers of shelterbelts were planted from Texas to North Dakota.

Besides decreasing soil erosion from wind, shelterbelts prevent snow in fields from being blown away, thus increasing soil moisture. This in turn reduces the

FIGURE 10-14 Terracing. (a) This Iowa corn field is protected from erosion by the aid of terraces, small earthen embankments (in green) that reduce water flow across the surface of the hilly terrain. The corn is planted in the stubble of last year's crop. **(b)** In Asia, terracing has been used in mountainous regions to grow rice.

Photo a courtesy of Tim McCabe/NRCS USDA; photo b © John Bill/Shutterstock, Inc.

FIGURE 10-15 Shelterbelts. Rows of trees planted along the margin of farm fields protect farmland in Michigan from the erosive effects of wind.

Courtesy of Erwin Cole/NRCS USDA.

demand for irrigation water during the growing season and helps replenish groundwater supplies. Shelterbelts can also improve irrigation efficiency by reducing the amount of water carried away from sprinklers by wind. In addition, shelterbelts provide habitat for animals, pest-eating insects, and pollinators. They also protect citrus groves from wind that blows fruit from trees. Planted around farmhouses, shelterbelts also reduce heat loss on windy winter days, increasing comfort and saving farmers money.

Overcoming the Economic Obstacles to Soil Erosion Controls   Soil erosion control measures like those just discussed are some of the most important measures needed to build a sustainable system of agriculture. In many less developed nations, however, farmers struggle to meet their basic needs. They may have neither the time nor the means to care properly for the land. Few see the benefits of soil conservation because the gains tend to materialize slowly.

Economics and short-term thinking impair soil-erosion control efforts in the more developed nations as well. Caught between high production costs and low prices for crops, farmers often ignore soil erosion and offset any losses in production by applying synthetic fertilizers. These additives artificially help farmers maintain yield, despite the loss in soil and valuable soil nutrients. Despite the importance of soil erosion controls to sustainable agriculture, many farmers are reluctant to invest in these practices. Such a view, while understandable, ignores the long-term cost of permitting unsustainable loss of soil—an erosion of the productive capacity of farmland.

Governments can promote soil conservation in a variety of ways. One landmark attempt to end the devastating loss of U.S. topsoil was the **1985 Farm Bill.** This law created a land conservation program (the **Conservation Reserve Program**) that directs the federal government to pay farmers to retire their most highly erodible cropland from production for 10 years and plant trees, grasses, or cover crops to stabilize and rebuild the soil. In 2011, 12.6 million hectares (31.1 million acres) were in the program at a cost of $1.8 billion per year.

The 1985 Farm Bill also established a federal program that called on farmers to develop soil erosion plans in exchange for eligibility for federal crop insurance, subsidies, and other benefits. So far, 1.5 million farmers have signed up for the program and have worked with the Natural Resource Conservation Service to develop plans for 134 million acres, about 25% of U.S. farmland. These efforts could cut soil erosion on some of the most productive soils in the country. The 1996 Farm Bill provided additional support for these efforts by providing money to assist farmers technically and financially with conservation measures. Half of the money is earmarked for livestock operations. This program is known as the **Environmental Quality Incentives Program** and was funded at 1.75 billion in 2012. Facing ever-increasing water shortages much of the money is used to conserve water—in example, by installing water-efficient irrigation systems. Cost–benefit calculations showed that for every

dollar invested in the program, the nation saved two dollars in reduced erosion and pollution.

Preventing Desertification   Protecting soil from desertification is also needed to preserve cropland and rangeland. Several of the measures designed to prevent soil erosion will be helpful in achieving this vital goal. For example, shelterbelts protect soil from the drying effects of wind. More direct actions are also possible. In China, for instance, agricultural officials have embarked on an ambitious program to plant a 4,480-kilometer (2,800-mile) "green wall" of vegetation to stop the spread of desert in the northern region (see **Spotlight on Sustainable Development 10-1**). Contour farming and terracing both increase soil moisture content and combat desertification. In Australia, huge semi-circular banks of soil are created in the windswept plains to catch seeds and encourage regrowth in areas denuded by livestock. Better land management—for example, controls on the number of livestock on rangeland—is also necessary. Pollution controls to slow down global warming are also vital to this effort.

Reducing Farmland Conversion   Efforts to prevent the spread of cities, the proliferation of highways, and other nonfarm uses of arable land are also vital to protecting farmland and creating a sustainable future. Slowing the growth of the human population is essential. Careful city planning and new zoning laws could help reduce farmland conversion by ensuring that homes, roads, airports, and businesses are not built on agricultural land.

Saving Irrigated Cropland/Using Water More Efficiently   As you learned earlier in this chapter, irrigated cropland produces a large portion (40%) of the world's food supplies, but huge expanses of irrigated land are being

10-1   The Green Wall of China: Stopping the Spread of Desert

China is a nation in trouble. With a growing population of nearly 1.4 billion people, China's land is falling into ruin. Centuries of overgrazing, poor agricultural practices, and deforestation have resulted in severe erosion and rapidly spreading deserts that gobble up the once-productive countryside.

The spread of deserts affects the lives of millions of peasants in China. In the highlands of northern China, for example, the land is cut with gullies, some hundreds of meters deep. Erosion from the raw, parched Earth is an astounding 30–40 metric tons per hectare (12–16 tons of topsoil per acre) per year—far above replacement level. In all, some 1.6 billion metric tons (1.45 million tons) are carried into the Yellow River annually, making it one of the muddiest rivers in the world.

According to one estimate, an area larger than Italy has turned into desert or semidesert in China. Although most of the desertification is occurring in northern China, few areas are immune to this problem.

In 1978, the Chinese government launched a 73-year reforestation project to stem the tide. By planting trees, shrubs, and grasses in semiarid regions, they hope to form a giant green "wall" across the northern reaches of the nation. When completed, the wall will extend more than 4,480 kilometers (2,600 miles) and will be 400 to 1,700 kilometers (250 to 1,000 miles) wide. This ambitious project is designed to return much of the land now falling into ruin to productive use.

To date, over 25 million hectares (62 million acres) have been replanted, and 6 million hectares (15 million acres) were naturally regenerated in mountainous and sandy areas. By 2050, China plans to replant another 10 million hectares (25 million acres). Unfortunately, as noted shortly trees don't grow well in arid regions. They are not part of the ecosystem. Recent studies have shown that manually planted tree seedlings and tree seeds fare poorly in the arid climate. Massive losses have been experienced in many areas over years or decades. The trees have depleted local soil moisture and died en masses, according to a study published in *Earth Science Reviews*.

The Yulin district is one of China's success stories. Before 1949, more than 400 villages and six towns had been invaded or completely covered by sand. Today, four major tree belts have been planted in the area, decreasing the southward push of the desert by 80% (**FIGURE 1**). Towering sand dunes now peep through poplar trees, and rice paddies sparkle in the sunshine. Grain production has been replaced by a diversified agricultural system, including animal husbandry, forestry, and crop production. The trees provide shade and help reduce wind erosion that causes the sand dunes to shift. Shrubs and grasses now thrive on land once stripped of its rich vegetation. Trees and shrubs grow in gullies, and grasses carpet slopes. Revegetation helps to hold the soil in place and reverse the local climate change.

Local residents have built a diversified economy in what was once an unproductive desert. Juice from the cherry trees, which thrive in the desert climate, is rich in vitamin C and amino acids; it is used to produce soft drinks, preserves, and beer. Twigs of the desert willow are used to make wicker baskets and trunks, which earn local residents $2 million in U.S. currency every year.

Despite the encouraging signs in China, a report by the Shanghai-based *World Economic Tribune* says that while nearly 10 million hectares (4,000 square miles) were planted each year, twice that amount is still being lost. In the northern province of Heilongjiang, home of China's largest concentration of virgin forest, loggers

FIGURE 1 Greenbelt. These trees planted in China are part of the effort to stop the spread of desert.
© Forrest Anderson, Rouviere Media.

reduced the tree cover from 50 to 35% in 30 years. Government pricing policies promote overcutting.

The reforestation project has been plagued by a shortage of money and technical expertise. In addition, survival rates of trees has been low, except in a few locations. Single-species forests are extremely vulnerable to disease, lack of water, and soil nutrient depletion, combined with disease have resulted in a massive die-off of many reforested areas, thus reducing the number of acres officially designated as "replanted." Some Chinese scientists have found better success by simply fencing areas off, thus restricting human use. This helps restore native grasses and could be a much more successful way of stopping the spread of deserts.

Reforestation is an attempt to restore the Earth's vegetative surface, which is vital to building a sustainable future. Efforts such as China's are needed on most continents to help reverse centuries of land abuse that have spawned the spread of deserts. Replanting forests not only slows the spread of deserts; it also helps restore wildlife populations and can help reduce global warming, which now threatens the world climate.

Adapted from: L. Ming (1988). Fighting China's Sea of Sand. *International Wildlife* 18(6):38–45, with permission.

salinized and waterlogged. Groundwater supplies are also declining very seriously in key agricultural areas such as the midwestern United States. You also learned that the growth in irrigated cropland is not keeping pace with population growth. One of the reasons for this is a lack of available surface water and groundwater.

One way to solve all of these problems is to use existing water more efficiently. The efficient use of resources, of course, is a key to building a sustainable society. Farmers, for instance, can improve irrigation efficiency through many simple and cost-effective measures. Lining irrigation ditches with cement or plastic can cut water losses by 30 to 50%, thus freeing up tremendous amounts of water to irrigate other land. Transporting water in pipes can result in even greater savings. Subsidies that help support these costly efforts can assist farmers all over the world. It is taxpayer money well spent, say supporters. Farmers can also use drip irrigation systems to deliver water directly to the roots of fruit trees and a few other crops (**FIGURE 10-16**). Conventional center pivot irrigation systems that once sprayed water into the air (with tremendous losses to evaporation) are modified to spray water downward, creating considerable savings (**FIGURE 10-17**). Computer systems can help farmers monitor soil moisture so they apply water only when it is needed and in the amount required by crops. Applying irrigation water at night or early and late in the day when evaporation is lowest can reduce demand by 50% or more. Improvements in soil organic matter through application of manure and other measures can also help because organic matter acts as a sponge, holding water in the soil. Efforts to protect aquifer recharge zones, areas in which groundwater supplies are recharged, can also ensure farmers a reliable supply of water.

KEY CONCEPTS

Numerous water efficiency measures can help reduce waste and free up water to expand irrigated cropland.

FIGURE 10-16 Increasing the efficiency of irrigation. Trickle systems deliver water to roots, cutting evaporation losses substantially. Trickle systems can be used only for certain crops, among them grapes (shown here) and fruit trees.

Courtesy of Lynn Betts/NRCS USDA.

Preventing Salinization and Waterlogging All water efficiency measures help increase irrigation water and expand cropland under irrigation, but certain measures can reduce salinization and waterlogging. This, in turn, could reduce the annual loss of farmland. By using computerized sensors that measure soil moisture, for example, farmers can apply only the amount of water needed by crops. This not only frees up water for other crops, it reduces salinization and waterlogging. Special drainage systems can also be installed to draw off excess water and prevent the buildup of salt and the potential for waterlogging, although such measures are quite costly. Unfortunately, water drained from farm fields may carry high levels of potentially toxic substances. Special care must be taken to avoid solving one problem (salinization) while creating another (surface water pollution). Government programs could also discourage irrigation in soils susceptible to these problems.

KEY CONCEPTS

More frugal application of irrigation water to crops and special drainage systems can reduce salinization and waterlogging in soils susceptible to the problem.

Soil Enrichment Programs

Soil erosion controls help preserve farmland from being washed or blown away. Because soil contains nutrients that are essential to plant growth, soil-erosion control -methods also protect soil fertility. As any good farmer will tell you, protecting or even enhancing the fertility of the soil is vital to maintaining or increasing its productivity.

(a)

(b)

FIGURE 10-17 Center pivot irrigation. (a) The standard device sprays water into the air, but much of the water evaporates before it hits the ground on hot days. **(b)** By turning the spray nozzles downward, much more water makes it to the crops and reduces agricultural water consumption.

Photo a © Muriel Lasure/ShutterStock, Inc.; photo b courtesy of Gene Alexander/NRCS USDA.

Soil fertility can be enhanced by the use of fertilizers and crop rotation. In many locations throughout the world, agricultural soil fertility is maintained by applying human wastes from homes or sewage treatment plants—an activity that returns the nutrients of foods we eat to their origin.

Organic Fertilizers One of the most effective means of replenishing soil fertility is to apply organic fertilizers to cropland and pasture. **Organic fertilizers** can be waste materials such as cow, chicken, and hog manure and human sewage. All of these replenish the soil's organic matter and add important soil nutrients such as nitrogen and phosphorus.

There are many sources of organic fertilizer. Currently, there are hundreds of millions of sheep, cattle, hogs, and other animals that produce billions of tons of waste. Putting it to good use—making cropland more productive—makes great sense. It helps recycle nutrients and prevents pollution of water supplies.

Organic fertilizers also include **green manure**—plants grown in a field that, rather than being harvested, are plowed under. Especially valuable are the leguminous plants such as alfalfa and clover, which are often grown during the off-season and plowed under before food crops are planted.

Soil enrichment with organic fertilizers of all sorts provides many benefits vital to building a sustainable agricultural system. First and foremost, it increases or helps to maintain soil fertility and crop yield. Because organic matter acts like a sponge in the soil, organic fertilizers increase the water-retention capabilities of the soil. Organic matter also provides an environment conducive to the growth of bacteria necessary for nitrogen fixation. Organic fertilizers help prevent shifts in soil acidity, and they tend to prevent the leaching of minerals from the soil by rain and snowmelt. In addition, careful application of human wastes on farmland helps to reduce water pollution by municipal sewage treatment plants.

Organic wastes have been successfully applied in some countries, such as China and India, for many years, but this practice is not without its problems; one of these is the cost of transporting waste to farms by pipelines or trucks. Another problem is that organic waste

GO GREEN

Grow some of your own food or to help your parents grow some of their food by starting an organic container garden on your deck or a small organic garden in your backyard. Many vegetables grow very well on sunny decks. This reduces visits to the grocery store, pesticide use, and energy required to deliver vegetables to local stores.

from municipal sewage treatment plants may be contaminated with pathogenic (disease-causing) organisms such as bacteria, viruses, and parasites. Theoretically, some of these organisms could be taken up by crops and therefore enter the human food chain. In industrialized nations, municipal waste may be contaminated with toxic heavy metals such as mercury and lead coming from factories connected to the plant. Better controls at sewage treatment plants or at the factories that produce these materials could alleviate the problem.

You can help promote the use of organic fertilizer by buying organically produced food from your grocery store or by purchasing from local growers either through farmer's markets or through a community-supported agricultural program that is described in **Spotlight on Sustainable Development 10-2.**

Synthetic Fertilizers In the more developed countries such as Canada, Australia, and the United States, farmers apply millions of tons of **synthetic fertilizer** to their land each year to boost crop production. Without these fertilizers, world food production would fall 40% or more, according to the Worldwatch Institute.

Synthetic fertilizers contain three nutrients: nitrogen, phosphorous, and potassium. Because of this, synthetic fertilizers only partially restore soil fertility. They do nothing to replenish organic matter or micronutrients (nutrients required by plants in very small quantity) necessary for proper plant growth and human nutrition. On land that is fertilized solely with synthetic fertilizers, soil fertility slowly declines over time. Moreover, excess fertilizer may be washed from the land by rains and end up in streams, causing a number of problems. To prevent the gradual depletion of nutrients and to help develop a more sustainable society, many agricultural experts call for much wider use of organic fertilizers.

Crop Rotation In modern agriculture, synthetic fertilizers and pesticides have allowed farmers to grow the same crops year after year on the same plots. This way, farmers can concentrate their efforts on crops that they know well. However, this process is generally viewed as unsustainable because it often depletes soil nutrients, increases soil erosion, and creates serious problems with pests and crop pathogens.

Crop rotation is a practice in which farmers alternate the crops they plant in their fields from one season to the next; for example, corn may be planted for 1 or 2 years, followed by alfalfa, a cover crop. The cover crop reduces soil erosion and replenishes soil nitrogen. If the cover crop is plowed under, it helps to replenish organic matter and return a variety of valuable nutrients to the soil. Therefore, this simple practice helps build the soil. Corn is commonly rotated with soybeans, a legume that adds nitrogen to the soil. Crop rotation has two additional benefits: It reduces pest damage and reduces the need for costly and potentially harmful chemical **pesticides** (substances that kill damaging insects, weeds, and other species). Properly planned and executed, crop rotation can boost yields by 10 to 15%.

> ### KEY CONCEPTS
>
> Crop rotation, alternating crops planted in a field one season after another, offers many benefits. Planting the proper crops can help replenish soil nutrients. It also helps reduce erosion, pest damage, and the need for costly and potentially harmful pesticides.

Increasing the Amount of Land in Production

There are several ways of meeting increasing demand for food. The previous section described a preventive approach—strategies such as erosion control and organic fertilizers. The other strategy is to develop new supplies. In the past, virtually all nations solved the problem of rising food demand by opening up new lands to the plow. Today, however, that option is quickly vanishing. In most parts of the world, potentially arable land is in short supply. In most of the major industrial nations, for example, farmland reserves are relatively small. In Canada, less than 5% of the land is capable of supporting crops, and virtually all of that land is in production. In Southeast Asia, 92% of the potential agricultural land is being farmed. In southwestern Asia, more land is currently being used than is considered suitable for rain-fed agriculture. Consequently, per capita food production in Asia has begun to fall as the population continues to grow.

For those countries that have little farmland reserve, efforts to protect soils, manage urban growth, and reduce population growth offer the greatest hope for meeting future food demand.

Africa and South America have large surpluses of land that could be farmed. Some experts believe that the nations of these continents should develop this land. However, much of this land is currently covered by tropical rain forests. As you may have learned elsewhere, although the tropical rain

forests are rich in plant and animal species, the soils they grow on are poor in nutrients. When stripped of vegetation, these soils also are prone to erosion in the intense tropical rains and may become hardened when exposed to sun. The potential for expansion on these continents is therefore not as great as some would lead you to believe.

Tapping unfarmed grasslands may be an option in some areas. These soils are rich and productive, but this strategy could severely deplete wildlife populations and disrupt many of the free services provided by nature—for example, flood control and local climate control.

> ### KEY CONCEPTS
>
> Grasslands and forests can be converted to farmland to meet the rising demand for food. In many parts of the world, though, and especially in the more developed nations, these farmland reserves are small. Even in countries where there is an abundance of reserve land, much of this land is covered with poor soils. Furthermore, the ecological cost of converting wild land to farmland would be enormous.

Increasing the Productivity of Existing Land: Developing Higher Yield Plants and Animals

Another way to increase food supplies is to develop new, higher yield plants and animals. New, high-yield varieties of rice and wheat developed during the Green Revolution, for example, produce three to five times as much grain as their predecessors. New varieties of plants are created by breeding closely related plants to combine the best features of the parents; these are called **hybrids.**

As the new hybrids were introduced into many poor nations, the hopes of the Green Revolution dimmed, however. Farmers soon found that the hybrids required large amounts of water and fertilizer, which were unavailable in many areas. Without these inputs, farm yields were not much higher than those of local varieties. In some cases, they were lower. Another problem was the high cost of the new varieties, which prevented many small farmers from buying them. Moreover, new plants were often more susceptible to insects and disease.

These setbacks stimulated new research to produce varieties that would increase productivity without huge inputs such as fertilizer. Today, plant breeders throughout the world are attempting to develop crops with a high nutritional value and greater resistance to drought, insects, disease, and wind. Plants with a higher photosynthetic efficiency are also in the offing. Efforts are even under way to incorporate the nitrogen-fixing capability of legumes into cereal plants such as wheat—a change that would decrease the need for fertilizers and reduce nitrogen depletion.

One exciting improvement is a new variety of corn, a staple for 200 million people worldwide. Because corn is such an important source of calories and protein, researchers spent

10-2   Community-Supported Agriculture

Community-supported agriculture (CSA) "is an innovative and resourceful strategy to connect local farmers with local consumers to develop a regional food supply and strong local economy," according to the University of Massachusetts Extension service, whose website contains a wealth of information on this idea.

The idea began in Japan in the very early 1970s and has since spread to other parts of the globe, including North America. No one knows the exact number of CSAs in the United States, but the number could be as high as 6,500, according to one source. Although each group is unique, community-supported agriculture generally involves two parties: a local farmer—typically a small farmer who produces crops and sometimes beef, pork, or chickens organically—and a group of residents in a nearby city or town who constitute the members. The members purchase produce directly from the farmer. In most groups, says author Sarah Milstein in an article in *Mother Earth News*, "Members pay ahead of time for a full season with the understanding that they will accept some of the risks of production." If the cucumbers produce poorly, so be it. You don't get cukes from the CSA. If zucchini grows particularly well, which is almost a given, you'd better learn how to make zucchini bread. "In other groups, members subscribe on a monthly basis and receive a predetermined amount of produce each week," she adds.

Produce is either picked up at the farm by members of the group or delivered to a central location by the farmer. Members of each group generally receive eight or more different types of vegetables each week—starting in the spring and continuing well into the fall. However, "Some groups offer fruit, herbs, flowers, bread, cheese, eggs, yogurt, beef, honey, maple syrup, and most anything else you can produce on a farm," says Milstein.

CSA programs vary in size. Slack Hollow Farm in Argyle, New York, has a dozen members. They cover only a small percentage of the farm's operating budget. The rest of the produce from their 7-acre farm is sold through a local food co-op. Across the Hudson River from Slack Hollow Farm in New York is Roxbury Farm, where growers cultivate 25 acres and sell to 700 members whose purchases cover about 90% of the farm's annual operating expenses (**FIGURE 1**).

FIGURE 1 Shopping for produce at a community-supported agriculture farm.
Courtesy of The Food Project (www.thefoodproject.org).

Community-supported agriculture is a win-win situation. Farmers acquire upfront funding, so vitally needed before the growing season starts, and they cultivate relatively secure markets. By growing a diverse array of crops, farmers reduce pest problems, increase soil fertility, and ensure a decent harvest. You, the consumer, receive an abundance of healthy food at a good price—provided the weather cooperates.

The environment benefits, too. Food produced without pesticides ensures healthy soil and healthy neighboring ecosystems. The birds that frequent the hedge rows surrounding farm fields live healthy lives, gobbling down tons of insects they harvest from the field and feeding the rest to their offspring.

In addition, because food is grown only a few miles from members' homes, rather than on distant farms hundreds or thousands of miles from the dinner table, very little energy is required to transport food to market. The less energy, the less pollution.

Most, if not all, community-supported agriculture operations are initiated by farmers looking for a secure local market for their products. They recruit members through word-of-mouth, brochures, flyers, media coverage, and other methods. If you're interested in such a program, ask around, and do some research on the web. Attend local farmer's markets and ask whether any of the participating farmers also engage in CSA or know farmers in the area who do.

Americans interested in finding a local CSA program can also check with the local U.S. Department of nearly two decades developing a new variety, quality-protein maize (QPM). Studies show that only about 40% of the protein in common corn is digested and used by humans. In contrast, roughly 90% of QPM's protein can be digested and used. In addition, QPM also produces 10% more grain. In areas where corn is a staple, such as Africa and Mexico, QPM could help curb hunger. QPM could help the residents of these areas become more self-sufficient in food production. Efforts are already under way to increase dramatically the use of this type of corn.

Some researchers are also exploring the use of perennial crops for agriculture. A **perennial** is a plant that grows from the same root system year after year—such as grasses. Today, most agricultural crops are **annuals,** plants that only last one season. Annuals are grown from seeds each year. Preliminary research suggests that productivity from perennials may be equal to or slightly lower than conventional annuals such as wheat, but the benefits from soil conservation, soil-nutrient retention, and energy savings may overwhelmingly favor them. Just as new varieties of plants help increase yield, so do fast-growing varieties of fowl and livestock.

KEY CONCEPTS

Numerous efforts are under way to safely increase the yield of plants and the growth rate of animals to increase food production.

Selective Breeding and Genetic Engineering Efforts are being made to improve plants and livestock by **selective breeding.** In selective breeding, organisms containing valuable characteristics are bred in hopes of acquiring offspring with these characteristics. This technique, used for hundreds of years, is effective but rather slow. Another more recent development being used to improve livestock and crop species is genetic engineering.

Genetic engineering is a complex process designed to mechanically transfer desirable genes into the genetic material of an organism. Genes that make cattle grow faster, for example, can be transferred into the ova of cattle. If the genes are incorporated into the DNA of the ova, the offspring would then carry this gene and pass it on to their offspring.

Genetic engineering may be used in other ways that improve agriculture. A new strain of bacteria developed by scientists at the University of California, for example, inhibits the formation of frost on plants, which may provide farmers with a way of reducing crop damage from early frost. Other scientists are working on genes that give plants resistance to herbicides used to control weeds. A group of scientists at the Monsanto Company has developed a strain of bacteria that grows on the roots of corn and other plants. When eaten by insects, the bacteria release a toxic protein that kills the pest. Geneticists have also introduced certain genes that allow oats to thrive in salty soils, and researchers are experimenting with genetically engineered bacteria that could help plants absorb nutrients more efficiently, thus increasing crop yields. In Canada, genetic engineers have developed a new strain of potatoes that poison Colorado potato beetles. Animal geneticists are now working on ways to improve livestock, combining genes from one species with those of another to improve efficiency of digestion, weight gain, and resistance to disease.

Genetic engineering was once touted as a savior for world agriculture. However, researchers are finding that it is more difficult to apply to agriculture and much slower than proponents once thought. Nevertheless, the successes of genetic engineering have fostered extraordinary enthusiasm in the business community. Dozens of new companies have formed in recent years, and billions of dollars have been invested in the fledgling industry. Still, safety questions remain. Will the genetically engineered bacteria escape into the environment, upsetting the ecological balance? Experts agree that, once unleashed, a new form of bacterium or virus would be impossible to retrieve. Controlling it could prove costly and damaging.

Some individuals have criticized genetic engineering as a means of tinkering with the evolutionary process. Deliberate genetic manipulations, such as the transfer of chromosomes from one species to another, are different from anything that ordinarily occurs during evolution. Is it right, critics ask, to interfere with the genetic makeup of living organisms? Will these intrusions alter the evolution of life on Earth?

Recent research suggests that the dangers of genetic engineering have been blown out of proportion and that genetically engineered bacteria are not generally a threat to ecosystem stability. Most scientists agree.

At least two studies now indicate that genetically engineered bacteria that are applied to seeds and take up residence on the roots migrate very little from the site of application in the short term. Critics, however, are concerned with long-term consequences.

Many ecologists want to see the industry properly monitored and advocate careful testing before release. Genetically altered species, they say, are analogous to alien species introduced into new environments. The history of such introductions has been fraught with difficulties. In Canada, the **Food Inspection Agency** has assumed responsibility for screening genetically modified foods. They review applications and determine whether new products can be used or subjected to testing.

Genetic engineering stands at a threshold. Still in its infancy, it offers an unparalleled opportunity for improving agriculture and animal husbandry. At the same time, this revolutionary science carries with it an unknown potential for environmental harm that concerns many of its critics.

KEY CONCEPTS

Geneticists can improve plant and animal strains by selective breeding and genetic engineering. Selective breeding has been used for hundreds of years. Genetic engineering is the deliberate transfer of genes from one organism to another. This technology has an enormous potential but poses many ethical questions and may create some serious environmental problems.

Protecting Wild Plant Species: Habitat Protection and Germ Plasm Repositories Tropical rain forests and other valuable habitats are being destroyed in record numbers to accommodate growing human populations. The loss of wildlands wipes out wild species, including the ancient relatives of modern crop species such as corn. The loss is not only aesthetic but also economic, for wild species from which modern crop species were derived may possess genes that confer resistance to drought, insects, and disease. These genes could be transferred from ancient plants to modern crop species with relative ease. Such genetic boosts are extremely important to the success of modern agriculture. In fact, in a period of 60 years corn harvests climbed from 20 bushels per hectare to 100 to 250 bushels—because of genetic improvements. Consequently, plant breeders are always looking for species that could help produce hardier, more resistant plants.

In 1979, a team of researchers from the United States came across a weedy grass species in Mexico that gave rise to modern corn. Only a few patches were left. This species contains genes that could provide modern corn plants with resistance to several costly diseases. Some researchers are also using the plant to develop a strain of corn that grows in fields much like grass—year after year from the same root structure. As noted earlier, plants such as this are called *perennials*. If successful, this perennial would save farmers the cost of plowing, sowing, and cultivating—and would reduce soil erosion. The potential economic and environmental benefits are enormous. This example points out the importance of protecting rain forests and other valuable habitats—and their immense genetic reservoir.

Because much of the habitat of many organisms is bound to be destroyed by bulldozers and chain saws, biologists have been searching through forests and fields to gather seeds for cold storage in genetic repositories (**FIGURE 10-18**). There they can be held for future study and possible use. The U.S. Department of Agriculture currently runs the National Germ Plasm System, which has (as of January 2011) over 12,000 species in stock with 535,000 samples from 13,300 species. They're adding 10,000 to 20,000 new samples per year.

In 1985, the less developed nations belonging to the U.N. Food and Agriculture Organization (FAO) voted to establish a worldwide system of storing seeds, cuttings, and roots from native plants, to be available to all countries. This system was created in large part to thwart genetic imperialism by the more developed nations, whose companies collect plants and seeds from LDCs and extract genes and important medicinal drugs, sometimes reaping huge profits while the less developed nations receive little or no benefit.

However important genetic repositories are to the immediate goal of protecting the world's species, this strategy has some major drawbacks. First, despite storage at low tempera-ture and humidity, many seeds decompose and must be replaced. Others undergo mutations when stored for long periods and are no longer useful. Finally, storage systems will not work for potatoes, fruit trees, and a variety of other plants.

FIGURE 10-18 Genetic repository. This room in a government facility is home to cuttings and seeds from thousands of plant species that are being preserved for future use.

Courtesy of Scott Bauer/USDA.

Developing Alternative Foods

Another strategy that may help expand food supplies is the development of alternative food sources—that is, new plants and animals taken from the wild, cultivated on farmlands, or grown in captivity.

Native Species As Sustainable Food Sources Many native plants and animals not currently eaten by humans could help meet the rising demand for food. The winged bean of the tropics, for example, could become a valuable source of food because the entire plant is edible: Its pods are similar to green beans. Its leaves taste like spinach. Its roots are much like potatoes, and its flowers taste like mushrooms. Food scientists are looking for other plants with similar potential.

Native animals may also provide an important, sustainable food source in years to come. In Africa, for instance, native grazers are already becoming a major source of protein. Native grazers are far superior to cattle introduced from Europe and America because they are genetically resistant to local diseases and rarely overgraze grasslands, unlike cattle. Native grazers also generally convert a higher percentage of the plant biomass into meat and may be cheaper to raise.

Fish from the Sea and Aquaculture Fish provide about 17% of the total animal protein consumed by the world's population and 6.5% of all protein (plant and animal), according to the U.N. Food and Agriculture Organization. Although three-quarters of the fish catch is consumed in the more developed nations, fish protein is important to the people of many poorer countries, in many cases supplying 40% of the total animal protein consumed.

During the 1970s, the world catch stabilized between 66 and 74 million metric tons a year. In the 1980s, it began to climb again, peaking at 96 million metric tons in 2000. It has remained around 90 million metric tons since 2001 (**FIGURE 10-19**).

The future of the world's fisheries is in jeopardy. Of 15 major oceanic fisheries, 11 are in decline. Continued fishing will inevitably result in dramatic population decreases in many commercially important fish species. Depletion of stocks,

FIGURE 10-19 World fish catch. (a) Freshwater fish production including aquaculture. **(b)** Marine fish catch. Note that marine fish catch has slightly declined, while freshwater production has steadily increased primarily as a result of aquaculture.

Data from Food and Agriculture Organization of the United Nations. *The State of the World Fisheries and Aquaculture, 2012.* www.fao.org/fishery/sofia/en.

called **overfishing**, results when commercial fishing interests deplete the breeding stock so that a natural fishery cannot be maintained. Protecting current stocks—so vital to maintaining a sustainable harvest—will require global cooperation.

Each year, millions of tons of fish that are netted by commercial fishing interests are thrown overboard because they are the wrong species or sex or are undersize. International fishing regulations require this to protect various species and populations. Undersize fish represent the future reproductive stock. Nontarget species are called *by-catch* and can amount to 15% of the total catch.

Unfortunately, most of the fish are dead before they hit the water. Because of this, efforts are under way to allow commercial fishing interests to keep by-catch and put it to good use. Fish discarded by Alaskan fishing companies would provide 50 million meals a year.

Although this may sound like a reasonable policy, consider the other side of the issue. The regulations were drawn up in an attempt to prevent trawlers from circumventing fishery management regulations designed to help sustain and rebuild fisheries. If by-catch is permitted to be kept, unethical companies could fish with abandon, pulling in whatever they wanted, potentially wiping out important fisheries. The answer may be in the development of better fishing gear that eliminates undersize fish.

Barring any substantial increases in fish from the sea, many observers believe that one of the world's greatest hopes for increasing fish and shellfish production is **fish farms,** commercial endeavors where fish are raised in ponds and natural bodies of water. Fish farms are already common in many parts of the world, and have dramatically increased the availability of fish as shown in **FIGURE 10-20**. Fish farms are forms of aquatic agriculture, called **aquaculture** in freshwater and **mariculture** in salt or brackish water.

Two basic strategies are employed in fish farming. In the first, fish are grown in ponds; population density is high and is maintained by intensive feeding, which is costly and therefore not always suited to the LDCs. Fish (and shellfish) can also be maintained in enclosures or ponds where they feed on algae, zooplankton, and other fish that occur naturally in the aquatic ecosystem. This system requires little food and energy and is quite suitable for less developed countries.

Worldwide, fish farms produced 52.5 million metric tons of food a year in 2008. Intensified efforts could double or triple this amount, providing additional food for the growing population.

KEY CONCEPTS

Most of the world's commercially important saltwater fish populations are in decline and in danger of being seriously depleted. The decline in wild fish populations has forced many countries to grow fish commercially in ponds, lagoons, and other water bodies.

Eating Lower on the Food Chain

According to one estimate, current food supplies would feed only approximately 2.5 billion people—slightly more than one-third of the world population—if everyone ate like Americans. If, however, everyone ate a subsistence diet, getting only as many calories as needed, the annual world food production could supply an estimated 6 billion people, still short of the current world population.

Armed with statistics such as these, some people propose that wealthier citizens of the world can contribute to solving world hunger by eating less and eating lower on

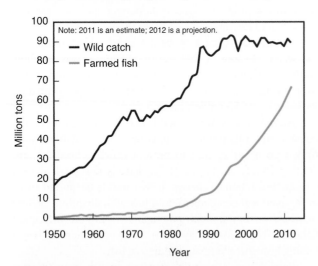

FIGURE 10-20 Fish farming on the rise. Fish farming has, in recent years, dramatically increased the supply of fish protein, helping offset the plateau in wild fish catch.

Data from Food and Agriculture Organization of the United Nations. *The State of the World Fisheries and Aquaculture, 2012.* www .fao.org/fishery/sofia/en.

the food chain. The logic behind this idea is that by eating less and consuming more grains, vegetables, and fruits—and less meat—citizens of the more developed nations would free grain for the less developed nations. A 10% decrease in beef consumption in the United States, for instance, would release enough grain to feed 60 million people in the LDCs. (Many more people could be sustained on a vegetarian diet.)

Although such a diet would be much healthier and would help reduce obesity and numerous health problems, which now afflict a large portion of American children and adults, the problem with this answer to world hunger, say critics, is that sacrifices on the part of the wealthy would most likely not translate into an increase in availability abroad. Another problem is that a large portion of land suitable for grazing is often not arable.

This is not to say that a vegetarian or a meat-conservative diet isn't desirable. It is. It is healthier and consumes far fewer resources. If there is a lesson to be learned from this issue, it is that to feed their people, the LDCs should concentrate on grain production rather than meat.

KEY CONCEPTS

Efforts to feed the world's people should focus on food sources low on the food chain—plants and plant products. Far more people can be fed on a vegetarian diet than on a meat-based one.

Reducing Pest Damage and Spoilage

Rats, insects, and birds attack crops in the field, in transit, and in storage. Conservatively, about 30% of all agricultural output is destroyed by pests, spoilage, and diseases. In the less developed nations, this figure may be much higher, especially in warm, humid climates where crops are grown year-round in conditions that are conducive to crop-damaging insects and disease.

Reducing the heavy toll of pests in the field could help increase the global food supply. To create a sustainable system of agriculture, however, pest control measures must be safe and sustainable. Numerous strategies for pest management fit these requirements.

Another measure of great importance is the improvement of storage and transportation. Inefficient transportation can delay food shipments while rats, insects, birds, or spoilage take their share (**FIGURE 10-21**). Spoilage can be prevented by improving transportation networks, refrigeration, and other steps. In the less developed nations, the supply of fish could be increased by 40% by improving refrigeration on ships, in transit, and in stores. Grain supplies in the less developed nations could be stored in dry silos or sheds to prevent the growth of mold and mildew and reduce rodent

FIGURE 10-21 Stopping losses. Better, more secure storage of grains can help reduce loss to rodents.

© SuperStock.

problems. Technical and financial assistance from the more developed countries could go a long way toward improving food storage and transportation, potentially increasing world food supply by 10% or more.

KEY CONCEPTS

Much of the world's food production is consumed by pests or rots in storage or in transit. Improvements in transit and storage, such as refrigeration, can greatly boost food supplies.

Creating Agricultural Self-Sufficiency in Less Developed Nations

Over the years, many less developed nations have gone from being self-sufficient in food production to becoming major importers of food. The reasons for this are many. Population growth, poor land management, and economics top the list. Continued growth of population raises the demand for food while poor land management decreases a nation's ability to produce food. As for economics, the main problem here seems to be that many people simply can't afford food. Poverty in the less developed nations is therefore one of the key causes of hunger and malnutrition. Trade policies also affect food supplies. Cheap grain from more developed nations, for example, makes it difficult for indigenous farmers to compete. Many have gone out of business as a result of cheap grain from countries such as Canada, Australia, and the United States, grain that was either donated as food aid or sold cheaply because of subsidies provided by the exporting nation's government. Staple crops in less developed nations have also been replaced by export crops, for example, tea, coffee, and bananas, reducing food supplies for local citizens. The force behind this shift has been economic. In an effort to pay back debts to more developed nations such as the United States, many

governments of the LDCs have encouraged the production of export crops. This generates tax income for the LDCs so they can pay back their debts. Unfortunately, it decreases food supplies for their people.

Improvements in all three arenas could go a long way toward increasing food supplies and helping countries become more self-reliant. Many of the provisions discussed earlier in this section can help improve land management. Population control strategies can help reduce pressure on limited food supplies. Changes in economic policies in both LDCs and MDCs and trade policies are required to promote self-sufficiency as well.

KEY CONCEPTS

Many LDCs have lost their ability to produce food as a result of overpopulation, farmland deterioration, and economic and trade policies. Reversing these trends could help nations become more self-reliant, which is vital for building a more sustainable future.

Legislation and New Policies: Political and Economic Solutions

Laws and policies that promote unsustainable farming and harvesting practices must be changed or abandoned—in all nations, rich and poor. The world's commercial fishing industry spends over $125 billion to catch $70 billion worth of fish. The remainder, $55 billion, came from governments (and hence, taxpayers) in the form of subsidies and low-interest loans. Such policies encourage overfishing and a depletion of this vital food source. Although in the short term, efforts to boost food production may be alluring, in the long run this artificial economic support may lead to ecological impoverishment and a sharp decline in food availability. This phenomenon of short-term production and profit at the expense of long-term production and profit crops up time and time again in environmental issues, and it must be carefully analyzed. Farmers, for instance, may find it costly to invest in soil conservation measures. By not investing in them, though, they destroy the productive capacity of their land. Sacrificing long-term sustainable production for short-term profit is a trade-off with serious social, economic, and environmental ramifications. Some advocates of sustainability who are opposed to subsidies argue nonetheless that there are times when they can be used to our benefit to prevent such trade-offs.

Voluntary efforts are also needed. For example, some U.S. farmers may find it advantageous to abandon crops that use large amounts of irrigation water in water-short regions. Shifting water-thirsty crops such as cotton and rice from the desert of California's Central Valley to more suitable climates could free up enormous amounts of water for crops requiring less water. Government policies that promote the cultivation of environmentally incongruous crops such as these should be examined and modified.

As noted earlier, changes in public policy in the less developed nations are also vital to creating a sustainable system of agriculture. For example, programs and policies that encourage the cultivation of cash crops such as coffee and bananas for export in place of staples needed to feed the people of a country deserve special attention.

Efforts are needed to encourage small-scale farming and better soil management in the less developed nations. Because many LDCs are poor, international aid can help promote sustainable agricultural practices including the many soil conservation measures described in this chapter.

Ending War

Throughout the world, dozens of civil conflicts are now raging, many of them in Africa. Besides the heavy toll war takes on people, these conflicts can seriously disrupt the production and distribution of food. Food supplies can be cut off when bridges are bombed or railroad yards are destroyed. Soldiers also damage crops or farmland to weaken an enemy's resolve. What is more, damage to farm fields and the distribution system can last for many years after a conflict has ended. Mines in farm fields, for instance, may prevent farmers from going back into them to begin planting. It goes without saying that putting an end to war is vital to helping provide not just a peaceful life, but adequate food required for health.

This chapter began by listing three challenges in agriculture: the near-term need to feed malnourished people, the long-term need to provide food for future generations, and the continuing need to protect and enhance the soil and the environment. It should be clear that there are many ways to meet these challenges. No one idea will work. Rather, these ideas must be integrated into a comprehensive policy to build a sustainable agricultural -system.

As we strive to create a system of food production that provides high-quality food at a reasonable price while protecting the environment, we must remember that even if we take care of the land and manage it sustainably, human population could very well outstrip our maximum sustainable food production. Population stabilization and long-term efforts to reduce the size of the human population are vital components of the long-term strategy to create a sustainable human presence.

To a man with an empty stomach, food is God.

—*Gandhi*

Exercise Analysis

There is a problem with the line of reasoning that says that because food production per hectare—or productivity—is on the rise, we shouldn't be worried about agriculture. Just because a system appears to be functioning well does not mean that it can be sustained over the long haul. World fish catch is at an all-time high, but the majority of the world's fisheries are in trouble.

To assess the sustainability of agriculture or any other system that relies on renewable resources, we must look beyond measures of current performance such as productivity to fundamentally more important indicators such as the amount of topsoil on farms, the nutrient levels in the topsoil, the amount of land taken out of production because of salinization and waterlogging, and the supply of water available for irrigation. These are better indicators of agriculture's sustainability. As this chapter points out, when you examine the statistics of agriculture, it becomes clear that current agricultural practices are unsustainable.

Why then do we see gains in agricultural productivity if the system is unsustainable? The answer is that gains in productivity largely result from offsetting measures such as irrigation and fertilizer use. Although the soil may be eroding away, farmers are able to maintain or increase productivity by adding more fertilizer. Sooner or later, this approach will fail.

This exercise requires four critical thinking rules. First, it requires us to question the conclusions—notably, that rising productivity means that agriculture is sustainable. It also requires us to define terms such as sustainability and productivity very carefully so we understand what we're talking about and think critically about it. Third, it calls on us to look for hidden assumptions. In this case, it is assumed that rising productivity signifies a healthy system, but what it really reflects is the use of offsetting measures. Finally, we must examine the big picture—that is, look at all of the factors that contribute to the sustainability of farming.

CRITICAL THINKING AND CONCEPT REVIEW

1. What percentage of the world's population is malnourished? What are the short- and long-range effects of malnutrition?
2. Using your knowledge of environmental issues and agriculture and your critical thinking skills, analyze the following statement: "Hunger is not an environmental issue."
3. What is desertification, what factors create it, and how can it be prevented? Does your lifestyle contribute to desertification? Why or why not?
4. Critically analyze this statement: "Soil erosion control is too expensive. We can't afford to pay for it because our crops don't bring in enough money."
5. What justifications can be made for a farmer or a country taking on the expense of soil erosion control and other sustainable farming practices?
6. Describe waterlogging and salinization of soils. How can they be prevented? Which of your proposed solutions might result in further problems?
7. Describe the decline in agricultural species diversity. How could this trend affect world agriculture? Give some examples.
8. Critically analyze this statement: "Technology can solve all of our food problems, so there is no need to slow population growth."
9. List and discuss the major strategies for solving world food shortages. Which are the most important? How would you implement them in this and foreign countries?
10. Critically analyze the statement, "Simply by practicing better soil conservation and replenishing soil nutrients, we can reduce the need for new farmland."

11. Describe the successes and failures of the Green Revolution. What improvements might be made?
12. Critically analyze the following statement: "The world fish catch climbed nicely during the 1980s, and there is no reason to believe that the oceans won't be a major source of new food."
13. Describe the pros and cons of policies designed to eliminate the discard of by-catch (fish thrown back into the sea because they are the wrong sex or species).
14. You have been appointed head of a U.N. task force. Your project is to develop an agricultural system in a poor African nation that imports more than 50% of its grain but still suffers from widespread hunger. Outline your plan, giving general principles you would follow and specific recommendations for achieving self-sufficiency.
15. Make a list of things you can do to help solve the world hunger problem and create a sustainable system of agriculture.
16. Can you determine ways that sustainable development principles applied to other human systems might influence agriculture and improve our prospects for feeding the world's people?

KEY TERMS

A horizon
accelerated erosion
annuals
aquaculture
B horizon
C horizon
climate
Conservation Reserve Program
conservation tillage
contour farming
crop rotation
D horizon
desertification
Environmental Quality Incentives
    Program
erosion
farmland conversion
fish farms
Food Inspection Agency

genetic engineering
green manure
Green Revolution
herbicides
humus
hybrids
infectious diseases
kwashiorkor
litter layer
marasmus
mariculture
minimum tillage
natural capital
natural erosion
1985 Farm Bill
O horizon
organic fertilizers
overfishing
parent material

perennial
pesticides
salinization
selective breeding
shelterbelts
soil
soil erosion
soil profile
strip cropping
subsidies
subsoil
sustainable agriculture
synthetic fertilizer
terraces
topography
topsoil
water table
waterlogging

© Don Mammoser/Shutterstock, Inc.

Preserving Biological Diversity

The worst sin toward our fellow creatures is not to hate them, but to be indifferent to them.

—George Bernard Shaw

LEARNING OBJECTIVES

After reading and studying the material in this chapter, students will be able to understand, knowledgably discuss, and, when appropriate, apply the information learned in this chapter, including:

1. The magnitude of species extinction
2. Why species go extinct and how humans contribute to extinction
3. The importance of keystone species
4. The many reasons why it is important to protect biodiversity
5. They many ways we can prevent extinction and protect biodiversity

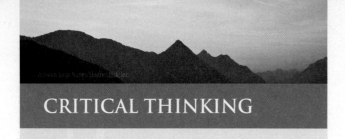

CRITICAL THINKING

Exercise

You own a guest ranch in western Colorado. Your ranch attracts people from all over the world who come to ride horses and view the abundant wildlife such as deer, elk, eagles, hawks, and an occasional black bear. You also raise sheep to supplement your family's income. One day, a neighboring rancher calls on you to ask for your help in killing coyotes, which he says are killing off his sheep and costing him lots of money. What he wants is permission to spread chunks of meat, containing a lethal poison called Compound 1080 on your property. What he is proposing is not legal, he freely admits, but it is done from time to time in remote parts of the West. He and other ranchers who want to control coyotes extract the poison from sheep collars—devices placed around the necks of sheep to selectively kill coyotes that attack them. Sheep collars are legal. Your neighbor will get the poison and even spread the scraps of meat on your property if you give him permission. He says it will also help protect your sheep herd.

What concerns would you have about this proposal? Make a list of questions you will need to answer before deciding whether to join your neighbor. What critical thinking rules will you use?

In 1996, a consortium of research institutes, including the Smithsonian Institution, received funding for a research project in the tropical rain forest of Nigeria and Cameroon. The researchers studied the local abundance, distribution, and life cycles of trees and shrubs with medicinal properties in a large segment of forest in Cameroon and in several smaller plots in Nigeria. Researchers searched for beneficial drugs by examining traditional medicines used by local villagers and by a mass screening of trees and shrubs. This project, funded by the U.S. Agency for International Development and several other U.S. government agencies, was an attempt to help develop alternatives to deforestation. It is believed that cultivation of medicinal trees could provide an important source of income to local communities. The field work, which ended in 2004, yielded numerous natural chemical compounds that could some day be used to treat tropical diseases, mental disorders, and chronic diseases.

This project is one of many efforts designed to save forests and the species that live within them. It is part of an effort to protect the natural environment and provide for the economic needs of people who live in such areas. As such, it is an example of the new philosophy of sustainable development.

This chapter examines biodiversity—the rich biological world—and new ways of protecting the planet's wealth of wild species. In the following pages, you will examine the causes of extinction and the countless benefits of plants and wildlife to human societies to help you fully understand the importance of saving species.

11.1 Biodiversity: Signs of Decline

By some estimates, as many as 500 million kinds of plants, animals, and microorganisms have made this planet home since the beginning of time. Today, scientists estimate that the world contains between 10 and 30 million species. Some put the estimate higher, up to 80 million. Only 1.9 million have been identified and named. Thus, 420 to 490 million species have become extinct over a period of 3.5 billion years.

This tells us that extinction is an evolutionary fact of life. Scientists refer to this as **natural extinction**. The natural occurrence of extinction is viewed by many as a justification for plant and animal extinctions occurring today as a result of human activities, known as **accelerated extinction**. Why not?

Natural extinction differs markedly from accelerated extinction for at least two important reasons. First, natural extinction represents a kind of evolutionary passing. That is to say, many millions of species have become extinct, but they did not disappear. They evolved into new species (**FIGURE 11-1**). Their kind may have vanished, but they gave rise to new species. Today, they are represented by their descendants. Modern extinctions, on the other hand, eliminate species entirely. They represent a dead end in evolution.

Second, the rate of extinction varies considerably. Even though some species did vanish in mass extinctions because of severe climatic changes or for other reasons, the rate of natural extinction—about one species every 1,000 years—is slow compared with today's accelerated extinction. Although no one can tell exactly how many species become extinct today, estimates based on loss of habitat and species diversity within these ecosystems suggest that it may be as many as 100 to 140 species every day or 36,500 to 50,000 per year. Harvard biologist E. O. Wilson thinks the number may be higher. As a result, many believe that we have entered into an era of extinction unparalleled in the history of the Earth.

Today, thousands of species are endangered or threatened. An **endangered species** is one that is in imminent danger of becoming extinct. A **threatened species** is one that is still abundant in its natural range but, because its numbers are declining, is likely to become extinct in time. Without concerted efforts to protect them, many threatened

FIGURE 11-1 Gone but not forgotten. This drawing illustrates the stages in the evolutionary history of the horse. All distinct species are extinct except for the modern horse. Like other species, horses evolved through intermediate stages that no longer exist.

species will become extinct. How serious is the problem? Today, three-fourths of the world's bird species are declining in number or threatened with extinction. Moreover, more than two-thirds of the world's 150 species of primates are threatened. Today, approximately 10,065 species of plants are threatened or endangered—2 of every 100 plant species on Earth according to the International Union for the Conservation of Nature. In the United States, approximately 1,519 species of plants and animals, according to the U.S. Fish and Wildlife Service, are threatened or in danger of extinction.

Scientists around the world are disturbed by an alarming disappearance of amphibians. Many species of frogs, toads, and salamanders are either experiencing steep declines in population size or have vanished. Amphibians are disappearing from a wide variety of habitats. The cause is still unknown, but most scientists believe that the decline is the result of a wide range of factors including habitat loss; environmental pollution, such as pesticides; exposure to ultraviolet radiation due to ozone depletion; disease; parasites; global warming; and others.

Unless we curb population growth, reduce pollution and habitat destruction, and manage our resources better, vast expanses of forests, wetlands, and grasslands will vanish over your lifetime. Millions of species may vanish as a result. The impact of habitat destruction and mass extinction will be felt worldwide.

 KEY CONCEPTS

Many species of plants and animals face extinction today as a result of human activities. Although extinction has occurred since the dawn of time, modern extinctions are occurring at a rate much faster than is biologically sustainable.

11.2 Causes of Extinction and the Decline in Biodiversity

Plant and animal extinction, like many other environmental problems, results from many factors. This section examines the two most important factors today: habitat alteration and commercial hunting/harvesting. It also examines other important factors, including (1) the introduction of alien and domesticated species, (2) pest and predator control, (3) the collection of animals for pets and research, (4) pollution, (5) ecological factors, and (6) the loss of keystone species.

 KEY CONCEPTS

Many factors contribute to the loss of species, but the two most important are the destruction and alteration of habitat and commercial harvesting.

Physical Alteration of Habitat

We humans have always altered the environment to meet our needs. Virtually every activity we undertake changes the biotic and abiotic conditions of the environment and, hence, the biological communities of aquatic and terrestrial ecosystems. Every product we buy comes with an environmental price tag. Extraction of the raw materials and production of finished products, even the transportation to stores or our homes, for example, causes damage. Some of the most dramatic environmental changes come from activities such as food production, timber harvesting, mining, road building, the manufacture and operation of automobiles, and the construction and maintenance of homes.

In recent years, even illegal immigration has taken a toll on wildlife and its habitat. Along the United States–Mexico border, thousands of illegal immigrants and law enforcement officials sent to curb this activity, have caused considerable damage to sensitive desert wilderness areas. One area that has been hard hit is the Cabeza Prieta National Wildlife Refuge in southwestern Arizona, a vast wilderness area that is home to many species including several endangered species. Mountains of trash and physical damage to the environment

are causing concern to wildlife officials. Drug smugglers have created roads and more than 1,300 miles of foot paths through these wilderness areas. Parts of wilderness and wildlife refuge fenced off to protect visitors from illegal activities are now being overgrazed by cattle from Mexican farmers.

Human activities tend to fragment the habitat of plants and animals. Human settlements, for example, become islands of human activity in the natural environment. As the human population grows, however, and as roads, highways, farms, and buildings increase in number, the pattern changes. Eventually that is left are a few islands of natural habitat in the human-dominated landscape (**FIGURE 11-2**). As the natural habitat fragments, species diversity declines.

Habitat alteration ranges from moderate to extreme and ranks as *the* most significant factor in the extinction of plants and animals the world over. It is no surprise to find that those areas of the world that are at risk are regions of the most intense human activity.

Nowhere is the loss of habitat more pronounced than in the tropical rain forests. Tropical forests house at least half of the Earth's species, perhaps more. Rain forests once covered 14% of the Earth's land surface. Over half of them have been cut down so they cover only 6% today. Countless species have perished as a result. At the current rate of destruction, the remaining rain forests could be gone by mid-century.

Coral reefs, wetlands, and estuaries (the mouths of rivers) are other critical habitats already greatly reduced and rapidly declining because of human development. Wetlands and estuaries, for example, are the home of many species but are also highly prized by humans for development. Damage to wetlands has been particularly severe in the industrial nations. New Zealand and Australia, for example, have lost more than 90% of their wetlands and countless species that lived in them. India, Pakistan, and Thailand have lost at least 75% of their mangrove swamps, a type of coastal wetland. Their ongoing destruction threatens the future of fish and waterfowl throughout the world.

FIGURE 11-2 Habitat fragmentation. Humans carve up the natural landscape to make room for cities, towns, farms, factories, housing developments, roads, shopping centers, and much more.

Courtesy of South Florida Water Management District.

Another ecosystem that has experienced serious losses is the tallgrass prairie of North America, which today is virtually nonexistent. Temperate rain forests, such as those in the Pacific Northwest, have also been subjected to intense harvesting. Worldwide, more than 56% of the temperate rain forests have been logged or cleared. No more than 10% of the original old-growth temperate rain forests remain in the United States. Canadian temperate rain forests have also been heavily cut. One of the many victims of this loss is a bird known as the *marbled murrelet* (**FIGURE 11-3**). The murrelet feeds on schooling fish that congregate near the shores of Canada, but nests on moss-covered branches of the temperate rain forest within 30 kilometers (18 miles) of the shore. Logging of old growth forests in British Columbia has dramatically reduced the habitat of this bird, causing equally dramatic declines in its numbers.

KEY CONCEPTS

Virtually all human activities alter the environment, changing the biotic and abiotic conditions and fragmenting habitat. Habitat alteration is the number one cause of species extinction. The most dramatic changes occur in biologically rich areas: tropical rain forests, wetlands, estuaries, and coral reefs.

Commercial Hunting and Harvesting

Humans have hunted and killed animals throughout history to provide food, fur, and other products. Several different types of hunting occur today. **Sport hunting** occurs for enjoyment and is generally well regulated. In fact, in many cases sport hunting benefits wild populations by helping to control numbers—so that they remain within the carrying capacity of the environment.

Another form of hunting is **subsistence hunting**, killing animals to provide food for indigenous people such as those who live in the tropical rain forests. Generally, such activities are carried out on a sustainable level and do not pose a threat to animals, although there are notable exceptions.

The third form of hunting is **commercial hunting**, or **harvesting**. This consists of large-scale efforts, such as the whale hunting of years past, and smaller operations, such as hunting African rhinos for their horns. Smaller operations may be legal or, in some cases, illegal.

Whale hunting is one of the most familiar examples of commercial hunting. In the 1700s and 1800s, commercial whalers hunted one species after another to the brink of extinction to provide oil, meat, and other products. The result has been a severe reduction in whale populations (**TABLE 11-1**). Thanks to efforts by the International Whaling Commission, commercial whaling has been greatly reduced. In its place is a new industry: whale watching, with annual revenues that exceed those of the commercial whaling industry itself.

Large-scale commercial hunting also doomed the passenger pigeon and greatly reduced the size of the bison herds of North America. The world's fisheries (fishing grounds) have also been heavily harvested for commercial gain, and many important fisheries have been eliminated. Overharvesting continues today in many of the world's remaining fisheries.

Commercial hunting may also occur illegally, on a much smaller scale. Illegal hunting is called **poaching**. Today,

FIGURE 11-3 The marbled murrelet. This bird and many others are in danger of extinction because of the destruction of temperate rainforests of North America.

© Glenn Bartley/age footstock.

TABLE 11-1

Whale Populations—Then and Now

Species	Number Before Commercial Whaling	Current Estimate
Blue	350,000	8,000–14,000
Bowhead	54,680	8,000
Fin	450,000	55,000–60,000
Gray	15,000–20,000	19,000–20,000
Humpback	119,000	30,000–40,000
Minke	250,000	1,000,000
Right	50,000	7,800
Sei (includes Bryde's)	350,000	144,000
Sperm	1,100,000	360,000

Data from International Whaling Commission, American Cetacean Society poachers continue to slaughter elephants, rhinos, tigers, and a variety of other endangered species because the economic benefits outweigh the risks of small fines or light jail sentences. A Bengal tiger coat, for example, sells for $100,000 in Japan. Poaching of ivory from tusks caused the near demise of African elephants, from an estimated 2.5 million in 1970 to an estimated 450,000 to 700,000 today according to the World Wildlife Foundation. Unfortunately, the illegal ivory trade has steadily increased since 2006 and is at an all-time high. In 2012, 35,000 African elephants were killed for their ivory tusks.

 KEY CONCEPTS

Commercial hunting and harvesting of wild species have occurred for centuries and represent the second largest threat to the world's animal species. This includes past activities, such as whale hunting, and present activities, such as commercial fish harvesting and poaching of endangered species.

The Introduction of Foreign Species

Foreign, or alien, species introduced accidentally or intentionally into new territories often do well in their new homes because conditions may be highly favorable to their growth and reproduction. One reason for this is that these species may face little environmental resistance—for example, there may be no natural predators to hold their populations in check—in their new home. As their populations expand, alien species often outcompete and eliminate native species. The English sparrow is an example. Deliberately introduced into this country in the 1850s, the sparrow quickly spread throughout the continent. It now competes for nesting sites and food once used by bluebirds, wrens, and swallows.

The starling is another intentional import from England that has had a similar effect on native North American bird species. Another introduced species that's raising havoc in North America is the grass carp, a native of China. Grass carp breed prolifically and are spreading through the rivers, squeezing out native species (**FIGURE 11-4**).

Native species can also migrate into new territories as a result of human activities with devastating effects. The barred owl, for instance, has spread across the continent, hopscotching from one patch of trees in cities and towns to another on the Great Plains, which was once a vast sea of grass that blocked the westward movement of the owl. It is now a resident of the forests of the Pacific Northwest. The barred owl harasses and preys on spotted owls and takes over their nesting sites. It may even interbreed and could further threaten this species, which is suffering because of the loss of vital old growth habitat.

Alien plants and animals have taken their toll on native species of Canada, too. European starlings have displaced many species. Raccoons that were released on Queen Charlotte Islands to provide additional income for local trappers have caused significant decreases in a type of sea bird

FIGURE 11-4 Grass carp introduced from Asia reproduce prolifically in many streams in the eastern United States and have become a major nuisance.

Courtesy of U.S. Fish & Wildlife Services.

known as an *alcid*. These burrow-nesting birds form huge vulnerable colonies. Raccoons raid the nests for food.

Nonnative plants also outcompete native species. Scotch broom, purple loosestrife, and crested wheatgrass have all been introduced into Canada and now compete aggressively with native species, sometimes wiping them out altogether. Today, 23% of Canada's wild plants are nonnative species. Islands are especially vulnerable to new species. In Hawaii, for example, 90% of all bird species have been wiped out by human inhabitants and by organisms (such as rats) introduced by humans.

Alien species often do well in warm climates such as Florida. In fact, Florida is a showcase of alien species gone wild. Australian pines, for example, were introduced as ornamentals and have spread rapidly along coastal beaches and canals. Their shallow roots are so dense that they destroy sandy beaches, where many sea turtles lay their eggs. Worst of all is a species called the *punk tree,* which grows in swamps, creating a dense tangle of vegetation impassable to many animal species. Boa constrictors released by hobbyists are now breeding prolifically in the Event glades of Florida.

The Great Lakes of North America have also been subject to numerous alien invasions, often with devastating economic and ecological consequences. Two of the latest are the zebra mussel and fishook water fleas, both accidentally introduced by ships carrying cargo from Europe. The fishook water flea feeds on phytoplankton, the main food source of small fish. Besides posing a threat to fisheries, the flea also has become a nuisance to commercial and sports anglers, clogging nets and lines. It is spreading rapidly throughout the Great Lakes and into surrounding waterways, just like the zebra mussel.

The alien invasion is nothing to take lightly. In the United States, a recent study that took 200 scientists 4 years to prepare estimates that 6,500 alien species of plants, animals, insects and arachnids (spiders), and disease organisms have gained a foothold in the United States. They have caused 315 native species to become endangered or threatened. Although no one knows for sure how much damage alien species cause, one study by researchers at Cornell University suggests that the damage in the United States caused by alien plants and animals may be over $120 billion per year! The main problems are predation on native species and habitat destruction.

Pest and Predator Control

Well-intentioned efforts to control species considered to be pests also affect wild populations of plants and animals, and they contribute to the steady decline in biodiversity on Earth. DDT and other pesticides, for example, have taken a huge toll on American wildlife. The peregrine falcon had disappeared in the eastern United States by the 1960s as a result of DDT. This pesticide caused eggshell thinning, which made eggs fragile and susceptible to breakage. The entire population of falcons east of the Mississippi River was wiped out and dramatic declines occurred in the West, too. Eagles and brown pelicans met a similar fate. Even the California condor, a massive vulture, suffered from eggshell thinning. Especially harmful to migratory birds that spend their summer in North America are those pesticides (such as DDT and related compounds) that have been banned in the United States but are still used in Latin America.

Studies suggest that nonylphenols, a nonactive ingredient of some pesticides, may have a deleterious effect on salmon. This near-ubiquitous water pollutant, also found in detergents, cosmetics, and plastics, enters waterways. In the Pacific Northwest, salmon migrating to the sea that are exposed to minute quantities of this chemical perish when exposed to salt water. Researchers believe that nonylphenols affect the endocrine system, making salmon unable to produce hormones needed to get rid of excess salt, an adaptation vital to the transition to oceanic life.

Predator control, once the cornerstone of wildlife management, has had a profound impact on native species such as wolves, bears, and mountain lions in North America. Killing off predators can also create severe habitat destruction, as the prey populations once controlled by these predators grow beyond the carrying capacity of their environment.

The Collection of Animals and Plants for Human Enjoyment, Research, and Other Purposes

Each year, millions of animals and plants are gathered from their native habitat throughout the world and exported to zoos, private collectors, pet shops, research institutes, and other places. The numbers are staggering. In 2012, for instance, nearly 182,000 live reptiles, and 310,000 reptile skins were legally imported into the United States according to the Convention on International Trade in Endangered Species (CITES) data base. That same year, nearly 2,350 live birds were imported into the United States. Tens of thousands of birds are also imported by Canada and Great Britain. For each bird or animal that makes it into someone's home, though, 10 to 50 may die along the way (**FIGURE 11-5**) Many perish after they are taken home, too.

Scientists throughout the world use a variety of wild caught primates such as monkeys for research on pressing medical problems such as AIDS. Taken from their homeland in Africa, as many as five chimpanzees die for every one that enters a laboratory. In 1975, the United States banned the importation of all primates for pets but allowed continued importation for zoos and research. Partly as a result, primate imports have dropped rapidly in the United States

FIGURE 11-5 Nature for sale. In the developing countries, local residents sell birds and animals they catch in nearby forests to dealers who export them, sometimes illegally, to the industrial nations. Many animals die along the way.

and elsewhere (**FIGURE 11-6**). Nonetheless, in 2012, the latest year for which data were available, the United States imported over 21,500 primates, down from 28,000 in 2006.

Some rare and endangered primates are still captured and sent to the United States. Because research animals often do not breed in captivity and because they have a high mortality rate, continual replenishment from wild populations is likely to continue. As noted earlier in the chapter, two-thirds of the world's 150 species of primates are threatened with extinction.

Go Green

Do not buy exotic pets such as birds or fish that come from the wild. Check the source very carefully to be sure that they are domestically raised.

Plants are also popular imports. In 2012, 294,500 live cacti (including some artificially propagated) were imported into the United States from 11 countries, down from 1.8 million in 2010. International trade in exotic flowers, especially those originating in the Mediterranean, is also a booming business and currently threatens many native plant species.

At home, collectors pillage the deserts in Texas and Arizona in search of salable cacti to adorn the lawns of customers. To reduce the ravaging impact of commercial cactus rustling, Arizona has made it illegal to remove 222 different plant species. With penalties up to $1,000 and jail sentences up to 1 year, Arizona has taken a small step to protect its native plants. With only two "cactus cops" to patrol the state, though, little progress can be made.

KEY CONCEPTS

Millions of plants and animals are taken from the wild and imported into developed countries for zoos, private collections, pet shops, and research, contributing to the worldwide loss of species.

Pollution

Pollution plays an important role in the decrease in the planet's biodiversity and is bound to play a larger role as the human population and economy expand. Today, pollution-caused problems such as global warming, acid deposition, and ozone depletion are creating major ecological changes. Pollution could combine with other extinction forces described in this chapter to greatly increase the threat to biodiversity. In fact, they could synergize creating a much greater threat to wildlife populations. For example, global warming caused by carbon dioxide pollution from the burning of fossil fuels combined with several other pollutants may be responsible for a massive die-off of the world's coral reefs, already besieged by sediment from onshore development, damage from ships, and chemical pollutants. Coral reefs contain an estimated 1 million species. By one estimate, 60% of the world's 230,000 square miles of coral reef could be lost by around 2040 if threats continue.

Global warming caused in large part by human pollutants and deforestation is also taking its toll on the polar bear, a species threatened already by numerous other factors, including chemical pollution, hunting, tourism, and oil drilling in its habitat. As Arctic Sea ice melts due to global warming, polar bears have difficulty accessing their food source: seals. Bears that are undernourished face difficulty surviving (**FIGURE 11-7**).

Pollution is also of concern on a local level. In the semiarid farmland of California's San Joaquin Valley, the heavy metal selenium—contained in waters draining from irrigated farm fields into specially built evaporation ponds (**FIGURE 11-8**) in the Kesterson National Wildlife Refuge—has caused massive biological effects. In 1985, for example, biologists found physical abnormalities in 42% of the chicks of waterfowl and wading birds at Kesterson—compared with an expected deformity rate of 1%. Problems included chicks without eyes, beaks, wings, and legs. Many embryos died. Adult birds and many other animals were also affected—as were crayfish, snakes, raccoons, and muskrats, which once flourished in the rich biological community but now have vanished. As one journalist put it, "The Kesterson Refuge

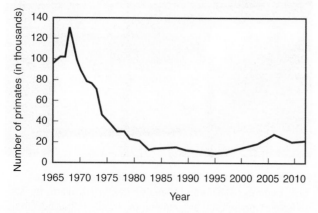

FIGURE 11-6 Decline in the number of primates imported into the United States.

FIGURE 11-7 Newly endangered. Polar bears are threatened by several pollutants, including greenhouse gases, which are causing global warming that is melting polar sea ice.

© Nik Niklz/Shutterstock, Inc.

FIGURE 11-8 Polluted evaporation ponds. Selenium and other toxic metals, leached from irrigated farmland and carried to nearby evaporation ponds (now closed) in the Kesterson National Wildlife Refuge in California's Central Valley, have resulted in a number of embryonic defects.

Courtesy of Gary Zahm/U.S. Fish & Wildlife Services.

had become a place that killed the animals it was supposed to protect." (In 1986, the evaporation ponds at Kesterson were filled with dirt.) Federal officials note similar problems in refuges in Utah, Wyoming, Texas, Nevada, Arizona, and other parts of California. In nonrefuge sites, they're proposing costly water treatment facilities to reduce selenium levels, followed by periodic removal of contaminated sediments from the bottom of the ponds in some locations.

Pollution affects many species the world over. One of them is the already-threatened beluga whales of Canada's St. Lawrence estuary. Scientists have found that older whales are dying of cancer and that young calves are ingesting poisons from their mothers' milk. Autopsies of dead whales show they are heavily contaminated with 25 different toxic chemicals. Although this is not the only threat to the beluga whale, whose population numbers around 200,000 worldwide according to the American Cetacean Society, it could be a major one.

KEY CONCEPTS

Pollution alters the physical and chemical nature of the environment in ways that impairs the survival of many species. Pollution and climate change (caused largely by pollution) is altering the health of the world's coral reefs and may cause widespread losses if something is not done to reverse the trend—and soon.

Pollution, combined with other forces such as habitat loss, introduction of alien species, and climate change could have a synergistic effect, resulting in much higher rates of extinction.

Biological Factors that Contribute to Extinction

Not all species are created equal when it comes to extinction. Some species, for instance, are quite adaptable and seem to do well in the human-dominated world.

Unfortunately, they're not always the most desirable species. Pigeons and cockroaches are two examples. Others do not fare so well. Some of these species have peculiarities in behavior or lifestyle that make them more vulnerable to extinction. Others produce so few young that they are not very resilient.

Consider the passenger pigeon (**FIGURE 11-9**). At one time, the passenger pigeon inhabited the eastern half of the United States in flocks so large they darkened the sky. Probably the most abundant bird species to ever live, the passenger pigeon is now extinct, in large part because of widespread commercial hunting and habitat destruction. Between 1860 and 1890, countless pigeons were killed and shipped to the cities for food. In 1878, the last colonial nesting site in Michigan was invaded by hunters. When the guns finally fell silent, over 1 billion birds had been killed. By this time, only about 2,000 remained. Broken into flocks too small to hunt economically, the birds were finally left alone. Did they recover? No. The number of birds dwindled year after year, until in 1914, the last bird held captive in the Cincinnati Zoo died.

This story illustrates an important point: Some species have a critical population size below which survival may be impossible. The passenger pigeon needed large colonies for successful social interaction and propagation of the species. Two thousand birds were simply not enough.

FIGURE 11-9 Museum specimens. The passenger pigeon, a once abundant species whose flocks darkened the skies, is extinct in large part because of commercial harvesting.

© Stephen J. Krasemann/Science Source.

Scientists know very little about the critical population size for many species. Unfortunately, overhunting, habitat destruction, and other activities discussed in this chapter can do irreparable damage before we realize it.

Another trait that influences an organism's survivability is its degree of specialization. Animals may be categorized as specialists or generalists. Specialists tend to be more vulnerable than generalists. Why? Generalists can exploit more food sources and can live in more habitats than specialists. They are more adaptable and, consequently, less vulnerable to habitat destruction and other forces.

Animal size also contributes to extinction. Larger animals such as the rhinoceros are easier (and often more desirable) prey for human hunters. Moreover, they are more likely to compete with humans for desirable resources such as grazing land. Larger animals also generally produce fewer offspring, making it more difficult for reduced populations to recover. The California condor, for example, lays a single egg every other year. Young condors remain dependent on their parents for about 1 year but are not sexually mature until age 6 or 7. Combined, these factors give the condor little resiliency to withstand pressures from human populations or natural disasters.

Another factor in extinction is the size of an organism's range. The smaller the range, the more prone the organism is to extinction.

FIGURE 11-10 Gopher tortoise. This slow-moving tortoise is a keystone species in the southeastern United States. Its burrow is home to many other species.

© Rose Thompson/ShutterStock, Inc.

KEY CONCEPTS

Many biological characteristics of organisms determine how vulnerable they are to human impacts on the environment, such as the number of offspring they produce, the size of their range, their tolerance for people, and their degree of specialization.

The Loss of Keystone Species

According to ecological theory, in some ecosystems the extinction of one critical species, known as a **keystone species**, may lead to the loss of many other species. For example, the gopher tortoise in the southeastern United States is a keystone species (**FIGURE 11-10**). Gopher tortoises are important because they dig long burrows in the sand, up to 400 feet long and 30 feet deep, that many other species share. The Florida mouse becomes a permanent resident, as does the gopher frog. Opossums, gray foxes, and indigo snakes frequent the burrows as well. The gopher tortoise is so important, in fact, that in areas where it has been eliminated, 37 species of invertebrates have disappeared.

Fig trees are a keystone in the tropical rain forests of Peru. Studies of plant communities in these forests show that three-fourths of the birds and mammals of the Amazonian rain forest rely on fruits. Most fruits, however, are available only 9 months of the year. During the remaining period, monkeys, peccaries, parrots, and toucans live on figs. Their loss would be devastating.

In the marine ecosystem in the U.S. Pacific, the sea otter is a keystone species. Sea otters are voracious eaters, feeding on sea urchins, abalone, crabs, and molluscs that inhabit kelp beds. The otter helps control sea urchin populations, which feed on kelp and seaweed. In locations where sea otters have been eliminated, kelp beds have disappeared because sea urchin populations increase in the otter's absence. When sea otters repopulated these regions, urchin populations fell and the kelp and seaweed recovered. Moreover, the seaweed provides a habitat for fish and other fish-eating creatures such as harbor seals. Thus, when the sea otter returns to an area, so do other species.

Scientists believe that many little-known animals may be keystone species and that their loss could have profound effects on natural systems and crops. Bats, for example, are a keystone species in the tropical rain forest because they pollinate flowers and disperse seeds of many plants. In fact, many tropical plant species are entirely dependent on bats. So are many important food crops such as avocados, bananas, cloves, cashews, dates, figs, and mangos.

Ironically, the lion's share of conservation funds is dedicated to nonkeystone species. The loss of these species, although tragic, would not be as devastating as the loss of unobtrusive, rare, or little-known plants and animals whose survival is vital to many other species.

KEY CONCEPTS

Keystone species are organisms on which many other species in an ecosystem depend. The loss of a single keystone species may have a devastating effect on other organisms.

A Multiplicity of Factors

The foregoing discussion shows that there are many factors that contribute to the loss of plants, animals, and other organisms. Several factors may operate simultaneously on a single species. California condor populations, for instance, were greatly affected by DDT, which caused eggshell thinning and a devastating decline in offspring (**FIGURE 11-11**). Habitat destruction caused by road and housing construction also played a major role in the decline of this species—taking away the condors' habitat. Even well-intentioned fire control efforts affected this giant bird, which has a wingspan up to 3 meters (nearly 10 feet). Fire suppression in the hills of southern California permitted brush to grow in the condors' habitat. This, in turn, eliminated takeoff and landing areas needed by these giant scavengers.

The loss of salmon populations in the Pacific Northwest is another example of the way in which many factors combine to eliminate a species. Most readers have heard about the many dams along the rivers that halted the upstream migration of adult salmon and the downstream migration of young salmon. Even with special salmon ladders, migrations are less than optimum. Young salmon, for instance, who migrate to the sea to mature, often get lost in reservoirs and never locate the specially constructed ladders. Salmon have also been the victim of years of overfishing. Logging and agriculture have increased sediment deposition in salmon spawning grounds. Agriculture has also taken its toll on salmon by reducing stream flows resulting from irrigation withdrawals. Agricultural chemicals and sediment eroded from farm fields have both added to the salmon's decline from historical levels of 10 to 16 million to 1.5 million today—three-quarters of which are raised in and released from hatcheries.

Not only do many factors contribute to the loss of species, but some observers are concerned that these factors may combine to produce an effect much greater than currently anticipated.

FIGURE 11-11 California condor. Like other species, the condor was threatened by a loss of habitat and the pesticide DDT.

© PhotoPalle/iStockphoto.

11.3 Why Protect Biodiversity?

Why should disappearing beetles, plants, or birds concern us? The answer to this question varies, depending on one's ethics and one's grasp of ecology and on one's understanding of the importance of other living things to the functioning of the planet and the human economy. In this section, we examine four reasons: (1) aesthetics and economics; (2) food, pharmaceuticals, scientific information, and products; (3) protecting free services and saving money; and (4) ethics. The first three categories can be categorized as *instrumental* or *utilitarian values*—that is, they are based on humans' obtaining some form of benefit. The fourth, based on ethical reasons to save endangered species and biodiversity, could be categorized as a nonutilitarian or an *inherent value* argument, for humans obtain no direct benefit.

Aesthetics and Economics

Wildlife expert Norman Myers once wrote, "Every time a species goes extinct, we are irreversibly impoverished." Wildlife and their habitats are a rich aesthetic resource. The sight of a trumpeter swan on a pond in Montana, the eerie cry of the loon at night, and the graceful dive of the humpback whale enrich our lives. Wild places provide solace to battle the stress of the modern world.

Saving wild places and other species protects the beauty around us. It can also be an intelligent economic choice. Tourist towns and resorts the world over depend on natural beauty to attract visitors. Natural beauty also attracts residents to an area and can be a major benefit to those who want to relocate.

Another byproduct of the preservation of biodiversity is a rapidly expanding industry called **ecotourism**, which caters to the demands of bird-watchers, wildlife lovers, kayakers, and others who want to travel the globe discovering the beauties that abound in distant countries. Ecotourists spend thousands of dollars each to glimpse a lion or cheetah in Africa or a toucan in Central America or to kayak remote rivers in Costa Rica (**FIGURE 11-12**). They sometimes stay in lodges that were built with little impact on the environment.

Go Green

Support organizations like the Nature Conservancy that buy and protect wildlife habitat.

The importance of ecotourism is that it provides a relatively sustainable alternative for nations and local villagers—people who once viewed wild species as competition for food and space or who regarded undeveloped land solely as a resource to provide wood products and other materials for human benefit.

FIGURE 11-12 Ecotourism. Tourists spend billions of dollars each year to visit faraway places, creating an economic boon to many developing nations and considerable incentive to protect native species and undeveloped land. Ecotourism has a downside, though: Hordes of tourists in tour vehicles or on the backs of animals can also cause severe environmental damage.

© Isifa Image Service S.P.O./Alamy Images.

The problem with ecotourism is that to travel to the far reaches of the globe requires substantial amounts of energy. And people often crowd ecologically sensitive areas. Trucks carrying sightseers in Africa are cutting up the savannah to offer a glimpse of wildlife.

 KEY CONCEPTS

One utilitarian argument for saving wild places and other species is that they are a source of beauty and pleasure. In addition, this can provide an economic benefit through activities such as ecotourism and bird watching.

Food, Pharmaceuticals, Scientific Information, and Products

"From morning coffee to evening nightcap," writes British environmentalist Norman Myers whose books focus on biodiversity, "we benefit in our daily life-styles from the fellow species that share our One Earth home. Without knowing it, we utilize hundreds of products each day that owe their origin to wild animals and plants." Virtually everything in our lives comes from the Earth's crust or the plants and animals that inhabit grasslands, oceans, lakes, rivers, and forests. From the oceans, we harvested millions of tons of fish worth more than $102 billion a year in 2010. From rain forests, we harvested a wide assortment of edible fruits and nuts and numerous plant products, such as rattan, which is used to make wicker furniture and baskets. Tropical rain forest plants also yield medicines to fight disease. As noted elsewhere in this book, important plant and animal genes needed to improve domestic crops and livestock come from nature. Wildlife provides a wealth of enjoyment for hunters, anglers, and nature lovers, and wild species are a huge source of scientific information that provides valuable insights into our world.

The economic benefits of wild species are huge. By some estimates, 40% of all prescription and nonprescription drugs are made with chemicals derived from or originally extracted from wild plants. Worldwide, the commercial value of the 7,000 or so drugs prescribed by doctors was estimated to be about $962 billion in 2012. The USDA estimates that each year genes introduced into commercial crop species yield over $1 billion worth of food. Similar gains can be documented for other major agricultural nations. Internationally, the most widely traded nontimber product is rattan, which is estimated to be worth more than $7 billion a year.

Wild species provide a wealth of scientific information that could be of great practical value. Norman Myers, in fact, asserts that "wild species rank among the most valuable raw materials with which society can meet the unknown challenges of the future." Many products from wild species now loom on the horizon and may offer us considerable financial gains and healthier lives. For example, the adhesive that barnacles use to adhere to rocks and ships may provide humankind with a new glue to cement fillings into teeth. A chemical derived from the skeletons of shrimps, crabs, and lobsters may help prevent fungal infections in human beings.

Scientists are also studying a wide range of species to learn new, economical, and environmentally friendly ways of producing a wide range of useful products. The abalone, for instance, produces a ceramic-like material much harder than those humans produce—and they do it in sea water using naturally occurring chemicals. Human ceramic production requires toxic chemicals and high-temperature furnaces. This effort is part of a movement called *biomimicry*.

 KEY CONCEPTS

Wild plants and animals are a valuable economic resource. They could provide new food sources to feed the growing human population, genes that could improve crop species, new medicines to combat disease, scientific knowledge, and an assortment of useful products to us.

Protecting Free Services and Saving Money

Ecosystems provide us with many invaluable services free of charge. For example, birds control insect pests. Plants produce oxygen and absorb carbon dioxide. Forests help maintain local climate. Vegetated land helps to replenish groundwater supplies, reduce flooding, and control erosion. Wetlands help purify water.

Although these services are free to us, they are extremely costly to replace with engineered designs. Consider an example. In 1989, a study of the economic impact of clear-cutting in the watershed of the Cedar River, from which the city of Seattle, Washington, acquires its drinking water, showed that continued harvesting would increase sediment in the river. To purify the water, the city would have to build a $120 million water treatment plant that would cost millions of dollars a year to operate.

It would replace a service that is provided free by the relatively intact forest ecosystem.

Collectively, wetlands, forests, grasslands, and other natural systems provide billions of dollars worth of services that most of us take for granted. But this does not mean that humans must create a hands-off policy. Many wildlands can sustain some level of harvest of natural resources without threatening biodiversity and environmental services. Mangrove swamps, for instance, are a nursery ground for commercially important fish and shrimp, but they also reduce coastal flood and storm damage and help filter sediment from waterways (**FIGURE 11-13**). If properly managed, mangroves can provide timber for construction, pulpwood for paper, and charcoal for energy. They can also provide food for livestock, shellfish for human consumption, and a number of other products. In Matang, Malaysia, for example, well-managed mangroves produce fish and wood products worth more than $1,000 per hectare ($400 per acre) per year. One job is provided for every 3 hectares (7.5 acres). If all of Southeast Asia's mangroves were managed sustainably, they would provide about $25 billion annually to the economy and create about 8 million new jobs. If exploited carelessly, this valuable economic and ecological asset, part of our natural capital, could be lost forever.

KEY CONCEPTS

Protecting natural systems helps preserve many extremely valuable ecological services such as flood control and natural water pollution purification. These services are very costly to replace with engineered solutions.

Ethics—Doing the Right Thing

To some people, protecting plants and animals is simply the right thing to do. What right, some individuals ask, do humans have to drive another species to extinction? Other organisms have a right to live, too. To these people, preserving other forms of life is part of our responsibility. In fact, most major religions instruct their followers to protect other species. Of course, not all people agree with this view. To them, human beings are the most important life-form, and thus, the needs and rights of other species take a backseat to ours.

KEY CONCEPTS

To many people, preservation of other species is ethically appropriate. Protecting them honors their right to exist and is therefore ethically correct.

11.4 How to Save Endangered Species and Protect Biodiversity—A Sustainable Approach

Protecting endangered species and biodiversity, like all environmental challenges, requires a number of strategies. Some will be stopgap in nature. These measures help us protect species in immediate danger of extinction. Others are more long term and preventive in nature. Both types of effort are essential to sustainability—as vital as both emergency rooms and preventive medicine are to protecting human health. Before we examine these measures, let us consider the question of priorities.

KEY CONCEPTS

Protecting endangered species and preserving the world's dwindling biodiversity will require many actions—ranging from short-term protective measures to long-term preventive efforts.

A Question of Priorities: Which Species Should We Protect?

Endangered pandas, blue whales, rhinos, and chimpanzees generally make the headlines because they are the most appealing or visible victims. Most preservation money is spent on these species. Interest in less appealing species is often difficult to stir, but many less conspicuous species are important components of natural systems, even keystone species. Many inconspicuous species are vital to human welfare. An adult frog, for example, can eat its weight in insects every day. In India, sharp declines in the frog populations may be partly responsible for higher rates of insect damage on crops and for an increase in malaria, a debilitating disease transmitted by mosquitos, a main component of the frog's diet. Losing species, therefore, is not just an aesthetic tragedy. It can have profound environmental, economic, and health consequences.

FIGURE 11-13 Mangrove swamp. This wonderfully dense thicket of trees along many coastlines provides many ecological services (including protection from storm surges) and can be sustainably harvested without loss of these services.

Protecting species, regardless of how appealing they are, is vital to sustainability.

KEY CONCEPTS

Although many species are endangered, most resources are expended on the most appealing or most visible ones. Many ecologically important species could vanish if efforts are not broadened.

Stopgap Measures: First Aid for an Ailing Planet

This section describes two important stopgap measures designed to protect endangered species and preserve biodiversity: the Endangered Species Act and captive breeding programs.

The Endangered Species Act In 1973, the United States Congress passed the **Endangered Species Act**. This act, which has become a model for many other countries, requires the U.S. Fish and Wildlife Service to designate and list endangered and threatened species in the United States. Endangered and threatened species are classified on the basis of biological criteria, primarily their population size and rate of population decrease. A species that is officially listed by the agency is then protected by law. It cannot be legally hunted, trapped, or harassed. Violators face prosecution, fines, and jail sentences. The presence of an endangered species on federally owned land or private property slated for development with federal monies is enough to stop a project—unless a plan can be worked out to relocate the species or develop the land without affecting the species.

The Endangered Species Act also requires the U.S. Fish and Wildlife Service to develop recovery plans for endangered species, including measures to protect their habitat. It also provides money to purchase this habitat. Finally, the law enables the United States to help other nations protect their endangered and threatened species by banning the importation of these species and by providing technical assistance.

Protection formally begins with the listing of an endangered or threatened species. In 2014, 469 animals and 720 plants were designated as endangered in the United States according to the U.S. Fish and Wildlife Service. An additional 176 animals and 154 plants were listed as threatened.

Since the Endangered Species Act went into effect, thousands of projects on federally owned land or privately owned land with federal funding have been reviewed for their impact on endangered species. In most cases, differences have been worked out amicably. Thus, the Endangered Species Act is not an impediment to development as some critics assert.

The most renowned exception was the case of the snail darter and the Tennessee Valley Authority's (TVA) Tellico Dam on the Little Tennessee River, but it ended with a successful resolution that benefited both people and this endangered fish (**FIGURE 11-14**). Problems began in 1975, when

(a)

(b)

FIGURE 11-14 The snail darter. (a) Measuring only 8 centimeters (3 inches), the snail darter created a great controversy between environmentalists and the TVA. **(b)** The impending destruction of the snail darter by the Tellico Dam brought the multimillion-dollar project to a standstill. After years of debate, Congress ordered the dam to be completed.
Photo a courtesy of U.S. Fish & Wildlife Services; photo b © Photos.com.

a federal court ordered a halt to the construction of a multimillion-dollar dam (already 90% completed) that would have flooded what was thought to be the fish's only breeding habitat. The order was upheld in the U.S. Supreme Court, and Congress established a committee to review a request for an exemption to the law. In 1979, the committee refused to grant an exemption, saying that the project was of questionable merit. The TVA applied more pressure on Congress, however, and later that year Congress authorized the completion of the dam. The story did not end in tragedy for the snail darter, though, for it was transplanted to several neighboring streams where it is doing well. Additional populations were discovered in several nearby streams.

A more recent example has been the battle in Oregon and Washington over protecting old-growth forests and the spotted owl, which was listed as threatened in 1989.

The Endangered Species Act is one of the toughest and most successful environmental laws in the United States. The success of the act, says Bob Davison, a National Wildlife Federation biologist, is best demonstrated by the fact "that there are species around today that would not have survived if the law had not forced agencies to consider the impacts of what they're doing while allowing development to proceed…. To a large extent, the law has succeeded in continually juggling those two competing interests."

As noted earlier, the U.S. Fish and Wildlife Service lists endangered species in other countries as well and forbids their importation. Governments throughout the world have enacted similar measures. An international ban on ivory trade, for instance, has nearly halted the illegal poaching of African elephants. In many countries, however, inadequate funding makes enforcement a joke. Inspectors can be paid off by illegal traffickers in endangered species. Governmental agents patrol only a small fraction of the poachers' range, and the courts have routinely been lenient on poachers.

The Endangered Species Act has been under attack for many years. When considering placing a species on the list of threatened or endangered species, critics want economic factors taken into account. They say that the economic impacts of listing a species as endangered should be considered during the process. It is their contention that if it is too costly to protect a species or if it would eliminate jobs, a species should not be listed. Critics also point out that efforts to save species often intrude on private property rights. They ask for compensation when, to protect an endangered species, a person's land can no longer be used. Restriction of land use is often called a "taking." For a debate on the takings issue, see **Point/Counterpoint 11-1**.

A report by the National Research Council, which provides scientific advice to Congress, outlined numerous changes in the act to make it more scientifically sound and economically responsible. For example, it called for faster development of recovery plans and guidelines to avoid provisions that are scientifically and economically unjustified. Such plans, they said, should spell out which human activities are likely to harm recovery and which ones are not—a step that would allow for better economic planning in and around protected areas.

The committee that prepared the report also recommended that a core of "survival habitat" be established as an emergency, stopgap measure when a species is first listed as endangered. This habitat would be able to support the population for 25 to 50 years. After more careful study, scientists could determine the exact dimensions of the critical habitat needed for the species to recover. This might result in either a downsizing or an increase in the protected habitat.

Numerous attempts were made by the Bush administration to rewrite the Endangered Species Act, making changes that would reduce the effectiveness of this legislation. Unable to get changes through Congress, the Bush administration weakened the Act through a number of regulatory decisions. For example, the administration approved a regulatory change that allows the EPA to approve new pesticides without considering their impact on wildlife. The administration also relaxed restrictions on the trade of endangered species to allow U.S. trophy hunters and wildlife traders to import more endangered species and body parts such as horns, antlers, and skins. During the early years of the Obama administration, the new president rescinded most of the damaging regulations.

The Convention on International Trade in Endangered Species International wildlife trade is big business made possible in part by improvements in transportation that make it feasible to ship live plants and animals from their homeland to almost anywhere in the world to satisfy the pet trade market. Today, the sale of animals, skins, and live plants is worth billions of dollars to those who engage in the legal and illegal trade of these items.

Unfortunately, legal and illegal wildlife trade has also caused massive declines in the numbers of many species of plants and animals. Recognizing the threat trade posed to the world's flora and fauna, the United Nations Environment Programme, The World Conservation Union, and the World Wide Fund for Nature launched an effort to ban commercial international trade of endangered or threatened species. The result was an agreement known as the **Convention on International Trade in Endangered Species of Wild Fauna and Flora**, also known as **CITES**. It went into effect on July 1, 1975, and currently has a membership of 179 nations.

CITES, the organization that administers the agreement, keeps a list of species (known as Appendix 1) threatened with extinction by international trade. CITES bans hunting, capturing, and selling of endangered and threatened species. In addition, the signatories meet every 2 to 3 years to assess successes and failures of the convention. They also address new concerns and pass resolutions to protect listed species of wildlife and plants. Meetings are attended by both member nations (signatories), nonmember states, and nongovernmental organizations, that is, nonprofit organizations interested in protecting the world's wildlife.

Important as it is, CITES has its shortcomings. Endangered and threatened species continue to be smuggled out of countries, especially in the less developed nations where wildlife populations are disappearing rapidly. Documents are falsified by traders so that it appears as if the species were captured or collected in areas where they are still relatively abundant. Species may also mislabeled indicating a nonendangered species. Enforcement and investigational staff responsible for monitoring international wildlife trade are small in number and sometimes corrupt. Enforcement is left to each country and is often lax. Where enforcement exists, it is often weak. Penalties are mild. The result: Illegal trade in plants and wildlife whose future is in question continues.

Captive Breeding and Release Programs Zoos are another frontline component of the global effort to save species from extinction. Today, many of the 1,200 zoos and aquariums throughout the world are now involved in breeding endangered species in captivity to prevent them

Compensate Landowners for Restrictions on Property Rights

Richard L. Stroup

Richard L. Stroup is a Professor of Economics at Montana State University and a Senior Associate of the Political Economy Research Center (PERC) in Bozeman, Montana.

To understand the takings issue, a good starting point is the Fifth Amendment to the United States Constitution, which includes the clause: "… nor shall private property be taken for public use, without just compensation." For most of the nation's history, this "takings" clause referred to eminent domain, the right of a government agency to take property to construct a road, a public building, or for some other public purpose. The clause confirms the government's right to take property but requires the government to compensate the former owner.

Compensation is important for two reasons. First, when resources are taken to provide a public good or service, the cost should be shared by the public. Taking a tract of land, a crop, or a tree from an owner without compensation would put an unjust burden on that owner. Second, compensation forces politicians and government officials to recognize the cost of what they are doing. The cost of a school, for example, includes resources such as the bricks used and the land on which it stands. To ignore the cost of these resources is to invite their overuse and waste. Compensation puts the resources into the budget process where, quite properly, it competes for access to tax dollars against other uses.

What has changed is that today's takings may occur not just through eminent domain but through regulations. Of course, regulation has long been used to protect the property and the person of individuals from harmful behavior. Preventing a polluter from wrongfully harming a downwind neighbor, for example, is an exercise of the government's police power. Courts agree that this type of regulation is not a taking, and compensation is not appropriate.

However, the expanding scope of environmental regulations has led the courts, including the Supreme Court, to consider the purpose of a regulation more carefully. If the purpose is to produce a public good, such as public access to a beach or habitat for wildlife, rather than to protect someone downwind from pollution, a court may rule that forcing a landowner to provide the service is a taking, and when a taking reduces the value of property, the government must compensate the property owner.

A 1992 decision by the Supreme Court made clear that when a public good is the purpose of the regulation, rather than preventing the violation of the rights of others, courts may find that a taking has occurred, and compensation is due. David Lucas, a Charleston, South Carolina developer, had purchased two oceanside lots in 1986 with the intention of building homes on them. In 1988, before Lucas began construction, the government of South Carolina enacted the Beachfront Management Act, which prohibited building on beachfront lots. Although his neighbors had built homes on their beachfront property, Lucas was not allowed to build on his. His property, for which he had paid $975,000, lost virtually all its value. Lucas was awarded compensation for the near-total taking.

Unfortunately, some activists seeking preservation regardless of cost tend to confuse the two kinds of regulation—stopping a harm and providing a public service. Another term that they have championed, "givings," also causes confusion. A regulatory taking may actually give value to the owner. For example, when a portion of a field is taken for a highway exit, the value of the remaining property may end up higher than the value of the entire tract prior to the taking. In this case, a portion of the property is taken by the state, but no compensation is required. The owner gains, and the government gains not only from the land use, but also by receiving more revenue from property taxes, higher taxes paid on income from the land, and, if the property is later sold, higher capital gains taxes.

Appropriate compensation encourages good management and cooperation. Today, the Fish and Wildlife Service simply dictates the use of land to carry out its Endangered Species Act mandates. If the agency paid the cost of those restrictions, its staff would seek cheaper land and use economical habitat enrichment techniques. Responding to these restrictions, landowners tend to be uncooperative. Some preemptively modify their land to keep endangered animals away. Case studies and statistical evidence showing the importance of this problem are accumulating.

Both logic and evidence suggest that treating the rights of landowners with greater respect, while continuing to hold them responsible for any wrongful harms they cause, would lead to more effective environmental protection. Until these rights are respected, the "takings" issue will remain.

Critical Thinking Questions

1. After you have read both essays, summarize the main points of each author in this debate; then list supporting information

2. Identify any arguments you would have liked expanded and not why.

3. Which viewpoint best corresponds to yours? Why?

In the early 19th century, British textile workers wrecked newly introduced machines out of fear that technological advances threatened their employment. Ever since, these so-called "luddites" have been a symbol of misguided opposition to human progress.

Modern "takings"—or "property rights"—advocates are best understood as latter day luddites. In the face of overwhelming scientific proof of growing environmental stresses, and of the necessity of a new system of property rights and responsibilities in order to protect our environment, takings advocates tenaciously cling to an outmoded, environmentally destructive conception of property rights.

The central contribution of 20th century environmental science has been a new recognition that, in a literal sense, everything *is* connected to everything else. We now know that burning of fossil fuels, once viewed as relatively benign, contributes to higher levels of atmospheric carbon and threatens to produce worldwide warming. We now know that wetlands destruction, the promotion of which was once government policy, increases flooding and pollution levels scores if not hundreds of miles downstream. And we know now that traditional logging practices, unless modified, threaten to wipe certain species out of existence.

Until only a few decades ago, it was hardly farfetched to think that a property owner's rights—as well as an owner's responsibilities—ended at the boundary of his or her property. Today we know better, that the security of the environment, and the long-term value of property itself, depends on careful attention to the effects of an owner's activities on the surrounding neighborhood and society as a whole.

Takings advocates dispute the idea that society can and should evolve to reflect this new scientific understanding. They reject the idea of new responsibilities to protect the environment as both unwanted and unfair. They assert that if society as a whole insists on these changes, the public should pay property owners for shouldering the new responsibilities. There is simply no legitimate support for this position.

Takings advocates seek to ground their argument in the Constitution of the United States, but it is a myth that anyone has a constitutional right not to follow public regulations designed to protect the environment. The Takings Clause of the Fifth Amendment provides that the public must pay "just compensation" when it seizes private property for use in building a road or a school, for example. But, as the U.S. Supreme Court has affirmed, this provision of the Constitution was never intended to prevent

Rethinking Property Rights and Wildlife Protection

John Echeverria

John Echeverria is Director of the Environmental Policy Project, Georgetown University Law Center.

Courtesy of John Echeverria.

regulation of uses of private property to protect society as a whole.

There is nothing surprising nor unprecedented about the idea that society's thinking about private property changes over time. In colonial Massachusetts, a law prohibited the building of a home more than easy walking distance from the meeting house; while this restriction probably strikes most modern American citizens as onerous, the rule apparently suited the society of that time and place. Likewise, in the early 20th century, the idea that government can set the minimum wage or fix maximum hours seemed an affront to private property rights, but the practice is well accepted now.

Also, in demanding public payment as a condition of refraining from harming our environment, takings advocates seek unfair windfalls at public expense. After all, reasonable environmental standards benefit everyone, including private property owners. In addition, much of the value of private property is created by nearby public investments, and it is hardly fair for owners to demand payments from the public based on speculative value created in part by the public in the first place.

This argument for the legitimacy of public regulation of the uses of private property suggests no disrespect for private property as an institution or its importance to the proper functioning of our economy. One of the key purposes of government is to safeguard property owners against theft and fraud. But the public's role in safeguarding private property rights goes hand in hand with the public's authority to ensure that those rights are exercised responsibly.

Nor does this viewpoint disparage the value of our market economy in producing an efficient mix of products and services. Quite the opposite. The familiar principle of "polluter pays" reflects a recognition that, in order for the market economy to function efficiently, firms whose activities produce external costs must be required to internalize these costs. In a competitive market, competing timber firms, for example, are driven to seek to maximize

(continued)

the return from the trees on their land, without regard to the negative costs of logging in terms of degraded fisheries, wildlife, or public drinking supplies. In the absence of regulation, these external costs would go unaddressed, and the companies would operate at less than an optimal level of efficiency from the standpoint of society as a whole.

Takings advocates would, in effect, exacerbate the problem of "market failure" by encouraging firms and individuals to ignore the external costs of their actions or, what amounts to the same thing, by forcing the public to pay them not to produce these external costs. In either case, the ultimate result is inefficient from an economic standpoint.

Fortunately, the passage of time solves many problems, and this will likely be the case with the takings issue. At bottom, the controversy over property rights reflects the painful process of change as we adapt our property concepts to reflect our new scientific knowledge. There is ground for hope that, as environmental education advances, firms and individuals will make investment decisions in harmony with the new environmental realities rather than in opposition to them. If so, today's takings luddites can be expected to pass into history along with their 19th century counterparts.

from disappearing from the face of the Earth. Canada's zoos are home to a dozen endangered or threatened animal species such as the whooping crane and ferruginous hawk. Botanical gardens are also lending a hand. In fact, today, about half of the world's 2,803 botanical gardens have developed programs to conserve rare plant species. Canada's 102 botanical gardens house at least 1,077 endangered, rare, or threatened plant species. More recently, zoos have been trying to establish captive-bred populations to provide animals for release into suitable habitat once the captive population was well established. One recent project was launched by zoos in the mainland United States to rescue three rare species of birds threatened by tree snakes on the island of Guam. The survival of the California condors was ensured by the San Diego Wild Animal Park and the Los Angeles Zoo, who raised condors in captivity for release into suitable habitat.

Many primates used for research are also being raised in captivity by commercial animal supply houses in an effort to reduce the drain on wild populations. Yet another way of reducing demand on wild populations is the use of laboratory tests—for example, cell cultures—rather than tests on live animals. Toxicity tests, for example, are being performed on cells grown in tissue culture. Actions such as these could stop the flow of animals from their natural habitat and help prevent the extinction of many primates.

The release of animals into the wild to reestablish populations that have been driven to extinction is an essential component of captive breeding programs. After all, what good is species preservation if the species must remain in captivity for eternity? Gray wolf populations, for instance, have been reestablished in and around Yellowstone National Park in Wyoming and in the Northern Rockies (**FIGURE 11-15**). The Mexican wolf was reintroduced into Arizona and New Mexico. They came from the Endangered Wolf Center in

FIGURE 11-15 Gray wolf. U.S. Fish and Wildlife Service Biologist Alice Whitelaw carries a sedated gray wolf, soon to be released in Yellowstone National Park, to an awaiting veterinarian for a checkup.

Courtesy of Pedro Ramirez, Jr./U.S. Fish and Wildlife Services.

eastern Missouri. Release programs have been sparked by major disagreements between environmentalists who seek to reestablish populations and ranchers who are concerned about predation by wolves on sheep and cattle. To address this problem, the Defenders of Wildlife established a Wolf Compensation Trust donated by private sources to pay for losses. (For a discussion of both sides of the wolf introduction, see **Point/Counterpoint 11-2**.)

The Mexican Wolf reintroduction into southwest New Mexico and southeast Arizona has been in progress for 8 years. In those years, the project has had more than its share of failures. Those failures never seem to affect the project or the forces that drive it.

Agencies in charge of the Mexican Wolf restoration project and environmental organizations that support it will undoubtedly claim that restoring lobos into the southwest will renew biodiversity in a region that supposedly has none.

What is biodiversity? Is it necessary to have a predator, as lethal to other species as wolves have historically been proven to be, in order to have biodiversity? Or is biodiversity just another buzzword brought into the midst of land management policies to further a political agenda?

Environmental organizations across the U.S. have given their support to wolf reintroduction, but only in agricultural communities. Wolves are never restored where people do not inhabit the area for the purpose of raising market livestock.

Supporters of wolf restoration efforts back up the U.S. Fish and Wildlife agency by saying the wolves were here first, and if necessary, those dependent on the land for livestock production can move aside for the animals. Naturally, ranchers see that as fighting words. When agencies and organizations cannot show scientific data that substantiates the need for another predator, supporters explain their compulsion to accomplish wolf reintroduction as the "it's the right thing to do." Who is it right for, certainly not the communities surrounding the wolf release sites, certainly not the game species, certainly not the ranchers and area hunters. How many times have we as individuals made an absolutely immoral decision and altered our lives or hurt someone else because, at the time, the bad decision seemed like the right thing to do?

In reality, the U.S. Fish and Wildlife agency is mandated by law to restore species to their former habitats. The Endangered Species Act (ESA), does not place upon the agency the requirement of good science to list a species. Science seldom has anything to do with it at all. A species can be listed as endangered by licking a stamp, putting it on an envelope, and mailing out a request. The Endangered Species Act has been up for reauthorization for 6 years but has never been reauthorized due to political pressure from environmental extremists.

The ESA has become inefficient and burdensome. The agencies that carry out the act seldom comply with the intent of the act to recover species. Very few species are ever removed from the list. The ESA supports enormous bureaucratic agencies, has given these agencies the power to manipulate people, whole communities and states, without having to show that any species they choose to

Mexican Wolf Reintroduction Debacle

Laura Schneberger

Laura and her husband Matt own a working cattle ranch that depends on a federal land grazing permit in the Gila National Forest in southwest New Mexico. She has three children and writes editorials and feature articles that have been published in many New Mexico newspapers and a few national publications.

Courtesy of Laura Schneberger.

recover is ever recovered sufficiently. Some grassroots organizations even argue that the ESA is so inefficient and out of date, it is no longer legal as it is now written.

The areas in the Southwest where government agencies and environmentalists have demanded the lobo be returned have diverse game populations that have not only survived intact, but have grown enough to provide tremendous hunting and tourism income for the states that they occupy. If lobos are successfully introduced into these regions, it is expected that prey species numbers will plummet to pre-1920s levels. The excess game that are now harvested and provide jobs and income in these states will no longer exist. The deer and elk herds in the state belong to the state to manage, not to the federal government.

Since the removal of the Mexican wolf from the Southwest, herds of elk, antelope, and deer have improved, increased, and been managed for the benefit of the herds and the communities. When lobos roamed the Southwest, very few deer could be found for game to support growing communities, livestock were brought in to supply meat, and wolf packs turned to livestock for their prey base.

Cattle still inhabit the areas that are being put to use as wolf habitat. Cattle and other prey species for wolves exist together in a complementary manner and cannot be separated as distinct prey species though wolf proponents try to do so. All ungulate herds in the Southwest are prey for wolves though none are owned by those hoping to introduce wolves. Either individuals or the state own the prey being set aside for reintroduced lobos. The ESA has no provisions for reimbursing individuals or communities for losses to endangered animals. Wolves will migrate into areas occupied by cattle when they follow elk and deer to the fresh grasses brought up by grazing cattle. The wolves will stay once they learn cattle are more easily preyed upon than deer or elk who have the ability to strike back at attackers with their forefeet.

Historically, Mexican lobos were responsible for the demise of hundreds of thousands of dollars in livestock in

(continued)

Arizona, New Mexico, and Texas. One study conducted from 1920 to 1922 showed that the stomach contents of 200 lobos were either empty, or contained beef, burro, horse, pig, or sheep. Only 17% contained antelope, deer, or rabbit.

In another historical account, a 3-year-old girl disappeared from a mining camp in Hillsboro, New Mexico for 4 days. She had been playing with a pack of semiwild burros living in the area of the camp. Trackers followed the burro herd for days trying to find the child. Her tracks were interspersed with those of the burros and those of a wolf pack. When found, she was alive, hungry, and exhausted; the burros had surrounded her and kept her moving while defending her from the attack of the wolves. The searchers found areas where tracks showed the burros had surrounded the girl and fought off the lobos. They even left one wolf carcass in the trail. Ironically, the wolves and the burros saved the child's life. The burros kept her moving and kept wolves off her, and the wolves kept the burros moving; if they hadn't, she would have suffered from hypothermia and died.

Just the fact that this story is still told sheds some light on how the people of the Southwest felt about wolves and why they were removed from the area. It isn't the only story left over from a bygone era that shows the Mexican lobo reintroduction to be a senseless endeavor.

Wolf reintroduction is seen by inhabitants of the rural Southwest as an attempt to control the use of land in poor states. It is being orchestrated by organizations that subsist on government and foundation money and bloated government bureaucracies. The welfare of wolves and their species is subordinate to the goal of land and people control.

COUNTERPOINT

Why Wolves?

Craig Miller

Craig is the Southwest Director of Defenders of Wildlife and lives in Tucson, Arizona, with his wife and two sons. All are eager to share their world with the big, good wolf.

Courtesy of Craig Miller.

Perhaps no animal has been shrouded in more controversy, misinformation, and emotion than the wolf. While just 50 years ago a federal campaign sought to eliminate wolves, a popular federal program now seeks to reestablish them. This shift reflects society's emerging understanding of the diverse benefits of wolves and our growing sense of responsibility to right the wrongs of the past when possible. Ensuring the survival of wolves is not only possible, but desirable and necessary. Wolf recovery is important to conserve the wolf itself, to maintain important ecological processes, and to provide spiritual, recreational, economic, and esthetic benefits to people. While wolf eradication was encouraged by myth, misinformation, and misunderstanding, wolf conservation is supported by ecology, evolution, economics, education, and ethics.

The influence of wolves runs deep. For tens of thousands of years they have been an important evolutionary force shaping and maintaining the landscape of North America and its inhabitants. The alertness of deer, the herding of ungulates, and the lodge-building of beaver are all, in part, attributed to the wolf. Wolves have similarly influenced humankind. Evolutionary biologist John Allman asserts that *Homo sapiens*, through their close alliance with wolves, were able to outcompete neanderthals. It is also believed that humans adopted aspects of wolf family structure and behavior to improve our own efficiency and chances for survival. It is ironic that man is responsible for the recent campaign to exterminate the animal, which through centuries of coexistence became "man's best friend."

The systematic destruction of wolves began in the early 17th century with the arrival of British Colonists and their livestock to North America. As settlements were established on top of territories already occupied by native people and native wildlife, policies of assimilation and extermination quickly became the law of the land. Domestic livestock rapidly displaced native populations of deer and other preferred prey of wolves. Overhunting by humans also severely reduced and in some cases entirely eliminated prey populations. Wolves turned to the larger, slower moving, and more abundant livestock. In the West stockmen faced many new challenges. Not knowing how to deal with drought, overgrazing, overhunting, or the general diminished carrying capacity of the land, ranchers targeted wolves with a vengeance. Wolf killing soon became a federal campaign carried out by professional trappers encouraged with bounties, and wolves were slaughtered by the millions. They were shot, trapped, poisoned, burned, and demonized to near extinction. By the mid-1950s wolf populations had been completely extirpated from 90% of their historic range in the lower 48 states.

The removal of wolves brought on abrupt and drastic changes in ecosystems. Animals such as deer and elk now have few natural checks and balances and die by the thousands during unnatural population explosions. As their numbers grow out of control they overgraze the landscape, destroy natural vegetation, and contribute to erosion, watershed degradation, and other problems. It is ironic that wolves were extirpated because of a fear of economic loss, because the ecological instability caused by their removal is now severely undermining the profitability of ranching and other resource-dependent industries. The landscape without wolves is biologically impoverished. Learning from history, a few western ranchers are now finding it possible to reinvigorate land productivity and profitability by mimicking the herding and regulatory role of wolves. These ranchers, however, agree that such management is labor intensive and that wolves would be less expensive and more effective (Jim Winder, personal communication).

Wolves also check other predators. Since the reintroduction of wolves in Yellowstone's Lamar Valley, biologist Robert Crabtree has found that wolves have reduced the coyote population by 50%. Conversely, in areas where wolves have been eliminated coyote populations have exploded. The greater density of coyotes has resulted in increased predator problems for livestock producers, as well as for other small to mid-size animals.

By looking for weaknesses in their prey, wolves usually kill vulnerable, sick, or starving animals. This process bolsters the genetic vigor and overall health of prey populations. Additionally, wolves create ecological hotspots when they leave behind carcasses. These carcasses attract a variety of species in high numbers, including eagles, wolverines, fishers, ravens, and more. Beetles and other invertebrates also prosper and attract other birds often carrying fleas, ticks, lice, etc. Carcass decomposition then benefits plant life with renewed fertilization, yielding thicker greener grasses and forbs to support the next generation of ungulates. Simply put, wolves are engineers of biodiversity. Their ability to maintain productive landscapes is the result of over 12,000 years of experience. The wise will embrace this understanding; others will reject it.

Critics of wolf restoration claim that the goal of wolf recovery is to control land use and to force ranchers out of business, but such claims ignore important facts. First and foremost, wolves kill livestock infrequently, especially when native prey is available. In wolf-occupied territories across the U.S. and Canada, less than one half of one percent of livestock present in wolf-occupied regions has been lost to wolves. To share the economic responsibility of wolves, Defenders of Wildlife reimburses ranchers at fair-market value for all verified losses. Since 1987, $115,000 has been paid to 112 ranchers. Defenders has also helped to minimize wolf–livestock conflicts by purchasing portable electric fencing, hiring herdsmen, purchasing hay, providing guard dogs, and disposing of carcasses. Defenders also developed a "wolf country" labeling program that creates a market premium for beef products produced in a manner compatible with wolf conservation. It is no coincidence that Arizona "wolf country" ranchers Will and Jan Holder have turned their first profit in 25 years in the presence of wolves. Despite rhetoric from traditional antagonists, opportunities exist to heal the wounds between conservation and agricultural communities as well as between humans and wildlife.

The crusade to exterminate wolves reflected the values of a culture then obsessed with taming wilderness. Today that mindset seems primitive considering that scientists are now predicting that within the next 50 years the only large mammals likely to remain on earth will be those which humans have decided to keep around. Wolves and other wildlife were placed here by a power higher than the livestock industry or the federal government. Right or wrong, humans are now determining the existence or nonexistence of these and other life forms. This serious responsibility warrants the thoughtful consideration of facts and the generous use of caution. Moreover, the overwhelming majority of the public recognizes the importance of maintaining a broad diversity of life and believes that species like the wolf are worth keeping. This sentiment is reflected in our national policy, the Endangered Species Act. It is our legal, biological, and moral obligation to prevent the senseless loss of wolves and other imperiled wildlife. Failing this would be a reckless abandonment of our responsibility to society, to our children, and to ourselves.

Critical Thinking Questions

1. List the main arguments of each author.
2. Does each author do a good job of supporting his or her arguments?
3. Which viewpoint corresponds to yours?

Long-Term Preventive Measures

Captive breeding programs and laws such as the Endangered Species Act are a kind of first aid measure in global efforts to protect species. They can be thought of as emergency measures needed to preserve the planet's biodiversity. They're similar to the kinds of treatment a heart attack patient receives in an emergency room of a hospital. An emergency room physician may save a patient suffering from a heart attack, but unless the patient diets, exercises, quits smoking, and learns to handle stress, he or she is not likely to survive very long. By the same token, emergency measures are not enough to ensure a diverse, biologically rich world. To achieve this goal, preventive measures are needed. Like other environmental problems, this one can be addressed by slowing, even halting, the growth of the human population and a fundamental redesign of human systems such as agriculture, energy, industry, and waste management.

Restructuring Human Systems for Sustainability To ensure a biologically rich world, we must address the root causes of species extinction and the loss of biodiversity. We need to tackle such issues as habitat destruction, overharvesting, and pollution as well as the fundamental driving forces behind these problems, notably unsustainable economic and population growth. To create sustainable solutions that confront these causes will require changes in human systems, from economics to transportation to waste management. This book outlines the need to restructure such systems by applying the operating principles of sustainability—especially conservation, recycling, renewable resource use, and restoration—to human systems. Such efforts go a long way toward eliminating many of the pressures that endanger our planet and the many species that live on it with us. **Spotlight on Sustainable Development 11-1** gives an example of what some ranchers and environmentalists are doing to change practices to help protect wildlife.

© hugolacasse/ShutterStock, Inc.

>> SPOTLIGHT ON SUSTAINABLE DEVELOPMENT

11-1 Predator Friendly Wool and Wolf Country Beef: You Must Have Read the Label Wrong

Imagine going to the store to the mall to buy a new sweater. One particular brand catches your eye. It's made from Predator Friendly Wool. After you pay for your purchase, you stop in at the restaurant next door to buy a salad made from Salmon Safe vegetables, while the gentleman at the next table enjoys a hamburger made from Wolf Country Beef.

Thanks to the efforts of far-seeing environmentalists, farmers, and ranchers, this scenario is becoming a reality. In Oregon, the Pacific Rivers Council is spearheading an innovative program to encourage farmers to grow crops in ways that help protect salmon fisheries. Farmers who join the program are seeking ways to reduce runoff from farm fields into streams, which

brings with it sediment and pollutants that destroy salmon habitat. Runoff can be reduced by planting buffer zones along streams to trap sediment running off from nearby land or by contour farming and other techniques. Cover crops, such as clover or rye, may be planted on cropland between seasons or in fallow fields, to protect them from the pernicious effects of rain.

Although the Salmon Safe program doesn't make pesticide prohibitions part of their certification, it does recognize the importance of reducing pesticide use in protecting fish and wildlife. The program's goal is to complement organic certification programs that focus on chemical use. Combined, the two programs provide optimal levels of protection for the nation's fish and wildlife.

In Montana and Idaho, sheep ranchers are trying their own brand of animal protection thanks to the efforts of the Growers' Wool Cooperative. Its Predator Friendly Wool Certification program seeks the coexistence of sheep and native predators, rather than eradication of predators. Ranchers use natural methods to keep coyotes away from their sheep, including llamas, which are quite protective, and guard dogs like **Great Pyrenees** (**FIGURE 1**). Guard dogs are fierce protectors of poultry and livestock, including sheep and cattle. They are especially effective at protecting against wolves, coyotes, and other dogs. These nonlethal measures reportedly make Predator Friendly Wool a popular item among some buyers.

In Arizona and New Mexico, the Defenders of Wildlife, a national organization, is working with cattle ranchers to help promote reintroduction of the Mexican wolf. The Mexican wolf or "El Lobo" is a subspecies of the gray wolf. It once roamed freely through much of Arizona and New Mexico. As with other predators, ranchers and government agents took a dim view of this predator, shooting them, poisoning them, and catching them in traps. By 1950, the Mexican wolf population was nearly gone. The last remaining wild wolf was shot in 1970.

Defenders of Wildlife, which has taken an active role in the reintroduction of the gray wolf in Montana, Idaho, and Wyoming, has crusaded hard to reintroduce the Mexican wolf to its native habitat using captive-bred animals (**FIGURE 2**). In 1998, the first wolves were introduced into the wild in Arizona, from a captive-breeding program of the Endangered Wolf Center in Eureka, Missouri, about 20 miles southwest of St. Louis. Released by the U.S. Fish and Wildlife Service in the Apache National Forest bordering New Mexico and Arizona, the wolves had a rough go of it. To date, 92 captive-bred Mexican wolves have been released, but the wild population is now (2012) only about 75 with only 3 breeding pairs, according to the U.S. Fish and Wildlife program. Defenders of Wildlife cite three major reasons for this discrepancy. To date, at least 23 wolves have been shot by ranchers. Ten wolves have been killed for killing cattle, and at least 10 have died after being recaptured.

Because wolves cannot be expected to remain in the National Forest, Defenders initiated a fund, known as the Wolf Compensation Trust, to reimburse ranchers in nearby areas for any losses incurred by wolves. The Fund compensates ranchers in Montana, Wyoming, and Idaho, too, and in 23 years has made payments in excess of $1.4 million to ranchers for loss of sheep, cattle (mostly calves), colts, and guard dogs killed or presumed killed by gray wolves released in Yellowstone and the northern Rockies. It also covers losses from wolves that have naturally migrated from Canada into northwestern Montana. The goal of Defenders of Wildlife is "to shift economic responsibility for wolf recovery away from the individual rancher and toward the millions of people who want to see wolf populations restored." Money for the fund comes from private donations. "When ranchers alone are forced to bear the cost of wolf recovery, it creates animosity and ill will toward the wolf," say Defenders. Federal legislation shifted responsibility for payments to states and Indian tribes. This frees up resources of Defenders to

FIGURE 1 Guardian dogs. The author's two Great Pyrenees, Badger (back), and Soleil, his mother (shown here), keep predators at bay. They've virtually eliminated losses of chickens by warding off coyotes, raccoons, foxes, wild dogs, opossum, and hawks. Prior to their arrival, predators would kill 20 to 30 chickens a year. Mortality is down to a few birds per year taken by occasional stealthy hawks and weasels.
Courtesy of Dan Chiras.

FIGURE 2 The endangered Mexican gray wolf. At this 60-acre facility in Missouri, Mexican gray wolves and red wolves are bred and raised for release into the wild.
© The Endangered Wolf Center, St. Louis, MO./Ashley Brown.

work with ranchers to prevent wolf predation in livestock, although the organization does continue to donate money to the states. The state of Wisconsin, for example, pays ranchers for livestock loss presumably caused by wolves. Unfortunately, there is some concern that ranchers are defrauding the state by making false claims for losses. Payments by other states are not reading available.

It should be pointed out that wolves are responsible for only a tiny percentage of all cattle deaths in the United States. Respiratory diseases, digestive problems, complications giving birth, and other diseases are major causes of death. Coyotes kill 22 times more cattle than wolves and domestic dogs five times more than wolves.

Defenders has also launched proactive efforts to prevent conflict between humans, livestock, and wolves and other predators. This program funds nonlethal deterrants that prevent bears and wolves from preying on livestock and even honey bees. Fencing to protect bees from bears, guard dogs to protect livestock, and fences to protect sheep are just a few of their privately funded efforts.

This program takes the economic sting out of wolf reintroduction. It demonstrates that conservation can be achieved through cooperation and that people need not be hurt economically by efforts to create a sustainable future.

In 1990, the Earth Island Institute initiated another innovative effort, the Dolphin Safe Tuna program. They later launched a Turtle Safe Shrimp program. Participants got bragging rights and marketing advantages for acting responsibly. It's an arrangement that works for people, the environment, and the economy.

More specifically, the sustainable strategy calls on people the world over to use only those resources they need and to use them with efficiency. All systems should be restructured to be maximally efficient. Such efforts would greatly reduce both our demand for resources and the habitat destruction caused by resource extraction (such as timber harvesting and mining). Recycling similarly reduces our need for both resource extraction and landfills, which destroy habitat. Recycling also uses less energy than making products from raw materials, thus reducing the amount of pollution we produce. All systems should maximize recycling by ensuring the return of usable waste to the production–consumption cycle and by incorporating recycled materials in new products and system components. Turning to renewable energy resources such as sunlight and wind would have an equally beneficial impact on wild species and natural habitats. Transportation, housing, industry, and other sectors can increase their dependence on renewable energy, which would help reduce some of our most serious pollution problems such as global warming, acid deposition, and oil spills. The restoration of natural habitats is also needed both to save wild species and to protect the free services nature provides. In the Pacific Northwest, for instance, five dams on the Columbia/Snake River watershed, which collectively produce only 5% of the region's electricity, are responsible for wiping out 90% of the wild salmon in 25 years. Accordingly, some environmentalists are calling for the destruction of these dams to ensure the salmon's future. In instances where dams have become obsolete and were torn down, salmon populations have rebounded.

As noted earlier, essential to these efforts are measures to stabilize human population growth and, perhaps, reduce human numbers over the long haul. Finally, growth management strategies are needed to stop the spread of human populations. These efforts, when combined, could make

KEY CONCEPTS

Protecting biodiversity will be best achieved by efforts that address the root causes of the crisis of unsustainability—our inefficient use of resources, continued population growth, our reliance on fossil fuels, our failure to recycle extensively, and our lack of attention to restoration. Addressing these issues will protect plants and animals and bring many other benefits to society.

tremendous inroads into the problem of species extinction, and they could offer many other benefits to society as well.

Setting Aside Biologically Rich Regions Buying habitats of endangered species and instituting growth management strategies in cities and towns to reduce the loss of habitat from urban sprawl are vital to the protection of endangered species and the maintenance of biodiversity—but they're not enough. Another preventive effort of extreme importance is the establishment of biodiversity reserves—protected areas characterized by a high biodiversity. The world map in **FIGURE 11-16** shows numerous high-priority areas in need of protection. Advocates of this approach argue that protecting these regions is the best economic investment in plant and wildlife diversity we can make, for it safeguards areas with extremely rich plant and animal life.

As you can see from the map, most of these regions are located in LDCs that lack the financial resources needed to protect them. Ironically, though, the wealthy, biologically poor MDCs will probably benefit the most from preserving these genetic resources. Therefore, many people argue that the more developed nations of the world should share the cost of preserving these areas. A 1% tax on internationally traded oil, for instance, could net billions of dollars a year

FIGURE 11-16 Biological hot spots. High-priority (dark red) and priority (light red) regions in need of protection to preserve important plants, animals, and other species.

and would go a long way toward establishing and maintaining large reserves in high-priority areas. An innovative way more developed countries can help in this effort is presented in **Spotlight on Sustainable Development 11-2**.

Many LDCs have made important efforts to protect their lands. China, which houses approximately 10% of the world's species, recently set aside 44 million hectares (110 million acres) of forest and tundra—or about 4.5% of its land area—as nature reserves to protect biodiversity. The government of Colombia has turned over 18 million hectares (45 million acres) of rain forest land, an area approximately three-fourths the size of Great Britain, to tribal peoples who have lived on the land sustainably for hundreds of years. The lands will be held in common by the people and cannot be sold unless they have the agreement of three-fourths of the adults in the affected tribes.

The Colombian government is also purchasing mountain land, where settlers of European descent have lived for almost 200 years, and is turning it back to its indigenous people. Combined with rain forests, some 26 million hectares (65 million acres) are now in the hands of local tribes. The success of this program has prompted other developing countries to follow suit. In 1991, Venezuela gave permanent title to a region of forest about the size of Austria to a native tribe.

Many areas in the more developed nations also need permanent protection. Although efforts are underway by private, nonprofit organizations such as the Nature Conservancy and by various governmental agencies, much

more can be done. A study of protected and biologically rich areas in Idaho showed that most of the protected areas—for example, areas set aside as federally designated wilderness—were rather low in biodiversity, whereas the high-biodiversity regions were in private hands and were not in any kind of program to ensure their protection. This type of analysis is called *gap analysis*, because it shows a gap between protection efforts and areas needing protection. Many other states are finding the same result. Encouraging landowners to protect these areas is vital to protecting biodiversity.

KEY CONCEPTS

Setting aside high-biodiversity areas for permanent protection will help to protect species from extinction and preserve biodiversity. Unfortunately, the majority of the most biologically diverse areas are located in the less developed nations, which lack the financial resources to protect them.

Buffer Zones and Wildlife Corridors: Protecting and Connecting Vital Areas Existing amid crops, cities, pastures, towns, and mines, protected areas (set aside for wildlife) have been only marginally successful in safeguarding biodiversity, much to the surprise of many. American ecologist William Newmark studied the loss of mammal species in national parks and reported an alarming decline in the number of species in all but the very largest parks. Utah's Bryce Canyon National Park, which is one of the

11-2 Debt-for-Nature Swaps

In 1984, Dr. Thomas E. Lovejoy, then with the World Wildlife Federation, proposed a new idea to help less developed nations set aside biologically rich areas. It is called the **debt-for-nature swap**. What is this?

Many LDCs have enormous debt—money they owe wealthier nations of the world who helped them finance oil development, agricultural equipment, dams, and other projects. Many of these nations are having trouble paying off their debt. What Lovejoy proposed was a trade of sorts. He suggested that the nations who owed money might agree to set aside land in exchange for some debt relief (**FIGURE 1**). By reducing debt, countries could free up money badly needed for other goals—among them literacy, health care, and sustainable development.

In a debt-for-nature swap, some of a less developed country's foreign debt can be exchanged for a commitment to invest in local conservation programs. A debt-for-nature swap typically involves three parties: a country in debt, a nation or bank to which the debt is owed, and an entity (an international nongovernmental organization, or NGO) that purchases the country's debt, usually at a significant discount. In a typical debt-for-nature swap, Country A, which owes a bank or a government a large sum of money (say,

FIGURE 1 Debt-for-nature swaps are another means of setting aside land to protect biodiversity. Photo taken along the Manú River in Peru.

©Alexey Stiop/Shutter Stock, Inc.

$10 million), sells its debt to an NGO at a considerable discount (for example, 10 to 20 cents on the dollar). In this scenario, the NGO would pay the creditor $1 to $2 million, and the entire debt would be forgiven. In exchange for the NGO's purchasing the debt, the debtor nation would enter into an agreement with the NGO to invest a certain amount of money in wildlife parks or other conservation measures.

Three international environmental NGOs based in Washington, D.C., have played a major role in promoting these swaps: Conservation International, the Nature Conservancy, and the World Wildlife Fund. The first ever debt-for-nature swap initiated by Conservation International occurred in 1987 in Bolivia. The agreement between Bolivia and Conservation International resulted in a commitment on the part of the Bolivian government to protect an area of biological importance in return for a small debt reduction. Conservation International purchased $650,000 of debt at an 85% discount. In exchange for the debt, the Bolivian government agreed to provide $250,000 in local currency to help manage the newly established Beni Biosphere Reserve, to create several new protected areas nearby, and to ensure that the forest in a vast tract of adjacent land would be managed sustainably.

Since the first debt-for-nature swaps in Bolivia, similar deals have been struck in nearly 40 countries including Ecuador, Costa Rica, Mexico, Poland, Egypt, Jamaica, the Philippines, and Vietnam. During the late 1980s and 1990s approximately five swaps were made annually. Debt-for-nature swaps have cropped significantly since 2000 (to about two per year) because of international initiatives that allow LDCs to restructure, even eliminate, debt at a much lower cost.

Debt-for-nature swaps are not a panacea. They represent one of many ways we can protect biodiversity. Existing programs have sparked a great deal of enthusiasm and hope, and they have infused conservation and environmental efforts with badly needed money.

nation's smallest parks, lost 36% of its species. California's Yosemite, which is nearly 20 times larger than Bryce, lost 23% of its species. Ecologists refer to such isolated patches of protected habitat as **ecological islands**.

The decline in species in ecological islands occurs because they often do not contain enough habitat for all of the species that live within them. Wide-ranging animals are especially affected. Small islands of habitat also reduce populations, sometimes below levels needed for successful reproduction. Tiny fragments of habitat often lead to interbreeding among members of a small population. This may result in inferior offspring, which are less fit. Small, isolated patches of habitat may also have different abiotic conditions from larger patches, making them less hospitable to species. Finally, human activities on the margins may affect the biotic and abiotic conditions.

Nonetheless, isolated patches can offer many benefits. Central Park in New York City, for example, is listed as one of the nation's 14 best bird watching sites. Why? Because it contains a variety of habitats and is strategically located along East Coast migration routes. It is one of the few places a migrating bird can stop and rest along the heavily populated Northeast coast.

To date, more than 12,000 sites on land and at sea similar to this in 167 nations have been classified as Important Bird Areas in countries such as Russia, Paraguay, South Africa, and the United States. Despite this designation, only 40% have received any form of protection, according to Birdlife International, the nonprofit organization that has spearheaded this effort protection.

The success of protected areas can be enhanced by establishing buffer zones around them. A **buffer zone** is a region in which *limited* human activity is allowed—for example, timber cutting or cattle grazing.

Another potentially important new idea in wildlife protection is the **wildlife corridor**, strips of land that connect habitats set aside to protect species. Wildlife corridors allow species to migrate from one habitat to another, breeding with members of other subpopulations. This, in turn, increases genetic diversity within a species, ensuring a better chance of survival. Connections between islands also open up new habitat. Food shortages encountered in one region, for example, can be offset by migrating to another area.

Wildlife corridors have been established in the United States, Europe, Central America, and Asia. The state of Florida and the Nature Conservancy are currently developing a wildlife corridor to protect the endangered Everglades panther, which has been relegated to a few patches of land widely dispersed throughout the state. If efforts to connect them are successful, the panther may someday roam much more freely in search of food and mates, hopefully living in relative harmony with the human population. Students can go online to view a map of this ambitious project.

KEY CONCEPTS

Islands of habitat are vital to protect species, but they are not enough to prevent species loss. Buffer zones between human activities and protected areas may provide an additional measure of protection. Wildlife corridors, areas that permit wildlife to move from one protected area to another, are also proving vital in our efforts to protect species diversity. Debt-for-nature swaps have been successful in many parts of the world.

Extractive Reserves Another way of protecting vital habitat in less developed nations is the **extractive reserve**, land set aside for native people to use on a sustainable basis (**FIGURE 11-17**). Huge tracts of rain forest in tropical countries, for example, are being preserved for sustainable harvesting of rubber, nuts, fruits, and other products. While providing a sustainable income, the reserves also help protect native species.

Interestingly, several studies show that income from extractive reserves actually exceeds revenues generated from agriculture, grazing, and lumber on the same land. In fact, government officials in the African nation of Niger halted a project to create a plantation in the rain

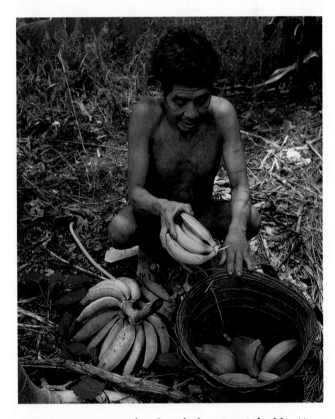

FIGURE 11-17 Harvesting the rain forest sustainably. Many products like fruit, rubber, and nuts can be harvested from the world's remaining rain forests without creating any damage to the plants, wildlife, and its long-term productivity. This helps to protect natural resources and preserve indigenous populations.

© Victor Englebert/Science Source.

forest. They decided to let the rain forest regrow after economic analysis showed that the economic and social benefits of a sustainable harvest from the intact forest outstripped those expected from the proposed project.

The world's very first extractive reserve was established in the Brazilian Amazon, and since that time, over 70 additional reserves have been established, protecting 18.5 million hectares (47.6 million acres) of rain forest. Proponents eventually hope to establish reserves on 10% of the Amazon's rain forest, about 40 million hectares (100 million acres). Encouraged by Brazil's success, several other less developed countries have established forest and lake reserves.

Although extractive reserves are an important component of the global plan to protect native species, it should be underscored that sustainable harvesting is vital to their success. Interestingly, scientists are finding huge tracts of jungle throughout the world that are inhabited by hunters and gatherers that are depleted or nearly depleted of animal species by native peoples in search for food, giving rise to a new term, **defaunation** (loss of fauna in an otherwise healthy ecosystem).

KEY CONCEPTS

Protected lands can be sustainably harvested by indigenous peoples to protect biodiversity. These extractive reserves can create a long-term source of food and income for native peoples that often exceed the economic benefits of timber harvesting and other short-term, environmentally destructive measures. The key to the success of this approach, however, is sustainable harvesting.

Improving Wildlife Management Besides establishing protected areas, buffer zones, and connecting corridors, efforts are needed to improve the way we manage fish and wildlife and the ecosystems they depend on. Most important are efforts to regulate harvests of fish and other commercially important species to avoid depletion. Steps are also needed to restore damaged wildlife habitat.

One of the most important ideas to come along in this field in recent years is the notion of **ecosystem management**. This calls for management of entire ecosystems—not just single species. It also instructs us in the practice of protecting all of the vital habitat a species requires—not just isolated sections of it.

KEY CONCEPTS

Saving species and protecting biodiversity will require many improvements in wildlife management—especially the adoption of ecosystem management, which takes a broader view of species protection.

The Biodiversity Treaty In 1992, the nations of the world met in Rio de Janeiro to sign a biological diversity treaty. Unlike other international treaties, this one focuses on biodiversity, not just threatened or endangered species. Currently ratified by more than 175 nations, the treaty calls on each member nation to develop a **national conservation strategy**. This is a detailed plan to manage and protect biological diversity.

Personal Solutions

As in all environmental issues, individuals can make important contributions. By practicing the principles of sustainability, you can become an important part of the solution. Here are some guidelines:

1. Use only what you need, and use all resources efficiently.
2. Recycle and buy recycled products.
3. Increase your reliance on renewable resources like wind energy and solar energy and support government programs aimed at increasing their use.
4. Help restore ecosystems; support groups that take an active role in these efforts.
5. Limit your family size; support private and government efforts throughout the world to provide family planning services and other means to help slow the growth of the human population.

You can also help educate others. Join groups and spread the word through educational campaigns, lobbying, television ads, posters, books, and pamphlets. Support organizations and politicians that promote energy efficiency and renewable energy and address population growth, habitat destruction, overharvesting of fish and other resources, and other environmental problems.

Progress in protecting biodiversity has been considerable. However, for most species, the situation is growing worse. The expanding human population and our growing demand for resources threaten to destroy hundreds of thousands of species in short order. The time for action is now. The next decade is extremely crucial—not just to the species that share this planet with us, but for ourselves.

KEY CONCEPTS

Saving species and protecting biodiversity require personal actions. We cannot wait for government or business to solve the problems for us.

To the conservationist, what is civilized in us is not music, literature, cinema, or architecture, although they represent tremendous accomplishment, but a compassion for all living things and a willingness to do more than simply care.

—Daniel D. Chiras

CRITICAL THINKING

Exercise Analysis

Before joining your neighbor, you would probably want to examine his basic premise—that coyotes are killing a large number of his sheep. Can you accept his assertion? Is he exaggerating? Is he biased by a hatred for coyotes? What is the true extent of the damage? Is he a reliable source? Do his losses merit the economic and environmental costs—or the legal risk—of the control measures he is proposing?

Next, you might want to know if poisoning coyotes would decrease wildlife populations on your ranch—and hence, your future income. Poisons that kill a coyote may be transferred to other animals that feed on the carcasses of poisoned predators. Furthermore, you would want to consider whether the killing of coyotes would increase the number of rabbits (which coyotes prey on) on your property—and thus decrease the amount of grass your sheep can eat. Moreover, do you really need to kill all of the coyotes to make the sheep safe? Are there other ways to control coyote predation that are less damaging to the environment and more economically sound?

CRITICAL THINKING AND CONCEPT REVIEW

1. Using your critical thinking skills, analyze the following statement: "Extinction is a natural process. Animals and plants become extinct whether or not humans are present. Therefore, we have little to be concerned about."

2. List and describe the factors that contribute directly to animal and plant extinction. Which ones are the most important? In what ways can factors that cause extinction synergize—that is, act together—to produce an even greater decline in species diversity.

3. Trophy hunters generally try to shoot the dominant males in a population. Natural predators, on the other hand, remove the sick, weak, and aged members of the population. Using your knowledge of ecology and evolution, in what ways are trophy hunting and natural predation different in their effects on the prey population?

4. Why are islands particularly susceptible to introduced species?

5. Discuss the ecological factors that contribute to species extinction.

6. Describe the concept of *keystone species*. What are its implications for the modern conservation movement?

7. You are placed in a high government position and must convince your fellow bureaucrats of the importance of preserving other species. How would you do this?

8. Outline a general plan for preserving species diversity. What measures are short term in nature, and what measures are long term? What measures are stopgap, and what measures are preventive?

9. Using your critical thinking skills and your understanding of ecology and human systems, analyze the following statement: "Setting aside land is all well and good, but aside from leaving pristine lands for the elite few who will be allowed to enjoy them, the benefits of this strategy are minimal, even in the long term. The loss of taxable land in this country impacts many, whereas the increase in biodiversity is barely measurable."

10. Define extractive reserve, biological corridor, and buffer zone. How can each one help protect biodiversity?

11. Fashion a statement to convince others why we should protect, even enhance, biodiversity. Which of your reasons are utilitarian? Which ones are not?

12. What measured can you and your family take to help protect and increase biodiversity? Do any of these measures help improve your life? Why or why not?

accelerated extinction
buffer zone
commercial hunting
Convention on International Trade in
Endangered Species of Wild
Fauna and Flora (CITES)
debt-for-nature swap
defaunation

ecological islands
ecosystem management
ecotourism
endangered species
Endangered Species Act
extractive reserve
harvesting
keystone species

national conservation strategy
natural extinction
poaching
sport hunting
subsistence hunting
threatened species
wildlife corridor

©Oleg Znamenskiy/Shutterstock, Inc.

Grasslands, Forests, and Wilderness: Sustainable Management Strategies

Our duty to the whole, including the unborn generations, bids us restrain an unprincipled present-day minority from wasting the heritage of these unborn generations.

—**Theodore Roosevelt**

 LEARNING OBJECTIVES

After reading and studying the material in this chapter, students will be able to understand, knowledgably discuss, and, when appropriate, apply the information learned in this chapter, including:

1. The important concept known as tragedy of the commons
2. The importance of rangelands, their condition, and techniques of sustainable rangeland management
3. The importance of forests, the extent of deforestation, and techniques to manage forests sustainably
4. The importance of wilderness and efforts being taken to protect it

CRITICAL THINKING

Exercise

One of the most dramatic changes on the planet in the last two decades has been the steady march of the world's largest desert, Africa's Sahara. In the 1970s and 1980s, researchers estimated that the desert spread southward at a rate of 5 kilometers (3 miles) per year. They attributed this expansion to drought, overgrazing, and agricultural land abuse in the semiarid grasslands bordering the desert. The 5 km/yr projection, however, was based primarily on measurements in a few locations, which researchers assumed were representative of the entire continent.

Using satellite observations of vegetation, however, scientists have found that the Sahara has advanced *and* retreated—largely in response to rainfall. From 1980 to 1984, for example, the desert's southern boundary moved 240 km (150 mi) south. Between 1984 and 1985, it moved north by 110 km (69 mi). In 1987 and 1988, it shifted northward again by 155 km (97 mi). In 1989 and 1990, however, the desert boundary shifted southward 77 km (48 mi).

Although the southern border of the desert in 1990 was 130 km (81 mi) farther south than in 1980, the researchers believe that the shift does not reflect a long-term trend, but rather differences in year-to-year rainfall.

Some critics of global warming use this data to argue that desertification caused by climatic shift is not occurring. They say that the shift of the desert is a natural phenomenon. How would you answer this claim? What critical thinking rules will you use?

In the early 1970s, scientists pored over satellite photographs of the drought-stricken African Sahel, a band of semiarid land that borders the southern Sahara. One of them noticed an unusually green patch of land amid the desert. Curious to find out the reason, Norman MacLeod, an American agronomist, flew to the site. There, surrounded by newly formed desert, was a privately owned ranch of 100,000 hectares (250,000 acres). Its grasses grew rich and thick even though vegetation in the surrounding fields had long since died, leaving the sandy soil unprotected. Why?

Stretching around the perimeter of the ranch was a fence that held out the cattle of the nomadic tribes that had overgrazed the surrounding communal property for decades. The ranch was divided into five sections. The ranch's cattle grazed each section on a 5-year rotation—a sustainable strategy in the semiarid land. This example shows us the devastation wrought by mismanagement but also the possibilities created by intelligent, ecologically based, and sustainable management.

This chapter examines the state of the world's grasslands and forests—extremely valuable sources of food, fiber, and numerous ecological services—and offers suggestions on sustainable natural resource management. It also discusses wilderness areas—vital sources of recreation that, if managed sustainably, could also become important means of preserving biodiversity. Before we begin, we must examine one of the reasons why lands are deteriorating.

12.1 The Tragedy of the Commons

The desert surrounding the ranch in the chapter opening was at one time lush communal grazing land. Tribespeople of the Sahel grazed their livestock on the land with abandon, eventually causing its destruction. As far back as ancient Greece, Aristotle recognized that communal property, shared by many people, often deteriorates severely. Early civilizations, for example, clear-cut communal forests and overgrazed their cattle on grasslands. History, however, shows that these civilizations paid dearly for their disregard. The skeletons of buildings from ancient cities stand out in deserts that were once the rich forest and grassland ecosystems of the Fertile Crescent. Much of Iran and Iraq, which is now barren desert, once supported cattle, farms, and rich forests (**FIGURE 12-1**). Historians believe that the decline of Greece and Rome may be partly the result of the rampant misuse of their lands.

FIGURE 12-1 Legacy of past abuse. View of barren hillsides of Iran. Once covered by rich forests, these lands have been ruined by centuries of abuse, starting with deforestation.

© iStockphoto/Thinkstock.

Economists have debated the fate of other common resources such as air, water, and land for decades. It was not until 1968, however, that professor Garrett Hardin exposed the cycle of destruction in a paper entitled "The Tragedy of the Commons."

In England, Hardin noted, cattle growers grazed their livestock freely on fields called **commons**, which fell into ruin as the users became caught in a blind cycle of self-fulfillment. Families found that they could enhance their personal wealth by increasing their herd size. Each additional cow meant more income at little if any additional cost because the farmer did not have to buy new land or feed. The commons provided them.

Hardin argues that the cattle growers were rewarded for doing wrong. They realized that increasing their herd would lead to overgrazing and deterioration of the pasture, but they also recognized that the negative effects of overgrazing would be shared by all of the community. Thus, each herdsman arrived at the same conclusion: He had more to gain than to lose by expanding his herd. This shortsighted thinking resulted in a spiraling decay of the commons. As each pursued what was best for himself, the whole was pushed toward disaster. "Freedom in a commons brings ruin to all," wrote Hardin.

The logic that compels people to abuse communal holdings has been with humankind as long as common property has. Today, however, the process has reached epic proportions. The overgrazed communal property of the Sahel, discussed in the introduction of this chapter, is one of the most recent examples. This tragedy has been worsened by unsustainable government policies. In the 1960s, for example, loans were given to nomadic people to drill for water. With a steady supply of water, previously nomadic tribes no longer migrated south during the dry season. The land around the wells and human settlements deteriorated even further as herds exceeded the carrying capacity of the land for cattle.

The tragedy of the commons provides partial explanation for many of the world's problems, from air pollution to the rapid increase in human population. In the rich, industrial nations as well as in the poor, developing countries, for example, fertility is driven largely by personal desires. In India, children provide security in old age or improve the social status of women. In Mexico, childbearing may be motivated by male pride; in the United States, by personal satisfaction. Individual self-serving acts of procreation create a global environmental tragedy that befalls all of humanity.

Hardin's analysis of the tragedy of the commons, although important, gives the impression that common resources are unique in their mismanagement. In many places, privately held resources are also abused (**FIGURE 12-2**). Erroneous frontier notions of the Earth as an unlimited supply of resources, ignorance of the impacts of human actions, and a short-term view of economics dictate unsustainable management strategies on a wide variety of privately held resources.

Today, many lands are gripped by the tragedy of overexploitation. Short-term exploitation may have been permissible at one time, when the human population was small in relation to the Earth's resources. Today, too many people

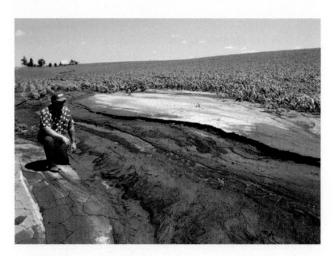

FIGURE 12-2 Private abuse. This privately owned farmland is deteriorating because of poor land management, which has led to severe soil erosion.

Courtesy of Lynn Betts/NRCS USDA.

depend on the biosphere for food, water, and other resources, and the cumulative effect of many small insults has become staggering. Local problems have spawned regional calamities. Regional problems are now spreading to create global concerns.

KEY CONCEPTS

Communal resources—that is, resources held in common by people, such as land, air, and oceans—often deteriorate as individuals become caught up in a cycle of self-gratification. Personal gain dictates actions that have negative effects shared by all who use communal property. Privately owned lands also deteriorate as a result of ignorance, greed, ill-advised management, and other factors.

12.2 Rangelands and Range Management: Protecting the World's Grasslands

Cattle and other livestock range over half the Earth's surface, feeding on grassland vegetation. So important are they to human nutrition that in many parts of the world domesticated animals outnumber humans three to one.

The grasslands on which many of the world's livestock depend are known as **rangelands** and are a vital component of global food, leather, and wool production. When properly managed, grasslands and the livestock they support can provide useful products *ad infinitum*. Regrettably, the vast majority of the world's rangeland is unsustainably managed. This section examines the problems facing rangelands and shows how they can be managed sustainably to ensure a steady supply of food and other resources while protecting the natural systems, even improving their ecological health. Before we begin, a few words on rangelands are in order to broaden your understanding.

Rangelands, grasslands on which livestock graze, are an important element of the global agricultural system. When properly managed, they can be a sustainable food source.

Grasses form the base of the food chain on rangelands. These hardy species are adapted to periodic drought, fire, and grazing. Care must be taken to protect the metabolic reserve of the plant, which ensures that the roots remain alive.

An Introduction to Rangeland Ecology

Dozens of plants, mostly grasses, inhabit the grassland biomes of the world. Grasses, like all other plant species, capture sunlight energy and use it to produce organic matter. Livestock and native grazers feed on grasses, acquiring the nutrients, both organic and inorganic, they need to survive. People, in turn, consume the meat of these animals or drink their milk. Thus, organic plant matter and its energy are passed up the food chain.

Grasses are unique among plants in that they can withstand a considerable amount of grazing, drought, and fire without suffering. This is possible because they grow from the **basal zone,** the lowermost portion of the leaf. (It's for this reason that our lawns withstand periodic mowing.) Most other plants grow from the top, so periodic grazing would kill them.

Even though grasses are adapted to grazing pressure, there are limits to how much grazing a grass species can withstand. As a general rule, it is safe to remove the top half of the exposed part of a grass plant. The rest constitutes a **metabolic reserve**—so named because it provides the photosynthetic capability needed to provide food for the roots of the plant, which may extend 2 meters (6 feet) or more into the ground (**FIGURE 12-3**). Without this reserve of nutrients, the plant is likely to die or be severely weakened.

Rangelands start to deteriorate when there is little or no recovery time between grazings. As a result, plants are so heavily grazed that their metabolic reserve is removed.

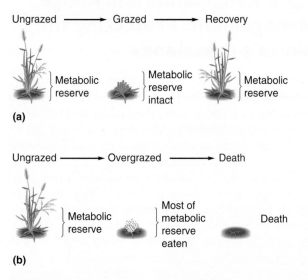

Ungrazed ⟶ Grazed ⟶ Recovery

Metabolic reserve

Metabolic reserve intact

Metabolic reserve

(a)

Ungrazed ⟶ Overgrazed ⟶ Death

Metabolic reserve

Most of metabolic reserve eaten

Death

(b)

FIGURE 12-3 The anatomy of grass. (a) Grasses are generally hardy species adapted to periodic grazing. **(b)** The lowermost portion, the metabolic reserve, however, cannot be removed without weakening or even killing the plant.

The Condition of the World's Rangeland

In the United States, 85% or over 110 million hectares (276 million acres) of grassland, much of it public land managed by the Bureau of Land Management, have undergone slight to moderate desertification in the last 200 years—equivalent to an area nearly twice the size of Texas! The Navajo Indians, for example, live on a 6 million-hectare (15 million-acre) reservation in Arizona, New Mexico, and Utah. Theirs is a sun-parched land, dusty and dry. To feed and clothe their people, the Indians have gradually increased the size of their sheep herds. Today, the herd size exceeds the carrying capacity by at least four times. Baked by the hot summer sun and swept by fierce winter winds, the overgrazed reservation is quickly becoming an arid dust bowl.

Northwest of Albuquerque, New Mexico, is another tract of parched desert land, the Rio Puerco Basin. In the late 1800s, its lush grasslands supported huge herds of cattle. A century later, however, the land has deteriorated under the strain of overgrazing. Erosion has formed deep gullies. Wind and rain erode the soil 5 to 10 times faster than it can be replenished. According to the **Bureau of Land Management (BLM)**, which manages most of the federally owned rangeland, 54.5% of the rangeland acreage assessed in 2012 meets all standards for good rangeland or is making significant progress toward meeting standards down 7 percentage points from 2006.[1] Some conservationists argue that current standards for good rangeland may not represent truly sustainable conditions. The remaining 46% do not meet standards of good rangeland. On much of this land, conditions appear to be deteriorating.

Worldwide, the condition of rangeland also appears to be less than optimum. The United Nations Environment Programme estimates that nearly 75% of the world's rangeland is at least moderately desertified, with a reduction in carrying capacity of at least 25%. The major culprit in this deterioration is overgrazing. In addition to desertification, this all-too-common practice results in severe erosion, declines in groundwater, loss of wildlife, and invasions of weeds.

Forests are also damaged by livestock. In some regions, branches are cut from trees for animal food. Entire forests are removed to make way for cattle ranches. In India, for instance, grasslands do not provide enough food for the nation's 196 million cattle, which forces people to turn to state forests and other lands.

[1] From the BLM's online Rangeland Inventory, Monitoring, and Evaluation Reports.

Rangeland Management: A Sustainable Approach

Grasslands have been sustainably grazed by native species for thousands of years with no apparent ill effect. In fact, grazing benefits grasslands. That's because grazers such as deer and buffalo spread seeds and fertilize the soil. The hooves of these animals pulverize rock and make soil. They also drive seeds into the ground.

Sustainable grazing management can, if properly carried out, protect and even improve the quality of rangeland. The island within the Sahel, discussed in the introduction to this chapter, provides living proof that humans can manage rangeland sustainably. What are the components of sustainable rangeland management?

Range management involves a number of basic techniques that are often employed simultaneously: (1) controlling the number of livestock on land, (2) deferred grazing, (3) controlling the distribution of livestock, (4) restoration, and (5) rangeland improvements.

Controlling Livestock Numbers

The first and most important step in managing rangeland sustainably is to control the number of animals and the duration of grazing on a piece of land. This measure honors one of the major principles of sustainability: living within the limits.

The key to proper range management is maintaining livestock populations within the carrying capacity of the ecosystem. The most important variable in determining the carrying capacity is the weather—especially the rainfall and temperature. In dry periods, the carrying capacity may be half that of a normal year. Accordingly, the number of livestock allowed to graze on a piece of land should be cut in half. In wet years, it may be double the normal year's carrying capacity.

Truly effective range management requires a willingness to cooperate with nature, one of the key principles of sustainable ethics. Many ranchers, however, are unwilling or unable to adjust their herd size with the weather. The cost they pay may be a gradual deterioration of their land. The irony of this is that in the long term they may be lowering the carrying capacity and ultimately destroying the land's productive capacity and their livelihood.

Rotational and Intensive Grazing

The ranch mentioned in the opening paragraph of this chapter was successful because its owners shifted cattle from one field to the next each year. This process, called **rotational grazing**, is beneficial because it reduces the grazing pressure on each section and ensures a healthy crop of grass. This technique is growing in popularity and is practiced by many cattle growers throughout the world. This technique reduces costs for feed, equipment, and veterinary care and is especially helpful to smaller operations that would be unable to compete with larger dairy herds.

Rotational grazing patterns must be tailored to the local conditions and ecosystem. Some farmers practice short, intensive rotational schemes; others may require more complicated schemes. One way to improve rangeland and pasture is **intensive grazing**, intentionally intensively grazing land for short periods, then rotating cattle on to other pastures. While this technique may seem as if it would damage land, properly controlled intensive grazing improves the

GO GREEN

If you're a meat eater, consider buying natural, grass-fed organic beef.

quality of pasture and rangeland. It forces cattle and other livestock to eat all grasses, not just the most delicious ones. If left on their own in a pasture for extended periods, livestock will graze on just the most tasty species, leaving less palatable grasses uneaten. As a result, the cattle tend to overgraze the most valuable plants. This often causes the most desirable—the most nutritious species—to die, leaving only less palatable species and weeds, which take over, decreasing the nutritive value and health of the grassland. By forcing cattle to eat all species, then moving them into another pasture, a rancher ensures the long-term health and viability of all grasses

On the author's farm in east-central Missouri, he has divided his pasture into 14 subpastures. He runs cattle in one at a time, forcing them to intensively graze the land (**FIGURE 12-4**). Cattle may remain in a small section for 2 to 5 days, depending on its size. Once that land is grazed down, he moves them to another pasture, and then another. Within 2 to 3 weeks the first pasture is ready for grazing. It looks lush and green like a lawn.

Rotational grazing in its various forms is a simple yet highly effective and sustainable strategy that benefits ranchers, the economy, and the environment. It increases the vigor

FIGURE 12-4 Part of the author's small herd of belted galloways on his Missouri farm are all-natural, grass-fed beef that are rotated from pasture to pasture after short, but intensive grazing periods.

Courtesy of Dan Chiras.

of the plants and enhances their nutritional value. It can even increase the carrying capacity of the land over the long run, permitting even greater profit. Furthermore, it helps to ensure a more diverse ecosystem beneficial to native species. Rotational grazing is a system of food production that helps human society meet its current needs in ways that ensure future generations a plentiful supply of food.

 KEY CONCEPTS

In rotational grazing, cattle are shifted from one pasture to another, permitting intensive grazing. This method enhances the condition of rangeland and can increase the carrying capacity in the long run.

Controlling the Distribution of Livestock

On rangeland, livestock often aggregate around water sources—stock ponds and natural water sources. Because of this, parts of the range may be severely overgrazed while other parts are ignored. In heavily grazed areas, more palatable and nutritious plants may perish, leaving behind less palatable and less nutritious species. In some cases, grasses may be destroyed entirely. Denuded areas, especially stream banks, often suffer serious erosion.

To promote more uniform grazing and protect the land from overgrazing, ranchers can construct fences around areas they want to protect. A more economical way of ensuring a more uniform grazing pattern is to locate water sources and salt licks strategically in areas that are underused.

Progressive cattle growers often divide their pastures into smaller pastures by installing temporary or permanent fencing in larger fields. At my farm in Missouri, Evergreen Four-Season Farm, our grass-fed beef cattle are moved from subpasture to subpasture, that is, rotating from one small field to the next as soon as one is thoroughly grazed. This practice forces our cattle to graze each pasture intensively. If cattle are allowed to range over a large area they tend to consume the tastiest grass species, leaving less desirable species uneaten. These species, in turn, may proliferate and come to dominate the pasture, decreasing the quality of the grass.

Cattle grazed less intensively in a large pasture also damage grassy areas with their hoofs as they move from one location to the next. Their paths often permanently destroy vegetation. Heavily grazing an area, therefore, not only helps to maintain species diversity and helps preserve the most nutritious grasses, it helps prevent permanent damage to grassy areas.

Intensive grazing also helps build soil, as cattle deposit manure over the small areas. The manure replenishes the organic matter as well as inorganic nutrients of the soil. Intensive but short-term hoof action also drives seeds into the ground, helping pastures remain productive over the long haul.

 KEY CONCEPTS

Fencing and careful distribution of water sources and salt licks can help promote a more uniform use of rangeland and pasture and protect some areas from degradation.

Restoring and Improving the Quality of Rangeland

Sustainable rangeland management also requires steps to restore deteriorated land. Restoration includes such measures as fertilizing land and replanting barren areas with hardy native grass species. We've planted clover to help improve our pastures. To improve the success of this measure, ranchers often run cattle over the land to force the seeds into the soil. Reseeded areas reduce soil erosion and often support a greater number of cattle than unseeded areas. They also help restore groundwater and protect streams from sediment, by reducing erosion. We've applied cow manure on denuded areas of our farm, caused by mismanagement of the previous owner, to help re-establish vegetation.

The productivity of the land can also be improved by periodic controlled burns. This returns nutrients to the soil and thus promotes the growth of grasses. The healthier the grass crop, the faster cattle grow to marketable size. In addition, controlled burns help kill undesirable plant species—for example, woody shrubs and prickly pear cacti.

Restoration and general improvement measures that increase productivity help to create a healthier ecosystem. Besides providing food, such lands are more stable and less prone to erosion and pests. Interestingly, numerous studies have shown that grasshoppers, rabbits, and even weedy plant species such as sagebrush are much more numerous in over-grazed areas than in healthy rangeland.

Revamping Government Policies

As in other sectors of the agricultural community, government policies that are not based on sound scientific understanding of rangeland carrying capacity can facilitate the rapid deterioration of this valuable resource. In Australia's Northern Territory, for example, ranchers who fail to keep a minimum number of cattle on public land are fined by the government. No maximum limits are set. Such policies promote the deterioration of the land and need to be replaced with sustainable grazing limits based on carrying capacity.

In 1978, the **Public Rangelands Improvement Act** was passed by the U.S. Congress as a means of promoting better range management. This law calls for improvements on public grasslands managed by the BLM and the U.S. Forest Service (USFS). The law required these agencies to develop guidelines for proper management. Whereas the Forest Service guidelines were designed by range management specialists and others, the BLM's policy was formulated by ranchers. Critics argue that because of this, the BLM policy is too lenient and contributes to the deterioration of these federally owned lands.

The Public Rangelands Improvement Act requires the BLM and Forest Service to reduce grazing on public land where damage is evident. This strategy is often unpopular among ranchers, who either cannot see the benefits of improving range conditions or dispute the claims that they are overgrazing the land. One of the chief weaknesses of the Public Rangelands Improvement Act is that it does not pertain to Indian lands in the West, where grazing reductions are desperately needed.

As a result of private and federal actions, U.S. rangelands seem to be gradually improving, but much more work is needed.

Sustainable Livestock Production

In many countries, cattle and other animals like pigs are raised in highly confined quarters. Cattle come from area farmers who raise them on grass (and mother's milk) until around 6 months of age. The cattle are then sold at auction and shipped off to **feedlots** also known as **confined feeding operations** (CFOs), as shown in **FIGURE 12-5**. In CFOs, the cattle are fed a diet rich in corn, which causes them to grow and fatten quickly. Cattle often live in filthy conditions. Air quality in and around CFOs is awful.

Unfortunately, cattle are grazers that thrive and produce their healthiest meat on a diet of grasses. They are not adapted to a corn diet. As a result, meat from cattle raised in CFOs, most of what's available to shoppers, is high in fat and low in heart-healthy nutrients (omega-3-fatty acids) that grass-fed beef provide. To prevent illness and promote growth in CFOs, feed is treated with antibiotics that can be passed on to consumers.

Although raising cattle in CFOs does not affect grasslands directly, it does have a tremendous environmental impact. For example, pen-raised cattle produce huge amounts of manure in limited spaces. Much of it is dumped in landfills. At one time, farmers used the manure their animals produced to fertilize their cropland. Today, however, many farmers produce only livestock and no longer grow crops—so they have less access to farmland on which to apply manure.

FIGURE 12-5 Confined feeding operations (CFOs). Most cattle are raised in crowded conditions where they are fattened prior too slaughter.

© Thoma/ShutterStock, Inc.

Another problem with livestock raised in enclosures is that they require enormous quantities of grain, mostly corn and sorghum. Ironically, most farmers who supply grain to feed cattle use artificial fertilizer on their land. If applied in excess, the nutrients may wash into nearby waterways, causing pollution. Artificial nutrients also fail to replace all of the nutrients removed by plants.

Hogs also tend to be raised in confined quarters where they are sustained on a steady supply of grain. Such operations also produce huge amounts of waste, which is often stored in waste lagoons, where the liquid evaporates or percolates into the ground, polluting groundwater. When waste lagoons break or flood, they can cause severe water pollution problems. Realizing the enormous economic and environmental costs of this technique, a small number of hog farmers allow their hogs to range freely in pastures, feeding on natural vegetation. The farmers provide Quanset hutlike shelters to protect the hogs but move the houses from time to time, to more evenly distribute grazing and to prevent the deposition of huge amounts of manure in one place. This technique appears to be working. It reduces odors created by most hog farms, and it spreads fertilizer evenly on the land, nourishing plants hogs will eat. This reduces the demand for grain, which saves farmers money. All in all, this technique costs about 70% less than confinement facilities and it helps protect the environment.

In the less developed nations, the wealthy class is the primary consumer of meat. Because livestock are fed grains or are sometimes produced on land that could grow food crops, meat production for the wealthy reduces overall food supplies. These practices also make food more costly for the poor.

Livestock production in confined spaces can be made more sustainable. One step to improve the sustainability of meat production is to return manure to cropland to provide nutrients. A sustainable system therefore might require a reintegration of livestock and crop production.

Meat production may have to be decreased in the long run to accommodate the growing population. Many more people can be fed on a vegetarian diet than on a meat-based diet. Central to all strategies are efforts to slow and perhaps stop the growth of the human population.

KEY CONCEPTS

In many countries, livestock are raised in pens and fed grains that could be used to feed large numbers of people.

12.3 Forests and Forest Management

Covering slightly more than one-quarter of the Earth's land surface, forests are a valuable asset to humankind. The most notable direct benefits are an estimated 5,000 commercial products—such as lumber, paper, turpentine, and others—worth tens of billions of dollars a year. Each year, primary and secondary products from U.S. forests sell for more than $448 billion, according to the USFS Forest Products Laboratory.

Primary products are roundwood (logs), and secondary products are lumber and paper. Tertiary products such as furniture and turpentine add billions more to the value of the nation's forests. The U.S. forestry and timber products industry employed an estimated 1.7 million people in 2008, according to the U.S. Forest Service. Forests are important to Canada's economy as well. In Canada, 320 communities and more than 433,000 people are economically dependent on forests and the $82 billion-dollar-a-year forestry industry. One of every 15 jobs in Canada is derived from timber, wood, paper products, and related industries. Moreover, forest products are the single most important export product the nation has to offer. In many poorer nations, forests are a source of wood and charcoal for cooking and heating.

Forests also provide refuge from hectic urban life and opportunities for many forms of recreation. They are home to many of the world's species. Forests protect watersheds from soil erosion, thus keeping rivers and reservoirs free of silt. They also reduce the severity of floods and facilitate aquifer recharge. In addition, forested lands assist in the cycling of water, oxygen, carbon, and other nutrients.

The United States has about 300 million hectares (740 million acres) of forestland (**FIGURE 12-6**). Canada has nearly 1 billion acres (397 million hectares) of forest land, about 10% of the world's total. According to estimates of the Worldwatch Institute, world forests cover an area 15 times greater than that of the United States—approximately 4.2 billion hectares (10.6 billion acres). About two-thirds of these forests have already been logged or disturbed in some way by human activities. In the United States, 95 to 98% of all forests have been cut at least once since the settlement by Europeans.

Despite the great benefits of forests, only about 13% of the world's forestland is under any kind of management. In addition, only about 2% of the world's forests are protected in forest reserves.

KEY CONCEPTS

The world's forests provide many social, economic, and environmental benefits. A large portion of the world's forests have been logged or disturbed. Very little forested land is under permanent protection.

Status Report on the World's Forests

Since the advent of agriculture, nearly one-half of the world's forests or about 3 billion hectares (7.5 billion acres) have been cleared and converted to other uses, mostly farms and human settlements. In East Africa, over 90% of the original moist forest is gone. In Brazil, a country with one of the world's largest forests, 38% of its trees have vanished. In the Philippines, 97% of its forests are gone. To date, the United States has lost about one-third of its forests. Europe has lost 70%. Much of this land has been converted to other uses such as farming, or has deteriorated severely after timber harvesting.

According to the United Nations (Framework Convention on Climate Change), the conversion of rain forest to agricultural land is the main cause of tropical rainforest deforestation.

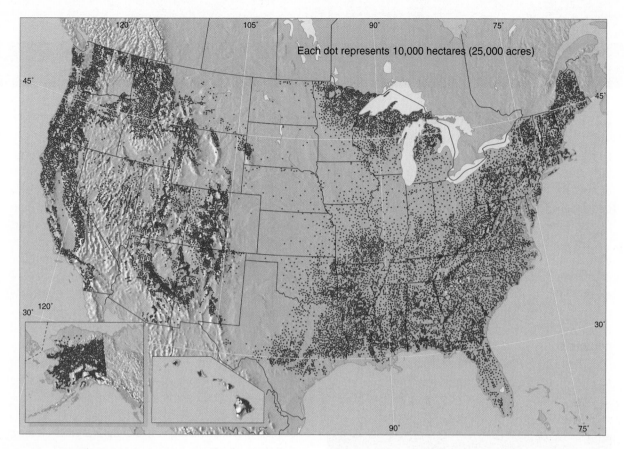

Each dot represents 10,000 hectares (25,000 acres)

FIGURE 12-6 U.S. forestland. This map shows the distribution of forestland in the United States.

Subsistence farming accounts for nearly half of all losses (88%). Commercial agriculture, such as growing soybeans and palm oil, is responsible for about one-third of the annual deforestation (28%), and logging accounts for about 14%. Finally, fuel wood harvest is responsible for about 5% of the losses.

Deforestation continues today although it is difficult to determine the exact amount. According to the Rainforest Action Network, an environmental organization, we deforest *and* alter 31 million hectares, or 78 million acres of tropical rain forest each year, an area larger than Poland. This number takes into account clear cutting and less devastating, though ecosystem altering, activities. The World Resources Institute claims that the rate of tropical deforestation is about 13 million hectares per year. For reference, 13 million hectares (32 million acres) of forests covers an area about the size of the state of North Carolina or Alabama.

Although most of the news regarding deforestation describes the rapid deforestation occurring in the rain forests in South America, Central America, Asia, and Africa, deforestation is also occurring in temperate deciduous forests and northern coniferous forests. In fact, Canada, Russia, and China are centers of timber cutting that many believe is occurring at unsustainable rates.

The heavy use of forests might not be so bad if efforts were made to replant trees at a rate commensurate with cutting. In less developed countries, for every 10 trees cut down, only 1 tree is replanted. In Africa, the ratio is 29 to 1.

Besides destroying habitat for many species, deforestation decreases sustainable fuel supplies needed for cooking and home heating. According to estimates about half the world's population—the poorest half—cook over wood fires. A very large percentage of them (two-thirds) are facing fuel wood shortages. In rural Kenya, shortages mean that some women must spend up to 24 hours a week in search of wood for cooking their families' meals. (**Spotlight on Sustainable Development 12-1** discusses deforestation in tropical rain forests.)

KEY CONCEPTS

Worldwide, about one-half of the world's forests have been cut. The land they once occupied has already been converted to other uses, mostly farming, or undergone severe deterioration. Deforestation continues today in tropical rain forests, northern coniferous forests, and temperate deciduous forests at a rapid pace and threatens the long-term sustainability of human civilization.

Root Causes of Global Deforestation

Deforestation throughout the world stems from numerous factors. A frontier mentality has pervaded many cultures. In the United States, colonists cleared eastern deciduous forests to make room for farms and settlements. Often the

12-1 Saving the World's Tropical Rain Forests

Rain forests are found in the tropics. Forming a thick, lush carpet of vegetation with a stunningly diverse array of species, the tropical rain forests of the world are being logged at a feverish pace. In 2006, the annual rate of deforestation was estimated to be about 13 million hectares (32 million acres) per year. Historically, other disturbances to forests caused by road building and other activities is similar to deforestation rates. Some experts think that tropical rain forests in all but a few places could be virtually obliterated early in this century if we do not enact strict measures to protect them (**FIGURE 1**).

The loss of tropical rain forests is one of the most serious problems facing the world today. It is a major cause of extinction because the tropical forests contain about half to two-thirds of the world's species.

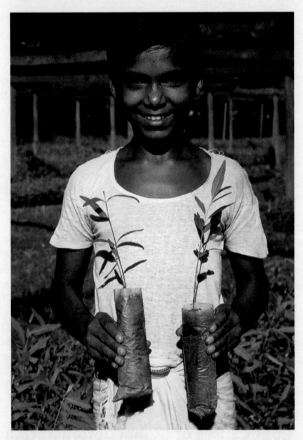

FIGURE 1 This boy in Nepal is holding trees that will be planted to restore a lost rain forest.
© Takeshi Takahara/Science Source.

The late Al Gentry, a researcher who studied tropical forests before his untimely death in an airplane crash, said that the "loss of so many species is not only a tragic squandering of the Earth's evolutionary heritage but also represents depletion of a significant part of the planet's genetic reservoir, a resource of immense economic potential." Genes from the tropical rain forest help boost agricultural production. Wild species are also a source of new medicines and a host of other new products.

Tropical forests play an important role in global cycling of oxygen and carbon dioxide. By various estimates, global deforestation accounts for about 13% of the world's annual increase in carbon dioxide. In less developed nations, deforestation results in 35% to 65% of annual emissions because of the relatively low levels of fossil fuel use and high rates of deforestation. It may, therefore, be a major factor in global warming.

Tropical rain forests are home to indigenous tribes that have lived sustainably for thousands of years. They are vast reservoirs of plants that may have medicinal value.

Clear-cutting tropical rain forests exposes the soil to intense rains that wash away the soil and fill nearby streams and rivers with sediment. These problems are commonplace in Pakistan, India, Thailand, and the Amazon Basin of South America. In Thailand, a devastating flood in 1988 was caused by massive deforestation by commercial timber companies, which ultimately led to a nationwide ban on timber cutting.

Much of the rain falling on intact tropical rain forests evaporates to become clouds again, only to rain down on the neighboring downwind forest. When forests are cut, surface runoff increases, and rainfall decreases by 50% or more downwind from clear-cut regions. As the area of deforestation spreads, rainfall declines over large regions, and once lush areas turn into barren tropical deserts. Forest fires, almost unheard of in tropical forests, become a problem due to the drying effect. Some tropical soils bake to a hard, brick-like consistency when exposed to sunlight.

Ironically, tropical soils are generally nutrient poor. In fact, approximately 95% of the nutrients are in the biomass (live plants) and only 5% are in the soil, just the opposite of a temperate forest. Cutting and burning the forests to make room for ranches and farms releases enough plant nutrients to support crops for a few years, but because the soil has little reserve and nutrients are quickly taken up by crops or washed away, the land quickly deteriorates, becoming barren and unproductive.

Although clear-cutting of large tracts is widespread in the tropics, efforts to harvest trees by selective cutting often alter the forest so much that they destroy as many species as clear-cutting.

Because of these problems, many nations have enacted laws to protect their forests. In 1988, Brazil established its first extractive reserve in the Amazon Basin, in the state of Acre. Set aside for harvesting rubber, nuts, fruit, oils, and other products, the 40,000-hectare (100,000-acre) reserve allows people to reap sustainable economic benefits while protecting the biodiversity of the forest. All told, 46 extractive reserves have been established in the Amazon, by Brazil and other countries, with hopes of more soon.

Extractive reserves can be managed sustainably and represent a long-term solution to economic and environmental problems facing less developed countries. To understand why, contrast cattle ranching with the sustainable harvest of rain forest products. In the state of Rondonia, Brazil, ranchers and farmers have cut down or disturbed at least 35% of the rain forest, creating millions of hectares of wasteland. On most

soils, ranchers can graze one cow per hectare for 5 years. In the next 2 years, because of declining soil fertility, they can graze only one cow per 4 hectares (10 acres). After that, the land is often destroyed.

In contrast, large tracts of rain forest set aside for sustainable harvest can produce a variety of products (among them latex, nuts, and fruits) indefinitely. Studies show that the income potential of sustainable harvest far exceeds cattle ranching and plantations.

One of the most important strategies for protecting existing tropical rain forests and providing wood for present and future generations is replanting the land already harvested and abandoned by timber companies or settlers. Small teak plantations are cropping up in Costa Rica in an effort to reestablish forests and produce valuable wood sustainably.

Although replanting does not duplicate the incredible diversity of forests, commercial plantations with a modest degree of diversity could be developed to provide wood and other products. This could dramatically reduce pressure on primary forests.

Although positive steps are being taken, rain forests are being destroyed much faster than they are being protected. Individuals can help by joining groups that are working to protect the forests; by avoiding wood products derived from rain forests (teak, disposable chopsticks, parquet floors, and many hardwoods); and by writing governmental representatives and urging action.

land eroded under poor management, so settlers picked up and moved into new territory. Seemingly unlimited forests merited little protection in the frontier days. Forests were viewed as an impediment to human progress, too. There was, of course, no understanding of the ecological importance of forests and the free services they provide to humans.

Unfortunately, many nations still view vast stands of timber as valuable only if it is harvested or cut down and converted to some other use. And, as nations run out of lumber, corporations expand into new territories, cutting trees in remote locations. Huge tracts are laid barren as a result.

In less developed countries (LDCs), deforestation occurs for several main reasons. Forests are leveled to provide fuel wood and charcoal for cooking and heating homes. According to the UN FAO, nearly half of all the wood (tropical and nontropical) cut each year is harvested to supply these needs.

Go Green

To protect forests, avoid disposable products like paper cups, paper plates, and disposable chopsticks.

Deforestation also results from poverty, unsustainable population growth, and landlessness. Countries often open up forests to resettlement to ease pressure in cities. The poor people who move into the forest cut trees and try to eke out a living. The faster the population grows, the faster the forest falls. Today, approximately 25% of the annual tropical deforestation is attributed to humans in search of sustenance.

Land ownership patterns in less developed nations also contribute to the loss of forests. In El Salvador, 2% of the landowners have title to 60% of the land. With the good land in the hands of a few, poor rural peasants often enter forests and cut down trees to carve out plots to farm. Many of these people exploit fragile and hilly terrain, semiarid grasslands, or rain forests. Because they are relegated to marginal lands, their farms often fail, and they must move on—often going deeper into the forest to repeat the cycle.

Industrial wood production, wood cut for paper making, lumber, and wood products, accounts for much of the remaining deforestation. This wood is destined for markets in MDCs and LDCs. About half of all industrial timber harvesting is carried out to supply the needs of Europe, North America, and Japan.

Deforestation results from many factors, including frontier thinking, a lack of knowledge of the importance of forests, population growth, poverty, and inequitable land ownership.

Unsustainable Government Policies Robert Repetto, an economist at the World Resources Institute, puts much of the blame for global deforestation on ill-advised government policies that influence how a nation's forests will be used. This is especially noticeable in nations where governments control the majority of the nation's forest reserves. The government of Canada, for instance, owns 94% of its forest land. The Indonesian government controls 74% of its forest land. Even governments that are committed to conservation often have policies that promote unsustainable forestry practices. Consider some examples.

Governments the world over have typically sold timber below market value to logging companies. Studies have shown that the U.S. Forest Service has lost money on most, if not all, of its national forests for years because it routinely auctions off timber rights below the cost of building roads (which the Forest Service—the taxpayer—pays for), surveying, paperwork, conducting auctions, and restoring landscapes and watersheds. Some environmentalists argue that below-cost timber sales are a form of public subsidy to the lumber industry; that amounts to about $330 million a year perhaps twice that amount. They also note that by lowering the cost of wood products, this subsidy discourages conservation by companies, builders, and individuals.

The United States is not alone in selling off timber below cost. Many other countries, among them Canada, Ghana, Indonesia, and the Philippines, let their timber go cheaply. Canada, for instance, has routinely sold timber at a loss—some suggest at losses greater than those of the United States. Much to the chagrin of environmentalists, and often without public consultation, provincial governments have leased huge tracts of land to private corporations from the United States, Korea, Hong Kong, Japan, and other countries. Some of this land is in provincial parks and wildlife reserves.

Many economic policies encourage an unsustainable exploitation of forest resources. Numerous less developed countries, for example, restrict the export of raw wood by international companies to create jobs at home and encourage economic development of domestic wood-processing industries. Bans or taxes that limit raw wood exports, they think, will increase the export of finished wood products (for example, furniture), netting higher revenues than the sale of raw wood. Unfortunately, says Repetto, many of the small mills are highly inefficient and use 50% more logs than the industry standard to achieve a given output of milled products. Such policies consequently result in a higher rate of deforestation.

Another problem is short-term contracts. In many less developed nations, 35 years or more are required for a stand of trees to recover from logging, but contracts are written for 20 years. Because companies have no long-term interest in their concession, such contracts discourage them from protecting forests. Long-term contracts might encourage them to harvest forests sustainably.

Heavy borrowing from international banks and industrialized nations has created enormous debt in the developing world. To pay back the loans, countries often lease their land at bargain rates to timber companies from other countries, which come in, clear the forests, then move on. Some countries promote timber harvesting and conversion of forest land to grow export crops. Export crops are also grown on prime farmland, forcing peasants to turn to forests and other fragile ecosystems to make a living.

Government tax policies also encourage deforestation. For instance, in Brazil, the government once offered huge income tax credits to investors in cattle ranches, once a leading cause of deforestation in the Amazon.

In the developing world, governments often promote settlement of forests by giving away land or by building roads into forests. The urban dispossessed then move in, clear the land, and attempt to eke out a living there.

The Canadian government has long encouraged deforestation with little regard for the environment. In British Columbia, where forests are falling at a rate far greater than the estimated annual sustainable yield, there are practically no institutional channels by which citizens can influence forest management on public lands. Citizen participation in decisions to log is extremely limited. As a result of recent deregulation, logging decisions are left largely to logging companies. In addition, forest laws that do exist expressly prioritize logging above all other considerations, making it difficult if not impossible to challenge environmentally harmful logging.

Canada's logging industry is a powerful force in its economy and political scene. Here, and elsewhere, logging companies reportedly violate rules and regulations regarding logging, cutting trees right to the banks of rivers; cutting on steep, erosion-prone slopes; making clear-cuts much larger than allowed by law; and failing to protect areas set aside for wildlife. Two-thirds of the coastal temperate rainforest has been degraded because of logging and development. In British Columbia, 140 stocks of salmon have been driven to extinction and 624 are at high risk, largely as a result of poor logging practices. Because of this, Canada is often considered the "Brazil of the North." What is more, the government seems to do little to stop lumber companies from such practices. In recent years, Canada's government has made significant progress in protecting its forests. The government has set aside large tracts for permanent protection and now requires sustainable timber harvest on hundreds of thousands of square kilometers.

When conservationists argue for controls on deforestation, the timber industry responds with the threat of lost jobs. In British Columbia, however, as in the American Pacific Northwest, wood products jobs have been steadily declining because of automation for years, while the annual cut has risen sharply. The timber companies wield an enormous amount of power and use the job issue as a smokescreen, say critics. Their power and influence throughout the world are additional factors responsible for widespread deforestation.

In less developed nations, violations and illegal cutting are also major problems. "Illegal logging occurs in all types of forests, across all continents, from Brazil to Canada, Cameroon to Kenya, and from Indonesia to Russia, destroying nature and wildlife, damaging communities and distorting trade," according to the World Wildlife Fund Global's website. It even occurs in the forests of Europe. Illegal logging accounts for 25% to 80% of all logging occurring in many nations, which greatly reduces tax revenues. To read more about the social, economic, and environmental impacts, you may want to visit their website.

The nations have few resources to monitor logging practices and enforce laws. Logging companies harvest beyond legal boundaries of logging concessions, cut in sensitive areas, and falsify records, taking more timber than allowed.

Especially troublesome is illegal timber harvesting of reserves set aside to protect native cultures and biodiversity, known as **extractive reserves**. Miners and loggers have moved in, despite official designation. In many less developed nations, governments have stripped rights of native peoples to the land, opening it up to exploitation.

KEY CONCEPTS

Ill-advised government policies including below-cost timber sales contribute to widespread deforestation and unsustainable forest management. These policies are often promoted by powerful economic interests that stand to gain from lenient timber-harvesting practices.

An Introduction to Forest Harvesting and Management

To understand the kinds of reform necessary, we must first look at the way trees are currently harvested. Trees are commercially harvested by four basic methods: clear-cutting, strip-cutting, selective cutting, and shelter-wood cutting.

KEY CONCEPTS

Trees are harvested primarily in four ways: clear-cutting, strip-cutting, selective cutting, and shelter-wood cutting.

Clear-Cutting **Clear-cutting** is a standard practice that is used primarily for softwoods (conifers) that grow in large stands containing relatively few tree species. Those species best suited to clear-cutting have seedlings that grow in open, sunny plots. Clear-cutting is also the method of choice on hardwoods such as aspen. It is the preferred method of harvest in tropical rain forests, which typically have many different tree species.

In clear-cutting operations, loggers remove all the trees in 16- to 80-hectare (40- to 200-acre) plots. They then burn the remaining material. This returns nutrients to the soil, facilitates regrowth, and reduces the threat of fires that could damage the regenerating forest. As the new stand grows, trees are thinned to prevent overcrowding.

Clear-cutting is one of the fastest and cheapest methods of harvesting trees. Clear-cutting may also increase **surface runoff**, the flow of water over the ground's surface, which enhances stream flow. This may increase the supply of water to cities, farms, and industry. Clear-cutting increases suitable habitat for some species, such as deer and elk, which benefits hunters. However, small clearings are better for elk than larger ones because they generally avoid open spaces larger than 8 hectares (20 acres). Elk prefer to remain at the edge of meadows, so they can escape into nearby forests should a predator arrive. Thus, large square or rectangular blocks are less advantageous for elk than smaller, irregular cuts. Another factor that determines whether a clear-cut increases or decreases elk habitat is the location of the cut. Winter range is a limiting factor in elk populations. Thus, clear-cuts in winter range, which make more food available, are more beneficial than cuts in the more abundant summer range. In the Rocky Mountain states, however, clear-cuts are generally made in elk summer range, high in the mountains.

Clear-cuts create unsightly scars that may take years to heal (**FIGURE 12-7**). If not replanted or reseeded naturally, soil erosion may become severe, especially on steep terrain. Eroded sediment fills streams and lakes, destroying fish habitat and increasing the cost of water treatment. Sediment also reduces the water-holding capacity of lakes and streams, which increases flooding, already more likely because of

FIGURE 12-7 Clear-cutting. This clear-cut, on steep slopes in the South Tongass National Forest of Alaska, is not only an eyesore but also increases soil erosion, impairing forest regrowth and polluting nearby lakes and streams. Notice the paths credited by dragging logs to the access road. These create channels that can accelerate soil erosion.

© Photodisc/Getty Images.

the elevated surface runoff. Erosion in clear-cut areas may deplete the soil of nutrients, thus impairing or even preventing revegetation.

Large open patches in mountainous terrain may accelerate a process called **sublimation**, the conversion of snow to water vapor. When this occurs, snowmelts actually decrease and stream flow declines.

Routine burning in clear-cuts can damage soils by destroying nutrient-cycling bacteria. Burning also volatilizes soil nitrogen, robbing nutrients from the soil. Burning may also destroy mychorrizal fungi (MIE-koe-RYE-zl FUN-gee), which grow in soils and attach to plant roots, greatly increasing their uptake of water and nutrients.

Clear-cutting fragments wildlife habitat, creating ecological islands. These islands are exposed to wind and more pronounced changes in temperature, humidity, and light that greatly affect indigenous species. In the Pacific Northwest, studies suggest that for each 10-hectare (25-acre) clear-cut, an additional 14 hectares (35 acres) will be degraded.

Clear-cutting destroys habitat and can contribute to the decline of many species. In the Pacific Northwest, heavy cutting of old-growth forests—ancient forests more than 250 years old, with many sections from 500 to 800 years old—threatens the spotted owl and dozens of other species dependent on this habitat (**FIGURE 12-8**). Soil erosion caused by excessive cutting of old-growth forests in the past century has devastated valuable salmon runs in Washington, Oregon, and California.

Canadian forests are also threatened. Old-growth forests are especially threatened by timber cutting. One of the hardest hit areas is Vancouver Island in British Columbia. Besides decimating native plants and animals, clear-cutting in Canada impinges on indigenous peoples who have hunted and harvested the forest sustainably for many years.

(For a debate on old-growth forests, see this chapter's Point/Counterpoint.)

In tropical forests, clear-cutting often has devastating results. Soils become baked in the sun and too hard to support growth; others wash away in torrential rains. For more on this subject, see Spotlight on Sustainable Development 12-1.

New regulations by the U.S. Forest Service are helping to reduce the impact of clear-cutting in national forests, but on private lands it is largely unregulated. Large commercial tree farms operate much like agribusinesses. Seedlings are planted, fertilized from airplanes, doused with herbicides to control less desirable species, and sprayed with pesticides to reduce losses. When the trees reach the desirable size, they are cut down, and the cycle begins again.

KEY CONCEPTS

Clear-cutting removes entire forests quickly and efficiently. Some tree species such as pines, which grow in open sunny fields, are best harvested in clear-cuts. Clear-cuts may benefit some species of wildlife but tend to destroy and fragment the habitat of many others. Clear-cutting creates ugly scars and can cause considerable environmental damage such as increased soil erosion.

Strip-Cutting Arguing for smaller clear-cuts on private and federal land, E. M. Sterling, an expert on forest management, notes that Austria harvests as much wood from its forests as does the Pacific Northwest. Yet Austrian forests show little evidence of clear-cutting because of strict forestry laws that apply to public as well as private lands. Austrian law, for

(a)

(b)

FIGURE 12-8 A bird and its forest. (a) The spotted owl is just one of many species that are adapted to **(b)** old-growth forests. When its habitat is destroyed, the owl disappears.

Photo a courtesy of John and Karen Hollingsworth/U.S. Fish & Wildlife Services; photo b courtesy of Dan Chiras.

instance, forbids clear-cutting on all steep, erodible land. It also limits the size of clear-cuts. A private landowner may cut 0.6 hectare (1.5 acres) without permission but must obtain a permit for larger clear-cuts. Seldom do clear-cuts exceed 2 hectares (5 acres). Most clear-cuts are narrow strips that blend in with the terrain and reseed on their own. This technique is called **strip-cutting**, and it leaves intact forests that will naturally reseed the small clear-cuts. It also reduces erosion and other environmental impacts of clear-cutting. U.S. clear-cutting can be improved by making cuts smaller and by blending them with the terrain.

Selective Timber Cutting   **Selective cutting**, is the removal of a limited number of trees from a forest. It takes place in multispecies (diverse) forests such as those of the northeastern United States. In selective cutting, foresters remove desirable tree species such as maple or beech trees. They generally remove the mature trees but will cut down deformed trees to get rid of them.

The object of selective cutting is to reduce visual scarring and to preserve species diversity in forests, which helps to protect forests from disease and insects. This method minimizes soil erosion and the destruction of wildlife habitat.

Selective cutting may sound like the answer to clear-cutting. Unfortunately, it has several major disadvantages, cost and time among them. This practice also removes the genetically superior trees, whose seed is needed to keep the forest healthy. As a result, a forest may slowly degenerate, producing lower quality wood.

Selective cutting in tropical forests can also be quite damaging if done with heavy equipment. A study in the eastern Amazon showed that when only 3% of the trees were removed, 54% had been uprooted, crushed, or damaged during the construction of roads and logging operations. Road building can also accelerate soil erosion, and—because larger tracts of forest must be harvested to achieve the same output as a clear-cut—more roads may need to be built. Selective cutting is also not suited to trees whose seedlings grow in sunny locations. If properly carried out, however, this technique produces little scarring, causes little or no erosion, and does little damage to wildlife habitat.

Go Green

Get off junk mail lists by writing (send a form letter) directly to those who send you junk mail or you can get off numerous lists at once by contacting the Mail Preference Service, Direct Marketing Association PO Box 643, Carmel, NY 10512. You can find a mail-in form at this web site, http://www.privacyrights.org/letters/jm1a.htm.

Shelter-Wood Cutting   **Shelter-wood cutting** is a kind of selective harvesting that addresses some of the concerns of critics. In this technique, poor quality trees are first removed from mixed timber stands, leaving the healthiest trees intact. These trees reseed the forest and provide shade for their seedlings. Once seedlings become established, loggers remove a portion of the commercially valuable mature trees, leaving enough in place to provide shade for the seedlings. Finally, when the seedlings become saplings, the remaining mature trees are harvested.

Shelter-wood cutting has many of the advantages of selective cutting. It leaves no unvegetated land except for roads, minimizes erosion, and increases the likelihood that the forest will regenerate. It also reduces habitat destruction and ensures a healthy seed source. However, it is more costly than either clear-cutting or selective cutting.

Shelter-wood and selective cutting can be economically competitive with clear-cutting in **second-growth forests**, ones that have been cut previously, if logging roads are already present. Even in low-diversity forests containing only one or two tree species, these techniques can be economically competitive. Because shelter-wood and selective cutting prevent the scarring of the land, they provide additional economic and aesthetic advantages to regions that rely heavily on tourism.

Creating a Sustainable System of Forestry

With the world population growing by about 83 million people per year and the demand for wood for fuel and wood products rising with it, steps are needed to create a sustainable wood and wood products production system. At least four major strategies are needed. First, efforts are needed to reduce demand for wood and wood products. Second, measures must be implemented to ensure a more sustainable harvest of timber. **Sustainable forestry** optimizes yield on forest land while protecting the long-term health and diversity of forest ecosystems. Third, efforts are needed to protect and perhaps even expand untapped forests throughout the world—for example, by establishing forest preserves that protect native species and provide ecological services such as flood control. Fourth, efforts are required to replant forests.

Four measures are required to create a sustainable system of wood production: (1) reductions in demand for wood and wood products, (2) sustainable management, (3) establishment of forest preserves, and (4) restoration of forest land.

Reducing Demand   Reducing the growth of the human population could help to decrease pressure on the world's forests and protect them from further destruction. Demand can also be reduced through efficiency measures. How?

Wood can be saved by using thinner saw blades in saw mills, which reduce the kerf (wood removed by the blade), and improved machines that do a better job of processing logs for plywood along with a host of other technologies. Special training for workers could also help reduce wood waste. According to the Worldwatch Institute, by training workers, improving equipment maintenance, and managing forests better, companies could produce the same amount of wood from one-third less land.

New homes could be built with about 10% less lumber by spacing studs (vertical framing members) a little wider and by other efficiency measures that would not compromise the structural integrity of the house. Building smaller homes that use 20 to 30% less wood could also reduce pressure on the world's forests.

Individuals can also reduce consumption by using fewer paper products, avoiding overpackaged items, and using the backs of scrap paper rather than new paper for homework and notes. The constant bombardment of printed advertising material can be stopped by writing companies and asking them to take you off their lists.

Another way of reducing the demand for trees is by recycling paper. Increased recycling of newsprint and other paper products could reduce the amount of forest cut down each year. Individuals, companies, colleges, and governments can help by purchasing paper products made from recycled paper.

The use of alternative building products can also ease pressure on the world's forests. One interesting innovation is the wooden I-beam, which is used to build ceilings, walls, and floors (**FIGURE 12-9**). These are made of plywood and oriented strand board, both of which are made from small-diameter, fast-growing trees. Forty to 60% less wood is needed to make a wooden I-beam to provide the same service as a 2″ × 10″ or 2″ × 12″ board cut from an old-growth tree. Alternative building materials such as straw bales can be used to build homes to reduce our demand for wood. Many other options are available to us at work and at home.

Demand for wood and wood products can be greatly reduced by controlling growth of the human population, using wood and wood products more efficiently, using alternative building materials, and recycling paper and wood.

Managing Forests and Tree Farms Sustainably
Protecting forests and the soils that support them is essential to creating a sustainable society. In the United States, forest management began in the late 1800s. In 1905, Republican President Theodore Roosevelt established the U.S. Forest Service. Its first head, Gifford Pinchot,

(a)

(b)

FIGURE 12-9 The wooden I-beam. This product, made from wood scraps and small trees, uses 40% to 60% less wood than a solid beam and thus greatly reduces our impact on old-growth forests. It also uses smaller, faster-growing trees, rather than large, old-growth trees.

Photos a and b copyright © Weyerhaeuser. Used with permission.

promoted careful use of forests over strict preservation. He promoted the multiple-use concept and sustained yield—using the forest for many purposes, from timber harvesting to recreation, but managing it scientifically to sustain its valuable services.

Pinchot's notion of multiple and sustained use has persisted. Scientifically based management has long been viewed as a key to managing forests to ensure a sustained yield. However, studies show that past knowledge of forest ecology was limited and that some practices were not sustainable.

In more traditional forest management, foresters tend to focus solely on yield. Timber production is paramount and forests are treated more like farms. In a sustainable forest management (SFM) program, forests are managed to protect and preserve a diverse set of goods and services, including recreation and ecological services (flood control). In addition, forest managers seek to protect the health of the soil and the condition of streams and lakes. In short, forests are managed as complex ecosystems, not as simplified farm production designed to produce one species maximally at the expense of the health of the soil and water. To achieve these ends, SFM often seeks to mimic natural cycles. Clear-cuts, for instance, are patterned after natural fires, which hopscotch across the landscape leaving patches of green between areas of devastation. They provide trees to reseed the area and furnish habitat for species dependent on the forest.

GO GREEN

Buy recycled paper products—notebook paper, pads, paper towels, toilet paper, etc.—and be sure to recycle all your waste paper.

Managers may protect dead trees, too, which are home to many bird species. Sustainable forest managers may also attempt to promote forests containing a mix of species, ages, and/or sizes of trees. Tree cuttings are done on longer cycles, rather than shorter cycles dictated by financial considerations. SFM can be applied to all types of forests, although it must be adjusted to climate, terrain, and type of tree.

Many other steps can be taken to ensure the health of forest ecosystems. One of the most important is to eliminate clear-cuts on steep terrain. This decreases erosion and protects streams and other surface waters. In the tropics, smaller clear-cuts help protect biodiversity and ensure reseeding. In the United States and other countries, reseeding by logging companies should be monitored more carefully, especially on public lands.

Protecting forests from natural hazards—including diseases, insects, fires, droughts, storms, and floods—is also important to creating a sustainable system of forestry. Diseases, insects, and fires account for most of the damage. Sound management that seeks to maintain trees and forests in a healthy state can minimize these problems. Just as in rangeland, the healthier the forest ecosystem, the fewer the pest problems. Maintaining a healthy forest may require periodic thinning of trees to reduce competition for water, which weakens trees and makes them more susceptible to pest damage. Maintaining genetic diversity in forests

also reduces damage from insects and disease organisms. Insect- and disease-resistant trees could be developed for reseeding on public and private land. Finally, all imported trees and lumber should be carefully inspected to avoid accidentally introducing pests.

Fire accounts for a large amount of U.S. forest destruction, leveling about 4 million hectares (9 million acres) of forest each year, according to the U.S. Forest Service Update: Wildfires in 2012, 88% of all forest fires are caused by human beings—either accidentally or deliberately. The remaining 12% are ignited by lightning, but these fires account for nearly 65% of the annual forest damage.

To protect watersheds, timber, and recreational opportunities, the Forest Service and state governments attempt to reduce forest fires by posting fire danger warnings and sponsoring television and radio announcements. In addition, the Forest Service has an active fire surveillance program that seeks to pinpoint fires and, if necessary, put them out as quickly as possible.

In the United States, efforts to protect forests from fire began in the early 1900s. Although strict fire control has saved billions of dollars worth of timber, ecologists and foresters now realize that it can be detrimental to forests. As our understanding of forest ecology has improved, it has become clear that some fires benefit forests. Small fires, for instance, burn dead branches that have accumulated on the ground and return nutrients to the soil. Most animals can escape minor ground fires, and larger living trees are generally unharmed by them. Periodic ground fires also prevent intense, destructive fires that occur if ground litter accumulates.

In forests protected from fires for long periods, the ample fuel supply caused by the buildup of dead branches may cause a fire to burn uncontrollably once it starts, spreading from treetop to treetop as a crown fire. Huge areas are destroyed in firestorms so hot that the soil is charred and wildlife perishes. Trees may be so severely burned that they die.

Periodic fires not only protect forests from devastating fires, they also foster their renewal. Why? Some forest species require occasional fires for optimal growth. The cones of the jack pine, for instance, open up and release their seeds during fires, as do those of the Douglas fir, sequoia, and lodgepole pine. Fires remove brush that shades seedlings and help replenish soils. Fires also help reduce disease and control potentially harmful insects.

Recognizing the benefits of periodic ground fires, forest managers now let many naturally occurring forest fires burn, provided they are not a threat to human settlements. The Forest Service also deliberately sets hundreds of fires each year (**prescribed fires**) to remove underbrush and litter (**FIGURE 12-10**). This practice reduces the chances of potentially harmful crown fires. It also improves wildlife habitat, soil fertility, timber production, and livestock forage.

Another way to ensure sustainable forestry is through certification. Timber companies practicing sustainable forest management can now be certified as sustainable by independent organizations that inspect their operations. Certification ensures retailers and customers that the wood was produced and processed in an environmentally acceptable manner. To

Courtesy of Ralph Saperstein.

Old Growth, Spotted Owls, and the Economy of the Pacific Northwest

Ralph Saperstein

Ralph Saperstein is the Public Policy Manager for Boise Cascade Corporation. He has spent the past 18 years tracking forest issues for the forest products industry.

The public debate over national forests in the Pacific Northwest focuses on the volatile passions generated by the images of ancient forests and endangered wildlife. Unfortunately, these images are far from reality. The forests of the Pacific Northwest have always been a mosaic of ages and ecosystems. Natural fires, volcanoes, floods, windstorms, insects, and disease have all played a major role in the evolution of our present forests. These forests were never all old growth, and the wildlife that they support have evolved in a wide variety of age classes and habitat types.

Congress designated national forests to provide the American people a stable source of timber, water, grazing, minerals, recreation, fish, and wildlife. The principle of multiple use has guided the management of national forests for nearly 100 years. Comprehensive land-use management plans developed under congressional directive and presidential administrations have assigned millions of acres of forestland to nontimber uses. In fact, before the listing of the northern spotted owl as a threatened species, only 30% of national forests in the Pacific Northwest were available to grow trees to produce lumber and paper products for millions of people. Protection plans for the spotted owl and other species set aside nearly 90% of our federal forest lands for uses that preclude timber harvesting.

Restrictions on forest management to protect old-growth and other forests have drastically reduced the supply of timber from our national forests. Literally dozens of companies, and the communities they support, were established on the promise of a sustained yield of timber from our national forests. Close to 100 towns in the Pacific Northwest are dependent on a lumber mill for their economic lifeblood. When the forest products industry gets a cold, the local communities catch pneumonia.

Trees are a valuable resource that can be planted and grown to produce lumber and paper products for future generations. Over 40% of the lumber used in the United States to build homes is manufactured in the Pacific Northwest. Stopping the scientific management of our national forest will lead consumers to turn to countries with little or no environmental restrictions for their forest products. All forests harvested in the Pacific Northwest are promptly replanted to grow trees for the future, as they have been for over 50 years. Federal and state laws require that streams are protected, wildlife habitat is created, and productive soils are cared for.

The forest products industry supports protecting old growth and wildlife habitat. The issue that must be addressed is one of balance. Over 1.6 million hectares (4 million acres) of old-growth forest are already preserved and will never be harvested. One must recognize that preserving forestland does not assure that it will never be susceptible to fire, wind, insects, and disease. These natural forces and the normal progression of forest growth will eventually drastically alter today's old-growth forests.

The spotted owl, followed by several fish species, has grabbed the attention of the entire nation. Special interest groups use the spotted owl and other fish and wildlife species as surrogates to eliminate needed management on millions of acres of forestland. Despite sympathetic court rulings and administrative actions, special interest groups have not demonstrated that scientifically based forest management cannot be compatible with fish and wildlife habitat enhancement or restoration. In fact, neither the government nor special interest groups have proven in court that harvesting trees harms a species that enjoys Endangered Species Act protection.

The spotted owl and fish issues have raised important philosophical issues about humans' role in the natural environment. The Endangered Species Act requires protection of all threatened or endangered species without regard to the social and economic costs of such protection. What once was a universally acceptable premise is now being scrutinized at all levels of society. There are literally thousands of species that threaten to stop development, agriculture, fishing, transportation, and power generation in all 50 states. The elevation of wildlife above people's needs is a philosophical change that will shake the foundations of modern society. The answer cannot be nature versus people. The solution to these conflicts must come from the wise management of our natural resources in balance with social and economic values.

For years, the tobacco companies would trot out their research scientists who announced the results of their latest studies "conclusively proving" that cigarette smoking was not linked to lung cancer. Today, against all evidence that logging the public's national forests is an economic and an ecological disaster, the timber industry continues to indulge in its own version of chronic denial. To listen to industry spokespeople, there are no problems in our national forests that more cutting will not solve.

Let's be clear: 95% of the original native forests that once covered most of our nation are gone. The 5% that remains is primarily in the Northwest, much of it badly fragmented by clear-cuts—the practice of cutting, then burning every living thing in 40- to 80-acre increments. What is left resides almost exclusively on public lands as part of the system of national forests. These forests belong to present and future generations of Americans, not to the timber industry.

The mere fact that in 150 years—a short time in the life of a forest—we have managed to dispatch all but 5% of this once dominant ecosystem attests to unsustainable forestry practices. We have, in fact, been hacking down our native forests at twice Brazil's rate. Ironically, everyone seems to be in agreement that Brazil, which still has 80% of its original forests intact, should stop cutting.

Typically, the industry argument is framed in terms of jobs versus owls: a narrow and inaccurate characterization. The two primary reasons for decreased timber employment are automation and exports. Tellingly, in Oregon, during the decade *before* the emergence of the spotted owl, while the total cut from national forests *increased* 15%, employment in the timber industry *decreased* 16%. The industry exports *more timber annually than the entire cut from all national forests.* According to Oregon Congressman Peter DeFazio's office, one-fourth of all the trees cut in the Northwest are exported. If you add minimally processed timber, exports account for up to 60% of all timber cut. It is clear we do not need federal timber for domestic consumption. Like a third-world colony, we export our raw materials—and our jobs—to foreign nations who then sell us back finished goods. If we simply stopped exports, we could cease logging national forests and experience no timber shortage.

Standing national forests offer much wider and more essential values than timber alone. Forests provide us with clean air and pure water. They abate flooding, moderate the climate, and deter desertification. Forests provide wildlife habitat and abundant fisheries. They are a source of medicines. The bark of the Pacific yew tree contains a chemical called taxol, a potent anticancer substance

Owls, Lies, and Taxpayer Waste

Victor Rozek

Victor Rozek is the general manager of the Native Forest Council.

remarkably effective against ovarian cancer. Tragically, for decades, the yew has been cut and burned as a weed species.

Standing forests also act as a vast carbon storehouse. Once cut, they release enormous amounts of carbon into the atmosphere, hastening global warming. And, of course, forests offer us recreation and inspiration. Yet, incredibly, when preparing timber sales, the government attributes no value to standing trees. In Alaska, the U.S. Forest Service sold 400-year-old trees for $1.48 each!

A congressional study showed that the U.S. Forest Service lost $5.6 billion in direct taxpayer subsidies over the past decade on timber sales. We are asked to pay for the destruction of our own heritage for the sake of temporary employment and short-term profits that will disappear with the last native forests.

As for the spotted owl, the disregard for existing federal laws has been so blatant that in May of 1991, U.S. District Judge William Dwyer issued a temporary injunction against timber sales in the Northwest. Judge Dwyer observed: "More is involved here than a simple failure by an agency to comply with its governing statute. The most recent violation of the National Forest Management Act exemplifies a deliberate and systematic refusal by the Forest Service and the Fish and Wildlife Service to comply with the laws protecting wildlife." It is absurd to suggest that endangered species are "threatening" to stop development or that the Endangered Species Act represents an "elevation of wildlife against people's needs." To the contrary, it is an act of last resort. It is precisely a century-long *imbalance* of placing human needs and economics ahead of everything else that has brought the forests to the point of near total ruin. The owl is an indicator species by which scientists judge the health of the entire forest ecosystem. And science tells us that the ecosystem is in deep trouble. As to balance, there are some 1,500 pairs of spotted owls and 5.5 billion people. Clearly, humans are not endangered. They can even be restrained, while owls cannot. Are we so impoverished that we need to kill the last handful?

The time has long passed for "compromise" or "balance." It's time to stop managing our national forests like a private social welfare program and return them to

(Continued)

present and future generations of Americans. We would not think of hiring displaced quarry workers to fill in the Grand Canyon. We would be far more foolish to sacrifice the enormous value of standing national forests for the benefit of wasteful employment that consumes much more than it produces.

Critical Thinking Questions

1. In your view, which author makes the most compelling case? Why?
2. What factors influence your view on this matter? Do you have certain biases or philosophies that affect your viewpoint?

ensure uniform standards, a consortium of environmental groups, foresters, timber producers, indigenous groups, and various independent certifiers formed the Forest Stewardship Council, which drew up guidelines for forest management and wood production. They now accredit independent certifiers, creating a uniform globally consistent system to ensure sustainable forest management and wood production. The Forest Stewardship Council has certified timber operations on 181.5 million hectares (445 million acres) in more than 80 countries as of February 2014. Today, several thousand

wood products are sold each year using Forest Stewardship Council-certified wood. Expect that to change, however, in the near future. Home Depot and Lowes sell some Forest Stewardship Council certified wood.

One promising development in the management of forests is the **Model Forest Program** launched by Canada in 1991. It seeks to involve a large number of stakeholders (organizations and individuals) in the management of Canada's forests, 94% of which are on public land—that is, land owned by provincial, territorial, or the federal governments. According

(a)

(b)

(c)

FIGURE 12-10 Benefits of forest fires. (a) Dense undergrowth in an Oregon pine stand results from the control of forest fires. **(b)** Controlled burning removes the undergrowth. **(c)** Periodic burning prevents disastrous fires, returns nutrients to the soil, and increases forage and wildlife.

Photo a © Jupiterimages/Comstock/Thinkstock; photo © Kirsanov/ShutterStock, Inc.; photo c © G. Perkins/Visuals Unlimited.

to their website, "The Canadian Model Forest Network (CMFN) represents 14 member organizations nationally." They create a process in which diverse forest interests can meet to reach agreement on proper forest management. They have involved over 500 organizations and 200 communities in important forest management decisions. According to their website, "These people are from aboriginal and non-aboriginal communities, industry (forestry and other land uses), government (municipal, provincial, and federal), nongovernment organizations, schools (elementary to university), and researchers. This allows us to openly discuss a range of topics. These are important issues to the forest and the community who depends on it." The goal is to create sustainable forest management that benefits the communities, the economy, and the environment. Management plans for hundreds of projects, ranging from 100,000 to 2.5 million hectares (250,000 to 6.2 million acres), are based on consensus among loggers, conservationists, scientists, and government officials. They seek to protect valuable ecological resources while establishing sound harvesting strategies. Other nations, including Mexico and Russia, have adopted similar programs.

Private efforts are also needed. In Alberta, conservation groups started a program to protect the grizzly bear, whose population has declined precipitously. This species, now in danger of extinction, may be saved by voluntary efforts of these organizations and timber companies. Spray Lakes Sawmills in Alberta, for example, has become a financial sponsor of the project and is voluntarily developing ways to harvest trees that have less impact on grizzly habitat.

Black bears in British Columbia could benefit from wide-ranging efforts to protect and better manage old-growth forests. Studies show that most black bears den in hollows in trees that are at least 500 years old. But much of British Columbia's old-growth forests are being cut and will be harvested every 80 to 100 years, seriously jeopardizing the bear's habitat.

KEY CONCEPTS

Better management of existing forests, based on sound scientific principles, including tree thinning, prescribed burns, and replanting, helps to create a more diverse and healthier forest that is less susceptible to disease and insects. Certification programs can help promote sustainable forest management.

Saving Primary Forests/Creating Forest Preserves Some forestry experts believe that we need to protect many of the remaining uncut forests, **primary forests**. To do so will require states and nations to set aside large tracts of primary forest throughout the world. Such measures, say proponents, will ensure continued delivery of the many free services forests provide, such as flood control. They will also help protect biodiversity and, in the tropics, will safeguard the homes of many indigenous peoples who still depend on forests.

In 1986, the U.S. Congress passed a law that prevents the U.S. Agency for International Development (AID) from funding projects such as dams and roadways that will destroy tropical forests in less developed nations. The law also directs the agency to help countries find alternatives to forest colonization and requires it to support preserves and other measures to save forests and promote biological diversity.

Primary forest preserves house insect predators (birds and other insects) and thus act as physical barriers to the spread of pests from harvested forests. Old-growth forests in the Pacific Northwest, for instance, house 100 times more insect predators than neighboring tree plantations—and thus tend to limit pest outbreaks.

KEY CONCEPTS

Saving uncut or primary forests helps preserve biodiversity but also protects nearby harvested forests from outbreaks of pests.

Restoring Forestland Restoration is a key principle of sustainability and is vital to efforts to create a sustainable system of timber production. Prior to 1500 AD, tropical rain forests covered 15.6 million square kilometers (6 million square miles)—a land area more than two times the size of the United States. Today, nearly 60% of the rain forests have been cut down.[2] At the current rate of destruction, rain forests could be gone by around 2050, well within most readers' lifetimes. Although some of the land has been converted to human use, at least two-thirds of it is unoccupied and available for replanting. These lands could be replanted with a variety of trees and managed sustainably. This would greatly reduce pressure on the world's remaining tropical rain forests. Planting only 5% of the rain forest land already cleared could nearly double the supply of commercially harvested wood. Tree farms that support a mixture of species and different-age trees can supply wood and nonwood products while providing wildlife habitat.

KEY CONCEPTS

Building a sustainable system of forestry will require efforts to replant millions of acres of forestland that has been cut and never replanted.

12.4 Wilderness and Wilderness Management

Wilderness, as defined by U.S. law, is "an area where the Earth and its community of life are untrammeled by man, where man is himself a visitor who does not remain." Why is it important? Why should we set aside large tracts of wilderness?

---

[2] Rain forests once covered 14% of the Earth's land surface; today they cover only about 6%, a loss of 58%.

Why Save Wilderness?

Wilderness provides an escape from modern society (**FIGURE 12-11**). Joseph Sax, author of *Mountains without Handrails,* writes that nature "seems to have a peculiar power to stimulate us to reflectiveness by its awesomeness and grandeur." It helps us understand ourselves and the world we live in, awakening us to the forgotten interdependence of living things. "Our initial response to nature," Sax writes, "is often awe and wonderment: trees that have survived for millennia; a profusion of flowers in the seeming sterility of the desert.... [It] is also a successful model of many things that human communities seek: continuity, stability and sustenance, adaptation, sustained productivity, diversity, and evolutionary change."

Wilderness, especially forested regions, is vital to preserving biodiversity and nature's free services. Many of the reasons for protecting wildlife also pertain to wilderness protection.

Historically, however, wilderness has been viewed by many people as something to exploit for short-term gain. In early colonial and postcolonial times, American lands represented untapped wealth—an unequaled opportunity to sustain a young, growing nation. To some frontierspeople and particularly early farmers, many natural resources were perceived more as obstacles than as assets. Forests needed to be cleared to permit farming. Marshes needed to be drained. Today many people view wilderness as simply a playground for an upper-middle-class elite—people who fight to protect these lands to the detriment of others who could reap economic benefits from mining and timber harvesting.

Differing views create enormous controversy. Today, many environmentalists lobby for more wilderness to be set aside. The mining and timber industries, both powerful lobbying forces, generally oppose wilderness designation, fearing that it will lock up valuable resources and hinder the economic development of the nation. Because only slightly over 16% of all government land holdings (excluding military bases and land on which public buildings are built), or about 4% of all land in the United States, has been afforded wilderness protection, environmentalists argue that the mining and timber industries' claims that wilderness is locking up the Earth's riches are unfounded. Locally, however, wilderness tracts can tie up huge parcels of land, threatening the economic well-being of communities that have long made a living by timber harvesting and mining. Wilderness advocates, though, note that wilderness designation can create a more sustainable economy by promoting businesses in rural areas that service hunters, anglers, hikers, backpackers, cross-country skiers, and others.

Should the United States set aside wilderness if it contains oil, natural gas, or minerals that could be used today? Many environmentalists believe that wildlands are more valuable than these finite resources because there are no substitutes for wilderness once it has been destroyed (Figure 12-11). They also note that there are many more sustainable ways to meet our demands for resources such as water, energy, minerals, and wood.

FIGURE 12-11 More than just a pretty place for backpackers. Wilderness restores us. It also offers numerous free ecological services, from climate control to watershed protection.

© Photos.com.

Preservation: The Wilderness Act

The earliest efforts at wilderness preservation in the United States began in the 1860s. John Muir, founder of the Sierra Club and a longtime wilderness advocate, is credited with much of the early interest in saving wilderness for future generations. Further advances came in the 1930s, when the U.S. Forest Service began to set aside large tracts of forest-land, called **primitive areas**, for protection. Between 1930 and 1964, the Forest Service established over 3.7 million hectares (9.1 million acres) of primitive areas in the national forests.

In 1964, Congress passed the **Wilderness Act**, establishing the **National Wilderness Preservation System**. Under this law, the Forest Service's primitive areas were renamed **wilderness areas**. The Wilderness Act directed the Forest Service, the Fish and Wildlife Service, and the National Park Service to recommend additional land within their jurisdictions for wilderness designation. By 2013, Congress had established more than 700 areas in 44 states, totalling 144.3 million hectares (109.5 million acres).

Fifty-four percent of the wilderness acres are in Alaska. Even so, more than half of American Wilderness Areas and within a day's drive from America's largest cities such as Seattle, Portland, Denver, Chicago, Phoenix, Los Angeles, and New York City according to the U.S. Forest Service.

In addition to setting aside land for wilderness designation, the Wilderness Act established rules and regulations for human activities in wilderness. For example, it forbids timber cutting, motorized vehicles, motorboats, aircraft landings, and other motorized equipment—except to control fire, insects, and diseases or where their use was already established.

Although the Wilderness Act sought to create an "enduring wilderness," many unwilderness-like activities were allowed to continue—notably livestock grazing and mining for metals and energy fuels—if claims were filed before the end of 1983. Wilderness areas throughout the United States are also riddled with private inholdings, property owned by individuals and companies who control the mineral and water rights on the property.

According to the U.S. Forest Service, more than 12 million people visit wilderness areas each year to hike, camp, backpack, raft, canoe, ski, ride horses, hunt, fish, bird watch, and photograph. Although it has largely been a success, the Wilderness Act has some weaknesses. For example, it did not provide a means of designating wilderness on the 180 million hectares (450 million acres) of land held by the Bureau of Land Management, mostly in the western states and Alaska. To correct this, the U.S. Congress passed the **Federal Land Policy and Management Act** (1976), which calls on the BLM to submit recommendations on the wilderness suitability of its land. In 2014, BLM Wilderness Areas contained 3.5 million hectares (8.7 million acres), 8% of the nation's total wilderness.

KEY CONCEPTS

The United States has a long history of wilderness preservation that continues today through the Wilderness Act. This law directs federal agencies to establish wilderness areas and stipulates the type of human activities that are permitted on these lands.

Sustainable Wilderness Management

Lured by the thought of quiet and solitude, backpackers pour into some U.S. wilderness areas only to be dismayed by crowds and special camping restrictions aimed at protecting lakes and streams from pollution. Wilderness crowding can result in severe environmental degradation. Grasses near favorite camping spots get trampled and eventually die, leaving behind only topsoil. Human waste deposited around campsites washes into streams. Streams are polluted by soap from dishwashing. Garbage often litters favorite sites and trails. Fragile areas get trampled. Wildlife are displaced.

Hiking trails become deeply rutted. Horses soil trails and overgraze popular areas.

Overcrowding and the environmental decay that accompanies it often occur near large metropolitan areas—in Colorado and California, for example. To wilderness proponents, these are signs of the need for more wilderness—especially if the U.S. population is to increase by more than 84 million between now and 2050.

Crowding and the environmental degradation from overuse can be reduced or eliminated by better management: (1) educating campers on ways to lessen their impact, (2) restricting access to overused areas, (3) issuing permits to control the number of users, (4) designating campsites, (5) increasing the number of wilderness rangers to monitor use, (6) disseminating information about infrequently used areas to divert campers from overused areas, and (7) improving trails to promote use of underutilized areas.

Globally, interest in wilderness protection is growing. Extractive reserves, which, if properly managed, could help protect biodiversity. Although some countries—such as Costa Rica, Brazil, and Colombia—have set aside large parcels of land for protection, many others have little or no protection whatsoever. Where they do exist, protected areas are often understaffed and overused.

Countries struggling to feed their people and meet the demands of rapidly growing populations often have little concern for protecting wilderness, even though it may be in their best long-term interest. Ecotourism provides an economic incentive to protect wilderness in less developed nations. Interest in potential pharmaceuticals from plants and other sources also provides an economic rationale for protecting lands. Finally, the growing realization that extractive reserves can be much more lucrative and environmentally sound than traditional land-use patterns (such as plantations and ranches) has provided additional impetus for saving land from development. Although wilderness management is generally based on the preservation ideology, these compatible uses bode well for the fate of the wilderness in less developed nations.

Wilderness, grasslands, and forests are vital to our future. They cater to many different needs: eating, shelter, relaxation, and escape. A world without them is almost unimaginable to those of us who love nature. A group of scientists peering through the glass of their space shuttle 100 years from now will see the evidence of our actions today. Whether they see patches of ancient desert within rich, productive land or just the opposite depends on actions and decisions we make today.

We're not so poor that we have to spend our wilderness or so rich that we can afford to.

—Newton Drury

Exercise Analysis

Critical thinking rules instruct us to question the methods by which information is obtained. In this example, early estimates of desertification were based on studies that looked only at isolated regions; that is, their sample size was quite small. This is a common error in scientific research. Conclusions based on small sample size—a few isolated observations or small numbers of test animals—should always be regarded with caution.

This is not to say that the Sahara may not be expanding. It may be; it may not be. Looking at the new evidence suggests that although the desert may ebb and flow, overall it is marching southward. The study cited in this exercise suggests that the desert marches southward in low-precipitation years, then recedes in years when precipitation is heavier. However, the trend has been an expansion of the desert. Clearly, further research is needed.

At this point, it is necessary to exercise one of the most paradoxical of all critical thinking rules—to tolerate uncertainty. Only time will tell whether the desert is indeed marching southward as a result of global warming and other problems mentioned in the introduction. Many global issues such as ozone depletion and global warming take place amid natural cycles. To discern real trends requires a decade or more of data.

CRITICAL THINKING AND CONCEPT REVIEW

1. What is the tragedy of the commons? Have you seen any examples of it? Is reproductive freedom—the right to have as many children as one wants—contributing to the tragedy of the commons?

2. Critically analyze the concept of the tragedy of the commons. Is it a valid phenomenon? Why or why not? Does the inadequacy of resource management on privately owned land negate the validity of this concept?

3. How can the operating principles of sustainability discussed earlier in the book help reshape the livestock and forestry industries? Give specific examples of ways these principles can be applied.

4. What are the major problems facing the world's rangelands? How can we create a sustainable system of providing meat for human society?

5. Define the following terms as they relate to forest management: *sustained yield*, *multiple use*, and *clear-cutting*.

6. What is meant by *sustainable forestry*? How is it different from the present method of cutting trees, which is based principally on sustained yield?

7. List and discuss ways to satisfy the growing need for wood and wood products in the coming years. Which of your ideas are the most ecologically sound?

8. In what ways can you reduce paper and wood waste and increase recycling?

9. Critically analyze this statement: "Wilderness is not essential to humanity. We should not be concerned with preserving wild areas. For the vast majority of the world's people, they are of no value."

10. Do you agree or disagree with the following statement? "The expanding U.S. population suggests the need for more wilderness designation." State reasons for your position.

basal zone
Bureau of Land Management (BLM)
clear-cutting
commons
confined feeding operations (CFOs)
extractive reserves
Federal Land Policy and
  Management Act
feedlots
intensive grazing
metabolic reserve

Model Forest Program
National Wilderness Preservation
  System
prescribed fires
primary forests
primitive areas
Public Rangelands Improvement
  Act
rangelands
range management
rotational grazing

second-growth forests
selective cutting
shelter-wood cutting
strip-cutting
sublimation
surface runoff
sustainable forestry
wilderness
Wilderness Act
wilderness areas

© BMJ/ShutterStock, Inc.

Water Resources: Preserving Our Liquid Assets and Protecting Aquatic Ecosystems

A river is more than an amenity—it is a treasure.

—**Oliver Wendell Holmes**

LEARNING OBJECTIVES

After reading and studying the material in this chapter, students will be able to understand, knowledgably discuss, and, when appropriate, apply the information learned in this chapter, including:

1. The hydrological cycle
2. Global water supply issues
3. Social, economic, and environmental impacts of groundwater overdraft, dams, and reservoirs
4. The many ways we can create a sustainable system of water supply
5. The causes and impacts of flooding
6. Techniques that allow us to sustainably control flooding
7. Threats to and the importance of wetlands, estuaries, coastlines, and rivers
8. Ways to protect wetlands, estuaries, coastlines, and rivers

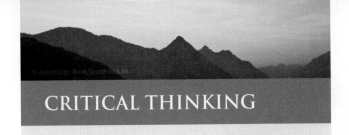
In Kansas, Nebraska, Colorado, and other western and midwestern states, groundwater from the huge underground reservoir, the Ogallala aquifer, is being depleted much faster than it can be replaced. An aquifer is a body of water located underground, usually in a permeable substrate such as sandstone. One proposed solution is to divert 14 million acre-feet (an acre-foot is one acre one foot deep) of water from the Missouri River at St. Joseph, Missouri, which is 260 meters—850 feet—above sea level, and pump it in canals and pipelines to agricultural areas in western Kansas, which is 900 to 1,000 meters—3,000 to 3,500 feet—above sea level. Is this a sound proposal?

To begin, you may want to prepare a list of all of the economic and environmental costs. I'll give you some numbers you can use to calculate pumping costs in the analysis section at the end of the chapter.

ater. We drink it. We wash with it. We cook with it. We play in it. We irrigate our crops with it, and we use it to power industrial plants, factories, and food processing plants. The abundance or lack of water often determines where we live and how well off we are.

Despite its importance to humans and other organisms, water is squandered and polluted by industry, agriculture, and many other systems. To live sustainably on the Earth, we must do a much better job of protecting and managing water resources.

This chapter discusses three components of sustainable water resource management: It examines water supply problems and ways to solve them. It explores flooding and ways to reduce it. Finally, it examines aquatic ecosystems such as wetlands, rivers, and oceans and ways to protect and restore them. The goal in all of this is to find sustainable ways to meet our needs for water now and in the future. Before looking at the problems of water supply, however, let's examine the water cycle.

13.1 The Hydrological Cycle

Water is part of a global recycling network known as the **hydrological cycle**, or **water cycle**. The hydrological cycle runs day and night, free of charge, collecting, purifying, and distributing water. Along its way, it serves humans and other living creatures in a multitude of ways.

The water cycle is driven by evaporation and precipitation. **Evaporation** occurs when water molecules escape from surface waters, soil, and plants, becoming suspended in air (**FIGURE 13-1**). When water molecules depart, they leave behind impurities. Thus, evaporated water is free of contamination until it mixes with atmospheric pollutants from human and natural sources.

Evaporation ensures the continuation of the water cycle, but also serves important biological functions. In plants, for example, the evaporation of water from leaves draws water and dissolved nutrients up from the roots through tiny vessels in the roots, stems, and leaves, much like sucking on a straw draws water from a glass. This helps deliver water and soil nutrients into the cells of plants so they can grow and reproduce. The evaporative loss of water from leaves is called **transpiration**. The loss of water from the soil and leaves is called **evapotranspiration**.

The evaporation of water is also helpful for animals. In humans, for instance, water evaporating from the skin helps keep our bodies from overheating on hot days.

In the atmosphere, water is suspended as fine droplets known as *water vapor*. Although we're not usually aware of it, you can detect its presence on a cold day when moisture beads up on cold windows or on a warm day when moisture collects on the outside of a glass of ice-cold lemonade.

The amount of moisture (water vapor) air can hold depends on the temperature of the air. The warmer the air, the more moisture it can hold. Atmospheric moisture content can be expressed in one of two ways. The first is the **absolute humidity**, which is the number of grams of water in a kilogram of air. If you extracted a sample of air and removed its water, this would be its absolute humidity. Another measure is the relative humidity. The **relative humidity** is the amount of moisture present in air compared with the amount it could hold if fully saturated, at any given temperature. If the local TV station's meteorologist says that the relative humidity is 50%, she means that the air contains 50% of the water vapor it can hold at the present temperature. If the relative humidity is 90%, the air has 90% of the moisture it can hold at that temperature. If the relative humidity is 100%, the air is said to be **saturated**. When the moisture content of a mass of air exceeds the saturation point, clouds, mist, and fog form—that is, water vapor becomes visible to us!

Water that has evaporated eventually returns to Earth through precipitation, thus continuing the hydrological cycle. Here's how it happens. On a warm summer day, sunlight warms the Earth. This causes water to evaporate from water bodies, the land, and plants. Warm, moisture-laden air then rises. As it rises, the air expands because atmospheric pressure decreases. As it expands, it cools. As it cools, its relative humidity of the air mass increases. When the air mass exceeds its saturation point, clouds form. This same phenomenon occurs when moisture-laden air is pushed upward by mountain ranges.

Clouds form when the relative humidity becomes slightly greater than 100%. When this occurs, water molecules begin to attach onto various small suspended particles, called **condensation nuclei**, in the air. Condensation nuclei are always present in the atmosphere; they may be salts from

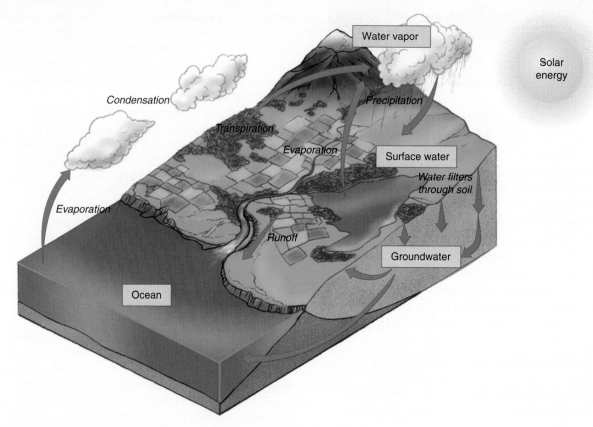

FIGURE 13-1 The hydrological cycle. Water is a renewable resource, purified and distributed throughout the planet via the hydrological cycle.

the sea, dusts, or particulates from factories, power plants, and vehicles.

Clouds and fog contain several hundred water-laden condensation nuclei per cubic centimeter of air. In tropical and semitropical regions, fine cloud droplets collide and coalesce, forming raindrops. Over a million fine water droplets must come together or **condense** to make a single drop of rain. Some evidence suggests that electrical activity in clouds may increase the rate of collision and thus enhance the formation of raindrops.

In temperate climates and at the poles, the temperature of the air in clouds is often well below the freezing point, even in the warmer months of the year. As a result, minute ice crystals form in the clouds. When the crystals reach a certain critical mass, they fall from the sky as snowflakes. In the spring, summer, and fall, the snowflakes generally melt as they fall, producing rain.

Clouds move about on the winds, generated by solar energy, and deposit their moisture throughout the globe as rain, drizzle, snow, hail, or sleet. This process, called **precipitation**, returns water to lakes, rivers, oceans, and land.

Water that falls on the land may evaporate again, or it may flow into lakes, rivers, streams, or groundwater. Eventually the water returns to the ocean, from which it may once again evaporate. The vast majority of the water in clouds and rain comes from surface waters, primarily the oceans.

On average, over 15 trillion liters (4 trillion gallons) of precipitation fall on the United States every day. Two-thirds of all this precipitation evaporates. Thirty-one percent finds its way into streams, lakes, and rivers, and 3% recharges groundwater (**FIGURE 13-2a**).

At any single moment 97% of the Earth's water is in the oceans. The remaining 3% is freshwater. Of this, most (99%) is locked up in polar ice, in glaciers, and in deep, inaccessible aquifers, underground reservoirs usually in some permeable material such as sandstone. This leaves a paltry 0.003% available for use by humans, but most of this water is hard to reach and much too costly to be of any practical value. At present, human civilization and all terrestrial forms of life are maintained by only a tiny fraction of the Earth's water supply, a fact that underscores the importance of treating our freshwater supplies with care.

KEY CONCEPTS

Water is a renewable resource, purified and distributed in the hydrological cycle, which is driven by solar energy, which causes evaporation and precipitation. Although the Earth is endowed with an enormous quantity of water, only a small fraction of the planet's water is available for human use, a fact that underscores the importance of managing it wisely.

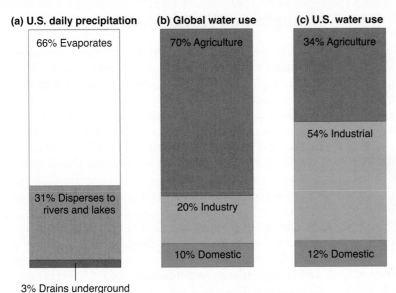

(a) U.S. daily precipitation

66% Evaporates

31% Disperses to rivers and lakes

3% Drains underground

(b) Global water use

70% Agriculture

20% Industry

10% Domestic

(c) U.S. water use

34% Agriculture

54% Industrial

12% Domestic

FIGURE 13-2 Where does the water go?
(a) Fate of precipitation in the United States,
(b) global water use, and **(c)** water use in the
United States.

13.2 Water Issues Related to Supply and Demand

Water issues are of great importance in the study of our environment. In this section, we examine issues related to supply and demand. We look at water shortages and impacts of supplying water to the world's people. Before we begin, however, let's take a look at where our water comes from and who uses it.

Where Does Water Come from and Who Uses It?

The world population withdraws enormous quantities of freshwater from lakes, streams, and rivers each year. Approximately 70% of this water is used for crop and livestock production. Industrial use accounts for about 20%, and domestic use makes up the remaining 10%, according to the Worldwatch Institute (Figure 13-2b). Of course, these averages vary from country to country. In the United States, for example, 33% of the freshwater is withdrawn for agriculture and 54% is for industry. Eleven percent is withdrawn for residential use (Figure 13-2c). In Asia, 85% is used for irrigation. Where does this water come from?

In the United States, surface waters supply the bulk of the freshwater we need each day. In fact, about three-fourths of our daily water supply comes from lakes, rivers, and streams, known as **surface waters**; the remainder comes from **groundwater**.

KEY CONCEPTS

Globally, agriculture and industry are the major users of water. Most water comes from surface water supplies—rivers, streams, and lakes.

Water Shortages

Although the Earth is endowed with a generous supply of freshwater, water shortages are becoming more and more common. Virtually every nation in the world experiences shortages, often extreme shortages. The consequences can be devastating. In many less developed countries (LDCs), women spend a large part of their waking hours fetching water, often walking 15 to 25 kilometers (10 to 15 miles) a day to get it—frequently from polluted streams and rivers. According to the World Health Organization, three out of every five people in the less developed nations do not have access to clean, disease-free drinking water.

To many, it may seem as if the world is running out of water. Far from it. Today's freshwater supply is the same as it was when civilization began. As you just learned, water is a renewable resource, constantly recycled by the hydrological cycle. Interestingly, enough drinkable water falls on the land each year to flood it to a depth of 86 centimeters (33 inches). On first glance, one would think that this amount should provide sufficient quantities to meet our needs several times over. Unfortunately, precipitation is not evenly distributed across the face of the Earth. Tropical rain forests are drenched with rain, receiving 250 centimeters or more (100 inches). Deserts receive under 25 centimeters (10 inches) a year. People living in areas that receive small amounts of precipitation, such as the desert, often suffer severe water shortages.

Water shortages result from lack of rain and are therefore most common in dry, arid climates, less so in semiarid regions, and uncommon in rainy areas. But rainfall patterns are just part of the equation. Demand also plays a big role. Any time demand exceeds supply, water shortages can occur. A stream and groundwater may be sufficient for a small town, but when that town burgeons to become a city, shortages may result.

Go Green

Take shorter showers, and never brush your teeth in the shower.

Water shortages occur even in areas with seemingly abundant supplies when demand exceeds supplies. Canada, for example, has 9% of the world's freshwater and ranks third among the world's nations in freshwater resources. Yet water shortages are common in many parts of Canada, especially the southern parts of the prairie provinces. Part of the reason for this is that Canada's indoor water use is second highest in the world, about 340 liters (88 gallons) per person per day. The other reason is that 90% of Canada's population is concentrated in the lower portion of this country and much of the nation's water resources are located in the North.

One thing is certain. As population increases, water shortages will become even more prevalent. Demand for agriculture and industry and personal use will rise significantly, causing even greater shortages than now exist. According to the Worldwatch Institute, between 1950 and 2050, the amount of water available per person will fall by 74%. Billions of people could suffer from serious water shortages. They and others believe that water scarcity may be the most underrated resource issue in the world today.

KEY CONCEPTS

Water shortages occur virtually everywhere. They are most prevalent in areas that naturally receive small amounts of precipitation but can occur in any region in which demand exceeds existing water supplies. Climate change, population growth, agricultural expansion, and demand for water by the industrial sector are likely to make water shortages even more prevalent in coming years.

Drought and Water Shortages

Drought also contributes mightily to water shortages. A **drought** exists when rainfall is 70% below average for a period of 21 days or longer. A severe drought results in a decrease in stream flow, sometimes a complete cessation of flow, and a drop in the level of lakes, streams, and reservoirs. This, in turn, reduces water available for irrigation, domestic use, and industrial production. In 2007, Georgia and many neighboring states of the southeastern United States suffered from prolonged drought that severely depleted water supplies (**FIGURE 13-3**). Drought struck the midwest in 2012.

Drought also results in a drop in the **water table**, the upper level of the groundwater. Droughts cause devastating losses to agricultural crops. They may also cause a loss of wildlife, especially aquatic organisms, and reduction in range production and increased stress on livestock. Drought years are also characterized by an increasing number of forest and grass fires as witnessed in California and many other western states in the past two decades.

Whether you live in New York, Toronto, Tulsa, or Bird City, Kansas, you've probably felt the impacts of water shortages at least once in your lifetime. Most people think of droughts as natural events. Although this is true, droughts can also be caused by human activities. Overgrazing and deforestation, for instance, decrease rainfall in downwind areas. The devastating fires in Indonesia and neighboring countries in 1999 were caused in large part by deforestation that led to a drying of neighboring intact forests. As you shall see, global climate change caused by carbon dioxide and other greenhouse gases from human activities is affecting rainfall patterns, creating excessive flooding in some areas and drought in others.

FIGURE 13-3 Depleted reservoir in Buford, Georgia. This reservoir, like many others in the southeastern United States in 2007 and 2008, is nearly empty as a result of prolonged drought. In 2012, widespread drought in the midwest created similar results in states including Texas, Kansas, Missouri, and Oklahoma.

© Chris Rank, Bloomberg News/Landov.

Droughts occur naturally, but also result from human actions such as deforestation, overgrazing, and climate change. Droughts reduce water supplies and create significant social, economic, and environmental problems.

Impacts of the Water Supply System

There is no doubt that water shortages cause severe problems, but before we rush headlong into a plan to develop more water supplies to meet our needs, we must carefully study the water supply system and its impacts. Such an understanding could help us find better—more sustainable—ways to meet our needs.

Water is withdrawn from lakes, rivers, and streams to serve human needs. In the case of lakes, water is generally removed by offshore pipes and pumped to water treatment plants. There it is filtered and chlorinated and then distributed through underground pipes to users (**FIGURE 13-4**). Rivers and streams, however, are generally dammed to create reservoirs that store water for year-round use. The water from the reservoir is drawn off as needed, filtered, disinfected, and distributed to users. In some areas, groundwater provides part or all of the water needed by humans. It is pumped from wells.

Although most of us take for granted the water that comes out of the tap when we turn the faucet, this water comes to us through an elaborate and costly system. The system has an enormous impact on our environment. The following section will examine some of these impacts.

Water comes from an elaborate and costly system that has a tremendous impact on the environment. Understanding the system and its impacts may help us design more sustainable systems of water supply.

Impacts of Excessive Water Withdrawals As a rule, small amounts of water can be removed from water supplies without any noticeable impact. When demand becomes excessive, however, impacts start to become evident. One of the reasons for this is that much of the water removed from streams and aquifers to serve humans never returns to its source. For example, over 80% of the agricultural water applied to crops evaporates. Consequently, many rivers that provide irrigation water flow at a fraction of their natural rates during high-use seasons. Some are completely dried up. The Colorado River in the United States, for example, is siphoned off by farms and cities so that by the time it reaches the sea, it is reduced to a mere trickle. The Yellow River in northern China has been running dry since 1985, and each year the dry period becomes longer. In 1997, it failed to deliver any water to the sea for 226 days. Even the mighty Nile, which flows into the Mediterranean, rarely has any water left in it by the time it reaches the sea.

FIGURE 13-4 Water systems. This diagram shows the components of the water supply system of a city.

No saltwater intrusion

- Main supply well
- Water table
- Pond
- Coastal well
- Fresh ground water
- Ocean
- Salt groundwater

Saltwater intrusion

- Pond has less water
- Water table
- Fresh groundwater
- Ocean
- Salt groundwater

Further saltwater intrusion

- Pond is dry
- Water table
- Fresh groundwater
- Ocean
- Salt groundwater

FIGURE 13-5 Saltwater intrusion into groundwater. Removal of excess groundwater in coastal regions causes freshwater aquifers to shrink, allowing saltwater to penetrate deeper inland underground. Groundwater overdraft not only causes saltwater to enter the freshwater aquifer but also causes surface water in the pond to decline.

Reductions in water flow and depletion reduce populations of aquatic species and recreation activities. It can cause neighboring habitat, especially wetland, to disappear. The vast wetland at the mouth of the Colorado River, for example, has been reduced to a tiny remnant of its former self. Water that is returned to surface water supplies is often polluted with agricultural pesticides or wastes from homes and factories.

Overuse also causes aquifers to dry up completely, driving farmers out of business. Groundwater is a vast resource. Despite the enormity of groundwater, supplies are on the decline worldwide. Groundwater overdraft—the removal of groundwater at a rate faster than it is replenished—has severe long-term implications. In coastal regions, for example, groundwater overdraft can lead to **saltwater intrusion**, the movement of saltwater from marine aquifers into freshwater aquifers (**FIGURE 13-5**). Saltwater then contaminates previously freshwater wells, destroying this resource.

Ponds, bogs, and streams mark the intersection of aquifers and the land surface. Many think of a pond as "exposed groundwater." Because of this link, groundwater overdraft also drains swamps and ponds, at times drying them up completely. Fish, wildlife, and recreation are often adversely affected.

Excessive withdrawal of groundwater threatens the long-term prospects for irrigated agriculture in the United States (**FIGURE 13-6**). As the water table drops, farmers, cities, towns, and homeowners often drill deeper wells to ensure an adequate water supply. This is a costly measure and a short-term answer. As groundwater levels continue to fall, even deeper wells must be drilled. Excessive groundwater withdrawals are causing the water table to drop in many less developed nations, including parts of China and India, the two most populous countries in the world. In both LDCs and more developed countries (MDCs), irrigated farmland must be retired due to a back of water.

Groundwater fills pores in the soil and thus supports the ground above the aquifer; therefore, when the water is withdrawn, the soil compacts and sinks, a process called **subsidence**. The most dramatic examples of subsidence have occurred in Florida and other southern states, where groundwater depletion has created huge **sinkholes** that may measure 100 meters (330 feet) across and 50 meters (165 feet) deep (**FIGURE 13-7**). Subsidence has occurred over large areas in the San Joaquin Valley of California, damaging pipelines, railroads, highways, homes, factories, and canals. Southeast

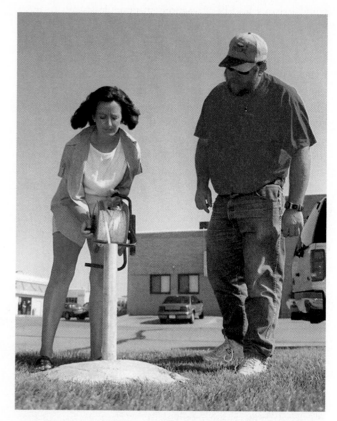

FIGURE 13-6 Declining groundwater. This teacher and his student are measuring the water levels in a well sunk into the Ogallala aquifer and are finding that water levels are dropping rapidly due to groundwater overdraft.

© Justin Wambold, *The North Platte Telegraph*/AP Photos.

of Phoenix, over 300 square kilometers (120 square miles) of land has subsided more than 2 meters (6 feet) because of groundwater overdraft. Huge cracks have formed, some 3 meters wide, 3 meters deep, and 300 meters long.

KEY CONCEPTS

Excessive withdrawal of surface water can cause streams to dry up with profound effects on fish and wildlife. Groundwater overdraft results in many problems, too, among them depletion of water levels in connected surface waters, subsidence, and loss of irrigation water for cropland.

FIGURE 13-7 One dramatic effect of groundwater overdraft. This large sinkhole in Florida developed quickly, swallowing a street. It was caused by depletion of groundwater, as explained in the text.

© AP Photos.

Impacts of Dams and Reservoirs Because water flow in streams and rivers varies with the seasons, dams are often constructed on them to retain water needed by cities, towns, farms, and industry. This ensures a steady supply year-round. In some cases, dams are built to generate electricity. Reservoirs behind dams provide recreational opportunities such as fishing, waterskiing, and boating. Dams help to control floods.

Despite their social and economic benefits, dams can have a profound impact on fish and wildlife. They flood streams and eliminate fish species that live and reproduce there. They interrupt migration. The once-plentiful salmon of the Pacific Northwest have been greatly reduced in number in large part because of dams that impair migration to the sea. Young salmon (called *smolts*) migrating to the ocean often lose their way in reservoirs. Why? Reservoirs eliminate current, which helps smolts orient. Thus, many young salmon never reach the sea, where they mature. Some salmon are killed in turbines that produce electricity. By one estimate, 8% to 15% of the young deep swimming smolts are killed at each dam.

Dams also interfere with the migration of salmon returning from the sea to spawning grounds. Salmon runs on many streams in the Pacific Northwest and on the East Coast have been decimated by dams built without the "ladders" that allow fish to bypass them. Even when ladders are built, there's no guarantee that fish populations will thrive.

Go Green

Turn the shower off while you soap up. Installing a water-efficient showerhead with a shut-off valve makes this easier.

Dams also flood the habitat of other species, such as bear and deer. Construction roads and noise also disturb wildlife.

Dams impact human activities and economies as well. New dams, for example, threaten recreational rivers, and they place towns that depend on income from rafting and kayaking in direct conflict with sprawling cities and suburbs that want the water.

In less developed countries, dams built for irrigation, hydropower, and drinking water often flood valuable farmland on which people have lived sustainably for hundreds of years. Pakistan's Tarbela Dam displaced 85,000 people. China's mammoth Three Gorges Dam and reservoir on the Chang River (formerly the Yangtze River), completed in 2009, displaced 1.4 million people in 20 towns. It destroyed nearly 100 square kilometers (40 square miles) and inundated about 41,000 hectares (100,000 acres) of farmland in the world's most heavily populated and agriculturally productive river valley. People displaced by dams and reservoirs must often move onto marginal land that deteriorates rapidly under heavy use.

Water withdrawals from reservoirs have many downstream impacts. For example, reduced water flows may allow saltwater to intrude into estuaries during high tide, upsetting the ecological balance in these important life zones. This is a serious problem in Everglades National Park in Florida. (See **Spotlight on Sustainable Development 13-1**.)

Reduced water flow also decreases the flow of nutrient-rich sediments to estuaries. This interrupts the natural flow of nutrients to floodplains, river deltas, and coastal waters. Rich farmland along riverbanks once replenished by periodic flooding must be abandoned or fertilized, often at a high cost.

Sediment that collects in reservoirs slowly fills them, making them useless. In the United States, well over 2,000 small reservoirs are totally clogged with sediment. Many larger reservoirs throughout the world will meet a similar fate. The $1.3 billion Tarbela Dam on the Indus River in Pakistan took 9 years to build, but because of upstream soil erosion, the reservoir could fill with sediment in 20 years. Lake Powell on the Colorado River will be filled in 100 to 300 years.

Dams generally release the coldest water from the bottom of the reservoir. This makes stream water exceedingly cold year-round and decreases the spawning of native species. The Glen Canyon Dam on the Colorado River, for instance, releases chilly water from the reservoir's bottom and has converted what was once a warm-water fishery to a rainbow trout fishery. The cold water now threatens several native species, such as the humpback chub (**FIGURE 13-8**). To avoid changing stream temperature, some existing dams have been retrofitted with devices that combine warm water from the reservoir's surface with cooler deep waters to ensure proper downstream water temperature.

Because up to 80% of the annual runoff of some streams may be withdrawn, wildlife habitat and good recreational sites may be lost. In addition, these projects usually transfer water from rural areas of low population density (often agricultural regions) to cities. This practice creates bitter conflicts between urban and rural residents.

Water withdrawals in reservoirs can also affect water quality. Diversion of clean, high mountain water in

13-1 Undoing the Damage to Florida's Kissimmee River

Between 1964 and 1970, the Army Corps of Engineers hacked away at the winding Kissimmee River in Florida, hauling up mud and dumping it along the banks. When they had finished this huge flood control project, a river that had once lazily meandered nearly 150 kilometers (100 miles) through the Florida marshes had been reduced to a canal 64 kilometers (40 miles) long, 60 meters (200 feet) wide, and 10 meters (30 feet) deep. The canal was designed to drain water quickly from the northern reaches of the watershed. On the heels of the huge dredgers that had converted the river into a canal came contractors who threw up concrete and earthen dams and locks every 16 kilometers (10 miles), creating huge reservoirs along the river's previous course. These were designed to control flooding downstream.

Once a rich habitat for bald eagles, deer, fish, waterfowl, and alligators, the Kissimmee River became a sterile tribute to our tireless efforts to control flooding. Most scientists condemned the channelization as a major environmental catastrophe, which destroyed three-fourths of the original 16,000 hectares (40,000 acres) of marsh, once a major habitat for dozens of species of waterbirds. Secondary canals built by landowners along the main canal drained another 80,000 hectares (200,000 acres). Soon after the canal's completion, the vast flocks of ducks that once rained down from the skies were gone. Gone, too, were the wading birds. By Florida Game and Fresh Water Fish Commission estimates, 90% of the waterfowl and 75% of the bald eagles vanished from the region, as did the largemouth bass that once attracted anglers from all over the nation.

Two years after this enormous project had been completed, Florida biologists began noticing changes in Lake Okeechobee (pronounced OAK-eh-CHO-bee), into which the Kissimmee's clean waters once flowed. Dead fish and dying vegetation were the most blatant signs that something was awry in the lake. It didn't take biologists long to determine that the loss of marshlands, which purify waters and hold back sediment, was the reason for Lake Okeechobee's sudden deterioration. The loss of the natural cleansing provided by wetlands—plus a heavy load of pesticides, fertilizer, animal wastes, and sediment from cattle ranches and farms that sprang up along the river's banks—created a

monumental water quality problem for the lake, which provides drinking water for Miami and coastal cities.

The waters of the Kissimmee River (which flow south) once fanned out across southern Florida to nourish the huge, multimillion-hectare wetlands known as the Everglades. On the southernmost tip of Florida is Everglades National Park. To make room for farms, much of the Everglades has been drained, and the water from Lake Okeechobee and the Kissimmee River basin has been shunted via canals to the coast. Reduced flows have disrupted the ecology of the Everglades, seriously threatening many species, some already endangered. Reduced water flows also produced an ironic backlash. Because of a lowered water table, farmland once prized because of its rich soil began to sink at a rate that could hinder farming in the region. Lower water flows have also resulted in saltwater intrusion into surface water and groundwater.

Ironically, studies made after the canal was completed indicate that it provides little or none of the expected flood control. Making matters worse, the canal is now seen as a major threat to downstream areas. After heavy rains in central Florida, for instance, a slug of water travels rapidly southward along the canal, wiping out nesting waterfowl and drowning unsuspecting wildlife.

Less than 2 years after the Army Corps of Engineers had trucked in the last load of cement, a special governor's conference committee released a report calling on the state to reflood the marshes that it had just drained. The report concluded that channelizing the river had been a big mistake. With a price tag of $30 million, it had also been a costly one. Even the Army Corps of Engineers commissioned a study to reevaluate the project and prepare recommendations for returning the river to its original state.

In 1983, Governor Robert Graham and supporters took steps to reverse the damage. In 1984, the Kissimmee River restoration began. However, because no federal funds were available to reclaim the river, funding had to come from another source, property taxes on residents in southern Florida. Financed by property taxes, the South Florida Water Management District (which was put in charge of the project) built three small dams to divert water from the canal back into the old river channel (**FIGURE 1**). This flooded the marshes along 20 kilometers (12 miles) of the river and cost $1.5 million.

These steps helped wetland vegetation, waterfowl, and fish populations recover, but full recovery will require other measures—especially steps to restore more normal water flows. In 1992, the U.S. Congress approved legislation that began a project to restore over 40 miles of river and 40 square miles of associated wetlands and flood plains. This was to be accomplished by filling in 35 kilometers (22 miles) of the 90 kilometer (56 mile) long canal and removing two of the five water control structures. The water district also purchased land along the river. Their goal was not to optimize one or a few valuable species, such as bass, but to restore the ecosystems badly damaged by channelization. They also worked upstream to help prevent runoff from farms and dairies. Restoring the Kissimmee River is part of a major ecological experiment aimed at saving Florida's fast-vanishing wetlands. According to the Florida Center for Environmental Studies, "Flora and fauna that disappeared when the system was channelized have returned in great numbers and are now thriving in the newly restored system." They add, "To date there have been three back-filling phases completed, one water control structure has been removed, and continuous water flow has been reestablished to 24 miles of the historic meandering Kissimmee River. The final stage of the project…began in December 2009 with the River Acres Flood Reduction Project and is scheduled to be completed by 2015."

Someday, the complex wetlands of Florida will function more like they did a hundred years ago. But at this very moment, engineers and construction companies are hard at work draining wetlands the world over. One of the largest lies along the Nile River. In the home of countless birds and wildlife, huge dredgers are now busily sucking up the mud and straightening the channel.

FIGURE 1 Small dam on the channelized portion of the Kissimmee River in Florida helps to restore wetlands by diverting water back into the original stream bed (left).
Courtesy South Florida Water Management District.

FIGURE 13-8 The humpback chub. This is one of several fish species of the Colorado River threatened by dams that release cold water into what was once a fairly warm-water fishery.
Courtesy of John N. Rinne, Rocky Mountain Research Station.

tributaries of the Colorado River, for instance, diminishes the flow and quality of the river, a source of water for more than 30 million people. The lower sections of the river now carry a heavy burden of sediment and dissolved salts because of the loss of clean upstream water. By the time the river reaches Mexico, its salt concentration is over 800 parts per million (violating a U.S.–Mexico treaty), compared with 40 ppm at its headwaters. Why is this of concern?

Water with a salt concentration over 700 parts per million (ppm) is toxic to many plants.[1] It cannot be used for agriculture unless it is diluted or desalinated (has the salt removed). Drinking water must have a salt concentration of 500 ppm or less. A bill passed by the U.S. Congress in 1973 authorized the construction of three desalination plants along the Colorado River to remove over 400,000 tons of salt from the river each year. Only one plant was

---

[1] For comparison, saltwater in the ocean has a concentration of 35 parts per *thousand*.

built. This facility in Yuma, Arizona, cost $250 million to build and costs about $15 to $20 million per year to operate.

Sediment and dissolved solids are a problem in agricultural runoff in Canada, too. In the South Saskatchewan River basin, irrigation water returning to streams contains levels of these pollutants that are sometimes twice what they were in the original source. They're also polluted by a host of other contaminants such as pesticides and nitrates.

The diversion of water can also have impacts on people and local economies. For example, excessive water withdrawal from two rivers that once fed the Aral Sea in the former Soviet Union has completely dried up one river and reduced the other to a mere trickle. The sea has dropped 12 meters (40 feet) in 30 years. Two-thirds of its water has evaporated now that it is no longer replenished. Towns once situated on the shore are now 48 kilometers (30 miles) inland. Because freshwater inflow has all but stopped, the concentration of salts in the sea has risen, killing off 20 of the 24 commercial fish species in the sea. Dozens of fishing towns have been abandoned (**FIGURE 13-9**).

KEY CONCEPTS

Dams and reservoirs have many benefits, such as flood control and water storage, but they also have profound impacts on people, wildlife, and the economy. They can eliminate native fishes, block fish migration routes, flood wildlife habitat and farmland, displace people, destroy recreational opportunities, reduce nutrient flow to estuaries, and alter water quality in streams. Ultimately, all dams have a finite life span because their reservoirs eventually fill with sediment.

FIGURE 13-9 The Aral Sea. This former inland sea in the former Soviet Union has been reduced by massive withdrawal of water to irrigate farmland. Most species of commercial fish have disappeared, and fishing villages have been abandoned. These fishing vessels and the fishing village have been left high and dry.
© ITAR-TASS/Sovfoto.

Creating a Sustainable Water Supply System

Fortunately there are numerous ways to make the present system of water supply and use sustainable. They require changes in both habits and technologies—most notably water-saving technologies. Many of these technologies are inexpensive and easy to install, and they provide enormous social, economic, and environmental benefits. The operating principles of sustainability are a useful guideline for revamping and rebuilding the water supply system.

KEY CONCEPTS

To avert future water shortages, we need a sustainable water policy and management strategy based on four of the operating principles of sustainability: conservation, recycling, restoration, and population control.

Population Stabilization Given the limited amount of freshwater available to humanity and the importance of ground and surface waters, creating a sustainable system of water supply will require measures to stabilize and perhaps stop the growth of the human population. Especially important are measures to control growth in water-short areas, such as the desert Southwest of the United States, northern Africa, the Middle East, and parts of the Indian subcontinent. Special attention must be paid to the placement of new housing and other development so that it is not built over aquifer recharge zones. See **Spotlight on Sustainable Development 13-2** for a stellar example of ecologically sound development.

KEY CONCEPTS

Efforts to slow and perhaps stop the growth of the human population are vital to living within the planet's capacity to supply freshwater.

Water Conservation: Using What We Need and Using It Efficiently For years, water shortages were addressed by expanding supply: damming rivers and streams and pumping water out of lakes or aquifers. As the cost of these options rose and concern over the environment increased, however, more and more attention has been focused on ways to use water more efficiently. One of the main benefits of water conservation is that it involves measures that can be brought online quickly and at costs well below those incurred by building new dams and reservoirs. Conservation measures also offer numerous environmental benefits.

Because agriculture and industry are the biggest water users in all countries, efforts to conserve water in these two sectors should be given the highest priority. Substantial

SPOTLIGHT on Sustainable Development

13-2 Ecological Solutions to Flooding and Water Supply Problems

What do The Woodlands, Texas, and Boston, Massachusetts, have in common, besides the distinguishing accents of their people? The answer is that both places have made important decisions in land-use planning that will save millions of dollars by reducing flood damage.

Boston boasts a fine city park system that stretches from the center of the city into the outlying suburbs. Few realize that the park system, with its meandering streams, was built in large part to help control flooding (**FIGURE 1**). After heavy rains, excess water flows into this basin, where it is slowly released to the sea, reducing flooding and property damage.

The city also purchased large wetlands in an outlying area, rather than letting developers drain and build on them. The wetlands were set aside to reduce flooding and also to provide valuable habitat for fish and wildlife. The project costs one-tenth as much as a dam to control flooding.

In 1971, landscape architect Ian McHarg wrote a landmark book entitled *Design with Nature*. In it, he suggests that builders consider wildlife needs, natural flooding, soil stability, and a half dozen other factors when constructing homes. By careful site analysis and design with nature, builders lessen their impact on the land, air, and water.

Texas developer George Mitchell decided to build a new town called The Woodlands using McHarg's ideas. On his forested tract north of Houston, Mitchell envisioned a city in harmony with the forces of nature. He and his staff of planners first analyzed the region and found that they could leave the natural drainage system as open space, which would carry water away more effectively and more cheaply than a storm sewer system. That step alone saved $14 million in construction costs. Roads were built on high ground, and buildings were restricted from aquifer recharge zones, thus protecting the groundwater that supplies Houston. In 1979, rainwaters drenched the site. The streams swelled by 55%. In neighboring towns built with little regard for nature, water flows increased 180%. Those towns suffered considerable flood damage, whereas The Woodlands managed well.

The Woodlands is an attractive community (**FIGURE 2**). Most of its trees still stand. The floodplains that were set aside for natural drainage and aquifer recharge harbor numerous birds and mammals, including bobcats and white-tailed deer. This community stands as a testament to the benefits of designing with nature. This approach to development permits nature to direct the design of human settlement and helps people to live in harmony with it—so important in building a sustainable future.

FIGURE 1 Boston, Massachaselfs. Aerial view of the park system of Boston, showing flood control aspects.

© Alex S. MacLean/The Picture Cube/Photolibrary.

FIGURE 2 The Woodlands, Texas. Note the large amount of wooded area surrounding the houses.

Courtesy of Dan Chiras.

gains in agricultural water efficiency can be achieved by lining the open, dirt-lined ditches that deliver water to crops (and that lose up to 50% of their water) with concrete or plastic (**FIGURE 13-10**). Lined ditches are 80 to 90% efficient. Pipes are even more efficient. Sprinklers, which waste up to half of their water, can be replaced with drip irrigation systems, which lose only 5%. However, drip irrigation can be used for only a limited number of crops, such as fruit trees, grapevines, and some vegetables. For the vast fields of wheat and corn, more conventional methods must be used. Modifications in sprinkler systems for these crops can cut water losses by 30 to 50%. Farmers can irrigate at night, early in the morning, or late in the day, when evaporation is lowest. In warmer climates, however, nighttime irrigation promotes fungal growth that reduces productivity, so this strategy won't work everywhere.

For many years, there has been little incentive to conserve water on farms in many countries, including the United States, because water projects were subsidized by taxpayers. In the words of one observer, the public often subsidizes the private use of water to grow taxpayer-subsidized crops. In the San Joaquin Valley of California, for example, irrigation water costs farmers about $5 per acre-foot (an acre of water one foot deep) thanks to publicly funded projects. Water is also cheap because the economic costs of the damage created by water

Go Green

Don't brush your teeth or shave with the water running.

projects (for example, the loss of fish) are not included in the cost. California's once-lucrative salmon fishing industry has collapsed, in large part, because of dams. Subsidized water projects are not unique to more developed nations. In less developed countries 80 to 90% of the cost of irrigation projects is subsidized by either the government or more developed countries.

FIGURE 13-10 Lined irrigation ditch. Concrete-lined ditches reduce water losses by 50%.

© Deyan Georgiev/Thinkstock.

Water conservation today makes great sense. According to the nonprofit group Environmental Defense Fund, recycling water and scheduling water application according to crop and soil needs costs about $10 per acre-foot. Switching from irrigated crops, such as cotton, to dryland crops costs farmers about $40 per acre-foot in lost crop yield. Water conservation by more efficient drip and sprinkler systems costs about $175 per acre-foot. Water from new dams, however, costs up to $500 per acre-foot.

By investing in water conservation, farmers can cut water demand and save money. They can also sell excess water to willing buyers. California cities and industries, for instance, currently purchase water for about $200 per acre-foot per year. If a farmer could invest $50 per acre-foot to save water and sell the excess for $200, he would reap a nice profit. Because farmers would be making a profit on publicly subsidized water, some critics say that it seems only fair that the public be reimbursed as well; farmers should not be allowed to reap the full profit of a publicly subsidized resource. No matter what the outcome of the debate, water conservation makes good sense in the long run as well as the short run. Worldwide, a 10% decrease in agricultural water use could double the amount of water available for cities and industries.

In southern California, the Metropolitan Water District pays for water conservation measures in the nearby Imperial Irrigation District. By lining irrigation canals and other measures, farmers free up substantial amounts of water for use in Los Angeles, San Diego, and neighboring cities—at a cost far below that of new dams and diversion projects.

Many sustainable farming practices reduce the demand for water. Increasing the organic content of the soil, for example, reduces water requirements because organic matter retains soil water. In addition, measures that reduce surface runoff (contour farming, terracing, and strip cropping) or that reduce wind blowing across farm fields (shelterbelts) result in a substantial reduction in irrigation water demands. In India, farmers plant rows of grass along the contours of hilly terrain to retard surface runoff, reduce soil erosion, and increase soil moisture content. Yields can increase by 50%. Such simple measures can be extremely cost-effective in large regions, making costly irrigation dams unnecessary. Shifting water-intensive crops such as cotton from regions of low moisture (California) to areas of higher precipitation (Alabama) could also result in substantial water savings.

Water conservation by industry can be achieved by redesigning existing processes and facilities to make them more efficient. The use of water by water-cooled electric power plants, one of the most water-consumptive industries in the world, could be cut one-fourth by using dry cooling towers (**FIGURE 13-11**), although these require more energy and are more expensive to operate than wet cooling towers.

Improvements can be made in cities and towns as well. Municipal water systems throughout the United States lose on average 12% of their water. Replacing leaky pipes can reduce water loss enormously, saving streams and the wildlife that depend on them. Such measures also provide important employment opportunities.

FIGURE 13-11 Wet and dry cooling towers. Water from electric power plants is cooled before being reused. Dry towers cost more to operate but use less water.

You and I can do many things to cut down on water use. One very simple and cost-effective solution is the high-efficiency showerhead, which uses about 2 to 2.5 gallons per minute. Using 70% less water than a traditional showerhead, the new models also cut down on energy demand and can save a family of four $200 to $250 a year in energy bills—all for a $12 to $20 investment in a new showerhead and 5 minutes of installation time.

Individual savings can add up. The installation of water-efficient showerheads in 1,300 homes would save 22,000 liters (85,000 gallons) of water a day, or about 8 million liters (31 million gallons) a year! Imagine the savings if everyone in the United States used one.

KEY CONCEPTS

Water conservation is a relatively fast and highly cost-effective means of meeting demands for water. Some of the largest gains can be made in agriculture and industry, although domestic water conservation measures can also save enormous amounts of water.

Restoring Watersheds and Wetlands to Protect and Enhance Our Water Supplies Slowing, even stopping population growth, and water conservation are the cornerstones of a sustainable water system. Other methods can also help us maintain, even expand, our water supply while protecting the environment. One of those is the restoration of vegetation in watersheds—replanting forests and grasslands, for instance. Revegetation reduces sedimentation in streams and reservoirs, extending their lifetime. Reductions in sediment load not only protect reservoirs, they reduce costs for water purification—providing an economic benefit—and protect fish and aquatic organisms—an ecological benefit. Revegetation also helps reduce surface runoff and thus enhances the infiltration of water into the soil, increasing groundwater supplies. Restoring wetlands increases natural filtering.

Another benefit of revegetation of denuded landscapes is that it may help restore normal rainfall patterns in areas where they've been disrupted by deforestation. In such regions, streams that have dried up may once again flow with clean, potable water, benefitting people and wildlife alike.

Preventive measures are also needed. Much greater care should be taken by lumber companies, farmers, and ranchers. Measures to protect streams during logging operations, for instance, help to maintain water quality and quantity. Erosion controls and sustainable rangeland practices reduce sedimentation and protect vital water supplies. These measures also encourage groundwater recharge.

KEY CONCEPTS

Restoring and protecting watersheds help to prevent siltation in reservoirs, enhance groundwater recharge, and can help restore rainfall.

Water Recycling In nature, water is recycled over and over in large part because of evaporation and precipitation. In the process, it is also purified and readied for reuse. Humans can also recycle water. In fact, in the coming decades, water recycling may become an integral part of a sustainable system of water supply. Wastewater from factories, for example, can be purified and returned to productive use instead of being released into surface waters.

Elaborate biological, chemical, and mechanical filtration systems can remove a wide assortment of wastes, from toxic metals such as lead to toxic organic compounds to acids.

In Tokyo, for example, Mitsubishi's 60-story office building has a fully automated recycling system that purifies all of the building's wastewater to drinking-water purity. Cities could also convert their existing sewage treatment plants to water recycling facilities. Sewage treatment plants remove many impurities from wastewater from homes, offices, and factories, but the final product in most plants, which is released in surface waters, is hardly potable (drinkable). However, sewage treatment plants can be designed to purify water so that it is safe to drink.

Another form of recycling, which lets nature do the work, is the release of wastewater onto **aquifer recharge zones**, regions where rain and snowmelt replenish aquifers. Nutrients in the wastewater are taken up by plants and degraded by bacteria in the soil so that as wastewater percolates through the soil it is cleansed. It then enters the aquifer in a clean state and can be reused. Effluents from sewage treatment plants, irrigation runoff, storm water runoff, and cooling water from industry could all be used to replenish water supplies in existing aquifers. Although aquifer recharge can help increase groundwater supplies, it is costly because many cities are far from aquifer recharge zones.

Water recycling not only helps reduce our demand on freshwater supplies, it also eliminates discharge into streams and thus helps cut water pollution—also vital to a sustainable future. Although building and maintaining large-scale recycling plants can be expensive, they are often environmentally and economically cheaper than new water diversion projects and dams. Many relatively inexpensive, low-tech, biological solutions are also available. In rural settings, for example, sewage from individual homes can be piped into artificial wetlands. The wetlands clean the water and make it usable for flushing toilets or watering gardens or fruit trees.

In some cases, it may be advantageous not to fully purify water. Nutrient-rich urban wastewater, for example, can be applied to pastures, cropland, golf courses, and college lawns. This practice returns valuable nutrients to the soil and reduces pollution of surface waters. In Colorado Springs, partially treated wastewater is used to irrigate lawns at the prestigious Colorado College. In Israel, where freshwater is limited, large amounts of municipal wastewater is used to irrigate cotton.

KEY CONCEPTS

Water can be purified and reused in industry, on farms, and in our own homes. Water recycling reduces pressure on surface water and groundwater supplies and reduces water pollution. Wastewater can be used to recharge groundwater supplies or to irrigate fields, golf courses, and parks.

Changing Government Policies to Create a Sustainable Water Supply System Revamping government policy can have a substantial impact on efforts to create a sustainable water supply system. One of the most important policy changes is economic: eliminating subsidized water projects (described earlier). Many observers argue that federal and state governments should require those who benefit from water projects to pay their full costs. Making users pay the full cost might encourage them to focus on least cost options, particularly conservation. The cost of a water project should also include the costs of correcting environmental problems, a step that would make conservation and recycling even more attractive.

Governments also establish building codes that include requirements for the type of plumbing fixtures in commercial and residential settings. Revised building codes that favor conservation by requiring water-conserving toilets and fixtures can promote substantial reductions in domestic and commercial water consumption in new and remodeled buildings. Mandating water-efficient toilets and showerheads as well as water-efficient lawns could cut water use in new homes by half. Today, as a result of federal regulation that went into effect in 1996, all states require that toilets installed in new and remodeled homes use no more than 5.8 liters (1.6 gallons) of water per flush—about one-half as much as current models and one-third as much water as toilets in older homes. Toilets use 30% to 40% of the indoor water consumed by a family, so water-conserving models can generate significant savings.

Considerable efforts are needed in arid regions to replace water-thirsty grasses in lawns with water-thrifty species of grass, trees, and shrubs. In Marin County in northern California, for example, one-third of the county's water supply is used to irrigate lawns. One town's government offers sizable subsidies to homeowners who **xeriscape** (ZER-eh-scape) their lawns—that is, remove water-intensive vegetation and replace it with water-conserving species (**FIGURE 13-12**). Each converted lawn saves about 460 liters (120 gallons) of water per day during peak months.

FIGURE 13-12 Xeriscaping. This lawn in Denver, Colorado, is composed of plants that require very little water, unlike conventional lawns in the area.

© David Winger/Alamy.

Smaller lawns planted with low-water grasses and water-conserving drip and root-zone irrigation systems can be encouraged. Government programs can also encourage the installation of water-conserving measures in existing homes. Some water departments, for example, now offer free water audits, free water-efficient showerheads, and sizable rebates to homeowners who replace existing toilets with low-water models.

Another highly overlooked but potentially important way to help expand water supplies is the **rain catchment system**, also known as a **catchwater system**. Water and snowmelt collected from a roof can be stored in large tanks and used to irrigate gardens and lawns. Passed through filters, this water is even clean enough to drink.

Gray water systems could also play a major role in water-short areas. Gray water is water from showers, sinks, and washing machines. It constitutes about 80% of the wastewater from a home. Gray water is suitable for lawn and garden irrigation. It can also be purified and reused to flush toilets (**FIGURE 13-13**). However, gray water requires extra plumbing because it must be kept separate from **black water** (from toilets). California recently adopted the United States' first gray water codes. Other states will soon follow.

Governments can control water prices, too. In most locations, the more water one uses, the cheaper it is. This, of course, does not encourage water conservation. If users are charged more, not less, for water over a certain sustainable level, water users often respond favorably, cutting back on demand.

Finally, municipal water agencies could adjust water rates by time of day, charging users more for water used when evaporation is greatest and less when evaporation is low. This incentive could help cut down on water evaporation from lawns and save huge amounts of water. (In warmer, moist climates, however, nighttime lawn irrigation may enhance fungal growth that damages lawns.)

KEY CONCEPTS

Many government policies, especially subsidies, contribute to the unsustainable use of water. Changes in these policies can greatly increase the efficiency of water use and promote more sustainable water supply strategies, such as conservation, rain catchment, recycling, gray water systems, abolishment of volume discounts, and implementation of time-of-use payments.

FIGURE 13-13 Gray water recovery. Gray water from sinks, showers, and washing machines can be collected by a separate set of pipes and reused for a variety of purposes. Gray water systems such as this one help to reduce household water demand.

Education: Learning to Use Water Wisely The battle to ensure an adequate water supply can be aided by massive public education. Teaching adults and children throughout the world ways to use this precious resource efficiently may be as important as the math or history they learn. A study of water conservation programs in the western United States, found that water conservation programs in cities are most successful if changes in pricing policies (which increase the cost of water the more someone uses it) are complemented by educational programs.

KEY CONCEPTS

Education is a key to creating individual responsibility and sustainable water-use patterns among citizens.

Tapping New Sources of Water: Dams and Desalination Plants The measures outlined previously represent the first line of attack. They are generally economical and easy to effect, and many are good for the environment, too. Conservation, for example, reduces environmental impact while supplying us with water at a lower cost than most other options. But there may be times when these measures won't provide enough water. Further supplies may be required. This means more dams and diversion projects or drilling more wells into groundwater.

Because the economic and environmental costs of dams and reservoirs are substantial, discretion is advised when undertaking such projects. In less developed nations, small-scale irrigation projects combined with efforts to use water more efficiently can eliminate the need for large and costly dams. Smaller projects help build local self-reliance so vital to sustainability.

Another way to increase water supply in communities near the ocean is by **desalination**, removing salts from seawater. Through various methods, seawater can be purified for drinking and irrigation.

Because 97% of the water on Earth is in the oceans, desalination might seem like the best answer to water shortages. Unfortunately, water produced by desalination has been 4 to 10 times more expensive than water from conventional sources. Recently, however, new technologies have greatly reduced the cost of desalination. Although the world's desalination capacity has increased dramatically, desalination produces only a tiny proportion of the freshwater consumed by humans. In the United States, desalination plants produce less than 1% of our daily freshwater requirement.

Desalination plants expand the Earth's carrying capacity, with potentially serious ecological impacts. Population growth in the Florida Keys, permitted in part by a new desalination plant, threatens coral reefs and wildlife such as the crocodile. Construction of houses and condominiums causes erosion that pollutes coastal waters. New residents produce an increasing amount of sewage and other pollutants that decrease water quality.

KEY CONCEPTS

As a last resort, rising demand can be met by developing new water supplies. New dams, diversion projects, and wells to tap into groundwater are options, though their costs are enormous. Desalination plants are a likely candidate for coastal communities but they are costly to operate.

13.3 Flooding: Problems and Solutions

Ironically, after shortages, the next major water problem encountered in many nations is flooding. Despite years of flood control work, floods cause billions of dollars worth of damage each year in the United States and throughout the rest of the world. Moreover, flooding appears to be on the rise in many places because of shifting rainfall patterns and more violent and more frequent storms that many atmospheric scientists believe result in part from global climate change and alteration of natural systems by human activities that reduce vegetative cover that increases surface runoff and, hence, flooding.

KEY CONCEPTS

Flooding is a major problem in many areas of the world and appears to be on the rise, in part as a result of climate change and alteration of the land such are deforestation.

Causes of Flooding

A simple correlation can be drawn between floods and their apparent causes: heavy rainfall and snowfall. The Red River in Canada, for example, has a long history of natural flooding. The river, which flows from south to north, often floods when winter snows in the southern reaches of the watershed melt before ice in the northern portion of the basin melts. Ice forms a natural dam. Closer examination, however, reveals many factors that go unnoticed. **FIGURE 13-14** lists factors that affect percolation, that is, water seepage into the ground, and runoff, water that flows over the surface of the land. All in all, precipitation that does not evaporate must either run off or percolate into the soil. Whether it flows across the land and empties into rivers or sinks quietly into the soil to become groundwater is largely a matter of the surface features, especially the vegetative cover. For example, a well-vegetated landscape covered by forests and grasses reduces surface runoff and promotes percolation into the soil. Put another way, heavily vegetated watersheds act as sponges. On the other hand, sparsely vegetated areas (for instance, deserts, overgrazed pastures, or clear-cuts) experience increased surface runoff and, hence, flooding.

The fate of rainfall in many cases rests not so much in the hands of nature as in the hands of farmers, urban planners,

Factors that increase percolation
- Slow rate of rainfall
- Small amount of rain
- Porous surface
- Flat land
- Loose soil
- Abundant vegetation: forests, marshes, grasses

Left side of watershed:
75% percolation
25% runoff

Right side of watershed:
25% percolation
75% runoff

Factors that increase runoff
- Rapid rate of rainfall
- Large amount of rain
- Nonporous surface
- Steep slope
- Compacted soil
- Lack of vegetation
- Clearing (clear-cuts), deserts

FIGURE 13-14 Percolation–runoff ratio. Numerous factors increase and decrease surface runoff and absorption (percolation).

developers, ranchers, and logging companies, who often strip vegetation from the land. This practice increases surface runoff. Water flowing rapidly over the surface of denuded land flows into streams and often spills over their banks, flooding nearby areas. Such water also often transports a substantial amount of soil from the land. Eroded sediment fills rivers and lakes and reduces their holding capacity. This makes flooding more likely, even after moderate rainfall.

Flooding in urban areas often results from highways, airports, shopping centers, office buildings, and homes, which greatly increase the amount of impermeable surface. Instead of soaking into the ground, rainwater washes off the surface in torrents.

Destruction of wetlands, swamps along rivers, is also responsible for increased flooding. Wetlands act as sponges, too, holding water and releasing it slowly into nearby waterways. Their loss eliminates this benefit and is thus partly responsible for the growing severity of floods in many countries.

FIGURE 13-15 shows the regions of the United States that are susceptible to flooding. Many of these areas are located along the Mississippi River and its tributaries. Several large rivers in Canada (not shown on the map) also experience periodic flooding, among them the Red River (in Manitoba), the Oldman and South Saskatchewan Rivers (in Southern Alberta), and the Saguenay River (in Quebec). The **floodplains** of these rivers, regions along the flanks of rivers naturally subject to flooding, are popular sites for cities, towns, and farms. Because cities and towns are often located along the banks of rivers, and because watersheds are often denuded, and storms tend to be growing in intensity, humankind can expect a long future of flooded basements, property damage, crop damage, and lost lives.

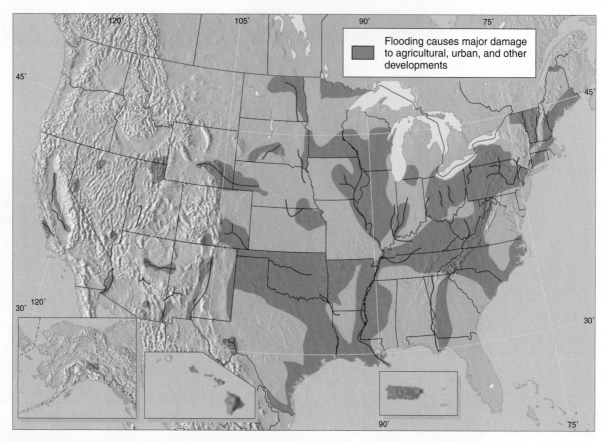

FIGURE 13-15 Flood-prone areas in the United States. Boundaries of the states are shown in black.

Floods result from natural variations in rain—but also from changes in the environment caused by humans, notably climate change and reduction in vegetative cover that increase runoff.

Controlling Flooding

Flood prevention measures in most countries have typically taken the form of dams and levees. **Levees** are embankments constructed along the banks of rivers to hold floodwaters back. As is so often the case, dams and levees are stopgap measures: They often treat the symptoms of a serious and costly problem, not the underlying causes.

Another technique for reducing flooding that is falling out of favor is **streambed channelization**, which is a kind of streamlining of rivers and streams, carried out by cutting or bulldozing vegetation along a stream's banks and deepening and straightening river channels. This creates a glorified ditch that may be lined with concrete or rock.

Streambed channelization in the United States was begun seriously in 1954, when the **Watershed Protection and Flood Prevention Act** authorized the Soil Conservation Service to drain wetlands along rivers to make more farmland and to reduce flooding. Between 1940 and

1971, 35,000 kilometers (22,000 miles) of U.S. streams were channelized by the Army Corps of Engineers and the Soil Conservation Service (according to American Rivers, a national river protection organization). More recent data are not available from any of the major players in this arena, although channelization is still occurring. Critics argue that many projects are of dubious merit and should not be undertaken. Why?

Streambed channelization generally eases flooding in the immediate vicinity, but this minor benefit is frequently outweighed by habitat destruction (loss of streambank vegetation), increased erosion of stream banks, loss of river recreation, and, most importantly, increased flooding downstream. (For more on this, see Spotlight on Sustainable Development 13-2.) Like dams and levees, streambed channelization is an unsustainable approach to a serious problem. What can be done to reduce flooding?

One of the most promising and sustainable approaches used to reduce flooding is sustainable **watershed management**. Sustainable watershed management includes steps to reduce deforestation and overgrazing. As you just learned, vegetation within a watershed increases the percolation of water into the ground, thus greatly reducing surface runoff and helping to control floods. This strategy also ensures more groundwater. A revegetated landscape also has streams that flow year-round, rather than drying up in dry months, as water is released from soils in a more controlled fashion.

Sustainable watershed management also includes measures to protect and restore wetlands in watersheds. Wetlands act as giant sponges, reducing surface flows and flooding. In fact, the devastating floods along the Mississippi River in 1993, which caused billions of dollars worth of damage to crops and homes, could have been averted had a relatively small percentage of the natural wetlands in the river's massive watershed been left intact. (Wetlands are discussed in more detail later in the chapter.) Wetlands along rivers absorb water and help reduce flooding.

Two general approaches can be taken to protect existing watersheds. Better regulating development in watersheds is vital. Mining, farming, and housing development, for example, can be regulated to ensure minimal runoff. Land can be developed in ways that protect forests and meadows needed to absorb storm water. Second, in areas where the landscape has been altered by human activities, efforts can be made to restore vegetation. Replanting trees, shrubs, and grasses on denuded hillsides can greatly reduce flooding, while also restoring wildlife habitat and making a more aesthetically appealing environment.

Watershed management also includes measures to redesign urban environments to slow down the rush of water during rainstorms. One effective means of achieving this important goal is the construction of holding ponds that absorb storm water runoff (**FIGURE 13-16**). Lakes in city parks can be situated in such a way as to trap much of the surface flow in a rainstorm. This water can then be released slowly to rivers to prevent flooding, or it could be used to water lawns or recharge aquifers. Holding ponds can also be developed near large parking lots or shopping malls. Some developers are even developing holding ponds under parking lots to absorb runoff and slowly release it into nearby streams. They use a special high-strength porous plastic called Rainstone to create a chamber to temporarily store water after rainstorms (**FIGURE 13-17a**).

Storm sewers can also be designed so as to divert water to special holding ponds or water tanks that serve factories. Individual homeowners can divert gutter water to underground holding tanks to be used later for watering lawns and gardens and for washing cars. (For examples of ways to avoid flooding, see Spotlight on Sustainable Development 13-2.) Finally, floodplain zoning laws can help restrict building and other human activities in flood-prone areas. Driveways and walkways can be constructed of materials called porous pavers that permit water to soak into the ground, thus reducing surface runoff (Figure 13-17b).

As in other areas, efforts to slow or even stop the growth of the human population are vital to solving this growing problem. More people mean more habitat destruction and more flooding. Zoning regulations can be helpful as well, restricting development in floodplains and sensitive areas of watersheds where home construction and roadways are ill-advised.

(a)

(b)

FIGURE 13-17 (a) Workers install Rainstone, a porous plastic product that holds rain runoff from a nearby parking lot, slowly releasing it into groundwater or a nearby stream to prevent flooding. **(b)** Porous pavers allow snow melt and rain to percolate into the ground, reducing surface runoff.

Photo a courtesy of Invisible Structures, Inc. (www.invisiblestructures.com); photo b © TMI/Alamy.

FIGURE 13-16 Water retention pond. This pond receives rainwater from a subdivision. The water is flowly released into a nearby stream.

Courtesy of J. Keizers.

Chapter 13: Water Resources: Preserving Our Liquid Assets and Protecting Aquatic Ecosystems   **289**

After years of devastating floods, some cities and towns are taking a new approach: They're "depopulating" river banks. After repeated devastating floods, many cities and towns are relocating to higher ground, out of harm's way. The federal government, which has spent billions of dollars rebuilding flooded towns through flood insurance programs and emergency aid, is helping in these new ventures, secure in the notion that aggressive buyouts of flood-prone land and relocation of homes saves lots of money in the long term. The Federal Emergency Management Agency, which assists after floods, and the U.S. Army Corps of Engineers, a long-time flood fighter and dam builder, have both engaged in an aggressive program to buy property from owners who have filed multiple disaster claims and return the lands to natural uses. The depopulated areas are cleared of human structures and returned to nature, creating ecological and recreation benefits.

Flood control can also be achieved by reducing our dependence on fossil fuels and curtailing other activities that produce greenhouse gases, which are believed to be changing rainfall and snowfall patterns that may be leading to more frequent and severe flooding throughout the world.

FIGURE 13-18 The Chesapeake Bay receives water containing pollutants from its large watershed. Pollution has damaged ecologically and economically important life forms including fish and shellfish.
Courtesy of GSFC/NASA.

KEY CONCEPTS

Many solutions to flooding—including dams, levees, and streambed channelization—are unsustainable; they treat the symptoms, not the root causes of flooding. Preventive measures—controlling population growth, protecting watersheds, and reducing global climate change—combined with restorative measures such as replanting trees and restoring wetlands, are far more effective in the long run.

13.4 Wetlands, Estuaries, Coastlines, and Rivers

From certain vantage points, Chesapeake Bay on the eastern seaboard of the United States resembles a vast ocean. It is not. It is an estuary, a region where freshwater from inland streams meets the ocean (**FIGURE 13-18**). For years Americans have treated this estuary and its vast wetlands with considerable disrespect. As a result, the rich abundance of organisms is diminishing, threatened by pollution, overfishing, and other activities.

The bay and its surrounding wetlands are home to a variety of economically important fish and shellfish, including blue crabs, oysters, and striped bass. Properly managed, the bay could provide enough food for nearly half the U.S. population. Chesapeake Bay is much more than a food source for the more than 16 million people who live near it, though. It is a popular habitat for all sorts of water fowl, too. It is also a source of recreation for hunters, anglers, boaters, and nature enthusiasts, all of which benefit the local economy.

Although it is only 310 kilometers (195 miles) long, the bay's shoreline measures nearly 13,000 kilometers (8,100 miles)—twice the entire U.S. Pacific coastline! The bay contains an abundance of species—nearly as many as the entire Atlantic Ocean. The abundance of fish and wildlife is

a result of many factors. The bay's shallowness, for instance, ensures adequate sunlight penetration, which enhances biological productivity. The nutrients this massive estuary receives from rivers that empty into it and the presence of numerous microhabitats (very distinct habitats within the bay) also contribute to its biological diversity.

Old-timers claim that the bay's waters once contained wall-to-wall oysters. Oyster populations have declined markedly, though, falling by 99% since 1870, with most of the decline occurring in recent times. Striped bass populations have also fallen off alarmingly. Various strategies have been tried to reverse the downward trend in the bass population, among them a moratorium on striped bass fishing in January 1985. In recent years, striped bass were produced by two federal fish hatcheries to supplement the declining populations. These efforts have resulted in an increase in the striped bass population. Because bass populations have recovered somewhat, limited commercial fishing has been permitted since 1990.

Environmental Protection Agency (EPA) studies show that the bay's submerged vegetation, so vital to fish, had dropped by 76% in the late 1900s, partly as a result of large algal blooms (bursts of algal growth resulting from inorganic pollutants) that block sunlight and impair the growth of submerged aquatic vegetation. Further damage to the bay is caused by aerobic (oxygen-requiring) bacteria, which deplete the oxygen supply when they decompose dead algae and organic pollutants in the water.

Population growth and commercial development—and the pollution they produce—are the biggest threats to the bay. The population is expected to reach 19 million by 2030. Most of the pollution comes from oil spills, sewage, toxic chemicals, heavy metals, and runoff from the bay's extensive drainage system, an area slightly smaller than Missouri. One of the worst pollutants is nitrogen from commercial duck farms, municipal sewage, and farms.

The story of Chesapeake Bay is an example of what is happening to many important surface waters in the United

States and elsewhere. It shows us how we are destroying a vital resource and ruining the habitat of other species that are often a valuable source of food for humankind. It reminds us that living sustainably on the planet requires much more than protecting human civilization from floods and providing adequate amounts of water. It requires efforts to protect wetlands, estuaries, coastlines, and rivers. Such actions protect biodiversity, but they also protect valuable food, income, and water sources for human populations.

Wetlands

One of the most endangered ecosystems on the planet is its wetlands. **Wetlands** come in many forms. Swamps are a good example. But so are wet meadows and certain woodlands along streams. What do these have in common? One thing is that the soils are typically saturated with water, either part of the year—as in the case of flooded woodlands (called *bottomlands*)—or year-round—as in the case of swamps.

Wetlands are supplied by surface and groundwater and support plants that are adapted for life in saturated soils. Biologists recognized two types of wetlands: inland and coastal, both of which are shown In **FIGURE 13-19**.

Inland wetlands are found along streams, lakes, rivers, and ponds; they include bogs, marshes, swamps, and river overflow lands (like bottomlands) that are wet at least part of the year.

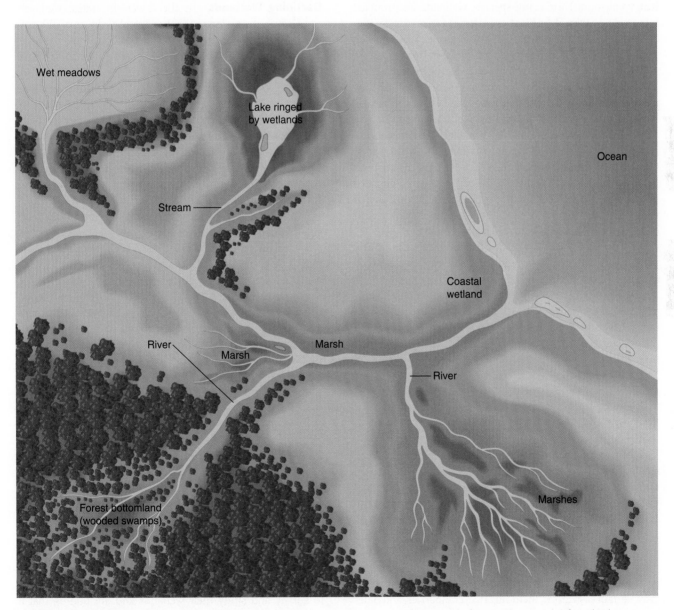

FIGURE 13-19 Wetlands. Wetlands are found in many locations, along rivers or at the mouths of rivers. They are also found in low depressions where groundwater intersects with the land surface or around lakes and ponds.

Coastal wetlands are wet or flooded regions along coastlines, typically associated with estuaries, the mouths of rivers where freshwater mixes with saltwater. Coastal wetlands include mangrove swamps, salt marshes, bays, and lagoons. Mangrove swamps in the United States are restricted to southernmost biomes of Florida, Louisiana, and Puerto Rico. Hawaii also has mangrove swamps. Wetlands along the rim of the Great Lakes are also considered coastal wetlands. They're typically associated with the mouths of rivers.

Wetlands are an extremely valuable and productive habitat for many animal and plant species, including many rare and endangered species. In fact, 45% of the species listed as threatened or endangered by the U.S. Fish and Wildlife Service rely directly or indirectly on wetlands for their long-term survival. In addition to mink, beaver, muskrats, and otters, wetlands are home to many shellfish, amphibians, reptiles, birds, and fish. While they serve as **primary habitat** (year-round) for many species, wetlands also provide important **seasonal habitat** (part-time) for migrating species such as ducks and geese. These species obtain food, shelter, and water during their perilous migrations.

The loss of wetlands can have a devastating effect on species. The Dusky Seaside Sparrow became extinct in 1987. The main reason for its disappearance was the destruction of wetlands around Merritt Island and St. John's Island in Florida. The Ivory Billed Woodpecker was thought to have met a similar fate, largely because of clearing bottomland hardwood forests. In 2005, however, more than 60 years after the bird was believed to be extinct, field biologists caught the bird on video in eastern Arkansas. The Louisiana Black bear, also known as the swamp bear, is now listed as threatened, primarily because of the loss of its wetland habitat. The list goes on.

Aside from their importance to wildlife, wetlands also play an important role in regulating stream flow, as noted earlier. Wetlands within and upstream from urban areas are especially important in flood control. Loss of wetlands decreases their ability to absorb peak flows and reduce flooding. The sponge effect has the added benefit of recharging groundwater supplies. Because streams are fed by groundwater, wetlands help to maintain year-round base flows. Wetlands also remove sediment in surface runoff and thus reduce sedimentation in streams. This helps keep streams and rivers clean. In addition, wetland plants absorb nitrogen and phosphorus, two common pollutants derived from heavily fertilized land, thus reducing water pollution. Coastal wetlands also help protect human settlements by absorbing **storm surges**, high waves accompanying high winds during hurricanes and other violent storms. Damage from Hurricane Katrina would have been greatly reduced had coastal wetlands not been reduced. Wetlands are used to grow certain cash crops, including rice, cranberries, and blueberries. Wetlands also have other economic values. Canada's remaining wetlands, for instance, annually generate between

Go Green

Water lawns and gardens conservatively—for instance, during the coolest part of the day.

$5 and $10 billion to the nation's economy from such activities and natural services as commercial fishing, sport fishing, hunting, trapping, recreation, groundwater recharge, and flood control, according to Environment Canada. Because their usefulness is not always apparent, wetlands are often filled in or dredged to make way for farms, housing, recreation, and industry.

KEY CONCEPTS

Wetlands are highly endangered ecosystems. They serve as habitat for aquatic organisms, birds, and mammals. They help control flooding, purify water, recharge groundwater, and buffer cities from storm surges. Unfortunately, many people fail to understand their economic and ecological value.

Declining Wetlands In the lower 48 states, wetlands once covered an area twice the size of California (90 million hectares, or 220 million acres). Today, half the wetlands are gone (**FIGURE 13-20a**). The greatest losses have occurred in California (91%), Ohio (90%), and Iowa (89%). Figure 13-20b shows the distribution of existing wetlands and losses by state.

Wetland destruction has slowed in recent years (**FIGURE 13-21**). Residential development and urban growth have been major contributors to wetland losses (**FIGURE 13-22**).

The remarkable decrease in the loss of wetlands is due to many factors. For one, many prime wetlands in agricultural states have been filled in and are now being farmed.

Wetland loss is a global phenomenon. Some areas have suffered greater losses than others, however. In California, New Zealand, and Australia, development has claimed more than 90% of the wetlands. Canada has lost huge amounts of wetlands largely to agriculture.

Coastal wetlands are victims of dredgers that scoop up muck from streams and bays to increase navigability. Dredging drains adjoining swamps. Cities often fill in swamps to make room for homes, recreational facilities, roadways, and factories. Farmers fill in swamps on their land to expand their arable land. About half of the world's coastal wetlands have been destroyed.

Mangrove swamps (located along the coastlines in semitropical and tropical regions) are one of the most valuable forms of wetland and the most threatened, suffering heavy losses in Asia, Latin America, and West Africa. In Ecuador, nearly half of the mangrove swamps have been cleared to build shrimp ponds. India, Pakistan, and Thailand have lost three-fourths of their mangroves.

Wetlands are also suffering from declines in environmental quality, primarily caused by pollution. Excess sediment deposition, pollutants that stimulate plant growth, and pesticide contamination are key problems, as is the reduction of water flow into wetlands. Today, many states have begun to survey and study their wetlands and are finding that a large percentage of them are impaired because of these or other problems including the invasion of exotic species.

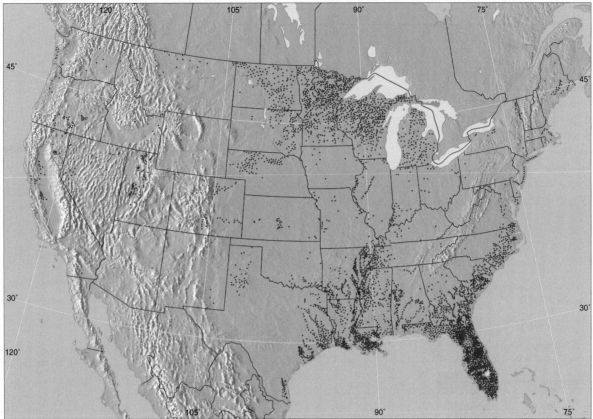

(a)

(b)

FIGURE 13-20 Map of wetlands in the United States. These maps show **(a)** where remaining wetlands are found in the lower 48 states and **(b)** the percentage loss in each state.

Wetlands are declining virtually everywhere, and massive numbers of them have already been destroyed in the United States and abroad. Fortunately, the rate of destruction in the few remaining U.S. wetlands has slowed considerably thanks to wetland restoration and the creation of new wetlands. Areas covered by wetlands in the United States are slowly increasing. Nonetheless, many remaining wetlands are impaired because of pollution, invasion of exotic species, and lack of adequate water flow.

Protecting Wetlands Concern for the loss of wetlands has stirred many governments into action. For example, Florida passed legislation in 1972 to regulate all wetland development. In 1988, New Jersey passed a law requiring buffer zones of 30 meters (100 feet) around important wetlands. Although a plan does not ensure protection, it is a step in the right direction.

The U.S. government has also assumed an increasing role in wetland protection through executive orders that prohibit federal agencies from supporting construction in wetlands when a practical alternative is available. The federal

FIGURE 13-21 Wetland losses in the United States. Wetland losses have declined dramatically over the past 50 years.

Data from T. E. Dahl, *Status and Trends of Wetlands in the Conterminous United States 2004–2009*; Frayer et al; Dahl and Johnson 1991; Dahl 2000; 2006.

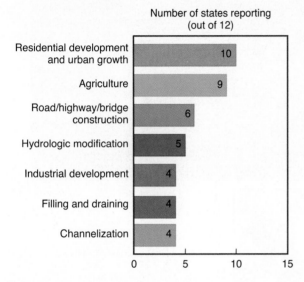

FIGURE 13-22 Source of wetland losses. These data show states reporting major causes of wetland losses. The numbers are the number of states reporting the problem.

Data courtesy of EPA.

Coastal Zone Management Act (1972) calls on states to develop plans to protect coastal wetlands. Virtually all states have either federally approved plans or state plans to regulate their wetlands. However, only 21 states have measures to protect inland, freshwater wetlands according to the Environmental Law Institute.

The federal government can purchase wetlands and set them aside as part of the **National Wildlife Refuge System**. The system today contains more than 9.7 million hectares (24 million acres) of wetlands. The U.S. Fish and Wildlife Service also buys wetlands to set aside. States own additional wetland acreage. All told, a little over one-fourth of our existing wetlands are protected.

Another important action aimed at protecting wetlands was the 1985 Farm Bill. It included important *swampbuster* provisions—rules that deny federal benefits (low-interest loans and crop insurance) to farmers who drain and farm their wetlands. In 1996, the swampbuster provisions were tightened. The law also established a **Wetlands Reserve Program**, a voluntary program to restore and protect wetlands on private farmland. Farmers are financially reimbursed for the land they enroll in the program.

Unfortunately, many wetland protection laws and regulations are like toothless watchdogs. In many instances, little is done to enforce them. Individuals can write their state representatives to find out what laws exist and how they are being enforced. Local action groups can help stimulate stronger enforcement when necessary. Personal action is needed more than ever in the face of deep cuts in the federal budget.

On an international scale, nations can cooperate to preserve wetlands. In 1971, representatives of many nations met in Iran to discuss the plight of wetlands and agreed to protect those lands within their jurisdiction. The agreement, known as the Ramsar Convention after the city where it was signed (Ramsar, Iran), seeks to stem the loss of wetlands. The list of protected areas—wetlands of international importance—includes 2,122 sites encompassing 205 million hectares (507 million acres), up from 1,021 sites in 2000. Currently 168 nations have signed the agreement. In 1986, the United States ratified the agreement. Four U.S. wildlife refuges were added to a list of wetlands of international importance.

Unfortunately, U.S. wetlands are facing a new round of threats—too numerous to cover in detail here. A 2001 ruling by the U.S. Supreme Court now allows certain types of wetlands (isolated wetlands) to be filled in—without any requirement to enhance wetlands elsewhere or create new wetlands. This ruling has already allowed companies to destroy valuable wetlands.

KEY CONCEPTS

Numerous efforts are under way at local, state, national, and even international levels to preserve and protect existing wetlands. Although these measures have not ended the destruction, they have greatly slowed it down.

Estuaries

Estuaries are the mouths of rivers, where saltwater and freshwater mix. Chesapeake Bay, described earlier in the chapter, is a prime example of an estuary. Like wetlands, estuaries are critical habitat for fish and shellfish. Together, coastal wetlands and estuaries make up the **estuarine zone**. Two-thirds of all fish and shellfish depend on this zone during some part of their life cycle. Because fish are an important source of food, protection of the estuarine zone is vital to feeding the world's people. It has an economic benefit as well.

The estuarine zone gets its richness from the land—eroded land. Eroded sediment, rich in nutrients, is carried in streams to the ocean, where it supports an abundance of aquatic organisms—especially algae, the base of the aquatic food web.

KEY CONCEPTS

Estuaries are the mouths of rivers. Combined with coastal wetlands, they form extremely biologically rich and economically important estuarine zones.

Damaging This Important Zone The estuarine zone is vulnerable to a variety of assaults: pollutants from sewage treatment plants or industries; sediment from accelerated erosion that buries rooted estuarine vegetation; oil spills; and dams that cut off the life-giving flow of nutrients from the land. Cities may withdraw so much freshwater upstream that rivers and estuaries run dry. In Texas, for example, drought and heavy water demands in past years have critically reduced water flow into estuaries. Further west, the Mexican delta of the Colorado River is a remnant of its former self because of excessive water withdrawal along its course (**FIGURE 13-23**). The Nile River estuary has

been similarly affected. Freshwater inflows are critical to maintaining the proper salt concentration in coastal wetlands. High salt levels in the estuarine zone reduce populations of shellfish.

In addition, pollutants from the ocean can concentrate in estuaries. Particles can carry heavy metals, organic pollutants such as PCB, and insecticides such as DDT—all of which poison aquatic organisms that depend on the estuarine zone.

Many organisms inhabit the estuarine zone, but the most sensitive to pollution are clams, oysters, and mussels. Pollution is generally thought to be one of the major factors responsible for the decline in mollusc harvests. Molluscs concentrate toxic heavy metals, chlorinated hydrocarbons, and many pathogenic organisms, including those that cause typhus and hepatitis. These pathogens may not affect the molluscs' survival, but make them unsafe for human consumption.

The estuarine zone is also subject to dredging and filling. Compounding the damage is the widespread problem

(a)

(b)

FIGURE 13-23 A delta dies. (a) Not more than 80 years ago the Colorado River flowed unhindered from northern Colorado through Utah, the Grand Canyon, Arizona, and Mexico before pouring out into the Gulf of California. But as one can see in this satellite image of the Colorado River Delta taken on September 8, 2000, all of its water is removed by farmers and cites before the river reaches its former destination. The former course of the Colorado River can be seen running diagonally from top left to bottom right. **(b)** A nine-year-old member of the Cocopah Indians of northern Mexico plays in dried up mud in an area his people fished and farmed for about 2,000 years.

Photo a courtesy of NASA/GSFC/MITI/ERSDAC/JAROS, and U.S./Japan ASTER Science Team; photo b © Gregory Bull/AP Photos.

of overharvesting of fish and shellfish. It is generally agreed that the decline in U.S. oyster production after 1950 was largely the result of overharvesting. Clams in the Northeast have likewise been severely overharvested.

Over 40% of the U.S. estuarine zone has been destroyed. The most severe damage has occurred in California, along the Atlantic coast from North Carolina to Florida, and along the entire Gulf of Mexico. Despite state and federal laws to protect this zone, destruction continues.

KEY CONCEPTS

Because of its location downstream from many human activities, the estuarine zone is seriously endangered. Pollution, decreased water flows, high salt levels, dredging, filling, and overharvesting of commercially important species, especially shellfish, in this zone results in enormous ecological and economic costs.

Protecting the Estuarine Zone Protecting estuaries and coastal wetlands is part of an overall sustainable management strategy whose beneficial effects ripple through the biosphere just as the adverse effects do now. Many efforts to solve other problems will also help protect the estuarine zone. Population stabilization, for example, which eases pressure on natural systems and reduces pollution, is vital to the long-term future of estuaries. Improved water pollution control is also essential. Pollution prevention efforts by businesses can help enormously. Watershed management, including revegetation and other erosion control measures, is also important. Water conservation programs on farms, factories, cities, and towns can reduce water withdrawals from streams and enhance freshwater flow into estuaries. Restrictions on dredging and filling and ending the overharvesting of fish and shellfish will also help.

Protecting the estuarine zone presents a unique challenge in the United States because 90% of the coastal land in the 48 coterminous states is privately owned. In 1972, Congress responded to the plight of the estuaries and coastal wetlands by passing the Coastal Zone Management Act (described earlier). This law set up a fund that provided the 35 coastal and Great Lakes states with assistance in developing their own laws and programs. It also provided them with money to purchase estuarine zones and estuarylike areas in the Great Lakes states. These regions are to be set aside for scientific study; as of February 2014, national estuarine research reserves had been established covering more than 0.52 million hectares (1.3 million acres) of estuary and coastal land in 22 states and territories.

The Coastal Zone Management Act also allows for the establishment of national marine sanctuaries. As of February 2014, 13 national marine sanctuaries and one marine national park have been set aside off the coasts of Washington, California, Hawaii, Maine, Massachusetts, North Carolina, Georgia, Florida, Texas, and American Samoa. These sanctuaries are established to protect vital habitat including coral reefs, open ocean, mangrove swamps, and islands to protect a variety of marine mammals and fish. Several are also home to historic ship wrecks and are thus popular diving spots. Some also support whale watching, sports fishing, and commercial fishing.

Half of the U.S. population lives in counties that are at least partly within an hour's drive of a coast, and most of the major cities are coastal. Discoveries of offshore oil, gas, and minerals pose new problems that require immediate answers. An abundant supply of cooling water makes the coastal zones prime candidates for new power plants and oil refineries.

Unfortunately, some states either have not adopted coastal protection programs or enforce their programs poorly, leaving their coastal waters vulnerable to further damage.

Fifteen additional federal laws have been passed to promote coastal zone management, but they are only as good as their enforcement. Lackadaisical enforcement is almost as bad as no enforcement at all.

The estuarine zone is being destroyed and severely damaged throughout the world, too. Today, over half the world's people (3.6 billion people) live and work within 120 miles of a coastline, occupying 10% of the Earth's land mass. Two-thirds, about 4.75 billion, live within 250 miles of the coastline. Without serious efforts to improve coastal zone management in the United States and abroad, we will almost certainly lose more of this ecologically and economically important habitat.

KEY CONCEPTS

Protecting the estuarine zone requires many efforts; these include establishing protected areas and carrying out numerous preventive measures—such as pollution control and soil erosion controls in watersheds of rivers and their tributaries that empty into estuaries.

Barrier Islands

The Atlantic and Gulf coast states are skirted by a chain of **barrier islands**, narrow sand islands separated from the mainland by lagoons, bays, and coastal waterways (**FIGURE 13-24**). An estimated 250 barrier islands lie along the Atlantic and Gulf coasts. Most of them are popular sites for recreation. Some have been designated as national seashores by the U.S. government and are used for recreation and wildlife habitat; however, most barrier islands are under private ownership.

Many of these islands have been developed for vacationers. Summer homes, roads, stores, and other structures have replaced the grass-covered dunes.

Barrier islands and their beaches are part of a river of sand that migrates down these coastlines. The islands grow and shrink from season to season and year to year in response to two main forces: First, waves, which tend to arrive at an angle to the beach, erode the beaches. As shown in **FIGURE 13-25a** waves create **beach drift**, a gradual movement of sand along the beach. In addition, the wind creates **longshore currents** parallel to the beach, which also move sand along it (Figure 13-25b).

(a)

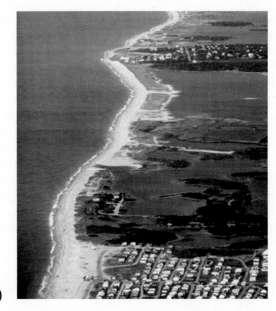

(b)

FIGURE 13-24 Barrier islands. (a) Map showing location of barrier islands off the coast of Texas. **(b)** Barrier island of Westerly, Rhode Island.

Photo courtesy of Hope Alexander, EPA Documeria/NOAA.

(a)

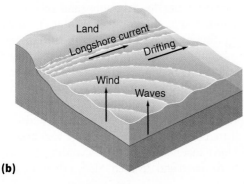

(b)

FIGURE 13-25 Forces that shape a beach. (a) Beach drift is caused by waves approaching obliquely. **(b)** Longshore currents are formed by offshore oblique winds and waves.

Combined, beach drift and **longshore drift** (movement of sand by longshore currents) cause the barrier islands to move parallel to the main shoreline, shortening on one end and elongating on the other. Homes built on the upcurrent side of the island may collapse into the sea (**FIGURE 13-26**). Second, winter storms tend to wash over the barrier islands, robbing them of sand and destroying houses, roads, and other structures.

KEY CONCEPTS

Barrier islands are offshore islands made of sand that are popular sites for housing and other forms of development. Barrier islands are subject to continuous erosion from waves and longshore currents and, therefore, are very unstable. Buildings can also be heavily damaged by hurricanes.

FIGURE 13-26 Losing ground. This home built on a shifting barrier island will soon be engulfed by the sea as the island "migrates" south.

Courtesy of Richard B. Mieremet, Senior Advisor, NOAA OSDIA.

Protecting Barrier Islands—Policy Options Federal actions and various relief programs have, in the past, encouraged development on barrier islands. When erosion and storm damage occurred, for example, the government provided disaster relief funds to rebuild. Federal flood insurance paid for the damage, and federally subsidized construction projects often helped rebuild roads and stabilize the islands. A few years later, storms would devastate the islands again, starting the rebuilding cycle all over at considerable expense. Recognizing the economic unsustainability of this act, Congress passed the **Coastal Barrier Resources Act** in 1982 to prohibit the expenditure of federal money for highway construction and other development on barrier islands. Despite denial of federal subsidies, development continues on many islands.

KEY CONCEPTS

U.S. government policy has promoted the development of barrier islands until quite recently.

Coastal Beaches

Coastal beaches, like barrier islands, are eroded by longshore currents. In fact, U.S. beaches are like great rivers of sand kept in constant motion by the major coastal currents. Sand lost in one area is replaced by sediment carried to the sea by rivers. Consequently, dams that trap sediment diminish the natural replacement of sand on coastal beaches. According to one estimate, dams hold back nearly 40% of the sediment that once reached the mouth of the Santa Clara River north of Los Angeles. This robs California beaches of 15 million metric tons (13.5 million tons) of sand each year. From New Jersey to Texas, the story is the same. According to one source, 70% of the world's beaches are eroding at a rapid rate.

Some communities erect barriers called **jetties** to prevent erosion by longshore currents. These structures only slow down the process. Jetties are sometimes built to maintain navigable passageways in coastal harbors. In 1911, for instance, two 300-meter (1,000-foot) jetties were built on the New Jersey coast north of Cape May to prevent sand from filling in the harbor. Although they performed admirably, the jetties have had a disastrous effect on downcurrent beaches. The beaches at Cape May grew thinner and thinner. By the 1920s, the town was fighting back by building small jetties to keep the remaining sand from being washed away and to trap the sand flowing in the longshore currents. To the townspeople's dismay, their efforts were fruitless. Beaches

KEY CONCEPTS

Beaches are constantly being eroded by longshore currents. Sand lost to erosion is replaced by sediment naturally eroded from the land. Dams on rivers trap the sediment and, therefore, rob beaches of their natural replenishing sand.

retreated by 6 meters (20 feet) a year. Lighthouses fell into the sea. The ocean threatened to swallow the airport. After years of anguish the town has turned to an expensive pumping system that draws sand from above the two large jetties and moves it down to the beaches, a costly replacement for a service nature used to provide for free.

Protecting Coastlines To protect coastlines, the delicate balance between sediment flow and erosion must be maintained. Dams should be built only as a last resort. Other sustainable options to acquire water or control flooding (discussed earlier in this chapter) can eliminate the need for new dams.

Some scientists argue against the construction of jetties so that beaches can be left to grow and shrink with natural cycles. Realizing that it makes more sense to cooperate with nature, in fact, some government officials on the Atlantic coast argue against any repair work on jetties and buildings destroyed after hurricanes, thus returning shorelines to their natural state. This step would begin a retreat from the shorelines, where nature probably intended humankind to be only a visitor.

KEY CONCEPTS

Protecting coastlines may require measures to ensure a steady stream of sediment from rivers and a hands-off policy toward building levees in vain attempts to prevent beach erosion.

Wild and Scenic Rivers

More and more people are flocking to rivers to embark on a variety of sports: fly fishing, kayaking, rafting, inner tubing, and canoeing. Undammed rivers offer much more than excitement for the river runner. Like wilderness, they provide opportunities for relaxation and reflection. They are home to fish and other forms of wildlife. Their canyons offer opportunity for geological study. "Rivers are more than an amenity," said former Supreme Court Justice Oliver Wendell Holmes, "they are a treasure." Unfortunately, for much of modern history rivers have been viewed in a utilitarian light. They've been perceived as sources of water and repositories for waste. They've been viewed and used as a means of transporting goods. They have also been viewed as a wild force that needed taming. Accordingly, many rivers have been dammed and diverted. Many others are slated for development.

KEY CONCEPTS

Rivers are a great source of recreation, but they provide many other benefits. Unfortunately, for many years people have viewed them solely as a valuable source of water or a troublesome source of floodwater that should be dammed.

Protecting Rivers   Foreseeing the need to protect rivers for recreation and habitat for fish and other species, the U.S. Congress passed the **Wild and Scenic Rivers Act** in 1968. This law made it possible to protect segments of rivers from dams, water diversion projects, and other forms of undesirable development. Congress established a three-tiered classification scheme: **Wild rivers** are rivers or sections of rivers that are relatively inaccessible and "untamed." They are free of dams and relatively unpolluted. **Scenic rivers** are free of dams, largely undeveloped, and of great scenic value. They are accessible in places by roads. **Recreational rivers** are rivers or sections of rivers that are readily accessible by roads and that may have some dams along their course or development along their shores. They offer important recreational opportunities. By 2011, 203 river segments in 38 states had been included in the system, totaling 20,270 kilometers (12,598 miles). Although that may sound like a lot, it's only one-quarter of 1% of all rivers. No matter what their classification, rivers protected by the "wild and scenic" designation are administered with the goals of ensuring nondegradation and of enhancing the values that caused them to be designated in the first place.

Wild or scenic river designation is often marked by controversy because so many interests vie for a river's benefits: municipal water consumers, paper manufacturers, farmers, ranchers, anglers, and white-water boaters. Competing interests make compromise difficult or impossible. A dammed river provides water for a new paper mill, waterskiing, and boating, but it also irretrievably floods the kayakers' rapids and the fly fishers' favorite pools.

KEY CONCEPTS

Many rivers or segments of rivers in the United States have been protected from development by the Wild and Scenic Rivers Act. Designating a river for protection is often fraught with controversy.

Sustainable Approaches to River Protection   The fight to keep free-flowing U.S. rivers safe from development continues today. Much of the pressure to dam American rivers, however, has been reduced because of economic and legal forces. After years of paying for questionable dams, Congress has found that many projects return only a few cents for every dollar invested. As a result, federally subsidized water projects, often handed out as political favors, have fallen into disfavor.

Further conflicts can be avoided by finding alternative means of supplying water, as discussed earlier in this chapter. Conservation is essential to this effort.

Many conservationists argue that valuable recreational rivers must be preserved just as endangered species are. When a scenic river gorge is dammed, it is gone forever.

A river is a vital resource to people and to a host of other species. Dammed and diverted to water-hungry, often wasteful consumers, a river becomes a symbol of unsustainable systems design. Our goal should be to manage our rivers and all other water resources wisely and efficiently, minimizing waste and damage, ensuring future generations the use of treasures we now enjoy and too often take for granted.

KEY CONCEPTS

In the United States the construction of new water projects has declined, in large part because of a cutback in federal subsidies for such projects and the realization that alternative sources of water, such as water conservation, are far cheaper and much better from an environmental perspective.

Let him who would enjoy a good future waste none of his present.

—Roger Babson

My list of costs for this project includes the following: (1) construction costs for the pipeline, canal system, and pumping stations; (2) pumping costs; (3) canal and pipeline maintenance; (4) interest on loans; and (5) maintenance and repair. Environmental costs would include (1) reduced water flow in the Missouri River, (2) possible effects on fish and wildlife, and (3) increasing pollution levels in the river as water is withdrawn.

The benefits of this project are many. The project would create jobs and boost the agricultural economy. It would help promote a continuation of farming in affected areas, which provides jobs and supports local economies. As farms prosper, so do their communities.

Although we can't examine all of the costs and benefits, let's take a look at the cost of pumping water and see how that compares to the value of the crops grown on the irrigated land. To begin with, it costs about $25 to pump one acre-foot of water 57 meters (188 feet) in elevation from Missouri to western Kansas. If 14 million acre-feet of water were pumped each year, how much would it cost?

To make this calculation, note that you'll be pumping water from 260 meters above sea level to about 900 meters above sea level, or about 640 meters.

Thus, a single acre-foot of water will cost about $280 to pump. Now multiply that by 14 million acre-feet. By my calculation, the annual price tag is nearly $4 billion. That's just electrical cost, though. In reality, repayment of loans for construction, interest, maintenance, and personnel will cost another $4 billion a year, according to my sources.

Now compare this to one economic benefit. Fourteen million acre-feet of water will irrigate about 3.4 million hectares (8.4 million acres) of corn, yielding about 754 million bushels of corn that would sell at $3 per bushel. That's about $2.3 billion worth of corn.

Thus, it would cost about $8 billion to transport water to produce $2.3 billion worth of corn. Remember that the $2.3 billion price tag is gross sales—not profit. It does not take into account costs for labor, machinery, fuel, fertilizer, pesticides, and so on. The profit would be considerably lower than $2.3 billion if these factors were taken into account.

This exercise shows one way of determining the feasibility of a project. We didn't have to work very hard to see that this project, although appealing in many respects, would be a very costly venture indeed.

CRITICAL THINKING AND CONCEPT REVIEW

1. What is the hydrological cycle? Draw a diagram showing how the water moves through the cycle. Why is this cycle important to you? How does human society alter the cycle? How do you affect it?

2. Define *transpiration, evaporation, relative humidity, absolute humidity, saturation,* and *condensation nuclei.*

3. Which sector is the largest water user in the world? In the United States?

4. Define the following terms: *groundwater, water table, saltwater intrusion,* and *subsidence.*

5. Discuss the problems caused by overexploitation of ground and surface water.

6. List the pros and cons of dams.

7. A dam is being proposed for a popular fishing and canoeing river near your home. The dam is designed to provide flood control and water for your community. You are a member of a local organization that opposes the dam. Prepare a list of options—ways to control floods and to supply drinking water to meet additional demand—that will not require a costly dam. What criteria would you use to evaluate each option?

8. Using your critical thinking skills and the knowledge you have gained, analyze the following statement: "Desalination of seawater is an important way of helping to meet future water demands."

9. Describe ways in which you and your family can help conserve water. Calculate how much water your efforts will save each day. How much will they save in a year?

10. Using your critical thinking skills, analyze the following statement: "Water is a renewable resource. Therefore, we can never run out."

11. What are the connections between flooding and water shortages? How can people solve both problems simultaneously?

KEY TERMS

absolute humidity
aquifer recharge zones
barrier islands
beach drift
black water
catchwater system
Coastal Barrier Resources Act
coastal beaches
coastal wetlands
Coastal Zone Management Act
condensation nuclei
condense
desalination
drought
estuaries
estuarine zone
evaporation
evapotranspiration

floodplains
groundwater
hydrological cycle
inland wetlands
jetties
levees
longshore currents
longshore drift
National Wildlife Refuge System
precipitation
primary habitat
rain catchment system
recreational rivers
relative humidity
saltwater intrusion
saturated
scenic rivers
seasonal habitat

sinkholes
storm surges
streambed channelization
subsidence
surface waters
transpiration
water cycle
water table
watershed management
Watershed Protection and Flood Prevention Act
wetlands
Wetlands Reserve Program
Wild and Scenic Rivers Act
wild rivers
xeriscape

© JVrublevskaya/Shutterstock, Inc.

CHAPTER

14

Nonrenewable Energy Sources

That human beings are fallible has been known since the beginning of time, but modern technology adds new urgency to the recognition.

—Garrett Hardin

LEARNING OBJECTIVES

After reading and studying the material in this chapter, students will be able to understand, knowledgably discuss, and, when appropriate, apply the information learned in this chapter, including:

1. The sources of energy in more developed and less developed nations
2. What energy is and the laws of energy
3. Our heavy dependence on fossil fuels
4. Energy fuel cycles and the importance of analyzing the impact at each stage of the cycle
5. The social, economic, and environment impacts of our use of fossil fuels
6. The prospects of fossil fuels from the standpoint of supply
7. Radioactivity and radiation
8. How nuclear reactors operate, including fission reactors, breeder reactors, and fusion reactors
9. The pros and cons of nuclear energy
10. Priorities for establishing a sustainable energy supply

CRITICAL THINKING

Exercise

Global climate change has led many proponents of nuclear energy in the United States to lobby the public and Congress for renewed support of nuclear energy to lessen America's dependence on foreign oil and coal. Advertisements in prominent magazines still tout the benefits of nuclear power. In such ads, proponents of nuclear power note that this technology has the added benefit of not contributing to global warming, a problem worsened by the combustion of fossil fuels—especially coal, oil, and oil byproducts such as gasoline, jet fuel, and diesel.

Is this thinking valid? Why or why not?

E nergy comes in many forms. In most industrial nations, coal, oil, and natural gas are the predominant fuels. These nonrenewable forms of energy are the lifeblood of modern industrial societies, but they are also a potential Achilles' heel. If you cut the supply off for even a brief moment, industry would come to a standstill. Agriculture and mining would halt. Millions would be out of work. Automobiles would vanish from city streets. Almost everything in our homes would cease to operate. In addition, as the Gulf of Mexico oil spill in 2010 showed, our dependence on these fuels can be extremely costly!

Many of the less developed nations have pinned their hopes for economic progress on their ability to tap into oil, coal, natural gas, and to a lesser extent nuclear power, which have fueled the industrial transformation of the wealthy nations. China, for example, is hoping to get much of its future energy from its abundant supplies of coal and from oil it imports from other countries. In recent years, China has built hundreds of coal-fired power plants to supply electricity to its vast and continually growing population, which is experiencing rapid economic growth.

Is the industrial world's dependence on coal, oil, natural gas, and nuclear energy sustainable? Can less developed nations achieve success by following in our footsteps?

This chapter examines the sustainability of the predominant nonrenewable energy fuels. It looks at their impacts and their abundance. It ends with some guidelines on creating a sustainable energy future.

14.1 Energy Use: Our Growing Dependence on Nonrenewable Fuels

U.S. and Canadian Energy Consumption

Throughout most of the 1800s, Americans had few choices for energy (**FIGURE 14-1**). Wood, a renewable resource, was the main form of energy. Today, the nation's options are many: coal, oil, natural gas, hydropower, geothermal energy, solar power, nuclear power, and wind.

American energy options began to expand in the late 1800s as wood, which once fueled the nation's factories, became depleted. Coal began to be used in factories, but coal was a dirty, bulky fuel that was expensive to mine and transport. When oil and natural gas were made available in the early 1900s, coal use began to fall. The new, cleaner-burning fuels were easier and cheaper to transport.

Today, the United States depends primarily on three fossil fuels: oil, natural gas, and coal. In 2012, oil accounted for nearly 38% of total energy consumption (**FIGURE 14-2A**). Natural gas provided nearly 28%, and coal provided 18% of our energy. All told, fossil fuels account for nearly 85% of our energy use. Nuclear power, another nonrenewable fuel, provided 8.5%. Renewable sources—solar, geothermal, wind, biomass, and hydropower—supplied a nearly 8%.

FIGURE 14-2B breaks down energy consumption by user in the United States in 2012. As it shows, the industrial sectors consume about 32% of the nation's energy. Transportation consumes about 28%, residential use accounts for about 21%, and the commercial sector (copy shops, restaurants, grocery stores, etc.) consume about 18.5%.

KEY CONCEPTS

Energy use in the United States has shifted considerably over the years. Today, the United States depends on a variety of fuel sources. Fossil fuels provide the bulk of the energy. Industry and business consume the majority of the fuel. Transportation is another major energy consumer.

Global Energy Consumption

Virtually all industrial nations get the energy they need from nonrenewable energy sources. On average, they receive about 85% of their energy from fossil fuels, 5% from nuclear power, a type of nonrenewable energy, and 10% from renewables such as solar and wind energy, although the renewable energy contribution is growing rapidly in many countries such as the United Kingdom, Germany, Japan, Spain, and Denmark. In the less developed countries, renewable energy sources such as biomass (wood and

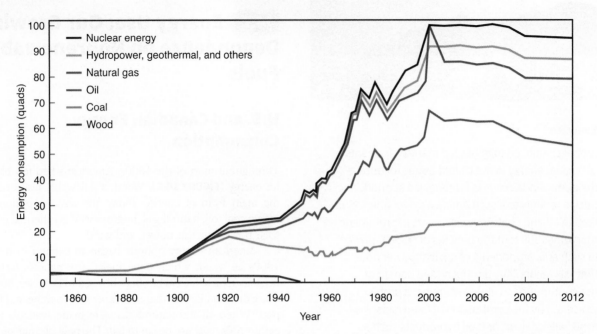

FIGURE 14-1 Changing options. Energy consumption in the United States by fuel type from 1850 to the present. As this graph shows, U.S. energy dependence has shifted over the years from wood to oil, coal, and natural gas. (Quad = quadrillion BTUs.)

Data from U.S. Energy Information Office, 2012.

cow dung, for example) play a much more important role in supplying demand, satisfying about 40% of their energy requirements. Nonrenewable fossil fuels supply about 60% of the total energy. Of nonrenewable energy fuels used in these countries, oil supplies the largest share. Coal and natural gas supply the rest. Nuclear power contributes only a tiny fraction of their energy demand, in large part, because of its extremely high cost.

Worldwide, the biggest users of energy are Americans, who make up about 4.4% of the world's population but consume about 18% of its primary energy. On a per capita basis, Americans consume more than twice as much energy as the people of Japan and Western Europe and about 16 times more per capita than the people of less developed nations. Canada is also a major consumer of energy, using more per capita than any other nation except for Luxembourg. With less than 0.5% of the world's population, Canada uses 2.6% of the world's energy. The reasons for this are many. It is a large country, situated in a cold climate. It has an energy-intensive industrial economy with logging, mining, agriculture, and energy production as the chief sources of income. Historically, energy prices have been low and Canadians tend to use energy inefficiently.

(a) Energy sources

(b) Energy consume

FIGURE 14-2 The U.S. energy profile. (a) This pie chart shows the major energy sources in the United States in 2012. Oil, natural gas, and coal are the three most commonly used fuels. **(b)** Major energy consumers in the United States. Industry and transportation lead the pack.

Data from U.S. Energy Information Administration, 2012.

14.2 What Is Energy?

Energy is all around us, but it is sometimes difficult to define. But what exactly is it?

Energy Comes in Many Forms

Let's begin by making a simple observation as a way to help define this term: Energy comes in many forms. For example, humans in many countries rely today on fossil fuels such as coal, oil, and natural gas. Some use nuclear energy to generate electricity. In other countries, wood and other forms of biomass are primary forms of energy. (Biomass includes a wide assortment of solid fuels, such as wood, and liquid fuels, such as ethanol derived from corn, and biodiesel, a diesel fuel made from vegetable oils.) And don't forget sunlight, wind, hydropower, and the geothermal energy—energy produced in the Earth's interior. Even a cube of sugar contains energy! Touch a match to it, and it will burn, giving off heat and light energy, two additional forms of energy.

Energy Can Be Renewable or Nonrenewable

Energy in its various forms can be broadly classified as either renewable or nonrenewable. Renewable energy, as noted earlier, is any form of energy that's regenerated by natural forces. Wind, for instance, is a renewable form of energy. It is available to us year after year, thanks in large part to the unequal heating of the Earth's surface. When one area is warmed by the sun, for instance, hot air is produced. Hot air rises, and as it does, cooler air moves in from neighboring areas. As the cool air moves in, it creates winds of varying intensity.

Renewable energy is everywhere and is replenished year after year, providing humankind with a potentially enormous supply… if only we're smart enough to tap into it! Nonrenewable energy, on the other hand, is finite. It cannot be regenerated in a timely fashion by natural processes. Coal, oil, natural gas, tar sands, oil shale, and nuclear energy are all nonrenewable forms of energy. Ironically, although most of these sources of energy were produced by natural biological and geological processes early in the Earth's history, and, although these processes continue today in some parts of the world, these fuels are not being produced at a rate even remotely close to our consumption. Coal, for instance, may be forming in some swamplands around the world. But its regeneration is taking place at such a painfully slow rate that it is meaningless. Put another way, contemporary production can never replenish the massive supplies that were produced over long periods of time many millions of years ago. Because of this, coal, oil, natural gas, and others are essentially finite. When they're gone, they're gone.

Energy Can Be Converted from One Form to Another

Another characteristic of all forms of energy is that they can be converted from one form to another. Natural gas, for example, when burned is converted to heat and light. Coal, oil, wood, biodiesel, and other fuels are also converted to heat and light during combustion. Heat can be used to generate steam that is used to spin a turbine attached to a machine called an alternator to produce yet another form of energy: electricity. Visible light contained in the Sun's energy can be converted to heat and electrical energy. Even wind can be converted to electricity or to mechanical energy to drive a pump to draw water from the ground.

Energy Conversions Allow Us to Put Energy to Good Use

When energy is converted to other forms, it produces valuable services. Coal, by itself, is of little value to us. It's a sedimentary rock and fun to behold, but it is the heat and electricity produced when coal is burned in power plants that are of value to us. Sunlight is pretty, too, and it feeds the plants we eat; but in our homes and factories, however, it is the heat that the sun produces and the electricity we can generate from it that is of primary value to us.

In summary, then, it is not raw forms of energy that we need. Not at all. It is the byproducts of energy that are unleashed when we "process" them in various energy-liberating technologies that meet the complex needs of society.

Energy Can Neither Be Created nor Destroyed

Another thing you need to know to deepen your understanding of energy is that energy cannot be created nor can it be destroyed. Physicists call this the **First Law of Thermodynamics** or, simply, the **First Law**.

The First Law says that all energy comes from pre-existing forms. Even though you may think you are "creating" energy when you burn a piece of firewood in a woodstove, all you are doing is unleashing energy contained in the wood—specifically, the energy locked in the chemical bonds in the molecules that make up wood. It, in turn, came from sunlight. The sun's energy came from the fusion of hydrogen atoms in the sun's interior.

Energy Is Degraded When It Is Converted from One Form to Another

More important to us, however, is the **Second Law of Thermodynamics**. The **Second Law**, says, quite simply, that anytime one converts a form of energy to another form—for example, when you convert natural gas to heat—it is degraded. Translated, that means energy conversions transform high-quality energy resource to low-quality energy. Natural gas, for instance, contains a huge amount of energy in a small volume; it's locked up in the simple chemical bonds that attach the carbon atom to the four hydrogen atoms of the methane molecules. When these bonds are broken, the stored chemical energy is released. Light and heat are the products. Both light and heat are less concentrated forms of energy, or lower quality forms of energy. Hence, we say that natural gas, a concentrated form of energy, is "degraded." In electric power plants, only about 50% of the energy contained in natural gas is converted to electrical energy. The rest is "lost" as heat that is dissipated into the environment. This leads to another important fact about nonrenewable fuels; once they've burned or reacted in the case of nuclear fuels their energy is gone forever. The heat radiates endlessly into outer space, heating the universe as it were.

Renewable energy resources, on the other hand, can be regenerated year after year after year. If we're going to persist as a society in the long term, it is renewable energy resources we'll need to rely on. Unlike fossil fuel energy and nuclear energy, renewable resources can return again and again, making our lives bright and cheery and comfortable so long as the sun continues to illuminate the daytime sky. With these important points in mind, let's define this thing we call energy.

No Energy Conversion Is 100% Efficient, Not Even Close to It!

This leads us to another important fact about energy: No energy conversion is 100% efficient. When coal is burned in an electrical power plant, only about one-third of the energy contained in the coal is converted to useful energy, in this case, electricity. The rest is lost as heat and light. The same goes for renewable energy technologies. One hundred units of solar energy beaming down on a solar electric module won't produce the equivalent of 100 units of electricity. You'll only get around 12 to 15% conversion on the most popular modules on the market today.

Energy is lost in all conversions. One hundred units of electrical energy won't produce 100 units of light energy in a standard incandescent lightbulb. In fact, most conventional lightbulbs in our homes (incandescent lights) convert only about 5% of the electrical energy that runs through them into light. The rest comes out as heat! (That's why the production of these light bulbs has been banned in the United States.)

Each conversion in a chain of energy conversions loses useful energy, as shown in **FIGURE 14-3**. Don't forget that. One implication is that to get the most out of our primary energy sources, we must reduce the number of conversions along the path.

But let's get something straight. Some of you may be wondering whether all of this discussion of "energy losses" is violation of the First Law, which states that energy cannot be created nor destroyed.

The truth be known, the "energy losses" I've been talking about during energy conversion are not really losses in the true sense of the word. Energy is not really destroyed; it is released in various forms, some useful and others, such as heat, not so useful. Chemical energy in gasoline that runs a car, for instance, is converted to mechanical energy of moving parts that propel us forward along the highways. Some is also lost as heat that radiates off the engine. This waste heat is of little value except to use on cold winter days when captured, at least in part, to warm the car's interior.

Eventually, however, all heat produced by a motorized vehicle escapes into outer space. It is not destroyed, per se; it escapes into space and is no longer available to us. Hence, the conversion results in a net loss of useful energy.

Energy Is the Ability to Do Work

To a physicist, energy is defined as "the ability to do work." More accurately, says engineer John Howe, "Energy is that elusive something that allows us to do work." We and our machines, that is.

Any time you lift an object, for example, or slide an object across the floor, you are performing work. The same holds for our machines. Anytime a machine lifts something or moves it from one place to another, it performs work.

FIGURE 14-3 Different forms of energy. Energy can be changed from one form to another; however, with each change a certain amount of energy is lost as heat.

Adapted from D. D. Chiras, et al. *Management for a Sustainable Future*, Ninth edition. Pearson Education (2005): Upper Saddle River, NJ.

According to physicists, work is also performed when the temperature of a substance, for example, water, is raised. Therefore, your stove or microwave is working when it boils water for hot tea or soup.

Energy, therefore, is valuable because it allows us to perform work. It powers our bodies. It powers our homes. It powers our society. We cannot exist without energy.

KEY CONCEPTS

There are many forms of energy. Energy is important because it allows us to perform work. Energy can be renewable or non-renewable. Raw energy is not as important to us in our homes as are the useful byproducts such as electricity, light, or heat. Energy can neither be created nor destroyed. It can only be converted from one form to another, and no energy conversion is 100% efficient, not even close.

During conversions useful energy decreases. Put another way, all conversions lose energy as heat that is dissipated into outer space.

14.3 Fossil Fuels: Analyzing Our Options

Energy is the lifeblood of modern society, but it does not come cheaply. In addition to its economic cost, huge environmental costs are posed by many forms of energy. As you shall soon see, the impacts of many of our current energy resources lead many to conclude that the current energy system is unsustainable. When analyzing the sustainability of the world's energy system, one must take into account available supplies as well as the impacts to the environment, climate, and human health. This section examines those impacts. Before we can understand them, though, we must first understand the many steps required to deliver energy to our homes, factories, and gas stations.

FIGURE 14-4 presents a diagram of some of the major steps involved in energy production and consumption. This chain of events constitutes an **energy system** and is composed of six major phases: exploration, extraction, processing, distribution, storage, and end use. Take a moment

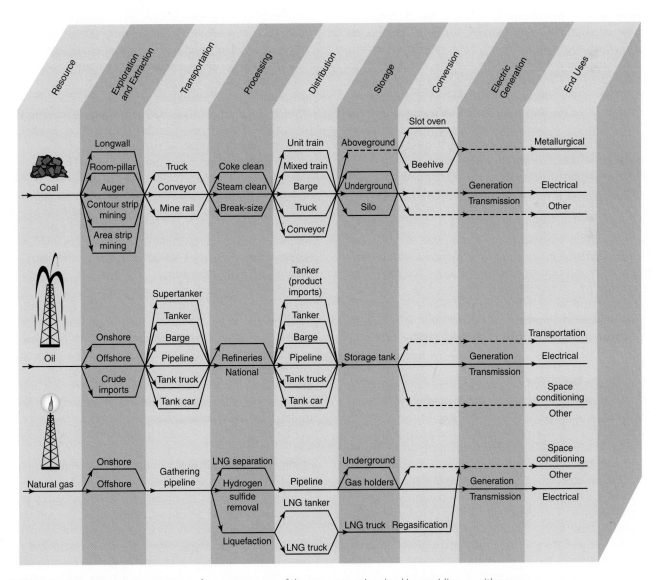

FIGURE 14-4 Energy systems. Most of us are unaware of the many steps involved in providing us with energy.

to familiarize yourself with these steps. As a rule, the most notable environmental impacts occur at the extraction and end-use phases.

Understanding energy systems and the ecosystem impacts along the way makes it clear that a simple flick of a light switch or a press on the gas pedal of an automobile creates a trail of environmental damage. It shows our personal connection to the long list of environmental problems facing the world.

KEY CONCEPTS

Energy does not come cheaply. In addition to the economic costs, society pays a huge environmental price for its use of nonrenewable energy in damage to human health and to the environment. These impacts arise at every phase of the energy systems. The most significant impacts arise from extraction and end use.

Crude Oil

Crude oil or **petroleum** is a thick liquid containing many combustible hydrocarbons (organic compounds made of hydrogen and carbon). Petroleum is created naturally from photosynthetic organisms (mostly algae) in the world's oceans. They settled to the bottom of the oceans. Over time, thick masses of dead algae were covered with sediment. Pressure and heat caused the organic materials to react forming at least 2,000 different organic chemicals that comprise crude oil. Found in deep deposits in the seafloor and on land (former seafloor), crude oil often occurs with natural gas. Geologists locate oil deposits through a careful study of geology—that is, the type of rocks found in various regions—and by other fairly low-impact means. However, to explore for oil in remote areas, roads must often be built so that exploration crews can enter. In Alaska and Canada, plans are in the offing to move into additional pristine northern areas as far north as the Arctic. Oil companies, supported by government subsidies, hope to explore and eventually drill for oil in these regions. One technology that could reduce the impact of oil development in such areas is **horizontal oil well drilling** (**FIGURE 14-5**). A typical oil well is drilled from the top. A horizontal oil well approaches the oil reservoir from the side, often some distance away. Using this technique, oil companies believe they can tap into oil reserves without entering extremely sensitive ecosystems. Even so, the impacts could be devastating. Numerous roads, oil processing facilities, power plants, landing strips, parking lots, worker housing, and other buildings must be erected at the well heads located in environmentally sensitive areas.

Once located, crude oil is extracted via wells drilled into deposits. Oil flows into the well and is pumped to the surface. Oil companies once thought that the yield from oil wells would be greatly increased by injecting water, steam, and carbon dioxide into the ground, which forces additional crude oil to the surface. Today, geologists realize that these techniques do not total increase yield. This has lead to a newer technique called **hydraulic fracking** or simply **fracking**.

FIGURE 14-5 Horizontal oil well drilling. Although it is challenging, oil can be tapped from drill rigs located several miles from the oil deposit, as shown here. This may help to reduce the impact that oil drilling and oil extraction can have on sensitive ecosystems.

Fracking is so named because it creates fractures in underground rock formations containing oil and natural gas. This is achieved by injecting a mixture of sand and chemicals into naturally occurring cracks under extremely high pressure (**FIGURE 14-6**). This causes the cracks to widen. This, in turn, allows oil and natural gas locked in deposits to flow out of the rock into the drilled well from which it is pumped to the surface. This procedure in the United States has helped expand domestic supplies of natural gas and oil.

According to opponents, fracking poses many unacceptable risks. This has led some countries to ban or strictly limit or regulate the industry. One impact comes from numerous toxic and carcinogenic chemicals that are pumped into the ground to liberate oil. They may migrate into groundwater supplies, contaminating drinking water for cities and towns. Scientists are concerned about potential health effects on those who consume the water.

Fracking may cause natural gas to leak into ground water supplies, contaminating drinking water as well. Some residents who live near fracking operations have found that combustible natural gas is emitted from their faucets. A match held near the water stream of a tap will ignite the gas, posing a severe safety risk (**FIGURE 14-7**). Opponents point to other potential environmental effects including the release of radioactive radium from water pumped into wells, then released into streams and rivers. Drilling waste also contains radioactive wastes.

Fracking also causes many small unnoticeable and larger noticeable earthquakes. Fracking requires large quantities of water, which could result in depletion of water supplies, especially in arid regions.

When the crude oil is extracted, it is transported by pipeline, truck, or ship to refineries for processing. Transportation of oil by ships and pipelines is a source of considerable environmental damage. Ships sometimes run aground and spill their contents into lakes and oceans, with devastating environmental consequences. Pipelines can break, spilling oil onto ecologically sensitive land. Oil drilling can be a major source

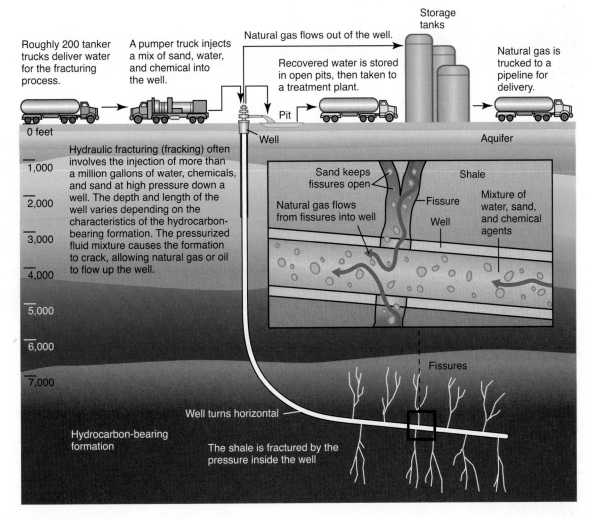

Roughly 200 tanker trucks deliver water for the fracturing process.

A pumper truck injects a mix of sand, water, and chemical into the well.

Natural gas flows out of the well.

Recovered water is stored in open pits, then taken to a treatment plant.

Storage tanks

Natural gas is trucked to a pipeline for delivery.

Pit

0 feet

Well

Aquifer

1,000

Hydraulic fracturing (fracking) often involves the injection of more than a million gallons of water, chemicals, and sand at high pressure down a well. The depth and length of the well varies depending on the characteristics of the hydrocarbon-bearing formation. The pressurized fluid mixture causes the formation to crack, allowing natural gas or oil to flow up the well.

2,000

3,000

4,000

5,000

6,000

7,000

Sand keeps fissures open

Natural gas flows from fissures into well

Shale

Fissure

Well

Mixture of water, sand, and chemical agents

Fissures

Well turns horizontal

Hydrocarbon-bearing formation

The shale is fractured by the pressure inside the well

FIGURE 14-6 Fracking.

of water pollution. Small leaks at offshore wells, for example, can result in oil contamination. Major leaks, such as the well blowout at the BP oil platform in the Gulf of Mexico, can also occur (**FIGURE 14-8**). The "spill" in the Gulf of Mexico occurred 5,000 feet below the surface and affected sea water and organisms from seafloor to the ocean's surface. The environmental and economic effects of the spill were enormous, and could persist for decades. For a discussion of the potential impacts of oil development in the Arctic, see **Spotlight on Sustainable Development 14-1.**

FIGURE 14-7 The dangers of fracking. Flammable methane is in this homeowner's well water very likely caused by fracking near her home.

© Les Stone/Corbis.

FIGURE 14-8 Oil spill in the Gulf. An explosion in an oil well in the Gulf of Mexico in 2010 resulted in the release of millions of gallons of oil for several months. The oil spread throughout much of the Gulf, affecting beaches, tourism, and countless wetlands. This is just one of many of the impacts of our dependence on oil.

Courtesy of U. S. Coast Guard.

14-1 Controversy Over Oil Exploration in the Arctic National Wildlife Refuge

In the far northeastern corner of Alaska lies the Arctic National Wildlife Refuge, or ANWR. Set aside to protect wildlife, this refuge encompasses 7.7 million hectares (19 million acres). It has been described as the last great American wilderness. Since 1977, however, the refuge has been the center of controversy that pitted oil companies against environmentalists. Oil companies and some politicians would like to open parts of the refuge to oil drilling; environmentalists and many others oppose the idea.

The delicate Arctic tundra of the coastal plain, a region 210 kilometers (115 miles) long and 50 kilometers (30 miles) wide, is the only section of land along the entire 1,800-kilometer (1,100-mile) north coast of Alaska that is closed to oil exploration and drilling. The coastal plain is the annual calving ground of several large caribou herds. It is also home to polar bears, grizzlies, wolves, moose, wolverines, and numerous small mammals. Many thousands of birds spend the summer there, raising their young and feeding on insects.

In 1991, after public protest over the tragic 1989 oil spill in Prince William Sound subsided, President George H. W. Bush and the oil companies introduced legislation that would allow drilling in the ANWR. Because of a massive outpouring of public protest, the bill was defeated, but the oil companies continued to push legislation that would permit them to explore for oil within this magnificent wildlife refuge.

In 2005, a couple of major oil companies reportedly withdrew support, after having conducted secret exploration in the region. Despite their having concluded that oil development in the area would be uneconomic, the George W. Bush Administration and Republican leaders unsuccessfully pushed for legislation that would open up a small but significant portion of ANWR to oil development. Although they hope to minimize damage by special drilling techniques, many conservationists fear that the pristine region will eventually be turned into a maze of roads, oil platforms, waste ponds, buildings, and gravel pits (**FIGURE 1**). Full-scale development would lead to construction of four airfields and 50 to 60 oil platforms. A power plant and several oil-processing

FIGURE 1 An oil drilling pad in Prudhoe Bay, Alaska. The delicate Arctic tundra will be damaged to supply more oil to satisfy America's needs.
© Hugh Rose/Accent Alaska.

plants would be built on the tundra, and hundreds of kilometers of roads and pipelines would crisscross this delicate land. The rising temperatures are threatening the Arctic tundra, which could dramatically alter plans to drill for oil in the Arctic. Currently, oil operations are permitted only when the Arctic tundra is frozen at least 30 centimeters (12 inches) and covered with at least 15 centimeters (6 inches) of snow. Because of global warming, the "window of opportunity" has shrunk from 200 days per year to only 100.

While proponents think that oil development and wildlife can coexist in harmony, critics predict a 20 to 40% decline in one of the major caribou herds. They expect over half of the musk oxen to perish. Populations of grizzlies, polar bears, wolverines, and other animals are also likely to decrease substantially. Air pollution, water pollution, and hazardous wastes are expected to have major impacts on wildlife and the long-term ecological health of the region. One exploratory well alone requires 35,000 cubic meters of gravel, which would be excavated from streambeds to make pads and roads. Oil spills on the tundra and careless waste disposal, common in nearby Prudhoe Bay, would change this delicate landscape, whose short growing seasons and harsh winters greatly impair the natural healing that normally takes place in the wake of human interference.

Proponents argue that we need the oil to cut reliance on foreign sources and that we need to drill for oil in ANWR now because of the 10- to 15-year lead time required to fully develop the area. Geologic evidence, some say, indicates a high potential for oil discovery, and oil is needed to help bolster Alaska's economy. They say that environmental regulations will ensure minimal impacts on wildlife and the environment.

Those in favor of preserving the refuge intact argue that the environmental costs are too high. Experience in nearby Prudhoe Bay shows that oil companies often ignore environmental regulations. Frequent violations have been cited. The Department of Interior has recorded over 17,000 oil spills since 1973 in the Arctic. In 2007, a record spill of 1,033,290 liters (267,000 gallons) occurred from a pipeline in a caribou crossing area in Alaska's North Slope. Where the oil saturates the soil, vegetation

fails to recover. Opponents of exploration also argue that because the entire north coast of Alaska is already open to oil exploration, the ANWR should not be—now or ever. If there's oil there, let it lie. Let us find alternatives and let the wildlife live in peace. Furthermore, they point out that slight improvements in automobile gas mileage could easily "provide" as much oil as the ANWR could generate over its lifetime—and at a much lower cost. Moreover, critics like President Barack Obama say that the amount of economically recoverable oil is about one-half the United States' annual oil consumption and will, therefore not lower gasoline prices.

Few of us will ever visit the ANWR, but many of us take comfort in knowing there are places where wildlife are free of the intrusion of modern industrial society. Those places, say opponents, may not last. For more on the debate see Point/Counterpoint 14-1 on the following page.

In a refinery, the oil is heated and distilled. This separates out many of the different hydrocarbon molecules in crude oil—including those that make up gasoline, diesel fuel, heating oil, and asphalt. Many small organic compounds are also extracted during this process and are used to make medicines, plastics, paints, and pesticides (**FIGURE 14-9**). Refineries use considerable amounts of energy and are major sources of air and water pollution.

The combustion of crude oil derivatives such as gasoline and diesel fuel produces enormous quantities of carbon dioxide (a greenhouse gas) and other pollutants such as sulfur dioxide and nitrogen dioxide, both of which are converted into acids in the atmosphere. In fact, many of the most significant local and global environmental problems stem from the combustion of oil and its byproducts.

KEY CONCEPTS

Oil is extracted from deep wells on the seafloor and on land; it is often found in association with natural gas. After it is extracted, crude oil is heated and distilled, a process that separates the components of oil, which produces useful fuel and nonfuel byproducts. The major impacts of the oil energy system come from oil spills and from combustion of oil and its byproducts.

Natural Gas

Natural gas is a mixture of low molecular weight hydrocarbons, mostly methane CH_4. It also contains small amounts of propane and butane that can be extracted, compressed, and liquefied. Natural gas is used to heat buildings, cook food, heat domestic water, and generate electricity. It is used to produce artificial fertilizer. Butane is used in grill lighters; propane can be used to heat homes, cook, and heat water for domestic uses.

Other 0.3 gallons
Kerosene 0.2 gallons
Lubricants 0.5 gallons
Feedstocks 1.2 gallons
Asphalt/road oil 1.3 gallons
Petroleum coke 1.8 gallons
Still gas 1.9 gallons
Liquefied gases 1.9 gallons
Residual fuel oil 2.3 gallons

Jet fuel 4.1 gallons

Distillate fuel oil
9.2 gallons

Gasoline
19.5 gallons

FIGURE 14-9 Products from a 44-gallon barrel of oil. Total is more than 44 because of "processing gain."

Data from the American Petroleum Institute.

Burned in homes, factories, and electric utilities, natural gas is often described as an ideal fuel because it contains few contaminants and burns cleanly. Substituting natural gas for coal in electric generating power plants, for instance, reduces carbon dioxide emissions by 50 to 70%. But precautions must be taken so methane in natural gas doesn't leak. Methane is 30 times more powerful as a greenhouse gas, so losses of only 3 to 4% of the methane could offset any benefit. Natural gas plants also frequently burn

Drilling for Oil in ANWR Is a Good Idea

Adrian Herrera

Adrian Herrera is an analyst for Arctic Power, a private lobby group from Alaska that monitors the ANWR debate on Capitol Hill. He has lived in Alaska 33 years and has worked in the Prudhoe Bay oil fields both as a field hand and as part of the environmental monitoring team.

Courtesy of Adrian Herrera.

Debate over oil exploration in the 10-02 area of the Arctic National Wildlife Refuge (ANWR) has been going on since the late 1980s and the topic has turned into more of a political football and fundraising tool rather than a rational discussion on energy consumption and care for the environment. The oil industry has largely abandoned the debate, leaving Congress to argue back and forth, year after year, getting little done.

Politically the problem lies in the need for 60 votes to overcome a filibuster in the Senate. ANWR bills are continually threatened with filibusters. Meanwhile we continue producing less oil, while consuming more.

Alaskans support drilling in ANWR overwhelmingly. Our economic future and livelihoods depend on it. Eighty-seven percent of our economy is based on North Slope oil, and with the Trans-Alaska Pipeline less than half full and decreasing fast, Alaskans are getting worried. Without oil taxes and royalties, the economy of the state would nose-dive. Expect major job losses and a decline in social infrastructure. The effect locally on the Inupiut native populations of the Arctic would be even more extreme, as their economy is near 100% reliant on tax revenue from oil. The local clinic, the school, the village fire truck, the sewer and water system would not exist—and did not exist—without oil tax revenues.

Many Native Alaskans see opposition to oil development in ANWR as a quiet attempt to force them out of existence. Opponents are denying the right of a people to use their private lands as they see fit (over 100,000 acres [40,500 hectares] of the 10-02 is private property).

ANWR is an issue of self-determination, human rights, and imperialism. The Alaskan Natives have walked and used the land for 40,000 years and are the true environmentalists. They should be allowed to live productive lives, access what is rightfully theirs, and build their futures like every other American. To drive this point home, the Alaska Federation of Natives, which represents all native peoples in the state, has supported development in the 10-02 with yearly resolutions since the debate started.

The fact is America needs oil and will for 100 years to come. Oil provides thousands of products we use everyday as well as raw energy that's turned into fuels. Oil lubricates wind turbine gears, oil produces polymers and resins used to produce components of hybrid cars, oil produces synthetic rubber used for tires, and bitumen for the roads they roll on. Natural gas also comes from oil wells and allows us to produce fertilizers to grow biofuel crops. From paint, to medicine, to fire retardants, to detergents and ink, oil is vital to our current lifestyle.

If you have to have oil, which we do, you have a choice to get it from home where you can control its production and benefit from the jobs and infrastructure or buy your oil from abroad where you have no control and are, in an economic sense, just burning dollar bills (we currently spend $37,750,000 per hour importing oil).

Pollution knows no boundaries and importing oil turns a blind eye to the environmental practices of producers. An oil spill caused by careless production in Nigeria is just as bad as one in Alaska. In Alaska, however, we have the strictest rules and regulations on Earth to mitigate and prevent catastrophes. A dozen federal, state, and local agencies monitor activity in the Alaskan Arctic 24/7 every day of the year. We do the best job humanly possible to protect our environment. Remember Alaskans live in Alaska because they prefer the wild nature and open spaces. Alaskans find it ironic that others believe we would do anything to destroy the very reason we moved there in the first place.

Much of the debate over the 10-02 area of ANWR centers around the porcupine caribou herd. Outside "environmentalists" claim there will be caribou armageddon and all the wildlife will be destroyed and land irreparably damaged. Nothing could be further from the truth. These same concerns were raised by groups back in the early '70s during the controversy over the development of oil in Prudhoe Bay, America's largest oil field, not 50 miles from ANWR. The 5,000 original caribou that made up the central arctic herd that migrate directly through the center of the oil field has grown and is now over 32,000 animals strong. Some armageddon!

All facilities are elevated to allow caribou to walk under them. No facility or pipeline is allowed to be built that will hinder migration. If it is found to do so, it must be removed or changed. This is the case throughout the Arctic and would be in the 10-02. Land use in the 10-02 is limited to 2,000 acres—less than 0.05% of ANWR's area. There is a good chance roads would not be used, and most of the production facilities already exist at Prudhoe, so few would be needed in the 10-02.

Drilling for Oil in ANWR Is a Good Idea (cont.)

Alaskans find it strange that many Americans have no problem building huge shopping malls to service increasing urban sprawl, laying thousands of miles of roads, running over deer and raccoons every night, everywhere all over the Lower 48, yet on a flat as a pancake stretch of distance coastline in the Arctic with the strictest environmental 24/7 regulation on Earth we cannot use 3 square miles of land to supply America with the energy we desperately need.

The 10-02 area is not "pristine." It is inhabited and hunted and used daily. It is not "special" or "scenic" as the rest (92%) of ANWR is, which is permanently off limits to oil development. It is flat, barren, freezing cold, and wind-swept every day of the year. It is ice covered for 8 months of the year and under total 24-hour darkness for 4 months of the year. Most animal life inhabits the plain only 6 weeks a year in summer when no oil production is allowed.

With the current energy crisis looming we have to take action. Alaskans support this issue and have the knowledge and ability to do this right. From the 10-02 we can supply America with a million barrels of oil a day for decades, providing jobs, security, and growing the national economy. Opening ANWR makes total sense and needs to happen now

Americans have always treasured our wild and special places. Iconic images of the Grand Canyon, Yosemite, and Yellowstone not only help define our nation and our natural heritage, they represent our democratic and egalitarian values. Our parks and wildlife refuges have been set aside for all Americans to enjoy and belong to every American and to subsequent generations. Instead of closing off areas for single uses such as oil and gas drilling or logging and handing lands over to multinational corporations, we work to keep our wild and special places open for everyone to enjoy.

The coastal plain of the Arctic National Wildlife Refuge is one of those wild and special places. The area is America's greatest wilderness, protecting more abundant and diverse wildlife than any other protected area above the Arctic Circle. The Arctic Refuge supports large populations of caribou, polar bears, musk oxen, and hundreds of migratory birds.

The coastal plain of the Arctic Refuge is also important to the Gwich'in, Alaska Natives whose name means "people of the caribou." The Gwich'in believe they evolved from the caribou. To the Gwich'in, the Porcupine (River) caribou herd

The Arctic Refuge: A Wild and Special Place That Should Be Protected for Future Generations

Athan Manuel
Athan Manuel is the director of the Lands Protection Program for the Sierra Club, where he oversees campaigns to protect public lands, national forests, and parks. He also defends wildlife and the Endangered Species Act and works to stop new offshore oil and gas drilling.

Courtesy of Athan Manuel.

that visits the coastal plain of the Arctic Refuge is sacred and a link to their traditional way of life and subsistence culture. The Gwich'in, along with the Alaskan Intertribal Council, are officially opposed to drilling in the Arctic because of the disruption of caribou migration patterns that would result from drilling threatens their subsistence way of life.

The annual migration of the caribou herd is one of our continent's most awe-inspiring wildlife spectacles. Each year, the caribou migrate 1,400 miles across Alaska and Canada, typically arriving on the Arctic Refuge in late May where the females give birth. According to some scientists, caribou birth rates could drop by as much as 40% if drilling begins in their habitat.

Unfortunately, oil and gas companies and their pro-drilling allies in Alaska and Washington, D.C. continue to push for drilling in this wild and special place, the vast majority of which is public land shared by all Americans.

Though no one proposes damming the Grand Canyon for hydroelectric power or generating geothermal energy from Old Faithful, oil companies like Exxon and Shell keep pushing for drilling in one of America's last wild places, the coastal plain of the Arctic Refuge.

Oil drilling would industrialize—and ruin—the Arctic Refuge. One need look no further than Prudhoe Bay in Alaska, one of America's largest oil fields, to see the damage done by drilling. Prudhoe Bay averages more than 500 oil spills a year—that's more than one oil spill per day. Four years ago, British Petroleum (BP) had to shut down their entire operation in Prudhoe Bay after they discovered spills caused by corroded pipelines and equipment. BP later revealed that the pipes that leaked had not been inspected in 11 years.

We simply cannot allow that kind of development and damage—and negligence—to occur in an ecosystem as fragile and as important as the coastal plain of the Arctic Refuge

We know what drilling will do to the Arctic Refuge—ruin one of America's wildest and most special places forever. We also know what drilling won't do. It won't solve our nation's energy problems. There is simply not enough oil in the Arctic

Refuge to reduce gas prices for consumers or to make America energy independent. At current rates of consumption, oil from the Arctic Refuge could provide at best a few months, not years, worth of energy. And more drilling leads to more burning of fossil fuels, which only accelerates the most pressing environmental threat of our day: global warming.

It does not make sense to ruin one of America's last wild places for a short-term supply of oil and gas, especially when we could save much more oil and gas through energy efficiency or by using renewable energy.

It's time to use America's technological know-how and innovation to reduce our dangerous dependence on fossil fuels like oil and gas. Congress took the first step in 2007 when members voted on—and President Bush signed into law—a bill to increase the fuel economy of our cars and light trucks.

In addition to fuel economy, we need to encourage the use of renewable energy like wind and solar power and adopt other existing energy-saving technologies that cut pollution, curb global warming, and create good jobs. Such a strategy will decrease our dependence on oil, reduce pollution, and protect wild places like the Grand Canyon, Old Faithful, and the coastal plain of the Arctic National Wildlife Refuge.

Critical Thinking Questions

1. Summarize each author's main points and supporting information.
2. Using your critical thinking skills, analyze each viewpoint.
3. Do you agree with one viewpoint? Which one? Why?

---

off excess gas, a process called **flaring**. This process releases hydrogen sulfide gas and a variety of pollutants, including carbon dioxide, carbon monoxide, and volatile organic compounds that are harmful to people and the environment. The Alberta (Canada) Research Council identified 200 chemical compounds in flare gas. In Alberta, 1.4% of the natural gas drawn from wells is flared.

Like oil, it is easy to transport natural gas within countries via pipelines. It is much more difficult to transport overseas, however. To do so, it must be compressed or compressed and liquified. Liquification requires large quantities of energy. It is then transported by ship. Natural gas is also fairly economical and requires relatively little energy to extract.

Natural gas is often associated with oil deposits and coal beds though it can be found by itself as well. In the eastern United States, natural gas is found in deep deposits of shale, a sedimentary rock formed from mud. Natural gas can become trapped in low-permeability shale. In recent years, scientists and energy companies have grown excited about a potentially large deposit

in the eastern U.S. known as Marcellus Shale. It covers 104,000 square miles and is located in Pennsylvania, West Virginia, eastern Ohio, and western New York. Fracking of these deposits is required to liberate natural gas, but as just noted, fracking can have serious environmental and health impacts.

Natural gas comes from wells as deep as 10 kilometers (6 miles). On land, drilling rigs generally have minimal impact unless they are located in wilderness areas, where operations destroy valuable land and roads and noise from heavy machinery and construction camps can disturb wildlife. However, natural gas extraction can cause

Go Green

Wearing more clothing, even insulating underwear and sweaters, indoors in the winter greatly increases comfort. Dress in several layers so you can find just the right combination to create comfort. By wearing warmer clothing, you can turn the thermostat down, saving energy and the environment.

subsidence in the vicinity of the well. One notable example is in the Los Angeles-Long Beach harbor area, where extensive oil and gas extraction began in 1928 and has caused the ground to drop 9 meters (30 feet) in some areas.

Natural gas is generally safe to transport in the gaseous form in pipelines, but as noted earlier, to transport it across oceans, it must be compressed. In the liquid form, natural gas is unstable and highly flammable. A ship containing liquefied natural gas could burn intensely.

KEY CONCEPTS

Natural gas is a combustible gas extracted from deep onshore and offshore wells. Natural gas is used to heat homes, cook food, and heat water. It is also used by factories to produce heat and some power plants to generate electricity. It is easy to transport within countries or from one country to its terrestrial neighbors. Natural gas is a clean fuel compared to oil and coal.

Coal

Coal is a solid organic fuel extracted from mines. It is the most abundant fossil fuel on Earth and is therefore relatively inexpensive. However, the production and consumption of coal are more environmentally disruptive than for any other major fossil fuel. Significant impacts occur at virtually every step in the process. (**TABLE 14-1** summarizes many of these impacts.)

KEY CONCEPTS

Coal is the most abundant fossil fuel, but its extraction and use result in enormous social, economic, and environmental costs.

Impacts of Coal Production: Surface Mining Coal is extracted by surface and underground mines, depending on the terrain and the depth of the seam. In hilly terrain where coal seams lie near the surface, as in the eastern United

States, bulldozers strip the rock and dirt lying over them (the **overburden**). The overburden is hauled away and eventually replaced after the coal supply has been depleted.

This type of mining is called **surface mining**, a term that refers to any mine that accesses its mineral from the surface. Several types of surface mines exist. Surface mines that follow the contour of the land are appropriately called **contour strip mines** (**FIGURE 14-10**). Contour strip mining defaces the landscape. Exposed hillsides can erode during rainy periods if proper precautions are not taken. Contour strip mining in the hilly terrain of Kentucky, for instance, has been shown to increase erosion from 0.4 metric tons per hectare (0.18 ton/acre) in undisturbed watersheds to nearly 150 metric tons per hectare (67 tons/acre). Access roads can also erode. During heavy rains, sediment from the roadways washes into nearby streams. Sediment also fills streams, reducing their water-carrying capacity. When rain falls, water spills over the streams' banks, flooding farms and communities.

Another mining technique in hilly or mountainous terrain is **mountaintop removal** (MTR) **mining**, which occurs predominantly in West Virginia and eastern Kentucky. In this technique, entire mountain tops are removed by dynamite and bulldozers to expose coal seams. The overlying material is often bulldozed into neighboring valleys, burying streams and forests. Although it is prohibited by law, this technique was given a boost by the George W. Bush administration in 2002, which rewrote Clean Water Act regulations. The changes made it legal to fill in valleys and streams by such operations. In mountaintop mining, forests are first clear-cut. The rock and dirt over the coal seam is scraped away destroying vegetation that can lead to flooding and landslides. The overburden is also dynamited to make it easier to remove. Explosions can damage foundations of homes and water wells. "Fly rock" may rain off mountains, endangering resident's lives and homes.

According to Mountain Justice, a nonprofit that opposes mountaintop mining, coal companies have buried over 1,920 kilometers (1,200 miles) of biologically crucial Appalachian

TABLE 14-1			
Major Environmental Impacts of Fossil Fuels			
Fuel	**Extraction**	**Transportation**	**End Uses**
Coal	Destruction of wildlife habitat, soil erosion from mine sites, sedimentation, aquifer depletion and pollution, acid mine drainage, subsidence, black lung disease, accidental death	Air pollution and noise from diesel trains	Air pollution from power plants and factories, especially acid pollutants and carbon dioxide; thermal pollution of waterways
Oil	Offshore leaks and blowouts causing water pollution and damage to fish, shellfish, birds, and beaches; subsidence near wells	Oil spills from ships or pipelines	Air pollution similar to that from coal
Natural gas	Subsidence and explosions	Explosions, land disturbance from pipelines	Fewer air pollutants than coal and oil

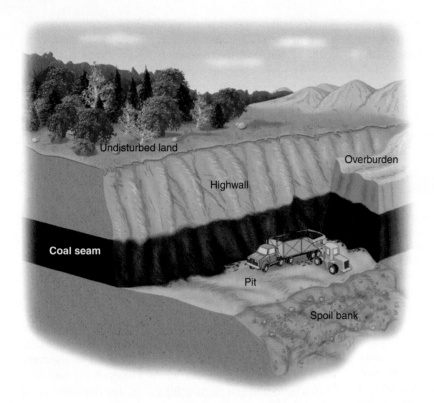

FIGURE 14-10 Contour strip mining. Contour strip mining is common in hilly terrain—for example, in the eastern U.S. coal fields. For many years, overburden was dumped on the downslope, destroying vegetation and creating serious erosion problems. Today, overburden is hauled away and stored. When the coal seam is exhausted, the overburden is replaced, graded, and planted.

headwater streams. Although companies are supposed to reclaim the land, this organization claims that "all too often mine sites are left stripped and bare" (**FIGURE 14-11**).

Liquid waste, usually generated by MTR coal-cleaning operations associated with the mining, is stored in large ponds behind manmade dams. This waste contains many toxic chemicals. Breaks in dams that often follow heavy rains release the waste into streams, causing extensive damage. One such spill released 306 million gallons of waste that polluted 75 miles of stream. Another dam failure in West Virginia killed 125 people and injured 1,100. Four thousand people were left homeless.

FIGURE 14-11 Mountaintop coal mine in West Virginia. Mines like this are extremely destructive to the environment and technically illegal, a fact that hasn't stopped some mining companies from engaging in this practice.

© Orjan F. Ellingvag/Dagens Naringsliv/Corbis.

In the relatively flat terrain of the West and Midwest, coal is usually mined in **area strip mines** (**FIGURE 14-12A**). In this type of surface mine, the topsoil is first removed by scrapers and is set aside for reapplication (**FIGURE 14-12B**). Next, the overburden is dynamited and removed by huge shovels called **draglines** to expose the coal seam. The coal is removed and hauled away, and another parallel strip is cut. The overburden from the new cut is then placed in the previous one (Figure 14-12B). As required by the **Surface Mining Control and Reclamation Act** (1977), the overburden must be regraded to the approximate original contour and the topsoil replaced. Seeds are sown, increasing the likelihood that the area will revegetate.

Area strip mines create eyesores, destroy wildlife habitat and grazing land, and may increase erosion. Proper reclamation can restore wildlife habitat and grazing land and eliminate the visual impact. Surface mines can also disrupt and pollute groundwater supplies in the West because many aquifers are located in or near coal seams. In Decker, Montana, extraction of coal from an aquifer seam resulted in a drop in the water table of 3 meters (10 feet) or more within a 3-kilometer (2-mile) radius of the mine. Some of the residents who depended on the aquifer were forced to drill deeper wells or find new water supplies.

KEY CONCEPTS

Coal is removed by surface and underground mines, both of which result in many environmental impacts. Surface mining is especially damaging to the environment; proper controls can greatly reduce impacts, though, and mine reclamation can help companies return lands to their original use.

(a)

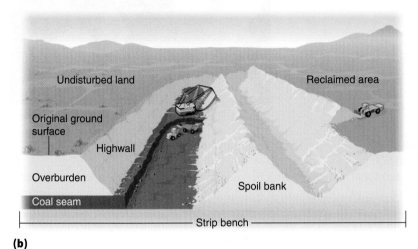

(b)

FIGURE 14-12 Area strip mine. (a) Aerial view of a strip mine. **(b)** In flat terrain, coal is extracted by strip mining. A dragline removes the overburden to expose the coal seam. Overburden is placed on the previously excavated strip and eventually regraded and replanted. If land is not carefully recontoured and replanted, it could be permanently ruined.

Photo courtesy of H. E. Malde/USGS Photo Library, Denver, Colorado.

Impacts of Coal Production: Underground Mines

In mountainous terrain where coal seams lie deep below the surface, coal is extracted via underground mines. Underground coal mines are notorious for explosions and cave-ins, and underground coal mining is ranked as the most hazardous of the major occupations in the United States. Between 1900 and 2003, 104,721 Americans were killed in underground coal mines, and over 1.63 million were permanently disabled. Because of stricter safety regulations, deaths from coal mining have dropped substantially since 1950 with the most dramatic decreases occurring since 1985. Fortunately, coal mining deaths have fallen dramatically in recent years. "While annual coal mining deaths numbered more than 1,000 a year in the early part of the 20th century," notes the Mine Safety and Health Administration (MSHA) of the U.S. Department of Labor, "they decreased to an average of about 451 annual fatalities in the 1950s, and to 141 (per year) in the 1970s." The MSHA report goes on to note that "From 2006–2010, the yearly average number of fatalities in coal mining decreased to 35." In 2011,

21 coal miners were killed in accidents. In 2012, 19 coal miners were killed in accidents.

While deaths in coal mines has declined substantially in the United States, other countries face serious challenges. China is the world's leading coal-producing nation. In 2007, a gas leak in an underground coal mine killed 35 workers in China. Earlier that year, 181 coal miners died when two coal mines flooded as a result of heavy rains. Before 2006, China's coal mines averaged 13 deaths a day as a result of fires, explosions, and floods, largely because of lax safety standards. That is nearly 5,000 deaths every year! Since then, coal mine deaths have dropped nearly by half, according to China's State Administration of Work Safety.

Underground coal mines also cause **black lung disease**, or **pneumoconiosis** (NEW-moe-cone-ee-OH-sis), a progressive, debilitating disease caused by breathing coal dust and dirt particles (**FIGURE 14-13**). Victims have difficulty getting enough oxygen because the tiny air sacs (alveoli) in the lungs break down. Exercise becomes difficult, and death is slow and painful. Despite safety improvements, one-third of all U.S. underground coal mines still have conditions conducive

(a)

(b)

FIGURE 14-13 Black lung disease. (a) Cross-section of a normal lung.
(b) Cross-section of a lung from a deceased coal miner who suffered from
black lung disease. The black material is carbon from coal dust.

Photo a © Donna Beer Stolz, Ph.D., Center for Biologic Imaging, University of Pittsburgh Medical School; photo b © Ken Eward/Science Source/Science Source.

to black lung. Between 1968 and 2010, black lung disease killed 76,000 U.S. miners, an average of more than 900 per year, according to the National Institute of Occupational Safety. In recent years, the death rate has increased due to lax regulations.

Collapsing mines also cause subsidence, a sinking of the surface. Cracks form on the surface, ruining good farmland. In some cases, streams vanish into the fissures. Over 800,000 hectares (2 million acres) of land has subsided in the United States from underground coal mining. For every hectare of coal mined in central Appalachia, over 5 hectares (12.5 acres) of surface becomes vulnerable to subsidence.

Another problem with underground mines is fires. In Pennsylvania, for example, some fires have burned for years, producing considerable amounts of pollution.

Many abandoned coal mines in the East leak sulfuric acid into streams. **Acid mine drainage**, as it is called, consists of sulfuric acid formed from water, air, and sulfides (iron pyrite) in the mine. A bacterium (*Thiobacillus thioxidans*) facilitates the conversion, making the waters highly acidic. Acid mine drainage kills plants and animals and inhibits bacterial decay of organic matter in water, allowing large quantities of organic matter to build up in streams. Sulfuric acid also leaches toxic elements such as aluminum, copper, zinc,

and magnesium from mines and mine wastes and carries them to streams.

Acid can render water unfit for drinking and swimming. Municipal and industrial water must be chemically neutralized before use. Acid also corrodes iron and steel pumps, bridges, locks, barges, and ships, causing millions of dollars of damage each year.

Active and abandoned U.S. mines produce several million metric tons of acid a year. Acid mine drainage pollutes over 15,530 kilometers (9,709 miles) of U.S. streams, most of which are in Appalachia (**FIGURE 14-14**).[1]

It also has polluted 73,000 hectares (180,000 acres). of reservoirs. Cleaning up the abandoned mines could take decades and billions of dollars, and further increases in coal production could increase acid mine drainage although awareness of the problem has resulted in efforts to reduce it.

Underground mines produce enormous quantities of wastes, which are transported to the surface and dumped near the mouth of the mine. These wastes, called **mine tailings**, contain heavy metals, acids, and other pollutants that wash into streams during rains. Coal-cleaning plants, which are designed to crush the coal and wash away impurities, also produce enormous quantities of waste. These wastes are often stored in holding ponds that sometimes leak and pollute nearby streams.

After the coal is extracted, it is loaded onto trucks or rail cars and shipped to power plants. Diesel trains, which move most of the nation's coal, pollute the air with

[1]Some estimate that as many as 19,200 kilometers (12,000 miles) of stream.

(a)

(b)

FIGURE 14-14 Acid mine drainage. (a) Map showing streams affected by acid mine drainage in the eastern United States. Acid mine drainage is prevalent in regions where rock contains the sulfur-containing compound known as iron pyrite. **(b)** Stream severely damaged by acid mine drainage.

Photo courtesy of D. Hardesty/USGS.

particulates and other potentially harmful pollutants such as sulfur dioxide.

Impacts of Coal Combustion   About 90% of U.S. coal is burned to generate electricity; most of the rest is burned by industry to generate heat or to produce steel and other metals. Power plants require enormous amounts of coal. For example, a 1,000-megawatt coal-fired power plant, which produces electricity for about 1 million people, burns about 2.7 million metric tons (3 million tons) of coal each year. Tens of thousands of tons of pollutants are released into the atmosphere during coal combustion, even with pollution control devices in place. These include particulates, sulfur oxides, nitrogen oxides, carbon monoxide, and carbon dioxide. These pollutants are largely responsible for two of the world's most significant environmental problems, acid deposition and global climate change. They cause billions of dollars in damage to fish, lakes, buildings, and human health.

Coal combustion also produces enormous quantities of solid waste. During combustion, a fine dust known as fly ash is produced. **Fly ash** is mineral matter that makes up 10% to 30% of the weight of uncleaned coal. It is carried up the smokestack with the escaping gases and, if captured by pollution control devices, becomes a hazardous solid waste. Sulfur dioxide gas is also emitted from the stack but can be removed by a pollution control device known as a **smokestack scrubber**. Scrubbers produce a toxic sludge, containing fly ash and sulfur compounds. Some mineral matter that is too heavy to form fly ash remains at the bottom of the coal-burning furnace as **bottom ash**. It, too, is a hazardous waste that must be disposed of. Wastes from coal-burning furnaces and their pollution control devices are generally buried in landfills, but toxic substances from these sites may leak into groundwater supplies, polluting them.

A 1,000-megawatt power plant may produce 180,000 to 680,000 metric tons (200,000 to 750,000 tons) of solid waste each year, including fly ash, bottom ash, and sludge from scrubbers. In 2012, coal-fired power plants in the United States produced about 109 million metric tons (121 million tons) of solid waste. Approximately 47% of the waste was used to make road surfaces and other products; the rest ended up in landfills.

Oil Shale

Oil shale is a sedimentary rock containing a combustible organic material known as **kerogen**. Oil shale is found in large deposits underlying much of the continental United States, with the richest ones in Colorado, Utah, and Wyoming. Similar deposits are also found in Canada, Russia, and China.

Kerogen can be extracted from oil shale by heat. This process produces a thick, oily substance known as **shale oil**. Like crude oil, it can be refined to produce a variety of fuels and useful chemicals.

Oil shale is rather abundant, but its extraction and processing are energy intensive and therefore costly. **Net energy analysis**, which determines how much energy it takes to extract and process an energy resource in relation to the amount produced, shows that shale oil production creates only about one-eighth as much energy as conventional crude oil extraction for the same energy investment. Its **net energy yield** (energy produced – energy invested = net energy yield) is therefore very low.

Oil shale development could result in serious environmental impacts as well. Strip mining to remove oil shale disturbs large tracts, causing erosion and reducing wildlife habitat. Mined shale is then crushed and heated in a large vessel, called a **surface retort**. Surface retorting produces enormous amounts of solid waste, spent shale. A small operation producing 50,000 barrels per day would generate about 19 million metric tons (21 million tons) of waste per year. Since the shale expands by about 12% on heating, not all of it could be disposed of in the mines from which the raw ore came. Dumped elsewhere, the spent shale may be leached by water, producing an assortment of toxic organic pollutants that could contaminate underground and surface waters.

Oil shale retorts require large quantities of water as well, but oil shale deposits are typically found in arid country. Retorts also produce significant amounts of air pollutants unless carefully controlled.

To bypass the solid waste problem, oil shale companies once experimented with *in situ* (IN SEE-two) **retorting**. In this process, shale deposits are first fractured by explosives. A fire is then started underground in the oily rock and is forced through the deposit. The heat from the fire forces some of the oil out of the rock. The oil is then collected and pumped to the surface.

In situ retorts eliminate many impacts caused by surface mining, but they have not worked well because it is difficult to fracture shale evenly (a step needed to promote uniform combustion) and because groundwater often seeps into them and extinguishes the fires.

In the early 1990s, the only oil shale plant operating in the United States, which cost $1 billion to build, closed its doors because it was barely covering operating costs and was producing shale oil at twice the cost of conventional crude oil.

Today, Shell oil company is experimenting with a new process that they believe could make oil shale development cost-effective. They drill holes in the oil shale and insert electric heaters into them. The oil shale is heated for 2 to 3 years until it reaches 343° to 371°C (650° to 700°F). Oil is released from the rock. It flows into collection wells in the shale and is then pumped to the surface. To contain the oil, they also freeze a large 6 to 9 meter (20 to 30 foot) region around the extraction zone to contain the oil and to prevent water from entering it. Both processes require lots of energy. Although Shell predicts that the final cost will be low, only time will tell. Huge amounts of coal-fired electricity would be needed to run the operation. To develop oil shale, numerous new power plants will need to be constructed and operated in oil shale regions. These plants would require huge amounts of water for cooling towers that would be difficult to obtain in the arid regions in which much of the United States' oil shale is located.

KEY CONCEPTS

Oil shale is a sedimentary rock containing an organic material (kerogen) that can be extracted by heating the rock. In a liquid state, this thick, oily substance, called *shale oil*, can be refined to make gasoline and a host of other chemical byproducts. Although it is abundant, oil shale is economically and environmentally costly to produce and is currently an insignificant source of oil.

Tar Sands

Tar sands, or **oil sands**, are sand deposits containing a combustible organic material called **bitumen**. They are another fossil fuel. Although they are found throughout the world, the largest deposits are in Alberta, Canada; Venezuela; and Russia. In the United States, six states have large deposits.

Tar sands are strip mined and then treated with hot water to extract the bitumen. *In situ* methods similar to those used in oil shale extraction have also been tested.

Tar sands are plagued with many of the problems that face oil shale. They are expensive to mine. In addition, oily sand sticks to machinery, gums up moving parts, and eats away at tires and conveyor belts. Tar sand expands by 30% after processing. As with oil shale, its production requires large amounts of water, which becomes badly polluted in the process.

The most significant barrier to tar sand development is a low net energy yield, and thus high cost. At least 0.6 of a barrel of energy is consumed per barrel produced. Worse, world reserves of tar sand oil are insignificant compared with world oil demands. Locally, however, tar sands can make a huge difference. Canada's oil sands, which is the largest deposit in the world, contain an estimated 1.7 trillion barrels of oil, of which some experts say 300 billion could be recovered. That's about 13 years worth of global oil demand, but enough to supply Canada at its current oil demand for 200 years. As noted earlier, oil sands are costly and environmentally damaging. The Canadian government has poured billions of dollars into research and development. By 2020, total investment is expected to be about $30 billion. Environmentalists argue that the federal and provincial governments have paid little attention to the potential impacts of tar sand development.

KEY CONCEPTS

Tar sands are sand deposits impregnated with a petroleum-like substance known as *bitumen*. Although there are large deposits, they are economically and environmentally costly to extract.

Coal Gasification and Liquefaction

The abundance of coal in the United States and abroad has stirred interest in coal gasification and liquefication. In both technologies, coal reacts with hydrogen to form **synfuels**, combustible organic fuels. **Coal gasification** produces combustible gases that can supplement natural gas. **Coal liquefaction** produces an oily substance that can be refined to make gasoline.

Coal gasification produces numerous air pollutants and requires large quantities of water, making it a dirty and resource-intensive alternative to natural gas. The economic cost of synthetic natural gas is high. An additional problem is the low net energy production. Synthetic gas produced from surface mines is about 1.5 times more expensive than natural gas, and synthetic gas from coal taken from underground mines is 3.5 times more expensive.

Coal liquefaction, the production of synthetic oil from coal, is similar to coal gasification. Although there are at least four ways of making a synthetic oil from coal, each involves the same general process: adding hydrogen to the coal. The oil produced by liquefaction must be purified to remove ash and coal particles.

Coal liquefaction could provide liquid fuels, but it would be costly. It produces air and water pollutants and requires large amounts of energy. Like coal gasification, it might be preempted by cheaper, cleaner, and renewable energy sources.

Coal gasification and liquefaction plants are currently in operation but produce relatively small amounts of liquid fuel.

Coal can be converted to gaseous and liquid fuels to replace oil and natural gas, but the processes are energy-intensive, inefficient, produce a lot of pollution, and are quite expensive.

14.4 Fossil Fuels: Meeting Future Demand

The sustainability of a fuel source must be judged on the basis of its social and environmental costs as well as its economic costs. An analysis of supplies is equally important. However, determining how much fossil fuel is available and how long it will last is fraught with difficulties. First, no one knows how much economically recoverable fossil fuel lies within the Earth's crust. Moreover, some countries have grossly exaggerated supplies, especially oil reserves. Second, future demand is not entirely predictable. How much will the former Soviet Union's and China's transformations to capitalistic economies dramatically increase fossil fuel consumption? How much will efforts in India, which houses over one-sixth of the world's people, increase? Or will the nations of the world adopt energy-efficient, renewable technologies on a large scale, cutting energy demand?

This section examines the projected lifetime of three fossil fuels: oil, natural gas, and coal. When examining the prospects for energy supplies, keep in mind the uncertainties outlined previously.

Oil: The End Is Near

Some experts predict that when the last drop of crude oil has been burned, the world's oil fields will have yielded between 1,800 and 2,200 billion barrels of oil. This figure is called the **ultimate production**. About half of this has already been used. So what's the problem? At the current rate of consumption, all remaining oil would last about 50 years.

The problem is that oil consumption is increasing, on average, approximately 2.5% per year. If oil consumption continues to grow at this rate the remaining oil will be gone much sooner.

Long before the last drop is burned, though, signs of shortage will be evident. You can understand this by studying **FIGURE 14-15**, which shows graphs of petroleum production in the United States and the world. As illustrated, world oil production is expected to rise to a peak and then begin to fall as reserves are depleted. When? No one knows for sure, but it has not yet occurred as of 2014. Optimistic predictions forecast a peak—very soon—after 2020. While global oil production has probably not yet peaked, many countries have already experienced peak production including the United States (1971), Iran (1988), Indonesia (1994), the United Kingdom (1999), Norway (2001), and Mexico (2004). As oil reserves decline, oil companies will have to work much harder to maintain production, which will raise the price of oil considerably. In addition, as domestic

(a) United States

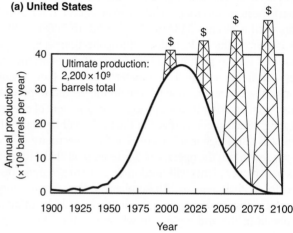

(b) World

FIGURE 14-15 The fate of U.S. and global oil supplies. Petroleum production curves for **(a)** the United States and **(b)** the world. U.S. oil production has already peaked. The oil derricks represent the increasing cost of drilling to extract declining supplies.

oil reserves fall, many nations will turn to the Middle East, which currently houses about two-thirds of the world's proven oil reserves (**FIGURE 14-16**). Christopher Flavin and Nicholas Lenssen of the Worldwatch Institute liken the developed world's dependency on oil to an addiction, note, "Not only is the world addicted to cheap oil, but the largest liquor store is in a very dangerous neighborhood."

Increasing our dependence on a fuel from a politically volatile region of the world can have devastating consequences. For example, the 1973 oil embargo imposed by the **Organization of Petroleum Exporting Countries** or **OPEC** and a second embargo by Iran in 1979 drove the price of oil from $3 a barrel to $34 in a short period, stimulating rapid inflation that brought the industrial world to its knees. The 1991 Persian Gulf War was another unfortunate consequence. This war cost several more developed nations $61 billion, a price paid primarily by the United States, Japan, and Saudi Arabia. Many observers believe that America's military involvement in the Middle East—which has costs hundreds of billions of dollars—is largely due to our desire to secure oil supplies.

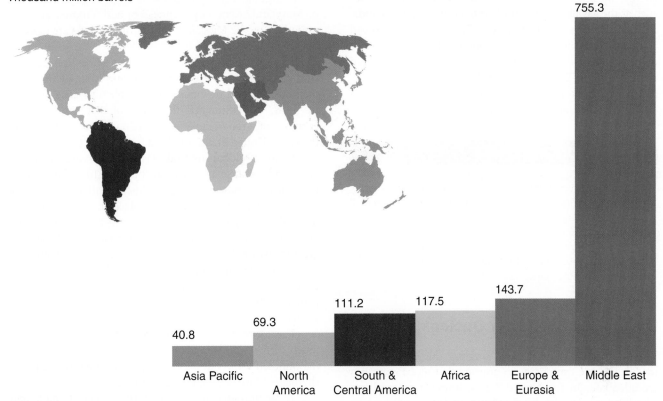

Proved reserves at end 2007
Thousand million barrels

Asia Pacific: 40.8
North America: 69.3
South & Central America: 111.2
Africa: 117.5
Europe & Eurasia: 143.7
Middle East: 755.3

FIGURE 14-16 Where will the industrial countries get their oil? This map shows where the world's oil lies.

Since 2004, oil and gasoline prices have increased dramatically upward as a result of many factors, including rising demand in the United States and China, a lack of oil refinery capacity, and insufficient supply. Some oil analysts pinned much of the blame on insufficient supply in relation to demand.

If demand for oil outstrips supply worldwide, exorbitant oil prices and global inflation could occur. If alternative energy sources are not available, many countries could experience severe economic turmoil: inflation, economic stagnation, and widespread unemployment. By 2050, oil production will be a small fraction of current levels.

Go Green

To save energy, slow down. Driving 65 miles per hour uses about 12.5% more gasoline than driving 55 mph. Driving 75 mph uses 25% more fuel than driving 55 mph.

Can't we simply find more oil? Although there's still more oil to find, it's unlikely that oil companies will make any major new finds to relieve the situation. U.S. oil fields are largely depleted. About 75% of our oil supplies have been used up (Figure 14–15a). Huge oil deposits in Texas and Alaska, for instance, are producing less and less oil each year. In fact, Texan and Alaskan oil production have dramatically declined since 1972 and 1988, respectively. World oil supplies are better but we may be at the half way point—that is, we may have consumed half of the global ultimate production. But what about future discoveries? Oil companies haven't found a really large oil deposit since 1962. Of the 40,000 oil fields in operation today, only about 40 have been huge. Today, the world consumes about 26.5 billion barrels of crude oil per year, yet annually discovers only about 4 billion new barrels of oil. Our prospects are not good. Many of the world's large oil fields are on the decline. As **TABLE 14-2** shows, oil production has already peaked and is now declining in many countries such as Mexico, Iran, and Russia. Even the mammoth oil fields of Saudi Arabia are thought to have peaked.

TABLE 14-2

Selected Nations in Which Oil Production Has Peaked

Country	Year Oil Production Peaked
United States	1970
Iran	1973
Russia	1988
Indonesia	1994
United Kingdom	1999
Norway	2001
Mexico	2004

Data from Association for the Study of Peak Oil.

Rising fuel prices could stimulate a massive shift toward energy efficiency—for example, new more efficient vehicles such as the gas-electric hybrid Toyota Prius, plug-in hybrids, and electric cars (**FIGURE 14-17**). To prevent widespread economic turmoil, however, nations must act soon to use oil more efficiently and to find replacements for oil—which powers our transportation system, heats some of our homes, and provides the raw materials for asphalt, plastics, medicines, cosmetics, and much more.

KEY CONCEPTS

Global oil supplies appear to be quite large; however, when one calculates how long they will last (based on the historical increases in energy consumption), it appears that oil supplies may last only 20 to 40 years. Long before the last drop of oil is removed, however, oil production will peak, resulting in a dramatic increase in prices that could have serious social and economic repercussions.

Natural Gas: A Better Outlook

The outlook for natural gas is brighter than the outlook for oil. For the time being, it appears that global supplies are adequate. In the United States, however, natural gas supplies are running low (**FIGURE 14-18**). As illustrated in Figure 14-18, natural gas production peaked in 1973, declined, then plateaued for many years. As you can see from the graph natural gas production has risen in recent years. This largely is due to a dramatic increase in gas drilling. The number of wells has increased from 342,000 to 482,000 from 2000 to 2012, according to the U.S. Energy Information Administration. Some relief may result from natural gas found in the Marcellus Formation in the eastern U.S., mentioned earlier. This rock is a black, low density,

organic rich shale that occurs beneath much of Ohio, West Virginia, Pennsylvania, and New York. Small areas of Maryland, Kentucky, Tennessee, and Virginia are also underlain by the Marcellus Shale. All locations are close to major population areas.

Some estimate that huge amounts of natural gas can be extracted from these deposits. Using some of the same methods that had previously been applied to shale deposits in Texas, it is estimated that 10% of that gas might be economically recoverable, which is only enough to supply the entire United States for about 2 years.

Because global natural gas supplies are relatively large and because it is a clean-burning fuel that can be used in cars, power plants, industry, and homes, many believe that natural gas could replace the more environmentally damaging fossil fuels, oil and coal. In fact, sustainable energy proponents believe that natural gas is a great transition fuel. It will help wean us from coal and oil, the most environmentally unsustainable fossil fuels, and provide us time to develop a sustainable energy system primarily based on conservation and renewable energy sources.

There are problems however, with this strategy, as much of the world's remaining natural gas supplies are in the Middle East. As domestic supplies decline, prices could skyrocket. Can't we merely import more?

To be imported from all countries except Canada and Mexico, natural gas must be compressed and shipped by tanker. Today, the United States imports only 2% of its natural gas this way and is not equipped to handle much more. Increasing imports would require numerous ports to develop facilities to handle ships and highly explosive gas, and many cities have rejected plans to do so. Even if we could, some experts predict that global natural gas production will peak between 2015 and 2025.

KEY CONCEPTS

Natural gas is a clean and rather abundant fossil fuel that could provide a source of energy as the world makes the transition to a more sustainable energy supply system.

Coal: The Brightest Outlook, the Dirtiest Fuel

Coal is the world's most abundant fossil fuel. Geologists estimate that the world's recoverable coal reserves are around 800 or 900 billion tons. While that seems like a lot of coal, at the current extraction rate, coal reserves would last only 132 years. The rate of coal consumption has been at 2 to 3% per year for many years. If the rate continues at 2.5% per year, the world's recoverable coal reserves will last 56 years—within the lifespan of most readers.

The United States has 27% of the world's coal supplies. Recoverable coal reserves are estimated to be about 257 billion short tons (a short ton is 2,000 pounds)

FIGURE 14-17 Toyota Prius, Gen II. This five-passenger car owned by the author is the second generation of the world's first mass-produced gasoline-electric hybrid vehicle. This car is powered by a gasoline-fueled internal combustion engine and a clean, quiet electric motor. Its low emissions earn its title as a super ultra-low-emission vehicle.

Courtesy of Toyota Motor Sales.

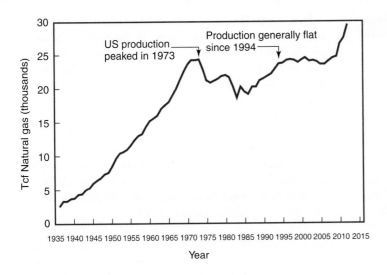

FIGURE 14-18 U.S. marketed natural gas production from gas wells, oil wells, and coal bed wells, 1930–2012. .

Adapted from U.S. Energy Information Administration, U.S. Natural Gas Withdrawals.

or 206 billion metric tons, according to the U.S. Energy Information Administration; the U.S. consumes about 889 million short tons a year. At that rate, U.S. coal supplies will last 290 years. U.S. coal supplies could last a very long time, even at increased levels of consumption.

Although growth rates in use would greatly reduce the life span of coal, the large supply suggests that coal could be with us for many years to come. Unfortunately, evidence strongly suggests that coal is an environmentally unsustainable fuel. In fact, with the exception of oil shale, coal is the dirtiest of all fossil fuels. Its combustion produces enormous amounts of ash, toxic sludge, particulates, and gaseous air pollutants. Although cleaner and more efficient ways to burn coal are being developed, progress has been slow. Air pollution control technologies and new clean-coal technologies do little to prevent one of the world's most potentially troubling environmental problems: global climate change caused in large part by carbon dioxide emissions from fossil fuel burning facilities, especially coal-fired power plants.

> **KEY CONCEPTS**
>
> Coal supplies in the United States and many other countries are abundant, but coal extraction and combustion make it an environmentally unsustainable fuel.

14.5 Nuclear Energy

Nuclear energy provides a growing percentage of the world's energy demands. In Canada, for instance, nuclear power provides approximately 7% of the total energy. In the United States, it satisfies 8.5% of our total energy demand. The nuclear reactors in use today in most nuclear power plants are fueled by naturally occurring uranium-235 (U-235). To understand how nuclear fuel works and some of its dangers, you must understand the atom and radioactivity.

Atoms: The Building Blocks of Matter

All matter is composed of tiny particles called *atoms*. Atoms, in turn, join to form molecules. The **atom** is composed of two parts: a centrally placed nucleus and an electron cloud (**FIGURE 14-19**). The **nucleus** contains two tiny subatomic particles, protons and neutrons, and constitutes 99.9% of the mass of an atom. Protons are positively charged particles. Neutrons have no charge.

Encircling the nucleus is a comparatively large area, the **electron cloud**. It contains much lighter particles of an atom, the **electrons**. These negatively charged particles orbit around the positively charged nucleus. To give you an idea of the relative size of the electron cloud and nucleus, a simple analogy will suffice. If an atom were the size of Mt. Everest (about 8,800 meters or 29,000 feet), the nucleus would be about the size of a football. The rest of the atom would be made of the electron cloud.

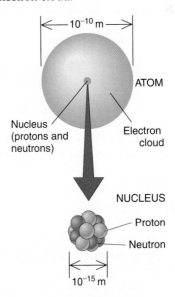

FIGURE 14-19 Structure of the atom. Atoms consist of a dense central region, the nucleus, containing protons and neutrons. Electrons are found in the much larger electron cloud, which surrounds the nucleus.

Radioactive Atoms

If you have ever studied chemistry, even a brief course in high school, you learned that atoms form elements, such as carbon, nitrogen, gold, and uranium. An **element** is the purest form of matter. Each element contains one type of atom. Pure carbon, for instance, contains only carbon atoms. Pure gold contains only gold atoms. Even so, not all atoms in an element are identical. Although carbon contains only carbon atoms, some carbon atoms are slightly different from others. These differences are small. They don't affect the chemical properties of the atoms, that is, how they react, but they can have profound consequences that cause some atoms to emit radioactivity.

How do atoms differ? Although all atoms of a given element contain the same number of protons in their nuclei, some contain slightly different numbers of neutrons. For example, all uranium atoms contain 92 protons, but some have 146 neutrons and others 143. These alternate forms are called **isotopes**. To distinguish one isotope from another, scientists add up the protons and neutrons and tack the sum of these two onto the name of the element. For example, the isotope of uranium that contains 146 neutrons is called uranium-238 because it contains 92 protons and 146 neutrons.

Excess neutrons in some isotopes make them unstable. To reach a more stable state, they emit **radiation**, high-energy particles or bursts of energy. Unstable, radioactive nuclei are called **radionuclides** (ray-de-oh-NEW-klides). They occur naturally or can be produced by various physical means. For the most part, the naturally occurring radionuclides are isotopes of heavy elements, from lead (82 protons in the nucleus) to uranium (92 protons in the nucleus).

KEY CONCEPTS

Elements are the purest form of matter. Although all atoms of a given element have the same number of protons and electrons, some atoms contain slightly more neutrons. Those atoms with excess neutrons tend to be unstable and emit radiation.

Radiation

Radionuclides emit three types of radiation: alpha particles, beta particles, and gamma rays. **Alpha particles** are the heaviest of the three forms of radiation. Each alpha particle consists of two protons and two neutrons, the same as a helium nucleus. Alpha particles are positively charged because they contain protons.

Because alpha particles have the largest mass of all forms of radiation, they travel only a few centimeters in air. They can

easily be stopped by a thick sheet of paper, so it is easy to shield people from this form of radiation. In the body, alpha particles can travel only about 30 micrometers (30 millionths of a meter, or about the width of three cells) in tissues. They cannot penetrate skin and are therefore often erroneously assumed to pose little harm to humans. But if alpha emitters enter body tissues—say, through inhalation—they can do serious, irreparable damage to nearby cells and their chromosomes.

Beta particles are negatively charged particles emitted from nuclei. They are equivalent to electrons, except they are more energetic. Beta particles are ejected from nuclei when neutrons in the nucleus are converted into protons, a process that helps stabilize radionuclides.

The beta particle is much lighter than the alpha particle and can travel much farther. It can penetrate a 1-millimeter lead plate and can travel up to 8 meters (27 feet) in air, but only 1 centimeter (0.4 inch) in tissue. Beta particles from some radionuclides have enough energy to penetrate clothing and skin but generally do not reach underlying tissues. They can, however, damage the skin and eyes (causing skin cancer and cataracts).

Gamma rays are a high-energy form of radiation with no mass and no charge, much like visible light but with much more energy. Gamma rays are emitted by nuclei to achieve a lower-energy, more stable state. They are often emitted after a nucleus has ejected an alpha or beta particle because even the loss of these particles does not always allow the nucleus to reach its most stable state. Some gamma rays can travel hundreds of meters in the air and can easily penetrate the body. Some can penetrate walls of cement and plaster or a few centimeters of lead.

Another common form of radiation is the X-ray. Unlike the three previously discussed forms, X-rays do not originate from naturally occurring unstable nuclei. Rather, X-rays are produced in X-ray machines. An X-ray machine contains a source of electrons, which are emitted when the machine is turned on, and a tungsten collecting terminal in a vacuum tube (**FIGURE 14-20**). Electrons strike the collecting terminal or target, collide with tungsten atoms, and are then rapidly brought to rest. The energy they carried in is released in the form of X-rays, which behave like gamma rays but have considerably less energy. Because of this they cannot penetrate lead.

All the forms of radiation described above are called **ionizing radiation** because they possess enough energy to rip electrons away from atoms, leaving charged atoms called **ions**. Ions are the primary cause of damage in tissues. Ionizing radiation differs from electromagnetic radiation such as radiowaves and magnetic fields, which are weak and unable to ionize materials.

KEY CONCEPTS

Physicists have identified three major types of ionizing radiation released by radioactive atoms: alpha particles, beta particles, and gamma rays. They vary in mass and ability to penetrate. All can cause electrons to be removed from molecules they strike. The X-ray is an artificially produced form of radiation that resembles the gamma ray.

FIGURE 14-20 Anatomy of an X-ray machine. Electrons strike a tungsten filament (red), causing it to release high-energy X-rays.

Understanding Nuclear Fission

With this background material, we continue our exploration of nuclear energy. As noted earlier, nuclear fuel contains U-235. Nuclear fuel does not burn like fossil fuels, but rather undergoes fission. **Fission** is the splitting of atoms. Atoms don't split spontaneously, though. Rather, inside a nuclear reactor U-235 atoms undergo fission when they are struck by neutrons. When they split, uranium atoms give off enormous amounts of energy (**FIGURE 14-21**). In fact, the complete fission of 1 kilogram (2.2 pounds) of U-235 could yield as much energy as 2,000 metric tons (2,200 tons) of coal!

Atomic fission takes place inside special devices, **fission reactors**, designed to contain the energy and radioactivity released during fission. As illustrated in **FIGURE 14-22**, U-235 is housed in long **fuel rods** located in the reactor core. U-235 atoms naturally emit neutrons that bombard other uranium atoms, causing them to split. The heat released during fission is used to produce steam, which drives a turbine that generates electricity. The steam is then cooled and the water is used again. Most nuclear plants are cooled by water and are called **light-water reactors (LWRs)**.

Fission Byproducts When a U-235 nucleus splits, it produces two smaller nuclei, called **daughter nuclei** or **fission fragments**. Over 400 different fragments can form during uranium fission, many of them radioactive. When U-235 nuclei split, they also release neutrons, which strike other nuclei in the fuel rods, creating a chain reaction. However, the chain reaction is controlled to keep it from

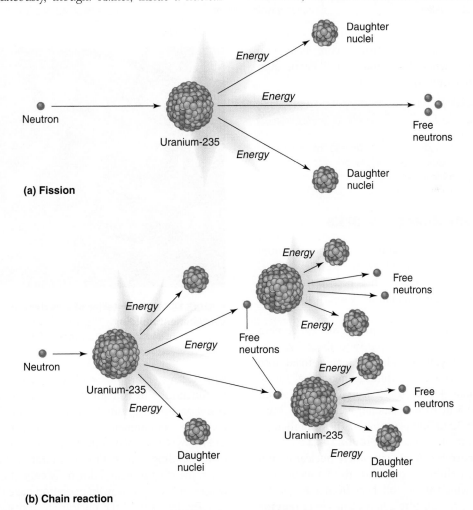

(a) Fission

(b) Chain reaction

FIGURE 14-21 Nuclear fission. (a) In a fission reaction a uranium-235 nucleus struck by a neutron splits into two smaller nuclei. Neutrons and enormous amounts of energy are released. **(b)** A chain reaction is brought on by placing fissile uranium-235 into a nuclear reactor. Neutrons liberated during the fission of one nucleus stimulate fission in neighboring nuclei, which in turn release more neutrons. Thus, the chain reaction can be sustained.

FIGURE 14-22 **Anatomy of a pressurized nuclear power plant.** As shown here, nuclear fission reactions in the reactor core heat up water to generate steam. The steam runs a turbine that generates electricity, as in conventional power plants. The water surrounding the reactor core circulates in a closed system. Heat is transferred to another closed system that generates steam. This double water-heating system helps prevent the escape of radioactivity from the system.

producing so much heat that the reactor core melts. (An atomic explosion in such a case would be unlikely because the fuel is not sufficiently concentrated.)

 KEY CONCEPTS

Splitting uranium-235 results in the release of massive amounts of energy including heat that can be used to generate steam to run a turbine to produce electricity.

How a Nuclear Reactor Works

Nuclear fuel is extracted from naturally occurring uranium ore. Uranium ore is a mixture of uranium-235, which is fissionable, and uranium-238, which is not. In the United States, most uranium ore is found in the western states such as Colorado and Utah.

To create fuel for nuclear power plants, fuel enrichment plants first concentrate the uranium-235 to about 3%. The enriched fuel, called **yellow cake**, is then formed into pellets. These are inserted into thin, stainless steel tubes, the fuel rods. About 4 meters (12 feet) long and about 1.2 centimeters (0.5 inch) in diameter, fuel rods are bundled together, with 30 to 300 per bundle. They can then be lowered into the core of the reactor (**FIGURE 14-23**).

The **reactor core** provides a site where nuclear fission can occur and be carefully controlled to avoid a meltdown. Bathed in water, which draws off heat from the nuclear fission reactions, the reactor core is housed in a large **reactor vessel** (Figure 14-23). It is made from 15-centimeter (6-inch) thick) steel (Figure 14-23).

In most countries, the reactor vessel is usually further housed in a **containment building**, a domelike structure

FIGURE 14-23 **The interior of a nuclear power plant containment building.** The reactor core is in the middle of the containment vessel, under 10.7 meters (35 feet) of water.
Courtesy of Mark Marten/U.S. Department of Energy.

with 1.2-meter (4-foot) walls of concrete (**FIGURE 14-24**). Its function is to contain the radiation in case the reactor vessel or pipes rupture. Unfortunately, not all reactors have complete containment buildings. The nuclear reactor at Chernobyl (described in the text) had only a partial containment building, which proved entirely inadequate in 1986 when the reactor melted down. Twenty of the former Soviet Union's 44 reactors now in operation have the same design.

The intensity of the nuclear chain reaction is controlled primarily by neutron-absorbing **control rods**. By absorbing neutrons, the control rods limit the rate of fission. Frequently made of cadmium or boron steel, the control rods fit in between the fuel rods and can be raised or lowered to control the rate at

FIGURE 14-24 The containment building of a nuclear power plant.

© Corbis/age footstock.

which fission takes place. They are also used to start and stop a reactor. Raising them, for instance, starts the chain reaction and lowering them back into place can shut down a reactor.

The energy released from nuclear fission in the fuel rods heats the water bathing the reactor core as shown in Figure 14-21. To avoid radioactive contamination, heat is transferred to water circulating in a separate set of pipes located in the wall of the reactor vessel. In a *pressurized nuclear reactor*, one of four basic designs, the superheated water is converted to steam that spins a turbine. The turbine is attached to an electricity generator by a shaft. As the turbine spins, the inner components of the generator spin, producing electricity, just as in coal-fired power plants.

In *boiling water reactors*, another type of reactor, water flows in a closed pipe through the reactor vessel. This superheated water boils inside the pipes and produces steam that drives a turbine directly, as shown in **FIGURE 14-25A**.

Another type of reactor, known as a *gas-cooled reactor*, is cooled by carbon dioxide gas. This gas circulates around the reactor core, as shown in **FIGURE 14-25B**. The heat it picks up is used to heat water in a separate closed loop. This water is converted to steam that drives a turbine connected to an electrical generator. This reactor uses graphite in its control rods.

The fourth and final type of nuclear fission reactor is a *light-water reactor* containing graphite control rods. Like others, the heated water is used to generate steam to run a turbine that spins an electrical generator. Of the reactors in place, most are pressurized and boiling water reactors. The third most common type is the gas-cooled reactor.

Besides the reactor vessel and containment building, nuclear reactors are equipped with a host of safety devices, among them backup power units for the controls, redundant electrical wiring in case one circuit burns out, and redundant pipes in case one bursts. Perhaps the most important is the emergency cooling system, which provides cold water to the reactor core should it become overheated. The result is a technology whose complexity exceeds most.

Nuclear Power: The Benefits

When nuclear power was first proposed, proponents claimed that its energy would be so cheap that it wouldn't be cost-efficient to meter houses. That dream has failed to materialize. In the United States, for instance, nuclear power typically costs about two to three times the cost of electricity from coal and wind farms. Building a nuclear power plant is two to six times more expensive than building an equivalent coal-fired power plant. Nevertheless, in 2012, the United States' 100 commercial nuclear power plants provided approximately 769 billion kilowatt-hours of electricity—8.5% of the nation's energy and 19% of the nation's electricity. Nuclear power also fits well into an electrical grid system that provides electricity to large numbers of people.

Perhaps the most convincing argument for using nuclear power, as opposed to coal and oil, is that it produces very little air pollution. Studies show that the release of radioactive materials into the atmosphere from nuclear power plants is insignificant under normal operating conditions. In fact, a coal-fired power plant may release more radioactivity than a normally operating nuclear power plant. Moreover, nuclear plants do not produce sulfur dioxide and nitrogen dioxide, which are released from fossil fuel combustion sources and are converted into strong acids in the air. Carried to the Earth in rain and snow, and on particulates, these acids can cause considerable damage to the environment. Nuclear power plants also produce no carbon dioxide. Another advantage of nuclear power is that it requires less strip mining than coal because the fuel is a much more concentrated form of energy. This results in less land disturbance and fewer impacts on groundwater, wildlife habitat, and so on to produce a given amount of electricity. Also, the cost of transporting nuclear fuel is lower than that for an equivalent amount of coal.

Nuclear Power: The Drawbacks

Despite its many advantages, nuclear power has some substantial drawbacks that have halted growth in the

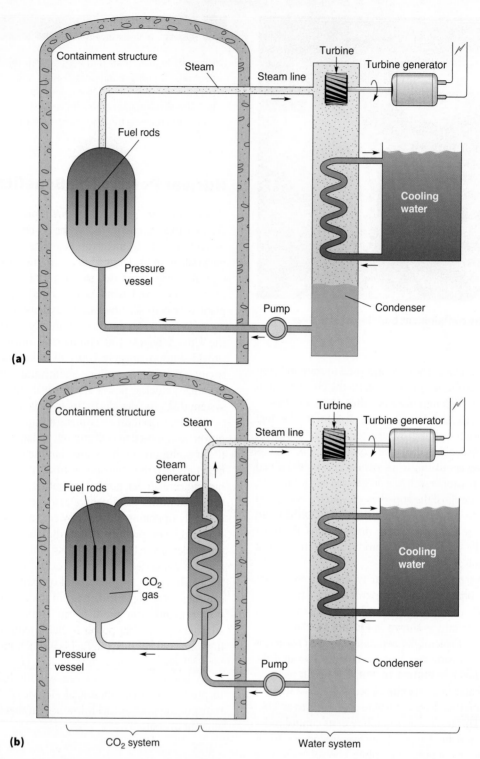

FIGURE 14-25 (a) Anatomy of a boiling water reactor and **(b)** a gas-cooled reactor.

industry in the United States: (1) waste disposal problems; (2) contamination of the environment with long-lasting radioactive materials from accidents at plants and during transportation; (3) thermal pollution from power plants; (4) health impacts; (5) limited supplies of uranium ore; (6) low social acceptability; (7) high construction costs; (8) a lack of support from insurance companies and the financial community; (9) vulnerability to sabotage;

(10) proliferation of nuclear weaponry from high-level reactor wastes; and (11) questions about what to do with nuclear plants after their useful life of 20 to 25 years is over. This section examines four major areas of concern: reactor safety, waste disposal, social acceptability and cost, and the proliferation of nuclear weapons. Before we examine these, however, a few words on the effects of radiation are in order.

Interest in nuclear power has declined substantially because of major problems, among them questions over reactor safety, unresolved waste disposal issues, low social acceptability, and high costs.

Health Effects of Radiation   All forms of radiation strip electrons from atoms in biologically important molecules in tissues. It is these charged molecules (ions) that damage and even kill cells and cause cancer. Radiation is like many other harmful substances. Its effect is determined by the dose—how much one receives—and the length of exposure. The higher the dose, the greater the effect. The longer the exposure, the greater the potential for damage.

Numerous studies have revealed some interesting generalizations about radiation: First, fetuses are more sensitive to radiation than children, who are in turn more sensitive than adults. Second, cells undergoing rapid division appear to be more sensitive to radiation than those that divide infrequently or not at all. Some of the most active—and thus most susceptible—tissues are those involved in the formation of blood cells and in immune protection. Epithelial (ep-eh-THEEL-ee-el) cells—those that line body organs such as the intestines—also undergo frequent division and are highly sensitive to radiation. In sharp contrast, nerve and muscle cells, which do not divide, have a very low sensitivity. Third, most if not all forms of cancer can be increased by ionizing radiation.

Health experts divide radiation exposure into two categories, high and low.

KEY CONCEPTS

The effects of radiation on human health depend on many factors, including the amount of radiation one is exposed to, the length of exposure, the age of the individual, and the type of cell.

Health Impacts of High-Level Radiation   The highest exposures to radiation cause death within a few hours or days. High-level doses that do not result in immediate death cause **radiation sickness**. Nausea and vomiting are the very first symptoms and develop soon after exposure. Several days later, diarrhea, hair loss, sore throat, reduction in blood platelets (needed for clotting), hemorrhaging, and bone marrow damage occur as a result of damage to fast-dividing cells. Although people survive sublethal exposures, many serious delayed effects result, including cancer, leukemia (a form of blood cancer), cataracts, sterility, and decreased life span. Pregnant women who are exposed to sublethal doses suffer from increased miscarriages, stillbirths, and early infant deaths.

Fortunately, high-level radiation exposures are rare. Such exposures would occur only in individuals working at or living very near nuclear power plants or munitions factories where a major accident has occurred. Nuclear war,

even on a limited scale, however, would also expose large numbers of people to dangerously high levels of radiation.

KEY CONCEPTS

High-level exposure to radiation leads to immediate death in the highest doses and radiation sickness at lower ones. Radiation sickness is survivable, but victims of it often suffer from serious delayed effects, including cancer and sterility. Pregnant women suffer from miscarriages.

Health Impacts of Low-Level Radiation   The effects of low-level radiation are often delayed and are therefore harder to determine with certainty. A growing body of evidence shows that low-level radiation increases the likelihood of developing several types of cancer, such as leukemia (**FIGURE 14-26**). Studies of uranium miners, workers who painted watch dials with radium early in this century, employees at nuclear weapons factories, radiologists, and X-ray technicians show that low-level exposure increases the incidence of many forms of cancer.

Studies suggest that low-level radiation is much more harmful than many scientists thought a decade and a half ago. Noted radiation biologist Dr. Irwin Boss, for example, estimates that low-level exposure is about 10 times more harmful than previously calculated. He and others have called for major revisions of the maximum allowable doses for workers.

KEY CONCEPTS

Numerous scientific studies show that exposure to low-level radiation has a measurable effect on human health, the most common effect being cancer.

Bioaccumulation   One problem with radionuclides is that they are often concentrated in certain tissues, where they can cause considerable damage. For example, iodine-131, which is released from nuclear power plants both during normal operations and in accidents, is deposited on the ground around a power plant. Fallout may be incorporated into grass eaten by dairy cows and passed to humans via milk. It is then selectively taken up by the human thyroid gland, where it irradiates cells and may produce tumors. Milk contaminated with I-131 is especially harmful to children.

Strontium-90 is released during atomic bomb blasts. It may also be released from reactors—in small amounts under normal operating conditions but in large quantities in accidents. Strontium-90 is readily absorbed by plants and may also be passed to humans through cow's milk. It seeks out bone, where it is deposited like calcium. With a half-life of 28 years, it irradiates the bone and can cause leukemia and bone cancer.[2]

Some radionuclides may be biologically magnified in the higher trophic levels of food chains.

---

[2]*Half-life* is the time it takes for a radioactive material to decrease to half its original mass, a consequence of radioactive decay.

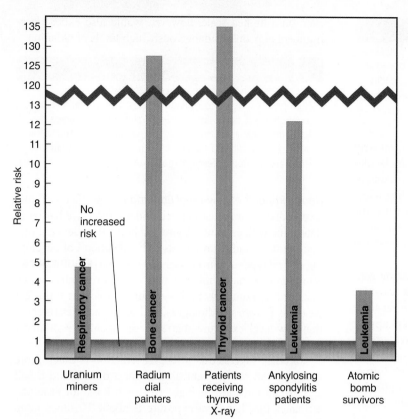

FIGURE 14-26 Cancer risk. This graph shows the relative risk of people exposed to radiation. Relative risk is a measure of the probability of developing cancer. As shown here, a uranium miner is four times more likely to develop respiratory tract cancer than someone who is not a miner. Radium dial painters are nearly 130 times more likely to develop bone cancer. Atomic bomb survivors, in general, are 3.5 times more likely to develop leukemia than nonirradiated people.

KEY CONCEPTS

Radionuclides often accumulate within tissues, where they irradiate cells. Thus, seemingly low levels in the environment may become dangerously concentrated in certain regions in the body.

Reactor Safety With this background information in mind, we turn to the issue of reactor safety. An accident at a nuclear power plant at Three Mile Island, Pennsylvania, in 1979 heightened awareness of the dangers inherent in this form of energy. This incident was caused by a malfunctioning valve in the cooling system, which triggered the worst commercial reactor accident in U.S. history. Radioactive steam poured into the containment building. Pipes in the system burst, releasing radioactive water that spilled onto the floors of two buildings. Some radiation escaped into the atmosphere, and some was dumped into the nearby Susquehanna River. The accident then took a turn for the worse. Hydrogen gas began to build up inside the reactor vessel, threatening to expose the reactor core, which would cause a meltdown. Although the gas buildup was slowly eliminated, thousands of area residents had to be evacuated. Photographs of the core showed that a partial meltdown had occurred.

The accident at Three Mile Island had many long-term effects. It cost the utility (and its customers) many millions of dollars to replace the electricity the plant would have generated. Even more money (over $1 billion) was needed for the cleanup.

Although utility officials claimed that the accident would cause few cancers, John Gofman and Arthur Tamplin, radiation health experts, estimated that the exposure to low-level radiation that residents received for 100 hours or longer would cause at least 300, and possibly as many as 900, fatal cases of cancer or leukemia. So far, evidence of elevated cancer rates has not been seen.

A 1975 study on reactor safety called the Rasmussen report (after its principal author, Norman Rasmussen of MIT) showed that the probability of a major accident at a nuclear power plant was about 1 in 10,000 reactor-years. (A reactor-year is a nuclear reactor operating for 1 year; for example, 10 reactors operating for 20 years are 200 reactor-years.) If 5,000 reactors were operating worldwide, as advocates once proposed for the year 2050, Rasmussen's prediction means that we could expect a major accident every other year! Each accident might cause between 825 and 13,000 immediate deaths, depending on the location of the plant. In addition, 7,500 to 180,000 cancer deaths would follow in the years after the accident. Radiation sickness would afflict 12,000 to 198,000 people, and 5,000 to 170,000 genetic defects would occur in infants. Property damage could range from $2.8 billion to $28 billion.

In 2012, 435 nuclear reactors were operating worldwide (**FIGURE 14-27**), with 62 under construction, according to the International Atomic Energy Agency. Today's

Go Green

Be sure to turn off your computer when not in use. Doing so can reduce electrical consumption required to run your computer by as much as 40% per year.

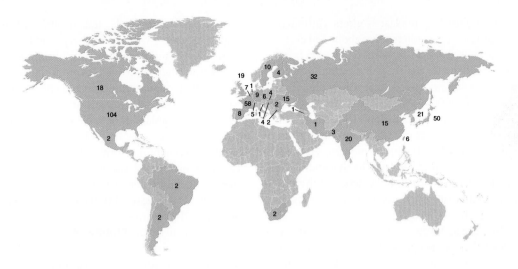

FIGURE 14-27 Global distribution of nuclear power plants.

Data from European Nuclear Society.

reactors produce about 12% of the world's electricity. If the Rasmussen report estimate is correct, we can expect a major meltdown every 20 years or so.

Because the Rasmussen study failed to include such possibilities as sabotage and human error, it has been discredited by the Nuclear Regulatory Commission, which argues that it underestimates the real risk. Human error could result in an even higher incidence of accidents. As a case in point, many experts believe that the accident at Three Mile Island was worsened by operator confusion, which turned a small problem into a disaster. Misjudgment and performance errors among personnel could negate technological improvements designed to make plants safer. Human errors and oversight can also occur during construction.

The accident at Chernobyl in the former Soviet Union in 1986 reinforced concerns about the possibility of human error (**FIGURE 14-28**). Releasing large amounts of radiation onto farms and cities throughout Europe, the accident was caused by plant operators who were running some tests on the reactor. The amount of cooling water flowing through the reactor core fell rapidly, and—because the operators had deactivated several backup systems—the temperature of the 200 tons of uranium in the reactor's fuel rods soared as high as 2,800°C (5,000°F), twice the temperature required to melt steel. An enormous steam explosion blew the roof off the building. Flames from 1,700 tons of burning graphite (a neutron-absorbing agent in the core) shot 30 meters (100 feet) into the air. While firefighters risked their lives to contain the disaster by spraying water down on the reactor core from the roofs of nearby buildings, the uranium fuel melted, spewing radioactive isotopes into the atmosphere.

Soon after the accident, 135,000 people living within a 30-kilometer (18-mile) radius of the plant were relocated. An additional 65,000 residents were evacuated in the months following the accident from areas as far as 330 kilometers (200 miles) from Chernobyl. Many of those living closest to the nuclear power plant will never be able to return to their homes.

FIGURE 14-28 The remnants of human error. This photograph shows the Chernobyl nuclear reactor soon after the 1986 explosion that blew off the top of the reactor building. This reactor and many others in the former Soviet Union were not built with suitable containment.

Courtesy of U.S. Department of Energy.

Besides losing their homes and their possessions, tens of thousands of residents may have been exposed to high levels of radiation. Estimates of human exposure near Chernobyl indicate that whole-body radiation for persons in the immediate area of the plant was 4 to 20 times higher than the allowable exposure for U.S. nuclear workers. Although no one knows for certain, it is predicted that about 1 of every 10 people, or about 15,000 people exposed to radiation from Chernobyl in the 30-kilometer radius, will die from cancer. The government of Ukraine asserts that Chernobyl has already caused 6,000 to 8,000 deaths, suggesting that the estimate of 15,000 deaths may be too low. The accident at Chernobyl also caused a number of immediate deaths from radiation poisoning—31 within the first 4 months. Workers who were saved by heroic procedures are likely candidates for cancer. New studies show that the incidence of thyroid cancer has increased dramatically in the Ukraine as a result of exposure to radioactive iodine.

Radiation also spread throughout Europe. One study shows that at least 13,000 people outside of the former USSR will die from cancer in the 50 years following the accident. A study by Dr. John Gofman of the Committee for Nuclear Responsibility suggests that the number may be much higher—about 500,000 to one million. The radiation was greatly diluted by the time it hit the United States, and estimates suggest that only 10 to 20 Americans will die from cancer from the Chernobyl accident. That's not a lot—unless, of course, you're one of the victims.

Besides exposing large numbers of people to potentially harmful radiation, the Chernobyl accident threatened crops, farmland, and livestock in Ukraine, the breadbasket of the former USSR. By some estimates, up to 150 square kilometers (60 square miles) of land is so contaminated that it cannot be farmed for decades. Agriculture outside the former USSR also suffered from the ill effects of radiation. Soon after

the accident, for example, Italian officials turned back 32 freight cars loaded with cattle, sheep, and horses from neighboring Austria and Poland because of abnormally high levels of radiation. Radioactive fallout from the accident also settled on Lapland, an expanse of land encompassing northern Sweden, Norway, Finland, and the northwestern part of the former USSR. Lapland is occupied by a seminomadic people who raise reindeer for food. The reindeer feed on lichen, which was contaminated by radiation after the accident. So heavily contaminated was the meat that Laplanders were forced to round up and slaughter their herds.

In 2011, a massive 9.0 magnitude earthquake and tsunami caused extensive damage to the Fukushima nuclear power plant in Japan (**FIGURE 14-29**). The accident caused a meltdown in three of the four reactors and release of massive amounts of radiation, equivalent to the 1986 Chernobyl accident. This accident caused the release of large amounts of highly radioactive material into the air and water. Radiation in the immediate vicinity will prevent citizens from ever returning to their homes. According to Japanese officials, the three reactor cores that melted down melted into the ground beneath the reactor buildings, where they remain. They could continue to release radiation into the environment for decades. Numerous highly radioactive fuel rods are also strewn about the site and radiation has been leaking into the water since the accident. Leaks may not be contained until 2015, according to authorities. For a detailed description of the accident and its impacts, students can check out the report by the World Nuclear Association, a professional organization representing the nuclear industry employees.

FIGURE 14-29 This four-reactor nuclear power plant in Japan was severely damaged by a massive earthquake and tsunami in March 2011. It has since released enormous amounts of radiation into the air and water.

Critics are also concerned with unforeseen technical difficulties. A hydrogen bubble at the Three Mile Island plant, for example, took the experts by surprise. Numerous backup systems in nuclear power plants are designed to prevent a meltdown and the release of radiation, but as the accident at Three Mile Island showed, unanticipated problems can and do arise.

Clouding the issue of reactor safety is the possibility of terrorism. Damage to the cooling systems of nuclear power plants could result in a meltdown with massive radiation leakage. Most plants are vulnerable to attack. Protection from ground and air assaults may be impossible. Even though security has been improved at many plants, the threat of well-planned terrorist actions cannot be ignored.

To address the question of reactor safety, the nuclear industry is promoting new, smaller designs that they claim are much safer than present models. Skeptics question whether this is true and point out that even if it is, making plants safer addresses only one of a dozen problems—among them, what to do with the enormous amounts of radioactive waste produced by a nuclear power plant.

Accidents at nuclear power plants are a major concern to those involved in the nuclear debate, but so are accidents at fuel processing plants—facilities that make nuclear fuel. In September, 1999, Japanese workers set off an uncontrolled chain reaction at a processing plant in Tokaimura when they accidentally poured seven times more uranium into a purification tank than they were supposed to. The reaction lasted about 20 hours and released large quantities of radioactivity into the plant and the neighboring environment. Sixty-three individuals were exposed to high levels of radiation and two are expected to die because of high-level exposure.

Accidents involving trucks and trains that transport uranium ore, yellow cake (processed ore), and uranium wastes also occur. Surprisingly common, accidents in the transportation sector generally result in the release of contaminants over a relatively small area. Special transport containers and other precautions are taken to minimize the potential for the release of nuclear wastes. As an added precaution, some cities ban trucks containing such wastes.

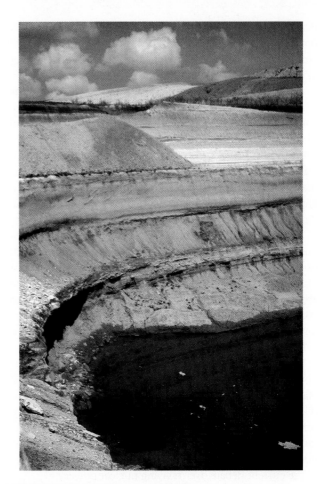

FIGURE 14-30 Inactive uranium mine. To acquire large amounts of uranium for nuclear power plants and weapons production, companies extract uranium ore from huge mines like this one. Enormous amounts of material must be removed. Radioactive waste can be washed from the site by rainwater or blown away by wind.

Courtesy of U.S. Fish & Wildlife Service.

KEY CONCEPTS

Several major accidents at nuclear power plants have raised awareness of the enormous damage a mechanical or human error, earthquake, or a tsunami can cause. Estimates suggest that many additional accidents are bound to occur in the future with costly social, economic, and environmental impacts.

Waste Disposal Nuclear power plants, nuclear weapons facilities, mines that extract uranium ore, and processing facilities that purify it produce enormous amounts of radioactive waste (**FIGURE 14-30**). For years, much of this waste has been improperly disposed of. In the early days of nuclear power, for example, millions of tons of radioactive waste were dumped on or near uranium mills, sometimes on the banks of rivers. Low-level wastes from power plants and other facilities (hospitals, laboratories) were buried in shallow landfills. In some instances, they were buried in boxes or steel drums that leaked radioactive materials into groundwater and surface waters. Until the 1960s, low- and medium-level radioactive wastes were often mixed with concrete, poured into barrels, and dumped at sea—a practice that continues today off the coast of England. In at least one location—Richland, Washington, home of the Hanford nuclear weapons facility—wastes were stored in underground tanks that have begun to leak high-level radioactive materials into the ground. Since it began operating, the plant has produced several hundred million liters of highly radioactive liquid waste. Since 1958, at least 2 million liters (500,000 gallons) of waste have leaked out of tanks into the soil and, possibly, the groundwater. Approximately 150 million liters (38 million gallons) of waste is still in storage tanks that could leak.

Low-level wastes are hazardous for about 300 years, and high-level wastes can be dangerous for tens of thousands of years. The half-life of one highly radioactive waste product of nuclear reactors, plutonium-239 (Pu-239), is 24,000 years. As a rule, it generally takes about eight half-lives for a material to be reduced to 0.1% of its original mass, at which point it is considered safe. For Pu-239, this is about 200,000 years. By comparison, the Egyptian pyramids are only about 4,500 years old!

A typical nuclear power plant produces 20 metric tons per year, and the U.S. nuclear reactors, therefore, produce about 2,000 metric tons per year. As a result, high-level nuclear wastes are increasing rapidly. In 1995, there were 28,000 metric tons (31,000 tons) of high-level radioactive waste being held at nuclear power plants, awaiting a permanent disposal site. By 2007, the amount had increased to approximately 50,000 metric tons (55,000 tons). In 1982, the U.S. Congress passed a law requiring the Department of Energy (DOE) to create a suitable site to build a disposal facility for such waste. After eliminating a number of candidates, DOE selected a remote site in Nevada (Yucca Mountain). The DOE spent over $2 billion studying the area, and many more billions developing the site, only to abandon it in 2010, because of major issues regarding its long-term safety. A fault line, for instance, was discovered below the site in 2007.

Meanwhile, high-level wastes are building up at nuclear power plants and weapons facilities throughout the United States. By 2015, this high-level nuclear waste is expected to total nearly 75,000 metric tons (83,000 tons). Clearly, something needs to be done about these wastes and the radioactive waste currently held at nuclear power plants. Point/Counterpoint 14-2 in this chapter presents both sides of the current debate over reviving nuclear power.

KEY CONCEPTS

Nuclear power has a long history in the United States. In the early years, many wastes from nuclear fuel production were carelessly disposed of. Wastes from nuclear power plants remain stockpiled at power plants because the U.S. nuclear industry lacks acceptable means of disposing the high-level wastes produced by reactors and uranium mills.

Social Acceptability and Cost   Two of the most important factors controlling the future of nuclear energy are its social acceptability and its cost. Several countries, including Sweden and Italy for instance, have voted to phase out nuclear power. Why is nuclear power so unpopular these days?

Besides safety concerns and concern over waste disposal, nuclear power plants are expensive to build and operate. A new nuclear power plant today costs between $2 and $8 billion, compared with $1 billion for an equivalent coal-fired power plant. Costs are high because of strict building standards, expensive labor, construction delays, and special materials needed to ensure plant safety. Because of the high cost and risk of damage, U.S. banks refuse to finance nuclear power plants, and utilities must turn to foreign investors.

The cost of operating and maintaining a nuclear power plant is also high. Repair costs, for example, are many times higher than those for conventional coal-fired plants.

Another factor in the cost equation is retirement costs. According to the Worldwatch Institute, one of six reactors already built has been retired. Their average life span was 17 years, compared with an industry estimate of 20 to 30 years. Dismantling these plants at the end of their life span may cost utilities about $0.5 to $1 million per megawatt, bringing the cost of decommissioning a 1,000-megawatt plant to $500 million to $1 billion. This cost will inevitably add to electric bills.

A rash of cancellations has also driven costs up. Even though utilities halt construction of nuclear plants, they must pay back the billions of dollars they borrowed to start construction. Customers inevitably foot most of the bill. Bondholders, who often finance such projects, can also be left in financial ruin.

High construction and maintenance costs, combined with other factors, make nuclear energy the most costly form of electricity currently available on a large scale.

A lack of public support and exorbitant costs, among other problems, have slowed the growth of the nuclear industry. In fact, the number of nuclear power plants worldwide has been declining in recent years. Although some countries like Canada and Russia still rely heavily on nuclear power and export technology and expertise through a government-owned business, the problems outlined in this section may be so insurmountable that even these nations will eventually phase out of nuclear energy.

In recent years, Russian companies backed by a major investment from the Russian government have begun to aggressively market nuclear power plants for sale or rent throughout the world, especially in many less-developed countries. Russians have developed floating nuclear power plants, that is, nuclear power plants that are assembled on barges in Russia, then transported to other countries, where they will be moored offshore, producing electricity. They have also developed a program by which cash-strapped nations can rent reactors designed, installed, and financed by Russian companies. Numerous reactors are now being installed. If successful, this program could increase the number of nuclear reactors and worldwide electricity production from them.

KEY CONCEPTS

Nuclear power has become a socially unacceptable form of electricity in part because of high costs during all phases of operation from construction to operation, repair, and retirement.

The Proliferation of Nuclear Weapons   At least 14 countries have or are suspected of having nuclear weapons. Some of these countries like North Korea and Israel are either politically unstable or are in volatile regions where war could easily erupt. Seven countries—the United States, China, the former Soviet Union, Britain, France, India, and Pakistan—have already tested nuclear weapons and several dozen more countries have the technical

In the early 1970s when I helped found Greenpeace, I believed that nuclear energy was synonymous with nuclear holocaust, as did most of my compatriots. Thirty years on, my views have changed, and the rest of the environmental movement needs to update its views, too, because nuclear energy may just be the energy source that can save our planet from another possible disaster: catastrophic climate change.

Look at it this way: More than 600 coal-fired electric plants in the United States produce 36% of U.S. emissions—or nearly 10% of global emissions—of CO_2, the primary greenhouse gas responsible for climate change. Nuclear energy is the only large-scale, cost-effective energy source that can reduce these emissions while continuing to satisfy a growing demand for power. And these days it can do so safely.

I say that guardedly, of course, just days after Iranian President Mahmoud Ahmadinejad announced that his country had enriched uranium. "The nuclear technology is only for the purpose of peace and nothing else," he said. Despite his assurances, there was widespread speculation that nuclear energy development in Iran was merely a cover for building nuclear weapons.

Although I don't want to underestimate the very real dangers of nuclear technology in the hands of rogue states, we cannot simply ban every technology that is dangerous.

Today, there are 103 nuclear reactors quietly delivering just 20% of America's electricity. Eighty percent of the people living within 10 miles of these plants approve of them (that's not including the nuclear workers).

Although I don't live near a nuclear plant, I am now squarely in their camp. And I am not alone among seasoned environmental activists in changing my mind on this subject. There are signs of a new willingness to listen, though, even among the staunchest antinuclear campaigners.

Here's why: Wind and solar power have their place, but because they are intermittent and unpredictable they simply can't replace big baseload plants such as coal, nuclear, and hydroelectric. Natural gas, a fossil fuel, is too expensive already, and its price is too volatile to risk building big baseload plants. Given that hydroelectric resources are built pretty much to capacity, nuclear is, by elimination, the only viable substitute for coal. It's that simple.

That's not to say that there aren't real problems—as well as various myths—associated with nuclear energy. Each concern deserves careful consideration:

- *Nuclear energy is expensive*. It is in fact one of the least expensive energy sources. In 2004, the

Nuclear Makes Sense

Patrick Moore

Patrick Moore, co-founder of Greenpeace, is chairman and chief scientist of Greenspirit Strategies, Ltd. He and Christine Todd Whitman are co-chairs of a new industry-funded initiative, the Clean and Safe Energy Coalition, which supports increased use of nuclear energy.

Courtesy of Patrick Moore..

average cost of producing nuclear energy in the United States was less than two cents per kilowatt-hour, comparable with coal and hydroelectric. Advances in technology will bring the cost down further in the future.

- *Nuclear plants are not safe*. In 1979, a reactor core meltdown at Pennsylvania's Three Mile Island nuclear power plant sent shivers of anguish throughout the country. What nobody noticed at the time, though, was that Three Mile Island was in fact a success story: The concrete containment structure did just what it was designed to do—prevent radiation from escaping into the environment. And although the reactor itself was crippled, there was no injury or death among nuclear workers or nearby residents. Three Mile Island was the only serious accident in the history of nuclear energy generation in the United States, but it was enough to scare us away from further developing the technology: There hasn't been a nuclear plant ordered up since then.

Although Three Mile Island was a success story, the accident at Chernobyl, almost 30 years ago [April 1986], was not. But Chernobyl was an accident waiting to happen. This early model of Soviet reactor had no containment vessel, was an inherently bad design and its operators literally blew it up. The multi-agency U.N. Chernobyl Forum reported that 56 deaths could be directly attributed to the accident, most of those from radiation or burns suffered while fighting the fire. Tragic as those deaths were, they pale in comparison to the more than 5,000 coal-mining deaths that occur worldwide every year. No one has died of a radiation-related accident in the history of the U.S. civilian nuclear reactor program. (And although hundreds of uranium mine workers did die from radiation exposure underground in the early years of that industry, that problem was long ago corrected.)

- *Nuclear waste will be dangerous for thousands of years*. Within 40 years, used fuel has less than one-thousandth of the radioactivity it had when it was removed from the reactor. And it is incorrect to call it waste, because 95% of the potential energy is still contained in the used fuel after the first cycle. Now that the United States has removed the ban on recycling used fuel, it will be possible to use that energy and to greatly reduce the amount of waste that needs treatment and disposal.

- *Nuclear reactors are vulnerable to terrorist attack.* The 6-feet-thick reinforced concrete containment vessel protects the contents from the outside as well as the inside. And even if a jumbo jet did crash into a reactor and breach the containment, the reactor would not explode. There are many types of facilities that are far more vulnerable, including liquid natural gas plants, chemical plants, and numerous political targets.
- *Nuclear fuel can be diverted to make nuclear weapons.* This is the most serious issue associated with nuclear energy and the most difficult to address, as the example of Iran shows. But just because nuclear technology can be put to evil purposes is not an argument to ban its use.

Over the past 20 years, one of the simplest tools—the machete—has been used to kill more than a million people in Africa, far more than were killed in the Hiroshima and Nagasaki nuclear bombings combined. What are car bombs made of? Diesel oil, fertilizer, and cars. If we banned everything that can be used to kill people, we would never have harnessed fire.

The only practical approach to the issue of nuclear weapons proliferation is to put it higher on the international agenda and to use diplomacy and, where necessary, force to prevent countries or terrorists from using nuclear materials for destructive ends. And new technologies such as the reprocessing system recently introduced in Japan (in which the plutonium is never separated from the uranium) can make it much more difficult for terrorists or rogue states to use civilian materials to manufacture weapons.

The 600-plus coal-fired plants emit nearly 2 billion tons of CO_2 annually—the equivalent of the exhaust from about 300 million automobiles. In addition, the Clean Air Council reports that coal plants are responsible for 64% of sulfur dioxide emissions, 26% of nitrous oxides, and 33% of mercury emissions. These pollutants are eroding the health of our environment, producing acid rain, smog, respiratory illness, and mercury contamination.

Meanwhile, the 103 nuclear plants operating in the United States effectively avoid the release of 700 million tons of CO_2 emissions annually—the equivalent of the exhaust from more than 100 million automobiles. Imagine if the ratio of coal to nuclear were reversed so that only 20% of our electricity was generated from coal and 60% from nuclear. This would go a long way toward cleaning the air and reducing greenhouse gas emissions. Every responsible environmentalist should support a move in that direction.

COUNTERPOINT

No Nuclear is Good Nuclear

Steve Clark

Steve Clark was director of Citizens for Clean Energy in the 1980s and has explored the technologies and policies of clean renewable energy and energy efficiency for 30 years. He is currently working on electric vehicles, energy-efficient lighting projects, and raising the energy IQ of the common person.

Courtesy of Steve Clark.

Humankind is faced with many questions concerning the paths that we should take to meet our ever-growing energy demand. It is important that we understand the implications of our energy choices and are not misled by false assumptions or misleading information. One example of the attempt to mislead the public on energy issues is the arguments made by the nuclear power industry.

Recently there has been a great deal of fanfare as several "greens" have announced their support of nuclear power. Like many others, they are rightfully alarmed by the environmental effects of fossil fuel energy use, specifically the emission of CO_2 that is leading to increased global warming and a potential climatic catastrophe. Looking at the predicted future of ever-increasing energy use, these new nuclear advocates have come to the conclusion that the only way to meet our ever-rising energy "demands" is by increasing supply. The best way, they claim, to increase supply of energy is through nuclear power.

The flawed assumption, that's rarely exposed, is that we need an ever-increasing supply of energy.

Energy expert Amory Lovins is one of many that has challenged this flawed assumption. Lovins and others have repeatedly pointed out that we humans waste a tremendous amount of our energy system. By using energy much more efficiently we can ensure the services we've grown accustomed to at a fraction of the current "supply." In other words, we can meet out needs by simply using less and live well, too.

Because energy has been cheap for so long, many of us have unquestioningly accepted inefficient energy technologies. We should, therefore, begin by questioning assumptions such as, is the less-than-10% efficient internal combustion automobile, the best way to provide personal mobility?

One previously "green" nuclear advocate, Patrick Moore, presents several misleading arguments in favor of expanding the role of nuclear power. One false assertion is that by falsely claiming renewable energy cannot meet the demand, he dismisses the one real option that will safely lead us to a sustainable energy future. "Nuclear energy," he asserts, "is the only large-scale, cost-effective energy source that can reduce these emissions while continuing to satisfy a growing demand for power."

According to a number of reputable energy analysts, renewable energy has many times the capacity to meet our needs. Solar energy striking 0.3% of the land area of the United States could theoretically meet all of the electrical needs of the entire country with current technology. Every hour, the sun radiates more energy onto the Earth than the entire human population uses in an entire year.

The current model of energy production and distribution is extremely inefficient. The only reason that we tolerate such inefficiency is that even with 60% wasted energy, the end-use energy is still cheap to the consumer. One reason for this is that we ignore much of the real cost of providing energy to consumers. We ignore

the environmental costs of producing energy, referring to problems such as global warming and nuclear waste as "externalities" and not including the costs of these and many other problems created by our current energy system in the price of energy.

The statement that nuclear energy is cost-effective is false. If this were the case, the people who finance the building of nuclear power plants would have invested in them. The reason no nuclear power plants have been built since the early 1970s is that the financial community pulled support. Mr. Moore claims nuclear, "…is in fact one of the least expensive energy sources." Without price guarantees, insurance against catastrophic failure (if they are safe, why do they need government guaranteed insurance against catastrophe?) and subsidized programs to deal with nuclear waste and the cost of decommissioning old power plants, nuclear power is not competitive, much less cheap.

The safety of nuclear power is interesting because it is really asking what level of risk and danger we are willing to accept. Mr. Moore tries to convince us that because machetes are used to kill people we should accept the risk of nuclear power. If someone uses a machete to kill, it's a failure to follow fundamental laws of social conduct. If someone suggests we overlook the hazards of nuclear power plants, it's a failure to understand fundamental ethical issues that are the basis of a sane society. E. F. Schumacher, author of *Small is Beautiful*, commented on this issue nearly 40 years ago: the technology of nuclear power may change, but fundamental issues of ethics do not. "No degree of prosperity could justify the accumulation of large amounts of highly toxic substances which nobody knows how to make safe and which remain an incalculable danger to the whole of creation for historical or even geological ages," he wrote. "To do such a thing is a transgression against life itself, a transgression infinitely more serious than any crime ever perpetrated by man." Schumacher went on to say, "The idea that a civilization could sustain itself on the basis of such a transgression is an ethical, spiritual and metaphysical monstrosity. It means conducting the economic affairs of man as if people really did not matter at all." Mr. Moore seems to have no sense of ethics concerning the misleading information he uses to promote nuclear power. He wrote, "The multi-agency U.N. Chernobyl Forum reported last year that 56 deaths could be directly attributed to the accident, most of those from radiation or burns suffered while fighting the fire." The U.N. task force suspect between 4,000 and 9,000 cancer deaths from the radiation released by the Chernobyl nuclear power plant.

Mr. Moore's former organization Greenpeace claims "that the true figures will exceed a quarter of a million cancer cases, with over 100,000 fatal cancers." According to the World Health Organization, the Chernobyl nuclear disaster will cause 50,000 new cases of thyroid cancer among young people living in the areas most affected by the nuclear disaster."

One of the very real problems with nuclear power is it leads directly to nuclear weapons proliferation. Mr. Moore states that the "only" solution to nuclear weapons proliferation is to "use force"—in other words, to wage war against anyone who tries to use the byproducts of the nuclear power plants for nuclear weapons. It doesn't seem to occur to him that not building the nuclear plants in the first place might be another practical approach to the problem.

"Wind and solar power have their place, but because they are intermittent and unpredictable they simply can't replace big baseload plants such as coal, nuclear, and hydroelectric." This is another false statement. Base load is another term for the need to keep coal and nuclear power plants running 24/7. This makes us slaves to the power industry's machines, which cannot be shut down when they are not needed.

While we do need reliable energy for many valid uses, renewable energy can meet all of these needs at all times of the day and night. Home photovoltaic systems have been doing this for many people for the last 20 years.

If renewable energy technology had received the vast research funding that has gone to nuclear energy, we would be living in a renewable energy society. Recent developments in solar power will resolve the problem once and for all if the "free market" is allowed to act and nuclear power is denied the subsidies needed of it to be an economically viable energy source. Combined with energy storage and energy efficiency, renewable energy can and will meet our energy needs cleanly, affordably, reliably, and safely.

A final point on the question of safe energy: No one has ever been killed by (or with) a solar panel.

Critical Thinking Questions

1. Summarize the views of both authors.
2. With which viewpoint do you agree? Why?

capability to produce nuclear weapons. The plutonium used in atomic bombs comes from special nuclear reactors as well as from conventional reactors. Critics of nuclear power argue that the spread of nuclear power throughout the world will make bomb-grade plutonium more widely available.

Nuclear fuels could be stolen by terrorist groups and easily fashioned into a crude but effective atomic bomb that would fit comfortably into the trunk of an automobile. Strategically planted, a terrorist bomb could prove disastrous. Concern for this danger further dampens public and political support for nuclear power.

KEY CONCEPTS

Countries with nuclear power plants can develop atomic bombs from waste products, notably plutonium-239.

Breeder Reactors

The world's supply of the uranium-235 used in light-water reactors is limited. In fact, only 10% of all naturally occurring uranium is U-235. Because uranium-235 supplies are limited, the nuclear industry has developed an alternative called the **breeder reactor**. Breeder reactors are fission reactors that are similar in most respects to the light-water fission reactor described earlier. However, besides producing electricity, breeder reactors convert uranium-238 (an isotope of uranium not used as a fuel in light-water reactors) into reactor fuel, plutonium-239.

In the breeder reactor, fast-moving neutrons from small amounts of plutonium-239 located in the fuel rods strike nonfissionable U-238 placed around the reactor core. This converts U-238 into fissile (fissionable) Pu-239 (**FIGURE 14-31**). Theoretically, for every 100 atoms of Pu-239 consumed in fission reactions, 130 atoms of Pu-239 are produced—hence, the name *breeder*.

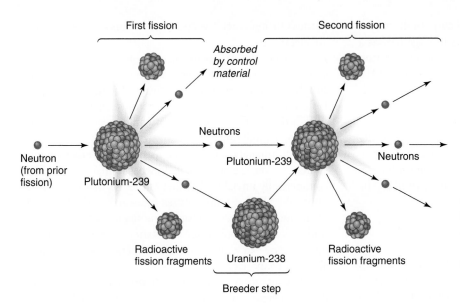

First fission

Second fission

Absorbed by control material

Neutron (from prior fission)

Plutonium-239

Neutrons

Plutonium-239

Neutrons

Radioactive fission fragments

Uranium-238

Radioactive fission fragments

Breeder step

FIGURE 14-31 Nuclear reactions in a breeder reactor. Neutrons produced during fission strike nonfissionable "fertile" materials such as uranium-238. U-238 is then converted into fissionable plutonium-239, which can be used as the reactor in the fuel.

The attraction of breeder reactors is that their fuel supply, U-238, is found in the wastes of uranium processing plants and in the spent fuel from conventional fission reactors. In the United States alone, the estimated supply of U-238 would last 1,000 years or more. In addition, breeder reactors could reduce the need for mining, processing, and milling of uranium ore. Moreover, fuel prices might remain stable because of the abundance of U-238. The breeder technology does not create chemical air pollution (if we ignore problems from mining and milling).

Breeder reactors have been under intensive development in the United States for well over 30 years. The most popular design is the **liquid metal fast breeder reactor**. It uses liquid sodium as a coolant rather than water. Heat produced by nuclear fission in the reactor core is transferred to the liquid sodium coolant, which transfers this heat to water. The water is then converted to steam that is used to generate electricity. Another type of breeder reactors the *thermal breeder*, details of where are beyond the scope of this book.

On the surface, the breeder reactor sounds like the answer to the world's electrical energy needs. Unfortunately, it has numerous problems, some so great that they may make the technology impractical. The most significant problem is that it takes about 30 years for the reactor to break even—that is, to produce as much Pu-239 as it consumes. The fate of breeder reactors hinges on drastically shortening the payback period. The second major problem is the cost: projected to be $4 to $8 billion. In addition, breeder reactors have many of the same problems as light-water reactors.

Another problem would be the large quantities of plutonium located at breeder reactors. Plutonium is long lived and extremely toxic if inhaled. It burns when exposed to air. Liquid sodium coolant is also dangerous. It reacts violently with water and burns spontaneously when exposed to air. A leak in the coolant system could trigger a catastrophic accident. Should the core melt down, a small explosion equivalent to several hundred tons of TNT might occur. Rupturing the containment building, the explosion would send a cloud of radioactive gas into the surrounding area.

In 1983, the U.S. Congress canceled funding for an experimental breeder reactor, the Clinch River project in Tennessee. Having spent nearly $2 billion for planning, Congress decided that further investments were unmerited. Numerous countries are pursuing breeder reactors, and many of them currently operate very small reactors. France, however, has two large breeder reactors in service today. Russia has launched an active breeder reactor program for domestic use and export.

KEY CONCEPTS

Breeder reactors are designed to produce energy and radioactive fuel from abundant nonfissile (nonfissionable) uranium-238 and could greatly increase the supply of nuclear fuels. Unfortunately, breeder reactors are costly to build and take 30 years to produce as much radioactive fuel as they consume.

Nuclear Resurgence

While nuclear reactor construction was halted by many countries in the 1970s, many countries have continued to develop nuclear power or have recently announced plans to build new nuclear reactors including Japan, China, India, South Korea, South Africa, Finland, France, and Russia.

New nuclear reactors are again being built in the United States, too, after a 20-year hiatus in large part as a result of support from the George W. Bush administration and the U.S. Department of Energy's New Nuclear 2010 Program. The Energy Policy Act of 2005 authorized federal loan guarantees for up to six nuclear reactors. These reactors known as Generation IV Very High–Temperature-Reactors produce electricity and hydrogen and, according to their designers, are safer than earlier designs. Four to six new nuclear reactors are expected to come on line by 2018, although older nuclear reactors will very likely be retired as well in coming years.

Nuclear Fusion

The sun is the source of virtually all forms of energy, even the fossil fuels. Although they consist of organic matter, they contain energy that was once sunlight. This energy is captured by plants that were later converted to coal. Oil comes from microorganisms as well, notably photosynthetic algae. The sun generates its energy by **nuclear fusion**, which occurs when two hydrogen nuclei join—or fuse—to form a slightly larger helium nucleus as shown in **FIGURE 14-32**. Fusion requires extremely high temperatures and pressure to overcome the electrostatic repulsion of the positively charged nuclei. When the hydrogen nuclei fuse, though, they emit large quantities of energy, mostly light and heat. Scientists hope to

harness fusion reactions and capture this energy to generate electricity.

If scientists can create fusion reactors, nuclear fusion could supply enormous amounts of energy because its fuel supply is also enormous: It's seawater. Energy analyst John Holdren estimates that at current rates of energy consumption in the United States, fusion would meet energy needs for up to 10 million years!

Unfortunately for proponents of this futuristic energy source, fusion has significant drawbacks that may make it commercially unviable. The first of these is that fusion reactions take place at temperatures measured in the hundreds of millions of degrees Celsius. The main obstacle, then, is finding a way to contain such an extremely hot reaction. No known alloy can withstand these temperatures; in fact, metals would vaporize.

To contain experimental fusion reactions, scientists have devised reactors that suspend tiny amounts of fuel in air within a metal reactor vessel. The most popular technique in experimental fusion reactors is **magnetic confinement** (**FIGURE 14-33**). In these reactors, a magnetic field suspends the superheated fuel. Heat given off by the fusion reaction is captured and used to boil water to make steam. Small-scale experimental reactors of this type have been developed and operated in the United States, Europe, the former Soviet Union, and Japan. Despite five decades of research, researchers consistently failed to reach the break-even point—that is, the point at which they get as much energy out of the system as they put into it. In 2013, researchers finally achieved this goal, the first step in a long, costly climb to commercialization.

In the proposed designs, heat released from the fusion reaction would be drawn off by a liquid lithium blanket, which transfers the heat to water. The lithium blanket would also capture neutrons, creating tritium (a hydrogen nucleus containing a proton and two neutrons), which could be extracted and used as additional fuel.

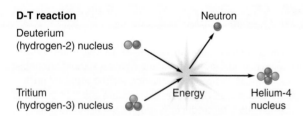

FIGURE 14-32 Nuclear fusion reactions. Hydrogen atoms can be forced together under intense heat and pressure to unite or fuse. This fusion reaction releases an enormous amount of energy mostly in the form of light and heat. Two potentially useful reactions could be used to create fusion reactors. The first (top) uses a form of the hydrogen atom that consists of one proton and one neutron, known as deuterium. (Normal hydrogen atoms contain only one proton and no neutrons.) The second (bottom) uses a hydrogen atom called tritium that contains one proton and two neutrons.

FIGURE 14-33 Fusion reactor. This drawing shows one type of fusion reactor, in which the fusion reactions occur suspended in an electromagnetic field.

Scientists estimate that a demonstration fusion reactor could be running by the year 2025. Because the U.S. Congress has cut funding, however, this goal is no longer attainable. The cost of a commercial fusion reactor cannot be accurately assessed at this time, but it could cost $12 to $20 billion, making its electricity unaffordable.

Another problem with this technology is the release of radioactivity. The deuterium–tritium fusion reactor is the most feasible type, but tritium is radioactive and difficult to contain. Because of the high temperatures in fusion reactors, tritium can penetrate metals and escape into the environment.

Fusion reactors would produce enormous amounts of waste heat. Fusion reactions would also emit highly energetic neutrons, which would strike the vessel walls and weaken the metal, necessitating replacement every 2 to 10 years. Metal fatigue could lead to the rupture of the vessel and the release of tritium and molten lithium, which burns spontaneously in air. A leak might destroy the reaction vessel and the containment facilities. Neutrons emitted from the fusion reaction would also convert metals in the reactor into radioactive materials. Periodic maintenance and repair of reactor vessels would be a health hazard to workers, and radioactive components removed from the reactor would have to be disposed of properly.

Despite the many problems, researchers worldwide are still seeking ways to develop fusion reactors.

KEY CONCEPTS

Enormous amounts of energy are produced when small atoms fuse. Fusion reactors could be powered by plentiful fuel sources and could supply energy for many millions of years. However, technical problems, costs, and environmental hazards present major obstacles to the commercial development of this form of energy.

14.6 Guidelines for Creating a Sustainable Energy System

The previous sections show the wide range of nonrenewable energy fuels and technologies, most of which pose significant environmental risks. In choosing options we must select those that are socially, economically, and environmentally sustainable. Creating a sustainable energy future requires us to consider a number of factors—many of which were introduced in this chapter. The following guidelines will help you judge all of our energy options.

First, energy resources should have a positive net energy yield—the higher the better. That is, energy we get out of an energy resource must exceed the energy we invest in its exploration, extraction, transportation, and so on.

Second, options should be chosen that minimize pollution and maximize public and environmental safety.

Third, options should be selected that are abundant and renewable. Abundant, renewable resources maximize returns on money invested in research, development, and commercialization.

Fourth, options should be pursued that are the most affordable both now and in the future. The cost of a fuel should include all costs: social, economic, and environmental.

Fifth, energy supplies should be matched with energy demand. Energy comes in a variety of forms. Each of these forms has a different **energy quality**, a measure of the amount of available work one can get out of it (**TABLE 14-3**). Oil and natural gas, for instance, are highly concentrated energy resources. When burned, they produce large amounts of heat useful for certain tasks, such as melting steel. Sunlight is a lower quality form. Streaming through the south-facing windows of a house, it produces heat appropriate for warming homes and offices. Using energy wisely means employing low-quality energy sources (sunlight) for tasks that call for it (home heating) and high-quality energy (oil) for appropriate tasks (running cars).

TABLE 14-3

Energy Quality of Different Forms of Energy

Quality of Energy	Form of Energy
Very high	Electricity, nuclear fission, nuclear fusion[1]
High	Natural gas, synthetic natural gas (from coal gasification), gasoline, petroleum, liquefied natural gas, coal, synthetic oil (from coal liquefaction)
Moderate	Geothermal, hydropower, biomass (wood, crop residues, manure, burnable municipal refuse), oil shale, tar sands

[1]Workable nuclear fusion reactors do not yet exist. Even if they were technically operable by the end of the first quarter of the 21st century, they would probably remain economically unfeasible.

KEY CONCEPTS

Although there are many energy options at our disposal, not all are sustainable. Creating a sustainable energy future will require a careful analysis of options for such factors as net energy yield, specific needs, efficiency, environmental impacts, abundance, renewability, and costs.

© hugolacasse/ShutterStock, Inc.

14-2 Coca-Cola Goes Green

As businesses expand, they have enormous opportunities to build facilities that use far fewer resources than conventional operations. In 2008, soft-drink giant Coca-Cola announced that its second largest bottler, Coca-Cola Hellenic Bottling Company, had announced plans to build 15 new combined heat and power plants by the end of 2009. These cogeneration facilities dramatically improve the efficiency of energy production and reduce emissions, including greenhouse gases.

The facilities will be built in Austria, Czech Republic, Greece, Northern Ireland, Poland, Slovakia, Nigeria, Serbia, the Ukraine, Italy, Russia, and Romania. These plants will reduce CO_2 emissions by at least 40% compared to conventional facilities.

Coca-Cola also announced that it is working with a pioneer in efficient alternative transportation, ZAP, to use 30 of the company's compact trucks for a new beverage distribution system in a city in Uruguay. The trucks use one-fifth the fuel of older models. If the tests are successful, the company announced that it may use them in larger cities around the world to combat rising fuel prices, worsening traffic congestion, air pollution, and shortages of parking.

14.7 Establishing Priorities

With the energy trends described in this chapter and the guidelines for a sustainable energy strategy just listed, we can establish a list of goals and actions needed to meet our future energy demands.

- **Action 1:** Greatly improve the energy efficiency of all machines, homes, appliances, buildings, factories, motor vehicles, airplanes, and so on. This action stretches fossil fuel supplies and gives us more time to make a transition to sustainable energy resources and technologies.
- **Action 2:** Install clean, renewable energy technologies to replace oil (which provides transportation fuel and raw materials to make plastics) because its supplies are limited. Also install renewable energy technologies to replace coal (primarily used to generate electricity) because it is such an environmentally costly fuel.
- **Action 3:** Install renewable energy technologies to replace natural gas (which is primarily used for heating and industrial processes). Domestic supplies are on the decline.

To create an energy system that is economically and environmentally sustainable is an enormously complicated task, made more difficult by current controversies and vested interest in maintaining the status quo, and the enormous political power of the 15 or so companies that control global energy production. To make wise choices, we must base our decisions on what is socially, economically, and environmentally sustainable. Our goal is to design a sustainable energy system that makes sense given the world's limited resources, one that fits into the economy of nature. Wise choices could help create broader prosperity, a better future, a stable economy, and a stronger, less vulnerable nation.

KEY CONCEPTS

Many experts on sustainable energy systems believe that, in the near term, efforts are needed to improve the efficiency of all energy-consuming technologies and to implement clean, sustainable, renewable, and affordable alternatives to natural gas, coal, crude oil, and their derivatives.

Most of our so-called reasoning consists in finding arguments for going on believing as we already do.

—James Harvey Robinson

CRITICAL THINKING

Exercise Analysis

To begin, you should first look at what nuclear power and oil are used for. In other words, be sure to dig deeply and define and understand all terms. As you learned in the chapter, nuclear power is a source of electricity. In the United States and most other nations, however, oil is used primarily to produce gasoline, diesel fuel, home heating oil, and jet fuel. It is also used to produce chemicals needed to make plastics, medicines, and other products. Very rarely is oil burned to generate electricity.

Increasing our reliance on nuclear power will do little if anything to reduce our dependence on oil. Some see this effort, which continues today in an expensive media campaign sponsored by the nuclear industry, as a ploy to dupe the public. Proponents of nuclear power are promoting a desirable goal (energy independence) but are offering a false solution (nuclear power). By defining terms more carefully, you can see that they're consciously or unconsciously hoodwinking the public.

On the issue of nuclear power and global warming, it is quite clear that what proponents say is true. A nuclear power plant produces no carbon dioxide and does not contribute to global warming. What about nuclear power's impacts, though? Would we simply be trading one problem for another?

CRITICAL THINKING AND CONCEPT REVIEW

1. Critically analyze the statement, "There is enough oil to go around for at least 65 more years."
2. Describe the term *energy system*. List all of the steps you can think of that are needed to produce gasoline from oil, starting with exploration; discuss some of the impacts on the environment associated with each step.
3. Critically analyze this statement: "Coal is our most abundant fossil fuel and, therefore, should be a major source of energy in the next hundred years."
4. What is oil shale? Discuss the benefits and risks of oil shale development.
5. Define *coal gasification* and *coal liquefaction*.
6. Describe how a light water fission reactor works. What is the fuel, and how is the chain reaction controlled?
7. What are the advantages and disadvantages of nuclear power?
8. Critically analyze this statement: "Nuclear energy is a clean source of energy and should be actively promoted."
9. What is a breeder reactor? How is it similar to a conventional fission reactor? How is it different? Discuss the advantages and disadvantages of the breeder reactor.
10. Write a 10-page research paper on the causes and impacts of the Fukushima nuclear power plant disaster.
11. What is nuclear fusion? Discuss the advantages and disadvantages of fusion energy.
12. You are studying future energy demands for your state. List and discuss the factors that affect how much energy it will be using in the year 2025.
13. List and discuss the guidelines for creating a sustainable energy future.

KEY TERMS

acid mine drainage
alpha particles
area strip mines
atom
beta particles
bitumen
black lung disease
bottom ash
breeder reactor
coal gasification
coal liquefaction
containment building
contour strip mines
control rods
crude oil
daughter nuclei
draglines
electron cloud
electrons
element
energy quality
energy system
First Law of Thermodynamics (First Law)
fission
fission fragments

fission reactors
flaring
fly ash
fuel rods
fracking
gamma rays
horizontal oil well drilling
hydraulic fracking
in situ retorting
ionizing radiation
ions
isotopes
kerogen
light-water reactors (LWRs)
liquid metal fast breeder reactor
magnetic confinement
mine tailings
mountaintop removal
natural gas
net energy analysis
net energy yield
nuclear energy
nuclear fusion
nucleus
oil sands

oil shale
Organization of Petroleum Exporting
    Countries (OPEC)
overburden
petroleum
pneumoconiosis
radiation
radiation sickness
radionuclides
reactor core
reactor vessel
Second Law of Thermodynamics
    (Second Law)
shale oil
smokestack scrubber
surface mining
Surface Mining Control and
    Reclamation Act
surface retort
synfuels
tar sands
ultimate production
yellow cake

© Mike Flippo/Shutterstock, Inc

Foundations of a Sustainable Energy System: Conservation and Renewable Energy

I'd put my money on the sun and solar energy. What a source of power! I hope we don't have to wait 'til oil and coal run out before we tackle that.

—Thomas Edison

LEARNING OBJECTIVES

After reading and studying the material in this chapter, students will be able to understand, knowledgably discuss, and, when appropriate, apply the information learned in this chapter, including:

1. The importance of energy conservation and energy efficiency in meeting our demands for energy
2. The social, economic, and environmental benefits of energy conservation and energy efficiency
3. Energy efficiency measures in homes and businesses
4. The many renewable energy options available to humankind, including solar, wind, biomass, green fuels, hydropower, geothermal, and hydrogen
5. The pros and cons of each renewable energy fuel option
6. The potential of renewable energy to meet human needs

347

CRITICAL THINKING

Exercise

An internationally known expert from the oil industry argues to a group of congressional representatives that solar energy is not an economically sound option. It's a great idea, he says, but the economics of solar energy aren't good enough to merit widespread use in the United States. He argues that the conventional fuels such as oil, coal, and nuclear energy are more economical and, therefore, more desirable options. If you were one of the senators on the panel listening to this testimony, what questions would you ask? How would you evaluate this advice? What critical thinking rules does this examination require?

(a)

(b)

FIGURE 15-1 A house for all seasons. Radial, that is. **(a)** Architect and builder Michael Reynolds builds homes out of used tires. Tire walls are covered by stucco and appear quite attractive. The interior walls absorb sunlight that streams in through south-facing glass in the winter months and keep the home very warm. **(b)** Planters can be used to grow food and flowers year-round.

Photos courtesy of Dan Chiras.

Michael Reynolds is a maverick in the construction industry. Based in Taos, New Mexico, Reynolds builds homes that embody many of the principles of sustainability. The walls, for example, are constructed of used automobile tires that otherwise would have ended up in landfills. The recycled tires are packed with dirt from the construction site, using a local resource. They're laid on top of one another like bricks to build thick walls. Cement stucco or earthen plaster is then applied to the tire walls, creating an appealing design (**FIGURE 15-1A**). Reynolds's houses, called *Earthships*, are generally built into the sides of hills, taking advantage of the Earth to shelter the house from summer heat and winter cold. With their thick walls well-insulated ceilings, and south-facing windows, Earthships are heated by the sun. They stay cool in the summer and warm in the winter.

Reynolds's homes are designed with interior planters that line the south wall, permitting residents to grow a variety of vegetables year round (**FIGURE 15-1B**). The plants are watered with wastewater from sinks and showers—commonly called *gray water*. In his most recent designs, Reynolds has devised a system to capture nutrients from toilet water. The waste is fed into specially lined outdoor planters, where it is broken down by bacteria and other microorganisms. The nutrients are used by plants growing in the planters.

Besides being heated by the sun, Reynolds's homes generate their own electricity from sunlight and are equipped with efficient lighting systems and appliances. His homes even capture and purify rainwater and snowmelt off the roof for cooking, drinking, bathing, washing dishes, and other uses.

Earthships are designed for self-sufficiency and environmental responsibility. They are unlike conventional homes, which Reynolds likens to patients in intensive care units that depend on outside support in the form of food, water, and energy. Without these inputs, a patient could not survive. Homes are equally as vulnerable.

Reynolds and a handful of other builders are part of a growing legion of citizens working to help make the transition to a sustainable society, one that supplies human needs while protecting and even enhancing the Earth's life-support systems. A key element of his design is the use of sunlight, a form of renewable energy. Conservation is also vital to the success of his design. This chapter examines two major "sources" of energy needed to create a sustainable global energy system: conservation and renewable energy. The final section of the chapter shows how these environmentally sustainable forms of energy could eventually replace what many consider to be unsustainable fuel sources: oil, coal, natural gas, and nuclear energy.

15.1 Energy Conservation: Foundation of a Sustainable Energy System

We humans need energy, and we need it badly. Yet at least half—perhaps as much as three-quarters—of the energy consumed in the world is wasted, largely because of inefficient technologies and wasteful practices. The **second law of thermodynamics** states that when energy is converted from one form to another, some energy is lost as heat. In other words, no energy conversion is 100% efficient; some waste is inevitable. The amount of energy wasted in the United States, however, far exceeds the inevitable loss.

Given our heavy dependence on nonrenewable energy supplies—and given their economic and environmental costs and eventual depletion—waste is economically, environmentally, and socially unsustainable. Writer Bruce Hannon put it best when he said, "A country that runs on energy cannot afford to waste it."

Go Green

To save energy in your apartment, dorm room, or home, close curtains, shades, and blinds at night on cold winter days. You can lose up to 20% of your heat through uncovered windows. If your domicile does not have window coverings, consider installing them.

KEY CONCEPTS

Energy is used wastefully in virtually all nations. Excessive waste is a sign of an unsustainable technology.

Economic and Environmental Benefits of Energy Conservation

Conservation is one of the biological principles of sustainability and is essential to the sustainable transition now under way. *Conservation* is used here to include frugality (using what you need) and efficiency (using it efficiently). Energy conservation offers numerous social, economic, and environmental benefits. In industry, energy conservation can reduce the cost of producing goods, giving companies an economic advantage in the marketplace. An Alcoa plant in Iowa, for example, found that with a minimum investment in energy-efficiency measures it could cut its $6 million annual energy bill by 25%, thus reducing energy costs by $1.5 million a year—a tidy sum that makes them more competitive and more profitable. Energy conservation also results in substantial reductions in pollution. Many factories, for example, produce their own electricity by burning coal. To reduce pollution, they've installed pollution control devices such as smokestack scrubbers on their power plants. These devices capture pollutants but produce a considerable amount of hazardous solid waste. By using energy more efficiently, companies can reduce the energy they need. Energy savings reduce fuel costs. They also result in economic savings by reducing the amount of hazardous waste that must be disposed of.

The economic savings from energy efficiency can be illustrated by comparing the cost of this strategy with the cost of supplying energy from conventional sources. For example, measures that improve the efficiency of machines and appliances cost about 1 cent per kilowatt-hour of energy saved, according to energy expert Amory Lovins. In contrast, electricity from coal-fired power plants costs 8 cents per kilowatt-hour (kWh). What this means is that for every penny a company invests in energy efficiency it will save 7 cents in electrical bills, if they're paying 8 cents per kWh. Similar calculations can be made for other fuels. Not surprisingly, the most energy-efficient companies and nations are also the most successful economically.

Besides saving money, energy efficiency addresses a number of environmental problems simultaneously. For example, reducing the combustion of fossil fuels reduces acid deposition, global warming, and urban air pollution. Cutting our demand for oil reduces the number of oil spills, thus saving aquatic species as well as the birds and mammals that depend on clean water. In addition, using less fossil fuel reduces the destruction of wildlife habitat by decreasing the extraction of fossil fuels. Besides saving money, efficiency measures can be rapidly deployed, so they begin saving individuals and businesses money very quickly. Finally, energy efficiency can help

KEY CONCEPTS

Energy-efficiency measures can reduce the world's dependence on fossil fuels. Energy-efficiency measures are easy to implement and cost a fraction of conventional energy while providing many environmental benefits.

us stretch limited supplies of fossil fuel, giving us more time to make a transition to a sustainable system of energy.

Energy Conservation Options

The range of energy conservation options available to businesses and homeowners is enormous. In some cases, efficiency can be achieved through minor changes in our actions. The Denver Marriott hotel, for instance, made substantial cuts in energy demand by asking its cleaning staff to raise the temperature setting in unoccupied rooms during the summer (so the air conditioning system didn't have to work so hard) and to lower it during the winter (to reduce heating costs). Around the home, shutting off lights, turning off televisions when they're not in use, and lowering the thermostat a little in the winter and raising it in the summer can result in significant reductions in energy consumption.

Every degree you raise the thermostat setting (for 8 hours) in the summer reduces your cooling costs approximately 8%. For commuters, driving within the speed limit, keeping cars tuned up, and being sure that tires are properly inflated could translate into millions of gallons of gasoline saved each year. Such measures require little or no initial investment and can save individuals a lot of money. When millions of people contribute to the cause, savings can be enormous.

Many devices that save energy are readily available for homes, businesses, and factories. Water-saving devices—especially showerheads, dishwashers, and washing machines—can save substantial amounts of energy in homes by reducing demand for hot water. High-efficiency lighting systems can cut electrical demand for lighting in homes and businesses by three-fourths. The compact fluorescent lightbulb (CFL), available in many stores today, can be substituted for standard incandescent bulbs and uses only one-fourth as much energy to produce the same amount of light (**FIGURE 15-2**). Although high-quality compact fluorescents cost more initially (about $1 to $2), they outlast 9 or 10 ordinary bulbs and save $30 to $50 over their lifetimes in reduced electrical demand. They also reduce carbon dioxide pollution by about one ton and reduce summer cooling costs because they produce little waste heat, unlike incandescent bulbs. Homeowners are recommended to install them in high-use areas.

Another longer-lasting, energy-efficient lightbulb, known as an LED, is now available (**FIGURE 15-3**). LED stands for light-emitting diode. These bulbs use about the same amount of energy as standard CFLs. Although they are still costly, they also outlast the long-lasting CFLs. An LED could last as many as 50,000 to 60,000 hours, saving several hundred of dollars over its lifetime. LEDs are now often used in taillights of trucks and cars, and even stoplights. The City of Denver replaced all of its stoplights with LEDs, which saves them $1 million a year in electric costs. LEDs can be used to replace outdoor floodlights and are also available for indoor use.

Some of the most substantial savings come from some of the easiest and least expensive measures. Caulking leaks in the building envelope—that is, the walls, foundation, and ceiling—of homes and businesses can dramatically reduce heat gain in the summer and heat loss in the winter. Applying weatherstripping around doors and windows can also reduce heat gain and loss. Sealing the leaks in a building saves lots of energy, slashes fuel costs, and dramatically increases comfort levels.

Substantial energy savings can also be achieved by installing a lot more insulation in homes, businesses, and factories. Doubling the insulation in a new home, for example, adds about 5% to the cost but pays for itself in 5 years in reduced energy costs.

Planting shade trees around homes can reduce summer cooling costs. One study showed that three trees planted near light-colored houses in a residential neighborhood in Phoenix could cut cooling costs by 18%. In Sacramento and Los Angeles, the savings are even greater—34 and 44%, respectively. Planting conifers on the north and west sides

FIGURE 15-2 The compact fluorescent lightbulb is one of many cost-effective means of saving energy. Over its lifetime, a $12 bulb can save $30 to $50 in electricity.

Courtesy of OSRAM SYLVANIA.

FIGURE 15-3 LED lights. While they are as efficient as CFLs, LEDs last 5 or 6 times longer.

© ludinko/iStock/Thinkstock.

TABLE 15-1

Individual Actions Count!

1. **Water Heating**
   - Turn down thermostat on water heater to 120°F.
   - Use less hot water (dishwashing, laundry, showers).
   - Install water-efficient showerhead.
   - Install faucet aerators.
   - Coordinate and concentrate time hot water is used.
   - Install an insulated water heater blanket.
   - Do full loads of laundry and use cooler water.
   - Hang clothes outside to dry.
   - Periodically drain 3 to 4 gallons from water heater.
   - Repair leaky faucets.

2. **Space Heating**
   - Lower thermostat setting at night and when away from your residence.
   - Insulate ceilings and walls.
   - Install storm windows, curtains, or window quilts.
   - Caulk cracks and use weatherstripping.
   - Use fans to distribute heat.
   - Dress more warmly.
   - Heat only used areas.
   - Humidify the air.
   - Install an electronic ignition system in furnace.
   - Replace or clean air filters in furnace.
   - Have furnace adjusted periodically.

3. **Cooling and Air Conditioning**
   - Increase thermostat setting at night or when away from your residence.
   - Use fans.
   - Cook at night or outside.
   - Dehumidify air.
   - Close drapes during the day.
   - Open windows at night.

4. **Cooking**
   - Cover pots and cook one-pot meals.
   - Don't overcook and don't open oven unnecessarily.
   - Double-up pots (use one as a lid for the other).
   - Boil less water (only the amount you need).
   - Use energy-efficient appliances (Crock-Pots).

5. **Lighting**
   - Cut the wattage of bulbs.
   - Turn off lights when not in use.
   - Use compact fluorescent bulbs whenever possible.
   - Use natural lighting whenever possible.

6. **Transportation**
   - Carpool, walk, ride a bike, or take the bus to work.
   - Use your car only when necessary.
   - Group your trips with the car.
   - Keep your car tuned and tire pressure at the recommended level.
   - Buy energy-efficient cars.
   - Recycle gas guzzlers.
   - On long trips, take the train or bus (not a jet).
   - Drive the speed limit.

---

of homes can cut wind and reduce winter heating costs. Tree planting near homes and businesses in the United States could cut energy demand by nearly 1 quad, or about 1/100th of our total energy demand.[1] Although it is not a lot, individual savings can be substantial.

Conservation does not mean freezing in the dark, as some would have you believe. For homeowners, adding attic insulation or installing storm windows reduces heat loss in the winter and retains cool air in the summer, resulting in a significant savings on utility bills and a more comfortable life. Both insulation and storm windows require small investments with sizeable returns on one's investment thanks to reductions in energy costs. (For additional suggestions, see **TABLE 15-1** Individual Actions Count!)

Energy efficiency can—and is being—designed into new products—for example, cars, appliances, buildings, and factories. Efficiency measures in new office buildings, for example, can cut energy for lighting, heating, and cooling by 75%! Energy-efficient homes can be built with heating bills of $100 a year or lower, even in cold climates!

One extraordinary means of reducing energy waste is **cogeneration**. In industrial cogeneration, waste heat from one process is captured and used in another, thus reducing waste

and improving the energy efficiency of a plant. For example, for years, many American industries produced their own steam on site for use in various processes. Steam generation requires natural gas or oil, which companies purchased along with electricity from local utilities. Today, many companies use waste heat from steam generation to produce electricity, thus boosting the efficiency of the system from 50%–70% to 80–90%—and driving down the cost of electricity.

Because of its favorable economics, cogeneration is an emerging source of energy. Growth in generating capacity has increased dramatically in the past 25 years—and continues to rise. In Germany, the United States, and other developed countries, small cogeneration technologies are being installed in restaurants, hotels, and apartment buildings. They supply space heat, hot water, and electricity at a much lower cost than conventional systems.

KEY CONCEPTS

There's no shortage of easy, economical ways to save energy. These include simple behavioral changes, such as turning off lights when one leaves a room, as well as many energy-efficient technologies, such as compact fluorescent lightbulbs and cogeneration facilities. Energy savings result in substantial social, economic, and environmental benefits.

---

[1]One quad is a quadrillion BTUs—about 170 million barrels of oil.

The Potential of Energy Efficiency

The United States and other industrial nations have made tremendous improvements in energy conservation, but the conservation potential has hardly begun to be tapped. Most experts agree: Although fossil fuels will run out and renewable energy supplies will take up the slack, energy efficiency could play a huge role in our future. Moreover, increased efficiency should not affect the level of services we receive. Much of our future energy demand can be met by freeing up energy currently wasted in three areas: transportation, buildings, and industry/business.

KEY CONCEPTS

Because energy is used so inefficiently, huge cuts in energy demand can be made by applying efficiency measures.

Saving Energy in the Transportation Sector Over one-fourth (28% in 2012) of the energy consumed in the United States is used in transportation. Energy consumption in the U.S. transportation sector has grown steadily since the 1960s and shows no sign of declining, a trend that many experts believe bodes poorly for our nation's prospects for creating a sustainable future (**FIGURE 15-4**).

Fortunately, dramatic improvements in the efficiency of vehicles are possible. Automobiles, the single largest source of fuel consumption in this sector, represent the greatest potential for reductions.

Worldwide there are over 110 car models that get over 40 miles per gallon, 70% of which are made by U.S. companies. Until very recently, only three of these vehicles were sold in the United States. In recent years, numerous auto manufacturers have begun producing gas-electric **hybrids**. A hybrid vehicle contains a gas and an electric motor that work in ways to greatly increase fuel mileage. Hybrids contain batteries that power the electric motor, saving on gas. The batteries are charged when the car brakes or slows down, so they never have to be charged. Three of the most efficient vehicles on the road in the United States are hybrid gas-electric vehicles, the Toyota Prius, the Honda Civic hybrid, and the Honda Insight.

More recently, several auto manufacturers including Ford and Toyota have introduced **plug-in hybrids**. These cars contain larger battery banks that allow them to travel farther on electricity and, thus, use less gasoline. They are charged at home or at the office when not in use. For more on hybrids and plug in hybrids, be sure to read **Spotlight on Sustainable Development 15-1**.

Go Green

When the time comes to buy a car, purchase the most energy-efficient model you can afford.

Recognizing the importance of boosting fuel economy, the Obama administration established much higher gas mileage standards for new cars and trucks that require 2016 model-year vehicles to be the average fuel mileage of 35.5 miles per gallon. For cars and light trucks, this is an increase of nearly 10 mpg over current standards. This is the first increase in 25 years. In 2011 President Obama announced an agreement with 13 auto manufactures to boost fuel economy of cars and light trucks to 54.5 mpg by 2025.

Energy-efficient vehicles are vital to address declining oil supplies, rising population, and increasing congestion on highways in cities. One of the most energy-efficient forms of transportation is mass transit—energy-efficient buses and trains that move large numbers of people with a fraction of the resources of single occupancy vehicles.

KEY CONCEPTS

Extremely energy-efficient vehicles are currently available. Many improvements in vehicles could increase efficiency even more, greatly cutting transportation energy consumption.

Saving Energy in Buildings About 40% of the energy consumed in the United States is used in buildings—office buildings, factories, and homes. The modern home and office building make many demands on the energy system. They require lighting, heating, and cooling. Many also contain various machines, appliances, and electronic equipment. Fortunately, very impressive savings can be made in all areas.

One encouraging development is the emergence of energy conservation companies, which provide innovative financing schemes for building owners who want to invest in energy efficiency. These companies have been successful in Europe for years and are becoming more common in the United States. They advise clients on ways to cut energy consumption and install various devices to achieve their goals. Some companies work out plans to be paid by their clients through energy savings accrued by the latter.

Another very encouraging development is the dramatic increase in **green building**—creating homes and

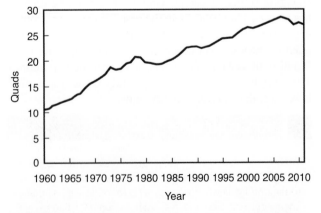

FIGURE 15-4 Energy consumption in the transportation sector. U.S. energy consumption in the transportation sector has increased consistently over the years, an unsustainable trend.

15-1 Reinventing the Automobile

Imagine a car that gets 150, maybe 200 miles per gallon, and can travel from the West Coast to the East Coast on one tank of gas. Imagine a car that comfortably seats five passengers but weighs only about 400 kilograms (900 pounds)—one-half to one-fourth the weight of a typical vehicle. Imagine also that this superefficient, lightweight car is also just as safe as a conventional vehicle and consists of a body and chassis that contain only 6 to 20 parts, compared with 250 or so in a conventional automobile.

Is this some creation of a science fiction writer? No. Far from it. This car is the brainchild of Amory Lovins of the Rocky Mountain Institute (RMI). Renowned for his work on energy efficiency, Lovins received funds from several foundations to develop designs for a superefficient car of the future to help drastically reduce the world's demand for fossil fuels and cut pollution.

Shown in **FIGURE 1**, this vehicle, called a *hypercar*, has a smooth underbody, and an incredibly aerodynamic form. These and other features could dramatically increase the car's efficiency without sacrificing either interior space, safety, or performance.

Lightweight composites (mixtures of materials) and other space-age materials would be used to construct the car. These materials are stronger and much lighter than those currently used by car manufacturers. Because composites absorb a lot more energy in crashes than the steel used in the bodies of a typical car, the lightweight hypercar would actually be safer. The vehicle's small engine, discussed shortly, provides for

an efficient performance and also leaves more "crumple zone"—area for the car to cave in during an accident. Force is absorbed by the materials, not the occupants.

The hypercar is a hybrid of sorts. Its propulsive unit, its engine, might consist of a small, fossil fuel-powered engine—only 20 horsepower or so—that is used to generate electricity. Other electricity-producing engines are also possible, for example, fuel cells (described in this chapter). Whatever the source, the electricity is sent to four small electric motors in the wheels. During braking, electric motors that drive the wheels function as small generators. They could theoretically capture up to 70% of the braking energy—energy that is lost in a typical automobile. This energy is stored in the battery for later use.

Further adding to the car's efficiency are special tires made of materials that reduce the friction on the road surface. This could, according to Lovins and his team at RMI, reduce the rolling drag by 65 to 80%. Together, the ultra-lightweight design, superb aerodynamics, and efficient hybrid propulsion system could translate into a 150 to 200 miles per gallon fuel rating—considerably higher than the 22.4 miles per gallon average achieved by cars and light trucks on the road in the United States. Air emissions from the vehicle would be 100 times lower than today's cars. In other words, 100 hypercars would produce as much pollution as one conventional vehicle.

Is this a wild dream? Apparently one student at Western Washington University didn't think so. He built a Corvette-sized two-seater light hybrid. In April 1994, the vehicle was tested by the Department of Energy and found to get 202 miles per gallon. In October 1994, a Swiss-built car that seats four achieved 150 miles per gallon.

As noted in the chapter, virtually all automobile manufacturers are now producing hybrid gas-electric cars that begin to resemble the hypercar with gas mileage between 40 and 70 miles per gallon. Toyota was the first to sell its car (Prius) in Japan, starting in 1999, and then in the United States in 2000. Honda came next with a model known as the Insight. It was first sold in the United States in 2000, but was discontinued in 2006.

FIGURE 1 The hypercar.
Copyright © 2005 Rocky Mountain Institute (www.rmi.org).

Numerous American manufacturers are now offering hybrid cars, trucks, SUVs, and vans. And all major U.S. auto manufacturers also have developed diesel hybrid cars that get 60 to 80 miles per gallon (at this writing, there are no plans to manufacture and sell these models).

How does a hybrid work? Hybrids have an electric motor and a small gas engine. In some models, like Toyota's Prius, the car initially operates on electric power; that is, when the car is turned on and begins to move, its electric motor provides the main propulsion. The gas engine kicks in at around 7 to 10 miles per hour. On the highway, however, the gas engine provides much of the thrust. The electric motor provides additional boost to pass or climb hills. When slowing down or going down a hill, however, the vehicle uses very little power at all. In fact, the gas engine may shut off entirely. The instantaneous gas mileage readout in the Prius, located on a computer screen in the center of the dashboard, records gas mileages of 100 miles per gallon, indicating that the car is coasting.

When the vehicle comes to a stop at a stoplight, the engine shuts off. When the operator presses on the gas pedal, however, the electric motor kicks in, followed by the gas motor. The car is off and running without hesitation.

Honda's hybrid Civic is similar to the Toyota Prius in many respects; that is, they contain a gas engine and an electric motor, but the electric motor is much smaller than that of the Prius. As a result, it can't be used to start the car from a standstill.

When a Honda Civic hybrid is turned on, the gas engine starts up and provides the power needed to move the car and most of the propulsion from that point on. The electric motor kicks in when additional power is needed, for example, when climbing a hill or passing another vehicle.

What about the batteries of a hybrid car? Do they need to be charged each night? No, they're charged continuously by the car during normal operation through an alternator, a device that provides electricity for lights and radios and the battery bank. As in the hypercar, batteries are also charged during braking. Electricity flows smoothly in and out of the lightweight battery bank (usually located behind the back seat) during various modes of operation, giving one a clean, energy-efficient ride! Like other functions in the car, it is all computer controlled. You drive; the car figures all of this out automatically.

Hybrids cost $2,000 to $3,000 more than non-hybrids. However, federal and state incentives have helped reduce the costs to consumers. Savings on fuel will help offset costs.

Expect to see a lot of hybrids on the road in coming years. Sales are going extremely well. The cars are pretty attractive. They're fast and relatively inexpensive, too. Although not reaching the full potential of the hypercar, they represent a dramatic shift in automobile technology with enormous benefit to the environment.

Another vehicle that holds promise is the plug-in hybrid electric vehicle, now under development. What's a plug-in hybrid?

Plug-in hybrid electric vehicles are hybrids, much like those just discussed, with a couple differences. First, they contain a larger battery bank. Second, unlike the hybrids just discussed, these cars are plugged in at night into a 120 or 240-volt outlet (like a dryer outlet). This charges the batteries. The electrical charge, in turn, powers the vehicle in most situations. According to the Institute for Analysis of Global Security, which researches transportation options that could help us free our dependence on foreign oil, "Plug-ins run on the stored (electrical) energy for much of a typical day's driving—depending on the size of the battery up to 60 miles per charge, far beyond the commute of an average American." They go on to say that when the electrical charge is used up, the car "automatically keeps running on the fuel in the fuel tank. A person who drives every day a distance shorter than the car's electric range would never have to dip into the fuel tank.

Because most of the energy used by plug-ins comes from electricity and not from gasoline and because electricity can be generated efficiently and cleanly from America's abundant renewable energy resources, especially solar and wind power, hybrid electrics may help us combat the coming shortages of oil without increasing global warming and a host of other environmental problems associated with fossil fuel use.

According to the Institute for the Analysis of Global Security, "The plug-in hybrid drive system is compatible with all vehicle models and does not entail any sacrifice of vehicle performance or driver amenities. A mid–size plug-in can accelerate from 0 to 60 miles per hour at less than 9 seconds, sustain a top speed of 97 mph and maintain 120 mph for about 2 minutes even with a low battery."

In 2014, five plug-in hybrids were available in the United States from Ford, Toyota, Honda, and Chevrolet. Unfortunately, the distance they can travel on battery power is fairly limited. The plug-in Toyota Prius, for instance, can only travel about 12 miles on battery power, according to the manufacturer, at which time the car reverts to the gas-hybrid mode. (In an EPA test, the car only went 6 miles on electric-only mode.) The Ford C-Max plug-in hybrid travels about 21 miles on battery power according to the manufacturer, before converting to the gas-electric mode. (In EPA tests, it switched over at 21 miles.) Even so, the combined fuel mileage of plug-in hybrids—that is, gas mileage when operating on gas and electricity—range from the equivalent of 95 to 115 miles per gallon.

One of the most unusual hybrids is the Chevy Volt (**FIGURE 2**). It is equipped with an electric motor and chargeable battery bank that allows the car to run entirely on electricity for about the first 20 miles. The car is also equipped with a small engine that can run on gasoline, biodiesel, or E-85 (an 85% mixture of ethanol and gasoline). The engine generates electricity that is used to power an electric motor. This car can achieve very high gas mileage. If you drove under 20 miles per day you could run entirely on electricity. If you drove 60 miles per day, you'd get an estimated 150 equivalent miles per gallon. For long trips the car gets around 40 miles per gallon.

Electric cars will also very likely become a major player in transportation throughout the world. Electric cars can be fast, powerful, and economical. Current models have a limited range, of 30 to 120 miles, making them ideal for commuting back and forth to work or school. (In the U.S., 90% of all automobile trips in a day are under 60 miles).

Electric cars are powered by electric motors. Electricity is stored in a battery bank that must be charged after use. Numerous electric cars appeared in 2010–2012, including the Nissan Leaf, which made its debut late in 2010, and the Ford Focus (**FIGURE 3**).

Electric motors are extremely efficient, and even when they use electricity generated at coal-fired power plants, they result in a substantial decrease in energy use and pollution.

Automobiles like RMI's proposed hypercar won't solve all of the world's transportation problems. If cars such as these became a popular alternative to conventional automobiles, they could drastically reduce fossil fuel consumption and air pollution, both impediments to the goal of creating a sustainable future. Super–fuel-efficient cars will not, however, solve such problems as urban sprawl and highway congestion during all hours of the day—and especially that time we mistakenly call "rush hour." Fortunately, there are ways to help solve these problems, notables high-speed mass transit.

FIGURE 2 The Chevy Volt.

FIGURE 3 Nissan Leaf. This is one of a dozen new electric vehicles that have been introduced by car makers in recent years.

other buildings that have as little impact on the environment as possible. One key element of green building is efficiency—making homes and other buildings as energy efficient as possible. Efforts are underway in Germany and the United States to spent produce *net-zero energy homes*—homes that are so efficient they can produce all the energy they need through solar and other renewable energy technologies. Green building is discussed in more detail in the next section.

Some experts believe that cost-effective energy-efficiency measures could cut total energy consumption in U.S. buildings by as much as 30% over the next two decades, despite a 15 to 20% increase in the number of buildings.

KEY CONCEPTS

Energy-efficiency measures in buildings can result in substantial energy savings in heating, cooling, lighting, and running appliances and electronic equipment.

Saving Energy in Industry   In the United States, industries consume one-third (28% in 2012) of the nation's energy. Over the past 40 years, industrial energy savings have been impressive. Energy required to generate one dollar of economic output has fallen by about 50% (**FIGURE 15-5**). Nevertheless, industries have many additional opportunities to cut energy consumption while increasing their economic strength and profitability. Many corporate executives now realize that an investment in energy efficiency is one of the most cost-effective ways of reducing expenses and boosting profits. Energy efficiency is not an impediment but a keystone to economic success.

Energy can be saved by installing more efficient lighting systems and insulating buildings to reduce heating and cooling costs. Energy use can also be cut dramatically by installing more efficient boilers, pumps, and motors. Many pumps, for example, operate at one speed. To control the flow of liquids, valves are used. Installation of variable speed motors has the same effect while dramatically cutting power demand.

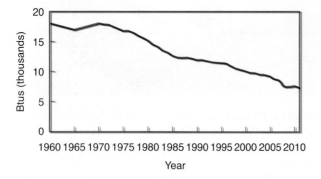

FIGURE 15-5 Energy vs. GNP. This graph shows that the energy required to generate a dollar of GNP (in year-2000 dollars) by the U.S. economy has declined over the years, largely as a result of an increase in efficiency.

Data from U.S. Energy Information Administration.

Recycling is also an important strategy for reducing industrial energy consumption. Roughly three-quarters of the energy consumed by industry is used to extract and process raw materials. Making metals from recycled scrap uses a fraction of the energy needed to make them from virgin ore.

KEY CONCEPTS

Industrial energy consumption accounts for a major portion of the world's energy demand but could be cut substantially, making companies more profitable and competitive.

Promoting Energy Efficiency

Energy efficiency is needed in all sectors of society. How can individuals, businesses, and nations be encouraged or compelled to use energy more efficiently?

Once again, many options are available. This section briefly discusses seven options: (1) education, (2) taxes, (3) feebate systems, (4) government-mandated efficiency standards, (5) voluntary programs, (6) changes in pricing, and (7) least-cost planning.

One of the most important steps is education. Educating people about the economic and environmental benefits of energy conservation is vital to the task at hand. Education can occur at many levels, for example, in schools, colleges, and on television. Government agencies such as the Environmental Protection Agency (EPA) and Department of Energy (DOE) can help educate businesses on the advantages of energy efficiency—and currently take an active role in this arena.

Some proponents of efficiency promote national taxes on fossil fuels—especially oil and coal—to promote efficiency. By raising the taxes on fossil fuels, making them more costly to consumers, governments could play an important role in stimulating energy conservation in all sectors of society.

Increased taxes on gasoline and other fuels in the United States and Canada could help cut energy consumption and generate revenues to build mass transit systems, promote conservation, and fund sustainable alternative transportation technologies. Because government currently subsidizes many fossil fuels through tax dollars from general revenues, proponents note, taxes would help it recover the money it already invests in providing us with artificially inexpensive fuel.

Another innovative measure is the *feebate* system. It consists of a fee (tax) paid by those who buy gas-guzzling cars (which would create a disincentive to consumers), and a rebate given to those who purchase energy-efficient autos. The rebate would be paid by the fees and would be an additional incentive to individuals. The United States had a gas guzzler tax for many years, but President Clinton abolished it in 1999.

A fourth means of stimulating conservation is through government-mandated efficiency standards. The **National Appliance Conservation Act** passed by the U.S. Congress in 1987, for instance, established energy-efficiency standards for all new appliances. It didn't spell out what changes needed to be made, only the amount of energy that appliances should be consuming. For example, the act required manufacturers of all major household appliances, such as refrigerators, to produce models that consume at least 20% less electricity than 1987 models. California passed a similar law, calling for a 50% cut in electrical usage by new appliances. The results were impressive. Today, homeowners can purchase many appliances that use a fraction of the energy consumed by earlier models. Efficiency standards could be applied to all new homes and factories as well.

Voluntary efforts are also vital. Many governments are starting to take a more active role in promoting energy efficiency in buildings through voluntary programs.

In the United States, the EPA and many states are promoting home energy efficiency through a variety of voluntary programs. EPA has its **Energy Star Homes program**. This program relies on builders to produce homes that use 30% less energy for heating, cooling, and so on than stipulated in the **National Model Energy Code**. When this goal has been verified by an independent certifier, the home receives an **Energy Star label**.

Many cities and states have **green building programs** that promote energy efficiency. By meeting certain criteria, home builders can apply for a green building certificate. This, in turn, helps builders market a home. A leader in commercial green building is the nonprofit group known as the U.S. Green Building Council. This organization created a commercial building rating system known as LEED—Leadership in Energy and Environmental Design—for new and existing buildings. Architects and builders must meet certain criteria to qualify for LEED certification, which dramatically reduces the energy and environmental impacts of new and existing buildings. They introduced LEED certification for homes in 2007.

Another encouraging development is the National Association of Home Builders' active role in researching and promoting energy-efficient homes. They released green home building guidelines in 2007. Some lending institutions are also providing loans for energy-efficient homes, although they tend to simply allow the lender to borrow more money, based on the belief that they will have more money in their pockets. Far better, say some critics, would be a program that gives lower interest loans to buyers of green built homes.

Yet another encouraging development is the net zero energy home. **Net zero energy** homes are designed to produce as much energy as they consume. Builders achieve this difficult goal by building superefficient homes—airtight homes whose walls and ceilings contain three to four times as much insulation as typical energy-efficient homes. The use of super energy-efficient windows, solar heat gain, and other measures result in homes that use 90% less energy than standard homes. Builders then install solar electric (**FIGURE 15-6**) and solar hot water systems to provide for the energy needs of the residents.

The net zero energy movement began in Germany thanks to the efforts of the Passive House (*Passiv Haus*) Institute. There is now a Passive House Institute headquartered in the United States that is carrying out this effort. The U.S. Department of Energy's Building America program has also played a very large role in promoting net zero energy buildings.

Conservation can also be stimulated by changes in pricing. For example, some utilities now charge customers more for electricity used during peak hours—times when demand is highest. Why? Meeting peak demand is very costly for utilities. It often requires utilities to buy electricity on the spot market, which can be very costly or build additional power plants that are needed for only a few hours a day. If utility companies can reduce peak demand through pricing, they won't have to build expensive new facilities.

Furthermore, Efficiency can also be stimulated by a process called **least-cost planning**. In the United States, utilities are regulated by the states. Thus, when a power company decides that it needs to build a new power plant to meet rising demand, it must first receive approval from a public utility commission. Currently, most states have requirements for least-cost planning or integrated resource

FIGURE 15-6 Author's solar array. This 4-kW solar electric array provides most of the electricity the author and his wife need to power their super energy-efficient, green-built home in east-central Missouri. The rest of his family's electricity is provided by a small wind turbine located in the same field.

Courtesy of Dan Chiras.

planning (choosing a variety of technologies, including renewables, and measures such as efficiency to meet energy needs). Least-cost planning requires power companies to select the least costly way of providing electricity, which almost always entails energy conservation programs. Some cost-saving options include improvements in generating efficiency, peak-pricing schemes, purchasing electricity from other companies (rather than building new power plants), cogeneration, and promoting energy conservation in homes, factories, and businesses. These steps are not only cheaper for utilities, they save customers money. In general, efficiency measures are two to three times cheaper than building new power plants. Moreover, the payback period of conservation is only 2 or 3 years compared to 20 for a new power plant. Unfortunately, efforts are currently under way in many states to eliminate requirements for least cost planning, and many states are not enforcing their requirements.

Finally, energy conservation can be stimulated by a combination of the measures listed previously here. The Canadian government, for instance, is trying to promote conservation through **The Efficiency and Alternative Energy program**, which entails 33 different initiatives to improve efficiency in all end-use sectors, including buildings, industry, transportation, and equipment. One of the keys of this program was the 1993 **Energy Efficiency Act**, which sets minimum efficiency standards for equipment. The **Motor Vehicle Safety Act** places stringent emissions standards on cars. All of these efforts are part of a strategy aimed at becoming more efficient and reducing greenhouse gas emissions to combat global warming.

Go Green

If your home or apartment is equipped with ceiling fans, use them. They help keep a home warmer in the winter and cooler in the summer, greatly reducing energy consumption and fuel bills. Ceiling fans can reduce cooling costs by up to 40% and heating costs by up to 10%.

KEY CONCEPTS

Not only are there many ways to use energy much more efficiently, there are also many ways to promote this strategy, including (1) education, (2) taxes on fossil fuels, (3) feebate systems (which levy a tax on those who choose energy-inefficient options and give rebates to those who opt for energy-efficient technologies), (4) government-mandated efficiency programs, (5) voluntary programs, (6) changes in pricing, and (7) least-cost planning.

Roadblocks to Energy Conservation

Despite the economic and environmental advantages of conservation, it still is not widely practiced. Why not? Part of the reason is that Americans, Canadians, and others have been blessed for many years with abundant energy and have had little incentive to use it efficiently.

Another reason is that fossil fuels and nuclear energy are subsidized by federal programs, which reduces the direct cost to the consumer. January 2014 oil was selling at $95 per barrel. If federal subsidies were added, the cost would be between $150 to $250 per barrel. Gas would be selling for $4.50 to $7.50 per gallon.

Still another reason is that high-efficiency products such as LED light bulbs often cost more than less energy-efficient ones, and many people neither calculate the long-term savings nor think about the environmental consequences of waste.

In addition, many governments have not been committed to energy conservation programs. The minimal investment is hard to reconcile, say some critics, when the government investment-to-savings ratio for conservation can be as high as 1:1,000—a thousand dollars saved for every dollar invested.

The lack of interest in energy conservation in the United States and other countries can also be attributed to the influence and power of energy companies, which are a dominant force on the political scene. Nonetheless, as Christopher Flavin and Nicholas Lenssen noted in a recent article, "Powerful economic, environmental, and social forces are pushing the world toward a very different energy system." One of the cornerstones of that system will be energy efficiency. Its potential is great, and many people, businesses, and nations are beginning to recognize that one of the greatest future sources of energy is our waste.

KEY CONCEPTS

Many roadblocks stand in the way of energy efficiency, including the illusion of abundance, federal subsidies that underwrite fossil fuels' true costs, higher initial costs for some energy-efficient products, and powerful political forces. Despite this, energy efficiency is becoming more popular in all sectors of society.

Renewable Energy Sources

Imagine a world powered by the sun, the winds, and other clean, renewable forms of energy. Although this may sound like a dream, 30 years from now you may well live in a world powered by a diverse mixture of clean, affordable renewable energy resources.

The shift to a renewable energy future has already begun. Brazil, for example, has turned to ethanol produced from sugarcane to power trucks and cars. California is turning to wind and geothermal (the Earth's heat) energy. Many other states are following suit. Colorado, Iowa, and South Dakota all produce 20% of their electricity from solar and wind. Several European nations, including Germany and Spain, are currently generating a substantial portion of their electricity from wind. Great Britain is following suit and Germany is gearing up for a 100% renewable energy future. Israel and other Middle Eastern countries are increasing their dependence on solar energy. Many Pacific Rim nations now get substantial amounts of energy from geothermal sources and plan to obtain substantially more in the future.

As world oil and natural gas reserves decline, as environmental problems caused by fossil fuels intensify, and as population spirals upward, more and more countries will shift to renewable energy resources. Each one will find many different clean, abundant, and affordable renewable fuels that are locally available. By increasing the use of these resources, nations can protect the environment and create more regionally self-reliant economies. By increasing their dependence on clean, affordable, and reliable renewable energy, they create a better future, greater economic and political security, and broader prosperity. The next section looks at many of the renewable energy options, discussing the pros and cons of each one.

 KEY CONCEPTS

Renewable energy will very likely become a major source of energy in the future; the transition to a renewable energy future has already begun in some nations.

Solar Energy Options

Oil, natural gas, oil shale, and coal all have limits. So does the sun, the origin of solar energy. In the sun's case, though, the supply of energy is expected to last at least 4 to 5 billion years. Although it is finite, solar energy is typically called a renewable resource.

Each day, about less than one billionth of the sun's energy strikes the Earth. Although this is a small fraction, it adds up to an impressive total. In fact, if all of the sunlight striking an area the size of Connecticut each year were captured and converted into useful energy, it could power the entire United States, including all homes, factories, and vehicles. Despite its enormous potential, solar energy provides a small fraction of U.S. energy needs. Contrary to popular misconception, this poor showing is not because solar energy is limited to a few areas. In fact, significant sources of solar radiation are available throughout the world, although some are better than others (**FIGURE 15-7**).

Four major solar technologies are in use worldwide: passive and active solar hot water, photovoltaics, and solar thermal electric. Understanding each one is vital to creating a sustainable energy future.

 KEY CONCEPTS

Solar energy is considered a renewable energy source, but it is really finite. Nonetheless, because it is so abundant and clean, it will very likely be a major contributor to future world energy supplies for hundreds of thousands of years.

Passive Solar Heating **FIGURE 15-8** shows a house designed to capture sunlight energy. In this structure, light from the low-angled winter sun streams through south-facing windows and is absorbed by interior walls and floors of brick, tiles, or concrete where it is converted to heat. The heat

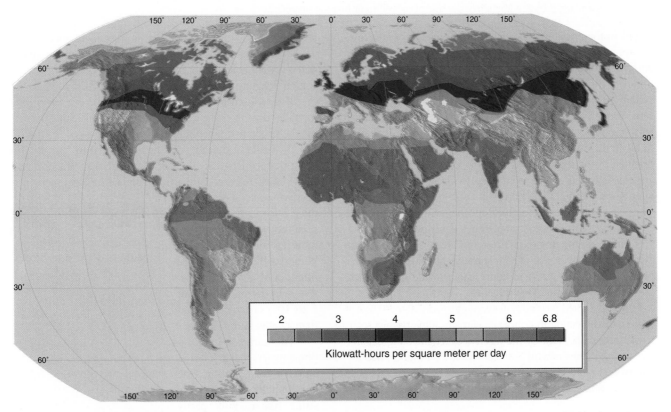

FIGURE 15-7 **Map of global solar energy availability.** The planet is bathed with sunlight energy, although some areas have considerably more than others.

FIGURE 15-8 A passive solar house. Sunlight streams into the house during the winter months, when the sun is low in the sky, and is kept out by the overhang during the summer months, when the sun is higher. Interior walls absorb sunlight energy and emit heat, keeping the house warm and cozy at a fraction of the cost of conventional heating systems.

stored in these materials, called thermal mass, radiates into the rooms, heating the air day and night. On cloudy days, passive solar homes are kept warm by residual heat that continues to radiate from thermal mass and by backup heating systems. This is an example of a **passive solar heating system**. Properly designed passive solar homes and buildings also require good insulation and insulated curtains or shades to block the outflow of heat at night. Passive solar homes are oriented to the south in the northern hemisphere. Windows are concentrated on the north side to allow the low-angled winter sun to enter. The homes are also built to be airtight. Overhangs are installed to block out the summer sun.

Passive solar energy is sometimes described as a system with only one moving part, the sun.[2] It can be added to an existing home by building a greenhouse. In addition to supplying winter heat, a greenhouse can provide a source of food most of the year (**FIGURE 15-9**).

Well-designed passive systems can provide 80 to 90% of a home's space heating in temperate climates. One passive solar home in Canada, built by the mechanical engineering department of the University of Saskatchewan, for instance, had an annual fuel bill of $40, compared with $1,400 for an average American home at the time of construction. The house was so airtight and well insulated that heat from sunlight, room lights, appliances, and occupants provided sufficient energy to maintain a comfortable interior temperature. The cost of this house was only slightly more than that of a tract home in a subdivision. Passive solar homes are also naturally cooled. In fact, everything a builder does to create a passive solar home like superinsulating the structure dramatically reduces cooling costs.

As a rule of thumb, solar houses cost 0 to 5% more to build than conventional houses of similar size. Because used home sales are based primarily on square footage, preowned passive solar homes are often a good buy. My first solar house cost about the same as conventional used homes, with a fraction of the utility bills! My second passive solar home was about 5 to 10% cheaper to build than a conventional home.

Thousands of American homeowners have selected another solar option, the **earth-sheltered, passive solar house**. Built partly underground to take advantage of the year-round constant temperature of the soil, properly designed earth-sheltered,

Go Green

When shopping for a new apartment or a new home, look for one whose long axis runs east to west and whose south side has plenty of shade-free windows to allow the low-angled winter sun to naturally heat the home. This will cut your annual heating and cooling bill by about 10%.

---

[2] Actually, the sun is stationary within our system. The Earth rotates around it.

(a)

(b)

FIGURE 15-9 Solar retrofit. (a) Homeowners can add solar greenhouses to existing houses. Greenhouses can be used to grow vegetables much of the year. **(b)** Sunlight penetrates the greenhouse glass and is stored in the floor or in water-filled drums. Warm air passes into the house. Fans often help facilitate air movement.

passive solar homes are well lighted, dry, and comfortable. They also require less external maintenance. Because they are sheltered by the earth, they stay warm in the winter and cool in the summer. They can easily be designed for passive solar heating. Combined, passive solar heating and earth sheltering can virtually eliminate the need for conventional heat.

KEY CONCEPTS

Buildings can be designed to capture solar energy to provide space heat. These are called passive solar homes. Properly designed structures can derive 80% to 90% of their heat from the sun.

Active Solar Hot Water (Thermal) Systems Active heating systems employ **solar collectors**, generally mounted on rooftops. Most collectors are insulated boxes with a double layer of glass on the sunny side (**FIGURE 15-10**). These are called **flat plate collectors**. Inside each collector is a black absorber plate that absorbs sunlight and converts it to heat. Pipes beneath the absorber plate circulate water or some other heat transfer fluid, and they absorb the solar heat. The heated water is piped to a superinsulated water storage tank usually located in the basement. After transferring its heat to the storage medium, the heat transfer fluid is returned to the collectors to be reheated. This process delivers a large amount of heat to the storage tank during the day, even in the winter so long as the sun is shining.

(a)

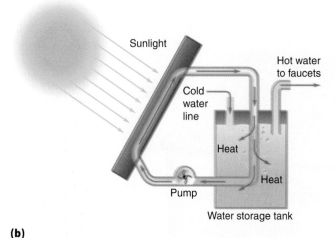

(b)

FIGURE 15-10 Active solar heating system. (a) Evacuated tube solar collector on the author's home. (b) Flat plate collectors like those shown here circulate a fluid through pipes in the collector; the fluid absorbs heat captured by the black interior of the panels. The heated fluid is then pumped to a storage tank, and the heat is transferred to some storage medium, usually water. The water can be used for space heating or heating water for a variety of uses.

Photo courtesy of Dan Chiras.

The hot water in the storage tank can be used for showers, baths, and washing dishes and can also be used to heat homes. There are even solar thermal systems to heat swimming pools and hot tubs. Specially designed active solar systems can be used in industry to provide hot water and steam, a major consumer of energy.

Solar collectors can be expensive to purchase, mount, and maintain. Leaks in systems can be costly to repair. In fact, a single repair can negate all energy savings for a 6- to 12-month period. Solar collectors are costly to add to a house after it is completed, too. And the extreme temperatures to which they are subjected can be hard on systems. Even so, solar hot water systems are cost competitive with electric and propane water heating systems. They can even compete with natural gas systems. Solar pool heating systems are much cheaper than gas or electric pool heating systems. However, some new and inexpensive models

are available with few moving parts and requiring little maintenance.

Solar Electricity: Photovoltaics Electricity can be generated from sunlight via solar electric (photovoltaic) modules, commonly referred to as panels. A photovoltaic (PV) cell consists of a thin wafer of silicon or some other materials that emit electrons when struck by sunlight. Numerous cells are wired together to form a PV module (**FIGURE 15-11A**). Electrons liberated from the material then flow out of the wafer, forming an electrical current (**FIGURE 15-11B**).

PV rocks attached to the modules are mounted on roofs or the ground. Some systems are designed to track the sun across the sky, optimizing electrical production. Solar manufacturers are also making roofing materials, a laminated solar material that is applied to metal roofing. Some manufacturers even apply PV material to glass for windows and skylights—so they perform double duty: they let light in and also produce electricity.

Electricity from photovoltaics currently costs about 10–12 cents per kilowatt-hour, or about the same as electricity from conventional sources. Fortunately, costs have fallen rapidly in the last four decades (about 95% since the 1970s) and are expected to decline substantially (around 75%) in the next decade. At this writing (January 2014), the cost of solar electric modules is at an all time low. Experts predict that new developments, for example, solar cells made from plastic, could soon make photovoltaics cheaper than electricity from conventional sources.

Photovoltaics are already in wide use in remote villages in the less developed nations. In such locations, photovoltaics are far cheaper than building transmission lines that carry electricity from distant power plants: Today, more than 6,000 rural villages in India rely on them. The governments of Sri Lanka and Indonesia have launched ambitious programs to install photovoltaics in remote areas. In the United States, it is often cheaper for a homeowner to install photovoltaics than to pay to string electric lines if the home is more than a few tenths of a mile from a power line. The federal government offers a 30% tax credit on solar electric systems, which will continue through 2016. This incentive makes solar electricity cheaper than electricity from conventional sources.

Today, photovoltaics provide only a tiny fraction of the world's electrical power, but sales are growing steadily, averaging a 16 to 17% per year increase in the 1990s.

(a)

Another exciting development is the increase in the efficiency of solar cells. Although they are not yet commercially available, solar cells with efficiencies over 40% have been produced by researchers at the National Renewable Energy Labs in Golden, Colorado. Solar cells currently on the market are one-fifth to one-third as efficient.

KEY CONCEPTS

Photovoltaics are thin wafers of material such as silicon that emit electrons when struck by sunlight, creating electricity. Photovoltaic module prices have fallen steadily so that solar electricity even without financial incentives is often cost competitive with conventional sources of electricity.

(b)

FIGURE 15-11 Solar modules. (a) Commonly referred to as solar panels, modules like these on the author's array, consist of numerous small solar cells wired together to produce direct current electricity. Glass protects the modules from rain and damage from hail. **(b)** Photovoltaic cells are made of silicon and other materials. When sunlight strikes the silicon atoms, it causes electrons to be ejected. Electrons can flow out of the photovoltaic cell through electrical wires, where they can do useful work. Electron vacancies in the silicon atoms are filled as electrons complete the circuit.

Photo courtesy of Dan Chiras.

Growth has increased even more in the 2000s, as shown in **FIGURE 15-12**. Japan, the United States, and the European Union have all adopted million roof policies, which seek to promote installation of solar electric and solar thermal systems on homes and businesses. For a case study on the use of solar electricity in business, see **Spotlight on Sustainable Development 15-2**.

Solar Thermal Electric Scientists and engineers have experimented for many years with ways to heat water with sunlight to generate steam that turns a turbine (a special device that drives an electrical generator). Most of their early schemes were large, costly structures. In southern California, one company has pioneered a less expensive alternative. It consists of a series of aluminum troughs that reflect sunlight onto small oil-filled tubes (**FIGURE 15-13**). The hot oil heats water, which is turned into steam that drives a turbine to generate electricity. This system produces enough electricity to supply 170,000 homes at a cost of 8 cents per kilowatt-hour—cheaper than nuclear power but more costly than coal (if you ignore the costly environmental impacts of electricity generated from coal). Numerous other companies have installed similar systems.

KEY CONCEPTS

Solar thermal electric facilities heat water using sunlight. Steam from this fairly cost-competitive process is used to generate electricity.

The Pros and Cons of Solar Energy One of the most notable advantages of solar energy is that the fuel is free. All we pay for are devices to capture and store it. As a result, solar energy and other renewables will provide a significant help against rising fuel prices (oil, coal, and natural gas prices will continue to rise well into the future). Solar energy is a huge energy resource available as long as the sun continues to shine. While the construction of solar technologies (flat plate collectors and photovoltaics) creates pollution and solid wastes, as does any manufacturing process, once a solar

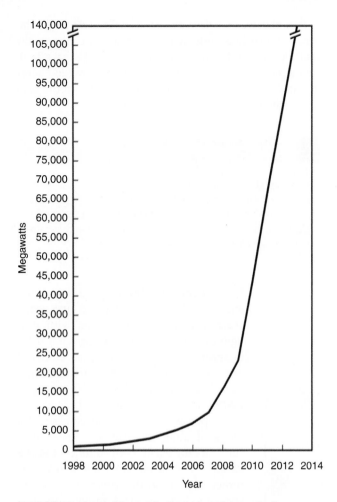

FIGURE 15-12 World cumulative solar photovoltaics installations, 1998–2013.

Data from European Photovoltaic Industry Association.

system is operating it is a very clean form of energy. It does not add to global warming, urban air pollution, and other environmental problems. Over their lifetimes, solar systems produce much more energy than is needed to make them. Years

of pollution-free operation offset the pollution created by production. In fact, it only takes 1 to 2 years for a PV system to generate all the energy it took to make the entire system. Also, most solar systems can be integrated with building designs and therefore do not take up valuable land.

Solar energy has many applications. It can, for example, be used to meet the low-temperature heat demands of homes or the intermediate- or high-temperature demands of factories. Solar energy can be used to cool buildings and generate electricity to power radios, lights, watches, road signs, stream flow monitors, space satellites, automobiles, and industrial motors.

No major technical breakthroughs are required either. Active solar water heating and passive solar space heating are well developed, although some improvements in design and lowered costs could enhance the appeal of others, such as photovoltaics and active solar space heating.

The major limitation of solar energy is that the source, the sun, is intermittent: It goes away at night and is blocked on cloudy days. Consequently, solar energy must be collected and stored. Photovoltaic systems, for example, require storage batteries, although most home owners simply send excess electricity onto the electrical grid during the day. They then draw electricity off at night at no charge. (The electrical grid becomes their storage medium.) Passive solar energy stores heat in thermal mass, but most solar users must install a backup system to provide heat during long cloudy periods.

For many years another disadvantage was cost. Although passive solar and solar hot water systems for domestic hot water and pool heating were often cheaper in the long run than conventional systems, a few forms of solar energy did not compete well economically with conventional sources—PVs or solar electricity, for example. As just noted, however, the cost solar electric systems has fallen dramatically, often making of them cost-competitive with—even cheaper than—conventional sources of electrical energy. This discussion of costs ignores the massive environmental and economic damage caused by conventional fuels and the huge subsidies that prop them up. When these two factors are taken into account, the economics of solar energy are even better.

KEY CONCEPTS

Solar energy technologies are well developed. Their advantage over other forms of energy production (such as nuclear or coal) is that they rely on a free, abundant fuel (sunlight) and are relatively clean systems to operate. Virtually all systems are economically competitive with conventional sources of electricity. Storing energy from intermittent sunlight remains one of their major drawbacks.

Wind Energy

About 2% of the sun's energy striking the Earth is converted into wind. (Wind is considered another form of solar energy.) Wind can be tapped to generate electricity, pump water, and perform mechanical work (grinding grain, for example). Electricity is produced by generators driven by blades (**FIGURE 15-14**). Large wind turbines are used for commercial

FIGURE 15-13 Solar thermal electric. These parabolic aluminum reflectors direct sunlight onto an oil-filled tube. The heat is then transferred to water, which boils and produces steam that drives small turbines to generate electricity.

Courtesy of Warren Gretz/NREL.

15-2 A Solar Giant Grows Taller

In 1998, the Japanese corporation Kyocera became the world's leading manufacturer of solar electric panels or photovoltaics. That year, the solar giant, long known for its commitment to environmental protection, became even taller with the completion of its new world headquarters, a state-of-the art office building (**FIGURE 1**). This 20-story building in Kyoto, Japan was designed to be the world's most environmentally friendly headquarters. What makes it so special?

As you can see from Figure 1, the building is equipped with nearly 2,000 PVs, on the roof and on the south side of the structure. This, in turn, reduces annual carbon dioxide emissions by 97 metric tons (107 tons), reducing global greenhouse gas emissions. The use of solar electricity also cuts sulfur and nitrogen oxide emissions, two pollutants that are responsible for acid deposition.

The Kyocera headquarters building acquires much of its electricity from a natural gas cogeneration unit. This system, which burns natural gas to make electricity, operates at about twice the efficiency of a centralized power plant.

Kyocera is air conditioned by ice—that is, by passing cool air over ice. The company makes ice at night using utility power, when demand for electricity is low. Spreading the load out over a 24-hour period makes optimal use of utility power and avoids costly expansion of power production to meet peak demands. It also helps to reduce brownouts during the day, when system-wide demand is high.

Special sensors in the building monitor outdoor and indoor temperature. When appropriate, outdoor air assists in cooling hot spots that develop around windows when the sun shines on them. In addition, the air flow is carefully monitored and controlled in each room to minimize cooling demand. Even the escalator is controlled by a sensor that monitors demand. When a person approaches, it turns on. When there's no activity, the escalator shuts down, saving energy.

Furthermore, Kyocera uses groundwater and rainwater captured on site to irrigate the landscape. High-efficiency lighting and heat-reflecting glass round out the energy features of this building.

Kyocera's headquarters is an environmentally friendly structure, yet aesthetically appealing. Designed to place the highest priority on environmental protection, it is a model for other companies, as is Kyocera itself, most of whose facilities worldwide are environmentally responsible operations. Their whole line of products—from eco-friendly fine ceramic components that have superior resistance to abrasion, heat, and corrosion to solar-power-generating systems with world-class efficiency—shows that companies can prosper while helping build a sustainable future.

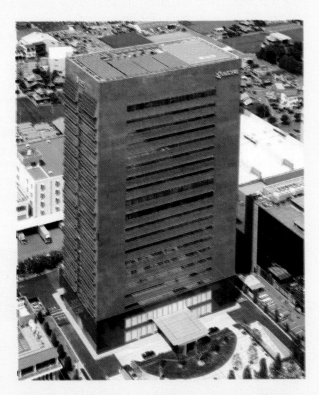

FIGURE 1 Kyocera's headquarters. This company is the world's leading manufacturer of PVs. This office complex is fitted with thousands of PV panels, which help provide energy.
Courtesy of Kyocera Corporation.

FIGURE 15-14 Wind farm. These windmill generators in Wyoming produce electricity costing about 5 cents per kilowatt-hour, about the cost of coal-generated electricity but without the adverse environmental impacts. The land can also be used for grazing.

Courtesy of Matt Reuer, Colorado College.

energy production. Smaller units can be used to power farms and households. Home owners who rely on wind often install photovoltaics to supplement the wind-generated power. My home, business, farms, and educational center in Missouri, for instance, are powered by five PV systems, a wind turbine, and two solar hot water systems (**FIGURE 15-15**).

 KEY CONCEPTS

Winds are produced by solar energy and can be used to generate electricity or to perform work directly, such as pumping water.

The Pros and Cons of Wind Energy Wind energy offers many of the advantages of solar energy: First, wind energy is an enormous resource. Wind resources are abundant in certain parts of every continent. Tapping the globe's windiest spots could provide 13 times the electricity now produced worldwide. The Worldwatch Institute estimates that wind energy could easily provide 20 to 30% of the electricity needed by many countries. In the United States, North

FIGURE 15-15 This solar array at the author's educational center in east–central Missouri provides most of the electricity he needs to power a home.

Courtesy of Dan Chiras.

Dakota, South Dakota, and Texas have enough wind to supply all the electrical demands of the entire country.

Wind energy is clean and renewable, uses only a small amount of land (**TABLE 15-2**), and is safe to operate. On a typical wind farm, only 10% of the land is used for roads and windmills; the remaining land can be grazed and even planted (Figure 15-14). Today, many farmers are leasing small portions of their land to companies to install wind generators. Large, commercial generators net them $3,000 to $5,000 *each* in rent and royalties, greatly boosting farm income.

Windmill technology is well developed, reliable, and efficient, and the fuel is free. Wind-generated electricity (from large windmills) is cost-competitive with other sources of electricity. Today, large wind farms are being built all over the world to provide electricity. China, Spain, Germany, Denmark, India, and the United States are leaders in wind power. Despite good wind resources, Canada has lagged behind other nations in developing wind energy. In the United States, new wind energy now costs on average about 5 cents per kilowatt-hour, clearly cost-competitive with coal. When one takes into account the environmental benefits, the economics of wind energy (like solar energy) become even more favorable. These factors, combined with the environmental benefit, make wind the fastest growing source of energy in the world (**FIGURE 15-16**). By the end of 2012, world wind power capacity was nearly 282,000 megawatts (a megawatt is equal to a million watts). Wind turbines generate electricity in over 70 countries.

There are, of course, disadvantages to wind. Critics argue that the wind does not blow all of the time, so backup systems and storage are needed. Contrary to popular misconception, wind is fairly reliable. Wind farms placed in good locations produce electricity 65 to 85% of the time. According to studies by the U.S. Department of Energy, however, a 1,000-megawatt wind farm requires only about 10 megawatts

Go Green

If you can afford it, buy green power (such as wind-generated electricity) from your local utility.

TABLE 15-2

Land Use of Selected Electricity-Generating Technologies, United States

Technology	Land Occupied[1]
Coal[2]	3,642
Solar thermal	3,561
Photovoltaics	3,237
Wind[3]	1,335
Geothermal	404

Data from L. R. Brown, C. Flavin, and S. Postel (1991). *Saving the Planet: How to Shape an Environmentally Sustainable Global Economy.* New York: Norton.
[1](Square meters per gigawatt-hour, for 30 year.)
[2]Includes coal mining.
[3]Land actually occupied by turbines and service roads.

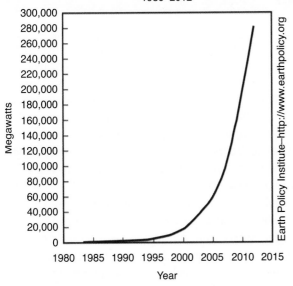

World Cumulative Installed Wind Power Capacity, 1980–2012

FIGURE 15-16 Wind energy. Wind has become the fastest growing source of energy in the world.

Data from EPI from GWEC, Worldwatch.

of backup power. As Mike Bergey of Bergey Windpower notes, "As a power source, wind energy is less predictable than solar energy, but it is also typically available for more hours during a given day." Another issue is visual impact. Individual wind turbines and wind farms are not visually appealing to everyone. Third, opponents of large-scale wind farms often cite bird kills as a significant concern. As shown in the accompanying viewpoint, bird mortality from commercial wind turbines does occur, but it is minor in comparison to other sources such as pet cats, automobiles, and cell phone towers.

Although winds do not blow all of the time in all locations, wind still could become a major source of electricity. It could supplement solar electricity, hydroelectricity, biomass, and conventional power sources. Surpluses produced in one area could be transferred elsewhere. Consistently windy locations like coastlines and the great plains of North America could also provide a steady stream of electricity to the national electrical grid, the high-voltage electric wires that criss cross the United State transporting electricity throughout the nation.

KEY CONCEPTS

Wind energy is clean, abundant, and fairly inexpensive, especially when one includes its low environmental costs. Wind energy could provide a significant percentage of our future energy demand.

Biomass

The organic matter contained in plants is called **biomass**. The energy found in this organic plant matter comes from sunlight. Useful biomass includes wood, wood residues left over from the timber industry, crops, crop residues, charcoal, manure, urban waste, industrial wastes, and municipal sewage. Some of these

fuels can be burned directly. Others are converted to methane (a gas) and ethanol (a liquid). The simplest way to get energy from biomass is to burn it. In some instances, though, it makes more sense to convert biomass to gaseous and liquid fuels such as ethanol. These can be burned in motor vehicles or used to produce raw materials for the chemical industry to manufacture drugs and plastics as oil and natural gas supplies decline.

Biomass (wood, for instance) supplies about 20% of the world's energy. In fact, it is the main source of energy for about half of the world's population, primarily those in the less developed countries. In sub-Saharan Africa, three-fourths of the energy comes from burning wood. In contrast, in the United States and other developed countries, biomass supplies only about 2.7% of our total energy needs. Wood-rich Canada currently supplies about 6% of its total energy demand from biomass, mostly wood.

Certain nonfood crops could also be grown to produce liquid fuel. For example, a desert shrub (*Euphorbia lathyris*) found in Mexico and the southwestern United States produces an oily substance that could be refined to make liquid fuel. In arid climates, the shrub could yield 16 barrels of oil per hectare (6 barrels per acre) on a sustainable basis. The copaiba tree of the Amazon yields a substance that can be substituted for diesel fuel without processing. Vegetable oils can also be used in place of diesel. According to one study, farmers could convert 10% of their cropland to sunflowers to produce all the diesel fuel needed to run their machinery. Eventually, the entire transportation system could be powered by renewable fuels.

Another potential source of energy is manure from livestock operations, especially cattle and pigs. Manure is fed into vats where it decomposes, giving off methane, a combustible gas. All over the world, plans are under way to use manure to produce a combustible fuel. This not only helps put a waste to good use, it helps reduce pollution.

KEY CONCEPTS

Biomass is organic matter such as wood, animal waste, or crop wastes that can be burned to produce heat and electricity or converted into gaseous or liquid fuels that can power factories, cars, trucks, and planes. It is a common fuel source in most developing nations but supplies only a fraction of the needs of people in the more developed nations.

The Pros and Cons of Biomass Biomass offers many benefits. It can help less developed nations reduce their dependence on nonrenewable energy resources. It also has a high net energy efficiency when it is collected and burned close to the source of production, and it has a wide range of applications. Unlike fossil fuels, biomass adds very little carbon dioxide to the atmosphere—as long as the amount of plant matter burned equals the amount of plant matter produced each year. Burning some forms of biomass, such as urban refuse, reduces the need for land disposal.

Several recent technological and scientific developments could improve the prospects for biomass. First is the use of high-efficiency gas turbines, similar to those in jet aircraft, to generate electricity from hot gases produced by

Bird Kills from Commercial Wind Farms: Fact or Fiction?

While commercial wind machines do kill birds, the problem has been blown way out of proportion. Studies show that bird kills from wind turbines pale in comparison to bird deaths from several common sources, among them plate glass windows, domestic cats, automobiles and trucks, tall buildings, smokestacks, and communication towers (see TABLE 1). Worldwide, hundreds of millions of birds—perhaps even billions—are killed each year by these sources. Commercial wind turbines, on the other hand, kill a miniscule number of the birds. So why has wind gotten such a bad rap?

Wind machines got a bad rap from one of America's oldest and largest wind farms: the Altamont Pass Wind Resource area in California. Located just east of San Francisco, Altamont Pass is home to a mind-boggling 7,000 wind turbines. It is also habitat for numerous raptors. Soon after the wind turbines were erected, the birds began to perch on the lattice wind towers in search of abundant prey (ground squirrels and other rodents) that live year-round in the grasses at the base of the towers.

Some scientists attributed this problem to a phenomenon called "target fixation." Fixed on their prey, the raptors descended from their perches and flew directly into the blades. In addition to those that perished as a result of collision with the blades, studies suggest that 8% died from electrocution (from power lines), and 11% died when they collided with electric wires. The causes of death in the remaining 26% is unknown. Some researchers speculate that many raptors died from poisoning by rodenticides, pesticides used to poison rodents. These chemicals, the scientists hypothesized, may have moved up the food chain, poisoning the unfortunate raptors.

Although many raptors have died at Altamont, a 2-year study of bird kill in the region showed that the highly publicized problem was indeed overblown. The researchers discovered only 182 dead birds in 2 years.

While any raptor death is of concern to those of us who cherish wildlife, the death rates at Altamont are insignificant compared to other factors. For example, scientists estimate that about 60 million birds die each year in the United States after being struck by motorized vehicles, according to the American Wind Energy Association's (AWEA) report "Facts About Wind Energy and Birds." Another 1.25 million die as a result of collision with tall structures such as buildings, smokestacks, and towers. Another 100 million to 900 million, perish after flying into plate glass windows, according to another report. Yet another 10 to 40 million birds perish after flying into communications towers and the guy wires that support them. Pesticides kill an estimated 67 million birds each year.

Cats are a lethal force in the lives of birds. Scientists who have studied bird deaths from cats in Wichita, Kansas, found that a single cat kills, on average, 4.2 birds per year. Other studies suggest even higher numbers. According to one study, a feral cat kills as many birds in 1 week as a large commercial wind turbine does in 1 to 2 years. Declawing a cat doesn't seem to help much. According to one estimate, the majority of cats (83%) kill birds, even declawed, and well-fed cats prey on wild birds. Neutering or spaying a cat does not seem to cut down on hunting, either. With more than 64 million cats in America alone, what's the total loss?

No one knows for sure, but if the situation in Wisconsin is indicative of the national toll, America's bird population is being decimated by our beloved feline companies. In Wisconsin alone, researchers estimate that cats kill approximately 39 million birds per year. Nationwide, the number is estimated to be around 270 million, very likely many many more. Two scientists who studied the issue in one small town in England, estimated that cats kill 20 million birds per year in Great Britain alone. "Even if wind were used to generate 100% of U.S. electricity needs, at the current rate of bird kills, wind would account for only one of every 250 human-related bird deaths," notes the AWEA. "So, if you want to save birds," says one wind energy expert who wished to remain anonymous, for reasons that will soon be clear, "put your cat in a blender, then sign up for wind-generated 'green electricity' from your utility."

TABLE 1

Estimated Annual Bird Death in the United States by Source

Activity/Source of Bird Motality	Estimated Annual Mortality
Killed by cats	270 million or more
Collisions with and electrocution by electrical transmission wires	130 to 170 million
Collisions with windows	100 to 900 million
Poisoning by pesticides	67 million
Collisions with motor vehicles	60 million
Collisions with communications towers	40 to 50 million

Data from American Wind Energy Association.

Clearly, the Altamont Pass wind farm is benign compared to a host of other lethal factors. Altamont also appears to be an isolated case. No other wind farm in the United States experiences mortality rates remotely close to the Altamont Pass facility. Why?

Acutely aware of the problem, contemporary wind developers select sites for new wind farms that are out of migratory pathways. Site selection has helped, but that's not all. Improvements in the design of commercial wind turbines have also helped to minimize bird kills at commercial wind farms. How?

Over the years, wind machines have gotten taller, blades have gotten longer, and the speed at which the blades rotate has declined substantially. Large blades help harness more energy from the wind, but large, slow-moving blades are also more easily avoided by birds. Ever-larger commercial wind machines currently under development could reduce the risk even more. (To learn more about efforts to reduce bird deaths even more, check out "Facts About Wind Energy and Birds" at the American Wind Energy Association's website, http://www.awea.org. Mick Sagrillo has written several articles on the topic, which you can check out at http://www.renewwisconsin.org. On the lower left, click on "Small Wind Toolboxes.")

Adapted with permission from Dr. Chiras' book, Power from the Wind, *published by New Society Publishers.*

---

the combustion of various forms of biomass. Second is the development of an enzymatic process that improves the efficiency of ethanol production from wood wastes—a procedure that has decreased the cost of ethanol from wood from $4 a gallon to $1.35 in the last decade. Further improvements could decrease costs to 60 cents a gallon. Wood wastes now dumped in landfills, urban tree trimmings, and sustainably managed tree farms could eventually form the base of a sustainable ethanol production to meet a portion of our transportation fuel needs.

Biomass is not a panacea, however. Although the world's reliance on this form of renewable energy will probably increase in coming years, it is doubtful that its use will increase as much as some proponents hope. Why not? Global warming could reduce water supplies and agricultural output in many areas, thus constraining biomass production. Further, some people believe that rising food demand may limit the amount of farmland available to produce biomass fuels. Biomass production could raise food prices, which is especially harmful in less developed countries. Already rising ethanol production has caused an increase in corn prices that has increased the cost of feeding cattle and the cost of feeding people whose main food sources are derived from corn. Furthermore, removing crop and forestry residues may reduce soil nutrient replenishment. Biomass can create large amounts of air pollution—for example, smoke from woodstoves. This smoke contains several toxic air pollutants, such as dioxins, and creosote. Studies show that children living in homes with wood-burning stoves have more respiratory problems than children in homes heated by other means. Finally, transportation costs for biomass are higher than traditional fossil fuels because biomass has a lower energy content per unit of weight.

KEY CONCEPTS

Biomass is a major source of energy in many less developed nations and could boost the energy supply of many more developed nations.

Biodiesel

Another biomass solution that holds great promise is biodiesel. **Biodiesel** is a renewable fuel made from an assortment of vegetable oils—for example, corn oil, canola oil, and soy oil. The liquid is a combustible, clean-burning fuel that could eventually power many of the nation's trucks, cars, buses, vans, ships, and trains, perhaps even jets. (There are already over 800 gas stations in the United States that sell biodiesel out of 170,000 stations.) You can locate gas stations at the Alternative Fueling Station Watch on the Internet.

Biodiesel is made by mixing vegetable oil with a solution of methanol containing sodium hydroxide (lye). The oil is usually heated. The methanol-lye mixture is then added to the vegetable oil. This solution is heated some more and then stirred for a period of time, usually about an hour. When the chemical reaction is complete, out comes biodiesel—long chain fatty acids that burn very nicely in diesel engines. The only waste product is glycerol, a dark thin oily substance that can be purified and used to make soap.

The Pros and Cons of Biodiesel   Biodiesel is a renewable fuel that could help us meet our needs for transportation fuels as the age of oil winds down. Biodiesel can help stimulate our nation's economy and reduce its reliance on foreign oil. It could also help stimulate rural economies. In the coming years, many farmers will be enlisted to produce "fuel grains" for the biodiesel market. Local biodiesel manufacturers could convert the grains to vegetable oil. Crop production and local manufacturing not only create stronger local and regional economies, they help forge the path to a more decentralized and sustainable system of fuel production.

According to Marc Franke, an Iowa-based proponent of biodiesel, the net energy efficiency of biodiesel production from soy oil is 3.2. The net energy production of biodiesel from canola or rapeseed oil is 4.3. What these figures mean is that for every unit of energy invested in biodiesel production, you get 3.2 units of energy output for soy oil and 4.3 units of energy from canola oil. According to Franke, it would take 7.0 acres of soybeans to supply soybeans to extract the oil needed to make biodiesel for a diesel car that travels 15,000 miles per year and gets 44 miles per gallon. It would take 2.7 acres of canola oil to do the same thing.

Seed crops are not the only potential source of biodiesel, however. Biodiesel can also be produced from vegetable oils discarded by restaurants—reducing disposal costs for them while providing a valuable, renewable liquid transportation fuel. U.S. restaurants, including all the fast-food chains, produce an estimated 3 billion gallons of waste vegetable oil per year, according to Franke!

There are a lot of other sources, too. For example, biodiesel can be manufactured from algae grown in ponds associated with sewage treatment plants, helping reduce pollution while generating liquid fuel for North America's transportation system. Diesel fuel can also be produced from the organic waste such as that generated at turkey farms using a process known as thermal depolymerization (TDP). Agricultural, organic wastes, says Franke, produce enough material to make 4 billion barrels of biodiesel each year!

According to Franke, biodiesel offers the same performance as regular diesel, but dramatically reduces tailpipe emissions. For example, biodiesel contains no sulfur; therefore, combustion of this fuel source eliminates sulfur oxide emissions that contribute to acid rain and snow—a big problem with conventional diesel vehicles.

There's no black smoke spewing from tailpipes of cars or trucks powered by biodiesel as they pass or power up a hill. Complete cycle carbon dioxide production from the manufacture of the fuel and use of biodiesel cars and trucks is 78% lower than vehicles powered by standard diesel fuel.

According to the U.S. Department of Energy, biodiesel is the fastest growing alternative fuel on the market today.

Yet another advantage of biodiesel is that the transition to this fuel won't require major changes in distribution systems. Biodiesel could be produced locally and sold at area filling stations using the same facilities that are used to dispense diesel fuel made from petroleum.

Biodiesel can also be run in space heaters and oil furnaces, even forklifts, tractors, and electric generators that currently use diesel. Your home someday could be heated with biodiesel. I use biodiesel in my lawn tractor and farm tractor at our farm located at The Evergreen Institute in east central Missouri (**FIGURE 15-17**).

Despite these benefits, biodiesel does have a few disadvantages. For one, it is not yet widely available. Second, in the United States, commercially manufactured biodiesel costs a bit more to produce than standard petroleum-derived diesel. This, of course, is largely due to the fact that biodiesel is only produced by small-scale production facilities. Increasing the scale of production could bring the cost down substantially. As oil prices rise and biodiesel production ramps up, costs could decline, making biodiesel the fuel of the future. Improving matters, starting in January 2004, producers received a federal tax incentive of up to $1 per gallon of biodiesel.

You can also make your own biodiesel at bargain basement prices in your basement. Biodiesel Solutions, a company in Fremont, California sells equipment you can set up in your basement or garage to manufacture biodiesel (**FIGURE 15-18**). They claim that it currently costs only about 70 cents per gallon to make, far lower than standard diesel, which is currently running about $3.60 per gallon.

Another problem with biodiesel is that soy biodiesel gels around 32° F, and, therefore can't be used at full strength when the weather gets cold. To get around this, biodiesel is ordinarily mixed with conventional diesel to which manufacturers add an antigelling agent to winter petroleum diesel.

Finally, large-scale biodiesel production will put increasing demand on farmland. Production could, if large enough, even reduce the exports of food crops. Although farmers won't suffer, those who rely on food from North America could find themselves in a bind. One way to lessen the impact of widespread biodiesel production would be to make it from algae and waste from turkey and hog farms.

FIGURE 15-17 Biodiesel-powered tractor. The author runs his farm equipment on biodiesel, including his tractor (shown here) with a load of (hand-gathered) loose hay on its way to the barn.

Courtesy of Dan Chiras.

FIGURE 15-18 Biodiesel reactor. Small biodiesel reactors can be used to manufacture biodiesel fuel in your garage.

© EvolutionBiodiesel

KEY CONCEPTS

Biodiesel is a liquid fuel that is made from vegetable oil derived from canola, soy, and other crops. It can be burned in all types of diesel vehicles without modifying them. This relatively clean-burning fuel reduces pollution, especial carbon dioxide emissions.

Vegetable Oil as a Fuel

Some people are burning vegetable oil directly in diesel cars, trucks, and buses. Some individuals acquire their fuel from unused vegetable oil. Others salvage vegetable oil from fast-food chains and other restaurants. How can vegetable oil be burned in a diesel car?

It may surprise readers to learn that the German inventor of the diesel engine, Rudolph Diesel, was able to run his engine on peanut oil. Modern diesel engines, however, can't run directly on vegetable oils. To run a vehicle on straight vegetable oil, which is thicker than biodiesel or conventional diesel fuel, car and truck owners must make some modifications to their vehicles. Several companies now manufacture conversion kits costing $300 to $800. Most of these kits include a separate tank for vegetable oil and heating elements for the tank, the fuel line, and the fuel filter. They heat the vegetable oil, which makes it lighter so it flows more readily.

Vegetable oil is an ideal fuel, but one needs to understand the pros and cons of this approach to compare it with other options, especially biodiesel. Let's start on the up side.

One of the biggest advantages of vegetable oil is that the fuel is widely available. You can stop and fill up anywhere in North America as long as there's a restaurant that deep fat fries its food in vegetable oil. (You'll have to filter the gunk out of the deep fryer oil first, however.)

Not only is the fuel abundant and widely available, so are diesel engine conversion kits. You can purchase them on the Internet.

Like other options discussed in this chapter, the fuel is renewable and can be picked up for free—for example, from fast-food restaurants whose owners are typically pleased to give it away, rather than pay for disposal!

Vegetable oil burns cleanly and thus helps solve another thorny issue—air pollution from conventional fossil fuels. Like biodiesel, vegetable oil use reduces carbon dioxide emissions. The fuel itself is often called *carbon neutral* but that's not entirely true. Although the amount produced during combustion equals the amount the plants take up during photosynthesis, it takes energy to make this fuel (gasoline to power a tractor, for instance). The consumption of energy, in turn, produces carbon dioxide. Even so, the fuel is light years ahead of conventional fossil fuels with respect to greenhouse gas production.

Vegetable oil fuels have some disadvantages. The conversion kits and installation of the kits cost money. In addition, the additional fuel tank takes up a considerable amount of trunk space. In many cars, the owner has to start the vehicle with ordinary diesel, then switch to vegetable oil after it has heated up. Five minutes before the car is turned off, the driver has to switch back to petroleum-based diesel to clean the lines of biodiesel. One German company has developed a conversion kit that uses the main fuel tank. Vehicles equipped with this can start on straight vegetable oil.

Vegetable oil can be burned in diesel vehicles and cars but requires a conversion kit. It can be produced from specially grown crops like corn, canola, and soy and can be derived from used cooking oil. Like biodiesel, it burns cleanly and can help us reduce our carbon footprint.

Hydroelectric Power

Humankind has tapped the power of flowing rivers and streams for thousands of years to run flour mills and, more recently, to produce electricity. River flow is made possible by two factors: sunlight energy, which causes water to evaporate. It is then deposited immediately as rain or sometime later as snow that flow into rivers. The second factor is gravity, which is responsible for the movement of water in streams.

Hydroelectric power is a renewable resource usually tapped by damming streams and rivers. Water in the reservoirs behind dams is released through special pipes. As it flows out, it turns the vanes of electric generators, producing electricity.

Hydropower supplies about one-fifth of the world's electrical demand. In Canada, it supplies about 12% of the nation's total annual energy demand. Hydropower creates no air pollution and is relatively inexpensive. Furthermore, the technology is well developed. Although dams and hydroelectric power plants are expensive to build, they often have lifespans 2 to 10 times greater than coal or nuclear powered plants.

There are some problems associated with hydropower. Reservoirs behind dams often fill with sediment, shortening the life span of a hydropower facility to 50 to 100 years. However, large projects with enormous reservoirs may last 200 to 300 years. Once a good site is destroyed by sediment,

FIGURE 15-19 Flooding by reservoir. The reservoir created by huge dams like this one flood thousands of acres of forest and farmland, sometimes even villages.

Courtesy of Master Sgt. Paul Gorman U.S. Air Force.

it is gone forever. Dams and reservoirs flood productive land, displace people, destroy stream fisheries, eliminate certain forms of recreation, upset nutrient replenishment in estuaries, and create many additional (**FIGURE 15-19**).

Many countries have a large untapped hydroelectric potential. In South America, for instance, hydroelectric generating potential is estimated at 600,000 megawatts. By comparison, the United States, the world's leader in hydroelectric production, has a present capacity of about 77,000 megawatts and an additional capacity of about 150,000 to 160,000 megawatts.

Estimates of hydroelectric potential can be deceiving because they include all possible sites, regardless of their economic or environmental costs. For example, half of the U.S. hydroelectric potential is in Alaska, far from places that need power. The potential for additional large projects in the continental United States is small because the most favorable sites have already been developed. Canada faces a similar shortage. Moreover, those developing sites that are available would result in massive environmental impacts and might be strongly opposed by the citizenry. In addition, the high cost of constructing large dams and reservoirs has increased the cost of hydroelectric energy by from 3 to 20 times since the early 1970s. Because of these factors, hydropower is projected to increase very slowly in the coming decades.

Two of the most sensible strategies for increasing hydropower may be to (1) increase the capacity of existing hydroelectric facilities—that is, add more turbines—and (2) install turbines on the many dams already built for flood control, recreation, and water supply. In appropriate locations, small dams could provide energy needed by farms, small businesses, and small communities. All projects, however, must be weighed against impacts on wildlife habitat, stream quality, estuarine destruction, and recreational uses.

In LDCs, small-scale hydroelectric generation may fit in well with the demand. In China tens of thousands of medium and small hydroelectric generators account for about one-fourth of the country's electrical output. Small instream generators that do not impede the flow of water can also provide

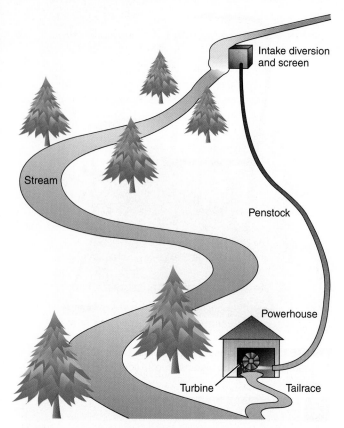

FIGURE 15-20 Small-scale hydropower. Flowing water from a nearby stream turns a turbine to generate electricity.

power to individual homes located near waterways. Water can also be removed, transported downstream in small pipes (to pick up speed), and then run through small generators (**FIGURE 15-20**). These can provide a remarkable amount of electrical energy 24 hours a day, 365 days a year! However, there aren't many suitable sites in most parts of the country for small-scale hydros, called *microhydros*.

KEY CONCEPTS

Hydroelectric power is renewable and operates relatively cleanly, but dams and reservoirs have an enormous impact on the environment. Although the potential hydropower sources are enormous, they are often far from settlements, and developing them would cause serious environmental damage.

Geothermal Energy

The Earth harbors an enormous amount of heat, or **geothermal energy**, which comes primarily from magma, molten rock beneath the Earth's surface. Geothermal energy is constantly regenerated, so it is a renewable resource. Geothermal resources fall into three major categories.

Hydrothermal convection zones are places where magma intrudes into the Earth's crust and heats rock that contains large amounts of groundwater. The heat drives the groundwater to the Earth's surface through fissures, where it may emerge as steam (geysers) or as a liquid (hot springs).

Geopressurized zones are aquifers that are trapped by impermeable rock strata and heated by underlying magma.

This superheated, pressurized water can be tapped by deep wells. Some geopressurized zones also contain methane gas.

Hot-rock zones, the most widespread but also the most expensive geothermal resource, are regions where bedrock is heated by magma. To reap the vast amounts of heat, wells are drilled, and the bedrock is fractured with explosives. Water is pumped into the fractured bedrock, heated, and then pumped out.

Geothermal energy is heavily concentrated in the so-called *Ring of Fire* encircling the Pacific Ocean and in the great mountain belts stretching from the Alps to China (**FIGURE 15-21**). It is also prevalent around the Mediterranean Sea and in East Africa's Great Rift Valley, which extends along the eastern part of the African continent. Within these areas, hydrothermal convection zones are the easiest and least expensive to tap. Hot water or steam from them can heat homes, factories, and greenhouses. In Iceland, for example, 65% of the homes are heated by the Earth's heat. Iceland's geothermally heated greenhouses produce nearly all of its vegetables; Russia and Hungary also heat many of their greenhouses in this way.

Geothermal is also being widely used to heat homes and other buildings in the United States and Canada. The heating systems, referred to as *ground-source heat pumps*, extract heat from the ground through pipe buried 6 to 8 feet below the surface. Using refrigeration technology, these systems commonly referred to as *geothermal systems*, pump heat from the ground into a building. In the summer, they can be run in reverse. That is, they can extract heat from a building and dump it back into the ground, thus cooling the building. Geothermal systems are about 25% more efficient at heating a building than conventional systems.

Steam from geothermal sources can be used to run turbines that produce electricity. Geothermal plants can produce electricity day and night and can provide electricity when wind or solar systems are not operating. They can also provide electricity in areas without sizable wind or solar resources.

Although still in the early stages of development in most countries, geothermal electric production is growing quickly in the United States, Italy, New Zealand, and Japan. El Salvador in Central America, however, currently generates 40% of its electricity from geothermal sources, and Kenya and Nicaragua acquire 11 and 10%, respectively.

Hydrothermal convection systems have several drawbacks. The steam and hot water they produce are often laden with minerals, salts, toxic metals, and hydrogen sulfide gas. Many of these chemicals corrode pipes and metal. Steam systems may emit an ear-shattering hiss and release large amounts of heat into the air. Pollution control devices are necessary to cut down on air and water pollutants. Engineers have also proposed building closed systems that pump the steam or hot water out and then inject it back into the ground to be reheated. Finally, because heat cannot be transported long distances, industries might have to be built at the source of energy.

KEY CONCEPTS

Geothermal energy is a renewable resource created primarily from magma, molten rock beneath the crust. Geothermal energy is used to generate electricity and to heat structures; it is a major source of energy in some countries.

Hydrogen Fuel

Hydrogen is another renewable fuel that many believe could be used to help replace oil, gasoline, and natural gas. It could someday be burned in our stoves, water heaters, furnaces, factories, and cars.

Hydrogen gas is produced by heating water to extremely high temperatures or passing electricity through it in the presence of a catalyst (a chemical that facilitates the breakdown of water into oxygen and hydrogen without being changed itself). The device used to generate hydrogen is known as an **electrolyzer**.

Hydrogen is a relatively clean-burning fuel. In fact, when hydrogen burns it produces only water and small amounts of nitrogen oxide (formed by the combination of atmospheric oxygen and nitrogen). Nitrogen oxides, which contribute to acid deposition, can be minimized by controlling combustion temperature and by installing pollution control devices. Unlike fossil fuels, hydrogen produces no carbon dioxide when burned.

Because hydrogen could be generated from seawater, it is an essentially limitless and renewable energy resource. Hydrogen is also easy to transport and has a wide range of uses. Unfortunately, at the present time it takes considerable

FIGURE 15-21 Global geothermal resources. Note the concentration of geothermal resources on the Pacific coastline—the so-called Ring of Fire.

energy to produce hydrogen. This low net energy yield makes it an expensive form of energy.

Hydrogen may also serve as a way of storing energy from hydroelectric, wind, solar, and other renewable energy sources. For example, when demand for these sources is low, the electricity they generate could be used to produce hydrogen from water. It would be stored for later use. When renewable sources are not available (say, on a calm day or during the evening), the hydrogen could be burned to produce electricity.

Although the immediate prospects for hydrogen are not spectacular, efforts to produce hydrogen more efficiently could go a long way to make this energy source more cost effective. Some energy analysts predict that one day areas rich in sunlight, wind, and water could become major sources of hydrogen fuel, which could be piped around the world in pipelines.

In 1996, the U.S. Congress passed the **Hydrogen Future Act**. This law supports research and development of hydrogen fuels and provides approximately $165 million to support these activities.

Fuel Cells   Hydrogen can also be used to produce electricity in a **fuel cell**. Many different types of fuel cells have been developed. **FIGURE 15-22** shows a simplified view of how a fuel cell works. As illustrated, this fuel cell looks a lot like a battery. It consists of an anode and a cathode separated by a membrane. Hydrogen is fed into the cathode where it reacts to form hydrogen ions and electrons. The electrons are drawn off by a wire, forming electricity, which powers lights, electronic equipment, and so on. Oxygen is introduced into the cathode, where it reacts with hydrogen ions that pass through the membrane from the anode and electrons returning from the circuit. The product is water.

Electricity produced by a fuel cell can be used to power electric motors in cars. In fact, in 1999 Daimler-Chrysler unveiled a hydrogen-powered car that can travel as fast as 90 miles per hour. It uses hydrogen and oxygen to run a fuel cell, which runs an electric motor to power the car. One company in Canada, Ballard Power Systems, is building hydrogen fuel cell powered buses, which are now being used experimentally in Chicago and British Columbia. During the public ceremony in Chicago, then Mayor Richard Daley, celebrated by drinking a glass of exhaust water collected from the tailpipe of an idling bus to underscore how safe the emissions are. If tests prove successful, Ballard hopes to enter full-scale production of fuel cell bus engines.

Iceland has committed itself to a hydrogen economy and will soon receive all of its energy from hydrogen and geothermal energy.

Although hydrogen-powered cars seem promising, remember that it takes energy (electricity) to make hydrogen.

Go Green

Combine errands to reduce your time, energy use, and pollution. When running errands, plan out your trip in advance so you can make a loop. Try to plan your trip so that you make only right turns. These means less time waiting to make left turns and less energy used. It works. UPS plans all its delivery routes so drivers make only right turns.

Studies show that it is far more efficient (three to four times) to use electricity directly to power an electric car than it is to use that electricity to generate hydrogen to run a fuel cell to produce electricity to run a car. Fuel cells use hydrogen and oxygen, but both hydrogen and oxygen also burn, so they can be burned directly in cars. For more on the pros and cons of hydrogen, see the Point/Counterpoint on the following pages. Today, a lot of fuel cell work is focused on using hydrogen-containing fuels such as methanol, natural gas, ethanol, or gasoline. The fuels will be fed into an onboard fuel processor that strips the hydrogen from the molecules. The hydrogen is then fed into the fuel cell. Although most fuel cells under development are powered by fossil fuels, the process is reportedly very clean and represents a major improvement over standard gasoline- and diesel-powered motor vehicles. They still produce significant amounts of

(a)

(b)

FIGURE 15-22 Fuel cell. Many people believe that fuel cells are the wave of the future. **(a)** Cells resemble batteries and produce electricity from hydrogen and oxygen. **(b)** Photo of a real fuel cell.

carbon dioxide. At this writing, Daimler-Chrysler, Toyota, Honda, Ford, and General Motors have fuel cell vehicles that could be brought to market soon.

KEY CONCEPTS

Hydrogen may become an important fuel in the future. Hydrogen can be produced by passing electricity through water, a renewable resource. When hydrogen burns, it produces heat, light, and water vapor. Fuel cells use hydrogen, either from water or organic fuels, to produce electricity. The electricity can be used to power cars, buses, and trucks.

15.3 Is a Renewable Energy Supply System Possible?

Enormous amounts of renewable energy are available and accessible with current technologies—far more than are available from fossil fuel reserves. This has led some scientists and environmentalists to propose a sustainable energy supply system far different from the one in existence today—energy efficiency and renewable energy resources. **FIGURE 15-23** shows one scenario for the projected shift in the world's energy supply system. As illustrated, renewables could take over for all of the nonrenewables. This transition will take many years, and thus, fossil fuels will be around for a long time.

Is this feasible? If so, how could renewable energy supplant the nonrenewables that currently power much of the world? Where will electricity come from? Where will the fuel come from to power the transportation system? What fuels will power industry?

Let's take these questions one at a time, remembering that energy efficiency will be a key to the success of any future energy system.

FIGURE 15-24 shows the distribution of three renewable energy fuels in the United States: wind, solar, and biomass. As you can see, the United States is blessed with renewable energy but wind, solar, and biomass are concentrated in different regions with some overlap. Proponents of a sustainable energy future believe that wind energy could provide massive amounts of electricity on windy days and nights because most wind farms are located in regions where winds are fairly consistent year round. As a result, wind could provide a fairly large and consistent amount of electricity. It can be transported throughout the United States via the electrical grid. Solar is also widely distributed and could provide a significant amount of electricity in sun-rich areas with surpluses shipped to other parts of the country.

As you can see from the map, biomass is concentrated in the northeastern quadrant of the country. It could meet local energy demands along with wind and solar energy from local sources as well as distant wind and solar farms. This view of our energy future relies on several key factors: the abundance of renewable energy and our ability to transport electricity from one region to another to meet our needs. Bear in mind that backup energy when the winds are not blowing or the sun is not shining could come from biomass as well as limited combustion of conventional fuel sources like natural gas and coal.

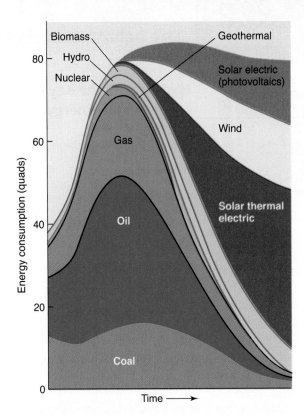

FIGURE 15-23 Transition to a sustainable energy supply system. One projection of possible energy resources in the United States. These figures are based on full commitment to renewable resources.

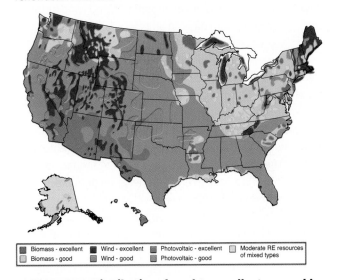

FIGURE 15-24 Distribution of good-to-excellent renewable energy resources.

Data from National Renewable Energy Laboratory.

TABLE 15-3 gives a further breakdown of energy demand by sector, showing what kind of energy is needed and how

KEY CONCEPTS

Not only is renewable energy abundant and accessible without major technological breakthroughs, it has many applications with sufficient potential to replace nuclear and fossil fuel supplies.

Chapter 15: Foundations of a Sustainable Energy System: Conservation and Renewable Energy   **375**

Hydrogen Is the Key to a Renewable Energy Future

© 2007 Schatz Energy Research Center.

Courtesy of Richard Engel.

Richard Engel
Richard Engel is a research engineer at the Schatz Energy Research Center in Arcata, California. His main interests are energy efficiency and renewable hydrogen energy systems.

As public concern about climate change and depletion of the world's fossil fuels escalates, renewable energy technologies are becoming more important and more widely adopted. Industry and government are working to implement large-scale wind and solar projects. Renewable energy is clean, reliable, and inexhaustible. Is it possible that renewable power and fuels will one day meet all of our energy needs?

Maybe so, but even if we invest in enough solar and wind-generating capacity to meet all our energy needs, there are some additional hurdles that need to be cleared. Renewable energy is abundant overall, but the sun goes down at night, streams dry up in summer, and wind turbines stop spinning on calm days. In other words, renewable energy sources are intermittent, while society's need for energy never lets up. If we start generating most of our electricity with renewables, the mismatch between energy supply and demand could become a significant problem.

What we need is a way to store energy, so the power generated on sunny or windy days can be used when it's dark or calm. Batteries are one familiar way to store energy. They convert electricity to chemical energy that can be converted back to electric energy on demand. For years, people who live in remote locations have used rechargeable batteries to store energy collected by their solar panels or wind turbines. Unfortunately, batteries are expensive, bulky, and toxic, and they gradually dissipate stored energy over time. Using huge numbers of batteries to store utility-generated energy for entire cities is neither practical nor economical.

This brings us to hydrogen, a practical way of storing electricity, again in chemical form. It is important to understand that hydrogen is not a *source* of energy. Instead, it's an energy *carrier*, a way of storing or transporting energy that comes from some other source.

Hydrogen is one of the most abundant elements here on Earth, but nearly all of it is bonded chemically to other elements, primarily in water and hydrocarbon molecules such as methane. These bonds must be broken to obtain the pure hydrogen that can be used as an energy carrier.

Hydrogen in water can be separated using electrolysis. This process can be powered by electricity from any source, including wind, solar, or other renewable power systems. The hydrogen can be burned like other fuels to provide heat or power. But more efficiently, it can be used in a fuel cell that combines hydrogen and oxygen electrochemically, a process that produces electricity, heat, and water. A renewable source of power, an electrolyzer, hydrogen storage tanks, and a fuel cell are all that's required to create a hydrogen energy system that provides reliable, around-the-clock power. The only emission is pure water.

Turning electricity into hydrogen also addresses another limitation of renewable energy. Hydrogen made using electricity effectively converts this energy into a portable fuel, something electric vehicle makers have struggled for years to do with batteries. Battery-electric cars offer great performance but have a limited driving range, typically less than 50 miles, because of the bulk and weight of the batteries. The average car is driven less than 50 miles a day, but what about longer trips? Hydrogen, used either to power fuel cells or as a clean-burning fuel in hybrid internal combustion/electric drive trains, can extend the driving range of electric cars to hundreds of miles.

Many people wonder why, with all its advantages, hydrogen energy does not yet play a major role in our energy economy. The main hurdles to fuel cell commercialization are high cost and product reliability, but these problems are gradually being resolved by the industry. Another obstacle that hydrogen, like any new fuel, faces is the so-called "chicken and egg" problem. How do you get manufacturers to make and sell fuel cell cars if there are no convenient places to fill them up? At the same time, how do you convince energy companies to build expensive new fueling stations if they're not convinced there will be vehicles coming in to fill up? This dilemma is also being overcome, mainly by government and large companies investing in mini-fleets of hydrogen vehicles and putting in their own fueling stations.

Vehicles, however, will probably turn out to be the last frontier for hydrogen fuel cell power, because conventional internal combustion engines and their fuels remain relatively inexpensive. In the meantime, we can expect to see fuel cells widely adopted as a replacement for batteries in small power applications, such as laptop computers and other portable devices. Fuel cells are also being used in remote, automated power applications, such as telecommunication repeater stations, and for backup stationary power in buildings.

Some people have made exaggerated claims that hydrogen is a silver bullet that will meet all our future energy needs. The truth is that we will need to use a diverse portfolio of many technologies in the future to generate, transport, and store energy. The special properties of hydrogen and fuel cells ensure they will play key roles among these many solutions. Hydrogen is safer and cleaner than the fuels we use today, and its ability to store energy makes it an ideal complement to the renewable power systems that will some day provide most of our electricity.

Hydrogen's role as a universal energy carrier is both well documented and long awaited. Although hydrogen's potential to provide clean, renewable electricity via fuel cells—those magical devices that convert chemical energy into electricity—warrants continued research, the same can be said for many other technologies including advanced battery systems. Despite predictions to the contrary, battery technology has advanced rapidly. New, more reliable and efficient batteries are now used to power cell phones, computers, and power tools. They are even being used in the newest generation of electric and plug-in electric hybrid vehicles or PHEVs—the very applications in which hydrogen fuel cells were predicted to dominate but have failed to appear. Likewise, questions about battery materials toxicity are becoming moot: Batteries are self-contained and can be completely recycled (the lead-acid battery industry has practiced recycling for over a century).

Why hasn't hydrogen enjoyed similar success? Here are a few reasons:

1. Generating, storing, and then producing electricity from hydrogen is a circuitous process. The roundtrip, *overall* efficiency (electric energy in vs. electric energy out) is seldom better than 25%. Battery systems score much higher with roundtrip efficiencies exceeding 80% in some cases.

Hydrogen's Role in Our Energy Future Is Dubious

Dominic Crea
Dominic Crea is a physicist, teacher, and an avid advocate of sensible renewable energy. He lectures on alternative transportation such as plug-in hybrid vehicles and has been instrumental in dispelling the myths of the hydrogen economy.

2. Hydrogen is difficult to store in quantity, even when compressed under enormous pressure. And while hydrogen gas packs more energy per unit of weight than any other fuel, it fares rather poorly when it comes to energy per unit volume. In an automotive context, fuel cell vehicles running on hydrogen rank about the same, with respect to mileage, as the latest generation of lithium-ion battery electric vehicles.

3. Fuel cell costs, reliability, durability, and standby losses also continue to foil the adoption of hydrogen as a universal energy carrier. In cars and trucks, fuel cell power trains haven fallen short of their goal; battery systems are leaping ahead with vehicles that can be purchased at a fraction of the price of their fuel cell counterparts. Furthermore, battery electric cars can be

recharged at home, work, and, increasingly, electric car-friendly businesses. In point of fact, car manufacturers are steering away from fuel cell vehicles and are now concentrating on PHEVs with lithium-ion batteries.

4. Hydrogen safety is still an issue of debatable concern. While the use of hydrogen in open air presents little danger, it can create a serious hazard when used in the home.

Admittedly however, the use of batteries to store electric energy makes sense only up to a maximum of perhaps 100 kilowatt-hours (about as much electricity as a normal household uses in 3 or 4 days). For significantly greater quantities of energy as might be required for air conditioning and heating, the bulk storage and delivery of hydrogen has been offered as a solution. There are, however, less expensive, more practical ways of achieving the same goal:

- The redistribution of renewable electricity to areas (and in forms) as needed is one solution that has failed to receive the attention it deserves. For example, surplus electricity generated from a rooftop solar electric panel in Ohio or from windmills along the Great Lakes, could be delivered to a bookstore parking lot in Michigan to recharge a PHEV while the owner sits inside the store sipping on coffee or socializing with friends.

    Conversely, this same vehicle represents a significant reservoir of readily dispatched electricity back into the grid—a two-way process—to help stabilize the supply of electricity during times of peak demand. This technology—V2G or "vehicle-to-grid—is a variant of "net metering." As such, it offers possibilities that are not available with hydrogen that is distributed via a universal pipeline (as proposed by hydrogen enthusiasts). To be specific, unlike the existing electric grid that allows for a bidirectional flow of electricity, pipelines are "one way" in operation; one cannot simply create hydrogen at home and send it "back down the pipeline" to other users as is done with net metering.

- Heating and air conditioning represent tremendous energy loads; neither batteries nor hydrogen offer reasonable solutions to this problem. In the case of air conditioning, surplus electricity can be converted directly into ice, circumventing the cost and complexity of a hydrogen fuel cell system. Many large buildings use this cost-saving technique—purchasing

power at night when it is cheap and refrigerating a large container of water into ice—to achieve cooling during the day. Likewise, heat can be stored directly in the form of chemical or physical changes rather than going through the wasteful processes of generating, storing, and reconverting hydrogen back into electricity.

- Finally, it might make more sense to combine electrolyzed hydrogen with carbon dioxide from the air to manufacture methane. This process, while less efficient than using the hydrogen onsite, offers the compelling benefit of creating a renewable, clean fuel that can be dispatched in our current natural gas (primarily methane) pipeline. (Hydrogen gas cannot be transported in existing pipes that deliver natural gas to homes and businesses.) Furthermore, renewable methane eliminates all the problems associated with storage, energy density, range, availability, and compatibility that are present with hydrogen. Renewable methane is being considered seriously; at three times the energy content per unit volume of hydrogen, it may prove to be the perfect fuel to be used in conjunction with the PHEV.

There is a bit of irony when it comes to the hydrogen economy: because so much has been promised over the years (and so little actually achieved), time, money, and interest that should have been devoted to other ideas has been lost. As a consequence, rather than unshackling us from our energy dependency, talk of a hydrogen economy has lulled us into complacency and inaction. In the meantime, we have become even more dependent on fossil fuels. Certainly, hydrogen will play a part in our energy future, but if we fail to recognize its limitations, if we fail to take a renewed look at other, more practical solutions, we may never get a chance to see hydrogen attain its true potential. Sadly, hydrogen's ultimate legacy may be the undoing of, rather than the salvation of, humankind.

Critical Thinking Questions

1. Summarize each author's main points and supporting information.
2. Using your critical thinking skills, analyze each viewpoint.
3. After studying the essays, do you think hydrogen will become a major fuel source in the future? Why or why not?

Table 15-3

Meeting Energy Needs of a Solar-Powered Society

Demand Sector	Sources	Application	Percentage of Total Energy Use
Residential and commercial	Passive and active solar systems, district heating systems	Space heating, water heating, air conditioning	20–25
	Active solar heating with concentrating solar collectors		
	Solar thermal, thermochemical, or electrolytic generation	Cooking and drying	~5
	Photovoltaic, wind, solar thermal, total energy systems	Lighting, appliances, refrigeration	~10
			Subtotal ~35
Industrial	Active solar heating with flat plate collectors and tracking solar concentrators	Industrial and agricultural process heat and steam	~7.5
	Tracking, concentrating solar collector systems	Industrial process heat and steam	~17.5
	Solar thermal, thermochemical, or electrolytic generation		
	Solar thermal, photovoltaic, cogeneration, wind systems		
	Biomass residues and wastes	Cogeneration, electric, drive, electrolytic, and electrochemical processes	~10
		Supply carbon sources to chemical industries	~5
			Subtotal ~40
Transportation	Photovoltaic, wind, solar thermal	Electric vehicles, electric rail	10–20
	Solar thermal, thermochemical, or electrolytic generation	Aircraft fuel, land and water vehicles	
	Biomass residues and wastes	Long-distance land and water vehicles	5–15
			Subtotal ~25
			Total 100

Data from H. W. Kendall and S. J. Nadis (1980). *Energy Strategies: Toward a Solar Future.* Cambridge, MA: Ballinger, p. 262.

it would be used. The point of this table is that there are options—many of them. They may seem farfetched right now, but as the world turns away from fossil fuels, many renewable energy advocates believe that the renewable alternatives will grow substantially and will become major sources of energy world wide.

Economic and Employment Potential of the Sustainable Energy Strategy

Numerous studies show that efficiency measures and most renewable energy technologies "produce" energy at the same cost or at a lower cost than conventional sources and eliminate many of the environmental impacts as well. They do this while employing more people than conventional energy supply strategies. The chief reason for this is that efficiency and alternative technologies tend to be more labor-intensive (requiring more people) than oil, coal, and nuclear energy, which are more capital-intensive (requiring huge investments in machines and fuels). Consider some examples.

A California-style wind farm generates electricity for about 5 cents per kilowatt-hour, one-half the cost of electricity from a nuclear power plant and without the serious environmental impacts. In addition, a wind farm that produces the same amount of energy as a nuclear power plant will employ over 540 workers, compared to 100 workers in the nuclear facility (**FIGURE 15-25**).

Similar gains can be made through energy efficiency. A study in Alaska, for instance, found that state expenditures on weatherization—home energy conservation—would create more jobs per dollar of investment than the construction of hospitals, highways, or new power plants. Conservation spending, for instance, would create three times as many jobs as highway construction.

FIGURE 15-25 Comparison of employment opportunities of various energy sources. This graph shows the difference in employment opportunities of different energy strategies, all producing the same amount of energy per year. Renewable resources turn out to be stellar job creators.

Weatherization of all homes in the United States would create 6 to 7 million job-years, according to Worldwatch Institute projections. (A job-year is one job for 1 year.) Six million job-years is the equivalent of 300,000 jobs over a period of 20 years! Not all jobs would be low-wage, either. Energy-efficiency companies would be run by well-paid personnel and would employ salespeople, managers, accountants, and installers.

A wise energy future is economically, environmentally, and socially sustainable. The current system fails on all of these criteria. Nonetheless, great barriers lie on the path to a sustainable system. One of the most significant is that industrial nations have been built on fossil fuels. Billions upon billions of dollars have been spent on the present fossil fuel-based system. Huge investments have been made in mining and drilling equipment, transportation networks, processing facilities, and power plants. Although the system is unsustainable, enormous resistance to change will come from many powerful political and economic interests: legislators from energy-producing states, coal and oil companies, and the power companies, which expect to profit from their investments.

Creating a sustainable system of energy will take many years. As fossil fuels and nuclear plants become obsolete, they can be replaced by renewable energy sources such as wind or solar energy. Some parts of the current energy system, such as electric lines, can be used to transport solar- and wind-generated electricity.

Shifting to renewable energy resources and conservation will also require massive amounts of investment capital to finance wind farms, geothermal plants, and so on. Although the shift to a sustainable energy strategy could create many new economic opportunities, jobs will be lost in some sectors. But many more jobs will open up than will be lost. Not only will there be more jobs, but they will undoubtedly be safer. Some workers will be able to turn their old skills to similar activities. Petroleum geologists and oil well crews, for example, might use their expertise to drill for geothermal resources.

The transition to a renewable energy system is occurring surely in many parts of the world. Many large corporations are becoming producers and consumers of renewable energy, including General Electric, BP, Shell, Microsoft, Google, and the Philadelphia Eagles football team. Individuals and businesses can help by buying green power, when it is available. Some utilities, for instance, offer electricity from wind sources to customers at a slightly higher cost. Governments can help, too, by providing tax breaks and grants for private research, and steps to increase demand for green power in the marketplace, and by using more renewable energy in its own facilities. The U.S. government has taken similar steps. The National Renewable Energy Lab, for instance, and many other government labs and facilities are now getting power from solar electricity and wind. Even the White House has been retrofitted with solar electric modules. Many states now have passed laws requiring an increase in renewable energy production (mostly wind) in the next decade. Colorado, for instance, requires major utilities to generate 30% of its electricity from renewable sources by 2020.

The length of the transition to renewable energy resources will depend on our political will and our willingness to change for the sake of the planet's future.

KEY CONCEPTS

Shifting to a sustainable system of energy will take many years. Most renewable energy technologies provide competitively priced electricity while creating more jobs than fossil fuels and nuclear energy and protecting the environment.

The great end of living is to harmonize man with the order of things.

—Oliver Wendell Holmes

CRITICAL THINKING

Exercise Analysis

Critical thinking warns us that broad statements—generalizations—such as those made by the supposed energy expert often fall apart on closer examination. A careful analysis of the issue reveals a somewhat different picture of reality—one much more favorable for solar energy.

Solar energy, as noted in this chapter, consists of at least four different technologies. One of them, solar thermal electric, is currently cost competitive with nuclear power. Solar thermal electric plants use sunlight to heat water or some other fluid, which is then converted into steam to run an electric turbine.

Another form of solar energy is solar electricity generated from photovoltaics (panels on rooftops that convert sunlight energy into electricity). Photovoltaics are not currently cost competitive with conventional fossil fuel resources in most applications in developed countries. It would cost a fortune for most homeowners, for instance, to add panels to their roofs to supply their homes with electricity.

In rural villages in the developing world, however, photovoltaics are quite cost-competitive because the expense incurred by running power lines to remote sites is astronomical. Even in developed countries such as the United States, though, photovoltaics can make sense. If a house is only half a mile from a power line, it is often more economical to install photovoltaics than to run a power line to the house.

Another solar technology is passive solar heating—that is, heating buildings with sunlight energy that streams in through south-facing windows. Passive solar often costs more, but it can be built economically. The author's house cost no more than a conventional home and is heated almost entirely by the sun, saving several thousand dollars a year.

This exercise also shows that it is important to define what is meant by various terms, such as *uneconomic*. Does the term refer only to conventional economics, which ignores costs such as damage to the environment? In this case, the oil industry spokesperson is talking only about the cost of producing conventional fuels. He's ignoring the enormous external costs from planetary warming, acid deposition, and urban air pollution.

Along that same line, does the cost of production include subsidies? As this chapter points out, oil in the United States is very heavily subsidized by the federal government. The nuclear and coal industries are also heavily subsidized. If subsidies were removed, they would be far less affordable, and many solar technologies would appear quite competitive. Thus, this exercise also shows the importance of considering the source of information that may be biased by economic self-interest.

CRITICAL THINKING AND CONCEPT REVIEW

1. In your view, is it imperative that we change to a sustainable energy system? Why or why not?

2. Make a list of all the energy-efficient, environmentally friendly vehicles available today. Describe each one and its pros and cons. Which type would make most sense for your lifestyle?

3. Make a list of all environmentally friendly transportation fuels, describe each one, and list their pros and cons.

4. Describe the types of solar energy technologies available. How does each one operate? What is it used for? Which ones are already cost competitive with current energy technologies?

5. What are the advantages and disadvantages of solar energy? How can the problems with the technologies be solved?

6. A person living in the Pacific Northwest argues that renewable energy is useless. The sun rarely shines in his neck of the woods. How would you answer this?

7. Describe the difference between passive and active solar systems. What features are needed in a home to make passive solar energy work?

8. What are photovoltaic cells? Why are they economical to use in rural villages in the developing nations.

9. Wind energy is cost competitive with conventional electricity from coal and cheaper than nuclear energy. Should we develop this energy resource in preference to nuclear power, coal, or shale oil? Why or why not?

10. What is biomass? How can useful energy be acquired from biomass?

11. How is geothermal energy formed? How can it be tapped? Describe the benefits and risks of geothermal energy.

12. Using your critical thinking skills, debate the following statement: "Hydroelectric power is an immensely untapped resource in the United States and could provide an enormous amount of energy."

13. What are the major problems facing hydrogen power? How could these be solved?

14. Using your critical thinking skills, discuss the following statement: "Conservation is our best and cheapest energy resource."

15. Discuss ways in which you could conserve more energy at home, at work, and in transit. Draw up a reasonable energy conservation plan for yourself and your family.

16. Using your critical thinking skills, debate this statement: "A sustainable energy strategy won't work. It will cost money and lose jobs."

17. List and describe the barriers to creating a sustainable energy supply system. How could they be overcome?

KEY TERMS

biodiesel
biomass
cogeneration
conservation
earth-sheltered house
The Efficiency and Alternative Energy program
electrolyzer
Energy Efficiency Act
Energy Star Homes program
Energy Star label
flat plate collectors

fuel cell
geopressurized zones
geothermal energy
green building
green building programs
gross national product (GNP)
hot-rock zones
hybrids
hydroelectric power
hydrogen
Hydrogen Future Act
hydrothermal convection zones

least-cost planning
Motor Vehicle Safety Act
National Appliance Conservation Act
National Model Energy Code
Natural Resources Canada
net zero energy
passive solar heating system
plug-in hybrid
second law of thermodynamics
solar collectors
Super Energy-Efficient Home program

CHAPTER

16

The Earth and Its Mineral Resources

Conservation is humanity caring for the future.

—Nancy Newhall

LEARNING OBJECTIVES

After reading and studying the material in this chapter, students will be able to understand, knowledgably discuss, and, when appropriate, apply the information learned in this chapter, including:

1. The types of mineral resources extracted from the Earth's crust and their importance to human society
2. Environmental impacts of mineral extraction and processing
3. Ways of dramatically reducing the impacts of our dependence on minerals
4. The many factors that affect the supply of minerals
5. Potential shortages of key minerals
6. Ways to increase mineral supplies

CRITICAL THINKING

Exercise

In the late 1980s and early 1990s, critics of recycling of metals and paper products often argued that recycling didn't make good economic sense. They pointed to the decline in price in recent years for recycled newspaper and other materials as a sign that the system should be scrapped. This position basically meant that we should retain the old ways: manufacturing aluminum and other metals from raw ore and making paper by cutting down trees to make pulp, which is then made into paper.

To strengthen their case against recycling, critics also pointed out that it is a rather dirty process. Using your knowledge of recycling and your critical thinking skills, how would you analyze this point of view?

Human societies depend on terrestrial and aquatic ecosystems and the Earth's crust for all our resources—from the food we eat to the steel from which we make a variety of products, including cars and trucks. Together, the ecosystems and the mineral-rich crust provide all the resources that make our lives and our economy possible. They are the infra-infrastructure of modern society.

In this, we will examine the minerals extracted from the Earth's crust (**FIGURE 16-1**), their economic importance, environmental impacts of their extraction, and ways to meet our need for minerals more sustainably.

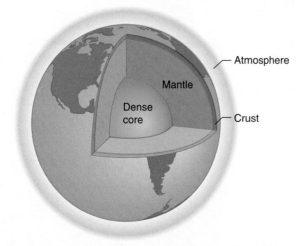

FIGURE 16-1 The Earth in cross section. The Earth's crust floats on the mantle. Beneath that is the fiery core.

16.1 The Earth's Mineral Riches

The Earth's crust contains many economically important minerals. **Minerals** are largely inorganic substances such as sand, gravel, aluminum ore, and iron ore. The term *minerals* also refers to a few organic substances such as coal and oil; these are technically referred to as *fuel minerals*. The rest are known as **nonfuel minerals.**

Nonfuel minerals are grouped into three basic categories: (1) **metal-yielding minerals,** such as aluminum and copper ore; (2) **industrial minerals,** such as lime; and (3) **construction minerals,** such as gravel and sand. Metal-yielding minerals must be processed, usually crushed and smelted (heated), to produce pure metals such as aluminum and copper. Industrial minerals and construction materials are used directly. They typically involve little processing.

Nonfuel minerals are made up of elements such as silicon and oxygen. The most abundant elements in the Earth's crust are oxygen, silicon, aluminum, iron, and magnesium. Other elements, such as gold and platinum, are also present, but are relatively rare.

Minerals typically occur in **rocks,** solid aggregates that usually contain two or more different types of mineral. Geologists divide rocks into three major classes: (1) **Igneous rocks,** such as basalt and granite, are those formed when molten minerals cool. (2) **Sedimentary rocks,** such as shale and sandstone, are formed from particles eroded from other types of rock. These particles are deposited elsewhere (for example, on the floor of the ocean) in huge quantities, where they build up in thick layers. (3) **Metamorphic rocks,** such as schist, are formed when igneous or sedimentary rocks are transformed by heat and pressure during mountain building.

Most metal-yielding minerals come from igneous rocks. These minerals are often concentrated in igneous rocks by geological processes. A concentrated deposit of minerals that can be mined and refined economically is called an **ore.** As just noted, most ores are mined and then crushed and processed (for example, heated) to yield their final product—for example, metals such as aluminum and zinc.

 KEY CONCEPTS

The Earth contains many valuable nonfuel minerals. Those metal-yielding minerals found in concentrated deposits called *ores* are mined and processed to produce metals.

Mineral Resources and Society

Few people realize how important minerals are to our lives. Metals derived from some ores, for example, are found in many products—among them buildings, computers, bicycles, glasses, and automobiles. Construction minerals are used to make roadbeds, schools, office buildings, and homes. Industrial minerals are used in fertilizers and concrete.

Minerals are so important to our welfare and our cultural evolution that scholars delineate the ages of human history by the chief minerals in use at the time: Stone, Bronze, and Iron. Although the Industrial Revolution is described in reference to the growing use of fossil fuels, minerals also played a key role in this transformation. In fact, fossil fuels were often used to process minerals into useful products.

Minerals are also vital to national economies. In the United States, mineral mining is big business. In the United States, the value of nonfuel mineral production in 2011 was $75 billion up from $57 billion in 2009. Processing and refining those ores boosted their value eight times—to $454 billion. Even recycling boosted the economy.

Today, more than a hundred nonfuel minerals are traded in the world market. These materials, worth billions of dollars to the world economy, are vital to industry, agriculture, and our own lives. Processing and refining these minerals in 2011 increased their value by nearly 10 times—to $704 billion, up from $454 billion in 2009. Even recycling minerals now contributes significantly to the U.S. economy. Nationwide, the value of metals and minerals recycled domestically was $30.4 billion. The United States, exports of scrap for recycling were worth nearly $20 billion.

Among metals, the most important are iron, aluminum, and copper. Iron is mainly used to make steel for automobiles, bridges, buildings, and a host of other products. Aluminum is used to build jet aircraft and to make beverage cans. Copper is mostly used to make electrical wire and water pipes. So important are the metal-yielding minerals that if any one of several dozen key minerals were suddenly no longer available at a reasonable price, industry and agriculture would suffer.

Who Consumes the World's Minerals?

The more developed countries are the major consumers of minerals. With approximately 20% of the world's population, they consume about 75% of its mineral resources. The United States, with about 4.4% of the world's population, consumes about 20% of its minerals. Japan, Canada, Europe, and Russia also consume large quantities of minerals.

In the past few years, mineral consumption by the industrial nations has leveled off and in some cases declined. One reason for this is that industrial nations have largely completed their **infrastructure**—bridges, highways, and buildings—and no longer need massive inputs of minerals. Another is that heavy manufacturing is on the decline in many countries, whereas nonmineral-intensive industries (among them high technology and services) are on the rise. Because the less developed nations are becoming industrialized, their share of the world's mineral consumption is on the rise.

Import Reliance

World mineral production is widely dispersed. However, some minerals are only found in commercially valuable quantities in specific countries. As a result, most nations are highly dependent on other countries (**FIGURE 16-2**). The vast majority of the United States's mineral resources come from politically stable countries. However, some vital minerals such as chromium and platinum come from politically volatile regions (**FIGURE 16-3**). To protect against embargoes or sudden declines in export from potential political upheaval in such nations, the United States stockpiles a 3-year supply of **strategic minerals** such as bauxite (aluminum ore), cadmium, and graphite, hoping that if political troubles arise in nations that supply minerals they will be resolved well before stockpiles are exhausted.

Industrial nations have for many years exploited less developed countries, extracting huge quantities of minerals at extraordinarily low prices. Less developed countries understandably feel cheated by the Western world, which converts the raw materials it acquires at a bargain into products that reap them trillions of dollars a year. To offset this problem, many governments of the less developed nations have taken steps to increase their share of profits—for example, by charging royalties, taxing raw ore exports, or encouraging the production of finished metals in their countries. The success of these efforts is reflected by statistics from the U.S. Bureau of Mines, showing that in 1982, the United States imported $4 billion worth of raw minerals and $25 billion worth of finished metals from foreign sources, mostly developing countries. In 2012, the imported raw materials were $8.1 billion, but the finished materials had risen to $147 billion.

Will There Be Enough?

The long-term outlook for minerals is mixed. Current estimates of world mineral reserves and projections of consumption suggest that about 75% of the economically vital

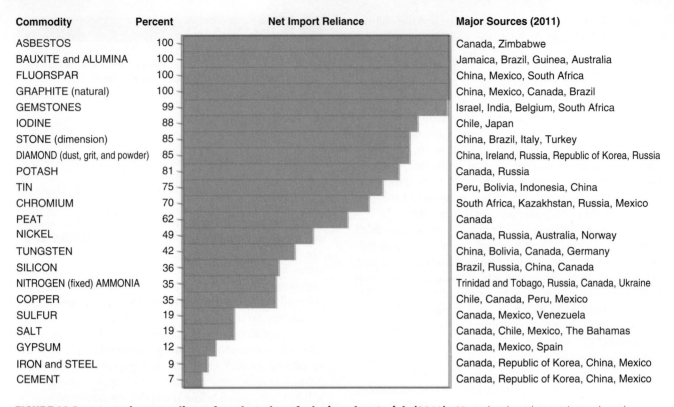

Commodity	Percent	Net Import Reliance	Major Sources (2011)
ASBESTOS	100		Canada, Zimbabwe
BAUXITE and ALUMINA	100		Jamaica, Brazil, Guinea, Australia
FLUORSPAR	100		China, Mexico, South Africa
GRAPHITE (natural)	100		China, Mexico, Canada, Brazil
GEMSTONES	99		Israel, India, Belgium, South Africa
IODINE	88		Chile, Japan
STONE (dimension)	85		China, Brazil, Italy, Turkey
DIAMOND (dust, grit, and powder)	85		China, Ireland, Russia, Republic of Korea, Russia
POTASH	81		Canada, Russia
TIN	75		Peru, Bolivia, Indonesia, China
CHROMIUM	70		South Africa, Kazakhstan, Russia, Mexico
PEAT	62		Canada
NICKEL	49		Canada, Russia, Australia, Norway
TUNGSTEN	42		China, Bolivia, Canada, Germany
SILICON	36		Brazil, Russia, China, Canada
NITROGEN (fixed) AMMONIA	35		Trinidad and Tobago, Russia, Canada, Ukraine
COPPER	35		Chile, Canada, Peru, Mexico
SULFUR	19		Canada, Mexico, Venezuela
SALT	19		Canada, Chile, Mexico, The Bahamas
GYPSUM	12		Canada, Mexico, Spain
IRON and STEEL	9		Canada, Republic of Korea, China, Mexico
CEMENT	7		Canada, Republic of Korea, China, Mexico

FIGURE 16-2 U.S. net import reliance for selected nonfuel mineral materials (2011). Many developed countries such as the United States are heavily dependent on other nations to supply minerals.

Data from U.S. Geological Survey, *Mineral Commodity Summaries* 2011.

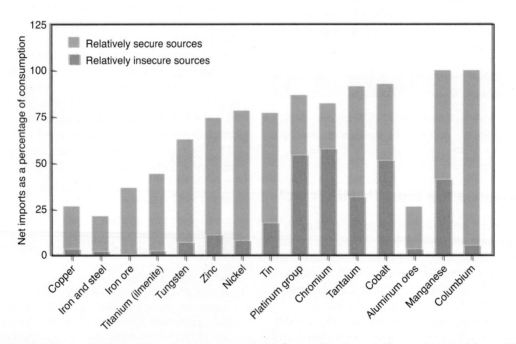

FIGURE 16-3 Risky business. The United States imports some minerals from politically unstable countries. Each bar in this graph shows the total imports as a percentage of U.S. consumption. The colored part (blue) of the bar indicates the percentage of the minerals that comes from insecure sources. The difference between the total and the colored bar (gold) represents imports from secure sources.

minerals are abundant enough to meet our needs for many years—or, if they are not, they have adequate substitutes. However, approximately 18 economically essential minerals will fall into short supply—some within a decade or two. Gold, silver, mercury, lead, sulfur, tin, tungsten, and zinc are among them.

Some experts warn us not to count on new discoveries and improved extraction technologies to meet our needs. Even if we could extract five times the currently known reserves of these materials, the 18 "endangered" minerals will be 80% depleted by or before 2040. Combined with declining oil supplies, the depletion of these important minerals could have serious global economic impacts.

 KEY CONCEPTS

Most minerals are present in adequate quantities or can be replaced by various substitutes, but some—about 18 minerals we use—are in danger of falling into short supply within the near future.

16.2 Environmental Impacts of Mineral Exploitation: A Brief Overview

Like fossil fuels, minerals are part of a production–consumption system that involves exploration, mining, processing, transportation, and end use. Many of these activities can cause significant environmental damage. In fact, mining and **smelting** (melting ore to extract metal) have created enormous amounts of environmental damage the world over. Copper production is a case in point. Copper is first extracted from the Earth in **open pit mines,** one type of surface mine (**FIGURE 16-4**). The rock (overburden) above the deposit is first excavated, and then the ore is removed. Next, the ore is crushed and run through a concentrator that removes impurities. The concentrated ore is then heated at high temperatures in a smelter to extract a crude metal, which is later resmelted to produce pure copper.

Each stage in this system produces major environmental impacts, even in the best regulated countries. For example, mining disturbs large areas. In the tropics, forests are leveled to make room for mines. Rock wastes from open pit mines are typically deposited on the land near the mine—burying vegetated areas (Figure 16-4). Mine wastes, called *tailings,* may also erode into nearby lakes and streams, clogging channels and causing a number of problems, such as flooding. Toxic metals in the mine wastes can contaminate nearby streams and reservoirs, killing aquatic life. Sulfur present in tailings may combine with water to form sulfuric acid, creating acid mine drainage, which is lethal to fish and numerous aquatic organisms.

Smelting produces enormous quantities of air pollution. Globally, copper and other nonferrous (noniron)

Go Green

To reduce mineral mining and pollution from smelting, recycle glass and all steel and aluminum cans.

FIGURE 16-4 Open pit mine. This aerial view of the Bingham copper mine in Utah shows the enormous impact of this activity. Waste from the mine is stockpiled around the site, creating further land disturbance (not shown here).

© Royce Bair/Hemera/Thinkstock.

smelters produce about 8% of the world's sulfur dioxide emissions. Sulfur dioxide combines with water in the atmosphere to produce sulfuric acid, which rains down on the land, creating severe ecological and economic problems. Smelters also release enormous amounts of toxic metals in the stack gases, among them arsenic, lead, and cadmium. Toxic metals and acids from smelters are responsible for huge *dead zones*—places where all vegetation has perished. Around the Sudbury smelter in Canada, 10,400 hectares (26,000 acres) have been turned into a barren moonscape because of air pollution from the smelter (**FIGURE 16-5**).

Other mineral extraction efforts rely on equally devastating techniques. In the Amazon, for example, gold is extracted by **hydraulic mining.** In this process, hillsides are blasted with high-pressure streams of water that wash away the soil and rock containing gold (**FIGURE 16-6**). The runoff is directed into sluices that separate out the gold. The sediment runs off into streams, killing fish and other aquatic life.

In North America, gold is sometimes extracted from piles of crushed ore or old tailings by a process called **heap leaching.** In this technique, miners spray a toxic cyanide solution on piles, letting it percolate down through the material. During its sojourn, the cyanide solution dissolves gold. The liquid is then collected, and gold is extracted. Cyanide collection reservoirs and contaminated piles endanger wildlife and groundwater. In 1990, for instance, 39 million liters (10 million gallons) of cyanide solution broke through a containment dam in South Carolina, flushing into a stream and killing at least 10,000 fish. Ducks and other waterfowl die by the thousands each year when they land in cyanide ponds near mines. Similar incidents have occured in Romania, Laos, Ghana (Africa), Alaska, and Costa Rica, killing fish and other forms of wildlife and posioning people.

FIGURE 16-5 Destruction from the Sudbury smelter. The dead zone around the copper smelter in Sudbury, Ontario shows the enormous environmental damage caused by emissions from the smelter.

© J.K. Enright/Alamy Images.

FIGURE 16-6 Hydraulic mining in Asia. Water sprayed on this hillside removes the soil and exposes subsoil containing gold, which is also removed by high-pressure water.

© Glow Images/age footstock.

In summary, mineral exploitation creates environmental damage on a scale matched by few other human activities. It is responsible for deforestation, soil erosion, water pollution, and air pollution. The environmental impacts are particularly severe in less developed countries, which produce a large portion of the world's minerals and have lax, if any, environmental regulations.

KEY CONCEPTS

The mineral production–consumption cycle produces extraordinary environmental impacts. The most significant occur in the mining and smelting phases.

16.3 Supplying Mineral Needs Sustainably

Humans need minerals, but they also need clean air, water, and healthy ecosystems. Can these two needs be met simultaneously?

Creating a Sustainable System of Mineral Production

Future demands for minerals and metals can be met in a more sustainable fashion in part by putting into practice the operating principles of sustainability, especially conservation, recycling, and restoration. Let's consider each one.

KEY CONCEPTS

Implementing the operating principles of sustainability—conservation, recycling, and restoration—can help human society create a more sustainable system of mineral production.

Recycling—Closing the Loop It is important to examine a few key concepts that explain why this strategy is so important in the effort to create a more sustainable system of mineral production. **Recycling** is a process in which valuable products such as waste metals are collected and returned to factories, where they are melted down and used to manufacture new products. Metals are infinitely recyclable.

Recycling helps increase the time a mineral or metal remains in use—that is, it increases its **residence time.** Recycling of materials helps to stretch limited mineral supplies, but it has many other environmental benefits. For example, it greatly reduces energy demand, slashes pollution, and reduces water use. Manufacturing an aluminum can from recycled aluminum, for example, uses only 5% of the energy required to make it from aluminum ore (bauxite). Less energy is required because aluminum scrap has already been processed and refined. Because less energy is involved, less air pollution is produced. Fewer tons of coal need to be mined, and that reduces the impact on the land.

In the United States and many other countries, recycling efforts have fallen short of their full potential, except in a few instances such as the automobile. Approximately 90% of all American cars are recycled. Nearly 100% of all lead acid batteries are also recycled. The recycling of aluminum, steel, and many other metals, however, could easily be doubled.

Meeting future demands sustainably will require a massive shift to recycling. Important as it is, recycling will not permit global exponential growth in mineral use to continue indefinitely, for several reasons. First, during the production–consumption cycle, some minerals and metals enter into long-term uses—for example, copper used for wiring homes and businesses. Second, some materials are lost through processing inefficiencies. Others are lost accidentally or are thrown away on purpose. Thus, it is impossible to recycle 100% of any given material. A practical goal would be 60 to 80%. Therefore, recycling can slow down the depletion of mineral supplies but cannot halt it. In theory, recycling can double our mineral resource base, but continually rising demand and inevitable losses will eventually deplete reserves.

Conservation—Decreasing Product Size, Increasing Product Durability

Conservation (using only what we need and using it efficiently) is another key element of a sustainable system of mineral supply. Conservation is often the cheapest, easiest, and quickest means of stretching mineral resources.

Conserving minerals may mean making smaller automobiles and finding ways to design many other products using less material. It also means making products more durable so that they last longer. Combined with recycling, conservation measures can greatly extend the lifetime of many valuable mineral resources. As **FIGURE 16-7** shows, continued exponential growth in mineral demand—the track we are now on—is the fastest route to depletion. Recycling will slow down the depletion but not eliminate it. Recycling and conservation measures combined will slow it down even more, giving us more time to develop new mining technologies and find substitutes. Conservation and recycling also help us to minimize our impact on the environment.

Individuals can also play a role by purchasing more durable products and by consuming less, meeting needs in other ways—for example, by borrowing or renting a tool rather than buying one. Individuals can also help by decreasing their demands for metal goods—for example, using mass transit in a city rather than buying a car.

Go Green

Set up a recycling bin in your dorm room or apartment, and, if your parents don't recycle, get them started.

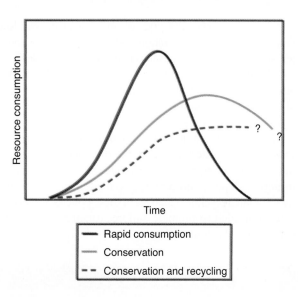

FIGURE 16-7 The future of the world's minerals. This graph shows three possible scenarios on a hypothetical time scale. The current path leads to the depletion of the world's mineral reserves. Conservation and recycling can greatly lengthen the life span of valuable minerals, helping us create a more sustainable economy. Which path will we take?

Promoting Conservation and Recycling Changes in existing laws are needed to stimulate recycling and conservation. For example, the **General Mining Act** (1872) permits miners to purchase federally owned lands for $5 per acre or less. All that is necessary is to file a claim and perform $100 worth of work on that claim each year. Because of this law, which was designed to promote mineral development, millions of acres of land have been transferred into private ownership for a fraction of their real value. Even foreign companies can purchase U.S. land. A Danish mining company, for instance, paid $275 for 108 acres of federally owned land that is estimated to contain $1 billion in minerals. The U.S. government does not collect royalties on minerals extracted from these lands—as it does for coal, oil, and natural gas. Because of the inexpensive price of this land and the lack of royalties, mineral prices are artificially low. The law of supply and demand correctly predicts that artificially low prices encourage demand. Low prices also discourage conservation and recycling efforts. At this writing (February 2014), efforts are under way to revamp the General Mining Act, and many observers are optimistic that such efforts will be successful, although they face an uphill battle against the powerful mining industry. So far, powerful mining companies have been successful in their efforts to block reforms. (For a debate on this law, see the Point/Counterpoint in this chapter.)

U.S. mining companies also receive a sizable tax exemption of about $500 million a year. Known as a **depletion allowance,** this exemption allows companies to deduct 5 to 22% of their gross earnings (depending on the mineral) from their taxes each year, even if the ore deposit is still producing. This law was intended to compensate mining companies for declining ore reserves. While it may

Chapter 16: The Earth and Its Mineral Resources **389**

do that, it also makes the price of minerals artificially low, again discouraging conservation and recycling. Many countries subsidize mineral development through tax laws or even through direct investment of government funds (tax dollars) in mineral development projects. Eliminating financial incentives for producing raw materials could go a long way toward promoting recycling and efficiency.

Governments could also help promote recycling and conservation by offering tax breaks or other incentives to companies that use recycled materials or that offer longer warranties. Deposit and refund systems for products could also stimulate recycling. Changes in these policies and others are badly needed to promote a more sustainable use of the world's mineral resources.

 KEY CONCEPTS

Reforming unsustainable laws, removing subsidies on raw materials, and giving financial incentives to companies that use resources efficiently and incorporate recycled materials in their products—will all help promote efficiency and conservation.

Restoration and Environmental Protection

New laws and tighter enforcement of existing laws could improve mining practices and reduce pollution from smelters in both industrial and nonindustrial nations. In the United States, mining and smelting operations are often exempt from existing laws. For example, mine wastes, which contain toxic metals that leach into nearby waterways, are currently not regulated by the **Resource Conservation and Recovery Act** (1976), a law that among other things places tight controls on hazardous wastes. This exemption, insisted on by the mining companies at the time the legislation was being considered, exists despite the fact that mines are the single largest producer of wastes in the United States. Moreover, smelters are not required to report their emissions to the Environmental Protection Agency (EPA), as other industries are. Efforts are currently under way to close these loopholes, but many observers are skeptical about their outcome.

In recent years, the Canadian government has begun to take steps to ensure the restoration and environmental protection of mined areas. Canadian mining companies are required to prepare an environmental impact statement for projects, which, among other things, outlines how federal and provincial regulations on mining will be honored. Although previous problems are enormous—for example, cleaning up abandoned mines in Canada is estimated to cost $9 billion—many companies have shown a willingness to cooperate to protect air, water, and wildlife. Dramatic improvements have been especially evident at Canada's large smelters.

 KEY CONCEPTS

Meeting mineral demands sustainably will require efforts to minimize environmental damage from mining and other operations. Tougher laws may be needed. Laws that exempt the mining industry from environmental protection need to be changed, too. Better efforts to restore the damage to natural systems caused by mining are also needed in many countries.

POINT · Reforming the 1872 Mining Act

Courtesy of Darrell Knuffke.

The General Mining Law of 1872: An Idea Whose Time Has Passed

Darrell Knuffke
Darrell Knuffke is retired as the western outreach coordinator for the Wilderness Society. Knuffke joined the Wilderness Society in 1985 and for 11 years directed its Central Rockies regional office in Denver. He lives in International Falls, MN.

Alan Septoff
Alan Septoff is Mineral Policy Center's Reform Campaign Director. Septoff joined MPC in 1997. He spends most of his time fighting anti-environmental initiatives supported by mining industry advocates in the U.S. Senate.

As latte shops thrive in the shadows of feed stores, no one disputes the notion that the American West is settled, the frontier closed.

As frontier disappeared, so did most of the 19th-century land laws meant to spur settlement. The lone statutory artifact from our expansionist youth is the General Mining Law of 1872. Congress enacted the law to promote mineral exploitation and development and to provide incentives to settlers by giving away mineral rights. Today, well over 125 years later, the law remains in full force and effect, unchanged in a new millennium. And there's nothing quaint about it.

Just by staking a claim, anyone can obtain the right to mine hard rock minerals—gold, silver, copper, and the like—on 239 million hectares (590 million acres) of the national land that all Americans hold in common. Once a claim is posted and registered (no limit to the numbers), the claimant can hold it indefinitely by paying $100 per claim per year to the federal government.

A claimant who discovers a "commercially viable" deposit can buy the land from the government through a process called patenting for $2.50 to $5.00 per acre. (There is currently a moratorium on patenting, which must be renewed by Congress each year.) More land, contiguous or not, is available for such things as milling and processing at $5 an acre, about which more will be said later.

If the claimant meets the doddering old law's terms, the government can't refuse the sale. After the sale, ownership of the land and its minerals is absolute. Owners may sell the land for any price or purpose—whether related to mining or not. Outside Phoenix stands a golf resort worth an estimated $60 million. The owner paid American taxpayers $170 for the mining claims it stands on.

Whether the land covered by a mining claim is patented or not, anyone who stakes a claim has a right to mine that claim. And that becomes a right protected by the takings clause of the Constitution. Because of this, the mining law is also a zoning law—all public lands not reserved for another purpose, like a National Park, are essentially zoned for mining. It doesn't matter if mining a claim would destroy a nearby town's water supply or if the land were better used for grazing livestock. . . . Once a valid claim is staked, the only foolproof way for the federal government to prevent mining is to buy back the claim.

The 1872 Mining Law offers the ultimate subsidy: We *give away* the minerals and the land with them. The Mineral Policy Center, a leading advocate for U.S. mining law reform, estimates that if the hard rock industry paid U.S. taxpayers an 8% royalty for minerals—the same rate that some coal and oil and gas industries pay for public resources—the Treasury would gain nearly half a billion dollars over 5 years. The industry claims it can't pay the royalty and still mine. But it pays that much and more on state and private land, where much of the nation's mining actually occurs. Of course, the industry claims it is *not* subsidized because it pays taxes on profits. By that lame reckoning, the public should provide at no cost the merchandise that every taxpaying business in the country sells.

One of the world's most valuable mines is the Stillwater, located on national forest land in Montana. Its platinum and palladium ore body is an estimated 26 miles long and contains over 225 million troy ounces of minerals worth $32 billion. What will taxpayers get for this monumental treasure? $10,180 in payment for the mining claims.

There are no environmental standards in the 1872 law. According to the U.S. Bureau of Mines, hard rock mining has contaminated more than 12,000 miles of our rivers and streams. The EPA lists over 60 mining related sites among its Superfund cleanup priorities. The industry argues that these are unfortunate relics of another time, that modern mining is different, responsible. Modern mining—especially gold mining—is, if anything, increasingly toxic and increasingly dangerous. Cyanide heap leach mining is responsible for the new gold rush in the American West—and for much else. In Colorado, taxpayers will have spent more than $100 million to clean up the poisonous stew a Canadian company left at the Summitville Mine when, after depleting a gold mine, it declared bankruptcy and walked away from its operation.

Although the 1872 Mining Law doesn't contain any explicit environmental protection provisions, it does effectively limit the amount of space that can be used to dump mining waste. For every 8 hectares (20 acres) of mining claims, the text of the law allows one 2-hectare (5-acre) claim (called a millsite claim) for ore processing and mine waste dumping. This wasn't an issue in 1872, when gold ore was rich—it didn't take much ore to get an ounce of gold. Nowadays, however, all of the rich ores have been mined out, and it takes many tons of ore to produce an ounce of gold. All of the ore that isn't gold has to be dumped as waste—on the millsite claims. In many cases, the law doesn't provide enough millsite acreage to dump all the waste produced by a modern mine.

Although clearly written into the law, it wasn't until March 1999 that the federal government began enforcing the millsite provision. The mining industry, which has for decades fought reasonable reform and defended as sacrosanct every provision of the old law, now finds it can't possibly live with the plain terms of *this* provision, and it is seeking relief where it has historically found it: in the U.S. Congress. Conservationists are girding to block a legislative rider that is sure to come, probably in the form of an appendage to an appropriations bill, that would deny the Interior Department any money to enforce the millsite limit that the law clearly mandates.

Reform advocates don't seek to bar mining on public lands. We do demand

- That federal land managers be able to prevent mining if it threatens resources of higher public value. The industry's mantra is "gold is where you find it." So are clean drinking water, biological richness, and healthy landscapes.
- Strong environmental safeguards where mining does occur and tough, enforceable standards for reclamation.

- Payment of a reasonable royalty to the public whose mineral resources these are.
- An end to patenting.

The 1872 Mining Law is a fiscal and environmental atrocity, yet it has survived every effort at reform. Its apologists are a few western members of Congress who think the region still lives on livestock, logging, and the mother lode. It doesn't and it hasn't for a very long time. The frontier is closed. So should be the most destructive provisions of this 128-year-old law.

COUNTERPOINT

The Case for Reasonable Reform

Stuart A. Sanderson

Stuart A. Sanderson is president of the Colorado Mining Association and is an attorney with nearly 25 years experience in mining and related issues, in both government and private practice.

Those who espouse repeal of the Mining Law contend that it amounts to a giveaway of valuable mineral resources. If that were so and if gold, silver, and other valuable mineral commodities were available in big, ready to mine "nuggets" at bargain basement prices, then why isn't the federal government completely "sold out?" Why don't you and I take advantage of the "free gold and other minerals" available? The facts belie such a conclusion. Of the more than 2.2 billion acres of public lands in the United States, only 3 million acres (less than one-quarter of one percent) have been transferred to private ownership through mineral patent during the entire 128 year history of the Mining Law. By comparison, more than 94 million acres have been transferred to railroads and 288 million acres conveyed to private ownership under agricultural homesteads. In addition, the mining industry has disturbed an amount of land that's less than the size of a county in Nevada.

"Minerals are where you find them." And minerals can be difficult to locate in commercial quantities; once they are found, it is very costly to extract and process them into a final product. That is why the Mining Law, a law that has been amended more than 20 times in its history, still provides appropriate incentives to reward those individuals who invest the time, capital, and resources to locate and produce minerals on public lands, minerals that are worth nothing in the ground unless someone assumes the risk of exploring for and developing them.

Mining is a capital-intensive industry, and the costs of mine development often run into hundreds of millions of dollars before the company even begins to recoup its investment. True, once the mining company has perfected a claim after location of a valuable mineral deposit, the company may apply for a mineral patent. But, as the Bureau of Land Management (BLM) points out, "the patenting of land under the [Mining Law] is not as easy as simply writing a check... In Nevada, mineral patent applications have contained from 1 to 500 claims per application. At a minimum, *it will cost the applicant $37,900 in direct costs to process a single claim or millsite.*" And these costs are based on the assumption that the claim will be uncontested. Companies are further required to pay annual maintenance fees in the amount of $100 per claim.

Contrary to what the opponents of mining have been saying, patent fees are not the only costs that the mining industry will incur in connection with the development of a claim and extraction of the mineral resource. In addition to the costs of mine development, mining companies add millions of dollars each year to local economies in the form of sales, property, and other taxes. The federal government also receives a large amount of revenue from mining. The Mining Law is no giveaway; it actually has generated billions of dollars in revenues to support local communities, schools, and economies. The mining industry shares its wealth for the benefit of its most valuable resource, the mine workers, who earn the highest average annual wages of all industrial workers, about $60,000 annually in Colorado. More than 320,000 people owe their livelihoods to the mining industry.

Opponents of mining falsely state that the Mining Law allows claimants to obtain patents on the cheap and convert such lands to residential, resort, or other valuable uses, reaping a windfall in the process. Federal regulations preclude such a practice and already require the forfeiture of a mining claim in the event of an unauthorized use. One often cited example of alleged abuse of the Mining Law involves land near Phoenix, Arizona, on which the claimant had obtained claims for sand and gravel. In the mid-1950s, the Congress excluded such common variety minerals from the operation of the Mining Law. Meanwhile, the city itself had grown and reached the proposed boundaries of the mining operation and the city government precluded the mine operator from developing the claim. Only then was the land later sold for commercial hotel and golf course development. Now you know the rest of the story....

These opponents also criticize the Stillwater Mine in Montana, which produces platinum and palladium. Antimining lobbyists argue that the operator's sole financial contribution is the amount paid for mineral patents. They conveniently ignore the massive costs the company has incurred in mine development, as well as the taxes and other economic contributions to Montana's economy. Stillwater is the largest private employer in Montana. According to the company's annual report, it paid its employees (95% of whom reside in the state) $57 million in salary and benefits in 1999. The company has spent millions developing the claim and pays on an annual basis more than $24 million payroll, property, and other taxes.

Does the mining industry support reform of the Mining Law? The answer is an emphatic "yes!" The industry has supported legislation in Congress to require the payment of a royalty. Industry is also willing to pay fair market value for the surface of the land in addition to any royalty for the minerals in the subsurface. Any unauthorized use would result in forfeiture of the mining claim. We have also supported legislation that would impose a permanent claim maintenance fee and would further earmark revenues for the cleanup of abandoned mines.

Any royalty imposed, however, must be reasonable and not discourage mineral production. Yet antimining special interests urge the imposition of a 12.5% royalty on all minerals, the same royalty paid by the coal industry. Such an argument, however, fails to account for the fact that the Mining Law applies to many different varieties of minerals, i.e., gold, silver, copper, palladium, and the like. Unlike coal, a one-size-fits-all approach would impose an inequitable royalty burden on these minerals, despite the fact that gold, copper, molybdenum, and other locatable minerals are priced differently and compete in different markets. Moreover, such a royalty would have a devastating impact on mineral production and would actually reduce the amount of revenues the federal and state governments receive in taxes and other fees. The accounting firm Coopers & Lybrand confirmed in a 1993 study that the imposition of even an 8% gross royalty would result in a minimum $422 million annual net loss to the Federal treasury. By contrast, the royalty proposed by the mining industry that would only tax net proceeds would actually increase revenues by $53 million.

Opponents of the Mining Law allege that additional one-size-fits-all federal reclamation standards are needed because the Mining Law contains no environmental standards. What they fail to point out is that mining activity is regulated by literally dozens of federal environmental laws, including the National Environmental Policy Act, the Clean Water Act, the Safe Drinking Water Act, the Endangered Species Act, and numerous others, not to mention hundreds of regulations. Mining operations on federal lands are required to meet the reclamation standards of the Federal Land Policy and Management Act, which direct mining operators to prevent degradation of such lands. Moreover, the states in which mining operations occur contain very comprehensive regulatory programs that govern the permitting of mining operations, and impose strict reclamation standards that require mine operators to return lands to a productive, post-mining land use. Colorado has one of the toughest such laws in the nation.

The idea that we need yet another set of federal reclamation standards is contrary to sound science and to the recommendations contained within the most recent study by the National Academy of Sciences. In a 1999 report entitled "Hardrock Mining on Public Lands," the Academy reviewed all existing federal and state regulations and concluded they were adequate, comprehensive, and well coordinated. The study further concluded that the federal government would be better served by improving the manner in which it implements and enforces existing laws, given the array of federal laws and remedies already on the books. We believe that the states, because of their greater familiarity with the terrain and other unique conditions within their borders, are best equipped to manage the environmental and reclamation plans for mineral activities. A new set of federal regulations would cost the taxpayer more and is no substitute for adequate budget, support, and enforcement of existing laws.

Mining is important to all Americans, adding nearly $523 billion annually to our nation's economy. The average American consumes more than 18,200 kilograms (40,000 pounds) of minerals each year, according to the National

(continued)

Mining Association. Your telephone contains more than 40 different minerals. Minerals are the salt in our food, the limestone in the sidewalk under our feet, the gypsum used in fire-resistant wallboard, the clay for the brick in our homes, the gold used in medical equipment and computers (not to mention jewelry and currency), and the molybdenum used in alloy steels. Modern life would not be possible without minerals. It is important, therefore, to fashion laws and regulatory programs in a manner that protect the environment without discouraging legitimate and responsible mineral resource development.

Critical Thinking Questions

1. Summarize each author's viewpoint, noting supporting data.
2. How do their positions overlap?
3. What are the points of difference?

16.4 Expanding Reserves

Given the continued growth of the world population, rising expectations of the residents of the less developed nations, economic expansion, and the inevitable loss of metals, recycling and conservation will very likely not satisfy 100% of our future demands. New deposits need to be discovered and mined.

Geologists use the term *reserves* to indicate deposits that they are fairly certain exist and that are feasible to mine at current prices. Several factors determine the size of available reserves and are worth studying as we consider ways of expanding reserves to meet future demand.

KEY CONCEPTS

 Future demand cannot all be satisfied by recycling and conservation efforts. Some new minerals must be mined.

Rising Prices, Rising Supplies

Many factors determine the size of mineral reserves. Economics is one of the most important. To understand this, we must look first at the law of supply and demand.

The law of supply and demand explains how both supply and demand affect the price we pay for raw materials, goods, and services. Basically, it says that when the demand for a raw material, good, or service exceeds its supply, prices tend to rise. A popular new car, for example, can be sold for more because demand is so high. In contrast, when demand is low in relation to supply, prices tend to be lower. However, that's not the end of the story. Prices also affect demand. As a rule, as prices rise, the demand for a product declines. As prices fall, the demand tends to rise. Supply and demand are elastic. That is, they change over time with changes in the price. Price is elastic as well, for it changes with supply and demand.

The law of supply and demand helps explain how businesses and our economy operate. It also helps to explain mineral reserves. Rising prices, for example, tend to make it economically feasible for companies to search for and produce more minerals. As a result, mineral reserves tend to expand as prices get higher, because it is now more economical to tap into lower grade ore deposits.

Over a decade ago, two rivals, Julian Simon (an economist who supports population growth) and Paul Ehrlich (an ecologist who thinks growth is a prescription for disaster), made a bet. Ehrlich bet $1,000 that the price of several key minerals such as copper and nickel would be higher at the end of the decade as a result of scarcity. Simon said they would not. Who won? Simon, the economist. Reserves of key minerals had expanded because of the many forces described in the previous and succeeding paragraphs.

Many people expect this cycle of rising price and expanding reserve to repeat itself *ad infinitum*, thus creating an endless supply of materials to meet human needs. While this line of reasoning may work in the short term, in the long term resource scientists remind us that mineral resources are finite. At some point, resources will simply run out or become so costly to mine that they will be unaffordable.

KEY CONCEPTS

Many factors determine the size of mineral reserves. One of the most important is the price. Reserves tend to expand as prices rise because companies are willing to spend more to extract lower grade ores, but ultimately, all mineral resources are finite.

16-1 Antarctica: Protecting the Last Frontier

Antarctica makes up one-tenth of the Earth's surface and is home to many animals, among them penguins, seals, and whales. Relatively untouched by humans, Antarctica has been the center of a huge controversy over minerals (**FIGURE 1**).

Between 1908 and 1943, Great Britain, France, Norway, Australia, New Zealand, Chile, and Argentina claimed 85% of Antarctica. The United States and the former Soviet Union, which had carried out their share of South Pole exploration and research, did not establish claims and refused to acknowledge the territorial claims of other nations. It was the position of these nations that it was preferable to retain an interest in all of the continent rather than to stake out a claim on one part.

In 1959, 16 nations signed a cooperative agreement, the **Antarctic Treaty System,** to maintain the continent for peaceful uses. The Treaty prohibited all military activities and weapons testing. The treaty also set aside arguments over land claims and permitted the nations that had signed it to carry out research unencumbered.

Since then, hints of Antarctic minerals and oil have threatened to destabilize the peace, pitting one nation against another. Part of the conflict arose between the haves—the seven claimant nations—and the have-nots—those who signed the treaty and have participated in Antarctic research and management but have no legal claim to the land or resources. The have-nots also include a growing number of less developed nations, who would like a share in the potential wealth of Antarctica. Added to the growing controversy was the involvement of several international environmental organizations, such as Greenpeace International and the International Union for the Conservation of Nature and Natural Resources. They are seeking to prevent development so that this vast, pristine wilderness will remain so.

The first real push for Antarctic mineral exploration came after the perilous climb in oil prices in the 1970s. Reports from the U.S. Geological Survey indicated that huge oil reserves might lie beneath Antarctica's continental shelf, but the evidence is sketchy. Sizable nonfuel mineral deposits may also lie beneath the ice and snow of Antarctica.

In 1988, the signatories of the treaty proposed regulations that would permit oil and mineral exploration. If approved, oil and mineral prospecting by seismic testing and other techniques with relatively minor environmental impacts could have begun. Full-scale development, such as mining and drilling, however, would have been barred until a new agreement was reached.

Many critics were concerned that development would threaten Antarctica's environment. Offshore oil rigs, for instance, would operate in some of the roughest seas in the world. Huge icebergs could rip apart drilling rigs. Ice floes could trap and crush ships. Mining would require huge amounts of energy to drill through the mile-thick ice cap.

Environmentalists were also concerned about the potential impacts of resource development on the rich marine ecosystem. Oil spills could be carried ashore, wiping out huge breeding colonies of penguins, birds, and seals. Oil slicks might also kill algae, the base of the aquatic food chain. Caught under the ice and in the chilly waters, oil could persist a hundred times longer than in warmer waters. Finally, cleaning up an oil spill would be impaired by the short "warm" season. If a well blew out, winter might set in before workers could plug it up.

Nonfuel mineral extraction would require enormous amounts of energy, create pollution, damage the ice pack, and severely disrupt the land surface. Most damaging to wildlife would be the use of ice-free coastal areas for processing and shipping ores.

FIGURE 1 Antarctica. This vast land may house coal, oil, and minerals. But is it worth risking this pristine area to boost natural resource supplies?

Experience had shown that research, generally a relatively benign activity, had already created significant air and water pollution. Further human activity on the scale required to extract minerals and oil, critics argued, would create an environmental disaster.

In October 1991, though, after considerable political pressure from environmentalists, 24 industrial nations signed an agreement that banned environmentally damaging oil and mineral exploration in Antarctica for at least 50 years. Japan signed the agreement in 1991 but has not ratified it. South Korea signed in 1992.

Besides prohibiting oil, gas, and mineral development, this treaty includes provisions that will protect the environment from the impacts of other human activities. Although the agreement still leaves the possibility of oil and mineral development open many years down the road, for now, Antarctica is saved from development. Protecting areas of the world from human development—especially environmentally fragile areas with slow recovery rates—is an essential element of global efforts to create a more sustainable human presence.

Technological Advances Expand Reserves

Technological improvements in the mining and processing of ore—especially those that make it more economical to mine reserves or tap into lower quality ores—have increased global reserves of several key minerals since the 1970s. Scientists and engineers are currently developing techniques to improve mining efficiency more, thus allowing companies to exploit even lower grade ores. One group of researchers, for example, discovered that certain algae bind gold ions. These algae could permit companies to extract gold from mine wastes or perhaps even from natural waters.

Although technological advances can increase mining efficiency, remember that technology is cost driven. When mineral supplies fall below a certain point, the costs of technologies required to access and mine the lower grade ores could conceivably become prohibitive—too great for consumers to bear. Mining would be curtailed. Thus, although technology can help, it is only part of the solution.

KEY CONCEPTS

Technological improvements make it feasible to mine less concentrated ores, which helps expand reserves.

Factors That Reduce Supplies

Higher prices and new technologies can expand reserves (economically recoverable minerals) within limits. However, many factors have an opposite effect. Rising labor costs, for example, tend to reduce reserves. Higher interest rates on money that companies borrow to operate also increase costs and make exploration, mining, and production more costly.

Energy prices also play an important role in determining when mining a particular mineral becomes uneconomical. As the concentration of an ore decreases, the amount of energy required for mining and refining is fairly constant up to a certain point, beyond which it increases dramatically (**FIGURE 16-8**). At this point, reserves become too expensive to mine; their costs exceed what the market will bear.

Rising energy prices combined with declining ore deposits could bring an end to some of our important mineral resources within 30 to 40 years, say some analysts. Runaway inflation could grip the world economy unless something is done.

A final factor is environmental costs. Mining lower grade ores produces greater environmental damage than mining higher grade ores. Why? Larger surface mines are needed to produce the same amount of ore. More material will be transported to smelters for processing. More waste will be

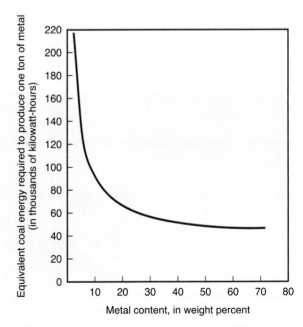

FIGURE 16-8 Energy and mineral concentration in ores. This graph shows that as the concentration in an ore body declines, the energy required to recover the mineral increases dramatically.

produced at the mines and smelters, and more air and water pollution will result. This could cause environmental protection costs to escalate, adding to the cost of production and accelerating the economic depletion of some minerals.

The prospects for expanding mineral reserves are mixed. As noted earlier, reserves for some minerals appear to be quite limited. Although supplies of others may be huge, environmental constraints and energy costs could put a damper on expansion, making conservation and recycling even more attractive.

Minerals from the Sea

The mineral deposits on land are finite, and on many continents, they have been heavily exploited. As a result, some companies are looking toward two untapped frontiers—Antarctica and the floor of the world's oceans—as potential sources for new minerals. Superficially promising, these options face serious economic, environmental, and social barriers. **Spotlight on Sustainable Development 16-1** examines the mineral potential of Antarctica. This section describes the seafloor deposits in the ocean.

The ocean is a vast resource of minerals, many of which are dissolved in the water itself. However, the concentrations of most dissolved minerals are generally too low to be of economic importance. More important are mineral deposits on the seafloor. The most important of these minerals are small lumps called manganese nodules (**FIGURE 16-9**). Abundant in the Pacific Ocean, manganese nodules contain several vital minerals, notably manganese (24%) and iron (14%), with smaller amounts of copper (1%) and cobalt (0.25%).

Several mining companies have explored the possibility of dredging the seabed to mine nodules. Preliminary studies show that seafloor mining may be technically feasible as well as profitable, but little is known about its environmental impact. At this point, the biggest impediment appears to be a political one. To date, the world is in a quandary over the issue of ownership. Mining companies in the West have the wealth to exploit the seafloor and argue that they should be allowed to do so. However, the less developed nations contend that the seabed belongs to all nations and that they should receive a portion of the proceeds.

At a United Nations Conference on the Law of the Sea, the less developed countries proposed an international tax on seabed minerals that could provide them with millions of dollars a year for agricultural and economic development. A comprehensive Law of the Sea treaty, worked out with U.S. negotiators during the Carter administration, would have included this plan, but the United States under Presidents Reagan, Bush, and Clinton refused to ratify the treaty—as did Great Britain and West Germany. Today, many people believe that a seabed taxation scheme is fair and necessary. Others see it as an unjust way of cutting into corporate profits. Consequently, progress toward mining the seabed has come to a halt.

16.5 Finding Substitutes

Historically, the substitution of one resource for another that has been depleted has been a useful strategy for industrialized nations. Shortages of cotton, wool, and natural rubber, for example, have been eased by synthetic materials made from oil.

Substitution will unquestionably play an important role in the future, too. It could help us find alternatives to some minerals that fall into short supply. It could help us replace materials whose production is environmentally damaging. For example, fiber-optic cables are made of thin glass fibers made from silica (sand). They are currently replacing many uses of copper, a mineral whose extraction and processing are environmentally harmful, even under the best of circumstances. Digital cameras have largely replaced film cameras that rely on silver, a product used to make photosensitive emulsion on film. But is substitution a crutch we can lean on forever?

Critics argue that substitutions have created unreasonable faith among the public in the ability of scientists to come up with new resources to replace those that are being depleted. Some resources may

FIGURE 16-9 Manganese nodules. Lying on the ocean's floor, these nodules range from the size of peas to that of oranges.

© Science VU/Visuals Unlimited.

 Go Green

You can help reduce mining and pollution from smelters by buying durable and used products.

not have substitutes. For instance, it may be impossible to find substitutes for the manganese used in desulfurizing steel, the nickel and chromium used in stainless steel, the tin in solder, the helium used in low-temperature refrigeration, or the tungsten used in high-speed tools. Scientists also note that many substitutes have limits themselves. Plastics have replaced many metals, for example, but the oil from which plastics are made is a limited resource.

Finding substitutes is a race against time. The time of economic depletion of some minerals is fast approaching, perhaps more quickly than substitutes can be found. A wise strategy would be to identify those resources that are nearest to economic depletion and then promote widespread conservation and recycling. Research to find substitutes should begin immediately, or if research is already under way, it should be greatly accelerated.

16.6 Personal Actions

Tolstoy wrote that "everyone thinks of changing the world, but no one thinks of changing himself." We're quick to point out all of the things government or business or other people can do but we are often remiss when it comes to taking personal action. As noted in previous chapters, personal actions are essential to building a sustainable future . They go a long way toward reducing demand and the impacts caused by meeting those demands, and they have the added advantage of saving money.

KEY CONCEPTS

Individual action is vital to building a sustainable future. Buying durable products, buying less, recycling, and choosing recycled materials are three steps people can take.

Tomorrow's growth depends on the use we make of today's materials and experiences.

—Elmer Wheeler

KEY CONCEPTS

Substitution of one resource for another that has become economically depleted has been a useful strategy in the past, but it may not work in all cases. Substitutes have limits, and some materials have no suitable alternatives.

CRITICAL THINKING

Exercise Analysis

In the late 1980s and the early 1990s, a dramatic increase in the number of U.S. recycling programs glutted the market with recycled newspapers and other products such as steel cans and plastic pop bottles. Consider paper recycling as an example: Recycling mills in the United States were flooded with newspaper, but their capacity to recycle this material increased only slightly during that time. This created a huge bottleneck in the system. Because of the glut of recyclables, prices for newsprint dropped considerably, sometimes making this endeavor rather uneconomical. Similar problems arose in plastic bottle and steel can recycling.

During this dark era for recycling, analysts predicted that these gluts would be transient. New plants would open up to take advantage of these materials. Were they right?

Absolutely. By 1993, paper recycling was back on its feet. New mill capacity had eased the paper glut. Plastic pop bottle recycling followed a similar pattern, as did steel can recycling. The problem with the critics was that they assumed a static condition—that the market would remain the same. Gluts were viewed as permanent. The critics failed to recognize that with millions of tons of relatively cheap material (recycled newspaper, plastic, and steel) on hand, businesses were bound to notice and avail themselves of the opportunities.

Assumptions can be dangerous, especially in dynamic systems that respond to abundance and price, like the economy. Critical thinking reminds us to examine assumptions carefully—our own and those of others.

Critics also point out that recycling is a dirty process. Indeed, recycling produces pollution. But digging a little deeper, you find that as a rule recycling produces much less pollution than manufacturing goods from raw materials. The energy savings alone can be staggering and fossil fuel energy is a leading producer of air pollution. Cut back on energy consumption, and you automatically reduce air pollution.

CRITICAL THINKING AND CONCEPT REVIEW

1. What are nonfuel minerals? Be specific. Why are they so important to our society?

2. Why is a 100% recycling rate very likely not achievable?

3. Describe the environmental impacts of mining and processing minerals.

4. Critically analyze this statement: "Economic forces will ensure a continual supply of mineral resources. As prices rise, we'll find new resources, develop new technologies, and find substitutes for minerals currently used. So there is nothing to worry about."

5. What are reserves? Describe the factors that cause mineral reserves to expand and contract.

6. Outline a plan to meet the future mineral needs of our society sustainably. Describe each element of your plan. What are the most important steps? What barriers would you expect to the implementation of each idea?

7. What legislative changes would help promote a sustainable system of mineral production and consumption?

8. Make a list of ways in which you can cut down your resource consumption.

9. Describe ways in which recycling and resource conservation can be promoted in your community. Be sure to list a variety of options, ranging from compulsory to voluntary. Which ideas do you like best? Why?

10. Using your critical thinking skills, analyze the following statement: "Outer space offers a vast supply of mineral resources. We will never run out of resources as long as we have access to these minerals." Do you agree with it or not? Why?

KEY TERMS

Antarctic Treaty System
construction minerals
depletion allowance
General Mining Act
heap leaching
hydraulic mining
igneous rocks
industrial minerals

infrastructure
metal-yielding minerals
metamorphic rocks
minerals
nonfuel minerals
open pit mines
ore
recycling

residence time
Resource Conservation and Recovery Act
rocks
sedimentary rocks
smelting
strategic minerals

© Hannamariah/ShutterStock, Inc.

Creating Sustainable Cities, Suburbs, and Towns: Sustainable Community Development and Environmental Protection

The optimist proclaims that we live in the best of all possible worlds, and the pessimist fears this is true.

—**J.B. Cabell**

LEARNING OBJECTIVES

After reading and studying the material in this chapter, students will be able to understand, knowledgably discuss, and, when appropriate, apply the information learned in this chapter, including:

1. The types of systems that support our lives and reasons why they are not sustainable
2. Why reforming human systems is vital to creating a sustainable future that protects and enhances the environment
3. How land use planning can help create more sustainable cities and towns
4. Ways to create a sustainable system of transportation

CRITICAL THINKING

Exercise

Hydrogen is a fuel that receives a lot of attention from the media these days, and many people, including politicians, are touting hydrogen as a potential fuel source for automobiles, trucks, and busses. They point out that hydrogen is made from an abundant and renewable resource, water, by simply running electricity through water. This causes the water molecules to split. Hydrogen can then be burned or introduced into a fuel cell that uses the hydrogen to generate electricity. When this occurs, the hydrogen combines with oxygen in the air to form water once again. The process produces no pollution, say proponents.

Analyze hydrogen as a fuel, and present your thoughts on the wisdom of pursuing hydrogen as a transportation fuel. You may want to do some research on hydrogen to learn more about it. How much energy does it take to make it? Would it be more efficient to use the electricity needed to split water molecules directly to power cars?

This book presents a systems approach to environmental protection. One of its goals is to show how principles of sustainability derived from the study of ecology and other areas can be used to restructure human systems such as energy, transportation, waste management, and housing to protect the natural systems that support our lives and to ensure a long, prosperous human presence. Sustainable development and redevelopment are, therefore, a vital strategy of environmental protection.

This book also presents many ideas on ways to create nature-compatible designs to permit humans to thrive within the limits of the natural world. These new designs could allow human civilization to prosper without disrupting nutrient cycles, climate, wildlife, and natural environments. In fact, they could even enhance the environment, our life support system, and the source of all human wealth.

TABLE 17-1 lists the human systems. Although restructuring human systems and the global economy are essential to protect the environment and build a sustainable future, the task will not be easy, inexpensive, or quickly executed. This chapter presents some additional ideas on sustainable systems, looking at land use and transportation.

TABLE 17-1
Human Systems

Energy
Transportation
Waste management
Water
Industry
Agriculture

17.1 Cities and Towns as Networks of Systems

Think of the city or town you live in. It consists of people and the built environment—streets, shopping centers, office buildings, and schools. Look a little harder, though, and you will see the city or town as a network of interdependent human systems. There's a transportation system, consisting of roads and highways, gas stations, automobiles, buses, trucks, and airports. There's an energy system, consisting of gas stations, power plants, underground pipes that carry natural gas, and power lines that transmit electricity. There's most likely an elaborate system of water supply, consisting of deep wells or dams and reservoirs, water treatment plants, and an extensive set of pipes under the streets to transport water to homes, factories, businesses, schools, government offices, car washes, and parks. There's a waste management system as well. It consists of thousands of generators of waste—homes, factories, office buildings, and copy shops that produce millions of tons of waste each year. It also consists of ways to deal with waste, such as recycling centers, waste-to-energy plants that burn garbage, landfills in which waste is buried, sewage treatment plants, and hazardous waste facilities.

These networks of systems are designed to meet our needs for raw materials, finished goods, and services. They also get rid of the mountains of waste produced by human society. Some futurists believe that making cities sustainable will require a restructuring of these systems. Virtually every aspect of city or town life will need to be rethought and redesigned, although not everyone agrees that this is necessary or, if they do, about how this should be accomplished.

KEY CONCEPTS

Cities and towns consist of numerous systems, such as energy, housing, and transportation, that many experts think are largely unsustainable. Making our living environment sustainable will require us to redesign human systems such efforts will help protect natural systems that support us.

The Invisibility of Human Systems

Although there has been a lot of progress in the last four decades in environmental protection, many problems are worsening. As pointed out earlier, the vast majority of the trends are leading us away from a sustainable existence. It may be hard for people to accept the assertion that human society and the systems it depends on are unsustainable. Throughout this book you have seen many statistics that support this conclusion. As has been pointed out in other chapters, pollution, species extinction, global warming, and a host of other environmental problems aren't mere surface wounds that can be fixed with Band-Aids; they're symptoms of deeper problems, most notably overpopulation and unsustainable human systems.

Why is it hard to grasp the problem with human systems? For much of the past 60 years, environmentalists, policymakers, teachers, and researchers have focused most of their attention on solutions that address the symptoms of the environmental crisis. Very little attention has been focused on human systems and their fundamental unsustainability. The basic assumption has been that if we solve the immediate problems, we'll be all right. A little change here and there will cut pollution and reduce the negative effects on people and other living creatures.

Another important factor is that for most people systems are invisible. That is, they were designed to operate with minimal bother to us (**FIGURE 17-1**). The fact that we don't

Go Green

One way to reduce your impact on the environment is to grow your food—or some of it—after you graduate or right now. You can even help your family start a vegetable garden.

think about systems very much is a tribute to the engineers who designed them. In fact, it's generally only when a system breaks down that we notice it exists.

We turn on a light switch, and electricity surges through the wires. We turn on the faucet, and out comes drinkable water. We go to the grocery store, and the shelves are packed with food. We pull up to the gas station, and there's plenty of fuel. We're aware of light switches, faucets, and gas pumps, but we're fairly ignorant about the rest of the systems. If we barely recognize their existence, how can we be concerned about them?

KEY CONCEPTS

Most efforts to solve environmental problems have focused on treating symptoms rather than on rethinking and revamping the systems that are at the root of the problems. Most people are unaware of the systems that support our lives until they break down.

Performance Versus Sustainability: Understanding a Crucial Difference

Another obstacle in the way of understanding the premise that human systems are unsustainable has to do with the distinction between performance and endurance. For most of us, the systems work well. They seem fine, so what's the problem? The problem is that just because a system is supplying us with the services we need doesn't mean that it is sustainable. For example, just because the world's fishing fleet is producing over 90 million tons of fish each year does not mean that this amount is a sustainable harvest level. In fact, this level of fish catch is severely depleting key fisheries, and we are headed for major shortages in the near future. The same is true with virtually all human systems.

KEY CONCEPTS

Just because a system such as energy or manufacturing appears to be functioning well does not mean it is sustainable in the long run.

FIGURE 17-1 The kitchen is more than a place to cook. The kitchen is an integral part of several key systems. Can you name them? How do natural systems support the human systems?

Why Are Human Systems Unsustainable?

Although it is difficult to determine the carrying capacity of the planet—that is, how many people it can support—it is easy to find evidence that we are exceeding it. Global warming, species extinction, land devastation, soil erosion, desertification, food shortages, and other problems are signs that we are exceeding the capacity of the Earth to support human life—and other life-forms as well.

These problems are partly a result of the massive size of the world population, but also a result of the design of the systems that support our lives. A system of energy based on efficiency and clean, renewable sources, for example, would have a fraction of the environmental impact of the present system. A system of agriculture based on minimum tillage, crop rotation, soil conservation measures, and natural pesticides would be able to feed the world's people with much less impact on water, wildlife, and soils. Both could endure for many centuries, very likely forever.

The present systems are unsustainable because they produce waste and pollution in excess of the Earth's capacity to absorb them. They end up poisoning us and other species. They change the climate. They adversely change the chemistry of rainfall. They've contributed to the present depletion of the ozone layer. They deplete both renewable and nonrenewable resources upon which our future depend.

Why do they do all of these things? As noted elsewhere, most systems are inefficient. They do not recycle or use recycled materials. They depend primarily on nonrenewable energy. They destroy but do not restore.

Go Green

Recycling saves energy and resources. Recycle everything you can. Buy recycled goods, like recycled paper, too.

KEY CONCEPTS

Human systems are unsustainable because they exceed the carrying capacity of the Earth. They produce pollution in excess of the planet's ability to absorb it, use renewable resources faster than they can be replenished, deplete nonrenewable resources, and fail to repair the damage created by their operation.

The Challenge of Creating Sustainable Cities and Towns

Cities and towns are a lot like the bird colonies that dot the rocky coastlines of many continents (**FIGURE 17-2**). Bird colonies are nesting and resting sites where large numbers of birds of the same species aggregate. They are also sites of enormous waste production and enormous food consumption—just like cities. But food for the colony and the city does not come from the immediate vicinity. It comes from neighboring ecosystems.

To be sustainable, cities and towns must have a lasting supply of resources. Proper resource management is therefore essential to the survival and sustainability of this pattern of habitation. Many changes can also be made in cities to ensure their sustainability. These changes are all designed to lessen resource demand and waste production while ensuring we meet our needs.

The challenge facing humankind is twofold. First, we must revamp existing infrastructure to make it sustainable. Second, we must build new infrastructure in a sustainable fashion. Consider the task of revamping existing infrastructure first—the so-called *redevelopment strategy*.

One of the first steps in reducing resource demand and pollution is to make all

Go Green

To save water and energy and reduce pollution, take shorter showers.

(a)

(b)

FIGURE 17-2 Colonies and cities. (a) Bird colony. **(b)** City. Cities resemble bird colonies in many ways. They are areas of intense activity where resources from outlying areas are consumed. They are also areas of concentrated waste production.

Photo a © Creatas/Alamy Images; photo b © Robert Glusic/Photodisc.

FIGURE 17-3 A sustainable home. This solar home utilizes three solar technologies: solar electric panels on the roofs of the garage and house to generate electricity; solar hot water panels on the roof of the garage (left) to provide domestic hot water; and passive solar, south-facing windows that transmit sunlight from the low-angled winter sun for interior space heat.

Courtesy of Department of Energy.

buildings much more efficient in their use of energy—electricity, fuel oil, and natural gas. We can make buildings much more efficient by sealing air leaks in the walls, adding insulation, and by replacing worn-out heating systems with newer, more efficient ones. We can also find ways to increase their dependence on renewable energy. For example, we can add solar electric systems to generate electricity or add solar hot water systems for domestic hot water and space heating (**FIGURE 17-3**). We can add low-flush toilets or, better yet, composting toilets and water-efficient showerheads, faucets, and appliances to reduce resource demand. We can use recycled paint or nontoxic paints. The list goes on.

Similar actions must be taken in transportation, housing, agriculture, waste management, and other systems—and are detailed in this book. Especially important are efforts to reduce **urban sprawl,** the continual expansion of human communities into farmlands and wildlands that provide vital resources required for our long-term economic health and survival.

Redevelopment is a vital but often overlooked strategy for sustainable development, in large part because so much infrastructure is already in place and because billions and billions of dollars have been invested in it. It is imperative that we make this infrastructure as sustainable as possible. Tens of millions of single-family homes and apartment buildings and millions of businesses are currently in use and are prime targets for sustainable redevelopment.

The second challenge—designing and building new infrastructure to be as sustainable as possible—takes place when systems need replacement or require expansion. To expand a transportation system in an urban environment, for example, cities can install light rail systems rather than adding new lanes to existing highways for cars. New housing could be constructed from recycled products and would be

equipped with a host of resource-saving devices so that new houses use only a quarter of the raw materials of existing houses. Rather than adding a landfill, a city could expand its recycling facilities to handle waste.

17.2 Land-Use Planning and Sustainability

A city or town is home to many different activities, some that conflict with one another: for example, housing and factories. For many years, cities and towns the world over have engaged in **land-use planning** to ensure that incompatible activities are kept apart. Land-use plans, for example, set up industrial zones and residential zones. Besides determining where people can live and do business, land-use plans establish sites for water pipes, electrical lines, roads, and shopping malls. Most land-use planning has little to do with ensuring a sustainable future. In some countries, land-use planning has been fashioned in a way to prevent or reduce sprawl and protect farmland and other ecologically important land.

 KEY CONCEPTS

Land-use planning helps cities establish the locations of various structures and activities and helps keep incompatible uses apart. As conceived and practiced in most places, it doesn't do much for sustainability.

Sustainable Land-Use Planning: Ending Sprawl

Sustainable land-use planning seeks to accomplish the same goals as land-use planning, notably separating humans from unsightly, noisy, and potentially dangerous activities. But it also attempts to achieve more efficient use of the land—in other words, to create patterns of land use that minimize the conversion of farmland and wildlands to asphalt, concrete,

KEY CONCEPTS

Two challenges face existing communities: revamping existing infrastructure and building new infrastructure sustainably.

and lawn. That is, it strives to prevent or control sprawl. This helps to preserve farmland, recreational areas, wetlands, scenic views, watersheds, aquifer recharge zones, and wildlife habitat—retaining aesthetic values and ecological services.

Sustainable land-use planning also entails efforts to coordinate key uses such as housing, business, and transportation development to reduce pollution, resource consumption, and loss of valuable land. Ultimately, land-use planning seeks the best for people, the economy, and the environment. It can in fact improve economies by reducing the loss of productive farmlands, by reducing air pollution and costly health problems, and by making the provision of government services such as police and fire protection more efficient.

FIGURE 17-4 shows four major types of traditional development: dispersed, compact, satellite, and corridor. Let's examine each one very briefly and assess its potential for promoting sustainable development.

KEY CONCEPTS

Sustainable land-use planning and development seek to optimize land use and minimize the loss of economically and ecologically important lands. They offer other benefits as well, including more efficient mass transit, reduced air pollution, and reductions in the cost of providing water, sewage, and other services.

Dispersed Development **Dispersed development** occurs in many cities and is commonly referred to as *urban sprawl*. Urban sprawl is the steady outward expansion of urban/suburban areas that occurs as new housing subdivisions, highways, shopping malls, and other forms of development spring up on the perimeter of existing cities and towns, taking over farmland, forest, and grassland (Figure 17-4a).

(a) Dispersed development

(b) Compact development

(c) Satellite development

(d) Corridor development

Commuter rail transit   Light rail transit   HOV or bus lane   Major highways and freeways

Urban center   1990 Urban area   2020 Urban area

FIGURE 17-4 Four development patterns. These maps of the Denver metropolitan area show four different development paths: **(a)** dispersed development, or urban sprawl; **(b)** compact development, the most sustainable alternative; **(c)** satellite development; and **(d)** corridor development. Each one has its benefits and its costs.

In many cities, land is already zoned for this type of development. Dispersed development, however, ranks low on the sustainability scale. In fact, it is the least desirable of all of the alternatives because it consumes lots of land. Farmland, forests, wetlands, and grasslands are all victims of sprawl. Because cities are often located near prime farmland, sprawl decreases a nation's long-term ability to produce food and displaces wildlife. Loss of natural habitat and replacement with paved surfaces often increases flooding. Furthermore, sprawl results in a rather haphazard pattern of settlement, with poor aesthetic appeal.

Because housing spreads out inefficiently on the landscape, dispersed development increases vehicle travel, adding to commuting time and increasing energy consumption and air pollution. Each of these has a cost. Increased commuting time means more time away from families and friends and less free time. Increased energy consumption, caused by the need to travel farther to work each day, costs more in gas and wear and tear on the automobile. Air pollution takes its toll on people and the environment.

Sprawl has other costs. Providing highways, mass transit, sewers, water, and other forms of infrastructure generally costs more than the alternative forms of development because more miles of water and sewer lines, telephone lines, and power lines must be run to supply homes and businesses strewn out inefficiently across the landscape. Police and fire protection in a sprawling urban environment is often more costly.

GO GREEN

To save energy and reduce pollution, live close to where you go to school and work. Walk or ride a bike to work, if you can.

KEY CONCEPTS

Dispersed development or urban sprawl is the most environmentally and economically unsustainable form of urban/suburban development.

Compact Development   Figure 17-4b shows an alternative development pattern, compact development. Ranking highest on the sustainability scale, **compact development** is a denser or more compact form of development. It requires the least amount of land to accommodate people and the services they require. Because of this, it helps to preserve open space, farmland, forests, and other ecologically and economically important lands surrounding a city.

Many measures can be implemented to ensure compact development: smaller lots, more efficient placement of houses in subdivisions, an increase in the number of multifamily dwellings such as apartments and condominiums, and more compact placement of services (shops and the like). Smaller lots mean that more houses can be located in an area.

Some people argue that compact development means crowding and unpleasant living conditions, but others note that this doesn't have to be so. For example, in a suburb, placing houses closer to the street so that there is less front lawn

and more back lawn gives people more usable outdoor space. Backyards can open into a common area that serves many homes: Community swing sets and community gardens could be added to create a sense of community and save resources. Narrowing the streets and using the saved space to create larger parks can provide playgrounds and baseball diamonds for neighborhood children. Existing zoning laws could be changed to allow homeowners to add small apartments in the tradition of the carriage house to existing homes or to convert unused space into apartments for renters. Condominiums and townhouses can accommodate far more people per hectare of land than single-dwelling units. These are ideal for singles and married couples without children or with small families. Paying attention to soundproofing, views, privacy, access to lawns/playspace to make such dwellings more livable can help make them a desirable option for people.

One promising development, which began in Denmark in the 1970s, is **cohousing,** a residential living arrangement that strives to create a community. Cohousing developments are small neighborhoods that offer many social and economic benefits, described in **Spotlight on Sustainable Development 17-1**.

In many cities and towns, businesses are strewn along highways in highly dispersed fashion. Aggregating services close to residences can cut down on driving time and make them accessible by foot or bicycle.

Compact development not only minimizes land use, it also reduces vehicle miles traveled and can substantially reduce energy consumption and air pollution. Perhaps more important, dense settlement patterns are much more amenable to mass transit systems, which, as discussed in the next section, are far easier to implement and more efficient if people live closer together. Mass transit systems use much less energy and require much less space to move people than automobiles do.

Compact development is also more efficient from an economic perspective. First and foremost, it is much cheaper to provide people living in a more compactly designed city with services such as fire protection, water, sewers, and other amenities than it is to provide them if homes and businesses are dispersed over a wider area.

Compact development, very common in Europe, creates a clear line between city and suburb and the outlying lands. In many U.S. cities, making compact development occur will require changes in existing zoning and other tools such as urban growth boundaries, described shortly.

KEY CONCEPTS

Compact development is a denser settlement pattern that offers many environmental benefits, such as reduced land use and air pollution and increased efficiency of mass transit.

Corridor and Satellite Development   The last two options are satellite development and corridor development. **Satellite development,** shown in Figure 17-4c, involves the development of outlying communities connected to the metropolitan area (the city and the surrounding suburbs) by highways and rail lines. Although this permits people to live outside the city and suburbs in small towns—and can create a very desirable quality of life—it is not as advantageous from many perspectives as compact development. It results in the conversion of a fair amount of open space and farmland. Although satellite development does not require as much land as dispersed development, it increases commuting time, energy consumption, and air pollution. Of the four options, it ranks number two in sustainability.

Corridor development, concentrating housing and business growth along major transportation corridors, is even less desirable (Figure 17-4d). It requires much more vehicle use to get to work and to services. It uses more energy and produces more pollution than compact development. It is, however, amenable to mass transit.

Clearly, of all of the options described in this section, compact development offers the best benefits and is the most sustainable form. When properly executed, land-use planning can help us achieve a more sustainable relationship with the Earth. Land-use plans can protect the foundation of tomorrow's civilization: renewable resources such as farmland, pastures, forests, fisheries, and wild species.

 KEY CONCEPTS

Corridor and satellite development, concentrating development along major traffic routes or in distant communities connected to the metropolitan area, respectively, are more desirable than urban sprawl but less sustainable than compact development.

Land-Use Planning and Building

Proper land-use planning is also be carried out on a smaller scale—for example, the development of a subdivision. When applied to specific sites, land-use plans take into account the slope of the land, soil quality, water drainage, location of wildlife habitat, and many other features. This permits planners and developers to design with nature, rather than redesigning nature. In The Woodlands, Texas, for example, a developer placed the homes outside the natural drainage areas, saving millions of dollars that would otherwise have been required to construct storm sewers. This also left an attractive open space rich in wildlife that local residents enjoy. Protecting natural drainage, wildlife areas, aquifer recharge zones, and other important land features can reduce development costs, increase the quality of life, and help protect valuable ecological services.

 KEY CONCEPTS

Many steps can be taken when building homes to protect valuable ecological assets. These can save developers money and reduce cost.

Statewide and Nationwide Sustainable Land-Use Planning

The Japanese provide a model that many countries could adopt to protect their land. In 1968, the entire country was placed under a nationwide land-use planning program. Lands were divided into urban, agricultural, and "other" classes. Several years later, the zoning classifications were expanded to include forests, natural parks, and nature reserves. The success of the Japanese plan lies in protecting land from the market system, which, left on its own, appropriates land irrespective of its long-term ecological value.

Many European nations have adopted similar programs. In Belgium, France, the Netherlands, and the former West Germany, national guidelines for land-use planning were established in the 1960s. Administered by local governments, they protect farmland, prevent urban sprawl, and help to establish **greenbelts,** undeveloped areas in or around cities and towns (**FIGURE 17-5**). The Netherlands has one of the best programs of all. Its national planning program also governs water and energy use.

Land-use planning at the federal level in the United States is rudimentary at best. Except for establishing national parks, wilderness areas, national forests, and wildlife preserves, the federal government has done little to systematically protect its land. Most zoning occurs on the community level. On the local level, planners concern themselves primarily with restrictions on land use for commercial purposes—housing developments and industrial development. Because of the United States' reliance on community-level planning, states are a patchwork quilt of conflicting rules and regulations. Some people believe that statewide land-use planning is needed.

Statewide land-use planning is an idea that is slow in coming. Oregon passed such a program in the early 1970s, and noteworthy programs now exist in Maine, New Jersey, Florida, Vermont, Washington, and Hawaii.

FIGURE 17-5 The greenbelt. This aerial photograph of Boulder, Colorado shows the greenbelt that surrounds the town, preserving valuable farms and open space and creating a more aesthetic living environment.

Courtesy of Dan Chiras.

© hugolacasse/ShutterStock, Inc.

17-1 Cohousing: Building a Community

In the past, many people lived in villages or in tightly knit urban neighborhoods. Families were familiar with people's past histories, their talents, and weaknesses. There was a sense of community that is often lacking in modern neighborhoods. Community provided a sense of belonging, a sense of security. It was a more practical way to live, too. People helped each other, sharing the burden of everyday life.

In modern society, community has withered and died. Designers of modern suburbs where many of us live take special care to ensure privacy, rather than promote closeness. In many neighborhoods, for example, fences line our back yards, walling us off from contact with neighbors. In fact, few of us see our neighbors any more. If we do, it is usually just to glimpse them coming and going—driving in and out of the garage. At night, many people returning from work simply slip into the garage without a trace. Although people do socialize, it is often with friends who live far from their homes. Given the distance that separates us, many people complain that they have to make appointments to see their friends. No one just drops in to say hello. Because there isn't much of a community in the neighborhoods we live in, many people begin to feel isolated. Life becomes a chore.

Disenfranchised with the lack of community in modern society, people throughout the world are banding together to recreate a sense of belonging through cohousing.

Cohousing generally consists of houses or apartments clustered around a common area and a common house. Usually consisting of around 30 homes or apartment dwellings, cohousing is occupied by people who planned their "development" and now manage the community they've developed. Individuals own or rent their homes, but share in the ownership of the commons. They also share in some of the work. One person, for example, might lead a committee that plans social events. Another might organize weekend work projects. Still another might coordinate day care.

In cohousing, many resources are shared, too. For example, in Golden, Colorado, Harmony Village's 27 families share a single lawn mower. The common house also offers a tremendous opportunity to share. In many common houses, there's a shared laundry facility. Two washers often serve an entire community. A television room and places for kids to hang out, to play without disturbing adults, are features of many common houses. The common house also typically has a large kitchen where community meals are prepared, sometimes every night of the week. The preparation of community meals is a shared responsibility. Imagine coming home from work to have a meal on the table and only being responsible for cooking once or twice a month.

Because of the sharing of resources, individual units are typically downsized. Kitchens, for example, tend to be much smaller, more like those found in apartments. If you are planning a big meal for friends, you book the common house and use the larger kitchen facility.

The common house may also have a room or two for guests, so you don't need a spare bedroom for that once-a-year visit by mom and dad. Mechanic shops, greenhouses, gardens, and playgrounds are also shared by residents, greatly cutting down on expenses and reducing personal living space.

Common meals and activities help knit a community together. But the initial weaving of the social fabric comes during the development of the community. In most instances, a core group of people does the planning from day 1 and sees the project through. This involves numerous meetings and discussions to reach consensus. Although this is difficult at first, with practice, residents say it becomes much easier. In fact, many people claim that the skills they learned in cohousing benefit them in both their home and work lives.

FIGURE 1 shows the layout of cohousing. As you can see, houses face a central area, a common area, where kids can play freely. Adults socialize in the commons, too. Cars are typically relegated to a peripheral position, usually garages, parking lots, or carports along the outside of the community. Although the car is never more than a block or so away from your home, you generally can't park next to your home, except perhaps to deliver groceries. The walkways are made for people, not cars. As a result, cohousing is kid and pedestrian friendly.

The common house is usually in view of all the units in cohousing. It is the heart of the community. Some cohousers have organized day care in the common house so they can leave for work and not have to transport their children to another location, saving time and money.

Privacy is designed into all cohousing so that people can satisfy this need too. In many instances, private patios or gardens are located outside the back door. You're not forced to socialize if you need alone time.

Cohousing is found in rural and urban settings. It is a great place for kids to grow up, and adults, too. It provides a rich social atmosphere.

Cohousing is not a new idea by any stretch of the imagination. It began in Denmark in the 1970s. The pioneers in cohousing sought to create a child-friendly atmosphere and opportunity to share some daily functions. Since that time, hundreds of new communities have sprung up in Europe, Canada, and the United States. Most are organized around the idea of community. However, a growing number of communities have been formed to create a more environmentally responsible way of living. In Nyland Cohousing in Lafayette, Colorado, the homes were designed and built to conserve energy and use solar energy for heating. Environmentally friendly building materials, were used, too. In Denmark, several cohousing developments have incorporated passive solar design and wind energy to generate electricity.

Cohousing is growing rapidly and is bound to increase in years to come. Being a more compact form of development, in which some space is often devoted to habitat preservation, cohousing is not only good for people, it is good for the planet.

Go Green

Take a break from your studies and visit a local cohousing community. You can find one near you by visiting the cohousing association web site.

Preliminary Sketch
Muir Commons CoHousing Community
Davis, CA

26 townhouses and
3600 sq. ft. common house on 2.83 acres

By   The CoHousing Company
     Berkeley, CA
     Feburary 1990

FIGURE 1   An architectural sketch of the layout of a typical cohousing community.

McCamant & Durrett Architects. Nevada City, California, 95959.

KEY CONCEPTS

Many states and nations have land-use planning regulations that minimize urban sprawl to protect farms, open space, and wildlife habitat.

Beyond Zoning

The main tool of land-use planning for years has been **zoning regulations,** which classify land according to use. Zoning can protect farmland and other lands from urban development. In rural Black Hawk County in Iowa, for instance, zoning laws prohibit housing developments on prime farmland, but permit them on lands with lower productivity.

New approaches are also being adopted to protect valuable land, especially farmland. One public policy tool is the **differential tax rate.** This allows city officials to tax different lands at different rates. Farmland is taxed at a lower rate than housing developments, making it more economical to farm—and encouraging farmers to hold onto their lands rather than sell them to developers.

Another technique that helps farmers keep their land instead of selling it to developers is the purchase of development rights. A **development right** is a fee paid to a farmer to prevent the land from being "developed," that is, being converted to housing. To determine the cost of a development right, two assessments of the land are made, one of its value as farmland and one of its value for development. The difference between the two is the development right. State or local governments may buy the development rights from the farmer and hold them in perpetuity. From then on, the land must be used for farming, no matter how many times it changes hands.

Still another way to reduce the spread of human populations onto valuable land is by **making growth pay its own way.** This idea calls on developers and new businesses to pay the cost of new schools, highways, water lines, sewer lines, police protection, and other forms of infrastructure needed as communities expand. The rationale is that the cost of new development should be passed on to those who profit from it, not to existing residents. A new home thus comes with a development fee attached, which the new homeowner pays.

This not only keeps local taxes from rising to subsidize new development, it encourages builders to locate new housing closer to existing schools, water lines, and highways. This, in turn, reduces sprawl and reduces the destruction of farmland and other ecologically important sites. Development fees might also encourage developers to install water-efficient fixtures and pay for water efficiency measures in existing homes and businesses, thus preventing an increase in demand.

Open space—fields, forests, and other valuable lands—can also be purchased and set aside for wildlife or designed for mixed use—that is, as parks in which people and wildlife can coexist. Open space acquisition is practiced in many areas. In some counties, a small sales tax raises money to purchase land. In others, real estate transfer fees (a tax paid when houses and land are bought and sold) are used to buy open space. Although these methods raise much money, open-space lands often command a premium price.

KEY CONCEPTS

Many techniques can help promote sustainable land-use patterns, including zoning, differential tax rates, purchase of development rights, making growth pay its own way, and open space acquisition.

Land-Use Planning in the Less Developed Nations

Urban sprawl is a major problem in the less developed nations. In fact, millions of hectares of farmland are destroyed each year by expanding urban centers. Land-use planning is therefore as essential in the less developed countries as in the industrial nations.

In some areas, land reform is badly needed. Wealthy landowners in many Latin American countries, for example, graze their cattle in rich valleys while peasants scratch out a living on the erodible hillsides. Hilly terrain that should be protected from erosion is being torn up by plows and washed away by rain. Some argue that sensible land use hinges on reform of these outdated landholding systems.

These are just a handful of ideas that could help reshape government policy to foster sustainability. Combined with many other ideas that promote sustainable ethics, revamp economics, and reshape unsustainable systems, they could form a national framework for dramatically realigning human systems to steer us onto a sustainable course.

KEY CONCEPTS

Land-use planning and land reform are also essential to creating sustainable land-use patterns in the developing nations.

17.3 Shifting to a Sustainable Transportation System

In 1950, the global automobile fleet numbered only 50 million; in 1998, the number was 508 million. In 2012, it had climbed to around 710 million.[1] Americans are among the most avid automobile users on the planet.

In the United States, there were over 111 million automobiles on the road in 2012. This impressive total results from a number of factors, among them inefficient land-use planning (that results in urban sprawl) and inadequate mass transit systems. Affluence and the expansive

---

[1] This number does not include SUVs, light trucks, and minivans.

nature of the country also contribute to America's having the highest per capita auto ownership on the planet. In the United States, automobile travel accounts for 90% of the motorized passenger transport. In Europe, where mass transit is much better developed and where cities are more compact, the auto still accounts for 78% of the passenger transport.

Air travel is another major component of modern transportation. World air travel has grown rapidly in the past five decades as shown in **FIGURE 17-6**. Besides carrying people, jets also carry huge amounts of freight. Unfortunately, the rising demand for air travel comes at an increasingly high environmental cost. Aviation consumes at least 5% of the world's oil each year. It is also the most polluting form of transport per kilometer traveled.

Today, more than 30% of the energy Americans consume is used by the transportation sector, and much of that powers our automobiles and jet aircraft. But are automobiles and jets sustainable forms of transportation?

Many observers believe that the answer to this question is no. The planet cannot absorb the pollutants produced by them. As the number of cars and jets expands and as the number of miles traveled each year increases, air pollution is bound to worsen. Oil spills could increase as the amount of oil being shipped increases. Atmospheric carbon dioxide levels are bound to rise as the combustion of fossil fuels increases. Acid deposition and urban pollution, both caused in large part by pollutants in automobile exhaust, could expand and threaten the health of people and ecosystems. Urban congestion, already at headache levels in many cities, is bound to worsen. Declining oil supplies also make the automobile and jet aircraft unsustainable. Road construction and maintenance are also quite costly. And tens of thousands of people are killed and injured each year in automobile accidents just in the United States.

Clearly, even though they provide us with great joy and mobility and are an important aspect of our economy, automobiles come with a huge price tag. So do jets. What can be done to shift to a more sustainable transportation system? Will the automobile continue to play a predominant role? Will jet aircraft travel continue to rise? The shift to a sustainable system of transportation will very likely occur in phases.

Phase 1: The Move Toward Efficient Vehicles and Alternative Fuels

The first step in the transition to a sustainable transportation system will be dramatic increases in efficiency, a change that has already begun. Improving fuel efficiency has several benefits essential to sustainability. First, it helps stretch oil supplies, providing time for the development of sustainable fuels and sustainable modes of transportation. Second, it reduces air pollution. The more efficient a vehicle is, the less pollution it produces. Third, it reduces the need to drill for and transport oil. Important as they are, efficiency gains could easily be offset by both the rising number of vehicles on the road and the increasing number of miles people are driving each year.

Go Green

To promote renewable energy and reduce pollution, consider buying a car that can run on ethanol or biodiesel or straight vegetable oil.

What is happening to fuel efficiency in the United States? In 1982, the average new American automobile got about 9 kilometers per liter of gasoline (22 miles per gallon). By 1992, the average fleet mileage had edged up to 11.9 kilometers per liter (about 28 mpg). But progress stagnated in the 1990s and 2000s because of heavy pressure on Congress from auto manufacturers interested in selling minivans, trucks, and sport-utility vehicles. Sales of these relatively inefficient, but highly popular vehicles, are more profitable than energy-efficient vehicles. Today, minivans, trucks, and sport-utility vehicles account for about two-thirds of all new car sales in the country. Because of this trend, in 2007, the average gas mileage of new vehicles in the United States fell to 19.8 mpg. In 2011, the average fuel economy for new vehicles was 24.1 mpg. The Environmental Protection Agency (EPA) and Department of Energy (DOE) under an Obama Administration directive has dramatically increased the average fuel economy of cars and light trucks to 35.5 mpg by 2016. In 2012, the president, working with major auto manufacturing, finalized even higher standards that will increase the average fuel

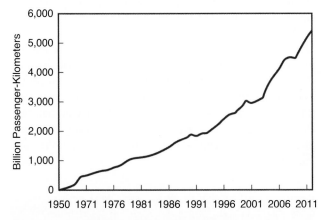

FIGURE 17-6 World air travel. More and more people are traveling by air these days, one of the most energy-inefficient modes of transportation.

Data from ICAO.

economy of all new cars and light-duty trucks to 54.5 mpg by 2015.

Creating More Fuel-Efficient Cars and Better Highway Systems Improvements in gasoline mileage can be achieved by many different techniques that can be combined to produce safe, lightweight, and extremely efficient vehicles. Engine redesign, for instance, can improve fuel efficiency. Improvements in aerodynamics—how the air flows over a moving vehicle—can also improve efficiency, as can lightweight materials such as the new foams and plastics. Space-age materials and air bags can increase the safety of the smaller, more energy-efficient vehicles. The alleged dangers of smaller cars could also be mitigated by tougher drunk-driving laws, enforcement of speed limits, and better driver education.

California has been a catalyst for many of the changes occurring in automobile design.

Some auto manufacturers have responded admirably to the challenge of producing energy-efficient vehicles. Numerous manufacturers are now selling hybrid gas-electric cars, including Toyota, Honda, and Ford. These vehicles have small gasoline-powered piston engines combined with electric motors, which are controlled by an on-board computer. In the Toyota Prius, the electric motors are used to power the vehicle at slow speeds (**FIGURE 17-7**). The gas engine is used for higher speeds. Hybrid vehicles typically get around 40 to 50 miles per gallon, but further improvements could greatly increase the gasoline mileage.

Go Green

When traveling long distance, consider taking the train or bus. Both options use less energy and produce less pollution per passenger mile than an automobile.

FIGURE 17-7 Toyota Prius. This five-passenger hybrid vehicle boasts outstanding gas mileage and low emissions. It runs on gasoline and electricity but never needs to be plugged in. Hybrid trucks, sedans, vans, and SUVs are gaining in popularity and could help us bide time to develop alternative transportation fuels and modes of transportation.

© Motoring Picture Library/Alamy.

Two promising developments are the plug-in hybrid and electric car. As described elsewhere, they operate primarily on electricity for short trips like commuting to work, and switch automatically to the gas-electric mode on longer trips.

Besides increasing efficiency of vehicles by improving engines and automobile aerodynamics, engineers are studying ways to change the way cars are driven. Imagine climbing into your car, punching your destination into the computer, and then sitting back to read the newspaper. This is the dream of some technologists who are working on cars that drive themselves. The experimental models are equipped with on-board computers that sense the sides of the road and traffic and keep the car in its lane. One of the chief benefits of this strategy would be that traffic would move more smoothly and more efficiently, reducing commuting time and pollution.

Another innovation is the use of computers to unjam congestion. Currently, engineers are using computers to understand and eliminate some of the factors that cause traffic congestion. These analyses are also helping cities to design highways to reduce congestion. Better controlled traffic lights and electronic signs that provide information to drivers to guide them toward better choices could reduce commuting time and prevent congestion.

KEY CONCEPTS

More efficient cars are part of the first phase of the transition to a sustainable energy system. Improvements in engines, the use of gas and electric motors, and better automobile aerodynamics are key elements of this effort. Computer systems that operate cars automatically, monitor traffic and signal congestion, and permit designers to create more efficient highways are also key to the success of these efforts.

Creating More Efficient Aircraft Aircraft manufacturers have made significant strides in improving fuel efficiency. Concerns over the impact of aviation and desires to reduce costs have led manufacturers such as Boeing/McDonnell Douglas and General Electric to take measures to improve efficiency in jet engines. In large part because of their efforts, today's new jets use about half as much fuel as those manufactured in the mid-1970s. Aircraft manufacturers are continuing to improve the efficiency of their jets. Even with these advances, air travel is growing so quickly that other means may be needed to reduce environmental impacts.

KEY CONCEPTS

Aircraft manufacturers have made impressive strides in improving fuel efficiency.

Alternative, Clean-Burning Fuels Alternative fuels are also part of the immediate transition to a more sustainable system of transportation. Cars, trucks, and buses can be powered by a variety of fuels, including hydrogen and ethanol, an alcohol produced from renewable sources such as corn, sugar cane, and wood. Besides being renewable, these fuels burn very cleanly. Interestingly, many new cars are now designed to operate on gasoline and a 85% ethanol–gasoline blend (called E-85). They are known as flex fuel cars.

Mentioned previously, another technology that holds promise is the electric vehicle. Electric vehicles are much cleaner than standard gasoline-powered engines.

Another technology for electric cars that some people think holds tremendous promise is the fuel cell. This device produces electricity used to run electric motors in cars, home generators, and other applications. Fuel cells are powered by hydrogen that can be derived from renewable resources (such as water) and nonrenewable fuels (gasoline, natural gas, and methanol). Unfortunately, it takes energy to make hydrogen. It is three to four times more efficient to use that energy (electricity) to power an electric vehicle.

Yet another fuel that holds great promise is biodiesel. Derived from vegetable oil, biodiesel burns in conventional diesel engines and produces a fraction of the air pollution of a conventional oil-based diesel. Vegetable oil is another fuel that can be burned in diesel engines with only minor modifications. For more on green cars and green fuels, you may want to read one of my newest books, *Green Transportation Basics,* published by New Society Publishers in 2010.

> ### KEY CONCEPTS
>
> Alternative fuels that burn cleanly and are renewable could also help reduce many problems created by the gasoline-powered automobile.

Phase 2: From Road and Airports to Rails, Buses, and Bicycles

The shift to clean-burning cars is essential, but like improvements in efficiency, it may be only a stopgap measure. By the year 2013, there were about 500 cities in the world that contained over a million people, and 30 of these housed over 10 million people. As cities grow, large numbers of commuters will very likely have to shift to **mass transit**—buses, commuter trains, and light rail (single-car trains)—to reduce highway congestion.

For a case study on one city's successful use of mass transit, see **Spotlight on Sustainable Development 17-2**.

Besides reducing highway congestion by taking cars off the road, mass transit is inherently more efficient than the automobiles. A diesel bus used to transport people, for instance, only gets 4 miles per gallon, but with 40 people on board the passenger miles per gallon of fuel increases to 160. Light rail, commuter trains typically run on electricity, perform extremely well, too.

Go Green

When commuting to school or work, take a bus or light rail. Both use much less energy than a single-occupancy vehicle.

Another option is carpooling and vanpooling—encouraging people to ride together. Carpooling is much more efficient than driving alone. For instance, the average new car on the road today gets only about 24 miles per gallon. If there's one person on board, that's 24 passenger miles per gallon. With two people, it jumps to close to 48 passenger miles per gallon (considering the additional weight being carried). Although that's an improvement it is not as efficient as light rail or bus transport. With three on board, it increases to nearly 72 passenger miles per gallon. If there are four passengers, it jumps to 96 passenger miles, still half that of the bus.

Given the relative efficiency of mass transit—and given declining fuel supplies, congestion, and pollution—many urban residents may within the next few decades give up their second and third automobiles. They will turn to more efficient and less polluting forms of transportation, among them commuter trains, light rail, and buses. Some mass transit users may even join car-share programs. Participants in commercial car-share programs lease cars by the hour for special trips but rely on mass transit for most of their daily trips. To learn more about car-sharing programs, see **Spotlight on Sustainable Development 17-3**.

The shift to mass transit is inevitable over the coming decades, but cities will have to improve their systems, making them much more rapid and convenient. With declining automobile traffic, cities may be able to convert highway lanes to light rail lines. Median strips could be converted to light rail systems serving surrounding suburbs. Fast, efficient buses could carry commuters from their homes to outlying rail stations, where people board high-speed trains that transport them rapidly to urban centers.

To be profitable, high-speed rail requires high-participation, high-density population in outlying areas, and a large central business district. In order to achieve this, it may be necessary to *densify* new suburbs—that is, to foster a more compact form of development, as described in the previous section.

Another idea gaining interest is in making new subdivisions more like small towns—that is, making them more self-contained. Known as new urbanist developments or *neotraditional towns*, these new communities are being built so that residents live within walking or biking distance of shops, stores, and small office buildings—which are often part of the new development itself. At this writing, nearly 100 are in various stages of development in the United States. They are also called new urbanist communities.

Urban centers can also densify by converting empty parking lots (produced by the decline in automobile use in cities) into office buildings. This makes mass transit more

17-2 Curitiba, Brazil—A City with a Sustainable Vision

Most modern cities have grown up around the automobile. City planners have literally shaped their cities around major transit corridors. Thus, the location of subdivisions, industrial facilities, and services has been largely determined by roads and highways, access routes primarily traveled by people in automobiles.

Proving that there is an alternative path is a city that's gaining wide recognition: Curitiba, Brazil. Lying near the east coast in the southernmost part of Brazil about 800 kilometers (500 miles) south of Rio de Janeiro, Curitiba is a showcase of wise planning and sustainable design principles that have served the city and the planet well.

Since 1950, Curitiba's population has grown from 300,000 to over 2.1 million. For most cities, this rate of growth, combined with poor planning, would have been a prescription for disaster. Poverty, pollution, crime, and highway congestion would have been the inevitable results—not so in Curitiba.

The city was blessed with a visionary mayor in the 1970s, Jaime Lerner, who adopted proposals first made in the 1960s to plot a future for the city based on mass transit, ecological design, appropriate technology, and public participation—all essential elements of sustainability. Lerner was an architect and planner of extraordinary vision.

As witness to his foresight, consider this: Today, 1.3 million commuters travel into Curitiba each day to go to work. Three-quarters of these people travel by bus. This remarkable feat is made possible by an extensive, privately operated bus system that transports people in and out of the city with remarkable speed.

In most cities, bus systems are notoriously slow. Although they move large numbers of people, they bog down in heavy traffic on city streets. In Curitiba, however, buses move passengers into and out of the city at such rapid speeds for several reasons. First, the city has constructed five major roadways that penetrate into the heart of the city. Each of these roadways has two lanes designated for buses. In addition, bus stops are equipped with special devices called raised tubes, which allow passengers to pay before they get on the bus (**FIGURE 1**). This greatly speeds up the boarding

process that slows down many a bus in the United States and other countries. Extra-wide doors also contribute to the speed of boarding. Double- and triple-length buses increase the system's capacity. Together, these innovations reduce the transit time by one-third.

Curitiba has made it possible to move in other directions as well. Many smaller bus routes connect residential areas with the main transit corridors so that one can travel about freely. Commuters can take a bus to the main route, hop on an express bus, and be downtown in record speed.

This system of transit not only makes good sense from an environmental standpoint—because buses transport people with fewer resources and much less pollution than the automobile—it also makes sense from an economic standpoint. It's much cheaper than a subway system.

So that the less fortunate, Curitiba's poor, can gain access to the system, the city purchased land along major corridors, which was developed for low-income families.

Curitiba has more to boast about than its mass transit system. It has established an extensive network of parks along natural drainages. This not only provides residents with someplace to escape from the buildings and roadways, it reduces damage from flooding. Prior to the establishment of this system, developers often built homes and other structures in drainage areas.

FIGURE 1 Raised tube bus stop in Curitiba.
© John Maier, Jr./Peter Arnold, Inc.

When floods came, many a home was damaged. Today, this system of parks with specially constructed ponds has nearly eliminated flooding and saved the city millions of dollars in engineering and construction costs. Low-tech solutions work and save money.

Curitiba promotes participation and cooperation, too. The city recognizes that solutions require the participation of many sectors, including business, government, community groups, and others. Fortunately, other cities are beginning to learn from Curitiba.

efficient and affordable. Unfortunately, the economics of mass transit is currently skewed by massive subsidies to the automobile and gasoline industry. According to national statistics, the automobile is subsidized to the tune of about $300 billion a year—or about $1,500 per car. This subsidy includes expenses for police protection, traffic control, city-paid parking, and other things—expenses that are paid out of general tax revenues. Oil is also heavily subsidized. Because of this, sustainable strategies such as mass transit generally have a difficult time competing with the automobile. Some critics say that the playing field is tilted heavily in favor of the automobile. Removing the hidden subsidies from oil and automobiles could make mass transit compete more favorably with the automobile.

Jet travel could shift in the coming decades as well. The Swiss have, for example, levied an extra tax on aircraft that do not meet air pollution standards. In 1997, Denmark introduced a $15 fee on all domestic flights that has encouraged people to travel by train instead. High-speed rail, which is growing in popularity in Europe and Japan, could help reduce air travel and shift passengers to a much more efficient form of transit.

The Bicycle For many urban dwellers, the bicycle is not a viable commuting option. Streets are too hilly, or workers live too far from work for practical commuting. In other places, cold winter climates prevent commuting except during the spring, summer, and early fall.

In some cities, however, the bicycle could play a significant role in transporting people. Where climates are mild, streets are not too crowded, and people live relatively close to work, the bicycle can transport surprisingly large numbers of people. Investments that promote bicycle commuting represent one of the cheapest options available to cities and towns.

For decades, the bicycle has been a major means of transportation in many European and Asian countries. In Beijing and other major Chinese cities, bicycles once outnumbered cars many times over, although bikes are being squeezed out by automobiles as China's economy improves. In some Asian cities, half of all trips are made by bike. Europe is also relatively bicycle friendly. In some cities, bicycles account for 20 to 30% of all trips. In the United States, in contrast, bicycles account for less than 1% of all trips. In Portland, Oregon; Austin, Texas; and other cities, however, bikes are becoming more popular thanks to a "fleet" of city-owned bikes that are left on the sidewalk for anyone who wants them (**FIGURE 17-8**). If you need to travel five blocks and don't want to walk, you hop on a bike and pedal to your next meeting or to lunch. Anyone who wants to use the bike you leave by the side of the road can have it.

Following the example set by Europe and Asia, some cities in the United States have laid out extensive bike paths for commuters. Davis, California, is a leader in promoting bicycle transportation. Today, 30% of all commuter transport within the city is by bicycle. Some streets are closed entirely to automobile transport, and 65 kilometers (40 miles) of bike lanes and paths have been established.

Bicycles won't replace cars, buses, and trains, but they can augment them—in some places more than others. Unfortunately, bicycle sales are declining in Asia because of an increased interest in motorized transport, mostly motorbikes and cars. In some countries, official policies are discouraging bicycle transport in an effort to make travel by car easier.

FIGURE 17-8 THE YELLOW BIKES OF AUSTIN, TEXAS. These bicycles are left on the street for anyone who wants to use them. Riders take them to their destination and leave them for someone else to use.
© AP Photos.

Economic Changes Accompanying a Shift to Mass Transit

The automobile industry is the world's largest manufacturing endeavor and supports a number of other economically important industries. Manufacturers of rubber, glass, steel, radios, and numerous automobile parts will also feel the impacts of the shrinking automobile market. So will the service sector: gas stations, automobile dealerships, and repair services. The shift to more efficient forms of transportation is likely to lead to significant shifts in the world's economy.

Today, 20 cents of every dollar spent in the United States is directly or indirectly connected to the automobile industry and its suppliers. Eighteen cents of every tax dollar the federal government collects comes from automobile manufacturers and their suppliers.

Although shifting toward a sustainable transportation system could result in a dramatic shift in our economy, experts point out that some of the steel and glass now destined for autos will be used for buses and trains. Many automobile workers will very likely find jobs in plants that produce buses and commuter trains. Many mechanics could shift as well to service the new fleet of more efficient vehicles. Some workers, however, may be forced to find employment in new areas. Helping them adjust to the changes is an important task.

Studies suggest that the employment potential of mass transit, like other sustainable strategies, exceeds that of the current automobile-based economy. A study in Germany showed that spending $1 billion on highways yields 24,000 to 33,000 (direct and indirect) jobs. The same amount spent on mass transit produces 38,000 to 40,000 jobs.

A sustainable transportation system is possible, but it will require a significant restructuring of the current, unsustainable system. Making that transition will require foresight and considerable political will. Many cities are already

© hugolacasse/ShutterStock, Inc.

SPOTLIGHT on Sustainable Development

17.3 Car Sharing: On the Road—Cheaper and Greener

Do you own your car—or does it own you? If you're like most people, you'll answer "yes" to both questions. You own a car that gets you where you want to go, but you're a slave to it, spending considerable amounts of money (and time) supporting your driving habit.

According to the American Automobile Association, each month Americans pay an average of $700 to own and operate a car. Payments for a new car can easily run $300 to $500 per month. Insurance adds from $75 to $150 per month to the cost of car ownership, and then there's gasoline, costing another $100 to $200, and maintenance expenses.

Fortunately, if you want the convenience of a car without the expense of ownership, there's an option for you—it's called car sharing.

Popular in Europe for more than three decades, car sharing has gained momentum in the United States. One company, Zipcar, offers car share programs in numerous cities across the country, including Boston, New York, Chicago, Denver, Los Angeles, Seattle, Washington, D.C., and Portland, Oregon. You can log on to their website to see the impressive list (www.zipcar.com). Hundreds of universities also participate, saving students money and helping curb campus traffic congestion and parking problems.

Most car share programs require an application and annual membership fee, which averages about $75. Once you join, vehicles are available for 1 hour to several days. (Special arrangements can be made to rent cars for long trips.) Businesses, families, and individuals can all participate. Car share programs screen applicants using age and traffic violation criteria to eliminate risky clientele. Qualified members are covered by comprehensive and liability insurance when behind the wheel.

Urban car share programs place their cars conveniently throughout the city, in reserved parking lots or spaces. Members pay a small hourly fee to use the car, typically $8 to $10, and a per-mile charge. Some programs give members a certain number of free miles before charging for mileage.

Car share companies use online and phone reservation systems, which allow you to reserve a car in a few quick keystrokes or with a phone call. A computer tells you where the car is and its license plate number. You show up at the site, hold your membership card next to the windshield, where it is read by a scanner, and the doors unlock. The car's onboard computer sends a signal to company headquarters, indicating your rental period has begun, and activating a billing record.

When you're done, you return the car to its parking space, lock it, and leave. Your credit card is billed monthly for usage. The program pays for gas, although members are responsible for filling the tank when the gauge drops below the one-quarter mark—using a company credit card. And what if you have a fender-bender? According to former Flexcar spokesperson John Williams, members pay half of the deductible ($500) if they caused the accident; otherwise, there's no penalty. (Policies vary among different organizations; be sure to inquire first.)

Car sharing programs make it easy to choose a vehicle to meet your needs by offering a wide range of vehicles, from small, efficient commuter cars such as the hybrid-electric Honda Civic or Toyota Prius to larger vehicles for special uses, such as Ford pickups and SUVs. According to Zipcar, more than half of their members say that they tried the service for the opportunity to get behind the wheel of many different makes and models of cars, including hybrid-electric vehicles.

How much you save depends on your driving habits and needs. Studies show that customers save $200 to over $400 a year.

To determine whether a car sharing program makes economic sense for you, Zipcar offers an online savings calculator. Just click on "Run the Numbers" on their website and enter the current or projected costs of your vehicle, including monthly payments, insurance premiums, fuel costs, parking, and maintenance costs. The program calculates the monthly and annual costs of car ownership and computes car sharing savings.

Beyond your pocketbook, car sharing also offers some environmental savings. Personal vehicles produce a large portion of the nation's annual emissions of carbon dioxide and other pollutants, contributing to global warming and localized smog. Accommodating the growing fleet of vehicles as we pave the planet also results in the loss of huge amounts of open space, farmland, and wildlife habitat.

Before Flexcar and Zipcar merged Zipcar had been responsible for removing more than 10,000 cars from U.S. city streets and highways. Their survey also showed that car share members drive an average of 4,000 fewer miles (6,437 km) each year, compared with their habits before joining the program. Most people drive less when they have to pay a per-hour or per-mile fee, and studies have shown that car sharing folks are more likely to combine trips, take mass transit, bike, or walk when it's convenient. Combined with cities that have good mass transit systems, car share programs are highly effective at reducing congestion and pollution in urban settings.

If car sharing hasn't come to your town, consider setting up a program yourself. Across the country, groups of friends, neighbors, and colleagues have established their own car share programs by using cars already owned by individual members of the group or by purchasing cars together.

To make a community car sharing program run smoothly, members should consider providing convenient locations to park the vehicles, and draft agreements on buying fuel, accessing keys, and servicing and insuring the vehicles. A booking system should also be established. The Eugene, Oregon BioCarShare program uses Online Resource Scheduler free software as their scheduling tool. Members log on at the website with their user name and password to reserve a car. Set up as a cooperative, BioCarShare requires members pay a joiner's fee, which is refundable, and a small monthly membership fee. Like most programs, they also require that drivers pay in proportion to their use, per hour or per mile. Car clubs such as these usually operate with standard insurance coverage—as long as the group or any of its members makes no profit. In most states, car clubs apply for insurance in the club's name and can list four or five people on a single policy.

Car sharing is an idea that offers the best of both worlds. It ensures people access to transportation while saving them considerable sums of money, and, it is good for the environment.

Adapted with permission from Dan Chiras, *EcoKids: Raising Children Who Care for the Earth,* Gabriola Island, British Columbia: New Society, 2004 making changes needed to effect this transition. Denver, Colorado, for instance, has already installed one successful light rail system and is building a second one to serve western suburbs.

Creating sustainable cities and towns is part of the challenge outlined in this text. It will take many years. The technology and knowledge required to make the changes are not barriers, but the political will and the costs of such endeavors surely stand in the way of such a massive shift. Incrementally, however, cities, businesses, and citizens are beginning to make the changes required to create a more sustainable human presence. Automobile manufacturers are making and selling tens of thousands of more efficient hybrid cars and trucks. The hypercar is currently under development. Biodiesel is currently sold at 10,000 gas stations in the United States, and numerous cities such as Denver are dramatically expanding their light-rail systems. Growth management strategies are popping up in the most unlikely places. The green living tips (Go Green boxes) in this and other chapters list personal actions that you can take to do your part in creating a more sustainable system of transportation and a better future.

KEY CONCEPTS

A shift away from the automobile will have serious repercussions on the global economy, but much of the slack could be taken up by a shift to the manufacture and maintenance of alternative transportation modes such as buses.

© hugolacasse/ShutterStock, Inc.

SPOTLIGHT on Sustainable Development

17.4 Masdar City: A Solar Oasis

In an effort to create a sustainable future, communities throughout the world are looking for ways to reduce their consumption of energy, water, land, and other natural resources. Entirely new green communities are being built, and some communities are being rebuilt green in the wake of disasters like the city of Greensburg, Kansas, that is building green after a devastating tornado ripped through their town in May, 2007.

The latest effort to build a green community from the ground up is occurring in a most unlikely place, the Persian Gulf. Groundbreaking for the new solar, car-free city took place in February, 2008. This nearly self-contained small desert city, called Masdar, is designed to be home for 50,000 people and is being built in the desert next to Abu Dhabi's international airport. The city will serve as a "hub for academic and corporate research on nonpolluting energy technologies," according to Andrew Revkin, *New York Times* science writer.

Revkin continues, "The 2.3-square-mile community, set behind walls to divert hot desert winds and airport noise, will be car free, according to the design by Foster + Partners, the London firm that has become a leading practitioner of energy-saving architecture."

Car-free Masdar City will feature narrow pedestrian streets shaded by canopies made of photovoltaic panels. The city will produce all of its electricity from abundant sunlight.

Water will be supplied by a solar-powered desalinization plant that extracts salts and other impurities from abundant seawater. Food will be grown in greenhouses. Waste will be composted and recycled.

Phase 1 of Masdar City is expected to be completed by the end of 2015 and include a graduate-level academic research center associated with the Massachusetts Institute of Technology. It will also include extensive commercial and residential buildings. The city will house 7,000 residents.

The world of politics is always twenty years behind the world of thought.

—**John Jay Chapman**

CRITICAL THINKING

Analysis

Hydrogen is nearly an ideal fuel. It can be made from an abundant resource, water, and is renewable. It produces little, if any, pollution when burned or used to power a fuel cell. In your research, however, you may have discovered that it takes a lot of energy to make hydrogen from water or to strip hydrogen atoms off other molecules such as methane (in natural gas). In fact, it takes a lot of electrical energy. Where does that energy come from? Burning coal or nuclear fuels. The combustion of coal and the use of nuclear fuels both create serious environmental and health problems.

Would it be better to use the electricity directly to power a car or truck, rather than use it to split water to make hydrogen to power a fuel cell to make electricity to run a car or truck? Actually, yes.

You may have found studies showing that using electricity directly—using it to power an electric vehicle—is three to four times more efficient than using electricity to create hydrogen to create electricity in a fuel cell to drive an electric motor in a car.

Bear in mind, too, that there are no sources of free hydrogen we can tap into. So, we'll always need to make it from other molecules, a process that takes a lot of energy.

CRITICAL THINKING AND CONCEPT REVIEW

1. List all the human systems you have relied on since you woke up this morning.
2. Where does the water you drink come from? Where does your garbage go? Is it recycled? Is it landfilled? Is it incinerated? Where does the wastewater from your domicile go?
3. Using your critical thinking skills and the knowledge you have gained in your coursework and reading, critically analyze the following statement: "We must restructure human systems to make them compatible with natural systems."
4. What is meant by the statement "Just because a system is functioning well doesn't mean that it is sustainable"? Give some examples.
5. List and describe several reasons why most if not all human systems are currently unsustainable.
6. When most people think about creating a sustainable future, they think about designing anew—that is, creating new superefficient homes and autos. Is this sufficient?
7. Critically analyze the following comment: "Our transportation system is just fine. My commute hasn't changed very much. Air pollution levels are down because of greater efficiency in automobiles and pollution control devices. What's everyone so concerned about?"
8. What are the traditional functions of land-use planning? How is sustainable land-use planning different?
9. Compare and contrast compact development and dispersed development patterns according to the following criteria: use of land, cost of infrastructure (roads, bridges, and so on), feasibility of mass transit, and air pollution.
10. Describe corridor and satellite development. What are the advantages and disadvantages of each?
11. Imagine that you are a developer. You are about to draw plans to develop a 80-hectare (200-acre) piece of property. Make a list of ways to make the development as environmentally sustainable as possible. Try doing this exercise by addressing one system at a time—for example, energy, waste, water, and transportation.
12. How can differential tax rates, the purchase of development rights, and making growth pay its own way (ending the subsidy for growth) be used to promote more sustainable land use?

13. Using your critical thinking skills, analyze the following assertion: "Land reform in the developing nations will help create a sustainable future."
14. The text outlined a simplified version of a plan to shift the developed nations such as the United States to a more sustainable system of transportation. Describe this plan and point out its strengths and weaknesses. How could the weaknesses be eliminated?
15. Debate the following statement: "A shift to a transportation system based in large part on mass transit will devastate the global economy and put thousands of people out of work."

KEY TERMS

cohousing
compact development
corridor development
development right
differential tax rate

dispersed development
greenbelts
land-use planning
making growth pay its own way
mass transit

satellite development
urban sprawl
zoning regulations

LEARNING TO LIVE WITH THE EARTH'S CARRYING CAPACITY

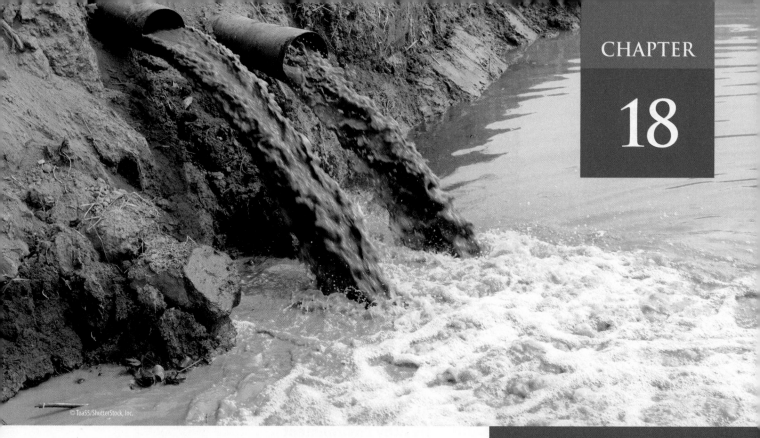

Principles of Toxicology and Risk Assessment

Life is a perpetual instruction in cause and effect.

—Ralph Waldo Emerson

LEARNING OBJECTIVES

After reading and studying the material in this chapter, students will be able to understand, knowledgably discuss, and, when appropriate, apply the information learned in this chapter, including:

1. Key principles of toxicology
2. How toxic chemicals affect living organisms, including humans
3. How toxic chemicals enter our bodies
4. Factors that affect the toxicity of a chemical substance or element
5. The causes of mutations, cancer, and birth defects and their effects on humans
6. How toxic chemicals can affect human reproduction
7. Sources of exposure and effects of exposure to asbestos and electromagnetic radiation
8. Ways to reduce or eliminate our exposure to toxic chemicals
9. Risk and risk assessment

CRITICAL THINKING

Exercise

You are an environmental manager for a major corporation. Your boss tells you that the company needs to cut back on toxic pollutants emitted into the air and water. If you don't, the company could face heavy fines. You turn to your staff and ask them to come up with a plan to reduce pollution. In a week, they come back with a proposal that calls for the addition of a $20 million pollution control device. You present the proposal to your boss, who says that the company can't afford it. He argues that it would be cheaper to continue releasing toxic pollutants and be fined than to comply with the law. What would you do? Would critical thinking help you out of this mess?

All across the globe, companies are taking a new approach to the problem of hazardous chemical wastes. Many companies are finding ways to reduce the amount of hazardous chemical wastes they produce by simple modifications of their processes. Others are finding ways to recycle hazardous wastes, generating profit at the same time. This approach, called **pollution prevention,** is like energy efficiency. It's a preventive action that is not only good for the environment, but also great for a company's bottom line. This encouraging shift has been stimulated in part by costly regulations, but also by business leaders who realize that industry is partly responsible for the deteriorating condition of the planet and that they can play a significant role in solving the problems we face. It has also been inspired by a realization that it costs less—much less—to prevent pollution than to generate pollution and then try to treat it to render it harmless. Despite progress, American industries released 4.09 billion pounds of hazardous materials into the air and water and onto land in 2011 according to the Environmental Protection Agency's (EPA's) Toxic Release Inventory.

This chapter deals with toxic substances, laying the groundwork for our study of pollution. It examines principles of **toxicology,** the study of toxic chemicals or **toxicants**, chemicals that adversely affect living organisms. It discusses the effects of toxic chemicals and several means of reducing exposure of people and the environment to these substances. It concludes with an introduction to a growing field, risk assessment, which helps us

18.1 Principles of Toxicology

Approximately 100,000 chemical substances are sold commercially in the United States. These chemicals are used to produce a long list of products, including cosmetics, food additives, and pesticides to which humans are routinely exposed. In the United States, chemical production and use skyrocketed after World War II. Of the commercially important chemicals, however, only a small number—perhaps 2%—are known to be harmful. Nonetheless, this small percentage amounts to hundreds of potentially dangerous chemicals that can pose a threat to workers and to the general public.

Go Green

Use environmentally friendly cleaning agents to reduce your exposure to toxic chemicals and to reduce water pollution.

Perhaps the biggest problem with toxic substances is our lack of knowledge about their effects. The National Academy of Sciences notes that fewer than 10% of U.S. agricultural chemicals (mostly pesticides) and 5% of food additives have been fully tested to assess long-term health effects. Testing potentially harmful substances is a costly and time-consuming task, made more difficult by the 700 to 1,000 new chemicals entering the marketplace each year.

KEY CONCEPTS

Many thousands of chemicals are produced each year in industrialized nations, but only a small percentage pose a risk to humans. Unfortunately, only a small fraction of these toxic chemicals has been fully tested to assess their toxicity.

The Biological Effects of Toxicants

People are exposed to toxic substances at home (**TABLE 18-1**), at work, or in the out-of-doors. Toxicants are present in the food we eat, the water we drink, the air we breathe, the furniture we buy, and even in the clothes we wear. These substances come from natural and human sources. Toxic substances also affect a wide range of plants and animals. This chapter deals primarily with effects on humans, although the principles you will learn apply to all living things.

In many cases, people have little control over exposure. Polluted air from nearby power plants or highways, for instance, exposes them to dozens of potentially harmful substances. In other cases, however, we intentionally expose ourselves to harmful substances, such as toxic fumes from cleaning agents.

Bear in mind, though, that exposure to a toxic substance does not necessarily mean that one will be adversely affected. The effect a toxic substance has on us, if any, depends on

TABLE 18-1

Common Toxicants Around the Home

Ibuprofen and Tylenol

Gasoline

Rubbing alcohol

Bleach

Mineral spirits

Many cleaning agents

many factors. Before we examine those factors, let's study some key characteristics of toxic effects.

KEY CONCEPTS

Humans are exposed to potentially toxic substances in virtually every aspect of our lives, but exposure does not necessarily mean that one will be adversely affected.

Immediate and Delayed Toxicity Toxicants produce a wide variety of effects. Some stimulate immediate effects. Some immediate effects are subtle, such as a slight cough or headache from urban air pollution. Others can be pronounced, such as the violent convulsions induced by exposure to certain insecticides. As a rule, immediate effects disappear shortly after the exposure ends; they are generally caused by fairly high concentrations of chemicals that result from short-term exposures.

Other toxic substances produce delayed effects such as cancer or birth defects. They may occur months to years after exposure and usually persist for years, as in the case of emphysema caused by cigarette smoke and air pollution. Delayed effects often result from low-level exposure over long periods (chronic exposures). It is important to note, though, that short-term exposures may also have delayed effects. A one-time exposure to certain cancer-causing agents, for example, can cause cancer that appears many years later.

KEY CONCEPTS

Toxic substances may produce immediate effects, ranging from slight to severe, or delayed effects, with a similar range of severity.

Local and Systemic Toxic Effects Still another important distinction regarding toxic chemicals is their site of action. Some toxicants exert local effects, most often at the site of contact. Certain industrial chemicals,

for instance, cause skin rashes. Other toxic substances get into the body and circulate to many different sites, where they exert their effects, often on entire organ systems, such as the brain and spinal cord (central nervous system). These effects are described as systemic and are the most common type. But how do toxicants cause their effects?

KEY CONCEPTS

Toxic substances can exert their effects locally, often at the site of exposure, or systemically—that is, in an organ system or even throughout much of the body.

How Toxicants Work

Toxic substances exert their effects at the cellular level in three major ways: First, some toxicants bind to enzymes, the cellular proteins that regulate many chemical reactions in the body. A disturbance of enzymatic activity can seriously alter the functioning of an organ or tissue. As examples, mercury and arsenic both bind to certain enzymes and block their activity. Second, some toxic chemical substances bind directly to nonenzyme molecules in the body, upsetting the chemical balance. Carbon monoxide, a pollutant from automobiles, gas stoves, leaky furnaces, and other sources, for example, binds to hemoglobin in the red blood cells. This, in turn, blocks their ability to bind to oxygen. This decreases the transport of oxygen throughout the body and can lead to death if levels are high enough. Third, some toxicants interact with the genetic material of cells, causing mutations, potentially harmful changes in the structure of the DNA (described shortly). As you will learn in a later section, mutations can lead to cell death, cancer, and birth defects.

KEY CONCEPTS

Toxic substances may bind to enzymes and other important molecules, including the genetic material, DNA. Binding to these chemicals alters cellular function, sometimes in profound ways.

Factors That Affect the Toxicity of Chemicals

Predicting the harmful effects of chemicals is no easy task. Age, sex, health, and a variety of other factors contribute to the final outcome. Consider the case of a family of six living near a Canadian lead and zinc smelter that released large amounts of lead. Although each member was exposed to high levels of lead, the symptoms varied. For example, the father and a 4-year-old boy suffered from acute abdominal pain and pancreatitis. The mother developed a nervous system disorder. Two other children experienced convulsions, and the last developed diabetes. Differences in age, sex, and genetic composition lead to such variations in toxic effect. They also make studies of the health effects of toxic agents very difficult to perform and can make the results difficult to interpret.

Complicating matters further, individuals can develop tolerance to certain toxicants. In other cases, they can become **hypersensitized** to them— that is, extremely sensitive to tiny doses. People exposed to formaldehyde, a chemical released from resins found in furniture, made from particleboard, plywood, and other products, are often tolerant of the chemical at high levels at first, but over time, they become sensitive to extremely low levels.

Go Green

Avoid products containing formaldehyde like particle board used to make book shelves and other types of furniture.

Numerous factors influence the effects of a given chemical, the four most important being (1) dose, (2) duration of exposure, (3) biological reactivity of the chemical in question, and (4) route of exposure.

KEY CONCEPTS

The toxicity of a chemical (how toxic it is) is determined by a variety of factors, such as the route of entry, dose, length of exposure, and age, sex, and genetic makeup of the individual.

Chemical Reactivity, Dose, and Duration of Exposure Chemicals vary considerably in their reactivity—how readily they react with biological molecules. Highly reactive chemicals generally have a greater effect on the body than those that are less reactive. In addition, the effect is also determined by the dose, the amount to which an individual or animal or plant is exposed. In general, the greater the dose, the greater the effect.

To study the effect of dosage, toxicologists often expose laboratory animals to varying amounts of a substance, determining the response at each level. The resulting graph is called a **dose-response curve** (**FIGURE 18-1**). To compare one chemical with another, toxicologists often determine the dose that kills half of the test animals—that is, the lethal dose for 50% of the test animals or **LD_{50}**. By comparing LD_{50} values, scientists can judge the relative toxicity of two chemicals. For example, a chemical with an LD_{50} of 200 milligrams per kilogram of body weight is half as toxic as one with an LD_{50} of 100 milligrams. Thus, the lower the LD_{50}, the more toxic a chemical.

The **duration of exposure** is the amount of time an individual or laboratory animal is exposed to a toxic substance. Exposures generally fall into two categories: **acute,** or short-term exposures, generally last less than 24 hours; **chronic,** or long-term exposures, last more than 3 months. Obviously, these are very broad categories, and many intermediate exposure possibilities exist. Studies show that for many toxicants, acute exposures result in very different effects than those resulting from repeated (chronic) exposure. An acute exposure to benzene, for instance, may result in a transient bout of depression. Repeated, chronic exposure may result in leukemia, a cancer of the white blood cells.

KEY CONCEPTS

The effect a chemical has on the body is determined by its reactivity, dose, and the duration of exposure. Humans are exposed to many different toxic chemicals at different doses and for different lengths of time.

Routes of Exposure The toxicity of a chemical is also dependent on the route of exposure—that is, how it enters our bodies—or the bodies of any animals. Three routes

Approximate Acute LD_{50}s of a Variety of Chemical Agents	
Agent	**LD_{50} (mg/kg)**
Ferrous sulfate	1500
Morphine sulfate	900
Phenobarbital sodium	150
DDT	100
Picrotoxin	5
Strychnine sulfate	2
Nicotine	1
d-Tubocurarine	0.5
Hemicholinium-3	0.2
Tetrodotoxin	0.10
Botulinus toxin	0.00001

FIGURE 18-1 Dose-response graph for two chemicals with differing toxicities. The LD_{50} is the amount of chemical that kills one-half of the experimental animals within a given time. The higher the LD_{50} value, the less toxic the chemical is. See the chart at right.

Chapter 18: Principles of Toxicology and Risk Assessment **425**

of exposure are most common: inhalation, ingestion, and dermal exposure (**FIGURE 18-2**).

Inhalation exposure, as the name implies, results from breathing a chemical vapor—that is, a chemical that is suspended in the air we breathe. Smokers inhale many potentially toxic substances, as do workers in chemical factories or commuters in heavy traffic. Toxic substances that are inhaled enter the bloodstream via the lungs and quickly become dispersed throughout the body. Inhalation is the most rapid route for chemicals to enter the body.

Ingestion, the next major route of exposure, occurs when toxic substances enter our bodies in the food we eat and the water or liquids we drink. Ingestion is the next most rapid form of exposure. Toxicants, however, can be neutralized by acids in the stomach and enzymes in the small intestine. Those that survive this harsh environment can be absorbed into the blood vessels in the lining of the intestinal tract and may be distributed throughout the body.

The slowest and least effective route of exposure is the skin. Toxic substances spilled on the skin may penetrate this rather thick layer and enter the blood vessels. As a general rule, the more readily a toxic substance is absorbed, the more effect it has.

Age and Health   Although we think of youngsters of all species as resilient, young, growing organisms are generally more susceptible to toxic chemicals than are adults. For example, two common air pollutants, ozone and sulfur dioxide, affect young laboratory animals two to three times more severely than they affect adults. Among humans, infants and children are more susceptible to lead and mercury poisoning than adults because their nervous systems are still developing (**FIGURE 18-3**).

Health is determined by many factors, among them one's nutrition, level of stress, and personal habits such as smoking. As a rule, the poorer a person's health is, the more susceptible he or she is to a toxicant.

Chemical Interactions   Modern society is dependent on thousands of chemicals, and most of us are exposed to many different substances in many different ways. Predicting the effect of these chemicals is therefore very difficult. Further adding to the difficulty of this task is the fact that chemical substances interact in a variety of ways. Some chemical substances, for example, team up to produce an **additive response**—an effect that is the sum of the individual responses (for example, 2 + 2 = 4). Others may produce a superadditive effect, also known as a **synergistic response**—that is, a response that is greater than the sum of the individual ones (2 + 2 = 6). One of the most familiar examples of synergism is the combination of barbiturate tranquilizers and alcohol; although neither taken alone in small amounts is dangerous, the combination can be deadly. Pollutants can also synergize. For instance, sulfur dioxide gas and particulates (minute airborne particles) inhaled together can reduce airflow through the lung's tiny passages. The combined response is much greater than the sum of the individual responses. The synergistic effect of smoking and asbestos is discussed later.

Another fascinating interaction is **potentiation,** which occurs when a chemical with no toxic effect combines with a toxic chemical and makes the toxicant even more harmful. This response can be represented by the equation 0 + 2 = 6. Isopropyl alcohol (rubbing alcohol), for instance, has no effect on the liver, but when combined with carbon tetrachloride, it greatly boosts the toxicity of the latter.

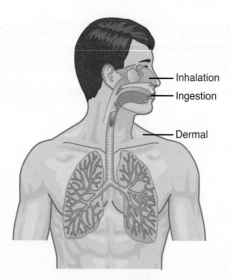

FIGURE 18-2 Three routes of exposure. This illustration shows the three major ways that toxic chemicals can enter the human body.

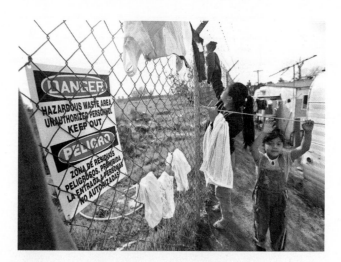

FIGURE 18-3 Children are more susceptible to poisoning.
© Gary Kazanjian/AP Photos.

Certain chemicals can also negate each other's effects, a phenomenon called **antagonism.** In these cases, a harmful effect is reduced by certain combinations of potentially toxic chemicals. This can be represented by the equation 2 + 4 = 3. In mice exposed to nitrous oxide gas, mortality is greatly reduced when particulates are present. Scientists are uncertain of the reasons for this phenomenon.

The many variables mentioned in the previous sections—such as age, nutrition, personal habits, and chemical interactions—often make it difficult to study the effects of pollutants on people. **Epidemiologists,** scientists who study the effects of chemicals and other disease-causing agents on people, must be careful to eliminate as many of these variables as possible so that they don't cloud the interpretation of the results.

Bioaccumulation and Biological Magnification Two additional factors that profoundly influence toxicity are bioaccumulation and biological magnification. **Bioaccumulation** is the buildup of chemicals within the tissues and organs of the

body. It results from at least three factors: (1) selective uptake, (2) resistance to chemical breakdown, and (3) long-term storage. Consider each one, beginning with selective update.

Iodine is actively absorbed from the bloodstream by the thyroid gland of humans. This is called *selective uptake*. What's the problem? Radioactive iodine, present in the milk of cows who feed on grass contaminated by an accident at a nuclear power plant, is selectively concentrated in the thyroids of people who consume the milk (**FIGURE 18-4**). The radioactive iodine remains in the gland and irradiates the cells, sometimes producing tumors.

Radioactive iodine released into atmosphere during accidents

Deposited on grass and eaten by cows

May lead to thyroid cancer

Passed to humans in milk

FIGURE 18-4 Bioaccumulation. Radioactive iodine released from accidents at nuclear power plants is taken up by cows and passed to humans in cow's milk. In humans, it accumulates in the thyroid gland in the neck.

Top photo © Photodisc; middle photo © WizData, Inc./ShutterStock, Inc.; bottom photo © Photodisc.

Heavy metals bioaccumulate because they resist chemical breakdown in the body. Further aiding the bioaccumulation of heavy metals, some of them bind strongly to proteins. Chlorinated hydrocarbons such as DDT are also resistant to breakdown and, because they are fat soluble, tend to concentrate in body fat. Here, they remain for many years.

The accumulation of toxic substances within an organism can have serious effects on an organism, as in the case of radioactive iodine. They may also have adverse effects on an organism that eat them. Consider an example. Scallops and other molluscs feed on material suspended in water. They also selectively take up certain toxic elements (heavy metals) from seawater, such as zinc, copper, cadmium, and chromium. The level of cadmium in scallops in polluted waters may be 2.3 million times higher than that of seawater. While such concentrations may not always cause problems to the organism, they can be toxic to organisms that eat them.

Toxic chemicals can move through food webs and can, in some instances, increase in concentration in the higher-level consumers. This phenomenon is known as **biological magnification**.

Biological magnification is defined as the progressive increase in the concentration of a chemical substance in the organisms of a food chain. As shown in **FIGURE 18-5**, the pesticide DDT is present in water and is taken up by **zooplankton,** single-celled organisms that live in water. Small fish ingest the fat-soluble DDT when they feed on zooplankton, and the persistent insecticide is then concentrated in their body fat. DDT also accumulates in the fat of fish-eating birds such as the osprey. As illustrated, tissue concentrations increase substantially at each level of the food chain.

Biological magnification occurs because DDT is a fat-soluble chemical that is stored in body fat; it is not readily broken down and is not excreted. Because of this, the more fish an osprey eats, the higher its DDT levels become. The concentration of DDT, in fact, may be several million times greater in fish-eating birds than it is in water (Figure 18-5). For humans, magnification that occurs in our food chain may be as much as 75,000 to 150,000 times. Synthetic chemicals such as DDT, some lead and mercury compounds, and even some radioactive substances are all biomagnified.

Biological magnification exposes organisms high on the food chain to potentially dangerous levels of persistent toxicants. Therefore, the presence of this phenomenon is important to keep in mind when judging the risk that a chemical poses to people and to the many species that live among us. Levels that may appear safe can actually be quite dangerous.

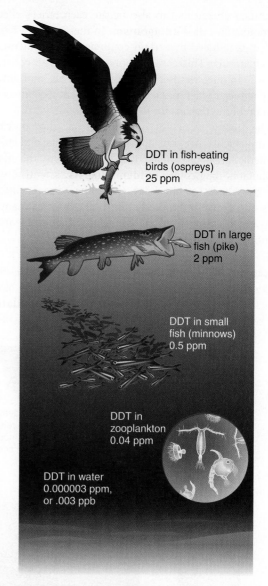

FIGURE 18-5 **Biological magnification.** Concentrations of some toxic metals and fat-soluble organic compounds such as the pesticide DDT increase at higher levels in a food chain. This phenomenon is known as *biological magnification* and results from the fact that fat-soluble molecules tend to be stored in the fat of organisms, rather than being broken down. As a result, they are passed with nearly 100% efficiency from one level of the food chain to the next.

18.2 Mutations, Cancer, and Birth Defects

Much of the concern about toxic chemicals today stems from their effects on the genetic material and from two likely offshoots of these effects: cancer and birth defects.

Mutations

The hereditary material of the cell is contained in a molecule called **DNA,** short for **deoxyribonucleic acid.** This

KEY CONCEPTS

Toxic substances may concentrate in certain tissues and organs, causing local effects. Some may also increase in concentration in the food chain, being the highest in top-level consumers. Because of these phenomena, ambient concentrations of a toxicant may be an insufficient means of predicting a chemical's toxic effects.

material is housed in **chromosomes** found in each cell in the body. The DNA not only passes on traits from parents to offspring, it also controls how cells grow and develop and how they function. DNA can be a target for various chemical and physical agents. Those agents that cause changes, or **mutations**, in the hereditary material are known as **mutagens.** The term *mutation* actually refers to three possible alterations in the hereditary material: (1) changes in the DNA molecule itself, (2) alterations of chromosomes that are visible by microscope (for example, deletion or rearrangement of parts of the chromosome), and (3) missing or extra chromosomes.

Mutations can be caused by chemical substances, such as benzene in gasoline, or by physical agents, such as ultraviolet radiation, X-rays, and other high-energy radiation. In humans, mutations can occur in normal body cells, or **somatic cells,** such as skin and bone. Such mutations occur quite frequently, but they are usually repaired by cellular enzymes. If a mutation is not repaired, and if it affects important genes, the cell may die or become cancerous (described shortly).

The reproductive cells, or **germ cells,** in the male and female gonads are also susceptible to mutagens. Unrepaired germ cell mutations may be passed to offspring. If an ovum with a genetic mutation, for example, is fertilized by a normal sperm, the mutation may be passed on to the embryo. The defective gene may prove lethal, killing the embryo, or it may manifest itself as a birth defect (described shortly), a metabolic disease (a biochemical disorder), or childhood cancer. Some germ cell mutations may not be expressed in the first generation but may appear in the second and third generations. This delayed effect makes it hard for scientists to pinpoint the causes of some diseases.

Genetic mutations are present in about 2 of every 100 newborns. The causes of mutations in humans are not well understood. Abnormal chromosome numbers, responsible for some diseases such as Down syndrome, are related to maternal age (**FIGURE 18-6**). Broken and rearranged chromosomes are also related to maternal age. As women enter their 30s, their chances of having a baby with an abnormal number of chromosomes increase; after age 40, the chances skyrocket. Geneticists once hypothesized that the older a woman is, the greater the chance that she has been exposed to mutagens; hence, the greater the chance that her child will have a mutation. New research suggests that the increased incidence of birth defects in babies of older women may result from the fact that older mothers are, for unknown reasons, more likely to carry a defective fetus to term than younger women.

Other diseases, associated with structural defects in the DNA molecule itself, seem to increase in incidence as the *father* gets older and are not related to the *mother's* age (**FIGURE 18-7**). These defects may be caused by mutagens.

KEY CONCEPTS

Chemical and physical agents can alter the hereditary material creating mutations that can occur in body cells or in reproductive cells. Those occurring in body cells may kill the cells or lead to uncontrolled growth, a cancer. Nonlethal changes in reproductive cells can be passed on to one's offspring resulting in embryonic death, birth defects, cancer, or metabolic diseases.

Cancer

One of the most serious consequences of mutations in cells is cancer (defined shortly). In the United States, recent statistics indicate that one of every two men and one of

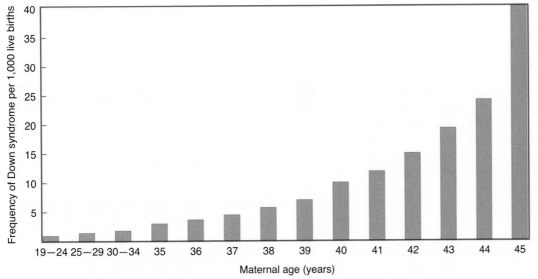

FIGURE 18-6 Down syndrome. The incidence of Down syndrome and several other abnormalities caused by abnormal chromosome numbers is related to the mother's age. The incidence of Down syndrome is about 5 in 1,000 at age 35, but 40 in 1,000 by age 45—about the time of menopause, when women stop ovulating (releasing ova).

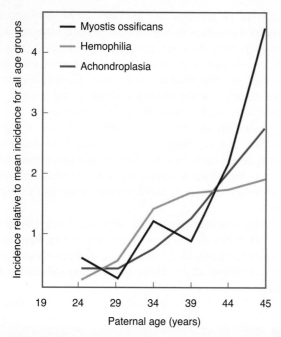

FIGURE 18-7 The incidence of three diseases in newborns: myositis ossificans, affecting bone; hemophilia, affecting blood; and achondroplasia, affecting cartilage. All are caused by DNA damage related to the father's age.

every three women will develop some form of cancer. In the United States, cancer kills about 577,000 people a year, according to the American Cancer Society. Worldwide, the number of deaths is estimated to be about 6 million per year, according to the World Health Organization.

According to two Oxford University scientists, 8,000 Americans die each year of cancer caused by environmental factors such as air pollution. Another 8,000 cancer deaths are attributed to food additives and industrial products, such as pesticides used around the house; 16,000 deaths result each year from occupational exposure to harmful substances. The researchers note, by comparison, that tobacco causes about 87% of the lung cancer deaths in the United States each year, or at least 173,000 deaths.

What is cancer, and how does it form? **Cancer** is a disease produced when a single cell or a group of cells somehow "escape" from mechanisms that control their growth. The unchecked division of these cells results in the formation of a **tumor** or **neoplasm**. In some instances, the tumor is limited to a single expanding mass of cells. This is referred to as a **benign tumor.** Benign tumors can create problems, for example, by compressing nerves or other vital tissues. If a tumor continues to grow, spreading into tissues and organs surrounding it, it is called a **malignant tumor.** Cells may break off from the site of formation, the **primary tumor,** and travel in the blood and other body fluids to distant sites. The spread of cancerous cells is called **metastasis** (meh-TASS-tah-siss). In distant sites the cancerous cells may form **secondary**

tumors. Certain types of cancers spread (metastasize) in characteristic ways. Breast cancer, for example, tends to spread to the bones. Lung cancer typically spreads to the brain.

Malignant tumors are dangerous because they continue to enlarge, demanding huge amounts of nutrients. They also often invade neighboring areas and destroy vital tissues and organs.

Cancers occur most commonly in the rapidly dividing cells of the body, such as those of the skin, bone marrow, lungs, and lining of the intestines. As a rule, nondividing cells such as nerve cells and muscle cells rarely become cancerous. Nonetheless, many different types of cells can form tumors; there are over 100 different types of cancer. Not only are there different types, but many types behave differently, a fact that has complicated efforts to find a cure for this disease. The most likely scenario is that we will discover numerous cures, many of them specific to a particular type of cancer.

What Causes Cancer? Cancer can be caused by a variety of factors—chemicals in the food we eat, physical agents such as X-rays, and biological agents such as viruses. Of these carcinogenic (cancer-causing) agents, the ones of greatest concern to most people are toxic chemicals. Chemicals that cause cancer are called carcinogens (car-SIN-oh-gins).

Carcinogens usually require repeated exposures for many years to stimulate tumor production. Thus, smoking a single cigarette (or even a whole pack) will not increase your chances of contracting lung cancer very much. However, smoking a pack a day for several years will greatly increase the likelihood that you will develop lung cancer. Moreover, the longer and more you smoke, the greater your chances are of getting cancer.

DNA-Reactive and Non–DNA-Reactive Carcinogens Chemicals that cause cancer by altering the DNA of cells are called **DNA-reactive**, or **genotoxic** (jean-oh--TOX-ick) **carcinogens**. A growing body of evidence, however, indicates that some chemical carcinogens may cause cancer without altering DNA directly. Such substances are referred to as **epigenetic** (ep-eh-gin-ET-ick) **carcinogens**.

Some of the epigenetic carcinogens, for instance, are thought to cause hormonal imbalances that lead to rapid cell proliferation. For example, toxicologists have discovered natural chemicals (certain plant-derived substances) and synthetic chemicals (for example, the pesticide DDT) that mimic the actions of the sex hormone estrogen. Estrogen stimulates cell division in the breasts and wombs (uteri) of women. It is widely believed to play an important role in the development of some types of cancer in women, particularly cancer of the breast and uterus. Toxicologists speculate that exposure to estrogen-like chemicals in our diet and the environment may also increase cancer risk.

Go Green

Whenever possible, avoid beverages in plastic bottles. Select drinks in glass bottles to avoid exposure to plasticizers.

Epigenetic carcinogens may act in other ways, too. For example, some chemicals may alter the function of the immune system, severely impairing its ability to recognize and destroy precancerous cells. Other epigenetic carcinogens result in persistent tissue injury. Asbestos fibers, for instance, may end up in cells, where they slice dividing chromosomes.

In recent years, growing evidence has shown that genes can be influenced by chemical modifications of proteins associated with DNA and by chemical modification of the DNA itself—specifically the addition of methyl groups on the DNA. Both of these epigenetic mechanisms can have profound effects on heredity but could also adversely alter cell function, causing disease, perhaps even cancer. Only additional research will tell us how significant these modifications will be.

KEY CONCEPTS

Most carcinogens react with the growth control genes of cells, causing mutations that lead to cancer. Evidence suggests that some other chemical carcinogens cause cancer through mechanisms that don't directly involve the DNA.

Biotransformation, Emotions, and Intrinsic Factors

Interestingly, many cancer-causing chemicals must be chemically modified by enzymes in the body in order to be able to react with the DNA. Nitrites, for example, are found in processed meats such as bologna (as a preservative) and are fairly benign chemicals. However, the liver converts nitrites into carcinogenic nitrosamines. Although the liver usually detoxifies chemical substances and protects us from harm, in this instance it converts a relatively harmless substance into a carcinogen with potentially lethal consequences. This process is called **biotransformation**.

Some studies indicate that emotions may also play an important role in the development of cancer (and other diseases), possibly by acting through the immune system. Researchers at the Johns Hopkins University, for example, studied the incidence of cancer in medical students who took a personality test between 1948 and 1964. The research showed that students who suppressed emotions were 16 times more likely to develop cancer later in life than students who vented their emotions. More research is needed to determine if the cause-and-effect relationship between mental health and cancer is real.

Other factors may also influence one's susceptibility to cancer. For example, nutritional deficiencies may promote cancer of the liver and esophagus. Excessive alcohol intake may promote cancer of the larynx, especially in heavy smokers.

Not every individual who is exposed to a potentially carcinogenic agent develops cancer. Scientists think that the reason for this is that certain intrinsic factors—for example, one's genetic makeup, immune system, and sex—may influence one's susceptibility to carcinogens.

KEY CONCEPTS

Most chemical carcinogens are not directly mutagenic; they must be chemically altered by enzymes in the body to be able to react with DNA. This process is called *biotransformation*.

Birth Defects

A **birth defect** is a physical (structural), biochemical, or functional abnormality. The most obvious defects are the physical abnormalities such as cleft palate, lack of limbs, or spina bifida (incomplete development of the spinal cord, often resulting in paralysis). According to some estimates, about 7% of all U.S. newborns have a birth defect. Others believe that the incidence of birth defects may actually be higher, between 10 to 12%. The reason for this discrepancy is that many minor defects escape detection at birth, among them mental retardation and certain enzyme deficiencies.

Agents that cause birth defects are called **teratogens** (tear-ah-toe-gens); the study of birth defects is **teratology** (from *teratos*, Greek for "monster"). Teratogenic agents include drugs, physical agents such as radiation, or biological agents such as the rubella virus, which causes German measles (**TABLE 18-2**). No one knows for sure what percentage of birth defects are caused by chemicals in the environment.

KEY CONCEPTS

Birth defects include structural and functional defects in newborns, which may be caused by chemical, physical, and biological agents called *teratogens*.

TABLE 18-2

Some Known and Suspected Teratogens in Humans

Known Agents	Possible or Suspected Agents
Progesterone	Aspirin
Thalidomide	Certain antibiotics
Rubella	Insulin
(German measles)	Antitubercular drugs
Alcohol	Antihistamines
Radiation	Barbiturates
	Iron
	Tobacco
	Antacids
	Excess vitamins A and D
	Certain antitumor drugs
	Certain insecticides
	Certain fungicides
	Certain herbicides
	Dioxin
	Cortisone
	Lead

Timing of Exposure Embryonic development can be divided into three parts: (1) a period of early development immediately after fertilization, the pre-embryonic period; (2) a period when the organs are developing, (the embryonic period); and (3) a growth phase, a period during which the organs have formed and the fetus primarily increases in size, the fetal period (1) (**FIGURE 18-8**). Teratogens exert their effects during organogenesis. As Figure 18-8 shows, organs are most sensitive early in organogenesis, and it is during this period that the most noticeable effects occur. Thus, the effects of a teratogen are related to the time of exposure.

Teratogenic effects are also determined by the type of chemical involved. Certain chemicals affect certain organs; for example, methyl mercury damages the developing brains of embryos. Other chemicals such as ethyl alcohol can affect several systems; children born to alcoholic mothers exhibit numerous defects, including growth failure, facial disfigurement, heart defects, and skeletal defects. Like most toxicants, teratogens usually exhibit a dose-response relationship: the greater the dose, the greater the effect.

Fetal development, occurring after organogenesis, may also be affected by harmful physical and chemical agents in our homes and places of work. Some toxic chemicals, for instance, stunt fetal growth and in higher quantities may even kill fetuses, resulting in a **stillbirth** or **spontaneous abortion.**

FIGURE 18-8 Embryonic development and teratogenesis. Schematic representation of human development, showing when some organ systems develop. Sensitive periods are early in development, indicated by the blue part of the bar. Exposure to teratogens during these times will almost certainly cause birth defects. Exposure to potentially harmful chemicals after organogenesis may result in physiological defects, minor anatomical changes, or death, if levels are sufficiently high.

Humans are not the only species whose offspring are altered by environmental chemicals. Numerous studies have shown that environmental pollutants such as selenium and **polychlorinated biphenyls (PCBs)** alter development in bird embryos.

KEY CONCEPTS

Birth defects may occur when an embryo is exposed to a teratogen during the critical period of organ development, organogenesis. Chemicals vary in their effects. Some affect specific organ systems. Others affect a number of systems. Exposure to some toxic chemicals after organ development can stunt growth and kill fetuses, resulting in a stillbirth.

18.3 Reproductive Toxicity

The effects of environmental chemicals on reproduction in humans and other species have become a major concern in recent years. Studies have shown that male factory workers become sterile when exposed on the job to DBCP, a soil fumigant that kills nematodes, a kind of round warm that damages crops. It is no longer used in the United States. Men who routinely handle various organic solvents often have abnormal sperm, unusually low sperm counts, and varying levels of infertility. A wide number of chemicals such as boron, cadmium, diethylstilbestrol (DES), methyl mercury, and many cancer drugs are toxic to the reproductive systems of men and women.

Reproductive toxicants may exert their effects long after exposure. Two researchers from Laval University in Quebec, for example, examined the records of 386 children who had died of cancer before the age of 5. Their study showed that at the time the children were conceived, many of the children's fathers had been working in occupations that exposed them to high levels of hydrocarbons. Some were painters exposed to paint thinners; others were mechanics exposed to car exhaust. This study suggests that hydrocarbons had entered the bloodstream of the fathers at work and may have traveled to the testes, where they damaged the germ cell DNA. The resulting mutations were passed to the offspring.

Go Green

When painting, select low, or better yet no, VOC paints, stains, and finishes.

In 1970, seven women (ages 15 to 22) were diagnosed with clear-cell adenocarcinoma (a type of cancer) of the vagina and cervix, a disease never seen before in women below the age of 30 and usually only seen in women over 50. A study of these women showed a correlation between their cancer and maternal ingestion of a drug known as DES. DES was widely used from the mid-1940s to 1970 in the United States and was administered to women who either had a history of miscarriages or had begun to bleed during pregnancy. Because bleeding is an early symptom of miscarriage, DES was given in hopes of preventing it. (It is now known that DES cannot prevent miscarriage.) Years later,

vaginal cancers and other abnormalities began to appear in the daughters of DES-treated women. Additional research has uncovered reproductive abnormalities in approximately one-fourth of male offspring exposed to DES through their mothers. Problems include small testes, cysts, and low ejaculate volume. No evidence of testicular tumors has been uncovered.

KEY CONCEPTS

Reproduction is a complex process, involving many steps. Chemical and physical agents may interrupt any of these complex processes, interfering with reproduction.

18.4 Environmental Hormones

Hormones are chemicals produced by certain cells in the body. They travel in the bloodstream to other parts and regulate numerous body functions, from reproduction to body temperature to growth. Hormones also help maintain internal constancy, or homeostasis.

The secretion of various hormones can be influenced by certain pollutants such as dioxins. Dioxins are a group of structurally similar chemicals found in some herbicides (chemicals used to control weeds) and in paper products. At high levels, dioxins are thought to be carcinogenic. At low levels, however, dioxins may suppress the immune system by binding to hormone receptors in the plasma membranes of body cells. As a result, some researchers classify dioxins as **environmental hormones.** Environmental hormones include a growing list of chemical toxicants. They can alter a variety of hormonal systems, creating a wide range of effects. For example, studies suggest that a number of common herbicides (the thiocarbamates) may produce thyroid tumors by altering normal hormonal balance.

Environmental hormones alter reproduction and other functions in nonhuman animals, too. Nonylphenols or chemicals that break down to form them, for example, alter the endocrine (hormonal) systems of salmon. Nonylphenols are present in dishwashing detergent, pesticides, and even some contraceptives (spermicides) and end up in waterways. New studies suggest that they can evaporate and travel long distances in the atmosphere, depositing via rain in other areas. Nonylphenols alter fish reproduction and cause other abnormalities. One major problem is that they appear to impair internal hormonal changes that are required by salmon as they migrate from freshwater to saltwater. Unable to adapt to the high concentrations of salt, many of these fish weaken and die. Some ecologists suspect that hundreds of environmental hormones may be responsible, in part, for the decline in numerous species of wildlife—fish, birds, and mammals. Effects may be subtle and difficult to demonstrate.

Another environmental hormone is a group of chemicals called phthalates (pronounced thall-lates). These chemicals are softening agents added to plastics—IV bags, beach balls, plastic milk jugs, and pop bottles, among others. Phthalates act as estrogens and may be partly responsible for the earlier onset of sexual maturity, including early breast development in girls. Exposure of embryos to moderate-to-high levels of four common plasticizers and other chemicals have recently been shown to cause abnormalities in the reproductive organs of newborn boys. Newborns of mothers who had moderate-to-high levels in their blood were more likely to exhibit reproductive organ abnormalities, including smaller penises and undescended testes—testes that fail to enter the scrotum during fetal development. Signs of demasculinization such as these have been observed in laboratory animals exposed to phthalates under controlled conditions.

Scientists believe that phthalates in the bloodstream of mothers may reduce the production of the male sex hormone such as testosterone by human fetal testes. This, in turn, leads to abnormal development of the sex organs. The effects, they postulate, could lead to further detrimental effects later in life, including lower sperm production in adulthood and an increased risk of testicular cancer.

Scientists in Spain have recently discovered that a plasticizer known as bisphenol-A may be responsible for rising levels of late-onset diabetes in developed countries like the United States. This chemical is added to polycarbonate plastic. This plastic is used to line aluminum and steel cans and is used to make many other products like hard plastic water bottles, certain microwavable containers, and baby bottles.

Earlier studies have shown that bisphenol-A leaches from the walls of food containers and is ingested by adults and children. Bisphenol-A is also commonly detected in the human blood where it has the same effects as the female sex hormone estrogen. Early exposure to bisphenol-A could also be at the root of the rising epidemic of obesity and may contribute to diabetes that occurs during pregnancy in some women.

Many species of wildlife could also be affected by these chemicals—even species that live far from human civilization. In a recent study, researchers found that male codfish that live in the open ocean far from sources of human pollution produce an egg-yolk protein normally found only in females. Their research suggests that the stimulation of the production of this female protein in males was the result of exposure to estrogen-like plasticizers. Exposure to such pollutants, researchers fear, may ultimately interrupt male sperm production.

The presence of the protein in fish isolated from pollution sources suggests, according to researchers, that the pollutants may be working their way through the food chain. Because of problems such as this, the EPA now requires companies to screen all new chemicals for potential hormonal effects.

KEY CONCEPTS

Certain pollutants in the environment enter the bodies of organisms and can alter the normal release of hormones, upsetting reproduction and other vital functions.

18.5 Case Studies: A Closer Look

A look at two case studies illustrates how complex human health and pollution issues are and the importance of good science.

Asbestos: How Great a Danger?

Asbestos is the generic name for several naturally occurring silicate mineral fibers. Asbestos is useful because of its flexibility, its great tensile strength, and its resistance to heat, friction, and acid. In the United States and other industrial countries, asbestos was once added to cement to make it more resistant to weather. Asbestos was also applied as a heat insulator on ceilings and pipes in factories, schools, and other buildings. In addition, asbestos was sprayed on walls and ceilings to make them more soundproof and fireproof. It was also once used in the manufacture of acoustic ceiling tiles, vinyl wallpaper, vinyl flooring, brake pads, brake linings, hair driers, patching plaster, and a multitude of other products.

Asbestos fibers can be easily dislodged. Floating in the air, these fine particles may be inhaled into the lungs, where they are neither broken down nor expelled. Remaining in the lungs for life, they produce three disorders: pulmonary fibrosis, lung cancer, and mesothelioma.

Pulmonary fibrosis, or **asbestosis,** is a buildup of scar tissue in the lungs that occurs in people who inhale asbestos on the job or in buildings with exposed asbestos insulation. This disease makes breathing nearly impossible and takes 10 to 20 years to develop.

Exposures to asbestos at low levels, even for short periods, can cause lung cancer. The death rate from lung cancer among asbestos insulation workers is four times the expected rate (**FIGURE 18-9**). Interestingly, the incidence of lung cancer in asbestos workers who smoke is 92 times greater than in asbestos workers who don't, providing a striking example of synergism.

Asbestos is the only known cause of **mesothelioma** (mez-oh-theel-ee-OH-ma), a cancer that develops in the lining of the lungs. This highly malignant cancer spreads rapidly and kills victims within a year from the time of diagnosis. Although the incidence of other asbestos-related diseases is clearly related to dose, mesotheliomas have been observed in individuals with only brief exposure.

Scientists have long wondered how asbestos causes cancer. Studies have shown measurable amounts of DNA in tissue fluids surrounding cells of the body. Research suggests that asbestos fibers may attach to this DNA and then pierce the cell membrane, carrying the DNA inside. Inside the cell this DNA may disrupt or turn off the genes that control a cell's growth. With the control mechanism altered or completely shut down, the cell begins to duplicate wildly. Another possibility is that the DNA may carry cancer-causing genes into the cell. Once inside, the genes become activated, triggering the cell to divide. Still other research suggests that asbestos

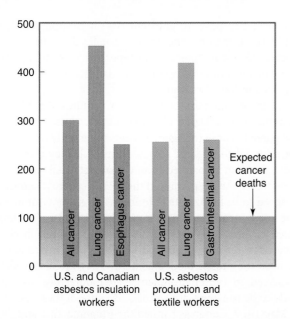

FIGURE 18-9 **Incidence of cancer in asbestos workers in the United States and Canada.** (Ratio of the number of observed to the number of expected deaths times 100.) Asbestos workers are much more likely to develop cancer than the general public.

fibers inside cells may damage chromosomes during cell division. This, in turn, may cause harmful mutations.

An estimated 8 to 11 million American workers have been exposed to asbestos since World War II. Studies of these workers show that over a third had lung cancer, mesothelioma, or gastrointestinal cancer. Between 1990 and 2020, some health experts expect about 2 million people, mostly workers, to die from exposure to asbestos.

Based on these and other statistics, the use of asbestos in the United States for insulation, fireproofing, and decorative purposes was banned in 1974. In 1979, the EPA began to help states and local school districts identify asbestos crumbling from pipes and ceilings. In 1989, the EPA ordered bans on almost all other applications, which went into effect in 1997 and will, with the previous ban, eliminate almost 95% of all asbestos use in the United States.

Companies that produce asbestos argue that removing asbestos from school buildings will not protect public health. Workers, they point out, have been the main victims of asbestos-related diseases. Furthermore, workers who smoke are at a much higher risk. Because young children do not smoke, their risk is very small. A far more economical way of dealing with the problem would be to stabilize the asbestos layer around pipes and in ceilings—say, by painting it.

KEY CONCEPTS

Asbestos is a naturally occurring silicate mineral fiber with many practical uses. Unfortunately, it produces three disorders: pulmonary fibrosis (buildup of scar tissue in the lungs), lung cancer, and mesothelioma. It is especially dangerous to asbestos workers and to individuals who also smoke.

Electromagnetic Radiation: A Hazard to Our Health?

In 1979, two researchers from the University of Colorado reported a link between high-current electric power lines and the incidence of childhood leukemia. Their study suggested that extremely low-frequency (ELF) magnetic fields produced when electricity flows through wires may be the cause of the increased incidence of cancer in children living nearby. The researchers found that the death rate from cancer in these children was twice what was expected in the general public.

ELF magnetic fields are virtually everywhere. Moreover, they easily penetrate building walls and the human body. Magnetic fields are also found around power stations, welding equipment, subways, and movie projectors.

In 1986, researchers from the University of North Carolina announced the results of a study that supported the Colorado research. The study showed a fivefold increase in childhood cancer (particularly leukemia) in residents living near the highest magnetic fields, 7 to 15 meters (25 to 50 feet) from wires carrying electricity from power substations to neighborhood transformers. Adding to the concern, a researcher from Texas recently found that magnetic fields increase the growth rate of cancer cells. In addition, cancer cells exposed to magnetic fields are 60 to 70% more resistant to the body's naturally occurring killer cells.

In 1986, another group of scientists reported that magnetic fields from electric waterbed heaters and electric blankets increased the likelihood of miscarriage. In the group that used electric heaters in their waterbeds, 61% of the miscarriages occurred from September to the end of January. By comparison, women using neither a waterbed heater nor an electric blanket had a 44% miscarriage rate during this same period.

ELF magnetic fields may also be a cause of birth defects in humans. Scientists know that magnetic fields affect fetal development in pigs, chickens, and rabbits.

Many scientists and industry representatives are skeptical about the link between cancer and magnetic fields. Eleanor Adair, environmental physiologist and senior research scientist at Yale University, notes that of the more than 30 epidemiological studies, about half show a weak correlation between magnetic fields—and the other half do not. She finds many serious problems with the design of virtually all epidemiological studies, further shedding doubt on their validity. In most studies, she notes, scientists did not measure the strength of the magnetic fields. Instead, they relied on measures of the proximity of homes to power lines or other sources and measured the size of the wires carrying electric current to homes. She also notes the failure of such studies to take into account other variables, such as exposure to smoking and pollution, that might be responsible for reported cancers. Other scientists, including David Carpenter from the University of Albany School of Public Health, disagree and feel that the studies probably underestimate the risk of cancer. An analysis of over 500 studies on the subject led the prestigious National Research Council to conclude that the bulk of the evidence suggests that exposure to ELF magnetic fields does not cause cancer, neurological or behavioral problems, or damage to reproductive cells or the developing fetus. They and others point out that electromagnetic radiation is not an ionizing radiation, like that emitted from radioactive materials, and thus probably can't cause cancer. Still others say that although ELF magnetic fields may not cause cancer, they may accelerate the growth of cancer cells.

KEY CONCEPTS

Some studies suggest that ELF magnetic fields produced when electricity flows through wires may increase the incidence of cancer in children living nearby. To date, however, results of studies on the effects of ELF magnetic fields have yielded inconclusive results.

18.6 Controlling Toxic Substances: Toward a Sustainable Solution

In 2011, the United States produced 34.3 million tons of hazardous wastes according to the EPA. Many laws have been passed to regulate chemical pollutants and toxic chemicals; however, most have called for end-of-pipe controls (**TABLE 18-3**). Although important, many experts realize that we also need long-term approaches, for example, efforts to prevent hazardous waste production in the first place. This section will look at both stop-gap and long-term, sustainable solutions. See **Spotlight on Sustainable Development 18-1** for a discussion of efforts to reduce exposure to lead.

KEY CONCEPTS

Many laws have been passed in the United States and other countries to reduce the release of toxic chemicals into the environment, but most of these relied on end-of-pipe controls. More recent efforts are aimed at preventing pollution—that is, eliminating hazardous waste production in the first place.

Toxic Substances Control Act

In 1976, the U.S. Congress passed the **Toxic Substances Control Act (TSCA).** The TSCA consists of three parts. The first is **premanufacture notification,** which requires all companies to inform the EPA 90 days before they import or manufacture a chemical substance not currently in commercial use. The agency then has 90 days to approve the chemical or, if necessary, place restrictions on its use to reduce exposure to people and the environment. To do this, EPA scientists review existing toxicity data on the chemical or data on similar chemicals. If the new chemical

is believed to pose insignificant risk, it is approved. If it could be hazardous, the agency usually asks the manufacturer to test its toxicity and report back. At this point, many manufacturers drop the candidate because of the cost involved in toxicity testing.

The second part of the TSCA requires the EPA to examine chemicals that were in commercial use before the law passed. Those the agency deems potentially hazardous are required to undergo toxicity testing. If they prove too risky, they can be banned from use or restricted.

Finally, the act permitted the EPA to control chemicals believed to be hazardous to humans and the environment. The most stringent controls were placed on polychlorinated biphenyls (PCBs), an insulating fluid used in electrical transformers. PCBs are stable in the environment because they resist biodegradation. They are also widely dispersed in the environment, can biomagnify, and are fairly toxic to laboratory animals. Because of these factors, in May 1979, Congress banned their manufacture and distribution except in a few limited cases.

KEY CONCEPTS

The Toxic Substances Control Act seeks both to prevent the introduction of chemicals that will be harmful to people and the environment and to eliminate those already in use that pose an unacceptable risk.

Market Incentives to Control Toxic Chemicals

In 1988, one of the most controversial environmental laws in the United States went into effect. The state of California's **Safe Drinking Water and Toxics Enforcement Act** set in motion a market strategy to reduce dramatically the exposure to toxic chemicals in foods and consumer products. Created as a result of a citizen-sponsored initiative, Proposition 65, which was passed overwhelmingly by voters in 1986, this law directs the state to set standards (acceptable levels) for potential toxicants in a variety of consumer products and foods. The law requires that manufacturers who violate these standards print warnings on their products, noting that the amount of the chemical in the product exceeds the state's safe level. Manufacturers are not required to meet standards, only to print the fact on the packages. The proponents of this law believe that consumers will shun products such as these, which will encourage manufacturers to reduce levels of toxicants in their products.

Numerous other market-based incentives for controlling toxic pollution are also available. Two of the more promising are pollution taxes and tradable permits. Basically, **pollution taxes,** or **green taxes,** are fees levied on environmentally unsound activities—for example, the use of toxic chemicals or the production of hazardous wastes. Taxes make it more costly to do things wrong and

TABLE 18-3

Federal Laws and Agencies Regulating Toxic Chemicals

Statute	Year Enacted	Responsible Agency	Sources Covered
Toxic Substances Control Act	1976	EPA	All new chemicals (other than food additives, drugs, pesticides, alcohol, tobacco); existing chemical hazards not covered by other laws
Clean Air Act	1963, 1967, 1970, amended 1977, 1990, 1997	EPA	Hazardous air pollutants
Federal Water Pollution Control Act	1972, amended 1977, 1978, 1987	EPA	Toxic water pollutants
Safe Drinking Water Act	1974, amended 1977, 1996	EPA	Drinking water contaminants
Federal Insecticide, Fungicide, and Rodenticide Act	1948, amended 1972, 1973, 1988	EPA	Pesticides
Act of July 22, 1954 (codified as § 346(a) of the Food, Drug, and Cosmetic Act)	1954, amended 1972	EPA	Tolerances for pesticide residues in food
Resource Conservation and Recovery Act	1976	EPA	Hazardous wastes
Marine Protection, Research, and Sanctuaries Act	1972	EPA	Ocean dumping
Food, Drug, and Cosmetic Act	1938	FDA	Basic coverage of food, drugs, and cosmetics
Food additives amendment	1958	FDA	Food additives
Color additive amendments	1960	FDA	Color additives
New drug amendments	1962	FDA	Drugs
New animal drug amendments	1968	FDA	Animal drugs and feed additives
Medical device amendments	1976	FDA	Medical devices
Wholesome Meat Act	1967	USDA	Food, feed, and color additives; pesticide residues in meat, poultry
Occupational Safety and Health Act	1970	OSHA	Workplace toxic chemicals
Federal Hazardous Substances Act	1966	CPSC	Household products
Consumer Product Safety Act	1972	CPSC	Dangerous consumer products
Poison Prevention Packaging Act	1970	CPSC	Packaging of dangerous children's products
Lead-Based Paint Poison Prevention Act	1973, amended 1976	CPSC	Use of lead paint in federally assisted housing
Hazardous Materials Transportation Act	1970	DOT (Materials Transportation Bureau)	Transportation of toxic substances generally
Federal Railroad Safety Act	1970	DOT (Federal Railroad Administration)	Railroad safety
Ports and Waterways Safety Act	1972	DOT (Coast Guard)	Shipment of toxic materials by water
Dangerous Cargo Act	1952	DOT (Coast Guard)	Shipment of toxic materials by water

NOTE: CPSC = Consumer Product Safety Commission   FDA = Food and Drug Administration

DOT = U.S. Department of Transportation   OSHA = Occupational Safety and Health Administration

EPA = U.S. Environmental Protection Agency   USDA = U.S. Department of Agriculture

Data from the Council on Environmental Quality.

18-1 Getting the Lead Out: Steps to Reduce Lead Exposure in the United States

Lead is one of the most useful metals in modern industrial societies. It is found in ceramic glazes, batteries, fishing sinkers, solder, and old pipe. Lead was once added to gasoline to help reduce engine knocking.

Unfortunately, lead is also a highly toxic poison, entering the body primarily through inhalation and ingestion. Although it affects many organs, lead has a special affinity for bone, brain, and kidneys.

High-level exposure in certain factory workers has been linked to neurological symptoms: fatigue, headache, muscular tremor, clumsiness, and loss of memory. If exposure is discontinued, patients may slowly recover, but residual damage such as epilepsy, idiocy, and hydrocephalus (fluid accumulation in the brain) often lingers. Continued high exposure may lead to convulsions, coma, and death. Some scientists believe that lead drinking vessels and lead pipes in water systems may have caused a decline in birth rates and increased psychosis in ancient Rome's ruling class, contributing to the fall of the Roman Empire.

Organic lead (alkyl lead in gasoline, for example) has been linked to a number of psychological disorders—including hallucinations, delusions, and excitement—and may lead to death.

Lead also damages the kidneys, causing a disturbance in the mechanisms that help us conserve valuable nutrients (such as glucose and amino acids) that might otherwise be lost in the urine. Prolonged, high-level exposure causes a progressive degeneration of the filtering mechanism that removes wastes from the bloodstream.

Lead has a profound effect on reproduction. Numerous reports show that the rate of spontaneous abortion is much higher in couples when one or both has been exposed to high levels of lead in the workplace. Recent studies show decreased fertility and damaged sperm in male workers with high-to-medium levels of lead in their blood. According to one study, exposure of a pregnant woman to high levels of lead in household drinking water nearly doubles her risk of having a retarded child.

Today, because of better controls on lead in the workplace and in commercial products, acute lead poisoning is rare in most countries. Nonetheless, many people throughout the world are regularly exposed to low levels of lead, with potentially serious consequences. Blood levels in humans in industrial nations today are about 100 times greater than before the advent of lead pollution. However, even residents of Nepal, a nonindustrialized country, have levels 10 times higher than those estimated to be present before the widespread use of lead—attesting to the global distribution of lead in the atmosphere.

Recently, scientists have begun to learn about the effects of low-level exposure, especially on the central nervous system. Herbert Needleman and his colleagues studied over 3,000 first-and second-grade children with different levels of lead in their bodies. Children with high lead levels (caused by exposure to low levels of lead over time) had significantly lower IQ scores than those with low levels. Attention span and classroom behavior were also significantly impaired. Several other studies have shown similar results. A recent study in England showed that at an early age even marginally elevated levels of lead may have lasting adverse effects on intelligence and behavior.

In 1990, Needleman reported that childhood exposure to lead levels once considered moderate or low can seriously and permanently alter the intelligence of adults. Studies show that elevated lead levels in the blood of men cause hypertension and increase the risk of heart disease.

Lead is found in our food, water, air, and soils. For most Americans, food is the number one source of lead exposure. Lead emitted by power plants, smelters, and boilers that burn used motor oil is frequently deposited in the soil, where it is taken up by crops.

Although food is the major source of lead, most of the concern for lead exposure in humans has focused on automobile exhausts. Regulations to eliminate lead from gasoline have markedly decreased concentrations in and around our cities.

In 1986, a major study by the EPA revealed that lead levels in drinking water in many cities exceeded federal standards, potentially threatening the health of millions of Americans. Lead in drinking water comes from lead-based solder and from lead pipes often found in older homes.

Lead victims are primarily children and, among them, mostly children of poor, black families who tend to live in regions where pollution from automobile exhaust is prevalent (**FIGURE 1**). Children may also eat dirt contaminated with lead from vehicles in previous years (before lead was removed from gasoline). Children living in old, neglected buildings often ingest flakes of lead-based paint, which was applied before a ban was enacted in the 1940s.

Children are more susceptible to lead in part because they absorb more in the intestines. Adults absorb about 8% to 10% of the lead they ingest, whereas the absorption rate in children may be as high as 40%. In addition, children are more sensitive to lead than adults. The developing brain seems to be the most sensitive organ. Furthermore, the toxicity of lead is increased in malnourished and iron-deficient children, who often come from poor, urban families.

A new study suggests that lead poisoning may be responsible for the higher incidence of cavities in children in economically depressed areas. A slight increase in lead concentrations in the blood (of just 5 micrograms per deciliter of blood) was attributed to an 80% increase in cavities.

Alarmed by the mounting evidence regarding the effects of lead on children, the U.S. EPA in 1973 began a progressive restriction of the lead content of gasoline. Today, it has declined to nearly zero (there's a nearly undetectable amount in unleaded gas). Blood lead levels have fallen as well. Other countries have also taken an active role in reducing lead emissions.

Cities in less developed nations are farther behind. In these countries, the lead content of the gasoline is, on the average, twice that of the more developed countries. Malnutrition and high lead levels in the air will almost certainly have serious effects on the children.

As noted previously, food is the major source of lead in the United States. In 1979, the Food and Drug Administration took steps to reduce the intake of lead from lead-soldered cans. Lead has been virtually eliminated in this application.

In 1986, Congress also banned the use of lead solder in pipes. Because drinking water accounts for about 20% of the lead exposure to Americans, the EPA adopted regulations that will reduce lead in public drinking water supplies, which could lower lead exposure in drinking water for about 138 million people. One regulation requires public water suppliers to treat their water with alkaline additives if lead is a problem or if the water is slightly acidic. This measure is expected to reduce lead leaching from pipes. The new EPA regulations also lower allowable levels of lead in the drinking water to about 5 parts per billion (ppb), down from 50 ppb. This measure is in all likelihood not going to result in a 10-fold decrease because the current limit of 50 ppb is for water measured at the tap, whereas the new standards measure lead levels in water leaving treatment plants. Because most lead enters water *after* it leaves the treatment plant, the new ruling may not lower public exposure very much. Only time will tell.

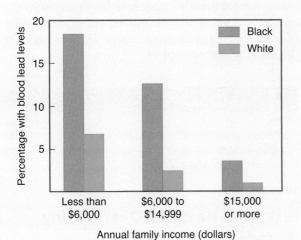

FIGURE 1 High lead levels in U.S. children from 6 months to 5 years old, according to parents' income and race (1976–1980).

thus provide an incentive to switch to more environmentally sustainable practices.

Tradable permits are permits that allow new or existing companies to release toxic chemical pollutants but at a low rate. In a given region, permits are granted to numerous companies so that total emissions in the region are acceptable. If a company can find a way to reduce emissions below its permitted levels economically, it can sell its remaining emissions to companies that are unable or unwilling to lower emissions. Companies that lower emissions can, therefore, make a profit on the sale of their emissions. This system encourages innovation and cost-effective means of reducing or even eliminating the production and release of toxic substances.

Market-based policies are designed to produce monetary incentives for companies to reduce pollution and regulatory burden. They also provide more freedom of choice for businesses to select methods that suit their unique situations. Accordingly, market-based actions are gaining popularity among businesses, government regulators, and environmentalists.

KEY CONCEPTS

Many efforts are under way to harness market forces, rather than impose regulations on companies, to encourage the use of nontoxic or less toxic products. These mechanisms promote innovation, freedom of choice, and cost-saving solutions that many businesses view as an acceptable way to reduce pollution.

The Multimedia Approach to Pollution Control: An Integrated Approach

The EPA has historically regulated toxic chemicals by controlling the emissions into various media (air, water, and soil) separately. In addition, each medium has been monitored and regulated by a separate branch of the EPA. Wastewater inspectors looked only at the pollutants in the wastewater. The air pollution inspectors examined what goes out of the smokestacks. So long as a company was obeying the necessary regulations, the inspectors were happy.

The shortsightedness of this method became evident in the early 1990s when EPA officials found that companies often disposed of wastes in the medium with the weakest regulations. Refinery wastes that would normally be shipped out in barrels and thus were subject to hazardous waste rules, for instance, might end up in the wastewater stream, which is less strictly regulated.

In 1991, William Reilly, then administrator of the EPA, called for a **multimedia approach** to prevent this environmentally costly toxic shell game. In the regional offices of the EPA, Reilly set up multimedia branches that trained inspectors in several media—for example, air and water releases. Their job is to determine if companies are in compliance with all pertinent environmental regulations, not just regulations in a specific medium—hence, the term *multimedia* (not to be confused with modern audiovisual productions).

Go Green

One of the most important things we can all do to reduce toxic pollution is to curb consumption as the production of virtually all products results in the production of hazardous wastes.

The EPA hopes that multimedia inspections will help the agency identify companies that are illegally or improperly disposing of wastes. Such a tightening in the regulatory approach could promote pollution prevention (preventive actions) and reduce emissions to the environment. Why? Increased agency awareness of the activities in a facility decreases a company's ability to hide pollution and avoid regulation by transferring pollution from one medium to another.

Controlling toxic substances also requires individual actions. By reducing or eliminating the purchase of toxic chemicals, you can help create a more sustainable world.

KEY CONCEPTS

Pollutants from factories exit via one of several avenues, such as wastewater or air pollution. For years, each medium has been regulated separately. Efforts are now under way to regulate and monitor several media simultaneously to keep companies from dumping potentially toxic substances in the least regulated medium.

18.7 Risk and Risk Assessment

Ralph Waldo Emerson once wrote, "As soon as there is life there is danger." Every day of our lives we face many dangers—some obvious, some hidden. The assessment of the risks (potential harm) of modern technological societies has become important in policymaking and is clearly vital to efforts to build a sustainable society. This section looks at risk and how it is assessed.

KEY CONCEPTS

Understanding and quantifying the risks posed in modern society is essential to creating socially, economically, and environmentally acceptable policies.

Risks and Hazards: Overlapping Boundaries

Two types of hazard are broadly defined by risk assessors: anthropogenic and natural. **Anthropogenic** (an-throw-poe-GEN-ick) **hazards** are those created by human beings. **Natural hazards** include events such as tornadoes, hurricanes, floods, droughts, volcanoes, and landslides. Be careful, however, events that are typically classified as natural hazards may actually be caused or worsened by human actions. Take flooding as an example. Although most of us think of floods as natural phenomena, the severity and frequency of flooding is often the result of such actions as channelizing streambeds or deforestation.

Three Steps in Risk Assessment

Hazards befalling human society exact an enormous price. They may kill people or destroy property. They may cause a deterioration in our health or lead to the extinction of plants and animals. How do we go about assessing the hazards and creating policies that minimize or eliminate them in socially and economically acceptable ways?

In the mid-1970s, a new science called **risk assessment** began to take root. Its goal was to help people understand and quantify risks posed by technology, our lifestyles, and our personal habits (smoking, drinking, and diet).

Risk assessment involves two interlocked steps: hazard identification and estimation of risk. **Hazard identification** involves steps to identify potential and real dangers. **Estimation of risk** generally involves two processes (**FIGURE 18-10**). The first determines the probability of an event or occurrence. This process answers the question, "How likely is the event?" The second stage determines the severity of an event, answering the question,

"How much damage could be caused?" To assess the risk of a toxic chemical, we must estimate the number of people (or other organisms) exposed, the levels and duration of exposure, and other complicating factors such as age, sex, health status, personal habits, and chemical interactions (described earlier). This information helps us determine what the impact might be—that is, how severe the effects might be.

Why Assessing the Risk of Toxic Substances Is So Difficult Estimating risk is fraught with uncertainty, especially in the area of toxic substances. Several factors contribute to the difficulties. First, our knowledge of the effects of toxic substances on humans is incomplete. As noted earlier in the chapter, modern societies use thousands of potentially toxic chemicals. Testing each one is expensive and time consuming. In addition, our knowledge of the toxic effects of pollution on humans is limited because it is neither practical nor ethical to test toxic chemicals on human beings. As a

FOUNDATION TOOL

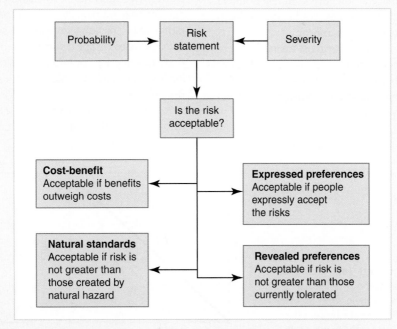

FIGURE 18-10 Determining the acceptability of a risk. After the probability and severity of a risk are determined, the next step is to determine its acceptability. Cost–benefit analysis is the most common method of determining risk acceptability, but three other methods can also be used, as shown here.

result, toxicologists must often rely on tests on laboratory animals (rats, mice, and rabbits) to estimate human toxicity. The results of experiments on laboratory animals cannot always be extrapolated to humans, though. As my graduate advisor once noted, "Contrary to popular belief, the human is not a large rat." Lab animals often react differently to chemicals than humans do; they may be able to break them down better, or they may not be able to break them down as well. Physiological differences between humans and lab animals therefore make it difficult to predict if a chemical harmful to an animal will be injurious to us.

Second, our lack of knowledge about toxic effects also stems from the fact that we humans are often exposed to many potentially harmful chemicals. For practical reasons, most toxicity tests are performed on one substance at a time. Because of chemical interactions described earlier, extrapolating the results from these tests to the real world can be misleading.

A third problem is that most animal tests of toxicity, especially those for cancer, are performed at high exposure levels. It is assumed that if a chemical is harmful at high levels, it will also be harmful at the lower levels. However, the fact that a large dose of a chemical induces cancer in a lab animal does not necessarily mean that the chemical will cause cancer at the low doses to which humans are typically exposed. Most toxicants have a **threshold level**—that is, a level below which no effects occur (**FIGURE 18-11**). Thus, extremely low levels of certain chemicals may be completely safe. The reason for this may be that at low levels, protective mechanisms in the body can cope with the chemical—either inactivating it or excreting it fast enough to prevent damage. Tissues may also be able to repair damage at low levels. At levels above the threshold, these mechanisms may not be able to keep up, and noticeable effects may then appear.

Although thresholds may exist for most toxicants, there are some exceptions, among them asbestos and radiation. Many scientists believe that even the smallest exposure has an effect and that damage caused by repeated low-level exposures accumulates over time.

If scientists can't make accurate extrapolations from high doses to low doses, why do they perform their experiments that way? Researchers use high doses to speed up their experiments. As a general rule, the entire process from exposure to manifestation takes about one-eighth of the life span of an animal. In humans, the time required to develop a noticeable cancer is 5 to 30 years after exposure. Any measure that speeds up the process, such as a high dose, helps cut costs. To test for low-level effects, scientists would need very large numbers of experimental animals to generate statistically valid results. High-dose studies, therefore, reduce the number of lab animals needed and can cut time and costs—a significant factor because cancer studies can cost $500,000 to $1 million per chemical. (For a debate on the validity of testing animals for cancer, see the Point/Counterpoint essays in this chapter.)

Risk Acceptability

No matter what we do—whether it is screwing in a light bulb or flying cross-country in a jet—we put ourselves (and possibly our environment) at risk. Nothing is safe—that is, entirely free from harm. The science of risk assessment recognizes that human life is haunted by hazards. Therefore, rather than speaking in terms of safety, which is absolute, the risk assessor speaks in terms of risk, which is relative. Activities that we commonly consider safe are referred to as *low-risk functions*. "Unsafe" activities are better labeled *high-risk functions*.

Knowing the relative risk of a technology is one thing. Knowing whether the risk is acceptable to the general

FIGURE 18-11 Threshold or not? **(a)** Hypothetical dose-response curves indicating the absence of a threshold level, a level below which no effect occurs. Some responses are linear: For every increase in dose, there is a corresponding increase in the response. In others, the response is nonlinear; that is, there is not a 1:1 correlation between dose and response, although the response does increase with increasing dose.
(b) A hypothetical dose-response curve showing a threshold level.

(a) No threshold

(b) Threshold

public is another story. **Risk acceptability,** in fact, is one of the trickiest issues facing modern society. Why? One reason is that our awareness and perception change from time to time. Thus, what appears safe one day becomes suspect the next, after a widely publicized accident. Irrational fears crop up and frighten us away from relatively low-risk activities. Jet crashes, for example, are widely publicized and may give the impression that this form of travel is quite risky. The fact of the matter is, you have a 1 in 5,000 chance of being killed in an automobile accident in a given year and only a 1 in 7 million chance of being killed in a jet crash.

The **perceived harm,** the damage people think will occur, heavily influences our assessment of risk acceptability. In general, the more harmful a technology and its byproduct are *perceived* to be, the less acceptable they are to society.

The acceptability of risks is also influenced by **perceived benefit**—how much benefit people think they will get from something. In general, the higher the perceived benefit, the greater the risk acceptability (**FIGURE 18-12**). As an example, the risks of a new steel mill might be acceptable to a community with high unemployment. Automobile travel provides the most telling example of the way in which perceived benefits affect our decisions. The risk of dying in an automobile accident in the United States is 1 in 5,000 in any given year. Over your lifetime, the risk of dying in a car accident is higher if you don't wear seat belts, about 2 in 100, than if you do, 1 in 100. Meanwhile, substances believed to be far less hazardous than driving are banned from public use because their benefit is not so highly valued. Chemical substances that pose a risk of one cancer death in a million people, for example, are banned.

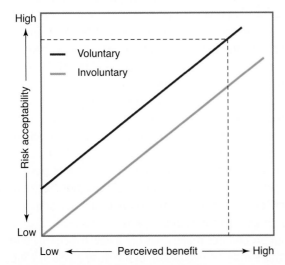

FIGURE 18-12 The acceptability of a risk increases as the perceived benefit rises. Voluntary risks—that is, ones people agree to—are generally more acceptable than risk imposed upon us without consent.

How Do We Decide If a Risk Is Acceptable?

Decisions on modern sources of risk—technologies, personal habits, and pollution—are becoming more and more commonplace. One of the most commonly used techniques is **cost–benefit analysis,** a process in which one analyzes the many costs and benefits and weighs one against the other (Figure 18-10).

One of the most important benefits of a sustainable strategy—including efforts such as energy efficiency, pollution prevention, recycling, and use of renewable resources—is that it offers many ways of having both a clean, healthy environment and economic vitality. It makes the environment and economy complementary, not antagonistic, forces. For now, however, many decisions must be made about risk acceptability. In this technique, one enumerates costs and benefits and then tries to determine if the benefits outweigh the costs. Although that may sound simple, it isn't. As a rule, benefits are generally easily measured: financial gain, business opportunities, jobs, and other tangible items. Many of the costs, however, are less tangible. External costs, costs incurred to society by pollution and other forms of environmental damage, are among the most difficult to quantify. In addition, researchers find it difficult to assign a dollar value to human health effects, environmental damage, and lost species. Moreover, the costs may be long term and borne by future generations.

Another problem with cost–benefit analysis is that the benefits, most importantly economic gain, often accrue to a select few who have inordinate power to influence the political system and sway people's opinions. American companies that design and build hazardous-waste incinerators, for instance, spend extraordinary sums of money trying to convince people in rural areas of the benefits of siting an incinerator near them.

In sum, cost–benefit analysis suffers because many costs are poorly documented, spread out, and unquantifiable, whereas the benefits are often obvious and readily quantifiable. Recent efforts by economists to assign a dollar value to environmental and health costs may help improve the process. In addition, the efforts of scientists to measure the impacts of technologies and their byproducts on wildlife, the environment, recreation, health, and society in general may also help.

KEY CONCEPTS

Several techniques are used to determine if the risk(s) posed by a technology or activity are acceptable. The most common is cost–benefit analysis, which weighs the costs against the benefits. Sustainable development strategies minimize social, economic, and environmental costs and maximize the benefits.

Animal Testing for Cancer Is Flawed

Philip H. Abelson

Philip H. Abelson was the Deputy Editor of *Science*. This article is copyright (© 1990) by the American Association for the Advancement of Science.

Courtesy of Philip H. Abelson.

The principal method of determining potential carcinogenicity of substances is based on studies of daily administration of huge doses of chemicals to inbred rodents for a lifetime. Then by questionable models, which include large safety factors, the results are extrapolated to effects of minuscule doses in humans. Resultant stringent regulations and attendant frightening publicity have led to public anxiety and chemophobia. If current ill-based regulatory levels continue to be imposed, the cost of cleaning up phantom hazards will be in the hundreds of billions of dollars with minimal benefit to human health. In the meantime, real hazards are not receiving adequate attention.

The current procedures for gauging carcinogenicity are coming under increasing scrutiny and criticism. A leader in the examination is Bruce Ames, who with others has amassed an impressive body of evidence and arguments. Ames and Gold summarized some of their recent data and conclusions in *Science* (31 August 1990, p. 970). Three articles in the *Proceedings of the National Academy of Sciences* provide an elaboration of the information with extensive bibliographies. The articles also provide data about other pathologic effects of natural chemicals.

A limited number of chemicals tested, both natural and synthetic, react with DNA to cause mutations. Most chemicals are not mutagens, but when the maximum tolerated dose (MTD) is administered daily to rodents over a lifetime, about half of the chemicals give rise to excess cancer, usually late in the normal life span of the animals. Experiments in which synthetic industrial chemicals were administered in the MTD to both rats and mice resulted in 212 of 350 chemicals being labeled as carcinogens. Similar experiments with chemicals naturally present in food resulted in 27 of 52 tested being designated as carcinogens. These 27 rodent carcinogens have been found in 57 different foods, including apples, bananas, carrots, celery, coffee, lettuce, orange juice, peas, potatoes, and tomatoes. They are commonly present in quantities thousands of times as great as are the synthetic pesticides.

The plant chemicals that have been tested represent only a tiny fraction of the natural pesticides. As a defense against predators and parasites, plants have evolved a large number of chemicals that have pathologic effects on their attackers and consumers. Ames and Gold estimate that plant foods contain 5,000 to 10,000 natural pesticides and breakdown products. In cabbage alone some 49 natural pesticides have been found. The typical plant contains a total of 1% or more of such substances. Compared with the amount of synthetic pesticides we consume, we eat about 10,000 times more of the plant pesticides.

It has long been known that virtually all chemicals are toxic if ingested in sufficiently high doses. Common table salt can cause stomach cancer. Ames and others have pointed out that high levels of chemicals cause large-scale cell death and replacement by division. Dividing cells are much more subject to mutations than quiescent cells. Much of the activity of cells involves oxidation, including formation of highly reactive free radicals that can react with and damage DNA. Repair mechanisms exist, but they are not perfect. Ames has stated that oxidative DNA damage is a major contributor to aging and to cancer. He points out that any agent causing chronic cell division can be indirectly mutagenic because it increases the probability of DNA damage being converted to mutations. If chemicals are administered at doses substantially lower than MTD, they are not likely to cause elevated rates of cell death and cell division and hence would not increase mutations. Thus, a chemical that produces cell death and cancer at the MTD could be harmless at lower dose levels.

Diets rich in fruits and vegetables tend to reduce human cancer. The rodent MTD test that labels plant chemicals as cancer causing in humans is misleading. The test is likewise of limited value for synthetic chemicals. The standard carcinogen tests that use rodents are an obsolescent relic of the ignorance of past decades. At that time, extreme caution made sense, but now tremendous improvements of analytical and other procedures make possible a new toxicology and far more realistic evaluation of the dose levels at which pathological effects occur.

The vast majority of the scientific community endorses the conduct of experimental studies in order to try to identify those materials that could cause disease and prevent harmful exposures. In the typical toxicologic study of 50 rodents, each animal is a stand-in for 50,000 people. Because rodents live about 2 years on average, they are usually exposed to amounts of the suspect agent that approximate what a human would encounter in an average lifetime of 70 years.

Some have argued that the use of the maximum tolerated dose (MTD) in these studies produces tissue damage and cell proliferation, which in turn lead to cancer. This is biological nonsense that ignores a fundamental characteristic of cancer biology that has been known for more than 50 years: Cancer is a multistage disease, with multiple causes, which arises by a stepwise evolution that involves progressive genetic changes, cell proliferation, and clonal expansion. Thus, swamping of tissues with high doses alone may well kill an animal or damage its tissues, but high doses alone are not sufficient to cause cancer.

A 2-year study at the National Institute of Environmental Health Sciences (NIEHS) by David Hoel and colleagues provides good evidence on this point. They looked for signs of damage in tissues taken from rats and mice used in cancer studies. Cancers occurred in organs that did not show apparent damage, and some damaged organs were completely free of tumors.

In addition, most compounds tested do not cause cancer only in the highest dose group tested but typically produce a dose-response relationship, where the amount of cancer developed is proportional to the dose administered. Some chemicals cause toxicity only; others cause only cancer. Not all those that cause cancer do so through organ toxicity. In fact, almost 90% of the substances shown to cause cancer in the National Toxicology Program do so without producing any increased cellular toxicity, and they also cause cancer at both lower and higher doses.

Animal studies are evolving and being further refined, as is our understanding of differences between species that need to be taken into account in conducting these studies. Every compound known to cause cancer in humans also produces cancer in animals, when adequately tested. For 8 of the 54 known human cancer-causing agents, evidence of carcinogenicity was first obtained in laboratory animals; in many cases, the same target organs and doses have been involved in producing cancer in both animals and humans.

Current Methods of Testing Cancer Are Valid

Devra Davis

Devra Davis is the author of National Book Award Finalist, *When Smoke Ran Like Water* (http://www.whensmokeranlikewater.com), and Director, Center for Environmental Oncology, University of Pittsburgh Cancer Institute, and Professor of Epidemiology, Graduate School of Public Health.

The NIEHS has carried out nearly 400 long-term rodent studies, which have been published following public peer review by specialists in the field. Chemicals nominated for testing usually represent a sample of potentially "problematic" materials. About 40% of the "suspect" pesticides evaluated to date have been found to cause cancer. As to the role of so-called natural pesticides, rodent diets are also loaded with many of these materials. Nevertheless, a number of test compounds added to these diets markedly increase the amount of tumors produced. Thus, animals are more sensitive to certain synthetic compounds than to the background level of natural materials. Humans are also likely to have acquired some resistance to these persisting, natural materials throughout our evolution.

To date, only about 20% of all synthetic organic chemicals have been adequately tested for their potential human toxicity. Those who must set public policy on the use of chemicals need a rational basis on which to stake their actions. Continued advances in animal testing provide an important contribution to environmental health sciences and to public health efforts to predict, and then prevent, the development of environmentally caused disease.

In summary, the current system should be used until it can be replaced by a demonstrably better one. There is no scientific basis for rejecting the MTD as capable of inducing cancer.

Critical Thinking Questions

1. Summarize each author's main points and supporting data. Do you see any inconsistencies or examples of faulty logic in either essay? Where?
2. Why is it that two scientists can disagree on an issue such as this?
3. Given the disagreement, what course of action would you recommend?

Actual Versus Perceived Risk The main purpose of risk assessment is to help policymakers formulate laws and regulations that protect human health, the environment, and other organisms. Ideally, good lawmaking requires that the **actual** risk, or the amount of risk a hazard really poses, be equal to the **perceived risk**, the risk perceived by the public. When actual and perceived risk are equal, public policy yields cost-effective protection.

When the perceived risk is much larger than the actual risk, costly overprotection occurs. In contrast, when the perceived risk is much smaller than the actual risk, underprotection occurs (**FIGURE 18-13**). It, too, can be costly to present and future generations.

KEY CONCEPTS

Risk assessment is designed to facilitate decision making by ensuring that the risk we perceive to be posed by any factor is equal to the actual risk.

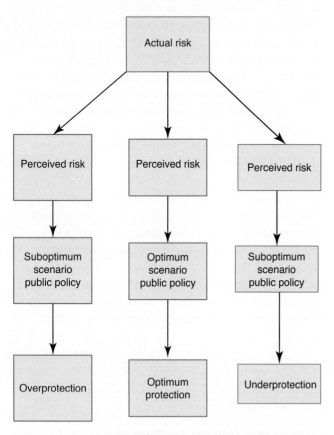

FIGURE 18-13 Matching risk. Matching the actual risk and the risk that a society perceives is essential to the formulation of good public policy. Perceived risk and actual risk do not always match, however, as shown. Policy decisions may result in overprotection or underprotection when the actual and perceived risk do not match. Optimal protection occurs when actual and perceived risk are equal.

The Final Filter: Ethics and Sustainability We hear over and over again that if it doesn't make economic sense, people won't do it. This is not always the case. Many decisions we make are based on our ethics, the values we hold—or, more simply, what we view as right and wrong. Values that affect our decisions come from our parents, relatives, friends, enemies, teachers, religious leaders, and politicians. These values shift over time, sometimes subtly, sometimes dramatically. They often change as we become older and as our priorities shift. Although our ethics are often never explicitly stated, they play an important role in our lives. They determine how we think, how we vote, how we treat one another, how we carry on business affairs, how we treat the Earth, and how we treat future generations.

Because values influence the way we think and the decisions we make, they often play an important role in decisions regarding risk acceptability. Most of us, however, assign different ethical priorities to different things. Some people care more about the environment than about economics. Others care more about economics than the environment. In sustainability, the goal is to place economics, the environment, and society on an equal level and to make decisions that serve all three simultaneously. This requires creative thinking, but the options are many.

KEY CONCEPTS

Our values, what we perceive as right or wrong, come from many sources and often affect our decisions. Sustainable ethics calls for value placed on economics, the environment, and society, simultaneously.

Space-Time Values and Sustainability Sustainable development requires a long-term perspective. Unfortunately, most of us are rather narrow in our outlook. **FIGURE 18-14** is a hypothetical graph of where our values lie. One axis plots our concern in space, the other in time. Space refers to self, family, friends, community, state, and so on, expanding outward. Time, obviously, ranges from the present to the long-term future.

As the scatter diagram shows, individual interest can be identified by a single point that denotes one's space and time concerns. These are called **space-time values.** Most people's interests lie toward the lower end of the scales, tending toward self-interest and immediate concerns. Some people call this selfishness, but it can also be considered a natural biological tendency to be concerned with the self. Among animals, awareness of the needs of others is a feature only of social creatures such as

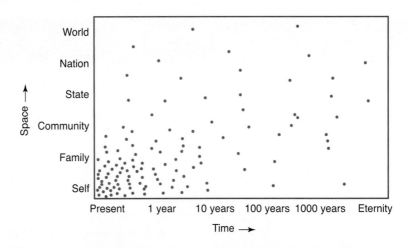

FIGURE 18-14 Space-time values. Graph of people's hypothetical spatial and temporal interests, indicated by the points. Most individuals tend toward the lower end of the scales, being concerned primarily with self and the present. This is a natural tendency, but the rapid pace of global environmental change makes it necessary to shift space-time values to reflect long-term needs and to stimulate actions needed to build a sustainable society. Unfortunately, most political leaders are mired in the immediate—given their human nature, the short terms of office, and reelection concerns.

monkeys and lions; however, concern for the uppermost end of the space-time graph is a distinguishing feature of the human animal. The unique human ability to ponder the consequences of actions is fortunate because humans have reached a position of unprecedented power as molders of the biosphere. Our power to change the world to our liking has never been greater, nor has our power to destroy it.

One of the things you have learned about sustainable development is that it takes a long-range view. It requires us to push the limits in space and time.

Chance fights ever on the side of the prudent.

—Euripides

CRITICAL THINKING

Exercise Analysis

Pollution control devices cost money, and they usually remove pollutants from a waste stream and create a concentrated waste that must be disposed of elsewhere. Pollution prevention, on the other hand, involves adjustments to the production system that reduce or even eliminate hazardous emissions. Several options are available under the pollution prevention strategy: modifying production processes, finding substitutes for toxic chemicals, recycling wastes, and a number of other cost-effective measures. Not only will these help you comply with the emissions standards and save your company huge sums of money, but you may actually exceed the standards—that is, reduce pollution emissions below those stipulated by state and federal laws and regulations.

Pollution has been dealt with after the fact for many years. In this exercise, all it took was a look at the bigger picture (alternatives to pollution control) and a little digging into the various options to find some viable solutions. This exercise is a good reminder to you to beware of conventional strategies for solving problems—not just in the environmental arena, but in all areas. Often, solutions that people propose reflect old ways of thinking—and therefore overlook creative and cost-effective measures that could help save money and build a sustainable society.

CRITICAL THINKING AND CONCEPT REVIEW

1. Define the terms *toxicant, carcinogen, teratogen*, and *mutagen*.
2. Using your critical thinking skills and your knowledge of toxicology, analyze the following statement: "All substances are toxic. It's the dose that makes the poison."
3. Describe how toxicants affect the cells of our bodies.
4. What is cancer? Discuss the steps in its formation.
5. Using your critical thinking skills, analyze the following statement: "If a chemical is found to be carcinogenic in a laboratory animal, it is likely to be carcinogenic in humans."
6. List some of the possible consequences of somatic and germ cell mutations in humans.
7. What is teratology? Do teratogenic chemicals always create birth defects when given during pregnancy? Why or why not?
8. Make a list of factors that influence the toxicity of a chemical in a given individual.
9. Define the terms *synergism, antagonism*, and *potentiation*. What is the difference between synergism and an additive effect?
10. A research study shows that a certain chemical is toxic to laboratory mice. The researchers believe that it may also be harmful to humans. What information would you need to support controls on human exposure?
11. Define the terms *bioaccumulation* and *biological magnification*. Based on your knowledge gained in this chapter, what factor(s) can be used to predict whether a chemical will be biologically magnified?
12. What is meant by a threshold level? Explain the reasons why threshold levels exist for many toxicants.
13. Describe the major provisions of the Toxic Substances Control Act.
14. What are the two major types of risk? Give examples.
15. Describe the steps involved in determining the level of risk posed by technology
16. What factors determine whether a risk is acceptable to a population? What is the difference between voluntary and involuntary risks? What is the difference between actual and perceived risks?
17. Many more people die in Montana and Wyoming from falls while hiking than are killed by grizzly bears. Why, then, are people so concerned about being killed by a bear when their chances of being killed in a fall are much greater?
18. What are space-time values? In general, where do your concerns lie in space and time?
19. Critically analyze the market-based incentives for controlling toxic substances that were discussed in this chapter.
20. The following statement was made at a public hearing regarding the contamination of a lake by effluents containing hazardous wastes generated by a chemical manufacturing plant. Using your critical thinking skills and your knowledge of toxicology, analyze the statement: "Our facility will produce hazardous substances, but because we are using state-of-the-art pollution control devices, releases of mercury and other toxicants will be minimal. Levels in the lake receiving our treated wasthewater will be rather small and should pose no threat to people drinking the water."

KEY TERMS

actual risk
acute
additive response
antagonism
anthropogenic hazards
asbestosis
benign tumor
bioaccumulation

biological magnification
biotransformation
birth defect
cancer
carcinogens
chromosomes
chronic
cost–benefit analysis

deoxyribonucleic acid (DNA)
DNA-reactive carcinogens
dose
dose-response curve
duration of exposure
environmental hormones
epidemiologists
epigenetic carcinogens

estimation of risk
genotoxic carcinogens
germ cells
hazard identification
hypersensitized
LD$_{50}$
malignant tumor
mesothelioma
metastasis
multimedia approach
mutagens
mutations
natural hazards
neoplasm
perceived benefit

perceived harm
perceived risk
pollution prevention
polychlorinated biphenyls (PCBs)
pollution (green) taxes
potentiation
premanufacture notification
primary tumor
pulmonary fibrosis
risk acceptability
risk assessment
Safe Drinking Water and Toxics
    Enforcement Act
secondary tumors
somatic cells

space-time values
spontaneous abortion
stillbirth
synergistic response
systemic
teratogens
teratology
threshold level
Toxic Substances Control
    Act (TSCA)
toxicants
toxicology
tradable permits
tumor
zooplankton

© De Visu/Shutterstock, Inc.

Air Pollution and Noise: Living and Working in a Healthy Environment

Not life, but a good life, is to be chiefly valued.

—Socrates

LEARNING OBJECTIVES

After reading and studying the material in this chapter, students will be able to understand, knowledgably discuss, and, when appropriate, apply the information learned in this chapter, including:

1. Natural and human sources of air pollution
2. Types of air pollution, including primary and secondary pollutants
3. Why air pollution is a sign of unsustainable human systems
4. How weather and topography affect air pollution levels
5. The many effects of air pollution including health effects and impacts on other species and materials
6. The many ways we can reduce or eliminate air pollution
7. How noise affects us and ways to reduce noise pollution
8. The importance of reducing indoor air pollution, sources of indoor air pollution, and ways to reduce our exposure to these toxic chemicals

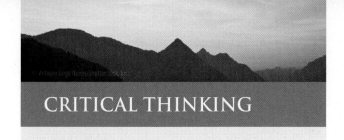

CRITICAL THINKING

Exercise

In a speech to a group of business executives, one opponent of the U.S. Clean Air Act noted that natural sources of air pollution such as volcanoes often exceed human sources. Because of this he argued that air pollution control legislation was misguided. He went on to say that he supported efforts in the U.S. Congress to weaken air pollution legislation because it was not only unnecessary but also hurt the economy. Regulations, he claimed, cost jobs and reduced the nation's productivity. Analyze this premise, using your critical thinking skills and your knowledge of air pollution.

At the Ford automobile manufacturing plant in Cologne, Germany, plant managers modernized the paint-spray line, cutting pollution generated by their operation by 70%. This change reduced the cost of painting a car by about $60 and will net the company hundreds of thousands of dollars in the next decade. Another German company that makes plastic films cut emissions by instituting controls that capture 90% of solvents previously lost to the atmosphere. The savings will pay for the technological improvements and are expected to save the company considerable sums of money in years to come. These two examples are part of a trend among industries designed to reduce operating costs while protecting the air we breathe.

This chapter discusses ambient air pollution—that is, pollution in the outside environment—and indoor air pollution. It looks at traditional strategies and sustainable approaches that make sense economically and environmentally. It also examines noise pollution.

19.1 Air: The Endangered Global Commons

Air is a mixture of gases, primarily nitrogen (78%) and oxygen (21%). It also contains small amounts of carbon dioxide (0.04%) and even smaller amounts of several inert substances, among them argon (almost 1%), helium, xenon, neon, and krypton. Suspended in this gaseous mixture is water vapor, which exists in varying amounts (depending on the atmospheric humidity). Air also contains various amounts of pollutants, chemicals that can have adverse effects on animals, plants, and materials.

The air we breathe is a renewable resource that is replenished by natural processes. Oxygen, for example, is replenished by plants and algae in fresh water and salt water. Air is also cleansed by other natural processes—for example, rain. Rain scours many pollutants from the sky, washing them into lakes and into our soils, sometimes with adverse effects.

Transparent and nurturing, air is a global resource, a global commons that benefits all of humanity and all the species that share this planet with us. Owned by no one and used by all of us, it is a source of oxygen vital to animals and carbon dioxide needed by plants. Industrial societies, however, have traditionally used the air as a waste dump for hundreds of pollutants. Like other commons, no one has sole responsibility for protecting our air, and even those who do not pollute it suffer from the disregard of others.

 KEY CONCEPTS

Air is a renewable resource cleansed by natural processes and regenerated by plants and algae. This global resource is used by many and protected by few. It suffers from the tragedy that befalls many commons.

Sources of Air Pollution

Pollutants in the atmosphere arise from two major sources: natural and anthropogenic. It may be surprising to learn that globally, the largest sources of many air pollutants are natural. Natural events such as volcanoes, dust storms, and forest fires produce huge quantities of air pollution each day (**TABLE 19-1**). Thus, in sheer quantity, natural pollutants often outweigh the products of human activities, the **anthropogenic pollutants,** pollutants from human sources. Nevertheless, anthropogenic pollutants generally create the most significant long-term threat to the biosphere. Why? Natural pollutants, except those from volcanoes, come from widely dispersed sources or infrequent events and therefore generally do not raise the ambient pollutant concentration very much. As a result, they have little effect on biological systems. In contrast, power plants, automobiles, factories, and other human sources emit large quantities in restricted areas, making a significant contribution to local pollution levels. The higher the concentration, the greater the effect.

 KEY CONCEPTS

Pollutants arise from natural and human or anthropogenic sources. Human-generated pollution is generally of greatest concern because it is produced in localized regions so that concentrations reach potentially dangerous levels.

TABLE 19-1	
Natural Air Pollutants	
Source	**Pollutants**
Volcanoes	Sulfur oxides, particulates
Forest fires	Carbon monoxide, carbon dioxide, nitrogen oxides, particulates
Wind storms	Dust
Plants (live)	Hydrocarbons, pollen
Plants (decaying)	Methane, hydrogen sulfide
Soil	Viruses, dust
Sea	Salt particulates

Anthropogenic Air Pollutants and Their Sources

Take a deep breath. If you live in a city or even a large town, chances are you have just inhaled tiny amounts of dozens of different air pollutants, most in concentrations too small (we think) to be harmful. If you are indoors, you may have inhaled even greater amounts because indoor air is often more contaminated than outdoor air, a subject discussed shortly.

This chapter concerns itself primarily with six major pollutants: carbon monoxide, sulfur oxides, nitrogen oxides, particulates, hydrocarbons, and ozone. Lead, another

important air pollutant, as well as carbon dioxide and chlorofluorocarbons are discussed. Although all of the pollutants are important, the U.S. Environmental Protection Agency (EPA) only regulates six: carbon monoxide, nitrogen oxides, sulfur oxides, particulates, ground-level ozone, and lead.

The criteria air pollutants are generated by three principal sources: transportation (autos, trucks, jets, and trains), fuel combustion at stationary sources (mostly power plants and factories), and various industrial processes.

Air pollutants arise from vaporization (or evaporation), attrition (or friction), and combustion. Combustion is by far the major producer. Among combustion sources, the burning of fossil fuels (coal, oil, and natural gas) and products refined from them (gasoline, diesel, jet fuel, and home heating oil) pose the most significant threat to the environment. In fact, the more developed nations' heavy reliance on fossil fuels, which have fueled most progress, is one of the root causes of the crisis of unsustainability.

Fossil fuels consist primarily of carbon and hydrogen atoms linked by chemical bonds. When they are ignited, the initial source of heat—say, a match—breaks some of the chemical bonds holding the atoms together. This releases energy in two forms: light and heat. Heat released in the process breaks other bonds, permitting combustion to occur until the fuel runs out. During combustion, oxygen in the air reacts with the carbon and hydrogen atoms of the fossil fuel to form carbon dioxide (CO_2) and water (H_2O). More commonly, combustion is not 100% complete. Incomplete combustion produces carbon dioxide, carbon monoxide (CO) gas, and unburned hydrocarbons (**FIGURE 19-1**).

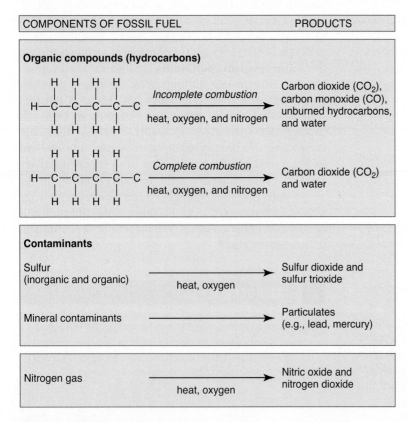

FIGURE 19-1 The chemistry of pollution. Products of fossil fuel combustion.

Many fuels are contaminated by heavy metals, such as mercury or lead. These unburnable contaminants may be carried off by hot combustion gases, escaping into the air as fine particulates. Other contaminants are also present. One of them is sulfur. It is especially prevalent in coal. Sulfur reacts with oxygen at high combustion temperatures, forming sulfur oxide gases, notably sulfur dioxide (SO_2) and sulfur trioxide (SO_3) (Figure 19-1). In the absence of pollution control devices, these gases escape with the other smokestack gases.

Combustion must take place in the presence of air, which provides oxygen. But air also contains nitrogen. During high-temperature combustion, nitrogen (N_2) reacts with oxygen to form nitric oxide (NO). NO is quickly converted to nitrogen dioxide (NO_2), a brownish orange gas seen in the air of many cities. Nitrogen dioxide is a key reactant in the formation of photochemical smog, described shortly. Nitrogen dioxide can also react with water in the atmosphere to produce nitric acid. For a review of the sources and effects of the criteria air pollutants, see **TABLE 19-2**.

Primary and Secondary Pollutants

The atmosphere contains hundreds of air pollutants from natural and anthropogenic sources. These pollutants are called **primary pollutants** because they are produced directly by various sources. Primary pollutants often react with one another or with water vapor. These reactions, which are often powered by energy from the sun, produce a whole new set of pollutants called **secondary pollutants.** For example, sulfur dioxide gas is a primary pollutant released from a variety of sources, such as coal-fired power plants, diesel trucks, and automobiles. In the atmosphere, SO_2 reacts with oxygen and water to produce sulfuric acid (H_2SO_4), a toxic and corrosive secondary pollutant with far-reaching effects.

Toxic Air Pollutants

Health officials and environmental activists have long been concerned about the effects of criteria air pollutants. Many are also concerned about hundreds of other potentially

TABLE 19-2

Major Air Pollutants: Their Sources and Health Effects

Pollutant	Major Anthropogenic Sources	Health Effects
Carbon monoxide	Transportation (cars, trucks, vans, SUVs, buses, and planes)	Acute exposure: headache, dizziness, decreased physical performance, death Chronic exposure: stress on cardiovascular system, deceased tolerance to exercise, heart attack
Sulfur oxides	Stationary combustion sources such as coal-fired power plants, industry	Acute exposure: inflammation of respiratory tract, aggravation of asthma Chronic exposure: emphysema, bronchitis
Nitrogen oxides	Transportation, stationary combustion sources such as coal-fired power plants	Acute exposure: lung irritation Chronic exposure: bronchitis
Particulates	Stationary combustion sources such as coal-fired power plants, industry	Irritation of respiratory system, cancer
Hydrocarbons	Transportation (cars, trucks, vans, SUVs, buses, and planes)	Unknown
Photochemical oxidants	Transportation, stationary combustion sources (indirectly through hydrocarbons and nitrogen oxides)	Acute exposure: respiratory irritation, eye irritation Chronic exposure: emphysema

toxic pollutants released into the atmosphere each year from factories and other sources. By some estimates, over 400 toxic air pollutants are released into the air in the United States. Toxic air pollutants can cause mutations, cancer, and possibly birth defects. Although generally emitted in much smaller quantities than the criteria pollutants, toxic air pollutants may be responsible for numerous cancer deaths each year. The U.S. EPA estimates that 45 of the toxic air pollutants may cause as many as 1,700 cases of cancer each year.

(a)

KEY CONCEPTS

Although much of the interest has focused on criteria air pollutants, many other potentially toxic chemicals are released into the air each year from industry and other sources.

Industrial and Photochemical Smog

Air pollution in cities generally falls into one of two categories, based on the climate and the type of air pollution. For years, older industrial cities such as New York, Philadelphia, St. Louis, Pittsburgh, and London belonged to a group of **gray-air cities** (**FIGURE 19-2A**). Newer, relatively nonindustrialized cities such as Denver, Los Angeles, Salt Lake City, Albuquerque, and Vancouver, Canada, belong to the group of **brown-air cities** (**FIGURE 19-2B**).

Gray-air cities were generally located in cold, moist climates. The major pollutants are sulfur oxides and particulates from factories. These pollutants combine with atmospheric moisture to form the grayish haze called **smog,** a term coined in 1905 to describe the mixture of smoke and fog that plagued industrial England. The gray-air cities depend greatly on coal and oil and are usually heavily industrialized. The air in these cities is especially bad during cold, wet winters, when both the demand for home heating oil and electricity and the atmospheric moisture content are high. Many of China's cities are gray-air cities due to the country's heavy reliance on coal.

Brown-air cities are typically located in warm, dry, and sunny climates and are generally newer cities with few polluting industries. The major sources of pollution in them are automobiles and electric power plants; the major primary pollutants are carbon monoxide, hydrocarbons, and nitrogen oxides.

In brown-air cities, unburned hydrocarbons and nitrogen oxides from automobiles and power plants react in the presence of sunlight to form a witch's brew of harmful secondary pollutants. Some of the worst are ozone, formaldehyde (the same chemical used in embalming fluid), and peroxyacylnitrate (PAN). The reactions that form these and other potentially toxic chemicals are called *photochemical reactions* because they involve both sunlight and chemical pollutants. The resulting brownish orange shroud of air pollution is known as **photochemical smog.** It's often referred to aix the brown cloud. Ozone (O_3) is one of the

(b)

FIGURE 19-2 A tale of two cities. (a) Gray-air smog in New York City. **(b)** Brown-air smog in Los Angeles.

Photo a © Dean D. Fetterolf/ShutterStock, Inc.; photo b © Chad Littlejohn/ShutterStock, Inc.

most prevalent chemicals in photochemical smog. This molecule is highly reactive. It erodes rubber, irritates the respiratory system and eyes, and damages crops, trees, and other plants.

Go Green

To reduce pollution, walk, ride a bike, or take a bus or train whenever possible.

In brown-air cities, early morning traffic provides the ingredients for photochemical smog, which usually reaches the highest levels in the early afternoon (**FIGURE 19-3**). Because the air laden with photochemical smog often drifts out of the city and because the reactions require sunlight and take time to occur, the suburbs and surrounding rural areas usually have higher levels of photochemical smog than the cities themselves. Major pollution episodes in brown-air cities usually occur during the summer months, when the sun is most intense. Pollution may extend for many miles. In Vancouver, British Columbia, photochemical smog may cover an area 50 kilometers (30 miles) downwind.

Today, the distinction between gray- and brown-air cities is disappearing. Most cities have brown air in the summer

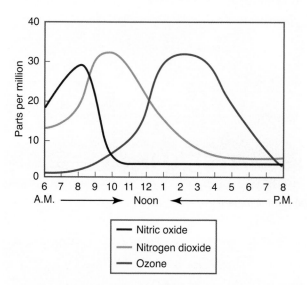

FIGURE 19-3 Photochemical smog. Nitrogen oxides and hydrocarbons (not shown here) react to form ozone and other photochemical oxidants. Because sunlight and time are required for the reactions to occur, maximum ozone concentration occurs in the early afternoon. Hydrocarbon levels would follow the same pattern as nitric oxide levels.

when sunlight and automobile pollutants are prevalent and gray air in the winter when pollution from wood stoves coal-fired power plants, and oil burners conspires with the moist, wet air to darken the skies.

Although anthropogenic pollutants are the most significant sources of air pollution, researchers have found some instances in which naturally occurring pollutants noticeably affect air quality. In Atlanta, for example, trees naturally emit a number of highly reactive hydrocarbons. These chemicals react with nitrogen dioxide from automobiles and other combustion sources to produce ozone. New research shows that hydrocarbons from trees are 50 to 100 times more reactive than hydrocarbons from human sources.

Pollution control often requires an analysis of both natural and anthropogenic sources. Such studies can keep us from making costly mistakes. For instance, the EPA once thought that the city of Atlanta could meet federal air quality standards for ozone by reducing the level of hydrocarbons from human sources, mostly automobiles, by 30%. New data, however, suggest that when the contribution from trees is added, the human-related sources would have to be cut by 70 to 100%. Thus efforts to cut nitrogen oxide were deemed more effective.

KEY CONCEPTS

Smog is a term applied to two different types of air pollution. In cities with drier, sunnier climates and minimal industrial activity, hydrocarbons and nitrogen oxides from motor vehicles and other sources react to form a brownish haze called *photochemical smog*. In older industrial cities with moister climates, particulates and sulfur oxides form *industrial smog*.

Air Pollution—A Symptom of Unsustainable Systems?

Criteria air pollutants come from a variety of sources—the main ones being motorized vehicles, factories, power plants, and homes. These, in turn, are components of basic human systems of transportation, industry, energy, and housing. To some observers, air pollution is viewed as an indication that human systems are unsustainable. This text focuses on ways to make these systems sustainable through ecological design, energy conservation, renewable energy, and a host of other measures. Such efforts are an attempt to create human systems that operate in ways that do not threaten the climate and other natural systems upon which we depend. They are, in short, a way to improve our prospects, create a more prosperous society, a healthier population, and a better future.

KEY CONCEPTS

Air pollution, like other forms of environmental deterioration, is a symptom of unsustainable systems of transportation, industry, housing, and energy production.

19.2 The Effects of Weather and Topography on Air Pollution

Many factors affect ambient air pollution levels. The following section examines some of the most important ones including wind and rain, topography, and temperature inversions.

The Cleansing Effects of Wind and Rain: Don't Be Fooled

Pollution is something of an enigma to many people. One day skies are clear. The next, the skies are filled with brown haze. Why?

Numerous factors contribute to this puzzle. Wind, for example, sweeps dirty air out of cities so that on a windy day the air above a city appears clean. Rain also cleanses the air in and around our cities. The folks in Seattle, for example, may brag about their clean air. The fact is, they're producing as much pollution as any other city of similar size; it just tends to be blown away or washed from the sky by rain.

In contrast to what many people think, air pollutants do not disappear. They are merely shifted elsewhere or deposited onto another medium—for example, surface waters or the soil or onto buildings. This transfer is known as **cross-media contamination**. Airborne pollutants can travel hundreds, sometimes thousands, of kilometers to other cities or to unpolluted wilderness, where they are deposited. In Canada, for example, ground-level ozone pollution in the Windsor-Quebec corridor and southern New Brunswick and western Nova Scotia comes not from Canadian sources, but largely from the United States.

Wind and rain both tend to cleanse the air above our cities, but pollutants do not disappear. They're either blown elsewhere or fall from the sky, ending up in waterways or in our soil or on buildings and other structures.

Mountains and Hills

Salt Lake City stretches out below a giant mountain range. Granite peaks and jagged cliffs make this one of the most scenic American cities—that is, when the mountains are visible through the air pollution. Residents of Salt Lake City would tell you that mountain ranges can be an asset, but they can also be a curse for they often block the flow of winds and trap pollutants for days on end. Mountains also block the sun, which helps disperse pollutants, as explained shortly. In Colorado and other mountainous states, ski resort towns nestled in the valleys among towering peaks often experience elevated levels of pollution as a result of their protected locations.

Mountains and hilly terrain can impede the flow of air, resulting in the buildup of pollutants in cities, towns, and industrialized areas.

Temperature Inversions

If you rode a hot-air balloon into the sky on a normal summer day, you would find that warm ground air rises. As it rises, it expands and cools. Thus, as you lifted into the sky, you'd find that the temperature decreases. If you had proper equipment to measure pollution, you would also find that pollution rises with the warm ground air. Because of this, the atmosphere is gently stirred, and ground-level pollution is reduced.

Atmospheric mixing is caused in part by sunlight. As sunlight strikes the Earth, it heats the rock and soil. This heat is transferred to the air immediately above the ground. The warm air then rises, mixing with cooler air above it. The temperature profile is shown in **FIGURE 19-4A**.

Under certain atmospheric conditions, a slightly different temperature profile is encountered. As **FIGURE 19-4B** shows, on some days air temperature drops as one gains altitude, but only to a certain point. After that point, the temperature begins to increase. This inverted temperature profile is called a **temperature inversion**. This bizarre temperature profile results in the creation of a warm-air lid over cooler air (Figure 19-4b). Because the cool, dense ground air cannot mix vertically, pollutants become trapped in the lower layers, sometimes reaching dangerous levels. How do temperature inversions form?

Temperature inversions occur in two ways. A **subsidence inversion** occurs when a high-pressure air mass (warm

(a) Normal pattern

(b) Thermal inversion

FIGURE 19-4 Temperature profiles. (a) During normal conditions, air temperature decreases with altitude; thus, pollutants ascend and mix with atmospheric gases. **(b)** In a temperature inversion, however, warm air forms a lid over cooler air, thus trapping air pollution.

front) slides over a colder air mass that failed to move out when the warm-air front arrived. Subsidence inversions may extend over many thousands of square kilometers.

A **radiation inversion,** the second type, is typically local and usually short lived. It is a phenomenon many of us witness on cold winter days. Radiation inversions begin to form a few hours before the sun sets. As the day ends, the air near the ground cools faster than the air above it. Thus, warm air lies over the cooler ground air, preventing it from rising and causing pollutants to accumulate. Making matters worse, radiation inversions also correspond with the evening commute, when traffic and pollution emissions peak. Fortunately, radiation inversions usually break up in the morning when the sun strikes the Earth and vertical mixing begins. Radiation inversions are common in mountainous regions—especially in the winter, when the sun is obstructed by the mountains and is therefore unable to warm the ground enough to stimulate vertical mixing.

Temperature inversions result in the formation of warm-air lids that form over cities and even large regions, impeding the vertical mixing of air. This, in turn, traps cooler pollutant-laden air below.

19.3 The Effects of Air Pollution

Despite significant progress in pollution control and declining emissions, over 116 million Americans—about 4 in 10 (38.5%) people—are breathing unhealthy air—air that violates federal air pollution standards for ozone at some point during the year.[1] According to the U.S. EPA, 195 counties in 24 states failed to meet federal air quality standards for ozone. Most counties are in the eastern third of the United States, although heavily populated California tops the list. The Los Angeles basin had the worst photochemical smog problem. Ozone poses the biggest problem. Particulates rank second, despite dramatic declines since 1970. All told, about 50 million Americans breathe air containing potentially harmful levels of particulates. According to a recent government report, approximately 60,000 Americans a year will die prematurely because of exposure to air pollution. Although 195 counties failed to meet federal air pollution standards, 2,905 counties met all standards. Eighteen states had all counties in compliance: Alaska, Florida, Hawaii, Idaho, Iowa, Kansas, Minnesota, Mississippi, Montana, Nebraska, New Mexico, North Dakota, Oklahoma, Oregon, South Dakota, Utah, Vermont, and Washington.

The American Lung Association (conservatively) estimates that air pollution costs Americans $50 billion a year in health costs, or about $200 per year for every man, woman, and child. Air pollution damages crops and buildings, costing us billions each year. Estimates put ozone crop damage at $5 to $10 billion a year. Many other costs cannot be calculated: the loss of scenic views, the destruction of a favorite fishing spot, the erosion of a valuable and important statue.

Air pollution causes extraordinary environmental and economic damage worldwide. In Europe, 20 to 25% of the forests are dead or damaged from factors such as disease, adverse weather, insects, and air pollution. Air pollutants weaken trees, making them more susceptible to a host of biotic and abiotic factors. The economic impact of such losses could be in the tens of billions of dollars.

Air pollution causes considerable damage in the less developed countries, which (with the exception of China) produce only a fraction of the world's air pollutants. Although no statistics are available, damage to health, forests, buildings, and statuary surely carries with it an enormous price tag.

KEY CONCEPTS

Air pollution adversely affects human health, damages the environment and the organisms that live in it, and damages buildings and a wide assortment of materials, costing billions of dollars a year in North America alone.

---

[1]Environmental Protection Agency, *Our Nations Air–status and Trends through 2010.* www.epaogov/airtrends/2011.

The Health Effects of Air Pollution

This section examines the impacts of pollution on human health, other organisms, and materials. These costs are typically referred to as *external costs*, as noted in previous chapters, because they are almost universally not factored into the costs of the activities (such as driving an automobile) that produce them. Air pollution has a variety of health effects, ranging from immediate to delayed and from slight irritation to potentially life-threatening conditions.

Immediate Health Effects Air pollutants may cause many immediate effects on human health, from shortness of breath to eye irritation to death (Table 19-2). Visitors to Los Angeles, for example, often complain of burning or itching eyes and irritated throats caused by photochemical smog. Residents in other cities report similar problems, especially in less developed nations where pollution laws are lax or nonexistent.

Some problems are not very obvious. Commuters in heavy traffic, for example, sometimes complain of headaches. Although they are often attributed to stresses and strains of work or highway congestion, they may also be due to carbon monoxide from automobile exhaust.

In 1966, researchers found that a sudden increase in levels of sulfur dioxide in New York City caused by adverse weather conditions resulted in a dramatic increase in the incidence of colds, coughs, rhinitis (nose irritation), and other symptoms. When the air cleared, the symptoms disappeared. More recent studies have shown that pollution episodes occurring in the summer months in the northeastern United States and Canada, which are characterized by an increase in sulfate and ozone, increase the incidence of hospital admissions for respiratory disease.

Far more serious health problems have been observed in industrial nations during pollution episodes—periods when atmospheric concentrations reach dangerous levels for short periods because of changes in the weather that impede the cleansing of the air. Several notorious episodes have been documented. The very first occurred in the Meuse Valley in Belgium in 1930; the second in Donora, Pennsylvania, in 1948; and the third in London in 1952.

Go Green

Donate used clothing to Goodwill or other similar organizations. Consider buying used clothing, too.

In each incident, pollution rose to dangerous levels as a result of temperature inversions that lasted 3 to 4 days. In addition, in each episode, the pollution came primarily from the combustion of coal in homes and factories. In Belgium, 65 people died. In Pennsylvania, 20 succumbed, and in London, 4,000 people may have died as a result of toxic levels of particulates and sulfur dioxide. In all locations, many other people became ill. Those who died or became ill were usually old or suffering from cardiovascular disease—and were therefore unable to cope with the added stress caused by the heavily polluted air.

Air pollution causes many immediate effects, such as shortness of breath, eye irritation, or upper respiratory tract irritation. In extreme cases, pollutants can become lethal.

Chronic Health Effects Exposure to some toxic substances may take place over many years and at low levels. This, in turn, may result in a variety of health effects.

In 2008, nearly 10 million Americans suffered from **chronic bronchitis,** a persistent inflammation of the bronchial tubes, which carry air into the lungs. Symptoms include mucus buildup, a persistent cough, and difficulty breathing. Cigarette smoking is a major cause of this disease, but urban air pollution is also a major contributing factor. Sulfur dioxide, nitrogen dioxide, and ozone are believed to be the major causative agents.

Emphysema is another chronic effect. Currently, nearly 3 million Americans suffer from this incurable disease, according to the American Lung Association. As these people become older, the small air sacs, or **alveoli,** in their lungs break down. This reduces the surface area for the exchange of oxygen with the blood. Breathing becomes more and more labored. When lung surface area is reduced by about 40%, victims suffer shortness of breath when exercising lightly.

Emphysema is caused primarily by cigarette smoking (about 80% of all cases) but may be caused by urban air pollution as well. One study, for instance, showed that the incidence of emphysema was higher in relatively polluted St. Louis than in relatively unpolluted Winnipeg (**FIGURE 19-5**). Studies in Great Britain showed that mail carriers who worked in polluted urban areas had a substantially higher death rate from emphysema than those who worked in unpolluted rural areas. Three air pollutants—ozone, nitrogen dioxide, and sulfur oxides—are all believed responsible for this disease.

More recently, several studies have shown that **bronchial asthma** attacks may be associated with elevated levels of pollution. Asthma is a chronic disorder, marked by periodic episodes of wheezing and difficult breathing. Most cases of asthma are caused by allergic reactions to common stimulants such as dust, pollen, and skin cells (dander) from pets. In some individuals, pollution may trigger asthma attacks. During such an attack, the passageways that carry air to the lungs (bronchi and bronchioles) fill with mucus, making breathing difficult. Irritants also stimulate the contraction of the smooth muscle cells in the walls of the smallest air-carrying ducts, the bronchioles, making it even more difficult to breathe. Periodic attacks can be quite disabling and may even lead to death. By one estimate, several thousand Americans die each year from severe asthma attacks. Victims are generally older individuals who are suffering from other diseases. Recent studies suggest that fine particulates are to blame for asthmatic attacks caused by air pollution.

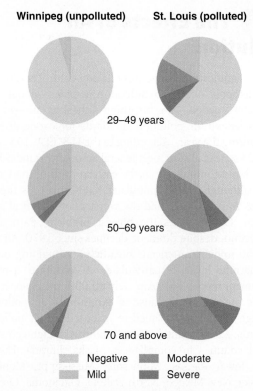

FIGURE 19-5 Urban air pollution and emphysema. Incidence of emphysema in Winnipeg and St. Louis. Note the increased incidence of emphysema in all three age groups in the more polluted urban environment of St. Louis.

A number of studies have shown that lung cancer rates are slightly higher among urban residents than among rural residents (even after the influence of cigarette smoking has been ruled out). Critical thinking recommends caution because health studies of human populations, or epidemiological studies, rely on statistical methods with some unavoidable shortcomings. In lung cancer studies, for example, researchers usually compare death certificates of urbanites with death certificates from rural residents. If the urban population shows a higher incidence of lung cancer, it is tempting to conclude that the cause was urban air pollution. But researchers must be careful to eliminate other causative agents, such as smoking and occupation. Most researchers eliminate smokers from the study, but not all take into account urban workplaces (factories), where men and women may have been exposed to high levels of cancer-causing pollutants. Thus, the slightly higher incidence of lung cancer in some urban settings may result from occupational exposure or some other factor. It could also result from pollution. Only time and further research will tell.

Long-term exposure to air pollution may result in a number of debilitating diseases, including bronchitis, emphysema, asthma, and lung cancer.

High-Risk Populations Not all individuals are affected equally by air pollution. Particularly susceptible are the old and infirm, especially people with preexisting lung and heart disorders. Carbon monoxide is especially dangerous to people with heart disease because CO binds strongly to hemoglobin, a protein in the red blood cells (RBCs). Hemoglobin normally binds to oxygen, which is transported through the bloodstream in the RBCs. The binding of CO to hemoglobin reduces the oxygen-carrying capacity of the blood. For sufficient oxygen to be delivered to the body's cells, the heart must pump more blood during a given period. This puts a strain on the heart and may trigger heart attacks in those whose hearts are already weakened.

Toxic substances often affect the young more severely than adults. Air pollution is no exception. In fact, some researchers estimate that the health risk from air pollution is six times greater for children than for adults. Several reasons may account for this difference. First, children may be more susceptible than healthy adults because they are more active and therefore breathe more. As a result, they may be exposed to more pollution. In addition, children typically suffer from more frequent colds and nasal congestion and thus tend to breathe more through their mouths. Air bypasses the normal filtering mechanism of the nose; thus, more pollutants enter the lungs.

 KEY CONCEPTS

Three groups are generally the most susceptible to air pollution: the young, the old, and the infirm (sick).

Effects on Plants and Nonhuman Animals

Air pollutants at levels people are exposed to have a wide range of effects on experimental animals. Unfortunately, very little research has been performed to study the effects of air pollution on wild species. We do know, however, that fluoride and arsenic released from smelters can seriously poison cattle grazing downwind. Acids produced from pollutants released by power plants, smelters, industrial boilers, and motor vehicles have been shown to be extremely harmful to wildlife, especially fish. In southern California, millions of ponderosa pines have been damaged by air pollution (mostly ozone) from Los Angeles.

Ozone, sulfur dioxide, and sulfuric acid are the pollutants most hazardous to plants (**FIGURE 19-6**). Ozone, for instance, makes plants more brittle. Farms in southern California and on the East Coast report significant damage to important vegetable crops. City gardeners also report damage to flowers and ornamental plants. Urban landscapers must plant pollution-resistant trees, notably locusts, along heavily traveled roads, but even these live only a fraction of their normal life span.

Sulfur dioxide damages plants directly, causing spotting of leaves. Recent studies show that air pollutants may

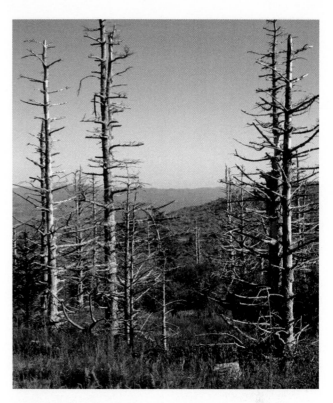

FIGURE 19-6 Dying forest. These trees in the Appalachian Mountains have been killed by acids in rain and snow, which are produced by pollution from automobiles, factories, and coal-fired power plants.

© Mary Terriberry/ShutterStock, Inc.

also make some plant species more desirable to leaf-eating insects. According to botanists from Cornell University, air pollution and other stresses cause plants to produce a chemical called *glutathione*, which protects leaves from pollution but also attracts insects that normally have no interest in these plant species.

One group of organisms that is extremely sensitive to air pollution is the lichens, which grow on rock, wood, or soil (**FIGURE 19-7**). Lichens are composite organisms, consisting of a fungus and an internal alga. Because they are amazingly hardy and resilient, lichens can endure a wide variety of climates, often living where few other life-forms exist. However, lichens obtain nourishment from the air, rain, and snow, which makes them extremely vulnerable to air pollution.

In the 1800s, scientists began to notice that lichens were being killed by pollution in many European cities. In the 1860s, 32 species of lichens were collected in the Jardin du Luxembourg in Paris. By 1896, not a single species survived. In the town of Mendlesham, England, 129 species were identified between 1912 and 1921. In 1973, only 67 remained. In England, researchers found that lichens were absorbing heavy metals and pollution; wherever pollution increased, the lichens died.

So sensitive are lichens to air pollution that air quality and the spread of pollution from an industrial source can be determined by mapping the presence or absence of them or by chemically analyzing the lichens in the area.

FIGURE 19-7 Lichens. This organism lives on rock and erodes away its surface. It is extremely sensitive to air pollution.

© Don Farrall/Photodisc.

Effects on Materials

Air pollutants can also severely damage metals, building materials (stone and concrete), paint, textiles, plastics, rubber, leather, paper, clothing, and ceramics (**TABLE 19-3**). The four most corrosive and harmful pollutants are sulfur dioxide, sulfuric acid, ozone, and nitric acid.

The Statue of Liberty, for example, required a costly ($35 million) face-lift after years of exposure to two secondary pollutants, sulfuric and nitric acids, produced from sulfur and nitrogen oxides. The Taj Mahal in India, like many other buildings in the world, is being defaced by sulfur oxide air pollution from local power plants.

Sulfuric and nitric acids not only cause cosmetic damage to metals, they also reduce their strength. In the Netherlands, bells that had been ringing true for 3 or 4 centuries have gone out of tune because of acid pollutants that have eaten away at them, lowering their pitch and rendering once familiar tunes indecipherable. Ozone cracks rubber windshield wipers, tires, and other rubber products, necessitating costly antioxidant additives. Particulates also cause damage. For example, particulates blown in the wind erode the surfaces of stone.

Society pays a huge price tag for cleaning sooty buildings, repainting pitted houses and automobiles, and replacing damaged rubber products. The damage can also be tragic, for air pollution attacks irreplaceable works of art such as the marble of the Parthenon in Athens, which has deteriorated more in 60 years than in the previous 2,000 because of air pollution.

TABLE 19-3

Damage to Materials from Air Pollution

Material	Damage	Principal Pollutants
Metals	Corrosion or tarnishing of surfaces, loss of strength	Sulfur dioxide, hydrogen sulfide, particulates
Stone and concrete	Discoloration, erosion of surfaces, leaching	Sulfur dioxide, hydrogen sulfide, particulates
Paint	Discoloration, reduced gloss, pitting	Sulfur dioxide, hydrogen sulfide, particulates, ozone
Rubber	Weakening, cracking	Ozone, other photochemical oxidants
Leather	Weakening, deterioration of surface	Sulfur dioxide
Paper	Embrittlement	Sulfur dioxide
Textiles	Soiling, fading, deterioration of fabric	Sulfur dioxide, ozone, particulates, nitrogen dioxide
Ceramics	Altered surface appearance	Hydrogen fluoride, particulates

19.4 Air Pollution Control: Toward a Sustainable Strategy

Air pollution exacts an enormous cost so air pollution control can result in many benefits. The economic benefits are surprisingly large. For example, an EPA report, *The Benefits and Costs of the Clean Air Act 1970 to 1990*, says that in 1990 Americans received about a $20 benefit for every dollar spent to control air pollution—and this included only health benefits.

How can air pollution be reduced to sustainable levels, though? We begin our study of pollution control with a look at some of the more traditional approaches included in the end-of-pipe strategy, which typically captures pollutants from smokestacks, sometimes converting them into other less harmful substances. We examine the pros and cons of this strategy. Then we turn to the alternative—and more sustainable—strategies.

Cleaner Air through Better Laws

Society's laws constantly evolve to fit the conditions of the present. This is true with U.S. clean-air legislation—now in its fifth decade of existence. Early clean-air laws were fairly weak and ineffective, but they laid the foundation for some of the most progressive environmental legislation in the world.

The federal **Clean Air Act** (1963) was our nation's first real attempt to reduce air pollution. But the law was weak and ineffectual. The first significant advances in protection came in 1970 when the Congress amended the law, adding (1) emissions standards for automobiles, (2) emissions standards for new industries, and (3) ambient air quality standards for urban areas. The ambient air quality standards established by the EPA covered six criteria air pollutants. These standards were designed to protect human health and the environment.

The 1970 amendments have reduced air pollution from automobiles, factories, and power plants. In addition, they stimulated many states to pass their own air pollution laws, some with regulations even tighter than federal ones. Despite these gains, the amendments created some problems.

For instance, in regions that exceeded ambient air quality standards, the law prohibited the construction of new factories or the expansion of existing ones. Predictably, the business community objected. In addition, some of the wording of the 1970 amendments was vague and required clarification. Of special interest were provisions dealing with the deterioration of air quality in areas that had already met federal standards.

Because of these and other problems, the Clean Air Act was amended again in 1977. To address the limits on industrial growth in areas that were violating the air quality standards, called **nonattainment areas,** lawmakers devised a strategy that allowed factories to expand and new ones to be built, but *only* if they met three provisions: (1) the new sources achieved the lowest possible emission rates; (2) other sources of pollution under the same ownership in that state complied with emissions-control provisions; and (3) unavoidable emissions were offset by pollution reductions by the same or other companies in the region.

The last provision, known as the **emissions offset policy,** forces newcomers to cut emissions in existing facilities or to request other companies in noncompliance areas to reduce their emissions. In most cases, the newcomers pay the cost of air-pollution control devices. The emissions offset policy has been used to reduce regional air pollution—because the air pollution emissions permitted from both the new and existing facilities are set below preconstruction levels. In 2003, the Bush administration changed regulations, allowing thousands of older power plants, oil refineries, and factories to expand operations without having to install pollution-control devices on existing (heavy polluting) facilities.

The issue of how to protect air quality in areas that were already meeting federal air quality standards sparked considerable debate in Congress. Environmentalists felt that the ambient air quality standards, in effect, gave industries a license to pollute up to permitted levels. That is to say, standards for the quality of air meant that clean-air regions could be polluted up to the standard. The 1977 amendments set forth rules for the **prevention of significant deterioration (PSD)** of air quality in **attainment regions,** regions where air quality meets federal standards. However, PSD requirements apply only to sulfur oxides and particulates. Many air pollution experts think that the PSD requirements should be expanded to include other pollutants, such as ozone.

The 1977 amendments also strengthened the enforcement power of the EPA. In previous years, when the EPA wanted to stop a polluter, it had to initiate a criminal lawsuit. Violators would often engage in protracted legal battles because legal costs were often lower than the cost of installing pollution control devices. Thanks to the 1977 amendments, the EPA is allowed to levy **noncompliance penalties** without going to court. These penalties are assessed on the ground that violators have an unfair business advantage over competitors that comply with the law. Penalties equal to the estimated cost of pollution control devices eliminate the cost incentive to pollute.

The Clean Air Act and additional regulations have helped reduce the release of six major pollutants, excluding carbon dioxide, despite a significant increase in economic output, vehicle miles traveled, energy consumption, and population (**FIGURE 19-8**). Additional gains in urban air quality have resulted from new regulations on wood-burning stoves. In 1990 and 1992, for instance, the EPA put into effect regulations requiring all new fireplace inserts and freestanding wood stoves to produce far less pollution. In fact, EPA-certified wood stoves emit 70 to 90% fewer particulates than older models. Many cities and towns have adopted regulations that prohibit wood burning during high-pollution days. Clean air legislation in Canada and other actions have resulted in significant declines in urban pollution as well. Canada's National Air Pollution Surveillance network showed marked declines in all pollutants except ground-level ozone since 1979.

FIGURE 19-8 Signs of success. (a) Despite dramatic increases in the U.S. gross domestic product, vehicle miles traveled, energy consumption, and population, emissions of six major pollutants (not carbon dioxide, however) fell thanks to more fuel-efficient vehicles and pollution control devices on factories, cars, and power plants. **(b)** Total pollution emissions of all major pollutants except carbon dioxide in the United States since 1990. Note that pollution concentrations have dropped significantly below national standard. Note also however, that the data shown here represents average of all monitors in the United States. **(c)** Lead concentrations.

Data for part a from U.S. EPA. *Our Nation's Air—Status and Trends Through 2010*. www.epa.gov/airtrends/2011/; data for part b from the Office of Air Quality Planning Standards, *Latest Findings on National Air Quality—Status and Trends through 2006*. Fig. 5. EPA-454/R-07-007, January 2008. www.epa.gov/airtrends/2007/f; data for part c from U.S. EPA. *National Emissions Inventory Air Pollution Emissions Trends* Data. www.epa.gov/ttnchie1/trends/.

In 1990, the Clean Air Act was amended once again to address inadequacies in the previous amendments. The new law, for example, set deadlines for establishing emission standards for 190 toxic chemicals, which had not been previously addressed. More important, it established a system of **pollution taxes** (market-based incentives) on toxic chemical emissions. These provide a powerful incentive for companies to reduce their releases.

The 1990 Clean Air Act amendments tightened emission standards for automobiles and raised the average mileage standards for new cars, a step that legislators hoped would improve automobile efficiency and help attack the pollution problem at its roots.

The 1990 amendments also established a market-based incentive program to reduce nitrogen and sulfur oxides, primary contributors to **acid deposition**. According to the law, the EPA was authorized to issue **tradable permits** to companies throughout the United States. These permits stipulate allowable emissions for nitrogen and sulfur oxides—lower than present emission rates—to improve air quality. Companies that find innovative and cost-effective ways to reduce pollutants below their permitted allowance can sell their unused credits to other companies at a profit. (These were first bought and sold on the commodities market starting in 1995.) This system encourages companies to develop cost-effective ways to prevent pollution. Thus, if a company can find inexpensive means of reducing its output of pollutants, it could benefit economically from the sale of its pollution credits. Finally, the 1990 amendments called for a phaseout of ozone-depleting chemicals.

Eliminating mercury from coal-fired industrial emissions may be one of the greatest challenges in the next 25 years. If new technologies that are currently being tested at coal-fired power plants are successful, electric utilities could possibly achieve upward of 90% reductions in this toxic heavy metal. These technologies being developed by the Department of Energy in conjunction with the nation's largest private utilities are designed to remove mercury from smokestack gasses by injecting fine carbon particles into them. Mercury binds to the carbon particles.

Although this technology is relatively new and has not (at this writing) been tested at full-scale power plants, experts say carbon-based sorbents promise to help the industry meet new mercury regulations imposed by the U.S. EPA and a growing number of state governments. Although such technologies may help, critics point out that mercury would become a solid waste that would need to be disposed of in a safe fashion.

KEY CONCEPTS

The Clean Air Act has been strengthened over the years to provide a comprehensive means of reducing air pollution. It consists of many parts, including standards for emissions from various sources, air quality standards, and several market-based incentives designed to reduce the emissions of pollutants.

Cleaner Air Through Technology: End-of-Pipe Solutions

Until recently, clean air laws and regulations have typically prescribed many technological controls to reduce air pollution. These devices either (1) remove harmful substances from emissions gases or (2) convert them into "harmless" substances. Such controls are called *end-of-pipe strategies*, for they address a pollutant after it has been produced.

End-of-pipe controls are much like first aid. They're essential in the short term and are thus a critical part of a sustainable strategy. Ultimately, however, many advocates promote preventive strategies, restructuring our energy system so that it produces very little air pollution in the first place. Before we examine such ideas, let's look at some of the end-of-pipe technologies.

KEY CONCEPTS

To date, many efforts to control pollution have relied on end-of-pipe solutions, mostly pollution control devices that capture pollutants or convert them into (supposedly) less harmful substances. Preventing pollution in the first place is a much more sustainable approach.

Pollution Control at Stationary Sources In many coal-fired power plants, bag filters separate particulate matter from the stack gases (**FIGURE 19-9A**). In these devices, smoke passes through a series of cloth bags that filter out particulates. Filters often remove over 99% of the particulates, but they do not remove gaseous pollutants such as sulfur dioxide.

Cyclones are also used to remove particulates, generally in smaller operations (**FIGURE 19-9B**). In the cyclone, particulate-laden air is passed through a metal cylinder. Centrifugal force causes the particulates to strike the walls and fall to the bottom of the cyclone, where they are removed. Cyclones remove 50 to 90% of the large particulates but few of the small- and medium-sized ones. The small particulates are the ones that cause the most harm because they can enter the lungs. Like bag filters, cyclones have no effect on gaseous pollutants.

Electrostatic precipitators, also used to remove particulates, are about 99% efficient and are installed on many U.S. coal-fired power plants (**FIGURE 19-9C**). In electrostatic precipitators, particulates first pass through an electric field, which charges the particles. The charged particles then attach to the wall of the device, which is oppositely charged. The current is periodically turned off, allowing the particulates to fall to the bottom.

The **scrubber,** unlike the other methods, removes particulates *and* sulfur dioxide gas (**FIGURE 19-9D**). In scrubbers, pollutant-laden air is passed through a fine mist of water and lime, which traps over 99% of the particulates and 80 to 95% of the sulfur oxide gases. Nitrogen oxides and carbon dioxide are not removed by these devices.

(a) Typical bag filter

(b) Basic cyclone collector

(c) Electrostatic precipitator

(d) Spray collector (scrubber)

FIGURE 19-9 Four pollution control devices used for stationary combustion sources.

Removing air pollution from stack gases helps clean up the air, but it creates a problem: hazardous wastes. Particulates from power plants trapped by the various pollution control devices, for instance, contain harmful trace elements such as mercury. Scrubbers produce a toxic sludge

KEY CONCEPTS

Pollutants can be filtered from smokestack gases, precipitated out, or even washed out. The problem with these techniques is that this approach is expensive and the pollutants must then be disposed of. Disposal in landfills can result in the pollution of groundwater.

rich in sulfur compounds. Improper disposal can create serious pollution problems elsewhere.

Mobile Sources Reducing pollution from mobile sources (cars, trucks, and planes) can be achieved by changes in engine design, especially those that improve efficiency. Such preventive measures are essential to sustainability. However, most new engine designs do not reduce emissions of carbon monoxide, nitrogen oxides, and hydrocarbons to acceptable levels. This makes it necessary to pass the exhaust gases through catalytic converters, special devices that are attached to the exhaust system (**FIGURE 19-10**). Catalytic converters transform carbon monoxide and hydrocarbons into carbon dioxide and water.

Catalytic converters once seemed like a logical strategy to reduce automobile emissions. Today, however,

FIGURE 19-10 The catalytic converter. In this device, exhaust gases pass over a hot catalytic surface that converts several pollutants into less harmful substances.

© Dorling Kindersley UI/age fotostock.

many people realize that they merely reduce one pollution problem (urban air pollution caused by carbon monoxide and hydrocarbons) while contributing to another (global warming caused by carbon dioxide). In addition, in many cities improvements in air quality brought about by catalytic converters are being negated by increasing numbers of motor vehicles on the road and increases in vehicle miles traveled. Catalytic converters in widespread use today also do nothing to reduce nitrogen oxides, a major contributor to acid deposition and photochemical smog. American auto manufacturers once argued that an affordable converter to do this job could not be developed. However, Volvo, the Swedish automobile manufacturer, introduced a catalytic converter in 1977 that lowered nitrogen oxide emissions to well below current U.S. automobile standards.

KEY CONCEPTS

Emissions reductions from automobiles and other mobile sources are achieved by catalytic converters, devices that convert pollutants in the exhaust of cars to less harmful substances, at least from a human health perspective.

New Combustion Technologies

The coal industry is actively seeking ways to make coal combustion more efficient and environmentally acceptable. These techniques are referred to as **clean coal technologies.** The U.S. government has launched a Clean Coal Power Initiative to seek ways to reduce sulfur, nitrogen, mercury, and carbon dioxide emissions from coal-fired power plants.

One technology for burning coal more efficiently—and therefore reduce emissions—is called **magnetohydrodynamics (MHD).** As shown in **FIGURE 19-11**, coal is first crushed and mixed with potassium carbonate or cesium,

substances that are easily ionized (stripped of electrons). The mixture is burned at extremely high temperatures and produces a hot ionized gas—a plasma—containing electrons. The plasma is passed through a nozzle into a magnetic field, generating an electrical current. The heat of the gas creates steam, which powers a turbine.

Magnetohydrodynamics is about 60% efficient, compared with the 30 to 40% efficiency of a conventional coal-burning power plant. MHD systems remove 95% of the sulfur contaminants in coal, have lower nitrogen oxide emissions, and produce fewer particulates than conventional coal plants; however, they release more fine particulates. Interest in MHD has declined because of technical difficulties.

Coal may also be burned more efficiently in **fluidized bed combustion (FBC),** a technology that is also more efficient and cleaner than conventional coal-fired burners. In FBC, finely powdered coal is mixed with sand and limestone and then fed into the boiler. Hot air, fed from underneath, suspends the mixture while it burns, thus increasing the efficiency of combustion. The limestone reacts with sulfur, forming calcium sulfate and reducing sulfur oxide emissions. Lower combustion temperatures in FBC reduce nitrogen oxide formation.

Several other more efficient and potentially cleaner coal-burning technologies are currently being tested or are in operation around the world. Some plants convert coal to combustible gases, carbon monoxide and hydrogen. In this process, called **coal gasification,** coal is brought into contact with steam and oxygen. The carbon monoxide and hydrogen produced in this process are then burned. Hot gases are used to turn a turbine to generate electricity. In the **integrated coal gasification combined cycle,** a plant's waste heat from the coal gasification process is used to produce steam to drive an additional turbine, boosting the efficiency of the plant.

Clean coal technologies also include pollution control devices like the mercury-capturing technique described earlier in the chapter. In addition, they include ideas on ways to capture carbon dioxide emissions from coal plants and sequester them, that is, remove them from the carbon cycle. For example, some proponents hope to capture carbon dioxide gas released from power plants and inject it into abandoned oil wells or into the deep ocean waters, a costly and difficult process that may not effectively remove them from the carbon cycle in the long run because deep waters often emerge at the surface in upwellings where they could release carbon dioxide into the atmosphere. This process is also fairly energy intensive and could increase coal combustion at electric power plants by 25% or more, according to some sources.

Efforts are also underway to develop technologies to utilize carbon dioxide emissions from coal-fired power plants, for example, by pumping them into nearby greenhouses where they can be used to promote plant growth. Chances are the leading technology will combine several technologies that not only capture CO_2 but also other pollutants or, better yet,

FIGURE 19-11 Magnetohydrodynamics. Coal is mixed with an ion-producing seed substance, such as potassium, and burned. A hot ionized gas is given off and shot through a magnetic field. The movement of the ionized gas through the magnetic field creates the electrical current. Air or water is also heated and used to run an electrical generator.

minimize their production. We must be careful to avoid higher energy use to achieve these goals as they could negate gains.

Clean coal technologies could play a key role in improving coal's environmental acceptability and performance, say proponents. However, critics still point out that this approach is a stopgap measure that could easily be offset by increasing dependence on coal.

Shifting to natural gas and using more efficient combustion technologies—in particular, natural gas turbines similar to jet engines—can also reduce pollution. Such efforts could dramatically reduce the emission of sulfur dioxide, particulates, and carbon dioxide from power plants and other facilities.

KEY CONCEPTS

Burning coal and natural gas more efficiently in improved burners helps reduce the amount of pollution emitted per unit of energy consumption and can help society meet energy demands more sustainably.

Economics and Air Pollution Control

The previous section describes some new technologies to burn fossil fuels more efficiently and a number of air-pollution control devices that are end-of-pipe solutions.

The end-of-pipe strategies can be very costly. Smokestack scrubbers and electrostatic precipitators, for instance, cost millions of dollars to install and operate. Catalytic converters on automobiles add $500 to the sticker price. Although air pollution controls generally save money, some controls may reap benefits far lower than their actual costs. One study, for instance, suggests that pollution controls on automobiles and other mobile sources cost about $15 billion a year but create benefits of under $1 billion a year. Although such analyses often grossly underestimate the economic savings of benefits (such as reducing the cost of global warming), they do suggest the need for more cost-effective alternatives. More important, what if one could reduce pollutants at little or no cost—or even better, while enhancing one's profits?

Another problem with end-of-pipe solutions is that they often result in the production of hazardous wastes that must be disposed of in another medium—the ground, for instance. In cases where pollution control devices are used, creative thinking and ingenuity can lead to strategies to profit from the waste. The chemical division of the Sherwin-Williams Company installed pollution control systems that capture solvents at a Chicago plant, saving that company $60,000 a year. In Germany and England, scrubber waste is used to make drywall used to finish walled in homed and offices (see **Spotlight on Sustainable Development 19-1**).

19-1 Germany's Sustainable Approach Pays Huge Dividends

In Iphofen, Germany, the Knauf gypsum manufacturing plant produces drywall from scrubber sludge from a nearby coal-fired power plant (**FIGURE 1**). This plant has been so successful that Knauf recently opened a second facility in England, which generates $48 million per year in revenue.

This example is just one of many environmentally sustainable economic projects in Germany, indisputably the world's leader in environmental protection.

In an article in *International Wildlife*, U.S. Senate Committee on Environment and Public Works counsel Curtis Moore writes that "the world's future begins in Germany." He goes on to say, "More than anywhere else on Earth Germany is demonstrating that the greening of industry, far from being an impediment to commerce, is...a stimulus." Germany is finding that efforts to use energy more efficiently (conserve), to recycle, and to tap into the Earth's generous supply of renewable resources—all vital components of the sustainable strategy—are creating enormous economic opportunities. Consider a few examples.

Germany passed legislation that requires all car manufacturers to take back and recycle old cars. To facilitate this process, new cars rolling off the assembly line contain bar-coded parts that identify the material used in their manufacture. German auto manufacturers have also made cars so that they can be more easily disassembled. Disassembly plants will be able to take a car completely apart in 20 minutes.

By 1994, take-back requirements went into effect for virtually all products, helping close the loop and reduce waste. By 1995, 72% of all glass and metals and 64% of all paperboard and plastic had to be recycled. Even old refrigerators are hauled free of charge to recycling centers, where ozone-depleting CFCs are extracted.

In the village of Neunburg stands a $38 million plant owned by a consortium of governments and industries. This facility, which produces hydrogen from solar-generated electricity, is really an experimental testing ground for new technologies. The Germans hope to be in the forefront when fossil fuel supplies run out or as nations look for alternatives to reduce global

FIGURE 1 Dried sludge ready to be converted to drywall.
Courtesy of Knauf Gips KG.

warming and other air pollution problems. If successful, Germany could power its entire economy by solar electricity and hydrogen.

Knowing that the future depends on clean, renewable energy, the German government initiated the Thousand Roofs program, which offers a 75% tax credit to homeowners who install photovoltaics on their roofs. The hope is that government funding will drive the costs downward and also provide valuable hands-on experience in manufacturing—experience that could position Germany well in ensuing years as photovoltaics come into wider use.

The Germans have found that even conventional pollution control strategies can pay off. For example, Germany adopted rules that required all power plants within its borders to cut sulfur oxide emissions by 90%, a goal achieved in 6 years. These rules stimulated considerable innovation on the part of Germany's industrial technologists. Today, technological innovators stand to make huge sums of money selling pollution control technology and know-how to Americans and others.

Adapted from Curtis A. Moore (1992). "Down Germany's Road to a Clean Tomorrow." *International Wildlife* 22 (5): 24–28.

Pollution control strategies are often overwhelmed by increasing activity. For example, catalytic converters cut pollution, but as the number of cars increases and as people drive more miles, the savings from the catalytic converter are often negated. From an economic standpoint, pollution control strategies cost business far more than is necessary. Pollution prevention strategies often achieve better reductions at a fraction of the cost and sometimes while generating additional profit.

KEY CONCEPTS

Air pollution control via end-of-pipe methods often adds substantial economic costs to various processes. It also produces waste products that, if not used for other purposes, end up in landfills.

Toward a Sustainable Strategy

Air-pollution control strategies have clearly improved the quality of the air. Although important in addressing immediate pollution problems, end-of-pipe solutions are falling into disfavor for reasons cited earlier. These and other problems with the conventional air-pollution control strategy have led many people the world over to seek alternatives—a sustainable approach that is based on several of the key principles of sustainability: conservation, recycling, renewable resources, restoration, and population control.

Conservation, recycling, and the use of renewable energy sources can produce substantial reductions in air pollution. Cutting energy demand by half through conservation measures, for instance, usually reduces air pollution output by half. Recycling aluminum rather than making it out of raw ore, however, cuts air pollution by 95%. Paper recycling reduces air pollution by 75%. Making steel from recycled scrap cuts air pollution by 30%.

Through individual, corporate, and government efforts in these areas, global demand for fossil fuel consumption can be substantially lowered—and along with it global pollution. Individuals can also help. Each unnecessary product we buy, each bottle or can we toss out, and each gallon of gas we waste contribute to air pollution. Individual responsibility is an effective means of cutting down on resource demand and pollution. When added together, individual contributions can make a big difference.

KEY CONCEPTS

Pollution prevention is a key element of a sustainable society. Pollution prevention results from enacting measures that promote conservation (frugality and efficiency), recycling, renewable resource use, and restoration of habitats.

19.5 Noise: The Forgotten Pollutant

Noise is one of the most widespread environmental pollutants. Few of us can escape it: Even when we are enjoying a backcountry camping trip, the silence is often broken by the roar of jets, chain saws, and off-road vehicles. Especially noisy are the cities and factories where many of us live and work. To understand noise pollution and its effects, let us first understand sound.

What Is Sound?

Sound is transmitted through the air as a series of waves. For a simple example, take the bass speaker element of a loudspeaker. When the music is loud, the speaker cone can actually be seen to vibrate, and a hand placed in front of it can feel the air move. If we could slow down the speaker cone to observe what was happening, we'd see that as it moves outward, it compresses the air molecules in front of it (**FIGURE 19-12**). As the speaker cone returns inward, the air it just compressed expands in both directions, much as a coiled spring would expand when released. This expansion compresses neighboring air molecules slightly farther away. These have the same effect that the outward-pushing speaker cone had (Figure 19-12). Thus, waves of expansion and compression are set up, transmitting the sound. Air molecules do not travel with the sound but only oscillate back and forth in the direction the sound is traveling. Sound is a train of high-pressure regions following one another through the air at about 340 meters per second (760 miles per hour).

Sound can be described in terms of its loudness and its pitch, or frequency. Loudness is measured in

FIGURE 19-12 The anatomy of sound. Sound waves, shown here, are compression waves. Sound vibrates molecules in the air, which in turn vibrate neighboring molecules, transmitting the sound from its source.

decibels (dB). The decibel scale encompasses a wide range of volume (**TABLE 19-4**). The lowest sound the human ear can detect is set at 0 dB. This is the threshold of hearing. In the decibel scale, each 10 dB increase results in a 10-fold increase in sound intensity. Thus, a 10 dB sound is 10 times louder than 0 dB sound. A 20 dB sound is 10 times louder than a 10 dB sound and 100 times louder than a 0 dB sound.

Pitch, or frequency, is a measure of how high or low a sound is. Bass notes played by a tuba have a low pitch, and a violin's treble notes have a high pitch. Pitch is measured in cycles (that is, waves) per second. This is the number of compression waves passing a given point each second. The higher the pitch, the larger the number of cycles per second. Cycles per second are commonly called **hertz (Hz),** after the German physicist Heinrich Rudolf Hertz.

TABLE 19-4

Sound Levels on the Decibel (dB) Scale, with Examples and Effects

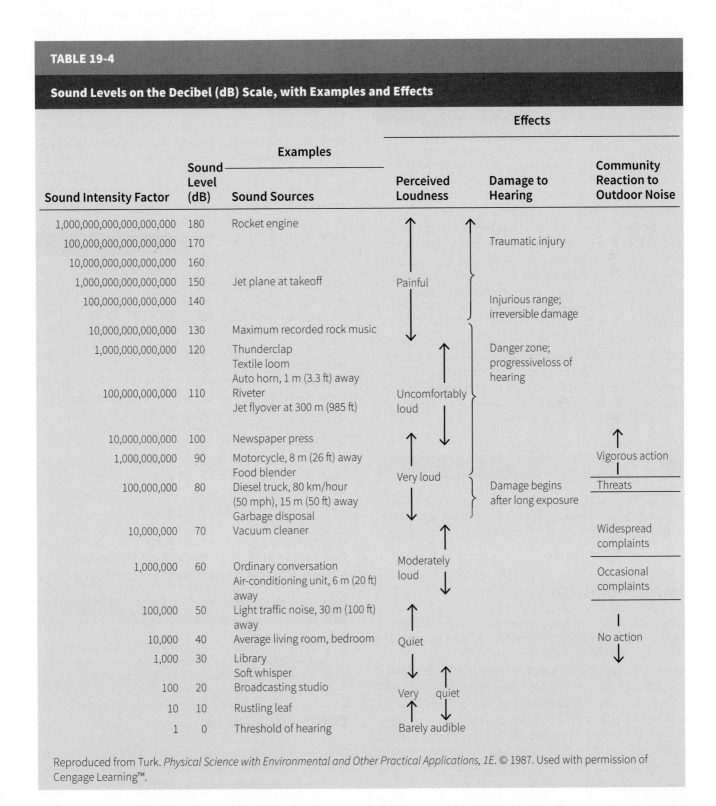

Reproduced from Turk. *Physical Science with Environmental and Other Practical Applications, 1E.* © 1987. Used with permission of Cengage Learning™.

The human ear is sensitive to sounds in the range of 20 to 20,000 hertz. Sounds below 20 hertz are not detected by the human ear and are described as *infrasonic*. Sounds above the audible range are called *ultrasonic*.

 KEY CONCEPTS

Sound waves are compression waves that travel through the air. Sound is characterized by loudness (measured in decibels) and pitch (how high or low it is).

What Is Noise?

Noise is any unwanted, unpleasant sound. What any individual considers noise depends on many variables, among them one's background, mood, and occupation. Location and hearing ability are also important determinants, as are the time of day, the duration and volume of a sound, and other factors. A sound may be pleasant when it is soft but noisy when it is loud, or it may be acceptable when you generate it but obnoxious when someone else does. On this people generally agree: The louder a sound is, the more annoying it becomes and the more likely people are to describe it as noise.

Go Green

Cover your ears when a fire engine or ambulance passes by. Avoid noisy venues and wear ear plugs at auto races.

Sources of noise fall into four categories: transportation, industrial, household, and military. As the world becomes more dependent on technology, noise pollution will grow worse. Of greatest concern are noises from off-road vehicles, construction, air traffic, home appliances, and surface transportation.

 KEY CONCEPTS

Noise is an unwanted, unpleasant sound. What individuals consider to be a noise depends on many variables, such as the time of day or loudness of the sound.

Impacts of Noise

As we grow older, our hearing often declines. Several factors seem to be at work here. The loss of hearing results from occasional infections in the inner ear and from the natural aging of the sound receptors. Exposure to noise also causes a deterioration of hearing. Men, for instance, generally suffer a greater loss of hearing with age than women do. This difference probably results from exposure to noise at work. If loud and persistent enough, noise can cause premature degeneration of the sensory cells inside the ear. According to one estimate, 1 in every 10 to 20 people suffers some hearing loss from anthropogenic noise.

Exposure to extremely loud noises, such as rock music, gunfire, and noisy machinery, can cause a momentary decrease in our ability to hear, which is called a **temporary threshold shift.** A permanent loss of hearing, or a **permanent threshold shift,** occurs after continued exposure to loud noise. Studies suggest that continuous, long-term exposure to noise levels as low as 55 dB can permanently damage hearing. Noise of this level is common in many factories and jobs, especially in construction and mining. Even city traffic noise can damage the hearing of people exposed on a regular basis. This explains why the average 20-year-old resident of New York City hears as well as a 70-year-old tribal resident of Africa's grasslands.

As a general rule, the higher the sound level, the less time it takes to induce a loss in hearing. Different frequencies produce differing amounts of damage. The lower frequencies, for example, do less damage than the higher-pitched sounds at the same loudness. Explosions (130 dB) and other extremely loud noises can cause instantaneous damage to the hair cells, resulting in deafness. Noise levels over 150 dB can severely damage the sensory cells, rupture the eardrum, and displace the tiny bones in the ear (the ossicles), which convey sound waves from the eardrum to the hair cells.

Workplace exposure to noise is slowly deafening millions of Americans. Large numbers of Americans, in fact, work at jobs where the noise levels are over 80 dB. Military personnel fall victim to noise from tanks, jets, helicopters, artillery, and rifles. Studies show that about half of the soldiers who complete combat training suffer so much hearing loss that they no longer meet the enlistment requirements for combat units. Bars, nightclubs, and traffic noise (especially diesel trucks and buses) are other important contributors to the deafening of America.

Go Green

To help reduce noise pollution, play your music at a lower volume. This will protect your hearing, save electricity, and reduce annoyance of others.

Noise interferes with family, friends, and coworkers, causing increased tension—not only because the sufferer can't understand what's being said, but also because deafened individuals may talk annoyingly loudly.

Noise also has profound effects on sleep. It may prevent us from falling asleep as soon as desired, or it may keep us from sleeping at all. Noise may wake us during the night or may alter the quality of sleep, leaving us irritable.

 KEY CONCEPTS

Noise affects us in many ways. It damages hearing, disrupts our sleep, and annoys us in our everyday lives. It interferes with conversation, concentration, relaxation, and leisure.

Controlling Noise

Noise pollution receives little attention worldwide compared with other forms of air pollution—in part because hearing loss is generally progressive. Because of this, victims are generally unaware of the gradual loss in auditory acuity. In fact, workers who are exposed to loud noises eventually become accustomed to them, partly because their hearing declines.

Noise control can be carried out on many different levels (**TABLE 19-5**). Individuals can take actions, even simple ones such as wearing ear plugs when operating chain saws or

---

TABLE 19-5

Controlling Noise

Design and Planning

Minimize air turbulence created by vehicles.

Minimize vibration from vehicles, machines, and home appliances.

Build railroads and highways away from densely populated areas.

Build better insulated subways, railroad cars, trucks, and buses.

Eliminate noisy two-stroke motorcycle engines.

Install better mufflers on motorized vehicles and equipment.

Build airports away from high-density population centers.

Restrict growth in the vicinity of existing airports.

Require low-noise home appliances, tools, and office equipment.

Pass laws to eliminate or block noise.

Control traffic flow to eliminate stop-and-start driving.

Barriers and Sound Absorbers

Build embankments along noisy streets.

Use smooth road surfaces.

Build level streets to cut down on engine noise generated on steep inclines.

Plant dense rows of trees along highways and around sources of noise.

Construct artificial noise barriers along streets and around factories.

Install better insulation in houses.

Use double- and triple-paned windows.

Add sound-absorbent materials in factories, offices, and homes.

Provide earplugs for ground personnel at airports and workers in factories and construction.

Personal Solutions

Use fewer power tools and appliances.

Time activities to minimize disturbance of others.

Use mass transit.

Refuse to buy noisy vehicles, tools, office equipment, and appliances.

Maintain vehicles to eliminate noise.

Wear ear guards when engaged in noisy activities.

Work with employer to reduce noise.

---

circular saws or other machinery. A nonprofit organization called Hearing Education Awareness for Rockers (H.E.A.R.), works with radio stations in the United States and Canada to protect audiences from loud rock music. In 1996, they gave out over 60,000 earplugs. Even some rock bands, among them Motley Crue, now sell ear protectors at concerts. Changing the design of machinery and other products can cut turbulence and vibration. Better urban planning can isolate people from noisy railroads, highways, factories, and airports. Legal controls on noise emissions can also be helpful.

Many countries have regulations to control noise from motorized vehicles, including Japan, Norway, the Netherlands, Sweden, Switzerland, the United Kingdom, Denmark, France, Italy, and Canada. In the United States, the **Noise Control Act** (1972) authorized the EPA to establish maximum permissible noise levels for motor vehicles and other sources. Individual cities and towns have also passed noise ordinances.

Noise in U.S. workplaces is controlled by the **Occupational Safety and Health Administration (OSHA).** OSHA's workplace standard is 90 dB for an 8-hour exposure, a level many authorities believe is too high. The agency can issue abatement orders for noise violations. These orders require the employer to first tackle the problem through engineering and design changes in the equipment. Should these prove infeasible, ear guards must be worn by workers, or workers must be moved regularly from noisy jobs to quieter jobs to reduce their overall exposure. Personal protective equipment is a less desirable solution because it is uncomfortable to wear and also blocks out important sounds or signals necessary for worker safety. Many workers find it uncomfortable and refuse to wear it.

The U.S. Environmental Protection Agency (EPA) also once played a role in noise control and abatement. Unfortunately, the EPA's noise control office, with a staff of about 800 people, closed its doors in 1982 when Congress refused to appropriate funds to support their work any longer. Before the EPA's noise office was eliminated, however, regulations on noise controls on interstate motor carriers, new trucks, motorcycles, and railroads had been created. These regulations are still on the books, as stipulated by the Noise Control Act, but nothing new has been added since then. New trucks generally comply with these regulations, as do cars, but motorcycles often fail to meet standards. Without a watchdog agency, there is little that can be done about this.

KEY CONCEPTS

Noise levels can be controlled by redesigning machinery and other sources, by sound insulating buildings, by separating sources of noise from people, and by other measures.

19.6 Indoor Air Pollution

Americans and other residents of industrial nations spend, on average, 50 to 90% of their time indoors, away from the harmful pollutants in the outside air. Well, not exactly.

Studies show that the air in our homes and offices contains high levels of potentially harmful substances known as **indoor air pollutants.** The U.S. EPA identifies four indoor air pollutants as the most dangerous: cigarette smoke, radon (a radioactive substance), formaldehyde, and asbestos.

Indoor air pollutants come from a surprising number of sources, among them cooking stoves, wood stoves, tobacco smokers, and heating systems. Furniture, wood paneling, plywood, and carpet also contribute to indoor pollution. Naturally occurring radioactive materials in the ground beneath a building may also release a potentially dangerous gas known as *radon*, which seeps into buildings. Outdoor air pollutants also enter buildings. Making buildings more airtight to reduce energy loss keeps these pollutants out, but it can also trap pollutants generated internally, resulting in higher levels. Special precautions must be taken when retrofitting a building for energy efficiency or constructing a new airtight, energy-efficient one to minimize or eliminate proper air exchange (more on this later). **TABLE 19-6** lists the four major indoor air pollutants, their sources, and their major effects.

KEY CONCEPTS

Indoor air pollutants come from combustion sources, from smokers, a variety of household and building products, and even from naturally occurring radioactive materials in the ground beneath a building.

How Serious Is Indoor Air Pollution?

The EPA estimates that indoor air pollutants may cause as many as 2,000 to 20,000 cases of lung cancer each year. Indoor air pollution, in fact, may cause more lung cancer than ambient air pollutants. EPA officials believe that toxic substances in our homes and offices are much more likely to cause cancer than outdoor air pollutants, for two main reasons. First, indoor levels are often much higher than outdoor levels, in some cases up to 100 times higher. Even in pristine rural areas, indoor air can be more polluted than outside air next to a chemical plant. Second, as noted earlier, people spend an enormous amount of time indoors.

As you might expect, the people with the highest risk are the young, elderly, and infirm—that is, individuals who are sick or who have heart and lung disease. Pregnant women are also at greater risk, as are office or factory workers who spend inordinate amounts of time indoors in polluted buildings.

The EPA estimates that one-fifth to one-third of all U.S. office buildings contain unhealthy levels of indoor air pollutants. These are often classified as **sick buildings.**

The EPA estimates that 10 to 20 million Americans suffer from **sick building syndrome,** an assortment of debilitating health effects caused by polluted indoor air. These individuals display a variety of health problems, among them chronic respiratory problems, sinus infections, sore throats, and headaches. People may also complain of dizziness, rashes, eye irritation, and nausea. These symptoms are largely caused by exposure to formaldehyde from furniture, plywood, and other sources in buildings. The EPA also estimates that 100,000 to 200,000 workers die prematurely each year as a result of exposure to indoor air pollution at work—either in offices or, more likely, in factories.

Go Green

To help reduce indoor air pollution, run an exhaust fan when you cook over a gas stove and run a bathroom exhaust when you shower. This will reduce moisture levels that promote mold.

TABLE 19-6

Major Indoor Air Pollutants

Pollutant	Source(s)	Effects on Human Health
Tobacco smoke—carbon monoxide, particulates, sulfur dioxide, nitrogen dioxide	Cigarettes, pipes, and cigars	Carbon monoxide—shortness of breath and heart attacks; particulates—lung cancer; sulfur and nitrogen dioxide—emphysema and chronic bronchitis. Effects seen in smokers and nonsmokers.
Radon	Naturally occurring radioactive material (radium) in the soil. Radon leaks through cracks in the foundation	Lung cancer
Formaldehyde	New furniture, particleboard, plywood, chipboard, and wood paneling	Irritation of the eyes, nose, and throat; nasal cancer and lung cancer
Asbestos	Insulation around pipes in old homes, school buildings, and factories	Lung cancer and buildup of fibrous tissue in the lungs, which makes breathing difficult

Although estimates of the effects of indoor air pollution are crude, they do suggest the need to take action. In order to understand our options, we examine three of the major indoor air pollutants: products of combustion, radon, and formaldehyde.

Why Is Indoor Air in Buildings So Polluted?

Indoor air pollution is not a recent phenomenon. It has been around since houses were built. Early homes, heated by fires, for example, could become choked with smoke. Indeed, this is the case in many less developed nations where people cook on open fires inside their homes.

Today, indoor air pollution exists, in part, because of efforts to tighten houses to reduce air infiltration and save energy. Cracks are sealed and many buildings are wrapped with a wind-proof material to reduce leakage. But the problem is worsened by the introduction of many new building materials containing pollutants. Carpets, furniture, plywood, oriented strand board, and a host of other products are made with glues that contain formaldehyde or other potentially toxic substances. These are trapped inside homes by weatherproofing measures.

Controlling Indoor Air Pollutants

Indoor air pollutants can be reduced or eliminated by a variety of techniques. Let's take each of the major indoor air pollutants and consider some strategies, starting with tobacco smoke.

Tobacco smoke is difficult to remove from a room, so one of the easiest solutions is to ask smokers to smoke outside. Better ventilation systems can remove the smoke-laden air and replace it with cleaner outside air. Bans on smoking in offices or special zones set up for smokers can also help reduce exposure.

Radon can be removed by special devices called energy recovery ventilators that pump stale, polluted air out of a building and draw new air in. During the process,

much of the heat from the outgoing air is transferred to the incoming air to save energy. Radon can also be controlled by placing plastic on the ground under the cement slab a house is built on and by installing pipes under the slab to draw radon out of the earth immediately below it—before it can seep into a building (**FIGURE 19-13**). Pipes can be installed in new construction and can be retrofitted in existing structures.

Formaldehyde is much more difficult to control because it is found in so many products. Fortunately, new building products such as formaldehyde-free oriented strand board are being made available for builders. One manufacturer is producing a wood sheathing for making floors and exterior walls out of straw. It performs well and is glued with a nontoxic material. Some builders are using natural materials such as adobe or straw bales to build walls, avoiding the use of glue-laden wood products altogether. Formaldehyde is also present in furniture, especially inexpensive products, which are often made with particleboard or other formaldehyde-containing wood products. These products should be avoided. Natural carpeting can help, too.

Asbestos can be removed from buildings, although it is costly. Alternatively, asbestos in pipes and ceilings can be stabilized with special sprays that coat the product and prevent fine fibers from flaking off.

Paints, stains, and finishes can also add lots of indoor air pollution. Latex paints contain small amounts of volatile organic compounds (VOCs), substances such as benzene, toluene, ethanol, and other compounds, that evaporate as the paint dries. These can cause a variety of health problems.

FIGURE 19-13 Radon protection. This house is protected from radon by pipes that draw radon from underneath the slab.

FIGURE 19-14 Milk paint? Yes, you read correctly. Paints were made out of milk long before pigments started being used. These paints, while expensive, are environmentally safe.

Courtesy of the Old Fashioned Milk Paint Company, Inc.

Stains and finishes have much higher concentrations of VOCs and outgass (give off) these materials for months after application.

To avoid these problems, some builders are using low toxic or nontoxic paints, stains, and finishes, containing reduced amounts of VOCs or none whatsoever. All major paint manufacturers now sell produced no-VOC paints. One company, for example, makes a water-soluble paint from milk protein, which is popular in hospitals and schools (**FIGURE 19-14**).

Solving indoor air pollution is not easy, but it can be done. Ways to build more healthful homes are becoming more widely known and many builders are adopting these practices.

> The power of man has grown in every sphere, except over himself.
>
> **—Winston Churchill**

CRITICAL THINKING

Exercise Analysis

The first statement made by the person cited in this exercise is true. That is, natural sources of pollution such as volcanoes often do exceed human sources of pollution. There is an important distinction, though. Natural sources are often dispersed. Although natural sources produce more pollution than human sources overall, the fact that the pollution from natural sources is spread out over a large area greatly dilutes its effect on ambient levels. As a rule, most natural sources do little to increase concentrations, and they rarely, except in isolated events, produce levels of pollution that are harmful to humans and other species. Oil, for instance, seeps from the ocean floor in rather large quantity overall. However, because the sources are many and widely dispersed, large oil slicks such as those produced when an oil tanker runs aground are not produced by natural seepage.

In contrast, human sources tend to be fairly concentrated. The Ohio River valley, for instance, is home to numerous power plants that burn coal. The sulfur dioxide they produce results in a substantial increase in atmospheric levels, with effects on downwind sites. Because of the concentrated nature of human sources, controlling air pollution from them is vital to protecting human health and the quality of the environment.

The second part of the speaker's argument, that environmental protection reduces productivity, is true. The extent to which it does so is rather small, though. Moreover, environmental protection efforts have long-term economic benefits. They lower economic externalities—social and health costs passed on to you and me. An accurate financial reckoning of the costs of pollution control should, many people argue, take into account economic externalities. As noted in this chapter, air pollution reductions can have enormous economic benefits in reduced healthcare costs alone.

On this topic, it is important to note that many laws and regulations could be revamped to encourage more creative and cost-effective solutions. Rather than weakening environmental legislation and regulation, we might do well to strengthen them through provisions that encourage pollution prevention and other measures that enable us to both cut pollution and boost productivity.

1. Describe the pollutants produced by burning fossil fuels. How is each product formed?

2. Define the terms *primary* and *secondary pollutant*.

3. In what ways are brown-air and gray-air cities different? What are the major air pollutants in each city? Why does the distinction break down?

4. What is photochemical smog? How is it formed?

5. Describe some factors that affect air pollution levels in your community.

6. What is a temperature inversion?

7. What are the health effects of (1) carbon monoxide, (2) sulfur oxides and particulates, and (3) photochemical oxidants?

8. List the major chronic health effects of air pollution. Which pollutants are thought to be the cause of these effects?

9. Of all the impacts of air pollution, which are of the most concern to you?

10. Discuss the air-pollution control legislation enacted by the U.S. Congress, highlighting important features of the acts and amendments. How would you characterize the conventional approach to air pollution? How has this approach contributed to the prevalent notion that environmental regulation is bad for the economy?

11. Make a list of ways you can help reduce air pollution.

12. Describe and compare end-of-pipe solutions to air pollution and preventive measures. Which of the two are the most sustainable? Where are the most economical?

13. Critically analyze this statement: "Coal is an abundant fuel with a potentially long life span. New technologies can be used to burn coal more efficiently and represent a sustainable solution to air pollution."

14. Why do you think noise pollution is often overlooked in our everyday concerns for our health and well-being?

15. Describe a sound wave and define these terms: *decibel, pitch*, and *frequency*.

16. List and describe reasons why what we consider to be noise is so subjective.

17. Critically analyze this statement: "Noise is a subjective thing. No one agrees on it, so why bother regulating it? Besides, no one's dying of noise pollution."

18. Define the terms *temporary threshold shift* and *permanent threshold shift* and the reasons each one occurs.

19. List the major indoor air pollutants and describe the effects of each one.

20. How would you go about estimating the number of people in the United States who are exposed to unhealthy air in their office or workplace? Describe an experimental design to derive fairly accurate numbers.

KEY TERMS

acid deposition
alveoli
anthropogenic pollutants
attainment regions
bag filters
bronchial asthma
brown-air cities
catalytic converters
chronic bronchitis
Clean Air Act
clean coal technologies
coal gasification
cross-media contamination
cyclones
decibels (dB)
electrostatic precipitators

emissions offset policy
emphysema
fluidized bed combustion (FBC)
gray-air cities
hertz (Hz)
indoor air pollutants
integrated coal gasification combined cycle
magnetohydrodynamics (MHD)
noise
Noise Control Act
nonattainment areas
noncompliance penalties
Occupational Safety and Health Administration (OSHA)
permanent threshold shift

photochemical smog
pollution taxes
prevention of significant deterioration (PSD)
primary pollutants
radiation inversion
scrubber
secondary pollutants
sick building syndrome
sick buildings
smog
subsidence inversion
temperature inversion
temporary threshold shift
tradable permits

© FloridaStock/Shutterstock, Inc.

Global Air Pollution: Ozone Depletion, Acid Deposition, and Global Climate Change

For 200 years we've been conquering nature. Now we're beating it to death.

—Tim McMillan

LEARNING OBJECTIVES

After reading and studying the material in this chapter, students will be able to understand, knowledgably discuss, and, when appropriate, apply the information learned in this chapter, including:

1. The importance of the ozone layer
2. The causes and social, economic, and environmental impacts of ozone depletion
3. International efforts that have dramatically reduced ozone depletion
4. The natural and anthropogenic sources of acid rain and snow
5. The social, economic, and environmental impacts of acid deposition
6. Efforts that have effectively reduced acid deposition, especially sustainable approaches
7. Global energy balance and the greenhouse effect
8. The relationship between global warming and climate change
9. Natural and anthropogenic factors that cause global warming and climate change
10. Why global warming and climate change are caused primarily by human activities
11. The social, economic, and environmental impacts of climate change
12. Ways to reduce and eliminate global warming and climate change
13. The social, economic, and environmental benefits of controlling global warming and climate change

CRITICAL THINKING

Exercise

A business magazine article notes that "on the issue of global warming the scientific community is divided." In support of this assertion, it quotes two prominent scientists. One says that he's "convinced the world is in a human-induced warming phase," and another claims that "there's simply not enough evidence to support such a conclusion." The author of the article goes on to say that because of the uncertainty among the scientific community, it makes no sense to launch a global effort to reduce carbon dioxide emissions. This position is supported by some segments of the business community (especially the oil and coal companies) but not by others (such as the insurance and natural gas industries). Can you detect any problem in this reportage? After you have finished, list the critical thinking rules that were helpful to you in this exercise.

In June 1992, representatives from nearly 180 nations met in Rio de Janeiro to hammer out the final language of Agenda 21, a massive global blueprint for sustainable development. They also negotiated final language for various agreements on global climate change, biodiversity, forest protection, and other subjects. This meeting was held in large part because of widespread recognition that human society is destroying the life-support systems of the planet.

In the 1960s and 1970s, most environmental issues were local or regional in scale. Today, however, many problems have reached global proportions. Solving them will require global cooperation such as that witnessed during the Earth Summit. This chapter looks at three global issues—ozone depletion, acid deposition, and global climate change—and sustainable solutions.

20.1 Stratospheric Ozone Depletion

Encircling the Earth is a layer of ozone gas (O_3), which screens out most of the sun's harmful ultraviolet (UV) radiation (**FIGURE 20-1**). Called the **ozone layer**, this protective zone occupies the inner two-thirds of the stratosphere, which extends 15 to 45 kilometers (10 to 30 miles) above

the Earth's surface. The ozone layer blocks 97–99% of the Sun's ultraviolet radiation and, thus, protects all terrestrial organisms from UV radiation's harmful effects. Without the ozone layer, terrestrial life would all but vanish. Some animals would suffer serious burns and would develop cancer and lethal mutations. Humans would be especially vulnerable to these changes. Unable to cope with the intense influx of ultraviolet radiation, many plants would perish—and with them, millions of species that depend on them for nutrition. How does the ozone layer work?

When ultraviolet radiation strikes ozone molecules, it causes them to split:

$$UV + O_3 \rightarrow O + O_2$$

The products, however, quickly reunite, reforming ozone. In this reaction, UV radiation is converted into heat as shown in the reaction:

$$O + O_2 \rightarrow O_3 + heat$$

Because ozone is regenerated, the ozone layer is a renewable form of protection.

  KEY CONCEPTS

The ozone layer is a portion of the stratosphere with a slightly higher concentration of ozone molecules. It forms a protective shield that filters out harmful ultraviolet radiation.

Activities That Deplete the Ozone Layer

Human civilization has and continues to threaten the ozone layer through two principal activities: (1) the use of a class of chemical compounds called *chlorofluorocarbons* and (2) jet travel through the stratosphere. Lets begin with chlorofluorocarbons.

The Use of Chlorofluorocarbons In 1951, manufacturers introduced a promising new product on the market in the United States, spray cans that contained a chemical substance known as *Freon-11*. Because it was under pressure it served as a propellant. Freon-11 and a similar compound, Freon-12, belong to a group of chemicals commonly referred to as **chlorofluorocarbons (CFCs)**. **TABLE 20-1** provides a few details on these CFCs. As just noted, **Freon-11** was used as a propellant in spray cans; **Freon-12** is used in refrigerators, air conditioners, and freezers. As you can see from the chemical formulas, CFCs contain two to three chlorine atoms (indicated by the symbol Cl) one fluorine atom, and a central carbon atom—hence the name chlorofluorocarbons.

CFCs have been used in other ways as well. They were once widely used as foam *blowing agents*. A blowing agent is a gas mixed with a liquid polymer such as polystyrene and blown or extruded into molds, to produce

A

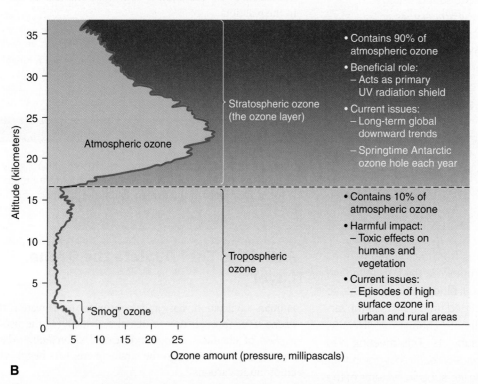

B

FIGURE 20-1 The ozone layer. (a) Ozone molecules absorb ultraviolet radiation striking the Earth's atmosphere and convert it into heat. **(b)** Ozone concentrations in the atmosphere and stratosphere.

TABLE 20-1

Commonly Used Freons

Generic Name	Use	Chemical Name	Chemical Formula
Freon-11	Spray can propellant	Trichloromonofluoromethane	Cl | Cl—C—Cl | F
Freon-12	Coolant in refrigerators, freezers, and air conditioners	Dichlorodifluoromethane	Cl | F—C—Cl | F

insulated cups or foam insulation boards for building homes. CFCs form bubbles in the foam, making it a good insulator. One class of CFCs was also used to clean electronic equipment, precision ball bearings, and medical equipment.

CFCs were attractive substances because they were believed to be chemically unreactive—or inert. Until the early 1970s, chemists believed that CFCs released from spray cans or escaping from refrigerators simply diffused into the upper layers of the atmosphere. There, they were presumed to be partially broken down by sunlight, a process that liberated some of the chlorine atoms. It was thought that this process, called *photodissociation* (breakdown in sunlight), would have little if any effect on the upper atmosphere.

In the early 1970s, two U.S. scientists, Mario J. Molina and F. Sherwood Rowland, began to question this assumption. They pointed out that **chlorine free radicals** (highly reactive chlorine atoms produced by the breakdown of CFCs) might react with stratospheric ozone. Shortly after their announcement, three research teams reported that a single chlorine free radical could react with and destroy ozone molecules in the stratosphere. The reactions are shown in **FIGURE 20-2**. Because of the nature of these reactions, a single chlorine free radical from a CFC molecule can destroy 100,000 molecules of ozone, thus depleting the ozone layer.

In addition to CFCs, scientists found that several chlorine and bromine compounds could deplete the ozone layer. Carbon tetrachloride, for example, is an ozone-depleting solvent that was widely used at the time in fire extinguishers and as a cleaning agent.

Like chlorine-containing compounds, bromine substances also diffuse into the stratosphere, where they photodissociate. Bromine then reacts with ozone, causing its concentration to fall. However, bromine free radicals are far more destructive of ozone than are chlorine free radicals and destroy hundreds of times more ozone molecules than chlorine atoms.

KEY CONCEPTS

Chlorofluorocarbon molecules were once used as propellants, refrigerants, blowing agents, and cleaning agents. CFCs diffuse into the stratosphere, where they break down, releasing chlorine atoms that react with ozone molecules, destroying them. Other chlorine and bromine-containing compounds have also been used widely and are known as *ozone depleters*.

Fluorine —
Carbon —
Chlorine —

UV radiation

Chlorofluorocarbon (CFC) molecule

(a)

Ultraviolet radiation from the sun strikes the CFC molecule and causes a chlorine atom to break away.

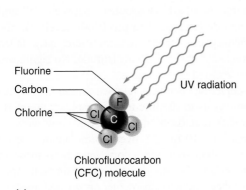

Chlorine free radical

Ozone molecule made up of three oxygen atoms (O_3)

Chlorine oxide

Oxygen (O_2)

(b)

The chlorine atom, known as a free radical, reacts with an ozone molecule to form chlorine oxide and diatomic oxygen.

Oxygen atom

Chlorine oxide

Oxygen (O_2)

Chlorine free radical

(c)

When a free atom of oxygen reacts with a chlorine oxide molecule, diatomic oxygen is formed, and the chlorine atom is released to destroy more ozone.

FIGURE 20-2 The chemistry of CFCs and ozone depletion. (a) Chlorofluorocarbons are dissociated by ultraviolet radiation in the stratosphere. This produces a highly reactive chlorine free radical. **(b)** The free radical can react with ozone in the ozone layer, forming chlorine oxide. This reaction reduces the ozone concentration. **(c)** A single molecule of Freon gas can eliminate many thousands of molecules of ozone because chlorine oxide breaks down, reforming the chlorine free radical. Note too that chlorine oxide also reacts with ozone molecules, destroying them.

High-Altitude Jets High-flying aircraft, such as military jets that routinely travel through the stratosphere, also destroy ozone. How? Jet engines release a pollutant called nitric oxide (NO). Nitric oxide gas, in turn, reacts with ozone to form nitrogen dioxide and oxygen:

$$NO \text{ (nitric oxide)} + O_3 \text{ (ozone)} \rightarrow$$
$$NO_2 \text{ (nitrogen dioxide)} + O_2 \text{ (oxygen)}$$

The threat to the ozone layer from high-flying jets is much lower than that from CFCs. The Concorde, a supersonic jet that once traveled between the United States and Europe in the stratosphere, was taken out of service in 2004 (**FIGURE 20-3**). In 1971, the U.S. Congress killed a proposal to subsidize the construction of 300 to 400 supersonic transports. Had they approved this proposal, jets would have become a major depleter of stratospheric ozone. Ordinary commercial jets also produce nitric oxide. Their effect on the ozone layer is still in question.

KEY CONCEPTS

All jets release nitric oxide gas, a pollutant that reacts with ozone. Military jets that travel through the stratosphere have the greatest impact. Jet travel poses a lower risk than the use of CFCs.

Ozone Depletion: The History of a Scientific Discovery

Evidence in support of the claim that CFCs were depleting the ozone layer began to be published in scientific journals in the 1980s. In May 1985, for example, British scientists reported a large decrease in the ozone layer above Antarctica, which has since been dubbed an *ozone hole* because of the severity of the decrease. In 1986 and 1987, intense study of the ozone hole strongly suggested that it was caused in large part by CFCs. Studies showed that the Antarctic ozone hole was also caused by several natural climatic conditions.

FIGURE 20-3 Supersonic speed: at what price? This supersonic transport once flew through the stratosphere, releasing nitric oxide, destroying ozone.

© Graham Bloomfield/ShutterStock, Inc.

One such weather phenomenon was a vortex of wind (a whirlpool in the atmosphere) that circles the South Pole during the winter months, blocking warmer air from penetrating the region. Studies showed that the vortex contributes to the formation of polar stratospheric clouds. CFCs and other ozone-depleting chemicals adhere to the surface of the ice crystals in these clouds. After winter, when sunlight returns to this frigid region, CFCs are released and broken down by sunlight. Ozone levels plummet as a result. During the 1990s, reductions of 40% to 50% were common, but reductions of up to 70% have been reported. After the vortex breaks down, huge masses of ozone-depleted air move northward, where they hover over nearby landmasses such as Australia, New Zealand, and the southern tips of South America and Africa.

In 1988, researchers discovered that a similar hole was forming over the Arctic. Ozone levels there, however, were not as severely depleted because the vortex in the Arctic is not as strong as it is in the south. Invading winds can break through, keeping the Arctic air much warmer and reducing ozone destruction. Although the wintertime ozone depletion is only 10 to 30%, less than in Antarctica, the Northern Hemisphere is more heavily populated. When the Arctic ozone hole breaks up, it also releases huge masses of ozone-depleted air that may linger over highly populated areas of North America and Europe (**FIGURE 20-4**).

The decline in ozone throughout the atmosphere and especially at the poles has resulted in a 3° decline at mid-latitudes in the Northern Hemisphere between 35° and 65° North latitude (compared to pre-1980s, levels). In the Southern Hemisphere, ozone concentrations have fallen by 6% between 35° and 65° South latitude. No significant changes have occured in the tropics.

Ozone-depleting chemicals released on the Earth's surface take many years to rise into the stratosphere; therefore, scientists predict that ozone depletion will grow worse—eventually reaching 10 to 30% over North America—even though, as you shall soon see, efforts have been made to eliminate the release of ozone-depleting substances. In April 2011, for instance, Arctic ozone depletion reached a record low, causing a massive increase in ultraviolet radiation.

KEY CONCEPTS

Studies of the ozone layer show substantial declines over the globe, with the highest level of depletion in the Southern Hemisphere and especially Antarctica.

The Many Effects of Ozone Depletion

Do the declines in the concentration of ozone gas in the ozone layer result in an increase in ultraviolet radiation striking the Earth?

Extensive studies show that UVB radiation striking the Earth has increased significantly in the Northern and Southern Hemispheres from about 40° N and 40° S.

(a)

(b)

FIGURE 20-4 Mapping ozone depletion. (a) This map of North America shows the ozone depletion at different latitudes between 1969 and 1988. **(b)** Computer-generated images of the changing size of the ozone hole over the Arctic.

Photo b Courtesy of NASA/JPL/Agency for Aerospace Programs (Netherlands)/Finnish Meteorological Institute.

Denver, Colorado is near 40° North latitude. These increases in UV radiation affect the northern U.S., Canada, most of Europe, Russia, and portions of Argentina, Chile, Australia, and New Zealand.

What impacts could ozone depletion have? Like many natural components of our environment, ultraviolet light is beneficial. In small amounts, ultraviolet radiation tans light skin and stimulates vitamin D production in the skin. However, excess ultraviolet exposure can cause problems. In humans, it can cause serious skin burns, cataracts (clouding of the eye's lens), skin cancer, and premature aging. Increased exposure to ultraviolet light may also suppress the human immune system, making us more susceptible to infectious diseases.

U.S. Environmental Protection Agency (EPA) researchers estimate that a 1% depletion of the ozone layer would lead to a 0.7 to 2% increase in ultraviolet light striking the Earth. This, in turn, would lead to an increase in skin cancer of about 4%. Over the next five decades, the EPA estimates that ozone depletion will result in approximately 200,000 cases of skin cancer in the United States. Worldwide, the number will be much higher. Especially hard hit will be countries such as New Zealand and Australia. In November 1991, a panel of scientists convened by the United Nations released a report on the predicted effects of global ozone depletion. They estimated that a 10% decrease in stratospheric ozone

concentrations will annually cause 300,000 additional cases of skin cancer and an additional 1.6 million cases of cataracts (loss of opacity) worldwide.

Studies of skin cancer show that light-skinned people are much more sensitive to ultraviolet radiation than more heavily pigmented individuals. In addition, studies show that some chemicals that are commonly found in drugs, soaps, cosmetics, and detergents may sensitize the skin to ultraviolet radiation. Thus, exposure to sunlight may increase the incidence of skin cancers among light-skinned people and users of many commercial products.

Land- and water-dwelling plants could also suffer from increasing ultraviolet radiation. Intense ultraviolet radiation is usually lethal to plants; smaller, nonfatal doses damage leaves, inhibit photosynthesis, cause mutations, or stunt growth. Declining ozone and increasing ultraviolet radiation could, therefore, cause dramatic declines in commercial crops such as corn, rice, and wheat, costing billions of dollars a year. It may also affect certain commercially valuable tree species.

In a hearing before the U.S. Senate in November 1991, Susan Weiler, head of the American Society of Limnology and Oceanography, testified that studies in Antarctica by other scientists have shown that phytoplankton (algae and other free-floating photosynthetic organisms) populations decrease about 6 to 12% when stratospheric ozone concentrations over the region drop by 40%. Because phytoplankton form the base of the aquatic food chain, damage to them could cause widespread ecological problems. One scientist thinks that ozone depletion and subsequent effects on the food chain may be the reason why two species of penguin are declining in Antarctica.

Finally, ultraviolet light is harmful to many products. Paints, plastics, and other materials deteriorate when exposed to ultraviolet light. Losses from further decreases in the ozone layer could cost society enormous amounts of money.

KEY CONCEPTS

Ozone depletion has resulted in an increase in ultraviolet radiation striking the Earth, especially in unpolluted areas. Ultraviolet radiation causes skin cancer, cataracts, and premature aging. It could also seriously damage ecosystems, crops, materials, and finishes.

Banning CFCs and Other Ozone-Depleting Chemicals: A Global Success Story

In the 1970s, fears caused by early projections of ozone depletion moved several nations, including the United States, Sweden, Finland, Norway, and Canada, to cut back on CFC emissions. In 1978, for example, the United States banned CFC use in spray cans. Freon-12—the refrigerant, coolant, and blowing agent—was not affected by the ban.

Additional scientific evidence on the decline of the ozone layer made it evident a decade later that worldwide cooperation was needed. As a result, in 1987, the United Nations sponsored negotiations aimed at reducing global CFC production. In September of that year, 24 nations signed a treaty called the **Montreal Protocol,** which would cut production of five CFCs in half by 1999 and freeze production of halons (used in fire prevention systems) at 1986 levels. (In halons, bromine atoms replace some or all of the chlorine atoms.) Although halons are used in much smaller quantities, as noted earlier, bromine is far more effective in destroying ozone than chlorine from CFCs and other sources.

This agreement paved the way for a gradual decline in CFC production in the industrial nations, but like so many other pollution control strategies, it would only slow the rate of destruction, not stop it. EPA computer projections showed that an 85% reduction in CFC emissions was needed to stabilize CFC levels in the atmosphere.

Before the Montreal Protocol went into effect, something unusual happened. In March of 1988, an international panel (described earlier) announced that ozone levels had fallen throughout the world. Two weeks later, DuPont, a major producer of CFCs, called for a total worldwide ban on CFC production—when only 2 weeks earlier it had said that it would not support a ban.

In June 1990, 93 nations signed an agreement in London that called for the complete elimination of CFCs and halons by the year 2000, if substitutes were available by then. The signatories also agreed to phase out other ozone-depleting chemicals including hydrochlorofluorocarbons (HCFCs), a class of chemicals (described shortly) once thought to be an excellent substitute for CFCs.

The news about the ozone layer continued to worsen. In 1992, the nations of the world met in Copenhagen to sign another agreement calling for an acceleration of the phaseout of CFCs, carbon tetrachloride, and other ozone-depleting chemicals within 4 to 9 years. The success of the global efforts to phase out CFCs is shown in **FIGURE 20-5**, a graph of projected concentrations of ozone-depleting compounds. Although progress has been impressive, police are finding that CFCs are being illegally imported into the United States (and presumably other countries as well) in massive quantities. Stopping this flow will be required to reach the goals of the various international treaties.

KEY CONCEPTS

As scientific evidence on ozone depletion accumulated, the nations of the world tightened restrictions on the production of ozone-depleting chemicals. Three international treaties have already been signed to eliminate the production of ozone-depleting compounds, and progress toward meeting these goals has been very impressive.

Substitutes for Ozone-Destroying CFCs

As the previous section showed, scientific reports led to the signing of the three treaties to ban CFCs and other ozone-depleting chemicals. The development of replacement chemicals has also played a big role in the dramatic change. It gave industry options and, in some cases, opportunities to profit from the shift to less harmful means of refrigerating

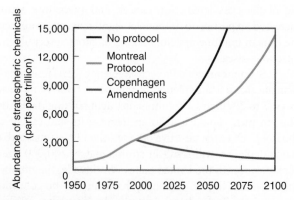

FIGURE 20-5 Graphing success. This graph shows the predicted concentrations of ozone-depleting chemicals in the atmosphere under various scenarios—no action, the Montreal Protocol, and the Copenhagen Amendments.

and of propelling liquids from spray cans. Many companies also led the way. AT&T, for example, was the first U.S. company to set a goal of phasing out the use of CFCs, which they succeeded in achieving in 1993.

CFCs are extremely important to modern society. Every refrigerator, freezer, and air conditioner once used them. To continue to provide these desirable services *and* protect the ozone layer, manufacturers pursued two routes: the use of less stable CFC compounds, which break down before they reach the stratosphere, and the production and use of non-CFC chemicals as substitutes.

Consider the first option, the production of less stable CFCs. This is achieved by substituting a hydrogen atom for one of the chlorine atoms of a CFC molecule. These molecules are known as **hydrochlorofluorocarbons** or **HCFCs**. HCFCs can be used in place of CFCs as refrigerants, cleaning agents, blowing agents, and so on. They work well and are affordable. Because they are less stable than CFCs, HCFCs break down in the lower atmosphere before they can ascend into the stratosphere, home to the ozone layer. Nonetheless, some molecules of HCFC reach the stratosphere where they photodissociate and destroy ozone. All in all, though, they cause much less damage to the ozone layer than CFCs—about one-fifth as much as CFCs.

Because HCFCs cause the destruction of ozone and are greenhouse gases, scientists and policymakers quickly came to view them as an interim solution. The HCFCs were scheduled for phase out so that by 2020, the production and consumption of HCFCs will be virtually eliminated.

The second option for replacing CFCs is the production of non-ozone-depleting substances—chemicals that perform the same functions as CFCs and HCFCs but do not deplete the ozone layer. These are called *zero ozone-depleting compounds*. Over the years, chemists have found numerous ozone-safe substitutes. Many of these chemicals are now being used in air conditioners and freezers, as cleaning agents, and so on. HFCs, hydrofluorocarbons, are the predominant example (**FIGURE 20-6**)

The Good News and Bad News about Ozone

The ozone story is one of humankind's greatest success stories. It not only illustrates how scientific knowledge can be used by society for the common good of humankind and all other species; it also shows how quickly changes can be made to bring about an end to the destruction of the global environment. It further illustrates the power of corporate responsibility. Many of the changes were brought about voluntarily without the need for governmental actions.

Already, studies are showing a decline in the concentration of ozone-depleting chemicals in the stratosphere. Unfortunately, many millions of tons of CFCs have already been released into the atmosphere and some CFCs last 100 years in the atmosphere. Because of this, scientists predict that the ozone layer will recover very slowly. Estimates on full recovery vary. National Oceanic and Atmospheric Administration (NOAA) scientists predict that if international agreements are adhered to, recovery to pre-1980 levels will be achieved by 2050. Others' predictions are less optimistic. They believe that at least 100 years will be required to return the ozone layer to 1985 levels. Another 100 to 200 years may be needed for full recovery.

Many people will contract skin cancer in the interim. The EPA notes that skin cancer in the United States has already reached epidemic proportions. One in five Americans will develop skin cancer in their lifetime and one American currently dies from skin cancer every hour. Melanoma, the most serious form of skin cancer, which has been linked to early exposure to excess ultraviolet light, is one of the fastest growing forms of cancer in the United States.

Nonetheless, many will be spared from skin cancer by the phaseout. The EPA estimates that the successful phaseout of CFCs will result in 295 million fewer cases of skin cancer over the next 100 years than would have occurred without these important changes.

Finally, many ozone-depleting chemicals (and their replacements) are greenhouse gases and thus contribute to global warming and climate change. They will accelerate the warming of the Earth's atmosphere, but their eventual decline may, when combined with other efforts, help us combat global warming.

KEY CONCEPTS

The concentration of CFCs and other ozone-depleting compounds in the atmosphere has begun to decline. Despite this progress, the ozone layer will take many years to recover. Many people will contract and die from skin cancer.

20.2 Acid Deposition

In the 1960s, forest ranger Bill Marleau built the cabin of his boyhood dreams on Woods Lake in the western part of New York's Adirondack Mountains. Isolated in a dense forest of birch, hemlock, and maple, the lake offered Marleau excellent fishing. Ten years after Marleau finished his cabin, however, something bizarre happened: Woods Lake, once a murky green suspension of microscopic algae and zooplankton, teeming with trout, began to turn clear. As the lake went through this mysterious transformation, the trout stopped

FIGURE 20-6 Hydrofluorocarbon. R410A refrigerant has virtually eliminated the use of HCFCs.

biting and soon disappeared altogether. Then the lily pads began to turn brown and die; soon afterward, the bullfrogs, otters, and loons disappeared.

What had happened to Woods Lake? What destroyed the web of life at this small, isolated lake, far from any sources of pollution? Scientists from the New York Department of Environmental Conservation say that Woods Lake is "critically acidified." As a result, virtually all forms of life in and around it have perished or moved elsewhere. The lake became acidified from acids and acid precursors deposited from the skies in several forms.

Acid deposition—the deposition of acidic substances generated from pollution largely human in origin—is commonplace today, as are lakes like Marleau's. In fact, Woods Lake is only one of about 375 lakes and ponds in the western Adirondacks turned acidic and hazardous to virtually all forms of life by acid deposition (**FIGURE 20-7**). In eastern Canada, 100 lakes have met a similar fate. In Scandinavia, the death count is 10,000. Across the globe thousands of lakes will die unless something is done, and quickly.

Acid deposition is a phenomenon not just of environmental interest, but of grave social and economic importance. Studies show that acid deposition turns lakes acidic, kills fish and other aquatic organisms, damages crops, destroys forests, alters soil fertility, and destroys statues and buildings. Moreover, scientists are finding that acid precipitation is more widespread than once thought and is taking a large toll on our environment and pocketbooks. Recent reports indicate that it poses a universal threat, affecting the more developed countries (MDCs) as well as many less developed countries (LDCs). Earthscan, an international environmental group, reports that acid deposition is already damaging soils, crops, and buildings in much of the developing world. Rapidly growing urban centers, with their poorly regulated industry and traffic congestion, are largely the culprits. Ironically, tough pollution laws in MDCs have given multinational corporations incentives to set up operations in LDCs, whose pollution laws (if they have any) are certainly much weaker.

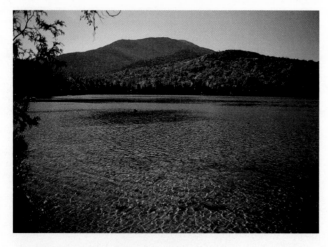

FIGURE 20-7 This crystal clear lake in the Adirondacks is too acidified to support aquatic life.

© Lynn & Sharon Gerig/Tom Stack & Associates, Inc.

What Is an Acid?

Understanding acid deposition requires a brief explanation of acids. An acid is a chemical substance that adds hydrogen ions to a solution. Hydrochloric acid (HCl), for example, dissociates into hydrogen and chloride ions and is therefore an acid.

Acids come in varying strengths. Strong acids add many hydrogen ions; weak acids add fewer. The degree of acidity of a solution, then, is related to the concentration of hydrogen ions in it. Chemists and others measure acidity on the **pH (potential hydrogen)** scale, which ranges from 0 to 14 (**FIGURE 20-8**). Substances that are acidic, such as vinegar and lemon juice, have low pH values—that is, less than 7. Basic (alkaline) substances, such as baking soda and lime, have high pH values—greater than 7. Basic substances have very low hydrogen ion concentrations. Neutral substances, such as pure water, have a pH of 7.

The pH scale is logarithmic, like the decibel scale, which measures loudness, or the Richter scale, which measures earthquake intensity. This means that a change of one unit on the scale—in this instance, 1 pH unit—represents a 10-fold change. Therefore, rain with a pH of 4 is 10 times more acidic than rain with a pH of 5, 100 times more acidic than rain with a pH of 6, and 1,000 times more acidic than rain with a pH of 7.

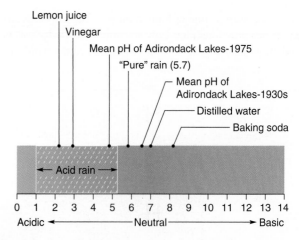

FIGURE 20-8 The pH scale. A pH scale is used to denote levels of acidity.

What Is Acid Deposition?

In an unpolluted environment, rainwater is slightly acidic, having a pH of about 5.7. The normal acidity of rainwater is created when atmospheric CO_2 is dissolved in water in clouds, mist, or fog and is converted into a mild acid (carbonic acid). Acid precipitation is rain and snow with a pH below 5.7.

Acid deposition includes two broad categories: wet deposition and dry deposition. **Wet deposition** refers to acids deposited in rain and snow. These acids are formed when two air pollutants, the sulfur and nitrogen oxides (which are both gases), combine with water in the atmosphere. Sulfur oxides form sulfuric acid; nitrogen oxide gases react with water to form nitric acid. Sulfuric and nitric acid are two of the three strongest acids known to science.

Produced in the atmosphere, these acids may accumulate in clouds and fall from the sky in rain and snow. Even coastal fogs may contain droplets of acid that, when deposited on buildings or plants, can cause noticeable damage. One study shows that moisture droplets in low-lying clouds (fog) tend to contain higher concentrations of acid than rain or snow that falls from them. Therefore, fog and clouds may bathe trees in highly acidic water. Making matters worse, recent studies suggest that the evaporation of recently deposited cloud water from forest canopies may result in acid concentrations on leaf surfaces much higher than those found in the cloud droplets themselves.

Sulfate and nitrate particulates are also present in the atmosphere. These pollutants may settle out of the atmosphere much like fine dust particles. This process is one form of **dry deposition**. Settling onto surfaces, these particulates can combine with water to form acids. Sulfur and nitrogen oxide gases may also be adsorbed onto the surfaces of plants or solid surfaces, where they, too, combine with water to form acids. This is another type of dry deposition.

Where Do Acids Come From?

Acid precursors, the primary pollutants that give rise to acids, come from natural and anthropogenic sources. Before we look at them, let us review how these pollutants are generated. As noted elsewhere in this text, fossil fuels such as coal and oil contain sulfur impurities. When these fuels are burned, sulfur reacts with oxygen in the intense heat and forms sulfur oxide gases. These gases are released into the atmosphere where they may be converted into sulfuric acid. The combustion of any organic matter in air, with its high concentration of nitrogen gas, produces nitrogen oxide gases. These gases are converted into nitric acid in the atmosphere after reacting with water.

Several natural events can produce significant amounts of sulfur dioxide. The natural sources of sulfur oxides include volcanoes and forest fires (**FIGURE 20-9a**). Bacterial decay of organic material such as plants also produces hydrogen sulfide gas, which can be converted into sulfuric acid in the atmosphere. Anthropogenic sources of this pollutant are of major concern, however, because they are often concentrated in urban and industrialized regions, causing local levels to be quite high. Two-thirds of all anthropogenic sulfur dioxide comes from electric power plants, most of which burn coal. Smelters, devices that melt mineral ores to extract pure minerals such as copper and aluminum, also release huge quantities of sulfur dioxide (**FIGURE 20-9b**).

Like the sulfur oxides, the nitrogen oxides arise from a wide variety of sources. Forest fires, for example, can release millions of tons. The two most important anthropogenic sources are electric power plants and motor vehicles. American factories, cars, and power plants in 2012 produced approximately 4.2 million metric tons (4.7 million tons) of sulfur dioxide and about 12 million metric tons (13.4 million tons) of nitrogen oxides a year. Since 2000, sulfur oxide emissions have plummeted by 70%. Nitrogen oxides have fallen by 40%.

The Transport of Acid Precursors

Acid precursors and acids formed in the atmosphere can remain airborne for 2 to 5 days and may travel hundreds, perhaps even thousands, of kilometers before being deposited. Studies have shown that acids falling from the sky in southern Norway and Sweden largely come from England and industrialized Europe. In the United States, acids falling in the Northeast come from the industrialized Midwest, primarily the Upper Mississippi and Ohio River valleys. Indiana and Ohio are the two major producers. Moving eastward, the mass of pollutants tends to converge on New York State and New England.

As noted earlier, acid deposition is a global phenomenon. It is found in the United States, Canada, the Amazon Basin, Europe, and the Netherlands, among other places. All

(a)   **(b)**

FIGURE 20-9 Sources of sulfur dioxide. (a) Volcanic eruptions, such as the Mount St. Helens blast shown here, is one natural source of sulfur dioxide. **(b)** Coal-fired power plants like this one produce most of the anthropogenic sulfur dioxide.

of these places are downwind of heavily polluted areas. In Europe and Scandinavia, rain and snow samples frequently have pH values between 3 and 5. In the White Mountains of New Hampshire, for example, the average annual pH of rainfall is about 4 to 4.21, nearly 100 times more acidic than normal precipitation. Rain samples collected in Pasadena, California in the 1970s had an average pH of 3.9. One of the lowest pH measurements was made in Kane, Pennsylvania, where a rainfall sample with a pH of 2.7 was recorded—rain as acidic as vinegar. The grand prize for acidic rainfall, however, goes to Wheeling, West Virginia, where a rainfall sample had a pH of 2—stronger than lemon juice. More recent studies of the pH of fog downwind from the Los Angeles basin, however, revealed levels as low as 1.7.

In southern Norway and Sweden and in the northeastern United States, two ominous trends were observed until recently. First, acid precipitation fell over a wider and wider area; second, the areas over which the strongest acids were falling were expanding (**FIGURE 20-10**).

KEY CONCEPTS

Acid precursors can be transported hundreds of kilometers from their site of production to their site of deposition often in valuable forest land and recreational regions and in wildlife habitats. Acid deposition occurs downwind from virtually all major industrial and urban centers. Acid deposition is increasing in strength (acidity) and expanding in geographic range.

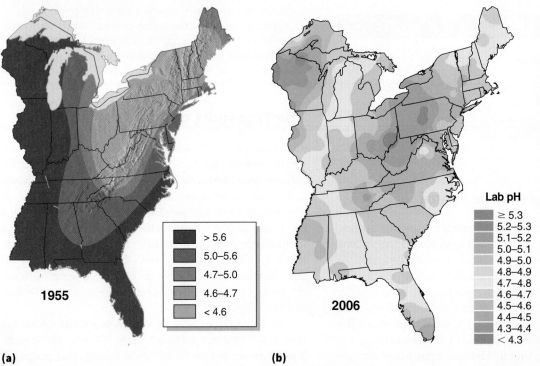

FIGURE 20-10 Acid precipitation: a growing problem. As shown here, acid precipitation in the eastern United States increased between (a) 1955 and (b) 2006. Note the worsening of acid rain and the wider area experiencing it. Thanks to regulations, emissions of acid precursors have fallen and acid deposition has begun to decline.

Data from National Atmospheric Deposition Program [NRSP-3], 2007.

The Social, Economic, and Environmental Impacts of Acid Deposition

Acid deposition has many impacts on the environment, our economy, and society. This section describes some very real impacts and some possible ones.

Acidification of Lakes Throughout the world, thousands of lakes and rivers and their fish have fallen victim to acid deposition. In the 1930s, for example, scientists surveyed lakes in the western part of the Adirondacks. Sampling the pH of 320 lakes, they found that most had pHs ranging from 6 to 7.5. In a 1975 survey of 216 lakes in the same area, a large number of them had pH values below 5, a level at which most aquatic life perishes (**FIGURE 20-11**). Of the acidified lakes, 82% were devoid of fish life. A more recent study of 1,500 lakes in New York's Adirondack Park found that 25% of the lakes are so acidic that fish no longer live in them. Another 20% of lakes are acidic enough to be endangered.

In the United States, eastern lakes had been particularly hard hit. One of every five lakes in Massachusetts, New Hampshire, New York, and Rhode Island was acidic enough to be harmful to aquatic life. As acid deposition continues to fall, these lakes could become lethal to virtually all forms of life.

Studies show that 4% of the lakes and 8% of the streams in acid-sensitive areas in the United States are chronically acidic. This includes lakes in the mid-Appalachians, the Adirondack Mountains, New England, the Atlantic coast, Florida, and the upper Midwest. Although the percentages may seem small, they translate into thousands of lakes and streams—once valuable wildlife habitat and fishing waters.

Studies have shown that acid deposition also occurs widely over the northern and central portions of Florida, with precipitation 10 times more acidic than normal. One-third of the lakes in Florida are now acidic enough to be harmful to aquatic life.

In the mountains of southern Scandinavia, acidification of surface waters has occurred at a rapid rate. In Sweden, approximately 20,000 lakes are without or soon to be without fish. Swedish authorities estimated that about 17% of the nation's lakes are acidified from acid deposition. Salmon runs in Norway have been eliminated because of the impact of acid precipitation on egg development, putting an end to inland commercial fishing in some areas.

In Canada, nine of Nova Scotia's famous salmon-fishing rivers have already lost their fish populations because of acidity. Eleven more are teetering on the brink of destruction. In a survey of 8,500 lakes in southeastern Canada, 13% (1,100 lakes) had pH values below 5.5.

KEY CONCEPTS

Acid deposition has acidified lakes throughout the world. Hundreds of lakes no longer support aquatic life, and thousands are on the verge of ecological collapse.

Why Are Some Lakes More Susceptible Than Others? All lakes are not created equal. Those that lie downwind from major industrial centers are most vulnerable, but location is not a sufficient predictor of a lake's future. Those found in watersheds with thin soils with little **buffering capacity** fall victim to this insidious form of pollution. Some lakes also contain **buffers**, chemical substances that allow aquatic systems to resist changes in pH. When hydrogen ion (H^+) levels increase, buffers combine with the free ions and eliminate them. When levels fall, they release them, thus maintaining a constant pH. In general, the higher the buffering capacity of a lake and its surrounding watershed, the less vulnerable it is. **FIGURE 20-12** shows acid-sensitive areas in the United States and Canada.

KEY CONCEPTS

The buffering capacity of the soil and surface waters—their ability to resist changes in pH—plays an important role in determining if a lake will be damaged by acid deposition.

How Does Acid Deposition Affect Aquatic Ecosystems? Many species of fish die when the pH drops below 4.5 to 5.0 (**FIGURE 20-13**). Falling pH is only part of the reason fish die. Scientists have found that acidic rainwater or snowmelt dissolves toxic elements such as aluminum from soil and rocks. The acidic waters carry the metals to streams and lakes. Aluminum irritates the gills of brook trout, causing a buildup of mucus and, ultimately, death by asphyxiation (**FIGURE 20-14**).

Spring poses a special threat to fish and other aquatic organisms. When the snow begins to melt, the surface melts first. This water drains through the unmelted snowpack and

FIGURE 20-11 Changing pH. These graphs show the pH level in Adirondack lakes. In the 1930s most lakes had a pH of 6.0 or higher, but in the 1970s a large percentage had a pH of less than 5.

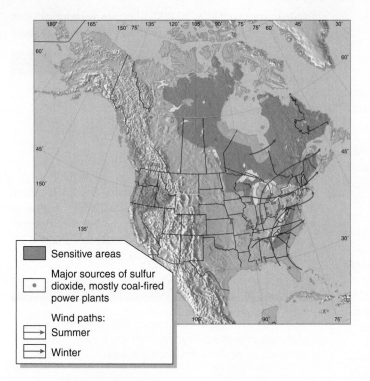

Sensitive areas

Major sources of sulfur dioxide, mostly coal-fired power plants

Wind paths:
Summer
Winter

FIGURE 20-12 Sensitive areas. This map of North America shows areas experiencing acid deposition and the most geologically vulnerable regions.

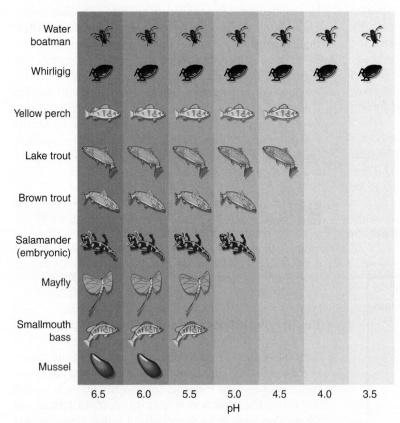

FIGURE 20-13 Differing sensitivity. The sensitivity of fish and other aquatic organisms to acid levels varies. The figure indicates the lowest pH (highest acidity) at which the various organisms can survive. The yellow perch, for example, can withstand a pH of 4.5, but populations plummet if the pH falls any further. Smallmouth bass and mussels are more acid sensitive, perishing at levels below 5.0 and 5.5, respectively.

leaches out most of the acids. In fact, the first 30% of the meltwater contains virtually all of the acid and typically has a pH of 3 to 3.5, which is toxic to eggs, fry, and adult fish as well. Thus, when snow begins to melt, the concentration of acid in nearby lakes and streams rises rapidly. This surge of acids coincides with the sensitive reproductive period for many species of fish.

KEY CONCEPTS

Acidity kills aquatic organisms outright, but it also impairs growth and reproduction. Acidity also leaches heavy metals, which are toxic to fish, from the soil. The spring snowmelt results in a sudden influx of acids and heavy metals that is highly damaging to aquatic life.

Widening the Circle of Destruction Many other species are affected by acid deposition. Songbirds living near acid-contaminated lakes in Scandinavia, for instance, lay eggs with softer shells than birds feeding near unaffected lakes. Scientists have found elevated levels of aluminum in the bones of affected birds and hypothesize that this comes from eating aquatic insects living in acidified waters. The aluminum interferes with normal calcium deposition, resulting in defective (soft) egg-shells and fewer offspring.

Acidification of surface waters may also be partly responsible for a nearly 50% decline in the population of black ducks on the East Coast from the 1950s to the early 1980s when populations reached an all-time low. Today, populations remain steady. Although other factors have accounted for part of this decline, it appears that acids are killing the aquatic insects needed by female ducks and their offspring.

Spotted salamanders are also adversely affected by acidity. In the laboratory, exposure to water with pH of 5 prevents normal embryonic development and results in gross deformities that are usually fatal. In one study, it was found that the mortality of fertilized eggs was 60% at pH 6 but only 1% at pH 7.

In the wild, spotted salamanders breed in temporary ponds created by melted snow. These ponds are likely to be highly acidic in regions where acid precipitation is prevalent; as a result, the fate of the spotted salamander is bleak.

The spotted salamander is as important as birds and small mammals in the food chain. A drastic change in its population would very likely have serious repercussions in the entire ecosystem. Many other amphibian species have disappeared or severely declined in their natural habitat, largely as a result of acid deposition.

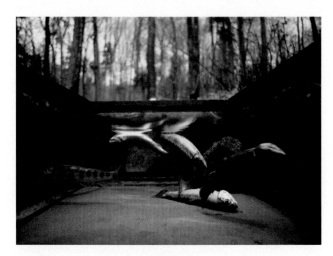

FIGURE 20-14 Death by asphyxiation. These fish were confined to a cage in a stream affected by acid rain. They died of asphyxiation caused by acid leaching aluminum from the soil. Aluminum irritates the gills and causes mucus buildup, which blocks oxygen influx and kills the fish.

© Ted Spiegel/National Geographic Creative.

FIGURE 20-15 Forest die-off. This forest in Germany is now largely dead because of many years of acid deposition.

© Marin Siepmann/imagebroker/Alamy Images.

 KEY CONCEPTS

Acid deposition affects birds living near lakes and aquatic species such as salamanders, which are a key element of the food chain.

Forest and Crop Damage Researchers estimate that 120,000 hectares (300,000 acres) of forest has been destroyed in the former Czechoslovakia by pollution, mostly acid precipitation. In Germany, 500,000 hectares (1.25 million acres) of forest is dead or dying (**FIGURE 20-15**). Even the famous Black Forest is now severely damaged by acidic pollutants from industry and automobiles. In Vermont's Green Mountains, half the red spruce, a high-elevation tree, have died from acid precipitation and acid fog. Lower-elevation sugar maples are also on the decline. The death of trees exacts huge ecological and economic costs. Forest-dwelling species perish as their forests die. The loss of timber reduces revenues. Swiss scientists believe that damage to trees on mountain slopes may increase the likelihood of avalanches because trees help retain snow on steep mountainsides. Further loss of the "barrier forests" endangers the safety of mountain residents, skiers, and highway travelers.

Numerous studies show that acid deposition damages forests both directly and indirectly. Consider direct damage first. Acid deposition damages leaves of birch and needles of pine trees. It also impairs germination of spruce seeds. It erodes protective waxes from oak leaves and leaches nutrients from plant leaves. In 1988, Robert Brock, a forest epidemiologist, reported findings from studies on Mt. Mitchell in North Carolina. In his study, he found that low-lying clouds containing acids that often bathe spruce and fir trees on the mountain are considerably more acidic than vinegar. Two days after a 2-day cloudy period, Brock found that needle

tips looked singed. Needles and leaves damaged by acid develop brown spots. Photosynthesis, the process by which needles and leaves make foodstuffs, declines—causing the tree to suffer nutritionally. Badly damaged leaves and needles fall off.

Indirect effects can also be important. Acids, for example, dissolve nutrients and valuable minerals from the soil trees grow in. These important substances are flushed from the soil, thus depriving plants of vital components essential for growth and reproduction. Acids also cause the release of toxic substances in soil, such as naturally occurring aluminum, which is normally chemically bound within the soil and thus not a problem. Recent evidence shows that aluminum damages cells in the water-transporting tubules of trees, closing off water transport. Trees die from thirst. The release of these substances can damage trees and other plants.

Go Green

Install compact fluorescent lightbulbs in your dorm room or apartment to save energy and reduce pollution.

In 1988, Professor Lee Klinger from the University of Colorado proposed another hypothesis to help explain why the world's forests are dying. One of the chief culprits, he says, may be an acid-loving moss that grows on the forest floor. Klinger has studied 100 regions in 30 states where forests are dying. In each one he has found a thick layer of moss carpeting the forest floor.

These mosses act as sponges, holding so much water that the surface soils become saturated. In affected areas, the feeder roots of the trees and the trees themselves die, for the same reason that a houseplant dies when it is overwatered:

Water eliminates air from the soil. Plants literally suffocate. Mosses may also kill mycorrhizal fungi that help trees absorb nutrients. Mosses also acidify the water passing through them. Acidic water dissolves toxic trace metals such as aluminum found in the soil, which can also kill the root system. It is likely that direct foliar damage and root damage may combine forces.

Acid deposition does not kill trees directly, according to the EPA. It is more likely to weaken them, by damaging their leaves, limiting the nutrients available to them, or poisoning them with toxic substances slowly released into the soil. Weakened trees are more susceptible to diseases, insects, drought, and other pollutants such as ozone that ultimately kill them. Combined, these forces may be responsible for the massive reduction in forest growth noticed by foresters in the eastern United States. It reminds us how important it is to consider the interaction of many changes we humans have created.

Just like trees, crops are affected directly and indirectly by acids. The direct effects include damage to leaves and buds. Acids falling on crop plants in the spring may impair growth at a very important time of year. In addition, acids may inhibit photosynthesis, the process by which plants produce carbohydrates and other important chemicals.

Acids damage plants indirectly—by altering the soil. For example, they may leach important elements from the soil, resulting in reduced growth. Acids impair soil bacteria and fungi that play an important role in nutrient cycling and nitrogen fixation, both essential to normal plant growth.

Concern for agriculture has also been raised by numerous researchers, but the results of many studies are inconclusive. Some researchers have reported that simulated acid precipitation decreases crop productivity, but others have found increases. Still others have found no effect. On balance, the EPA concludes that food crops are not seriously affected by acid deposition.

 KEY CONCEPTS

Acid deposition damages forests in many parts of the world and affects crops as well. Trees and other plants are damaged directly by acids but also indirectly, through changes in the soil chemistry and soil-dwelling organisms. Acids may not kill trees outright but will seriously weaken them, making them susceptible to disease, insects, and other air pollutants.

Acids: Fertilizing Effect Aquatic and terrestrial plants require sulfur and nitrogen to grow. In some instances, acid rain may enhance soil fertility and improve crop growth.

On balance, acids probably do more damage than good. Direct damage to growing plants and damage to the soil offset the fertilizing effect. Studies have also shown that nitrogen in the form of nitrates or nitric acid deposited in Chesapeake Bay actually stimulate excessive growth of algae and aquatic plants.

Plant overgrowth in Chesapeake Bay caused by nitric acid impairs navigation. In addition, plants growing on the surface block sunlight needed for photosynthesis in plants and algae in deeper layers. These deeper plants help maintain oxygen levels; without them, other forms of aquatic life may perish. In addition, when aquatic vegetation dies in the autumn, it decays. The bacteria that break down this organic matter rob the water of oxygen, killing many aquatic organisms.

In a recent study of the effects of nitrogen fertilization on the prairies of Minnesota (published in *Science*), researchers found that additional nitrogen benefited weedy species that overtook native prairie grasses. Native species require less nitrogen and were choked out by weedy species. This not only altered the species composition, it resulted in a greater emission of carbon dioxide, with implications for global warming. The excess carbon dioxide emissions result from the fact that weedy species decompose much more rapidly than native prairie grass species.

 KEY CONCEPTS

The sulfur and nitrogen in sulfuric and nitric acid promote plant growth, but their negative effects (such as direct damage and changes in the soil chemistry) typically outweigh any benefits resulting from their fertilizing effect.

Damage to Materials

Acid precipitation also corrodes humanmade structures. It has taken its toll on some of special importance, such as the Statue of Liberty, the Canadian Parliament Building in Ottawa, Egypt's temples of Karnak, and the caryatids of the Acropolis—not just architectural works but works of art, priceless treasures (**FIGURE 20-16**). Acid rain may also damage house paint and etch the surfaces of automobiles and trucks. A U.S. report claims that acid precipitation causes an estimated $5 billion a year in damage to buildings in 17 northeastern and midwestern states. The price tag includes the cost of repairing mortar, galvanized steel, and stone structures as well as the cost of repainting. It does not include damage to automobile paint, roofing materials, and concrete, potentially adding billions of dollars to the cost.

KEY CONCEPTS

Acids cause billions of dollars of damage to priceless statues, buildings, and materials.

Solving a Growing Problem— Short-Term Solutions

The release of acid precursors has dropped significantly since the 1990s in large part because of regulations on coal-fired power plants. Acid deposition has also declined,

FIGURE 20-16 Lady Liberty gets a face-lift. The Statue of Liberty was recently given a multimillion-dollar face-lift because of damage from air pollution, including acid deposition. The repair work cost well over $35 million.

© JP Laffont/Sygma/Corbis.

FIGURE 20-17 Liming rivers. Millions of dollars are spent by the Swedish government to lime rivers each year to offset the influx of acids from the sky.

© Martin Bond/SPL/Science Source.

The first significant U.S. governmental action came with the passage of 1990 amendments to the Clean Air Act. It contained provisions for substantial reductions in sulfur oxide emissions but more modest cuts in nitrogen oxides by the year 2010. Although the goals were not met, there were substantial reductions in the release of acid precursors.

In 1991, the United States and Canada signed an agreement to reduce emissions of sulfur dioxide and nitrogen dioxide from power plants, smelters, and other stationary sources. Canada, highly motivated to protect its lakes from further deterioration, met its goal by 1993. Huge reductions occurred at some facilities such as the world-famous Inco nickel smelter in Sudbury, Ontario, where 90% reductions were achieved. The United States met its goals but a little later.

International efforts to control sulfur oxide emissions generally rely on three strategies: (1) the installation of scrubbers on new and existing coal-fired power plants, (2) the combustion of low-sulfur coal or natural gas in utilities, and (3) the combustion of desulfurized coal—that is, coal that has had most of its sulfur removed. Nitrogen dioxide levels are more difficult to control because nitrogen oxide gases come from the nitrogen in air that reacts with oxygen in high-temperature furnaces. Changes in the temperature of combustion, however, can help reduce nitrogen oxide emissions.

Another approach to the problem involves treating the lakes themselves. In 1977, for example, the Swedish government embarked on an expensive program to neutralize acidic lakes by applying lime to thousands of lakes and rivers (**FIGURE 20-17**). These actions improved the water quality in many lakes, saving fish populations, but cost tens of millions of dollars per year.

Critics argue that liming is a short-term, stopgap solution, a little like CPR administered to a heart attack victim. In Canada liming costs $120 per hectare ($50 per acre). Treating a single lake can cost between $4,000 and $40,000. In 5 years, however, treated lakes turn acidic again.

Stopgap measures are essential, as noted earlier in the book. They are the first line of defense against acid deposition, but many much less costly—indeed, profit-generating—options are available to us.

and rivers are also steadily improving. Although statistics on recovery aren't available, many lakes too acidic to support fish are being restocked.

Long-Term Sustainable Strategies

Liming lakes and other approaches outlined in the previous section generally treat the symptoms of the problem while ignoring the root causes—in this case, our heavy dependence on and inefficient use of fossil fuels. To solve this problem sustainably, we must find ways to address the root causes. Increasing energy efficiency and reducing our dependence on fossil fuels by using solar and wind energy are key elements of a sustainable strategy.

Recognizing the importance of these alternative measures, several countries—including Germany, Great Britain, Japan, Norway, and Sweden—are actively developing alternative fuel sources of energy and increasing the efficiency with which they use fossil fuels. In Mexico, utilities, government, and industry have teamed up in hopes of making energy efficiency a cornerstone of the country's development. In India, private businesses have launched a program to install thousands of energy-efficient compact fluorescent lightbulbs in homes and businesses.

Population stabilization, growth management, and recycling are also essential elements of our prevention strategy. All of these reduce our demand for energy and, therefore, our production of pollutants. These strategies help solve more than acid deposition—they'll help reduce habitat destruction, the loss of species, urban air pollution, water pollution, and other environmental ills.

KEY CONCEPTS

Fuel efficiency, renewable fuels, recycling, population stabilization, and growth management are key elements of a sustainable design strategy to help prevent the production of acid precursors—and hence reduce acid deposition.

Are Controls on Acid Deposition Working?

The Acid Deposition provisions of the 1990 Clean Air Act called for a dramatic reduction of sulfur dioxide emissions to 9 million metric tons (10 million tons) per year by 2010—over a 50% reduction. In 2013, sulfur oxide emissions were at 4.2 million metric tons. It also required reduction, albeit a modest one (about 10%), of 1.8 million metric tons (2 million tons) per year in nitrogen oxide emissions. In 2013, nitrogen dioxide emissions had dropped from 22.7 to 11.3 million metric tons!

The 1990 Clean Air Act also directed the **National Acid Precipitation Assessment Program (NAPAP)** to develop market-based controls on acid deposition—that is, strategies that encouraged electric utilities to cut emissions through innovation and other methods that made more sense economically than old command and control regulations. One idea was the tradable permit. Furthermore, the NAPAP was directed to monitor progress and economic costs and benefits of controls.

As demonstrated by the dramatic decreases in acid precursors, market-based controls were very effective in cutting sulfur emissions. In response to the decreases in sulfur dioxide emissions, concentrations of sulfate and sulfuric acid in precipitation have shown a significant decline in many parts of the United States. Decreases in wet sulfate deposition have averaged more than 30% in the Eastern U.S. (**FIGURE 20-18**). Nitrate concentrations in rain, as expected, have shown less change (**FIGURE 20-19**). Many lakes and streams also reflect changes in the chemical makeup of the rain. There is now evidence of recovery of acidified lakes in New England as well as badly damaged lakes of the Adirondacks.

KEY CONCEPTS

Market-based strategies, in particular tradable permits, have proven successful in reducing sulfur dioxide emissions in the United States, with corresponding changes being seen in the acidity of and snow as well as many lakes and streams. Reductions in acid precursors are one of the greatest environment success stories in the United States.

20.3 Global Climate Change

Scientists have long known that air pollution can affect local weather. For example, smoke from factories can substantially increase rainfall in areas downwind. Air pollution can also affect weather globally. Evidence for this phenomenon has been growing steadily for many years. In 1995, a group of 2,500 atmospheric scientists concluded that the bulk of the evidence showed that human activities, especially emissions of carbon dioxide, were having a discernible effect on global climate that could have profound effects on people, the economy, and the environment. They contended that human activities were increasing pollutants that caused **global warming,** an increase in average global temperature. Global warming, in turn, was altering the Earth's climate, that is, causing **global climate change.** Climate is the average weather conditions. Their announcement was met with skepticism by a handful of scientists and by industry, especially the oil and coal industries.

In December 1999, the head of the U.S. National Oceanic and Atmospheric Administration and the head of the British Meteorological Office, agencies that normally do not venture into the political arena, agreed with this assessment. They published a joint letter in a London newspaper, *The Independent*, in which they wrote, "Our climate is changing rapidly and it's important that we take action now.... we believe the evidence is almost incontrovertible, that man has an effect (on climate) and therefore we need to act accordingly."

To understand this issue, we first look at the science of global energy balance, the basis for the planet's climate.

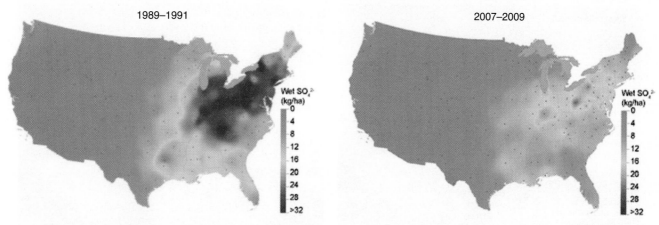

FIGURE 20-18 Acid deposition on the decline. These maps show a significant decrease in wet sulfate deposition in the U.S.

Data from Environmental Protection Agency 2009 Environmental Results. http://www.epa.gov/airmarkt/progress/ARP09_3.html

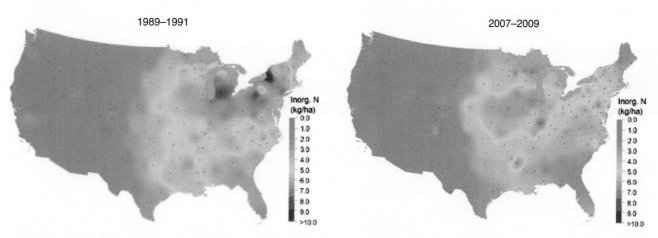

FIGURE 20-19 Acid deposition on the decline. These maps show the decrease in wet nitrate deposition resulting from reductions in atmospheric nitrogen oxide emissions.

Data from Environmental Protection Agency 2009 Environmental Results. http://www.epa.gov/airmarkt/progress/ARP09_3.html

The Science Global Energy Balance and the Greenhouse Effect

Each day the Earth is bathed in sunlight. Approximately one-third of the sunlight striking the Earth and its atmosphere is reflected back into space. The rest is absorbed by the air, water, land, and plants. As anyone who has sat in the sun knows, sunlight is absorbed by surfaces and converted into heat or, more technically, infrared radiation. As shown in **FIGURE 20-20a**, this heat is slowly radiated back into the atmosphere and eventually escapes into outer space. As a result, energy input is balanced by energy output.

Scientists have long known that certain chemical substances in the atmosphere alter this balance. These include water vapor, carbon dioxide (CO_2), nitrous oxide (N_2O), methane (CH_4), and chlorofluorocarbons (**TABLE 20-2**). How do these gases affect the Earth's energy balance? Consider carbon dioxide. Carbon dioxide molecules in the air absorb infrared radiation escaping from the Earth's surface and radiate it back to Earth, acting much like the glass in a greenhouse (**FIGURE 20-20b**). Other molecules act similarly.

This phenomenon, known as the **greenhouse effect,** is essential to life on planet Earth, for it helps maintain the Earth's surface temperature. Without it, the Earth would be at least 30°C (55°F) cooler than it is today—and inhospitable to most life-forms.

The Earth's surface temperature can be altered by changes in the concentrations of CO_2 and other chemical substances previously mentioned. An increase in carbon dioxide levels, for instance, causes more heat to be radiated back to the Earth, which warms the planet. Decreases in carbon dioxide or other similar compounds have the opposite effect.

The chemical substances that increase the Earth's surface temperature are called **greenhouse gases** because they act somewhat like the glass in a greenhouse. You've observed the greenhouse effect in a car or house heated by the sun. The glass permits sunlight to enter the car's interior, where it is absorbed by interior surfaces and converted to heat. The glass prevents heat from escaping. On Earth, sunlight is absorbed by plants, dirt, buildings, parking lots, and roadways and is converted into heat. The heat or

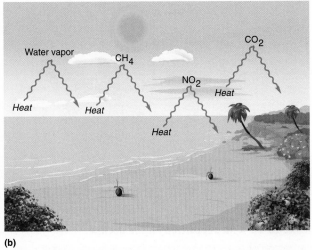

FIGURE 20-20 Global energy balance. (a) This drawing shows the influx of solar radiation and its fate. **(b)** Carbon dioxide and other greenhouse gases radiate heat back to Earth, causing the Earth's surface temperature to increase.

TABLE 20-2

Major Greenhouse Gases and Their Characteristics

Gas	Atmospheric Concentration (ppm)	Annual Increase (%)	Life Span (Years)	Relative Greenhouse Efficiency (CO$_2$ = 1)	Current Greenhouse Contribution (%)	Principal Sources of Gas
Carbon dioxide (CO$_2$) (from fossil fuels)	380	0.4	x[1]	1	57 (44)	Coal, oil, natural gas, deforestation
Carbon dioxide (from biological sources)					(13)	
Chlorofluorocarbons (CFCs)	0.000225	5	75–111	15,000	25	Foams, aerosols, refrigerants, solvents
Methane (CH$_4$)	1.774	1	11	25	12	Wetlands, rice, fossil fuels, livestock
Nitrous oxide (N$_2$O)	0.319	0.2	150	230	6	Fossil fuels, fertilizers, deforestation

Data from World Watch Institute, U.S. EPA, and *Journal of Geophysical Research*.

[1]Carbon dioxide is a stable molecule with a 2- to 4-year average residence time in the atmosphere.

infrared radiation then rises, but some of it is trapped in the atmosphere by greenhouse gases.

Studies of the global energy balance show that the Earth's surface temperature is influenced by two key factors: those that affect the amount of sunlight striking the Earth and those that alter the amount of heat lost or retained. In other words, the Earth's temperature is affected by the influx of solar energy and loss of heat. Both natural and anthropogenic factors influence these processes.

Natural Factors That Influence Global Temperature

The average temperature of the Earth and, as a consequence, the Earth's climate, changes naturally over time. In fact, the Earth cycles between cold spells (ice ages) and interglacial warm periods. We're currently in an interglacial warming period.

These natural changes in climate, while not completely understood, are the result of at least five natural phenomena: (1) changes in the Earth's orbit around the sun, (2) changes in the Earth's tilt, (3) increases or decreases in solar activity (sunspots), (4) increases or decreases in volcanic activity, and (5) chaotic interactions of the climate system.

As you may recall from early science classes, the Earth orbits around the sun in an elliptical path. Scientists have found that the Earth's orbit changes over time, sometimes bringing the Earth slightly closer to the sun, causing warming, sometimes moving the Earth slightly farther away from the sun, causing cooling.

You may also recall from earlier studies that the Earth is tilted at approximately a 23.5-degree angle. Surprisingly, the Earth's tilt can also change very slightly, altering the average temperature and either heating or cooling the planet.

Solar activity is also inconstant. In fact, the sun's output goes through an 11-year cycle. Increased output results in natural planetary warming.

Changes in volcanic activity also play a role in climate. Increased volcanism, for instance, results in the release of ash that reduces solar gain, cooling the Earth. According to scientists, volcanic activity seems to play an important role in year-to-year climate variability, and there is not much evidence to suggest that prolonged periods of volcanic eruptions cause long-term changes.

The last cause of long-term natural climate change is potentially chaotic interactions within the climate system. Complex and inherently unpredictable interactions take place between major components of the Earth's climate, for example, the oceans and the atmosphere. Such interactions could, say some scientists, cause the climate system to change over time, resulting in either a period of cooling or warming that could last for decades to centuries. The continued action of these effects could even restore conditions to something resembling the earlier state.

As shown in **FIGURE 20-21**, the Earth's temperature and the concentration of two naturally occurring greenhouse gases, methane and carbon dioxide, also change cyclically—approximately every 100,000 years or so. What causes these changes? Scientists believe that natural causes—for example, changes in the Earth's tilt or solar activity—stimulate

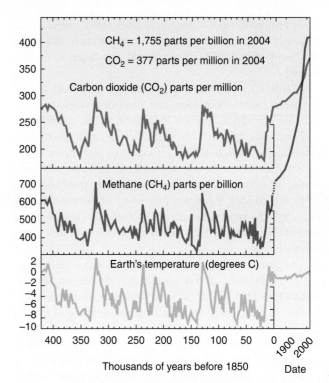

FIGURE 20-21 Natural cycles. The Earth's temperature (bottom panel) cycles approximately every 100,000 years. Methane and carbon dioxide levels also undergo long-term cycles, as explained in the text. Notice that today, carbon dioxide and methane levels far exceed those found in the previous 450,000 years. What effect will this have on average global temperature?

Reproduced from Springer and *Climatic Change*, vol. 68, 2005, pp. 269–279, A slippery slope: How much global warming constitutes "dangerous anthropogenic interference?" J. E. Hansen, and is given with kind permission from Springer Science and Business Media.

increases in temperature. It in turn increases forest and grassland fires and the decay of organic material, which increase carbon dioxide and methane levels. These greenhouse gases, in turn, increase global average temperature up to a point.

Anthropogenic Factors That Increase Global Temperature

Human activities also affect global average temperature. Two activities are most important: the release of greenhouse gases and deforestation. Greenhouse gases, such as carbon dioxide and methane, come from natural and anthropogenic sources. Although the release of greenhouse gases from natural sources has remained fairly constant over the past 100 years, emissions from human sources have increased dramatically. Annual carbon dioxide emissions from the combustion of fossil fuels, for instance, climbed from a mere 534 million tons (of carbon) to about nearly 9,700 million tons per year in 2012 (**FIGURE 20-22a**) in the past 100 years.

Not all greenhouse gases have a natural source. Chlorofluorocarbons and their replacements, the hydro-chlorofluorocarbons, for example, have no natural source.

KEY CONCEPTS

The Earth's climate shifts naturally as a result of many factors, including changes in solar activity and volcanic activity, as well as changes in the Earth's orbit and the tilt of the Earth.

Their production and release have risen rapidly since 1950, although CFCs release has declined dramatically as a result of a use ban and HCFCs release will also begin to decline as well.

Table 20-2 lists the four greenhouse gases and several important facts about each one, including their sources, the rates at which they are increasing in the atmosphere, and their life spans. It also lists their relative greenhouse efficiency—that is, how they compare as a greenhouse gas to carbon dioxide. As shown, one molecule of CFCs is equivalent to 15,000 molecules of carbon dioxide, partly because CFCs last so long. Pay special attention to the contribution of each to the greenhouse effect in column 6. This takes into account the relative effect of each one and the amount in the atmosphere.

As you can see, carbon dioxide is the major greenhouse gas and emissions of this greenhouse gas are on the rise (**FIGURE 20-22a**). Not all carbon dioxide comes from the combustion of fossil fuels. Deforestation and torching of cleared forests, for example, also produce a significant amount of carbon dioxide. How does deforestation result in an increase in atmospheric carbon dioxide? As you may recall from studying ecology, all plants, including trees, take up carbon dioxide from the atmosphere, which they convert into plant matter. As trees are stripped from the land, the Earth's capacity to absorb carbon dioxide decreases. Atmospheric levels increase.

Are Global Warming and Global Climate Change Occurring?

While many people believe that the Earth's oceans and our atmosphere are warming, others are not so sure. What is happening?

As shown in **FIGURE 20-22c**, the Earth's temperature is indeed increasing and the climate is shifting as a result. The temperature increase is determined from data obtained from thousands of temperature sensors distributed over land and sea throughout the Earth's surface. Temperature readings show a steady increase in global average temperature since the 1880s with only a small dip in temperature from 1940 to 1960. Scientists believe that this transient decline may have been caused by particulate pollution from factories during the post-war surge in economic activity. Particulates reduce sunlight striking the Earth. Starting in the 1960s, air

(a)

(b)

(c)

FIGURE 20-22 Global carbon dioxide emissions and global temperature. (a) Annual carbon emissions from fossil fuel burning, 1751–2012. **(b)** Annual mean carbon dioxide levels in the atmosphere have risen dramatically since 1960. **(c)** Graph of average global temperature since 1950.

Part a data from Earth Policy Institute (http://www.earth-policy.org/indicators/carbon_emissions_2013); Part b data from Scripps Institute of Oceanography; part c data from Earth Policy Institute (http://www.earth-policy.org/indicators/C51).

pollution laws reduced particulate pollution, eliminating this cooling effect.

While rising average global temperature shows that the Earth is warming, there's an abundance of additional evidence that change is underway. Here is a summary of evidence for global warming and global climate change:

Rising sea level. In the past century, sea level has risen 17 centimeters (6.7 inches), as a result of melting glaciers in mountainous areas and in Antarctica, according to the National Aeronautics and Space Administration.

Rising ocean temperatures. Studies show that the oceans are warming to a depth of 3,000 meters (10,000 feet). The oceans currently absorb about 80% of the heat added to the climate system by global warming. As the temperature of water increases, water expands. This is thought to be part of the reason for rising sea level and changing intensity of storms, especially hurricanes. Heat from oceans adds energy to hurricanes, increasing wind speeds, and the damage they cause.

Melting glaciers. Global warming is evidenced by rapidly melting glaciers throughout the world.

Breakup of Antarctic ice sheet. Huge blocks of ice, often the size of small New England states, are breaking off the Antarctic ice sheet, signaling a melting and break-up of this massive expanse of ice.

Dramatic decrease in Arctic polar sea ice and land-based ice in Greenland (**FIGURE 20-23**). A research ship once took 2 years to traverse the North Pole; in 2000, the ship made the trip in a few months because of the lack of ice.

FIGURE 20-23 Loss of arctic sea ice. The yellow line is the average boundary of the ice pack from 1979 to 2004. This photo was taken in 2005, showing a dramatic decrease in the extent of the ice.

Courtesy of Josefino Comiso and NASA/Goddard Space Flight Center Scientific Visualization Studio.

Heat waves and drought on the rise. Heat and drought are taking their toll on agriculture worldwide, causing severe economic problems in some areas. In the early 2000s, severe drought plagued many parts of the United States. They have continued into the next decade. Heat waves are also responsible for numerous deaths. In 2003, an estimated 40,000 people died in Europe as a result of a record-breaking heat wave.

Tornadoes and violent storms are on the rise. Heat energy stored in oceans and in the atmosphere spawns more violent storms. A recent study of the severity of

© hugolacasse/ShutterStock, Inc.

SPOTLIGHT on Sustainable Development

20.1 Intel Pledges Huge Purchase of Green Energy

In 2008, computer chip giant Intel announced that the company would purchase more than 1.3 billion kilowatt hours a year of renewable energy certificates (RECs) from Sterling Planet to offset carbon dioxide emissions from their operations. The electricity will be generated by wind, solar, small hydroelectric facilities, and biomass. According to the U.S. EPA, this purchase makes Intel the single-largest corporate buyer of green power—electricity generated from clean, renewable resources—in the United States.

Besides offsetting their emissions, Intel officials hoped that their purchase would help stimulate other companies to follow suit and, hence, boost the market for green power. Their efforts, combined with those of other companies, could help lower the generating cost of electricity from renewable resources such as wind.

To learn of other companies, check out EPA's Green Power Partners list online. At this writing, Intel takes the number one spot, and PepsiCo is number two. Companies are not the only entities that have made major commitments to purchase green power. The Commonwealth of Pennsylvania, the city of Houston, and the city of Dallas are all major purchasers, as are Sprint Nextel, IBM, and the U.S. Department of Veterans Affairs.

storms in the United States showed that violent downpours are on the rise. Studies also show that the occurrence of tornadoes is on the rise in the United States (**FIGURE 20-24**). In the 1980s, 600 to 800 tornadoes occurred in the United States. In the 2000s, the number had increased to 1,000 to 1,700 per year. In 2008, the United States experienced nearly 1,700 tornadoes that killed over 1,700 people; at this writing (May 2011) there were already over 1,000 reported tornadoes. Worldwide, the incidence of major weather events has also increased. In 1999, there were four times as many major storms and floods as there were in the 1960s, and damage was seven times higher. Plagued by devastating floods in Europe in recent years that many experts think were caused by global climate change, the European insurance industry has become a major supporter of international efforts to curb the emission of greenhouse gases.

Severity of hurricanes is on the rise. Studies show that although the number of hurricanes has not increased, their severity has (**FIGURE 20-25**). Hurricane Katrina is a good example. It killed 1,836 people, and displaced 1 million residents in Louisiana and Mississippi. Katrina destroyed 200,000 homes, costing the insurance industry $25.3 billion. It destroyed 500,000 hectares, (1.3 million acres) of forest, costing $5 billion. The total economic impact of this event is estimated to be around $150 billion.

Forest fires on the rise. During the 16-year period from 1987 to 2003, fires burned 6.5 times as much area per year as they did between 1970 and 1986. While the increase in forest and grass fires typically has been blamed on fire suppression and grazing policies, it is more likely the cause of rising temperature. Severe forest fires have continued throughout the western United States and Australia in recent years.

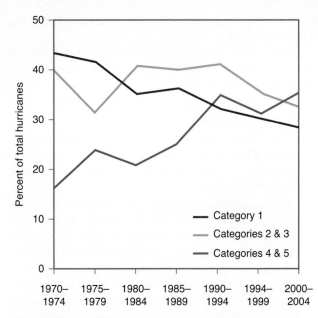

FIGURE 20-25 Category 1, 2, and 3 hurricanes are on the decline while more violent and destructive category 4 and 5 hurricanes are on the rise.

Adapted from P. J. Webster, et al., *Science* 309 (2005): 1844–1846.

KEY CONCEPTS

Abundant evidence clearly shows that global warming is occurring and that the climate is changing as a result. Evidence includes: rising global temperature, rising sea level, increasing ocean temperatures, melting glaciers, loss of polar ice, heatwaves, droughts, more violent storms, and an increase in forest fires.

Are Human Activities Causing Global Warming?

To most objective observers the evidence is pretty straightforward: the Earth's temperature—and its climate—are changing dramatically. Even though these potentially devastating changes are occurring, critics say that this does not mean that humans are the cause? Natural factors could be playing a role.

In a report of the International Panel on Climate Change, researchers presented sound evidence to suggest that while natural factors are indeed warming the planet, they pale in comparison to human causes. As shown in **FIGURE 20-26**, scientists determined the rate of heating (or cooling). They expressed these values, known as radiative forcing, in watts per square meter, that is, the amount of heat per square meter of the Earth's surface. As illustrated, some human activities increase temperature; others decrease global temperature. On balance, though, human activities are increasing the Earth's temperature. But how do they compare to natural factors?

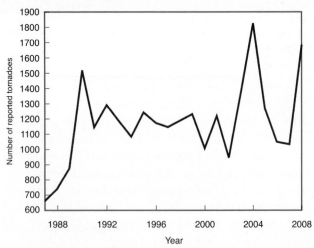

FIGURE 20-24 Tornadoes. The number of tornadoes each year is on the rise.

FIGURE 20-26 Radiative forcing (RF) components. This graph shows the warming and cooling effects of human pollutants and natural forces, measured as radiative forcing. The natural effects are dwarfed by human factors. (LOSU = level of scientific understanding.)

Reproduced from *Climate Change 2007: The Physical Science Basis*. Working Group I Contribution to the Fourth Assessment Report of the Intergovernmental Panel on Climate Change. Cambridge University Press.

As shown in the figure, natural radiative forcing is only a small fraction of human-induced warming. Computer studies bear out this contention. Using complex and increasingly more reliable computer models of global climate, scientists plugged in data on natural and anthropogenic factors that affect global climate to estimate the average temperature of various continents as well as future temperature. The results are shown in **FIGURE 20-27**. In this graph, the blue-shaded area indicates the average temperature that would be expected if only natural factors were involved in global temperature. The pink shaded areas indicate the predicted temperature if natural and human factors were involved. The black lines represent the actual average temperature based on observations. As you can see, there's a very good correlation between actual and predicted temperature if both natural and human factors are taken into consideration.

 Go Green

Recycle all waste, and use recycled products to reduce our nation's energy demand and environmental pollution.

KEY CONCEPTS

While natural forces affect global warming and global climate change, human activities appear to be the main driving force.

Predicting Future Effects of Global Warming and Global Climate Change

While scientific evidence strongly suggests that global warming and global climate change are occurring and that human activities are the main cause, what will the future bring? How much change will occur? How will future releases of greenhouse gases and deforestation affect our climate, our economy, our society, and ecosystems?

Predicting future impacts is fraught with difficulty. Many factors will determine the level of greenhouse gases in the future. Population growth and the rate of economic growth and industrialization will be primary determinants, however, industrialization alone is not a sufficient predictor. What is most important is the amount of fossil fuel that will be burned to spur economic growth and industrial development.

Public policy will also have a profound effect on future concentrations of greenhouse gases. Measures to promote renewable energy and efficient energy use could reduce the growth in the emissions of carbon dioxide. International treaties are also important.

To predict the effects of different levels of atmospheric greenhouse gases, scientists use **global climate models,**

FIGURE 20-27 Actual and predicted temperature. The blue-shaded areas show the predicted temperature of various regions based on natural factors. The pink-shaded areas show predicted temperature if both natural and anthropogenic factors were included. The black line represents the actual temperature. Notice that there's a very good correlation between the pink-shaded area and the black line.

Modified and based on *Climate Change 2007: Synthesis Report.* Contribution of Working Groups I, II and III to the Fourth Assessment Report of the Intergovernmental Panel on Climate Change, Figure SPM.4. IPCC, Geneva, Switzerland.

computer programs that contain mathematical equations that attempt to simulate the various aspects of our climate and how they change. Scientists feed data into the computer regarding projected greenhouse gas levels, and the computers use this information to predict possible climatic effects.

Global climate models have been criticized by some as being inaccurate. Although early climate models were crude, considerable efforts have gone into fine-tuning models to make them much more valuable tools for predicting future climate. With this in mind, let's take a look at some of the potential impacts of rising greenhouse gas emissions and deforestation.

Rising Temperatures, Rising Sea Levels The UN-sponsored Intergovernmental Panel on Climate Change (IPCC) predicts a 1.1° to 6.4°C increase in average global temperature by year 2100—that's a 2° to 11.5°F increase. How much temperature increases depends on emissions. The higher the emissions, the greater the temperature increase. As you can see in Figure 20-21a, emissions are rising rapidly. Although a 1.1° to 6.4°C increase may seem insignificant, it could drastically alter global climate and sea level. IPCC scientists predict that sea level

could increase 0.2 to 0.6 meters (8 to 24 inches) by 2100 as a result of the melting of glaciers and the land-based Antarctic ice pack and an expansion of the seas resulting from warmer temperatures.

Rising sea levels would threaten coastal cities throughout the world. Today, over half the world's population lives in coastal cities and towns. In fact, over 40 of the world's largest cities are in coastal regions. Among cities most at risk are Miami, New Orleans, Bangkok, Hamburg, London, St. Petersburg, Shanghai, Sydney, Alexandria, and Dhaka. In the United States, more than half of the population lives within 83 kilometers (50 miles) of the ocean. Even a modest increase in sea level would flood coastal wetlands, low-lying fields, and cities. The rise in sea level would also worsen the damage from storms. Waves produced during hurricanes and other storms would sweep further inland, damaging more homes and cities than they do today. Many people would have to relocate. Cities may be forced to build levees to hold back the seas or move to higher ground.

The inland creep of the ocean usurps farmland and wildlife habitat and create more crowding as people compete for a limited land base. The changes would not come cheaply, and they would not be easy.

Less-developed nations would also suffer enormously if the oceans rise. In Asia, for instance, rice is produced in many low-lying regions that might be reclaimed by the sea. Storm surges could carry saltwater onto some of the remaining fields, killing crops and poisoning the soil.

Low-lying island nations, many of them tropical resort places, would also be dramatically affected by rising sea level. Many island nations in the South Pacific are already beginning to feel the effect as ocean waters rise and rob them of valuable beaches and shoreline (**FIGURE 20-28**). In fact, a coalition of island nations is one of the leading proponents in global climate change negotiations.

FIGURE 20-28 A victim of global warming. This tiny island, Funafuti, in Tuvalu, is losing ground to rising sea levels caused by global warming. It won't be long before the island disappears. Residents are already fleeing to make new homes.

© Ashley Cooper/Alamy Images.

KEY CONCEPTS

Scientists predict a climatically significant increase in temperature over the next 100 years that could result in a rise in sea level with potentially devastating effects on coastal populations and ecosystems.

Rising Temperatures, Changing Rainfall Patterns

Global temperature is expected to continue to increase after 2100 even if the emission of greenhouse gases ceases. According to some computer models, not only will temperature and sea level increase, rainfall patterns will change (**FIGURE 20-29**). Much of Europe, Africa, and South America, for example, are already experiencing decreased rainfall. Such decreases reduce agricultural production, causing food shortages. Many areas, however, are experiencing increased rainfall. Although that may seem beneficial, it too can disrupt farming and lead to decreased food production. That's because rain-soaked fields often can't be planted on time.

KEY CONCEPTS

Global temperature increases could shift rainfall patterns, increasing precipitation in some areas and decreasing it in others. Too little rain in some areas and too much in others could have a profound effect on food production and the economy.

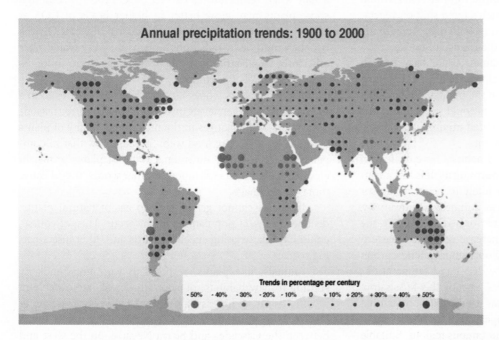

Annual precipitation trends: 1900 to 2000

Trends in percentage per century

- 50% - 40% - 30% - 20% - 10%   0   + 10% + 20% + 30% + 40% + 50%

FIGURE 20-29 Changing rainfall patterns. This map shows measured changes in global rainfall over the past 100 years.

Reproduced from *Climate Change 2001: Synthesis Report*. A contribution of Working Groups I, II and III to the Third Assessment Report of the Intergovernmental Panel on Climate Change. Cambridge University Press. Used with permission of IPCC.

Rising Temperatures, Violent Storms Computer models predict that global warming could spawn bizarre and increasingly more violent weather, notably tornadoes and hurricanes, such as those witnessed in growing numbers in recent years. Warming seas impart more energy to the atmosphere, which can greatly intensify a hurricane as it did in the case of Hurricane Katrina, which struck Louisiana and Mississippi in 2005. An increase in devastating storms could have potentially serious economic impacts as well, by destroying crops and making food more expensive. Property insurance would rise.

KEY CONCEPTS

Computer models and current evidence suggests that rising temperatures could result in more violent and costly storms, including tornadoes and hurricanes.

The Ecological and Health Impacts of Global Climate Change

In 2000, a report by the National Research Council predicted a temperature increase in North America of 3° to 6°C (5° to 10°F) over the next 100 years. In this report, scientists predicted a major shift in climate zones in North America. The authors suggest that the tropical climate of equatorial regions will shift northward into the lower tier states. Their climate, in turn, will shift to the mid-tier states and the climate of the top-tier states would move into Canada. The effect of such a rapid shift on vegetation and wildlife could be devastating, with many trees dying as temperatures climb outside of their range of tolerance.

Go Green

Combine trips to save energy and reduce pollution—or walk or ride your bike to do errands if you can.

A great many plants and animals could face difficult times as the planet warms and its climate shifts. In fact, if the change in temperature continues at the predicted rate, many species could become extinct. Others could suffer enormous declines in their populations. A limited number will be able to adapt or migrate to suitable habitat.

Interestingly, most species and habitats have dealt with changing conditions for eons, so warming itself is less a concern than the rate at which it is likely to occur. Professor Margaret Davis of the University of Minnesota shows why, in a computer simulation she performed to predict the effects of a global temperature increase on several eastern tree species. She found that if global atmospheric carbon dioxide concentrations double by 2050 and temperatures rise as predicted, hardwood trees east of the Mississippi River would face tremendous losses. Her studies predict that beech trees would probably disappear from the southeastern United States, except in a few mountainous regions. Suitable beech tree habitat would shift north to New England and southeastern Canada, which is now the extreme northern limit of the tree's range.

Although species can shift their range, most would not be able to move anywhere as fast as required. At the end of the last ice age, for instance, beech trees "migrated" northward by dispersing seeds at a rate of about 10 kilometers (6 miles) per 50 years—far slower than the 500-kilometer (300 miles) migration needed to avoid destruction from global warming taking place within the next 50 years. Many other trees in the United States would face a similar fate. Over time, these trees will die out, possibly being replaced by less desirable heat-resistant species.

Practically every ecosystem on Earth would be affected by global warming. Some of the most important and most threatened are coastal ecosystems, notably mangrove swamps and coastal marshes. The future of these areas and the services they provide—among them protecting coastal regions from erosion and providing habitat for commercially important food species—would not be bright. As sea level rises, most mangrove swamps would be lost.

Some studies suggest that plants might thrive in a CO_2-rich world. After all, carbon dioxide is essential to plant growth. Higher levels of carbon dioxide are used to increase growth of plants such as tomatoes in commercial greenhouses. Some studies show that plants use water more efficiently when exposed to higher levels of carbon dioxide. Some individuals have interpreted this to suggest that a carbon dioxide-rich world would be a greener world.

A closer examination of the issue, however, suggests that the benefits of increasing levels of CO_2 are overstated. In fact, studies have shown that elevated CO_2 levels benefit certain plants but harm others—among the latter corn, sugar cane, and many grasses. In addition, studies show trees in tropical regions grow more slowly when exposed to higher temperatures. Moreover, photosynthesis grinds to a halt when temperatures exceed 38°C (100°F). Although some plants may grow better in higher levels of CO_2, at higher temperatures that will accompany higher CO_2, the benefit would be lost. Although plants may use water more efficiently, this feature won't help them in severe droughts spawned by global warming.

Making matters worse, elevated levels of CO_2 result in a reduction in nitrogen levels in virtually all plants, for reasons not understood. Changes in the nutritional quality of plants could affect the entire food web. Studies show that to compensate for the lower nutritional value of plants grown in elevated CO_2, insects eat more. In other words, they'd cause more crop damage.

If insects cannot get enough to eat of natural plants, declines in insect populations might occur. This, of course, could have devastating effects on birds and other insect-eating species.

Animals, like plants, respond to warming trends by shifting to new habitats, but not all animals are capable of moving as far as necessary. Stanford University researcher Dennis Murphy studied the Great Basin Mountains, lying between the Cascades and Sierra Nevadas on the west and the Rockies on the east. His studies indicated that 44% of the mammal species, 23% of the butterfly species, and a smaller percentage of bird species would disappear by a 3°C (5.4°F) increase in global temperature. A study published in the prestigious journal *Nature* says that global climate change could drive as many as 1 million species to extinction by 2050.

Some species may actually thrive amid global warming— for example, organisms that are responsible for infectious diseases. Robert Shope of the Yale University School of Medicine predicts that some diseases now restricted to tropical areas could invade new territory as the planet warms. For example, one form of rabies, now transmitted by vampire bats, could spread northward from Mexico and could result in damage of about $1 billion a year to the Texas cattle industry. Some scientists believe that the recent outbreaks of the deadly Hanta virus, which is transmitted by rodents in the desert Southwest, may be a result of rising global temperature.

Warmer climates are already causing the spread of insects that carry malaria and dengue fever to higher altitudes and higher latitudes. In Africa, studies show that malaria is spreading to higher altitude regions in five countries, which scientists attribute to the warming climate. In Mexico and Costa Rica, dengue fever is spreading into the highlands as well. "The control issue looms largest in the developing world, where resources for prevention and treatment can be scarce," writes Dr. Paul Epstein, Associate Director of the Center for Health and the Global Environment at Harvard Medical School in an article in *Scientific American*. "But the technologically advanced nations, too, can fall victim to surprise attacks—as happened last year (1999) when the West Nile virus broke out for the first time in North America, killing seven New Yorkers. In these days of international commerce and travel, an infectious disorder that appears in one part of the world can quickly become a problem continents away if the disease-causing agent, or pathogen, finds itself in a hospitable environment."

KEY CONCEPTS

Many plant and animal species and entire ecosystems could be profoundly influenced by global climate change, especially if the rate of change occurs faster than their ability to adapt, which seems inevitable. Disease could also "migrate" into new territory.

Cooling and Changing Ocean Currents

While most of the discussion of global climate change focuses on the effects of higher temperatures, many scientists believe that global warming may plunge some areas of the planet into a period of intense and catastrophic cooling. Global warming could bring on ice-age conditions in Great Britain, Europe, and Russia. How?

As noted, seawater circulates throughout the oceans. For example, warm water from tropical regions circulates northward in the Atlantic Ocean via the Gulf Stream. This makes Great Britain and Europe (located on the same latitude as northern Canada) much warmer than they would otherwise be. Warm saltwater circulating northward cools as it moves toward the North Pole. The cooler, denser water then sinks and flows back toward the equator, only to rise again. This gigantic current forms a huge conveyor shown in **FIGURE 20-30** that distributes heat from the equator to the poles. It makes the poles warmer and the equator cooler than they would be otherwise, but global climate change could alter all of this, and abruptly. How?

Scientists believe that melting glaciers and increased rainfall in the northern latitudes could infuse the ocean with less dense, cool fresh water that could shut down the conveyor belt. The cool less-dense fresh water would essentially put a halt to an essential driving force of the global conveyor belt—the sinking of dense, cool, salt water. Studies of ocean currents and climate suggest that such changes could occur abruptly, and soon, and may take decades, even centuries, to reverse. The result? Areas like Europe could become much colder, while the equator becomes hotter. As you shall soon see, scientists have already detected significant changes in this system, which has caused considerable concern in Europe and stimulated major efforts to curb greenhouse gases.

KEY CONCEPTS

Global warming could slow down or halt oceanic currents that deliver warm water to northern regions, causing significant and costly cooling.

National and International Security

As the previous material illustrates, climate change could have extremely serious economic and environmental consequences throughout the world. The U.S. Department of Defense noted in a classified report that global climate change could result in a global catastrophe that would cost millions of lives through warfare and recurrent natural disasters. The report warns that major European cities could sink beneath rising seas as Great Britain is plunged into a Siberian climate. It also warns of nuclear conflict, serious droughts, famine, and widespread civil unrest that could erupt throughout the world.

The document predicts that abrupt climate change could cause many nations to develop nuclear weapons to defend and secure dwindling food, water, and energy supplies. Some experts who have viewed the classified document argue that the threat of terrorism is currently eclipsed by the potential global instability that could result from global climate change. "Disruption and conflict will be endemic features of life," concludes the Pentagon. "Once again, warfare would define human life." Climate change "should be elevated beyond a scientific debate to a U.S. national security concern," say the authors, Peter Schwartz, CIA consultant

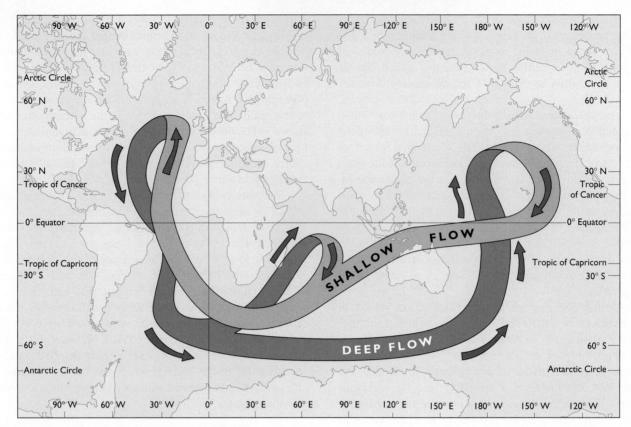

FIGURE 20-30 **Surface and deep water exchange.**

and former head of planning at Royal Dutch/Shell Group, and Doug Randall of the California-based Global Business Network.

 KEY CONCEPTS

> Global climate change is not just an economic and environmental issue, it has potentially serious implications for global peace.

Scientific Uncertainties: What We Don't Know

Although there is strong evidence that the oceans and atmosphere are warming, that climate is changing, and that human activities are the primary cause of these trends, there are some scientific uncertainties worth considering.

One significant uncertainty is the potential for changes that could cause the Earth's temperature to rise more quickly. These factors are not taken into account in the projections of global warming by the IPCC, leading some scientists to view the IPCC projections as conservative. Because they do not take into account factors that could result in a rapid deterioration of climatic conditions, climate could deteriorate much more quickly than anticipated.

Evidence of this is already being seen. **FIGURE 20-31**, for instance, shows the predicted melting of arctic sea ice, using computer models. The red line shows that observed

FIGURE 20-31 This graph shows the computer projections of sea ice (blue) and actual measurements (red), indicating that the computer models underestimate melting.

© 2008 University Corporation for Atmospheric Research, illustration by Steve Deyo.

melting, indicating a much faster rate of disappearance than predicted.

Changes now in motion that could stimulate dangerous positive feedbacks are shown in **FIGURE 20-32**. Many of these changes could lead to dangerous positive feedbacks. A **positive feedback** occurs when one factor leads to an increase in another. In nature, positive feedback mechanisms are rare. Homeostatic mechanisms, tend to operate through

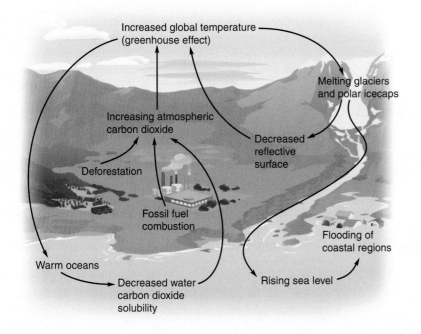

FIGURE 20-32 Dangerous feedback. If the planet continues to warm, many changes could occur. Some could accelerate the increase in atmospheric carbon dioxide levels and result in an acceleration in the rate of global warming.

negative feedback mechanisms that maintain more or less constant conditions. In natural systems altered by humans, positive feedback mechanisms are dangerously common. To understand this phenomenon, consider some examples of positive feedbacks.

The oceans currently serve as a major reservoir for CO_2. In fact, they store 60% more CO_2 than the atmosphere does. Without the oceans, CO_2 levels in the atmosphere would be considerably higher than they are today. As the Earth's temperature rises, however, the ocean's ability to dissolve and hold CO_2 will very likely decline (Figure 20-32). (Warm liquids, like a warm soda, hold less CO_2 gas.)

If the amount of CO_2 dissolved in the oceans declined, the oceans would release CO_2 they have absorbed in the past 100 years. This would cause atmospheric CO_2 levels to increase much more rapidly, accelerating global warming and global climate change. Rising temperatures would further reduce the amount of CO_2 held in ocean waters, creating a dangerous positive feedback cycle. Research suggests that rising temperatures are already reducing the ability of oceans to hold CO_2.

Several additional positive feedback mechanisms may also be initiated. Melting glaciers, for instance, decrease the reflective surface of the Earth, resulting in an increase in the absorption of sunlight. This would result in a further warming of the atmosphere, for reasons explained at the beginning of this section. Rising temperatures, in turn, would accelerate the melting of glaciers. Rising temperatures would also increase the amount of energy people use to cool their houses—which adds more CO_2 to the atmosphere, once again accelerating the rise in temperature.

As noted earlier, the rise in atmospheric CO_2 concentrations is partly the result of deforestation—the rapid loss of trees, especially in the tropics and in parts of the United States, such as the Pacific Northwest. Trees absorb CO_2, which they use to produce food molecules and structural

tissues. Worldwide, forests are being cut much faster than they regrow. As a result, deforestation "contributes" approximately one-fourth of the annual global increase in CO_2 and, as noted in Table 20-2, about 13% of the annual rise in global temperature. Trees may also be destroyed by climatic change (rising temperature and drought) and by fires. This loss could also become part of a wildly accelerating positive feedback mechanism that causes a rapid increase in CO_2 in the atmosphere and a more rapid shift in global temperature. An increase in forest fires, already witnessed, could also contribute increasing CO_2 levels, resulting in further warming.

Although most potential feedbacks are positive, there are a few factors that could counteract these forces, either negating the increase in temperature or reducing it. The oceans, for instance, tend to act as a temperature buffer, keeping temperatures from rising as rapidly. In addition, rising temperatures may result in an increase in evaporation of water from lakes and oceans, increasing cloudiness. This could reduce sunlight penetration, reducing the climb in global temperature. However, some forms of clouds trap heat. Increased particulates in dusty air from denuded, parched lands could also reflect incident solar energy.

Current evidence suggests that the depletion of the ozone layer will have a cooling effect on global climate. Other information suggests an adjustment in the other direction. A study of 17 computer models published in *Science* in 1991 also suggests that global warming may affect snowfall and average temperature in unanticipated ways. As a rule, climate modelers think that global warming will result in a net decrease in planetary snow cover. This, they predict, would increase absorption and increase global warming. However, five of the world's best climate models show that a reduction in snow cover would cool the atmosphere and counteract a small part of the greenhouse warming. The cooling effect may result from an increase in cloud cover or an increase in heat radiation into space. In sum, many climate modelers acknowledge that current models don't do a very good job of simulating cloud behavior.

How these factors play out is anyone's guess at this time, but many scientists believe that, overall, the positive feedback loops—factors that increase temperature—may dominate.

KEY CONCEPTS

Many factors could complicate predictions of future climate. Some factors could result in a runaway increase in atmospheric CO_2 levels, which would greatly accelerate warming. Others could reduce warming.

Solving a Problem in a Climate of Uncertainty: Weighing Risks and Benefits

Global climate is a very complex system, and our understanding, while growing, is still not complete. Many uncertainties still exist with regard to the behavior of the CO_2 sinks such as the oceans and forests. Future levels of greenhouse gases and global warming activities such as deforestation are unknown.

With this much uncertainty, some people prefer to take a wait-and-see approach. Some companies, and some politicians, oppose measures to cut greenhouse gas emissions, arguing that the threat is not real and that such changes will hurt the nation's economy. Critics of this viewpoint argue that the stakes are much too high to take no action. They note the economic and environmental benefits of cutting fossil fuel combustion are so significant that we should make the conversion despite the uncertainty. In short, they argue, the cost of accepting the greenhouse hypothesis and acting accordingly is far lower than the potential cost of rejecting it.

Today, many nations and companies like Microsoft, Google, Toyota, BP, and Wal-Mart, to name a few, are taking steps to reduce CO_2 emissions. Germany plans to develop a 100% renewable energy economy. Great Britain has pledged to reduce its CO_2 emissions by 60%. The U.S. government under the leadership of President Barack Obama have taken many measures to reduce our nation's CO_2 emissions.

Fortunately, many state and local governments have taken the lead in reducing greenhouse gas emissions by taking steps to reduce their carbon footprint. Many of their efforts, like energy efficiency, could have profound economic benefits, too, and could help spawn a renewable energy industry vital to the world's long-term future.

KEY CONCEPTS

Some uncertainty exists on global climate change, which has slowed progress. Even though there are some uncertainties, many people believe that the costs of reducing or even eliminating greenhouse gases are outweighed by the potential social, economic, and environmental costs of global climate change.

Solving the Problem Sustainably

Humans produce over 9.7 billion tons of carbon (in the form of CO_2) a year—40% from the less developed nations and 60% from the more developed industrial nations. To create a sustainable society, free from the impacts of global warming, the Intergovernmental Panel on Climate Change believes that CO_2 emissions need to be cut by about 75%, down to about 2 billion tons per year. Unfortunately, the world is poised to increase emissions. Emissions from the less developed nations are projected to quadruple in the next half century. China's emissions now exceed those of the United States, the leader in greenhouse gas emissions. By 2020, the less developed nations could be producing as much CO_2 as the industrial nations. Emissions from the more developed nations are projected to increase by 30%.

Further adding to the dilemma, tropical deforestation and burning of forests could greatly accelerate CO_2 emissions and global climate change. The fires that occurred in Asia in 1997 and 1998 released more CO_2 than western Europe produces in a year. These fires were intentionally ignited in many cases. In others, however, extraordinarily dry rain forests caught fire in lightning storms. The dry conditions were caused by deforestation in upwind areas, which tends to reduce rainfall.

Global warming is ultimately a symptom of overpopulation and unsustainable human systems and technologies designed to meet the needs of people. Especially noteworthy in the group of human systems are energy, transportation, housing, industry, and timber production. The energy system, for example, relies heavily on fossil fuel, the combustion of which produces huge quantities of CO_2.

The good news is that numerous solutions are available to reduce, even eliminate, greenhouse gas emissions. Bans on ozone-depleting and greenhouse-enhancing CFCs, discussed earlier in the chapter, are one example. Other sustainable solutions, those that strike at the roots of the problems, include measures to reduce population growth and eliminate the harmful emissions of CO_2, methane, CFCs, and nitrous oxide. This section discusses some of the most effective measures. Not surprisingly, the most effective approaches turn out to be applications of the operating principles of sustainability conservation, recycling, renewable resource use, restoration, and population stabilization.

KEY CONCEPTS

Redesigning human systems according to the principles of sustainability could help alleviate global warming and help create a prosperous, safe future.

Population Stabilization and Restoration To reduce atmospheric CO_2 levels and stave off or stop the inevitable increase in global temperature will require a massive reforestation of the Earth. Australia recently announced that it was embarking on an ambitious program to plant 1 billion trees, partly to offset global warming. A few other countries are following suit, among them China, although efforts are below what's required to have any appreciable effect.

Norman Myers, an international expert on tropical forests, argues that replanting 2.6 million square kilometers (1 million square miles) of tropical rain forest would reduce annual emissions of CO_2 by 2.25 billion metric tons (2.5 billion tons), or about 33%. Although this ambitious project would cost approximately $100 billion, Myers argues that it is a small price to pay, especially when one takes into account the potential economic damage caused by climate change. A bad hurricane can easily cause several billion dollars worth of damage if it hits a metropolitan area. According to the EPA, protecting the U.S. east coast from rising sea waters could cost the nation $75 to $110 billion. Global warming would necessitate costly modifications of irrigation systems and hydroelectric dams estimated at $100 billion. Replanting could also bring direct economic benefits to less developed nations through reduced soil erosion and sustainable harvest of forests.

Individuals can help plant trees in clear-cuts, roadsides, abandoned fields, and backyards. If you really want to offset the carbon dioxide produced by your lifestyle, you would have to plant 400 trees. A family of four would need to plant 2.5 hectares (6 acres) of fast-growing trees to offset its lifetime carbon dioxide production. Obviously, not everyone can replant trees, but you can help by supporting public and private reforestation projects.

KEY CONCEPTS

Stabilizing population growth can help reduce humanity's need for fossil fuels and other greenhouse-enhancing activities such as deforestation. Restoring forests, especially in the tropics, and restoring other carbon sinks like wetlands, could have a profound effect on global carbon dioxide levels.

Recycling, Energy Efficiency, and Renewable Energy

Individuals and businesses throughout the world can also help reduce global CO_2 levels by recycling and reusing all materials to the maximum extent possible. **TABLE 20-3** shows a comparison of energy use in manufacturing cans and bottles. It shows that recycled glass and aluminum require far less energy to produce—and, thus, produce less CO_2 air pollution—than if they were made from raw ore and used only once. Refillable bottles use much less energy even if they are only used 10 times.

Energy efficiency is another means of reducing CO_2 and methane emissions. According to energy experts Amory Lovins of the Rocky Mountain Institute and Christopher Flavin of the Worldwatch Institute, global emissions could be cut by 2.7 billion metric tons (3 billion tons) per year within two decades by cost-effective and profitable technologies. No technical breakthroughs are needed, either. Combined with reforestation efforts outlined above, these efforts could nearly eliminate anthropogenic global CO_2 emissions.

Individuals can cut energy demand by walking, bicycling, or riding a bus to school or work; by building smaller,

TABLE 20-3

Energy Consumption per Use for 12-Ounce Beverage Containers

Container	Energy Use (BTUs)
Aluminum can, used once	7,050
Steel can, used once	5,950
Recycled steel can	3,880
Glass beer bottle, used once	3,730
Recycled aluminum can	2,550
Recycled glass beer bottle	2,530
Refillable glass bottle, used 10 times	610

Data from L. R. Brown, C. Flavin, and S. Postel (1991). *Saving the Planet: How to Shape an Environmentally Sustainable Economy.* New York: Norton.

more energy-efficient homes; by insulating existing homes; by recycling; and by using efficient appliances or doing some things by hand (for example, mixing by hand rather than using an electric mixer, or drying clothes on a line rather than using a dryer). When buying appliances, choose the most energy-efficient ones available. New energy-efficient refrigerators, the leading user of electricity in homes, can cut electrical demand from 1,200 kilowatt-hours per year to 300 to 400.

Renewable energy technologies can dramatically reduce CO_2 and methane emissions. Many options are available, and several are quite cost competitive.

How effective are these and other measures outlined in this section? According to a study by the American Solar Energy Society (**FIGURE 20-33**), widespread implementation of energy efficiency, wind energy, biofuels, biomass, solar electricity, and other renewable energy technologies could reduce the United States CO_2 emissions by 60 to 80%, well within the range of what is needed to reach climatic stability. Their studies show that these measures would actually improve the economy. Efficiency measures especially would save individuals and companies huge sums of money.

KEY CONCEPTS

Recycling and energy efficiency greatly reduce energy demand and cut greenhouse gas emissions. Renewable energy technologies can provide us with much-needed power with little or no impact on the global climate.

International Cooperation to Halt Global Warming

This chapter opened with a discussion of the Earth Summit, the highly acclaimed global conference that addressed such pressing problems as forest protection and global warming. One outcome of the meeting was an agreement calling on

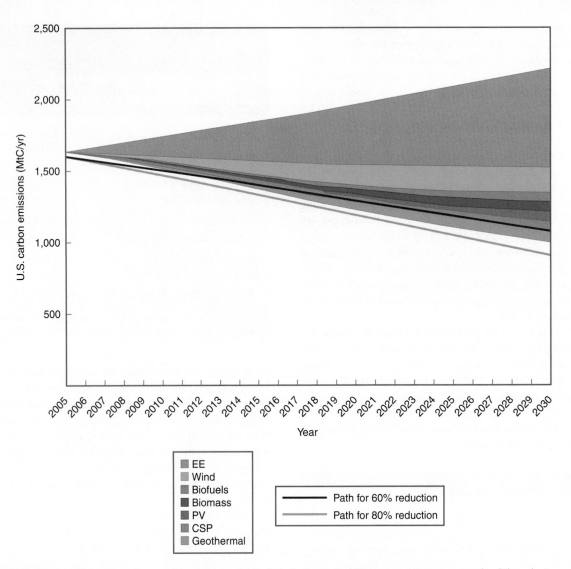

FIGURE 20-33 Energy efficiency and renewable energy could easily help the United States meet its energy needs while reducing carbon dioxide emissions by 60% to 80%. (EE = energy efficiency, PV = solar electricity, CSP = concentrated solar power.)

Reproduced from American Solar Energy Society, *Tackling Climate Change in the U.S.: Potential Carbon Emissions Reductions from Energy Efficiency and Renewable Energy by 2030*. January, 2007.

nations to reduce CO_2 emissions. The nations also signed an agreement to protect forests.

Although both agreements are important, they fall seriously short of the job at hand. Further, the global climate agreement is largely voluntary. By 1997, little progress had been made in reducing CO_2 levels. During that year, the United Nations sponsored a meeting in Kyoto, Japan, to negotiate an international agreement on global climate change. Known as the 1997 **Kyoto Protocol** and signed by 174 nations (as of November 2007), this agreement calls for a slow, steady decline in greenhouse gas emissions. It went into effect in 2005. This agreement commits the industrial nations and the former Eastern Bloc nations to cut CO_2 and other greenhouse gas emissions 5.2% below 1990 levels between 2008 and 2012. The signatories have met twice since then, but so far success in achieving these goals has been poor unfortunately. Many countries have increased greenhouse gas emissions, most noteworthy are China and

India (**FIGURE 20-34**). The fastest growth in greenhouse emissions has occurred in the less developed nations, with growth rates of about 40%, in large part because of industrialization (Figure 20-34). There is some good news, however, notably in the United Kingdom, France, and Germany, where greenhouse gas emissions are falling as a result of energy-efficiency policies.

Many observers believe that, as was the case with the first treaty on ozone protection (the Montreal Protocol), nations will see the need to take more drastic steps to curb or even stop the rise in greenhouse gas, which the vast majority of the world's atmospheric scientists believe is responsible for record-hot temperatures in the past two decades and a host of other changes, including a rise in sea level. Continuing bad news—hot years, devastating floods, and violent storms—could spur further agreements that eventually lead us to a more sustainable system of energy production and a climate we can all live with.

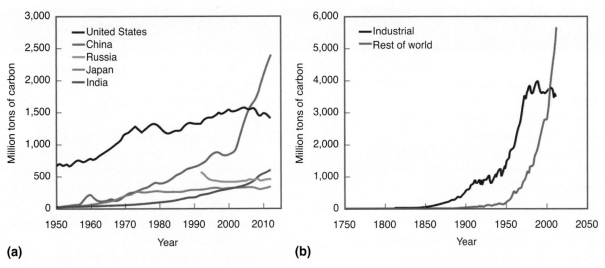

(a)　　　　　　　　　　　　　　　　**(b)**

FIGURE 20-34 Carbon dioxide emissions. (a) This graph shows the release of carbon dioxide by country. Note that U.S. emissions have leveled off while China's emissions continue to rise at a rapid rate. **(b)** Note the rapid increase in carbon dioxide emissions in developing nations.

Data from Earth Policy Institute (www.earth-policy.org/indicators/C52/carbon_emissions_2013).

CRITICAL THINKING

Exercise Analysis

Critical thinking rules encourage us to question sources of information and their conclusions. In this example, we find that the author of this article is grossly in error when he claims that the scientific community is divided on the issue. The key word is *divided*, which makes it sound as if there's a 50–50 split of opinion.

In truth, about 99% of the United States' 700 atmospheric scientists believe that global warming is a reality; only a handful embrace the opposite view. The evidence the author introduces to support his assertion—a quote from each side of the issue—is terribly misleading.

This type of reporting is quite common in newspapers, television, and magazines. In journalism, it is customary to get both sides of the issue, but one must not mistake the notion of two sides for an equal split of opinion. Statements such as the one we're analyzing here, although intended to give a balanced view of issues, in reality provide an extremely unbalanced view.

What would have been a more accurate statement? The author might have noted opposing views but also noted that, although not all scientists agree that global warming is occurring, the vast majority do.

That said, it is still important to note that even though the majority of the world's atmospheric scientists agree that global warming is happening, it doesn't mean they're right. For many centuries, scientists held that the Earth was at the center of the solar system.

This exercise shows that digging a little deeper and finding out more often throws conclusions into question.

CRITICAL THINKING AND CONCEPT REVIEW

1. What is the ozone layer? Why is it important to life on Earth?

2. Describe the major activities and pollutants that deplete the ozone layer.

3. What measures have been taken to reduce the destruction of the ozone layer? Will they reverse the decline quickly? Why or why not?

4. HCFCs, replacements for CFCs, also deplete the ozone layer, but to a lesser degree. Consequently, they are considered interim solutions. What other reason accounts for their interim-solution status?

5. Some critics of the environmental movement suggest that stratospheric ozone depletion is a hoax. Look up some of their writings and, if possible, the work on which they base their claims. What do you find? Are the conclusions based on good scientific evidence? With your knowledge of the issue, debate the main points they make.

6. Critically analyze the following statement: "We must balance the harm done by ozone depletion and maintaining CFCs for refrigeration with the potential loss of refrigeration in developing countries, which could result in many lost lives due to foodborne disease."

7. Define the terms *acid deposition, acid precursor, wet deposition*, and *dry deposition*.

8. How are acid precursors formed?

9. Using your critical thinking skills, analyze the following statement: "Natural events such as volcanoes produce far more acid precursors than human activities, so we shouldn't worry about controlling anthropogenic sources. Such controls won't do anything to affect acid levels."

10. Using your critical thinking skills and your knowledge of biology, how would you assess current efforts aimed at reducing acid deposition or counteracting its impact (for example, smokestack scrubbers, liming lakes, and breeding acid-resistant fish)?

11. Outline key elements of a sustainable strategy for reducing acid deposition.

12. What is the greenhouse effect? What gases are responsible for it? Which ones are natural? Which ones are anthropogenic?

13. Describe the potential social, economic, and environmental impacts of the continued rise of greenhouse gases. Describe how environmental impacts could lead to social and economic impacts.

14. Describe the factors that could accelerate greenhouse warming and those that may lessen it.

15. Outline the short-term and long-term solutions that have been proposed for addressing global climate change and critically analyze each one. Which ones would be the most effective? Which ones would be the most politically acceptable?

16. Compare the difference between heeding warnings about global warming and ignoring them. What action do you recommend? Why?

KEY TERMS

acid deposition
acid precursors
buffering capacity
buffers
chlorine free radicals
chlorofluorocarbons (CFCs)
dry deposition
Freon-11
Freon-12

global climate change
global climate models
global warming
greenhouse effect
greenhouse gases
hydrochlorofluorocarbons (HCFCs)
Kyoto Protocol
Montreal Protocol

National Acid Precipitation
  Assessment Program (NAPAP)
ozone layer
pH (potential hydrogen)
positive feedback
wet deposition

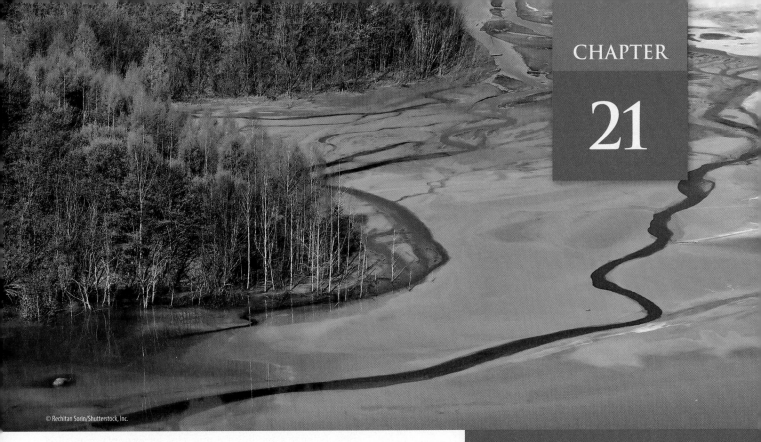

© Rechitan Sorin/Shutterstock, Inc.

Water Pollution: Sustainably Managing a Renewable Resource

It is a crime to catch a fish in some lakes, and a miracle in others.

—Evan Esar

LEARNING OBJECTIVES

After reading and studying the material in this chapter, students will be able to understand, knowledgably discuss, and, when appropriate, apply the information learned in this chapter, including:

1. Sources of surface water pollution, including point and nonpoint sources
2. Types of water pollutants, their sources, and their impacts
3. Ways to sustainably address water pollution
4. Sources of ground water pollution and their impact
5. Ways to sustainably reduce and eliminate ground water pollution
6. Pollution of the ocean, especially oil and plastics
7. Impacts of pollutants in the oceans

CHAPTER OUTLINE

CRITICAL THINKING

Exercise

A family living along Lake Superior argues that they don't need to conserve water. There's plenty of water in the lake, so why should they sacrifice? Do you agree with this assertion? When you have finished, list the critical thinking rules you used to analyze this issue.

Author Loren Eiseley once wrote, "If there is magic in this planet, it is in water." Water has a practical side, too. Water covers 70% of the Earth's surface and makes up two-thirds or more of the weight of most animals and up to 95% of the weight of plants. It plays an important role in metabolism, the chemical reactions in the cells of the bodies of all organisms. It is vital to agriculture and industry, too. Despite its crucial role in our lives, water is one of the most badly abused resources. We've covered water resources and watershed protection elsewhere. This chapter covers water pollution: where it comes from, how it affects living organisms, and finally, ways to reduce it. Like other chapters, this one outlines some components of a sustainable response to this problem.

21.1 Surface Water Pollution

Water exists in many forms. It is found in the atmosphere as mist or clouds. It is locked in glaciers as solid ice. It is found on the Earth's surface in ponds, lakes, streams, and rivers, which are collectively referred to as **surface waters.** Water is also stored underground in aquifers. Such deposits are part of the planet's vast **groundwater** holdings. This chapter deals with water pollution occurring in groundwater and surface waters.

Water pollution is any physical or chemical change in water that adversely affects organisms. Like many other problems, it is global in scope, but the types of pollution vary according to a country's level of development. In poor, nonindustrialized nations, water pollution is predominantly caused by human and animal wastes, pathogenic organisms from this waste, pesticides, and sediment from unsound farming and timber practices. The rich industrial nations also suffer from these problems, but their more opulent lifestyles and numerous industries create an additional assortment of potentially hazardous pollutants: heat, toxic metals, acids, pesticides, organic chemicals, and an assortment of pharmaceuticals (excreted in peoples' urine). In between these two extremes are numerous countries with various levels

of industrialization. They often have inadequate laws—or none at all—to combat water pollution, or if they do have good pollution control laws, they frequently lack adequate funding to enforce them. Their waters are therefore often badly polluted with an assortment of industrial and municipal wastes.

KEY CONCEPTS

Nations produce many different types of water pollutants. The water in all countries is plagued with pollutants from human and animal wastes, but in the industrial nations numerous toxic chemicals also contribute to water pollution.

Sources of Water Pollution

Like air pollutants, water pollutants come from numerous sources, both natural and anthropogenic. As a rule, anthropogenic sources are the most important because they tend to be localized and thus contribute significantly to the deterioration of local waterways or groundwater required by humans.

Like air pollutants, water pollutants respect no boundaries. Pollutants produced in one city, for example, may affect water quality in another city located downstream. They may even flow from one country to another, ending up in the water supply of downstream neighbors, or they may flow into oceans and seas. For example, for many years the rivers flowing into the Mediterranean Sea and the sea itself were viewed as a

Go Green

To reduce water pollution, compost organic kitchen waste rather than using the garbage disposal.

convenient dump for domestic and industrial wastes. Along Italy's coast, the waters became a cesspool. Fortunately, internal actions and negotiations among the nations that border the Mediterranean have spawned a region-wide plan that is successfully reversing the years of decline.

Pollutants also enter surface waters from the air and surrounding land. This problem is called **cross-media contamination** and refers specifically to the movement of a pollutant from one medium (such as air) to another (such as water). Pesticides sprayed on crops, for example, may evaporate, becoming airborne. They may then be transported in the air to nearby lakes. They may then flow to the oceans. Volatile organic chemicals from factories may evaporate and later be washed out of the sky by rains.

KEY CONCEPTS

Water pollutants arise from natural and anthropogenic sources. They travel freely from one location to another through rivers, streams, and groundwater. They are also transported from one medium (land or air) to another (water).

(a)

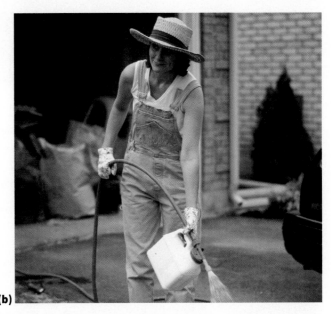

(b)

FIGURE 21-1 Water pollution sources. (a) This pipe dumping factory waste into a nearby river is an example of a point source. **(b)** Urban streets, lawns, and gardens are nonpoint sources. Heavy rains often wash pollutants from these sources into nearby waterways.

Photo a courtesy of U.S. Environmental Protection Agency, Great Lakes National Program; photo b © Photos.com.

Point and Nonpoint Sources

When we ponder the sources of water pollution, we generally think of factories, power plants, mines, oil wells, and sewage treatment plants that release tons of potentially toxic chemicals into sewers and lakes and rivers (**FIGURE 21-1A**). These **point sources,** so named because they are discrete locations, are relatively easy to identify and control. Although they are obvious sources for which controls can be effected, pollution control can be costly. Abandoned metal mines in Canada's Arctic region, for example, leak a variety of pollutants (mostly heavy metals) into nearby streams and will cost billions of dollars to clean up.

In the past decade, scientists have come to realize that point sources are only half the problem. The other half are the **nonpoint sources**—less discrete sites such as farms, forests, lawns, and urban streets (**FIGURE 21-1B**). Rainwater washes pollutants from these sources into nearby streams, rivers, and lakes. You may want to take a moment to study nonpoint sources listed in **TABLE 21-1**.

Together point and nonpoint sources can result in huge problems. Many Canadian waters are polluted by direct discharge of pollutants from industrial and municipal point sources and by indirect discharge from numerous nonpoint sources. The Frazer River in Canada, for instance, receives incompletely treated sewage and a variety of toxic chemicals from wood treatment and wood pulp and paper mills. It is polluted by chemical leaching from nearby landfills.

In the United States, nonpoint sources release nearly two-thirds of the pollutants that end up in our waterways (**FIGURE 21-2**). These substances include dust, sediment, pesticides, asbestos, fertilizers, heavy metals, salts, oil,

grease, litter, and even air pollutants washed from the sky by rain. Because the sources are many and spread out, control has proven difficult. It is also far easier for governments to force a company to install a pollution control device on a factory than to try to force tens of thousands of city dwellers to get oil leaks in cars fixed or to apply pesticides to lawns more conservatively. Although there are numerous sources of nonpoint water pollution, **FIGURE 21-3** shows that agriculture is the predominant nonpoint source in the United States, affecting nearly 60% of the nation's streams.

The remaining portion of this section covers the major types of point and nonpoint water pollution.

KEY CONCEPTS

Water pollution arises from identifiable point sources, such as factories, and from diffuse nonpoint sources, such as farm fields and streets. Point sources are much easier to control.

Organic Nutrients

In more developed countries, large quantities of organic chemicals are released into surface waters from point sources such as feedlots, sewage treatment plants, and various industries such as paper mills and meat-packing plants.[1] In many cases, these wastes are treated prior to release into streams. Even so, large quantities of organic waste still enter surface waters. Flooding can also cause severe problems. In 1999, for example, a series of hurricanes and tropical storms hit the southeastern United States. In North Carolina, 20 municipal sewage treatment plants were flooded, releasing enormous amounts of untreated human waste into streams and other water bodies. Making matters worse, many of the

---

[1] An organic compound consists primarily of carbon and hydrogen atoms.

TABLE 21-1

Major Nonpoint Pollution Sources in the United States

Activity	Explanation
Silviculture	Growing and harvesting trees for lumber and paper production can produce large quantities of sediment.
Agriculture	Disruption of natural vegetation leads to increased erosion; pesticide and fertilizer use, coupled with poor land management, can pollute neighboring surface water and groundwater.
Mining	Leaching from mine wastes and drainage from mines themselves can pollute surface and groundwater with metals and acids; disruption of natural vegetation accelerates sediment erosion.
Construction	Road and building construction disrupts vegetation and increases sediment erosion.
Salt use and ground-water overuse	Salt from roads and storage piles can pollute groundwater and surface water; saltwater intrusion from groundwater overdraft pollutes ground and surface water.
Drilling and waste disposal	Injection wells for waste disposal, septic tanks, hazardous waste dumps, and landfills for municipal garbage can contaminate groundwater.
Hydrological modification	Dam construction and diversion of water both can pollute surface waters.
Urban runoff	Pesticides, herbicides, and fertilizers applied to lawns, as well as residues from roads, can be washed into surface waters by rain.

state's 4,000 hog farms were flooded out by heavy persistent rainfall. Waste on the ground and in lagoons, ponds where the waste is held for treatment, flowed into streams (**FIGURE 21-4**). Hundreds of millions of gallons of hog wastes poured into wetlands, creeks, and rivers.

In less developed countries (LDCs), organic pollutants come primarily from human sewage and animal wastes. Raw sewage often enters water without treatment.

In surface waters, organic pollutants are consumed by naturally occurring bacteria. With an abundant food source, the bacterial populations proliferate. As bacteria consume the organic matter, they help to purify the waters and are therefore part of nature's system of restoration. Unfortunately, there's a catch.

The degradation of organic pollutants by bacteria utilizes oxygen, which is dissolved in water (**FIGURE 21-5**). Excess organic matter therefore causes oxygen levels to drop; as a result, fish and other aquatic organisms may perish. When oxygen levels become very low, anaerobic (non–oxygen-requiring) bacteria take over, breaking down what's left.

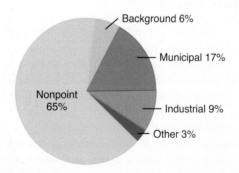

FIGURE 21-2 Major sources of U.S. stream pollution. This pie chart shows the contribution of point and nonpoint sources to total water pollution in the United States. Municipal, Industrial, and Other are point sources (Background is natural pollution.). Nonpoint sources are by far the greatest polluter of U.S. waters.

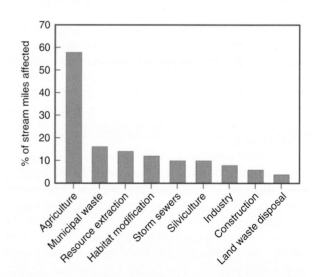

FIGURE 21-3 Sources of nonpoint water pollution affecting streams. This bar graph lists nonpoint water pollution sources in the United States by their impact—that is, the percentage of miles of streams they affect in the United States. Agricultural activities and municipal wastes (sewage) turn out to be the number one and two contributors.

In the process, though, they produce foul-smelling and toxic gases (methane and hydrogen sulfide).

Oxygen depletion in rivers and streams occurs more readily in the hot summer months because stream flow is generally low and organic pollutant concentrations are higher. In addition, increased water temperatures speed up bacterial decay.

Because streams are flowing systems, oxygen levels may return to normal if the pollution is a one-time event. However, when numerous sources of organic pollutants are found along the course of a river—or when organic input exceeds the ability of the aquatic system to restore conditions through bacterial decomposition and water flow—recovery may be impossible. Lakes can also recover from organic pollutants, but usually much more slowly than rivers.

Go Green

Use low- or nonphosphate detergents to reduce water pollution.

The organic nutrient concentration in streams is estimated by a test that determines the rate at which oxygen is depleted from a test sample. In this test, polluted water is saturated with oxygen and held in a closed bottle for 5 days; during this period, bacteria in the water degrade the organic matter and consume the oxygen that was added to it. The amount of oxygen remaining after 5 days gives an indication of the organic matter present; the more polluted the sample, the less oxygen is left. This standard measurement is called the **biochemical oxygen demand,** or **BOD**.

KEY CONCEPTS

Organic nutrients come from a variety of sources, primarily treated and untreated waste (human and other animals) accidentally and intentionally released into waterways. Organic compounds stimulate bacterial growth, which depletes oxygen levels in water bodies, killing off oxygen-dependent species such as fish. Oxygen levels can return to normal levels, but only if the influx of organic materials ceases.

Inorganic Nutrients—Nitrates and Phosphates

Inorganic nutrients are chemical compounds produced by a variety of sources, many of which also release organic nutrients. Two of the most common inorganic nutrients are nitrogen and phosphorus. Unlike organic nutrients, which stimulate the proliferation of bacteria, inorganic nutrients tend to promote the growth of algae and aquatic plants.

In terrestrial and aquatic ecosystems, nitrogen exists primarily in the form of ammonia and nitrates. Phosphorus is found in phosphates. In aquatic ecosystems, nitrogen and phosphorus are often limiting factors for populations of algae and other aquatic plants, that is, they are found in limited amount. Consequently, if levels of these pollutants increase, algal and plant growth increase dramatically. Lakes and rivers become choked with thick mats of algae or dense

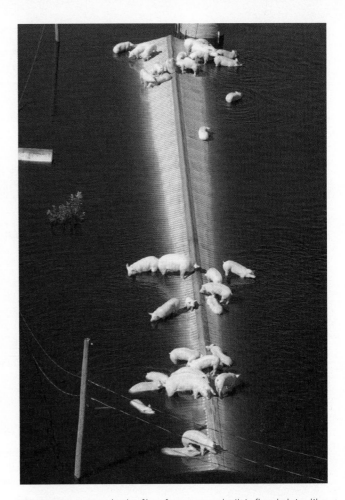

FIGURE 21-4 Hundreds of hog farms, many built in flood plains like this one were flooded by Hurricane Floyd in 1999, drowning tens of thousands of animals. Sewage lagoons overflowed and waste from the thousands of farms flowed into rivers, lakes, and the ocean.
© Reuters NewMedia, Inc./Corbis Images.

growths of aquatic plants. In freshwater lakes and reservoirs, phosphate is usually the limiting nutrient for plant growth; marine waters are usually nitrate limited.

Excessive plant growth impacts people and the environment. For example, during the summer the proliferation of plants often impairs fishing, swimming, navigation, and recreational boating. In the fall, most of these plants die and are degraded by bacteria. When large amounts of plant matter decay, dissolved oxygen levels fall, killing other aquatic organisms. As oxygen levels drop, anaerobic bacteria resume the breakdown and produce the noxious products noted earlier. Thus, inorganic nutrients create many of the same problems that organic nutrients do. Where do inorganic nutrients come from?

Inorganic fertilizer from croplands is the major anthropogenic source of plant nutrients especially nitrates in fresh waters. When highly soluble fertilizers are used in excess, as much as 25% may be washed into streams and lakes by the rain. Applying fertilizer in smaller amounts and implementing measures to reduce surface runoff of farmland, such as terracing and strip cropping, could greatly reduce the amount of phosphates and nitrates entering surface waters.

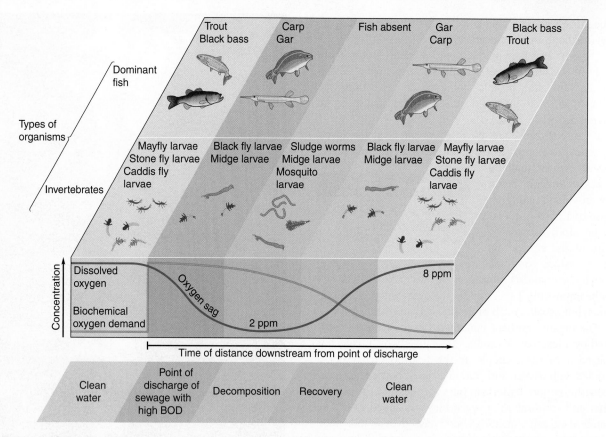

FIGURE 21-5 The oxygen sag curve. Oxygen levels and biochemical oxygen demand in a stream are shown downriver from a point source of organic nutrients. Oxygen levels drop immediately below the pollution source as a result of bacterial decay of organic matter. Oxygen levels recover downstream after the organic matter is removed from the water.

Fertilizers also pollute groundwater with nitrates. In Canada's Lower Frazer River Valley and in southern Ontario, nitrate pollution in groundwater has been traced to the application of manure and artificial fertilizer and from septic tanks.

Laundry detergents are the second most important anthropogenic source of inorganic nutrient pollution in the United States and other developed countries. Many detergents contain synthetic phosphates called tripoly-phosphates (TPPs). These chemicals cling to dirt particles and grease, keeping them in suspension until the wash water is flushed out of the washing machine.

Low and nonphosphate detergents are currently widely available, but many people don't need phosphate-based detergents at all. In fact, nearly 60% of the U.S. population lives in soft-water regions where soap-based cleansing agents work as well as detergents. In the hard-water regions, harmless substitutes for TPPs can be used. For example, lime soap-dispersing agents have been used in bar soaps for years and can be used for laundry detergents.

Nitrates and nitric acid can also enter surface waters from the atmosphere. According to the Environmental Defense Fund, about 25% of the nitrogen polluting the Chesapeake Bay comes from wet and dry acid deposition. Nitrates and phosphates are also contained in animal waste—humans and livestock being the principal sources. These pollutants may seep into groundwater under intense livestock operations, polluting the drinking water of local residents.

Eutrophication and Natural Succession

In undisturbed ecosystems, lakes naturally pick up plant nutrients from surface runoff and rainfall. The natural accumulation of nutrients in lakes is called **natural eutrophication** (u-TROH-feh-KAY-shun). Given sufficient time, natural eutrophication and natural soil erosion can transform shallow lakes into swampland and then into dry land, a process called **natural succession**. As shown in **FIGURE 21-6**, inorganic nutrients that accumulate in lakes stimulate plant growth. The plants eventually die and contribute organic sediment to the lake's bottom. This sediment combines with silt from erosion, gradually filling in a lake. Of course, shallow lakes are many times more vulnerable than deeper lakes.

Humans can accelerate the decay of lakes and ponds. Accelerated erosion, caused by human activities, and nutrients entering from farms, feedlots, and sewage treatment

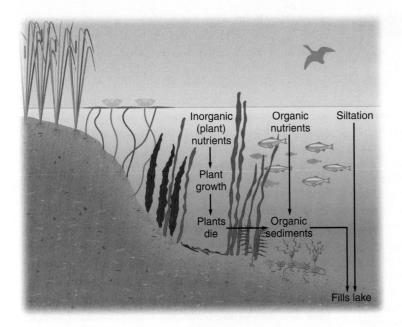

FIGURE 21-6 Eutrophication and succession.
This drawing shows the contributions of inorganic and organic nutrients and sediment to the succession of a lake into swampland. Inorganic nutrients contribute to eutrophication. Combined with sediment from natural or anthropogenic sources, they can cause a lake to fill in.

plants—called **cultural eutrophication**—both contribute to the deterioration of lakes. Inorganic nutrients may cause good, productive lakes to become choked with vegetation. In temperate climates, the plants rot in the fall. Sediment from erosion and organic debris caused by excessive plant growth can eventually fill the lake, destroying it.

The fate of lakes overfed with nutrients from sewage treatment plants and farms, however, is not as dim as scientists once believed; if nutrient inflow is greatly reduced or stopped, a lake may make a comeback. Lake Washington near Seattle, for example, became a foul-smelling, eutrophic eyesore after decades of abuse during which millions of gallons of sewage were dumped into its waters. In 1968, local communities began to divert their wastes to Puget Sound, an arm of the sea with a greater capacity to assimilate the wastes. Lake Washington began a slow recovery.

Of course, the diversion of sewage to Puget Sound has had some negative effects. Although the sound cleanses itself more quickly than the lake, certain toxic substances in the waste and in surface runoff (nonpoint pollution) are having a harmful effect on marine life. The effects are especially noticeable in areas where industries have discharged their wastes for decades. Today, sediments in the sound contain high levels of toxic organic wastes and heavy metals, which may explain the high incidence of tumors in sole, a fish that lives on the ocean's bottom. Shellfish beds have been closed to prevent people from harvesting contaminated shellfish. Efforts are now under way to minimize sewage and industrial waste discharged into the sound.

The story of Puget Sound reminds us that when it comes to pollution there is no "away." The diversion of waste from one location to another is often not a remedy, just a quick fix that transfers a problem from one location to another. It also

Go Green

Use nontoxic biodegradable cleaning agents to reduce water pollution.

illustrates the importance of preventing problems in the first place. Minimizing or eliminating the discharge of pollutants into the environment is a far more sustainable approach. The Puget Sound case also illustrates the economic impact of human actions. Contamination of shellfish beds by human wastes not only threatens natural systems and organisms, it threatens human health and the economic well-being of those who make a living from the sea.

Eutrophication is the most widespread problem in U.S. lakes, reservoirs, and ponds according to the Environmental Protection Agency (EPA). Nearly all receive wastes from industry and municipalities, but even if these sources were eliminated, many lakes would probably not improve significantly because of continued pollution from nonpoint sources.

KEY CONCEPTS

The accumulation of nutrients in lakes, from both natural and human sources, is called *natural* and *cultural eutrophication,* respectively. Combined with the deposition of sediment from natural processes and human activities, cultural eutrophication causes lakes to age prematurely.

Infectious Agents

Water may be polluted by pathogenic (disease-causing) organisms, including bacteria, viruses, and protozoans. The major infectious diseases transmitted via water include viral hepatitis, polio (viral), typhoid fever (bacterial), amoebic dysentery (protozoan), cholera (bacterial), schistosomiasis (parasitic worm), and salmonellosis (bacterial). These diseases are especially harmful to the young, old, and infirm. The major sources of infectious agents are (1) untreated or improperly treated human sewage, (2) animal wastes in fields and feedlots located near waterways, and (3) meat-packing and tanning plants that release untreated animal wastes into water.

Waterborne infectious diseases are a problem of immense proportions in the less developed nations of Africa, Asia, and Latin America killing more than 100,000 people per year. This results from the fact that human wastes often end up in surface waters from which many people receive their drinking water.

Infectious agents were once the major water pollutants of now-developed nations, too, before sewage treatment plants and disinfection of drinking water became commonplace. Despite these improvements, problems remain. In many cities, sewage and storm runoff travel through the same pipes. Both end up in the sewage treatment plant. However, when heavy storms occur, excess water containing both sewage and stormwater bypasses the sewage treatment plants, ending up in rivers and lakes. Flooding also causes sewage treatment plants located in flood plains (a common practice) to overflow.

Despite sophisticated wastewater and drinking water treatment facilities in more developed countries (MDCs), infectious agents can cause enormous problems. Outbreaks can also occur result of eating contaminated food such as eggs, tomatoes, and spinach.

Measuring the level of pathogenic organisms would be costly and time consuming for cities and towns. However, by measuring levels of a naturally occurring intestinal bacterium, the **coliform bacterium,** water quality personnel can determine how much fecal contamination has occurred **(FIGURE 21-7)**. The higher the coliform count, the more likely the water is to contain some pathogenic agent from fecal contamination. EPA studies show that about 15% of U.S. rivers assessed violate standards for coliform bacteria. Rivers in many LDCs are very heavily polluted with human wastes. A river in New Delhi, India, for example, has a coliform bacteria count of 7,500 per 100 milliliters, which is extremely high, before it enters the city. The count jumps to 24,000,000 per 100 milliliters after leaving the city. Disease and mortality rates are understandably very high.

Monitoring coliform bacteria in water is not by itself sufficient to protect human health. Several studies show that swimmers have developed gastrointestinal disorders and flu-like symptoms after swimming in seawater containing levels of coliform bacteria considered safe by government standards. Researchers have found that these symptoms are related to exposure to another noncoliform bacteria, a group known as **fecal streptococcal bacteria**. Some water quality experts believe that they are a more sensitive indicator of human fecal contamination. Others argue that, to protect human health adequately, two or more microorganisms ought to be monitored, including viruses.

KEY CONCEPTS

Numerous infectious agents are found in surface waters, especially in LDCs where they cause considerable amounts of illness and death. Sewage treatment facilities and drinking water purification have greatly reduced the incidence of disease in MDCs, although outbreaks do occur. To monitor infectious agents, officials generally measure fecal coliform levels, a harmless bacterium itself, but an indicator of the presence of fecal contamination and potentially pathogenic microbes.

Toxic Organic Water Pollutants

About 100,000 synthetic organic compounds are in use today in the industrialized nations. These include solvents, pesticides, and a host of others. Some of these chemicals find their way into surface waters directly

FIGURE 21-7 Fecal coliform bacteria.
This nontoxic bacterium is an indicator organism. Drinking water quality personnel measure it to determine the level of fecal contamination in water supplies. Its presence suggests the possible presence of pathogenic organisms.

© Andrew Syred/Science Source.

from factories.[2] Others, such as pesticides, enter waterways in surface runoff. Still others, such as toxic cleaning agents, enter sewage from homes. Concerns over toxic organic pollutants are many: (1) Many toxic organic compounds are nonbiodegradable or degrade slowly, so they persist in the ecosystem; (2) some are biomagnified in the food web; (3) some bioaccumulate; (4) some chemicals cause cancer in humans and aquatic organisms, whereas others are converted into carcinogens after reacting with the chlorine used to disinfect water; (5) many are toxic to fish and other aquatic organisms; and (6) some are mere nuisances, giving water and fish an offensive taste or odor.

Go Green

Use natural and nontoxic cleaners to help reduce water pollution.

Unfortunately, our knowledge of the effects of synthetic organics, which are often found in low concentrations, is rudimentary. Reports of diseases traceable to a single chemical are few, but some experts worry that cancer and genetic damage may result from long-term exposure to the wide array of toxic substances found in the waterways of the world; even if only present at low concentrations.

KEY CONCEPTS

Numerous toxic chemicals enter the waterways from factories, homes, farms, lawns, and gardens. They may have many different effects on people and the environment. As a rule, concentrations in surface waters are low.

Toxic Inorganic Water Pollutants

Toxic inorganic water pollutants encompass a wide range of chemicals, including metals, acids, and salts. Some of them have become major water pollutants. In most states, for example, toxic metals such as mercury and lead are found in levels that are considered unsafe for people and other organisms. Metals come from many sources, among them industrial discharge, urban runoff, sewage effluents, and mining. They are also derived from air pollution fallout. This is an example of cross-media contamination. Mercury emitted from coal-fired power plants, for instance, is washed from the sky and enters rivers, lakes, and streams. Recent surveys of U.S. drinking water show that some heavy metals such as lead come from pipes; others may come from groundwater supplies.

KEY CONCEPTS

Numerous inorganic pollutants such as acids and heavy metals like mercury make their way into the surface and groundwater of industrial nations, usually from industrial sources. Some, such as lead and mercury, can have serious health effects.

---

[2] Toxic organic pollutants differ from organic nutrients discussed earlier because they do not nourish plants. They are harmful to both plants and animals because they alter metabolism.

Mercury One of the more common and potentially most harmful toxic metals is mercury. In the 1950s, mercury was thought to be an innocuous water pollutant, although it was known to have been hazardous to miners and to 19th-century hatmakers exposed to mercury during the manufacture of hat brims. Both miners and hatters frequently developed tremors, or "hatter's shakes," and lost hair and teeth.

GO GREEN

Use energy efficiently to reduce mercury release from coal-fired power plants. Energy efficiency also reduces other air pollutants that eventually end up in surface waters.

In the 1950s, an outbreak of mercury poisonings in Japan raised awareness of the hazard. Residents who ate seafood from Minamata Bay, which was contaminated with methyl mercury, developed numbness of the limbs, lips, and tongue. They lost muscle control and suffered from other neurological defects, among them deafness, blurring of vision, clumsiness, apathy, and mental derangement. Of 52 reported cases, 17 people died and 23 were permanently disabled.

Mercury is a byproduct of manufacturing the plastic vinyl chloride, from which beach balls, toys, and other products are made. It is also emitted in aqueous wastes of the chemical industry and from incinerators, coal-fired power plants, research laboratories, and even hospitals.

In streams and lakes, inorganic mercury is converted by bacteria into two organic forms. One of these, dimethyl mercury, evaporates quickly from the water. The other, **methyl mercury,** however, remains in the bottom sediments and is slowly released into the water, where it enters organisms in the food chain and is biologically magnified. The mercury poisoning that affected those eating seafood from Minimata Bay resulted from biological magnification of methyl mercury. In Minnesota and other northern states, mercury levels in fish are so high that wildlife officials publish booklets that warn anglers to limit their intake of fish. Mercury even contaminates saltwater fish like swordfish, making some hazardous to eat, especially for young children and pregnant women.

KEY CONCEPTS

Mercury is emitted from many sources, among them vinyl factories and coal-fired power plants, and is one of the most common and most toxic inorganic pollutants, primarily exerting its effect through the nervous system.

Nitrates and Nitrites As discussed earlier, nitrates and nitrites are common inorganic pollutants of water. Nitrates in surface and groundwater come primarily from animal waste and artificial fertilizers. The sources include septic tanks, barnyards, sewage treatment plants, and heavily fertilized crops. Besides stimulating algal and plant growth, as noted previously, nitrates are also converted to toxic nitrites in the intestines of humans. Nitrites combine with the hemoglobin in red blood corpuscles and form methemoglobin, which has a reduced oxygen-carrying capacity. Nitrites can be fatal to infants. Although rare today, nitrite poisonings usually occur in rural areas where drinking water is contaminated by septic tanks, farmyards, or intensive livestock operations such as pig farms.

Salts Sodium chloride and calcium chloride are used on winter roads, driveways, and sidewalks to melt ice and snow. However, meltwater from ice and snow can transport these salts into nearby lakes, streams and groundwater. Salts kill sensitive plants such as the sugar maple. Apple and peach trees are also affected.

In surface waters, salts may kill salt-intolerant organisms, allowing salt-tolerant species to thrive. However, the fluctuations in water flow lead to varying salt concentrations. This is a condition that neither salt-tolerant species (which thrive in high salt concentrations) nor salt-intolerant organisms can survive.

Many cities and towns are now adding a solution of magnesium chloride to roads before snowstorms. Magnesium chloride prevents snow from sticking to the road and prevents dangerous black ice, making winter driving safer. It is also used on driveways, parking lots, and sidewalks. This liquid works like antifreeze—it lowers the freezing point of water. This prevents ice from forming a strong bond on the road and helps keep roads from becoming slippery, thus reducing accidents. Magnesium chloride also replaces the sand-salt mix once applied to roads in cities and towns throughout the country. Does magnesium chloride use pose any environmental problems?

Environmental studies conducted in the 1980s showed magnesium chloride to be less toxic than salt. Tests have shown that the proper application of magnesium chloride produces no negative effects on groundwater, surface water, or vegetation. Further research, however, shows that magnesium chloride, far from being safe, can cause extensive damage to roadside vegetation. It appears to be toxic to trees and inhibits water absorption by their roots.

According to the New Hampshire Department of Environmental Services, chloride that enters surface waters can be toxic to forms of aquatic species such as fish, insects, and amphibians. They note, "Elevated chloride levels can… pose a risk to species survival, growth, and/or reproduction." They also note that chloride's toxic effects increase when magnesium is present. These chemicals can reduce species diversity. "The presence of salt also releases toxic metals from sediment. When these chemicals are released into the water

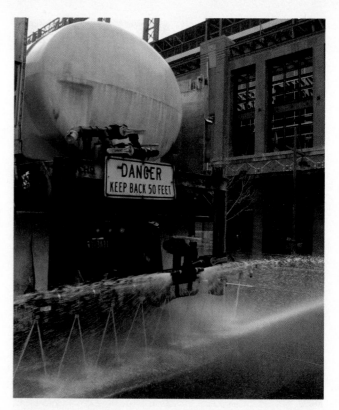

FIGURE 21-8 Deicing. Spraying salt solutions on roadways can damage nearby waterways.
© John Leyba/The Denver Post/Getty Images.

they can negatively affect nutrients and dissolved oxygen that aquatic species require" (**FIGURE 21-8**).

Chlorine Chlorine is a highly reactive inorganic chemical. It is commonly used (1) to kill bacteria in drinking water; (2) to destroy potentially harmful organisms in treated wastewater released from sewage treatment plants into streams; and (3) to kill algae, bacteria, fungi, and other organisms that grow inside and clog the pipes of the cooling systems of power plants. Chlorine and some of the products it forms in water are highly toxic to fish and other organisms.

Chlorine reacts with organic compounds to form **chlorinated organics.** These chemicals may show up in drinking water downstream from sewage treatment plants and other sources. Many of them are known carcinogens and teratogens. However, medical studies indicate that the rates of certain cancers (liver, intestinal tract) are only slightly elevated in populations consuming water contaminated by these compounds.

Sediment

Sediment is the leading water pollutant in the United States in terms of volume. Consisting of dirt particles and sand, sediment is largely a byproduct of a number of poorly planned and executed practices such as logging, agriculture, mining, and construction of roads and buildings. Agriculture, for example, increases erosion rates on average four to eight times above normal, sometimes much more. Poor construction and mining may increase the rate of erosion by 10 to 200 times. Sediment is also a major problem in the LDCs, where poor farming practices, overgrazing, and careless logging practices commonly occur.

Sedimentation, the deposition of sediment in surface waters, has many social, economic, and environmental impacts. Sediment, for instance, destroys spawning and feeding grounds for fish. It also smothers fish eggs and fry and thus reduces fish and shellfish populations. Suspended sediments decrease light penetration in water bodies, which may impair the growth of algae and aquatic plants, the base of the food chain. Each of these environmental impacts exacts a social and economic cost. As rivers and lakes are destroyed, for instance, they reduce recreational opportunities such as fishing; this not only robs many individuals of satisfying entertainment, it also reduces revenues to local bait shops and other segments of the local economy dependent on fishing.

The deposition of sediment in lakes speeds up natural succession. Sediment deposited in streams, called **streambed aggradation** (ah-gra-DAY-shun), results in a gradual widening of the channel (**FIGURE 21-9**). As streams become shallower, they become wider. Water temperature may rise, lowering the amount of dissolved oxygen and making streams more vulnerable to organic pollutants that deplete oxygen. Lowered dissolved oxygen also wipes out species such as trout that require higher levels. Streambed aggradation also makes streams more susceptible to flooding. In addition, sediment can fill shipping channels, which must then be dredged at considerable expense. Sediment fills in agricultural and municipal reservoirs throughout the world, reducing their lifespan. Moreover, hydroelectric equipment associated with dams may be worn out by sediments. Finally, some pollutants—such as pesticides, nitrates, phosphates from agricultural fertilizers, and pathogenic organisms—bind to sediment. This extends their lifetime and biological impacts. Sediment pollution can be checked and even eliminated, however, by good land management and restoration.

FIGURE 21-9 Streambed aggradation. This streambed is slowly becoming wider and shallower by deposition of sediment.

Courtesy of Rosgen, D.L. (2008). *River Stability Field Guide*. Fort Collins, Colorado: Wildland Hydrology.

Thermal Pollution

Unknown to many people, industry also pollutes water by dumping warm or hot water into cooler surface waters. Rapid or even gradual changes in water temperature can disrupt aquatic ecosystems with profound effects. Where does all this hot water come from?

Power plants and factories often must use water to cool liquids or machinery. They obtain this water from nearby lakes, oceans, and rivers. The U.S. electric power industry is a major contributor (**FIGURE 21-10**). It uses nearly 90% of all cooling water in the United States. Steel mills, oil refineries, and paper mills also use large amounts of water for cooling.

Small amounts of heat have no serious effect on the aquatic ecosystem, but large quantities can shift conditions beyond the range of tolerance of aquatic organisms, discussed elsewhere in the text. Heat-intolerant organisms perish. Elimination of heat-intolerant species may allow heat-tolerant species to take over. These are usually less desirable species.

Thermal pollution lowers the dissolved oxygen content of water, at the same time increasing the metabolic rate of fish and aquatic organisms. **Metabolism** consists of the chemical reactions occurring in the cells of organisms. Because metabolism requires oxygen, some species may be eliminated entirely if the water temperature rises 10°C (18°F). As an example, at the Savannah River nuclear power plant, the number of rooted plant species and turtles was at least 75% lower in ponds receiving hot water than in ponds at normal temperature. The number of fish species was reduced by one third.

Sudden changes in water temperature can cause **thermal shock,** the sudden death of fish and other organisms that cannot escape into cooler neighboring waters. Thermal shock is frequently experienced when power plants begin operation or when they temporarily shut down for repair.

Chapter 21: Water Pollution: Sustainably Managing a Renewable Resource   **521**

FIGURE 21-10 Thermal pollution. This schematic drawing shows a cooling system of a nuclear power plant and its effect on the temperature of surface waters of a nearby stream into which the heated water is released. Water from the river is used to cool water that circulates through the reactor. Heat picked up from this process by the river water is then dumped, along with the water, back into the stream (or lake), causing a dramatic increase in water temperature that is harmful to aquatic life.

The latter event can devastate heat-tolerant species that inhabit artificially warmed waters.

Fish spawn and migrate in response to changes in water temperature, and thermal pollution may interfere with these processes. Water temperature also influences the survival and early development of aquatic organisms. For instance, trout eggs may not hatch if water is too warm. Thermal pollution can also increase the susceptibility of aquatic organisms to parasites, certain toxic chemicals, and pathogens.

Thermal pollution can be controlled by constructing artificial ponds for collecting and cooling water before it is released into natural waterways—lakes and streams. Cooling towers are another way to dissipate heat.

 KEY CONCEPTS

Water from rivers and lakes is used to cool many industrial processes, electric power production being one of the major ones. Heat generated by these processes is often discharged directly into surface waters, where it kills organisms outright or shifts the composition of the aquatic system.

21.2 Groundwater Pollution

Aquifers are underground water bodies located in porous rock such as sandstone. They supply nearly one-fourth of the annual water demand in the United States and drinking water for over half the population. In rural areas, groundwater supplies 98% of the drinking water. Aquifers are also a major supplier of water in many other countries. In Canada, a country containing seven of the world's largest lakes and with nearly 8% of its land area covered by lakes and rivers, groundwater still contains 37 times more water than surface waters. According to one estimate, approximately one in four Canadians relies on groundwater for domestic uses. In some areas, however, such as Prince Edward Island, groundwater supplies 100% of the domestic demand.

Scientists are now reporting that U.S. and Canadian groundwater is increasingly threatened by pollution. In fact, many pollutants are present at much higher concentrations in groundwater than they are in most contaminated surface supplies. Also, many contaminants are tasteless and odorless at concentrations thought to threaten human health.

Go Green

Use organic fertilizers and natural pesticides on lawns and gardens. Be sure to follow application directions so you don't apply too much.

About 4,500 billion liters (1,185 billion gallons) of contaminated water seeps into the ground in the United States every day from septic tanks, cesspools, oil wells, landfills, agriculture, and ponds holding hazardous wastes. Unfortunately, very little is known about the extent of groundwater contamination. The EPA estimates that 1% of the drinking water wells in the United States have contaminants that exceed the standards designed to protect human health. Although that may seem small, 1% of hundreds of thousands of wells is a large number. In fact, one study reported that at least 8,000 private, public, and industrial wells in the United States are contaminated.

Groundwater is an important source of drinking water in many nations and may be heavily contaminated in numerous industrialized nations by toxic waste pits, septic tanks, oil wells, and landfills. Groundwater in some rural areas may also be contaminated by agricultural chemicals, notably pesticides and fertilizer.

Effects of Groundwater Pollution

Thousands of chemicals, many of them potentially harmful to health, turn up in water samples from polluted wells. The most common chemical pollutants are chlorides, nitrates, heavy metals, and various toxic organics such as pesticides, degreasing agents, and petroleum products.

Low-molecular-weight organic compounds, for example, from leaking underground gasoline storage tanks, are particularly worrisome because many of them are carcinogenic. Concern among medical experts is great because some fear that there is no threshold level for some of these compounds—that is, there is no level free from risk of cancer or other problems. Others fear that many chemicals may act synergistically, making a potentially difficult problem much worse.

With so many different chemicals in groundwater, it is not surprising that the effects vary widely. Nonetheless, some effects are more prevalent, according to Beverly Paigen, an Oakland, California, researcher renowned for her studies at Love Canal in Niagara Falls, New York. In a summary report of health studies of Americans exposed to groundwater pollutants, Paigen noted that the most common problems include miscarriage, low birth weight, birth defects, and premature infant death. Adults and children suffer skin rashes, eye irritation, and a whole host of neurological problems, including dizziness, headaches, seizures, and fainting spells. In a widely publicized case in San Jose, California, pollutants from a leaky underground storage tank owned by the Fairchild Camera and Instrument Company are thought to have doubled the rate of miscarriage in pregnant women and tripled the rate of heart defects in newborns. In Woburn, Massachusetts, contaminated groundwater is blamed for a doubling in the childhood leukemia rate.

Thousands of chemicals may be found in a nation's groundwater. Many of them are potentially harmful to human health, causing problems for unborn children (miscarriage, birth defects, and premature infant death) and adults (rashes and neurological problems).

Cleaning Up Groundwater

Many people think of groundwater as underground rivers. Nothing could be farther from the truth. Groundwater typically moves slowly—from 5 centimeters (2 inches) to 64 centimeters (2 feet) a day. Because groundwater moves so slowly, it can take years for water polluted in one location to appear in another. Additionally, once an aquifer is contaminated, it can take several hundred years for it to cleanse itself.

Detecting groundwater pollution is expensive and time consuming. Numerous test wells must be drilled to sample water and determine the rate and direction of flow. Despite intensive drilling, health officials can easily miss a tiny stream of pollutants that flows through one portion of a large aquifer. Liquids that do not readily mix with water, for example, may travel along the top or bottom of the aquifer in thin layers and are often difficult to detect.

Preventing groundwater pollution is generally the cheapest way to protect this vital resource and is an essential element in creating a sustainable water supply system. How can we prevent this problem? Cutting down on the use of potentially toxic chemicals and taking measures to reduce or eliminate the production of hazardous wastes are important first steps. Individuals can help by using natural or biodegradable cleaning agents. Hazardous waste recycling and improvements in the ways we dispose of wastes would also help.

Because many aquifers are already contaminated by potentially toxic substances, efforts are needed to reclaim them. It may, for example, be necessary to pump contaminated water from polluted aquifers to the surface and purify it. The water can be pumped back into the ground at that site or used for a variety of purposes. This process, although feasible for some contaminants, is extremely expensive. Many other approaches are being tested. For example, steam, various solvents, and surfactants (substances that reduce surface tension of liquids) can be injected into aquifers to improve recovery rates of chemicals.

New techniques are also being developed to use naturally occurring bacteria in soil and groundwater to clean up some contamination; this is called **bioremediation**. For instance, hydrocarbons (such as crude oil, gasoline, and creosote) that have leaked from storage tanks or are spilled from vehicles have polluted more groundwater used for drinking than any other class of chemicals in the United States. Microbiologists have known since the late 1970s that certain bacteria can digest or break down hydrocarbons in the soil and groundwater, converting them into carbon dioxide and methane gases. Bacteria in the soil, however, as a rule can degrade only about 1% of the hydrocarbon pollution flowing past them. Why? Quite simply, they lack key chemical nutrients needed for metabolism. By supplying these nutrients, researchers may be able to accelerate the bacterial decomposition of hydrocarbons.

Groundwater moves slowly and takes many years to cleanse itself. Preventing groundwater pollution is essential to creating a sustainable water supply. Equally important are efforts to clean up groundwater supplies already contaminated by potentially toxic chemicals.

21.3 Ocean Pollution

"When we go down to the low-tide line," Rachel Carson wrote, "we enter a world that is as old as the Earth itself—the primeval meeting place of the elements of Earth and water, a place of compromise and conflict and eternal change." Today this compromise, conflict, and eternal change have taken on a new meaning as humankind forges out into the oceans in search of food, fuel, and minerals. In our use of the seas, we spill many toxic substances, such as oil. Ocean pollutants also arise from land-based activities, though. The Kansas farmer and the Minnesota factory, for instance, have an impact on this vast body containing more than 1.3 billion cubic kilometers of water. Many inland pollutants are washed from the land or dumped into rivers, in which they are transported to the world's oceans.

This section deals with five crucial challenges: (1) oil pollution, (2) plastic pollution, (3) medical wastes, (4) red tide (algal blooms), and (5) sewage.

KEY CONCEPTS

The oceans are polluted by chemicals spilled into them directly and by pollutants washed from the lands and transported to them by rivers.

Oil in the Seas

Oil spills from oil tankers receive a considerable amount of attention and lead many to believe that oil-tanker accidents are the major source of oil in the seas. The massive release of oil from the damaged offshore oil well in 2010 in the Gulf of Mexico tops all spills (**FIGURE 21-11**).

On April 20, 2010, a deep-water offshore oil rig owned by British Petroleum (BP) experienced a devastating explosion that killed 11 workers. Over the next 108 days, the leaking well released an estimated 5 million barrels of oil into the Gulf of Mexico. Oil flooded pristine coastal beaches that are the heart of much of the region's economy. The oil slick was estimated to be more than 208 kilometers (130 miles) long and 110 kilometers (70 miles) wide and affected the coastlines of four states: Louisiana, Alabama, Mississippi, and Florida.

Oil invaded fragile ecosystems, including wetlands and bayous, on which many species of wildlife depend. Biologists believe that the oil from the leaking drill rig could cause long-term damage to these ecosystems and could, in fact, irreversibly alter some coastal ecosystems in the region.

Some of the nation's most valuable wildlife refuges and conservation areas are located in the area affected by the spill. The Gulf of Mexico is home to hundreds of species. Each year, an estimated 5 million migratory birds such as ducks and geese pass through the region. The Gulf provides winter refuge or a resting spot on long migrations to 70% of the nation's waterfowl. Many endangered species rely on Gulf waters, among them the North Atlantic bluefin tuna, four species of sea turtles, six whale species, sharks, and dolphins.

While the price tag of this disaster may never be known, it could be many billions of dollars. Especially hard hit has been the fishing industry in coastal communities, such as Louisiana's oyster and shrimp industry. In Louisiana, the annual retail sale of seafood comes to $1.8 billion. Recreation fishing nets another $1 billion in sales, and saltwater sport fishing generates about $757 million each year. Thousands of jobs depend on the Gulf water.

Although huge oil spills do occur and can pose a threat to coastlines, coastal wetlands, and fish and wildlife, nearly half (47%) of the oil entering our oceans comes from natural seepage from offshore oil deposits, according to the National Research Council (NRC). According to their report, *Oil in the Sea III,* natural seepage contributes about 4.3 million barrels (180 million gallons) of oil per year.

The second largest source of oil pollution in the sea comes from oil runoff from paved areas in urban and suburban areas, including streets and parking lots. Oil dripping from cars or dumped down sewer drains runs into rivers after rainstorms and makes its way to the sea. Oil is also

FIGURE 21-11 Oil flowed for days from this offshore oil well owned by British Petroleum (BP), in the Gulf of Mexico that was also severely damaged by an explosion.

Courtesy of U. S. Coast Guard.

released from boats and jet skis that often burn a mixture of gas and oil. Together, these and several other sources, contribute the vast majority of petroleum introduced to the environment through human activity. On average, they release approximately about 3.3 million barrels per year or about 140,000,000 gallons worldwide each year.

The third largest source is oil transportation, a category that includes oil spills and releases from refineries. Together, theses sources release slightly more that 1 million barrels (44,000,000 gallons) of crude oil or refined products into the world's oceans, or about 12% of the total.

Oil and gas exploration and associated activities release about 264,000 barrels or 11,000,000 gallons worldwide, each year, according to the NRC. That's under 3% of the total.

KEY CONCEPTS

Half of the oil polluting the oceans comes from natural seepage; the rest comes from human sources, including tanker accidents, accidents at oil wells, and inland disposal. The largest source is inland disposal.

Biological Impacts of an Oil Spill What happens to oil spilled from a tanker or a well? You may be surprised to learn that about 25% of the chemicals making up crude oil are volatile—that is, they evaporate readily. Within 3 months, one-fourth of the chemical substances in oil evaporate, becoming air pollutants. Relatively nonvolatile compounds that are lighter than water float on the surface, where they are broken down by bacteria over the next few months. Nearly 60% of the oil spill is destroyed in this way. The remaining 15% consists of heavier compounds that stick together and sink to the bottom in huge globs. In cold polar waters, oil decomposes very slowly. In some cases, it may become incorporated into sea ice and be released very slowly for years afterward.

Before oil's chemical components can evaporate or be broken down, they can have many harmful effects. Oil kills plants and animals in the estuarine zone. Especially hard hit are the barnacles, mussels, crabs, and rock weed (a type of algae). Their recovery after a major spill may take 2 to 10 years. Oil also settles on beaches and kills organisms that live there. It settles to the ocean floor and kills benthic (bottom dwelling) organisms such as crabs. Those organisms that survive may accumulate oil in their tissue, making them inedible. Oil also poisons algae and may disrupt major food chains and decrease the yield of edible fish. It also coats birds, impairing flight or reducing the insulative property of feathers, thus making the birds highly vulnerable to hypothermia (**FIGURE 21-12**). Gallant efforts are often made to rescue birds sullied by oil. After cleaning off the oil, the birds are kept in captivity, nursed back to health, and then released. Two studies published in 1996, however, suggest that such efforts (which are quite costly) are fairly ineffective.

GO GREEN

Never dump oil down a sewer—ever! Change oil through companies that recycle or reuse oils.

FIGURE 21-12 Oil-covered duck. This bird was beyond the help of volunteers who attempted to save many oil-covered animals after the 1989 oil spill in Prince William Sound (Alaska).

Courtesy of John S. Lough/Exxon Valdez Oil Spill Trustee Council.

One extensive study of more sensitive species of rehabilitated seabirds found that they lasted, on average, only 4 days after release. A study of brown pelicans showed that half of the birds released into the wild died within 6 months. Only 12 to 15% of the rehabilitated birds were still alive after 2 years, far lower than the 80 to 90% two-year survival rate of birds not exposed to oil. Why?

Birds immersed in oil inhale and ingest (swallow) toxic substances. These birds commonly suffer from anemia, immune system suppression, tissue damage, and hormonal imbalances. Researchers hypothesize that rehabilitated birds, although apparently healthy upon release, fare poorly because their immune systems cannot protect them adequately under the stressful conditions of life in the wild.

Oil also endangers fish hatcheries in coastal waters and can contaminate commercially valuable fish.

The amount of damage caused by oil pollution depends partly on the amount of oil spilled and the direction in which it is carried by the wind and ocean currents. If slicks reach land, they damage beaches and shorelines, coastal wetlands, recreational areas, and marine organisms. Oil may be driven over portions of the continental shelf, a highly productive marine zone; there it can poison clams, scallops, flounder, haddock, and other important food species. Oil driven out to sea has fewer environmental consequences because there life is less abundant in the open waters of the ocean.

The damage resulting from an oil spill also depends on when it occurs, too. For example, the devastating spill off the coast of Alaska in 1989 (Prince William Sound) occurred only 2 weeks before hundreds of thousands of ducks and other waterfowl migrated through the sound or came to nest. Had the spill occurred after the migration, wildlife losses would have been much lower.

Oil pollution of the oceans poses less of a threat to the overall marine environment than was once feared, a National Research Council committee has concluded. Oil can have serious local effects, the group noted, and these can persist for decades, as in the case of Prince William Sound. Overall, however, the marine environment has not suffered irreversible damage from oil. The committee was quick to point out that scientists have only limited knowledge of the potential damage of oil in tropical and Arctic regions, where much of the current oil development is occurring. Studies of the 1989 oil spill in Alaska, for example, showed that the hydrocarbons in the oil are more stable and more persistent in the cold waters than in warmer seas. This could make their impact even greater.

KEY CONCEPTS

Much of the oil spilled from human sources evaporates or is broken down by naturally occurring bacteria or sinks to the bottom. Before it is eliminated, however, it can cause serious environmental damage to aquatic plant and animal species as well as marine mammals and birds. The extent of the damage depends on the amount spilled, the location of the spill, the prevailing weather conditions, and the season.

Reducing the Number of Oil Spills   Thanks to public outcry since the 1970s, the number of oil spills has fallen dramatically. **FIGURE 21-13** shows that the number of large spills fell from an average of 25 per year in the 1970s to 3.3 per year in the 2000s to 2 per year from 2010 to 2013. Tougher governmental standards for new oil tankers went into effect in 1979. New safety standards for older tankers were phased in between 1981 and 1985. Dual radar systems, backup steering controls, collision avoidance aids, and improved inspection and certification were instrumental in reducing spills. In addition, under these regulations, crude oil must be cleaned before it is pumped on board to eliminate sludge buildup in storage tanks. This sludge was once rinsed out at sea. New regulations also require tankers to

have separate ballast tanks. These tanks are filled with salt water to help ships keep their balance when returning after discharging their cargo. In older ships, empty oil tanks were once filled with water for ballast. When the ship arrived at port, the oil-contaminated water was dumped into the sea.

The serious effects of oil washing and combined ballast-oil tanks must not be underestimated. About 1.3 million metric tons (1.4 million tons) of oil were once released each year during tank purging and ballast tank discharge—over six times the amount released by tanker spills.

Although the number of oil spills has decreased worldwide since the 1970s, the quantity of oil released fluctuates (**FIGURE 21-14**). In 2004, for example, 4.6 million gallons were spilled from ships and barges; in 2002, oil spills from these sources was 4.5 greater, about 21 million gallons.

In 1990 the U.S. Congress passed the **Oil Pollution Act,** which established a $1 billion fund to be used to clean up oil spills and pay for damages. The money comes from a 3-cents-per-barrel tax on domestic and imported oil and thus passes the costs of cleanup and damage on to the consumer. Under the bill, oil companies would ultimately be responsible for cleanup costs, but only up to a point because the law sets strict limits on their financial liability. A company responsible for a spill, for instance, would pay only $10 million for cleanup. If the spill occurred near an onshore facility where damage is much greater, the company would pay $350 million for each spill. The fund would be used to pay the additional costs. Many critics are disappointed with the liability limits because the costs of cleaning oil are likely to be far greater. Cleanup in Prince William Sound to date has cost Exxon $2.5 billion and the federal government $154 million. In addition, in 1992, Exxon agreed to pay an additional $1.25 billion in criminal fees, restitution, and civil

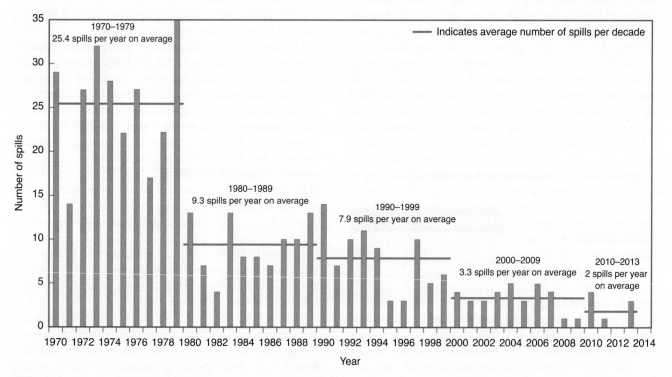

FIGURE 21-13 Number of spills over 700 tonnes.

Data from International Tanker Owners Pollution Federation Ltd., ITOPF.

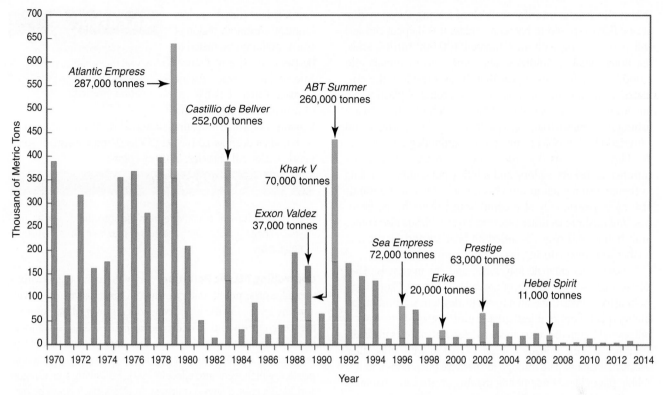

FIGURE 21-14 Oil spills from 1970 to 2009. Note the dramatic decline in recent years thanks to higher safety standards. (Tonnes = metric tons.)

Data from International Tanker Owners Pollution Federation Ltd., ITOPF.

recovery (to pay damages to salmon fishing companies and other parties), although this money has never been paid.

Although the law requires new tankers to be equipped with double steel hulls or effective double containment features to reduce the likelihood of spills, existing oil tankers are not required to be retrofitted. However, existing tankers must be escorted by two towing vessels in high-risk areas, and single-hull vessels started being phased out in 1995. By 2015 all tankers with single hulls are to be banned from use.

In addition, the law establishes regional oil spill response teams that can be deployed immediately after a spill to coordinate cleanup efforts. It also allows states to set stricter guidelines. A state, for instance, could mandate double hulls for use in its harbors.

KEY CONCEPTS

Many efforts are under way in the United States and other countries to reduce oil spills, including new standards for the construction and operation of oil tankers. As a result, the number of oil spills from tankers has decreased dramatically since the 1970s.

Sustainable Solutions to Oil Pollution Many of the steps outlined in this text to make human systems more sustainable will have a profound impact on water quality. This section describes how key principles of sustainable development—conservation, recycling, renewable energy use, restoration, and population control—can be applied to reduce oil pollution, beginning with conservation.

Conservation is one of the most important principles of sustainability. Energy conservation—using what we need and using it efficiently—is one of the most important areas to practice this principle. For example, energy efficiency in automobiles and increased use of mass transit will decrease the world's dependence on crude oil and will reduce the number of oil wells that leak oil into the ocean. It will also decrease oil spills occurring each year. Because much of the oil entering the oceans comes from disposal of oil on land, recycling of oil, rather than disposing of it in sewers, will also help protect the oceans. Renewable energy resources such ethanol and biodiesel cut pollution of water and air by reducing our dependence on fossil fuels.

KEY CONCEPTS

Conservation, recycling, and renewable resource use can dramatically reduce our dependence on oil and reduce oil pollution in the seas.

Plastic Pollution

A young seal swims playfully in the coastal waters of San Diego Bay. Floating in its watery domain is a piece of a plastic fishing net that has drifted with the currents for months. The seal swims around and around curiously and then plunges through an opening in the net, only to be entrapped. At first, the net is just a mild nuisance, but as the seal grows, the filament begins to tighten around its neck.

Eventually, it cuts into the seal's skin, leaving an open ring of raw flesh exposed to bacteria. Unless it is helped, the seal will perish, along with an estimated 100,000 turtles, seals, sea lions, whales, dolphins, and other sea mammals and 1,000,000 sea birds each year that die as a result of the estimated 10 million metric tons (9 million tons) of plastics. No one knows the number of fish that die each year from nylon fishing nets, plastic bags, six-pack yokes, plastic straps, and a myriad of other objects made from nonbiodegradable plastic. Plastic waste is discarded into the ocean annually by commercial fishers, sailors, and military personnel according to the environmental group Greenpeace. Tens of thousands of tons of plastic may also come from private boats, factories, and onshore facilities (**FIGURE 21-15**). Plastic also comes from bottles and bags discarded on beaches and plastic that washes into rivers during heavy rainstorms.

Plastic nets entangle fish, birds, and sea mammals. They may strangle, starve to death, or drown their victims. Plastic bags, looking like jellyfish, are eaten by sea turtles. One scientist pulled enough plastic out of a leatherback turtle's stomach to make a ball several feet in diameter. In such cases, starvation is the usual result because the animal's stomach is packed with indigestible plastic that cannot pass through its digestive tract. Birds and fish gobble plastic beads resembling the tiny crustaceans that are a normal part of their diet. They may become poisoned and die. Discarded plastic eating utensils, when swallowed, may cut into an animal's stomach lining, causing it to bleed to death.

Plastic is found throughout the world's oceans, but in 1997, a California sailor, surfer, and volunteer environmentalist and a friend, who were heading back from a sailing race in Hawaii, headed into an area known as the doldrums. Known technically as the North Pacific Subtropical Gyre, it is a region of perennial high pressure. It forms a huge, slowly spiraling vortex of warm equatorial air. It draws in winds and turns them gently until they expire, hence the nickname, the doldrums. Several major sea currents also converge in the gyre, bringing with them floating debris from the Pacific coasts of Southeast Asia, North America, Canada, and Mexico.

It is here that the sailors found an enormous region of plastic pollution estimated to be twice the size of France. Known as Great Pacific Garbage Patch, it is likened to a marine soup, the main ingredient of which is floating plastic debris. Plastic garbage can be found as deep as 10 meters (33 feet) and consists of small plastic flecks and particles, in many colors.

KEY CONCEPTS

Millions of tons of plastic are dumped into the ocean each year, killing hundreds of thousands of marine mammals, fish, and birds.

Controlling Plastic Pollution   As with many other environmental issues, public outcry created a groundswell of activity at many levels. On the local level, concern over plastic spurred the Oregon Fish and Wildlife Department to sponsor annual beach cleanups to remove plastic that might be washed out to sea (**FIGURE 21-16**). Italy placed a ban on all nonbiodegradable plastics, which went into effect in 1991. Following suit, Oregon and Alaska passed laws requiring that all six-pack yokes be biodegradable. In 2007, San Francisco became the first city to ban plastic shopping bags in grocery stores and large-chain pharmacies. Numerous countries have banned or taken action to reduce the use of plastic bags including Australia, Bangladesh, Ireland, Italy, South Africa, and Taiwan.

On a larger scale, the U.S. Congress passed the **Plastic Pollution Control Act** (1988), which makes it unlawful for

FIGURE 21-15 Victim of plastic pollution. A young hawksbill sea turtle is caught in a plastic fishing net cut loose by commercial fishing interests.

© Jeff Rotman/Alamy Images.

FIGURE 21-16 Trash on a beach in California. National and local initiatives are helping to reduce the amount of plastic and other trash like this on our beaches and in our oceans.

© T. O'Keefe/PhotoLink/Photodisc/Getty Images.

any U.S. vessel to discard plastic garbage into the ocean. This legislation also requires all manufacturers of six-pack yokes to use degradable plastic if technically feasible. In 1988, the U.S. Senate approved an international treaty banning the disposal of plastics in the ocean. The agreement was signed by 28 nations and went into effect in December 1988.

Despite these steps, millions of fish, birds, and sea mammals will perish in coming years. Without stricter controls, international cooperation, and private efforts, the ever-increasing use of plastic will bring about the unnecessary death of innumerable sea creatures. The use of biodegradable plastics could help, but they are not a panacea. In fact, the term *biodegradable plastic* is a bit of a misnomer. Biodegradable plastics are plastics mixed with starch or some other material that breaks down, dispersing small particles of plastic into the environment.

KEY CONCEPTS

Many steps have been taken to reduce the disposal of plastic into the ocean, but huge amounts are still being disposed of each year, posing a serious threat to sea life.

Medical Wastes and Sewage Sludge

In the summer of 1988, many Americans were shocked to learn that medical wastes were being illegally dumped into the ocean. Bloody bandages, sutures, vials of AIDS-infected blood, and used syringes washed up onto the eastern coast of the United States as well as onto the shores of Lake Erie.

Because there was no way to track the wastes to their source, the U.S. Congress passed the **Medical Waste Tracking Act** (1988). It established a 2-year program that covered 10 states and required people generating, storing, treating, and disposing of medical wastes to keep records that were accessible to the public. If wastes showed up on shorelines, they could be traced back to their source so that responsible parties could be brought to justice. Unfortunately, this law expired in 1990 and has not been reauthorized, leaving controls up to the states. All states have some rules regarding medical waste disposal.

A far more important concern has been the dumping of sewage into the ocean from sewage treatment plants. In 1988, Congress passed the **Ocean Dumping Ban Act,** which prohibited the dumping of sewage sludge (organic material from sewage treatment plants, discussed later) into the ocean. This longstanding practice of many coastal cities officially ended on December 31, 1991. Even though it is no longer legal to dump human wastes into the ocean, sewer overflows which occurs during heavy rainstorms, result in a considerable amount of raw sewage entering the ocean each year.

KEY CONCEPTS

Medical wastes and sewage have long been dumped into the ocean, but steps have been taken to eliminate both practices in the United States. The direct dumping of sewage has stopped entirely, but millions of gallons of sewage enter the sea each year from sewage treatment plants located near the coast.

Red Tide

Another growing problem occurring in the oceans of the world is known as *red tide*. **Red tide** is a term that refers to reddish algal blooms in coastal waters. Reports of sometimes harmless, sometimes noxious, and sometimes deadly red tides appear to be on the rise. At one time, they were found mainly in the waters off the coasts of Europe and North America. Today, they are also reported in coastal waters of Asia and South America. Scientists are not certain whether there is an actual increase in the incidence of such outbreaks or simply more awareness of the problem and more frequent reporting.

Red tides are only occasionally red. They may be brown or green or orange. Some evidence suggests that they may be caused by elevated levels of plant nutrients, such as algal blooms in freshwater. The problem with red tides is that the organisms that discolor the water contain toxic substances that protect them from predators. These organisms may be ingested by shellfish, however, which themselves become toxic to people.

Scientists have also discovered an entirely new family of microorganisms (pfeisteria) that release an extremely toxic substance into the water. The substance is so toxic that scientists must take very special precautions when working with it—similar to those taken when studying the AIDS virus in the laboratory. First identified in coastal waters of North Carolina, this organism causes massive fish kills. In the laboratory, the organism is nourished by levels of phosphate similar to those found in polluted rivers.

Some ecologists believe that toxic red tides represent a serious problem worldwide and are urging states and nations to step up efforts to reduce pollution that may be feeding these organisms.

KEY CONCEPTS

Outbreaks of microscopic and often highly toxic algae appear to be on the rise worldwide and may be caused by an increase in inorganic nutrient pollution from agriculture, industry, and sewage treatment plants.

21.4 Water Pollution Control

TABLE 21-2 offers a quick summary of the different sources of water pollution and shows the major pollutants from each one. This information is helpful in plotting control strategies. You may want to take a few minutes to study the table. The challenge is to devise solutions that prevent the problem in the first place. Also bear in mind the need to address both point and nonpoint sources.

As with other issues, water pollution control has traditionally focused on end-of-pipe measures: pollution control devices that remove the pollutants from a waste stream. Although these are important, the waste must go somewhere. Ultimately, the most sustainable approaches are those that prevent the pollutant from being generated in the first place.

TABLE 21-2

Common Water Pollutants

Sources	Bacteria	Nutrients	Ammonia	Total Dissolved Solids	Acids	Toxins
Point sources						
Municipal sewage treatment plants	•	•	•			•
Industrial facilities						•
Combined sewer overflows	•	•	•	•		•
Nonpoint sources						
Agricultural runoff	•	•		•		•
Urban runoff	•	•		•		•
Construction runoff		•				•
Mining runoff				•	•	•
Septic systems	•	•				•
Landfills						•
Forestry runoff		•				•

Data from U.S. Environmental Protection Agency.

KEY CONCEPTS

Reducing water pollution requires efforts on two levels—those that capture wastes emitted from various sources end-of-pipe solutions) and those that prevent waste production and pollution. The most sustainable are those that prevent waste.

KEY CONCEPTS

Legislation to address water pollution has historically focused on point sources—primarily factories and sewage treatment plants. Gains made in controlling such sources, however, have often been offset by increasing levels of pollution from nonpoint sources such as city streets, lawns, and farm fields.

Legislative Controls

The cornerstone of U.S. water pollution policy is the **Clean Water Act.** While important, it has focused primarily on point sources, notably sewage treatment plants and factories. In fact, the Clean Water Act has provided funding to construct and improve thousands of U.S. sewage treatment plants that handle municipal wastes (**FIGURE 21-17**). It has also required factory owners to install water-pollution control equipment to reduce the discharge of industrial waste into surface waters (**TABLE 21-3**). For a success story, see **Spotlight on Sustainable Development 21-1**.

Unfortunately, in many cases nonpoint pollution from expanding cities offsets the gains from sewage treatment plants. Pesticides from lawns; oil on driveways, parking lots, and streets; nitrate fertilizers from farm fields; and dozens of other nonpoint sources all contribute millions of tons of pollution each year to U.S. waterways. Ironically, despite intense efforts to control point sources, in many regions

FIGURE 21-17 Sewage treatment plant. This photograph of a sewage treatment plant shows the large, circular trickling filters that are used for secondary treatment.

© Graham Prentice/Dreamstime.com.

TABLE 21-3

Major Provisions of the U.S. Clean Water Act

Planning

The states receive planning grants from the EPA to review their water pollution problems and to determine ways to solve them by reducing or eliminating point and nonpoint water pollutants.

Standards Development

The states adopt water quality standards for their streams. These standards define a use for each stream and prescribe the water quality needed to achieve that use.

Effluent Standards

The EPA develops limits on how much pollution may be released by industries and municipalities. These limits are developed at the national level based on engineering and economic judgments. The EPA or the states are required to make the discharge limits for individual plants more stringent if necessary to meet the state water quality standards.

Grants and Loans

The EPA provides financial assistance to state water programs for the construction of sewage treatment plants, processing of permit applications, monitoring of water quality, and enforcement.

Dredge and Fill Program

The EPA develops environmental guidelines to protect wetlands from dredge and fill activities. These guidelines are used to assess whether permits should be issued.

Permits and Enforcement

All industries and municipal dischargers receive permits from either the EPA or the states. The EPA and the states regularly inspect these dischargers to determine whether they are in compliance with the permit and take appropriate enforcement actions if necessary.

Data from U.S. Environmental Protection Agency.

---

experiencing rapid population growth water quality has not improved or has improved only slightly.

Controlling Nonpoint Pollution

Recognizing that nonpoint sources are a major source of water pollution, the U.S. Congress amended the Clean Water Act in 1987. Under the amended law, the states were required to identify nonpoint sources and draft plans to control them. The EPA currently provides $200 million per year in grant money to states to address nonpoint sources. States are required to partially match federal funds as well. This program has resulted in numerous successes.

Some state and local governments have taken actions on their own by passing zoning laws aimed at reducing agricultural and urban runoff. Maryland, for instance, limits the amount of pavement accompanying development in watersheds of Chesapeake Bay. In northern California, loggers are prohibited from clogging streams with silt. In addition, local soil conservation districts help to identify trouble spots and work with farmers to decrease erosion and other problems. North Carolina has instituted some innovative approaches to control nonpoint sources. For example, it currently permits industry to work with farmers to implement nonpoint pollution controls—for example, erosion controls. Such measures are often cheaper than building treatment facilities and offer multiple benefits.

Despite these gains, much work is needed to reduce nonpoint water pollution to sustainable levels. Because agricultural runoff is the largest source of nonpoint water pollution, affecting between 50 and 70% of all surface water and groundwater in the nation, laws that address pollution from farms are important. New regulations that require terracing of steep roadbanks, revegetation of denuded land, and use of mulches to hold soil in place while grasses take hold along newly constructed highways and housing sites, are also essential. Laws that require sedimentation ponds to collect runoff before it can reach streams and that require porous pavements or porous pavers that soak up rainwater could also help reduce erosion and the influx of pollutants into nearby water bodies.

KEY CONCEPTS

In the United States, efforts to control nonpoint water pollution are still in their infancy but are gaining in popularity because they are often economical solutions that offer other benefits as well. Especially important are efforts to control agricultural runoff.

Preventing Groundwater Pollution Several federal laws give the EPA the authority to prevent and control sources of groundwater pollution and to clean up polluted groundwater. In 1984, the EPA adopted a formal groundwater protection strategy that, among many things, established an Office of Groundwater Protection. This strategy focuses on building state capacity—that is, state regulation and control of groundwater. Under the plan, the EPA provides technical assistance in analyzing problems and gives advice needed by states to establish their own groundwater protection programs. Many states have adopted their own groundwater standards. Some have taken preventive measures by mapping out aquifers and banning industrial development over them. Today, the vast majority of the states have developed programs that at least minimally regulate discharges to groundwater. Clearly, much improvement is needed, especially in groundwater pollution prevention.

In 1996, Congress toughened the U.S. drinking water law by adding regulations that protect groundwater. Congress passed legislation that requires the EPA to set standards for

© hugolacasse/ShutterStock, Inc.

21-1 Cleaning Up the Great Lakes

Carved by ancient glaciers, the five mammoth Great Lakes hold 90% of America's fresh surface water. Approximately 36 million people live within their drainage basin. The Great Lakes provide drinking water for 30 million people and is an enormous source of recreation, jobs, and revenue as shown in **TABLE 1**. Like so many of the continent's waters, they have suffered years of abuse from pollution and poor land management. Especially hard hit were Lakes Ontario and Erie.

Cultural Stress: The Death of a Lake?

The story of Lake Erie serves as a reminder of the immense impact of human civilization and as a lesson on ways to prevent future deterioration of the world's waters. Lake Erie was once surrounded by dense forest. The healthy forests reduced erosion and streams ran clean. Today, more than 13 million people live in the lake's watershed. The dense woodlands that once protected the soil were cut down to make room for farms, homes, industries, and roadways. As a result, large quantities of topsoil washed into the rivers and the lake, clogging navigable channels and destroying spawning areas essential to the lake's once rich fish life. In the early 1900s, many of the wetlands along the lake's shore were drained, and hundreds of small dams were built to provide power to mills, blocking

the upstream migration of fish such as the walleye and sturgeon.

By the 1960s, the lake's waters were polluted with a wide assortment of organic and inorganic nutrients. Raw sewage floated on the water's surface, and algal blooms (rapid growth of algae) were commonplace (**FIGURE 1**). Dissolved oxygen levels frequently dropped to low levels, especially in the profundal zone that occupies the large central basin of the lake. Blue-green algae proliferated in the warm summer months in the shallow western end of the lake, creating a foul-smelling, murky green water. Lead, zinc, nickel, mercury, and other toxins from industry polluted harbors and built up in nearshore sediments. In 1970 and 1971, mercury levels in fish from Lake Erie often exceeded safe levels set by the U.S. Food and Drug Administration (FDA). Reductions in mercury discharges in 1975 resulted in a decline in mercury in fish, but violations of health standards continue.

In colonial days, numerous fish species inhabited the lake and its tributaries. Largemouth and smallmouth bass, muskellunge, northern pike, and channel catfish were common in the tributaries of the lake. Lake herring, blue pike, lake whitefish, lake sturgeon, and others lived in the open waters. By the 1940s, however, blue pike and native lake trout had vanished. Sturgeon, lake herring, whitefish, and muskellunge managed to hold on, but in reduced numbers.

TABLE 1

Income and Jobs Generated by the Great Lakes

Activity	Number of Jobs or Participants	Annual Revenue
Recreational boating	4 million recreational boats	$16 billion
Tourism	9.2 million anglers	
	4.6 million hunters	$30 billion
	2.9 million bird watchers	
Shipping	226,000 jobs	$48 billion

FIGURE 1 Plant nutrients from sewage treatment facilities caused massive algal blooms in the Great Lakes.
Courtesy of T. Joyce/NOAA.

Lake Erie suffered from overfishing, introduction of alien species, pollution, and destruction of shorelines and spawning grounds. Algal blooms, beach closings, thick deposits of sludge, oxygen depletion, taste and odor problems, and contaminated fish were the legacy of years of disregard and mismanagement.

By the late 1950s, large areas of Lake Erie's central basin were anoxic (without oxygen) for weeks on end during the summer. Until the late 1970s, anoxia spread cancerously. The lake was pronounced dead; many feared that the other lakes would follow suit.

The Joint Cleanup Program—Not Enough

Alarmed by the condition of Lake Erie and the other lakes, in 1972, the United States and Canada agreed to restrict the discharge of pollution into the Great Lakes. The Great Lakes Water Quality Agreement, updated in 1978, called for controls of point and nonpoint pollution sources. It demanded that releases into the lakes of "any or all persistent toxic substances" be "virtually eliminated." With cooperation from industrial and municipal polluters, the lakes began to show signs of recovery. Lake Erie, the shallowest and fastest flowing of the lakes, made a quick recovery. Today, it is teeming with fish. Gone are the massive algal blooms and the raw sewage discharges that once discolored the waters. The other lakes also show signs of improvement. In a recent report to the U.S. Congress, in fact, Great Lake states reported on a survey that included nearly all of the nearshore waters. This report showed that most of the waters are now safe for swimming and other recreational activities, and much of it is clean enough for drinking water.

Despite these efforts, 43 areas failed to meet the standards established in the U.S.–Canadian accord a decade or so later. In 1987, a Buffalo-based environmental group, Great Lakes United, published a report noting that while highly visible and odoriferous pollutants such as sewage had been significantly reduced, many toxic substances not visible to the naked eye continue to pour into the Lakes. For instance, PCBs and pesticides still persist at unacceptable levels. Officials administering the 1978 agreement conceded that the U.S. and Canadian governments had not fully lived up to its terms.

Toxicants enter the lakes from factories and sewage treatment plants; from nonpoint pollution, including farmland and urban runoff; from toxic fallout—that is, pollutants deposited from the atmosphere; and from resuspension of substances contained in the bottom sediments, caused in part when harbors and river mouths are dredged.

One of the most significant avenues of entry for toxins is atmospheric deposition. Trace metals, pesticides, phosphorus, nitric acid, nitrates, sulfates, sulfuric acid, and organic compounds are all deposited from the air. Approximately 60 to 90% of all PCBs entering Lakes Superior and Michigan come from the atmosphere. An estimated 14,000 metric tons (15,400 tons) of aluminum is deposited in Lake Superior from the atmosphere, and nearly 29,000 metric tons (31,900 tons) annually falls from the skies into Lake Michigan. The atmosphere is also a major source of phosphorus, providing an estimated 59% of the total input to Lake Superior. Without controls on atmospheric deposition, it will be impossible to end the discharge of toxins into the Great Lakes.

What does the continued pollution assault mean? First, commercial fishing, once an economic mainstay in the region, has virtually ceased in all of the lakes except Superior. Even there, the commercial fishers operate under continual uncertainty, never knowing when their catch will exceed safe limits for contaminants and be declared unsafe by the FDA. Second, introduced Pacific salmon and reintroduced lake trout survive in the lakes, but the salmon population must be restocked each year. Lake trout do not reproduce successfully either, except in Superior. Restocking costs millions of dollars a year. Third, continued pollution has forced states to issue warnings advising women who are pregnant, lactating, or of childbearing age not to eat certain fish. Parents are also advised not to feed their children lake-caught fish. This warning came after a scientific study on chronic low-level exposure, which showed that infants of women who had eaten PCB-contaminated fish two to three times a month were smaller, more sluggish, and had slower reflexes than infants of women who had not eaten contaminated fish. Children also accumulate the toxicants, which may impact future fertility. Acceptable levels for most of the 800 pollutants found in the lakes simply have not been established, primarily because of a lack of information on health effects. Currently all of the Great Lakes are under fish consumption advisory—meaning the fish consumption is inadvisable or should be limited to prevent health problems. **FIGURE 2** shows the contaminents of greatest concern. PCBs are the main culprit. Aquatic life in these waters is also in jeopardy from both toxic chemicals that accumulate in the food chain and habitat loss or degradation and competition and predation by nonnative species such as the zebra mussel and sea lamprey. You can learn more online at the EPA's Great Lakes monitoring page and at Healthy Lakes.org.

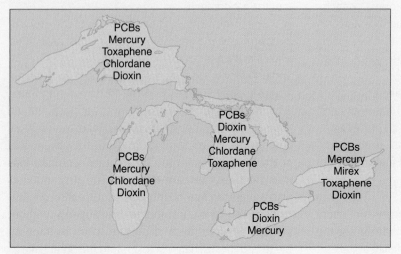

FIGURE 2 Toxic chemical threats in the Great Lakes.

Data from Environmental Protection Agency.

Reducing the problems posed by toxic pollutants is complicated. Eight states, two provinces, numerous tribal councils, and two nations share an interest in managing the waters of the Great Lakes. The result is often conflict that holds up the important steps needed to bring the Great Lakes back to a full, productive, and healthy life.

In recent talks, Canada and the United States agreed on ways to control airborne pollutants. They also agreed to remove contaminated sediments from the lakes and to better control groundwater pollution that contributes to the degradation of the Great Lakes. At least 15 to 30 years of rehabilitation are needed, but improvements are already beginning to be seen. Clearly, there's a long way to go.

drinking water quality for 85 additional chemical substances, including pesticides and various industrial chemicals.

The new law requires drinking water suppliers to test for these chemicals and to maintain the drinking water standards. It also requires them to monitor drinking water supplies for other substances that might pose a threat to human health. Even though the EPA is now monitoring these chemicals, hundreds more could be monitored.

 KEY CONCEPTS

The U.S. government and many states have done a better job of addressing groundwater pollution than nonpoint water pollution because groundwater is an important source of drinking water for many communities.

Water Pollution Control Technologies: End-of-Pipe Approaches

The first sewage treatment plant in the United States was built in Memphis, Tennessee, in 1880; today, there are over 14,770 of them. Sewage entering a treatment plant often contains pollutants from homes, hospitals, schools, and industries. It contains human wastes, paper, soap, detergent, dirt, cloth, food residues, microorganisms, and a variety of household chemicals. In some cases, water from storm drainage systems is mixed with municipal waste drainage systems to save the cost of building separate pipes for each, which can be exorbitant. Combined systems generally work well, but as noted earlier, during storms inflow may exceed plant capacity. Consequently, some untreated storm runoff and sewage pass directly into waterways, raising the coliform count in downstream waters and rendering them

unfit for swimming. Combined sewage overflows, as they are called, annually amount to 5 trillion liters (1.3 trillion gallons) according to the EPA.

 KEY CONCEPTS

Sewage treatment plants receive waste from many sources within cities, including homes, office buildings, factories, hospitals, and, in some cases, stormwater drainage systems.

Primary Treatment Sewage treatment can take place in three stages: primary, secondary, and tertiary treatment. Primary treatment physically removes large objects by first passing the sewage through a series of grates and screens. Sand, dirt, and other solids settle out in grit chambers (**FIGURE 21-18**). The solid organic matter, or sludge, settles out in a settling tank.

 KEY CONCEPTS

Primary treatment is a means of filtering out large objects and solid organic matter.

Secondary Treatment The secondary stage destroys biodegradable organic matter through biological decay (Figure 21-18). Sludge from primary treatment enters a large tank, where bacteria and other organisms decompose the agitated waste. Another common way is to pass the liquid sludge through a trickling filter (**FIGURE 21-19**). Here, long pipes rotate slowly over a bed of stones (and sometimes bark), dripping wastes on an artificial decomposer food web consisting of bacteria, protozoa, fungi, snails, worms, and insects.

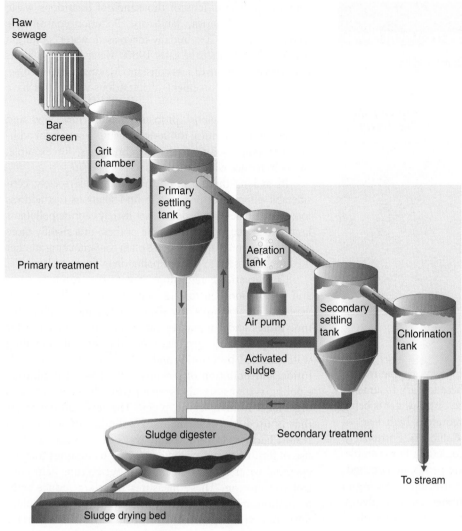

FIGURE 21-18 **Sewage treatment.** Primary and secondary sewage treatment facilities. The secondary treatment facility contains all of the components of the primary system but has an additional aeration tank (or trickling filter) to further decompose organic matter.

FIGURE 21-19 Trickling filter. A decomposer food chain consisting of bacteria and other microorganisms in the rock or bark bed of this system consumes organic matter, nitrates, and phosphates in the liquid sewage.

© Photos.com.

The bacteria and fungi consume the organics and, in turn, are consumed by protozoans. Snails and insects feed on the protozoans. Some inorganic nutrients—notably nitrates and phosphorus—are also removed during secondary treatment.

Secondary treatment is completed by a secondary settling basin, or clarifier, which removes residual organic matter. In most municipalities, liquid remaining after secondary treatment is chlorinated to kill potentially pathogenic bacteria and protozoans and is then released into receiving streams, lakes, or bays.

The efficiency of primary and secondary treatment is shown in **TABLE 21-4**.

 KEY CONCEPTS

Secondary treatment further destroys organic matter and removes much of the nitrogen and phosphorus in sewage by employing bacteria and other decomposers.

TABLE 21-4
Removal of Pollutants by Sewage Treatment Plants

Substance	Percentage Removed by Treatment	
	Primary	Primary and Secondary
Solids	60	90
Organic wastes	30	90
Phosphorus	0	30
Nitrates	0	50
Salts	0	5
Radioisotopes	0	50
Pesticides	0	0

Tertiary Treatment

Many methods exist for removing the chemicals that remain after secondary treatment. Most of these tertiary treatments are costly and therefore are rarely used unless water is being released into bodies of water that require a high level of purity—for example, Lake Tahoe in California. Fortunately, some cheaper options are gaining recognition. For example, effluents can be transferred to holding ponds after secondary treatment, where algae and water hyacinths growing in the water consume the remaining nitrates and phosphates. Certain aquatic plants such as duckweed absorb dissolved organic materials directly from the water.

Aquatic plants grown in sewage ponds can be harvested and converted into food for humans or livestock. In Burma, Laos, and Thailand, duckweed has been consumed by farmers for years. The protein yield of a duckweed pond is six times greater than that of an equivalent field of soybeans. One of the problems with this approach, however, is that water hyacinths and duckweed also absorb toxic metals from the water. Therefore, consumption by humans and livestock must be carefully monitored. Another approach is the use of artificial wetlands, either indoors in greenhouses or outdoors.

 KEY CONCEPTS

Tertiary treatment brings sewage wastewater near to drinking water quality but is rarely used because it is so costly.

Sustainable Solutions for Water Pollution

Despite an outpouring of laws to control pollution, the United States and other countries have hardly come to grips with the problem. Tens of thousands of hazardous waste sites litter the American landscape. Pollution control laws passed in the 1970s initially decreased water pollution nationwide, but since the early 1980s, water pollution levels have remained more or less constant. Making matters worse, regulation and enforcement of hazardous-waste laws have been lax.

Solutions to global problems require new laws and tighter controls. Critical thinking demands a search for additional systemic solutions. New technologies, for example, can help reduce waste.

In making the transition to a sustainable society, considerable effort must be made to find solutions that address root causes. These strategies do not merely capture pollutants bound for waterways and dispose of them in a slightly more acceptable manner, in the process often contaminating another medium; they seek to prevent pollution in the first place.

Conservation, recycling, renewable resource use, and population control are four forms of pollution prevention. For example, by using energy more efficiently, individuals and companies can reduce their demand for electricity. This, in turn, decreases the amount of cooling water needed by utilities and cuts back on thermal pollution and pollution of streams and lakes with chlorine. Cooling systems are also treated with chemicals to keep organisms from clogging pipes. The less cooling water that is used, the less chemical pollution of waters will result. Farmers can find many ways to cut back on their use of fertilizers and pesticides. Many techniques that are essential to creating a sustainable agriculture help control soil erosion. These measures therefore reduce sedimentation and the release of pesticides, inorganic plant nutrients, and organic chemical pollutants into surface and groundwaters.

TABLE 21-5 shows the dramatic reductions in water pollution possible from recycling aluminum, steel, paper, and glass. Because recycling these commodities also reduces air pollution—and because air pollution can become water pollution through cross-media contamination—this strategy pays a double dividend. A third payoff comes from the reduction of mining wastes, which often make their way into streams.

In contrast to fossil fuel power plants, most renewable energy sources such as photovoltaics and wind energy require little if any water. Thus they produce little if any air and water pollution. Population stabilization, of course, is the ultimate pollution prevention strategy. Each avoided birth means that much less waste going into waterways.

The sustainable strategy also seeks to weave human actions into the cycles of nature. That is, it attempts to return nutrients to their site of origin or to utilize them for other purposes—for example, by using municipal sewage to fertilize crops. This approach is known as **land disposal.**

In ancient times, land disposal of human sewage was commonplace; it is still practiced in many LDCs, such as China and India. As the populations of LDCs grew and

TABLE 21-5

Environmental Benefits Derived from Substituting Secondary Materials for Virgin Resources

Environmental Benefit	Reduction of Use (percent)			
	Aluminum	Steel	Paper	Glass
Energy use	90–97	47–74	23–74	4–32
Air pollution	95	85	74	20
Water pollution	97	76	35	—
Mining wastes	—	97	—	80
Water use	—	40	58	50

Data from R. C. Lechter and M. T. Sheil. 1986. "Source Separation and Citizen Recycling." In *The Solid Waste Handbook.* W. D. Robinson, ed. New York: Wiley.

became more urbanized, this natural method of recycling wastes was gradually phased out. Land disposal of sewage can be used to recharge aquifers. This technique relies on surface vegetation, soil, and soil microorganisms to filter many potentially harmful chemicals. Sewage can be piped to pastures, fields, and forests (**FIGURE 21-20**). Organic matter in the effluent enriches the soil and improves its organic content and ability to retain water. Nitrates and phosphates serve as fertilizers. The water supports plant growth and seeps into the ground, helping to recharge aquifers. Crops nourished by effluents from treated sewage show a remarkable increase in yield.

Land disposal of sludge from sewage treatment plants is not a perfect solution. First, treated sewage may contain harmful bacteria, protozoans, and viruses that could adhere to plants consumed by humans or livestock or become airborne after the effluent dries. To get around this problem, the Swiss and Germans heat their sludge

to destroy such organisms before applying it to pastures and cropland. Alternatively, sewage sludge can be decayed in compost piles before application. The heat given off during composting kills all of the viruses, bacteria, and parasite eggs.

The second problem is that toxic metals found in some sewage may accumulate in soils and be taken up by plants and livestock. Metal-contaminated sewage usually comes from industries. By removing metals from their waste stream (an option called **pretreatment**) or by finding ways to prevent them from entering the waste stream entirely (pollution prevention), industries can eliminate the problem.

Third, transporting sludge to fields increases the cost of sewage treatment, limiting some of the incentive to use this method. Experts are quick to point out, however, that land disposal is 10 times cheaper than building and operating a tertiary treatment plant.

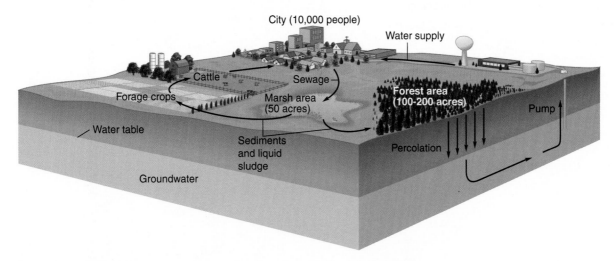

FIGURE 21-20 Closing the loop. Land disposal of sewage helps fertilize farmlands and forests, reduce surface water pollution, and replenish groundwater supplies. This strategy helps prevent the pollution of surface waters.

Scientists at the University of Maryland recently developed an unusual way to put sludge to good use. By combining it with clay and slate, they formed odorless *biobricks*, which look like ordinary bricks. Washington's Suburban Sanitary Commission recently put the idea to the test by building a 750-square-meter (8,300-square-foot) maintenance building in Maryland with 20,000 biobricks. If successful, biobricks could help reduce land disturbance resulting from mining materials used for brick making—and could help cut sewage disposal costs and environmental contamination. Biobricks are also being marketed as a fuel for wood stoves (**FIGURE 21-21**).

Restoration efforts can also help clean up water supplies. Restoration of wetlands can help reverse the decline of aquatic systems and purify surface water.

These are just a handful of steps needed to solve the water pollution dilemma sustainably. Of course, personal actions are also needed. Limiting family size and reducing the consumption of unnecessary goods are important actions we can each take. By reducing the purchase of goods, we can greatly reduce hazardous waste production at factories. As a general rule, each ton of garbage generated by consumers is preceded by 5 tons of manufacturing waste and 20 tons of waste produced at the source (the mine or forest where the raw materials were acquired). So every savings we make, by buying less or by buying more durable items, can effect large upstream savings in waste and pollutants that might otherwise make their way into our waterways.

Another effective way to cut your personal contribution to water pollution is to install a composting toilet. The composted wastes can be added safely to gardens, eliminating the need for synthetic fertilizer. If and when you become a homeowner, restrict your use of synthetic fertilizers, insecticides, herbicides, bleaches, detergents, disinfectants, and other household chemicals—or find safe alternatives. Select low-phosphate or nonphosphate detergents for washing your clothes. You may also contact local and federal officials to support further cleanup efforts.

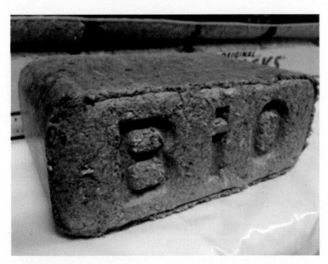

FIGURE 21-21 Biobrick. Compressed biomass (sawdust, nut shells, tree branches) formed into bricks can be used as fuel in wood stoves. The bricks are very dense with low moisture content; therefore, they burn slowly with high heat output.
© BioPellet

KEY CONCEPTS

Preventing pollution is a key to creating clean, healthy ground and surface waters, which are a prerequisite to living sustainably on the Earth. Reducing consumption of goods, recycling, using renewable resources, and stabilizing population growth will collectively reduce our production of water pollutants.

It is astonishing with how little wisdom mankind can be governed when that little wisdom is its own.

—**W.R. Inge**

Water conservation in the home is designed to save water, but it has other benefits as well. For example, it helps reduce energy demand because it takes energy both to pump water to treatment plants and to heat water for domestic use. The less water you use, the less energy is needed in your home and at the water treatment plant. This in turn results in numerous environmental savings, from reduced habitat destruction from coal mining to less air pollution.

Water conservation in the home also means less water going down the drains to sewage treatment plants. That saves energy at the sewage treatment plant. It also reduces the demand for chlorine, which is used to disinfect the effluent of sewage treatment plants. Can you think of any other benefits?

This exercise clearly shows how important it is to examine the big picture. Creating a sustainable future requires steps with multiple benefits. We need to conserve energy and reduce pollution wherever we can; so look at each solution as an opportunity to solve several problems. Many people think that conservation means sacrifice, when in fact water conservation measures in the home can be quite inexpensive and easy to implement. Simple changes in behavior—such as not flushing the toilet every time or watering your lawn less frequently—actually reduce your work. Conservation measures can save you money as well, which can be used to enhance your life.

CRITICAL THINKING AND CONCEPT REVIEW

1. List the major types of water pollutants found in developing and developed nations. What are the major differences? Why do these differences exist?
2. Define the terms *point source* and *nonpoint source*. Give some examples of each and explain why nonpoint sources of water pollution are often more difficult to control than point sources.
3. Describe where organic nutrients come from, what effects they have on aquatic ecosystems, and how they can be controlled.
4. What are inorganic plant nutrients, and how do they affect the aquatic environment? List some strategies for eliminating these pollutants.
5. Define the terms *natural eutrophication* and *cultural eutrophication*. Describe how inorganic and organic nutrients accelerate natural succession in a lake.
6. What are the major sources of infectious agents in polluted water? How can they be controlled?
7. What are some of the major inorganic water pollutants? How do they affect the aquatic environment and human populations?
8. Why is chlorine used in the treatment of human sewage and drinking water? What dangers does its use pose? How would you determine if the risks of chlorine use outweigh the benefits?
9. What are the major sources of sediment, and how can they be controlled? What are the costs and benefits of sediment control?
10. A factory that has been polluting a nearby stream with toxic wastes (volatile organic chemicals) has agreed to stop. The owners are proposing the use of evaporation ponds, where the wastes would sit until they evaporated. Using your critical thinking skills, your systems-thinking abilities, and your knowledge of water pollution, critically analyze this proposal.
11. What are the major sources of groundwater pollution? How can groundwater pollution be reduced or eliminated? What can you do?
12. Discuss the sources of oil in the ocean and ways to reduce oil contamination.
13. Critically analyze the following statement: "Oil naturally seeps into the ocean. In fact, about half of the oil in the ocean comes from natural sources. Because of this, it makes little sense to worry about spills from tankers and other sources."

14. Describe primary and secondary wastewater treatment. What happens at each stage, and what pollutants are removed?

15. A housing development is being built in your town. The developers say that this project will not increase pollution in the river that flows into the town because they will pay to expand the town's sewage treatment plant. Do you agree or disagree with their assertion? Why?

16. Pollution control devices are often viewed as stopgap measures. Why are such measures essential for long-term sustainability?

17. List and critically analyze all of the options available to us to prevent pollution, such as population stabilization. Why are these approaches essential to the task of building a sustainable society?

KEY TERMS

biochemical oxygen demand (BOD)
bioremediation
chlorinated organics
Clean Water Act
coliform bacterium
cross-media contamination
cultural eutrophication
fecal streptococcal bacteria
groundwater
inorganic nutrients

land disposal
Medical Waste Tracking Act
metabolism
methyl mercury
natural eutrophication
natural succession
nonpoint sources
Ocean Dumping Ban Act
Oil Pollution Act
Plastic Pollution Control Act

point sources
pretreatment
red tide
sediment
sedimentation
streambed aggradation
surface waters
thermal shock
water pollution

© B. Brown/Shutterstock, Inc.

PESTS AND PESTICIDES: GROWING CROPS SUSTAINABLY

What we do for ourselves resides with us. What we do for others and the world remains and is immortal.

—**Albert Pine**

LEARNING OBJECTIVES

After reading and studying the material in this chapter, students will be able to understand, knowledgably discuss, and, when appropriate, apply the information learned in this chapter, including:

1. Evolution of chemical pesticide control
2. Types of pesticides and how they kill pests
3. Impacts of chemical pesticide, on humans and the environment
4. How genetic resistance is making it more difficult to control pests
5. Traditional and sustainable approaches to controlling or eliminating pesticides, especially integrated pest management

CRITICAL THINKING

A newspaper reported that a pesticide commonly found in small amounts on fruits and vegetables caused cancer in mice. Food safety advocates were quoted in the article as saying that the pesticide should be banned. Environmental Protection Agency (EPA) officials argued that the pesticide was safe because their calculations showed that it would only cause one additional case of cancer per million people in the population each year. How would you analyze this issue? What critical thinking rules are necessary?

Ron Rosmann is not a revolutionary; he's an Iowa farmer. Like many of his cohorts, though, he is sowing the seeds of a revolution in farming. For years Rosmann, like other corn and soybean farmers, believed that without chemical pesticides on his farm, he'd be out of business. **Pesticides** are chemical substances that kill insects, weeds, and a whole assortment of organisms that reduce crop output.

In the 1980s, Rosmann visited a couple of farmers that grew crops without pesticides. Much to his surprise, their fields looked great. There were few weeds and insects. Encouraged by what seemed an impossibility, Rosmann began to experiment with alternatives to pesticides the next year. On his farm, weeds were the biggest problem, and **herbicides** (chemicals used to kill them) were a big expense. By using techniques he had seen the year before, Rosmann found that he could virtually eliminate herbicides. In the 9 years that followed, he used herbicides only once, and then in small amounts because he had mistakenly let weeds get out of control. Today, Rosmann saves $4,000 to $5,000 a year on herbicides. Much to his surprise, his crop yields have gone up. In other words, he's growing more food per acre at a lower cost.

Rosmann's efforts are a prime example of sustainable development. They are good for people, the environment, and the economy. Unlike Rosmann, farmers in the industrial countries collectively spend billions of dollars on chemical pesticides. To them, weeds and other pests are an impediment to efficient farming. They reduce production and profits.

Each year weeds, insects, bacteria, fungi, viruses, birds, rodents, mammals, and other organisms—pest species—consume or destroy an estimated 52% of the world's food production (this includes both preharvest and postharvest losses). The highest rate of destruction occurs in the tropics and subtropics, where because of favorable year-round climate, as many as three crops are grown each year on the same field. The warm weather and long growing season create optimal conditions for crop-eating insects.

Crop destruction from pests is high even in the more developed nations, despite elaborate and costly control strategies. In North America, for instance, preharvest losses are estimated to be about 31%, and postharvest losses amount to 9% of what is left. Together, about 40% of annual U.S. production is lost to various pests. Some of the most common insect pests in the United States and their ranges are shown in **FIGURE 22-1**.

Similar losses are reported in other more developed countries. In a world where about 20% of the people are malnourished, such losses are tragic. The silver lining behind this ominous cloud, however, is that efforts to reduce losses through pest control could help increase food supplies. As you will see in this chapter, pest control must be embarked on judiciously. It requires a careful analysis of the social, economic, and environmental costs of all -strategies—and the use of sustainable strategies.

To help set the stage for the discussion of sustainable pest management, this chapter begins with a discussion of the conventional pest control strategies and the many environmental problems that make them unsustainable. It ends with a discussion of the many options, including those pursued by Ron Rosmann, for controlling pests economically and in a socially and environmentally sustainable manner.

22.1 Chemical Pesticides

In China over 3,000 years ago, farmers controlled locusts by burning infested fields. In the ancient Middle East, open ditches were used to trap immature locusts. In 1182 CE, Chinese citizens were required to collect and kill locusts in an effort to control an outbreak. Such measures involving direct human intervention are often called cultural controls.

Over the years, many farmers have also used various toxic chemicals. Early chemical pesticides, known as **first-generation pesticides**, were simple but sometimes highly toxic preparations made of ashes, sulfur, arsenic compounds, ground tobacco, or hydrogen cyanide. Lead, zinc, and mercury compounds were also used. In Greece, for instance, farmers used arsenic, sulfur, caustic soda, and olive oil to control crop pests. Today, few of the first-generation pesticides remain in use. Most proved to be too toxic to people or ineffective. Many were environmentally persistent. Many of the compounds still contaminate soils.

KEY CONCEPTS

Pest control measures have been used for centuries. Earliest techniques included cultural controls, such as burning fields to kill locusts, and chemical controls, the use of toxic chemicals to kill insect pests. Early chemicals have been abandoned because they were toxic, ineffective, or environmentally persistent.

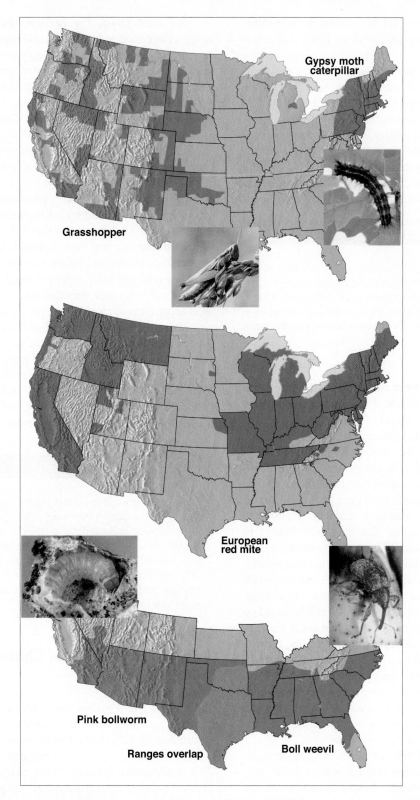

FIGURE 22-1 The geography of major pests. These maps of the United States show the most prominent insect pests and where they are found.

Gypsy moth caterpillar photo © WizData, Inc./ShutterStock, Inc.; grasshopper photo courtesy of USDA ARS; pink bollworm photo Jack Kelly Clark. Used with permission of the University of California Statewide IPM Program; boll weevil photo courtesy of USDA ARS.

Modern Chemical Pesticides

In 1939, the Swiss chemist Paul Müller discovered the insecticidal properties of a synthetic organic compound called **DDT (dichlorodiphenyltrichloroethane)**. This discovery ushered in a new era of chemical control. DDT became the first in a long line of **second-generation pesticides**, synthetic organic pesticides. For the 25 years following its development, DDT was viewed by many as the savior of humankind, for it was quite lethal to insect pests and thus increased crop yield. Since DDT was relatively inexpensive to produce, its use became widespread. In 1944, Müller was awarded a Nobel prize.

Over the years, thousands of new chemical pesticides have been synthesized and tested. Today, according to the U.S. EPA, 900 different substances are used in over 33,000 commercial formulations of herbicides, insecticides, fungicides, miticides, and rodenticides, with one goal in mind: to reduce pests to tolerable levels. Some pesticides, including DDT, are **broad-spectrum pesticides** that attack a wide variety of organisms; others are **narrow-spectrum pesticides**, used in controlling a few specific pest species.

KEY CONCEPTS

The first synthetic organic chemical used as a pesticide was a chemical called DDT, whose discovery ushered in a new era of synthetic chemical pesticides. Chemical pesticides are either broad-spectrum substances that kill a variety of pest species or narrow-spectrum chemicals that are effective against one or a few pests.

Types of Chemical Pesticides Synthetic chemical pesticides fall into three chemical families: (1) chlorinated hydrocarbons (organochlorines), (2) organic phosphates (organophosphates), and (3) carbamates.

The first group to be developed, the **chlorinated hydrocarbons**, is a high-risk group that includes DDT, aldrin, kepone, dieldrin, chlordane, heptachlor, endrin, mirex, toxaphene, and lindane. All of these have been banned or drastically restricted because of their ability to cause cancer, birth defects, neurological disorders, and damage to wildlife and the environment. Chlorinated hydrocarbons are extremely resistant to breakdown, persist in the environment, are passed up the food chain (biomagnified), and remain for long periods in body fat. As you will see later in this chapter, a pesticide banned in the United States may still be found on food imported from countries where no bans are in place.

The second group, the **organic phosphates**, consists of chemicals such as malathion and parathion. These pesticides, while still toxic, break down much more rapidly than chlorinated hydrocarbons after being sprayed on a crop. The chlorinated hydrocarbon DDT, for example, has a half-life in the environment of a couple of years; parathion, an organic phosphate, has a half-life of a couple of days.[1]

The organic phosphates are water soluble and are excreted in the urine. Because they are water soluble, they are less likely to bioaccumulate. Despite their benefits, the organic phosphates are still of great concern, for humans exposed to even low levels may suffer an assortment of health effects such as drowsiness, confusion, cramps, diarrhea, vomiting, headaches, and difficulty breathing. Higher levels cause severe convulsions, paralysis, tremors, coma, and death.

The third group, the **carbamates**, are widely used today as insecticides, herbicides, and fungicides. One of the most common is carbaryl (commonly known as Sevin). As a group, the carbamates are, like the organic phosphates, less persistent than the chlorinated hydrocarbons, remaining a few days to 2 weeks after application. They are also water soluble and do not bioaccumulate. Carbamates are nerve poisons, as are the chlorinated hydrocarbons and organic phosphates. Although they are environmentally safer than their predecessors, carbamates have been shown to cause birth defects and genetic damage.

Organophosphates and carbamates have one key advantage over the chlorinated hydrocarbons— their lack of persistence. This means that they kill their pests and then disappear within a few days or weeks. Unfortunately, they are much more toxic to humans than the chlorinated organics. They enter the body rapidly through the skin, digestive system, and lungs and therefore can be harmful to people who apply them to crops and to people who live nearby.

Although chlorinated hydrocarbons, organic phosphates, and carbamates are the three main types of pesticides in use today, they are not the only ones. Two others worth noting are the **triazines** (TRY-ah-zeens), a relatively harmless family of chemical herbicides, and the **pyrethroids** (PIE-rith-roids), a group of natural and synthetic insecticides, also relatively nontoxic to humans.

KEY CONCEPTS

Three main types of synthetic organic chemical pesticides have been developed over the years—chlorinated hydrocarbons, organophosphates, and carbamates. These are all neurotoxins. Organic phosphates and carbamates are less persistent in the environment or body fat of organisms than chlorinated hydrocarbons—and are thus less likely to be biomagnified in the food chain. Unfortunately, both are much more toxic to people.

Growth in the Use of Chemical Pesticides

Approximately 2.9 million metric tons (3.2 million tons) of chemical pesticides were used throughout the world in 2007, approximately 22% in the United States and about 57% in Europe and other more developed nations. The remaining 21% is used primarily in less developed countries. The United States is a leading agricultural nation and a leader in pesticide use. In fact, about one-fifth of all pesticides used in the world each year are applied to U.S. crops. Of these pesticides, 10% are insecticides for insect control, 25% are herbicides for weed control, 14% are fungicides, and the remaining 26% include nematicides, fumigants, rodenticides, molluscicides, and fish/bird poisons, according to the U.S. EPA.[2]

---

[1] Half-life is the time it takes for half of the chemical to break down in the environment.

[2] Data from the U.S. EPA, 2006–2007 Pesticide Market estimates, the latest available data. These amounts are active ingredients only.

Overuse

According to University of Illinois entomologist Robert Metcalf, author of a standard college textbook on pest management, farmers apply more than twice the pesticide they need. Adding to the unnecessary environmental contamination are a cadre of misinformed homeowners who apply about one-tenth of all U.S. chemical pesticides on gardens, lawns, and trees in higher amounts per acre than farmers do. Rains wash excess pesticides into sewers, streams, and other water bodies. Soils exposed to heavy applications of pesticide often suffer because beneficial bacteria are destroyed. Without bacteria, grass clippings decay very slowly, creating a layer of undecomposed material called *thatch*. Thatch buildup on lawns is believed to be one byproduct of excessive use of pesticides.

Individuals not only apply pesticides in excess, they often take poor precautions to protect themselves when applying chemicals to gardens and lawns. That is, they rarely use or wear gloves and face masks. As a result, extremely toxic pesticides are often restricted—that is, not available for homeowners to use.

KEY CONCEPTS

Pesticides are often applied far in excess of what's needed by farmers and especially homeowners. Homeowners also often fail to take precautions to protect themselves from exposure.

Biological Impacts of Pesticides

Pesticides constitute a very small percentage (around 3%) of the commercial chemicals commonly used in the United States each year. Nonetheless, because they are released into the environment in large quantities and have the potential to alter ecosystem balance and threaten human health, their use has created widespread and often heated controversy.

In the 1940s, DDT and other new pesticides were applied to a variety of crops, with astonishing results. DDT was even used to delouse soldiers and civilians in World War II. DDT and other chemical pesticides proved to be fast and efficient in controlling insects, weeds, and other pests. In India, for example, before the use of DDT in the 1950s to control malaria-carrying mosquitoes, there were over 100 million cases of malaria a year. By 1961, the annual incidence had been reduced to 50,000. Since then, the national eradication program has suffered numerous problems and the incidence of malaria has increased dramatically. In 2011, for instance, there were 219 million cases but only 660,000 deaths, according to the World Health Organization (WHO).

Pesticides also allowed farmers to respond quickly to pest outbreaks, thus avoiding economic disaster. Pesticides were relatively cheap and easy to apply, and their use resulted in substantial financial gains as yields increased. Some insecticides, such as DDT and dieldrin, persisted long after application, giving extended protection. The persistence of these pesticides, in fact, was thought to be one of their major advantages.

The successes of the early phase led to a rapid expansion in pesticide production and use. During this period, researchers developed many new pesticides. In the ensuing years, however, many problems started to emerge.

KEY CONCEPTS

The initial success of DDT led to a rapid increase in pesticide use and in the number of new chemical pesticides. Problems soon began to emerge, however.

Destroying Natural Pest Controls and Beneficial Insects One of the most troubling problems is that pesticides, especially broad-spectrum chemicals, often destroy natural biological pest controls—that is, insect predators and parasitic insects that naturally help control pests and potential pests. The loss of these beneficial insects (predators, for example) may result in a proliferation of pest species. In fact, some insect species that were previously benign suddenly increase in number and begin creating a need for additional pesticides. Agronomists call this population explosion of new pests an **upset**.

A classic example of such an upset took place in California. Spider mites, once only a minor crop pest, have become a major pest because of the use of pesticides that killed off many of their natural enemies, which were more sensitive to the sprays. Today, mites cause twice as much damage as any other insect pest in California and cost farmers (in damage and control) five times what they cost before the widespread use of pesticides. Two of modern farming's most costly pests, the cotton bollworm and the corn-root worm, were minor problems before widespread pesticide use reduced their natural predators. In the United States, one-third of the nation's 300 most destructive insect pests are secondary pests—that is, insects that once caused little or no trouble at all.

Pesticides also destroy other beneficial insects, such as honeybees, which play an important role in pollination. About one-third of the food we eat comes from plants pollinated by insects. The honeybee is responsible for 80% of the pollination, according to the USDA (**FIGURE 22-2**). Honeybees pollinate 130 U.S. crops (mostly fruits, nuts, and vegetables) annually, worth an estimated $20 billion a year in 2013, according to the U.S. Department of Agriculture (USDA). Apple orchards are particularly hard hit. Honey produced by honeybees was valued at $150 million, down from $225 million in 2003. Captive honeybee colonies are

FIGURE 22-2 Honey bees in danger. Honey bees produce honey and pollinate crops, especially fruit and nut trees.

© Geoff Kuchera/iStock/Thinkstock.

FIGURE 22-3 Pesticide resistance. A tobacco budworm crawls through deadly DDT (white crystals) unaffected. Because of genetic resistance, many other species are unaffected by the toxic chemicals designed to kill them.

©1998 Fred Ward/Black Star.

on the decline. Since 1947, they have fallen from 5.9 million to 2.9 million in 2007. Between 2007 and 2010, the USDA estimates that one-third of U.S. honeybees died. The loss is refered to as *colony collapse disorder*. Numerous factors are likely responsible for the decline. Especially trouble-some may be a new class of pesticides known as neonicotinoids derived from nicotine. This pesticide may weaken bees making them more susceptible to other factors. Honeybee colonies are also being destroyed by parasitic mites that suck the blood from adult bees. Mites are transmitted by Africanized honeybees and other species. Losses are also attributed to habitat loss, and even habitat modifications. Researchers believe that additional losses may be the result of parasites, virus, a fungal parasite that attacks the gut (*Nosema*), pesticides—and very likely a combination of these factors. Bees are under more stress due to many factors like habitat loss and exposure to neonicotinoid pesticides, which could weaken their immune systems, making them more vulnerable to mites and other organisms.

KEY CONCEPTS

One major impact of pesticides is that they often kill natural pest control agents such as predatory insects that are more sensitive to pesticides. This unleashes the growth of pest populations. Pesticides also kill other beneficial insects such as pollinators.

Creating Genetically Resistant Pests   Another unanticipated effect of pesticide use has been the dramatic increase in **genetically resistant insects**. Because of genetic diversity, a small portion of any insect population (roughly 5%) is genetically resistant to a pesticide. That is, it is not killed by a normal application (**FIGURE 22-3**). Why are they resistant? They may contain enzymes, for example, that destroy or detoxify the poison. The presence of these enzymes is a product of their unique genetic makeup. Therefore, an initial application of pesticides will kill all but the genetically resistant members of a pest population. Although these survivors do little damage at first, over time they reproduce and form a sizable population, which can cause significant crop damage.

As farmers encountered genetic resistance, they began to try new approaches. The first strategy was to increase the amount of pesticide. Because of genetic resistance, this solution is effective only in the short term. Why? Even though the higher dose wipes out the majority of the insect pests, it invariably leaves behind a small subpopulation that is genetically resistant to the increased dose. They soon proliferate and cause farmers to apply even more, creating a vicious addictive cycle that many observers call the **pesticide treadmill**. When DDT and other insecticides were first introduced into Central America, cotton fields were sprayed eight times each growing season; today, because of genetic resistance, 30 to 40 applications are typical.

The second strategy was the development of new pesticides. However, scientists found that it was expensive to create pesticides and that the insects developed genetic resistance to these new chemicals almost as quickly as they were released.

Genetic resistance to DDT was first reported in 1947 by Italian researchers. In 2014, approximately 584 insect species were resistant to at least one form of pesticide up from 557 in 2007 (**FIGURE 22-4**). Over 20 of the world's worst pests are now resistant to most types of insecticide. Moreover, farmers are finding that crop diseases and weeds are also developing resistance to the very chemicals that are used to control them. Worldwide, hundreds of plant pathogens (organisms that cause disease) and 232 species of weeds have developed resistance to at least one pesticide.

Since the introduction of DDT, insects have never met a chemical pesticide they couldn't defeat. What is more, despite the increased application of chemical pesticides, annual losses due to pests have continued to climb. Although insecticide use has increased 10-fold since World War II, crop damage has doubled. Today, not only do insects

FIGURE 22-4 Pesticide resistance. This graph shows the number of species of insects, that are resistant to at least one pesticide.

take a larger percentage of the U.S. harvest than they did before the introduction of DDT, but damage from fungi and weeds is climbing as well.

 KEY CONCEPTS

Pesticide use results in the formation of genetically resistant pest species. Higher doses and new pesticides, used to combat them, only worsen the situation by creating even more resistant species.

Damage to Fish and Wildlife Many chemical pesticides have proved harmful to nontarget species: birds, fish, and other animals. Survival rates in newly hatched trout in

Go Green

Use organic and natural pest control measures on gardens and lawns.

upstate New York, for example, were depressed by DDT that had been applied to nearby forests. Insect- and worm-eating birds also perished in areas where aerial spraying of insecticide had occurred. As a result of widespread pesticide use, populations of many birds plummeted.

The manufacturers of pesticides argued that the chemicals were found in only minute concentrations in the environment and could not be the cause of declining populations of fish and wildlife.

However, studies showed that although DDT and DDE (a toxic breakdown product of DDT) levels in aquatic ecosystems were quite low, concentrations were higher in producers and still higher in consumers because of biomagnification. Fish-eating birds, the consumers at the top trophic level, had the highest concentrations of these chemicals. Although these levels were not lethal to adults, they impaired reproduction. DDT and DDE reduced the deposition of calcium in the eggshells of birds that feed on fish and of other birds. This list includes peregrine falcons, brown pelicans, cormorants, bald eagles, gulls, and ospreys. Of the two compounds, biologists discovered that DDE posed the greater physiological threat to birds, in part

because it persisted longer. Reduced eggshell calcium levels create a thinner, more fragile shell that cracks easily during incubation. As a result of widespread DDT contamination, many predatory birds populations were nearly wiped out. In one study of bald eagle reproduction, North Dakota State University zoologist James Grier showed that the number of young per nest in northwestern Ontario declined by about 70% between 1966 and 1974.

The newer pesticides are also harmful to fish and wildlife. One pesticide, diazinon (die-AS-eh-non), was once commonly used on sod farms and golf courses throughout the United States to control insects. In 1984, three fairways were treated with diazinon in Hempstead, New York. In the 2 days following treatment, 700 Atlantic brant geese, 7% of the state's entire population, died of acute diazinon poisoning.

Diazinon was also used to treat nine fairways at a golf course in Bellingham, Washington. After application, the area was irrigated to decrease pesticide concentrate on the surface. Nevertheless, 85 American wigeon ducks perished after eating grass on one of the treated fairways on the day of application.

In 1988, the EPA banned the use of diazinon on golf courses and sod farms because of similar incidents. It is still available for use in other outdoor areas. Although it continues to kill wildlife, especially birds, mortality rates are down now that it is no longer used on golf courses and sod farms.

Birds have also been dramatically affected by the use of granular carbofuran (car-bow-FUHR-an), which was developed in 1970 by the FMC Corporation of Philadelphia. Farmers throughout the United States use carbofuran to eradicate nematodes and insects from corn, rice, and other crops. Although carbofuran apparently poses no threat to humans when applied to crops, it is lethal to songbirds. In fact, a songbird can die after ingesting a single granule.

In 1989, EPA records showed that about 2 million birds were dying of carbofuran poisoning each year. In 1991, the state of Virginia banned the use of carbofuran, and since that time, it has been banned nationwide by the EPA. Though the ban will likely have a positive impact on birds in the United States, export of carbofuran continues. To learn more about this and other pesticides, you can log on to Beyond Pesticides' website (http://www.beyondpesticides.org).

 KEY CONCEPTS

Pesticides poison fish, birds, and other species outright and also biomagnify in the food chain. Thus, seemingly low concentrations in the environment can result in very high levels in organisms in the uppermost trophic level, which may impair reproduction.

Although the general public is exposed to pesticides in a variety of ways, it is farm and chemical workers who are exposed to the highest levels from direct exposure to pesticides on the job as well, indicating widespread misuse. Workers pick up pesticides on their clothing and skin through accidents and negligence or by prematurely entering sprayed fields. In most cases, workers are poorly protected (**FIGURE 22-5**). They are often given few instructions and no protective gear to minimize exposure—especially in the less developed countries, where poor worker safety standards and practices and widespread illiteracy are common. In the less developed countries, workers often complain of being sprayed with insecticides from helicopters or planes while they are working in fields and on plantations.

In less developed countries where farm labor is abundant and worker safety provisions minimal, workers are often viewed as an expendable commodity. Christopher Brady, who is active in development work in Latin America and Africa, writes, "As a result of increasing health risks to the sprayers, Tela (a Honduran subsidiary of Chiquita Brands, Inc., a U.S. company) has changed their hiring policy. Sprayers are now only hired on a 6-month contract." They are "let go and new ones hired before any serious health problems may be detected." Workers spraying herbicides on banana plantations are given a thin cotton mask as their only protective gear. Moreover, empty barrels once containing concentrated pesticide are often used by local villagers to hold drinking water.

Symptoms of poisoning are many. They include insomnia, nausea, and loss of sex drive. Some people exposed to pesticides complain of reduced powers of concentration, irritability, and nervous disorders. Spills on one's skin may cause rashes and a burning sensation. In severe cases, death will occur.

In the United States, at least 45,000 workers are seriously poisoned each year, but many experts believe that this figure grossly underestimates the number of serious poisonings. Surveys in California, for instance, show that three-fourths of all serious poisonings go unreported. Each year 200 to 1,000 people die from pesticide poisoning in the United States. Although there is no way to know the true extent of the problem, the World Health Organization estimates that worldwide there are at least 25 million pesticide poisonings annually, in LDCs. These result in an estimated 300,000 deaths each year, according to World Health Organization estimates. Pesticide poisoning also results in numerous chronic and fatal illnesses. What is more, the WHO expects the number of poisonings to increase if pesticide use increases in the less developed countries.

Although farm and chemical workers are the groups most heavily exposed to pesticides, residents of rural and even suburban areas are often exposed to potentially dangerous levels if they live near agricultural lands. People living near pesticide-treated fields may be exposed because one-half to three-fourths of the sprayed material never reaches the ground but is carried away by light winds (**FIGURE 22-6**). This phenomenon is known as **pesticide drift**. Recent studies have also shown that pesticide applied to one crop may vaporize under sunlight and drift to neighboring crops. Exposure may also occur through contaminated groundwater. Families living near fields sprayed with herbicides and pesticides may suffer from persistent headaches, cramps, skin rashes, dizziness, high blood pressure, chest pains, persistent coughs, internal bleeding, and leukemia. They could also develop cancer or other diseases over time.

Pesticides can also be transmitted in air currents hundreds or thousands of miles from their site of use—for example, as far north as the Arctic. Pesticides are washed from the sky by rain where they enter various food chains; eventually exposing humans to potentially harmful levels.

In an effort to control mosquitoes, many cities in the United States routinely spray insecticides in neighborhoods and nearby breeding areas. In Florida, for instance, various mosquito control agencies argue that controlling the insect is vital to real estate interests and tourism because it minimizes the risk of encephalitis, a potentially deadly brain infection that is caused by an organism carried by mosquitoes.

To protect residents from West Nile Virus, which is carried by mosquitoes, many cities have started or increased pesticide spraying. Trucks and aircraft spew out tons of pesticides while residents sleep. What long-term health impact this has, if any, is unknown. Florida wildlife officials, however, think that pesticide use is largely responsible for a 70% decline in the population of snook, a popular sport fish. Adding to the problem, city officials use a variety of pesticides in city parks, and lawn-care companies and individuals douse lawns and trees with a variety of toxic substances, often incorrectly and without warning neighbors.

Consumers may also become the victims of pesticide poisonings. Although such instances appear to be rare, millions of people could be consuming high levels of pesticides on fruit and vegetables, especially in poor countries such as India. Pesticide residues on these foods often exceed safe levels. While they may not cause immediate death they could contribute to long-term illnesses such as cancer.

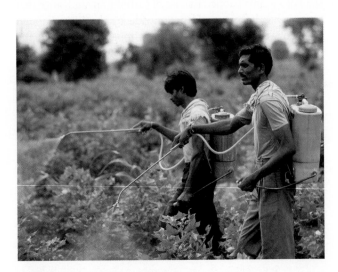

FIGURE 22-5 Farm workers at risk. Unprotected farm workers are exposed to the highest levels of pesticides.

© Joerg Boethling/Alamy.

FIGURE 22-6 Pesticide drift. Pesticide sprayed from planes contaminates the ecosystem because much of the pesticide drifts away. Various avenues for the dispersal of pesticides are shown. Average values for DDT concentrations are indicated in parts per million (ppm) and billion (ppb). Families living near sprayed fields are exposed to herbicides and pesticides.

Rain 0.1–0.3 ppb

Trade winds 0.1–0.3 ppb

Fat of man 6–12 ppm

Fat of cows 0.5 ppm

Rivers and lakes 0.001–0.2 ppb

Groundwater 0.001–0.2 ppb

KEY CONCEPTS

Pesticides contaminate many foods and have been found in human body tissues even in remote areas of the world, indicating that pesticides are globally distributed. Chemical and farm workers are frequently exposed to the highest levels, especially in the less developed nations. Residents living near farms and rural areas can also be exposed to pesticides as a result of pesticide drift. The effects of pesticide exposure range from mild neurological problems to death, depending on the exposure level and type of chemical.

The Economic Costs of Pesticide Use

Although pesticide use helps protect crops and saves us billions of dollars, pesticide use has an environmental impact. All environmental impacts have an economic cost. Human pesticide poisonings, described in the previous section, cause illness and death, both of which exact a huge economic cost. Some pesticides have been shown to damage crops. Crops have been destroyed by herbicides that were reportedly safe to apply. Organic farmers are finding that pesticides evaporate from neighboring farms, contaminating their crops. In 2003, the Bush administration adopted a policy that will make it impossible for farmers to sue manufacturers of pesticides that fail to live up to company claims.

Pesticides may also end up in groundwater in rural communities, causing a potential health risk and necessitating costly cleanup. The Monsanto Company, which produces herbicides, estimated that of 6 million wells in a surveyed area in the United States, 13% (770,000 wells) were contaminated with one or more of five herbicides; 6,600 wells

contained levels exceeding the EPA's maximum contamination level. Although the company insists that none of the herbicides poses a threat to health, it is offering well owners whose water is contaminated above specified concentrations up to $2,000 a well to make corrections.

KEY CONCEPTS

Pesticide use is not just a threat to human health and the environment, it causes considerable economic damage.

Herbicides in Peace and War

Forty-six percent of all chemical pesticides applied to crops are herbicides. Although there are over 180 types of synthetic herbicides on the market, four products are used in greatest quantity: atrazine, alachlor, butylate, and 2,4-D (2, 4-dichlorophenoxyacetic acid). Two herbicides have received most of the attention: 2,4-D and 2,4,5-T. These are nonpersistent synthetic organic compounds similar in function to plant hormones called **auxins**. When sprayed on plants, 2, 4-D and 2,4,5-T increase the metabolic rate of cells so much that plants cannot keep up with increased nutrient demands and literally grow to death.

KEY CONCEPTS

Plant auxins are hormones that stimulate plant growth. Two of the most widely studied herbicides are auxin-like compounds that cause plants to grow so fast they cannot keep up with nutrient demands.

Peacetime Uses: Pros and Cons 2,4-D and 2,4,5-T were once widely used to control brush and plants along roadways, power lines, and pipelines. They have also been used to control poison ivy and ragweed, eliminate unwanted trees in commercial tree farms and in national forests, kill aquatic weeds, and rid rangelands of brush and poisonous plants. Overall, three-fourths of these chemicals are used for weed control on farms.

The benefits of these and other herbicides are many:

1. They decrease the amount of mechanical cultivation needed to control weeds—and thus reduce labor costs.
2. They reduce weed damage when soils are too wet to cultivate, because crops can be sprayed by plane.
3. They help farmers reduce water usage because water escapes more rapidly from ground that has been tilled to control weeds.

However beneficial their use may be, herbicides also have many drawbacks:

1. Some weeds have become resistant to herbicides and have become more troublesome.
2. Some nonpest species proliferate after spraying. To get rid of them, additional herbicide or more toxic ones may be needed.
3. When herbicides are used, farmers often reduce tillage; the weeds killed by herbicides remain on the ground and provide food and habitat for insect pests. Herbicide use therefore may actually increase insect pest populations, increasing the need for insecticides. In addition, herbicides may decrease the farmer's incentive to rotate crops, an effective way of reducing insect pests.
4. Herbicides may make some crops more susceptible to insects and disease. For example, some herbicides reduce the waxy coating on plants, change their metabolic rates, and retard or stimulate plant growth; all of these effects may make plants more susceptible to insects and disease.
5. Some herbicides are toxic and may cause birth defects, cancer, and other illnesses in animals.

Critics argue that **integrated weed management** could reduce the use of herbicides. Such management would employ special equipment such as wick applicators that apply herbicide only on weeds between rows. Wick applicators consume much less herbicide than aerial spraying and create less environmental contamination. The reasonable use of herbicides could be complemented by mechanical cultivation, proper spacing of rows for healthy crops, biological weed controls, and crop rotation.

Controversy over Wartime Use of 2,4-D and 2,4,5-T Herbicides were used extensively in the Vietnam War in the 1960s and early 1970s as defoliants: to prevent guerrilla ambushes along roads and waterways; to deter the movement of soldiers through demilitarized zones and across the border of Laos; to destroy crops that might be eaten by the enemy; and to clear areas around camps. Three herbicide preparations were sprayed from planes, helicopters, boats, trucks, and portable units from 1962 to 1970.

The most effective and most controversial of the herbicides was **Agent Orange**, a 50-50 mixture of 2,4-D and 2,4,5-T (**FIGURE 22-7**). During the war, over 42 million kilograms (93 million pounds) of Agent Orange was sprayed on the wetlands and forests of Vietnam, decimating 1.8 million hectares (4.5 million acres) of countryside and at least 190,000 hectares (470,000 acres) of farmland. Over half of the mangrove vegetation of South Vietnam (1,930 square kilometers, or 744 square miles) and about 5% of the hardwood forests were destroyed. The forests alone represent about $500 million worth of wood, a supply that would last 30 years.

The prospect for these forests seems dim because hardy weed species such as cogon grass and bamboo have invaded the deforested zones. Ecologists fear that these species may greatly delay recovery or prevent it altogether. The National Academy of Sciences estimates that defoliated mangroves may take 100 years to recover because the destruction was so great that few seed sources remain. Because mangrove swamps protect coastlines from storms and because they are vital habitat for commercially important fish, numerous efforts are under way to replant the mangrove swamps in Vietnam.

Defoliants and numerous insecticides used to control mosquitoes created an ecological disaster in Vietnam,

FIGURE 22-7 Spraying Agent Orange. Vietnamese jungles were sprayed with Agent Orange to clear the vegetation.

© Department of Defense/AP Photos.

 KEY CONCEPTS

Herbicides provide many benefits but do not come without social and environmental costs. Fortunately, there are many ways to greatly reduce or eliminate herbicide use and minimize or eliminate the environmental damage.

resulting in the death of numerous fish and animals. In one survey of a heavily sprayed forest visited years after the war ended, Harvard University biologist Peter Ashton found 24 species of birds and 5 species of mammals, compared with 145 and 170 bird species and 30 and 55 mammal species in two nearby forests that had not been sprayed. The total impact of such actions will never be known.

Agent Orange may have been the cause of serious medical problems that developed in soldiers and villagers throughout Vietnam. Studies suggest that health effects attributed to Agent Orange probably resulted from contaminants in Agent Orange belonging to a group of chemicals known as dioxins. **Dioxins** cause birth defects and cancer in mice and rats. Dioxin is believed to be 100,000 times more potent than the tranquilizer thalidomide, which caused many birth defects in Europe and the United States. Soldiers from the United States and Australia who fought in herbicide-defoliated areas and were exposed to substantial amounts of the chemical developed severe headaches, nausea, diarrhea, internal bleeding, chlor-acne (a severe skin rash similar to acne), and depression.

In 1970, as a result of the public outcry in the United States, the government banned the use of Agent Orange in Vietnam. In 1985, 2,4,5-T was banned in the United States. Use of 2,4-D continues today to control hardwood trees that are considered weeds in commercial evergreen forests and along highways.

Vietnam veterans returning from the war also began to suffer unusual medical disorders. A high proportion of the men have fathered offspring that were born dead or aborted prior to term, as well as infants with multiple birth defects. Other veterans have developed lymphoma, leukemia, and testicular cancer.

Perhaps the most compelling evidence linking adverse health effects to Agent Orange came from a study carried out by Vietnamese doctors. They examined the rate of birth defects in the offspring of 40,000 Vietnamese couples and found that women whose husbands had fought in South Vietnam, where Agent Orange was sprayed, were 3.5 times more likely to miscarry or give birth to defective children than women whose husbands had remained in the north during the war.

Data recently released from an independent study of military veterans indicate an increased risk of elevated blood pressure, benign fatty tumors, miscarriage, visual and skin sensitivity to light, and depression. Other evidence has implicated dioxin as the cause of soft-tissue sarcomas (cancers); veterans exposed to dioxin had a rate of soft-tissue sarcomas seven times higher than normal.

In May 1984, manufacturers and Vietnam veterans reached an out-of-court settlement that established a $180-million fund for veterans and their families claiming injury from Agent Orange. In 1989, seven companies that were being sued by veterans for damages caused by Agent Orange agreed to pay them $240 million. Still unconvinced that the data showed a link between Agent Orange and these serious problems, the Veterans Administration (VA) has reluctantly agreed to treat all Vietnam veterans, a change in policy that may be too late for many of the soldiers.

KEY CONCEPTS

Extensive use of chemical defoliants (Agent Orange) during the Vietnam War has resulted in substantial environmental and health impacts in soldiers and their offspring. The health effects are attributed to dioxins that contaminated the defoliant.

22.2 Controlling Pesticide Use

Concern over the health and environmental effects of pesticide use has led to a number of actions to minimize impacts, including outright bans of harmful pesticides, registration, and establishing tolerance levels on produce. This section discusses these actions and their strengths and weaknesses.

Bans on Pesticide Production and Use

In the United States, caution regarding pesticide use began in 1962 with the publication of the late Rachel Carson's book *Silent Spring*, which pointed out many of the real and potential impacts of pesticide use (**FIGURE 22-8**). This widely read book is credited with touching off the controversy that resulted in increased public awareness and research on the effects of the indiscriminate use of pesticides. Concern reached a peak a decade later with the near extinction of peregrine falcons, brown pelicans, cormorants, bald eagles, and other bird species exposed to DDT.

Go Green

Lobby your college or university food service to use organic fruits and vegetables produced by local farmers, if possible.

Faced with the growing body of evidence that illustrated the many hazards of using chemical pesticides, environmentalists pushed for bans on certain pesticides. Scientists have found that such bans have greatly benefited U.S. wildlife. The ban on DDT in the United States, for instance, has resulted in increases in the number of endangered bald eagles, brown pelicans, ospreys, and other species. Recent studies show that DDE and DDT levels have dropped in wild populations and that normal reproductive rates have returned. Affected bird populations are recovering nicely and some may be moved off the endangered species list.

Bans on the production and use of pesticides in the United States, however, are only part of the answer. Continued use of pesticides outside the United States, as in the case of carbofuran, poisons migratory species such as songbirds that winter in Central America and Mexico but spend spring and summer in the United States. In addition, much of the produce imported into the United States has recently been found to be contaminated with pesticides, many of which were banned in the United States. Because one-fourth of the fruit and vegetables sold in the United States comes from foreign countries, global bans of harmful pesticides are needed to protect human health.

FIGURE 22-8 Antipesticide crusader. Scientist Rachel Carson first drew attention to the dangers of pesticides in her environmental classic Silent Spring.

© AP Photos.

Unfortunately, only 21% of all pesticides sold in the U.S. have been adequately tested for carcinogenicity. Less than 10% have been tested for their ability to cause genetic mutations and less than 40% have been tested for their potential to cause birth defects.

KEY CONCEPTS

Many environmentally harmful pesticides have been banned in the United States in the past 30 years. Some of these pesticides, however, continue to be used in other countries, where they poison wildlife and migratory birds and contaminate crops destined for local markets as well as markets in the United States and other more developed countries.

Registering Pesticides

Another way of controlling pesticides is the act of registration, described shortly. In the United States, pesticide registration was first required by the **Federal Insecticide, Fungicide, and Rodenticide Act (FIFRA)**, a law currently administered by the Environmental Protection Agency (EPA). The FIFRA is designed to protect farmers, farm workers, the general public, and the environment from new chemical pesticides. It does this by requiring manufacturers to test new pesticides for a range of effects before they can be registered for use. Pesticides introduced prior to the passage of the law in 1972, however, were originally not affected.

Registration is a kind of permitting process managed by the EPA. When a company develops a new pesticide that it wants to market, it must perform a number of tests on plants and animals to assess the potential toxicity of the chemical. Test plots are also sprayed with the pesticide to determine residues—that is, how long the chemical remains. These data are then submitted to the EPA scientists for their review. Taking into account the average American diet, residue levels, and toxic effects, the EPA then determines on which crops, if any, the pesticide can be used.

The critical thinking skills you have learned may suggest a problem right away: Is the average American diet accurately determined? If the truth be known, the average American diet includes a great deal of red meat, chicken, and other meat products. That automatically excludes vegetarians, however, who tend to consume large quantities of fruits and vegetables. Ironically, many vegetarians who select a special diet for health reasons are inadvertently exposed to the highest levels of pesticide—unless, of course, they consume organic produce. **Organic produce** includes fruits and vegetables grown without pesticides or synthetic fertilizers. Most grocery stores contain a small selection of organically grown produce. In a number of states large grocery stores such as Wild Oats specialize in healthy food, including organically grown fruits and vegetables, as do many smaller health food stores.

Pesticides are registered for general or restricted use. General use means that anyone can purchase and use them. Technically, restricted-use pesticides are to be used by licensed applicators—farmers and lawn-care companies, for instance.

The problem with this system is that pesticide use is largely regulated by an honor system. Labels on restricted-use products describe their legal uses, and only licensed applicators can buy them; other than those limitations, there's very little if anything to stop people from using pesticides any way they please.

In 1988, the FIFRA was amended to correct some of its weaknesses. One of the most significant gains was a plan to register many pesticides that had been introduced before the EPA took over the process in 1972. Pesticide registrations in the 1950s and 1960s were made with very little if any sound toxicological data, say critics. Registering the over 300 chemical pesticides will take many years to complete and will be funded by fees paid by the chemical companies. Some experts see this as a weeding-out process. They believe that many pesticides will be canceled as companies that are afraid their product won't be approved for health and environmental reasons won't invest the money needed to register them.

Despite this improvement, critics say that additional reform is necessary. For example, pesticide registration does not currently require manufacturers to test for neurotoxicity—toxic effects on the nervous system, the brain, spinal cord, and nerves. To many critics, this is a glaring omission because about half of the pesticides (especially the organophosphates) are insect neurotoxins that could also affect the human nervous system.

New research also shows that some pesticides damage the immune system, the body's defense against bacteria, viruses, and even cancer. Pesticide registration, however, requires no test of immune system effects—another omission, say critics. In addition, pesticide registration does not take into account possible synergism.

Perhaps the most glaring problem, though, is that the FIFRA provides virtually no monitoring functions. The end users are free to do as they please. They can apply as much pesticide as they want and can apply it wherever they want. Moreover, there is no one to see that solutions are mixed correctly or that equipment is working properly. As a result, farmers frequently apply much more than is needed.

Licensing and training are the only avenues available to address this problem. To be licensed, most states require farmers to take a test. In Colorado, for example, farmers must study a booklet provided by the EPA and then take an open-book test. If they pass, they're licensed to spray restricted pesticides on their fields. Critics argue that more rigorous education and testing are needed.

Governments, pesticide companies, farmers' groups, and universities could provide additional hands-on training and monitoring to be sure that pesticides are used more safely. As mentioned earlier, programs that teach farmers to become better at identifying pests and monitoring pest populations in the field could decrease insecticide use and costs. Special crop scouts, such as those now in use in Indonesia, could be trained to monitor pest populations and determine when they have reached the threshold level. They could also assist farmers in finding alternative ways of controlling pests. (See **Spotlight on Sustainable Development 22-1** for more on Indonesia's program.)

Establishing Tolerance Levels and Monitoring Produce

To help protect public health, the FIFRA also authorizes the EPA to set **tolerance levels** for pesticides on fruits, vegetables, and other foods. Tolerance levels are concentrations in or on foods that are believed to pose an acceptable health risk. For cancer, the EPA sets the concentration at a level it thinks will cause no more than one additional cancer death in one million people. This determination is fraught with difficulty as Interestingly, although the U.S. EPA sets tolerance levels for pesticides on foods, there is no limit to the number of different pesticides that can be present. Chemical pesticides can have synergistic effects at very low levels, and little research has been performed to assess potential effects.

Although the EPA sets tolerance levels, it is up to the U.S. Food and Drug Administration (FDA) and state agricultural agencies to monitor the nation's food supply and to enforce tolerance levels. Inspectors examine fruits and vegetables and have the authority to seize and condemn foods containing residues that exceed EPA levels or that contain illegal pesticides—but only when they are shipped from one state to another. Inside the states, the responsibility for monitoring food lies with state agencies.

Both the FDA and state agencies suffer from a common problem: a lack of funds and a shortage of inspectors. Given these problems and the massive amount of food consumed by the American public, only a small portion of our food is tested. Furthermore, examiners test only for the presence of a handful of the pesticides. Unless you grow your own food, you are probably consuming small amounts of many pesticides.

Similar problems occur in other countries. In Ireland, for instance, the Pesticide Control Service has only 11 scientists to test food for the entire country. Consequently, few farmers are inspected, and only 2,000 food samples are tested annually.

Currently, the EPA reported that at least 45 pesticides on food are thought to be carcinogenic at least to nonhuman animals. In 1987, the National Academy of Sciences issued a report concluding that 1 million Americans alive today will develop cancer as a result of pesticide contamination of their food—that's one of every 250 Americans. Add to that possible birth defects, miscarriages, mutations, neurological effects, and other milder symptoms, and it is little wonder that the EPA ranked pesticides in food as one of the nation's most serious health concerns.

Critics of pesticide regulations argued that the tolerance levels set for pesticide residues were inadequate because they failed to take into account the special diets of vegetarians and children.

In 1996, the U.S. Congress strengthened pesticide regulations by passing the **Food Quality Protection Act**. It requires the EPA to estimate risks of pesticides from all sources, including drinking water, food, and home uses. It also requires the EPA to determine risk to infants and children and build in protection for these age groups. Historically, however, critics point out that the EPA rarely revises tolerance levels when new scientific data about risks become available. Also, it rarely bans a pesticide if it is harmful to wildlife but not to people.

Some additional reforms in pesticide controls were also included in the recent legislation. For example, the EPA is directed to consider health risks other than cancer and to consider hormone-mimicking properties of pesticides. Some pollutants mimic naturally occurring hormones and can dramatically alter reproductive and other functions in fish, wildlife, and perhaps even humans.

22-1 Indonesia Turns to Biological Pest Control

Indonesia is a country of islands—nearly 14,000 of them—in Southeast Asia. In 1983, this rural nation, once the world's leading importer of rice, succeeded in growing enough rice to feed its own people. New strains of rice, fertilizers, and an intricate irrigation system deserved much of the credit for the success. In 1985, however, the notorious brown planthopper threatened the progress of the previous years. This insect causes rice to dry out, rot, and fall in the field. Infestations of the insect can cause enormous damage.

To combat the planthopper, in 1985, the government decided to try integrated pest management (IPM). It was advised to do so by an Indonesian entomologist, Dr. Ida Oka, who had received his Ph.D. from Cornell University under David Pimentel, a leading expert on natural insect control. Oka had been in charge of pest management in the late 1970s and early 1980s and had implemented sound IPM for rice and other crops. During that time pesticide use had been greatly reduced, and rice yields had soared. Oka had resigned, however, when a new minister of agriculture was appointed. This official's pro-pesticide policies, in which the government paid about 85% of the cost of pesticides to farmers, resulted in widespread use of chemical pesticides and clearly placed the farmers on the pesticide treadmill. This policy was largely responsible for the outbreak of the brown plant-hopper in 1985.

Indonesian scientists had found that the use of pesticides to control the planthopper and other insects killed many beneficial insects that preyed on the pest. One of the beneficial insects destroyed in the spraying was the wolf spider, which can devour 5 to 20 brown planthoppers a day. If left alone, the beneficial insects could often control the harmful ones. Researchers also found that farmers sprayed fields regularly, regardless of whether they needed it. The overuse of pesticides actually increased the severity of infestations, so much so that by 1986 the country was in danger of becoming a rice importer once again.

Convinced of the danger of the continuing use of pesticides, the Indonesian government asked the United Nations (UN) Food and Agricultural Organization (FAO) to help it promote an IPM program. In 1986, with the help of the FAO, the government embarked on a program to educate farmers on IPM and the dangers of pesticides. Experts ventured into the rice paddies, where they showed farmers how to diagnose problems, calculate the ratio of good bugs to bad ones, and decide how much damage the crop could stand without a decrease in yield. They then taught farmers ways to reduce spraying and methods to protect beneficial predatory insects.

Early results showed that IPM worked. IPM reduced pesticide use substantially. Trained farmers, for example, apply one-ninth as much pesticide as they did before training, with no decrease—sometimes even an increase—in crop yield. Farmers have also learned to discern more carefully insect damage from fungal damage, and this helps to reduce pesticide use.

The average yield on farms using pesticides was 4% lower than on fields controlled by IPM. Despite the high subsidies for insecticides, the farms using IPM proved more profitable than those sprayed more frequently. The government saves an estimated $120 million a year on pesticide subsidies. Indonesia's streams and wildlife also started showing signs of recovery.

The success of the pilot program in Indonesia convinced the government to adopt IPM as a national pest control strategy. The government, in fact, banned 56 of the 57 pesticides previously approved for farming in Indonesia, in order to help protect predatory insects. The government then launched a massive campaign to educate Indonesia's farmers in IPM. In 1992, the country hired 2,000 crop scouts to educate farmers on IPM and hoped that all farmers would soon be using this method.

Since then, integrated pest management has spread to Thailand, Bangladesh, Sri Lanka, Malaysia, India, and China. Farmers in Indonesia who were not introduced to the technique raised their voice. They wanted to be included in this effort to help the world move to a more sustainable, environmentally safe form of farming. Agricultural experts believe that IPM could be used on fields that provide 45% of the rice for people living in southern and southeast Asia and could save millions of dollars, preserve wildlife, and protect human health without endangering high crop yields.

Another problem is that the chemicals that pesticides are mixed with or dissolved in may also be toxic. In fact, studies of pesticides used indoors show that some of these substances reach rather high levels—and remain high indoors for a considerable length of time after a house has been sprayed for cockroaches, ants, or other insects.

The next section shows that there is a way to greatly reduce or even eliminate pesticide use. It is called integrated pest management. In the interim, individuals can avoid some produce containing pesticides by growing some of their own fruits and vegetables or by purchasing organically grown produce. Washing fruits and vegetables can help but won't eliminate all pesticide residues. Beyond that, individuals can exercise their democratic prerogatives by writing local, state, and federal officials and asking for tougher laws to regulate pesticide registration and use.

 KEY CONCEPTS

The EPA also sets tolerance levels—acceptable levels of pesticides on food—which are monitored by the U.S. Food and Drug Administration. Unfortunately, inadequate funding prohibits thorough sampling.

Weakening Protection

Federal pesticide regulations have been refined over the years, providing more and more protection to people, wildlife, and the environment. Some states have even increased protection by passing more stringent regulations. Other states, however, are working to weaken pesticide regulations. Congress is currently considering legislation that could result in more pesticide contamination of streams and rivers from commercial timber operations and a host of other actions.

22.3 Integrated Pest Management: Protecting Crops Sustainably

Integrated pest management (IPM) is a new strategy of pest control gaining popularity throughout the world. It depends on four sustainable means of pest control: environmental, genetic, chemical, and cultural. Two or more of these approaches may be used simultaneously to control pests, with little or no damage to the environment and human health. Studies show that they also offer important economic benefits. The central goal of integrated pest control: to reduce pest populations to levels that do not cause economic damage, while protecting human health and the environment. The goal is not to eliminate pest species entirely, which may be impossible anyway.

 KEY CONCEPTS

Integrated pest management entails a set of alternative strategies to control pest populations at manageable levels. It offers substantial social, economic, and environmental benefits and is the cornerstone of a sustainable system of agriculture.

Environmental Controls

Environmental control is a term used to designate a number of techniques that alter the biotic and abiotic conditions in crops, making them inhospitable to pests. They generally rely on knowledge more than on technology.

KEY CONCEPTS

Environmental controls seek to change both the biotic and abiotic conditions of crops in ways that reduce the growth of pest populations, while causing little if any damage to the crop itself.

Increasing Crop Diversity: Heteroculture and Crop Rotation Planting single crop species (monocultures) generally promotes pest and disease outbreaks. Crop diversity, on the other hand, reduces the amount of food available to any one pest and helps prevent

  Go Green

Buy organic fruits and vegetables, especially lettuce, beans, spinach, celery, peppers, apples, peaches, and pears, which are not directly sprayed with pesticides and not peeled.

such rapid population growth. Several techniques can increase crop diversity, among them heteroculture and crop rotation.

A farmer who plants several crops side by side in his field, rather than huge expanses of one crop covering, is practicing **heteroculture**. This simple but effective measure works because it provides environmental resistance to pests; that is, pest populations are often much smaller in heterocultures than in monocultures because there is less to eat. In addition, some crops harbor predatory insects that feed on pests in nearby crops. Corn and peanuts grown in adjacent fields, for instance, can reduce corn borers by as much as 80%. Part of the reason for this success may be that predatory insects that feed on the corn borer live in peanut crops.

One of the most recent and innovative ways of intermixing crops and **strip cropping** (**FIGURE 22-9**). In this technique, alternative strips of corn and soybeans (or other crops), each with a dozen or more rows, are planted side by side in the same field. This not only decreases insect pests but also increases yield because the corn protects

FIGURE 22-9 Strip cropping. Alternating strips of alfalfa with corn on the contour protect this crop field in northeast Iowa from soil erosion and increase productivity of the crops as explained in text.

Courtesy of Tim McCabe/NRCS USDA.

the soybeans from wind while the openness of the field provides more sunlight to the corn.

Heteroculture decreases pesticide use and also provides a means of diversifying farm production from year to year. In some ways, then, it is a form of insurance. Ron Rosmann, for example, plants half of his 200-hectare (500-acre) farm in corn and soybeans. The rest he devotes to hay, pasture, cattle, hogs, chickens, and a tree nursery to help pay for his three sons' college educations. A bad year for corn will not wipe him out.

Heteroculture can be practiced by home gardeners with great success. I have been growing vegetables in a pesticide-free garden for nearly 40 years and have had virtually no trouble with insects, in large part because I intermix species. Small patches of carrots are planted next to peas, which are next to spinach, and so on. I also interplant onions, marigolds, and other species that repel pests.

Crop rotation, a means of reducing soil erosion and increasing soil fertility, also helps control pests, for at least two reasons. First, the healthier the soil, the healthier and more resistant the plants are to insects and disease. Second, it also helps hold down pest populations because it reduces food available from one year to the next for specialized insects—those that feed on only one crop. For instance, wireworms feed on potatoes but not alfalfa. Therefore, if potatoes and alfalfa are alternated from year to year in the same field, wireworm offspring that hatch in the alfalfa patch will have little to feed on. Their numbers will decline. When potatoes are planted the next year, few wireworms will be around. Although the population may increase during the growing season, it will generally not reach a level that causes harm. The next year alfalfa is planted, and wireworm offspring once again perish. Gardeners can also practice crop rotation on a small scale to hold insects in check.

Altering the Time of Planting Some plants naturally escape insect pests by sprouting early or late in the growing season. A good example of this adaptation is the wild radish, which sprouts early in the season before the emergence of the troublesome cabbage maggot fly.

Agriculturalists can use their knowledge of an insect's life cycle to their advantage by coordinating plantings with the expected date of hatching. A slight delay in planting of wheat, for example, helps protect it from the destructive Hessian fly. In general, if a pest emerges early in the spring, planting can be delayed to avoid that pest (within the limits of the growing season). Without food, the pest will perish.

Home gardeners can also foil pests by planting seedlings instead of seeds. Certain crops such as lettuce can be mowed down by hungry slugs and other insects when they first emerge from the ground. If larger seedlings are planted, though, the plant has a better chance of surviving.

Altering Plant and Soil Nutrients The levels of certain nutrients in soil and plants can also affect pest population size. By regulating soil nutrients, then, a farmer may be able to control pests. Nitrogen is one of the important nutrients that insects and parasites derive from plants. Too much or too little nitrogen can alter the population size of various pests. For example, grain aphids reproduce better on grains high in nitrogen. Other insects, such as the greenhouse thrip and mites, do poorly on high-nitrogen spinach and tomatoes, respectively. Once again, knowledge reigns supreme. A little knowledge of pest nutrient requirements, soil nutrient levels, and plant nutrient content can become a useful ally.

Controlling Adjacent Crops and Weeds All crops are surrounded by forests, meadows, or other crops. Each of these houses harmful as well as beneficial insects. To reduce pest damage, farmers often avoid planting crops that provide food and habitat for harmful pests next to other crops. Those that provide habitat for beneficial insects are often encouraged. Farmers have even used low-value crops to attract pests away from more valuable crops. The former are called **trap crops**. Alfalfa is a good example. When planted adjacent to cotton, it lures the harmful lygus bug away from the cotton plants, which prevents serious damage to the cotton. Some farmers may even spray the alfalfa with pesticide to get rid of the bug, using far less chemical than would be necessary if the entire cotton crop had to be sprayed.

KEY CONCEPTS

Adjacent weeds or crops may harbor beneficial or harmful insects. Controlling what grows beside a given crop can therefore be a powerful tool in protecting crops with little or no pesticide use.

Biological Control: Introducing Predators, Parasites, and Disease Organisms In nature, thousands of potential insect pests never become pests because of natural controls exerted by predators, diseases, and parasites. These are biotic components of natural environmental resistance. Farmers can capitalize on this knowledge to manage a variety of pests including weeds, insects, and rodents.

To date, scientists have documented well over 500 examples of partial or complete control of insects and weeds through natural predators, parasites, or other biological agents. This technique is generally referred to as **biological control**. One classic example of the effectiveness of biological control is the control of the prickly pear cactus in Australia.

The prickly pear was introduced into Australia from its native Mexico. By 1925, over 24 million hectares (60 million acres) of land had been badly infested. Half of this land was abandoned because of the thick carpet of cactus. Farmers introduced a cactus-eating moth to Australia in 1925 to eradicate the pest, and 7 years later, much of the land had been cleared and could once again be used (**FIGURE 22-10**).

The predatory lady beetle was introduced into California from Australia in the 1880s to control an insect that destroyed citrus trees. Parasitic insects from Iran, Iraq, and Pakistan have been introduced to control the olive scale, an insect that once threatened the state's olive trees. Both lady beetles and the predatory insects now exert complete control on their prey, keeping their populations at manageable levels without the use of pesticides.

A few insect pests can also be controlled by birds, a natural control organism whose potential has been overlooked.

(a)　　　　　　　　　　　　**(b)**

FIGURE 22-10 Prickly pear cactus invasion. (a) A prickly pear cactus infestation in Queensland, Australia. **(b)** The same area after the introduction of the cactus-feeding moth.

Photos courtesy of Queensland Department of Agriculture, Fisheries and Forestry, Brisbane, Australia.

Brown thrashers can eat over 6,000 insects in one day. A swallow consumes 1,000 leafhoppers in 12 hours, and a pair of flickers can snack on 500 ants and go away hungry. In China, thousands of ducklings are driven through rice fields; in some places they reduce the populations of insects by 60 to 75%, allowing farmers to reduce insecticide use considerably. Their droppings also provide fertilizer for the crops. We use ducks and chickens to help reduce insects including ticks in our gardens and around our home (**FIGURE 22-11**).

Bacteria and other microorganisms can be brought to bear on pests. One common example is the bacterium *Bacillus thuringiensis* (bah-SILL-us thur-in-GEEN-siss) or BT, used to control many leaf-eating caterpillars. Cultivated in the lab and sold commercially, it is available as a powder that is either dusted on plants or mixed with water and then sprayed on plants. Caterpillars that eat the bacteria die because BT produces a toxic protein that paralyzes the digestive system of insects. Humans and other organisms are usually unaffected. I use BT routinely in my garden to control cabbage butterflies.

BT is used by organic gardeners with considerable success. It has been sprayed in China to control pine caterpillars and cabbage army worms. In California, it has been used for many years to control various crop-eating caterpillars, and it is currently applied in the northeastern United States to help control gypsy moths, which devastate forests. Another strain of BT has been employed in the battle against mosquitoes in Colorado and other states. The use of BT and other techniques has resulted in a measurable reduction in insecticide use in certain crops, especially almonds and tomatoes.

Researchers have also successfully inoculated corn plants with genetically altered bacteria containing the BT gene. The bacteria multiply in the corn plant as they grow, and they kill European corn borers that feed on the stalks. Studies show that the bacteria do not migrate into the corn kernels.

FIGURE 22-11 Free-range chickens. These are a handful of the author's 50 free-range chickens. They roam the property feeding on vegetation and insects like ticks, producing the healthy eggs, rich in heart-healthy omega-3-fatty acids. The author sells the eggs through several local farmers' markets.

Courtesy of Dan Chiras.

Viruses and fungi may be used similarly. In Australia, after years of fruitless efforts to control rabbits, scientists introduced a pathogenic myxoma (mix-OH-ma) virus, which eliminated almost all the rabbits within one year. Unfortunately, the virus has evolved to an nonvirulent form, and the rabbit has evolved resistance. Control is no longer as effective as it once was.

Biological control agents must be developed with caution to ensure that they do not pose a threat to humans, livestock, and natural ecosystems. One of the major concerns has to do with the introduction of alien species into new environments. Careful testing is necessary to be sure that an organism introduced to control a pest does not become a pest itself. In the early 1980s in sub-Saharan Africa, for example, an insect known as the mealybug became a major pest, attacking cassava plants. This plant produces an edible root that is a staple for about 200 million people. To find a control for this troublesome pest, researchers scoured the mealybug's South American homeland for natural enemies. They eventually located a small parasitic wasp that injects its eggs into the larvae of the mealybug. When the eggs hatch, they devour the larvae. Before the researchers could release the wasp, though, they had to perform extensive tests to see if the wasp would survive in its new home and if it would become a pest itself. Convinced that it wouldn't, the bug was first released in 1986 and today is providing protection in 24 African countries. So far, the wasp seems to be working fine.

Another undiscussed problem of biological controls is that target organisms can develop genetic resistance to them, as did the Australian rabbits mentioned earlier. Researchers in Kansas recently found that larvae of the Indian meal moth, which feed on grain stored in sheds and bins, develop genetic resistance to BT. In such cases

new controls could be introduced or alternated with BT. In some instances, biological control agents themselves may undergo genetic changes that offset the newly acquired resistance of the pest. This process is called **coevolution**. As yet, there is no record of such changes in biological control agents, but some scientists think that they are inevitable.

KEY CONCEPTS

Many natural biological control agents can be used to prevent the outbreak of pest species, including the introduction of natural predators, birds, bacteria, viruses, and fungi.

Genetic Controls

Integrated pest management includes two major genetic control strategies, the sterile male technique and the breeding of genetically resistant plants and animals. Both are important components and can be used in conjunction with other methods.

Sterile Male Technique   As the previous examples have shown, a little knowledge of biology can go a long way. Another example in which knowledge is brought to bear on pest control is the **sterile male technique**. As the name implies, this technique involves the introduction of sterile males of insect pests into the environment. Males are raised in captivity and sterilized by irradiation or by exposure to certain chemicals. The sterilized males are then released in large numbers in infested areas. The males far outnumber the fertile wild males and thus account for a large percentage of the matings with wild females. Because many insect species mate only once, eggs produced by such a union are infertile. Thus, if the population of sterilized males greatly exceeds that of the wild males, most of the eggs will be infertile. Insect populations can be brought under control swiftly.

The sterile male technique has been used effectively against several species of insect pests, including the screwworm fly in Mexico and the United States, the Mediterranean fruit fly in Capri, the melon fly on the island of Rota (near Guam), and the Oriental fruit fly on Guam.

The sterile male technique has not always succeeded. It has, for instance, proved unsuccessful in mosquito control. Scientists believe that the chief reason for its failure in this instance is the lower sexual activity of sterilized males compared with wild males. Other reasons may include an inadequate number of sterile males, the in-migration of additional pests, and impatience on the part of state agricultural agents. Some researchers also suggest that through natural selection a new strain of insects may evolve that recognizes and avoids sterile males.

Despite these problems, the sterile male technique is an important tool in integrated pest management. It is species specific, can be used with environmental controls, and can be effective in eliminating pests in low-density infestations.

Developing Resistant Crops and Animals Many species are naturally resistant to pests and disease organisms. However, crop species often have lost their resistance through years of special breeding programs aimed primarily at increasing yield. Many geneticists think that introducing genes that provide genetic resistance to crops can help reduce or eliminate pesticides. For example, researchers have found that certain oils in the skins of oranges, grapefruits, and lemons are highly toxic to the eggs and larvae of the Caribbean fruit fly, which lays its eggs in the skins of these fruits. The flies' larvae destroy the fruit, but scientists may now be able to breed citrus selectively to increase the amount of toxic oils in their peels. By creating natural resistance, this technique eliminates the need for pesticides.

Cornell University scientists are developing a potato plant whose leaves, stems, and sprouts are covered with tiny, sticky hairs that trap insects and immobilize their legs and mouth parts. Field tests show that this plant can reduce by half the infestation of green peach aphids, which (despite their name) also attack potatoes. The new variety was developed by crossing cultivated potatoes with a wild species with sticky hairs, which grows as a weed in Bolivia.

Other genetic research has led to insect-resistant wheat soybeans, alfalfa, cotton, and potatoes. Work on chemical factors that attract insects to plants may help scientists selectively remove them to make plants unappealing.

The Monsanto Company is working on another promising weapon in the fight against pests. Robert Kaufman and his colleagues have isolated the gene that gives BT its pesticidal action. The scientists have transferred that gene to another bacterium, *Pseudomonas fluorescens*, which lives on the roots of corn and several other plants. The transplanted gene renders the host bacterium lethal to insects and other organisms, such as the black cutworm, that feed on the roots of commercially important crop species. Simply by planting seeds that have been pretreated with *P. fluorescens* bearing the toxic gene, farmers may be able to provide long-term protection without many of the dangers of pesticides. However, widespread use of BT in corn and other crops may result in the development of resistance and loss of the BT control.

Monsanto hopes that more insecticidal genes can be added to *P. fluorescens* in the years to come, giving corn a wider range of protection, reducing chemical pesticide use

and in the process protecting wildlife from the harmful toxic pesticides.

Root-zone protection is not the only strategy that geneticists are developing. Numerous bacteria colonize aboveground plant parts; fitted with insecticidal genes from BT and other naturally occurring biological agents, these bacteria could create a protective barrier to ward off dozens of insect pests.

Chemical Controls

Chemical controls are also be a part of IPM. These chemicals are conventional pesticides and a new breed of natural (presumably nontoxic) substances. First, though, we look at the conventional pesticides.

Reducing the Use of Second-Generation Pesticides Even with wider use of biological control agents and other strategies of IPM, second-generation pesticides will likely remain a part of our pest control strategy for many years to come. However, several principles should guide their use: (1) They should be applied sparingly; (2) they should be applied at the most effective time to reduce the number of applications; (3) they should destroy as few natural predators, nonpest species, and biological control agents as possible; (4) they should not be applied near drinking water supplies; (5) they should be carefully tested for toxic effects; (6) they should be avoided if they are persistent and bioaccumulate; (7) they should be used in ways that reduce exposure to workers and nearby families; and (8) they should be used to reduce populations to low levels, with environmental, genetic, and cultural control measures used to maintain populations at low levels.

One way to minimize the use of insecticides is to spray only affected areas. Insects, for example, may infest a small portion of a crop. If possible, only that portion should be sprayed. A technique that is useful for herbicides is the use of special wick applicators, rather than sprayers, which deliver a small dose directly to the target species with minimum environmental contamination (**FIGURE 22-12**).

Timing of application also helps farmers reduce pesticide use. Consider an example. The red spider mite can be kept under control in apple orchards by applying insecticides early in the season, well before the mite's natural predators emerge. Throughout the rest of the season, the natural predators will keep the spider mites under control. No further pesticides are needed. By using a similar approach on cotton, researchers at Texas A&M have cut pesticide use by 70% while maintaining normal cotton crop production.

FIGURE 22-12 Wick applicator. Instead of spraying the entire crop with herbicide, farmers can use this device to apply pesticides only to the weeds growing between the rows.

Courtesy of Lynn Betts/NRCS USDA.

Better monitoring of crops also help reduce pesticide use. Some farmers spray their entire crop when they encounter any pests. They don't even bother to see how extensive the infestation is. In many cases, only a tiny portion of the crop needs spraying, a discovery that could result in considerable savings of time and money.

Others spray on a routine schedule, often provided by the pesticide salesperson, without even checking to see if insect pests are present in harmful numbers. Schedules may not reflect the needs of farmers' crops. Were farmers to check, they might find that the pests were under control—or weren't even there. A little monitoring might save lots of time, money, and environmental contamination.

Successful IPM requires a better understanding of insect biology, as well as better skills in recognizing and counting insects in a farmer's fields. By knowing what insects are present and where they are in a field, farmers can become wiser participants in the ecosystem they manage.

Finally, more developed nations have an important role to play by helping less developed countries develop a sustainable pest management program through technical and financial assistance.

KEY CONCEPTS

Conventional pesticides will remain in use, but they must be screened carefully to avoid harmful ones. Those that are used must be administered with caution—in amounts and at a time when they are most effective to minimize their environmental impact. This requires better monitoring of crops and infestations.

Third-Generation Pesticides This section discusses a whole new arsenal of chemicals that are produced in nature and could, with a little ingenuity, be applied to pest control on a large scale. These nontoxic agents, such as natural chemical repellants, could, in conjunction with measures outlined previously, displace potentially harmful second-generation pesticides. This new class of chemical compounds is sometimes referred to as the **third-generation pesticides.**

KEY CONCEPTS

Scientists and farmers are exploring a whole new group of naturally occurring chemicals to control insect pests. These could eventually become the cornerstone of pest management in a sustainable system of agriculture.

Pheromones Insects and other animals release chemicals called **pheromones** (FAIR-eh-moans). Pheromones provide a chemical means of communication. One well-known group of pheromones is the **sex attractants**, which are emitted by female insects to attract males at the time of breeding. Effective in extraordinarily small concentrations, pheromones draw males to females. This evolutionary adaptation ensures a high rate of reproductive success.

Some sex attractants are now produced commercially and are available for pest control. Some of these substances are used in **pheromone traps** that lure males. These traps may contain a pesticide-laden bait or a sticky substance that immobilizes insects (**FIGURE 22-13**). Pheromone traps of various sorts have been used to control at least 25 insect species and can be used with other IPM methods.

FIGURE 22-13 Pheromone trap. This trap contains a sticky substance to immobilize male gypsy moths, which are drawn to it in search of mates by chemical sex attractants known as pheromones.

© Nigel Cattlin, Holt Studios International/Science Source.

Pheromone traps can also be used to pinpoint the time when pests appear. By knowing precisely when insect pests emerge, farmers can time their pesticide applications for maximum effectiveness. This technique helps reduce the amounts of pesticide applied.

Another technique for controlling insects with pheromones is known as the **confusion technique**. In this method, pheromones are sprayed on crops at breeding time. Unsuspecting males are drawn by the pheromone in all directions. Fertile females hardly stand a chance against this flood of sex attractant pheromone and may never find a partner. One modification of this technique involves the release of wood chips treated with sex attractants. Males are attracted to the wood chips and may attempt to breed with them.

Finally, pheromones can be used to lure beneficial insects out of fields. Once they are gone, pesticide sprays can be applied.

The use of pheromones offers several advantages over second-generation pesticides. They are, for example, nontoxic and biodegradable and therefore not expected to have any significant environmental impacts. They can be used at low concentrations. Third, they are highly species-specific. The major disadvantage is the high cost of developing them.

juvenile and molting hormones disrupts an insect's life cycle, sometimes resulting in death. For example, larvae treated with juvenile hormone are prevented from maturing and eventually die. If given molting hormone, they will enter the pupal stage too early and die. Interestingly, some plants contain chemicals structurally similar to juvenile hormone, an evolutionary adaptation that helps protect them from hungry insects. When ingested by larvae, these chemicals prevent the larvae from pupating. This, in turn, prevents formation of the adult form that produces eggs and additional generations of larvae.

Insect hormones applied to crops offer many of the same advantages that pheromones offer, including biodegradability, lack of toxicity, and low persistence in the environment. Like pheromones, however, they are costly. Insect hormones act rather slowly, sometimes taking a week or two to eliminate a pest, by which time extensive damage may have been done. In addition, insect hormones are not as species specific as pheromones and therefore may affect natural predators and other nonpest species. The timing of application is also critical, for hormones are effective only at certain times in an insect's life cycle.

Researchers recently discovered a plant from Malaysia that produces juvenile hormone. They hope that the

KEY CONCEPTS

Pheromones are natural sex attractants released by insects. They can be used to draw insects into traps or can be sprayed on fields to confuse males so they cannot find females with which to mate. These substances are nontoxic and biodegradable.

Insect Hormones   The life cycle of many insects is shown in **FIGURE 22-14**. As illustrated, adult insects lay eggs, which develop into larvae (caterpillar stage). The larvae are voracious eaters and are often the most troublesome form of pest species. Eventually, the caterpillar spins a cocoon in which it undergoes an amazing change, transforming into a flying form such as a moth or butterfly—the adult form.

The entire life cycle of insects is regulated by two hormones, **juvenile hormone** and **molting hormone. Hormones** are chemical substances that are produced by specific cells in the body and travel through the bloodstream to distant sites, where they exert some effect. Altering the levels of

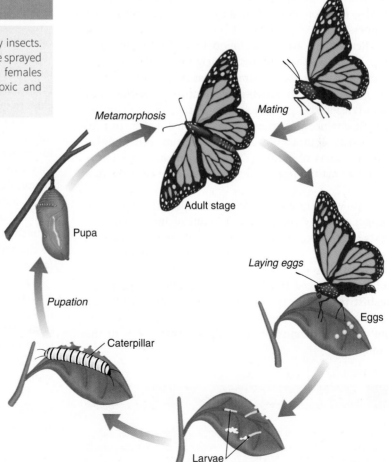

FIGURE 22-14 The insect life cycle.

genes responsible for the production of this hormone can be transferred to commercially important crops, offering another avenue of protection.

KEY CONCEPTS

Two hormones control the life cycles of insects and can be sprayed on crops to alter these cycles, leading to the death of pests. Unfortunately, to be effective, application must be precisely timed and these hormones take a fair amount of time to reduce pest populations.

Natural Chemical Pesticides Natives of the South American tropics have used the seeds and leaves of the neem tree for many years to control pests. Researchers have found that this tree produces chemicals that kill or repel a variety of insects. This extract may become useful in the control of larvae that feed on vegetables and ornamental crops.

Egyptian researchers found that flies ignored a species of brown algae left out on a counter to dry. Curious, researchers extracted a mixture of chemicals from the algae and found that they, too, repel a variety of insects that attack cotton and rice. Natural chemicals such as these may prove useful in years to come. Like other third-generation pesticides, they are biodegradable and nonpersistent.

Researchers have also found that some plants produce chemicals that alter insect metabolism. Petunias, for example, synthesize a chemical that dramatically stunts the growth of corn earworms. Scientists hope that they can transfer the genes to crop plants, through either genetic engineering or more conventional means, to provide a natural protection against pests, thus providing an on-site means of control. Nicotine, caffeine, and citrus oil are all natural insecticides and are under investigation today.

Gardeners often use a dilute solution of soap or hot pepper to spray their plants to control insects such as pesty white flies and aphids. I've used them for years on vegetables and ornamentals grown in my indoor planters. Gardeners also use pyrethrums, a naturally occurring substance derived from plants. It's an effective pesticide against many insects. We use a dilute solution of soap to spray our cattle to ward off flies and are experimenting with other natural chemicals to control flies in the summer.

KEY CONCEPTS

Plants have evolved many natural insect repellants that can be commercially produced and sprayed on crops to help control pests.

Cultural Controls

The final components of IPM are the **cultural controls**, any one of a dozen techniques to control pest populations that do not fall under the previous categories—environmental, genetic, and chemical. These methods include cultivation to control weeds, noisemakers to frighten birds, and manual removal of insects from crops (especially suitable for gardeners). Also included in this group are such measures as destroying insect breeding grounds; improved forecasting of insect emergence; quarantines on imported foods, notably fruits and vegetables, to prevent the spread of pests; and water and fertilizer management to ensure optimum crop health and resistance to pests. One of the most badly needed cultural controls is monitoring.

KEY CONCEPTS

Insect pests can be controlled by many techniques that do not require the use of chemicals. These are called *cultural controls* and include measures such as noisemakers to frighten birds from crops, manual removal of insects, and quarantines on imported food.

Educating the World About Alternative Strategies

According to a report by Worldwatch Institute's Peter Weber, "Farmers seem to see in pesticides...the illusion of a guarantee, which they can never get from the weather, markets, or politicians." Some banks even require farmers to use pesticides to qualify for crop loans.

With a little imagination and thought, farmers are finding that they can reduce pesticide use by 50% and still not lower harvests. As farmers like Ron Rosmann are finding, they can even find ways to curtail pest damage without chemicals.

Training the world's farmers in IPM is vital to efforts to build a sustainable society. This will require a massive educational effort. Schools and universities, extension services, and even agricultural magazines can help retrain farmers.

For many years, though, in virtually all countries, farmers have received most of their advice on pest control from the sales reps of the chemical manufacturers that produce pesticides—who have an obvious conflict of interest. In some universities, much of the research on pest control is sponsored by pesticide manufacturers, which may bias the system.

Fortunately, many farmers have become aware of the pitfalls of the indiscriminate use of pesticides. Cost consciousness has led to efforts to seek alternatives on the part of both small farmers and large corporate farms. In addition, several U.S. universities have sustainable agriculture programs that teach students about IPM, among other subjects, including Iowa State University and the University of California at Davis. State university extension programs also provide information to farmers and gardeners on alternatives to chemical pesticides.

Universities in the less developed countries are also beginning to train farmers and students.

In less developed countries, agricultural departments have trained farmers in IPM and encouraged them to educate fellow farmers. In Indonesia, the government trained and hired 2,000 crop scouts. These scouts work with farmers to monitor insect populations and also teach the farmers the techniques of IPM. Farmers can also work together without government support. Ron Rosmann, introduced at the beginning of the chapter, is a member of Practical Farmers of Iowa (PFI), a group of more than 1,500 farmers and friends of farmers interested not only in cutting costs but also in preventing soil erosion and using fewer chemicals—ultimately, farming more sustainably. Groups like PFI have been started in several other states as well. Farmers who belong to such groups are pioneering new ways to produce crops that could result in dramatic decreases in the use of pesticides while maintaining or increasing yields. By sharing their ideas with nonmembers, they can help spread the word.

Farmer groups are also beginning to form in less developed countries. Bolivian farmers, for instance, formed a group called the Association of Ecological Producers. A similar group was formed in Mexico and draws off a long tradition of pesticide-free agriculture. Mexico's more than 13,000 organic farmers export an estimated $300 million worth of food to the United States and Europe.

Nongovernmental organizations or **NGOs**—among them, consumer and environmental groups—can play a major role in educating farmers and government officials worldwide. A ban on several widely used and highly toxic pesticides in the Philippines, for instance, is the result of the efforts of numerous NGOs. Concerted efforts on their part resulted in a dramatic shift in official and public opinion away from using chemical pesticides toward safer alternatives. The success of these groups can be attributed in large part to the efforts of the NGOs to demonstrate alternatives available to farmers. NGOs have also publicized research showing that crop yields don't have to fall when crops are protected by IPM. Through their work, they have successfully promoted IPM, which has gained wider acceptance than chemical pesticides.

The shift to pesticide-free farming will require attitude changes on the part of consumers, too. To make sustainable agriculture successful will require buyers to choose organic produce. In some cases, organic produce is competitively priced. In most cases, though, it costs more. So, consumers must be willing to pay extra to support this industry and to protect their health and the health of the environment. Many consumers will have to rethink and change their buying habits. Some conventionally grown fruits and vegetables tend to have more pesticides than others. **TABLE 22-1** lists the Environmental Working Group's "dirty dozen" and "clean fifteen." To reduce your intake of pesticides, they recommend that when you buy fruits and vegetables that are on the "dirty dozen" list, you purchase organically grown products.

TABLE 22-1	
Environmental Working Group's Dirty Dozen and Clean Fifteen in 2013	
Dirty Dozen Plus	**Clean Fifteen**
Apples	Asparagus
Celery	Avocados
Cherry tomatoes	Cabbage
Cucumbers	Cantaloupe
Grapes	Sweet corn
Hot peppers	Eggplant
Imported nectarines	Grapefruit
Peaches	Kiwi
Potatoes	Mangos
Spinach	Mushrooms
Strawberries	Onions
Sweet bell peppers	Papayas
Kale, collard greens	Pineapples
Summer squash (zucchini and yellow crookneck squash)	Sweet peas—frozen Sweet potatoes

Educating farmers and others about sustainable pest management is imperative. Universities, agricultural agencies, farmers' groups, and nonprofit organizations are several of the avenues available for promoting pesticide-free farming.

Government Actions to Encourage Sustainable Agriculture

Governments can promote sustainable agriculture by providing low-cost crop insurance for farmers who make the transition to IPM. As a farmer shifts from chemical-intensive use to IPM, losses can be severe because the ecology of the farm may be severely out of balance. With insurance, farmers can wean themselves from pesticides and not go bankrupt in the process.

Governments can also help by developing **organic certification programs**. Developed nationally and in at least 20 states, these programs set standards farmers must meet to permit them to label their produce "organically grown." Certification will help consumers determine which produce is truly pesticide-free. It could also stimulate other farmers to think about switching to chemical-free methods of farming. To the delight of many people, numerous U.S. farmers have already made the switch. In 2011, there were nearly 13,000 certified organic growers in the United States according to the USDA. That's good news to some, but it is only a small fraction of the 2 million U.S. farmers.

You can help, too by purchasing organic produce and taking other steps suggested in the Go Green sidebars.

> Our doubts are traitors and make us lose the good we oft might win by fearing to attempt.
>
> **—William Shakespeare**

CRITICAL THINKING

As with many other issues, it is important to dig deeper and always consider the big picture. It is crucial to uncover biases that may enter into the arguments of some proponents as well. Scrutinize the experiment reported in the newspaper. Was it performed correctly, and are the results applicable to humans?

Suppose you found that the results were valid. You would then want to look at the next most important question: Should the pesticide be banned? In the public policy arena, this question pivots on another relatively simple question: Are the risks worth the benefits?

In order to answer this, it is necessary to seek more information and viewpoints. If you did, you might find that farmers and pesticide manufacturers would argue that this insecticide helps them prevent crop damage, which saves farmers millions of dollars a year. They also say that pesticides make food cheaper for consumers. The pesticide manufacturing industry also provides thousands of jobs. Farmers provide food for billions of people. In the developing world, this food keeps millions of people alive.

As for the costs, if EPA estimates are correct, 10 to 20 U.S. citizens will contract cancer each year, and many of them will die each year. The pesticide may also contaminate groundwater, kill fish and birds, and have other adverse environmental effects.

When you dig deeper, you would very likely find that there are alternative ways of controlling insects without using potentially harmful pesticides. These methods may have the added benefit of reducing groundwater contamination and mortality in birds, fish, and other wildlife. In some cases, alternative methods lower crop yield, but because of lower input costs, farmers make as much or even more money.

Can you think of any other costs and benefits? After analyzing both sides of the argument, what is your opinion? Should the pesticide be banned? Would your opinion change if you were one of the cancer victims? If you were a farmer?

CRITICAL THINKING AND CONCEPT REVIEW

1. List and discuss reasons why pest damage is high in developed nations despite the extensive use of chemical pesticides.
2. Critically analyze this statement: "Without pesticides, crop damage in the United States and other major food producers would be much higher."
3. Describe some of the environmental and health problems caused by the use of pesticides. How can they be avoided?
4. How are beneficial species affected by insecticide use? Give some examples.
5. What does the term *pesticide treadmill* refer to?
6. Why do DDT and other chlorinated hydrocarbons persist in the environment? Why do they cause problems even though they are found in low concentration in water?
7. Make a list of the major components of integrated pest management. What role do soil conservation measures play in pest control? What advantages does IPM strategy offer over current management techniques?
8. Explain why crop rotation and increasing crop diversity reduce pest populations.
9. Describe some of the biological control methods. Give examples. Using your knowledge of biology and evolution, would you expect pest species to develop resis-tance to biological controls?
10. Describe why the sterile male technique works.
11. Discuss some ways in which genetic engineering and selective breeding may be used to help cut down on pest damage.
12. You are appointed director of the state agricultural department. Outline a way to encourage farmers to minimize and ultimately eliminate pesticide use.
13. Using critical thinking, analyze this statement: "U.S. agriculture cannot go organic. We cannot eliminate pesticides without greatly reducing productivity and farmers' profits."

KEY TERMS

Agent Orange
auxins
biological control
broad-spectrum pesticides
carbamates
chlorinated hydrocarbons
coevolution
confusion technique
crop rotation
cultural controls
DDT (dichlorodiphenyltrichloroethane)
dioxins
environmental control
Federal Insecticide, Fungicide, and Rodenticide Act (FIFRA)

first-generation pesticides
Food Quality Protection Act
genetically resistant insects
herbicides
heteroculture
hormones
integrated pest management (IPM)
integrated weed management
juvenile hormone
molting hormone
narrow-spectrum pesticides
nongovernmental organizations (NGOs)
organic certification programs
organic phosphates
organic produce

pesticides
pesticide drift
pesticide treadmill
pheromones
pheromone traps
pyrethroids
registration
second-generation pesticides
sex attractants
sterile male technique
strip cropping
third-generation pesticides
tolerance levels
trap crops
triazines
upset

CHAPTER
23

Hazardous and Solid Wastes: Sustainable Solutions

LEARNING OBJECTIVES

After reading and studying the material in this chapter, students will be able to understand, knowledgably discuss, and, when appropriate, apply the information learned in this chapter, including:

1. The ways hazardous wastes, including nuclear wastes, enter our environment
2. The importance of leaking underground storage tanks to groundwater pollution and the extent of the problem
3. Traditional and sustainable approaches to hazardous waste
4. Solid waste production and ways to deal with solid waste, both traditional and sustainable

CRITICAL THINKING

Exercise

As a hazardous-waste manager of a chemical company, you are faced with a dilemma. Your company produces over 400,000 pounds of highly toxic waste annually. The cost of disposal is several million dollars. An official from a new waste disposal firm contacts you and says that he can dispose of the waste at half the cost. When you ask him where it is going, he says that it will be shipped to a developing nation, where it will be incinerated. He tells you that pollution controls on the incinerator where the waste will be burned are not as sophisticated as those on the one owned by the U.S. disposal company you've always used, but it doesn't really matter because there's so little toxic waste burned in the country anyway. What choice do you make? What factors will affect your choice?

The 3M Corporation, headquartered in Minnesota, is a $30 billion dollar-a-year corporation and a model of corporate ingenuity and foresight. In 1975, the company started a Pollution Prevention Pays program designed to reduce solid and hazardous waste generated by the company's many facilities. By substituting less toxic or nontoxic chemicals for solvents, by recycling everything they could, and by modifying manufacturing processes, 3M has made substantial cuts. From 1990 to 2011, 3M cut global emissions of carbon dioxide by 72%. They reduced the emission of volatile organic chemicals (VOCs) by 98% from 1990 to 2013 and continue to find ways to reduce emissions. VOCs contribute to photochemical smog and many are toxic to humans and other species. The company has reduced solid waste by 67%. They have made great strides in reducing water pollution, too. All in all, the company has eliminated the release of more than 3.8 billion pounds of pollutants. These and other actions taken by the company saved about $1.8 billion between 1975 and 2011. To date the company has completed 10,000 pollution prevention projects suggested by company employees. Moreover, the company continues to reduce its environmental footprint.

Although interest in reducing waste by recycling and using resources more efficiently is increasing, businesses are a long way from tapping into the full potential of this strategy. Many continue to discard millions of tons of perfectly usable material, including cardboard, wood, office paper, and metals. Moreover, some companies continue to illegally and irresponsibly dump hazardous waste into the air, water, and soil.

This chapter discusses solid and hazardous wastes. Like other chapters in this book, it shows how individuals, businesses, and governments have traditionally addressed the problem of waste and how limited these strategies are. It will also show more sustainable approaches, measures that make sense from social, economic, and environmental perspectives.

23.1 Hazardous Wastes: Coming to Terms with the Problem

Hazardous wastes are waste products of homes, factories, businesses, military installations, and other facilities that pose a threat to people and the environment. They are toxic, carcinogenic, or mutagenic. For many years, pollution like hazardous waste was seen as a sign of progress. Today, many individuals view pollution in general and hazardous waste in particular as signs of unsustainable technologies or unsustainable industrial systems (**FIGURE 23-1**).

KEY CONCEPTS

Hazardous wastes come from a variety of sources, among them factories and our homes. These toxic materials are now viewed as signs of unsustainable practices.

FIGURE 23-1 Toxic hot spot in Stratford, CT. Inactive industrial sites like this one are often contaminated by a variety of hazardous chemicals. Cleanup of such sites can easily cost several million dollars.

Courtesy of Bill Empson/U.S. Army Corps of Engineers.

Love Canal: The Awakening

The severity of the U.S. hazardous-waste problem caught the attention of the American public in the 1970s when toxic chemicals began to ooze out of a hazardous-waste dump known as Love Canal in Niagara Falls, New York. The story of Love Canal began in the 1880s, when William T. Love began digging a canal that would run from the Niagara River just above Niagara Falls to a point on the river below the falls. The canal was built to divert water to an electric power plant to supply future industrial facilities that would be built along its banks. Unfortunately for Love, the canal was never completed. Only a small remnant of the canal remained in the early 1900s. In 1942, the Hooker Chemical Company signed an agreement with the canal's owner to dump hazardous wastes into the abandoned canal. In 1946, Hooker bought the site, and from 1947 to 1952 it disposed of over 20,000 metric tons (22,000 tons) of highly toxic and carcinogenic wastes, including dioxin.

In 1952, the story took an ironic twist. In that year the city of Niagara Falls began condemnation proceedings on the property. This legal maneuver would allow the city to acquire the land to build an elementary school and residential community. Hooker sold the land to the city for $1 in exchange for a release from any future liability. Hooker insists that it warned against construction on the dump site itself, but it allegedly never disclosed the real danger of building on it. Before turning the land over to the city, Hooker sealed the pit with a clay cap and topsoil, once thought sufficient to protect hazardous-waste dumps.

Troubles began in January 1954, however, when workers removed the clay cap during the construction of the school. In the late 1950s, rusting and leaking barrels of toxic waste began to surface. Children playing near them suffered chemical burns; some became ill and died. Hooker said that it had warned the school board not to let children play in contaminated areas, but the company apparently made no effort to warn local residents of the potential problems.

Go Green

Recycle oil from your car engine, if you change it yourself.

The problem continued for years. Chemical fumes took the bark off trees and killed grass and plants in vegetable gardens. Smelly pools of toxins welled up on the surface. In the early 1970s, after a period of heavy rainfall, basements in homes near the dump began to flood with a thick, black sludge of toxic chemicals. The chemical smells in homes around the dump site became intolerable.

Tests in 1978 on water, air, and soil in the area detected 82 different chemical contaminants, a dozen of which were known or suspected carcinogens. In that same year, the state health department found that nearly one of every three pregnant women in the area had miscarried, a rate much higher than expected. Birth defects were observed in 5 of 24 children. Another study, released in 1979 by Dr. Beverly Paigen of the Roswell Cancer Institute, showed that over half of the children born between 1974 and 1978 to families living in areas where toxic chemicals were present had birth defects. In this study, the overall incidence of birth defects in the Love Canal area was 1 in 5, compared with a normal rate of less than 1 in 10. The miscarriage rate was 25 in 100, compared with 8 in 100 for women moving into the area. Asthma was four times as prevalent in contaminated areas as in uncontaminated areas in the region; the incidence of urinary and convulsive disorders was almost three times higher than expected. The incidence of nasal and sinus infections, respiratory diseases, rashes, and headaches was also elevated.

As a result of public outcry, the school was soon closed. The state fenced off the canal and evacuated several hundred families (**FIGURE 23-2**). President Carter declared the site a disaster area. In May 1980, a new study revealed high levels of genetic damage among residents living near the canal. An additional 780 families were evacuated from outlying areas.

In 1987, the Environmental Protection Agency (EPA) announced plans to clean the sewers and dredge two creeks in the Love Canal area to remove sediments contaminated with toxic chemicals. In 1988 and 1989, the creeks were diverted so they would dry. Bulldozers then removed the top 18 inches of mud, which was burned in a special incinerator built for this project. The removal of the sediments cost $13 million, and sewer cleanup added another $5 million to the price tag. Incineration of the sediment cost over $26 million. Love Canal cost the state of New York and the federal government approximately $272 million for cleanup, relocation of residents, and other expenses.

A 1980 study by the EPA showed that chemical contamination was pretty much limited to the canal area (the actual dump), an area immediately south of it, and two rows of houses on either side of the canal (**FIGURE 23-3**). The last group of residents to be evacuated, the report said, were probably moved out unnecessarily. The EPA study also showed that the dump had contaminated shallow groundwater but not deeper aquifers. The EPA

FIGURE 23-2 The price of pollution. Houses awaiting the bulldozer in the Love Canal area of Niagara Falls, New York. This tragedy not only cost the government (and taxpayers) millions of dollars, it also resulted in considerable disruption of people's lives as families were forced from their homes.
© Joe Traver/Time Life Pictures/Getty Images.

concluded that further migration of toxic chemicals was highly unlikely. Based on this study and other work, the EPA and the state of New York declared two-thirds of the evacuated Love Canal site "habitable" and proposed to sell the houses. Over 260 homes were renovated and sold to new owners. Approximately 150 acres east of the site have been sold for light industrial use. Lois Gibbs, the Love Canal resident largely responsible for drawing public attention to the disaster and getting the state and federal governments to take action, argues that the decision to resettle the area was improperly made. In fact, she claims that in assessing the habitability of the Love Canal site, the New York State Health Department compared it to only two other sites, both badly contaminated by industrial wastes, and deemed it suitable for resettlement. Gibbs warns that resettling Love Canal will put more people at risk. According to Gibbs, 18,000 metric tons (20,000 tons) of hazardous waste still remain at the site. It will take 20,000 years for these wastes to decompose. So far, there have been no medical concerns among new residents.

KEY CONCEPTS

Love Canal was a hazardous-waste disposal site in Niagara Falls, New York, similar to many others around the world. Leakage from the site caused serious health problems in residents living near it. The incident alerted the public and government officials to the problem of improper hazardous-waste disposal.

The Dimensions of the Problem

In the years following the Love Canal incident, health officials found that Love Canal was not an isolated incident, but just the tip of an enormous hazardous-waste iceberg. In 1989, the EPA announced that the number of hazardous-waste sites was approximately 32,000. The General Accounting Office estimated that the number of hazardous-waste sites could be much higher, perhaps 100,000 to 400,000. These estimates do not include the 17,000 hazardous-waste *hot spots* on U.S. military bases. Hazardous wastes have also been mixed with oils that are sprayed on dirt and gravel roads in rural areas to prevent dust. Tens of thousands, perhaps hundreds of thousands, of badly contaminated sites may exist in Europe, especially in the former Eastern Bloc nations and the nations that once were part of the Soviet Union. Canada is home to many contaminated sites, too. One of the worst of all is the Sydney steel plant in Nova Scotia. For nearly 100 years, the plant's operators dumped waste from its operations in a nearby creek. Today, 34 hectares (85 acres) of the tidal flats where the creek opens onto the ocean contain an estimated 500,000 metric tons (454,000 tons) of toxic coal-tar from the facility. Cleanup proved far more difficult and far more costly than anticipated. At this writing, the federal and provincial governments have spent over $52 million cleaning up the site and are still not finished. The estimated cost is $400 million. Findings such as these have fueled widespread concern about cancer.

Making matters worse, each year U.S. factories create an estimated 31 million metric tons (34.3 million tons) of hazardous waste from large facilities—about one

The Love Canal study area

FIGURE 23-3 Love Canal. Area within the green rectangle was closed off, and numerous families were evacuated after health studies showed elevated incidence of birth defects, stillbirths, and a variety of symptoms most likely attributable to toxic wastes from the hazardous-waste dump. Residents were also evacuated from an outer region (the declaration area); but, tests have shown that hazardous wastes have not migrated into this area and that these residents may have been unnecessarily evacuated.

metric ton for every man, woman, and child.[1] But the United States is not alone. European countries and many less developed nations also produce tens of millions of tons of hazardous waste each year. In 1985, 90% of the hazardous wastes in the United States were improperly disposed of, according to one estimate. This waste ended up in abandoned warehouses; in rivers, streams, and lakes; in leaky landfills that contaminate groundwater; in fields and forests; and along highways. No current estimates are available, but the percentage is now believed to be much lower.

Improper waste disposal has left behind a long list of costly effects: (1) groundwater contamination, (2) drinking well closures, (3) habitat destruction, (4) human disease, (5) soil contamination, (6) fish kills, (7) livestock disease, (8) sewage treatment plant damage, (9) town closures, and (10) difficult or impossible cleanups. Irresponsible and ill-conceived waste disposal continuing today will create a legacy of polluted groundwater and contaminated land that could persist for decades, perhaps centuries.

Most experts believe that the nation is in for a much longer, more difficult battle than once was anticipated. Why? There are many more sites that are far more difficult to deal with than anybody anticipated. The price tag could also be much higher than expected. The U.S. Office of Technology Assessment (OTA) estimates that it will cost $100 billion to clean up the 10,000 sites in the United States that pose a serious threat to health. Researchers at the University of Tennessee estimate the cleanup of all hazardous-waste sites in the United States by state and federal programs will cost $750 billion.

KEY CONCEPTS

The problem with hazardous wastes is twofold. First, tens of thousands of hazardous-waste sites are in need of cleanup in the United States and other countries. These sites contaminate groundwater and can affect the health of people who live nearby. Cleaning them up will cost many billions of dollars. Second, factories continue to generate millions of tons of hazardous waste each year.

LUST—It's Not What You Think

You feel dizzy. Your head spins. Your insides ache. You haven't been yourself for weeks. What may be ailing you is LUST—but not the usual kind. Your symptoms may be caused by the groundwater pollution from a nearby leaking underground storage tank, which EPA's top acronymists once dubbed LUST. Some time later, no doubt responding to complaints from citizens vigilant for political correctness, they dropped the L, leaving UST: underground storage tanks.

---

[1] This includes only facilities that produce over 1,000 kilograms (450 pounds) of hazardous waste per year. Thousands of smaller facilities also generate hazardous waste but are not required to report it.

Generally concentrated in heavily developed urban and suburban areas, underground storage tanks are used primarily to store petroleum products such as gasoline, diesel fuel, and fuel oil. They're found at gas stations, factories, and homes. Some are used to store hazardous wastes, too.

The problem with tanks is that moisture and soil acidity gradually corrode the steel they're made of, causing them to deteriorate over time and leak petroleum byproducts, toxic chemicals, and hazardous wastes. The main concern is the potential effect on groundwater and human health. Even a small leak can contaminate large quantities of groundwater. For example, 1.5 cups of hazardous liquid leaking out of a tank per hour can contaminate nearly 4 million liters (over 1 million gallons) of groundwater in a day. Contaminated groundwater is very difficult—sometimes impossible—and expensive to clean up.

Go Green

Take unwanted paint, stains, finishes, and other hazardous chemicals to recycling or disposal facilities.

Contaminated groundwater poses a problem to people and animals (such as livestock) that drink it. Contaminated drinking water used for baths and showers can also be dangerous. Benzene, a component of gasoline that can cause cancer, is absorbed through the skin when bathing. Showering generates dangerous vapors that can cause skin and eye irritation.

Although there are many sources of groundwater contamination, according to the U.S. EPA, leaking underground storage tanks are the number one cause of groundwater pollution, followed by landfills (**FIGURE 23-4**). A small leak can go undetected for a long time, causing considerable contamination. According to the EPA, the most significant effects are seen in the shallow aquifers, the ones domestic water is typically drawn from. Nationwide, 3 to 5 million underground storage tanks containing hazardous materials dot the United States. Since the underground storage program began, 514,000 leaky tanks have been discovered. As of September 2013, about 85% or more than 436,000 have been cleaned up, leaving nearly 78,000 to be addressed. To see how many leaking underground storage tanks there are in your state, log onto www.epa.gov/oust/wheruliv.htm.

Major oil companies have already spent millions to clean up polluted groundwater and soil and to install new tanks at gas stations. The cost of such actions can be exorbitant. Unfortunately, half of U.S. service stations are owned by independent dealers who generally are not financially able to replace the leaking tanks.

About 90% of the cleanup and replacement of leaking underground storage tanks is being financed and performed by private industry. The rest falls on the shoulders of the states. The EPA sets guidelines for cleanup and replacement and also provides financial assistance to help states. Today, all 50 states have created funds to pay for their part of the

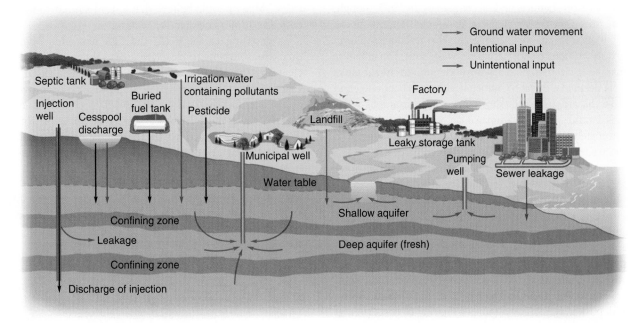

Legend:
→ Ground water movement
→ Intentional input
→ Unintentional input

Labels: Septic tank, Injection well, Cesspool discharge, Buried fuel tank, Irrigation water containing pollutants, Pesticide, Factory, Landfill, Leaky storage tank, Municipal well, Pumping well, Sewer leakage, Water table, Confining zone, Leakage, Shallow aquifer, Deep aquifer (fresh), Confining zone, Discharge of injection

FIGURE 23-4 Source of groundwater pollution. Groundwater is contaminated by a variety of sources such as septic tanks, farms, and abandoned industrial sites.

cleanup cost. State and federal funds are derived from taxes on gasoline.

> **KEY CONCEPTS**
>
> Hundreds of thousands of underground storage tanks have been installed in industrial nations and are used to store many potentially toxic substances such as gasoline, diesel fuel, hazardous waste, and heating oil. Over time, steel tanks corrode and begin to leak, contaminating groundwater used for cooking food, drinking, and bathing.

23.2 Managing Hazardous Wastes

Two hazardous-waste problems face virtually all industrial nations and, to a lesser extent, the less developed countries (LDCs). First, cleaning up existing hazardous-waste sites, leaking storage tanks, and polluted groundwater. Second, dealing with the enormous amounts of hazardous waste produced each year to avoid creating new sites and further contamination of groundwater.

The first problem—cleaning up toxic hot spots—requires immediate action. It's one place where the Band-Aid approach is appropriate. That said, all solutions that fall into this category must be sustainable; they must not merely shift the problem from one location (a contaminated factory site) to another (a landfill where the hazardous wastes are dumped). The second problem—dealing with waste that is currently being produced—calls for long-term preventive measures. Preventing pollution should be the primary goal.

> **KEY CONCEPTS**
>
> Addressing the issue of hazardous waste requires plans to clean up contaminated sites and preventive actions to reduce or eliminate hazardous-waste production.

The Superfund Act: Cleaning Up Past Mistakes

In June 1983, the 2,400 residents of the town of Times Beach, Missouri, agreed to sell their 800 homes and 30 businesses to the federal government for $35 million. Why? The roads in Times Beach, like those elsewhere, had been sprayed with oil containing hazardous wastes, including dioxin. How did the oil get contaminated?

As noted earlier in the chapter, unscrupulous hazardous-waste disposal companies had mixed toxic chemical wastes with waste oil and then spread it on dirt roads to control dust. In Times Beach, the dioxin levels in the soil were 100 to 300 times higher than levels considered harmful during long-term exposure. The town had to go, and the federal government bought it. Today, Times Beach is a ghost town bordered by a tall chain-link fence. Its only occupants are occasional EPA officials or scientists from companies that are exploring ways to detoxify the soil.

The $35 million purchase price for this contaminated piece of real estate came from a special fund known as the **Superfund.** It was created in 1980 by the **Comprehensive Environmental Response, Compensation and Liability Act, CERCLA** for short. Commonly called the **Superfund Act,** it and its two amendments (1986 and 1990) established a $16.3 billion fund financed by state and federal governments and by taxes on chemical and oil companies. In 1995, though, the tax on businesses expired and continued funding (about $1.2 billion/year) has come from general tax revenues and from companies responsible for hazardous-waste sites. The money is earmarked to clean up both leaking underground storage tanks that are deemed a threat to human health and abandoned and inactive hazardous-waste sites, including hazardous-waste dumps, landfills, and contaminated factories, mines, and mills. Unfortunately many regions still lack the money required to clean up sites, so a lot of cleanup work is on hold.

Under CERCLA, the EPA is authorized to collect fines from parties responsible for the contamination—totaling up to three times the cleanup cost. This law makes owners and operators of hazardous-waste dump sites and contaminated areas, as well as their customers, responsible for cleanup costs and property damage. Under the law, then, all businesses, hospitals, schools, cities, and other parties that deposited hazardous waste at a site are liable for a portion of the cleanup cost, based on the type and the volume of waste they deposited. CERCLA requires a sharing of costs even at licensed hazardous-waste disposal facilities where hazardous wastes were legally disposed of in previous years. As one industry representative put it, "You are liable for your waste forever." If a company or party is no longer in business, the remaining parties must pay the cost.

Superfund has clearly had an impact. As of 2013, 1,156 sites have been cleaned up, nearly 70% of the sites on its priority list.

KEY CONCEPTS

The Superfund Act provides money to clean up hazardous-waste dumps and other contaminated sites. This money comes from a tax on oil and petrochemicals and is replenished by fees charged to those responsible for the contamination, including the owners and operators of waste sites and the companies that paid to have their waste disposed of in them.

Problems with the Superfund Program   Despite these successes, the Superfund program has been riddled with problems. One of the most significant is cost. Stabilizing a leaking pond designed to hold hazardous wastes costs the EPA $500,000. A study to determine what chemicals are leaking from a site can cost $800,000.

A second problem is that it has created a legal nightmare, described as a "massive web of litigation between the EPA, waste depositors, and insurance companies." Of the funds paid in the early years the program has been operating, nearly 60% have been for legal fees. This money has largely been spent on identifying liable parties to get them to pay their fair share—not on cleanup.

The Superfund Act has also been criticized because, although it provides money for cleanup and financial compensation for property damage, it fails to provide avenues for victims of illegal dumping of hazardous wastes and their families to be compensated for personal injury or death. According to former senator George Mitchell, "Under the legislation, it's all right to hurt people, but not

trees." Many people believe that a fund similar to worker's compensation is needed for victims.

According to the OTA, the EPA has often opted for quick-fix solutions to clean up areas. Three-quarters of the cleanups, they say, are inadequate in the long term. In Love Canal, for instance, the EPA simply put a clay cap over the dump site and dug a ditch around it to collect hazardous wastes that escape. In other sites, contaminated soil was excavated and hauled off to another landfill. Incineration and biological destruction of the wastes might have been more long-lasting solutions.

KEY CONCEPTS

The cleanup of hazardous-waste sites under the Superfund Act has proven extremely slow, costly, and litigious. Much of the money initially spent went to legal fees. The Superfund Act has been criticized because it provides money for cleaning up property but no compensation for damage to human health. Many of the cleanups are considered inadequate.

Alternative Cleanup Funding Options   Some critics of CERCLA argue that the law has created an adversarial relationship among many parties that is counterproductive and inefficient. Too much money is spent on litigation and too little on cleanup. Although critical of CERCLA, these individuals do not suggest we abandon the program, only that we find options that put the money to better use.

One proposal calls for a federal hazardous-waste tax levied on each ton of hazardous waste disposed of, incinerated, or treated. This money would generate revenue for cleanup of the most heavily contaminated sites. There would be no need to try to assign responsibility for contamination and no need for the lengthy litigation that often results.

The American Association of Property and Casualty Insurers has proposed funding cleanup via a small fee on each new commercial insurance policy written in the United States. Their calculations suggest that this would generate more than $4 billion a year that could be used for cleanup. Both this and the previous alternative might allow the EPA to redirect its efforts to cleanup.

KEY CONCEPTS

Rather than spending millions of dollars to identify responsible parties, the Superfund program might work better with a no-fault policy—one that provides funds to clean up sites regardless of who is liable.

What to Do with Today's Waste: Preventing Future Problems

The high cost of cleanup strongly suggests the need for active preventive measures to avoid further contamination. Several possibilities exist.

The Resource Conservation and Recovery Act: Preventing Improper Disposal Cleaning up hazardous-waste sites is an essential first step in protecting health and creating a sustainable future. It should help reduce further contamination of groundwater. Efforts are also needed to prevent illegal and improper waste disposal. In 1976, the U.S. Congress passed the **Resource Conservation and Recovery Act (RCRA)**. This law is designed to monitor hazardous waste to eliminate illegal and improper waste disposal.

Under RCRA, the EPA was designated the nation's hazardous-waste watchdog. The EPA's first role was to determine which wastes were hazardous. RCRA also called on the agency to establish a nationwide reporting system for all companies handling hazardous chemicals. This requirement created a trail of paperwork that follows hazardous wastes from the moment they are generated to the moment they are disposed of—a so-called cradle-to-grave tracking. Congress believed that this requirement would make it difficult for waste generators to dump hazardous wastes improperly. RCRA also directed the EPA to set industry-wide standards for packaging, shipping, and disposing of wastes. Only licensed facilities could receive wastes.

Go Green

Use nontoxic cleaning agents, paints, stains, and finishes.

Unfortunately, RCRA's implementation has been slow. It was not until 4 years after Congress adopted the law that the EPA came up with its first hazardous-waste regulations. To the dismay of many, the regulations were full of loopholes, and about 40 million metric tons (44 million tons) of pollutants annually escaped control.

Because of public pressure, in 1984, Congress passed a set of amendments to eliminate RCRA's loopholes and ensure proper waste disposal. For example, under the original law, if a company produced under 1,000 kilograms (2,200 pounds) of hazardous wastes per month, it could dump them in a local landfill. The amendments changed the rules so that any individual or company that generated more than 100 kilograms (220 pounds) of hazardous waste a year must follow the same guidelines imposed on large waste producers.

The 1984 amendments also declared a national policy to reduce or eliminate land disposal of hazardous waste. Congress made it clear that land disposal technologies must be a last resort. The 1984 amendments gave preference to reuse, recycling, detoxification (such as incineration), and other measures discussed shortly. These approaches are bringing the United States closer to a sustainable waste management system.

The 1984 amendments to RCRA also addressed leaking underground storage tanks. After May 1985, for example, all newly installed underground tanks had to be protected from corrosion for the life of the tank. The lining of the tank must consist of materials compatible with stored substances. Furthermore, owners and operators must have methods for detecting leaks, must take corrective action when leaks occur, and must report all actions.

KEY CONCEPTS

The Resource Conservation and Recovery Act established a nationwide reporting system to monitor hazardous wastes from their production until their disposal. This provision seeks to eliminate illegal and improper hazardous-waste disposal. Recent amendments call for an elimination of land disposal of hazardous wastes and regulations to prevent leakage from underground storage tanks.

Weaknesses in RCRA Despite these changes, RCRA still has many loopholes. Some critics believe that its definition of hazardous wastes is too narrow. Michael Picker of the National Toxics Campaign, for example, thinks that municipal waste (garbage) should be classified as hazardous waste because it contains toxic chemicals such as pesticides, and heavy metals like lead (from batteries). Leachates (water containing contaminants) seeping from some municipal landfills are as toxic as those coming from regulated hazardous-waste facilities. According to John Young of the Worldwatch Institute, more than one of every five hazardous-waste sites on the U.S. Superfund cleanup list is a municipal landfill.

Picker also thinks that sewage and untreated wastewater handled by publicly owned sewage treatment plants should be considered a hazardous waste. Toxic chemicals in the sewer system, released by factories (legally and illegally) and homeowners, can escape into the air and into waterways. Agricultural wastes, mostly pesticides, are not regulated. Mill and mine tailings are also excluded from most regulatory control. By expanding the definition of what is toxic and by instituting better controls, the government could greatly cut back on the release of hazardous materials into the environment.

One lesson environmentalists have learned in the past four decades is that passing a law is not a guarantee of protection. Why not? For one thing, agencies responsible for administering and enforcing new laws don't always perform as they are instructed. Some drag their feet because they don't approve of the law. Additionally, agencies may be so underfunded and so overworked that they can't take on new responsibilities or, if they do, they do a poor job. The EPA is a case in point. Understaffed and underfunded, the EPA today struggles to implement RCRA and the handful of other laws aimed at protecting public health and the environment. Since the agency was formed, its workload has more than doubled, but funding has until quite recently remained at more or less the same level (adjusted for inflation).

KEY CONCEPTS

Additional steps may be needed to further reduce environmental contamination by hazardous wastes, among them broadening the definition of what is hazardous so that municipal waste, sewage, pesticides, and mine waste are included. Passing a law to protect the environment does not always result in immediate or expected gains for a number of reasons, notably a lack of funds or personnel to carry out the work.

Exporting Toxic Troubles Another lesson we've learned in the last few decades is that tough environmental legislation often has unanticipated effects elsewhere. For example, regulations that add to the cost of disposing of hazardous wastes have caused some companies to find ways to prevent waste production, but the very same regulations have caused unscrupulous companies to illegally dump their wastes or to export toxic wastes (including incinerator ash) abroad. In the 1980s, these wastes often ended up in cash-hungry LDCs or Eastern Bloc nations, neither of which had adequate laws requiring proper hazardous-waste disposal.

In 1986, Congress amended RCRA by establishing procedures to notify importing countries and obtain prior written consent. However, these regulations may be insufficient and EPA officials think that hundreds of tons of hazardous wastes are still being exported illegally.

Exporting hazardous waste to a nation without its full consent goes against principles of international law. Accordingly, numerous African nations have passed laws banning the import of hazardous wastes. In some countries, importing hazardous wastes is punished by stiff jail terms and multimillion-dollar fines. In Nigeria, an importer can be put to death. The Organization of Eastern Caribbean States and 22 Latin American countries have also joined forces to stop the dumping of hazardous wastes on their soils.

In 1990, the European Economic Community (EEC) (a coalition of European nations) agreed to ban exports of toxic and radioactive waste to 68 former European colonies. Many of the less developed nations who were part of the accord also agreed not to import hazardous wastes from non-EEC members. As of February 2014,

180 nations have signed an agreement (the Basel Convention) that bans the transfer of hazardous wastes to less developed nations, including Canada, Mexico, and numerous European nations. The United States has signed the agreement but our participation has not been ratified by Congress. Although these are important steps forward, many less developed nations are still open to imports, representing a potentially huge repository for hazardous wastes. Bear in mind also that signing an agreement will not stop the illegal flow.

KEY CONCEPTS

Tighter regulations for the disposal of hazardous wastes in the United States and other nations have led to efforts to reduce waste production by some companies, but also to illegal dumping by less scrupulous ones. Some companies export hazardous wastes to less developed nations with little or no oversight of such practices. Although many less developed nations ban such activities, many still accept wastes.

Dealing with Today's Wastes: A Variety of Options

In 1983, the National Academy of Sciences (NAS), a prestigious body composed of the nation's leading scientists, issued a report outlining options for handling hazardous wastes (**FIGURE 23-5**). This section discusses the various options and illustrates some of the most effective means of dealing with the problem, notably reducing or eliminating hazardous waste, showing that you don't have waste if you don't make it.

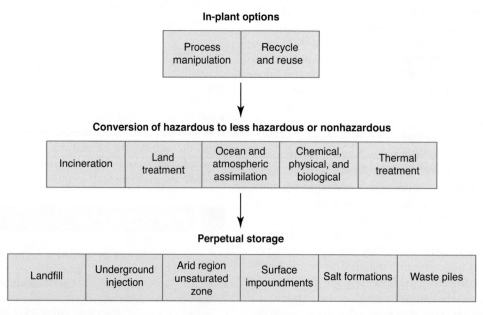

FIGURE 23-5 Hazardous-waste options. A three-tier hierarchy of options for handling hazardous wastes. The top tier includes in-plant options that reduce the production of hazardous waste in the first place. It contains the most desirable options. The middle tier converts hazardous materials into nonhazardous or less hazardous substances. Perpetual storage, the lowest tier, is the least desirable but often the cheapest alternative.

Many options exist for getting rid of waste. The most sustainable approaches involve steps that reduce or eliminate hazardous-waste output. You don't have waste if you don't make it.

Changes in manufacturing processes such as process manipulation and substitution are often the simplest and most cost-effective means of reducing or eliminating hazardous wastes. Waste output can also be dramatically reduced by recycling and reusing wastes.

Process Manipulation, Re-cycling, and Reuse

As illustrated in Figure 23-5, at the top of the NAS list are in-plant options, generally relatively simple and cost-effective changes that reduce hazardous-waste production. All qualify as highly sustainable practices because they reduce hazardous-waste output—avoiding the need to dispose of waste.

> **Go Green**
>
> Consume less! All products you buy result in the production of hazardous or solid wastes.

In-plant options include three general actions. The first, **process manipulation,** involves alterations in manufacturing processes to cut waste production. One alteration of manufacturing is known as substitution. **Substitution** involves the use of nontoxic or less toxic substitutes in manufacturing. Cleo Wrap, the world's largest producer of gift wrap, for example, switched to nontoxic inks and cut its annual production of hazardous waste by 140,000 kilograms (300,000 pounds). Industries can also change the chemical composition of their products, eliminating those that are harmful or that might produce harmful byproducts during manufacturing. Nontoxic household cleaners are a good example. Another type of process manipulation involves the monitoring of manufacturing processes to locate and fix leaks that are emitting toxic chemicals into the environment. Exxon Chemical, for example, installed floating lids over vats that contained volatile organic compounds, greatly reducing losses from evaporation. According to OTA, U.S. industries could reduce or prevent more than 50% of their hazardous-waste generation through process manipulation.

The second and third in-plant options are the **reuse and recycling strategies.** In some instances, companies can capture toxic wastes and, with little or no purification, reuse them to manufacture other products or sell them to other companies for reuse. In metal finishing, nickel that's left behind in rinse water can be recovered and reused or sold. For plants that cannot afford onsite technologies, shared facilities or third-party recyclers can provide an economical alternative. Regional waste exchanges—including both private and government-operated facilities that put waste producers in touch with companies that need their waste—can assist in the exchange of potentially hazardous wastes.

Reuse and recycling strategies help cut the output of waste by putting perfectly good materials to use. These strategies can save companies millions of dollars a year in materials costs. In addition, they eliminate the cost of waste disposal, cut down on potential environmental and health damage, and eliminate costly cleanups.

Conversion to Less Hazardous or Nonhazardous Substances

Process manipulation, waste reduction, recycling, and reuse can make a significant dent in waste production. Unfortunately, not all waste can be eliminated, reused, and recycled. Some waste will be produced. The NAS recommended that, where appropriate, remaining wastes be destroyed or detoxified—that is, converted to less hazardous or nonhazardous materials.

Detoxification can be accomplished for certain types of waste by **land disposal,** applying them to land. When mixed with the top layer of soil, some waste materials can be broken down by chemical reactions, by oxidation from sunlight, or by bacteria and fungi in the soil. Some nondegradable wastes may be absorbed onto soil particles and held there indefinitely (we think). Others may migrate into deeper layers. A word of caution: Land treatment is an expensive option, requiring care to avoid polluting ecosystems, poisoning cattle and other animals, and contaminating groundwater. Also, studies show that changes in conditions that bind toxins to soil particles may cause a sudden and unanticipated release into the environment. Plants can also remove toxic materials from contaminated soil, as explained in **Spotlight on Sustainable Development 23-1.**

Another option available for organic wastes is **incineration.** High-temperature furnaces at stationary waste disposal sites, on ships that burn wastes at sea, and on mobile incinerators can be used to burn toxic organic wastes such as PCBs, pesticides, and dioxin (**FIGURE 23-6**). In these facilities, oil and natural gas are used as fuels. Hazardous substances are injected into the furnace or mixed with the fuel before combustion.

Mobile hazardous-waste incinerators can be used to clean up contaminated sites or to deal with wastes stored in warehouses. Permanent hazardous-waste incinerators can also be established to deal with wastes from a variety of producers. Such facilities may also be used to provide energy for communities and factories. However, many communities object to hazardous-waste incinerators, fearing the release of toxicants from spills during transport or leaks at the plants. Incinerators may not always perform adequately, and operating personnel may bypass regulations. Low-level releases from smokestacks may also result in a long-term exposure to hazardous chemicals. Recent studies also show that when hazardous waste is burned, unburned waste and other chemical compounds in the exhaust may combine to form toxic air pollutants. Incinerators also require fuel and produce carbon dioxide, a greenhouse gas.

© hugolacasse/ShutterStock, Inc.

23-1 Plants Clean Up Toxic Chemicals

Several small rafts float in a pond near the Chernobyl nuclear power plant, the site of the world's most costly nuclear reactor accident. The raft contains sunflowers, whose roots dangle in the water. These rafts are not intended to make the pond more aesthetically appealing, but to decontaminate the radioactive water.

This is just one of dozens of similar efforts worldwide to use plants to clean up soils and waters contaminated by hazardous materials such as lead, selenium, oil, and radioactive substances.

Cleaning up contaminated sites with plants is technically referred to as *phytoremediation*. It relies on plants' abilities to absorb, store, and in some cases even detoxify toxic substances. The plants are often assisted by bacteria and single-celled fungi. These organisms help to break down some chemicals. The sunflowers in the radioactive pond near Chernobyl act as a kind of biological sponge. The roots absorb the radioactive cesium. Radioactive strontium contained in the water is taken up and is stored in the shoots. After 3 weeks or so, the plants are removed and discarded. This technique costs a fraction of what the more elaborate filtering mechanisms do.

Plants can also be used to remove radioactive cesium and strontium from contaminated soils. One species, known as *Indian mustard*, is particularly effective (**FIGURE 1**).

FIGURE 1 Indian mustard absorbs radioactive contaminants in the soil in this field.

Courtesy of Edenspace Systems Corporation.

Lead-contaminated soil can also be cleansed by plants, but because lead forms strong chemical bonds with soil components, scientists must first apply a special agent to release the lead so that it can be taken up by plant roots. After absorbing all of the lead they can, the plants are removed and properly disposed of. Eventually, scientists would like to see the lead such plants contain recycled. Plants could, for example, be added to lead smelters. The organic matter would burn off, leaving behind the lead for reuse.

Experiments are also underway to use plants to clean up contaminated groundwater. One of the most commonly encountered groundwater pollutants is a chemical known as TCE (trichloroethylene). This dry cleaning agent and degreaser was widely used before being banned.

Scientists have found that TCE can be removed by poplar trees whose roots extend 12 to 15 meters (40 to 50 feet) into the ground. Studies have shown that poplars can remove 95% of the TCE in experimental settings.

Another serious contaminant is selenium, a toxic substance that is leached from irrigated soils in California and other arid regions. Irrigation water draining from farm fields can have extremely high concentrations of this substance, too high to release the water safely into lakes and streams. To handle this waste, special evaporation ponds have been built as a kind of low-tech repository. In these ponds, the selenium-rich water evaporates, leaving behind the toxic metal. Over time, selenium concentrations can become quite high. Wildlife that inhabit the ponds can suffer enormously.

Scientists are now experimenting with artificial wetlands placed upstream from the evaporation ponds. These wetlands contain plants that remove selenium and could greatly reduce the selenium levels deposited in the evaporation ponds. A 36-hectare (90-acre) wetland built by Chevron Corporation in California currently removes up to 75% of the selenium from the 10 million liters (2.6 million gallons) of wastewater the company pumps through it every day.

Plants aren't the answer to hazardous waste cleanup, but they could be a welcome tool in the gigantic task of cleaning up the hundreds of thousands of contaminated aquifers and sites throughout the world.

FIGURE 23-6 Chemical waste incinerator. This incinerator in the United Kingdom is used to destroy toxic organic wastes from factories. High-temperature combustion and good pollution controls help to reduce emissions from the facility.

© Martin D. Vonka/ShutterStock, Inc.

Low-temperature decomposition of some wastes, including cyanide and toxic organics such as pesticides, offers some promise. In this technique, wastes are mixed with air and maintained under high pressure while being heated to 450° to 600°C (840° to 1100°F). During this process, organic compounds are broken into smaller, biodegradable molecules. Valuable materials can be extracted and recycled. This process is that it uses less energy than incineration.

Chemical, physical, and biological agents can also be used to detoxify or neutralize hazardous wastes. For example, lime can neutralize sulfuric acid. Ozone can be used to break up small organic molecules, nitrogen compounds, and cyanides. Many bacteria can degrade or detoxify organic wastes and may prove helpful in the future. New strains capable of destroying a wide variety of organic wastes may be developed through genetic engineering.

KEY CONCEPTS

Hazardous wastes can be converted to nontoxic or less toxic substances by chemical, physical, and biological means, such as neutralization, combustion, low-temperature thermal decomposition, and bacterial decay. Such measures, although effective, present some problems and are clearly not as desirable as preventive measures.

Perpetual Storage In-plant modifications and conversion techniques that destroy or detoxify wastes cannot rid us of all of our hazardous waste. By various estimates, 25 to 40% of the waste stream will remain even after the best efforts to reduce, reuse, recycle, and destroy it—although some companies have achieved far better reductions.

Residual hazardous waste could be stored by one of a half dozen options (Figure 23-5). For example, residual waste could be dumped in **secured landfills,** excavated pits lined by impermeable synthetic liners and thick, impermeable layers of clay. To lower the risk of leakage, landfills should be placed in arid regions—neither over aquifers nor near major water supplies. Special drains must be installed to catch any liquids that leak out of the site. These wastes should then be destroyed. Groundwater and air should be monitored regularly to detect leaks.

Growing public opposition to this strategy makes it more difficult for companies to find dump sites. Some observers have labeled this the **NIMBY syndrome:** Get rid of the stuff, but Not In My BackYard. Paradoxically, it seems that most of us want the products available in an industrial economy that inadvertently generates waste, but few of us want the wastes dumped (or even burned) nearby.

Even though the EPA has issued tough regulations for hazardous-waste landfills, critics argue that landfills are only a temporary solution. No matter how well constructed they are, they will eventually leak. In an attempt to reduce problems for future generations, the EPA has drawn up a list of chemicals that cannot be disposed of in landfills.

Landfills are one of the cheapest waste disposal options today and are therefore often favored by industry. The savings they offer today are very likely to be charged to future generations, though.

Other methods of perpetual storage include (1) use of specially built warehouses that hold wastes in ideal conditions and prevent any material from leaking into the environment; (2) deposition in geologically stable salt formations; and (3) deposition deep in the ground in arid regions where groundwater is absent and earthquakes are nonexistent.

KEY CONCEPTS

Not all waste can be eliminated by prevention, recycling, or detoxification. Perpetual storage remains the final, yet least sustainable, option.

Disposing of Radioactive Wastes

High-level radioactive wastes are some of the most hazardous of all wastes, but they have long been ignored by many countries. High-level radioactive waste is generated by commercial nuclear power plants and weapons production facilities. Lower-level wastes come from research laboratories and hospitals. Many radioactive wastes have a long lifetime. Others can concentrate in animal tissues. Virtually all pose a serious threat to human health.

High-level radioactive waste is not a problem that will go away, even though the nuclear power industry and nuclear weapons manufacturing are declining. The waste already produced and now stockpiled at nuclear reactors and the continued operation of the existing power plants, fuel processing facilities, and nuclear weapons facilities necessitates long-term, low-risk storage of nuclear wastes.

Recognizing the need to do something with nuclear wastes rather than continuing to stockpile them at power plants or weapons facilities, Congress passed the **Nuclear Waste Policy Act** in 1982. This law established a timetable for the Department of Energy (DOE) to select a deep underground disposal site for high-level radioactive wastes. The DOE focused its attention on a site in Nevada called Yucca Mountain. Located on federal land about 160 kilometers (100 miles) northwest of Las Vegas, Yucca Mountain was proposed to someday be home to a huge underground storage site excavated in the volcanic rock, costing an estimated $7 to $8 billion (**FIGURE 23-7**).

The DOE had hoped to open the site in 2010. In order for federal decision makers to approve the site, though, they had to be relatively sure that earthquakes and volcanic eruptions would not threaten the stability of the repository. In 2009, the Senate passed a bill to close Yucca Mountain after 25 years and $13.5 billion spent. As a result of concern over earthquakes, scientists discovered that the site lay over a major fault line.

Several other options are available to deal with high-level radioactive waste. Radioactive waste, for example, can be bombarded with neutrons in special reactors to convert some of it into less harmful substances. However, existing reactors do a poor job of altering cesium-137 and strontium-90, two of the more dangerous byproducts of nuclear fission.

Seabed disposal has been used in the past by the United States and European countries, but is now forbidden. Still, some scientists suggest that the seabed may provide a site for radioactive wastes; the effects are difficult to predict.

The problems of disposal suggest to some that nuclear power should be phased out. Cleaner, less costly measures to produce energy should be developed. As noted earlier, though, we need to establish cost-effective and low-risk methods to dispose of (safely store) the vast amounts of waste that have already been generated.

In the United States, low-level radioactive waste from hospitals and research laboratories is packaged and shipped to three disposal sites—in Nevada, Washington, and South Carolina—where it is buried in the ground. Medium-level waste from nuclear power plants and weapons facilities is another matter altogether. In the 1980s, the DOE began construction on a medium-level radioactive waste depository called WIPP (Waste Isolation Pilot Plant). WIPP is the first geological repository for the disposal of radioactive waste produced by the U.S. defense programs—nuclear weapons research and production. It also houses waste from the disassembly of nuclear weapons. The waste slated for disposal in the site consists mostly of protective clothing, tools, glassware, and equipment contaminated with waste. It does not house high-level radioactive waste or spent nuclear fuels.

FIGURE 23-7 An aerial view of Yucca Mountain, Nevada. This site was once slated to become home to much of the nation's radioactive waste. Questions about the adequacy of the site and the potential problems created by shipping wastes to it from all parts of the country and other issues caused the federal government to abandon plans to use the Yucca Mountain site in 2009.

Courtesy of U.S. Department of Energy.

WIPP is an experimental project of DOE monitored and regulated by the Environmental Protection Agency. This $700 million facility was carved out of a thick salt deposit 630 meters (2,100 feet) below the ground in southeastern New Mexico near Carlsbad. Radioactive waste is held in steel canisters.

In 1993, the DOE started receiving wastes to test the safety of the site. The facility will eventually hold 6.2 million cubic feet of radioactive waste. In 1998, the EPA gave final approval for the site, saying that it complies with disposal regulations and is safe to contain wastes for 10,000 years.

Although the site has been open for many years, many citizens are concerned about accidents that could occur as radioactive waste is shipped to WIPP from federal weapons facilities in California, Colorado, Idaho, Illinois, Nevada, New Mexico, Ohio, Tennessee, South Carolina, and Washington. Most of the waste disposed of at the site is generated as nuclear weapons are disassembled. To reduce the risk of transportation accidents, DOE requires wastes to be shipped in special casks certified by the Nuclear Regulatory Commission and subjected to a series of tests to demonstrate their ability to survive severe crashes and punctures followed by fires or immersion in water.

Unlike many other hazardous wastes, most nuclear wastes cannot be reused or recycled. Process modification can help reduce waste output, but for the most part the problem has to be addressed at the end of the pipe. The end of the cold war and the decline in the nuclear arsenals of the United States and the former Soviet Union have dramatically reduced production at nuclear weapons facilities. Nuclear waste from power plants could also decline as they are phased out in the United States and elsewhere. Nonetheless, huge volumes of waste already exist and must be dealt with somehow.

KEY CONCEPTS

Disposing of radioactive waste will be a problem for a long time, even though nuclear power and nuclear weapons production are on the decline. Safe, permanent repositories are needed to store huge amounts of waste produced by power plants and other facilities.

Some Obstacles to Sustainable Hazardous-Waste Management

Hazardous-waste production has dropped steadily since 1980 in the United States. This encouraging trend has resulted in large part because of efforts to reduce hazardous-waste production by process modification, recycling, and reuse.

For years, one of the main problems with hazardous waste was that much of it was highly diluted in water released by various industrial processes. This dilute waste stream was typically pumped into deep wells, from which hazardous substances may leak into groundwater. Besides contaminating drinking water, industrial wastes pumped into deep wells can increase the incidence of earthquakes. Two Ohio geologists believe that industrial waste injected into the ground by a chemical company in Ohio may have created underground pressure that triggered the 4.9-Richter-scale earthquake that shook northeastern Ohio and a nearby nuclear power plant in 1986. Their studies suggest that pumping waste into the sandstone increased fluid pressure in the pores of a bed of rock that lies directly above the crystalline bedrock, causing the earthquake along a fault in the bedrock.

The sheer volume of polluted water has made it hard to regulate. Removing the hazardous substances from the water is extremely costly, so most plant owners are unwilling to make the investment in wastewater treatment.

Today, underground injection of hazardous waste continues, but at a greatly reduced level, about 5% of the total release. By far, the most significant release occurs in the air. It accounts for nearly 30% of all toxic chemicals released into the environment.

The push to reduce hazardous waste release into the environment has stirred considerable interest in incinerators. Incinerators are equipped with pollution control devices and can remove many of the chemical pollutants. They can also be used to generate energy and reduce environmental pollution by hazardous materials.

One person's solution, however, often becomes another person's problem. Incinerators often require landfills to dispose of highly toxic residues. Leaks may pollute the site and seep into groundwater, which is used for irrigation and drinking. Although plants typically have state-of-the-art pollution control equipment, small amounts of toxic substances may be emitted and over time could accumulate in the soil, ecosystems, and crops downwind. An incinerator becomes a hazardous-waste magnet, drawing shiploads in by truck and train from neighboring cities and states. Accidents could cause spills, environmental damage, injury, and evacuations.

Although proponents of incineration think that the rural siting strategy may be shrewd business, it ultimately diverts attention from finding sustainable solutions, such as process modification, reduction, recycling, and reuse. These are environmentally more acceptable and, in the long run, are more sustainable strategies for dealing with a nation's hazardous wastes.

KEY CONCEPTS

Hazardous wastes have been released in dilute wastewater discharges and injected into deep wells for many years. Incinerators have also been used to destroy wastes. Although hazardous-waste production appears to be on the decline, companies still release large quantities into the air, water, and land.

Individual Actions Count

As with all environmental issues, individuals can participate in many ways. For example, each of us can contribute by recycling oil, not dumping it in sewers or in vacant lots. We can properly dispose of paints, paint thinners, and other potentially toxic wastes. Many cities and counties provide toxic roundups, periodic collection dates during which they will take toxic household products for disposal free of charge. Some of these products, such as paints, can be reclaimed and used again.

One of the best strategies is to avoid toxic chemicals such as pesticides and cleaning agents in the first place. You can cut back on potentially hazardous materials by purchasing environmentally safe cleaning products and insecticides. Reducing your consumption of nonessential goods, whose production invariably creates toxic wastes that poison our land and water, also helps.

You can also learn about waste sites in your area and become active in grassroots organizations working on reduction, recycling, and safer disposal methods. Together, millions of Americans using resources wisely can make significant inroads into the hazardous-waste problem.

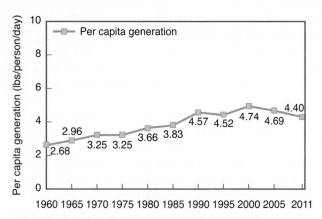

FIGURE 23-8 **Per capita municipal solid waste generation, 1960–2011.**

Data from Environmental Protection Agency (www.epa.gov/epawaste/nonhaz/municipal).

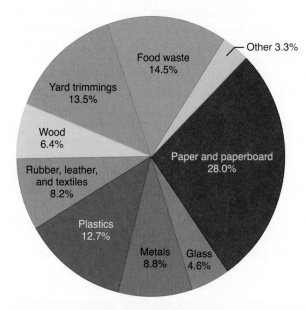

FIGURE 23-9 **Anatomy of U.S. waste, 2011.** This pie chart shows the composition of U.S. municipal solid waste by weight, after recovery of some recyclable and compostable materials.

Data from U.S. Environmental Protection Agency, 2011.

KEY CONCEPTS

Individuals can help reduce the hazardous-waste threat by properly disposing of hazardous materials, avoiding their use whenever possible, using nontoxic alternatives, and cutting down on nonessential consumption—because the production of goods often results in the generation of hazardous waste at factories.

Solid Wastes: Understanding the Problem

Each year, human society produces mountains of **municipal solid wastes:** discarded paper, metals, leftover food, and other items that come from businesses, hospitals, airports, schools, stores, and homes. The problems are especially acute in the more developed nations. In 2011, for instance, Americans generated 226 million metric tons (250 million tons) of municipal solid waste—enough garbage to fill the Superdome in New Orleans three times a day. On a more individual level, in 2010 Americans produced on average approximately 464 kilograms (1,021 pounds) per person—2 kilograms (4.4 pounds) per person per day (**FIGURE 23-8**). A city of 1 million people could fill the Superdome once every year. The composition of the U.S. municipal waste stream is shown in **FIGURE 23-9**.

Municipal solid waste production has increased sharply since 1980, but growth slowed in the 1990s and leveled off in the 2000s (**FIGURE 23-10**). The sheer volume of waste creates major problems in many cities, where land for disposal is growing scarcer and more costly by the day.

FIGURE 23-10 **Total municipal solid waste generation, 1960–2011.**

Data from Environmental Protection Agency (www.epa.gov/epawaste/nonhaz/municipal).

Garbage disposal is also of concern to those interested in building a sustainable future because it squanders the Earth's resources. The more that is thrown away, the more minerals that must be mined. The more we throw away, the more trees that must be cut. The more plastic we discard, the more oil wells that must be drilled. Each of these activities produces enormous waste itself and equally impressive amounts of environmental damage.

Science fiction writer Arthur C. Clarke once noted that "solid wastes are the only raw materials we're too stupid to use." In the United States, 53.7% of the municipal solid waste was buried in landfills, and 11.6% was burned in incinerators in 2011. Landfill disposal not only wastes valuable resources, it costs communities millions of dollars each year. For most local governments, the cost of trash disposal is usually exceeded only by the costs of education and of highway construction and maintenance.

Fortunately, many Americans recycle their waste. In 2011, 35% of all solid waste (our trash) was recycled and composted (**FIGURE 23-11**).

Like so many other problems, municipal solid waste is the product of many interacting factors: (1) large populations; (2) high per capita consumption; (3) low product durability; (4) our heavy dependence on disposable products; (5) low reuse and recycling rates; (6) a lack of personal and governmental commitment to reduce waste; (7) widely dispersed populations, in which producers of recyclable and reusable items are separated from those willing to purchase these materials; and (8) relatively cheap energy and abundant land for disposal.

KEY CONCEPTS

More developed nations produce enormous amounts of solid waste each year, much of which is burned or landfilled, squandering precious resources and creating an enormous and costly waste disposal problem, waste production is increasing in many countries such as the United States; recovery rates (recycling and composting) are growing much faster, a trend that bodes well for the future.

23.4 Solving a Growing Problem Sustainably

Actions to reduce our output of solid waste generally fall into three broad categories (**FIGURE 23-12**). The traditional response to solid waste is known as the **output approach.** It consists of ways to deal with trash flowing out of cities and towns. Most often, this means incinerating trash or dumping it in landfills. A more sustainable strategy is known as the **input approach.** This consists of activities that reduce the amount of materials entering the production–consumption cycle—for example, efforts to reduce consumption and waste, say, by increasing product durability. The third approach, also essential to building a sustainable society, is the **throughput approach.** It consists of ways to direct materials back into the production-consumption system, creating a closed-loop (cyclic) system akin to those found in nature. Reuse and recycling fall under this category.

KEY CONCEPTS

Solid waste management strategies fall into one of three categories: those that deal with waste after it has been produced, those that divert waste back into the production–consumption cycle, and those that prevent waste generation in the first place.

The Traditional Strategy: The Output Approach

The most widely used strategy throughout the world is the output approach, landfilling and incineration. Landfilling and incineration are end-of-pipe controls. Like many others discussed in this text, they are not sustainable in the long run.

KEY CONCEPTS

Worldwide, most trash is still dumped in landfills.

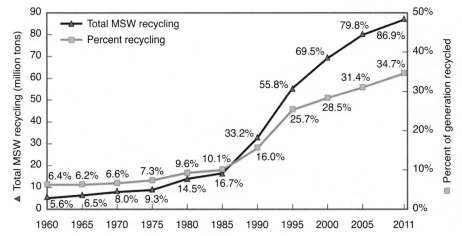

FIGURE 23-11 Municipal solid waste generation and recycling rates in the United States, 1960–2011.

Data from Environmental Protection Agency (www.epa.gov/epawaste/nonhaz/municipal).

FIGURE 23-12 Strategies for reducing solid waste. There are many ways of dealing with waste. The strategies fall within one of three groups: output, input, and throughput. In most cases, a combination of all three must be applied to alleviate the solid waste problem. Efforts should concentrate on the input and throughput solutions, which will go a long way in helping to create a sustainable society.

Dumps and Landfills Until the 1960s, garbage dumps were prevalent features of the American landscape. Public objection, however, to wafting odors, rat- and insect-infested midden heaps, and dark plumes of smoke that billowed out of burning dumps forced cities to look for other ways to deal with their growing trash problem. The federal government contributed to the demise of the dump by passing RCRA, discussed earlier. Besides addressing hazardous waste, RCRA also required all open dumps to be closed or upgraded by 1983.

The open dump has been replaced by the sanitary landfill. A **sanitary landfill** is a natural or humanmade depression into which solid wastes are dumped, compressed, and daily covered with a layer of dirt. Because solid wastes are no longer burned, as they were in many open dumps, air pollution is greatly reduced. Because trash is covered each day with a layer of dirt, odors, flies, insects, rodents, and potential health problems are eliminated or sharply reduced.

Despite their immediate benefits, landfills have some notable problems. First and most important, landfills require land. In the United States, the trash from 10,000 people in a year will cover one hectare (2.47 acres) about 1.2 meters (4 feet) deep. Around many cities, usable land is in short supply or is expensive. Second, landfills, like dumps, require a great deal of energy for excavation, filling, and hauling trash. Third, they can pollute groundwater. Toxic household wastes (paint thinner, pesticides, and other poisons) and feces (from disposable diapers, kitty litter, and backyard cleanup of Rover's messes) are discarded in municipal landfills, where they can leak into groundwater. Landfills are a major source of groundwater contamination and many retired landfills are listed as Superfund sites, as you learned earlier in the chapter. Fourth, they produce methane gas from the decomposition of organic materials. Methane can seep through the ground into buildings built over or near reclaimed landfills. Methane is explosive at relatively low concentrations. Fifth, landfills sink or subside as the organic trash decays, requiring additional regrading and filling. Buildings constructed on top of reclaimed landfills may suffer serious structural damage. Sixth, they have low social acceptability. Quite understandably, most people don't want

the noise, traffic, and blowing debris that come with even the best-managed landfills.

Although landfilling is ultimately an unsustainable strategy, there are many ways to make it more environmentally acceptable. Energy requirements, for example, can be cut by new methods of waste collection. Packer trucks now reduce waste volume by 60%, meaning that fewer trucks are needed to haul garbage to landfills. This saves on fuel consumption and pollution.

Vacuum collection systems can also be used to save energy in dense urban settings, especially apartment complexes. Solid waste is dumped into pipes that carry it to a central collection point. One such system is in operation in Sundbyberg, Sweden. Garbage is whisked away from wall chutes to a central collection facility, where the glass and metals are removed by an automated process. The burnables are incinerated, providing heat for the 1,100 apartments using the system. A similar system handles 45 metric tons (50 tons) of waste per day at Disney World in Florida. Today, over 400 such systems are in operation in Europe in hospitals, apartment buildings, and housing tracts.

Water pollution problems can be reduced by locating landfills away from streams, lakes, and aquifers. Test wells around the site can be used to monitor the movement of pollutants, if any, away from the site. Special drainage systems and careful landscaping can reduce the flow of water over the surface of a landfill, thus reducing the amount of water penetrating it. Impermeable clay caps and liners can reduce water infiltration and the escape of pollutants. In addition, pollutants leaking from the site can be collected by specially built drainage systems and then detoxified. The toxic seepage is shipped to hazardous-waste facilities.

Methane gas produced in landfills can be drawn off and sold as fuel, supplementing natural gas. Subsidence damage to buildings built on reclaimed sites can be reduced by removing organic wastes before disposal and by allowing organic decay to proceed for a number of years before construction.

Go Green

Recycle all wastes. Start a compost pile if you can— and buy recycled products whenever possible.

Open garbage dumps have been replaced by sanitary landfills, pits into which garbage is dumped and then covered daily with a thin layer of dirt to eliminate odors and pests. Although sanitary landfills are unsustainable, they can be made more environmentally acceptable by locating them away from groundwater supplies, collecting and treating toxic leachate, capturing methane gas, and other measures.

Ocean Dumping For years, many city officials have viewed the world's vast oceans as a huge garbage dump for wastes, including municipal garbage and human sewage. At one time, U.S. municipalities dumped their trash in 126 offshore dump sites. Although most of the solid waste dumped at sea was mud and sediment from dredging of harbors, estuaries, and rivers, considerable amounts of industrial hazardous wastes, radioactive wastes, municipal wastes, and sludge from sewage treatment plants were dumped at sea. Even today, city planners are trying to find ways to justify dumping some of their trash offshore. Some propose building offshore "islands of trash" from discarded automobiles, demolished buildings, and other solid wastes to support hotels and airports. Others talk of (and build) artificial reefs of discarded stone, toilets, cement slabs, and automobiles.

Ocean dumping of many wastes in U.S. waters declined as a result of the **Marine Protection, Research, and Sanctuaries Act** passed by Congress in 1972. The long-term goal of this act was to phase out all ocean dumping, especially of sewage and solid wastes. In 1977, the EPA issued regulations calling for an end by 1981 to all dumping of industrial wastes and sludge in the Atlantic Ocean, where 90% of the waste disposal was occurring. Although hazardous-waste and solid waste dumping declined sharply, sewage sludge disposal remained a significant problem because New York and other cities continued to dump sludge in the ocean for many years.

In 1988, the U.S. Congress passed the **Ocean Dumping Ban Act** (1988), which called for an end to sewage sludge disposal at sea by December 21, 1991. The EPA negotiated agreements with local jurisdictions that phased out ocean dumping of sewage sludge by June 1992 (**FIGURE 23-13**). Unfortunately, many other countries continue to view the oceans as a vast repository for their waste.

Oceans have long been viewed as a huge waste dump site for a variety of wastes, including radioactive materials. This practice is no longer legal in the United States, but it continues in other countries.

Incineration Incineration is another output control widely used in many industrial countries. In Denmark, 60% of the municipal solid waste is incinerated. The Netherlands

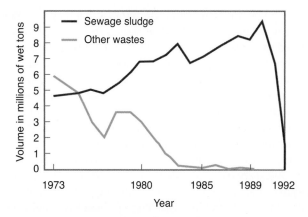

FIGURE 23-13 Signs of improvement. Ocean dumping of sewage sludge has declined significantly in recent years. Other waste dumping was largely brought to a halt in the early 1980s.

and Sweden both burn about one-third of theirs. The United States burns nearly 14% of its solid waste.

Burning trash can reduce the waste volume by two-thirds, cutting requirements for landfill space. Incineration can also be used to produce heat and electricity. Because of this, incinerators are often called **waste-to-energy (WTE) plants (FIGURE 23-14)**. One ton of garbage is equivalent to about one barrel of oil. But even with the energy gain, WTE plants cost more to build and operate than landfills. WTE plants are also much more expensive than recycling programs. Nationwide, recycling projects cost about one-third as much as incinerators. Each incinerator must be individually designed to accommodate the local mixture of burnable (leaves) and nonburnable refuse (steel cans). Operating these incinerators is made more difficult because the mixture varies from season to season. In the spring and fall, for example, yard and garden waste increases dramatically. But all yard waste does not behave the same. Wet leaves, for example, do not burn well. To avoid problems caused by wet leaves and other similar materials, municipalities may require homeowners to separate combustible material (dry branches) from wet organic matter (grass clippings).

FIGURE 23-14 WTE plant. This truck is dumping garbage that will be burned to produce electrical energy.

©Ton koene/age fotostock.

Chapter 23: Hazardous and Solid Wastes: Sustainable Solutions **583**

Another problem with incinerators is that they may emit toxic pollutants, especially when plastics are burned. Ash produced in garbage incinerators may be hazardous to human health. Test data show that the ash from incinerators contains dioxins and toxic metals, such as lead and cadmium, in concentrations that may be harmful to humans. Consequently, some scientists and environmentalists argue that ash from garbage incinerators should be reclassified as a hazardous substance and disposed of in hazardous-waste facilities, which is much more costly than dumping in ordinary landfills.

Ash disposed of in ordinary landfills may, over the years, begin to leak into groundwater, polluting public and private drinking water supplies. Cleaning up these sites could cost many hundreds of millions of dollars.

Many municipalities are opting for incinerators in which nonburnable materials such as glass and steel cans are removed before combustion, rather than for mass burn facilities, in which the entire waste stream is burned without separation. Separation of noncombustible wastes permits recycling and increases the combustion efficiency of incinerators. It is also much cleaner and is more efficient than mass burn, although it still produces hazardous materials.

WTE plants are a step in the right direction, but they are not as cost-effective and sustainable as waste management programs that rely on waste reduction, composting, and recycling. Incinerators also rank low on the social acceptability scale.

Incinerators may seem like a good way to solve the growing trash problem, but critical thinking suggests the need for a long-term view of the problem. The most important question today is not just how to reduce landfilling, but how to cut our waste; conserve valuable resources; protect our air, water, and land; and ensure vital wildlife habitat. In short, how do we create a sustainable system of waste management? Clearly, incinerators reduce trash but don't contribute significantly to the goals of sustainability.

KEY CONCEPTS

Garbage can be burned in incinerators, which greatly reduces trash. This option, however, even when linked to energy production, is viewed by many as an unsustainable way of dealing with municipal solid waste.

Sustainable Options: The Input Approach

Sustainable solutions are those that often approach problems by addressing their root causes—that is, by confronting them at their source. In solid waste, as in hazardous waste, the most effective and sustainable strategies are those that seek to reduce the amount of materials entering the production–consumption system in the first place. This strategy is called **source reduction.** The three main source reduction strategies include measures that (1) increase product life span, (2) reduce the amount of materials in goods and their packaging, and (3) reduce consumption (demand for goods).

KEY CONCEPTS

Source reduction techniques reduce the amount of waste entering the waste stream and represent the most sustainable waste management strategy.

Increasing Product Life Span: Making More Durable Products The trouble with today's products is that, as writer John Ruskin noted, "There is hardly anything in the world that some man cannot make a little worse and sell a little cheaper." In the long run, however, cheaply made goods end up costing consumers more than well-made and slightly more expensive goods. The rapid turnover may be profitable to businesses, but it is unsustainable from ecological and personal economic standpoints. Planned obsolescence of products destroys the air, water, and land. More durable toys, garden tools, cars, and clothing require less frequent replacement and thus decrease resource use. It's a great personal strategy and also can be a profitable business strategy because consumers who are fed up with shoddy products will gravitate to well-made ones.

KEY CONCEPTS

Higher quality, more durable goods last longer and consequently reduce the amount of waste. Manufacturers and consumers can play significant roles by making and buying more durable products.

Reducing the Amount of Material in Products and Packaging In the United States, packaging makes up 30% of all municipal solid waste, according to the EPA.

Of course, packaging is necessary, but much of it is superfluous and wasteful. (Consider the package containing the latest software you bought.) The Campbell Soup Company realized this and redesigned its soup cans; today, they use 30% less material than they did in the 1970s. Some beverage companies now package drinks in **aseptic containers**, boxes constructed of several thin layers of polyethylene, foil, and paper (**FIGURE 23-15**). The containers hold milk, juices, and wine and keep them fresh for several months without refrigeration. Aseptic containers require less energy, too. Canned drinks, for instance, must be pasteurized for 45 minutes, whereas the contents of aseptic packages are sterilized out of the package for only 1 minute. This not only greatly reduces energy demand but preserves flavor. Because beverages in aseptic containers like soy milk do not require refrigeration during transportation and storage, this lowers the energy demand. Being lighter than cans also helps cut down on transportation costs. Aseptic containers can be recycled, but access to recycling programs that handle them is fairly limited.

KEY CONCEPTS

Efforts to make products smaller and more compact can significantly reduce resource demand.

FIGURE 23-15
Aseptic container. Aseptic containers like this one are convenient and energy efficient but are fairly difficult to recycle.

© Jones & Bartlett Learning.
Photographed by Kimberly Potvin.

The Throughput Approach: Reuse, Recycling, and Composting

In natural systems, all waste is food material for other organisms. In short, there is no waste. Everything is recycled. Creating a sustainable society, say many experts, will require steps to greatly increase recycling efforts to emulate nature.

Recycling is part of the throughput approach. The second part of this approach is reuse. Both strategies remove useful materials from the waste stream and channel them back to manufacturers (in the case of recycling) or to end users (in the case of reuse) (**TABLE 23-1**).

Reducing Consumption H. W. Shaw once wrote, "Our necessities are few, but our wants are endless." Critics of modern-day society argue that ceaseless efforts to satisfy our endless wants are a big part of the solid waste dilemma the United States, Canada, and other countries face. By cutting back on consumption, individuals can help reduce solid wastes and many other problems described in this book.

Reductions in consumption will require shifts in our attitudes. The attitude that "new is always better" leads many consumers to purchase new goods when old ones still work. The fashion industry thrives on its ability to convince the public that new fashions are "in"—and that anyone wearing the old is "out of fashion." Advertisers capitalize on this strategy as well, and many consumers fall into the trap.

Another way to reduce consumption is through outright bans. Some cities, for example, have taken steps to reduce waste by banning objectionable materials—notably plastic bags—prompting the plastics industry to take measures to develop recycling facilities.

Each of us can make a personal effort to reduce consumption with little noticeable change in lifestyle. There are limitless possibilities; all we have to do is try.

The Reuse Option Reuse is the return of operable or repairable goods into the market system for someone to use. In most cities, organizations such as Goodwill and the Disabled American Veterans pick up used products, including clothes, shoes, silverware, plates, pans, books, tools, bicycles, and appliances. Many of them provide drop-off stations or pick up goods at your home. These products are cleaned and then resold to the needy and the frugal. Profits go to help the needy. For-profit secondhand stores also provide consumers with options for buying children's clothing and toys, furniture, appliances, plumbing supplies, and a host of other products. Check them out. You can often get great bargains!

Go Green

Donate old (usable) clothes and other items to Goodwill or similar organizations.

Packaging materials—such as cardboard boxes, bottles, and grocery bags—can also be reused, saving both energy and materials. Shopping bags can be reused by individuals. Reusable beverage containers can be sterilized, refilled, and returned to the shelf, sometimes completing the cycle as many as 50 times. Disposable and recyclable bottles and cans have virtually eliminated the reusable container from the market in more developed countries, although this technique is popular in less developed nations.

TABLE 23-2 compares the energy demand of refillable glass bottles (used 10 times) with various other packaging options. As illustrated, when it comes to energy demand, recycling is generally more efficient than one-time use (compare the energy required to make an aluminum can from raw ore versus recycled scrap). Reusing glass bottles only

TABLE 23-1

Reuse and Recycling of Solid Wastes

Material	Reuse and Recycling
Paper	Repulped and made into cardboard, paper, and a number of paper products. Incinerated to generate heat. Shredded and used as mulch or insulation.
Organic matter	Composted and added to gardens and farms to enrich the soil. Incinerated to generate heat.
Clothing and textiles	Shredded and reused for new fiber products or burned to generate energy. Donated to charities or sold at garage sales.
Glass	Returned and refilled. Crushed and used to make new glass. Crushed and mixed with asphalt. Crushed and added to bricks and cinder blocks.
Metals	Remelted and used to manufacture new metal for containers, buildings, and other uses.

Data from B. J. Nebel (1981). *Environmental Science*. Englewood Cliffs, NJ: Prentice-Hall, p. 297.

TABLE 23-2

Energy Consumption per Use for 12-Ounce Beverage Containers

Container	Energy Use (BTUs)
Aluminum can, used once	7,050
Steel can, used once	5,950
Recycled steel can	3,880
Glass beer bottle, used once	3,730
Recycled aluminum can	2,550
Recycled glass beer bottle	2,530
Refillable glass bottle, used 10 times	610

Data from L. R. Brown, C. Flavin, and S. Postel (1991). *Saving the Planet: How to Shape an Environmentally Sustainable Economy*. New York: Norton, Table 4-1, p. 71.

10 times, though, is four times more efficient than recycled glass. Unfortunately, there aren't many reusable products on the market other than canvas shopping bags and glass milk bottles available at some grocery stances.

Besides saving energy, the reuse option (1) reduces the land area needed for solid waste disposal, (2) provides jobs, (3) provides inexpensive products for the poor and the thrifty, (4) reduces litter, (5) decreases the amount of materials consumed by society, and (6) helps reduce pollution and environmental degradation.

KEY CONCEPTS

Reusing materials and products cuts down on resource demand and offers many other economic and environmental benefits. Citizens can participate in two ways: by donating used products and buying used goods.

The Recycling Option In human societies, **recycling** refers to the return of materials to manufacturers, where they can be melted down or chopped up, refashioned into the original finished material, and then incorporated into products. Paper, for instance, is shredded, immersed in water, de-inked, and then used to make more paper. Recycling conserves resources, alleviates future resource shortages, reduces energy demand, cuts pollution, saves water, and decreases solid waste disposal and incineration.

Consider some examples of the benefits: Each 1.2-meter (4-foot) stack of newspaper you recycle saves a 12-meter (40-foot) Douglas fir tree. Recycling a ton of newspaper saves 17 trees. Paper recycling uses one-third to one-half as much energy as the conventional process of making paper from wood pulp. Small savings can add up to make huge changes. If the United States, for example, increased paper recycling by 30%, we would save an estimated 350 million trees each year. Paper recycling has an added benefit of reducing the air pollution generated in making paper by 95%.

Aluminum recycling offers great benefits as well. Aluminum recycling requires 95% less energy than making aluminum from raw ore (bauxite). Thus, a manufacturer can make 20 aluminum cans from recycled metal with the same energy it takes to make one can from bauxite ore. Aluminum recycling produces 95% less air pollution as well. Similar environmental benefits are available from recycling other metals and plastics.

Japan is a world leader in recycling. Currently, about one-half of that nation's waste is composted and recycled. In the United States and many other countries, however, recycling efforts are a long way from achieving their full potential, although there are exceptions. One exception is the automobile. In the United States, approximately 90% of all cars are recycled. Lead in car batteries is another exception, with about 96% of all batteries making their way back into the manufacturing process. **FIGURE 23-16** shows current recycling rates of numerous materials in the

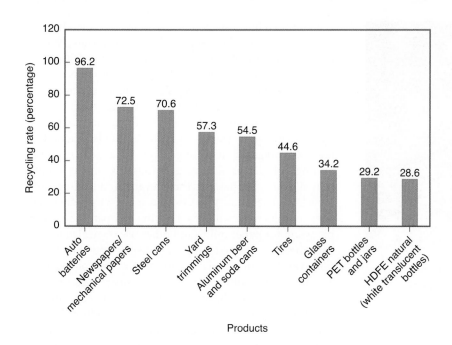

FIGURE 23-16 Recycling rates of selected products, 2011.

Data from Environmental Protection Agency (www.epa.gov/epawaste/nonhaz/municipal).

United States. Despite these impressive statistics, we recycle only 35% of all municipal solid waste. Recycling rates for many products could easily double or triple.

Recycling's Growing Popularity Recycling is widely practiced in Europe and Japan. Many major U.S. cities have developed recycling programs to reduce landfilling. In 2011, there were over 9,867 curbside recycling programs serving 161 million Americans, according to EPA's report *Municipal Solid Waste in the United States* (2011). Seven states—Minnesota, New Jersey, New York, Maine, South Carolina, South Dakota, and Virginia—report total recycling rates of 40% or more. Sixteen states report recycling rates of 30 to 39%. Impressive as this is, many other countries are doing much better. Japan, for example, currently recycles about half of its garbage. One suburb of Tokyo, for instance, currently recycles and composts 90% of its garbage.

Types of Recycling Programs Cities and towns offer a variety of recycling options. Some use drop-off sites, where residents can deposit their recyclables on the way to work or to the grocery store (**FIGURE 23-17**). In some locations, recycling companies will actually pay for some materials such as aluminum and copper. Drop-off centers can be successful, but generally only if containers are conveniently placed—for example, at train stations, near parks, or near heavy-use intersections. Many colleges and universities also offer recycling programs with recycling containers in the hallways of classroom buildings.

Curbside recycling, in which recyclables are periodically picked up by trash haulers at the curb, is by far the most successful type of recycling program (**FIGURE 23-18**). Participation rates as high as 60 to 80% can be expected if recyclables and trash are picked up on the same day.

Unsorted trash can also be collected and shipped to resource recovery centers, where it is separated by machines or people. This technique, called *end-point separation*, is generally more costly than source separation in which people sort recyclables at home or in the office. Moreover, complete separation may not be possible at these facilities, thus lowering the value of recyclable materials.

A growing number of recycling programs are being run by private haulers. Toronto, for example, is home to a program run by Waste Management, Inc.

FIGURE 23-17 Drop-off site. This roadside center is conveniently located so that people can drop off recyclables on the way to nearby shopping centers.

© Stephen Finn/ShutterStock, Inc.

FIGURE 23-18 Curbside recycling. Many cities have curbside recycling programs, which make recycling very convenient. Trash and recyclables are often picked up by different trucks.

© Marquisphoto/ShutterStock, Inc.

Obstacles to Recycling. From an energy and resource standpoint, recycling is generally not as good an option as reuse, but it is far better than burning materials and infinitely better than throwing them away. If recycling is such a good idea, why don't Americans, Canadians, and other people do more of it? The reasons are complex.

First, some industrial societies grew up with abundant resources. Steeped in the frontier notion that "there's always more," industrialists and political leaders have traditionally seen little

> **Go Green**
>
> At the end of the academic year, don't throw away perfectly usable items. Donate them through http://www.freecycle.org or to local charities. You can also contact http://www.dumpandrun.org. This organization collects discards and sells them to incoming students the next year, donating the proceeds to charity.

need for recycling, except perhaps in times of war when raw materials were in short supply. Given the seemingly inexhaustible supply of materials, factories were primarily set up to handle virgin material. The production–consumption system was built without recycling in mind. Changing this ingrained and wasteful system will not be easy.

Second, the nation's tax laws today support extraction and discourage recycling. As pointed out elsewhere, U.S. laws work against recycling. Even today, for example, mining companies receive generous tax breaks (depletion allowances) that give them an unfair advantage over recyclers. These tax breaks often make virgin materials cheaper than recycled ones. Logging companies that supply wood for paper mills (and other uses) are also heavily subsidized by the federal government—and thus by the taxpayer. Logging roads in national forests, for instance, are built with public money, and timber sales on public land have long been made below their cost, further benefiting virgin paper industry over recycling. Federal subsidies create unfair economics. Hard-rock mining companies can also purchase federal land at $5.00 an acre, thanks to the Mining Act of 1872. They are also not required to pay any royalty to the government for extracted minerals, which artificially reduces the price of minerals and metals.

The traditional extractive industries receive another hidden subsidy that's far more difficult to quantify. It is called an *economic externality*—a cost that is passed on to the public from pollution and other harmful effects of these activities. Because the traditional ways of making paper, steel cans, and other products produce more pollution, they have a bigger impact on our health and our environment than does recycling, which uses far less energy and produces considerably less pollution. The higher environmental cost of virgin materials, however, is not completely reflected in the price of the product. It is, instead, paid in federal taxes that go to clean up our air and water. It is paid in higher health bills and in dozens of other ways that few of us are aware of.

Another difficulty is the built-in transportation price differential, mandated by law, for scrap metal and ore. For example, ore travels more cheaply than metal bound for a recycling mill.

In some locations, such as the U.S. West, cheap landfill costs were once a barrier to overcome. In Colorado, for example, landfill tipping fees—the cost to dump a ton of trash in a landfill—once ranged from $4 to $7 per ton. Recycling costs $30 to $50 a ton. On the East and West Coasts, landfill tipping fees ranged from $40 to $100, making recycling far more profitable. As landfills were closed and new, more expensive sites were developed, however, tipping fees began to rise. In the northeastern United States, tipping fees now average $67 per ton. In most of the rest of the nation, where undeveloped land is more abundant, tipping fees now average $42 to $44 per ton. In addition, many interior states are a long way from the nation's paper recycling mills, making it more expensive and less profitable to recycle paper.

Recycling suffers from an image problem as well. In the 1970s, many recycled paper products were of inferior quality. Many people who used them were dissatisfied and soon returned to virgin materials. Since that time, however, recycled paper products have improved immensely. Recycled office paper, stationery, and computer paper are indistinguishable from virgin stock. Still, the notion that recycled paper is inferior sometimes persists.

Plastics pose a special problem for recycling. Most plastic is perfectly recyclable. The problem is that there are many different types of plastic used for packaging. Making matters worse, some packages contain two or more types of plastic, making it difficult to recycle. A plastic ketchup bottle, for example, has five layers of plastic, each one different, each one providing a special feature needed to make a perfectly squeezable bottle to deliver our ketchup.

One way plastic manufacturers have helped promote plastic recycling is by placing codes on plastic packaging so that individuals and recyclers can tell exactly what type of plastic it is made of (**FIGURE 23-19**). In most cities, only two or three of the most widely used plastics are recycled.

KEY CONCEPTS

Although recycling is a promising strategy for waste management, many obstacles hinder its full implementation including subsidies and cheap landfill costs. Two of the most important are federal laws and subsidies that give the raw materials industry an unfair economic advantage over recycling. The production of materials from virgin resources has a much higher external cost than recycling, but that's not reflected in prices.

FIGURE 23-19 Plastic code. The codes on the bottom of virtually all plastic containers tell the type of plastic the product contains, making it easier for individuals and recyclers to tell one from another.

© Jones & Bartlett Learning. Photographed by Kimberly Potvin.

Overcoming the Obstacles. Despite these barriers, U.S. recycling efforts are on the rise and are bound to increase substantially. In 1988, the U.S. Environmental Protection Agency announced a nationwide goal of reducing the solid waste stream by 25% through waste reduction and recycling by 1994, a goal reached a year early. In 2004, however, the rate had climbed to 29.7%. In 2009, it inched upward to 33.8%. In 2011, it had climber to 35%.

Lofty government goals are only part of the equation, though. Rising energy prices, decreasing landfill space, and depletion of high-grade ores will all increase recycling in coming years. Public awareness through education from schools, environmental groups, and the media can also help. Policy changes are also needed to eliminate preferential freight rates and subsidies for timber harvesting and mining.

To be successful, recycling programs must also be convenient and cost-effective. Recycling also requires individual commitment and action. Separating trash before collection requires a little effort, but not much more than the effort we expend to separate our white clothes from our colored clothes when we do our laundry. I like to think of source separation as the small price we have to pay for the riches the Earth gives us. It is a small sacrifice that pays off in a better world and a sustainable future.

To promote recycling among citizens, cities and towns can create several economic incentives. In Seattle, residents pay disposal fees based on the amount of trash they produce. Those who compost and recycle pay far less for trash removal than those who do not.

KEY CONCEPTS

Expanding the amount of garbage that is recycled will require removal of the subsidies that give benefits to companies that use virgin materials rather than recycled ones. Rising landfill costs, tipping fees, and fuel costs could help increase recycling rates. Making recycling programs convenient and economical can also stimulate this industry. Commitment on the part of citizens is also vital to making recycling a success.

Procuring Recycled Materials: Closing the Loop. Besides overcoming the barriers just described, to make recycling successful, we must also develop markets for recycled materials. Creating markets for recyclables causes an important shift in the production–consumption system and our impact on the Earth. Most human economies are linear in design, as illustrated in **FIGURE 23-20A**. They are dependent on resources extracted from the environment, which are then fashioned into products, used, and then disposed of. A sustainable industrial system based on recycling is cyclic (**FIGURE 23-20B**). It resembles the nutrient cycles in ecosystems. In such a system, materials flow cyclically through the system—from production facility to consumer and back for reuse a lot like carbon or nitrogen in ecosystems.

Although some raw materials will need to be extracted and some wastes will invariably be produced in a sustainable system of production and consumption, the goal of a sustainable economy is to maximize the amount of material that

(a) Linear production system

Wastes

Mines, forests, farms   Manufacturing and processing   Stores   Homes   Landfill

Wastes

Mines, forests, farms   Manufacturing and processing   Stores   Homes   Landfill

(b) Cyclic production system: The industrial ecosystem

FIGURE 23-20 Systems of production and consumption. By reducing waste, reusing, composting, and recycling, individuals and businesses can convert **(a)** the linear production system into **(b)** a cyclic one. Jobs will shift away from the ends of the linear system to the center, where recycling and composting are done. Jobs at all levels of income will be created in recycling and composting industries.

In November 1990, Germany's minister for the environment issued a decree calling for an 80% reduction in the amount of packaging entering the waste stream. To meet these goals, retailers are required to collect packaging materials and recycle them. This puts pressure on manufacturers to reduce excess packaging wherever possible and perhaps to help develop markets to use the recyclables. To promote public participation, the government is also levying a small deposit (equivalent to about 30 cents in U.S. currency) on most packages. Retailers can avoid the deposit and take-back requirements if industry establishes alternative collection and recycling systems that meet the established goals.

Laws requiring deposits on beverage containers, commonly known as *bottle bills*, can also dramatically increase recycling rates. At this writing, eleven U.S. states have bottle bills. In such states, over 90% of the glass and aluminum containers are returned for recycling. Container deposit legislation is also on the books in the Netherlands, Scandinavia, Russia, Canada, Japan, parts of Australia, and a handful of less developed nations.

is recycled within the economy. As you shall soon see, this will mean a shift away from employment at both ends of the system and toward the middle, where the recycling occurs (both pickup and remanufacturing).

Building an industrial ecosystem—that is, a system of commerce that is an analog of the natural ecosystems—means even greater increases in the amount of material that is returned by end users; but this is not enough. To create a truly cyclic economy, companies must be willing to use the materials picked up in curbside programs or drop-off sites, as noted above. Financial incentives and disincentives can be used to promote remanufacturing. For instance, Colorado offers a tax credit to companies that purchase equipment that allows them to incorporate recycled materials into the manufacturing process. In addition, people like you and I must be willing to purchase recycled paper and other products to help support manufacturers. If recyclers cannot find a market for their materials, all the recycling in the world is of no use.

Go Green

Help promote recycling by purchasing products made from recycled materials. You'll find there are many available once you start looking.

Recycling programs in the United States are currently hindered by small markets for some materials. For example, in the past decade, municipalities starting many new recycling programs have discovered that there are no markets for some materials or that the payments they will receive are inadequate to support their programs. When New York State began recycling plastic soft drink bottles, it couldn't find a buyer; the recycling program became an expensive trip to the landfill. Over time, though, markets developed, and the soft drink bottles are now ground up and used to make carpeting, filler for pillows, jackets, and other products. In the late 1980s, many cities found that newsprint recycling efforts were crippled by a dramatic decline in newsprint prices. Over the past two decades, gluts of recyclable materials have always seemed to end, as markets develop in response to rising secondary material supplies.

Recycled newsprint can be used to make a variety of useful products besides new newsprint. For instance, it can be used to make egg cartons, cereal boxes, map tubes, drywall, ceiling tiles, animal bedding, and insulation. The insulation filling the rafter spaces in my environmentally sustainable home is

100% recycled newsprint and cardboard. Cities and towns can encourage companies to manufacture these products from waste, thus putting locally available wastes to good use and helping build stable, self-reliant regional or state economies.

Individuals can help expand markets by purchasing recycled products. Governments can also help. When added together, local, state, and federal governments account for an estimated 40% of the U.S. **gross national product**—that is, the nation's total annual output of goods and services. Government is the single most important purchaser in the U.S. economy. Think of the impact that local, state, and federal governments could have if they all started buying recycled paper and other materials. The demand they created would greatly increase production, lower costs, and create supplies for the rest of us. For a summary of federal programs, check out the EPA's online Database of Environmental Information for Products and Services.

Today, all 50 states and the District of Columbia have procurement programs, purchasing millions of reams of recycled office paper and paper products. New York, for example, allows state agencies to purchase recycled paper if it comes within 5% of the cost of virgin stock; California allows purchase if it comes within 10%. These are called **price preference policies.** In these two states, about one-fourth of all the office paper, tissue, paper towels, and cardboard purchased by state agencies are made from recycled stock. On average, the states pay only about 2% more for office paper. Who knows what they saved in reduced pollution and health bills? To see a list of states with mandated recycled product procurement, check out Practice Green Health's website. You'll be amazed at the number of programs.

In October 1991, President George H. W. Bush signed an executive order that required all federal departments and agencies to purchase products made with recycled materials whenever possible. This order also requires federal agencies to name recycling coordinators to increase recycling of discarded items by the nation's 3 million federal employees.

Unknown to many, the Resource Conservation and Recovery Act called on federal agencies to purchase recycled materials in 1976, but only if the products were reasonably priced.

Unfortunately, "reasonable cost" provisions were interpreted to mean the lowest cost. Given federal transportation policy, the smaller markets, and the failure of current pricing systems to reflect external costs (among other problems), recycled products are often slightly more expensive. Colorado officials, however, have found that some items are cost-competitive or even cheaper. Furthermore, money saved on one product, can be used to pay the slightly higher price of another, thus keeping state spending constant.

As recycling becomes more common, more and more manufacturers are bound to shift to recycled materials. Recycling facilities are bound to open in many states, creating jobs and economic opportunities. In the not-too-distant future, recycling could well be the chief source of materials to make consumer products.

It may be surprising to many, but recycling is much more prevalent in the less developed countries than in the industrialized nations. The poor raid the dumps for food, clothing, and materials for shelter; they also seek out discarded metals and other goods that they can sell. The more developed nations can help promote recycling as they assist less developed nations in their efforts to increase their economic well-being. Information on the energy and material savings, as well as the technologies for recycling, could help convince these countries and businesses to adopt recycling policies.

KEY CONCEPTS

One of the most important boosts to recycling is purchasing products made from recycled materials, an activity in which governments, businesses, and individuals can all engage. Governments can help strengthen the markets by providing incentives to companies that use recycled materials or by requiring their agencies and departments to buy products made from recycled materials.

Composting. Another valuable throughput strategy that reduces waste and recycles materials is **composting**, the process in which nutrients from organic wastes such as leaves, grass clippings, cardboard, and paper are returned to the soil. Composting is a form of nutrient recycling.

In most large-scale operations, organic matter is collected from various sources, stockpiled, mixed with some dirt, and then allowed to decompose. The resulting product, **compost,** is a nutrient-rich, organic material that can be used to build soil fertility. Composting may occur in backyards, in neighborhood facilities, or in large municipal operations.

In 2011, 3,090 municipal composting operations were in business in the United States, according to the EPA. In Seattle, for example, zoo officials compost all of the manure produced by the zoo's many animals. When decomposed, the manure forms a rich, organic soil supplement they call *ZooDoo*, which is sold to gardeners and homeowners for use around the yard. Yard waste composting programs exist in virtually every state in the United States. Despite this success, most large-scale composting programs are in Europe—in the Netherlands, Belgium, England, and Italy—and in Israel.

Widespread composting, combined with recycling, can result in substantial reductions in municipal waste. Nationwide, organic wastes such as leaves, grass clippings, and kitchen scraps constitute about 25% of the garbage dumped into landfills or burned in incinerators, During the fall, compostable waste can make up 75% of the waste stream. Composting these wastes would not only reduce landfilling but could save considerable amounts of energy needed to transport wastes to landfills. Composting is often much cheaper—30 to 60% lower—than landfilling and can help recycle nutrients to farmland, closing the loops of an extremely important nutrient cycle.

Composting can be practiced very successfully at home. Gardeners can make their own compost piles of leaves, grass clippings, and vegetable wastes

Go Green

Start a compost pile at your home or your parents' residence.

from the kitchen. By mixing these materials with a little soil (which contains the bacteria that do the breakdown) and by watering the pile from time to time to keep it moist, homeowners can produce a nutrient-rich soil supplement for lawns and gardens—eliminating the need for artificial fertilizer. A commercially available container or a simple wooden enclosure helps keep neighbors' dogs or neighborhood raccoons and skunks out of the pile. Home composting also eliminates the need to haul wastes to central facilities.

Another waste product of cities and towns is sewage sludge. As noted, sludge is typically dumped in landfills, wasting very valuable soil nutrients. Fortunately, sludge can be combined with municipal solid waste (leaves, paper, grass clippings, and so on) and composted. This process, called **co-composting,** destroys viruses and bacteria in the sludge, so the product can be sold for use on farms or on gardens and lawns. Co-composting is expected to become increasingly popular in the United States and other countries because it costs less and is more ecologically sound than landfilling.

 KEY CONCEPTS

Composting is a way of returning the nutrients contained in organic matter such as yard waste and food scraps to the soil. This strategy not only reduces the need for landfilling, it helps nourish soils, creating a more closed-loop system.

The Economic Benefits of Recycling, Reuse, and Composting John Young of the Worldwatch Institute once wrote that "more efficient use of materials could virtually eliminate incineration and dramatically reduce dependence on landfills. It could also substantially lower energy needs….Taken together, source reduction, reuse, and recycling cannot only cut waste but also foster more flexible and self-reliant economies. Decentralized collection and processing of secondary materials can create new industries and jobs."

In fact, numerous studies show that recycling creates far more jobs than the traditional strategies of landfilling and incineration. A study in New York showed that recycling 10,000 tons of trash through curbside recycling programs produced, on average, 36 new jobs. Landfilling or incinerating that trash produced only about one job each. While some jobs will be lost in the front end of the production–consumption system—that is, at the mines and in the forests where raw materials are extracted—many more jobs will be opened up in the middle, the recycling portion. According to a 2011 study by the Scrap Recycling Industries, recycling in the United States is a $90-billion-dollar-a-year industry. It employs nearly 460,000 workers with wages of $26 million a year. In a 2011 report, the Tellus Institute estimated that doubling recycling rates in the United States would create 1.5 million new jobs in the next 20 years. As Young points out, while the economic health of nations is often measured by the amount of materials they consume, prosperity doesn't have to be so tightly linked to consumption. We can live well and live sustainably.

 KEY CONCEPTS

Recycling is not only good for the environment, it often costs less than other strategies and creates many more jobs.

We are poisoning ourselves and our posterity.

—Barry Commoner

CRITICAL THINKING

Exercise Analysis

This is a clear-cut case of dollars versus environmental protection that hinges on ethics—values. You are fairly sure that the incinerator in the developing country will produce toxic emissions and that spills along the way are also possible. Lax rules may result in considerable environmental contamination. Furthermore, waste from the incinerator will very likely be dumped in an open pit, where it could leak into the groundwater. In essence, your economic savings will be passed on as an economic liability to unsuspecting residents of the receiving nation.

A savings of a million dollars is fairly substantial, however. Your boss will no doubt take notice and may even give you a sizable raise for saving the company so much money. How do you decide? Which is more important: protecting the environment of another country or saving your company money and maybe setting yourself up for a raise? Do you feel a sense of intragenerational equity? What other options are available to you?

CRITICAL THINKING AND CONCEPT REVIEW

1. Summarize the major events occurring at Love Canal. Who was to blame for this problem? What might have been done to avoid it?

2. You are appointed to head a state agency on hazardous-waste disposal. You and your staff are to make recommendations for a statewide plan to handle hazardous wastes. Draw up a plan for eliminating dumping. Which techniques would have the highest priority? How would you put your plan into effect?

3. Discuss the major provisions of the Resource Conservation and Recovery Act (1976) and the Comprehensive Environmental Response, Compensation, and Liability Act (1980), the so-called Superfund Act. What are the weaknesses of each?

4. Describe the pros and cons of hazardous-waste management strategies, including process modification, recycling and reuse, conversion to nonhazardous or less hazardous materials, and perpetual disposal.

5. Debate the following statement: "All hazardous wastes should be recycled and reused to eliminate disposal."

6. A hazardous-waste site is going to be placed in your community. What information would you want to know about the site? How would you go about getting the information you need? Would you oppose it? Why or why not?

7. List personal ways in which we can each contribute to lessening the hazardous waste problem.

8. Discuss some of the options we have for getting rid of radioactive wastes. Which ones seem the most feasible to you? Why?

9. Debate the following statement: "Victims of improper hazardous-waste disposal practices should be compensated by a victim compensation fund developed by taxing the producers of toxic waste."

10. Describe the three basic approaches to solving the solid waste problem. Give examples of each one. Which is (are) the most sustainable? Why?

11. Describe the pros and cons of landfilling, incineration, source reduction, composting, reuse, and recycling.

12. What shifts in activities would result from changing the linear production–consumption system to a cyclic one? How will this affect jobs? In your opinion, are the proposed shifts necessary? How can we soften the blow of such a change?

13. Critically analyze this statement: "Recycling is not as great as advocates would have you think. Recycling takes energy, for example, to pick up materials and reprocess them. It produces waste and pollution."

KEY TERMS

aseptic containers
co-composting
compost
composting
Comprehensive Environmental Response, Compensation and Liability Act (CERCLA)
detoxification
gross national product
hazardous wastes
incineration
input approach

land disposal
low-temperature decomposition
Marine Protection, Research, and Sanctuaries Act
municipal solid wastes
NIMBY syndrome
Nuclear Waste Policy Act
Ocean Dumping Ban Act
output approach
price preference policies
process manipulation
recycling

Resource Conservation and Recovery Act (RCRA)
reuse
reuse and recycling strategies
sanitary landfill
secured landfills
source reduction
substitution
Superfund
Superfund Act
throughput approach
waste-to-energy (WTE) plants

ETHICS, ECONOMICS, AND ENVIRONMENT

Environmental Ethics: The Foundation of a Sustainable Society

Modern man is the victim of the very instruments he values most. Every gain in power, every mastery of natural forces, every scientific addition to knowledge has proved potentially dangerous, because it has not been accompanied by equal gains in self-understanding and self-discipline.

—Lewis Mumford

LEARNING OBJECTIVES

After reading and studying the material in this chapter, students will be able to understand, knowledgably discuss, and, when appropriate, apply the information learned in this chapter, including:

1. Unsustainable (frontier) ethics and how they adversely affect the environment, our society, and our lives
2. Sustainable ethics and how they can favorably affect the environment, our society, and our lives
3. How sustainable ethics can be promoted
4. Obstacles to the adoption of sustainable ethics and ways to overcome them

© wavebreakmedia/Shutterstock, Inc.

Exercise

Many people assert that the only way to get others to act reasonably to protect the environment is through economic incentives or disincentives. "People just won't act unless they see that it will help them financially," they say. "Either penalize them or give them some economic incentive if you want action." What is your opinion of this statement? Does it stand up to critical thinking?

This text began with an outline of today's environmental crisis—really a crisis of sustainability. Each part has explored a facet of it and has explained the role of individuals, businesses, and government in creating problems and solving them. Many chapters outlined traditional strategies, which address problems but often fail to address their root causes. These chapters also included a set of sustainable remedies—solutions that confront the root causes and thus hold great promise for creating lasting solutions and long-term prosperity. Included in the sustainable approach are ways to promote conservation, recycling, renewable resource use, restoration of renewable resources, population stabilization, and population growth management.

In order to bring about desired change, we also need a set of values that support a sustainable society. Without them, changes in our practices will be slow in coming. Earlier in the text, I described such a set of values, called *sustainable ethics*. They are based on the realization that our current way of life is not sustainable. This chapter reexamines the subject, highlighting key points already made and expanding the discussion to deepen your appreciation of our ethical options. We begin with a review of frontier ethics. Next, we explore in more depth sustainable ethics and discuss some ways to develop and implement a global sustainable ethic.

24.1 The Frontier Mentality Revisited

To understand sustainable ethics, we first examine the predominant ethic of today's industrial society, the frontier ethic (**TABLE 24-1**). The **frontier ethic** is characterized by three tenets: First, the Earth has an unlimited supply of resources for exclusive human use. In other words, "There is always more, and it's all for us." Embodied in this notion is the belief that the Earth has an unlimited capacity to assimilate pollution from human activities. Second, humans are apart from nature, rather than a part of it. In other words, we can survive without natural systems and are immune to the natural forces and ecological laws that affect all other organisms. Third, human success is best achieved through the domination and control of nature.

TABLE 24-1

Frontier and Sustainable Ethics Compared

Frontier Ethics	Sustainable Ethics
The earth is an unlimited bank of resources for exclusive human use.	**The earth has a limited supply of resources used by all species.**
When the supply runs out, move elsewhere.	**Recycling and the use of renewable resources will prevent depletion.**
Life will be made better if we just continue to add to our material wealth.	Life's value is not simply the sum total of our bank accounts.
The cost of any project is determined by the cost of materials, energy, and labor. Economics is all that matters.	The cost is more than the sum of the energy, labor, and materials. External costs such as damage to health and the environment must be calculated.
The key to success is through domination and control; nature is to be overcome.	**We must understand and cooperate with nature.**
New laws and technologies will solve our environmental problems.	Individual efforts to solve the pressing problems must be combined with tough laws and new technologies.
Humans are apart from nature; we are above nature, somehow separated from it and superior to it.	**We are a part of nature, ruled by its rules and respectful of its components. We are not superior to nature.**
Waste is to be expected in all human endeavors.	**Waste is intolerable; every wasted object should have a use.**

The frontier mentality has been a part of human thinking for tens of thousands of years. It may have begun to emerge in the hunting and gathering societies and was clearly present in agricultural societies. It flourishes today in the industrial world. The damage this outmoded way of thinking creates has reached enormous proportions, in large part because human numbers have increased dramatically and many of our technologies that are used to meet the needs of the nearly 7 billion Earth residents create enormous, now life-threatening, changes in the environment.

A prime example of the frontier mentality can be seen in the settling of North America. Early frontierists cut down forests to make room for farms and grew crops on the soil until the nutrients had been depleted or the soil eroded away. They then moved on, forging into new territory to start the cycle over again. Modern society displays the same brand of careless frontierism in its quest for wealth. By cutting down forests without replanting them; by depleting ocean fisheries with little concern for the long-term well-being of fish populations; and by polluting the air, water, and soil, modern society has created the potential for widespread ecological collapse.

The frontier ethic is not unique to the Western world. Similar views are evident in Latin America, Africa, and Asia. In Panama, for example, to promote economic development, past leaders have called for "the conquest of the forest." In 1940, Brazil's president wrote of the need to "conquer the land, tame the waters, and subjugate the jungle." In Asia, Japan has become a world leader in exploiting the global environment. Thailand and Indonesia have suffered mightily under resource development. In the 12th century, the king of Sri Lanka wrote, "Let not a single drop of water that falls on the land go into the sea without serving the people."

Perhaps the most troubling aspect of frontier ethics is the disregard for the Earth it fosters. Why worry about soil erosion or water pollution? Why worry about overfishing? Why worry about tropical forests? There is always more, and it's ours for the taking. Such thinking can lead us to pursue resources in high-risk areas, such as the Gulf of Mexico. The BP oil spill in 2010 demonstrated the enormous economic and environmental impacts of our all-out pursuit of resources, in this case oil.

Another aspect of the frontier notion that humans succeed best through the subjugation of nature is that this strategy often backfires on us, creating severe ecological backlashes such as pest resistance, increased flooding, wildlife extinction, global climate change, and severe soil erosion. Ecological backlashes affect our economic well-being and our own survival, underscoring an important but often overlooked fact: The future of people and nature is closely intertwined.

Today, people are learning to cooperate with nature—to weave human activities into the economy of nature. Biological pest control, watershed protection to control flooding, solar

Go Green

Hold on to your cell phone. Resist the temptation to upgrade every two or three years.

energy, growth management to protect valuable farmland, and a host of other measures are good examples of ways we can meet human needs without destroying the biosphere, our life-support system, and foreclosing on future generations. Refraining from building homes on barrier islands or in the floodplains of rivers and grazing cattle at the carrying capacity of grasslands are additional means of living and thriving within natural cycles and the limits inherent in them. They all pay huge dividends in the long term and represent a kind of insurance policy for the future. That is, they will allow us to prosper well into the future. The Spotlights on Sustainable Development included in many chapters in this book illustrate the ways people are learning to fit into nature's grand scheme, abiding by immutable ecological laws.

Despite these gains, the frontier mentality is still the dominant social paradigm of modern society. If you listen to politicians or newscasters, you will hear the talk of unlimited possibilities—for continued growth, bold new frontiers, and vast, virtually limitless resources. So deeply imbedded is the frontier mentality that it affects how most of us view our problems and how we go about solving them. What is more, nearly all of the social and political institutions function to maintain the chief goal of frontierism: continued growth. This deep entrenchment makes the frontier mentality difficult to dislodge.

KEY CONCEPTS

The frontier ethic has long historical roots and is deeply embedded in modern civilization. This notion of unlimited resources for exclusive human use and of humans being apart from nature, succeeding by dominating it, profoundly influences the way we meet our needs and how we view and solve environmental problems of great urgency.

24.2 Sustainable Ethics: Making the Transition

As entrenched as our belief system may seem, it can be changed. Colonialism and slavery and the beliefs that supported them, for instance, have fallen by the wayside. The Communistic Soviet Union fell dramatically and almost without warning, despite (or maybe because of) 60 years of rule. New paradigms have replaced these outmoded systems when it became clear that they did not serve humanity well. Although their demise was not without turmoil, they did fall (**FIGURE 24-1**). The same may hold true for the frontier ethic, which a growing number of people are recognizing has outlived its welcome. Change may occur as the evidence of global environmental decline and the social and economic repercussions of it become more evident. The current worldwide concern for global warming may be a catalyst for a global shift in ethics. What will replace frontier ethics?

FIGURE 24-1 Communism fell dramatically in the former Soviet Union despite 60 years of domination.

© Michael Euler/AP Photos.

Leopold's Land Ethic: Planting the Seed

One of the most influential proponents of sustainable ethics was the late Aldo Leopold, a wildlife ecologist best known for his book *A Sand County Almanac*. Leopold carried on the battle of John Muir, the founder of the Sierra Club and a long-time crusader for wilderness. Leopold described the need to include nature in our ethical concerns—more specifically, to extend our concerns beyond people. He called his ethic a **land ethic.** The land ethic teaches us to respect the land and ecosystems. It instructs us, in Leopold's own words, to enlarge "the boundaries of the community to include soils, water, plants and animals, or collectively: the land." In so doing, Leopold thought that the role of *Homo sapiens* would shift from that of "conqueror. . .to plain member and citizen of it."

Leopold first suggested the need for a land ethic in 1933. His book, which was published in 1949, took the message further. Charles E. Little, author and founder of the American Land Resource Association, calls the land ethic "one of the most important ideas of the century."

Leopold's view was considerably more encompassing than the view of Theodore Roosevelt–style conservationists, who protected resources principally because they were valuable to humans. The land ethic has helped to change the thinking of many people the world over.

Leopold suggested caution and deferred rewards in our use of natural resources. For Leopold, conservation required equal amounts of reflection and action. He called for individual responsibility in maintaining the health of the land,

but his guidelines revolved mostly around resource management. The land ethic, while planting the seeds of the ethical revolution needed to build a sustainable society, falls short of the complex challenges facing the world today.

A New View to Meet Today's Challenges: Sustainable Ethics

The main tenet of sustainable ethics is that "there is *not* always more." The Earth has a limited supply of nonrenewable resources such as metals and oil. Even renewable resources can be depleted if harvested at a rate that exceeds the replacement rate, so it is safe to say that many renewable resources have limits, too. Furthermore, the Earth has a limited capacity to absorb our waste. Truly, there are very real biophysical limits.

To develop a sustainable society, we must recognize that infinite growth of material consumption in a finite world is an impossibility. We must also accept the fact that ever-increasing production and consumption in a world of limits could destroy the life-support systems of the planet, upon which we depend.

Sustainable ethics also holds that humans are not *apart from* but rather *a part of* nature. We are dependent on natural systems for a variety of goods and services. This principle was called *dependence*. Sustainable ethics also holds that humans are subject to the laws that govern all life on Earth.

Just as humans depend on natural systems, so too these systems depend on us. Human society has an enormous impact on the future of the biosphere. The damage we create in the biosphere often has profound influences on people. This principle is called *interdependence*. The notions of dependence and interdependence are eloquently articulated in a quote once attributed to the Native American Chief Seattle who watched as European settlers usurped the Native American lands of the Pacific Northwest:

> You must teach your children that the ground beneath their feet is the ashes of our grandfathers. So they will respect the land, tell your children that the Earth is rich with the lives of our kin. Teach your children what we have taught our children—that the Earth is our mother. Whatever befalls the Earth, befalls the sons of the Earth. If men spit upon the ground, they spit upon themselves.
>
> This we know. The Earth does not belong to man; man belongs to the Earth. This we know. All things are connected like the blood which unites one family. All things are connected.
>
> What befalls the Earth befalls the sons of the Earth. Man did not weave the web of life; he is merely a strand in it. Whatever he does to the web, he does to himself.

Sustainable ethics also maintains that the key to success is through cooperation rather than domination and control. In short, we must learn to fit the human economy within the economy of nature. Put another way, we must live within the Earth's carrying capacity. Just as all aircraft must obey certain laws of aerodynamics, human actions must obey the laws of nature, respecting limits and protecting the life-support systems upon which our economy and our lives depend.

The core value of sustainable ethics is respect and care for the community of life. This value reflects a duty to one another and to other life-forms, both now and in the future. The core value implies that human activity should not occur at the expense of other species or other people. This is known as intergenerational equity (fairness to future generations), intragenerational equity (fairness to others who are alive today), and ecological justice (fairness to other species). As you have seen many times in this text, managing our activities so that they do not threaten the survival of other species or eliminate their habitats is as much a matter of ethics as it is of practicality.

One outgrowth of a change to a sustainable ethic is restraint. In regard to technology and development, the ability to say "we can" would not inevitably be followed by "we will." Instead, new questions would be asked: "Should we?" Should we build this dam? Should we build more nuclear weapons? Should we have another child? Thus, "we can" would be followed by this important question: "What are the environmental consequences?"

Sustainable ethics turns us away from self-centered thinking and favors good for the whole of society and the Earth. Restraint is exercised because it benefits all people alive today, future generations, and the many species that share the planet with us. Restraint could help create a high-synergy society in which the individual parts function for the good of the whole.

Sustainable ethics can be manifest by seven principles by which society can operate. They are: conservation, recycling, renewable resources, restoration, sustainable management, population control, and adaptability. These operating principles could help us reshape all human activities and systems, from the way we acquire food to the way we transport ourselves to and from work.

Sustainable ethics and practices are growing worldwide, as witnessed in 1992 by the Rio Conference and in 1997 by the follow-up conferences, both of which brought leaders from virtually every nation of the world together for a common goal.[1] Although the success of these efforts is less than expected, they did initiate many important programs.

Go Green

Wash with cold water using cold-water detergents. They dramatically reduce energy use.

The sustainable ethical system is a new paradigm that lays the foundation for a sustainable society. However different a sustainable society may be, it does not necessarily renounce all technology, all growth, or all material goods. Instead, it advocates a thoughtful look at the long-term

---

[1]The 1997 conference was judged by some critics as a limited success.

health of the planet and an evaluation of the consequences of technology, population growth, and materialism. All decisions are passed through the sustainability filter to determine if they enhance our present and our future.

Toward a Humane, Sustainable Future

In 1992, the late Donella Meadows, Dennis Meadows, and Jörgen Randers, three of the four authors of the groundbreaking 1972 book *The Limits to Growth* (which forecast social and environmental disaster as a result of trends in population growth, resource use, and pollution), published a sequel entitled *Beyond the Limits*. In it, the authors noted that many trends have worsened. They wrote, "The human world is beyond its limits. The present way of doing things is unsustainable. The future, to be viable at all, must be one of drawing back, easing down, healing." Sustainable ethics can provide an ethical framework for these changes.

Because ethics are the foundation of human behavior, changes in the way we think could begin to alter the ways we act. If our actions are based on a set of principles conducive to a humane, sustainable existence, our future could shift similarly. But how does one go about changing the way the world thinks?

24.3 Developing and Implementing Sustainable Ethics

The daunting task of changing the way people think and act worldwide elicits great pessimism. Nonetheless, examples of the shift in ethical values abound. This section describes several ways by which we can change the way the world thinks. The Viewpoint in this chapter discusses the rationale behind a future-centered perspective.

"Why should I feel obligated to future generations? We're inevitably separated by time and space. My presence here on Earth now will have no influence on someone living 200 years from now."

You may have heard this opinion expressed by your friends—perhaps you even hold it yourself. But if you have ever explored a wilderness preserve, used a library, or visited a historical monument, you already have some reasons for being responsible. Much of what we value in our family, our society, and our world has been provided by our predecessors, sometimes at considerable cost and effort on their part.

In today's world we face a number of issues that will affect future generations even more profoundly than they affect us now. Exploding world populations, shrinking nonrenewable resources, and plant and animal organisms threatened with extinction all add up to one thing—an ailing environment.

These are not isolated problems. Each stems from a common perception of our relationship to the world and our future. This perception can be characterized by a description of people and things as unique, immediate, individual, and separate from everything else. Any solutions that we might use to resolve our problems would have to start by challenging this perception.

Let me propose four basic considerations that may suggest a new paradigm for understanding our relationship to the world and our future. I believe our moral responsibility to coming generations will follow directly from these.

1. Future generations will be essentially the same as we are. They may have different wants and priorities, but they will manifest the same basic needs for food, water, air, and space. In addition, they will have the same basic physical and mental capacities with which to interact with their environment. Once born, they too will claim a right to life and protection from life-threatening conditions, such as extreme temperatures, toxins, famine, and disease. To give them life without also providing the basic means to sustain and enhance life would be cruel. If we expect the species to continue, we are obliged to leave a hospitable environment for those still to come.

2. One is born into a given generation by historical accident. None of us chose when to be born or to

Why We Should Feel Responsible for Future Generations

Robert Mellert
The author teaches philosophy and future studies at Brookdale Community College in Lincroft, New Jersey. He has published numerous articles on process philosophy, ethics, and future studies.

Courtesy of Robert Mellert.

whom. Because we have no special claim to the time and place of our birth, justice would require that we have no more rights over the world and its resources than anyone else.

3. Our survival as a species is more important than our individual survival. This is confirmed in nature every day; parents, whether they be rabbits, wolves, whales, or humans, spend their energies to reproduce and care for their young before they themselves die. Many will even risk their own lives for their offspring. This is because life is not ours to keep, but to share with others.

4. Even after we die, the effects of our life continue. We will be present in the memories of others and in the habits and traditions we shared with them. Our ideas will continue to enlarge the range of options for others; even when these memories and ideas are no longer consciously a part of the future, they will ripple onward, actively influencing the course of future events and people. What has been can never die. We are what we have been given and what we have chosen to make of these "gifts." In short, we are the product of our ancestors—all that they died for and believed in—and we are the product of our decisions. Future generations will be the result of what we are now and how they use what we leave them.

If we accept these four simple ideas, it is easy to see why we have an obligation to future generations. Our obligation is based on the truth that we are more than unique and separate individuals, living only in the immediacy of the now. We are, rather, parts of a much larger whole, one that transcends space and time. As John Locke, the great English philosopher, once said, we owe the future "enough and as good" as we received from the past.

Promoting Models of Sustainability

The Spotlights on Sustainable Development in this text highlight examples of people, businesses, and governments that have taken significant actions based on their commitment to the Earth and to building a sustainable world. Models such as these are extremely important catalysts in the sustainable transition (**FIGURE 24-2**). Similar examples are often highlighted in books, television news reports, newspaper articles, documentaries, magazine articles, government reports, and other media. They not only inspire other people to take action, they offer practical guidance on the ways to achieve similar goals. In addition to inspiring hope, models illustrate that humans can prosper within limits. The successes of models are also broadcast by word-of-mouth and via the Internet. For a look at an inspiring model, see the story of one of my other publishers, New Society Publishers, **Spotlight on Sustainable Development 24-1**.

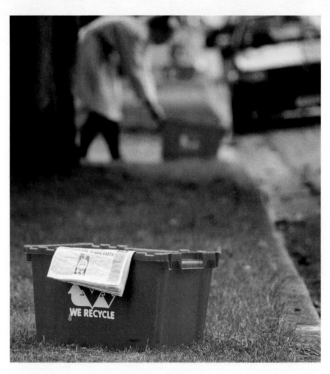

FIGURE 24-2 Models of sustainability. What people do has a profound influence, especially if it provides a sound model that others can study and emulate.

© Photodisc/Getty Images.

KEY CONCEPTS

Publicizing models of sustainable action offers inspiration and practical examples of ways individuals, businesses, and governments are building a sustainable future.

© hugolacasse/ShutterStock, Inc.

SPOTLIGHT on Sustainable Development

24-1 First Publisher in North America to Go Carbon Neutral

New Society Publishers is a progressive publisher on Gabriola Island, British Columbia that puts out about 25 books a year. As their name implies, their books are designed to help people understand the many pressing problems of our times and take actions to build a more just and economically and environmentally sustainable future.

In May, 2005, New Society Publishers became the first publisher in North America to become carbon neutral. What does carbon neutral mean?

It means that they have taken steps to ensure that their operations do not add to the carbon dioxide emissions to the atmosphere. New Society Publishers (NSP) achieved this goal, in part, by purchasing 213 metric tons of carbon offsets; these offsets, which I'll explain shortly, have neutralized the effect of the 213 metric tons of emissions that the company released into the

atmosphere during 2003 as a result of the publisher's use of paper, fuel, and electricity and their production of garbage.

"This is a great way to play our part, as we try to find ways to conduct our business in a more environmentally friendly manner," said Chris Plant, co-owner of NSP (**FIGURE 1**). "We see it as a fundamental component of being a socially responsible business and walking our talk."

Becoming "carbon neutral" is a new expression that is gathering momentum around the world. The growing list of companies and organizations that have become carbon neutral includes the HSBC Bank, with 10,000 offices in 76 countries; the City of Newcastle, UK; and Volvo's fleet sales team. If you know where to look, it is possible to buy carbon-neutral flights, cars, holidays, concerts, conferences, and houses.

FIGURE 1 Chris and Judith Plant, owners of New Society Publishers, were the first North American publishers to go carbon neutral. They have published ten of the author's books on renewable energy, green building, and related topics.

Courtesy of Dave Scougal and New Society Publishers.

In 2003, Canada's Premiers' Conference on Prince Edward Island became carbon neutral through the purchase of 850 trees planted by Tree Canada to offset the emissions from the ministers' flights to PEI.

New Society's 213 metric tons of carbon dioxide emissions came from its use of paper (140 metric tons), flights and couriers (65 metric tons), trucking and vehicle travel (7 metric tons), fuel for heat and power (0.75 metric tons), and the garbage it produced (0.14 metric tons).

Half of the publisher's emissions were offset by a contribution to Tree Canada, which will plant 182 trees to absorb 106 metric tons of NSP's emissions. The other half (for 107 metric tons) was offset by a contribution to the Solar Electric Light Fund (SELF), based in Washington, D.C. SELF brings solar electric lighting to villagers in countries such as South Africa and Bhutan, replacing the use of kerosene, which is a dangerous and expensive carbon-producing fossil fuel.

NSP's emissions analysis was done by climate specialist Guy Dauncey, author of the award-winning book *Stormy Weather: 101 Solutions to Global Climate Change* (New Society Publishers, 2001). Mr. Dauncey is himself carbon neutral, offsetting his personal emissions each year, a habit he shares with Hollywood stars Leonardo DiCaprio, Brad Pitt, and Jake Gyllenhaal. "It takes a bit of time and effort, and some number crunching," said Mr. Dauncey, "but it's nothing that any other business need be shy of."

Mr. Dauncey has also created a template that NSP can use in future years and a paper called *Going Carbon Neutral—A Guide for Publishers*, which can be used by any publisher in the world to help them do the same. This is available for free downloads on NSP's website at http://www.newsociety.com.

The Gabriola Island–based publisher, founded in the late 1970s, has a staff of eight people and, as noted above, publishes over 25 titles a year, including ten of my books. Its goal as a company, says Plant, is "to contribute in fundamental ways to building a more ecologically sustainable and just society."

NSP has a history of environmental firsts. In 2001, they became the first publisher in North America to print all of their books on ancient forest–free paper (100% postconsumer recycled, nonchlorine bleached). Since then, NSP has saved thousands of trees with this strategy, receiving awards from both the publishing community (Publisher of the Year) and Canadian Businesses for Social Responsibility (Ethics in Action). Mr. Dauncey's analysis shows that by switching to postconsumer recycled paper NSP prevented 112 metric tons of CO_2 from entering the atmosphere in 2003, because recycled paper has a smaller environmental footprint than virgin paper. Along with other decisions, using forest-free paper has helped NSP's annual emissions to be 119 metric tons fewer than they would otherwise have been.

Judith Plant, NSP's other co-owner, said, "Our staff has been really behind us on this one. As soon as we knew what our emissions were, we sat down and discussed how to offset them. We plan to continue doing this every year."

NSP is the second publisher in the world to become carbon neutral. The first was the small British publisher Snowbooks, which was launched in 2004.

Education

Educators—at all levels of the educational system—can play an important role in forging environmentally sustainable attitudes and actions among children, teenagers, and adults the world over. Recognizing the importance of environmental education, some states (such as Wisconsin) have made environmental education mandatory. Some colleges, such as Prescott College in Arizona and Colorado College in Colorado Springs, Colorado, where I teach, include environmental education or environmental sustainability in their mission statement.

Care must be taken in developing an environmental education program. Steve Van Matre, head of the Institute for Earth Education, has proposed an alternative to issues-based environmental education for schoolchildren. He argues that environmental education for school-age children is often disorganized and hit-or-miss. It falls down in three areas: (1) It fails to develop a true understanding of the importance of the environment in supporting our lives. It also tends to overlook (2) how we individually impact natural systems and (3) what we can do personally to change our habits. By focusing on these three areas and developing strong connections with the environment, Van Matre believes, we can make important gains in environmental education. David Orr, director of the environmental studies program at Oberlin College, argues for the need to revamp college curricula and the ways colleges operate, making them models of sustainability. The environmental studies building Orr developed with students and experts—and now teaches in—is a model of sustainable building (**FIGURE 24-3**). It uses about one-fifth as much energy as a conventional teaching facility and is heated by passive solar energy and powered in part by solar electricity. It can be naturally ventilated and recycles water. It even uses an electrolyzer to produce hydrogen gas for heating and hot water. Numerous other colleges have followed suit, building classrooms and offices with many environmentally friendly features (**FIGURE 24-4**).

KEY CONCEPTS

All levels of education can be enlisted in the effort to foster an understanding of ethics and actions that will contribute to a sustainable future.

Churches

Churches can also become an important force in creating global change needed to foster a sustainable future. Religious leaders who are not already involved, for example, may want to explore a broader sustainable ethic with their parishioners and suggest ways to change. Many churches sponsor adult education classes that deal with social, political, and environmental issues. These also can be an excellent route for raising awareness and promoting a new Earth ethic. Some churches have even gone so far

(a)

(b)

FIGURE 24-3 Practicing what they teach. (a) Oberlin College's environmental studies program is taught in this state-of-the art classroom building on campus. **(b)** Workers install photovoltaic panels on the roof of the building during its construction.

Photo a courtesy of Oberlin College; photo b © John M. Walsh, The Morning Journal/AP Photos.

FIGURE 24-4 The Tutt Science building at Colorado College, where the author is a visiting professor, is LEED-certified with many green features, including energy-efficient design, energy-efficient lighting, water-efficient bathrooms, and a host of environmentally friendly building materials.

Courtesy of the Tutt Science Center at Colorado College.

FIGURE 24-5 Solar-powered church. This large solar electric array located in the parking lot of the Wesley United Methodist Church in Sedalia, Missouri, helps power the facility.

Courtesy of Dan Chiras.

as to install solar electric systems to produce their electricity (**FIGURE 24-5**). Reverend Larry Rice who resides in Missouri not only preaches environmental responsibility, his ministry offers numerous classes, publications, and DVDs on renewable energy and other similar topics (**FIGURE 24-6**).

 KEY CONCEPTS

Churches are an established forum for education whose efforts could be partially dedicated to the exploration of sustainable ethics and sustainable lifestyles.

Declarations of Sustainable Ethics and Policy

Nations can help redirect human values and actions on a broad scale by adopting declarations of sustainable development and by ratifying constitutional amendments that make the environment a leading consideration. In the United States, for instance, numerous states have passed laws or adopted official language embracing the notion of sustainable development. Many businesses have done the same. Although talk can be cheap, such acts can have a solidifying effect—galvanizing people around key concepts and goals. In 1992, over 170 nations adopted the **Rio Declaration**, a kind of international declaration for sustainable development (see below).

KEY CONCEPTS

Official declarations by states, nations, and groups of nations can help promote widespread adoption of sustainable ethics and actions. However, such proclamations need to be followed by actions to be truly meaningful.

A World Organization Dedicated to Sustainable Development

Participants in the 1992 Earth Summit in Rio created an organization known as the **United Nations Commission on Sustainable Development (UNCSD).** UNCSD consists of high-level government representatives—heads of governments' environmental protection branches—with the power to make decisions. The commission uses its powers of publicity and peer pressure to encourage the implementation of **Agenda 21**, a massive blueprint for sustainable action. It also promotes the Rio Declaration, a statement of 27 principles regarding environment and development, and

FIGURE 24-6 Reverend Larry Rice. A minister and tireless advocate of renewable energy and sustainable living.

© BILL GREENBLATT/UPI/Landov.

the agreement on forest protection. The commission could become an international watchdog, overseeing promises made at the Rio conference, "shining the spotlight on countries that renege and pushing through stronger commitments as necessary," as Worldwatch Institute's Hilary French says.

Equally important, the commission helps coordinate the UN's many programs and agencies that deal with the environment—among them the Development Program, the World Bank, the Food and Agriculture Organization, and the International Monetary Fund. All UN entities will report to the UNCSD on what they are doing to integrate the sustainable policies adopted at the 1992 Earth Summit.

This effort could go a long way toward instilling sustainable beliefs and practices among important players in the world community. At this writing, however, the commission's success has been rather limited.

 KEY CONCEPTS

The United Nations Commission on Sustainable Development is an international organization dedicated to promoting sustainable development.

A Role for Everyone

One of the recurring themes of this text is that sustainable development will require efforts on the part of all of us. Without widespread participation and cooperation based on a new environmental ethic, sustainable development is unlikely to succeed. Political leaders can play a major role in re-educating people through speeches and television advertisements. Parents can help by teaching their children and by engaging in environmentally responsible behaviors like recycling. Children can help by teaching their parents. Businesses can work with their employees, and employees can help instill a sustainable ethic in businesses they work in. The opportunities are limitless.

Surely, promoting a global sustainable ethic will be slow and uphill work. Many people do not see the need for change, and others will resist change because it threatens personal interests. Clearly, many obstacles exist. Overcoming those obstacles requires an understanding of them.

KEY CONCEPTS

Participation by all sectors of society is essential to building a sustainable society.

24.4 Overcoming Obstacles to Sustainability

The Industrial Revolution did not occur overnight—nor did the Agricultural Revolution. The Sustainable Revolution will very likely take a century or more. The reason for the slow

pace is simple: There are many obstacles to the adoption of sustainable ethics and sustainable practices. This section explores some of these obstacles and ways to overcome them.

 KEY CONCEPTS

Many obstacles lie on the road to sustainability, and the process will likely take a century or more.

Faith in Technological Fixes

Many people express extreme optimism in the power of technology to solve our problems, a phenomenon labeled **technological optimism.** The late economist Julian Simon for instance, said that there are no limits. As long as we exercise our ingenuity, we can develop new technologies to expand our reserves or find substitutes. So why change our views? The world is unlimited!

In their zeal to find high-tech solutions, technological optimists often overlook logical, inexpensive, and Earth-friendly low-tech solutions, such as conservation. As you have seen in other chapters, conservation is a far more effective way of reducing global carbon dioxide emissions and is far cheaper than nuclear energy, carbon sequestration, even renewable energy technologies, or other options. A dollar invested in conservation, in fact, reduces carbon dioxide emissions seven times more than a dollar invested in nuclear power. A dollar invested in energy efficiency and conservation reduces carbon dioxide emissions three to five times more than a dollar invested in a residential solar electric system. Recycling is a far cheaper solution than high-tech trash incinerators. Nonetheless, many municipalities are still building incinerators.

Unfortunately, humanity and often unqualified optimism in technology blinds us to the obvious solutions at our feet. Spurred by thoughts of limitless resources, many applied scientists, technologists, politicians, and businesspeople promise an unlimited future. Believing that these experts can't be wrong, many people place their faith in their proposed solutions and resist changing their views.

Today, more and more people are beginning to understand that technological fixes won't save the day and are calling for simpler, more effective solutions that strike at the root causes of the crisis of unsustainability. This is not to say that technology has no role in creating a sustainable world. Far from it. Technology will play an important part in the future. The solar system that is powering my computer is a good example of the positive benefits of technology. The point is that technology is not enough. A new ethic that fosters

 KEY CONCEPTS

The belief that technology can solve all of the world's problems hinders the development of a sustainable ethic and the implementation of simple, cost-effective, sustainable practices.

personal responsibility and action is a large part of the equation. Education can help illustrate the limits of technology and the many positive examples of ways that sustainable ethics and personal actions can help forge an enduring human presence.

Apathy, Powerlessness, and Despair

Many people are apathetic about the course of modern society. Involved in their own lives, they see resource limitations and pollution as problems for which someone else must take responsibility. Apathy is effortless, noncontroversial, and cheap.

Where does apathy come from? In part, many of us are taught or conditioned to be apathetic—not to rock the boat or make waves. Someone bulldozes a favorite forest of ours to put up a shopping mall, and someone else calms our indignation by saying, "That's the price of progress."

Apathy may also stem from our early education. Government and history teachers, for example, instruct students about democracy but may fail to drive home the important principle that democracy relies on participation. To many of us, democracy means freedom, but democracy also comes with certain responsibilities.

Go Green

Wash only full loads of laundry or adjust the water level on washing machines to save water and energy.

First and foremost, we have a responsibility to participate: to vote, to write congressional representatives, to sit on citizen advisory committees. Many people, however, don't even vote.

Another cause of apathy is powerlessness. In the United States and other countries, people are often paralyzed by a sense of powerlessness—feelings of insignificance. Big corporations influence government decisions, steamrolling citizen interests. If you can't do anything, why bother?

This pervasive feeling has a powerful effect on our own lives and on the environment. Feelings of insignificance create many of the environmental problems facing the world and keep us from solving them. I call this the **Paradox of Inconsequence**. Here's how it goes: Being just one of the more than seven billion people on Earth is an excuse many of us use for continuing to do what we have always done, living unsustainable lifestyles. What difference, we ask, does it make if you drive 10 miles per hour over the speed limit and waste a little gasoline? You're just 1 of 316 million Americans and more than 7 billion Earth residents. You're just a small fraction of the problem.

The trouble with this thinking is that millions of people think the same thing and thus go about their affairs as if their roles were of no consequence. Their actions can add up to a great deal of waste and pollution, though. The many problems we have are, in fact, the result of millions of individuals' actions.

Ironically, feelings of insignificance also keep us from solving many problems. What difference does it make if you drive the speed limit, recycle aluminum and paper, and keep your thermostat at a reasonable temperature? Because you are only 1 of over 316 million Americans or 7 billion world citizens, your contribution to resource depletion and pollution is insignificant. So why bother acting responsibly?

Feelings of insignificance create many of the problems we face and keep us from solving them. But all of this can be turned around. Millions of people acting responsibly can have huge impacts. It is a powerful mathematics. For example: Ascend a marble stairway that tens of thousands have ascended before you. Feel the grooves worn by their feet. Although each one would say he or she had no effect, together all have had an enormous impact. If we can have a negative impact, we can also have a positive impact through our actions. Education can go a long way to convince individuals of the collective benefit of responsible living.

FIGURE 24-7 shows a hypothetical curve that plots apathy and its close cousin despair against empowerment—taking control of your life and taking action. It suggests, and with good reason, that as people take control of their lives, get active, and work toward change, apathy and despair tend to decrease. A good example of what individuals can do is presented in **Spotlight on Sustainable Development 24-2.**

KEY CONCEPTS

Apathy and feelings of powerlessness, insignificance, and despair hinder progress toward sustainability. Many examples show that individuals can be truly powerful in changing the world.

The Self-Centered View

The average man and woman on the street often take a self-centered view. As a rule, their actions are driven not by concern for sustainability but by their own welfare. Author and social critic Tom Wolfe coined the phrase *the me-generation* to describe the Americans of the 1980s who seemed intent on self-indulgence. The 1990s proved to be much the same. Many people of the me-generation buy

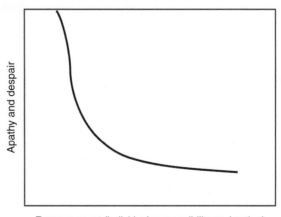

FIGURE 24-7 Apathy and despair. Getting active can help reduce one's feelings of despair. When an individual's actions are combined with those of others, people can make a big difference.

© hugolacasse/ShutterStock, Inc.

24-2 Students Put Rotting Garbage to Use Heating School

Science students in Maryland Heights, Missouri, launched an ambitious program that could start a trend in the United States and abroad. Their project: to capture methane, a combustible gas released from a nearby landfill, to heat their school.

Today, thanks to the efforts of a handful of students and a local businessperson, all 117 classrooms and two gymnasiums at Pattonville High are heated by methane from rotting garbage. Pattonville High is believed to be the first school in the United States that is heated by methane from rotting garbage.

The methane, piped to the school by an 1,100-meter (3,600-foot) pipeline from the 35-hectare (85-acre) landfill, is burned in the furnace in place of natural gas. What makes this all the more exciting is that the landfill operators were burning off the waste gas. Even more exciting is that this project, which cost $175,000, saves the school $40,000 a year in heating bills. The savings would pay the capital costs in less than 5 years. The principal of the school, Tom Byrnes, said, "The methane gas is there. It's a matter of burning it off or using it productively."

The school is not entirely heated by methane. Some natural gas is needed for the cafeteria and other operations, but the use is minimal.

The U.S. Environmental Protection Agency (EPA) estimates that 750 landfills in the United States could be tapped to generate gas. If only half of them were tapped, it would cut the emissions of thousands of tons of carbon dioxide each year. According to the EPA, this would be equivalent to removing 12 million cars from the road. Obviously, landfill waste gas is not the answer to creating a sustainable future, but it can help save money and reduce emissions.

New regulations require that landfills with more than 2.5 metric tons (2.75 tons) of trash be tested for gaseous emissions. If emissions exceed a certain level, the new Clean Air Act amendments require the operator to install a system to collect and burn off the gas, to prevent possible explosions caused when the gas reaches certain levels. If methane can be put to good use, say proponents, it should be.

---

what they can afford, giving little thought to the effects of their consumption on the environment. Replacing the self-centered approach to life with a global environmental perspective is the main thrust of sustainable ethics.

Ironically, for many years, environmental groups failed to promote individual responsibility and action among their members and the general public. Much of their educational outreach was through direct mail. Their appeals generally outlined the problems we faced and presented proposals to solve them through legislative action. Individuals were asked to help, but individual involvement was generally limited to writing a check. Wendell Berry, philosopher and author of *The Unsettling of America*, writes that "the giving of money has. . . become our characteristic virtue. But to give is not to do. The money is given in lieu of action, thought, care, time." I call this phenomenon the *cash conscience*. Some groups have realized the shortcomings of this approach and have taken steps to promote more individual action.

Go Green

Hang clothes on a line to dry. It can save lots of energy.

Environmental groups have also spent much of their time pointing accusatory fingers at lawmakers, inept government regulators, and business executives. Although these parties are indeed part of the equation, so are individuals. This lopsided approach nurtures the idea that the blame for the environmental crisis lies almost entirely in someone else's hands, a phenomenon I call *blame shifting*. In addition, it fosters the idea that problems can be solved only by regulating someone else's activities or by applying new and more efficient technologies to control someone else's pollution. It nurtures the idea that money alone can solve our problems. The legislative and legal work of environmental organizations such as Environmental Defense, the Natural Resources Defense Council, Greenpeace, the Sierra Club, and others is vital to our efforts to improve the environment, but individual action is equally, if not more, important.

KEY CONCEPTS

Concern for the self, the cash conscience, and blame shifting hinder efforts to foster individual action and responsibility.

Ego Gratification

According to psychologists, as we pass from infancy to adulthood we develop a sense of the self, of our own separate identity. Psychologists call this the *derived self*. Once the derived self becomes established, it also needs reaffirmation. We can get that from our parents, friends, teachers, and ourselves. If not, we may seek it in new clothes, fast cars, and extravagant homes, self-rewarding behavior that builds our egos or reaffirms that we are important. After all, important people live in big houses and drive expensive cars, don't they? Reaffirmation may be at the roots of materialism and overconsumption. They lead to the accumulation of material possessions, which depletes essential resources, causes pollution, and reduces wildlife habitat. What can a society do to build feelings of self-worth?

Adult education programs and better training in health classes can help us break patterns of child rearing and create new ones. Religious leaders can help as well.

 KEY CONCEPTS

Feelings of inadequacy are often offset by the accumulation of material possessions. Overconsumption is a key factor in the decline of the environment.

Economic Self-Interest and Outmoded Governmental Policies

Most of us live in countries whose socioeconomic systems are out of line with the natural world—that is, whose human systems are operating unsustainably. Greed, ignorance, and a host of other factors drive us further down the road to ecological ruin. Our governments and the economic systems they often serve are therefore a barrier to sustainability. Although other chapters have pointed out some of the flaws in both government policy and economics, the next three chapters go into more depth, showing where reform is needed to place these human activities back on a sustainable course.

 KEY CONCEPTS

Unsustainable economic systems and governments that support them are major barriers to sustainable development.

24.5 Sustainable Ethics: How Useful Are They?

Each day people are faced with dozens of decisions, many of which influence the environment. For example, should you ride your bicycle or drive your car to school or work? Should you turn up the heat or put on a warm sweater? The sum of the decisions made by many millions of individuals such as yourself adds up. Environmentally sustainable decisions result in positive outcomes. Environmentally unsustainable decisions create negative results.

Corporations and governments face many decisions as well. Single decisions at this level have impacts as profound as many millions of individual decisions. For example, should the government develop nuclear power or push for energy conservation? Should the government use recycled paper or continue buying virgin stock? Should a corporation produce disposable products?

Ultimately, most of our decisions are influenced by our values—what we view as right or wrong, desirable or undesirable. Values are learned from our parents, relatives, peers, teachers, religious leaders, politicians, writers, news commentators, reporters, books we read, and articles in magazines and newspapers. Values may shift over the years, sometimes drastically.

Many people today base their judgments on **utilitarianism**, a doctrine by which the worth of things is determined by their usefulness. This view tends to put human needs above most, if not all, others. Economics is the yardstick of many utilitarian decisions. Utilitarian resource management, for example, finds the fastest, cheapest ways of acquiring resources. It puts human needs above all else. Thus, forests may be reseeded, but not so much to provide habitat for animals as to provide wood for future generations.

At first glance, frontier ethics appears to be heavily utilitarian. It serves human needs. After closer examination, though, it is clear that it is anything but utilitarian, for it leads to destructive behavior that forecloses on our future. In the long run, frontier ethics is nonutilitarian.

At the opposite end of the spectrum is a new and controversial view of **natural rights.** It says that all living things have rights, irrespective of their value to human society. Wilderness should be set aside, not just so people can use it, but to protect the species that have lived on the land for thousands of years. Animals, plants, and insects should be preserved, not because they are of use to humans, but because they have a right to live.

Another value system is based on **divine law**, the word of God as written, for example, in the Bible. For the most part, divine law dictates personal behavior—interactions between people. The Ten Commandments instruct us on ways to treat one another: Thou shalt not kill. Thou shalt not commit adultery. One religious group has recently added an eleventh commandment: Thou shalt protect the Earth. Divine law does indeed call on us to protect the environment, God's creation. Virtually every religion has some teachings to advance this mandate.

These are some of the principal value systems that affect our decisions. As you think about them, you may realize that no one system determines your actions at all times. In some cases, you may act out of utility. In others, you may make a sacrifice for the good of the whole.

Sustainable ethics is another system of values. It calls on us to consider a new set of parameters in making decisions. It reminds us of our place in the world and implores us to act in cooperation and consideration, rather than in isolation and strict self-interest. At first this may seem anything but utilitarian. It is utilitarian, however. Sustainable ethics is a means of serving human needs and the needs of nature, which are the same. It serves people in the present and the future. In many ways, sustainable ethics combines aspects of natural rights and utilitarianism. Some religious scholars also contend that it incorporates divine law. Bottom-line, sustainable ethics and practices that manifest these ideas will help all countries ensure brighter futures, stronger economies, healthier populations, and even more efficient government.

KEY CONCEPTS

Values affect the way we act. To some people, the value of things is based on their usefulness to humans. Sustainable ethics may be one of the most useful sets of values because it serves people, the environment, and the economy much better than shortsighted, outmoded systems of belief.

A human being is part of the whole, called by us "Universe." He experiences himself, his thoughts, and feelings as something separated from the rest—a kind of optical delusion of his consciousness. This delusion is a kind of prison for us, restricting us to our personal desires and to affection for a few persons nearest to us. Our task must be to free ourselves from this prison by widening our circle of compassion to embrace all living creatures and the whole of nature in its beauty.

—**Albert Einstein**

CRITICAL THINKING

Exercise Analysis

You, your friends, and your parents probably do many things to help protect the environment that don't benefit you economically. For example, you may recycle even though you probably don't earn enough from the bottles and cans you return to make it worth your while economically. So do millions of other people in the United States and other countries. You and your family are therefore living proof that some people do act for unselfish reasons.

The cynic who asserts that people respond only when they're financially benefited is making a broad generalization that fails to stand up to scrutiny. A more accurate statement might be that some people will take actions on their own, whereas others seem to respond best when they're financially affected. People who make statements such as the one that launched this exercise may be motivated by money themselves. Consequently, they may think that everyone else is the same. This bias clearly affects one's view of the world. Beware of powerful generalizations; they can become crippling thought stoppers.

1. Describe the three major tenets of frontier ethics.
2. For one week, when you listen to the news, watch TV, read articles, and listen to people talk, note examples of the frontier mentality. Make a list of them.
3. Is your personal ethic closer to a frontier or a sustainable ethic? Do your actions reflect your personal beliefs? How could you improve?
4. Discuss the tenets of sustainable ethics. Indicate which of these tenets coincide with your personal beliefs. Which ones don't? Why?
5. Critically analyze the tenets of sustainable ethics. Do you see any flaws in them?
6. Describe modes by which sustainable ethics can be promoted worldwide. Would you approve of such an activity?
7. Describe the roles that apathy, self-centeredness, feelings of insignificance, and technological optimism play in creating environmental problems and blocking change. How can each one be changed?
8. "Animals and plants have rights," says a leading philosopher, "irrespective of their value to humans." Do you agree? Why or why not?
9. Critically analyze the following statement: "Sustainable ethics is, in many ways, a utilitarian set of beliefs."
10. Using your critical thinking skills, analyze the following statement: "All of the changes prescribed in this book won't be enough to create a sustainable society. We need to change attitudes and ethics. Without changes at the deep, personal level, we cannot create a sustainable future."

KEY TERMS

Agenda 21
divine law
frontier ethic
land ethic

natural rights
Paradox of Inconsequence
Rio Declaration
technological optimism

United Nations Commission on
  Sustainable Development (UNCSD)
utilitarianism

© Gordon Bell/Shutterstock, Inc.

Economics and the Environment

Thirty years of exploration into biology, physics, and human nature have brought me to the realization that humanity itself has, in fact, sharpened the sword that is potentially responsible for piercing it through the heart.

—Michael Reynolds

LEARNING OBJECTIVES

After reading and studying the material in this chapter, students will be able to understand, knowledgably discuss, and, when appropriate, apply the information learned in this chapter, including:

1. Micro and macroeconomics
2. Types of economic systems and how they affect the environment
3. How principles of economics like supply and demand operate and their implications on the condition of the environment
4. How economic factors affect pollution control and resource management
5. How cost-benefit analysis works and its shortcomings when it comes to environmental management
6. How modern economics can be changed to benefit the environment
7. Steps needed to create a sustainable, friendly economic system
8. Ways in which a sustainable economic approach can create jobs

Exercise

"In order to improve our economy we need to grow—that is, produce and sell—more goods," say many business economists. "If we can't do it here, we need to tap into global markets. Without economic growth, jobs won't be available for those who need them. We have to keep growing economically, even if it means lowering environmental standards."

This view is widely held in the United States and other countries. Critics point out, however, that it is flawed for a couple of reasons. Based on what you've learned so far from this book and your course, can you think of any reasons why? Can you think of any alternatives to economic growth to create a strong and sustainable economy? What critical thinking rules come into play?

I n southern India, people once made traps for monkeys by drilling small holes in coconuts, filling the shells with rice, and chaining them to trees. The success of this trap was attributed to the fact that the hole was large enough for a monkey to put its empty hand in but too small for it to pull a handful of rice out. As monkeys clung tenaciously to their rice, villagers threw nets over them. The monkeys were trapped by their own refusal to let go.

Many observers believe that humankind is caught in a similar dilemma. Clinging tenaciously to an unsustainable way of living and doing business, many among us ignore the critical environmental problems such as global warming that undermine our long-term future including the health of our economy. As a result, we often resist changes needed to build a sustainable human economy. This chapter begins with an overview of some key principles of economics, describes the economics of pollution control and resource management, and then discusses weaknesses in economic systems from the viewpoint of sustainability. It concludes with a discussion of ways to build sustainable economic systems in industrial nations.

25.1 Economics, Environment, and Sustainability

Manufacturing and trade began in earnest with the rise of towns and villages. Towns and villages, in turn, owe their origin to agricultural surpluses made possible by technological advancements. New implements and techniques permitted farmers to produce an abundance of food with less labor. Displaced farm workers and their families moved to cities and towns, where many took up trades to earn a living.

Although the economic system began a long time ago, the science of economics began only two centuries ago with the publication of the Scotsman Adam Smith's book *The Wealth of Nations.*

Economics is a science that seeks to understand and explain the production, distribution, and consumption of goods and services. Economics is generally divided into two broad disciplines, microeconomics and macroeconomics.

Microeconomics is concerned with economic behavior of individuals, households, and businesses—both large and small. It concerns itself with decisions that are made in markets where goods and services are being bought and sold. Microeconomics examines how our decisions and behavior affect the supply and demand for goods and services, which, in turn, determines prices. It also examines how prices affect the supply and demand of goods and services. Two of the main concerns of microeconomics are inputs and outputs. **Inputs** include the commodities that companies need to produce goods and services, including raw materials, labor, and energy. **Outputs** include goods and services. These include materials, goods, and services that companies produce for use by consumers or other companies.

Macroeconomics is a branch of economics that deals with the bigger picture, that is, the performance, structure, and behavior of national or regional economies. It deals with growth, inflation, and unemployment, and with national economic policies and governmental actions such as taxation, that influence these issues.

Economics, like ecology and environmental science, is concerned with relationships. It employs scientific tools to discover the laws that regulate economies. Economists recognize two broad disciplines within micro- and macroeconomics. The description of economic facts and relationships falls within the purview of the first discipline, known as **descriptive economics.** Descriptive economics is a relatively pure science. Its questions can be answered by research and facts.

Economics melds with political science and sociology when it attempts to answer value-laden questions. For example, should companies pay for pollution controls? Should the economy continue to grow? Such questions cannot be answered by empirical facts and figures. There are no right or wrong answers to them, for they require value judgments and are left to the political process and the economic players themselves. This realm of economics is called **normative economics.**

More and more, people are calling for a new kind of economics, an **ecological economics** or **sustainable economics**, one that concerns itself with supplying the needs of people while protecting the environment. In this chapter, we examine sustainable economics. As you ponder this new and exciting discipline, remember that it seeks ways to ensure that economic development serves people—and serves them equitably, while protecting, even enhancing the natural world. Sustainable economics, as envisioned by its many proponents, is a system that would supply goods and services in ways that honor limits and protect the carrying capacity of the planet. It is a system that would enhance human welfare not only in the present but also in the future. Before we examine what a sustainable economic system might look like and study ways to make this vision a reality, we must first examine the current economic system. To do so, we examine some of the basic precepts of economics.

Economic Systems

Economics is a discipline that helps businesses and societies solve three fundamental problems:

1. *What* commodities should it produce and in what quantity?
2. *How* should it produce its goods?
3. *For whom* should it produce them?

Of course, there are many ways for a society to solve these three basic problems. The type of economy influences the answers. Economists recognize two basic types of economic systems: command and market. A **command economy**, also known as a planned economy, is one in which the state or government manages the economy. Examples are the economies of North Korea, Cuba, and, until recently, mainland China. In command economies, the state or government controls all major sectors of the economy—from food production to the manufacture of household goods. In other words, government officials dictate production and distribution goals. They make all decisions about the distribution of income—who gets what—and *what* the economy produces and *how much* is produced.

Government decisions, in an ideal world, are supposed to be made to benefit all, but even in these systems based on egalitarian goals, corruption abounds and the ruling elite often live high on the hog while the majority of the population lives in poverty.

In a **market economy**, the production, distribution, and price of goods and services are made by private individuals—business owners—based upon their own and their customers' interests, rather than on an overarching macroeconomic plan as in command economies. That is to say, in market economies what companies produce and the quantity they produce each year are determined by consumer demand and businesses and the potential for profit. Although there may be a high demand for a particular product or service, if a company cannot make a profit, it will generally not produce it.

In market economies, like those of most nations of the world, governments (theoretically) take a back seat to the marketplace. In other words, the market operates on its own without government intervention. Companies produce the goods and services for which there is the highest demand or products that yield the highest profit, thus answering the first question: What goods and services and in what quantity? Profit ultimately dictates how goods are produced. In market economies, the least costly method of production yields the greatest profit.

In a market economy, the question "For whom?" is also determined by money. In general, whoever can afford a good or service can get it.

One of the key principles of economics is the **law of scarcity**, which states that most things that people want are limited (**FIGURE 25-1**). As a result, their sale is rationed. In a market economy, price is the principal rationing mechanism. For instance, few of us drive a Tesla, a sporty electric vehicle that costs around $110,000, even though many of us might like to. Why? Because the price greatly exceeds the ability of the vast majority of us to pay for this car. In a sense, then, price rations Tesla sales and purchases. In contrast, in command economies, governments ration most of the output, although prices do play a role.

FIGURE 25-1 A treat for the wealthy. High-end vehicles, such as this sporty Tesla electric car, are targeted to the world's wealthy.

© Olga Besnard/ShutterStock, Inc.

In truth, most nations' economies are mixed—that is, they contain elements of market and command economies. China's economy is becoming more and more a market economy because thousands of government-owned businesses were converted to private ownership. In the market economies of Great Britain, Canada, and the United States, for example, free enterprise is often touted as the guiding rule. Let the market determine who gets what. Nonetheless, in these and other countries, governments also influence economic behavior.

The market economy of the United States is highly influenced by public policy. The U.S. government, for example, often subsidizes certain activities, giving some businesses and some products an economic advantage over others. The subsidies to oil, coal, and nuclear energy, for instance, artificially lower their price. Mining companies, for instance, are given special tax breaks as they deplete their mineral reserves. The economic incentives are intended to provide money for further mineral exploration and development.

Regardless of whether you agree or disagree with the policy, it represents governmental interference with the free-market economy that alters the cost of producing goods. Consider one specific example: Each year the U.S. military spends an estimated $50 billion protecting oil tankers in the Persian Gulf, according to the Rocky Mountain Institute. This oil, which will be refined in the United States, is worth an estimated $10 billion. If oil companies paid the cost of protecting their oil themselves, the cost of fuel oil, diesel, jet fuel, gasoline, and hundreds of other products derived from crude oil would be much higher. Instead, taxpayers pay the bill. To be fair, government policies provide subsidies help to foster a sustainable future. Companies that install wind turbines to produce electricity in the United States are given a federal production tax credit—a government incentive for each kilowatt-hour of electricity they generate. There are numerous incentives for homeowners and businesses who want to install solar electric systems, too.

Governments also influence businesses through various laws and regulations—for example, those that require companies to control pollution. Laws that stipulate how much pollution a company can emit affect the price of goods and services. Outright government bans on dangerous products limit product availability and, therefore, dictate production and consumption, interfering in a free economy to protect health or the environment. (For a look at a creative way to regulate natural resources, see **Spotlight on Sustainable Development 25-1**.)

Another way of influencing the market is through federally mandated freight rates. Freight rates are set by the federal government. In the case of raw ore versus recycled materials, raw ore travels more cheaply than scrap for recycling—providing a great benefit to the mining industry and a deterrent to recycling.

State and federal governments also subsidize various activities by special tax breaks or by sponsoring or funding research. These and other subsidies create an uneven playing field, which benefits some and harms others. Last, but far from least, governments also impose tariffs (taxes on imports) to regulate the flow of goods into a country, thus stifling free international competition. For example, there's a tariff on ethanol imported from Brazil and a limit to how much can be imported—measures that favor domestic production of ethanol for fuel. Governments can also exempt products from tariffs. Mexico, for instance, charges no tariff on imported renewable energy technologies like solar electric systems in an effort to promote their adoption.

Regulations, bans, subsidies, and other policy instruments are levers through which governments influence market economies to protect natural resources, people, and businesses.

Although they're nearly extinct, command economies also generally consist of a mixture of market and command practices. China, for example, allows a great deal of free-market enterprise within an economic system once tightly controlled by the central government.

Finally, both types of economic systems have enormous impact on the environment. The command economies of the former Soviet Union, for example, produced enormous amounts of air and water pollution and have left behind a trail of toxic hot spots, not unlike those of the Western world, where free enterprise has been the rule.

KEY CONCEPTS

Two types of economic systems exist: command and market economies. Command economies are run in large part by governments. In market economies, decisions about production are largely determined by prices and by people's ability to pay. Nonetheless, governments affect market economies in many ways: regulations, bans on certain products or activities, taxes, and subsidies. Both types of economies have had enormous environmental impact.

The Law of Supply and Demand

In market economies, the three essential questions posed in the previous section are generally solved by price. The price of a good or service is, in turn, determined by two additional factors: supply and demand. **Supply** refers to the amount of a resource, product, or service that's available. **Demand** refers to the amount people want. The predictable interplay of price, supply, and demand constitutes the **law of supply and demand.**

To understand the relationship between these three factors, take a moment or two to study the graph in **FIGURE 25-2A**, which shows a demand curve for rice. On the vertical axis of the graph is the price (P) of rice per bushel. On the horizontal axis is the quantity (Q) that people will buy at each price. This graph shows that as the price falls, the demand increases and vice versa. The relationship between P and Q is inverse—that is, as one increases, the other decreases.

Most of us are familiar with the interplay of price and demand. We all know, for example, that popular rock bands sell their tickets for a higher price than up-and-coming bands.

25-1 A Small Business Park Unpaves The Way to a Sustainable Future

Umeå is a small city in northern Sweden, a country renowned for environmental stewardship. Among its many environmental features, Umeå is home to a remarkable small business park—a model that other cities could emulate. The business park is home to franchises of three multinational corporations: a Ford Motor Company sales and service dealership; a Statoil gas station, car wash, and convenience store; and a McDonald's fast-food restaurant.

Okay, you say, there's no way these three companies possibly contribute to environmental protection and a sustainable future.

Well, let's take a look . . .

To begin, all three businesses have green roofs—not roofs painted green, but rather specially constructed roofs that are planted in grass and other forms of vegetation. These roofs help give back to nature a little more green space, taken up by the development itself. They absorb moisture, which nourishes the plants, rather than letting it run off parking lots, picking up oil and other pollutants that then wash into surface waters. Green roofs also help keep the buildings cooler in the summer.

The parking lot at the Ford dealership is unlike other parking lots you'll find in industrial nations. This one is made from porous pavers, special blocks that allow moisture to seep into the ground where it replenishes groundwater. By letting water seep into the ground, these pavers reduce surface runoff, water pollution, and flooding.

The buildings are made of natural or recycled building materials, too. Moreover, the buildings are built to be efficient—so efficient that they use 60% less energy than similar structures. As an example, motion sensors in the buildings turn off lights when the rooms are unoccupied.

Energy demands are met from renewable energy sources, such as the sun and wind. Electricity comes from a wind generator on the coast about 9 kilometers (15 miles) away. The Ford dealership also has solar panels on one side that preheat incoming air, reducing energy demand of the heating systems. It also has skylights that reduce daytime electrical consumption by 60 to 70%. At the McDonald's, waste heat from various processes such as cooking grills and deep fryers and from refrigeration systems is captured and piped to parts of the complex that require heat.

The business park also uses water efficiently and captures and reuses all of the storm water onsite. It flows

FIGURE 1 Umeå business park in Sweden.
Courtesy of Ford Motor Company AB, Sweden.

into an onsite water garden. Ninety-nine percent of the water from the car wash is filtered and reused. Sewage from the facility is used for fertilizer in local farming operations.

In addition, the business park reuses or recycles all of the waste that it produces. Wood used to build the facility comes from local wood lots and was sustainably grown and harvested.

When building the small business park, the developer removed a house located on the site. Rather than demolishing the structure and hauling the waste to the landfill, he lifted it and trucked the building to a new location. Rather than cutting down a large oak tree on the site, it, too, was transplanted elsewhere. The developer left a pine tree that was home to an endangered beetle on the site, too, rather than cutting it down.

The list of environmental attributes of this remarkable facility does not stop here. The gas station, for instance, sells three types of fuels that replace environmentally unfriendly diesel and gasoline. The convenience store sells organic produce and other organic foods. They encourage customers to recycle food containers by including well-labeled receptacles.

The Ford dealership recycles all waste oils and other waste fluids from cars such as brake fluid and radiator fluid. They use vegetable oil, rather than a petroleum-based fluid, in their hydraulic lifts. They even work hard to make this the healthiest and safest possible workspace for their employees. All three businesses provide ongoing education for their employees to ensure the continuation of sustainable business practices.

Now, you say, sure this is all well and good, but the businesses must be suffering.

Wrong.

The fact is that during the first year of operation revenue from sales at the car dealership and service shot through the roof. The other businesses are thriving as well, indicating that this is clearly not a case of environmental philanthropy, but a sound, economically profitable business practice that provides a model for the rest of the world. Over a half a million people visit the park every year, and city officials are encouraging other similar development.

Imagine how the world would change if all businesses pursued a similar strategy?

Adapted with permission from Dan Chiras, *EcoKids: Raising Children Who Care for the Earth,* Gabriola Island, British Columbia: New Society Publishers, 2004.

The supply curve is of interest chiefly to producers (**FIGURE 25-2B**). It plots the relationship between price and the quantity that suppliers will produce. The supply curve illustrates an intuitively simple concept: The higher the price, the more producers are willing and able to produce. At lower prices, producers often have to reduce production. Numerous examples exist. The fall in oil prices in the early 1980s, for example, put many American oil drillers out of business and caused economic hardship throughout the world. What happened? As oil prices fell because of conservation (which reduced demand) and cheaper foreign oil, many American companies went out of business or shut down wells because they couldn't produce oil profitably at the lower prices.

The economy is a balancing act with two principal players, supply and demand. Supply and demand interact to determine the price of goods and services. Graphically, this is represented by the intersection of the supply and demand curves and is known as the **market price equilibrium point** (**FIGURE 25-2C**). The market price equilibrium point represents the price at which both consumers are willing to buy a product and producers can afford to produce it.

The law of supply and demand is a relatively simple idea that explains a great deal of economic activity of businesses and individuals—specifically the market relationship between buyers and sellers. The law of supply and demand says that the price of a good or service determines both the quantity supplied by manufacturers and the quantity purchased by consumers. That is, the price determines how much a manufacturer provides and how much is consumed.

Although you may not have studied the law of supply and demand, you are probably familiar with it. You may have seen instances when the demand for a resource exceeded the supply, leading to an increase in price. This is one of the reasons for rising oil prices in this decade.

(a) Demand curve

(b) Supply curve

(c) Market price equilibrium point

FIGURE 25-2 Supply and demand curves for rice. (a) The demand curve shows the relationship between the price (P) on the vertical axis and demand (Q) on the horizontal axis. This graph shows that a rise in prices reduces the demand. Falling prices increase demand. **(b)** The supply curve shows the relationship between the price and supply, or amount produced. The higher the price, the more farmers will produce. **(c)** The market equilibrium point is the intersection of the supply and demand curves. It's the price people will pay for rice and the amount farmers will produce at that price.

Imagine an economy in which literally millions of prices are set by this kind of interaction without interference by governments. That system would be a **free-market system.** However, as noted previously, in most countries, governments meddle in economic affairs, sometimes benefiting businesses through subsidies and sometimes thwarting free enterprise through regulations.

The free-market system may also be tampered with by business itself. Some businesses, for instance, push for protectionist trade policies that limit or eliminate imports from foreign producers, in an effort to keep markets to themselves. Others buy up competitors, creating monopolies. Monopolies eliminate competition and can result in prices that are set at any level the monopoly wants (provided the public will pay). Antitrust laws in the United States are aimed at protecting individuals from monopolies. Companies may also try to drive others out of business. For example, Walmart purportedly marks down prices below those of nearby competitors like small hardware stores. Outpriced by this corporate giant, the small businesses are eventually forced to close their doors.

KEY CONCEPTS

The law of supply and demand shows that prices affect both the supply of goods and services and the demand for them. The prices of goods and services in a market economy are largely determined by the interaction of supply and demand.

Environmental Implications Of Supply And Demand

The law of supply and demand has some very real implications for sustainability. First, consider the impact of declining supplies by looking at the ivory trade. For years, elephants were slaughtered by the thousands in Africa to support the profitable ivory trade. When populations began to decline, African countries made it illegal to shoot elephants, and many countries prohibited the importation of ivory. Nonetheless, poaching continued. Why? Bans on ivory export and the dramatic decline in wild elephant populations caused by overhunting increased the price of ivory. That, in turn, gave poachers a considerable economic incentive to continue the illegal slaughter, even at the risk of being killed by game wardens. The supply graph predicts such activity, showing that the higher the price, the more willing someone is to produce a given product. People become rich, and the elephant is pushed toward extinction. It's a simple line of cause and effect with devastating consequences.

Government regulations that put a stop to the sale of ivory products in many countries have helped to solve the problem. These bans caused the market for raw ivory to evaporate, making poaching unprofitable and brightening the future of the elephant. Now that the elephant populations are recovering, efforts are under way to allow limited ivory production—say, from animals killed to keep populations on game preserves from skyrocketing and ruining the environment.

Supply and demand economics also have considerable impact on conservation efforts. For instance, as the price of oil climbs many industrial nations are finding ways to improve the efficiency of factories, automobiles, and homes. Supply and demand can also spawn wasteful behavior, though. The fall in oil prices in the early 1980s, created by conservation and by increases in production by the United Kingdom and other countries, eroded many people's resolve to save energy. The U.S. government, in fact, lost virtually all interest in energy efficiency at that time.

As a general rule, abundant supplies (at least in the near term) lead to low prices, which often foster wasteful practices. "We've got plenty, so why conserve?" seems to be the attitude of many. Section 25-4 discusses some of the weaknesses of supply and demand.

KEY CONCEPTS

Economics has a profound impact on activities that affect the quality of our environment and the sustainability of our society. Rising prices can increase environmentally responsible behavior such as energy conservation but also increase environmentally destructive actions like poaching.

Measuring Economic Success: The GNP

Economists need ways to measure economic activity. The most widely used measure of a nation's economy is its gross national product. The **gross national product (GNP)** is the market value of the nation's output—in other words, the dollar value of all goods and services that a nation produces and sells (including government purchases) in a given year. GNP includes domestic and overseas operations of all businesses. (**Gross domestic product [GDP]** is a subset of the GNP and includes all economic activity within the borders of a nation.)

Real GNP is the GNP adjusted for inflation. Per capita GNP is defined as follows:

Per capita GNP = GNP ÷ Total population

Widely used to track economies, the GNP gives a general picture of the wealth of nations and the living standards of their people. As you will soon see, the GNP, like the law of supply and demand, is insensitive to environmental concerns and sustainability.

KEY CONCEPTS

The gross national product is a measure of the economic output of a nation, including all goods and services sold domestically and abroad, and is used as a means of tracking the success of economies.

25.2 The Economics of Pollution Control

The free-market economic system in the United States and other countries, as well as the command economies of socialist and communist nations, treated pollution with almost uniform disregard until the 1960s (**FIGURE 25-3**). (Some disregard continues today.) Pollution was something that issued from smokestacks and that symbolized progress and prosperity. The costs incurred from pollution were largely ignored and are considered an **economic externality**, a cost to society and the environment not paid directly by manufacturers or consumers. Economic externalities include damage to human health, fish and wildlife populations, vegetation, climate, and others. Numerous examples have been pointed out in this text.

In many instances, businesses were simply unaware of the external costs of their activities. Gradually, though, citizens throughout the world began suing polluters (if their system of government allowed it). Governments established pollution standards to protect people and the environment, and businesses began to curb pollution, usually by installing pollution control devices. Reducing pollution reduces external costs, but increases internal costs. As a result, pollution control became part of the cost of doing business—the costs were internalized. Such costs are passed on to the consumer.

U.S. pollution laws have forced significant controls on pollution and subsequent cost internalization. Among them are the Clean Air Act, the Clean Water Act, the Resource Conservation and Recovery Act, and the Surface Mine Control and Reclamation Act. Although these laws are important, they have not resulted in a full internalization of costs. In fact, despite billions of dollars spent on controlling pollution, environmental damage (externalities) still occurs at unsustainable levels.

The reason environmental destruction continues is largely one of policy. Most government policy designed to control pollution has relied on end-of-pipe controls. In most cases, such controls reduce only the output of pollution and lessen problems or slow their rate of development. To a large extent, economics determines the level of control—that is, just how much money we are willing to spend on controlling pollution.

FIGURE 25-3 Disregard for pollution. In the past, factories such as this one belched out huge amounts of smoke, which was considered a sign of progress. Little thought was given to the damage it created and the subsequent costs.

© AbleStock.

Cost–Benefit Analysis and Pollution Control

As you may recall from discussions of risk assessment elsewhere in the text, the chief goal of pollution control is to reduce pollution in ways that yield the maximum benefit at the lowest cost. This goal is made possible by cost–benefit analysis. To understand how this process operates, take a look at **FIGURE 25-4**. **FIGURE 25-4A** illustrates an intuitive concept: the higher the level of pollution, the higher the cost of damage. Thus, the more acid deposition, the more damage one can expect.

FIGURE 25-4B shows the relationship between control costs and the amount of pollution that can be removed from a smokestack. This graph indicates that a little bit of effort can remove large quantities of pollutants cheaply. As more and more pollutants are removed, however, the cost to remove each additional increment rises. In the initial phase, for example, removing 100 units of pollution may cost only 1 cent per unit. Later on, the cost of removing pollution increases rapidly—for example, to $1 per unit of pollution. The phenomenon is called the **law of diminishing returns**. It states that for each additional dollar invested, we get a smaller return.

If both graphs are placed together, the point where the two lines intersect is the break-even point—theoretically, the level at which costs and benefits are equal (**FIGURE 25-4C**). If a society wants even lower pollution levels, it will have to pay more (**FIGURE 25-4D**).

Although this sounds very easy, it isn't. That's because determining the cost of pollution is very difficult. To do so, one must first determine the amount of damage. For example, how many people will die from air pollution-induced cancer? How many fish will be poisoned? The second problem comes in assigning a value to lost lives, lost wilderness, polluted air, extinct species, or obstructed views. How much is a person's life worth? How much are the many free services (oxygen production and carbon dioxide trapping, for example) provided by nature really worth?

One way around the dilemma of the unpriceable good is to calculate mitigation costs, the cost of offsetting damage. How much would it cost to restore an eroded statue? How much would it cost to move an endangered species to a new habitat? These costs can be surprisingly high. Engineers estimated that the cost of replacing the free services like flood control of nearly 3,400 hectares (8,500 acres) of wetlands

in the Charles River Basin (near Boston, Massachusetts) was over $17 million a year. The Army Corps of Engineers instead opted to protect the wetlands, which provide natural flood protection, rather than build expensive flood control facilities along the river.

Reducing all life-forms to dollars and cents bothers many people. It strikes many as morally wrong to consider sacrificing people's lives so that society can have its endless supply of disposable pens, diapers, and razors. Others consider it a cruel form of business logic. How can people or species such as the spotted owl be sacrificed for profit? Clearly, it is not an easy issue.

Because of the difficulties in assessing damage and the elusive nature of determining value, modern society is bound to find itself torn between two factions. On the one side are those who would allow some acceptable damage as a trade-off for the benefits of modern technology. On the other side are those who argue

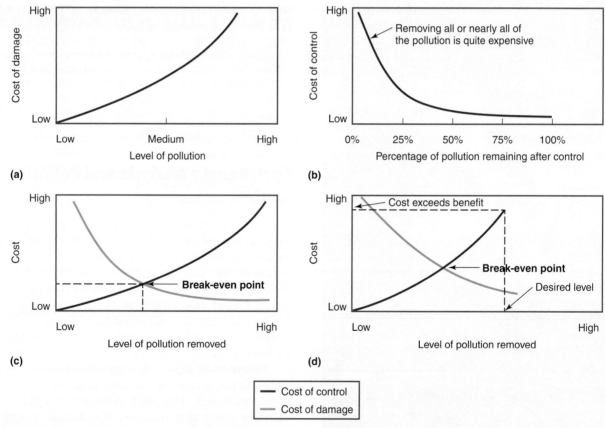

FIGURE 25-4 Cost of damage vs. cost of control. Optimizing pollution control efforts requires balancing the cost of cleanup against the economic benefits of control. **(a)** This graph shows that as pollution levels rise, the amount of damage increases, as reflected in rising costs. **(b)** Removing pollution requires an economic expenditure when pollution control devices are used. Removing 100% of the pollution is quite expensive. Removing 75% costs less per unit removed. This graph shows the relationship between cost of control and percentage of pollution remaining. **(c)** This graph shows the point at which pollution control costs equal economic damage. This is the break-even point. Unfortunately, determining the full costs of pollution is not easy, and estimates may be in serious error. **(d)** Society may wish to lower pollution levels past the break-even point. The desired level of pollution may technically cost society more than it benefits in reducing damage.

for eliminating harmful pollutants altogether. Torn between polar opposites, we muddle along, using the best scientific information available on environmental and health damage while battling one another in courts and legislatures over the proper levels of control. Interestingly, all this conflict could be greatly reduced by applying sustainable measures: implementing conservation, practicing pollution prevention, recycling, and turning to renewable energy technologies. These alternative solutions eliminate many adverse impacts because they eliminate the source of the problem—and they often do it at a fraction of the cost of the traditional end-of-pipe controls.

 KEY CONCEPTS

Cost–benefit analysis is a balancing act that allows one to determine how much should be spent to reduce pollution to a level at which the costs of control equal the benefits (reduced externalities). Sustainable approaches have the advantage over end-of-pipe solutions in that they can reduce pollution emissions at a far lower cost.

Who Should Pay for Pollution Control?

In today's world, pollution control devices remain the dominant means of dealing with pollution. One of the most frequent economic questions is, "Who should pay for pollution control?" Should corporations be required to pay for scrubbers and other devices to clean up the environment? In such cases the costs will ultimately be passed on to the consumers of industry's products. This strategy is known as the **consumer-pays option.** Another option, the **taxpayer-pays option,** places the financial responsibility on taxpayers. Government payments to coal miners who suffer from black lung disease are an example of this option.

Individuals who favor the consumer-pays option argue that the people who use the products that create pollution should bear the cost. The more one buys (and pollutes and depletes), the more one should pay. Frugal individuals should not have to subsidize the cost of cleanup through taxes. Passing the cost directly to the consumer, they add, could create more frugal buying habits—so essential to our long-term sustainability.

Those who support the taxpayer-pays option argue that through their elected officials, voters (and, hence, taxpayers) have allowed industry to pollute with impunity for years. Today, new standards are imposed on industry that place costly burdens on companies that have been operating lawfully for long periods. They think that when society changes its rules, society (citizen taxpayers) ought to be required to pay for controls. Advocates of the taxpayer-pays option also argue that society has elected and continues to elect officials who make deals with polluters to entice them and their polluting businesses into the community. In other cases, elected officials have overlooked flagrant violations of environmental laws. If society is responsible for elected officials who permitted pollution and other forms of environmental destruction, then the voting public (taxpayers) must bear at least some of the cost.

As in most controversies, there is validity to both arguments. In the case of old industries suddenly confronted with new laws, taxpayers might bear the economic burden of new controls. However, in new industries, pollution control costs should probably be borne by the corporation and consumers.

KEY CONCEPTS

Pollution controls and other environmental protection strategies are paid for by consumers if the costs are borne by businesses, or by taxpayers if the costs are shouldered by government.

Does Pollution Control Always Cost Money?

The previous discussion may leave the impression that reducing pollution always costs inordinate sums of money. Far from it. In some cases, pollution control devices can be installed to capture marketable products that might otherwise be dispersed into the air. These materials can be sold or reused, thus generating considerable profits. Pollution prevention strategies can be even more profitable. Redesigning chemical and industrial processes, for example, can sharply reduce energy and material demand and eliminate hazardous waste disposal. These reductions often save companies considerable sums of money. In other words, companies profit from such changes.

Nowhere is this case for pollution prevention more obvious than in the 1989 *Exxon Valdez* oil spill in Alaska or the Gulf oil spill in 2010, where a little prevention could have saved billions of dollars (**FIGURE 25-5**). Accidents such as this have three costs: direct costs, indirect costs, and repercussion costs. **Direct costs** from an oil spill are those that oil companies incur in the weeks and months following

FIGURE 25-5 Alaskan oil spill. This tragedy substantially boosted the GNP of Alaska for a while, but such expenditures are not a true measure of economic welfare.

Courtesy of Exxon Valdez Oil Spill Trustee Council/NOAA

an accident. These include the costs of lost oil, cleanup, waste disposal, and ship repairs. The oil lost from the *Valdez* was worth an estimated $4.8 million. Cleanup costs were $2.5 billion the first year. The company also agreed to pay an additional $1.25 billion in fines and penalties.

Indirect costs are those costs that are incurred by state and federal agencies. Oil companies are usually required to reimburse the government for some or all of these costs. Another indirect cost results from damage to the local economy: reduction in tourism, fishing revenues, and so on. Yet another indirect cost is damage to wildlife. The Clean Water Act holds companies liable for the cost of damage to natural resources from oil spills. The law permits federal agencies to collect money for lost sea otters, waterfowl, and eagles. All told, indirect costs in the *Exxon Valdez* case could come to several hundred million dollars.

Repercussion costs, the image problems arising from the spill, cause people to boycott the company or reduce their patronage. Adverse publicity may also result in more costly restrictions on oil tankers. Companies invariably balance the costs of control and prevention against the possible costs of an accident. If they don't incorporate the full costs, however, cost–benefit analyses are likely to be flawed and pollution may end up costing them hundreds of millions of dollars.

KEY CONCEPTS

Reducing or eliminating pollution can be a profitable venture that adds to the bottom line of companies. Pollution prevention and other techniques often save companies considerable sums of money, especially if all costs are calculated, including direct, indirect, and repercussion costs.

25.3 The Economics Of Resource Management

The previous section looked at the economics of pollution control as it is typically practiced. This section examines the economics of resource management—for example, the management of forests and soils—offering additional insights into the problems of modern society as well as some new solutions. Many decisions about natural resources are influenced by basic economic considerations, among them a factor called time preference.

KEY CONCEPTS

The management of natural resources is affected by numerous economic factors.

Time Preference

Time preference is a measure of one's willingness to postpone some current income for greater returns in the future. This is often determined by immediate needs. For example,

suppose your parents offer you $100 today or promise to invest it so it is worth $200 by the end of the year. If you are short on cash and need to pay for school books, you may take the money now. Your decision to accept the money now is based on your current needs, which in this case outweigh the benefits of waiting the year, even though you would be $100 ahead. Economists would say that your need for current income outweighs greater returns in the future.

Time preference is also influenced by uncertainty. In the previous example, how certain are you that your parents will be able to invest the money successfully so you get the $200? Another factor affecting time preference is the rate of return. The higher the rate of return, the more likely it is that you will wait for the income. For instance, if your parents said they had a sure thing they could invest in and that your money would grow to $1,000 a year from now, you'd probably wait. You could borrow $100 from a friend, if you needed the money, at 10% interest and pay her back at the end of the year with the $1,000 you earned—and you would have lots of money left over.

Inflation also affects time preference. In times of inflation, people are apt to buy now to avoid higher costs later. But inflation can also drive interest rates up, making savings more appealing.

Time preference applies equally well to the ways in which we manage many of our natural resources such as water, farmland, and forests. Take agriculture as an example. Farmers have two basic choices when it comes to managing their land. They can choose a depletion strategy, which permits them to acquire an immediate high rate of return for a short period. This might involve the use of artificial fertilizers, herbicides, and pesticides to maximize production. Soil erosion control and other techniques might be ignored. Alternatively, farmers may choose a conservation strategy, which includes techniques to conserve topsoil and maintain soil fertility. These actions require immediate monetary investments. They may cost the farmers more in the short run and cut into immediate profits. When one takes into account the possible loss of future income from soil erosion, however, the conservation strategy may make the most sense.

The choice of strategy in this example depends on the time preference. Will farmers choose the cheapest method of production, which gives the highest profit in the short term, or will they choose a slightly more expensive route, forgoing immediate high profits in favor of sustainable profits in the future? The economic needs of farmers determine, to a large extent, their time preference. For example, a young farmer looking forward to a productive career may opt for the conservation strategy. His immediate needs may be small. He may have no family and few debts. He can sacrifice income now for larger returns in the long run. However, an established farmer may have a family to support, kids in college, and excessive debt. He may, therefore, maximize his short-term profits through a depletion strategy.

Farmers' willingness to give up the potentially higher later income resulting from conservation may also result from uncertainty about future prices, the long-range prospects for farming, and interest rates. If the price of corn is high this year but might drop significantly in coming years, farmers may choose to make their money now. If the bottom falls out of the market in the next few years, they will have made the most of this short-term opportunity. If interest rates are likely to rise, short-term profit making may be the preferable choice. High interest rates on loans for machinery and seeds that farmers purchase tend to encourage the depletion strategy.

 KEY CONCEPTS

Resource management is influenced by time preference, one's temporal preference for earnings. Time preference is affected by current needs, uncertainty, and inflation.

Opportunity Cost

Another factor that greatly influences economic decisions regarding resource management is the **opportunity cost**—the cost of lost opportunities. For instance, the conservation strategy requires a monetary investment. The money put into conservation could have been invested in the stock market or a new business venture, possibly yielding more profit with less work than the conservation strategy. As a result, when opportunity costs are high, farmers are likely to choose options other than conservation.

Opportunity costs are also incurred when resources are wantonly destroyed. Tropical rainforests contain plants that produce chemicals that can be used as medicines. Losing these species via deforestation creates a significant opportunity cost—a loss of profit and potentially life-saving drugs. These losses suggest that a broader view of opportunities should be considered during resource management decisions and when making new laws and regulations.

 KEY CONCEPTS

Money can be put to many uses. Many people choose options that provide the highest returns. That may not include investment in wise resource management

Net Present Value

Economists rely on an economic calculations tool to determine the net present value of different profit-making ventures. The **net present value** is the dollar value of the stream of income generated by a strategy, while taking into account inflation and other economic factors. This technique can be used to compare very different economic strategies—for example, installing a solar electric system vs. paying the utility company for electricity they generate.

Through a complicated set of calculations, economists can determine which strategy makes the most sense based on the value of current dollars. The strategy with the highest net present value is generally chosen. To understand how it works, consider an example. Suppose you were contemplating installing a solar electric system on your property to generate all of the electricity you and your family consume. Being a smart buyer, you want to determine how this strategy compares to business as usual—buying electricity from the power company.

To do this, you would first project the cost of all of the electricity you'd buy from the local utility over a period of 30 to 40 years, the life of a solar system, and compare that to the cost of the solar electric system. To arrive at an accurate estimate of electrical costs from the utility, however, you'd have to take into account rising electrical costs and the declining value of the dollar due to inflation. (A dollar today is worth more than it will be 10 years from now due to inflation.)

To determine the value of the electricity from the solar system, you'd have to include the cost of the system, including installation, maintenance, and repair over the 30 to 40 year life of the system. You'd also need to calculate the cost of borrowing money to purchase it (interest on loans or interest lost if you took the money out of a savings account). You'd have to calculate inflation, too. When all of this is done, you can compare the economic cost of both strategies based on today's dollars—that is, the present value of both strategies. Whichever is less expensive makes the best economic decision.

This technique is also used by large corporations and organizations, such as the World Bank, which funds development in less developed countries. Big business uses this method all the time. A timber company, for instance, will use it to determine whether it makes more sense to clear cut a tropical rain forest or harvest its wood and other products sustainably over a long period. Unfortunately, it is often more rational from an economic standpoint to liquidate a resource than to harvest at a sustainable rate that would result in a lower value and a lower rate of return on the investment. The strategy, then, is to make your money and run, investing it in another high-rate-of-return venture. When this happens, economics and ecology clash.

Creating a sustainable system of resource management requires rethinking the way we calculate the value of different opportunities. Education can provide economists and resource managers with a new view of time preference, as well as alternative (sustainable) strategies to achieve respectable economic returns without depleting ecological resources. Bearing in mind that sustainable strategies provide a way of ensuring the long-term prosperity of our species, many critics argue that we must learn to factor in the long-term and hard-to-quantify benefits (like cleaner air) of sustainable strategies that might have lower net present value or yield lower returns. One means of fully valuing

Go Green

When staying in a hotel or motel room, throw all of the trash in one waste can. Plastic bags that line the receptacles are often thrown out each time a trash can is emptied, so dumping all of your waste in one trash can will help reduce this waste of resources.

such strategies is to calculate the replacement or mitigation cost of various actions. **Replacement cost** is an estimate of the true value of a resource to human society. How much would it cost to replant a tropical rain forest and reestablish the complex ecosystem and the benefits it provides such as lowering atmospheric carbon dioxide levels and maintaining rainfall patterns that are essential to agriculture?

By factoring these costs into the economic calculus used to determine net present value and opportunity costs, resource managers get a more accurate picture of the economics of various strategies. In this light, many profitable ventures will seem foolish. Sustainable strategies will appear much more attractive under a more ecologically sound means of accounting.

 KEY CONCEPTS

Calculating the long-term costs and benefits of different economic strategies based on the current value of money—or present value—is a technique often used in resource management. Decisions based on the present value tend to emphasize immediate returns, which results in the liquidation of natural resources rather than their sustainable harvest.

Ethics

To many people, money is a key driving force in society. You can't convince them that other people will act out of a sense of duty to future generations or other species. Others see this view as hopelessly pessimistic. They point out that many noneconomic factors influence economic decisions. One of the key factors is ethics. In some cases, ethics can be as powerful as—or even more powerful than—profit and other economic factors.

In building a sustainable society, writers, educators, business leaders, and government leaders can play an important role in creating a long-term view that seeks to ensure sustainability. This view, if widely held, could help foster wiser management of the Earth's resources. Most important, it could help shift time preference and encourage us to reconsider opportunity costs and discount rates.

 KEY CONCEPTS

Not all decisions about resource management are based on economics. Ethics can play a big role in determining actions, overriding other immediate concerns such as opportunity costs.

25.4 What's Wrong With Economics: An Ecological Perspective

Herman Daly, who has led the effort to create an Earth-friendly system of economics, studied three leading economics textbooks and found that not one of them mentioned pollution,

the environment, or natural resources. Some forward-thinking economists see this almost complete disregard for the environment as a fundamental flaw in their discipline. Worldwatch Institute's Sandra Postel writes, "While the environment and the economy are tightly interwoven in reality, they are almost completely divorced from one another in economic structures and institutions."

In my book *Lessons from Nature*, I outline four major flaws in economic thinking when viewed through the lens of sustainability: It is shortsighted. It is obsessed with growth. It promotes dependency, and it tends to exploit people and the environment. This criticism is not meant to be a denunciation of capitalism or a condemnation of those who are part of the economic system—which is all of us. It is intended to show how we can bring the human economy back into line with the economy of nature.

 KEY CONCEPTS

The economic system has several key flaws when viewed from an ecological perspective. Correcting these flaws can help us create a sustainable human economy.

Economic Shortsightedness

Earlier in this chapter you learned that the law of supply and demand governs modern economic transactions and profoundly influences our thinking. Proponents of sustainable economics, however, point out that this crucial law of economics is shortsighted, for it fails to take into account the finite supplies of many natural resources: oil, natural gas, and minerals. As a rule, supply and demand economics focuses on immediate supplies and is blind to long-term stocks and long-term demand. Current prices reflect short-term supplies. "Why worry about running out?" ask supply and demand proponents. As a resource is depleted, rising prices will stimulate exploration and more discovery, thus opening up new supplies. Falling supplies may also force us to find substitutes, permitting society to continue on the endless treadmill of production and consumption.

At some point, nonrenewable resources become economically depleted—that is, they fall into such short supply that they are no longer affordable. They're too expensive to extract. No one would pay the price. If substitutes are not available—and there appear to be a number of important minerals for which there are no substitutes—hard times are likely. Long before that point, though, the rising prices of declining resources could stimulate crippling global inflation.

Supply and demand economics as it is practiced today is a serious impediment to sustainability. To overcome it, supply and demand economics must be adjusted to reflect ecological realities. Key economic players, among them professors of business and economics, can assist by pointing out the limitations of supply and demand theory. Such activities could temper growth and stimulate efforts to recycle and use resources more efficiently.

Changes are also needed in public policy to help adjust economic activity to honor limits. The basic goal of these policy measures would be to adjust current prices to reflect long-term supplies. Several market tools are available, including a range of incentives and disincentives. One of the most important adjustment tools is **user fees**, or **green taxes**. User fees are taxes on raw materials, paid by producers and ultimately passed on to consumers. One example is the severance tax charged to coal companies on each ton of coal they mine. Carbon taxes are currently used in several European countries; they are another user fee.

User fees artificially increase the cost of raw materials and finished products. This, in turn, helps promote conservation and raises revenues that can be used to develop alternative supplies. For instance, money raised by a tax on coal could be used to promote energy conservation or renewable energy such as solar (**FIGURE 25-6**). Thus, a tax can help ensure future generations that they too will have access to resources needed for a healthy, productive life. User fees help us honor the rights of future generations.

(a)

(b)

FIGURE 25-6 User fees. (a) Even small taxes on coal consumption could help support more environmentally sound energy strategies **(b)** such as wind power.

Photo a © Dmitriy Kuzmichev/ShutterStock, Inc.; photo b © T.W. van Urk/ShutterStock, Inc.

Another means of instilling vision in the economic system is to revamp cost–benefit analysis, described earlier in the chapter. Efforts are needed to identify all environmental, social, and human health impacts—present and future—and to quantify them. As best they can, companies should ensure that all costs are incorporated into the price of their product or service. This is known as **full-cost pricing**. By incorporating the external costs into the cost of a good or service, we can narrow the gap between the market price and the real cost.

Full-cost pricing is an ideal to work toward. It could stimulate economic change. For instance, it might compel companies to find ways to prevent problems in the first place, through sustainable practices such as pollution prevention, energy efficiency, recycling, and renewable energy use.

Governments can play a role by requiring least-cost policies. Numerous states now require utilities to choose the least costly ways of producing new energy. To reflect the cost of various strategies more accurately, some states require utilities to add 15% to the cost of conventional strategies such as coal, oil, and nuclear power. This helps account for economic externalities and usually makes environmentally sustainable strategies more competitive.

KEY CONCEPTS

Economic systems and the participants in them often fail to take into account long-term supplies, a dangerous trend that results in an underpricing of many natural resources and leads to environmentally unsustainable activities. Several mechanisms are available to incorporate such concerns into economic decision making, including user fees, least-cost planning, and full-cost pricing.

Economics and Growth

In the 1967 edition of his popular economics textbook, Nobel prize-winning MIT economist Paul Samuelson defined economics as "the science of growth." Subsequent editions have dropped the wording, but the bias remains. In his 1984 edition, for example, Samuelson wrote, "Today, the ultimate measure of economic success is a country's ability to generate a high level of and rapid growth in the output of economic goods and services. Greater output of food and clothing, cars and education, radios and concerts—what else is an economy for if not to produce an appropriate mix of these in high quantity and fidelity?" Although some economists do not subscribe to this view, many do—especially those in business and government.

Our focus on growth is rooted in the frontier notion that "there is always more." For many decades, human civilization has been caught up in it and has been willing to pay almost any price for it.

Economic growth is based on an increasing consumption of goods and services. Such an increase generally arises from (1) an increase in population size and (2) an increase in the amount each of us buys—per capita consumption.

Because it means greater production and, presumably, greater economic wealth, population growth has traditionally been viewed as an asset to society. Each new baby is viewed as a potential new consumer.

The dangerous aspect of our nearly singular focus on growth is that it tends to equate economic growth with progress. The faster an economy is growing, many assume, the better off people are. Thus, a rising GNP is taken to mean that a country's health and people's lives are improving. Over the years, so many people have bought into this logic that economic growth has become a way of life. In fact, it is no exaggeration to say that economic growth has become the abiding principle of economics and business and the central focus of political campaigns and government policy. With the growth-is-essential philosophy so deeply embedded in our society, many people are blind to the outcome of continual growth. Few people today recognize that continuous economic growth is ultimately incompatible with the economy of nature, where very real limits exist.

KEY CONCEPTS

Economic growth is fueled by population growth and ever-increasing per capita consumption. Continual economic growth is the abiding principle of many economies and the measure of success, but it is ultimately incompatible with the finite world in which we live.

Growth and the GNP: Some Fundamental Flaws

Economic growth is measured by tracking the GNP of nations, but the GNP is believed by many observers to be a flawed index of a nation's welfare. Why? The GNP includes many goods and services that make no contribution to the welfare of the people. More to the point, the GNP fails to differentiate between revenue from different activities. For example, the GNP includes all expenditures on homes, books, parks, wildlife refuges, and food that improve the standard of living. However, it also includes expenditures on oil spill cleanup, cancer treatment, and air pollution cleanup. It includes hospital bills from cancer and the ever-increasing cost of insurance to protect our homes from flooding and storms worsened by climate change. Therefore, a country with filthy air and polluted water faced with rising cancer rates and frequent costly tornados and hurricanes will register a high GNP. In other words, as the GNP is structured, a dollar spent on a piano lesson for a future Mozart is just as important as a dollar spent on a pack of cigarettes.

By carefully defining terms, as is required by critical thinking, we can see the fallacy of overdependence on GNP as a measure of success. A telling example is the state of Alaska, whose economic output increased by $1 billion in 1989 because of the *Valdez* oil spill.

Another problem with the GNP is that it fails to account for the destruction of natural assets, a nation's natural capital.

Natural capital is a nation's ecological wealth. It includes topsoil, forests, grasslands, open spaces, farmlands, fisheries, and wild species—resources of great aesthetic and economic importance. In many ways, natural capital is like money in the bank. Nations draw on this capital to fuel their industries and to satisfy human needs for food, building materials, and other goods and services. But most nations are dipping into the bank account, taking not only the interest but the principal itself. Thus, a nation can be rapidly depleting its natural capital while posting a high GNP, and there is no way of knowing it. Herman Daly once wrote, "most nations are treating the Earth as if it were a corporation in liquidation." That is, most nations are selling off their natural assets such as their forests to make a quick buck. Although their GNPs may be impressive, sooner or later, they will have no productive capacity left. How does exploitive behavior such as this affect GNP?

Suppose that lumber companies in a country cut down the nation's forests to increase exports but failed to replant them. A study of the GNP of the country during the cutting phase would show that the economy is strong. If the destruction of the forests were taken into account, however, it would be clear that conclusions about economic strength would be meaningless. The GNP fails to show the dramatic decline in assets. Unfortunately, many countries the world over are depleting their natural resources. Bolivia, Colombia, Ethiopia, Ghana, Indonesia, and Kenya, for example, depend on primary resources (such as minerals, timber, and crops) for 75% or more of their exports, but they are rapidly depleting them (**FIGURE 25-7**). Their GNPs hide evidence of the decline. Soon there will be no assets and no source of income.

Yet another largely unrecognized problem with the GNP is that it is blind to accumulated wealth. Modern economies today compare one to another on the basis of their annual growth in GNP. This criterion, however, fails to take into account the accumulated wealth of a country—how much people already have. A newly developed nation may have a 5% annual growth in its GNP compared with a 2% growth rate in a developed country such as the United States. Does

FIGURE 25-7 Declining natural capital. The GNP fails to take into account declines in natural assets.

© Photodisc.

this mean the newly developed nation is doing better? Not necessarily. It only means that its economy is expanding more rapidly. Most likely, it has more room to expand, but its people are far less wealthy than those of the United States.

Finally, GNP also pays no attention to the distribution of wealth in a society. Many projects that improve the GNP of a nation fail to improve the lives of the people they are designed to help. A select few benefit, and in some cases, the masses may be harmed. (Some people call this the *trickle-up* theory.) In developed countries, the GNP hides frightening inequities. In the United States, for example—a country deemed rich by GNP standards—one-fifth of all children live in poverty, and until recently about 35 to 40 million Americans have had no health insurance. Looking at the GNP, one would be compelled to think that all was well within U.S. borders.

Author Wendell Berry calls the GNP the "fever chart of our consumption." It is much more than that. It is an indiscriminate measure of our consumption, our waste, and our disregard for environmental conditions. It has become a measure of our unsustainability. Even its inventors argue that the GNP is prone to overvaluing production and consumption of goods and not reflecting *improvement in human well-being*. Because the GNP fails to make this distinction, critics argue that it is a false measure of success. Clearly, something else is needed to measure progress.

KEY CONCEPTS

Continual economic growth is the abiding principle of many economies and the GNP is our measure of success. However, the GNP is an inaccurate measure of the welfare of a nation's people because it fails to distinguish economic activity that enhances our welfare from that which results in a decreased quality of life. It also fails to take into account resource depletion, accumulated wealth, and the distribution of wealth.

Alternative Indicators of Progress One promising development in recent years is the push to develop alternative measures of progress. The Yale economists William Nordhaus and James Tobin, for example, devised a measure that adjusts the GNP to make it a more accurate representation of the good that people receive from their nation's economic growth. This measure, called the **net economic welfare (NEW)**, subtracts the *disamenities* of an economy— the cost of pollution, the cost of medical care for victims of lung disease caused by urban air pollution, and other costs of unsustainable economic development—from the GNP. The NEW also *adds* the cost of certain activities, such as household services (cleaning and cooking) provided by men and women, that are not part of the traditional GNP calculations but that improve well-being.

FIGURE 25-8 shows the relationship between NEW and GNP. NEW is lower than the GNP as you might expect. Moreover, NEW is growing more slowly than GNP. In other words, as the nation's output grows, the economic benefits of growth fall behind, largely because of rising pollution and

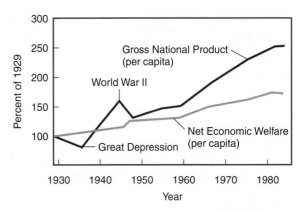

FIGURE 25-8 Net economic welfare vs. gross national product. The GNP is the market value of all goods and services produced by a country. U.S. per capita GNP has risen continuously for many years. The NEW, or net economic welfare, is a measure of beneficial goods and services. It is derived by subtracting the negative aspects of the GNP, which do nothing to improve the quality of life, such as damage from air pollution. As shown here, the per capita NEW is lower than the GNP and rises at a slower rate, suggesting diminishing returns from economic growth.

environmental destruction. We are spending more money but getting less for it.

Economist Herman Daly proposed an alternative measure, the **Index of Sustainable Economic Welfare (ISEW)**, which is now called the **Genuine Progress Indicator (GPI)**. Although the procedure for calculating the GPI is complex, the GPI is the outcome of an attempt to measure whether a country's economic growth, resulting from an increase in the production and consumption of goods and services have improved the health and well-being of the people in the country or caused a deterioration in them.

GPI takes into account *the enhancement of nature's ability to provide its many services*, for example, to provide clean air and water, among other things. These factors are, therefore, part of a more inclusive definition of progress. The GPI, therefore, is considered by proponents as better approximation of the nation's true economic welfare. Growth in GPI, in turn, can be compared to the GNP to determine whether economic progress, measured by growth in GNP is actually improving our lives or not.

When Daly compared the GPI to the GNP of the United States, he made a startling discovery. In the 1970s, the GNP grew annually by 2% per capita. In the 1980s, the per capita GNP grew by 1.8% per year. The GPI during the 1970s, however, climbed only 0.7% per capita (**FIGURE 25-9**), indicating that only about one-third of the annual growth in GNP improved the lives of Americans. Largely because of environmental degradation, the GPI declined by 0.8% per year in the 1980s. During the 1990s and the 2000s, the growth in the GNP continued but GPI leveled off. What does this mean?

Although our economic output is increasing, and economists and politicians would have us believe that we are better off, our lot in life is actually declining. In other words,

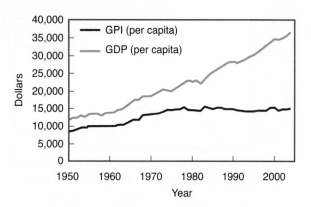

FIGURE 25-9 Alternative indicators of progress. This graph is part of a set of indicators that measure the well-being of the United States. Note that while GNP continues to rise, GPI falls.

Data from *The Genuine Progress Indicator, Redefining Progress: The Nature of Economics*, Washington, D.C. (2006).

despite a growing economy, which most people think is good news, we are actually worse off.

The United States is not the only country to show a discrepancy between the GNP and GPI, but it is one of the few countries for which the GPI continues to plummet. In Germany, the GPI is actually increasing faster than the GNP.

Fortunately, many efforts are under way to recalculate GNPs to take into account the depletion and deterioration of natural resources. Essential as these efforts are, it is important to note that aggregate measures such as the GPI hide important trends needed to assess a nation's progress. To get an even more accurate picture, over 75 U.S. cities, towns, and states are developing or have developed alternative measures of success—often called **benchmarks** or **indicators of sustainability**. These include data on a variety of social, economic, and environmental trends that allow communities to track crime, energy consumption, pollution, and a host of other factors. Indicators of sustainability tell communities where they are and where they're going. They provide valuable information needed to develop policies that address the most critical problems in an era of limited revenues. Trends data also show how effective policies are at solving problems.

Go Green

When staying in a hotel room, take the used soap with you and use it at home—otherwise it will be thrown out.

KEY CONCEPTS

Alternative measures of economic performance that take into account economic negatives (such as the depletion of natural resources) provide a more accurate picture of the welfare of nations. Even more precise information about the welfare of people comes from efforts to track key social, economic, and environmental trends—the indicators of sustainability.

Making the Economic System Work Better
Advocates of a sustainable future argue that nations should strive to reduce the difference between the GNP and GPI,

that is, we should make our economic systems work *for* us, not to our detriment. For every gain in economic output, we should receive a similar social and environmental benefit, not a shot of pollution that needs to be cleaned up. How can this be done?

One way of maximizing the social benefit of economic activity is to reduce the pollution, waste, and environmental destruction per dollar of GNP. This can be accomplished by applying the operating principles of sustainability, discussed in previous chapters. By becoming more efficient in our use of resources; by recycling and reusing all materials; by converting to clean, renewable energy supplies; and by restoring ecosystems including farmland and rangeland we can reshape our economic system so that economic activity translates into measurable improvements in our lives and the ecosystems upon which we depend.

Corporations, small businesses, and even individuals can promote these activities by adopting new approaches. As many previous examples in this text have shown, such changes can result in substantial economic savings. Governments can also assist by providing a range of incentives and disincentives, many of which have been discussed in previous chapters.

KEY CONCEPTS

Applying the principles of sustainability can help us ensure that economic activity translates into improvements in the quality of our lives in all ways—social, economic, and environmental.

Rethinking Growth: Focusing on Development
Many advocates of sustainable development assert that our economy as presently configured is much like a cancer, eating away at its host, the Earth. They remind us that sooner or later all untreated cancers kill their hosts. Put another way, continual economic growth in a finite world results in environmental degradation and resource depletion. Although some advocates of sustainability think we can grow continually and sustainably, others disagree.

Critics of economic growth argue that the future of industrial nations depends more on development than on continual economic growth. Continued economic growth requires a continual extraction of resources and ever-increasing production and consumption, which many observers believe cannot be sustained.

As Michael Kinsley of the Rocky Mountain Institute notes, "Development is very different from growth. After reaching physical maturity, we humans can continue to develop in many beneficial and interesting new ways—learning new skills, gaining deeper wisdom, and much more." Likewise, a nation can develop without increasing (growing) its material consumption. It can still create jobs and expand cultural and educational opportunities without expanding its demand for resources.

In the more developed nations, then, sustainable economic development must strive to create a higher quality of life without depleting resources or causing environmental impacts that undercut future generations and other species. Many of the strategies outlined in this chapter will help us reach this goal. Especially important are efforts to use resources more efficiently and to reduce our demand for unnecessary goods and services. Using renewable resources and recycling also meet our needs, at a fraction of the environmental costs of traditional strategies.

In the less developed nations, development and limited growth within ecological limits are probably required. Limited economic growth is essential to raise the standard of living of the poor. Such efforts are needed to improve the lives of the world's people and will help put an end to the cycle of poverty and environmental decay, say proponents of this viewpoint.

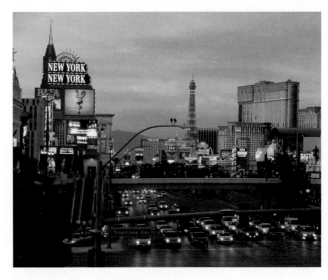

FIGURE 25-10 The price of interdependence. The city of Las Vegas, Nevada, is surrounded by desert. If its energy or water were cut off, the city could not exist.

© Steve Hoskinson/ShutterStock, Inc.

KEY CONCEPTS

Creating a sustainable future may require us to develop—to maintain and improve human well-being—without increasing economic throughput. Many sustainable strategies such as efficiency and recycling permit the attainment of this goal at a fraction of the environmental costs of continued growth.

Fostering Local and Regional Self-Reliance

The economic system of much of the world also tends to foster dependency among individuals, regions, and nations through trade. Economic interdependence, while desirable in many respects, creates very real problems from an environmental perspective. First and foremost, it tends to separate producers of goods and services from consumers. As a result, consumers are often blind to the environmental costs of satisfying their needs and desires.

Trade and interdependence allow human populations to flourish beyond the local and regional carrying capacity. Water imported from the Colorado River to the cities and towns of southern Arizona, for example, has resulted in explosive population growth. Food imports allow populations in regions with poor agricultural potential to flourish. Without infusions of fuel, food, and water, many regions could not support large human settlements (**FIGURE 25-10**). If resource supplies dwindle, many of these regions could suffer extreme economic and social hardship. In the U.S. West, for instance, studies of global warming suggest that water flow in the Colorado River, which currently serves an estimated 20 million people, could fall by 30%, with devastating effects on this population. This could cripple agriculture in arid southern California.

Despite these problems, business economists usually refer to the global marketplace (and interdependence in general) as a highly desirable goal. From an ecological viewpoint, however, the global marketplace and the rising interdependence that accompanies its development could be dangerous.

Some leading thinkers on the subject argue that from a sustainable vantage point, the future of human society lies not in globalization, but rather in local or regional self-reliance—that is, human communities living within the means of a bioregion. Food, water, energy, and other resources would come from local sources, not be imported from sources hundreds, even thousands of miles away. Greater regional and local self-reliance does not eliminate all imports and exports; it seeks *greater* self-sufficiency and reliance on resources that are immediately available—for example, sunlight and wind energy instead of imported oil. (For a discussion of sustainable rain forest management, see **Spotlight on Sustainable Development 25-2**.)

The transition to a sustainable society may ultimately require increasing regional self-reliance. For the United States, regional self-reliance means tapping into local energy supplies, preferably clean and renewable ones. For states, it may mean developing diverse economies that produce many of the goods and services needed by people on a day-to-day basis. In fact, many products could be manufactured locally from recycled materials gleaned from our wastes. Diversity in regional economies could provide an added benefit: It could help make communities and states more recession proof.

Greater local self-reliance, although in opposition to the dominant view of economic success, is beginning to emerge throughout the world. In Brazil, for instance, a program to produce ethanol for cars and trucks was created to reduce the nation's

Go Green

When leaving a hotel room, turn the heat down in the winter and turn the temperature up in the summer to save energy.

dependence on foreign oil. California's public utilities have developed an energy strategy that relies principally on energy efficiency and local supplies of renewable energy, such as solar, wind, and geothermal. Many states have programs to promote the purchase of goods produced within the state. Efforts to promote U.S. or state-produced goods are another sign of emerging self-reliance.

KEY CONCEPTS

For most of us, international trade is viewed as a highly desirable activity, but the interdependence it creates has many adverse effects on our long-term sustainability. Greater local and regional self-reliance, although controversial, may be essential to achieving a sustainable future.

25.5 Creating a Sustainable Economic System: Challenges in the Industrial World

University of Colorado economist Kenneth Boulding was one of the first to write about changes needed in the U.S. economy. In 1966, he coined the phrase *cowboy economy* to describe the present economic system, characterized by maximum production, consumption, resource use, and profit. Boulding suggested that the cowboy, or frontier, economy be replaced by a *spaceship economy*—an economic system that recognizes that the Earth, much like a spaceship, is a closed system wholly dependent on a fragile life-support system.

© hugolacasse/ShutterStock, Inc.

SPOTLIGHT on Sustainable Development

25-2 Cultural Survival: Serving People and Nature Sustainably

Throughout the tropics, efforts are under way to improve the lives of people who live in the rain forests without destroying the rich diversity of their biome. One way to protect a resource is to set it off-limits to people. Such reserves, however, stand little chance of remaining intact if local people are robbed of the resources they need to survive. Recognizing that in some instances land must be used in order to be preserved, some countries have established extractive reserves.

Extractive reserves are protected regions where local residents harvest the natural products of the forest, among them fruits and nuts, rubber, oils, fibers, and medicines. Because they are harvested in ways that do not harm the forest, extractive reserves can be productive sources of food and resources *ad infinitum*. In fact, indigenous peoples of the world have depended on a sustainable harvest of forest resources throughout history (**FIGURE 1**).

Although numerous countries have set aside huge tracts of land, the success of extractive reserves also depends on efforts to market forest products in ways that benefit those who do the work. Jason Clay, an anthropologist with a U.S.-based group known as Cultural Survival, is one of many volunteers helping to make this happen. Clay manages a rain forest marketing project. He helps rain forest inhabitants process and sell the nuts they harvest. Instead of receiving 4 cents

FIGURE 1 Nonprofit organizations are finding ways to help those closest to the land reap great financial reward from the sustainable harvest of forest products.
© Marcus Brown/ShutterStock, Inc.

per pound, the typical price for unprocessed nuts, processing and marketing raise the price to $1.00 per pound. Cultural Survival charges its commercial clients an additional 5%, which it uses to support the group's activities throughout the world.

One of Cultural Survival's biggest customers is a company called Rainforest Products, which currently markets two cereals in the United States, Rainforest Crisp and Rainforest Granola. Both cereals contain Brazil nuts and cashews harvested by indigenous Amazonians. Ben and Jerry's Ice Cream of Vermont once marketed a popular ice cream flavor called Rainforest Crunch, made with handpicked Brazil nuts and cashews from the Amazon.

"For the forests, it is a question of use it or lose it," says Cultural Survival's Jason Clay. "The value of the rain forest will have to be tested in the marketplace. But the point is to change the market, not the forest."

By that, Clay means that the rain forest needs to be viewed differently—as a source of sustainably harvested fruits, nuts, and other products, rather than as a source of wood and pasture land. Do such nonwood forest products have adequate economic value, though?

The World Bank argues that "the extractive reserves are *the* most promising alternative to land clearing and colonization schemes" (italics added). In 1989, Charles M. Peters of the New York Botanical Gardens' Institute of Economic Botany and his associates proved this to be true. These researchers evaluated a 1-hectare (2.5-acre) plot of Amazonian forest in Peru to determine the economic value of a variety of options. An inventory of plant life turned up 842 trees with diameters greater than or equal to 10 centimeters (4 inches). The trees included specimens from 275 species. Seventy two of the species yield products that are currently marketed by natives. Sixty species produce commercial timber, one species produces rubber, and 11 species produce edible fruits. Peters and his associates found that the 1-hectare survey area produced fruit whose annual net worth was about $400 and rubber worth more than $22.

The 1-hectare plot also contained 93.8 cubic meters of merchantable timber, which if clear-cut would be worth about $1,000. However, clear-cutting would destroy the fruit and rubber trees and would put an end to future timber production. The $1,000 net income would be a one-time profit.

The researchers also examined selective cutting of marketable trees, which could be compatible with fruit and rubber harvesting. They found that this option by itself yielded about $310 per year.

As pointed out in the chapter, economists often rely on the discount rate to determine the present economic value of different profit-making strategies. Net present value (NPV) is equal to the net revenue produced each year, divided by a discount rate. The equation is $NPV = V/r$, where V is the annual revenue and r is the discount rate. Peters and his colleagues used a 5% inflation-free discount rate to determine the net present value of the fruit and latex resources of all future harvests. They found that the NPV of sustainable fruit and latex harvests was $6,330 per hectare, assuming that 25% of the fruit crop was left in the forest for regeneration.

The researchers used another equation to determine the NPV of a perpetual series of sustainable timber harvests. Using that equation, they found that the NPV of sustainable harvest by selective cutting was $490. The fruit, rubber, and sustainable timber harvests of the 1-hectare plot have a combined net worth of $6,820. The value would increase even further if the revenues from medicinal plants, small palms, and other plants were included.

Clear-cutting has a much lower net present value because, as noted previously, the practice is not sustainable. Interestingly, even conversion of the forest to tree farms was found to have a lower net present value than sustainable harvest. Timber and pulpwood produced on a 1-hectare plantation of a commercially valuable tree called *Gmelina arborea* in Brazil was estimated at $3,184, less than half that of the natural forest. Calculations also showed that a fully stocked cattle ranch in Brazil has a present value of only $2,960 per hectare, even if the costs of weeding, fencing, and animal care are excluded.

This economic assessment clearly shows that the value of a standing forest is much higher than the most common alternatives. Except for a few cases, though, decisions about the forests seem to favor economically less productive and environmentally unsustainable options.

The goal of sustainable economics is to serve people and to protect, even enhance the environment, making extractive reserves an essential element of the new economy. Some advocates believe, however, that for such ventures to be truly successful, the people who do the work have to have more control in the market. By eliminating the middle persons, locals can generate more income for local collectors. If extractive reserves are successful, not only will lifestyles improve, but large tracts of rain forest and the species that live in them could be saved.

An economic system that seeks to meet human needs while protecting and enhancing the life-support systems of the planet is called a **sustainable economy** (TABLE 25-1). The notion of a sustainable economy was not widely accepted when Boulding first proposed it. Today, as more and more nations face limits, experience environmental deterioration firsthand, and ponder threats such as global warming, the need to reshape economic systems is becoming clear. How do we create a sustainable economy?

The previous sections have outlined numerous ways to make the economy more sustainable—among them user taxes, full-cost pricing, and better cost–benefit analysis. They outlined some efforts to create new measures of progress and described the need for more self-reliance. New laws and private initiatives that promote conservation (efficiency), restoration, and recycling were also discussed, as were efforts that encourage the use of renewable resources and promote population stabilization. These steps could help revamp major sectors of our society, such as agriculture, transportation, housing, and energy. No one of the changes listed in this paragraph or this text will work alone; all are needed, and each one is vital to the effort.

Because many private and government initiatives that promote these activities have been outlined in other chapters, in this section we examine four major ideas: harnessing market forces, corporate reform, green products and green seals of approval, and appropriate technology. Each of these is an important element in the complex equation of sustainability.

KEY CONCEPTS

Numerous changes can help us forge a sustainable economy, one that meets human needs without foreclosing on future generations by destroying the natural resource base that makes all economic activity possible.

TABLE 25-1

Characteristics of a Sustainable Economy

Participants take a long-term systems view that

- Recognizes the importance of natural systems to human well-being.
- Ensures that economic activities improve the quality of life of all, not just a select few.
- Seeks ways to ensure that economic activities maintain or improve natural systems.
- Uses all resources extremely efficiently.
- Promotes maximum recycling and reuse.
- Relies heavily on clean, renewable technologies.
- Restores damaged ecosystems.
- Promotes regional self-reliance.
- Relies on appropriate technology.
- Designs and builds using principles of ecological sustainability.

Harnessing Market Forces to Protect the Environment

Companies often argue that, left to their own devices, they could find ways to reduce pollution at a much lower cost than governmentally mandated controls. In response to this complaint and the growing cost of regulation, former Democratic senator Tim Wirth of Colorado and a Republican colleague, the late John Heinz of Pennsylvania, assembled a multidisciplinary, bipartisan team from businesses, colleges and universities, the environmental community, and government. Their goal was to propose ways that government could tap into economic forces to reduce pollution and manage resources sustainably. Such measures could supplement—even supplant—traditional laws aimed at regulating pollution and resource use.

Their report, issued in 1988, outlined a number of "marketplace solutions," including (1) economic disincentives, (2) economic incentives, (3) tradable and marketable permits, (4) laws that eliminate market barriers to efficient resource use, and (5) laws that remove subsidies for environmentally destructive activities.

KEY CONCEPTS

Many market mechanisms can be brought to bear on environmental problems, allowing businesses to innovate, save money, and reduce the burden of regulations, while reducing their environmental impact.

Economic Disincentives   Governments have many legal means to force companies to control pollution, reclaim damaged lands, or reduce resource consumption. In the United States and many other industrial countries, the mainstay of environmental control is a complex and costly set of rules and regulations.

Regulations are generally backed by legal recourse, among them fines and prison sentences. Fines are a type of economic disincentive to discourage environmentally destructive activities.

The 1988 report proposed a different set of economic disincentives, among them user fees and pollution taxes, described earlier. User fees and **pollution taxes** impose a fee on products that waste resources, pollute the environment, or damage ecosystems. A good example is the **gas guzzler tax**, a tax on cars that get poor gas mileage, which President Clinton rescinded through executive order. Carbon taxes are a good example of pollution taxes.

Economic disincentives can work. The 1990 Clean Air Act, for instance, imposed a steep tax on the use of ozone-depleting chlorofluorocarbons (CFCs) that provided considerable incentive for companies to find alternatives. Pollution taxes have long been used in Europe to reduce the emissions of harmful environmental pollutants, but have been fiercely opposed by business in the United States. Efforts are currently underway to impose a carbon tax on fossil fuels to be used to reduce emissions and to develop clean, carbon-free energy sources, although it is doubtful it will ever pass given the strong opposition of oil and coal companies and a strong public campaign they have mounted to convince citizens that this will cause severe damage to the economy. A more likely candidate would be tradeable permits, discussed shortly.

Hazel Henderson, a leader in rethinking economics, calls user fees and pollution charges green taxes and argues that they should be applied to a wide variety of products and activities, among them disposable goods, airplane travel, international tourism, and oil consumption. Interestingly, some European nations that embrace the idea that present generations have an obligation to future generations have instituted a number of green taxes. In such cases, green taxes have been found to generate a substantial source of revenue. Henderson notes that many business executives find green taxes more acceptable than bureaucracy and regulation. Surprisingly, most economists approve of the idea.

KEY CONCEPTS

Green taxes—levies on undesirable products or activities—create a disincentive to companies that spurs interest in finding environmentally sustainable alternatives.

Economic Incentives In recent years, government officials have also explored a variety of economic incentives that induce companies to comply with pollution laws. Tax credits can also be effective incentives. A tax credit works this way: A government gives a company a tax credit—say 10%—for investing in recycling equipment or for buying recycled material. This lowers the cost of business and encourages responsible practices. Tax credits can be given to individuals who invest in environmentally responsible products, for example, solar energy, wind energy, and conservation. The United States offered a 30% tax credit on solar electric and other renewable energy systems from 2004 to 2016.

Faced with budget shortfalls, governments are often wary of tax credits because they can lower tax revenues. Therefore, careful analysis is a must when considering offering tax credits.

KEY CONCEPTS

Tax incentives like tax credits to individuals and businesses can be used by governments to encourage sustainable practices and products.

The Permit System Another incentive is the tradeable or **marketable permit**. A marketable permit works this way: The EPA grants a permit to Company A, allowing it to emit 10,000 metric tons of sulfur dioxide each year, but the company finds an inexpensive yet effective way to cut emissions to 5,000 metric tons. It can then sell its permit for the 5,000 metric tons to Company B. For Company B, purchasing the permit may be cheaper than installing pollution control devices. In this instance, the total emission of pollution in the region remains the same. However, permit systems are generally designed to reduce pollution levels. In this example, the EPA would lower permitted levels of emissions for all companies in an area. Those that can reduce pollution below their permitted value—say, through pollution prevention—can sell unused permitted emissions.

In Colorado, a marketable permit system was used to control water pollution entering Dillon Reservoir, located in the Rocky Mountains 110 miles west of Denver. Nitrogen and phosphorus entering the reservoir from farms, sewage treatment plants, and other sources had begun to make the reservoir eutrophic. A study of various options to reduce phosphorus pollution showed that additional controls in treatment plants would cost the towns about $1,800 per kilogram ($820/pound) of phosphorus removed. Controlling nonpoint pollution would cost only about $150 per kilogram ($68/pound)—over 10 times less. The legislature and the EPA approved a marketable permit plan allowing the publicly owned sewage treatment plant to pay for nonpoint pollution controls on farmland, saving an estimated $1 million per year.

Marketable permits have been used in the United States to reduce carbon dioxide emissions in certain states, and efforts are underway to expand this program nationwide, although there are strong forces that oppose efforts to regulate carbon dioxide emissions—largely from oil and coal companies and utilities.

KEY CONCEPTS

Marketable permits provide companies with an economic incentive to reduce pollution. If they can find a cost-effective way to reduce emissions, they can sell their unused permitted emissions.

Removing Market Barriers Numerous laws and regulations create subsidies for environmentally destructive activities, as discussed in chapters on mining, forestry, and solid waste. They make it more difficult for environmentally desirable companies and products to compete. Subsidies, therefore, create market barriers. Preferential freight rates—mandated by federal regulations—make it cheaper to haul virgin materials than scrap bound for recycling. Subsidies to oil, gas, coal, and nuclear industries give them a considerable advantage over renewable fuels.

By removing the subsidies and shipping regulations, economically inefficient and environmentally unsustainable practices can be eliminated. Removing these barriers can help protect the environment, ensure sustainable practices, and reduce government spending.

KEY CONCEPTS

Unsustainable practices are often subsidized by governments or are given economic advantage over more sustainable practices. Removing these market barriers can create a level economic playing field, permitting sustainable activities and products to flourish.

Corporate Reform: Greening the Corporation

A sustainable economy depends in large part on the emergence of companies that operate sustainably. In 1989, a number of environmental groups and several managers of major pension funds in the United States, including those of California and New York, joined forces to form the Coalition for Environmentally Responsible Economies (CERES). The CERES proposed a set of guidelines for responsible corporate conduct. Called the Valdez Principles, they call on companies to do the following:

1. Protect the biosphere by reducing and eliminating pollution.
2. Promote the sustainable use of natural resources by ensuring sustainable management of land, water, and forests.
3. Reduce the production and disposal of wastes by recycling, waste minimization, and other measures; use safe disposal methods for wastes that cannot be handled otherwise.
4. Employ safe and sustainable energy sources and use energy efficiently.
5. Reduce risk to the environment and to workers.
6. Market safe products and services—those that have minimal environmental impacts—and inform consumers of the impacts of the products and services a company offers.
7. Restore previous environmental damage and provide compensation to persons who have been adversely affected by company actions.
8. Disclose accidents and hazards and protect employees who report them.
9. Employ environmental directors, with at least one member of the board of directors qualified to represent environmental interests; employ environmental managers, with a senior executive responsible for environmental affairs.
10. Assess and audit progress in implementing the Valdez Principles.

Take a moment to look back over them, and you will see that almost all of the principles of sustainability (and many other ideas discussed in this chapter) are embodied in them. Number 9 is particularly important. It instructs companies to appoint high-ranking executives to the task of environmental management. IBM, for example, appointed a vice president of environmental health and safety, with a staff of about 30 people, to ensure that corporate environmental policy is carried out. IBM has also upgraded the status of its environmental staff, creating a more powerful and autonomous group that ensures production goals are met while minimizing pollution and reducing environmental damage. The success of corporate programs depends in large part on leadership from CEOs or presidents of corporations who look favorably upon corporate environmental policy.

Although few businesses and governments adopted the Valdez Principles, many have begun to implement the principles and practices of a natural capitalism, described in **Spotlight on Sustainable Development 25-3**. **Natural capitalism** incorporates some of the key principles of sustainable economics described in this book and has been popularized by Amory and Hunter Lovins and Paul Hawkens in a book called *Natural Capitalism*. Individuals can help influence the direction of business by investing in socially and environmentally responsible companies and mutual funds. The New Alternatives Fund, for example, is a mutual fund that invests in companies involved in many environmentally sustainable activities, among them renewable energy, cogeneration, insulation, efficient light bulbs, and other forms of energy conservation. Offering a respectable rate of return, it and other environmentally responsible investments illustrate that good business and environmental protection can go hand in hand.

> ### KEY CONCEPTS
>
> Companies can be forced to become sustainable through regulations and market mechanisms, but corporate responsibility and action through voluntary efforts can be even more effective in reducing the environmental impact of business.

Green Products and Green Seals of Approval

Individuals can also influence corporate policy—and make an enormous impact on the environment—by avoiding environmentally unfriendly products (among them disposable diapers and pens) and purchasing green products, defined as environmentally friendly goods and services. **Green products** may be goods made from recycled materials—for example, books, toilet paper, and paper towels made from recycled paper. They also include nontoxic cleaners and high-efficiency showerheads or lightbulbs. They include a wide range of recyclable and reusable products, such as shopping bags.

Today, hundreds of green building products are available, so that a house can be built from the foundation to the roof using environmentally friendly materials (**FIGURE 25-11**). Many nontoxic, environmentally friendly paints, stains, and finishes are also available. Many of them are cost competitive with conventional products, too. Using the operating principles of sustainability—conservation, recycling, renewable resource use, and restoration—as guidelines for product purchase will help you select the most sustainable products.

Individuals can find out about a company's environmental and social record by consulting the book *Shopping for a Better World*, which is published by the Council on Economic Priorities, a nonprofit consumer/environmental group. It rates hundreds of products that people routinely buy. Companies are scored in 10 categories, including environmental policy, charitable giving, minority advancement, and women's advancement.

25-3 Principles of Natural Capitalism

The principles of natural capitalism provide the basis for a complete rethinking of business. They show how, in contrast to conventional wisdom, far greater profits are achieved through protecting and enhancing nature, culture, and community than by harming them.

Radically increase resource efficiency

Most companies can dramatically increase the productivity of almost any resource they use. Fourfold increases in resource productivity are now the basis of economic development policy for the European Union, and 10-fold increases or greater are possible with existing technologies. Increasing efficiency also encompasses the development of innovative business models that focus on meeting consumer needs in ways that require fewer manufactured products and reward companies for reducing their environmental footprint.

"Biomimicry"

Biomimicry calls for a shift away from conventional "heat, beat, and treat" manufacturing methods, which require enormous energy inputs and usually create toxic byproducts, to production based on models derived from nature. The benign productive processes of living things can guide industrial innovation and teach us how to eliminate waste and toxics. This principle, outlined in the book *Biomimicry*, by Janine Benyus, can be the source of dynamic industrial innovation.

Invest in restorative practices

All good capitalists reinvest in productive capital. Restoring the world's depleted natural and human capital is a critical foundation of sustainable wealth creation. This principle encourages businesses to behave in ways that restore the capacity of both the earth and society to sustain life by investing in human and natural capital.

The goal

No net loss of natural or social capital. No current balance sheet accurately captures the real economic value of natural or social capital. But no human system that systematically degrades either one can long endure. To achieve genuine prosperity and an economy worth sustaining, it is essential to ensure that neither form of capital is diminished.

FIGURE 25-11 Green building products. Virtually every part of a home can now be made from an environmentally friendly or green building material like bamboo flooring (shown here). Bamboo is a fast-growing species that can be harvested repeatedly from the same rootstalk.

Teragren SIGNATURE Bamboo Flooring, photo courtesy of Teragren LLC.

Another means of learning about products is government labeling programs. In the early 1980s, the government of the former West Germany instituted a product-labeling program known as the *Blue Angel* program. The Blue Angel label on a product assures the customer that the product is environmentally acceptable and was produced in an environmentally acceptable manner. Since its beginning, the program has scrutinized over 3,850 products and services.

The Blue Angel program offers even more complete analysis of the environmental acceptability of products and have had a profound influence on German industries. In some cases, companies whose products have been given poor ratings altered their manufacturing processes or made drastic changes in the products themselves to receive a seal of approval.

Ecolabeling programs have also emerged in Europe, Canada, Japan, and the United States. The Green Seal program in the United States, headed by Earth Day cofounder Denis Hayes, for instance, passes judgment on products using criteria similar to those used in

Canada (**FIGURE 25-12**). This program currently certifies over 1,000 products and services—everything from coffee filters to windows and doors to lodging.

Another popular label that can help individuals buy green is the Energy Star. It is found on 50 products from air conditioners to stereo receivers to computer screens to houses. It indicates that the product is one of the most energy efficient in its category. The Energy Star program is sponsored by the U.S. EPA and Department of Energy.

The EPA also recently initiated a WaterSense label for washing machines, showerheads, toilets, and more. Like the Energy Star labels, they help buyers identify the most water-efficient appliances and fixtures on the market.

A truly sustainable economy would require that virtually all products be produced in efficient, environmentally sustainable ways. In the meantime, individuals can send a strong signal to the marketplace by purchasing green products.

KEY CONCEPTS

Products vary in the degree to which they contribute to sustainability. Some promote the principles of sustainability, such as recycling, renewable resource use, and restoration. By purchasing such green products, individuals help promote a sustainable economy. Product labeling programs can help individuals select the most environmentally sustainable products.

Appropriate Technology and Sustainable Economic Development

In his classic book *Small Is Beautiful*, the late E. F. Schumacher popularized the term **appropriate technology**. Summarized in **TABLE 25-2**, appropriate technology puts people to work in meaningful ways. Compared with many modern forms of technology, it is efficient on a small scale. It uses locally available resources, takes less energy to run, and produces minimal amounts of waste. Some examples of appropriate technologies geared to a less exploitive lifestyle include passive solar heating for homes and businesses, wind generators and solar modules for making electricity, bicycles for commuting to work when the weather permits, and compost piles or compost worm bins to decompose yard wastes.

The use of appropriate technologies can help the less developed nations reduce their resource demand and impact on the environment. It could also free up resources for other people, making a better life possible for a larger number of people.

Appropriate technology, like other suggestions in this chapter, is not a panacea. It is one part of the solution. (See **TABLE 25-3** for a summary of the ideas.) The choice for appropriate technologies belongs to us and the business world.

KEY CONCEPTS

Appropriate technologies are an essential part of a sustainable future. They rely on local resources, are efficient, and produce little if any pollution.

A Hopeful Future

"Throughout most of its tenure on Earth," economist Herman Daly points out, "humanity has existed in near steady-state conditions. Only in the past two centuries has growth become the norm." As pointed out earlier, continual economic growth may be the norm today but is very likely not sustainable. If we are to create an enduring human presence, our economic system must conform to the design principles of the ecosystem. The human economy must operate

FIGURE 25-12 The Green Seal of approval. This seal affixed to a product indicates that it is a green product—one whose purchase causes less harm to the environment than competing products made by other manufacturers.

Courtesy of Green Seal (www.greenseal.org).

TABLE 25-3

Steps Essential to Building a Sustainable Economy

Goals

- Reduce population growth, then gradually reduce population size through attrition.
- Reduce resource consumption and waste by reducing demand, increasing product durability, and increasing efficiency.
- Recycle, reuse, and compost to the maximum extent possible.
- Develop a wide range of renewable energy resources.
- Protect and conserve renewable resources, such as farmland, fisheries, forests, grasslands, air, and water.
- Improve renewable resource management to ensure sustainability.
- Repair past damage to natural resources by replanting forests and grasslands, reseeding roadsides, restoring streams, cleaning groundwater, and reducing overgrazing.
- Increase national and regional self-sufficiency by using renewable resources whenever possible.
- Support sustainable development projects in developing nations.
- Develop sustainable ethics and promote individual and corporate responsibility and action.
- Improve social conditions by promoting democracy, justice, and a more equitable distribution of wealth.
- Work for global peace and cooperation.

Some Policy Tools

- Environmental education programs.
- Green taxes and full-cost pricing requirements for business.
- Alternative measures of progress, such as the NEW and GPI, that look at a wide range of economic, social, and environmental conditions.
- Laws that harness market forces, including (1) economic disincentives, (2) economic incentives, (3) tradeable and marketable permits, (4) laws that eliminate market barriers that promote inefficient resource use, and (5) laws that remove subsidies of environmentally destructive activities.

within the limits of the economy of nature. However, that does not mean that the economic future is bleak. The sustainable economy need not be a prescription for dull living and stagnation. In fact, it may entail growth in some sectors, such as solar energy and energy efficiency, while others, such as oil and steel production, are phased out. A sustainable economy based largely on renewable resources does not mean a retreat into the Dark Ages of drafty, cold domiciles. Solar heating and insulation could mean even more comfortable living spaces. Many nonrenewable resources (metals, for instance) would remain in use, recycling through the system many times. Some mining would be necessary to

replace what is lost or locked up permanently in structures. Renewable fuels such as ethanol and biodiesel could power our transportation system. Solar electric systems (such as the ones on my home and educational center, which supply 100% of my electricity) along with wind energy might provide the electricity we need.

Clearly, today's economy and that proposed by a growing number of people are worlds apart. One of the principal differences is the continued high level of consumption and waste in the former, which is bound to deplete our nonrenewable (oil and aluminum) and renewable (wood from the tropical rain forest) resources and possibly create a population crash. In contrast, a truly sustainable economy would achieve a level of production and consumption that could be sustained forever.

 KEY CONCEPTS

> A sustainable economy conforms to ecological design principles and is dynamic and full of opportunity.

25.6 Environmental Protection Versus Jobs: Problem or Opportunity?

During the 1992 presidential election campaign, then-President George Bush repeated a theme that arose many times during his campaign: We need environmental protection but not at the cost of jobs. He pointed out that we can't have higher-mileage cars because it would put tens of thousands of American autoworkers out of work. We can't save the spotted owl because it will cost jobs. Ignoring the loss of employment from automation and other factors (which cost the timber industry 12,000 jobs from 1977 to 1987), Bush and others hammered on the point over and over again, while Bill Clinton and Al Gore asserted that the choice between jobs and environmental protection was a false one.[1] Who's right in this debate? Do environmental regulations cost jobs? Do they cripple businesses?

Environmental regulations and permits can delay projects, such as dams and power plants. However, several studies show that delays are often the result of poor planning on the part of the companies and government agencies overseeing the work. Delays could be greatly reduced if corporations and governmental agencies were willing to invite the public to give input early on. As it is, citizens often become involved late in projects, and—because their concerns have not been incorporated—they oppose them.

---

[1] Another 35,000 jobs in the timber industry could be lost in the next 45 years as a result of mechanization, according to the Association of Forest Service Employees for Environmental Ethics.

Business leaders argue that environmental regulations also cut productivity and reduce income, resulting in loss of jobs. **Productivity** is the dollar value of goods per hour of paid employment. This figure is often used to determine how healthy an economy is. Businesspeople argue that environmental regulations divert workers from productive jobs (miner) to nonproductive ones (mine safety inspector), which raises the cost of doing business and lowers productivity. Careful analysis shows that environmental regulations do indeed diminish the output of industry. By various estimates, environmental regulations decrease productivity by an estimated 5 to 15%. However, they decrease output much less than other factors, such as high energy prices and the general shift to a service economy. Efforts to prevent pollution and other problems can actually increase productivity, as pointed out in previous chapters.

In an article in *International Wildlife*, Curtis Moore described the efforts of Germany to protect the environment. He wrote: "More than anywhere else on Earth, Germany is demonstrating that the greening of industry, far from being an impediment to commerce, is in fact a stimulus." He went on to say, "Precisely because it is so environmentally advanced, Germany is lean, competitive, and poised to dominate the global marketplace." Through end-of-pipe controls and numerous sustainable strategies that address the root causes of the problem, Germans have shown that cleaning up industry can stimulate business. How? For one, tough standards stimulate innovation, and nations with the most rigorous requirements for pollution control often become leading exporters of control technologies. In addition, environmentally responsible business can result in better employment opportunities.

Several studies show that far more jobs are created by environmental controls than have been lost. The Clean Air Act and Clean Water Act have created at least 300,000 jobs in pollution control. Studies by the EPA show that very few jobs are lost as a result of environmental regulation. Only companies that are hanging on for dear life are affected. Even the AFL-CIO admits that not one plant shutdown can be attributed to environmental regulations.

This is not to say that tighter controls on pollution, reductions in timber cutting, or other environmental regulations are always benign. They do affect communities, sometimes profoundly. Environmental protection measures, especially the kinds of changes needed to develop a sustainable system, will result in massive shifts in employment. Employment will very likely shift from the ends of the present linear system of production and consumption (mines, forests, and waste dumps) to the center of an efficient, cyclic system.

Numerous studies suggest that this shift will result in a net increase in the number of jobs. Thus, restructuring the economy and our way of life to rely more on energy efficiency, recycling, mass transit, and alternative fuels will create far more jobs than are lost. A large nuclear power plant employs 100 workers. A coal-fired power plant that generates the same amount of electricity employs 116 workers, whereas a solar thermal facility employs more than twice as many workers, about 250. However, a wind farm employs more than twice as many people again, 540 in all. Furthermore, wind and solar thermal energy are less expensive and more environmentally benign than nuclear power.

Studies in the United States show that energy efficiency also creates jobs. Michael Renner of the Worldwatch Institute estimates that weatherizing all U.S. households could create 6 to 7 million job-years—that's 300,000 jobs lasting 20 years each!

A 1985 European study that examined the employment potential of six energy conservation and renewable energy technologies in four countries (Great Britain, Denmark, France, and the former West Germany) found that these technologies could create 142,000 job-years—7,000 jobs lasting 20 years. A full-fledged program for all 12 European nations could create 530,000 job-years.

Michael Renner notes that "although a shift from fossil fuels to solar energy entails job losses in the oil, coal, and gas industries, there are overlaps among the kinds of supplies and skills required by the solar industry that will minimize overall job loss." Companies that produce the materials needed for solar panels are already in existence. Many of the skills needed to tap renewable energy supplies are similar to those required for conventional construction and heating system installation. Work opportunities could spring up in a variety of existing occupations, among them carpentry, plumbing, and construction.

Shifting to mass transit may also create more jobs than are created by our current system of automobile transit. A German study showed that spending $1 billion on highway construction yields only 24,000 to 33,000 direct and indirect jobs. Spending the same amount on railway and light-rail construction creates 38,000 to 40,000 jobs. Furthermore, the shift from cars and trucks to railroads, subways, light rail lines, and buses offers alternative job opportunities for much of the workforce.

Recycling offers similar employment opportunities. Alcoa estimates that at least 30,000 people in the United States are employed in aluminum recycling, nearly twice as many as in aluminum production. Vermont's recycling facilities employ 550 to 3,000 people for each 1 million tons of materials they handle. On the other hand, incinerators equipped to handle the same amount employ only 150 to 1,100 people, and landfills employ only 50 to 360.

Environmental protection can help save companies money. Pollution prevention measures, for instance, cut down on the production of hazardous waste and substantially reduce waste disposal costs. They also eliminate future liability for cleanup and may reduce or eliminate employee health effects and future claims for health impacts. Pollution prevention can reduce lawsuits and greatly slash the cost of compliance—installation and operation of pollution control equipment. Many examples cited in this text highlight companies that have saved considerable amounts of money by preventing pollution, using energy more efficiently, and recycling and reusing wastes.

Worldwatch Institute's Cynthia Pollock Shea notes that "businesses that protect the environment can make a healthy profit." As German and Japanese companies are showing us, environmental protection is a precondition for success.

Environmental protection, rather than being an impediment to economic progress, may be a stimulant. Many sustainable strategies result in cost-competitive goods and services that create as many or more jobs, with little environmental impact. A sustainable economy, however, will involve a major employment shift.

A penny will hide the brightest star in the Universe if you hold it close enough to your eye.

—Samuel Grafton

CRITICAL THINKING

Exercise Analysis

Two essential critical thinking rules are to examine one's biases and to consider the big picture. These rules are important in analyzing the assertions made in this exercise. Consider bias first.

Many people in positions of power are wedded to the notion that all growth is good. As noted in the chapter, however, much of the economic growth occurring in the United States has not improved our well-being. Growth results in extraordinary costs, such as lost wildlife, lost open space, polluted air, crowding, and so on. As pointed out in the chapter, a substantial portion of the growth of the U.S. GNP has resulted from "bad" economics—pollution cleanup, repair of damage from pollution, and unnecessary healthcare expenses. Statistics also show that economic growth tends to benefit the wealthiest members of societies; that is, economic benefits of the growing economy are not trickling down to the less fortunate, as they're supposed to.

Also noted in this chapter is the assertion that to improve an economy it is not always necessary to manufacture more goods. One of the fastest and most profitable ways of improving an economy is to use resources more efficiently. By using energy, water, and materials more efficiently, for instance, companies can produce goods more cheaply, make a higher profit, and/or compete more effectively. Pollution prevention is another strategy with extraordinary economic benefits. All of these strategies permit companies to produce goods while improving the bottom line and the environment.

CRITICAL THINKING AND CONCEPT REVIEW

1. Define the following terms: *command economy*, *market economy*, and *law of scarcity*.
2. Describe the law of supply and demand. What is the market price equilibrium? What are the major weaknesses of the law of supply and demand? How can they be corrected?
3. Define your own economic goals. Would you classify them as consistent with a frontier economy or a sustainable economy?
4. Using your critical thinking skills, analyze the following statement: "Economic growth is essential to a healthy economy."

5. In your view, is continued economic growth in the developed nations possible? If so, why, and for how long? If not, what are the alternatives?
6. Describe the gross national product (GNP) and its strengths and weaknesses.
7. Define the term *net economic welfare* (NEW). How does it differ from the GNP? Is it greater than the GNP, or less? Does it grow as quickly as or more slowly than the GNP? Is this good or bad?
8. Describe the index of sustainable economic welfare (GPI). How does it differ from the GNP?

9. Using your critical thinking skills, analyze the following statement: "The GNP is a fundamentally sound economic measure and a good indicator of the well-being of a nation's people because economic health means a higher standard of living for all."

10. Discuss how time preference and opportunity costs are related. How do they affect the ways in which people manage natural resources?

11. Define the term *economic externality* and describe how externalities can be internalized. What are the benefits of doing this?

12. How can economic externalities be avoided?

13. What is the economically optimal level of pollution control? Why is it impractical to consider reducing pollution from factories and other sources to zero via pollution control devices?

14. Describe the law of diminishing returns. How does it apply to pollution control? Can you think of any other examples where the law applies?

15. The analysis of the economic system and of economic thinking suggests a number of weaknesses when viewed through the lens of sustainability. What are they? Describe each one and discuss how each can be corrected.

16. Of what value are alternative measures of progress?

17. In what ways can local or regional self-reliance promote sustainability?

18. Describe market forces that can be used to promote environmentally responsible behavior in businesses.

19. Describe a sustainable economy. What are its main goals? In your view, is it a practical alternative to the current economic system? What are its strengths and weaknesses?

20. Using your critical thinking skills, analyze this statement: "Cleaning up the environment will put thousands of people out of work. We simply can't afford to do it."

KEY TERMS

appropriate technology
benchmarks
command economy
consumer-pays option
demand
descriptive economics
direct costs
ecological economics
economic externality
economics
free-market system
full-cost pricing
gas guzzler tax
Genuine Progress Indicator (GPI)
green products
green taxes

gross domestic product (GDP)
gross national product (GNP)
Index of Sustainable Economic
    Welfare (ISEW)
indicators of sustainability
indirect costs
inputs
law of diminishing returns
law of scarcity
law of supply and demand
macroeconomics
market economy
market price equilibrium point
marketable permit
microeconomics
natural capital

natural capitalism
net economic welfare (NEW)
net present value
normative economics
opportunity cost
outputs
pollution taxes
productivity
repercussion costs
replacement cost
supply
sustainable economics
sustainable economy
taxpayer-pays option
time preference
user fees

© Erika J Mitchell/Shutterstock, Inc.

Law, Government, and Society

Our environmental laws are not ordinary laws, they are laws of survival.

—Edmund Muskie

LEARNING OBJECTIVES

After reading and studying the material in this chapter, students will be able to understand, knowledgably discuss, and, when appropriate, apply the information learned in this chapter, including:

1. Forms of government and how they affect the environment
2. How government policies affect environmental protection for better and for worse
3. How democratic governments work and key players in the formulation of environmental policy
4. How environmental law has evolved to create broader and more sustainable environmental protection
5. Principles of environmental law
6. Measures that can help create more environmentally friendly governments worldwide

CRITICAL THINKING

Exercise

In 1991, many nations began to draft national reports in preparation for the United Nations Conference on Environment and Development in June 1992. The national reports were designed to assess what nations were doing to protect their environments and also to promote sustainable development. In the U.S. National Report, the authors frequently pointed to the billions of dollars that the government and the nation's businesses were spending on pollution control as a sign of our commitment to environmental protection and sustainable development. The authors also cited case studies that showed specific actions on the part of businesses and government agencies.

To some critics, the authors of the U.S. National Report seemed to be saying that because the United States is spending a great deal of money on environmental problems and is putting enormous effort into solving them, the nation is acting responsibly and with sufficient vigor. Is there a fallacy in this kind of reasoning?

We need a future we can believe in, one that is neither so optimistic as to be unrealistic nor so grim as to invite apathy or despair. In short, we need a future that is hopeful *and* attainable. This text has outlined one of many possible futures, a sustainable one in which human society thrives and prospers and yet protects, even enhances, the environment. Creating such a system requires an understanding of basic principles of ecology and an appreciation of how vital natural systems are to our lives. It will also require that we address key issues such as population growth and dramatically change the way we use and manage resources. To create a sustainable human presence we must also alter the way we deal with pollution. Building a sustainable future, say many observers, will require fundamental changes in human systems, from agriculture to transportation to industry. The changes needed depend on a shift in people's attitudes and changes in economic systems. Such modifications may come about in many ways—for example, through personal change and personal initiative through market forces (such as the demand for green products). One effective means of social change is through government.

This chapter examines government: the roles of government, the participants in government policy making, and the decision-making process. It then discusses environmental law. Next, it discusses some key obstacles to a sustainable future and ways to overcome them. Finally, it outlines important changes in international governance needed to make a sustainable future a reality.

26.1 The Role of Government in Environmental Protection

In another chapter, I described two economic systems: market economies, in which the marketplace determines the availability and price of goods and services, and command economies, in which the government controls these functions. Most have features of both economies. However, they tend to lean in one direction or the other.

Forms of Government

For the most part, governments and economic systems go hand in hand. Countries with market economies are generally **democratic nations** with representative governments—that is, governments largely staffed by people who are elected by the people. Representative governments operate by rules (laws) that are, technically, agreed on by the majority through their elected officials. In democratic nations, the means of production and distribution of goods and services are, for the most part, privately held and privately managed. Profit is a major driving force of this system, and because of this, many large corporations play a powerful role in influencing policy, often to the detriment of the people. Thus, the economic systems tend to be market driven. However, democratic nations have found that varying degrees of governmental control are needed to reduce unlawful activities.

Communist nations are based on the notion that all property is held in common. It is the antithesis of capitalism, with its privately held production capacity. Communist or socialist nations are politically organized around the social ownership of all goods and services. In other words, technically speaking, the people own and operate the production and distribution network. Such nations emphasize the requirements of the whole, rather than individual liberty. Profit is not a motive in such systems. The goal is that goods should be distributed equally, according to needs (although in practice this rarely occurs). In communist nations, the government manages the economy in the name of the betterment of people. History shows, however, that personal oppression is often rampant in such systems. Many thinkers believe that such systems are fundamentally unsustainable, and transformation of them is a key element of creating a sustainable future.

Just as economies are mixed, so are governments. New Zealand, Germany, and Sweden are all democratically governed, but in all three nations, the government provides health care and other services that are paid for through taxes. In China, one of the few remaining communist nations, the private sale of products for profit is currently permissible and growing rapidly. This chapter focuses primarily on democratic nations.

KEY CONCEPTS

Two basic forms of government exist: democratic and communist. Democratic nations are those in which most decisions are made by elected officials. The economies tend to be controlled by market forces. Communist nations are those designed around the notion of public ownership of the means of production and the distribution of goods and services according to need, not ability to pay.

FIGURE 26-1 High-speed commute. In Europe, high fuel costs and other factors help promote interest in energy-efficient mass transit.
© ollo/iStockphoto.

Government Policies and Sustainability

Although democratic nations generally support free-market economies, governments have found it necessary to regulate free enterprise to protect unbridled business interests from exploiting people and the environment. The tools for regulating or influencing development are many, but fall into three general categories: taxes, financial incentives and expenditures, and laws and regulations. The President is also given a special tool, known as the *executive order*, which we examine shortly.

Taxes are a means of generating funds for public services, among them highways, education, road maintenance, police protection, pollution control, and wildlife programs. Taxes are levied on personal property, commodities, services, and income (both personal and corporate). Taxes can be used to discourage undesirable activities. For instance, a sizable tax on gasoline in European nations discourages automobile use and promotes the purchase of energy-efficient automobiles. It's not surprising that some of the best mass transit systems are found in Europe (**FIGURE 26-1**).

Two other common tools are tax credits and income tax deductions. These are two types of financial incentives. A **tax credit** is a dollar amount of the purchase price that a person or company can deduct from income or corporate tax. In the United States in the late 1970s and early 1980s, for example, the federal government offered individuals and businesses sizable tax credits if they invested in energy conservation (insulation, weatherstripping) and renewable energy (solar panels and wind energy). Many states offered similar tax credits.

Income tax deductions work similarly. When a government offers a deduction for an activity, the amount of the purchase or some percentage of the purchase price is deducted from one's total income. This lowering of income lowers the amount of tax one pays. If your taxable income was $50,000, but you had a $5,000 tax deduction, you would only be taxed on $45,000. The U.S. government

currently offers income tax deductions for children. Couples with children receive a tax break for every child they have. Interest on loans for second homes was, until recently, also tax deductible. Tax policies that favor undesirable practices can also have significant impacts on our future.

Direct government expenditures can also have a profound effect on the environment and on sustainability. For instance, in the 1980s, government grants to communities for sewage treatment plant construction created thousands of jobs and helped reduce water pollution. Governments also fund research and development in many activities, such as energy conservation and renewables. Government-funded research helps reduce the cost of commercial development and thus provides an economic advantage to some technologies. Government procurement (purchase) programs for environmentally beneficial technologies and materials (such as photovoltaics and recycled products) help increase production and bring costs down, making new products more attractive to citizens.

Government expenditures on environmentally less desirable technologies (nuclear energy research, for instance) and activities (war and oil development) have the opposite effect. The most obvious reason, say critics, is that such expenditures direct society down a potentially unsustainable path of development. The farther we go along that path, the harder it will be to shift to a sustainable one. Such expenditures also divert money from other important tasks, such as energy efficiency and renewable energy development.

Cutting government expenditures on environmentally destructive activities could free up considerable sums of money that could be invested to promote sustainability. Even modest reductions in global military spending could finance a massive campaign to combat global climate change, for example.

The final governmental tool, and probably the most widely used one, is laws and regulations. In democratic societies, laws emanate from legislative bodies: congresses, parliaments, and town councils. They are generally designed to curb or ban those activities people have deemed undesirable. For example, the **U.S. Corporate Average Fuel Economy Act** is a law passed in 1975. It established a set of goals for gas mileage standards for automobile manufacturers to help curb the waste of fuel.

A nation's laws are only as good as their enforcement. Enforcement depends on having adequate interest, personnel, and funds. If an agency lacks funds or human power or even the will to enforce a new law, little will come of it.

The Corporate Average Fuel Economy Act (just cited) is an example of a law whose utility was greatly reduced by lack of interest on the part of the responsible government agency, the Department of Transportation (DOT). Unfortunately, the DOT it bowed to heavy pressure from Ford and General Motors, rolling back the mileage standards twice during the Reagan administration. In 1985, the average fleet mileage was far below Congress's goal of 11.9 kilometers/liter (27.5 mpg). When President George H. W. Bush took office, however, he pledged to raise standards, and by 1989, the mileage standard of the average new car did increase.

In 2010, the Obama administration set new fuel standards that will raise current fuel economy standards by nearly 10 mpg by the 2016 model year in an effort to help combat global climate change caused by the greenhouse gas, carbon dioxide. According to the new ruling, automakers will be required to meet a fleet-wide average of 35.5 mpg by 2016. This is the first increase in fuel economy standard in over 20 years. This could save car and truck owners as much as $3,000 over the life of a vehicle.

Like the president or chief operating officer of a corporation, the president of a nation can have a powerful impact on environmental protection. Unsympathetic leaders make it difficult to move forward. President Reagan, for instance, who was in office from 1981 to 1989, openly opposed the environmental protection goals of previous administrations. He actively sought to undermine protection by weakening existing laws and agencies, such as the Environmental Protection Agency (EPA). President George H. W. Bush also worked to undermine the nation's environmental laws and regulations.

President Clinton was much more sympathetic to the environmental issues. He and Vice President Gore took numerous steps to improve environmental protection, even greening the White House as described in **Spotlight on Sustainable Development 26-1**. President George W. Bush also worked to weaken environmental laws and regulations—far more actively and successfully than President Reagan. President Obama, on the other hand, has pushed for significant actions to protect the environment including cuts in greenhouse gases and huge increases in renewable energy production.

Go Green

To save gas and time, vote by mail.

The United States is not the only country in which environmental laws fail to accomplish their goals because of lack of funds, apathy, lack of enforcement, or political opposition. Before it disintegrated, the Soviet Union had some of the most stringent environmental standards in the world, but enforcement was virtually nonexistent. The country lacked the technical expertise and the financial resources to control pollution.

Environmental lawmakers usually set broad goals in their legislation, but the job of writing specific regulations to put those goals into effect lies in the hands of experts within various government agencies. In the United States, the EPA is the principal environmental rule maker. A host of other agencies also participate in this process when laws pertain to matters under their jurisdiction. The Forest Service, for instance, drafts regulations for management of forests. These agencies propose regulations, based on their interpretation of the law. The regulations are then subject to public comment. (This is where citizens and nonprofit organizations can play a role in helping to shape final regulations!) After receiving input from various stakeholders, the agency may modify its proposed regulations and put them into effect. The EPA is currently empowered by nine federal laws, among them the Clean Water Act, the Clean Air Act, the Superfund Act, and the Resource Conservation and Recovery Act (RCRA).

The **executive order** is yet another tool for effecting development and protecting the environment, but it is reserved for the President of the United States. An executive order is just that, an order issued by the President of the United States. President Nixon used an executive order to establish the Environmental Protection Agency. President Clinton used it effectively and often to reduce the environmental impact of the federal government. Executive Order 12843, for instance, implemented an early phaseout of ozone-depleting chemicals at federal facilities. Executive Order 12844 accelerated the acquisition of vehicles that use alternative (clean-burning) fuels by federal agencies. Executive Order 12845 required the federal agencies to purchase energy-efficient (EPA Energy Star) computers, monitors, and printers. Executive Order 12856 directed federal agencies to cut toxic emissions by 50%. President Clinton also signed Executive Order 12873 promoting the purchase of green products. The executive order can also be used to weaken environmental protection; President Bush, for instance, used it often in this fashion, even to override decisions of the U.S. Supreme Court.

Through the mechanisms outlined previously, governments around the world have made tremendous strides in protecting the environment, managing resources, and controlling population. Many chapters in this text have pointed out the most significant advances. Since 1958, the U.S. government has passed nearly two dozen environmental laws and amendments. Similar progress has been made in most other industrial nations. Although environmental conditions are better than they would have been in the absence of these measures, most governments have dealt with environmental problems after the fact; that is, they have typically relied on end-of-pipe solutions or regulations that merely slow down the destruction of global ecosystems. This chapter describes

26-1 Greening The White House

Shortly after he took office, President Bill Clinton (**FIGURE 1**) began the task of remodeling the White House. Although Presidents and their wives have long been known to spruce up the executive mansion to meet their individual tastes, this remodeling project was unlike any other president's in U.S. history. Far from the cosmetic changes of previous administrations, this was to be a "green makeover"—a series of environmental upgrades of the White House complex, consisting of the Executive Residence, where the President and his family live, the Old Executive Office Building (OEOB) where many White House staffers work, and the White House grounds.

Spurred on by Vice President Al Gore, long known for his commitment to environmental issues, this project was designed to achieve a number of important goals: to reduce pollution, slash energy demand, protect water quality, conserve water, improve the comfort of the structure, and save the government and taxpayers money. Actions were many and varied from reducing pesticide use to using soaker hoses to water plants to reduce water consumption to installing an energy-efficient refrigerator in the First Family's residence. Here's a quick overview of some of the many measures taken on behalf of the President:

- Installing more efficient cooling units, known as chillers, which provide cool air to rooms.

FIGURE 1 President William J. Clinton
Courtesy of Helene C. Stikkel/U.S. Department of Defense.

- Installing chillers that do not use ozone-depleting CFCs in air coolers, using HFC-134a, which contains no chlorine.
- Upgrading heating, ventilation, and air conditioning systems with energy management control systems that allow more precise control of room temperature to reduce overall heating and cooling demands.
- Capturing waste heat to preheat domestic hot water.
- Replacing window-mounted air conditioners with much more efficient ones—there are 1,000 air conditioners and about 10% are replaced every year. New units have timers and occupancy sensors.
- Rezoning steam heating system for better control of heating.
- Cleaning up the White House complex paint shops—they were a source of toxic VOCs (volatile organic solvents). Problems resulted from inadequate ventilation, numerous open cans, pans, and trays of paint, lacquer, thinner, solvents, etc.
- Painting off site at a state of the art painting facility maintained by the National Park Service.
- Implementing a no smoking policy in the Residence, in East and West Wings, and OEOB.
- Reducing pesticide use and training pesticide applicators.
- Reducing fertilizer use.
- Adjusting mowing practices to minimize water demand.
- Using mulching lawn mowers to grind up grass clippings, returning nutrients to soil.
- Irrigating early in the day between 6 and 7 AM to reduce water use.
- Using soaker hoses whenever possible for watering vegetation.
- Installing moisture sensors for irrigation system.
- Adjusting and replacing faulty sprinkler heads.
- Using integrated pest management to control insect pests.
- Installing compact fluorescent lighting wherever possible—for example in table lamps—to reduce electrical demand and pollution.
- Installing occupancy sensors for lighting.
- Upgrading outdoor lighting, cutting electrical consumption by 51%.
- Rehabilitating existing historical skylights on fifth floor of the OEOB to increase daylight—dirt and paint on glass had reduced light penetration. This will reduce electrical use in these areas by 20 to 75%.

- Purchasing energy-efficient computers and office equipment, cutting energy consumption by 50% or more.
- Installing water conserving devices in restrooms, kitchens, and other areas.
- Recirculating cooling water in chillers, rather than using fresh water to supply them. This eliminates 5 gallons of water per minute previously discharged into sewage system.
- Instructing staff to use green products as often as possible.
- Establishing a source reduction policy to reduce paper consumption, limit use of disposable products, and conserve office supplies.
- Making special events green by minimizing disposable items, using recycled materials, making bulk purchases, and making recycling containers visible.
- Recycling batteries.
- Increasing composting of yard waste.
- Recycling compact fluorescent light bulbs to reduce flow of heavy metals such as mercury to landfills.
- Recycling polystyrene plastic in the cafeteria.

- Replacing glass in third-floor solarium and greenhouse in Executive Residence with high-performance insulating glass.
- Retrofitting windows of OEOB with double-paned glass.

The greening of the White House is an inspiration to all of us. As President Clinton said at a speech on Earth Day, 1994, "We're going to identify what it takes to make the White House a model for efficiency and waste reduction, and then we're going to get the job done. . . . Before I can ask you to do the best you can in your house, I ought to make sure I'm doing the best I can in my house." That's called leadership by example.

But this project wasn't carried out by the President on weekends. The American Institute of Architects coordinated 100 national experts, who made recommendations. The Environmental Protection Agency and the Department of Energy played a huge role in the success of this project, as well. Tonight when the current President sits down to read in his study, he'll be working under the light of a compact fluorescent light bulb.

policies needed to confront the roots of the problems and create a sustainable way of life.

What about the less developed nations, where nearly 6 billion people now live? Many less developed nations have adopted policies to control population growth, save wildlife, and reduce pollution. However, progress has been hindered in many cases by a lack of money, rapid population growth, corruption, civil war, war with neighboring nations, and governmental apathy and ineptitude. In many cases, hunger and poverty are so severe that environmental matters are relegated to the back seat. Without population stabilization and programs to raise the standard of living and the availability of food, little progress can be expected.

KEY CONCEPTS

Democratic governments regulate activities, such as environmental protection, via three measures: (1) tax policy, (2) direct financial incentives and expenditures, and (3) laws and regulations. The President also wields enormous power through the executive order.

26.2 Political Decision Making: The Players and the Process

Now that we've seen the basic tools by which governments influence the environment and shape the future of a country (indeed, the entire world), let us take a closer look at the players and the process.

Government Officials

In communist and other authoritarian governments, policy is largely determined by a ruling elite. Ironically, as noted earlier in this chapter, in such systems the people whom the government is meant to serve generally have little or no input. In fact, they often live under severe oppression and poverty, struggling to obtain enough food. In contrast, in democratic nations a wide assortment of government officials make important decisions that affect the environment and the future of the nation. Some government officials (senators, for instance) are elected by the people; others (agency heads) are appointed by the president, governors, or mayors. Their role is to make laws and regulations and enforce them. In other words, they are the creators and guardians of public policy, although corruption and outside interests can undermine their responsibility.

Corruption is known to abound in communist and other authoritarian nations, but representative government, for all its merits, does not always operate as its founding fathers intended. Appointed officials in the EPA and state health departments, for example, have been known to overlook violations of environmental and health laws or weaken laws and regulations. Elected officials are often inclined to vote in favor of legislation that is supported by big-money interests that, not coincidentally, donate large sums to their campaigns. This is not to detract from the fact that dedicated, hard-working men and women at all levels of government have also made important contributions to bettering the environment.

Government officials are charged with the task of creating laws and regulations and enforcing them for the general good. In democratic nations, government officials are either elected or appointed to office. Theoretically, both types are accountable to the general public. Elected officials are generally more responsive, though, as their reelection depends on satisfying the electorate.

The Public

In a democratic society, voters generally have a say in policy making through their elected representatives (**FIGURE 26-2**). Letters, phone calls, attendance at town meetings, and responses to surveys are four lines of communication that permit us to let our opinions be known on a variety of issues.

Although citizens sometimes forget that, elected representatives are their employees. Their job is to interpret public preferences and find ways of enacting policies that satisfy the needs of the electorate. In many cases, however, the populace is divided on issues and elected officials make decisions based on their party's position (presumably the majority of a representatives constituency). Elected representatives may even vote against their constituencies' wishes. Public officials are also frequently influenced by special interests—business groups or environmental groups. For this and other reasons, many people have grown cynical about their influence on the government and about their own future. Many people feel victimized and disempowered by the political process, which leads to apathy and poor voter turnout at elections.

Apathy is counterproductive to the political process. Citizens can make a difference. A letter or email to a congressional representative may not change a vote, but many such communication could. As a rule, for each email or letter they receive, legislators estimate that there are 5,000 to 10,000 people who feel the same way. One thousand letters means there are 5 million people who think similarly, which is very compelling to an elected official.

Letters and emails to state legislators can have special impact because they rarely hear from the public. Twenty or 30 letters are cause for concern. In their minds, 40 or 50 letters indicate a disaster requiring action. So, write and vote. Let your voice be heard.

Public policy is forged by elected officials with input from the general public. Although many people are discouraged with the process and are apathetic, emails and other forms of communication to elected officials can help shape public policy.

Special Interest Groups

The **theory of public choice** states that politicians act in ways that maximize their chances of reelection. To do so, they must appease the voters, and, in recent years, a select group

of billionaries, often reacting on their own behalf. Special interest groups—organized forces of like-minded individuals with common goals (such as environmental organizations and industry groups)—often work through **lobbyists,** who meet with legislators and their aides to convince them of the merits of their views. Lobbyists often draft legislation. When Harry Truman was president, about 450 lobbyists worked the halls of Congress; today, an estimated 23,000 lobbyists frequent the nation's capital.

Special interests and billionaires reward politicians monetarily, providing funds needed to run costly political campaigns. To make an even larger impact, many special interests band together to form **political action committees,** or **PAC**s. A PAC is a consortium of individuals or organizations that pool their financial resources to create a single large donation.

According to Common Cause, a nonprofit citizens' lobbying group, PAC contributions pay off "in billions of dollars worth of government favors for the corporations and other special interest groups that make them." PAC contributions from business, for example, may help to defeat or to water down important environmental legislation.

Another new political and economic force is the 527 group. These political organizations accept huge donations from wealthy individuals and ordinary citizens. The donations are used to run ads supporting like-minded candidates

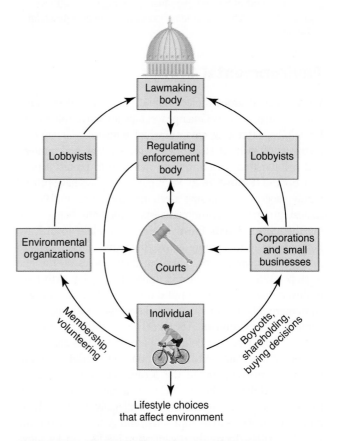

FIGURE 26-2 The many lines of connection among government, individuals, and corporations. Laws are passed by lawmakers. These, in turn, affect the behavior of individuals and businesses. Individuals, environmental groups, and businesses also influence the laws through lobbyists and the court system.

or political positions. They currently bypass federal election campaign contribution laws.

One of the most insidious political games played by special interests, and especially business interests, is known as the **Double-C/Double-P game.** Double-C/Double-P stands for "Commonize the Costs and Privatize the Profits." Here's how it is played: Business lobbyists support and sometimes even propose legislation that promotes government programs that could generate huge profits for the private sector. Thus, taxpayers end up paying for programs that support business. Publicly supported sports stadiums are a good example.

Although democracy can be distorted by special interest groups, especially business PACs, money does not always reign supreme. In the United States, powerful antipollution laws, auto safety standards, hazardous-waste laws, and other important environmental laws have been enacted despite strong opposition from some of the business community. A good measure of this success can be attributed to the efforts of environmental groups.

FIGURE 26-3 High-profile activism. Greenpeace activists attempt to prevent the slaughter of a minke whale by Japanese whalers.
© Kate Davison, HO/Greenpeace/AP Photos.

KEY CONCEPTS

Special interest groups and extremely wealthy individuals can wield an inordinate amount of political power through lobbyists and financial support of political campaigns. Nonetheless, citizen interests have, on many occasions, overcome powerful PAC interests.

Environmental Groups

Environmental groups affect public policy in several ways. Some groups, such as Greenpeace, take an active role on the front lines, protesting environmental injustices. Greenpeace members often meet face to face with whalers, seal hunters, and polluters, endangering their own lives to protest actions they oppose (**FIGURE 26-3**). Such public displays have proved highly successful in raising awareness on issues. In 1980, a radical group of environmentalists came into being, calling themselves **Earth First!** Under the leadership of Dave Foreman, they took more active, frequently illegal steps to protect the environment. They put their bodies on the line to stop bulldozers. They drove spikes into old-growth trees to thwart chain saws. They put dirt in the fuel tanks of heavy equipment. Their tactics were called **ecotage**—sabotage in the name of the environment. (For a debate on the pros and cons of direct action, see the Point/Counterpoint in this chapter.)

Some years later, Paul Watson, one of the original founders of Greenpeace, started the oceanic equivalent of Earth First! Called the Seashepherd Conservation Society, it roams the high seas, ramming ships that are illegally whaling, flying miniature airplanes to disrupt sonar, and generally disrupting illegal activities. Today, the organization has its own TV show.

Most environmental groups operate within legal means. Some groups, for instance, are involved in education. They may prepare educational materials for schools or deliver talks to public gatherings. Other groups work to preserve land to protect biodiversity. The Nature Conservancy, for instance, purchases land that is then set aside to protect endangered wildlife and plants, to provide future generations with opportunities to enjoy the out-of-doors. Other groups are involved in lobbying, writing new laws and getting them passed, and strengthening existing laws. Still others serve as watchdogs, making sure that polluters obey the rules and that government agencies do their jobs, such as the Trial Lawyers for Public Justice in Washington, D.C.

Many environmental groups have become involved in proactive planning, devising sustainable alternatives to environmentally harmful projects. Environmental Defense has been a leader in this arena. Its staff of economists, public policy experts, and scientists research issues and offer economically and environmentally sound alternatives. Environmental Defense worked with McDonald's Corporation, for example, to institute the company's recycling and waste reduction program.

Some nonprofit groups spend much of their time in court, suing governmental agencies when they are not doing their job or are doing it poorly, or suing businesses that violate environmental laws.

Many environmental groups operate on several levels. Environmental Defense (ED), for instance, began as a litigator, suing government agencies and businesses. Today, it works closely with business and government to find alternative practices that serve people and the environment. ED also publishes books and newsletters that influence public opinion and public policy.

Environmental groups can offset the disproportionate influence of business, but they are usually at a disadvantage. Corporations such as Exxon Mobil wield far more influence in the political process.

KEY CONCEPTS

Environmental organizations take many forms and work in many ways to protect the environment.

Environment and Law: Creating a Sustainable Future

The late Edmund Muskie, a longtime senator from Maine, once noted that "environmental laws are not ordinary laws, they are laws of survival." Today, more and more people are realizing that environmental protection is indeed essential to the long-term future, but they also recognize that environmental protection is not enough to achieve sustainability. The social and economic goals described in this text are also an essential component. They will help us build a better future, more effective and cost-efficient government, broaden prosperity, and a more stable world through mutual responsibility. This section, however, examines U.S. environmental laws.

Evolution of U.S. Environmental Law

How did the U.S. environmental laws and regulations evolve? The progression of laws followed a logical evolutionary route that began with ordinances imposed by local governments. Interested in protecting health and environmental quality, officials of cities and towns passed local laws to limit activities of private citizens for the good of the whole. For example, municipal ordinances regulated trash burning within city limits to reduce air pollution. These efforts began in the 1800s.

In the mid-1900s, however, it became clear that efforts to control problems such as air pollution through local ordinances were often inadequate, especially in densely populated regions. Disputes arose between neighboring municipalities with different laws. Regions with strict laws were hampered in cleaning up their rivers by upstream cities with lax pollution laws. Thus, states began drafting legislation to regulate water pollution.

Soon, state laws proved inadequate, too, because air and rivers flow freely across state borders (**FIGURE 26-4**). Thus, interstate conflicts over pollution replaced the conflicts between neighboring municipalities. In addition, state programs were often inadequately funded and lacked the technical expertise needed to set pollution standards. State agencies also found themselves powerless against large corporate polluters with political influence in the courts and legislatures. Because of these problems and the growing effectiveness of special interest groups, environmental controls shifted to the federal government, gradually in the 1940s and 1950s and then more rapidly.

Initially, the federal government restricted itself to research on health and pollution control technologies. This approach met little opposition from state and local governments. Next, the federal government began offering grants to fund pollution control projects and the formation of state pollution enforcement agencies.

With increasing pressure from environmental groups, citizens, and many businesses, the federal government began to take a larger role in enforcement. Today, much to the dismay of some, the federal government sets ambient pollution standards and standards for emissions from factories, automobiles, and other sources and can take strict enforcement actions if needed.

FIGURE 26-4 Environmental laws evolved from local to state level in large part because pollution knows no boundaries. Rivers like the Ohio River (shown here) flow from state to state. Laws and regulations protecting water quality in a downstream state are meaningless if not matched by similar protections in upstream states.
© Daniel Dempster Photography/Alamy Images.

Stiff fines and penalties are frequently used to coerce businesses into cooperating. Corporate executives have even been jailed.

The shift to federal control is based on at least two important principles of American federalism: (1) When it is important to maintain uniform standards, the federal government provides the best means. Uniform policies minimize interstate conflicts and create a fairer economic system for businesses. (2) The power of the federal government to tax is much stronger than a state's. To control pollution effectively requires expensive research, which the federal government can more easily afford. Furthermore, it would be costly, time-consuming, and redundant for each state to carry out this extensive research.

The shift to the federal level has some disadvantages, however. First, the federal government officials may not always understand the problems of the regions it regulates as well as local officials do. Another criticism is that states should have the right to do as they please with their own resources; in other words, federal control diminishes self-determination and self-governance. Without central controls, however, states impinge on each other's quality of life. For example, poor watershed management leading to erosion in the Rocky Mountain states could have long-term adverse impacts on downstream users to the east and west.

One way of addressing these problems is to develop federal standards but to allow the states to manage their own programs. Thus, the Clean Air Act, the Surface Mining Control and Reclamation Act, and the Resource Conservation and Recovery Act all permit the states to run their own programs

KEY CONCEPTS

U.S. environmental law began locally and evolved to higher levels of government as it became apparent that local problems spread to other areas and required a higher political authority to address them.

as long as they are at least as stringent as the one set up by federal law. These acts also provide money to assist the states in setting up their own programs.

The National Environmental Policy Act

One of the most significant advances in environmental protection of the past three decades was the **National Environmental Policy Act** (**NEPA**) (1969). NEPA is a brief, rather general statute with several key goals. First, it declares a national policy to "use all practicable means" to minimize the environmental impact of federal actions. More specifically, NEPA requires that decisions regarding federally controlled or subsidized projects (such as dams, highways, and airports) must be made after considering possible adverse impacts outlined in an **environmental impact statement (EIS).** Among other things, an EIS must describe (1) what the project is; (2) the need for it; (3) its environmental impact, both in the short term and the long term; and (4) proposals to minimize, or (in the language of bureaucrats) to mitigate, the impact, including alternatives to the project. Drafts of the EIS are available for public comment and review by federal agencies. Written comments from the public must be addressed in the final EIS, which is issued at least 30 days before undertaking the proposed action.

The EIS has been an effective way of getting businesses and governmental agencies to focus on the environmental impacts of their projects. Thus, it has helped promote vision in economic development strategies. The underlying goal of the EIS is that those who become aware of potential impacts of their plans will act responsibly to avert them as much as possible. In this sense, the EIS is a political carrot (an inducement) rather than a political stick (a punishment). Available early in the planning stage, it can help decision makers determine whether they are implementing their policies, programs, and plans in compliance with the national environmental goals expressed in the NEPA and other legislation.

NEPA also established the **Council on Environmental Quality (CEQ)** in the executive branch. The Council once published an annual report (*Environmental Quality*) on the environment and on environmental protection efforts of the federal government. It also developed and recommended new environmental policies to the President. The CEQ was all but abolished by President Clinton.

NEPA has significantly affected federal decision making. It has led to hundreds of lawsuits filed by environmental groups against the government, perhaps more than any other environmental statute. In addition, several states and nations throughout the world have passed laws or issued executive or administrative orders patterned after NEPA. France, Canada, Australia, New Zealand, and Sweden all require EISs. California passed an **Environmental Quality Act** in 1970 that requires EISs for all projects—private and public—that will affect the environment in a significant way.

The success of NEPA has been great, but the law is not without flaws. One of the most frequent criticisms is that EISs are too lengthy and deal with too many peripheral subjects. Reports are often ignored; they may show serious adverse impacts, but the project will be approved and carried out anyway, often without ameliorative actions. A common complaint from environmentalists is that reports are often based on inadequate information. Projections of environmental impact are difficult to make and often too subjective. Practically no work has been done to see if the projected impacts actually materialize; thus, Americans continue to be unprepared to make sound projections about impact. EISs may be "doctored" by agencies or private consulting firms that write them for federal agencies, to hide the real impacts. Some agencies can avoid writing EISs by simply stating that there will be no adverse environmental impact; it is then up to others to prove the need in court. Other critics of the EIS contend that it is too costly and often leads to delays in important projects. The paperwork and time involved seem excessive.

To answer some of these complaints, the Council on Environmental Quality issued streamlined procedures for preparing EISs in 1979. They call for (1) a maximum length of 150 pages, except for more involved projects; (2) a summary of no more than 15 pages that describes major findings and conclusions; (3) documentation and referencing of projected impacts; and (4) the use of clear, concise, and understandable language.

EISs are prepared after the fact; that is, they are designed to assess impacts of projects that have already been planned. A more appropriate method would be to require analyses of need and then seek sustainable ways to meet the needs. Some environmentalists believe that NEPA should require agencies to select the most environmentally benign and cost-effective approach, both in the short and the long terms.

Another unanswered problem is that although environmental groups can sue an agency that they believe should have filed an EIS or that has filed an inadequate EIS, they cannot recover attorney's fees from such suits. Recovery of attorney's fees would relieve the costly burden of forcing governmental agencies to heed the law.

Although it has its critics and still stands in need of improvement, NEPA is the cornerstone of U.S. environmental policy. Numerous federal agencies have reported important environmental benefits from it as well as economic savings from recently revised rules.

KEY CONCEPTS

The National Environmental Policy Act requires all projects on federally owned land and all federally subsidized projects to be examined for their environmental impact. It established the environmental impact statement as a means of assessing potential effects and suggesting ameliorative actions.

The Environmental Protection Agency

Another major environmental accomplishment in the United States was the establishment of the **Environmental Protection Agency (EPA)** in 1970. The EPA was founded by an executive order drafted by President Nixon, which called for a major reorganization of 15 existing federal agencies that were already working on important environmental issues.

Civil disobedience, ecotage, and other direct actions aimed at stopping ecological destruction have scored a few spectacular victories. In Australia, civil disobedience successfully halted construction of the economically and environmentally costly Franklin Dam. In the Pacific Northwest, direct action brought national attention to the spotted owl and other old-growth forest-dependent wildlife, which may yet lead to their permanent protection.

For the most part, however, direct action has proven largely ineffective, even counterproductive, at protecting the environment. Moreover, it raises important ethical questions that its advocates often ignore.

Of the two basic types of direct action—civil disobedience and ecotage—ecotage is the least effective because all it does is increase the cost to resource developers. If millions of dollars can be made from cutting old-growth forests or drilling for oil, adding a few thousand dollars to the costs will not stop the cutting or drilling. Ecotage may make the ecosaboteurs feel good, but it does little to protect the environment.

The most successful direct actions have all used civil disobedience—a human blockade in the case of Franklin Dam, tree-sitting in the case of old-growth forests. The goal of such actions is to gain media exposure so that people learn about environmental problems. Yet, civil disobedience suffers from two major weaknesses. First, it only works a few times. The first person who sits in a redwood tree is news. The hundredth person is boring.

More importantly, civil disobedience works because it polarizes people into action—letter writing, testimony, and other lobbying. This seems good at first because it encourages Congress or state legislatures to address environmental problems. In the long run, however, direct actions polarize people on both sides of an issue. They help opponents of environmental protection as much as they help supporters.

This is apparent in the West, which has recently seen the rapid growth of a counterenvironmental crusade known as the Wise Use Movement. Their followers are people upset by environmental tactics, government bureaucracy, taxes, and other problems. Direct action drives people into this camp as much as it gains environmental supporters.

Beyond the effectiveness of the direct action are questions of ethics. A few environmental problems may be life-and-death issues for the human race. But, to be honest, most are just questions of values: Is a fish worth more than a dam? Is an owl worth more than the houses that can be built from wood?

Direct Action Wastes Limited Resources of the Environmental Community

Randal O'Toole
As an economist for Cascade Holistic Economic Consultants, of Portland, Oregon, Randal O'Toole has examined management practices on over 60 national forests. He is the author of *Reforming the Forest Service*, a comprehensive package of reforms based on his research on the national forests.

Courtesy of Randal O'Toole.

Environmentalists sometimes turn to direct action out of frustration that the current system does not allow them to express their values. But too often, direct action attempts to forcibly or legislatively impose their values on others, which threatens the stability of our society.

Instead of trying to impose values on others, the environment is better protected by finding ways to express environmental values. Many resources, such as air, wildlife, and outdoor recreation on the public lands, are free or nearly free. As a result, they are overused and abused. When a forest manager compares timber values, which directly add to the manager's income, with recreation values, which may be real but contribute nothing to the manager's income, timber values nearly always win out.

The solution is to find institutions that allow us to express our appreciation for the environment. Some of these institutions already exist. Groups like The Nature Conservancy buy critical wildlife habitat. The Clean Air Act allows polluters to sell their pollution permits, thus giving them an incentive to reduce their pollution. Other institutions must be created. Recreation fees would give forest managers a reason to protect scenic beauty and other resources that recreationists value. Tradeable water rights would allow fishing groups like Trout Unlimited to buy irrigation water and leave it in the stream.

Like anyone, environmentalists have limited resources to spend on actions to protect the environment. Those resources must be divided between lobbying, litigation, research, and other activities. To be as effective—and ethical—as possible, environmentalists should concentrate on promoting institutions that protect our environment as well as the stability of our society.

COUNTERPOINT

Direct Action Fills an Empty Niche in Wildland Conservation

Howie Wolke

Howie Wolke is a wilderness backpacking guide, owner of Big Wild Adventures, based in Paradise Valley, Montana, just north of Yellowstone. A longtime wildland activist, he is the author of *Wilderness on the Rocks,* a critique of wildland conservation in America, and coauthor (with Dave Foreman) of *The Big Outside,* a descriptive inventory of the remaining big roadless areas in the United States.

Courtesy of Howie Wolke.

Civil disobedience and other direct actions catapulted old-growth forest destruction into the national consciousness. Direct action helped stop the infamous Vegas-Barstow dirt bike race in the fragile California desert.

Recently, the U.S. Forest Service cancelled the remainder of the Cove and Mallard timber sales in central Idaho's Salmon-Selway Ecosystem. Nearly a decade of direct actions had slowed logging to less than half of the original goal. Now, a new policy to protect roadless areas has given the uncut wildlands a major reprieve. And in Seattle, mass demonstrations and human blockades made the World Trade Organization's ability to veto environmental and labor laws of sovereign nations an international issue. Overnight Ecotage (monkey-wrenching) has also received lots of press, but the subject upsets many, and discussing it tends to obfuscate the biodiversity issue.

Traditional tactics (letters, lobbying, litigation, articles, slide shows, rallies, and boycotts) have also saved many wild places. The Grand Canyon remains undammed and the National Wilderness System is expanding, proving that traditional tactics work. Or do they?

Despite impressive victories, habitat destruction is rampant. Consequently, species, subspecies, and populations become locally, regionally, or entirely extinct as native wildland ecosystems collapse. Dozens of species and subspecies in the United States have become extinct since the Endangered Species Act was enacted in 1973. Since before World War II, the U.S. Forest Service has annually destroyed—mostly via logging and road building—about a million acres of unprotected wilderness, despite the growing National Wilderness System (mostly growing with rocky high-altitude additions).

Wildland ecosystems are under assault everywhere, by loggers, miners, cattlemen, power companies, dam builders, and subdividers. Urban sprawl abuts wilderness. Air pollution kills forests. Fences halt wildlife migrations. Poachers exterminate black bear and black

rhino for body parts used for mythical aphrodisiacs. Ignorant all-terrain vehicle enthusiasts crush desert tortoises, bird nests, insects, and small mammals—not to mention the monumental environmental threat of global warming.

Today's global ecological crisis is complex and unprecedented. The late Cretaceous extinction 65 million years ago pales when compared with today's galloping loss of biodiversity. Globally, over one species per hour is wiped out. The obvious concern, of course, is survival. How much biodiversity can we lose before the life-support web crumbles or before humans can no longer live comfortably? Or at all? In truth, we don't know and perhaps can never know in time to avert monumental disaster.

But the question of survival obscures another profound consideration. Life has intrinsic value. Some economists reduce all conflicts to opposing values: costs versus benefits. "Is an owl 'worth' more than the houses derived from wood?" they ask. Such reductionism opposes nearly all enlightened historic human thought. Do we value only human life? Are we appalled only when torture or genocide victims are human? I think not. The value of thriving diverse native life transcends the almighty buck. Those who resort to nonviolent civil disobedience-related direct actions are imposing their values on others no more than corporations that donate money to political campaigns or, for that matter, than those who threw the Boston Tea Party.

Three questions should always be confronted when considering direct action: (1) Is the action nonviolent toward life? (2) Will the action increase public awareness of a problem? (3) Will it slow or delay destruction so that long-term protection remains plausible?

The success of direct action is apparent. Besides some obvious victories, direct action brings new activists into wildland battles. It permits various mainstream environmental groups to take stronger positions without appearing "radical." Despite contrary claims, civil disobedience still draws attention. Yes, the first tree-sitter is news; the hundredth isn't. Unless she stays tree-bound for many months. Civil disobedience is effective when applied strategically: Save it for the worst atrocities. And use your imagination.

The bottom line is this: *Nothing* yet has stopped today's spiraling loss of habitat and biodiversity. We need better laws, better incentives, better news media, better representatives, and better institutions. But, we can't afford to wait. We must seize any opportunity to save habitat. Life is falling through the cracks of the well intentioned. In these complex times, simple solutions won't work. Wildland conservation must be eclectic.

Just as lobbying isn't for everyone, neither is civil disobedience. To expand the realm of effective conservation activism, few avenues of ecological defense should be closed. Specialization works. Land trusts protect valuable open space on private lands. Wilderness Watch lobbies and educates. The Earth-Justice Legal Defense Fund litigates. Other groups specialize in direct action. Ultimately, no human or nonhuman society can sustain itself in a wasteland. The challenge is to protect habitat. Sustainability and the joy of unharnessed life will follow.

Critical Thinking Questions

1. There are many forms of direct action to protect the environment. Which ones, if any, do you support? Why?
2. What values influence your answer to question 1?

The EPA was directed to carry out the Federal Water Pollution Control Act and the Clean Air Act. Today, it manages and enforces most of the environmental protection laws that issue from Congress. Current responsibilities of the EPA include research on the health and environmental impacts of a wide range of pollutants, as well as the development and enforcement of health and environmental standards for pollutants outside the workplace. The EPA is concerned with a variety of areas, including pesticides, hazardous wastes, solid wastes, toxic substances, water pollution, air pollution, radiation, and noise pollution.

The EPA can provide incentives to state and local communities through grants that help pay for water pollution control projects. Grants to universities have helped expand the research capability of the agency. The EPA carries on much of its own research at four National Environmental Research Centers, located at Cincinnati, Ohio; Research Triangle Park, North Carolina; Las Vegas, Nevada; and Corvallis, Oregon.

The EPA is often caught in the crossfire between opposing groups—for example, between environmentalists who seek tighter controls and the businesses the EPA regulates, which commonly complain that regulations are too stringent and costly.

Through polls, the American people continue to express a strong concern for a healthy environment; they support maintaining existing laws or even strengthening them. A growing population and an expanding economy create an ever-increasing burden on the environment and on the agencies that regulate environmental issues. Thus, many argue that the EPA and other agencies involved in environmental problems should be expanded, not merely maintained or even reduced.

KEY CONCEPTS

The U.S. Environmental Protection Agency manages and enforces most of the country's environmental laws. It sets standards for pollution and does research to expand our understanding of the effects of toxic substances and other pollutants on human health and the environment.

Principles of Environmental Law

In the United States, government's authority to protect the environment is conferred by the U.S. and state constitutions. Environmental protection is conferred by common law, federal and state statutes and local ordinances, and regulations promulgated by state and federal agencies. Statutory law and common law are the mainstays of environmental protection.

KEY CONCEPTS

Environmental law is based on statutes and local ordinances—that is, specific laws passed by Congress and other governing bodies—and on common law, principles derived from thousands of years of legal decisions.

Statutory Law Many examples of state and federal laws for environmental protection and resource management have appeared throughout this book. These are forms of **statutory laws**—that is, laws written by and agreed upon by legislative bodies. Statutory laws generally establish broad goals, among them the protection of health and the environment by reducing air pollution, or the judicious use of natural resources. However, as pointed out in this chapter, Congress and the state legislatures lack the time and expertise to determine specifically how these goals can be met. Thus, Congress assigns the technical details—the setting of standards, pollution control requirements, and resource management programs—to federal agencies such as the EPA.

Besides setting standards of acceptable behavior and calling on agencies to determine regulations, statutory laws also provide authority for various enforcing agencies to take legal action—to fine polluters or to take polluters to court to face criminal charges and possible jail sentences.

Go Green

Take an active role—become a letter writer. To save resources, email your letters to government offices.

Common Law Many environmental cases are tried on the basis of **common law,** a body of unwritten rules and principles derived from thousands of years of legal decisions. Common law is based on proper or reasonable behavior.

Common law is a rather flexible form of law that attempts to balance competing societal interests. As an example, a company that generates noise may be brought to court by a nearby landowner who argues that the factory is a nuisance. The landowner may sue to have the action stopped through an injunction. In deciding the case, the court relies on common-law principles. It weighs the legitimate interests of the company in doing business (and thus making noise) and the interests of society (which wants its citizens employed and wants to collect taxes from the company) against the interests of the landowner (who is trying to protect the family's rest, health, and enjoyment of property).

The court may favor the **plaintiff** (the one who files the lawsuit) if the damage (loss of sleep, health effects, and inconvenience) is greater than the cost of preventing the risk (costs of noise abatement, loss of jobs, and loss of tax revenues). However, the court might not issue an **injunction,** an order to cause the factory to shut down; instead, it may simply require the **defendant** (the one defending the case or whose actions are being contested) to reduce noise levels within a certain period, striking a balance between competing interests. Cases such as this one illustrate the **balance principle.**

Many cases involving common law are settled on two legal principles: nuisance and negligence.

Nuisance The most common ground for action in the field of environmental common law is **nuisance.** A nuisance is a class of wrongs that arises from the unreasonable, unwarranted, or unlawful use of a person's own property that obstructs or injures the right of the public or another individual. It may result in annoyance, inconvenience, discomfort, or hurt. This means that one can use one's personal property or land in any way one sees fit, but only in a reasonable manner and as long as that use of the property does not cause material injury or annoyance to the public or another individual.

Generally, two types of remedy are available in a nuisance suit: compensation and injunction. Compensation is a monetary award for damage caused. Injunctions are court orders requiring the nuisance to be stopped.

Nuisances are often characterized as either public or private. Until recently the two were distinctly different concepts. A **public nuisance** is an activity that harms or interferes with the rights of the general public. Typically, public nuisance suits are brought to court by public officials. A **private nuisance** is one that affects only a few people. For example, the pollution of a well affecting only one or two families is considered a private nuisance. A public nuisance would be pollution that affects hundreds, perhaps thousands, of landowners along a river's shores. The most common environmental nuisance is noise. Water pollutants and air pollutants such as smoke, dust, odors, and other chemicals are other major nuisances.

Historically, the distinction between private and public nuisance has hampered pollution abatement because the courts have traditionally held that an individual could bring a private nuisance suit only if the individual had suffered a unique injury—that is, one different from that suffered by others. For example, an individual would not be permitted to sue a company for polluting a river shared by many others. In this case, the only legal recourse would be a public nuisance suit brought by an official (the local health department, for instance). In such instances, though, public officials are often unwilling to file suit against local businesses that provide important tax dollars and jobs for the community (and possibly campaign support as well).

Increasingly, the distinction between private and public nuisance is fading; private persons can bring suit to stop a public nuisance. As a result, the private individual is gradually acquiring more power to stop polluters.

Several common defenses are used to fight nuisance suits. Because most nuisance suits are decided by balancing the rights and interests of the opposing parties, **good-faith efforts** of the polluter may influence the decision. For example, if a small company had installed pollution control devices and had attempted to keep them operating properly but was still creating a nuisance, the court might hold it liable but would probably be more lenient in damages or conditions of abatement. If, on the other hand, the company had made no attempt to eliminate a pollution nuisance, the court might generally be more severe.

The availability of pollution control must also be considered. If a company is using state-of-the-art pollution control and still creates a nuisance, the court may not impose damages or an injunction. In contrast, if the company has failed to keep pace with pollution control equipment, the court may order it to install such controls.

In states that still distinguish between public and private nuisance, a **class action suit** can be filed. Class action suits are brought on behalf of many people and seek remedy for damage caused to the entire group by a nuisance. In order for a federal class action suit for compensation to be permitted, however, each person named in the suit must have suffered at least $10,000 in damage. If not, the suit can be dismissed.

One defense against nuisance suits is that the plaintiff has "come to the nuisance." **Coming to the nuisance** occurs when an individual moves into an area where a nuisance—such as an airport, animal feedlot, or factory—already exists and then begins to complain. An old common-law principle holds that if you voluntarily place yourself in a situation in which you suffer injury, you have no legal right to sue for damages or an injunction. In most courts, however, even though you purchase property and know of the existence of a nuisance, you still have the right to file suit to abate it or recover for damages. This is based on another common-law principle: that clean air and the enjoyment of property are rights that go along with owning the property. Thus, if population expands toward a nuisance, it may be the responsibility of the party creating the nuisance to put an end to it.

According to environmental attorney Thomas Sullivan, "The courts are moving to strict liability for environmental nuisances, so that practically speaking, there are no good defenses. The solution is: do not create nuisances."

 KEY CONCEPTS

People can use their property or land in any way they see fit, but only if it does not cause injury or is an annoyance to others, which is called a *nuisance*. Many environmental lawsuits are based on the nuisance principle.

Negligence   A second major principle of common law is **negligence.** From a legal viewpoint, a person is negligent if he or she acts in an unreasonable manner and if these actions cause personal or property damage.

Negligence provides a basis for liability, just as nuisance does, but negligence is generally more difficult to prove. What is reasonable action in one instance may not be reasonable in another. Statutory laws and regulations help the courts determine whether behavior is reasonable. For example, regulations drawn up by the EPA specify how certain hazardous wastes should be treated. Failure to comply with those standards may be evidence of negligence. Negligence may be shown in instances in which a company fails to use common practices in the industry. For example, a company may be found negligent if it fails to transport hazardous wastes in containers like those used by other companies.

In a much broader sense, the courts may decide that a company is negligent simply if it fails to do something that a reasonable person would have done. For example, negligence might be demonstrated if a company failed to test its wastes for the presence of harmful chemicals when a reasonable person would have done so. Likewise, negligence may stem from action that a reasonable person would not have taken. In summary, then, negligence can result from either inaction or action that may be deemed unreasonable considering the circumstances.

The **concept of knowing** is fundamentally important to negligence suits. Briefly, negligence can be determined on the basis of what a defendant knew or should have known about a particular risk. The standard of comparison is what a reasonable person should have known under similar circumstances. For example, a man on trial may argue that he is not negligent because he did not know that a harmful chemical was in the materials that he dumped into a municipal waste dump and that subsequently polluted nearby groundwater. His argument will be valid if a reasonable person in his position could not have known about the wastes.

Interestingly, in some cases, determinations of liability for damage or harm need not be based on proof of negligence. For instance, in cases where the risk is extraordinarily large, one need prove only that injury or damage occurred, not that the operator was negligent or acted unreasonably.

Business interests often try to lessen their liability by getting Congress to pass laws that put ceilings on liability. The Price-Anderson Act, for example, frees utility companies from financial liability incurred by a nuclear power plant accident. The act requires the government to pay for all damages outside the plant, but only up to $560 million. The airlines have a similar law; at this writing, the oil companies have successfully pushed for similar limits on liability resulting from tanker spills.

 KEY CONCEPTS

Negligence is an unreasonable act that causes personal or property damage and can be a basis for liability in environmental lawsuits.

Problems with Environmental Lawsuits Legal actions to stop a nuisance or collect damages from a nuisance or act of negligence carry with them a **burden of proof:** In order to win, plaintiffs must prove that they have been harmed in some significant way and that the defendant is responsible for that harm. This is not always easy, for several reasons. First, the cause-and-effect connection between a pollutant and disease may not have been definitely established by medical research. If doubt exists, the case is weakened. Second, diseases such as cancer may occur decades after the exposure. This makes it extremely difficult to prove causation. As a rule, it is easier to link cause and effect with acute diseases. Third, it is often difficult or impossible to identify the party responsible for damage, especially in areas where there are many industries or where illegal acts such as midnight dumping of hazardous wastes have occurred. Any reasonable doubt about the party responsible for personal or property damage may weaken a lawsuit.

The **statute of limitations** limits the length of time within which a person can sue after a particular event. This provision creates problems in cases of delayed

Go Green

Support environmental groups like Environmental Defense, the Natural Resources Defense Counsel, and the Nature Conservancy. Supporting local groups is extremely important, too. They are often woefully underfunded.

diseases. Statutes of limitations are designed to reduce lawsuits in which evidence is unavailable or memories of potential witnesses may have faded. In latent-disease cases, though, they create a major obstacle to individuals seeking compensation for damage in states that apply the time limitations to the onset of exposure. This essentially makes it impossible for cancer victims to file suit. Other states start the judicial clock from the time the victim learns of the disease. This makes it a little easier to collect compensation for diseases such as cancer, black lung disease, and emphysema.

KEY CONCEPTS

The burden of proof needed to win environmental lawsuits is not always easy to achieve because of many factors, such as inconclusive data connecting pollution with disease and long lag times between exposure and effect.

Resolving Environmental Disputes Out of Court

An increasing number of disputes between environmentalists and businesses are being settled out of court, either privately or by dispute resolution, known as **mediation.** This approach often employs a neutral party who mediates the discussion between opposing parties. The mediator keeps the proceedings on track, encourages rivals to work together, and tries to resolve the dispute in a way that is satisfactory to both groups.

The benefits of mediation over litigation are many. Mediation is much less costly and time consuming. Perhaps most important of all, mediation may bring about more satisfactory resolutions. For instance, environmental lawsuits often hinge on specific points of law rather than on substantive issues, and thus, they may have little to do with what the plaintiff really wants. An environmental group might bring a suit over the adequacy of an EIS but, in reality, might want the government to ensure protection of a valuable species that would be affected by the project. In mediation, this will be the central issue. Mediation may therefore result in more appropriate solutions.

Mediation also tends to promote a more accurate view of problems. For instance, in lawsuits each party tends to bring up evidence that favors its goal and may ignore or dismiss unfavorable or ambiguous information that might weaken its stand. In mediation, both parties are encouraged to openly discuss the uncertainties of their positions, discovering many points of agreement and building a better understanding of opposing positions.

Mediation has its drawbacks, too. First, funding is inadequate. In the past, mediation has been financed largely by foundations. Federal, state, and local governments need to develop programs to fund mediation. Second, some groups fear they will lose their constituency if they enter into negotiations on certain issues because they will be giving the impression that they are failing at their stated goals and compromising with the "enemy." A third drawback is a lack of faith in the outcome of mediation. Unlike court orders, resolutions drawn up in mediation are not legally enforceable. Thus, months of discussion may produce nothing but a piece of paper that polluters will ignore.

Out-of-court settlements present another legal problem. Such settlements have hindered environmental law. Eager to avoid a costly settlement, a company may pay victims if they agree to dismiss the company from further liability. Out-of-court settlements may also benefit plaintiffs, saving them the time, headaches, and costs of environmental litigation.

Although advantageous to both parties, these settlements provide no precedents for environmental laws. In short, the fewer cases that make it to court, the fewer guidelines courts have to settle future cases. This lack of precedent may discourage attorneys and citizens from filing court cases. Without clear examples from the past, they may simply be unwilling to face costly, time-consuming legal battles.

Dispute resolution is growing in the United States, but it will not replace litigation. Still, it can play a valuable role in the future. Eight states have organizations that offer mediation services, and a growing number of private organizations have been formed to provide professional mediators with experience and knowledge in environmental issues. Thus, more and more disputes may be settled by this noncombative approach.

KEY CONCEPTS

Out-of-court settlements through mediation provide a less hostile approach to conflict resolution and can reduce or avoid costs of environmental lawsuits for both defendant and plaintiff.

26.4 Creating Governments That Foster Sustainability

From an environmental standpoint, democratic governments display the same weaknesses as the economy: They are generally shortsighted, obsessed with promoting growth, and exploitive of people and nature. They also tend to promote dependency and undermine self-reliance.

Why do governments and economies have the same flaws? In a nutshell, the frontier ethics in many countries is largely responsible for creating mutually supporting systems of government and economics. We create what we believe in.

To build a humane, sustainable society, governments and the economies they serve must be made responsive to the needs of future generations and other species that share this planet with us. First and foremost, governments must adopt the goal of sustainability as a central organizing principle. This section offers additional ideas on ways to correct the flaws in governments and economies.

KEY CONCEPTS

Most governments promote shortsighted, growth-oriented, exploitive policies and economies. By adopting sustainability as a key goal, governments could help us reshape our unsustainable economy and prevent environmental problems.

Creating Government with Vision

Responding to problems requires a consensus for action, which may be translated into appropriate governmental policies. In the United States, one of the key barriers to building a sustainable future is a lack of consensus about the severity of the world's problems and the need for long-term action. As humorist Herbert Prochnow once noted, "How can a government know what the people want when the people don't know?" How can we agree on the need to tackle a problem when there is so much conflicting evidence, much of it planted by public relations firms hired by large corporations or conservative think tanks funded by millions of dollars by donations from big business? How can we agree on a course for the future when few agree on what is best? This section discusses ways of increasing public awareness of problems and the need for long-term sustainable solutions—creating a society and a government with a vision.

KEY CONCEPTS

A long-term perspective is essential to create a sustainable government.

Increasing Public Awareness Through Research and Education The lack of consensus on environmental issues results in part from widespread ignorance about these issues among the general public. Often, people who speak out against environmentalist concerns have very little knowledge of the issues.

One solution to the problem of public ignorance is government support for scientific study of environmental problems and sustainable solutions. Research is also needed on ways values can be changed to achieve a sustainable society. Support for this work could come from governments and private foundations.

Information is important, but in order to make a difference it must be disseminated to the citizenry. Concerted efforts are needed to inform people the world over of the problems facing the global ecosystem. Well-informed teachers and public officials can help disseminate this knowledge, as can the media (**FIGURE 26-5**). Environmental groups and religious leaders can also play a role. Education can assist in bringing sustainability into immediate concerns of everyday citizens, elected officials, and businesspeople.

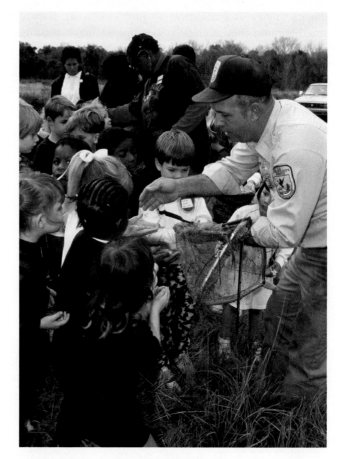

FIGURE 26-5 Outdoor classroom activity. Environmental education of children of all ages is a key to building a sustainable society.

Courtesy of John and Karen Hollingsworth/USFWS.

KEY CONCEPTS

Research and education are essential to raise the level of awareness of environmental problems and sustainable solutions among the public and elected officials.

Getting Beyond Crisis Management Public ignorance of issues is also a result of a tendency for public policymakers to practice **crisis politics,** which refers to the management of immediate crises at the expense of long-range problems. Henry Kissinger put it best when he said that, in government, the "urgent often displaces the important." In crisis management, immediate problems such as strikes, disaster relief, and pressing economic problems tend to get all of the attention. They reduce the amount of time policymakers spend thinking about and addressing long-range problems, especially environmental ones. When new problems do arise, governments ignore them. Too heavily embroiled in the present crisis, they defer decision making until a small—and probably solvable—matter becomes a monstrous problem that defies solution. In addition, governments are generally unwilling to devote money and resources to problems when they're small because they're already spending so much treating the latest list of crises. In so doing, they end up with a menu of very costly problems that could have been solved at a fraction of the cost. A good example is the savings and loan scandal. Government representatives were warned about the problem in 1984, but ignored it until 6 years later, when a minor problem had grown to a $500 billion headache.

When immediate issues take precedence over important long-term problems, governments lumber from crisis to crisis. It's not an efficient way to run a government or a society. Making matters worse, legislators often apply marginal corrections to the present-day crises. Thus, they often end up treating the symptoms and not the causes. In time, our problems grow worse.

In 1985, then Senator Al Gore introduced a bill to Congress called the **Critical Trends Assessment Act.** Had it passed, it would have required the federal government to establish an **Office of Critical Trends Analysis (OCTA)** to examine long-term trends, especially economic and environmental ones. The OCTA would collect and analyze vast amounts of data from a variety of government agencies. Using knowledge from many sources, the OCTA would then advise the President on policy needs to avert problems. State governments could also establish OCTAs, whose central purpose would be to forge a sustainable economic and environmental future.

OCTAs might help convince influential government leaders as well as business interests of the importance of many environmental issues, reducing ignorance and creating a consensus for change that helps improve our prospects. Combined with the many other changes described in this text, OCTAs could help shape a sustainable future.

Another approach would be the establishment of special commissions on the future to address critical trends. In 1977, for example, President Carter directed several key agencies in the federal government to study population, resource, and pollution trends and issue a report on their findings, which would serve as a foundation for longer-term planning. The resulting study, *The Global 2000 Report to the President*, was a gold mine of information about the future, on which many decisions could be based. This report could be updated and revised.

Another way of creating vision in government is to establish special study sections in various governmental agencies, such as the Departments of Agriculture, Energy, and the Interior. Reports from these agencies could be distributed to political leaders, teachers, and the public.

Internationally, long-range vision can be facilitated by the United Nations, through the Environmental Programme (UNEP) or specially established commissions. In the mid-1980s, for instance, the UN established the **World Commission on Environment and Development.** Members of the commission were asked to produce a global agenda for sustainable economic development. Their report, *Our Common Future*, called on the General Assembly of the United Nations to find ways to achieve sustainable development; it was the main stimulus for the United Nations Conference on Economic Development in Rio de Janiero in 1992. The UN's role is described more fully later in this chapter.

KEY CONCEPTS

In government, long-range planning and action (which are essential for sustainable development) are often sacrificed in the interest of addressing immediate problems, creating a phenomenon called *crisis politics.* Crisis management tends to apply marginal solutions and allows little problems to become larger and more difficult to solve. Long-term thinking and planning is vital to solving issues sustainably.

Getting Beyond Limited Planning Horizons Another barrier to sustainability is the limited planning horizon of many elected officials. Limited planning horizons result from several factors. First and foremost, politicians, like most of the rest of us, tend to be shortsighted. Reelection concerns also contribute to this phenomenon. Frequent reelection campaigns tend to force politicians to choose short-term solutions, rather than long-term answers that may be more difficult and painful—and consequently less popular. To get reelected, politicians need results, even if they treat only the symptoms of a crisis. Members of the U.S. House of Representatives, for instance, serve 2-year terms. Although this may have been appropriate 200 years ago when the nation's problems were few, today some observers think that it is inadequate. Why? During their 2-year terms, many House members spend inordinate amounts of their time collecting money, plotting reelection strategies, and campaigning—time that they could have devoted to their job.

One way to address this problem is to institute longer terms of office for members of the U.S. House of Representatives. Term limitations could also help. Fifteen states now have some form of term limitation—that is, laws that limit the number of years a state or federal official can serve. Colorado's governor, for instance, can serve no more than three 4-year terms. A 6-year nonrenewable term for the President, for example, might reduce the reelection pressures that lead to shortsighted political decisions and take valuable time away from the office. It would also reduce the amount of time spent on the campaign trail (at taxpayer expense!) while in office and increase the amount of work time. Longer terms of office, of course, could have damaging effects as well.

Shorter campaigns could help make democracies work better, too. In England, campaigning begins 6 weeks before general elections. Thus, incumbents spend less time on the campaign trail than U.S. politicians. In addition, they spend considerably less money.

Overcoming the barrier of limited planning horizons will also require more farsighted leaders, who are willing to implement policies that have implications well past the next election. Such policies may not favor the immediate electorate as much as the future electorate. Citizens must become willing to elect men and women with a vision of the future and with the skills to articulate and popularize their vision. If we expect visionaries, though, we must become more visionary ourselves and make tough choices.

Environmental groups at the national and state levels are helping support responsible leadership. You can help by supporting candidates who promote the efficient use of resources, recycling, renewable resource use, restoration and sustainable management of resources, growth management, and population stabilization. Citizens can write or visit their elected officials. One other effective way for citizens to make their voices heard is to join advocacy groups.

 KEY CONCEPTS

Longer terms of office for elected officials could reduce political shortsightedness and encourage solutions that contribute to sustainability, as would term limitations and shorter campaigns.

Becoming Proactive A government that lives and acts for today is by definition a **reactive government.** Its laws and regulations are often ineffective, and in the long run they can complicate matters, making truly effective solutions harder to reach. Many of the laws passed by reactive governments are **retrospective**—that is, they attempt to regulate something that has already gotten out of hand. For example, the Superfund Act provides money to clean toxic hot spots in the United States. Retrospective laws are part of crisis management. They're necessary like emergency rooms in hospitals but less desirable in the long run than long-term, sustainable solutions.

Sustainability relies in large part on **proactive laws,** laws that attempt to prevent problems in the first place or to confront them while they're still small and easily solvable. A good example is the Toxic Substances Control Act, with its provisions for screening new chemicals before they are introduced into the marketplace. Laws that promote solar energy, conservation, recycling, growth management, and population stabilization are additional examples. Ultimately, such laws could help us transition from the wasteful frontier economy to the efficient, sustainable society.

Western democracies today generally have a mix of proactive and retrospective (and reactive) laws, with a strong leaning toward a reactive approach. As we move further into the 21st century and as the world population continues to grow exponentially, proactive government becomes more necessary than ever before.

 KEY CONCEPTS

Creating a sustainable future requires short-term solutions to address immediate problems, but it ultimately hinges on a long-term, proactive approach that seeks to prevent problems.

Ending Our Obsession with Growth

Some of the steps outlined above could also help nations reduce their obsession with growth. Offices of critical trends analysis, for instance, could help show the potential outcome of current policies that promote continual growth in production and consumption at the expense of the environment. Education and research could have similar effects. As pointed out, new measures of progress could also temper our obsession with growth. Governments can help by promoting the use of these measures and by publicizing real progress toward improvement—rather than the illusion of progress suggested by the GNP.

In 1949, the U.S. Congress passed the **Full Employment Act,** which calls on the government and its various agencies to promote maximum production and consumption to promote full employment. Although full employment is a desirable goal, the notion that we can achieve it only through maximum production and consumption is erroneous and ecologically ill-advised. Repealing this law and replacing it with a **Sustainable Futures Act** might be a symbolic act, but it could be an important step in building a sustainable future.

The Sustainable Futures Act would call on Congress to actively promote sustainability through conservation (efficiency and pollution prevention), recycling, renewable resource use, restoration, growth management, and population measures. Moreover, it would require that all new policies be judged through the lens of sustainability. The act might also set up a special legislative body to review all existing policy and to suggest amendments or new laws that would bring human activities—from international trade to energy to manufacturing—in line with the dictates of sustainability.

 KEY CONCEPTS

Many measures could help temper the obsession with growth found in many nations, which is so detrimental to the goal of sustainability.

Reducing Exploitation and Promoting Self-Reliance

Reducing the exploitation of people and of natural resources is another goal of sustainable government. Interestingly enough, many of the changes mentioned above would contribute to this goal. Steps that reduce our obsession with growth and create vision, for instance, tend to make societies less resource-intensive and thus less exploitive. Other steps can also be taken to reduce our exploitive ways. Additional ideas are listed later.

KEY CONCEPTS

Countless measures that promote the goals of sustainability could help reduce society's exploitation of natural systems and of other people—and promote greater self-reliance, broader prosperity, and a better future.

Creating Offices of Sustainable Economic Development

One idea that may promote self-reliance and an end to our exploitative economy is the establishment of offices of sustainable economic development (OSED) in state governments or, perhaps, at the national level. All state governments currently have offices of business development or offices of economic development, which promote business by recruiting new business and helping existing business become more profitable. As a rule, little thought is given to matters of sustainability.

OSEDs could serve as a catalyst for change. First, they could help existing businesses become more sustainable—for example, by using energy and other resources more efficiently, by recycling and purchasing recycled goods, and by helping build industries that divert wastes to useful purposes. OSEDs could promote pollution prevention and renewable energy resources. Second, they could assist in promoting sustainable business opportunities—for example, helping identify locally available resources (wastepaper or aluminum scrap) around which new businesses could be developed. By focusing on sustainable economic progress, OSEDs would help build strong, stable, and regionally self-reliant economies.

Working with the state offices of critical trends analysis and elected officials, OSEDs could help create a comprehensive strategy for sustainability. They could also help promote better land-use planning, which is vital to sustainable economic development throughout the world. For a summary of ideas on creating a sustainable government, see **TABLE 26-1**.

TABLE 26-1

Summary of Recommendations for Sustainable Government

- Adopt a national sustainable development strategy that promotes conservation, recycling, renewable resource use, restoration, and population control.
- Repeal or modify existing laws that hinder sustainability.
- Increase support of research on environmental problems and sustainable solutions to them.
- Increase support of education on environmental issues and sustainable solutions.
- Pass a national Sustainable Futures Act.
- Establish offices of critical trends analysis at state and national levels.
- Establish offices of sustainable economic development in state governments.
- Evaluate all new and existing policies on the basis of sustainability.
- Adopt sustainable land-use planning at state and national levels.
- Implement policies and actions outlined in Agenda 21 and other international agreements.
- Support sustainable development in nonindustrialized nations.

Models of Sustainable Development

In any transition, people and nations often become models of desirable action. In the sustainable transition, several European countries have recently emerged as models. One of the best examples is Germany.

Germany probably has the highest environmental standards in the world. Part of the reason for this lies in the **Green Party,** a political party that advocates measures to protect the environment. The Greens support reductions in pollution and hazardous wastes, putting an end to nuclear power and nuclear weapons, efficiency measures, and the use of renewable resources. Rallying under the banner, "We are neither right nor left, but in the front," the Greens have been instrumental in transforming German society (**FIGURE 26-6**).

The Greens represent a major shift in political thinking. They take a long-term view of the future, calling for redirection of policy consistent with sustainable ethics. The Green Party has also spread to the United States. Colorado, Connecticut, Washington, and dozens of other states now have green parties, which promote social justice and a clean and healthy environment. Green Party members write letters and phone political leaders, stage peaceful demonstrations, and work to get their members elected to state legislatures and other offices.

Conventional political wisdom holds that in the United States popular ideas are absorbed by major political parties. They enter the mainstream of political thought and become

FIGURE 26-6 Reinhard Buetikofer, chairman of Germany's Green Party (right) and election campaign manager, Fritz Kuhn (left), take a ride in a bike taxi at the opening of the Green election campaign center in Berlin in 2005. The sign on the bike taxi reads "Take Part!"
© Fritz Reiss/AP Photos.

integrated into public policy. This has happened in Germany. Thus, no matter whether the U.S. Green Party survives or is simply absorbed into the mainstream, it may have begun a movement that could reshape American politics.

Sweden, Norway, and the Netherlands are also models of sustainability, whose policies and actions could inspire other countries to follow suit—especially when skeptical leaders see the economic benefits of pursuing sustainable options. Besides providing inspiration, model countries also become a source of new technologies and will enjoy a competitive advantage over those countries that are resisting change.

 KEY CONCEPTS

The Green Party has been instrumental in encouraging sustainable practices in Germany and other countries, which serve as models for other nations. Taking the lead in sustainable development also positions some nations to become leading exporters of knowledge and technologies.

26.5 Global Government: Toward a Sustainable World Community

Although some people oppose the idea of international governance, others are in favor of it. On the environmental front, environmental problems have grown to global scale and require international efforts to solve them.

 KEY CONCEPTS

Many environmental problems are global in scale and, therefore, require international cooperation.

Regional and Global Alliances

Cooperation and participation are key principles of sustainable development. These efforts are often fostered by various alliances. Several environmental alliances have been discussed in this text, among them the International Whaling Commission (IWC). Composed of members from all of the whaling nations, the IWC sets quotas on whale kills and enacts outright bans. The IWC, however, has no enforcement power; it relies principally on cooperation. Not all nations comply with the regulations. To help give the IWC a little muscle, the United States and other nations prohibit trade with countries that violate the agreements. International treaties have also helped protect the ozone layer.

Some of the most recent alliances emerged from the Earth Summit in June 1992 and a smaller follow-up conference in 1997. This section examines the various accords and discusses further measures needed to improve them.

 KEY CONCEPTS

Because the task of building a sustainable future is global in nature, regional and global alliances are needed to solve the problems that afflict many nations.

The Climate Convention Signed by 154 nations at the 1992 Earth Summit, this agreement calls on the world community to hold greenhouse gas emissions at 1990 levels by the year 2000. Unfortunately, this agreement did not create any stipulations that required actions. In the years that followed the signing of this accord, little was done. So, in 1997, the nations of the world met again in Kyoto, Japan to hammer out a more rigorous agreement. Industrial and former Eastern Bloc nations agreed to cut carbon dioxide and other greenhouse gas emissions to 5.2% below 1990 levels between 2008 and 2012. Unfortunately, global CO_2 emissions are rising rapidly because of economic growth of China and India.

In 1998, more than 160 countries meeting in Buenos Aires adopted further actions. In a move that stunned the world, British Petroleum, the third largest oil and gas company in the world, announced that it would reduce carbon dioxide emissions at its facilities by 10%.

In November, 2000, the signatories of the Kyoto Accord met in the Netherlands, but were unable to reach an agreement on ways to cut emissions in the United States. In 2004, Russia signed the agreement, thus putting it into effect. The United States has signed the agreement but is not seeking ratification by the Senate.

Although the Climate Convention is an important first step forward, it will—if followed—only slow the rate of climate change because greenhouse gas emissions already exceed the Earth's carrying capacity.

 KEY CONCEPTS

The Climate Convention is a first step in reducing global greenhouse gases, but success has been limited and much more is needed to solve the problem of global warming.

The Biodiversity Convention This agreement was signed by all participants except the United States, which refused because of language the Bush administration deemed unacceptable. President Clinton signed it in 1993 soon after taking office, however. This agreement asks nations to prepare inventories of species to be preserved and to develop strategies to conserve and use biological resources sustainably. Like the climate treaty, no deadlines were set. In addition, the language of the proposal is quite weak.

 KEY CONCEPTS

The Biodiversity Convention calls on nations to take steps to protect species.

Agenda 21 By far the most comprehensive document to emerge from the Earth Summit, **Agenda 21** is a massive work plan for national action and international cooperation aimed at achieving sustainable development. Its 39 chapters of nonbinding recommendations are a blueprint for a sustainable future. For instance, Agenda 21 urges nations to adopt national strategies for sustainable development. It calls for community-based sustainable development. It also urges countries to adopt national strategies that eliminate obstacles to the full participation of women in sustainable development. It recommends that countries eliminate subsidies for environmentally destructive activities and encourages them to ensure public input.

Despite these and other important recommendations, Agenda 21 has some serious weaknesses, according to some critics. Language that would commit more developed nations to donate a certain percentage of their annual GNP to sustainable development overseas was eliminated. In addition, Agenda 21 did not propose new ways of eliminating debt. Nor did it offer new ways to accelerate the transition to sustainable systems of energy (because of pressure from the United States, Saudi Arabia, and Kuwait). All references to full-cost pricing were eliminated. Moreover, the chapter on forestry does not recommend policies of sustainable forest management. The chapter on population fails to underscore the importance of population stabilization to sustainable development and avoids the term *family planning* altogether, reportedly because of pressure from the Vatican.

KEY CONCEPTS

Agenda 21 is a comprehensive list, containing an estimated 4,500 actions, to promote sustainability. Despite its size, the document misses very important elements of sustainability, such as population stabilization and the shift to renewable energy.

Forest Principles Another nonbinding agreement adopted by attendees of the Earth Summit was the **Authoritative Statement of Principles on the World's Forests.** Although in this instance the United States government pushed for stronger efforts to preserve native forests, 77 less developed nations, led by India and Malaysia, watered the agreement down. In fact, they eliminated language pointing out the value of forests to the global environment. In addition, Malaysia eliminated wording that encouraged efforts to promote international trade of products from sustainably managed forests. According to Gareth Porter of the Energy and Environment Institute, the result is a set of principles that underscore sovereign rights of nations to exploit their forests, legitimizing existing policies in those countries that are currently endangering the world's forests.

KEY CONCEPTS

By most accounts, the forest principles articulated at the Earth Summit did little to advance global efforts aimed at sustainable forest harvest and management.

Rio Declaration The final outcome of the Rio Conference was the **Rio Declaration.** It includes recommendations for a number of political principles vital to achieving sustainable development, among them public access to government information on the environment. It also calls on nations to use environmental impact statements and to exercise caution in their development plans. In addition, it encourages timely notification of activities that could exert their effects across a nation's boundaries. The Rio Declaration also recognizes the special responsibility of the more developed countries in achieving global environmental restoration because of their technological and financial capabilities, consumption patterns, and pollution production.

Although the outcome of the 1992 Earth Summit was less than many would have hoped, the progress made there was a good start, especially considering the complexity of negotiations (involving over 150 nations with very different goals and priorities). Because many environmental, economic, and social problems will inevitably worsen in the coming years, progress is bound to occur in areas that were ignored or deleted because of their political sensitivity.

KEY CONCEPTS

The Rio Declaration espoused important political principles essential to the goal of sustainable global development.

Strengthening International Government

Solving global environmental problems may require national sustainable development plans and international agreements. Efforts to strengthen existing international institutions, notably the United Nations, could also help (**FIGURE 26-7**). Some observers think that we may need some form of world government to tackle the problems facing humankind.

KEY CONCEPTS

Two schools of thought exist among those who believe in the need to strengthen international environmental protection and sustainable development. One advocates a strengthening of existing structures such as the United Nations; the other promotes the creation of a global government.

Strengthening the United Nations' Role in Sustainable Development The United Nations has several key programs, such as the Fund for Population Activities (UNFPA), the Food and Agricultural Organization (FAO), the United Nations Environment Programme (UNEP), and the United Nations Development Program. Over the years, these programs have helped scores of less developed nations in many ways. Conferences on population and its impact, sponsored by UNEP, have helped convince most

FIGURE 26-7 The United Nations. The UN has helped increase global cooperation in pollution control and other important issues. It could provide a forum for more progress on building a sustainable society.

© UN/DPI Photo.

less developed nations that something must be done to control rampant population growth. In 1987, UNEP facilitated negotiations that led to the Montreal Accord and two revisions that address depletion of the ozone layer, and in 1990, it sponsored negotiations that produced the groundbreaking London Agreement further reducing the release of ozone-depleting chemicals. In 1988, the UN Commission for Europe negotiated a treaty that calls for a freeze on nitrogen oxide emissions at their 1987 levels by using the best available control technology on new vehicles and power plants in Europe. The Rio Conference and the follow-up conference in 1997, sponsored by the UNEP, are additional examples. For a discussion of the follow-up conference, see **Spotlight on Sustainable Development 26-2**.

Despite its successes and its importance to the future of the world, UNEP operates on a meager $300 million annual budget. The 330 professional staff and 350 support staff are, according to most observers, much too small for the task at hand. Increasing the budget and the staff of UNEP could go a long way toward improving its effectiveness. It would allow UNEP to gather more data and disseminate information more widely. In addition, it would allow UNEP to expand efforts to promote alliances on key issues such as population growth, desertification, and deforestation.

A global sustainable society and government may seem like Utopia. As writer Rolf Edberg reminds us, though, "The utopia of one generation may be recognized as a practical necessity by the next." I think that the present generation is beginning to recognize that sustainability is a practical necessity.

Edberg also notes that achieving such goals depends on our ability to free ourselves from "ideas and emotions that once had a function in our battle for survival but have since become useless." Starving masses and the universal fear of environmental destruction may be the psychological forces that set the stage for a new world society governed by sustainable principles.

KEY CONCEPTS

Several agencies within the United Nations already work diligently toward the goal of environmental protection and sustainable development. Strengthening these organizations could greatly improve these efforts.

The legitimate object of government is to do for a community of people what ever they need to have done, but cannot do at all, in their separate and individual capacities.

—Abraham Lincoln

26-2 Earth Summit II: The Follow Up to Rio

Five years after the highly regarded Earth Summit in Rio de Janeiro, the United Nations sponsored a follow-up meeting, commonly called Earth Summit II. This week-long, follow-up conference held in New York City in June, 1997, was attended by more than a hundred presidents, prime ministers, and other top government officials and was designed to assess progress made since Rio.

What attendees heard was not encouraging. Vice President Al Gore gave the opening speech, noting that deforestation was occurring at rates only slightly less than those of the 1980s. Greenpeace International representative Clif Curtis summed up the grim message of the conference, noting that governments have done far too little to turn around the ongoing environmental decay. He noted that "environmental degradation is worsening at the global level" and pointed out that "we're going backwards on the road to sustainable development." For example, although 150 nations signed the Framework Convention on Climate Change at the 1992 Earth Summit, in which they agreed to reduce or prevent greenhouse gas emissions, in 2009, global carbon emissions reached a record high of 8.4 billion tons. The Framework Convention called on nations to curb emissions of carbon dioxide so that by 2000, emissions of this greenhouse gas would not exceed 1990 levels. However, U.S. carbon dioxide emissions had increased 8% since 1992 and were projected to be 13% above 1990 levels. Emissions in Japan and the nations of the European Union were an estimated 6% higher than 1990 levels. Moreover, many rapidly industrializing nations such as China, Brazil, and Indonesia had experienced 20 to 40% increases in emissions since 1990.

At Rio I, 158 countries signed an agreement to protect plant and animal species by preserving and restoring natural habitats. This agreement, known as the Convention on Biodiversity, went into effect in December 1993. Since that time, an estimated 100,000 species of plants and animals have become extinct, according to estimates of a nonprofit organization, Earth Summit Watch, which monitors progress of the Earth Summit and encourages governments to implement measures they agreed to. The loss of forest habitat, as noted earlier, is a key factor in this decline and is continuing at a rapid pace.

One bright spot is the global phaseout of leaded gasoline. But on balance, Earth Summit Watch notes that "a lot of international institutional development, dialogue, and debate" have taken place, but "not much concrete improvement" in environmental conditions. While many countries are addressing the complex problems, continued growth of the human population and the economy place extraordinary demands on the environment. Progress in environmental protection cannot keep pace with the destruction.

In December 1997, national leaders met to address global climate change, perhaps one of the most pressing issues of our times. Many proposals were offered by the United States, Europe, and small island nations that will suffer the most if the climate continues to warm and oceans rise. Although change will be slow, it is clear that global consensus is emerging on the topic of global climate change: Most nations recognize that something must be done to reduce the emissions of greenhouse gases.

CRITICAL THINKING

Exercise Analysis

One of the myths we propagate in our society is that results can be measured by the amount of money we spend on a problem. Thus, if we are spending a lot of money, we must be doing a good job. Unfortunately, results don't always parallel expenditures. For example, billions of dollars have been spent on hazardous-waste cleanup in the United States, but much of this money ends up in the pockets of attorneys, not in actual cleanup efforts. Spending on water pollution control at sewage treatment plants keeps sewage out of streams, but sludge ends up in landfills.

The U.S. National Report also fails to address the most important question of all: Is the money we are spending enough to create a sustainable way of life? In other words, is it spent on stopgap measures, or does it go into systemic solutions that foster a sustainable human presence? Far more important than dollars spent is the outcome—for example, how much pollution is actually removed, and is it enough to ensure sustainability?

The important lesson here is that when you dig a little deeper, you find that simple but appealing logic falls apart. The report fails to answer the most important question of all: Are solutions advancing the transition to a sustainable society or merely making marginal improvements?

CRITICAL THINKING AND CONCEPT REVIEW

1. What are the major tools by which governments control economic activity and environmental protection? Make a list and describe the pros and cons of each one, as you see it.
2. In what ways are democracies unrepresentative? Can you give some examples?
3. Look around your community. What aspects of it could have been planned better? Why?
4. This chapter lists a number of weaknesses in the current system of government. What are they? What proposals were given to overcome them? Evaluate each of the proposals and make a list of the benefits of each one. Critically analyze them, pointing out problems and ways to solve them.
5. Critically analyze the following statement: "Nations should be left on their own. Others should not meddle in their internal affairs, even in the name of global environmental protection."
6. List the pros and cons of increasing the power of the United Nations in matters of environmental protection.
7. Debate this statement: "We need an international democratic government to address social, economic, and environmental problems."

KEY TERMS

Agenda 21
Authoritative Statement of Principles
   on the World's Forests
balance principle
burden of proof
class action suit
coming to the nuisance
common law
communist nations
concept of knowing
Council on Environmental Quality
   (CEQ)
crisis politics
Critical Trends Assessment Act
defendant
democratic nations
direct government expenditures
Double-C/Double-P game
Earth First!

ecotage
environmental impact
   statement (EIS)
Environmental Protection
   Agency (EPA)
Environmental Quality Act
executive order
Full Employment Act
good-faith efforts
Green Party
income tax deductions
injunction
lobbyists
mediation
National Environmental Policy Act
   (NEPA)
negligence
nuisance
Office of Critical Trends Analysis (OCTA)

plaintiff
political action committees (PACs)
private nuisance
proactive laws
public nuisance
reactive government
retrospective
Rio Declaration
statute of limitations
statutory laws
Sustainable Futures Act
tax credit
theory of public choice
U.S. Corporate Average Fuel
   Economy Act
World Commission on Environment
   and Development

GLOSSARY

Abiotic factors Nonliving components of the ecosystem, including chemical and physical factors such as nitrogen, temperature, and rainfall.

Accelerated erosion Loss of soil due to wind or water in land disturbed by human activities.

Accelerated extinction Elimination of species due to human activities such as habitat destruction, commercial hunting, sport hunting, and pollution.

Acid deposition Rain or snow that has a lower pH than precipitation from unpolluted skies; also includes dry forms of deposition such as nitrate and sulfate particles.

Acid mine drainage Sulfuric acid that drains from mines, especially abandoned underground coal mines in the East (Appalachia). Created by the chemical reaction among oxygen, water, and iron sulfides found in coal and surrounding rocks.

Acid precursor Refers to sulfur oxides and nitrogen oxides, chemical air pollutants that can be converted into sulfuric and nitric acids, respectively, in the atmosphere.

Active solar Capture and storage of the sun's energy through special collection devices (solar panels) that absorb heat and transfer it to air, water, or some other medium, which is then pumped to a storage site (usually a water tank) for later use. Contrast with *passive solar*.

Actual risk An accurate measure of the hazard posed by a certain technology or action.

Acute effects In general, effects that occur shortly after exposure to toxic agents. Contrast with *chronic effects*.

Acute exposure Short-term exposure, usually to high levels of one or more agents. Symptoms generally appear soon after exposure.

Adaptation A genetically determined structural, functional, or behavioral characteristic of an organism that enhances its chances of reproducing and passing on its genes.

Adaptive management Experimental approach to resource management that allows officials to monitor and evaluate practices and change them as needed.

Adaptive radiation Evolution of several life-forms from a common ancestor.

Advanced industrial society Post-World War II industrial society characterized by a great increase in production and consumption, increased energy demand, and a shift toward synthetics and nonrenewable resources.

Age-specific fertility rate Number of live births per 1,000 women of a specific age group.

Agricultural land conversion Transformation of farmland to other purposes, primarily cities, highways, airports, and the like.

Agricultural revolution One of the dominant epochs of human cultural development characterized by the cultivation of plants for food and fiber and the domestication and use of animals for meat and other products.

Agricultural society A group of people living in villages or towns and relying on domestic animals and crops grown in nearby fields. Characterized by specialization of work roles.

Algal bloom Rapid growth of algae in surface waters due to increase in inorganic nutrients, usually either nitrogen or phosphorus.

Alien species (or foreign species) Any species introduced into or living in a new habitat. Also known as an *exotic*.

Alpha particles Positively charged particles consisting of two protons and two neutrons, emitted from radioactive nuclei.

Alveoli Small sacs in the lungs where exchange of oxygen and carbon dioxide between air and blood occurs.

Ambient air quality standard Maximum permissible concentration of a pollutant in the air around us. Contrast with *emissions standard*.

Annuals Plants that must be grown from seeds—for example, domestic corn and radishes.

Antagonism In toxicology, when two chemical or physical agents (often toxins) counteract each other to produce a lesser response than would be expected if individual effects were added together.

Anthropogenic hazard A danger created by humans.

Appropriate technology A term coined by the late E. F. Schumacher to refer to technology that is "appropriate" for the economy, resources, and culture of a region. It is characterized by small- to medium-sized machines, maximum human labor, ease of understanding, meaningful employment, use of local resources, decentralized production, production of durable products, emphasis on renewable resources (especially energy), and compatibility with the environment and culture.

Aquaculture Cultivation of fish and other aquatic organisms in freshwater ponds, lakes, irrigation ditches, and other bodies of water.

Aquatic life zone Distinct regions of the ocean akin to terrestrial biomes.

Aquifer Underground stratum of porous material (sandstone) containing water (groundwater) that may be withdrawn from wells for human use.

Aquifer recharge zone Region in which water from rain or snow percolates into an aquifer, replenishing the supply of groundwater.

Artificial selection Selective breeding to create new plant and animal breeds to bring out desirable characteristics.

Asbestos One of several naturally occurring silicate fibers. Useful in society as an insulator but deadly to breathe, even in small amounts. Causes mesothelioma, asbestosis, and lung cancer.

Asbestosis Lung disease characterized by buildup of scar tissue in the lungs. Caused by inhalation of asbestos.

Asthma Lung disorder characterized by constriction and excessive mucus production in the bronchioles, resulting in periodic difficulty in breathing, shortness of breath, coughing. Usually caused by allergy and often aggravated by air pollution.

Atmosphere Layer of air surrounding the Earth.

Atom A basic unit of matter consisting of a nucleus of positively charged protons and uncharged neutrons, and an outer cloud of electrons orbiting the nucleus.

Autotroph Organism, such as a plant, that produces its own food, generally via photosynthesis. See *producer*.

Auxins Plant hormones responsible for stimulating growth.

Bacteria A group of single-celled organisms, each surrounded by a cell wall and containing circular DNA. Responsible for some diseases and many beneficial functions, such as the decay of organic materials and nutrient recycling.

Barrier islands Small, sandy islands off a coast, separated from the mainland by lagoons or bays.

Beach drift Wave-caused movement of sand along a beach.

Beta particles Negatively charged particles emitted from nuclei of radioactive elements when a neutron is converted to a proton.

Big bang Theory of the Universe's formation. States that all matter in the universe was infinitely compressed 15 to 20 billion years ago and then exploded, sending energy and matter out into space. The matter was in the form of subatomic particles, which formed atoms as the Universe cooled over millions of years.

Bioaccumulation The buildup of chemicals within tissues and organs of the body; results in a resistance to breakdown and slow excretion rates.

Biochemical oxygen demand (BOD) Measure of oxygen depletion of water (largely from bacterial decay) due to presence of biodegradable organic pollutants. Gives scientists an indication of how much organic matter is in water.

Biodegradable Material that can be broken down by naturally occurring organisms such as bacteria in air, water, and soil.

Biogas A gas containing methane and carbon dioxide. Produced by anaerobic decay of organic matter, especially manure and crop residues.

Biogeochemical cycle Complex cyclical transfer of nutrients from the environment to organisms and back to the environment. Examples include the carbon, nitrogen, and phosphorus cycles.

Biological community An assemblage of plants, animals, and microorganisms.

Biological control Use of naturally occurring predators, parasites, bacteria, and viruses to control pests.

Biological extinction Disappearance of a species from part or all of its range.

Biological infrastructure Also known as infrainfrastructure. All that supports human life and the human economy: the air, water, soil, and living things.

Biological magnification Buildup of chemical elements or substances in organisms in successively higher trophic levels. Also called *biomagnification*.

Biomass As measured by ecologists, the dried weight of all organic matter in the ecosystem. In the energy field, any form of organic material (from both plants and animals) from which energy can be derived by burning or by bioconversion, such as fermentation. Includes wood, cow dung, agricultural crop residues, forestry residues, scrap paper.

Biomass pyramid See *pyramid of biomass*.

Biome One of several immense terrestrial regions, each characterized throughout its extent by similar plants, animals, climate, and soil type.

Biosphere All the life-supporting regions (ecosystems) of the Earth and all the interactions that occur among organisms and between organisms and the environment.

Biotic factor The biological component of the ecosystem, consisting of populations of plants, animals, and microorganisms in complex communities.

Biotic (reproductive) potential Maximum reproductive potential of a species.

Birth control Any measure designed to reduce births, including contraception and abortion.

Birth defect An anatomical (structural) or physiological (functional) defect in a newborn.

Black water Water from toilets and bidets contaminated with feces. Kitchen sink water containing food scraps is often placed in this category.

Bloom See *algal bloom*.

Breeder reactor Fission reactor that produces electricity and also converts abundant but nonfissile uranium-238 into fissile plutonium-239, which can be used in other fission reactors.

Broad-spectrum pesticide (or biocide) Chemical agent effective in controlling a large number of pests.

Bronchitis Inflammation of the bronchi caused by smoking and air pollutants. Symptoms include mucus buildup, chronic cough, and throat irritation.

Brown-air cities Newer, relatively nonindustrialized cities whose polluted skies contain photochemical oxidants (especially ozone) and nitrogen oxides, largely from automobiles and power plants. Tend to have dry, sunny climates. Contrast with *gray-air cities*.

Buffer zone Region around a protected area in which limited human activity is permitted.

Calorie Amount of heat needed to raise 1 gram of water 1 degree Celsius or 1.8 degrees Fahrenheit. One thousand calories is equal to one kilocalorie, commonly written as *calorie*.

Cancer Uncontrolled proliferation of cells in humans and other living organisms. In humans, includes more than a hundred different types, afflicting individuals of all races and ages.

Carbon cycle The cycling of carbon between organisms and the environment.

Carcinogen A chemical or physical agent that causes cancer to develop, often decades after the original exposure.

Carrying capacity Maximum population size that a given ecosystem can support for an indefinite period or on a sustainable basis.

Catalyst Substance that accelerates chemical reactions but is not used up in the process. Enzymes are biological catalysts. Also see *catalytic converter*.

Catalytic converter Device attached to the exhaust system of automobiles and trucks to rid the exhaust gases of harmful pollutants.

Cation Any one of many kinds of positively charged ions.

Cellular respiration Process by which a cell breaks down glucose and other organic molecules to acquire energy. Also called *oxidative metabolism*.

Chlorofluorocarbons Organic molecules consisting of chlorine and fluorine covalently bonded to carbon. CC13F (Freon-11) and CC12F2 (Freon-12) are common forms. Used as spray can propellants and coolants. Previously thought to be inert, but now known to destroy the stratospheric ozone layer. Also called *chlorofluoromethanes* and *freon gases*.

Chlorophyll Pigment of plant cells that absorbs sunlight, thus allowing plants to capture solar energy.

Chromosomes Genetic material of organisms, containing DNA and protein. Carries the genetic information that controls all cellular activity.

Chronic bronchitis Persistent inflammation of the bronchi due to pollutants in ambient air and tobacco smoke. Characterized by persistent cough.

Chronic effects In general, the delayed health results of toxic agents, for example, emphysema, bronchitis, and cancer. Contrast with *acute effects*.

Chronic exposure Exposure to one or more toxic agents generally over a long period. Symptoms (for example, cancer) usually appear long after exposure.

Chronic obstructive lung disease Any one of several lung diseases characterized by obstruction of breathing. Includes emphysema, bronchitis, and diseases with symptoms of both of these.

Clear-cutting Removal of all trees from a forested area.

Climate Average weather conditions, including temperature, rainfall, and humidity.

Climax community or ecosystem See *mature community*.

Closed system A system that can exchange energy, but does not exchange matter, with the surrounding environment. Example: the Earth. Contrast with *open system*.

Coal gasification Production of combustible organic gases (mostly methane) by applying heat and steam to coal in an oxygen-enriched environment. Carried out in surface vessels or *in situ*.

Coal liquefaction Production of synthetic oil from coal.

Coastal wetlands Wet or flooded regions along coastlines, including mangrove swamps and salt marshes. Contrast with *inland wetlands*.

Coevolution Process whereby two species evolve adaptations as a result of extensive interactions with each other.

Cogeneration Production of two or more forms of useful energy from one process. For example, production of electricity and steam heat from combustion of coal. Increases energy efficiency.

Coliform bacterium Common bacterium found in the intestinal tracts of humans and other species. Used in water quality analysis to determine the extent of fecal contamination.

Commensalism Relationship between two organisms that is beneficial to one and neither harmful nor helpful to the other.

Common law Body of rules and principles based on judicial precedent rather than legislative enactments. Founded on an innate sense of justice, good conscience, and reason. Flexible and adaptive. Contrast with *statutory law*.

Commons Any resource used in common by many people, such as air, water, and grazing land.

Community Also called a *biological community*. The populations of plants, animals, and microorganisms living and interacting in a given locality.

Competition Vying for resources among members of the same or different species.

Composting Aerobic decay of organic matter to generate a humuslike substance used to supplement soil.

Confusion technique (of pest control) Release of insect sex-attractant pheromones identical to pheromones released by normal breeding females to attract males for mating. Release in large quantities confuses males as to the location of the females, thus minimizing the chances of males finding females and helping to control pest populations.

Conifers Trees with needles for leaves that remain green (and photosynthetic) year round.

Conservation A strategy to reduce the use of resources, especially through increased efficiency, reuse, recycling, and decreased demand.

Conservation biology See *restoration ecology*.

Consumer (or consumer organism) An organism in the ecosystem that feeds on autotrophs and/or heterotrophs. Synonym: *heterotroph*.

Continental drift Movement of the Earth's tectonic plates on a semiliquid layer of mantle, forcing continents to shift position over hundreds of thousands of years.

Contour farming Soil erosion control technique in which row crops (corn) are planted along the contour lines in sloping or hilly fields rather than up and down the hills.

Contraceptive Any device or chemical substance used to prevent conception.

Control group In scientific experimentation, a group that is untreated and compared with a treated, or experimental, group.

Control rods Special rods containing neutron-absorbing materials. Inserted into a reactor core to control the rate of fission or to shut down fission reactions.

Convergent evolution The independent evolution of similar traits among unrelated organisms resulting from similar selective pressures.

Cosmic radiation High-energy electromagnetic radiation similar to cosmic rays but originating from periodic solar flare-ups. Possesses extraordinary ability to penetrate materials, including cement walls.

Cost-benefit analysis Way of determining the economic, social, and environmental costs and benefits of a proposed action such as construction of a dam or highway. Still a crude analytical tool because of the difficulty of measuring environmental costs.

Crisis of unsustainability Used to define the general state of affairs of our society living, as it were, in an unsustainable fashion.

Critical population size Population level below which a species cannot successfully reproduce.

Critical thinking The most ordered type of reasoning of which we're capable. Helps us to discern between beliefs (what people believe to be true) and knowledge based on facts.

Crop rotation Alternating crops in fields to help restore soil fertility and also control pests.

Cross-media contamination The movement of pollution from one medium, such as air, to another, such as water.

Crossing over Transfer of genetic material from one chromosome to another during the formation of gametes.

Crude birth rate Number of births per 1,000 people in a population at the midpoint of the year.

Crude death rate Number of deaths per 1,000 people in a population at midyear.

Cultural control (of pests) Techniques to control pest populations not involving chemical pesticides, environmental controls, or genetic controls. Examples: cultivation to control weeds and manual removal of insects from crops.

Cultural eutrophication Eutrophication (see definition) due largely to human activities.

Daughter nuclei Atomic nuclei that are produced during fission of uranium.

DDT (dichlorodiphenyltrichloroethane) An organochlorine insecticide used first to control malaria-carrying mosquitoes and lice and later to control a variety of insect pests, but now banned in the United States because of its persistence in the environment and its ability to bioaccumulate.

Debt-for-nature swap Arrangement made between a debtor nation, an intermediary, and a lending institution in which the intermediary buys up a debt of a debtor nation at a discount in exchange for funds or programs to protect vital habitats in the debtor nation.

Decibel (dB) A unit to measure the loudness of sound.

Deciduous trees Trees that lose their leaves each year during the late fall and remain leafless.

Decomposer An organism that breaks down nonliving organic material. Examples: bacteria and fungi.

Decomposer food chain A specific nutrient and energy pathway in an ecosystem, in which decomposer organisms (bacteria and fungi) consume dead plants and animals as well as animal wastes. Essential for the return of nutrients to soil and carbon dioxide to the atmosphere. Also called *detritus food chain*.

Deforestation Destruction of forests by clear-cutting.

Demographic transition A phenomenon witnessed in populations of industrializing nations. As industrialization proceeds and wealth accumulates, crude birth rate and crude death rate decline, resulting in zero or low population growth. Decline in death rate usually precedes the decline in birth rate, producing a period of rapid growth before stabilization.

Demography The science of population.

Depletion allowance Tax relief given to extractive industries as they deplete reserves. Intended to allow the companies to invest more in exploration. Gives extraction industries unfair advantage over recycling companies.

Desert Biome located throughout the world. Often found on the downwind side of mountain ranges. Characterized by low humidity, high summertime temperatures, and plants and animals especially adapted to lack of water.

Desertification The formation of desert in arid and semi-arid regions from overgrazing, deforestation, poor agricultural practices, and climate change. Found today in Africa, the Middle East, and the southwestern United States.

Detoxification Rendering a substance harmless by causing it to react with another chemical or destroying the molecule through combustion or thermal decomposition.

Detritus Any organic waste from plants and animals.

Detritus feeders Organisms in the decomposer food chain that feed primarily on organic waste (detritus), such as fallen leaves.

Detritus food chain See *decomposer food chain*.

Deuterium An isotope of hydrogen whose nucleus contains one proton and one neutron (a hydrogen atom has only one proton).

Development In this text, development is defined as ways of improving human well being.

Dioxins A large group of highly toxic, carcinogenic compounds containing some herbicides (2,4-D) and 2,4,5-T and Agent Orange. Once disposed of by mixing with waste crankcase oil that was spread on dirt roads to control dust.

Diversity A measure of the number of different species in an ecosystem.

DNA (deoxyribonucleic acid) A long-chained organic molecule that is found in chromosomes and carries the genetic information that controls cellular function and is the basis of heredity.

Dose-response curve Graphical representation of the effects of varying doses of chemical or physical agents.

Doubling time The length of time it takes some measured entity (population) to double in size at a given growth rate.

Ecological backlashes Ecological effects of seemingly harmless activities, for example, the greenhouse effect.

Ecological equivalents Organisms that occupy similar ecological niches in different regions of the world.

Ecological justice An ethical concept stating that other organisms have a right to live and prosper using the Earth's resources.

Ecological niche See *niche*.

Ecological system See *ecosystem*.

Ecology Study of living organisms and their relationships to one another and the environment.

Economic depletion Reduction in the supply of a resource to the point at which it is no longer economically feasible to continue mining, extracting, or harvesting it.

Economic externality A cost (environmental damage, illness) of manufacturing, road building, or other actions that is not taken into account when determining the total cost of production or construction. A cost generally passed on to the general public and taxpayers; external cost.

Ecosphere See *biosphere*.

Ecosystem Short for *ecological system*. A community of organisms occupying a given region within a biome. Also, the physical and chemical environment of that community and all the interactions among organisms and between organisms and their environment.

Ecosystem management A new approach to managing human and natural landscapes. Attempts to protect entire watersheds and biological communities, rather than small isolated parcels or individual species, and considers a wide range of factors, including abiotic and biotic factors of the environment.

Ecotone Transition zone between adjacent ecosystems.

Element A substance, such as oxygen, gold, or carbon, that is distinguished from all other elements by the number of protons in its atomic nucleus. The atoms of an element cannot be decomposed by chemical means.

Emigration Movement of people out of a country to establish residence elsewhere.

Emissions offset policy Strategy to control air pollution in areas meeting federal ambient air quality standards, whereby new factories must secure emissions reductions from existing factories to begin operation; thus the overall pollution level does not increase.

Emissions standard The maximum amount of a pollutant permitted to be released from a given source.

Emphysema A progressive, debilitating lung disease caused by smoking and pollution at work and in the environment. Characterized by gradual breakdown of the alveoli (see definition) and difficulty in catching one's breath.

Endangered species A plant, animal, or microorganism that is in immediate danger of biological extinction. See *threatened species*.

End-of-pipe controls Refers to strategies and technologies such as sewage treatment plants that attempt to deal with pollution after they are created. Generally, pollutants are removed from smoke stack gases or liquid effluents.

Energy The capacity to do work. Exists in many forms, including heat, light, sound, electricity, coal, oil, and gasoline.

Energy quality The amount of useful work acquired from a given form of energy. High-quality energy forms are concentrated (such as oil and coal); low-quality energy forms are less concentrated (such as solar heat).

Energy system The complete production-consumption process for energy resources, including exploration, mining, refining, transportation, and waste disposal.

Entropy A measure of disorder. The second law of thermodynamics applied to matter says that all systems proceed toward maximum disorder (maximum entropy).

Environment All the biological and nonbiological factors that affect an organism's life.

Environmental control (of pests) Methods designed to alter the abiotic and biotic environments of pests, making them inhospitable or intolerable. Examples include increasing crop diversity, altering time of planting, and altering soil nutrient levels.

Environmental impact statement (EIS or ES) Document prepared primarily to outline potential impacts of projects supported in part or in their entirety by federal funds.

Environmental phase (of the nutrient cycle) Part of the nutrient or biogeochemical cycle in which the nutrient is deposited or cycles through the environment (air, water, and soil).

Environmental resistance Abiotic and biotic factors that can potentially reduce population size.

Environmental science The interdisciplinary study of the complex and interconnected issues of population, resources, and pollution.

Environmental Protection Agency (EPA) Federal agency charged with creating and enforcing environmental regulations.

Epidemiological studies Studies of diseases based on analysis of populations.

Epidemiology Study of disease and death in human populations that attempts to find links between causes and effects through statistical methods.

Epilimnion Upper, warm waters of a lake. Contrast with *hypolimnion*.

Estuarine zones Coastal wetlands and estuaries.

Estuary Coastal region such as an inlet or mouth of a river where salt and fresh water mix.

Ethanol Grain alcohol, or ethyl alcohol, produced by fermentation of organic matter. Can be used as a fuel for a variety of vehicles and as a chemical feedstock.

Eukaryotes Cells whose genetic material is located in nuclei. Contain distinct organelles that carry out specific functions.

Eutrophication Accumulation of nutrients in a lake or pond due to human intervention (cultural eutrophication) or natural causes (natural eutrophication). Contributes to process of succession (see definition).

Evapotranspiration Evaporation of water from soil and transpiration of water from plants.

Evolution A long-term process of change in organisms caused by random genetic changes that favor the survival and reproduction of the organism possessing the genetic change. Through evolution, organisms become better adapted to their environment.

Exclusion principle Ecological law holding that no two species can occupy the exact same niche.

Experimental group In scientific experimentation, a group that is treated and compared with an untreated, or control, group.

Exponential curve See *J curve*.

Exponential growth Increase in any measurable thing by a fixed percentage. When plotted on graph paper, it forms a J-shaped curve.

Externality A spillover effect that harms others. The source of the effect (say, pollution) does not pay for the effect.

Extinction See *biological extinction*.

Extractive reserve A forest protected for the sustainable harvest of fruit, nuts, and other products, generally by local people.

Fall overturn Annual cycle in deep lakes in temperate climates, in which the warm surface water and cool subsurface water mix.

Fallout Deposition of radioactive materials produced during an atomic detonation.

Family planning Process by which couples determine the number and spacing of children.

Feedlot Fenced area where cattle are raised in close confinement to minimize energy loss and maximize weight gain.

First law of thermodynamics Also called the *law of conservation of energy*. States that energy is neither created nor destroyed; it can only be transformed from one form to another.

First-generation pesticides Earliest known chemical pesticides such as ashes, sulfur, ground tobacco, and hydrogen cyanide. Contrast with *second-* and *third-generation pesticides*.

First-law efficiency A measure of the efficiency of energy use. Total amount of useful work derived from a system divided by the total amount of energy put into a system.

Fission Splitting of atomic nuclei when they are struck by neutrons or other subatomic particles.

Fission fragments See *daughter nuclei*.

Floodplain Low-lying region along a river or stream, periodically subject to natural flooding. Common site for human habitation and farming.

Fly ash Mineral matter escaping with smokestack gases from combustion of coal.

Food chain A specific nutrient and energy pathway in ecosystems, proceeding from producer to consumer. Part of a bigger network called the *food web*. See *decomposer food chain* and *grazer food chain*.

Food web Complex intermeshing of individual food chains in an ecosystem.

Foreign species See *alien species*.

Fossil fuel Any one of the organic fuels (coal, natural gas, oil, tar sands, and oil shale) derived from once living plants or animals.

Freons See *chlorofluorocarbons*.

Frontier mentality A mind-set that views humans as "above" all other forms of life rather than as an integral part of nature and that sees the world as an unlimited supply of resources for human use regardless of the impacts on other species. Implicit in this view are the notions that bigger is better, continued material wealth will improve life, and nature must be subdued.

Fuel rods Rods packed with small pellets of radioactive fuel (usually a mixture of fissionable uranium-235 and uranium-238) for use in fission reactors.

Full-cost pricing Technique to include all costs, including the cost of ecological damage, into the price of a product or service.

Gaia hypothesis Term coined by James Lovelock to describe the Earth's capacity to maintain the physical and chemical conditions necessary for life.

Galaxy Grouping of billions of stars, gas, and dust, such as the Milky Way galaxy.

Gamma rays A high-energy form of radiation given off by certain radionuclides. Can easily penetrate the skin and damage cells.

Gasohol Liquid fuel for vehicles, containing nine parts gasoline and one part ethanol.

Gene Segment of the DNA that either codes for proteins produced by the cell (structural gene) or regulates structural genes.

Gene pool Sum total of all the genes and their alternate forms in a population.

Generalists Organisms that have a broad niche, usually feeding on a variety of food materials and sometimes adapted to a large number of habitats.

Genetic control (of pests) Development of plants and animals genetically resistant to pests, through breeding programs and genetic engineering. Also, introduction of sterilized males of pest species (see *sterile male technique*).

Genetic engineering Isolation and production of genes that are then inserted into bacteria or other organisms. Can be used to produce insulin and other hormones. May someday also be used to treat genetic diseases.

Geopressurized zone Aquifer containing superheated, pressurized water and steam trapped by impermeable rock strata and heated by underlying magma.

Geothermal energy Energy derived from the Earth's heat that comes from decay of naturally occurring radioactive materials in the Earth's crust, magma, and friction caused by movement of tectonic plates.

Global change General term that refers to all of the changes—climatological, ecological, environmental, perhaps even economic—caused by greenhouse gas accumulation in the atmosphere.

Global climate change Change in the Earth's climate caused by greenhouse gases.

Global warming Warming of the Earth's atmosphere caused by the build up of greenhouse gases.

GNP See *gross national product*.

Gradualism Theory of evolution holding that species evolve over long periods. Contrast with *punctuated equilibrium*.

Grasslands Biome found in both temperate and tropical regions and characterized by periodic drought, flat or slightly rolling terrain, and large grazers that feed off the lush grasses.

Gray-air cities Older industrial cities characterized by predominantly sulfur dioxide and particulate pollution. Contrast with *brown-air cities*.

Grazer food chain A specific nutrient and energy pathway starting with plants that are consumed by grazers (herbivores).

Green product General term referring to environmentally friendly products. They may be made from recycled materials or may be fully recyclable. They may be reusable or nontoxic or may help promote efficient use of resources.

Green Revolution Developments in plant genetics in the late 1950s and early 1960s resulting in high-yield varieties, producing three to five times more grain than previous plants but requiring intensive irrigation and fertilizer use.

Green tax General term applying to user fees, severance taxes, and pollution taxes.

Greenhouse effect Mechanism that explains atmospheric heating caused by increasing carbon dioxide. Carbon dioxide is believed to act like the glass in a greenhouse, permitting visible light to penetrate but impeding the escape of infrared radiation, or heat.

Greenhouse gas Any of several naturally occurring or anthropogenic gases that trap heat escaping from the Earth's surface, causing the atmosphere to heat up.

Gross national product (GNP) Total national output of goods and services valued at market prices, including net exports and private investment.

Gross primary productivity The total amount of sunlight converted into chemical-bond energy by a plant. This measure does not take into account how much energy a plant uses for normal cellular functions. See *net primary productivity*.

Groundwater Water below the earth's surface in the saturated zone.

Growth factor Any one of many biotic and abiotic factors that stimulate growth of populations. Contrast with *reduction factors*.

Habitat The specific region in which an organism lives.

Half-life Time required for one-half of a given amount of radioactive material to decay, producing one-half the original mass. Can also be used to describe the length of residence of chemicals in tissues. *Biological half-life* refers to the time it takes for one-half of a given amount of a substance to be excreted or catabolized.

Hazardous waste Any potentially harmful solid, liquid, or gaseous waste product of manufacturing or other human activities.

Herbicide Chemical agent used to control weeds.

Herbivore Heterotrophic organism that feeds exclusively on plants.

Heteroculture Agriculture in which several plant species are grown simultaneously to reduce insect infestation and disease.

Heterotroph An organism that feeds on other organisms such as plants and animals. It cannot make its own food-stuffs.

Homeostasis State of relative constancy in organisms and ecosystems. A kind of dynamic equilibrium.

Hot-rock zones Most widespread geothermal resource. Regions where bedrock is heated by underlying magma.

Humus Decaying organic matter that increases fertility, aeration, and water retention of soils.

Hunting and gathering society People who lived as nomads or in semipermanent sites from the beginning of human evolution until approximately 5000 B.C. Some remnant populations still survive. They gathered seeds, fruits, roots, and other plant products, ate carrion, and hunted indigenous species for food.

Hybrid Offspring produced by cross-mating of two different strains or varieties of plants or animals.

Hybrid car A new type of vehicle powered by gasoline and electricity (which it generates itself).

Hydrocarbons Organic molecules containing hydrogen and carbon. Released during the incomplete combustion of organic fuels. React with nitrogen oxides and sunlight to form photochemical oxidants in photochemical smog.

Hydroelectric power Electricity produced in turbines powered by running water.

Hydrological cycle The movement of water through the environment, from atmosphere to earth and back again. Major events include evaporation and precipitation. Also called the *water cycle*.

Hydrosphere The watery portion of the planet. Contrast with *atmosphere* and *lithosphere*.

Hydrothermal convection zone Rock strata containing large amounts of water heated by underlying magma and driven to the surface through cracks and fissures in overlying rock layers. Forms hot springs and geysers.

Hypolimnion Deep, cold waters of a lake. Contrast with *epilimnion*.

Hypothesis Tentative explanation for a natural phenomenon.

Immature ecosystem An early successional community characterized by low species diversity and low stability. Contrast with *mature ecosystem*.

Immigration Movement of people into a country to set up residence there.

Income tax deduction An expense by an individual or business that can be subtracted from one's gross earnings, thus lowering one's taxes. Results in a smaller overall reduction in one's taxes than a tax credit.

Indoor air pollution Generally refers to air pollutants in homes from internal sources such as smokers, fireplaces, woodstoves, carpets, paneling, furniture, foam insulation, and cooking stoves.

Induced abortion Surgical procedure to interrupt pregnancy by removing the embryo or fetus from the uterus. In the first trimester, generally carried out by vacuum aspiration. Contrast with *spontaneous abortion*.

Industrial Revolution Period in history marked by a shift in manufacturing from small-scale production by hand to large-scale production by machine. Occurred in England in the 1700s and in the United States in the 1800s.

Industrial smog Air pollution from industrial cities (gray-air cities), consisting mostly of particulates and sulfur oxides. Contrast with *photochemical smog*.

Industrial society Group of people living in urban or rural environments characterized by mechanization of industrial production and agriculture. Widespread machine labor causes high energy demands and pollution. Increasing control over natural processes leads to feelings that humans are apart from nature and superior to it.

Infant mortality rate Number of infants under 1 year of age dying per 1,000 births in any given year.

Infectious disease Generally, a disease caused by a virus, bacterium, or parasite that can be transmitted from one organism to another (example: viral hepatitis).

Infrared radiation Heat, an electromagnetic radiation of wavelength outside the red end of the visible spectrum.

Inland wetlands Wet or flooded regions along inland surface waters. Includes marshes, bogs, and river outflow lands. Contrast with *coastal wetlands*.

In-migration Movement of people into a state or region within a country to set up residence.

Inorganic fertilizer Synthetic plant nutrient added to the soil to replace lost nutrients. Components include nitrogen, phosphorus, and potassium. Also called *artificial fertilizer* or *synthetic fertilizer*.

Input approach A method of solving an environmental problem by reducing the inputs. For example, reducing consumption and increasing product durability can cut production of solid wastes, pollution, or hazardous wastes.

Insecticide One form of pesticide, used specifically to control insect populations.

Integrated pest management Pest control with minimum risk to humans and the environment through use of a variety of control techniques (including pesticides and biological controls).

Integrated wildlife or species management Control of populations through the use of many techniques, including the reintroduction of natural predators, habitat improvement, reduction in habitat destruction, establishment of preserves, reduced pollution, and captive breeding.

Intergenerational equity Key concept of sustainable ethics. A kind of fairness to future generations.

Interspecific competition Competition among members of different species.

Ion A particle formed when an atom loses or gains an electron.

Ionizing radiation Electromagnetic radiation with the capacity to form ions in body tissues and other substances.

Isotopes Atoms of the same element that differ in their atomic weight because of variations in the number of neutrons in their nuclei.

J curve A graphical representation of exponential growth.

Juvenile hormone (JH) Chemical substance in insects that stimulates growth through early life stages. Used with some success as an insecticide. When applied to infested fields, JH alters normal growth and development of insect pests, resulting in their death.

Kerogen Solid, insoluble organic material found in oil shale.

Keystone species Critical species in an ecosystem whose loss profoundly affects several or many others.

Kilocalorie One thousand calories. See *calorie*.

Kilowatt One thousand watts. See *watt*.

Kinetic energy The energy of objects in motion.

Kwashiorkor Dietary deficiency caused by insufficient protein intake; common in children 1 to 3 years of age in less developed countries. Characterized by growth retardation, wasting of muscles in limbs, and accumulation of fluids in the body—especially in feet, legs, hands, and face.

Lag effect The tendency for a population to continue growing even after it has reached replacement-level fertility. Caused by an expanding number of women reaching reproductive age.

Land ethic View that extends ethical concerns beyond humans to the ecosystem.

Landfill Depression in the ground or excavated site in which waste is deposited. Sanitary landfills are so named because trash is covered daily with soil to prevent odors and proliferation of rodents and flies. Can be used for hazardous wastes, but preferably as last resort. Must be lined with clay and impermeable liners to accept hazardous materials.

Land-use planning Process whereby land uses are matched with the needs of the community and environmental considerations—for example, need for open space and agricultural land and for control of water and air pollution.

Laterite Soil found in some tropical rain forests. Rich in iron and aluminum but generally of poor fertility. Turns bricklike if exposed to sunlight.

Least-cost planning Process in which demand for a resource (such as electricity) is met in the least costly way. Generally includes only economic costs.

Less developed country Term describing the nonindustrialized nations, generally characterized by low standard of living, high population growth rate, high infant mortality, low material consumption, low per capita energy consumption, low per capita income, rural population, and high illiteracy.

Life expectancy The number of years people in a society live, on average.

Light water reactor Most common fission reactor for generating electricity. Water bathes the core of the reactor and is used to generate steam, which turns the turbines that generate electricity. Contrast with *liquid metal fast breeder*.

Limiting factor A chemical or physical factor that determines whether an organism can survive in a given ecosystem. In most ecosystems, rainfall is the limiting factor.

Limnetic zone Open water zone of lakes through which sunlight penetrates; contains algae and other microscopic organisms that feed on dissolved nutrients.

Linear thinking A mode of thinking that oversimplifies complex issues. Often ignores complex networks of cause and effect and systems impacts.

Linearity Refers to linear thinking and linear systems design.

Liquefaction Production of liquid fuel from coal.

Liquid metal fast breeder Fission reactor that uses liquid metals such as sodium as a coolant.

Lithosphere Crust of the earth. Contrast with *hydrosphere* and *atmosphere*.

Littoral drift Movement of beach sand parallel to the shoreline. Caused by waves and longshore currents parallel to the beach.

Littoral zone Shallow waters along a lakeshore, where rooted vegetation often grows.

Macronutrient A chemical substance needed by living organisms in large quantities (for example, carbon, oxygen, hydrogen, and nitrogen). Contrast with *micronutrient*.

Magma Molten rock beneath the Earth's crust.

Malnourishment A dietary deficiency caused by lack of vital nutrients and vitamins.

Manganese nodules Nodular accumulations of manganese and other minerals such as iron and copper found on the ocean floor at depths of 300 to 6,000 meters. Particularly abundant in the Pacific Ocean.

Marasmus A dietary deficiency caused by insufficient intake of protein and calories; occurs primarily in infants under the age of 1, usually as the result of discontinuation of breastfeeding.

Mariculture Cultivation of fish and other aquatic organisms in salt water (estuaries and bays).

Mature community A community that remains more or less the same over a long period of time. Climax stage of succession. Also called a *climax community*.

Mature ecosystem An ecosystem in the climax stage of succession, characterized by high species diversity and high stability. Contrast with *immature ecosystem*.

Measure of economic welfare Proposed standard that takes into account the accumulated wealth of a nation.

Megawatt Measure of electrical power equal to a million watts, or 1,000 kilowatts. See *watt*.

Mesothelioma A tumor of the lining of the lung (pleura). Caused by asbestos.

Metalimnion See *thermocline*.

Metastasis Movement of cancer cells to another location where new tumors are formed.

Methyl mercury Water-soluble organic form of mercury, formed by bacteria in aquatic ecosystems from inorganic (insoluble) mercury pollution. Able to undergo biological magnification.

Micronutrient An element needed by organisms, but only in small quantities—such as copper, iron, and zinc. Contrast with *macronutrient*.

Migration Movement of people across state and national boundaries to set up new residence. See *immigration, emigration, in-migration*, and *out-migration*.

Mill tailings Residue from uranium processing plants. Spent uranium ore that is contaminated with radioactivity.

Mineral A chemical element (such as gold) or inorganic compound (such as iron ore) existing naturally.

Minimum tillage Reduced plowing and cultivating of cropland between and during growing seasons, to help reduce soil erosion and save energy. Also called *conservation tillage*.

Molecule Particle consisting of two or more atoms bonded together. The atoms in a molecule can be of the same element but are usually of different elements.

Monoculture Cultivation of one plant species (such as wheat) over a large area. Highly susceptible to disease and insects.

More developed country A convenient term that describes industrialized nations, generally characterized by high standard of living, low population growth rate, low infant mortality, excessive material consumption, high per capita energy consumption, high per capita income, urban population, and low illiteracy.

Municipal solid waste Refers to garbage from homes and businesses in cities and towns. Typically contains high percentage of recyclable and compostable materials.

Mutagen A chemical or physical agent capable of damaging the genetic material (DNA and chromosomes) of living organisms in both germ cells and somatic cells.

Mutation In general, any damage to the DNA and chromosomes.

Mutualism Relationship between two organisms that is beneficial to both.

Narrow-spectrum pesticide A chemical agent effective in controlling a small number of pests.

Natural erosion Loss of soil occurring at a slow rate, but not caused by human activities. A natural event in all terrestrial ecosystems.

Natural eutrophication See *eutrophication*.

Natural gas Gaseous fuel containing 50%–90% methane and lesser amounts of other burnable organic gases such as propane and butane.

Natural hazards Dangers that result from normal meteorologic, atmospheric, oceanic, biological, and geological phenomena.

Natural resource See *resource*.

Natural selection Process in which slight variations in organisms (adaptations) are preserved if they are useful and help the organism to better respond to its environment.

Negative feedback Control mechanism present in the ecosystem and in all organisms in which signals cause processes to shut down.

Net energy See *net useful energy production*.

Net migration Number of immigrants minus the number of emigrants. Can be expressed as a rate by determining immigration and emigration rates.

Net primary productivity Gross primary productivity (the total amount of energy that plants produce) minus the energy plants use during cellular respiration.

Net useful energy production Amount of useful energy extracted from an energy system.

Niche Also called an *ecological niche*. An organism's place in the ecosystem: where it lives, what it consumes, what consumes it, and how it interacts with all biotic and abiotic factors.

Nitrate (NO_3) Inorganic anion containing three oxygen atoms and one nitrogen atom linked by covalent bonds.

Nitrite (NO_2) Inorganic anion containing two oxygen atoms and one nitrogen atom. Combines with hemoglobin and may cause serious health impairment and death in children.

Nitrogen cycle The cycling of nitrogen between organisms and the environment.

Nitrogen fixation Conversion of atmospheric nitrogen (a gas) into nitrate and ammonium ions (inorganic form), which can be used by plants.

Nitrogen oxides Nitric oxide (NO) and nitrogen dioxide (NO_2), produced during combustion when atmospheric nitrogen (N_2) combines with oxygen. Can be converted into nitric acid (HNO_3). All are harmful to humans and other organisms.

Noise An unwanted or unpleasant sound.

Nonattainment area Region that violates EPA air pollution standards.

Nonpoint source (of pollution) Diffuse source of pollution such as an eroding farm field, urban and suburban lands, and forests. Contrast with *point source*.

Nonrenewable resource Resource that is not replaced or regenerated naturally within a reasonable period (fossil fuel, mineral).

Northern coniferous, boreal forest Same as *taiga*.

Nuclear fission Splitting of an atomic nucleus when neutrons strike it. Products are two or more smaller nuclei, neutrons (which can cause further fission reactions), and an enormous amount of heat and radiation energy.

Nuclear fusion Joining of two small atomic nuclei (such as hydrogen and deuterium) to form a new and larger nucleus (such as helium), accompanied by an enormous release of energy. Source of light and heat from the sun.

Nuclear power (or energy) Energy from the fission or fusion of atomic nuclei.

Nutrient Any substance that is required for the survival of an organism.

Nutrient cycle Same as *biogeochemical cycle*.

Off-road vehicle (ORV) Any vehicle used cross country, often in a recreational capacity (four-wheel-drive vehicles, dune buggies, all-terrain vehicles, snowmobiles, and trail bikes).

Oil See *petroleum*.

Oil shale A fine-grained sedimentary rock called *marlstone* containing an organic substance known as kerogen. When heated, it gives off shale oil, which is much like crude oil.

Old-growth forest Ancient forests with trees often 150 to 1,000 or more years old. Also called *primary forest*.

Omnivore An organism that eats both plants and animals.

Open system A system that freely exchanges energy and matter with the environment. Example: any living organism. Contrast with *closed system*.

Opportunity costs Costs of lost money-making opportunities (and potentially higher income) incurred when we make a decision to invest our money in a particular way.

Optimum range Biotic and abiotic conditions in which survival and reproduction of a given species is maximized.

Ore Rock bearing important minerals—for example, uranium ore.

Ore deposit A valuable mineral located in high concentration in a given region.

Organic farming Agricultural system in which natural fertilizers (manure and crop residues), crop rotation, contour planting, biological insect control measures, and other techniques are used to ensure soil fertility, erosion control, and pest control.

Organic fertilizer Material such as plant and animal wastes added to cropland and pastures to improve soil. Provides valuable soil nutrients and increases the organic content of soil (thus increasing moisture content).

Organismic phase The part of the nutrient cycle in which nutrients are located in organisms: plants, animals, bacteria, fungi, or others.

Out-migration Movement of people out of a state or region within a country to set up residence elsewhere in that country.

Output approach A method of solving an environmental problem by controlling the outputs. For example, composting or burning trash reduces the land requirements for solid waste disposal. Control devices reduce air and water pollution.

Overgrazing Excessive consumption of producer organisms (plants) by grazers such as deer, rabbits, and domestic livestock. Indication that the ecosystem is out of balance.

Overpopulation A condition resulting when the number of organisms in an ecosystem exceeds its ability to assimilate wastes and provide resources. Creates physical and mental stress on a species as a result of competition for limited resources and deterioration of the environment.

Overshoot The phenomenon occurring when a population of organisms exceeds the carrying capacity of its environment.

Oxidants Oxidizing chemicals (for example, ozone) found in the atmosphere.

Oxygen-demanding wastes Organic wastes that are broken down in water by aerobic bacteria. Aerobic breakdown causes the oxygen levels to drop.

Ozone (O$_3$) Inorganic molecule found in the atmosphere, where it is a pollutant because of its harmful effects on living tissue and rubber. Also found in the stratosphere, where it helps screen ultraviolet light. Used in some advanced sewage treatment plants.

Ozone layer Thin layer of ozone molecules in the stratosphere. Absorbs ultraviolet light and converts it to infrared radiation. Effectively screens out 99% of the ultraviolet light.

Paradigm A major theoretical construct that is central to a field of study. For example, the theory of evolution and the structure of DNA are two paradigms that are central to biological science.

Parasitism Relationship in which one species lives in or on another, its host.

Particulates Solid particles (dust, pollen, soot) or water droplets in the atmosphere.

Passive solar Capture and retention of the sun's energy within a building through south-facing windows and some form of heat storage in the building (brick or cement floors and walls). Contrast with *active solar*.

PCBs See *polychlorinated biphenyls*.

Perennial A plant that grows from the same root structure year after year (for example, rose bushes).

Permafrost Permanently frozen ground found in the tundra.

Permanent threshold shift Loss of hearing after continued exposure to noise. Contrast with *temporary threshold shift*.

Pesticide A general term referring to a chemical agent that kills organisms we classify as pests, such as insects and rodents. Also called *biocide*.

Petroleum A viscous liquid containing numerous burnable hydrocarbons. Distilled into a variety of useful fuels (fuel oil, gasoline, and diesel) and petrochemicals (chemicals that can be used as a chemical feedstock for the production of drugs, plastics, and other substances).

pH Measure of acidity on a scale from 0 to 14, with pH 7 being neutral, numbers greater than 7 being basic, and numbers less than 7 being acidic.

Pheromone Chemical substance given off by insects and other species. Sex-attractant pheromones released into the atmosphere in small quantity by female insects attract males at breeding time. Can be used in pest control. See *pheromone traps* and *confusion technique*.

Pheromone traps Traps containing pheromones, to attract insect pests, and either pesticide to kill them or a sticky substance to immobilize them. These traps may be used to pinpoint the emergence of insects, allowing conventional pesticides to be used in moderation.

Photochemical oxidants Ozone and a variety of oxygenated organic compounds produced when sunlight, hydrocarbons, and nitrogen oxides react in the atmosphere.

Photochemical reaction A chemical reaction that occurs in the atmosphere involving sunlight or heat, pollutants, and sometimes natural atmospheric chemicals.

Photochemical smog A complex mixture of photochemical oxidants and nitrogen oxides. Usually has a brownish-orange color.

Photosynthesis A two-part process in plants and algae involving (1) the capture of sunlight and its conversion into cellular energy and (2) the production of organic molecules such as glucose and amino acids from carbon dioxide, water, and energy from the sun.

Photovoltaic cell Thin wafer of silicon or other material that emits electrons when struck by sunlight, thus generating an electrical current. Also called *solar cell*.

Physical infrastructure Refers to humanmade environment or humanmade systems: buildings, bridges, factories, and so on.

Phytoplankton Includes single-celled algae and other free-floating photosynthetic organisms.

Pioneer community The first community to become established in a once-lifeless environment during primary succession.

Pitch (or frequency) Measure of the frequency of a sound in cycles per second (cps) hertz (Hz)—compressional sound waves passing a given point per second. The higher the cps, the higher the pitch.

Pneumoconiosis (black lung) A debilitating lung disease caused by prolonged inhalation of coal and other mineral dusts. Results in a decreased elasticity and gradual break-down of alveoli in the lungs. Eventually leads to death.

Point source (of pollution) Easily discernible source of pollution, such as a factory. Contrast with *nonpoint source*.

Pollution Any physical, chemical, or biological alteration of air, water, or land that is harmful to living organisms.

Pollution prevention Any one of several methods to reduce pollution production, such as process modification and substitution.

Polychlorinated biphenyls (PCBs) Group of at least 50 organic compounds, used for many years as insulation in electrical equipment. Capable of biological magnification. Disrupts reproduction in gulls and possibly other organisms high on the food chain.

Population A group of organisms of the same species living within a specified region.

Population control In human populations, all methods of reducing birth rate, primarily through pregnancy prevention and abortion. In an ecological sense, regulation of population size by a myriad of abiotic and biotic factors.

Population crash (dieback) Sudden decrease in population that results when an organism exceeds the carrying capacity of its environment.

Population growth rate Rate at which a population increases on a yearly basis, expressed as a percentage. For world population: GR = (crude birth rate – crude death rate) × 100. For a given country, population growth rate must also take into account the net migration rate.

Population histogram Graphical representation of population by age and sex.

Positive feedback Control mechanism in ecosystems and organisms in which information influences some process, causing it to increase.

Potential energy Stored energy.

Predator An organism that actively hunts its prey.

Prey Organism (such as deer) attacked and killed by predator.

Primary air pollutant A pollutant that has not undergone any chemical transformation; emitted by either a natural or an anthropogenic source.

Primary consumer First consuming organism in a given food chain. A grazer in grazer food chains or a decomposer organism or insect in decomposer food chains. Belongs to the second trophic level.

Primary forest See *old-growth forest*.

Primary pollutant Pollutant produced by combustion or other sources. Can be chemically modified after release, creating a secondary pollutant.

Primary succession The sequential development of biotic communities where none previously existed.

Primary treatment (of sewage) First step in sewage treatment, to remove large solid objects by screens (filters) and sediment and organic matter in settling chambers. See *secondary* and *tertiary treatment*.

Proactive government One that is concerned with long-range problems and lasting solutions. Contrast with *reactive government*.

Producer (autotroph or producer organism) One of the organisms that produces the organic matter cycling through the ecosystem. Producers include plants and photosynthetic algae.

Productivity The rate of conversion of sunlight by plants into chemical-bond energy (covalent bonds in organic molecules). See *gross* and *net primary productivity*.

Profundal zone Deeper lake water, into which sunlight does not penetrate. Below the limnetic zone.

Prospective law Law designed to address future problems and generate long-lasting solutions. Contrast with *retrospective law*.

Punctuated equilibrium A theory of evolution stating that species are fairly stable for long periods and that new species evolve rapidly over short periods of thousands of years that punctuate the equilibrium. Contrast with *gradualism*.

Pyramid of biomass Graphical representation of the amount of biomass (organic matter) at each trophic level in an ecosystem.

Pyramid of numbers Graphical representation of the number of organisms of different species at each trophic level in an ecosystem.

Quad One quadrillion (10^{15}) Btus of heat.

Rad (radiation absorbed dose) Measure of the amount of energy deposited in a tissue or some other medium struck by radiation. One rad = 100 ergs of energy deposited in one gram of tissue.

Radioactive waste Any solid or liquid waste material containing radioactivity. Produced by research labs, hospitals, nuclear weapons factories, and fission reactors.

Radioactivity Radiation released from unstable nuclei. See *alpha* and *beta particles* and *gamma rays*.

Radionuclides Radioactive forms (isotopes) of elements.

Rain shadow Arid downwind (leeward) side of mountain range.

Range of tolerance Range of physical and chemical factors in which an organism can survive. When the upper or lower limits of this range are exceeded, growth, reproduction, and survival are threatened.

Rangeland Grazing land for cattle, sheep, and other domestic livestock.

Reactive government A government that lives and acts for today, addressing present-day problems as they arise. Shows little or no concern for long-term issues and solutions. Contrast with *proactive government*.

Reactor core Assemblage of fuel rods and control rods inside a reactor vessel. Bathed by water to help control the rate of fission and absorb the heat.

Real price (or cost) The price of a commodity or service in fixed dollars—that is, the value of a dollar at an earlier time. Helpful way to determine whether a resource has experienced a real increase in cost or whether higher costs are simply due to inflation.

Reclamation As used here, the process of returning land to its prior use. Common usage: to convert deserts and other areas into habitable, productive land.

Recycling A strategy to reduce resource use by returning used or waste materials from the consumption phase to the production phase of the economy.

Reduction factors Abiotic and biotic factors that tend to decrease population growth and help balance populations and ecosystems, offsetting growth factors.

Relative humidity The amount of moisture in a given quantity of air, divided by the amount the air could hold at that temperature. Expressed as a percentage.

Rem (roentgen equivalent man) Measure that accounts for the damage done by a given type of radiation. One rad = one rem for X rays, gamma rays, and beta particles; but one rad = 10 to 20 rems for alpha particles because they do more damage.

Renewable resource A resource replaced through natural ecological cycles (water, plants, animals) or through natural chemical or physical processes (sunlight, wind).

Replacement-level fertility Number of children a couple must have to replace themselves in the population.

Reproductive age Age during which most women bear their offspring (ages 14–44).

Reproductive isolation Any of many mechanisms that prevent species from interbreeding or producing viable offspring.

Reserve Deposit of energy or minerals that is economically and geologically feasible to remove with current and foreseeable technology.

Residence time Length of time a chemical spends in the environment.

Resilience Ability of an ecosystem to return to normal after a disturbance.

Resource (as a measurement of a mineral or fuel) Total amount of a mineral or fuel on earth. Generally, only a small fraction can be recovered. Compare with *reserve*.

Resource (in general) Anything used by organisms to meet their needs, including air, water, minerals, plants, fuels, and animals.

Restoration ecology Study of restoring ecosystems to their natural state after human interference. Also called *conservation biology*.

Retorting Process of removing kerogen from oil shale, usually by burning or heating the shale. Can be carried out in surface vessels (surface retorting) or underground in fractured shale (*in situ* retorting).

Retrospective law One that attempts to solve a problem without giving much attention to potential future problems. Contrast with *prospective law*.

Reverse osmosis Means of purifying water for pollution control and desalination. Water is forced through porous membranes; pores allow passage of water molecules but not impurities.

Risk acceptability A measure of how acceptable a hazard is to a population.

Risk assessment The science of determining what hazards a society is exposed to from natural and human causes and the probability and severity of those risks.

Risk probability The likelihood that a hazardous event will occur.

Risk severity A measure of the total damage a hazardous event would cause.

Salinization Deposition of salts in irrigated soils, making soil unfit for most crops. Caused by rising water table due to inadequate drainage of irrigated soils.

Saltwater intrusion Movement of salt water from oceans or saltwater aquifers into freshwater aquifers, caused by depletion of the freshwater aquifers or low precipitation or both.

Sanitary landfill Solid waste disposal site where garbage is dumped and covered daily with a layer of dirt to reduce odors, insects, and rats.

Scrubber Pollution control device that removes particulates and sulfur oxides from smokestacks by passing exhaust gases through a fine spray of water containing lime.

Second law of thermodynamics States that when energy is converted from one form to another, it is degraded; that is, it is converted from a concentrated to a less concentrated form. The amount of useful energy decreases during such conversions.

Secondary consumer Second consuming organism in food chain. Belongs to the third trophic level.

Secondary pollutant A chemical pollutant from a natural or anthropogenic source that undergoes chemical change as a result of reacting with another pollutant, sunlight, atmospheric moisture, or some other environmental agent.

Secondary succession The sequential development of biotic communities occurring after the complete or partial destruction of an existing community by natural or anthropogenic forces.

Secondary treatment (of sewage) After primary treatment, removal of biodegradable organic matter from sewage using bacteria and other microconsumers in activated sludge or trickle filters. Also removes some of the phosphorus (30%) and nitrate (50%). See also *tertiary treatment*.

Second-generation pesticides Synthetic organic chemicals such as DDT that replaced older pesticides such as sulfur, ground tobacco, and ashes. Generally resistant to bacterial breakdown.

Second-law efficiency Measure of the efficiency of energy use, taking into account the unavoidable loss (described by the second law of thermodynamics) of energy during energy conversions. Calculated by dividing the minimum amount of energy required to perform a task by the actual amount used.

Secured landfill One lined by clay and synthetic liners in an effort to prevent leakage.

Sediment Soil particles, sand, and other mineral matter eroded from land and carried in surface waters.

Selective advantage An advantage one member of a species has over others by virtue of some adaptation it has acquired.

Selective cutting Selective removal of trees. Especially useful for mixed hardwood stands. Contrast with *clear-cutting* and *shelter-wood cutting*.

Sewage treatment plant Facility where human solid and liquid wastes from homes, hospitals, and industries are treated, primarily to remove organic matter, nitrates, and phosphates.

Shale oil Thick, heavy oil formed when shale is heated (retorted). Can be refined to produce fuel oil, kerosene, diesel fuel, and other petroleum products and petrochemicals.

Shelterbelts Rows of trees and shrubs planted alongside fields to reduce wind erosion and retain snow to increase soil moisture. May also be used to reduce heat loss from wind and thus conserve energy around homes and farms.

Shelter-wood cutting Three-step process spread out over years: (1) removal of poor-quality trees to improve growth of commercially valuable trees and allow new seedlings to become established, (2) removal of commercially valuable trees once seedlings are established, and (3) cutting remaining mature trees grown from seedlings.

Sigmoidal curve An S-shaped curve.

Simplified ecosystem One with lowered species diversity, usually as a result of human intervention.

Sinkhole Hole created by sudden collapse of the Earth's surface due to groundwater overdraft. A form of subsidence.

Slash-and-burn agriculture Farming practice in which small plots are cleared of vegetation by cutting and burning. Crops are grown until the soil is depleted; then the land is abandoned. This allows the natural vegetation and soil to recover. Common practice of early agricultural societies living in the tropics.

Sludge Solid organic material produced during sewage treatment.

Smelter A factory where ores are melted to separate impurities from the valuable minerals.

Smog Originally referred to a grayish haze (combination of smoke and fog) found in industrial cities. Also pertains to pollution called *photochemical smog*, found in newer cities. See *industrial smog*.

Social Darwinism The application (or misapplication) of the theory of evolution to social behavior.

Sociocusis Hearing loss from human activities.

Soil horizons Layers found in most soils.

Solar collector Device to absorb sunlight and convert it into heat.

Solar energy Energy derived from the sun. Used for heating water for domestic use, space heating, and producing electricity.

Solar system Group of planets revolving around a star.

Sonic boom A high-energy wake creating an explosive boom that trails after jets traveling faster than the speed of sound.

Spaceship Earth Metaphor introduced in the 1960s to foster a greater appreciation of the finite nature of Earth's resources and the ecological cycles that replenish oxygen and other important nutrients.

Specialist Organism that has a narrow niche, usually feeding on one or a few food materials and adapted to a particular habitat.

Speciation Formation of new species.

Species A group of plants, animals, or microorganisms that have a high degree of similarity and generally can interbreed only among themselves.

Species diversity Measure of the number of different species in a biological community.

Spontaneous abortion Loss of an embryo or fetus from the uterus, not caused by surgery. Generally the result of chromosomal abnormalities. Contrast with *induced abortion*.

Spring overturn Annual cycle in deep lakes in temperate climates, in which surface and subsurface waters mix.

SST (supersonic transport) Jet that travels faster than the speed of sound.

Stable runoff Amount of surface runoff that can be counted on from year to year.

Star Spherical cloud of hot gas, such as the sun, fueled by nuclear fusion reactions in its core.

Statutory law Law enacted by Congress or a state legislature. Contrast with *common law*.

Sterile male technique Pest control strategy whereby males of the pest species are grown in captivity, sterilized, and then released en masse in infested areas at breeding time. Sterile males far exceed normal wild males and mate with normal females, resulting in infertile matings and control of the pest.

Sterilization A highly successful procedure in males and females to prevent pregnancy. In males the ducts (vas deferens) that carry sperm from the testicles are cut and tied (vasectomy); in females the Fallopian tubes, or oviducts, which transport ova from the ovary to the uterus, are cut and tied (tubal ligation). Sterilization is not to be confused with castration in males (complete removal of the gonads).

Stratosphere Outer region of the earth's atmosphere, found outside the troposphere, extending 11 to 40 kilometers (7 to 25 miles) above the Earth's surface. Innermost layer of the stratosphere contains the ozone layer.

Streambed aggradation Deposition of sediment in streams or rivers, thereby reducing their water-carrying capacity.

Streambed channelization An ecologically unsound way of reducing flooding by deepening and straightening of streams, accompanied by removal of trees and other vegetation along the banks.

Strip cropping Soil conservation technique in which alternating crop varieties are planted in huge strips across fields to reduce wind and water erosion of soil.

Subsidence Sinking of land caused by collapse of underground mines or depletion of groundwater.

Succession The natural replacement of one biotic community by another. See *primary* and *secondary succession*.

Sulfur dioxide (SO_2) Colorless gas produced during combustion of fossil fuels contaminated with organic and inorganic sulfur compounds. Can be converted into sulfuric acid in the atmosphere.

Sulfur oxides (SO_x) Sulfur dioxide and sulfur trioxide, common air pollutants arising from combustion of coal, oil, gasoline, and diesel fuel. Also produced by natural sources such as bacterial decay and hot springs. Sulfur dioxide reacts with oxygen to form sulfur trioxide, which may react with water to form sulfuric acid.

Supply and demand theory Also known as the *law of supply and demand*. Economic theory explaining the price of goods and services. The supply of and demand for goods and services are primary price determinants. High demand diminishes supply, creating competition for existing goods and services, thus driving up prices.

Surface mining Any of several mining techniques in which all the dirt and rock overlying a desirable mineral (coal, for example) are first removed, exposing the mineral.

Surface runoff Water flowing in streams and over the ground during a rainstorm or snowmelt.

Surface water Refers to bodies of water visible from the Earth's surface, such as lakes, rivers, ponds, and marshes.

Sustainable economics Economic system that seeks to meet human needs while protecting the life-support systems of the biosphere.

Sustainable ethics (mentality) A mind-set that views humans as a part of nature and Earth as a limited supply of resources, which must be carefully managed to prevent irreparable damage. Obligations to future generations require us to exercise restraint to ensure adequate resources and a clean and healthy environment.

Sustainable society A society based on sustainable ethics. Lives within the limits imposed by nature. Based on maximum use of renewable resources, recycling, conservation, and population control.

Sustained yield concept Use of renewable natural resources, such as forests and grassland, that will not cause their destruction and will ensure continued use.

Sympatric speciation Formation of new species without geographical isolation. Common in plants.

Synergism Phenomenon occurring when two or more agents (often toxicants) together produce an effect larger than expected based on knowledge of the effect of each alone. Sometimes simply refers to two or more components acting together.

Synergistic response Refers to a biological response to two or more chemicals or physical factors such as radiation that is greater than the sum of responses to individual factors.

Synfuel See *synthetic fuel.*

Synthetic fertilizer Same as inorganic fertilizer.

Synthetic fuel Gaseous or liquid organic fuel derived from coal, oil shale, or tar sands.

Systems thinking Mode of thinking that takes into account complex webs of cause and effect. Contrast with *linear thinking.*

Taiga Biome found south of the tundra across North America, Europe, and Asia, characterized by coniferous forests, soil that thaws during the summer months, abundant precipitation, and high species diversity.

Tar sands Also known as *oil sands* or *bituminous sands.* Sand impregnated with a viscous, petroleumlike substance, bitumen, which can be driven off by heat, producing a synthetic oil.

Tax credit An expense by an individual or business that can be subtracted directly from one's annual tax liability.

Technological fix A purely technological answer to a problem. Also called a *technical fix.*

Technological optimism Undying faith in technological fixes.

Tectonic plates Huge segments of the Earth's crust that often contain entire continents or parts of them and that float on an underlying semiliquid layer.

Temperate deciduous forest Biome located in the eastern United States, Europe, and northeastern China below the taiga. Characterized by deciduous and nondeciduous trees, warm growing season, abundant rainfall, and a rich species diversity.

Temperature inversion Alteration in the normal atmospheric temperature profile so that air temperature increases rather than decreases with altitude.

Temporary threshold shift Momentary dulling of the sense of hearing after exposure to loud sounds. Can lead to permanent threshold shift.

Teratogen A chemical or physical agent capable of causing birth defects.

Terracing Construction of small earthen embankments on hilly or mountainous terrain to reduce the velocity of water flowing across the soil and thus reduce soil erosion.

Tertiary treatment (of sewage) Removal of nitrates, phosphates, chlorinated compounds, salts, acids, metals, and toxic organics after secondary treatment.

Thermal pollution Heat added to air or water that adversely affects living organisms and may alter climate.

Thermocline Sharp transition between upper, warm waters (epilimnion) and deeper, cold waters (hypolimnion) of a lake. Also called *metalimnion.*

Thermodynamics The study of energy conversions. See the *first* and *second laws of thermodynamics.*

Third-generation pesticides Newer chemical agents to control pests, such as pheromones and insect hormones.

Threatened species A species whose population is declining and could become extinct.

Threshold level A level of exposure below which no effect is observed or measured.

Throughput approach A method of solving an environmental problem by recycling and reuse. For example, recycling or reusing hazardous wastes reduces their output.

Time preference A measure of the value of an immediate gain in comparison with a long-term gain.

Tolerance level (for pesticides) Level of residue on fruits and vegetables permitted by EPA because it is considered "safe."

Total fertility rate Average number of children that would be born alive to a woman if she were to pass through all her childbearing years conforming to the age-specific fertility rates.

Toxicant A chemical, physical, or biological agent that causes disease or some alteration of the normal structure and function of an organism. Impairments may be slight or severe. Onset of effects may be immediate or delayed.

Tradeable permit Permit issued by government agency that allows companies to release certain amounts of pollution. Companies that find ways to reduce pollution can sell permits to other companies.

Transpiration Escape of water from plants through pores (stomata) in the leaves.

Tree farms Private forests devoted to maximum timber growth and relying heavily on herbicides, insecticides, and fertilizers.

Tritium (hydrogen-3) Radioactive isotope of hydrogen whose nucleus contains two neutrons and one proton. Can be used in fusion reactors.

Trophic level The position of an organism in a food chain.

Tropical rain forest Lush forests near the equator with high annual rainfall, high average temperature, and notoriously nutrient-poor soil. Possibly the richest ecosystem on Earth.

Tumor Any abnormal growth of cells. May or may not spread. Tumors that do not continue to grow or spread to other sites are known as *benign tumors.* Those that continue to grow and spread to other areas are called *malignant tumors.*

Tundra (alpine) Life zone found on mountaintops. Closely resembles the Arctic tundra in terms of precipitation, temperature, growing season, plants, and animals. Extraordinarily fragile.

Tundra (Arctic) First major life zone or biome south of the north pole. Vast region on far northern borders of North America, Europe, and Asia. Characterized by lack of trees, low precipitation, and low temperatures.

Ultimate production Total amount of a nonrenewable resource that could ultimately be extracted at a reasonable price.

Ultraviolet (UV) light or radiation Electromagnetic radiation from sun and special lamps. Causes sunburn and mutations in bacteria and other living cells.

Undernourishment A lack of calories in the diet. Contrast with *malnourishment.*

Variation Genetically based differences in behavior, structure, or function in a population.

Waste minimization Any one of several strategies to deal with hazardous and solid wastes such as recycling. Designed to minimize or reduce waste output.

Waste-to-energy plant Incinerator for rubbish that produces small amounts of electricity from heat given off by combustion.

Water cycle See *hydrological cycle.*

Water table Top of the zone of saturation.

Waterlogging Saturation of soils with water due to poor soil drainage and irrigation. Decreases soil oxygen and kills plants.

Watershed Land area drained by a given stream or river.

Watt Unit of power indicating rate at which electrical work is being performed.

Wave power Energy derived from sea waves.

Wet cooling tower Device used for cooling water from power plants. Hot water flows through rising air, which draws off heat. Cool water is then returned to the system.

Wetlands Land areas along freshwater (inland wetlands) and salt water (coastal wetlands) that are flooded all or part of the time.

Wilderness An area where the biological community is relatively undisturbed by humans. Seen by developers as an untapped supply of resources such as timber and minerals; seen by environmentalists as a haven from hectic urban life, an area for reflection and solitude.

Wilderness area An area established by the U.S. Congress under the Wilderness Act (1964), where timber cutting and use of motorized vehicles are prohibited. Most are located in national forests.

Wildlife corridor Protected piece of land that connects protected habitat, allowing animals to migrate.

Wind energy Energy captured from the wind to generate electricity or pump water. An indirect form of solar energy.

Wind generators Windmills that produce electrical energy.

Zero population growth A condition in which population is not increasing; the population growth rate is zero.

Zone of intolerance Range of environmental conditions that an organism cannot survive in.

Zone of physiological stress Upper and lower limits of range of tolerance, where organisms have difficulty surviving.

Zooplankton Nonphotosynthetic, single-celled aquatic organisms.

INDEX

Note: Page numbers followed by *f*, or *t* indicate materials in figures, or tables respectively.

A

A horizon layer, 188
Abelson, Philip H., 444, 444*f*
abiotic, 57
  components of ecosystems, 60
  conditions, 60
abiotic factors
  altering, 123
  ecosystem homeostasis and, 109, 110*f*
abortion
  Catholic Church and, 177
  induced, 169
  spontaneous, 432
absolute humidity, 271
abstinence, 169
abyssal plain, 101
abyssal zone, 102
accelerated erosion, 189
accelerated extinction, 216
acceptability of risk, determining, 441*f*
achondroplasia, in newborns, 430*f*
acid deposition, 44, 238, 463
  Adirondack lakes, 484, 484*f*, 487, 487*f*
  aquatic ecosystems, 487–488
  Canada, 487
  Chesapeake Bay, 516
  damage to materials, 490
  decline of, 493*f*
  defined, 484
  dry deposition, 485
  effect on plants, 489–490
  effect on wildlife, 487
  lakes, 483–484, 484*f*, 487, 487*f*
  North America, 488*f*
  Scandinavia, 484, 486, 487
  social, economic, and environmental impacts
    of, 487–490
  transport of acid precursors, 485–486, 486*f*
  wet deposition, 485
acid deposition control
  Acid Deposition provisions, 1990 Clean Air
    Act, 492
  effectiveness of controls, 492
  long-term strategies, 492
  National Acid Precipitation Assessment
    Program, 492
  short-term solutions, 490–491
acid mine drainage, 318, 319*f*
acid precipitation, 486*f. See also* acid deposition
  measurement, 486
acid precursors, 485–486, 486*f*
acid rain, 490. *See also* acid deposition; acid
    precursors

acidity, pH scale, 484*f*
acids, 484
  fertilizing effect, 490
  natural and anthropogenic sources, 485
acquired immune deficiency syndrome
    (AIDS), 178
acquisitiveness, 45
active solar hot water systems, 361–362, 362*f*
actual risk *vs.* perceived risk, 446, 446*f*
acute exposures, 425
Adair, Eleanor, 435
adaptations, 43, 120
additive response, 426
Adirondack lakes, 484, 484*f*, 487, 487*f*
advanced industrial age, 138–139
Advanced Industrial Society, 138
  fossil fuel combustion, 138
advertising, mass consumption culture and, 45
aesthetics, biodiversity protection and,
    225–226
Afghanistan, desertification in, 192
AFL-CIO, 638
Africa
  deforestation in, 253
  desertification in, 192
  deserts in, 90
  grasslands in, 89
  HIV/AIDS in, 178
  hunger in, 185
  land surpluses in, 205
  population growth in, 153, 178
  rural despair in, 150, 150*f*
  Sahel, 246
African elephants, poaching, 220, 229
African violets, 93
Africanized honeybees, 123, 125*f*, 546
agencies regulating toxic chemicals, federal laws
    and, 437*t*
Agency for International Development
    (AID), 265
Agenda 21, 477, 605, 662
Agent Orange, 549
Agricultural Revolution, 18, 138, 606
  population growth and, 148
agricultural societies, 134–137
  environmental impact of, 136
  features of, 133*t*
  mass-produced foods, 135
  plows and, 134
agricultural technologies, 43
agriculture, 514*t*, 521
  altering planting time, 556
  creation of self-sufficiency, 211
  genetic engineering, 207–208
  integrated pest management, 555–564
  intercropping, 185

  legislation and policies, 211–212
  perennials crops for, 207
  roots of, 134*f*
  sustainable, 185, 196
    governments and, 564
  swidden, 134, 136
  world, challenges facing, 187
agroforestry, 135–136, 135*f*
Ahmadinejad, Mahmoud, 337
AID. *See* Agency for International Development
AIDS. *See* acquired immune deficiency syndrome
air
  endangered global commons, 451–455
  global circulation of, 82*f*
air pollutants
  anthropogenic, 451, 455
  carbon monoxide, 452, 452*t*
  hydrocarbons, 452, 452*t*
  natural, 451
  ozone, 452
  particulates, 452, 452*t*
  photochemical smog, 455
  primary, 453
  secondary, 453
  sources and health effects, 453*t*
  sulfur oxide, 452, 457
  toxic, 453
air pollution, 139
  biomass, 369
  brown-air cities, 454
  damage to materials from, 460*t*
  Donora, PA, 457
  effects of, 457–460
  environmental and economic damage, 457
  fossil fuel combustion and, 138
  gray-air cities, 454
  health effects, 457–459
  impact of, 122*f*
  indoor, 471–474
  Meuse Valley, Belgium, 457
  mountains and hills, 456
  particulates, 496–497
  plants and nonhuman animals,
    effects on, 459
  redesigning systems and, 35*f*
  smelting and, 387
  smog, 454–455
  sources of, 451
  in Sunbelt cities, U.S., 159
  sustainable development and, 35
  urban, 150
  urban sprawl and, 406
air pollution control devices, 461–468
  bag filters, 463
  catalytic converters, 464, 465*f*
  economic benefits, 461